# ADVANCED MECHANICS OF MATERIALS

# ADVANCED MECHANICS OF MATERIALS

**Roman Solecki**
Emeritus, University of Connecticut

**R. Jay Conant**
Montana State University

New York Oxford
OXFORD UNIVERSITY PRESS
2003

Oxford University Press

Oxford New York
Auckland Bangkok Buenos Aires Cape Town Chennai
Dar es Salaam Delhi Hong Kong Istanbul Karachi Kolkata
Kuala Lumpur Madrid Melbourne Mexico City Mumbai
Nairobi São Paulo Shanghai Taipei Tokyo Toronto



Published by Oxford University Press, Inc.
198 Madison Avenue, New York, New York, 10016
http://www.oup-usa.org



**Library of Congress Cataloging-in-Publication Data**

Solecki, Roman.
Advanced mechanics of materials / by Roman Solecki, R. Jay Conant.
p. cm.
Includes bibliographical references and index.
ISBN 0-19-514372-8
1. Strength of materials. I. Conant, R. Jay. II. Title.

TA405 .S655 2002
620.1'12—dc21 2002025750

Printing number: 9 8 7 6 5 4 3 2 1

Printed in the United States of America
on acid-free paper

TO

Maria and Pat

# CONTENTS

CHAPTER THREE
## Displacement and Strain



CHAPTER FOUR
## Relationships Between Stress and Strain

CHAPTER FIVE
## Energy Concepts

CHAPTER SIX
## Numerical Methods I



CHAPTER SEVEN
## Numerical Methods II: Finite Elements

CHAPTER EIGHT
## Beams

CHAPTER ELEVEN
## Buckling and Vibration



CHAPTER TWELVE
## Introduction to Fracture Mechanics

## Appendixes

# PREFACE

The engineering of a successful product requires an ability to analyze the product as a whole, as well as its individual components and assemblies, in order to verify design decisions and to ascertain that the product will perform as intended without premature failure. The analysis invariably involves developing a model, or idealization, of the actual problem and then applying the principles of engineering to the model in order to predict its behavior.

Engineers need an understanding of the fundamental concepts and an ability to apply these concepts to the solution of engineering problems. Consequently, the developments in this text tend to evolve from *fundamental principles* such as equilibrium and conservation of energy. The basic concepts and principles are amply discussed and numerous examples are given throughout the text to help the reader grasp the ideas. We begin with theoretical and conceptual developments, presenting relatively simple examples aimed at illuminating certain aspects of the theory. We then apply the theory to specific types of problems such as beam bending, plate bending, vibration and buckling of beams and plates, and finite element analysis. The book contains general topics such as bending of curved beams, thick-walled cylinders, torsion of bars with noncircular cross sections, and finite element analysis. More specialized topics are also discussed, including fracture mechanics and piezoelectric behavior of materials.

## PREREQUISITES

This book is concerned with the development of analytical methods for solving problems in mechanics of materials that are generally considered beyond the scope of basic courses in the discipline. It is intended for advanced undergraduate and beginning graduate students in engineering.

The authors assume that the student has successfully completed an elementary course in *mechanics of materials* (or *strength of materials*). In addition it is assumed that the student has completed the usual undergraduate math sequence consisting of differential and integral calculus, multivariate calculus, and differential equations. Students will also find it helpful to have completed a course in machine design (mechanical engineering), structures (civil engineering), or aircraft structures (aerospace engineering) which covers yielding under combined loading, an introduction to fracture mechanics, and fatigue. Although not essential, familiarity with these topics will give the student experience with solving a fairly broad range of problems that will prove helpful in studying this text. Chapter 5 covers energy methods in mechanics using the fundamental concepts and definitions of thermodynamics as a starting point. Therefore students will benefit from familiarity with concepts concerning the first and second laws of thermodynamics as developed in undergraduate thermodynamics courses. Many examples and problems in the text lend themselves to numerical or symbolic solution on a computer. Hence some background in computer programming and in the use of computer algebra programs such as Mathcad will prove useful.

# 1 INTRODUCTION

This book is concerned with the development of analytical methods for solving problems in mechanics of materials that are generally considered beyond the scope of basic courses in the discipline. As such, the developments tend to evolve from fundamental principles such as equilibrium and conservation of energy.

In analyzing a particular physical problem, the analyst must develop a mathematical model that describes the problem. The mathematical model consists of equations that are solved to obtain formulas predicting the behavior of the mathematical model. It is imperative to keep in mind that what is being analyzed is a model of the physical problem rather than the physical problem itself. This is, perhaps, a subtle distinction, but it is of paramount importance. Thus, how well the results of the analysis predict the behavior of the physical problem depends on how well the mathematical model represents the physical problem at hand.

Any mathematical model is an idealization of the physical problem that the model represents. In developing the model, certain assumptions are made regarding the behavior of the physical problem in order to simplify the analysis. The goal of the analysis is to capture the essential behavior of the physical problem while ignoring behavior that is considered nonessential. For example, in developing the Bernoulli–Euler beam theory studied in mechanics of materials courses, among other things it is assumed that straight lines perpendicular to the neutral axis before bending remain straight and perpendicular after bending. The resulting equations describe the deflection of the beam due to bending but ignore the additional deflection due to the effects of shear. Thus the deflection due to shear is considered to be nonessential in this theory. The resulting mathematical theory yields a fourth-order ordinary differential equation that can be solved using techniques learned in an undergraduate differential equations course. The equation has been studied extensively, and solutions for a wide variety of loads and boundary conditions are tabulated in books such as Roark [1].

The value of the Bernoulli–Euler beam theory is that it leads to a simple differential equation that can be readily solved, and the results obtained are applicable in a variety of practical situations. Generally speaking, the use of simplifying assumptions enable results to be obtained with the use of fewer resources. For example, a solution may be obtained without the use of a computer, or with a computer program that executes in a few seconds instead of a couple of hours.

More accurate mathematical models generally result as the number of assumptions is reduced. However, this increased accuracy is usually accompanied by increased mathematical complexity. The analyst must carefully consider whether the additional complexity is warranted. One could develop, for example, a beam theory that includes large as well as small displacements, and that includes material behavior beyond the elastic limit. Such a theory would be quite complicated and would require substantially more time to apply to a particular problem than elementary beam theory. If the beams of interest undergo only small displacements and the stresses are required to be within the elastic limit of the material, it would be difficult to justify the additional time and effort needed to use a complicated theory.

Some design situations may demand that the most accurate theory available be used. For example, in the design of aerospace structures, weight considerations usually necessitate the use of small safety factors. Since part of the purpose of a safety factor is to account for uncertainty in the analysis, use of a very accurate theory is warranted. In other design situations, it may be known that a particular problem is outside of the scope of an elementary theory. Elementary beam theory, which neglects shear deflection, is generally considered to give accurate results for slender beams, where the length-to-thickness ratio of the beam exceeds 10. For design situations involving short, stubby beams a more advanced beam theory that includes the effects of shear deflection must be used. In still other situations it might be known or suspected that secondary behavior, which would ordinarily be neglected as nonessential, is important.

However, even in situations where very advanced mathematical models are required, simplified models can still be useful in order to quickly obtain a design that is in the ballpark, or to gain confidence in the results of a complicated analysis that may require months to complete. Obtaining the natural frequencies of the wing of a commercial airliner requires the use of sophisticated computer programs that require large amounts of input data and hence are prone to input errors. A relatively quick estimate of these frequencies can be obtained by computing the natural frequencies of a simple beam having the average properties of the wing. While the natural frequencies thus obtained will not be highly accurate, they will, nonetheless, provide ballpark estimates that are useful in gaining confidence in the computer solution. If substantial differences between the solution obtained from the simple model and the computer solution exist, these must be reconciled.

In a surprising number of situations, important aspects of the behavior of complicated systems can be explained with the use of simple models. Photographs of failed buildings and bridges from the recent earthquakes in San Francisco, CA, and Northridge, CA, indicate that although the structures were quite complicated, they acted much like single-degree-of-freedom spring-mass systems! Many of these structures underwent severe side-to-side motion that led to collapse of the supports (bridge support columns, for example) but the main structure remained essentially undamaged, suggesting that the main structure behaved as a rigid mass. The response curves (amplitude versus frequency) for contact piezoelectric transducers used in ultrasonic testing of materials are strikingly similar to the response curves for damped, harmonic oscillators subjected to a harmonic forcing function. Thus, knowledge of the behavior of the relatively simple damped, harmonic oscillator can provide useful insight into the behavior of piezoelectric transducers, which are inherently more complicated.

Generally speaking, Chapters 2 through 5 are devoted to theoretical and conceptual developments and contain relatively simple examples aimed at illuminating certain aspects of the theory. In the remaining chapters the theory is applied to specific classes of problems such as beam bending, plate bending, vibration and buckling of beams and plates, and finite element analysis.

In Chapter 2 the concept of stress is developed, along with the equilibrium equations. Transformation equations for representing the stress components in different coordinate systems are also developed. From these, principal stresses and their directions are obtained. Concepts useful in the study of the plastic behavior of materials, namely, octahedral stress, mean stress, and deviatoric stress, are developed.

Chapter 3 is devoted to the study of displacement and strain. Expressions are derived for normal and shear strains in terms of the displacement components. The concept of compatibility of strains is introduced, and the equations pertaining to compatibility are derived. Transformation equations for strain, analogous to those for stress, are derived. The concept of rotation is introduced, and it is shown that infinitesimal deformations can be represented as combinations of strains and rotations. From the transformation equations, principal strain expressions are obtained. The dilatation, an expression of volume change for infinitesimal deformations, is derived in this chapter, and the concept of deviatoric strain, analogous to deviatoric stress, is introduced.

The relationship between stress and strain for different material behavior is explored in Chapter 4. First the generalized Hooke's law for isotropic materials is developed. The reduction of the general equations to the important two-dimensional cases of plane stress and plane strain is carried out in this chapter. The general relations between stress and strain for anisotropic materials are developed, and specialization of these relations for orthotropic materials, used for fiber-reinforced composites, is carried out. The stress-strain relations for transversely isotropic materials (wood, for example) are also developed in this chapter. Attention is then turned to stress-strain relations for viscoelastic materials based on the Maxwell model and the Kelvin–Voigt model. This chapter continues by examining material behavior beyond the elastic limit. The maximum shear theory and the distortion energy theory for predicting yielding in multiaxial stress situations are developed and contrasted. The Prandtl–Reuss equations giving the relations between stress and strain rate for plastic behavior of elastic–perfectly plastic materials are then developed. This is followed by the development of stress-strain-temperature equations for problems in which the temperature is not uniform. Chapter 4 closes with the discussion of a very unusual and practically important, contemporary topic. Certain materials exhibit an interesting behavior, whereby an electric field is generated when the material is strained. It is called the *direct piezoelectric effect*. This is always accompanied by the converse behavior, called the *converse piezoelectric effect,* in which strain occurs when the material is placed in an electric field. Materials exhibiting these effects are called *piezoelectric materials*. Recent application areas that make use of piezoelectric materials are *active vibration control,* in which a sensor (which could be piezoelectric) is used to sense vibration and send an electric signal to a piezoelectric actuator that provides corrective antivibration, and *smart structures,* which are embedded piezoelectric sensors that can detect damage (such as delamination in a composite) or excessive deflection in a flexible structure and signal a piezoelectric actuator to provide stiffening.

Energy concepts are explored in Chapter 5. Fundamental concepts leading to the first and second laws of thermodynamics are developed and specialized for solids. The concept of strain energy is introduced for a linear elastic solid and specialized for beams.

Castigliano's theorem is then derived for beams. The principle of virtual work is established for deformable bodies, and the theorem of minimum total potential energy for elastic bodies is derived from this. The principle of minimum complementary energy is developed next. These principles are used to obtain Castigliano's first and second theorems for bodies of arbitrary shape composed of nonlinear elastic materials. The Rayleigh–Ritz method is introduced as an approximate method for solving problems concerning deformable bodies. Next, the Betti–Rayleigh reciprocal theorem is developed. It is used to obtain beam deflections via Green's functions. Finally, a general stress-strain equation is developed for linear elastic materials, and it is shown that a general anisotropic elastic material requires 21 independent elastic constants for its description.

Chapter 6 introduces the reader to numerical methods. The finite difference method is used to solve both ordinary differential equations and partial differential equations. The method of iteration is used for ordinary differential equations. Finally, the method of collocation is discussed in conjunction with the solution to ordinary differential equations.

Numerical methods are continued in Chapter 7, which is devoted to finite element analysis. The chapter begins with a development of the stiffness method of analysis for plane frame structures. Since most of the key ideas of finite element analysis are contained in the stiffness method, it provides a solid background from which to develop the finite element method. Energy methods, in particular Castigliano's theorems, are used in developing the stiffness method as well as the finite element method. The finite element material is, of necessity, brief and elementary. It does, nonetheless, cover all of the basics of the method and provide the reader with an introduction sufficient to enable subsequent study of specialized texts on the subject.

The analytical tools developed in earlier chapters are used to study the behavior of beams in Chapter 8. The chapter begins by considering the bending of continuous beams. The three moments theorem is developed, as well as the method of initial parameters. Castigliano's theorem is also used to solve these problems. Next, unsymmetric bending of straight beams is addressed. This is followed by the study of curved beams subjected to both out-of-plane and in-plane loading. Castigliano's theorem is used to develop Biezeno's theorem for circular rings with out-of-plane loads and supports. This is followed by the analysis of beams on elastic foundations. Straight beams, both infinite and finite, are considered first, and these are followed by curved beams. Green's functions for beams are introduced in the next section. This is followed by thermal effects in beams and composite beams. Next, the use of Fourier series in the solution to beam problems is considered. Also a discussion of approximate solutions to beam problems via the finite difference method and the Rayleigh–Ritz method is presented. The chapter ends with an extensive discussion of piezoelectric, also composite, beams. Here the material introduced in Chapter 4 is utilized. Several examples are solved.

Chapter 9 contains an introduction to two- and three-dimensional problems in elasticity. Appropriate equations from Chapters 2, 3, and 5 are combined to yield Navier's equations. This is followed by a discussion of boundary conditions in elasticity. Axisymmetric problems for thick-walled cylinders and circular disks are examined for both pressure and centrifugal loading due to rotation. Both elastic and plastic material behavior is considered. Composite disks and cylinders composed of orthotropic materials are considered as are disks of varying thickness. Next, Airy's stress function is introduced and is used to formulate the torsion problem for prismatic members with noncircular cross sections. The membrane analogy is discussed and applied to several example problems. This chapter closes with the application of the finite difference method to the solution to Navier's equations for a two-dimensional beam bending problem, and to the torsion problem for a square cross section.

The analysis of elastic plates is covered in Chapter 10. The assumptions employed in plate theory are introduced and used to reduce the three-dimensional elasticity equations to the two-dimensional plate equations. These are applied to the solution of several axisymmetric circular plate bending problems for both solid and annular plates. Rectangular plates are considered next, along with the appropriate boundary conditions. Solutions are obtained through the use of double trigonometric Fourier series. Solutions for laminated composite plates, with both isotropic laminae and orthotropic laminae, are obtained. Finally, approximate solutions to rectangular plate problems based on the finite difference method and the Rayleigh–Ritz method are developed.

Buckling and vibration are covered in Chapter 11. Mathematically, the equations for buckling and vibration are identical, so it is natural to discuss the two phenomena in a single chapter. The chapter begins by developing the equation of motion for a beam subjected to an axial compressive load. Solutions for harmonic motion that lead to the natural frequencies and mode shapes are obtained for various boundary conditions. The effect of an axial load on the natural frequencies is clearly seen. Next, the critical buckling loads and mode shapes are obtained for various boundary conditions. The discussion of beams ends with the application of the Rayleigh–Ritz method to obtain approximate buckling solutions. The chapter continues with a discussion of the buckling and vibration of thin-walled curved members such as rings. The approach taken is similar to that used for beams. The chapter concludes by considering the buckling and vibration of rectangular plates.

Chapter 12 deals with the influence of cracks on the behavior of a linearly elastic solid. We consider the solution of some idealized problems in which the crack is assumed to be straight, finite, or semi-infinite. Of particular interest is the stress distribution near the tip of the crack. Thus the concept of a stress intensity factor (SIF) is introduced. SIF helps in predicting the possibility of fracture damage. The concept of the $J$ integral is also introduced and its relation to the energy rate is determined.

## REFERENCE

[1] Roark, R. J., *Formulas for Stress and Strain,* 4th ed., McGraw-Hill Book Company, New York (1965)

# 2 STRESS AND EQUILIBRIUM EQUATIONS

In this chapter we are concerned with stress, its components, and the relationships between them. Once the stress components are introduced and their symmetry is established we derive the formulas for the stress acting on a plane inclined with respect to the Cartesian coordinate planes. The differential equations of equilibrium of an infinitesimal element are determined in both a Cartesian and a polar coordinate system. Principal stresses and stress invariants are also discussed.

## 2-1 CONCEPT OF STRESS

A solid* can be subjected to the action of two types of external forces. Concentrated or distributed loads applied to the surface of a solid belong to the first type. These so-called *surface tractions* result from the contact with other solids or fluids and are transmitted through the surface to the interior of the solid. The second type consists of forces acting directly on all particles of the solid. These forces, called *body forces,* are proportional to the mass contained in the volume element (e.g., gravity, electromagnetic forces). Let us cut through the solid. The contact forces from the portion of the body removed by the cut act on the cut surface as distributed loads. Both the magnitude of these loads and their directions will vary, in general, from point to point. Now isolate from the surface of the cut (Fig. 2-1) an arbitrary point $M$, surrounded by a small area $\Delta A$, and the corresponding surface load $\boldsymbol{F}$ acting on that area. In general, when $\Delta A$ changes, $\boldsymbol{F}$ changes. However, when the area shrinks to the point $M$, the ratio of this load to the area approaches a limit, called the *stress vector,* at point $M$ and denoted by $\boldsymbol{t}$,

$$\boldsymbol{t} = \lim_{\Delta A \to 0} \frac{\boldsymbol{F}}{\Delta A}$$

The stress vector is measured in units of force per unit area.

*Or any other continuous medium.

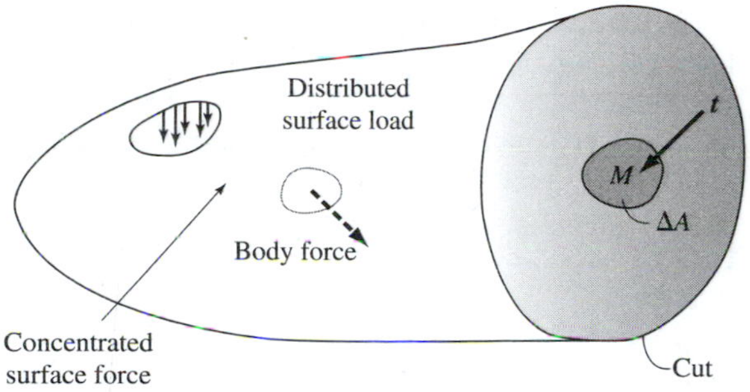


Figure 2-1

## 2-2 STRESS COMPONENTS AND EQUILIBRIUM EQUATIONS

### 2-2-1 Stress Components in Cartesian Coordinates—Matrix Representation

To better understand the stress behavior throughout the solid and to establish equilibrium conditions to be satisfied, we analyze an infinitesimal volume element cut around an arbitrarily chosen point. For convenience, this element is oriented so that one coordinate axis is perpendicular to each of its faces. We now introduce a Cartesian coordinate system $Oxyz$. Thus the element has the form of a rectangular solid oriented so that its three intersecting edges coincide with the $xyz$ axes (Fig. 2-2).

Since the stresses are distributed continuously throughout the solid, a continuous load is applied to each face of the element in the free-body diagram. For clarity let us show instead of this load its resultants. We note that since the variation of stress across an infinitesimal area is also infinitesimal, the resultant forces can be attached to the centers of the faces and the resultant moments neglected. We also observe that the

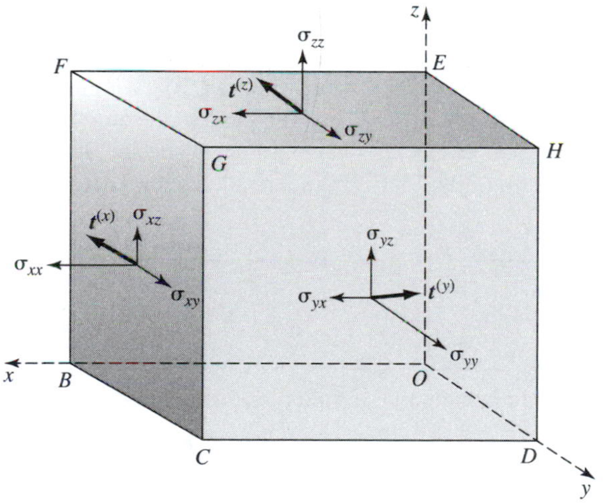


Figure 2-2 Stress components in Cartesian coordinate system

distance between two parallel faces is infinitesimal, implying that both the magnitude and the direction of the stress vectors acting on parallel faces differ infinitesimally.

To facilitate the distinction between stress vectors acting on various faces, let us assign the labels indicating the direction of the normal to the appropriate face. Thus we denote stress vectors acting on faces with normals pointing in the positive $x$, $y$, or $z$ directions (these are faces $BCGF$, $CDHG$, and $EFGH$, respectively) by $\boldsymbol{t}^{(x)}$, $\boldsymbol{t}^{(y)}$, and $\boldsymbol{t}^{(z)}$. Thus the stress vectors acting on faces with normals pointing in the negative $x$, $y$, or $z$ directions are $\boldsymbol{t}^{(-x)}$, $\boldsymbol{t}^{(-y)}$, $\boldsymbol{t}^{(-z)}$. But if we take, for instance, two faces of a cut with stress vectors $\boldsymbol{t}^{(x)}$ and $\boldsymbol{t}^{(-x)}$, respectively, then the consideration of the equilibrium across this cut leads to the conclusion that $\boldsymbol{t}^{(-x)} = -\boldsymbol{t}^{(x)}$, and so on (see also [5, p. 67]). Next we decompose each of the stress vectors into three component vectors parallel to the coordinate axes. The magnitudes of these vectors, called *stress components,* are denoted as follows:

$$\begin{aligned} \sigma_{xx} &= \boldsymbol{t}^{(x)} \cdot \boldsymbol{i}, & \sigma_{xy} &= \boldsymbol{t}^{(x)} \cdot \boldsymbol{j}, & \sigma_{xz} &= \boldsymbol{t}^{(x)} \cdot \boldsymbol{k} \\ \sigma_{yx} &= \boldsymbol{t}^{(y)} \cdot \boldsymbol{i}, & \sigma_{yy} &= \boldsymbol{t}^{(y)} \cdot \boldsymbol{j}, & \sigma_{yz} &= \boldsymbol{t}^{(y)} \cdot \boldsymbol{k} \\ \sigma_{zx} &= \boldsymbol{t}^{(z)} \cdot \boldsymbol{i}, & \sigma_{zy} &= \boldsymbol{t}^{(z)} \cdot \boldsymbol{j}, & \sigma_{zz} &= \boldsymbol{t}^{(z)} \cdot \boldsymbol{k} \end{aligned} \tag{2-1}$$

where $\boldsymbol{i}$, $\boldsymbol{j}$, and $\boldsymbol{k}$ are unit vectors along the $Ox$, $Oy$, and $Oz$ axes, respectively. Note that the first subscript identifies the direction of the normal to the face on which the component is acting, whereas the second index represents the direction of the component vector. Stress components are positive when they point in positive coordinate directions on faces with outer normals directed in positive coordinate directions, or in negative coordinate directions on faces with outer normals directed in negative coordinate directions. They are negative otherwise (Fig. 2-2). The stress components can be arranged into a square matrix,

$$\boldsymbol{S} = \begin{bmatrix} \sigma_{xx} & \sigma_{xy} & \sigma_{xz} \\ \sigma_{yx} & \sigma_{yy} & \sigma_{yz} \\ \sigma_{zx} & \sigma_{zy} & \sigma_{zz} \end{bmatrix} \tag{2-2}$$

known as the *stress matrix*. The matrix representation of the stress components and of other quantities is particularly important when numerical methods or symbolic algebra are applied (see Chapter 7 and Examples 2-7 and 2-8). The elements on the main diagonal of the matrix, $\sigma_{xx}$, $\sigma_{yy}$, $\sigma_{zz}$, are called *normal* stresses because they are the components normal to the surface on which they act. The remaining elements are called *shear* stresses, or *shearing* stresses, because they represent the shearing of the surface on which they act with respect to the neighboring surface. It will be shown in the next section that the matrix $\boldsymbol{S}$ has only six independent elements.

## EXAMPLE 2-1

A stress vector $\boldsymbol{t}^{(x)}$ acting at point $M$ on a plane perpendicular to the $Ox$ axis equals

$$\boldsymbol{t}^{(x)} = 2\boldsymbol{i} + \boldsymbol{j} + \boldsymbol{k} \tag{a}$$

Find the stress components.

**SOLUTION**

We find from Eqs. (2-1) that

$$\begin{aligned} \sigma_{xx} &= \boldsymbol{t}^{(x)} \cdot \boldsymbol{i} = (2\boldsymbol{i} + \boldsymbol{j} + \boldsymbol{k}) \cdot \boldsymbol{i} = 2 \\ \sigma_{xy} &= \boldsymbol{t}^{(x)} \cdot \boldsymbol{j} = (2\boldsymbol{i} + \boldsymbol{j} + \boldsymbol{k}) \cdot \boldsymbol{j} = 1 \\ \sigma_{xz} &= \boldsymbol{t}^{(x)} \cdot \boldsymbol{k} = (2\boldsymbol{i} + \boldsymbol{j} + \boldsymbol{k}) \cdot \boldsymbol{k} = 1 \end{aligned} \qquad (b)$$

where Eq. (*a*) has been used.

## 2-2-2 Symmetry of Shear Stresses

We begin by analyzing the conditions of equilibrium requiring that the sum of the moments of all applied forces with respect to the coordinate axes must equal zero. To this end let us consider again the rectangular solid element (Fig. 2-3). It is now necessary to take into account the body forces whose resultant $\boldsymbol{b}$ (per unit volume) acts at the geometrical center of the element. It will be more convenient to perform the analysis using

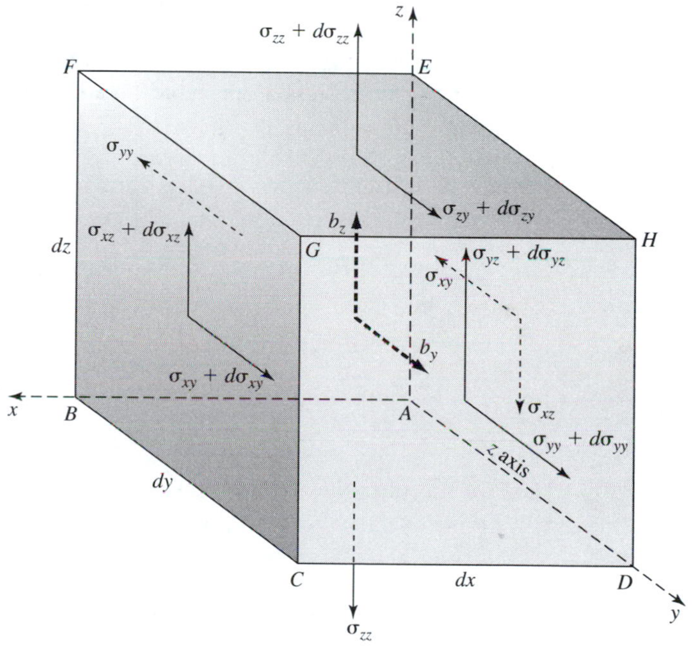


Figure 2-3 Calculating the moments of stress components about $x$ axis

TABLE 2-1

| Face | Moment About $x$ Axis |
|---|---|
| $BCGF$ | $\underbrace{(\sigma_{xz} + d\sigma_{xz})}_{\text{stress}}\,\underbrace{dy\,dz}_{\text{area}}\,\underbrace{\frac{dy}{2}}_{\text{moment arm}} - (\sigma_{xy} + d\sigma_{xy})\,dy\,dz\frac{dz}{2}$ |
| $CDHG$ | $-(\sigma_{yy} + d\sigma_{yy})\,dx\,dz\frac{dz}{2} + (\sigma_{yz} + d\sigma_{yz})\,dx\,dz\,dy$ |
| $ADHE$ | $\sigma_{yx}\,dy\,dz\frac{dz}{2} - \sigma_{xz}\,dy\,dz\frac{dy}{2}$ |
| $AEFB$ | $\sigma_{yy}\,dx\,dz\frac{dz}{2}$ |
| $ABCD$ | $-\sigma_{zz}\,dx\,dy\frac{dy}{2}$ |
| $FGHE$ | $(\sigma_{zz} + d\sigma_{zz})\,dx\,dy\frac{dy}{2} - (\sigma_{zy} + d\sigma_{zy})\,dx\,dy\,dz$ |

stress components rather than stress vectors. First consider the moment of these forces about the $x$ axis. In order not to obscure the figure when applying the condition

$$\sum M_x = 0 \tag{2-3}$$

only those stresses will be shown which are neither parallel to $Ox$ nor intersecting it. Therefore the stresses that do not contribute to Eq. (2-3) are not shown in Fig. 2-3. Recalling that the quantities appearing in Fig. 2-3 are actually forces per unit area or per unit volume, we must multiply them by the appropriate area or volume. The moments about the $x$ axis of the forces acting on all faces of the element are shown in Table 2-1.

Summing the expressions in Table 2-1 and adding moments due to the body forces, we obtain

$$\begin{aligned}
&(\sigma_{yz} + d\sigma_{yz})\,dx\,dy\,dz - (\sigma_{yy} + d\sigma_{yy})\,dx\,dz\frac{dz}{2} + (\sigma_{xz} + d\sigma_{xz})\,dy\,dz\frac{dy}{2} \\
&\quad - (\sigma_{xy} + d\sigma_{xy})\,dy\,dz\frac{dz}{2} - \sigma_{zz}\,dx\,dy\frac{dy}{2} + (\sigma_{zz} + d\sigma_{zz})\,dx\,dy\frac{dy}{2} \\
&\quad - (\sigma_{zy} + d\sigma_{zy})\,dx\,dy\,dz - \sigma_{xz}\,dy\,dz\frac{dy}{2} + \sigma_{yx}\,dy\,dz\frac{dz}{2} + \sigma_{yy}\,dx\,dz\frac{dz}{2} \\
&\quad - \sigma_{zz}\,dx\,dy\frac{dy}{2} - b_y\,dx\,dy\,dz\frac{dz}{2} + b_z\,dx\,dy\,dz\frac{dy}{2} = 0
\end{aligned}$$

Next we substitute these expressions into Eq. (2-3). On dividing this equation through by $dx\,dy\,dz$ and passing to the limit $dx \to 0$, $dy \to 0$, $dz \to 0$, we obtain

$$\sigma_{yz} = \sigma_{zy} \tag{2-4}$$

This procedure can be repeated to satisfy the equilibrium conditions

$$\sum M_y = 0, \qquad \sum M_z = 0$$

which lead to the following relations:

$$\sigma_{yx} = \sigma_{xy}, \qquad \sigma_{zx} = \sigma_{xz} \tag{2-5}$$

We see from Eqs. (2-4) and (2-5) that the stress matrix, Eq. (2-2), is symmetric and has the following final form:

$$S = \begin{bmatrix} \sigma_{xx} & \sigma_{xy} & \sigma_{xz} \\ \sigma_{xy} & \sigma_{yy} & \sigma_{yz} \\ \sigma_{xz} & \sigma_{yz} & \sigma_{zz} \end{bmatrix} \tag{2-6}$$

## EXAMPLE 2-2

A narrow, deep beam with rectangular cross section ($b < h$) is subjected to a uniform line load of intensity $p =$ constant per unit length (Fig. 2-4). The beam is supported at the ends by the vertical forces $V = pl/2$. The stress components in the beam have been found to be

$$\sigma_{xx} = -\frac{py}{5bh^3}[30x(x-l) - 20y^2 + 3h^2] \tag{a}$$

$$\sigma_{yy} = -\frac{p}{2bh^3}(y+h)(2y-h)^2 \tag{b}$$

$$\sigma_{xy} = \frac{3p}{2bh^3}\left(\frac{x-l}{2}\right)(4y^2 - h^2) \tag{c}$$

$$\sigma_{zz} = \sigma_{xz} = \sigma_{yz} = 0 \tag{d}$$

Determine and plot stress components along the planes $y = 0$, $x = l/4$, and $x = l/2$ assuming $h/l = 1$.

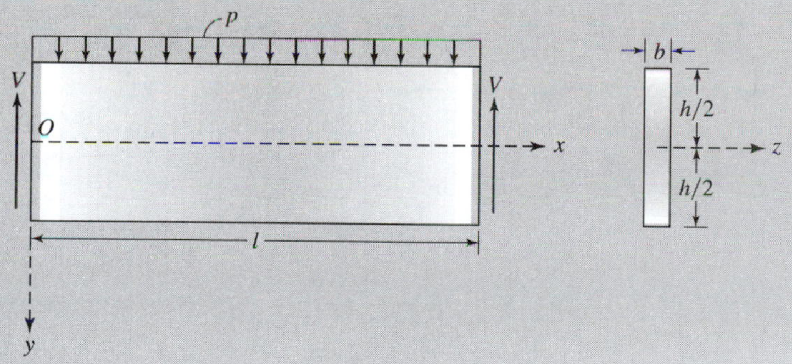


Figure 2-4

## SOLUTION

**1.** $y = 0$. This plane is normal to the $Oy$ axis, hence the only stresses present here are $\sigma_{yy}$, $\sigma_{xy}$, and $\sigma_{yz}$. From Eq. ($b$),

$$\sigma_{yy} = \frac{-p}{2b} = \text{constant}$$

From Eq. ($c$),

$$\sigma_{xy} = \frac{-3p(x - l/2)}{2bh}$$

From Eq. ($d$),

$$\sigma_{yz} = 0$$

The stresses $\sigma_{yy}$ and $\sigma_{xy}$ are plotted in Fig. 2-5$a$ and $b$.

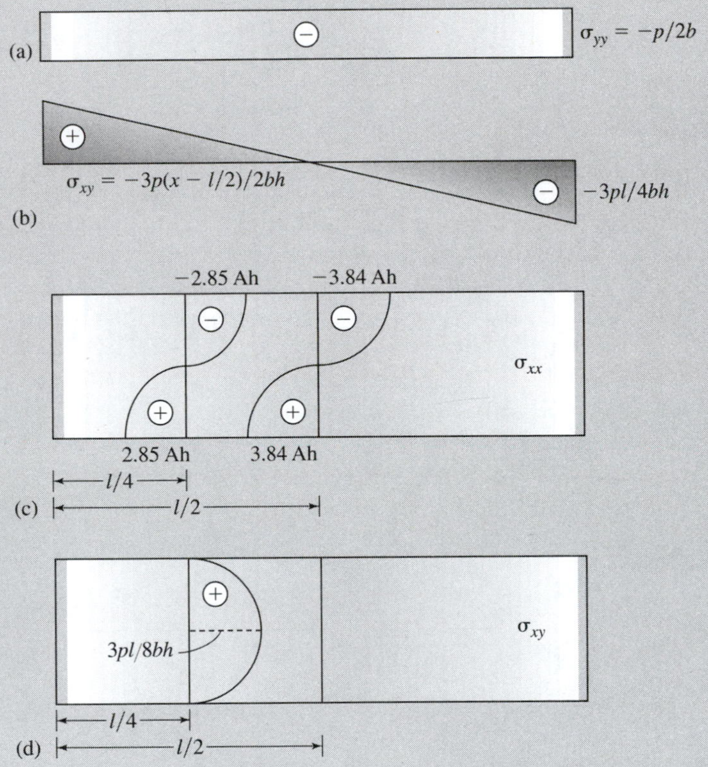


**Figure 2-5**

2. $x = l/4$. Since the $Ox$ axis is normal to this plane, the stresses appearing here are $\sigma_{xx}$, $\sigma_{xy}$, and $\sigma_{xz}$. From Eq. (*a*),

$$\sigma_{xx} = \frac{-py[-(15l/2)(3l/4) - 20y^2 + 3h^2]}{5bh^3}$$
$$= \frac{-py(-45l^2/8 - 20y^2 + 3h^2)}{5bh^3}$$

From Eq. (*c*),

$$\sigma_{xy} = \frac{-3pl(4y^2 - h^2)}{8bh^3}$$

From Eq. (*d*),

$$\sigma_{xz} = 0$$

The stresses $\sigma_{xx}$ and $\sigma_{xy}$ are plotted in Fig. 2-5*c* and *d*.

3. $x = l/2$. This case is analogous to case 2. From Eq. (*a*),

$$\sigma_{xx} = \frac{-py[-15l(l/2) - 20y^2 + 3h^2]}{5bh^3}$$
$$= \frac{-py[-(15l^2/2) - 20y^2 + 3h^2]}{5bh^3}$$

From Eq. (*c*),

$$\sigma_{xy} = 0$$

From Eq. (*d*),

$$\sigma_{xz} = 0$$

The stress $\sigma_{xx}$ is plotted in Fig. 2-5*c*.

## 2-2-3 Stresses Acting on an Inclined Plane

When analyzing stress we are usually interested in the largest values of its components. These do not occur exclusively on planes normal to $Ox$, $Oy$, or $Oz$. Then it is very often desirable to find the stress components acting on a surface whose normal is not parallel to any of the coordinate axes $Ox$, $Oy$, or $Oz$ and to relate them to the stress components defined in Section 2-2-1. Let us consider an infinitesimal tetrahedron as shown in Fig. 2-6*a*. It can be interpreted as a corner cut from the element analyzed in Section 2-2-2.

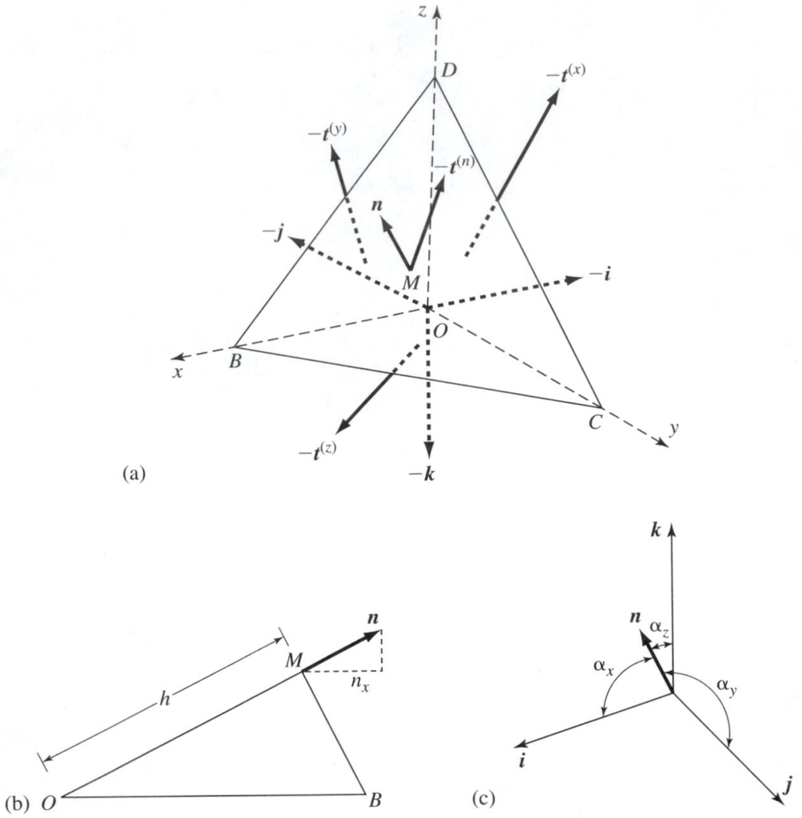


Figure 2-6 Stresses on an inclined plane

Recalling that a vector normal to any plane in space uniquely defines this plane, let us introduce a unit vector $\boldsymbol{n}$ normal to the inclined plane *BCD*. It is assumed that it is pointing away from the tetrahedron and is attached at a point *M* such that *OM* will be collinear with $\boldsymbol{n}$. The remaining faces of the tetrahedron are characterized by the unit outer normal vectors $-\boldsymbol{i}$, $-\boldsymbol{j}$, and $-\boldsymbol{k}$ with directions opposite to *Ox*, *Oy*, and *Oz*. To complete the geometrical description of the tetrahedron we denote by $n_x, n_y, n_z$ the cosines of the angles $\alpha_x, \alpha_y, \alpha_z$ (Fig. 2-6*c*),

$$
\begin{aligned}
n_x &= \boldsymbol{n} \cdot \boldsymbol{i} = |\boldsymbol{n}||\boldsymbol{i}| \cos \alpha_x \\
n_y &= \boldsymbol{n} \cdot \boldsymbol{j} = |\boldsymbol{n}||\boldsymbol{j}| \cos \alpha_y \\
n_z &= \boldsymbol{n} \cdot \boldsymbol{k} = |\boldsymbol{n}||\boldsymbol{k}| \cos \alpha_z
\end{aligned}
\tag{2-7}
$$

The quantities $n_x$, $n_y$, and $n_z$ identifying the orientation of the inclined surface *BCD* in terms of its outer unit normal vector $\boldsymbol{n}$ are known as *direction cosines*.

We also maintain that

$$\begin{aligned}\Delta A_1 &= \Delta A_{OCD} = n_x\,\Delta A\\ \Delta A_2 &= \Delta A_{BOD} = n_y\,\Delta A\\ \Delta A_3 &= \Delta A_{BOC} = n_z\,\Delta A\end{aligned} \tag{2-8}$$

where $\Delta A$ is the area of $BCD$.

To prove relations (2-8) we first note that $\measuredangle BMO$, $\measuredangle CMO$, and $\measuredangle DMO$ are right angles and that therefore [see also Eqs. (2-7) and Fig. 2-6*b*]

$$h = n_x(OB) = n_y(OC) = n_z(OD) \tag{2-9}$$

On the other hand the volume $\Delta V$ of the tetrahedron can be calculated in four different ways taking in turn each face as the base. Hence,

$$\Delta V = \tfrac{1}{3}h\Delta A = \tfrac{1}{3}(OB)\Delta A_1 = \tfrac{1}{3}(OC)\Delta A_2 = \tfrac{1}{3}(OD)\Delta A_3 \tag{2-10}$$

Eqs. (2-9) can now be substituted into Eqs. (2-10) to give

$$h\,\Delta A = \frac{h}{n_x}\Delta A_1 = \frac{h}{n_y}\Delta A_2 = \frac{h}{n_z}\Delta A_3 \tag{2-11}$$

and Eqs. (2-8) follow directly. Let $\boldsymbol{t}^{(n)}$ be the stress vector acting on $BCD$ and let $\boldsymbol{t}^{(-x)}$, $\boldsymbol{t}^{(-y)}$, and $\boldsymbol{t}^{(-z)}$ be the stress vectors applied to the remaining faces (see Section 2-2-1). The condition of equilibrium of the tetrahedron takes the following form:

$$\boldsymbol{t}^{(n)}\Delta A + \boldsymbol{t}^{(-x)}\Delta A_1 + \boldsymbol{t}^{(-y)}\Delta A_2 + \boldsymbol{t}^{(-z)}\Delta A_3 + \frac{h}{3}\Delta A\,\boldsymbol{b} = 0 \tag{2-12}$$

where the last term is the body force. In the limit, with $h \to 0$ and on substituting Eqs. (2-8) we find that

$$\boldsymbol{t}^{(n)} = \boldsymbol{t}^{(x)}n_x + \boldsymbol{t}^{(y)}n_y + \boldsymbol{t}^{(z)}n_z \tag{2-13}$$

To write this vector equation in the component form

$$\boldsymbol{t}^{(n)} = t_x^{(n)}\boldsymbol{i} + t_y^{(n)}\boldsymbol{j} + t_z^{(n)}\boldsymbol{k}$$

we dot it with $\boldsymbol{i}$, $\boldsymbol{j}$, or $\boldsymbol{k}$, obtaining the following expressions:

$$\begin{aligned}t_x^{(n)} &= \boldsymbol{t}^{(n)}\cdot\boldsymbol{i} = \sigma_{xx}n_x + \sigma_{xy}n_y + \sigma_{xz}n_z\\ t_y^{(n)} &= \boldsymbol{t}^{(n)}\cdot\boldsymbol{j} = \sigma_{xy}n_x + \sigma_{yy}n_y + \sigma_{yz}n_z\\ t_z^{(n)} &= \boldsymbol{t}^{(n)}\cdot\boldsymbol{k} = \sigma_{xz}n_x + \sigma_{yz}n_y + \sigma_{zz}n_z\end{aligned} \tag{2-14}$$

where relations (2-1) have also been used. Finally we can put Eqs. (2-14) in matrix form,

$$\boldsymbol{t}^{(n)} = \boldsymbol{S}\boldsymbol{n}$$

where $\boldsymbol{t}^{(n)}$ and $\boldsymbol{n}$ are column matrices,

$$\boldsymbol{t}^{(n)} = \begin{Bmatrix} t_x^{(n)}\\ t_y^{(n)}\\ t_z^{(n)}\end{Bmatrix}, \qquad \boldsymbol{n} = \begin{Bmatrix} n_x\\ n_y\\ n_z\end{Bmatrix} \tag{2-15}$$

and where $\boldsymbol{S}$ is defined by Eq. (2-6).

## 2-2-4 Normal and Tangential Stresses—Stress Boundary Conditions

Relations (2-13) and (2-14) have several applications. Let us first express the projection $\sigma_n$ of the stress vector $\boldsymbol{t}^{(n)}$ on the normal $\boldsymbol{n}$ in terms of the stress components (2-1). On taking the dot product of Eq. (2-13) with $\boldsymbol{n}$ we get

$$\sigma_n = \boldsymbol{t}^{(n)} \cdot \boldsymbol{n} = (\boldsymbol{t}^{(x)} \cdot \boldsymbol{n})n_x + (\boldsymbol{t}^{(y)} \cdot \boldsymbol{n})n_y + (\boldsymbol{t}^{(z)} \cdot \boldsymbol{n})n_z \tag{2-16}$$

It is, however, known from Eq. (2-7) that

$$\boldsymbol{n} = n_x\boldsymbol{i} + n_y\boldsymbol{j} + n_z\boldsymbol{k}$$

Substituting into Eq. (2-16) and using again Eq. (2-1) yields the following result:

$$\sigma_n = \sigma_{xx}n_x^2 + \sigma_{yy}n_y^2 + \sigma_{zz}n_z^2 + 2\sigma_{xy}n_xn_y + 2\sigma_{xz}n_xn_z + 2\sigma_{yz}n_yn_z \tag{2-17}$$

Thus Eq. (2-17) relates the stress component normal to an arbitrarily inclined plane at a point to the stress components parallel to the coordinate axes $Ox$, $Oy$, and $Oz$ at the same point.

When the loads are applied to a portion of the surface of a solid, then the left-hand sides of Eqs. (2-14) are known. These equations then relate applied stress to internal stress and are called *stress boundary conditions*. This will be illustrated in the next example.

### EXAMPLE 2-3

Consider a solid in the form of a very long rod (long in the $z$ direction) whose cross section is a quarter-circle, as shown in Fig. 2-7. The rod is resting on a rigid foundation as seen in the figure and is subjected to loads independent of $z$. The stress along $OB$ is unknown, hence it cannot be present in the boundary conditions on $OB$. Here the boundary conditions are imposed on displacements, unless one assumes that the rod can slide freely along $OB$ and $Oz$, in which case the shear stresses $\sigma_{xy}$ and $\sigma_{zy}$ must be zero. The load $p(y)$ on $OC$ varies linearly according to

$$p(y) = p_0\frac{R - y}{R} \tag{2-18}$$

Now let us consider any infinitesimal portion of $OC$ near an arbitrarily located point $M$ (Figs. 2-7 and 2-8). Let the face $BCD$ of the tetrahedron in Fig. 2-6 coincide with the vertical surface in Fig. 2-8, where a portion of $BCD$ is shown. Since the normal $\boldsymbol{n}$ must be directed as shown in the figure, we have [see Eq. (2-6)]

$$n_x = -1, \qquad n_y = n_z = 0 \tag{2-19}$$

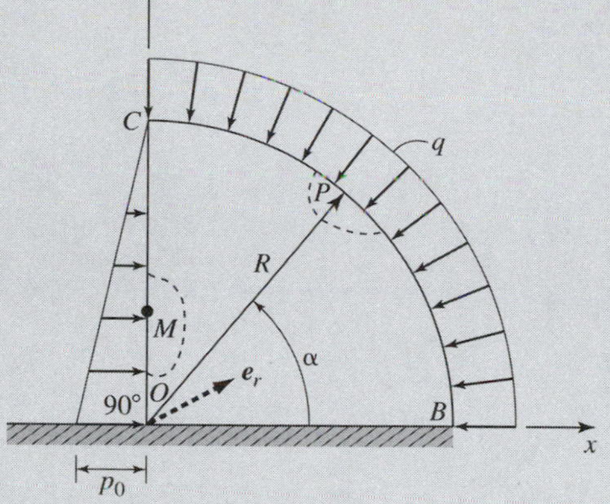


**Figure 2-7**

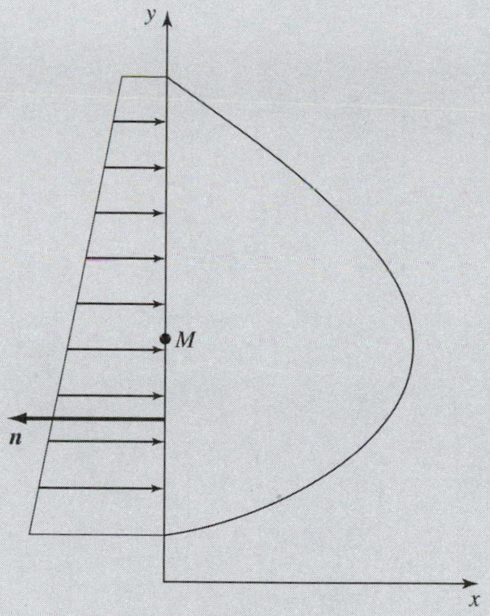


**Figure 2-8**

As the quantity $p(y)$ from Eq. (2-18) represents the magnitude of the applied stress vector $\boldsymbol{t}^{(n)}$, we have

$$\boldsymbol{t}^{(n)} = p(y)\boldsymbol{i}$$

Therefore the components of both sides of this expression are given by

$$t_x^{(n)} = p(y), \qquad t_y^{(n)} = t_z^{(n)} = 0 \tag{2-20}$$

When Eqs. (2-18), (2-19), and (2-20) are used in Eqs. (2-14), we obtain

$$\sigma_{xx} = -p_0\frac{R-y}{R}, \qquad \sigma_{xy} = 0, \qquad \sigma_{xz} = 0 \tag{2-21}$$

These are the boundary conditions to be imposed on $\sigma_{xx}$, $\sigma_{xy}$, and $\sigma_{xz}$ along $OC$. Proceeding in a similar manner with an infinitesimal segment of the circular arc $BC$ about a point $P$ (Fig. 2-9), we align the face $BCD$ of the tetrahedron with the cut-out portion of the arc. Due to its negligible dimensions it can be treated as completely flat. Let us introduce a local coordinate system $r$, $\theta$, $z$, denoting by $\alpha$ the angle between $OB$ and $OP$. This provides

$$n_x = \cos\alpha, \qquad n_y = \sin\alpha, \qquad n_z = 0 \tag{2-22}$$

Since now $\boldsymbol{t}^{(n)} = -q\boldsymbol{e}_r$, where $\boldsymbol{e}_r$ is the unit vector in the radial direction, we get

$$\begin{aligned} t_x^{(n)} &= \boldsymbol{t}^{(n)} \cdot \boldsymbol{i} = -q\boldsymbol{e}_r \cdot \boldsymbol{i} = -q\cos\alpha \\ t_y^{(n)} &= \boldsymbol{t}^{(n)} \cdot \boldsymbol{j} = -q\boldsymbol{e}_r \cdot \boldsymbol{j} = -q\sin\alpha \\ t_z^{(n)} &= \boldsymbol{t}^{(n)} \cdot \boldsymbol{k} = -q\boldsymbol{e}_r \cdot \boldsymbol{k} = 0 \end{aligned} \tag{2-23}$$

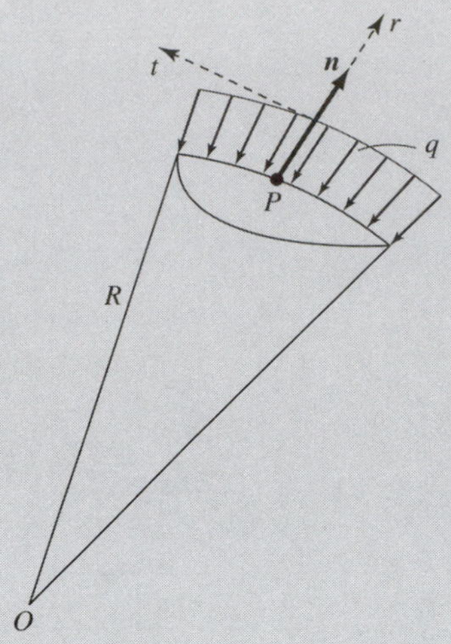


**Figure 2-9** Circular portion of boundary

Finally, on substituting Eqs. (2-22) and (2-23) into Eqs. (2-14) we obtain the following boundary conditions:

$$\begin{aligned} \sigma_{xx} \cos\alpha + \sigma_{xy} \sin\alpha &= -q \cos\alpha \\ \sigma_{xy} \cos\alpha + \sigma_{yy} \sin\alpha &= -q \sin\alpha \\ \sigma_{xz} \cos\alpha + \sigma_{yz} \sin\alpha &= 0 \end{aligned} \tag{2-24}$$

Equation (2-17) gives $\sigma_n$, the magnitude of the projection of the stress vector $\boldsymbol{t}^{(n)}$ on the normal $\boldsymbol{n}$ to an inclined plane, in terms of the direction cosines of $\boldsymbol{n}$ and the Cartesian components of stress. Another component of the vector $\boldsymbol{t}^{(n)}$, called the *tangential* stress, lies on the inclined plane. It is seen in Fig. 2-10 that

$$|\boldsymbol{t}^{(n)}|^2 = \tau^2 + \sigma_n^2$$

Therefore,

$$\tau = \left[\boldsymbol{t}^{(n)} \cdot \boldsymbol{t}^{(n)} - \sigma_n^2\right]^{1/2} \tag{2-25}$$

With Eq. (2-13) this becomes

$$\tau = \left[ \left(t_x^{(n)}\right)^2 + \left(t_y^{(n)}\right)^2 + \left(t_z^{(n)}\right)^2 - \sigma_n^2\right]^{1/2} \tag{2-26}$$

Equations (2-14) can be now substituted into Eq. (2-26) to give

$$\begin{aligned} \tau = \big[&(\sigma_{xx}n_x + \sigma_{xy}n_y + \sigma_{xz}n_z)^2 + (\sigma_{xy}n_x + \sigma_{yy}n_y + \sigma_{yz}n_z)^2 \\ &+ (\sigma_{xz}n_x + \sigma_{yz}n_y + \sigma_{zz}n_z)^2 - \sigma_n^2\big]^{1/2} \end{aligned} \tag{2-27}$$

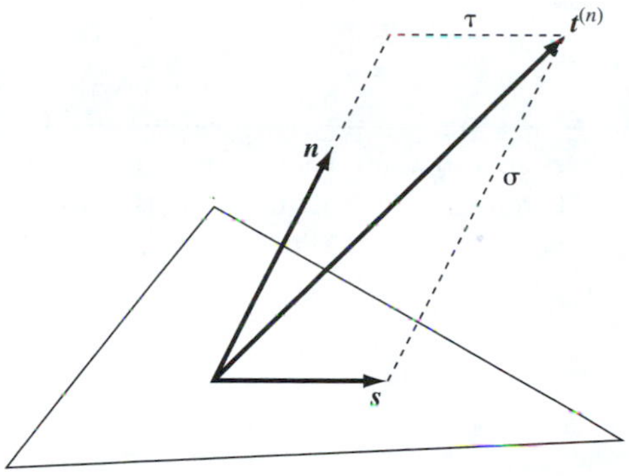


Figure 2-10 Tangential stress

The next step is to find the direction of the tangential component $\tau$. To this end we denote by $\boldsymbol{s}$ the unit vector acting in the direction of the projection of $\boldsymbol{t}^{(n)}$ on the inclined plane and we replace the vector $\boldsymbol{t}^{(n)}$ by its new decomposition

$$\boldsymbol{t}^{(n)} = \sigma_n \boldsymbol{n} + \tau \boldsymbol{s} \tag{2-28}$$

On substituting this into the equilibrium Eq. (2-13) we get

$$\sigma_n \boldsymbol{n} + \tau \boldsymbol{s} = \boldsymbol{t}^{(x)} n_x + \boldsymbol{t}^{(y)} n_y + \boldsymbol{t}^{(z)} n_z \tag{2-29}$$

Solving Eq. (2-29) for $\boldsymbol{s}$ yields

$$\boldsymbol{s} = \left(\boldsymbol{t}^{(x)} n_x + \boldsymbol{t}^{(y)} n_y + \boldsymbol{t}^{(z)} n_z - \sigma_n \boldsymbol{n}\right)/\tau \tag{2-30}$$

The direction cosines, which are the components $s_x$, $s_y$, and $s_z$ of $\boldsymbol{s}$, are readily evaluated by multiplying Eq. (2-30) by $\boldsymbol{i}, \boldsymbol{j}$, or $\boldsymbol{k}$. This generates the following expressions:

$$\begin{aligned} s_x &= (\sigma_{xx} n_x + \sigma_{xy} n_y + \sigma_{xz} n_z - \sigma_n n_x)/\tau \\ s_y &= (\sigma_{xy} n_x + \sigma_{yy} n_y + \sigma_{yz} n_z - \sigma_n n_y)/\tau \\ s_z &= (\sigma_{xz} n_z + \sigma_{yz} n_y + \sigma_{zz} n_z - \sigma_n n_z)/\tau \end{aligned} \tag{2-31}$$

Here we have used Eqs. (2-1), that is, $\sigma_{xx} = \boldsymbol{t}^{(x)} \cdot \boldsymbol{i}, \ldots,$ and Eqs. (2-7), that is, $n_x = \boldsymbol{n} \cdot \boldsymbol{i}, \ldots.$ Equations (2-31) can also be put in matrix form,

$$\boldsymbol{s} = (\boldsymbol{S} - \sigma_n \boldsymbol{I})\boldsymbol{n}/\tau \tag{2-32}$$

Here $\boldsymbol{s}$ is a column matrix whose elements are direction cosines of the vector $\boldsymbol{s}$, and $\boldsymbol{I}$ is the unit matrix,

$$\boldsymbol{s} = \begin{Bmatrix} s_x \\ s_y \\ s_z \end{Bmatrix}, \qquad \boldsymbol{I} = \begin{bmatrix} 1 & 0 & 0 \\ 0 & 1 & 0 \\ 0 & 0 & 1 \end{bmatrix} \tag{2-33}$$

The matrices $\boldsymbol{S}$ and $\boldsymbol{n}$ have been defined by Eqs. (2-6) and (2-15), respectively.

## EXAMPLE 2-4

Assume that the stresses in a solid are known functions of the Cartesian coordinates $x$, $y$, and $z$ (Fig. 2-11). Then their values at any point $M$ can be readily calculated by substituting the coordinates $x_M$, $y_M$, and $z_M$ of $M$ for $x$, $y$, and $z$. Let the stress matrix obtained for a specific point $M$ be

$$\mathbf{S} = \begin{bmatrix} 10 & 3 & 5 \\ 3 & 1 & 2 \\ 5 & 2 & 4 \end{bmatrix} \times 10^3 \text{ psi} \tag{a}$$

Consider also a plane through $M$ characterized by the direction cosines

$$n_x = \frac{\sqrt{3}}{2}, \qquad n_y = \frac{\sqrt{2}}{4}, \qquad n_z = -\frac{\sqrt{2}}{4} \tag{b}$$

1. Find the normal component of the traction $\sigma_n$ and its tangential component $\tau$ acting at $M$ on this plane.
2. Find the direction cosines of $\sigma_n$ and of $\tau$.
3. Find the magnitude of the traction vector $\boldsymbol{t}^{(n)}$ and its angle with the normal $\boldsymbol{n}$.

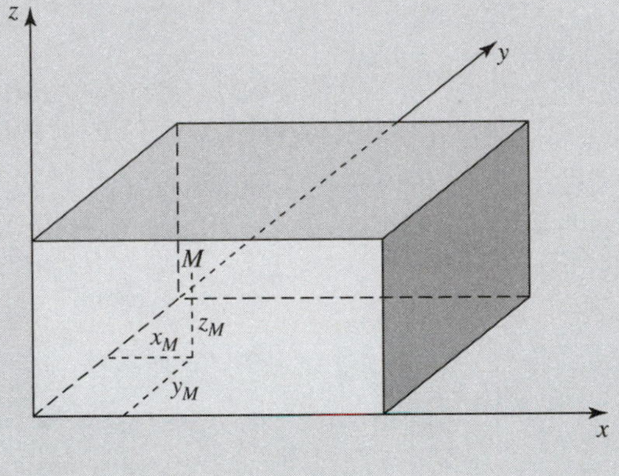


**Figure 2-11**

### SOLUTION

1. In order to calculate $\sigma_n$ we use Eq. (2-17), which gives

$$\sigma_n = \left[10 \times \frac{3}{4} + 1 \times \frac{1}{8} + 4 \times \frac{1}{8} + 2 \times 3\left(\frac{\sqrt{3}}{2}\frac{\sqrt{2}}{4}\right) - 2 \times 5\left(\frac{\sqrt{3}}{2}\frac{\sqrt{2}}{4}\right) - 2 \times 2\left(\frac{\sqrt{2}}{4}\frac{\sqrt{2}}{4}\right)\right] \times 10^3 \text{ psi}$$
$$= 6.401 \times 10^3 \text{ psi} \tag{c}$$

Next the magnitude of the shear stress vector $\tau$ is found from Eq. (2-27),

$$\tau = \left[\left(10\times\frac{\sqrt{3}}{2}+3\times\frac{\sqrt{2}}{4}-5\times\frac{\sqrt{2}}{4}\right)^2+\left(3\times\frac{\sqrt{3}}{2}+1\times\frac{\sqrt{2}}{4}-2\times\frac{\sqrt{2}}{4}\right)^2\right.$$

$$\left.+\left(5\times\frac{\sqrt{3}}{2}+2\times\frac{\sqrt{2}}{4}-4\times\frac{\sqrt{2}}{4}\right)^2-6.401^2\right]^{1/2}\times 10^3\text{ psi}$$

$$= 6.330\times 10^3\text{ psi} \tag{d}$$

**2.** The direction cosines of the shear stress vector $\tau$ are evaluated from Eqs. (2-31) and Eqs. (*c*) and (*d*),

$$s_x = \left(10\times\frac{\sqrt{3}}{2}+3\times\frac{\sqrt{2}}{4}-5\times\frac{\sqrt{2}}{4}-6.401\frac{\sqrt{3}}{2}\right)\Big/6.360 = 0.379$$

$$s_y = \left(3\times\frac{\sqrt{3}}{2}+1\times\frac{\sqrt{2}}{4}-2\times\frac{\sqrt{2}}{4}-6.401\frac{\sqrt{2}}{4}\right)\Big/6.360 = -0.011 \tag{e}$$

$$s_z = \left(5\times\frac{\sqrt{3}}{2}+2\times\frac{\sqrt{2}}{4}-4\times\frac{\sqrt{2}}{4}+6.401\frac{\sqrt{2}}{4}\right)\Big/6.360 = 0.925$$

The accuracy of these results can be verified by using the condition

$$s_x^2+s_y^2+s_z^2=1 \tag{f}$$

**3.** In order to calculate the magnitude of the vector $\boldsymbol{t}^{(n)}$,

$$|\boldsymbol{t}^{(n)}| = \left[\left(t_x^{(n)}\right)^2+\left(t_y^{(n)}\right)^2+\left(t_z^{(n)}\right)^2\right]^{1/2} \tag{g}$$

we use Eqs. (2-14). Thus,

$$|\boldsymbol{t}^{(n)}| = [(\sigma_{xx}n_x+\sigma_{xy}n_y+\sigma_{xz}n_z)^2+(\sigma_{xy}n_x+\sigma_{yy}n_y+\sigma_{yz}n_z)^2$$

$$+(\sigma_{xz}n_x+\sigma_{yz}n_y+\sigma_{zz}n_z)^2]^{1/2}$$

$$= \left[\left(10\times\frac{\sqrt{3}}{2}+3\times\frac{\sqrt{2}}{4}-5\times\frac{\sqrt{2}}{4}\right)^2+\left(3\times\frac{\sqrt{3}}{2}+1\times\frac{\sqrt{2}}{4}-2\times\frac{\sqrt{2}}{4}\right)^2\right.$$

$$\left.+\left(5\times\frac{\sqrt{3}}{2}+2\times\frac{\sqrt{2}}{4}-4\times\frac{\sqrt{2}}{4}\right)^2\right]^{1/2}$$

$$= 9.023\times 10^3\text{ psi} \tag{h}$$

The next step is to evaluate the angle between the traction vector $\boldsymbol{t}^{(n)}$ and the normal $\boldsymbol{n}$. To this end we use the formula

$$
\begin{aligned}
\cos(\boldsymbol{t}^{(n)}, \boldsymbol{n}) &= \left(t_x^{(n)}\boldsymbol{i} + t_y^{(n)}\boldsymbol{j} + t_z^{(n)}\boldsymbol{k}\right) \cdot \boldsymbol{n} / \left|\boldsymbol{t}^{(n)}\right| \\
&= \left(t_x^{(n)} n_x + t_y^{(n)} n_y + t_z^{(n)} n_z\right) / \left|\boldsymbol{t}^{(n)}\right| \\
&= [(\sigma_{xx} n_x + \sigma_{xy} n_y + \sigma_{xz} n_z) n_x + (\sigma_{xy} n_x + \sigma_{yy} n_y + \sigma_{yz} n_z) n_y \\
&\quad + (\sigma_{xz} n_x + \sigma_{yz} n_y + \sigma_{zz} n_z) n_z] / \left|\boldsymbol{t}^{(n)}\right| \\
&= \left[\left(10 \times \frac{\sqrt{3}}{2} + 3 \times \frac{\sqrt{2}}{4} - 5 \times \frac{\sqrt{2}}{4}\right)\frac{\sqrt{3}}{2}\right. \\
&\quad + \left(3 \times \frac{\sqrt{3}}{2} + 1 \times \frac{\sqrt{2}}{4} - 2 \times \frac{\sqrt{2}}{4}\right)\frac{\sqrt{2}}{4} \\
&\quad \left. - \left(5 \times \frac{\sqrt{3}}{2} + 2 \times \frac{\sqrt{2}}{4} - 4 \times \frac{\sqrt{2}}{4}\right)\frac{\sqrt{2}}{4}\right] \Big/ 9.023 \\
&= 0.7093 = 40.6^\circ \qquad (i)
\end{aligned}
$$

In arriving at Eq. ($i$), Eqs. (2-7) and (2-17) have been used.

## 2-2-5 Transformation of Stress Components—Stress as a Tensor

It is often necessary or convenient to change the coordinate system and then refer the stress components to this new system, as shown in the following examples.

1. *Sheet of fiber-reinforced composite.* Assume that we have a rectangular lamina reinforced by the parallel fibers in a direction not parallel to the sides of the lamina. The basic Cartesian coordinate system has axes along the sides of the lamina. To determine stresses in the fibers we have to use another Cartesian coordinate system with one axis in the direction of the fibers.
2. *Thin-walled pressure vessel.* Assume that we have a weld spiraling around a tank. To analyze the integrity of the tank we need to calculate the stresses tangent and normal to the weld.
3. *Glue joint in a wooden board.* Such a joint is in general inclined. We again are interested in the stresses in the joint, hence the need of an auxiliary coordinate system.

The first step toward the goal of changing the coordinate system was already made in Section 2-2-4. Let the direction of the normal $\boldsymbol{n}$ to the previously chosen inclined plane define one of the new coordinate axes, say, $Ox'$, and let us introduce the remaining two axes, $Oy'$ and $Oz'$, in directions perpendicular to $Ox'$. These lie therefore in a plane parallel to the face $BCD$ of the tetrahedron in Fig. 2-6. By merely changing some symbols, we can make Eq. (2-17) serve as a template for the normal stress in the direction $Ox'$. To this end we denote

$$\sigma_{x'x'} = \sigma_n, \qquad n_{x'x} = n_x, \qquad n_{x'y} = n_y, \qquad n_{x'z} = n_z, \qquad \boldsymbol{i} = \boldsymbol{n} \tag{2-34}$$

observing that $n_x, n_y$, and $n_z$ are direction cosines (see Section 2-2-3) of the angles between the new axes $x'$, $y'$, $z'$ and the old ones, $x$, $y$, $z$. We obtain immediately

$$\sigma_{x'x'} = \sigma_{xx}n_{x'x}^2 + \sigma_{yy}n_{x'y}^2 + \sigma_{zz}n_{x'z}^2 + 2\sigma_{xy}n_{x'x}n_{x'y} + 2\sigma_{xz}n_{x'x}n_{x'z} + 2\sigma_{yz}n_{x'y}n_{x'z} \tag{2-35}$$

The remaining stress components, $\sigma_{x'y'}$ and $\sigma_{x'z'}$, acting on the same plane, are readily evaluated following the procedure from Section 2-2-4. First introduce the unit vectors $\boldsymbol{j}'$ and $\boldsymbol{k}'$ along the axes $Oy'$ and $Oz'$. Next take the dot product of the vector

$$\boldsymbol{t}^{(n)} = t_x^{(n)}\boldsymbol{i} + t_y^{(n)}\boldsymbol{j} + t_z^{(n)}\boldsymbol{k} \tag{2-36}$$

with $\boldsymbol{j}'$ and $\boldsymbol{k}'$,

$$\boldsymbol{t}^{(n)} \cdot \boldsymbol{j}' = \sigma_{x'y'}, \qquad \boldsymbol{t}^{(n)} \cdot \boldsymbol{k}' = \sigma_{x'z'}, \qquad \boldsymbol{i} \cdot \boldsymbol{j}' = n_{xy'}, \qquad \boldsymbol{j} \cdot \boldsymbol{j}' = n_{yy'}$$

and so on. Finally, on substituting expressions (2-14) for $t_x^{(n)}, t_y^{(n)}, t_z^{(n)}$, using Eqs. (2-34), and simplifying the results, we obtain

$$\begin{aligned} \sigma_{x'y'} = {} & \sigma_{xx}n_{x'x}n_{xy'} + \sigma_{yy}n_{x'y}n_{yy'} + \sigma_{zz}n_{x'z}n_{zy'} + \sigma_{xy}(n_{x'y}n_{xy'} + n_{x'x}n_{yy'}) \\ & + \sigma_{xz}(n_{x'z}n_{xy'} + n_{zy'}n_{x'x}) + \sigma_{yz}(n_{x'z}n_{yy'} + n_{x'y}n_{zy'}) \end{aligned} \tag{2-37}$$

$$\begin{aligned} \sigma_{x'z'} = {} & \sigma_{xx}n_{x'x}n_{xz'} + \sigma_{yy}n_{x'y}n_{yz'} + \sigma_{zz}n_{x'z}n_{zz'} + \sigma_{xy}(n_{x'y}n_{xz'} + n_{x'x}n_{yz'}) \\ & + \sigma_{xz}(n_{x'z}n_{xz'} + n_{xx'}n_{zz'}) + \sigma_{yz}(n_{x'z}n_{yz'} + n_{x'y}n_{zz'}) \end{aligned} \tag{2-38}$$

Note that $n_{xy'} = n_{y'x}$, and so on. Also note that if the inclined plane were perpendicular to $Oy'$ or to $Oz'$, then the remaining stress components, $\sigma_{y'y'}$, $\sigma_{z'z'}$, $\sigma_{y'z'}$, could be obtained from Eqs. (2-36), (2-37), and (2-38) by replacing $x'$ by $y'$ or by $z'$, and so on. When written in matrix form, the relationship between the stress components in the $Ox'y'z'$ coordinate system and in the $Oxyz$ coordinate system becomes

$$\boldsymbol{S}' = \boldsymbol{N}^T\boldsymbol{S}\boldsymbol{N} \tag{2-39}$$

where $\boldsymbol{N}^T$ is the transpose of the matrix $\boldsymbol{N}$, and

$$\boldsymbol{S}' = \begin{bmatrix} \sigma_{x'x'} & \sigma_{x'y'} & \sigma_{x'z'} \\ \sigma_{x'y'} & \sigma_{y'y'} & \sigma_{y'z'} \\ \sigma_{x'z'} & \sigma_{y'z'} & \sigma_{z'z'} \end{bmatrix}, \qquad \boldsymbol{N} = \begin{bmatrix} n_{x'x} & n_{y'x} & n_{z'x} \\ n_{x'y} & n_{y'y} & n_{z'y} \\ n_{x'z} & n_{y'z} & n_{z'z} \end{bmatrix} \tag{2-40}$$

An inspection of Eqs. (2-35), (2-37), and (2-38) discloses a certain regularity in their structures. To make it more obvious let us use the following notation, known as *subscript notation*. The axes $x$, $y$, and $z$ will be denoted by $x_1$, $x_2$, and $x_3$, respectively, whereas the axes $x'$, $y'$, and $z'$ become $x_1'$, $x_2'$, and $x_3'$. We also change the notation for the stress components. The subscripts $x$, $y$, and $z$ are replaced by 1, 2, and 3, respectively, and so are the subscipts $x'$, $y'$, and $z'$. To distinguish between, say, $\sigma_{xx}$ and $\sigma_{x'x'}$, we denote the latter by $\sigma'_{11}$. Finally, the symbols used for the direction cosines are replaced by the symbols as follows:

| Axes | $x$ | $y$ | $z$ |
|---|---|---|---|
| $x'$ | $n_{11}$ | $n_{12}$ | $n_{13}$ |
| $y'$ | $n_{21}$ | $n_{22}$ | $n_{23}$ |
| $z'$ | $n_{31}$ | $n_{32}$ | $n_{33}$ |

Let us apply this notation to rewrite Eq. (2-35),

$$\sigma'_{11} = \sigma_{11}n_{11}^2 + \sigma_{22}n_{12}^2 + \sigma_{33}n_{13}^2 + 2\sigma_{12}n_{11}n_{12} + 2\sigma_{13}n_{11}n_{13} + 2\sigma_{23}n_{12}n_{13} \quad \text{(2-41)}$$

Similarly, Eqs. (2-37) and (2-38) become

$$\begin{aligned}\sigma'_{12} = \sigma_{11}n_{11}n_{21} + \sigma_{22}n_{12}n_{22} + \sigma_{33}n_{13}n_{23} + \sigma_{12}(n_{12}n_{21} + n_{11}n_{22})\\ + \sigma_{13}(n_{13}n_{21} + n_{23}n_{11}) + \sigma_{23}(n_{13}n_{22} + n_{12}n_{23})\end{aligned} \quad \text{(2-42}a\text{)}$$

$$\begin{aligned}\sigma'_{13} = \sigma_{11}n_{11}n_{31} + \sigma_{22}n_{12}n_{32} + \sigma_{33}n_{13}n_{33} + \sigma_{12}(n_{12}n_{31} + n_{11}n_{32})\\ + \sigma_{13}(n_{13}n_{31} + n_{11}n_{33}) + \sigma_{23}(n_{13}n_{32} + n_{12}n_{33})\end{aligned} \quad \text{(2-42}b\text{)}$$

The real advantage of this notation becomes apparent after we replace numerical indices 1, 2, and 3 by the characters $i$, $j$, $k$, assuming that they take the values 1, 2, or 3. It is further assumed that an index appearing twice demands summation. That is, for instance,

$$\sigma_{ii} = \sigma_{11} + \sigma_{22} + \sigma_{33} \quad \text{(2-43)}$$

Similarly,

$$\sigma_{ij}n_{jk} = \sigma_{i1}n_{1k} + \sigma_{i2}n_{2k} + \sigma_{i3}n_{3k}$$

and so on. We see by inspection that Eqs. (2-41) and (2-42) can be put in compact form,

$$\sigma'_{ij} = \sigma_{kl}n_{ik}n_{jl} \quad \text{(2-44)}$$

Any quantity that transforms in this manner is called a *Cartesian tensor of rank 2*. Therefore stress is a Cartesian tensor. We may incidentally point to the moment of inertia as another familiar example of a Cartesian tensor of rank 2. Comparison of the formula for transformation of the moment of inertia with Eq. (2-44) shows that they are identical.

## 2-2-6 Stress Components in Polar Coordinates

Polar coordinates are used quite often, particularly in cases dictated by geometry (e.g., circular disks, cylindrical shafts). The results from Section 2-2-4 can be easily adapted to derive formulas expressing stress components in polar coordinates from those in Cartesian coordinate systems (Fig. 2-12). Let us introduce the Cartesian coordinate system $Ox'y'z'$ such that the $x'$ axis coincides with the radial direction $r$, the $y'$ axis with the tangential direction $\theta$, and the $z'$ axis with the axial direction $z$. We note that

$$x = r\cos\theta = x'\cos\theta, \qquad y = r\sin\theta = x'\sin\theta, \qquad z = z = z' \quad \text{(2-45)}$$

is the coordinate transformation from $r$, $\theta$, $z$ to $x$, $y$, $z$, while

$$r = \sqrt{x^2 + y^2}, \qquad \theta = \tan^{-1}(y/x), \qquad z = z \quad \text{(2-46)}$$

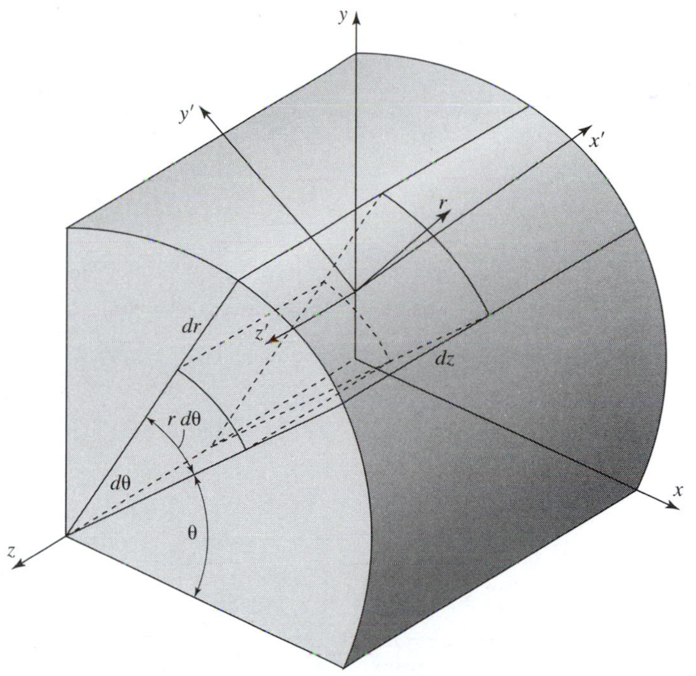


Figure 2-12 Stress components in polar coordinates

is the inverse transformation. In view of Eqs. (2-45) we see that the matrix of the direction cosines in Eq. (2-40) reduces to

$$\begin{bmatrix} n_{x'x} & n_{y'x} & n_{z'x} \\ n_{x'y} & n_{y'y} & n_{z'y} \\ n_{x'z} & n_{y'z} & n_{z'z} \end{bmatrix} = \begin{bmatrix} \cos\theta & -\sin\theta & 0 \\ \sin\theta & \cos\theta & 0 \\ 0 & 0 & 1 \end{bmatrix} \tag{2-47}$$

We also denote

$$\begin{aligned} \sigma_{rr} &= \sigma_{x'x'}, & \sigma_{\theta\theta} &= \sigma_{y'y'}, & \sigma_{zz} &= \sigma_{z'z'} \\ \sigma_{r\theta} &= \sigma_{x'y'}, & \sigma_{rz} &= \sigma_{x'z'}, & \sigma_{\theta z} &= \sigma_{y'z'} \end{aligned} \tag{2- 48}$$

Substituting Eqs. (2-47) and (2-48) into Eq. (2-39) and expanding the results, this becomes

$$\begin{aligned} \sigma_{rr} &= \sigma_{xx}\cos^2\theta + \sigma_{yy}\sin^2\theta + \sigma_{xy}\sin 2\theta \\ \sigma_{\theta\theta} &= \sigma_{xx}\sin^2\theta + \sigma_{yy}\cos^2\theta - \sigma_{xy}\sin 2\theta \\ \sigma_{zz} &= \sigma_{zz} \\ \sigma_{rz} &= \sigma_{xz}\cos\theta + \sigma_{yz}\sin\theta \\ \sigma_{r\theta} &= \tfrac{1}{2}(\sigma_{yy} - \sigma_{xx})\sin 2\theta + \sigma_{xy}\cos 2\theta \\ \sigma_{\theta z} &= -\sigma_{xz}\sin\theta + \sigma_{yz}\cos\theta \end{aligned} \tag{2-49}$$

The inverse relationships can be readily obtained. To this end we premultiply Eq. (2-39) by $\boldsymbol{N}$ and postmultiply by $\boldsymbol{N}^T$. By using the fact that

$$\boldsymbol{N}\boldsymbol{N}^T = \boldsymbol{N}^T\boldsymbol{N} = \boldsymbol{I} \tag{2-50}$$

we get the following expression for the stress matrix $\boldsymbol{S}$:

$$\boldsymbol{S} = \boldsymbol{N}\boldsymbol{S}'\boldsymbol{N}^T \tag{2-51}$$

Writing explicitly the elements of the matrix equation yields the desired expressions:

$$\begin{aligned}
\sigma_{xx} &= \sigma_{rr}\cos^2\theta + \sigma_{\theta\theta}\sin^2\theta - \sigma_{r\theta}\sin 2\theta \\
\sigma_{yy} &= \sigma_{rr}\sin^2\theta + \sigma_{\theta\theta}\cos^2\theta + \sigma_{r\theta}\sin 2\theta \\
\sigma_{zz} &= \sigma_{zz} \\
\sigma_{yz} &= \sigma_{rz}\sin\theta + \sigma_{\theta z}\cos\theta \\
\sigma_{xy} &= \tfrac{1}{2}(\sigma_{rr} - \sigma_{\theta\theta})\sin 2\theta + \sigma_{r\theta}\cos 2\theta \\
\sigma_{xz} &= \sigma_{rz}\cos\theta - \sigma_{\theta z}\sin\theta
\end{aligned} \tag{2-52}$$

## EXAMPLE 2-5

The Cartesian components of stress at a point $M$ are given by

$$S = \begin{bmatrix} 2 & 2 & 1 \\ 2 & 3 & 0 \\ 1 & 0 & 1 \end{bmatrix} \tag{$a$}$$

Let us introduce, in a plane parallel to $Oxy$, a local polar coordinate system with the origin at $M$ (see Fig. 2-13).

1. Show that $\sigma_{rr}$ must have at least one extremum at $M$.
2. Find the angle $\theta$ such that the radial stress component $\sigma_{rr}$ is maximum.
3. Find a direction $\theta$ (if any) such that $\sigma_{rr} = 0$ at $M$.

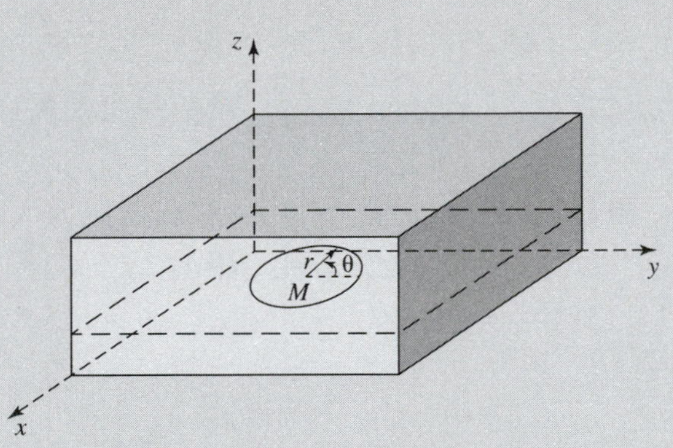


**Figure 2-13** Local polar coordinate system

**SOLUTION**

1. We draw a diagram of $\sigma_{rr}$ versus the angle $\theta$ at $M$. Without making any calculations we determine that as we move around point $M$, changing $\theta$ from 0 to $2\pi$, $\sigma_{rr}$ must return to its original value (see Fig. 2-14). Thus unless $\sigma_{rr}$ = constant, it must have at least one extremum. This proves our statement.
2. In order to find max $\sigma_{rr}$ we differentiate the first of Eqs. (2-49) with respect to $\theta$ and equate the result to zero. Thus,

$$\frac{d\sigma_{rr}}{d\theta} = -2\sigma_{xx}\sin\theta\cos\theta + 2\sigma_{yy}\sin\theta\cos\theta + 2\sigma_{xy}\cos 2\theta = 0 \qquad (b)$$

which, using Eq. (*a*), becomes

$$-4\sin\theta\cos\theta + 6\sin\theta\cos\theta + 2(\cos^2\theta - \sin^2\theta) = 0 \qquad (c)$$

or

$$2\sin\theta\cos\theta + 2 - 4\sin^2\theta = 0 \qquad (d)$$

We note that the left-hand side of Eq. (*d*) is periodic, with the period equal to $\pi$. Using Mathcad we found that Eq. (*d*) has two roots in the range $0 \leq \theta \leq \pi$,

$$\theta_1 = 0.9079, \qquad \theta_2 = 2.479 \qquad (e)$$

Now we are sure that $\sigma_{rr}$ has one maximum and one minimum. In order to determine where they are located, we examine the second derivative of $\sigma_{rr}$ with respect to $\theta$. With Eq. (*b*) we obtain

$$\frac{d^2\sigma_{rr}}{d\theta^2} = -2\sigma_{xx}\cos^2\theta + 2\sigma_{yy}\cos^2\theta - 4\sigma_{xy}\sin 2\theta = 2\cos 2\theta - 8\sin 2\theta \qquad (f)$$

Substituting here results (*e*), we get

$$\left.\frac{d^2\sigma_{rr}}{d\theta^2}\right|_{\theta_1} = -8.276, \qquad \left.\frac{d^2\sigma_{rr}}{d\theta^2}\right|_{\theta_2} = 8.245 \qquad (g)$$

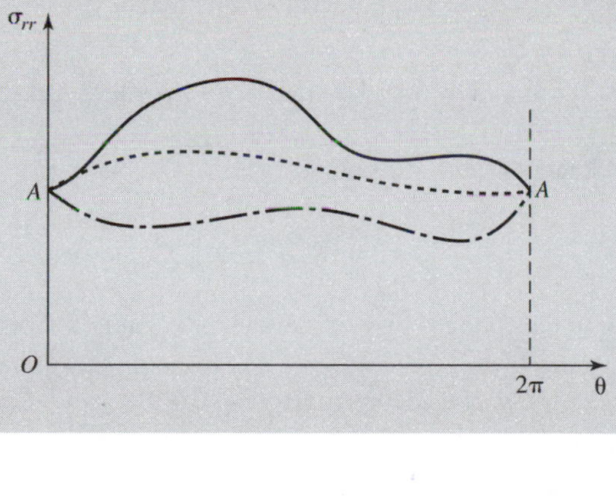


**Figure 2-14** Radial stress as a function of $\theta$

Thus $\sigma_{rr}$ is maximum at $\theta_1$ and minimum at $\theta_2$

$$\max \sigma_{rr} = 4.327, \qquad \min \sigma_{rr} = 1.445$$

**3.** The first line of Eq. (2-49) becomes

$$\sigma_{rr} = 2\cos^2\theta + 3\sin^2\theta + 2\sin 2\theta = 0$$

Using the trigonometric identities we reduce this to

$$2 + \sin^2\theta \pm 2\sin\theta\sqrt{1 - \sin^2\theta} = 0$$

or

$$5\sin^4\theta + 4 = 0$$

This equation does not have any real roots, which implies that there is no such point $M$ in which $\sigma_{rr} = 0$.

## 2-2-7 Equilibrium Equations in Cartesian Coordinates

In previous sections we derived expressions for determining stresses at a point in one coordinate system in terms of the stresses in another coordinate system. We still need, however, equations from which to obtain the variation of the stress distribution from point to point as a function of geometry, load, and the boundary conditions. The first step in this development will be made in the present section. In the following sections we shall see what other equations are necessary to perform stress analysis, what are the limitations, and possible simplifications. Let us consider again an infinitesimal element, as discussed in Section 2-2-1 (Fig. 2-15). It is now necessary to examine the conditions of equilibrium of the parallelepiped by projecting all forces acting on it on the coordinate axes $Ox$, $Oy$, and $Oz$. It is now also necessary, as it was in Section 2-2-2, to distinguish between forces acting on opposite parallel faces infinitesimally close to each other.

Let $d\mathbf{t}^{(x)}$ represent the differential change of $\mathbf{t}^{(x)}(x, y, z)$ as $x$, $y$, and $z$ change by differential amounts. By the chain rule,

$$d\mathbf{t}^{(x)} = \frac{\partial \mathbf{t}^{(x)}}{\partial x}\,dx + \frac{\partial \mathbf{t}^{(x)}}{\partial y}\,dy + \frac{\partial \mathbf{t}^{(x)}}{\partial z}\,dz$$

If $y$ and $z$ are kept constant, we obtain

$$d\mathbf{t}^{(x)} = \frac{\partial \mathbf{t}^{(x)}}{\partial x}\,dx$$

Similarly, due to the changes in $y$ or $z$ alone, $d\mathbf{t}^{(y)}$ and $d\mathbf{t}^{(z)}$ become

$$d\mathbf{t}^{(y)} = \frac{\partial \mathbf{t}^{(y)}}{\partial y}\,dy, \qquad d\mathbf{t}^{(z)} = \frac{\partial \mathbf{t}^{(z)}}{\partial z}\,dz$$

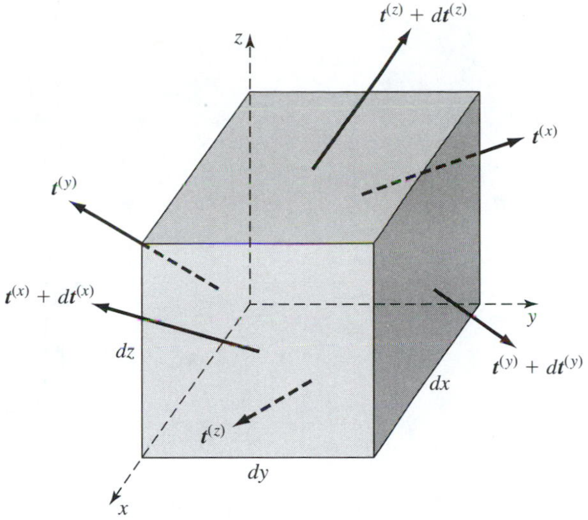


Figure 2-15

We now project on $Ox$ all forces acting on the element and sum these forces. This is accomplished by taking the dot product of each force vector with the unit vector $\boldsymbol{i}$. For equilibrium to hold, the sum must equal zero,

$$\left(\boldsymbol{t}^{(x)} + \frac{\partial \boldsymbol{t}^{(x)}}{\partial x}\,dx\right)\cdot \boldsymbol{i}\,dy\,dz - \boldsymbol{t}^{(x)}\cdot \boldsymbol{i}\,dy\,dz + \left(\boldsymbol{t}^{(y)} + \frac{\partial \boldsymbol{t}^{(y)}}{\partial y}\,dy\right)\cdot \boldsymbol{i}\,dx\,dz$$
$$- \boldsymbol{t}^{(y)}\cdot \boldsymbol{i}\,dx\,dz + \left(\boldsymbol{t}^{(z)} + \frac{\partial \boldsymbol{t}^{(z)}}{\partial z}\,dz\right)\cdot \boldsymbol{i}\,dx\,dy - \boldsymbol{t}^{(z)}\cdot \boldsymbol{i}\,dx\,dy + \boldsymbol{b}\cdot \boldsymbol{i}\,dx\,dy\,dz = 0$$

where we used the relations from Section 2-2-1 showing that $\boldsymbol{t}^{(-x)} = -\boldsymbol{t}^{(x)}$, and so on.

Projections of all forces on $Oy$ and $Oz$ are obtained in similar way,

$$\left(\boldsymbol{t}^{(x)} + \frac{\partial \boldsymbol{t}^{(x)}}{\partial x}\,dx\right)\cdot \boldsymbol{j}\,dy\,dz - \boldsymbol{t}^{(x)}\cdot \boldsymbol{j}\,dy\,dz + \left(\boldsymbol{t}^{(y)} + \frac{\partial \boldsymbol{t}^{(y)}}{\partial y}\,dy\right)\cdot \boldsymbol{j}\,dx\,dz$$
$$- \boldsymbol{t}^{(y)}\cdot \boldsymbol{j}\,dx\,dz + \left(\boldsymbol{t}^{(z)} + \frac{\partial \boldsymbol{t}^{(z)}}{\partial z}\,dz\right)\cdot \boldsymbol{j}\,dx\,dy - \boldsymbol{t}^{(z)}\cdot \boldsymbol{j}\,dx\,dy + \boldsymbol{b}\cdot \boldsymbol{j}\,dx\,dy\,dz = 0$$

$$\left(\boldsymbol{t}^{(x)} + \frac{\partial \boldsymbol{t}^{(x)}}{\partial x}\,dx\right)\cdot \boldsymbol{k}\,dy\,dz - \boldsymbol{t}^{(x)}\cdot \boldsymbol{k}\,dy\,dz + \left(\boldsymbol{t}^{(y)} + \frac{\partial \boldsymbol{t}^{(y)}}{\partial y}\,dy\right)\cdot \boldsymbol{k}\,dx\,dz$$
$$- \boldsymbol{t}^{(y)}\cdot \boldsymbol{k}\,dx\,dz + \left(\boldsymbol{t}^{(z)} + \frac{\partial \boldsymbol{t}^{(z)}}{\partial z}\,dz\right)\cdot \boldsymbol{k}\,dx\,dy - \boldsymbol{t}^{(z)}\cdot \boldsymbol{k}\,dx\,dy + \boldsymbol{b}\cdot \boldsymbol{k}\,dx\,dy = 0$$

Substituting here Eqs. (2-1), (2-4), and (2-5) and noting that, for example,

$$\left(\frac{\partial \boldsymbol{t}^{(x)}}{\partial x}\,dx\right)\cdot \boldsymbol{i} = \frac{\partial}{\partial x}\left(\boldsymbol{t}^{(x)}\cdot \boldsymbol{i}\right)dx$$

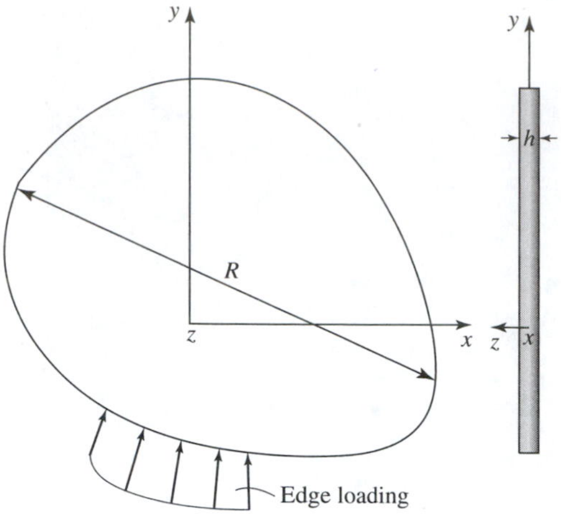


Figure 2-16 Plane stress

we obtain finally the following system of differential equations of equilibrium:

$$\begin{aligned}
\frac{\partial \sigma_{xx}}{\partial x} + \frac{\partial \sigma_{xy}}{\partial y} + \frac{\partial \sigma_{xz}}{\partial z} + b_x = 0 \\
\frac{\partial \sigma_{xy}}{\partial x} + \frac{\partial \sigma_{yy}}{\partial y} + \frac{\partial \sigma_{yz}}{\partial z} + b_y = 0 \\
\frac{\partial \sigma_{xz}}{\partial x} + \frac{\partial \sigma_{yz}}{\partial y} + \frac{\partial \sigma_{zz}}{\partial z} + b_z = 0
\end{aligned} \tag{2-53}$$

Let us consider, as a special case, a thin plate of uniform thickness, as shown in Fig. 2-16, where $h \ll R$, and $R$ is a typical dimension that characterizes the size of the plate. First we place the coordinate axes $Ox$ and $Oy$ on one of the stress-free faces (the large, flat, surfaces of the plate). We further assume that the external load is parallel to $Oxy$ and is applied to the edge of the plate. It is also assumed that the body forces act only in the $Oxy$ plane. In this situation the stresses $\sigma_{xz}$, $\sigma_{yz}$, and $\sigma_{zz}$ are zero at $z = 0$ and at $z = h$ (they must be zero because we assumed that the flat surfaces of the plate are stress free). For $0 < z < h$, the values of these stresses are very small in comparison to other stresses (because they are zero at the faces and the distance between them is very small), and they can therefore be neglected. Similarly, the stress gradients in the $z$ direction will be negligibly small. The resulting state of stress, called *plane* stress, is described by the following equations:

$$\begin{gathered}
\sigma_{xz} = \sigma_{yz} = \sigma_{zz} = 0 \\
\sigma_{xx} = \sigma_{xx}(x, y), \qquad \sigma_{xy} = \sigma_{xy}(x, y), \qquad \sigma_{yy} = \sigma_{yy}(x, y)
\end{gathered} \tag{2-54}$$

We then find that with Eqs. (2-54) the equilibrium Eqs. (2-53) become

$$\begin{aligned}
\frac{\partial \sigma_{xx}}{\partial x} + \frac{\partial \sigma_{xy}}{\partial y} + b_x = 0 \\
\frac{\partial \sigma_{xy}}{\partial x} + \frac{\partial \sigma_{yy}}{\partial y} + b_y = 0
\end{aligned} \tag{2-55}$$

When the condition of plane stress exists, at least in an approximate sense, the three-dimensional problem can be reduced to a simpler two-dimensional one.

### 2-2-8 Equilibrium Equations in Polar Coordinates

In some applications polar coordinates are more convenient than Cartesian coordinates. To obtain equilibrium equations we consider an infinitesimal element suitable for this system, such that along each of its faces one variable—$r$ or $z$—will be constant (Fig. 2-17). We now introduce unit vectors $\boldsymbol{e}_r$, $\boldsymbol{e}_\theta$, and $\boldsymbol{e}_z$ along the radial direction $r$, the tangential direction $\theta$, and the axial direction $z$, respectively (Fig. 2-18), pointing toward the positive coordinate directions. Noting, due to the curvature, the distinction between unit vectors $\boldsymbol{e}_r^{(+)}$, $\boldsymbol{e}_r^{(-)}$, $\boldsymbol{e}_\theta^{(+)}$, and $\boldsymbol{e}_\theta^{(-)}$, we see in Fig. 2-19 that

$$
\begin{aligned}
\boldsymbol{e}_r^{(-)} \cdot \boldsymbol{e}_r &= \cos\frac{d\theta}{2} \approx 1, & \boldsymbol{e}_r^{(-)} \cdot \boldsymbol{e}_\theta &= -\sin\frac{d\theta}{2} \approx -\frac{d\theta}{2} \\
\boldsymbol{e}_\theta^{(-)} \cdot \boldsymbol{e}_r &= \sin\frac{d\theta}{2} \approx \frac{d\theta}{2}, & \boldsymbol{e}_\theta^{(-)} \cdot \boldsymbol{e}_\theta &= \cos\frac{d\theta}{2} \approx 1 \\
\boldsymbol{e}_r^{(+)} \cdot \boldsymbol{e}_r &= \cos\frac{d\theta}{2} \approx 1, & \boldsymbol{e}_r^{(+)} \cdot \boldsymbol{e}_\theta &= \sin\frac{d\theta}{2} \approx \frac{d\theta}{2} \\
\boldsymbol{e}_\theta^{(+)} \cdot \boldsymbol{e}_r &= -\sin\frac{d\theta}{2} \approx -\frac{d\theta}{2}, & \boldsymbol{e}_\theta^{(+)} \cdot \boldsymbol{e}_\theta &= \cos\frac{d\theta}{2} \approx 1
\end{aligned}
\tag{2-56}
$$

Each of the stress vectors shown on the free-body diagram of the element (Fig. 2-20) can be decomposed into three components:

$$
\begin{aligned}
\boldsymbol{t}^{(r)} &= \sigma_{rr}\boldsymbol{e}_r + \sigma_{r\theta}\boldsymbol{e}_\theta + \sigma_{rz}\boldsymbol{e}_z \\
\boldsymbol{t}^{(\theta)} &= \sigma_{r\theta}\boldsymbol{e}_r + \sigma_{\theta\theta}\boldsymbol{e}_\theta + \sigma_{\theta z}\boldsymbol{e}_z \\
\boldsymbol{t}^{(z)} &= \sigma_{rz}\boldsymbol{e}_r + \sigma_{\theta z}\boldsymbol{e}_\theta + \sigma_{zz}\boldsymbol{e}_z
\end{aligned}
\tag{2-57}
$$

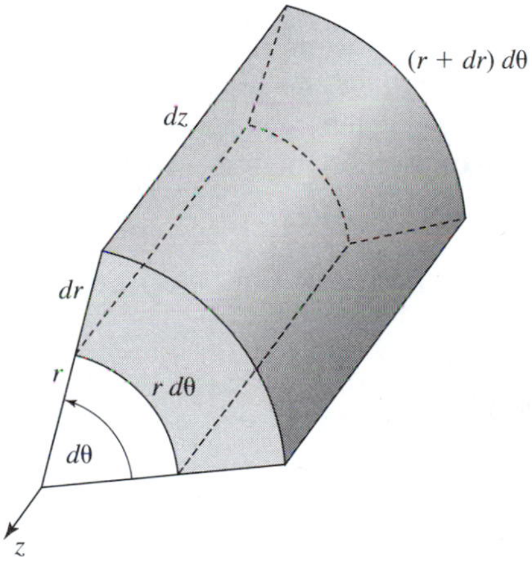


Figure 2-17

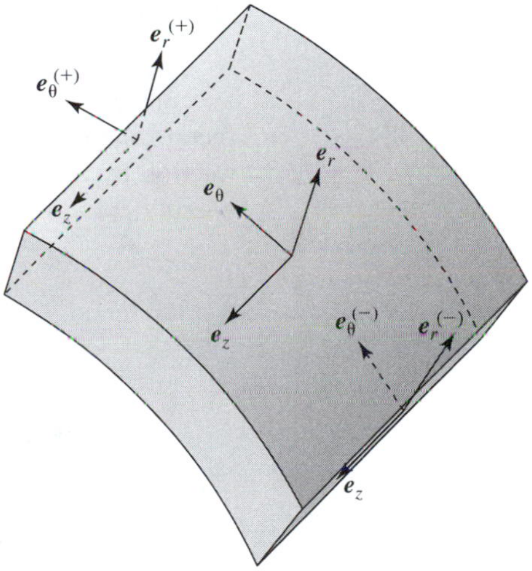


Figure 2-18 Unit vectors in polar coordinates

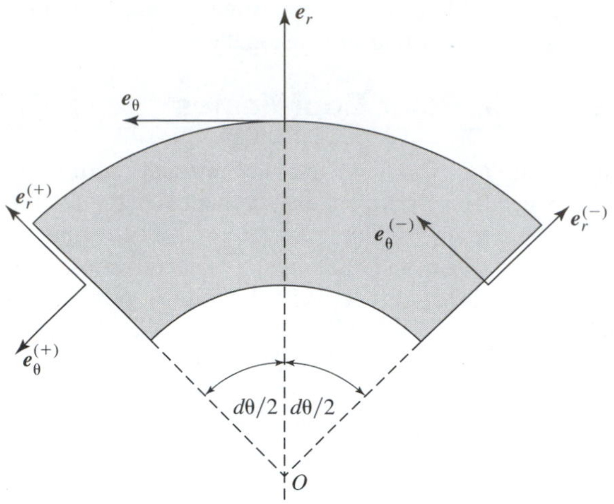


Figure 2-19 Unit vectors on two edges of an infinitesimal element

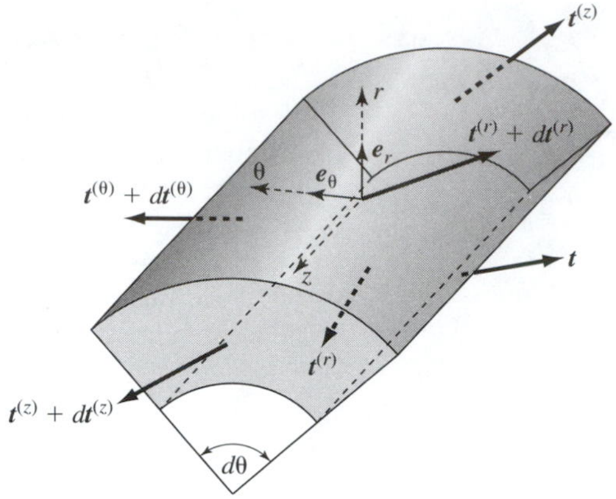


Figure 2-20

where the symmetry of the stress components ($\sigma_{\theta r} = \sigma_{r\theta}$) was used. The symmetry can be proved in the same way as was shown in Section 2-2-2 for the case of Cartesian coordinates. The unit vectors appearing in Eqs. (2-57) should be replaced either by $\boldsymbol{e}_r^{(-)}$ and $\boldsymbol{e}_\theta^{(-)}$ or by $\boldsymbol{e}_r^{(+)}$ and $\boldsymbol{e}_\theta^{(+)}$, as seen in Fig. 2-19. Forces acting on the element (Fig. 2-20) are now projected on the directions $r$, $\theta$, $z$. We shall also use the formulas

$$d\boldsymbol{t}^{(r)} = \frac{\partial \boldsymbol{t}^{(r)}}{\partial r}\, dr, \qquad d\boldsymbol{t}^{(\theta)} = \frac{\partial \boldsymbol{t}^{(\theta)}}{\partial \theta}\, d\theta, \qquad d\boldsymbol{t}^{(z)} = \frac{\partial \boldsymbol{t}^{(z)}}{\partial z}\, dz$$

to represent the change in the stress vectors associated with moving from one element face to the corresponding face on the opposite side of the element. Multiplying each traction by the area of the corresponding face and taking the dot product with $\boldsymbol{e}_r$, we obtain the following equation of equilibrium in the $r$ direction:

$$\left(\boldsymbol{t}^{(r)} + \frac{\partial \boldsymbol{t}^{(r)}}{\partial r}\, dr\right) \cdot \boldsymbol{e}_r (r + dr)\, d\theta\, dz - \boldsymbol{t}^{(r)} \cdot \boldsymbol{e}_r r\, d\theta\, dz$$

$$+ \left(\boldsymbol{t}^{(\theta)} + \frac{\partial \boldsymbol{t}^{(\theta)}}{\partial \theta}\, d\theta\right) \cdot \boldsymbol{e}_r\, dr\, dz - \boldsymbol{t}^{(\theta)} \cdot \boldsymbol{e}_r\, dr\, dz$$

$$+ \left(\boldsymbol{t}^{(z)} + \frac{\partial \boldsymbol{t}^{(z)}}{\partial z}\, dz\right) \cdot \boldsymbol{e}_r r\, d\theta\, dr - \boldsymbol{t}^{(z)} \cdot \boldsymbol{e}_r r\, d\theta\, dr + \boldsymbol{b} \cdot \boldsymbol{e}_r r\, d\theta\, dr\, dz = 0$$

Substituting here Eqs. (2-57) we find

$$\left[\left(\sigma_{rr}+\frac{\partial\sigma_{rr}}{\partial r}\,dr\right)\boldsymbol{e}_r+\left(\sigma_{r\theta}+\frac{\partial\sigma_{r\theta}}{\partial r}\,dr\right)\boldsymbol{e}_\theta+\left(\sigma_{rz}+\frac{\partial\sigma_{rz}}{\partial r}\,dr\right)\boldsymbol{e}_z\right]\cdot\boldsymbol{e}_r(r+dr)\,d\theta\,dz$$

$$-\left(\sigma_{rr}\boldsymbol{e}_r+\sigma_{r\theta}\boldsymbol{e}_\theta+\sigma_{rz}\boldsymbol{e}_z\right)\cdot\boldsymbol{e}_r r\,d\theta\,dz$$

$$+\left[\left(\sigma_{r\theta}+\frac{\partial\sigma_{r\theta}}{\partial\theta}\,d\theta-\sigma_{\theta\theta}\,d\theta\right)\boldsymbol{e}_r^{(+)}+\left(\sigma_{\theta\theta}+\frac{\partial\sigma_{\theta\theta}}{\partial\theta}\,d\theta+\sigma_{r\theta}\,d\theta\right)\boldsymbol{e}_\theta^{(+)}\right.$$

$$\left.+\left(\sigma_{\theta z}+\frac{\partial\sigma_{\theta z}}{\partial\theta}\,d\theta\right)\boldsymbol{e}_z\right]\cdot\boldsymbol{e}_r\,dr\,dz-\left(\sigma_{r\theta}\boldsymbol{e}_r^{(-)}+\sigma_{\theta\theta}\boldsymbol{e}_\theta^{(-)}+\sigma_{\theta z}\boldsymbol{e}_z\right)\boldsymbol{e}_r\,dr\,dz$$

$$+\left[\left(\sigma_{rz}+\frac{\partial\sigma_{rz}}{\partial z}\,dz\right)\boldsymbol{e}_r+\left(\sigma_{\theta z}+\frac{\partial\sigma_{\theta z}}{\partial z}\,dz\right)\boldsymbol{e}_\theta+\left(\sigma_{zz}+\frac{\partial\sigma_{zz}}{\partial z}\,dz\right)\boldsymbol{e}_z\right]\cdot\boldsymbol{e}_r r\,d\theta\,dr$$

$$-\left(\sigma_{rz}\boldsymbol{e}_r+\sigma_{\theta z}\boldsymbol{e}_\theta+\sigma_{zz}\boldsymbol{e}_z\right)\cdot\boldsymbol{e}_r r\,d\theta\,dr+\left(b_r\boldsymbol{e}_r+b_\theta\boldsymbol{e}_\theta+b_z\boldsymbol{e}_z\right)\cdot\boldsymbol{e}_r r\,d\theta\,dr\,dz=0$$

where $b_r$, $b_\theta$, and $b_z$ are polar components of the body force vector $\boldsymbol{b}$. Here we have used the relations $\partial\boldsymbol{e}_r/\partial\theta=\boldsymbol{e}_\theta$, and $\partial\boldsymbol{e}_\theta/\partial r=-\boldsymbol{e}_r$ (see also [6, p. 667]). Applying Eqs. (2-56), dividing by $r\,dr\,d\theta\,dz$, and evaluating for $dr\to 0$, $d\theta\to 0$, $dz\to 0$, leads to the first of the following equations:

$$\begin{aligned}\frac{\partial\sigma_{rr}}{\partial r}+\frac{1}{r}\frac{\partial\sigma_{r\theta}}{\partial\theta}+\frac{\partial\sigma_{rz}}{\partial z}+\frac{1}{r}(\sigma_{rr}-\sigma_{\theta\theta})+b_r&=0\\ \frac{\partial\sigma_{r\theta}}{\partial r}+\frac{1}{r}\frac{\partial\sigma_{\theta\theta}}{\partial\theta}+\frac{\partial\sigma_{\theta z}}{\partial z}+\frac{2}{r}\sigma_{r\theta}+b_\theta&=0\\ \frac{\partial\sigma_{rz}}{\partial r}+\frac{1}{r}\frac{\partial\sigma_{\theta z}}{\partial\theta}+\frac{\partial\sigma_{zz}}{\partial z}+\frac{1}{r}\sigma_{rz}+b_z&=0\end{aligned}\tag{2-58}$$

The last two of Eqs. (2-58) are obtained in a similar manner, that is, by taking the dot products of the $\boldsymbol{t}$ vector with $\boldsymbol{e}_\theta$ or $\boldsymbol{e}_z$. We note that when the stress distribution is independent of $z$ then every cross section perpendicular to the $Oz$ axis is subject to the same deformation. Let us now make three equidistant cuts $\alpha$, $\beta$, and $\gamma$ perpendicular to $Oz$, as seen in Fig. 2-21. Since the "slices" between $\alpha\beta$ and $\beta\gamma$ deform in the same way, there is no shear in cut $\beta$. Thus there are no shear stresses acting on this cut. Consequently, since the location of cut $\beta$ is arbitrary,

$$\sigma_{rz}=\sigma_{\theta z}=0\tag{2-59}$$

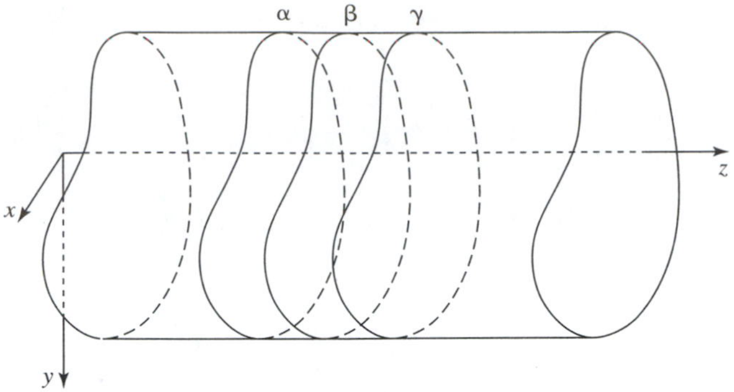


Figure 2-21 Two-dimensional state of stress

throughout. This two-dimensional state of stress occurs when we have, for instance, a prismatic bar with the applied load independent of $z$. Eqs. (2-58) reduce then to

$$\begin{aligned} \frac{\partial \sigma_{rr}}{\partial r} + \frac{1}{r}\frac{\partial \sigma_{r\theta}}{\partial \theta} + \frac{1}{r}(\sigma_{rr} - \sigma_{\theta\theta}) + b_r = 0 \\ \frac{\partial \sigma_{r\theta}}{\partial r} + \frac{1}{r}\frac{\partial \sigma_{\theta\theta}}{\partial \theta} + \frac{2}{r}\sigma_{r\theta} + b_\theta = 0 \end{aligned} \qquad (2\text{-}60)$$

In the case of axial symmetry there is no dependence of the stress components on $\theta$ and, as any plane axial cut is a symmetry plane, the shear stresses vanish there. This occurs when every cross section of a solid is circular and the applied load does not depend on $\theta$. Therefore,

$$\sigma_{r\theta} = \sigma_{\theta z} = 0 \qquad (2\text{-}61)$$

With Eq. (2-61), Eqs. (2-58) become

$$\begin{aligned} \frac{\partial \sigma_{rr}}{\partial r} + \frac{\partial \sigma_{rz}}{\partial z} + \frac{1}{r}(\sigma_{rr} - \sigma_{\theta\theta}) + b_r = 0 \\ \frac{\partial \sigma_{rz}}{\partial r} + \frac{\partial \sigma_{zz}}{\partial z} + \frac{1}{r}\sigma_{rz} + b_z = 0 \end{aligned} \qquad (2\text{-}62)$$

Finally, assuming that there is no dependence of the stress components on either $z$ or $\theta$, all shear stresses are zero and we arrive at the following equation of equilibrium:

$$\frac{d\sigma_{rr}}{dr} + \frac{1}{r}(\sigma_{rr} - \sigma_{\theta\theta}) + b_r = 0 \qquad (2\text{-}63)$$

## 2-2-9 Applicability of Equilibrium Equations

The equilibrium equations derived in Section 2-2-7 are very general and applicable to the stress analysis of any solid, independently of the material from which it is made. It must be understood, however, that these equations alone are not sufficient for stress analysis. This is evident from the observation that, in general, they represent three differential equations in six unknowns. Other equations describing deformation, material properties, and boundary conditions must be added to make the stress analysis a solvable problem. This will be discussed briefly in Chapter 9. Except for a limited number of cases, the system of equations to be solved cannot be handled analytically. Numerical methods, some of which are described in Chapters 6 and 7, must then be applied. It is sometimes possible to simplify the mathematical description by making additional assumptions. This is well known from the elementary strength of materials (i.e., simple beam theory). This approach will also be utilized in this text to discuss deflections of plates, beams on elastic foundation, and so on. There will be few instances, discussed in the following chapters, when equations derived in Section 2-2-7 will be applied with no simplifying assumptions. This refers in particular to the axisymmetric case described by Eq. (2-63) and applicable in the theory of circular disks and thick-walled cylinders (Chapter 9).

## 2-3 PRINCIPAL STRESSES AND INVARIANTS

### 2-3-1 Characteristic Equation

If an inclined plane with a unit normal $\boldsymbol{n}$ is rotated, then both the normal stress $\sigma_n$ [Eq. (2-17)] and the shear stress $\tau$ [Eq. (2-27)] will change. Let us examine whether it is possible to find a direction of the normal to the inclined plane such that the shear stress acting on that plane is zero. First we assume that at least one such direction exists, and then we evaluate the corresponding unit normal vector $\boldsymbol{n}$ and the normal stress $\sigma_n$.

To this end let us multiply both sides of Eq. (2-31) by $\tau$ and then set $\tau = 0$. This results in the following system of linear algebraic equations, homogeneous in $n_x$, $n_y$, and $n_z$:

$$\begin{bmatrix} \sigma_{xx} - \sigma_n & \sigma_{xy} & \sigma_{xz} \\ \sigma_{xy} & \sigma_{yy} - \sigma_n & \sigma_{yz} \\ \sigma_{xz} & \sigma_{yz} & \sigma_{zz} - \sigma_n \end{bmatrix} \begin{Bmatrix} n_x \\ n_y \\ n_z \end{Bmatrix} = \begin{Bmatrix} 0 \\ 0 \\ 0 \end{Bmatrix} \tag{2-64}$$

or, using previously introduced symbols for matrices,

$$(\boldsymbol{S} - \sigma_n \boldsymbol{I})\boldsymbol{n} = 0$$

Since $\sigma_n$ is unknown in the preceding system, there are three equations and four unknowns. A fourth equation is provided by the condition that $\boldsymbol{n}$ be a unit vector, and hence

$$n_x^2 + n_y^2 + n_z^2 = 1 \tag{2-65}$$

We have to solve Eq. (2-64), where the unknown $\sigma_n$ is a parameter. It is seen by inspection of Eq. (2-64) that the seemingly obvious solution, $n_x = n_y = n_z = 0$, violates the condition of Eq. (2-65) and is therefore unacceptable. It is known from algebra that a homogeneous system of linear algebraic equations possesses a nonzero solution if and only if the determinant of a matrix formed from the coefficients of the unknowns is zero (see [3, p. 467]). Thus it is required that

$$\det\begin{bmatrix} \sigma_{xx}-\sigma_n & \sigma_{xy} & \sigma_{xz} \\ \sigma_{xy} & \sigma_{yy}-\sigma_n & \sigma_{yz} \\ \sigma_{xz} & \sigma_{yz} & \sigma_{zz}-\sigma_n \end{bmatrix} = 0$$

or

$$\det[\boldsymbol{S} - \sigma_n \boldsymbol{I}] = 0 \tag{2-66}$$

Expanding Eq. (2-66) yields

$$(\sigma_{xx}-\sigma_n)\left[(\sigma_{yy}-\sigma_n)(\sigma_{zz}-\sigma_n)-\sigma_{yz}^2\right] - \sigma_{xy}[\sigma_{xy}(\sigma_{zz}-\sigma_n)-\sigma_{xz}\sigma_{yz}]$$
$$+\,\sigma_{xz}[\sigma_{xy}\sigma_{yz}-\sigma_{xz}(\sigma_{yy}-\sigma_n)] = 0$$

or

$$\sigma_n^3 - I_1\sigma_n^2 + I_2\sigma_n - I_3 = 0 \tag{2-67}$$

where

$$I_1 = \sigma_{xx} + \sigma_{yy} + \sigma_{zz} = \operatorname{tr} \boldsymbol{S} \tag{2-68$a$}$$
$$I_2 = -\sigma_{xy}^2 - \sigma_{yz}^2 - \sigma_{xz}^2 + \sigma_{xx}\sigma_{yy} + \sigma_{yy}\sigma_{zz} + \sigma_{zz}\sigma_{xx} \tag{2-68$b$}$$
$$I_3 = \sigma_{xx}\sigma_{yy}\sigma_{zz} - \sigma_{xx}\sigma_{yz}^2 - \sigma_{yy}\sigma_{xz}^2 - \sigma_{zz}\sigma_{xy}^2 + 2\sigma_{xy}\sigma_{yz}\sigma_{xz} = \det \boldsymbol{S} \tag{2-68$c$}$$

Here tr denotes the *trace of the matrix,* that is, the sum of the terms on the main diagonal of the matrix. The quantities $I_1$, $I_2$, $I_3$ are called *stress invariants* for reasons to be explained shortly. First we note that Eq. (2-67), the so-called *characteristic equation,* is a cubic equation. Its roots are those values of $\sigma_n$ that guarantee the existence of a plane on which $\tau = 0$.

We observe now that the presence of such a plane through a point is a physical phenomenon, that is, a change of the coordinate system cannot affect either the inclination of this plane or the value of the corresponding normal stress $\sigma_n$. Consequently the roots of the characteristic equation, Eq. (2-67), cannot depend on the coordinate system. This implies that the numerical values of the coefficients of $\sigma_n$ in Eq. (2-67) must also be independent of the orientation of the coordinate system. This justifies the term *invariant* for those coefficients. The invariance with respect to the coordinate transformation can also be verified directly. To this end the coordinates $x$, $y$, $z$ are changed into $x'$, $y'$, $z'$ and the stress components $\sigma_{xx}, \ldots, \sigma_{xz}$ are replaced by the expressions relating them to the new stress components, $\sigma_{x'x'}, \ldots, \sigma_{x'z'}$, Eqs. (2-35), ( 2-37), and (2-38). After tedious manipulations this results in

$$\sigma_{xx} + \sigma_{yy} + \sigma_{zz} = \sigma_{x'x'} + \sigma_{y'y'} + \sigma_{z'z'} \tag{2-69}$$

and similar relations for the remaining two cases. This confirms that the rotation of the coordinate system does not change the values of the invariants at a fixed point.

## 2-3-2 Principal Stresses and Principal Directions

Equation (2-67) has three roots: $\sigma_1, \sigma_2$, and $\sigma_3$. These can be shown to be real (see [3, p. 467]) and are usually ordered in the following way:

$$\sigma_1 > \sigma_2 > \sigma_3$$

On substituting these roots, one at the time, into Eqs. (2-64) a system of three equations in three unknowns, $n_x^{(i)}, n_y^{(i)}, n_z^{(i)}$, is obtained for each root. Here the superscript $i$ ($i = 1, 2,$ or 3) indicates with which of the three roots the normal $\boldsymbol{n}$ is associated. This normal, determining the *principal plane* on which the shear stress $\tau$ is zero, is known as the *principal direction*. There are obviously three principal directions, and they can be shown (see [2, p. 100]) to be mutually perpendicular. We note that now only two of the three Eqs. (2-64) are independent, the third being a linear combination of the remaining two. The "missing" equation is Eq. (2-65).

From the known properties of the algebraic equations it follows also that the stress invariants are directly related to the principal stresses,

$$\begin{aligned} I_1 &= \sigma_1 + \sigma_2 + \sigma_3 \\ I_2 &= \sigma_1\sigma_2 + \sigma_2\sigma_3 + \sigma_3\sigma_1 \\ I_3 &= \sigma_1\sigma_2\sigma_3 \end{aligned} \tag{2-70}$$

These are standard matrix eigenvalue problems, which can be solved readily using routines found in, say, Mathcad and also on advanced pocket calculators.

## EXAMPLE 2-6

For a given stress matrix representing the state of stress at a certain point,

$$\boldsymbol{S} = \begin{bmatrix} 1 & 2 & 3 \\ 2 & 2 & 0 \\ 3 & 0 & 2 \end{bmatrix}$$

find the stress invariants, the principal stresses, and the principal directions.

### SOLUTION

From Eqs. (2-68) we find

$$I_1 = 1 + 2 + 2 = 5$$
$$I_2 = -2^2 - 0^2 - 3^2 + 1 \cdot 2 + 2 \cdot 2 + 2 \cdot 1 = -5$$
$$I_3 = 1 \cdot 2 \cdot 2 - 1 \cdot 0^2 - 2 \cdot 3^2 - 2 \cdot 2^2 + 2 \cdot 2 \cdot 0 \cdot 3 = 4 - 8 + 18 = -22$$

Therefore Eq. (2-67) becomes

$$\sigma_n^3 - 5\sigma_n^2 - 5\sigma_n + 22 = 0 \tag{a}$$

A cubic equation can be solved by using formulas for roots, or applying available computer programs. The roots are

$$\sigma_1 = 5.14, \qquad \sigma_2 = 2.00, \qquad \sigma_3 = -2.14$$

Next we calculate the corresponding principal directions.

**1.** $\sigma_n = \sigma_2 = 2.00$. Substituting this and the values of the stress components into the first two of Eqs. (2-64) results in the following system of equations:

$$(-2)n_x^{(2)} + 2n_y^{(2)} + 3n_z^{(2)} = 0$$
$$2n_x^{(2)} + (2-2)n_y^{(2)} = 0$$

while from Eq. (2-65),

$$\left(n_x^{(2)}\right)^2 + \left(n_y^{(2)}\right)^2 + \left(n_z^{(2)}\right)^2 = 1$$

Solving the preceding system by eliminating, say, $n_x^{(2)}$ and $n_z^{(2)}$ from the first two equations and substituting the results into the third one, we obtain

$$n_x^{(2)} = 0, \qquad n_y^{(2)} = 0.832, \qquad n_z^{(2)} = -0.555$$

**2.** $\sigma_n = \sigma_1 = 5.14$. Proceeding as before, we get

$$(1-5.14)n_x^{(1)} + 2n_y^{(1)} + 3n_z^{(1)} = 0$$
$$2n_x^{(1)} + (2-5.14)n_y^{(1)} = 0$$
$$\left(n_x^{(1)}\right)^2 + \left(n_y^{(1)}\right)^2 + \left(n_z^{(1)}\right)^2 = 1$$

with the solutions

$$n_x^{(1)} = 0.657, \qquad n_y^{(1)} = 0.418, \qquad n_z^{(1)} = 0.627$$

**3.** $\sigma_n = \sigma_3 = -2.14$. Now,

$$(1+2.14)n_x^{(3)} + 2n_y^{(3)} + 3n_z^{(3)} = 0$$
$$2n_x^{(3)} + (2+2.14)n_y^{(3)} = 0$$
$$\left(n_x^{(3)}\right)^2 + \left(n_y^{(3)}\right)^2 + (n_z^{(3)})^2 = 1$$

with the solutions

$$n_x^{(3)} = -0.754, \qquad n_y^{(3)} = 0.364, \qquad n_z^{(3)} = 0.546$$

The three principal directions are

$$\boldsymbol{n}^{(1)} = n_x^{(1)}\boldsymbol{i} + n_y^{(1)}\boldsymbol{j} + n_z^{(1)}\boldsymbol{k} = 0.657\boldsymbol{i} + 0.418\boldsymbol{j} + 0.627\boldsymbol{k}$$
$$\boldsymbol{n}^{(2)} = n_x^{(2)}\boldsymbol{i} + n_y^{(2)}\boldsymbol{j} + n_z^{(2)}\boldsymbol{k} = 0.832\boldsymbol{j} - 0555\boldsymbol{k}$$
$$\boldsymbol{n}^{(3)} = n_x^{(3)}\boldsymbol{i} + n_y^{(3)}\boldsymbol{j} + n_z^{(3)}\boldsymbol{k} = -0.754\boldsymbol{i} + 0.364\boldsymbol{j} + 0.546\boldsymbol{k}$$

and their orthogonality can be verified by calculating the appropriate dot products of the unit normals with each other.

## 2-3-3 Plane Stress—Principal Stresses and Principal Directions

The plane state of stress, explained in Section 2-2-7, again deserves our special attention. First we place the $Oxy$ plane parallel to the flat faces of the plate (Fig. 2-22). Next we apply a special coordinate transformation consisting of rotation of $Oxy$ by an angle $\theta$ around $Oz$. By using the fact that $z' = z$ we get

$$\begin{aligned} n_{x'x} &= \cos\theta, & n_{x'y} &= \sin\theta, & n_{x'z} &= 0 \\ n_{y'x} &= -\sin\theta, & n_{y'y} &= \cos\theta, & n_{y'z} &= 0 \\ n_{z'x} &= 0, & n_{z'y} &= 0, & n_{z'z} &= 1 \end{aligned}$$

so that the matrix $\boldsymbol{N}$, [Eq.(2-40)] becomes

$$\boldsymbol{N} = \begin{bmatrix} \cos\theta & -\sin\theta & 0 \\ \sin\theta & \cos\theta & 0 \\ 0 & 0 & 1 \end{bmatrix} \tag{2-71}$$

From the product of matrices [Eq. (2-39)], or directly from formulas (2-35) and (2-37) [see also Eqs. (2-49)], we find

$$\begin{aligned} \sigma_{x'x'} &= \sigma_{xx}\cos^2\theta + \sigma_{yy}\sin^2\theta + 2\sigma_{xy}\sin\theta\cos\theta \\ \sigma_{y'y'} &= \sigma_{xx}\sin^2\theta + \sigma_{yy}\cos^2\theta - 2\sigma_{xy}\sin\theta\cos\theta \\ \sigma_{x'y'} &= (\sigma_{yy} - \sigma_{xx})\sin\theta\cos\theta + \sigma_{xy}(\cos^2\theta - \sin^2\theta) \end{aligned} \tag{2-72}$$

It is also convenient to represent these equations in the following form:

$$\sigma_{x'x'} = \tfrac{1}{2}(\sigma_{xx} + \sigma_{yy}) + \tfrac{1}{2}(\sigma_{xx} - \sigma_{yy})\cos 2\theta + \sigma_{xy}\sin 2\theta \tag{2-73a}$$

$$\sigma_{y'y'} = \tfrac{1}{2}(\sigma_{xx} + \sigma_{yy}) - \tfrac{1}{2}(\sigma_{xx} - \sigma_{yy})\cos 2\theta - \sigma_{xy}\sin 2\theta \tag{2-73b}$$

$$\sigma_{x'y'} = -\tfrac{1}{2}(\sigma_{xx} - \sigma_{yy})\sin 2\theta + \sigma_{xy}\cos 2\theta \tag{2-73c}$$

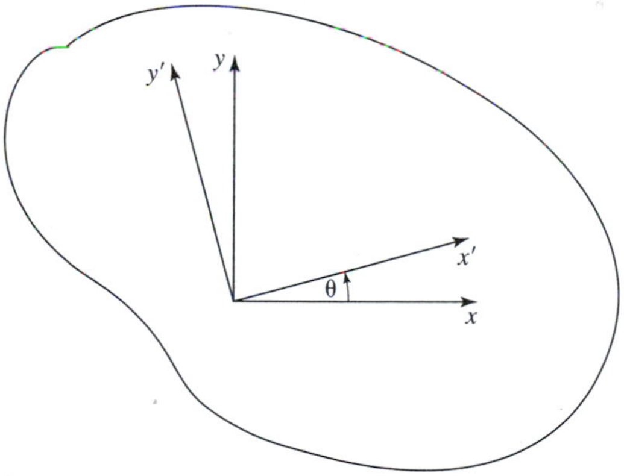


Figure 2-22

In going from Eqs. (2-72) to Eqs. (2-73) the trigonometric reduction formulas $\sin 2\theta = 2\sin\theta\cos\theta$ and $\cos 2\theta = \cos^2\theta - \sin^2\theta$ have been used. The principal stresses and the principal directions can be determined either as a special case of Eqs. (2-67), (2-64), and (2-65) or directly from Eqs. (2-73). Using the first method, we find from Eqs. (2-68) that

$$I_1 = \sigma_{xx} + \sigma_{yy}, \qquad I_2 = \sigma_{xx}\sigma_{yy} - \sigma_{xy}^2, \qquad I_3 = 0 \tag{2-74}$$

Thus Eq. (2-67) becomes

$$\sigma_n^3 - (\sigma_{xx} + \sigma_{yy})\sigma_n^2 + \left(\sigma_{xx}\sigma_{yy} - \sigma_{xy}^2\right)\sigma_n = 0$$

This yields the following principal values of stress:

$$\sigma_2 = 0, \qquad \sigma_{1,3} = \frac{\sigma_{xx} + \sigma_{yy}}{2} \pm \sqrt{\left(\frac{\sigma_{xx} - \sigma_{yy}}{2}\right)^2 + \sigma_{xy}^2} \tag{2-75}$$

where the subscripts are assigned consistent with the comments in Section 2-3-2.

Applying the second method, we conclude from Eq. (2-73*c*) that $\sigma_{x'y'} = 0$ when

$$\tan 2\theta_p = \frac{2\sigma_{xy}}{\sigma_{xx} - \sigma_{yy}} \tag{2-76}$$

This relationship determines the principal directions $\theta_p$. On substituting Eq. (2-76) into Eqs. (2-73*a*, *b*) we again obtain the values determined by Eqs. (2-75). One property of the principal values should be noted. When Eqs. (2-73*a*, *b*) are differentiated with respect to $\theta$, we readily obtain

$$\frac{\partial\sigma_{x'x'}}{\partial\theta} = 2\sigma_{x'y'}, \qquad \frac{\partial\sigma_{y'y'}}{\partial\theta} = -2\sigma_{x'y'} \tag{2-77}$$

where Eq. (2-73*c*) has been used. This indicates that for $\sigma_{x'y'} = 0$, $\sigma_{x'x'}$ and $\sigma_{y'y'}$ reach their extremal values. Therefore the principal values calculated from Eqs. (2-75) are either maximum or minimum. In the following we will assume that

$$\sigma_1 = \sigma_{\max}, \qquad \sigma_3 = \sigma_{\min}$$

Let us also find the maximum of $\sigma_{xy}$. Differentiating Eq. (2-73*c*) with respect to $\theta$, equating the result to zero, and denoting its solution by $\overline{\theta}$ yields*

$$\tan 2\overline{\theta} = -\frac{\sigma_{xx} - \sigma_{yy}}{2\sigma_{xy}} = -\frac{1}{\tan 2\theta_p} \tag{2-78}$$

The angles $\theta_p$ from Eq. (2-76) and $\overline{\theta}$ from Eq. (2-78) are related. Indeed, comparing Eq. (2-76) with Eq. (2-78) we see that

$$\tan 2\overline{\theta}\tan 2\theta_p \equiv \frac{\sin 2\overline{\theta}\sin 2\theta_p}{\cos 2\overline{\theta}\cos 2\theta_p} = -1$$

---

*In this way we obtain a local maximum that is not always the absolute maximum.

or

$$\frac{\cos 2(\overline{\theta} - \theta_p) - \cos 2(\overline{\theta} + \theta_p)}{\cos 2(\overline{\theta} - \theta_p) + \cos 2(\overline{\theta} + \theta_p)} = -1$$

and

$$\cos 2(\overline{\theta} - \theta_p) = 0$$

Therefore,

$$\overline{\theta} = \theta_p \pm \frac{\pi}{4} \tag{2-79}$$

Thus on substituting $\overline{\theta}$ given by Eq. (2-79) for $\theta$ appearing in Eq. (2-73*c*), we find the maximum value of $\sigma_{x'y'}$,

$$\max \sigma_{x'y'} = \sqrt{\left(\frac{\sigma_{xx} - \sigma_{yy}}{2}\right)^2 + \sigma_{xy}^2} = \frac{\sigma_{\max} - \sigma_{\min}}{2} = \frac{\sigma_1 - \sigma_3}{2} \tag{2-80}$$

In arriving at Eq. (2-80) the trigonometric formula

$$\cos^2 2\theta_p = \frac{1}{1 + \tan^2 2\theta_p}$$

has been used, where $\tan 2\theta_p$ is expressed by Eq. (2-76). Relation (2-79) indicates that maximum shearing stress, Eq. (2-80), occurs on a plane bisecting the angle between the principal planes.* We should remember that Eq. (2-80) was derived under the assumption that the element is rotated about the $Oz$ axis. Rotation about another axis may result in shear stresses larger than indicated by Eq. (2-80).

## 2-3-4 Plane Stress—Mohr's Circle

Mohr's circle is a geometric representation of the relations derived in the previous section between stress components in two different coordinate systems. This visualization is useful in determining both the stresses in the transformed coordinate system, Eqs. (2-73), and the principal stresses, Eqs. (2-75). First we plot a circle with the center at a point $C$ with coordinates $(\sigma_1 + \sigma_2)/2$, 0 and with the radius $R = \sigma_1 - \sigma_2$ in the $\sigma\tau$ plane (Fig. 2-23). Here $\sigma$ and $\tau$ stand for the normal and the shear stress components, respectively. Since $\sigma_{xx}$, $\sigma_{yy}$, and $\sigma_{xy}$ are known, we now find on this circle a point $M$ with coordinates $\sigma_{xx}$ and $\sigma_{xy}$. Next we draw the radius $CP$ making an angle $2\theta$ with the radius $CM$ and find the coordinates $\sigma_p$, $\tau_p$ of the point $P$. As shown in Fig. 2-23,

$$\begin{aligned}\sigma_p &= \frac{\sigma_1 + \sigma_2}{2} + R\cos(2\theta - 2\alpha) \\ &= \frac{\sigma_1 + \sigma_2}{2} + (\sigma_1 - \sigma_2)(\cos 2\theta \cos 2\alpha + \sin 2\theta \sin 2\alpha)\end{aligned} \tag{2-81}$$

$$\tau_p = R\sin(2\theta - 2\alpha) = R(\sin 2\theta \cos 2\alpha - \cos 2\theta \sin 2\alpha) \tag{2-82}$$

*Thus forming a 45° angle with either of the principal planes.

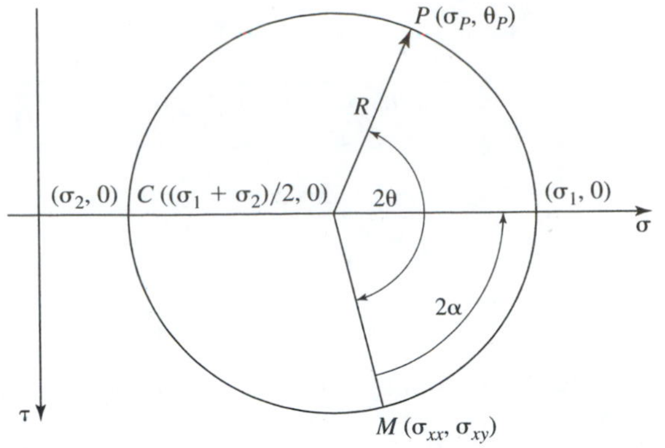


**Figure 2-23** Mohr's circle

But

$$\sin 2\alpha = \frac{\sigma_{xy}}{R}, \qquad \cos 2\alpha = \frac{1}{R}\left(\sigma_{xx} - \frac{\sigma_1 + \sigma_3}{2}\right)$$

and also, from Eqs. (2-75),

$$\sigma_1 + \sigma_2 = \sigma_{xx} + \sigma_{yy}, \qquad \sigma_1 - \sigma_2 = \sqrt{\frac{1}{4}(\sigma_{xx} - \sigma_{yy})^2 + \sigma_{xy}^2}$$

Then Eqs. (2-81) and (2-82) become

$$\sigma_p = \frac{\sigma_{xx} + \sigma_{yy}}{2} + \frac{\sigma_{xx} - \sigma_{yy}}{2} \cos 2\theta + \sigma_{xy} \sin 2\theta \tag{2-83}$$

$$\tau_p = \frac{\sigma_{xx} - \sigma_{yy}}{2} \sin 2\theta - \sigma_{xy} \cos 2\theta \tag{2-84}$$

Finally, comparing Eq. (2-84) with the relations (2-73) and (2-83) with (2-73*a*) we get

$$\sigma_p = \sigma_{x'x}, \qquad \tau_p = -\sigma_{x'y'} \tag{2-85}$$

Eq. (2-73*b*) is obtained from Eq. (2-73*a*) by replacing $\theta$ by $\theta \pm \pi/2$.

## 2-3-5 Octahedral Stresses

Once the principal directions have been found, we can introduce Cartesian coordinates $x$, $y$, $z$ collinear with them. Now all shear stress components vanish by the definition

$$\sigma_{xy} = \sigma_{yz} = \sigma_{xz} = 0 \tag{2-86}$$

whereas normal stress components become

$$\sigma_{xx} = \sigma_1, \qquad \sigma_{yy} = \sigma_2, \qquad \sigma_{zz} = \sigma_3 \tag{2-87}$$

Let us construct a set of planes, each of which is equally inclined to each of the coordinate axes (Fig. 2-24). The normals $\boldsymbol{n}$ to these planes form equal angles with the axes

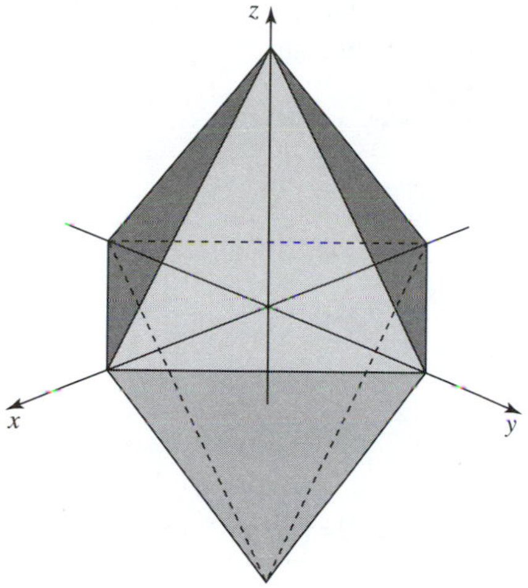


**Figure 2-24** Set of planes equally inclined to the coordinate axes

$Ox$, $Oy$, $Oz$. Thus,

$$n_x = \pm n_y = \pm n_z \tag{2-88}$$

The condition that $\boldsymbol{n}$ be a unit vector gives, using Eq. (2.88),

$$3n_x^2 = 1$$

from which it follows that

$$n_x = \pm\frac{\sqrt{3}}{3}, \qquad n_y = \pm\frac{\sqrt{3}}{3}, \qquad n_z = \pm\frac{\sqrt{3}}{3} \tag{2-89}$$

There are eight possible combinations of values of the components $n_x$, $n_y$, and $n_z$, each corresponding to one of the eight planes in Fig. 2-24. Substituting Eqs. (2-86), (2-87), and (2-88) into Eqs. (2-17) and (2-27) we find the normal and shearing stresses on these planes. These stresses, called *octahedral* stresses, play an important role in the theory of plasticity. They are denoted by $\sigma_{\text{OCT}}$ and $\tau_{\text{OCT}}$,

$$\begin{aligned} \sigma_{\text{OCT}} &= \tfrac{1}{3}(\sigma_1 + \sigma_2 + \sigma_3) = \tfrac{1}{3}I_1 \\ \tau_{\text{OCT}}^2 &= \tfrac{1}{3}\left(\sigma_1^2 + \sigma_2^2 + \sigma_3^2\right) - \sigma_{\text{OCT}}^2 \\ &= \tfrac{1}{3}\left(\sigma_1^2 + \sigma_2^2 + \sigma_3^2\right) - \tfrac{1}{9}(\sigma_1 + \sigma_2 + \sigma_3)^2 \end{aligned} \tag{2-90}$$

Therefore,

$$\begin{aligned} 9\tau_{\text{OCT}}^2 &= 3\left(\sigma_1^2 + \sigma_2^2 + \sigma_3^2\right) - (\sigma_1 + \sigma_2 + \sigma_3)^2 \\ &= (\sigma_1 - \sigma_2)^2 + (\sigma_2 - \sigma_3)^2 + (\sigma_3 - \sigma_1)^2 \\ &= 2I_1^2 - 6I_2 \end{aligned} \tag{2-91}$$

As we know, $I_1$ and $I_2$ do not depend on the coordinate system; therefore $\sigma_{\text{OCT}}$ and $\tau_{\text{OCT}}$ do not depend on it either. Thus, introducing an arbitrary coordinate system $Oxyz$, in which $I_1$ and $I_2$ are expressed by Eqs. (2-68), we obtain from Eqs. (2-90) and (2-91) the following expressions for the octahedral stresses:

$$\begin{aligned}\sigma_{OCT} &= \tfrac{1}{3}(\sigma_{xx}+\sigma_{yy}+\sigma_{zz})\\ 9\tau_{OCT}^2 &= 2(\sigma_{xx}+\sigma_{yy}+\sigma_{zz})^2 - 6\big(-\sigma_{xy}^2-\sigma_{yz}^2-\sigma_{xz}^2+\sigma_{xx}\sigma_{yy}+\sigma_{yy}\sigma_{zz}+\sigma_{zz}\sigma_{xx}\big)\\ &= (\sigma_{xx}-\sigma_{yy})^2+(\sigma_{yy}-\sigma_{zz})^2+(\sigma_{zz}-\sigma_{xx})^2+6\sigma_{xy}^2+6\sigma_{yz}^2+6\sigma_{xz}^2\end{aligned} \tag{2-92}$$

It can be shown (see [1, p. 229]) that

$$\frac{\sqrt{3}}{2} \le \frac{\tau_{OCT}}{\tau_{MAX}} \le 1$$

## 2-3-6 Mean and Deviatoric Stresses

When describing the material behavior of metals one concludes that in certain cases some stress components play a predominant role in comparison with other components. Plastic behavior of metals, in particular, is reported to be independent of the average normal stress $\sigma_M$ defined as follows:

$$\sigma_M = \tfrac{1}{3}(\sigma_1+\sigma_2+\sigma_3) = \tfrac{1}{3}(\sigma_{xx}+\sigma_{yy}+\sigma_{zz}) = \tfrac{1}{3}I_1 = \sigma_{OCT} \tag{2-93}$$

Let us construct a diagonal matrix $\boldsymbol{M} = \sigma_M \boldsymbol{I}$,

$$\boldsymbol{M} = \begin{bmatrix} \sigma_M & 0 & 0 \\ 0 & \sigma_M & 0 \\ 0 & 0 & \sigma_M \end{bmatrix} \tag{2-94}$$

which is called the *mean stress* matrix. We define now the *deviatoric* stresses,

$$\begin{aligned} &\sigma'_{xx} = \sigma_{xx}-\sigma_M, \quad &&\sigma'_{yy} = \sigma_{yy}-\sigma_M, \quad &&\sigma'_{zz} = \sigma_{zz}-\sigma_M \\ &\sigma'_{xy} = \sigma_{xy}, &&\sigma'_{xz} = \sigma_{xz}, &&\sigma'_{yz} = \sigma_{yz} \end{aligned} \tag{2-95}$$

Replacing $\sigma_M$ by Eq. (2-93), we construct the *deviatoric stress matrix* $\boldsymbol{D}$,

$$\boldsymbol{D} = \begin{bmatrix} \tfrac{2}{3}\sigma_{xx}-\tfrac{1}{3}\sigma_{yy}-\tfrac{1}{3}\sigma_{zz} & \sigma_{xy} & \sigma_{xz} \\ \sigma_{xy} & \tfrac{2}{3}\sigma_{yy}-\tfrac{1}{3}\sigma_{xx}-\tfrac{1}{3}\sigma_{zz} & \sigma_{yz} \\ \sigma_{xz} & \sigma_{yz} & \tfrac{2}{3}\sigma_{zz}-\tfrac{1}{3}\sigma_{xx}-\tfrac{1}{3}\sigma_{yy} \end{bmatrix} \tag{2-96}$$

or

$$\boldsymbol{D} = \boldsymbol{S} - \boldsymbol{M} = \boldsymbol{S} - \sigma_M \boldsymbol{I} \tag{2-97}$$

Deviatoric stresses play an important role in the theory of plasticity. It is accepted that, for the most part, they influence the yielding of ductile materials. We next calculate the principal stress deviations following the same procedure that was adopted in calculating the principal stresses. Hence we expand the determinant

$$\det[\boldsymbol{D} - \sigma'_n \boldsymbol{I}] = \det \begin{bmatrix} \sigma'_{xx} - \sigma'_n & \sigma'_{xy} & \sigma'_{xz} \\ \sigma'_{xy} & \sigma'_{yy} - \sigma'_n & \sigma'_{yz} \\ \sigma'_{xz} & \sigma'_{yz} & \sigma'_{zz} - \sigma'_n \end{bmatrix} \tag{2-98}$$

obtaining, in analogy to Eq. (2-67), the following characteristic equation:

$$(\sigma'_n)^3 - I_{1d}(\sigma'_n)^2 + I_{2d}\sigma'_n - I_{3d} = 0 \tag{2-99}$$

where $I_{1d}$, $I_{2d}$, and $I_{3d}$ depend on $\sigma'_{xx}, \ldots$ in the same way in which $I_1$, $I_2$, and $I_3$ depend on $\sigma_{xx}, \ldots$. Therefore [see Eqs. (2-68)],

$$\begin{aligned} I_{1d} &= \sigma'_{xx} + \sigma'_{yy} + \sigma'_{zz} \\ &= \left(\tfrac{2}{3}\sigma_{xx} - \tfrac{1}{3}\sigma_{yy} - \tfrac{1}{3}\sigma_{zz}\right) + \left(\tfrac{2}{3}\sigma_{yy} - \tfrac{1}{3}\sigma_{xx} - \tfrac{1}{3}\sigma_{zz}\right) \\ &\quad + \left(\tfrac{2}{3}\sigma_{zz} - \tfrac{1}{3}\sigma_{xx} - \tfrac{1}{3}\sigma_{yy}\right) = 0 \end{aligned} \tag{2-100}$$

$$\begin{aligned} I_{2d} &= -\sigma'_{xx}\sigma'_{yy} - \sigma'_{yy}\sigma'_{zz} - \sigma'_{zz}\sigma'_{xx} + (\sigma'_{xy})^2 + (\sigma'_{yz})^2 + (\sigma'_{zx})^2 \\ &= \tfrac{1}{2}[(\sigma'_{xx})^2 + (\sigma'_{yy})^2 + (\sigma'_{zz})^2] + (\sigma'_{xy})^2 + (\sigma'_{yz})^2 + (\sigma'_{zx})^2 \\ &= \tfrac{1}{6}[(\sigma_{xx} - \sigma_{yy})^2 + (\sigma_{yy} - \sigma_{zz})^2 + (\sigma_{zz} - \sigma_{xx})^2] + (\sigma_{xy})^2 + (\sigma_{yz})^2 + (\sigma_{zx})^2 \\ &= 3\sigma_M^2 - I_2 \end{aligned} \tag{2-101}$$

where $I_2$ is given in Eq. (2-68*b*). Lastly,

$$I_{3d} = I_3 - \sigma_M I_2 + 2\sigma_M^3 = I_3 + \sigma_M I_{2d} - \sigma_M^3 \tag{2-102}$$

Substituting now the relationship (2-97) into Eq. (2-98), we obtain

$$\det[\boldsymbol{S} - (\sigma'_n + \sigma_M)\boldsymbol{I}] = 0 \tag{2-103}$$

Comparing Eq. (2-103) with Eq. (2-66) we see that it can be obtained from the latter by replacing $\sigma'_n + \sigma_M$ by $\sigma_n$. Writing Eq. (2-99) in the same form as Eq. (2-67), we obtain its roots by subtracting $\sigma_M$ from the roots of Eq. (2-67),

$$\sigma'_1 = \sigma_1 - \sigma_M, \qquad \sigma'_2 = \sigma_2 - \sigma_M, \qquad \sigma'_3 = \sigma_3 - \sigma_M \tag{2-104}$$

This reasoning also proves that the principal directions coincide in both cases.

## 2-4 THREE-DIMENSIONAL MOHR'S CIRCLES

Once the principal stresses at a point $O$ are known, the stresses acting on an arbitrarily inclined plane through this point can be found using expression (2-39) given in Section 2-2-5. Specifically one can calculate the normal stress $\sigma_n$ and the tangential stress $\tau$

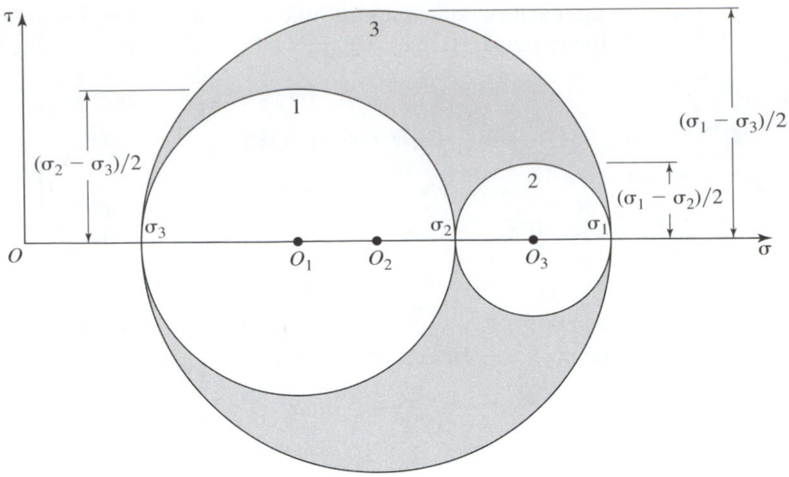


**Figure 2-25** Coordinate system $\sigma_n$, $\tau$

acting on this plane. An insight into the relation between the principal stresses $\sigma_1, \sigma_2, \sigma_3$ on one side and $\sigma_n$ and $\tau$ on the other can be gained from the geometrical construction devised by Otto Mohr* and known as three-dimensional Mohr's circles. First we introduce a coordinate system $\sigma_n$, $\tau$ (Fig. 2-25) and, assuming that $\sigma_1 > \sigma_2 > \sigma_3$, we mark these quantities on the $\sigma$ axis. Next we plot three circles with radii $(\sigma_1 - \sigma_3)/2$, $(\sigma_2 - \sigma_3)/2$, and $(\sigma_1 - \sigma_2)/2$, respectively.

It will be shown that for any possible rotation of the inclined plane the stresses $\sigma_n$ and $\tau$ acting on it will always be located within the shaded area, as seen in Fig. 2-25. To this end let us select arbitrarily a plane through $O$ with a unit normal $\boldsymbol{n}$. Hence,

$$n_x^2 + n_y^2 + n_z^2 = 1 \tag{2-105}$$

There is an orientation of the plane for which

$$\sigma_{xx} = \sigma_1, \qquad \sigma_{yy} = \sigma_2, \qquad \sigma_{zz} = \sigma_3$$

and

$$\sigma_{xy} = \sigma_{yz} = \sigma_{zx} = 0$$

Therefore, with Eq. (2-17) we get

$$\sigma_n = \sigma_1 n_x^2 + \sigma_2 n_y^2 + \sigma_3 n_z^2 \tag{2-106}$$

Similarly, from Eq. (2-27) we obtain

$$\tau^2 = \sigma_1^2 n_x^2 + \sigma_2^2 n_y^2 + \sigma_3^2 n_z^2 - \sigma_n^2 \tag{2-107}$$

*Otto Mohr (1835–1918), German civil engineer.

On solving Eqs. (2-105), (2-106), and (2-107) for $n_x^2$, $n_y^2$, and $n_z^2$, and noting that all these quantities are nonnegative, we obtain the following inequalities:

$$n_x^2 = \frac{\tau^2 + (\sigma_n - \sigma_2)(\sigma_n - \sigma_3)}{(\sigma_1 - \sigma_2)(\sigma_1 - \sigma_3)} \geq 0 \tag{2-108$a$}$$

$$n_y^2 = \frac{\tau^2 + (\sigma_n - \sigma_3)(\sigma_n - \sigma_1)}{(\sigma_2 - \sigma_3)(\sigma_2 - \sigma_1)} \geq 0 \tag{2-108$b$}$$

$$n_z^2 = \frac{\tau^2 + (\sigma_n - \sigma_1)(\sigma_n - \sigma_2)}{(\sigma_3 - \sigma_1)(\sigma_3 - \sigma_2)} \geq 0 \tag{2-108$c$}$$

Note that since the denominators in Eq. (2-108*a*, *c*) are positive and the denominator in Eq. (2-108*b*) is negative, the following is true:

$$\begin{aligned} \tau^2 + (\sigma_n - \sigma_2)(\sigma_n - \sigma_3) &\geq 0 \\ \tau^2 + (\sigma_n - \sigma_3)(\sigma_n - \sigma_1) &\leq 0 \\ \tau^2 + (\sigma_n - \sigma_1)(\sigma_n - \sigma_2) &\geq 0 \end{aligned} \tag{2-109}$$

Simple algebraic manipulations show that we may also write the inequalities (2-109) in the following form:

$$\left[\sigma_n - \tfrac{1}{2}(\sigma_2 + \sigma_3)\right]^2 + \tau^2 \geq \tfrac{1}{4}(\sigma_2 - \sigma_3)^2 \tag{2-110$a$}$$

$$\left[\sigma_n - \tfrac{1}{2}(\sigma_3 + \sigma_1)\right]^2 + \tau^2 \leq \tfrac{1}{4}(\sigma_3 - \sigma_1)^2 \tag{2-110$b$}$$

$$\left[\sigma_n - \tfrac{1}{2}(\sigma_1 + \sigma_2)\right]^2 + \tau^2 \geq \tfrac{1}{4}(\sigma_1 - \sigma_2)^2 \tag{2-110$c$}$$

Now let us consider an arbitrary point $P$ in the $\sigma_n$, $\tau$ plane. Because of the symmetry of Fig. 2-25, the choice of the sign of the coordinate $\tau$ is immaterial. Let us therefore choose $\tau > 0$. Note that since the coordinates of the center $O_1$ of circle 1 are

$$\sigma_{n1} = \frac{\sigma_2 + \sigma_3}{2}, \qquad \tau_1 = 0$$

the square of the distance $PO_1$ equals

$$\left(\sigma_n - \frac{\sigma_2 + \sigma_3}{2}\right)^2 + \tau^2$$

This is exactly the left-hand side of the inequality (2-110*a*), whose right-hand side represents the square of the radius of circle 1. Thus this inequality demonstrates that whatever the choice of the inclined plane, the stresses $\sigma_n$ and $\tau$ are such that the point $P$ is outside circle 1. Similar reasoning applied to inequality (2-110*b*) implies that point $P$ must be inside circle 2, whereas inequality (2-110*c*) shows that $P$ must be outside circle 3. Therefore the inequalities (2-110) prove that the magnitudes of the stresses $\sigma_n$ and $\tau$ are such that the point $P$ with coordinates $\sigma_n$ and $\tau$ will indeed be located within the shaded area of Fig. 2-25. From Fig. 2-25, or from Eq. (2-110*b*), it is also evident that

$$\max \tau = \frac{\sigma_1 - \sigma_3}{2}$$

## 2-5 STRESS ANALYSIS AND SYMBOLIC MANIPULATION

Most of the time the stress analyst will employ numerical solutions using commercially available packages or developing his or her own. Sometimes, however, the analyst may face the necessity of developing either exact or approximate formulas. This usually requires some intellectual effort, but it also involves many straightforward but tedious and error-prone algebraic, differential, or similar manipulations.

During the past few decades several programs were developed whose objective is to free engineers, physicists, and others from spending hundreds of hours on these manipulations. These packages, known as *symbolic manipulation programs,* let the computer operate on symbols rather than on numbers (they do not, however, exclude numerical calculations), effectively allowing the user to perform more challenging tasks. Symbolic manipulation packages are available for both main-frame computers and personal computers. In this text we have used Macsyma [4], Maple [6] and Mathcad [7] programs. Obviously the programs residing on large computers are more powerful than the programs adapted for personal computers, but still, the personal computer packages can perform an amazing assortment of tasks. A few of these are algebraic manipulations, solution of systems of linear algebraic equations, operations on matrices, differentiation, integration, vector and tensor analyses, and others. The following examples show some simple applications of Mathcad7. In other chapters, notably in Chapters 8, 9, and 10, we discuss additional applications.

### EXAMPLE 2-7

Knowing the elements of the matrix $\mathbf{S} - \sigma_n \mathbf{I}$ [Eq. (2-64)] derive the characteristic Eq. (2-67).

### SOLUTION

After the Mathcad work space appears, we define the matrices $\mathbf{S}$, $\sigma_n$, and $\mathbf{I}$:

$$\mathbf{S} - \sigma_n \mathbf{I} = \begin{bmatrix} \sigma_{xx} - \sigma_n & \sigma_{xy} & \sigma_{xz} \\ \sigma_{xy} & \sigma_{yy} - \sigma_n & \sigma_{yz} \\ \sigma_{xz} & \sigma_{yz} & \sigma_{zz} - \sigma_n \end{bmatrix}$$

We calculate the determinant of this matrix by clicking on the square matrix and then on SPACE. Then click on SYMBOLICS > MATRIX > DETERMINANT, and obtain the following result:

$$\sigma_{xx}\cdot\sigma_{yy}\cdot\sigma_{zz}-\sigma_{xx}\cdot\sigma_{yy}\cdot\sigma_n-\sigma_{xx}\cdot\sigma_n\cdot\sigma_{zz}+\sigma_{xx}\cdot\sigma_n^2-\sigma_{xx}\cdot\sigma_{yz}^2$$
$$-\sigma_n\cdot\sigma_{yy}\cdot\sigma_{zz}+\sigma_{yy}\cdot\sigma_n^2+\sigma_{zz}\cdot\sigma_n^2-\sigma_n^3+\sigma_n\cdot\sigma_{yz}^2-\sigma_{xy}^2\cdot\sigma_{xz}$$
$$+\sigma_{xy}^2\cdot\sigma_n+2\cdot\sigma_{xy}\cdot\sigma_{xz}\cdot\sigma_{yz}-\sigma_{xz}^2\cdot\sigma_{yy}+\sigma_{xz}^2\cdot\sigma_n=0 \qquad (a)$$

To calculate the coefficients of decreasing powers of $\sigma_n$, we select $\sigma_n$ in the preceding equation and then SYMBOLICS > POLYNOMIAL COEFFICIENTS. This results in

$$\begin{bmatrix} \sigma_{xx}\cdot\sigma_{yy}\cdot\sigma_{zz}-\sigma_{xx}\cdot\sigma_{yz}^2-\sigma_{xy}^2\cdot\sigma_{zz}+2\cdot\sigma_{xy}\cdot\sigma_{xz}\cdot\sigma_{yz}-\sigma_{xz}^2\cdot\sigma_{yy} \\ -\sigma_{xx}\cdot\sigma_{yy}-\sigma_{xx}\cdot\sigma_{zz}-\sigma_{yy}\cdot\sigma_{zz}+\sigma_{yz}^2+\sigma_{xy}^2+\sigma_{xz}^2 \\ \sigma_{xx}+\sigma_{yy}+\sigma_{zz} \\ -1 \end{bmatrix}$$

The top row in the preceding expression is the coefficient of $\sigma_n^0$. The second, third, and fourth rows are the coefficients of $\sigma_n$ to the power of 1, 2, and 3, respectively. Denoting

$$\sigma_{xx}\cdot\sigma_{yy}\cdot\sigma_{zz}-\sigma_{xx}\cdot\sigma_{yz}^2-\sigma_{xy}^2\cdot\sigma_{zz}+2\cdot\sigma_{xy}\cdot\sigma_{xz}\cdot\sigma_{yz}-\sigma_{xz}^2\cdot\sigma_{yy}=I_3$$
$$-\sigma_{xx}\cdot\sigma_{yy}-\sigma_{xx}\cdot\sigma_{zz}-\sigma_{yy}\cdot\sigma_{zz}+\sigma_{yz}^2+\sigma_{xy}^2+\sigma_{xz}^2=-I_2$$
$$\sigma_{xx}+\sigma_{yy}+\sigma_{zz}=I_1$$

we write Eq. ($a$) in the form of Eq. (2-67),

$$\sigma^3-I_1\sigma^2+I_2\sigma^1-I_3=0$$

## EXAMPLE 2-8

Given the matrices $\boldsymbol{S}$ [Eq. (2-6)], and $\boldsymbol{N}$ [Eq. (2-40)], find the elements of the matrix $\boldsymbol{S}'$ [Eq. (2-40)] using the transformation formula (2-39).

### SOLUTION

To define the matrices symbolically we use the boldface equal sign obtained by using CTRL+= and then use INSERT → MATRIX. From Eq. (2-6) where, for convenience we replaced $\sigma$ by $s$, we get

$$\boldsymbol{S}=\begin{bmatrix} s_{xx} & s_{xy} & s_{xz} \\ s_{xy} & s_{yy} & s_{yz} \\ s_{xz} & s_{yz} & s_{zz} \end{bmatrix}$$

and from Eq. (2-40), using INSERT > MATRIX, we get

$$N = \begin{bmatrix} n_{x'x} & n_{y'x} & n_{z'x} \\ n_{x'y} & n_{y'y} & n_{z'y} \\ n_{x'z} & n_{y'z} & n_{z'z} \end{bmatrix}$$

Now we must generate the matrix $\boldsymbol{N}^T$. To this end we click on the matrix and press SPACE until the matrix is between the editing lines. When we choose MATRIX > TRANSPOSE from the SYMBOLICS menu, we get

$$N^T = \begin{bmatrix} n_{x'x} & n_{x'y} & n_{x'z} \\ n_{y'x} & n_{y'y} & n_{y'z} \\ n_{z'x} & n_{z'y} & n_{z'z} \end{bmatrix}$$

Finally we must generate the matrix product $\boldsymbol{N}^T\boldsymbol{SN}$. To this end we first multiply the matrix $\boldsymbol{N}^T$ by $\boldsymbol{S}$, generating the product matrix $\boldsymbol{P}$ (use matrix palette and select a product of two vectors). Using the same procedure, we multiply the resulting matrix in turn by the matrix $\boldsymbol{N}$,

$$P = \begin{bmatrix} n_{x'x} & n_{x'y} & n_{x'z} \\ n_{y'x} & n_{y'y} & n_{y'z} \\ n_{z'x} & n_{z'y} & n_{z'z} \end{bmatrix} \cdot \begin{bmatrix} s_{xx} & s_{xy} & s_{xz} \\ s_{xy} & s_{yy} & s_{yz} \\ s_{xz} & s_{yz} & s_{zz} \end{bmatrix}$$

$$S' = \begin{bmatrix} n_{x'x} & n_{x'y} & n_{x'z} \\ n_{y'x} & n_{y'y} & n_{y'z} \\ n_{z'x} & n_{z'y} & n_{z'z} \end{bmatrix} \cdot \begin{bmatrix} s_{xx} & s_{xy} & s_{xz} \\ s_{xy} & s_{yy} & s_{yz} \\ s_{xz} & s_{yz} & s_{zz} \end{bmatrix} \cdot \begin{bmatrix} n_{x'x} & n_{y'x} & n_{z'x} \\ n_{x'y} & n_{y'y} & n_{z'y} \\ n_{x'z} & n_{y'z} & n_{z'z} \end{bmatrix}$$

$$P = \begin{bmatrix} n_{x'x}\cdot s_{xx} + n_{x'y}\cdot s_{xy} + n_{x'z}\cdot s_{xz} & n_{x'x}\cdot s_{xy} + n_{x'y}\cdot s_{yy} + n_{x'z}\cdot s_{yz} & n_{x'x}\cdot s_{xz} + n_{x'y}\cdot s_{yz} + n_{x'z}\cdot s_{zz} \\ n_{y'x}\cdot s_{xx} + n_{y'y}\cdot s_{xy} + n_{y'z}\cdot s_{xz} & n_{y'x}\cdot s_{xy} + n_{y'y}\cdot s_{yy} + n_{y'z}\cdot s_{yz} & n_{y'x}\cdot s_{xz} + n_{y'y}\cdot s_{yz} + n_{y'z}\cdot s_{zz} \\ n_{z'x}\cdot s_{xx} + n_{z'y}\cdot s_{xy} + n_{z'z}\cdot s_{xz} & n_{z'x}\cdot s_{xy} + n_{z'y}\cdot s_{yy} + n_{z'z}\cdot s_{yz} & n_{z'x}\cdot s_{xz} + n_{z'y}\cdot s_{yz} + n_{z'z}\cdot s_{zz} \end{bmatrix}$$

$$S' = \begin{bmatrix} n_{x'x}\cdot s_{xx} + n_{x'y}\cdot s_{xy} + n_{x'z}\cdot s_{xz} & n_{x'x}\cdot s_{xy} + n_{x'y}\cdot s_{yy} + n_{x'z}\cdot s_{yz} & n_{x'x}\cdot s_{xz} + n_{x'y}\cdot s_{yz} + n_{x'z}\cdot s_{zz} \\ n_{y'x}\cdot s_{xx} + n_{y'y}\cdot s_{xy} + n_{y'z}\cdot s_{xz} & n_{y'x}\cdot s_{xy} + n_{y'y}\cdot s_{yy} + n_{y'z}\cdot s_{yz} & n_{y'x}\cdot s_{xz} + n_{y'y}\cdot s_{yz} + n_{y'z}\cdot s_{zz} \\ n_{z'x}\cdot s_{xx} + n_{z'y}\cdot s_{xy} + n_{z'z}\cdot s_{xz} & n_{z'x}\cdot s_{xy} + n_{z'y}\cdot s_{yy} + n_{z'z}\cdot s_{yz} & n_{z'x}\cdot s_{xz} + n_{z'y}\cdot s_{yz} + n_{z'z}\cdot s_{zz} \end{bmatrix}$$

$$\times \begin{bmatrix} n_{x'x} & n_{y'x} & n_{z'x} \\ n_{x'y} & n_{y'y} & n_{z'y} \\ n_{x'z} & n_{y'z} & n_{z'z} \end{bmatrix}$$

For instance, the term $\boldsymbol{S}^1$ becomes

$$n_{x'x}^2 \cdot s_{xx} + 2n_{x'x}\cdot n_{x'y}\cdot s_{xy} + 2n_{x'x}\cdot n_{x'z}\cdot s_{xz} + n_{x'y}^2\cdot s_{yy} + 2n_{x'y}\cdot n_{x'z}\cdot s_{yz} + n_{x'z}^2\cdot s_{zz}$$

It is seen that the elements of the matrix $\boldsymbol{S}'$ are identical to the components (2-35), (2-37), and (2-38), and to the remaining stress components, which we did not list here.

## PROBLEMS

### Section 2-2

**2-1** Derive Eqs. (2-5).

**2-2** Show all steps in the derivation of Eq. (2-13).

**2-3** Derive Eq. (2-17), then write it as a product of three matrices.

**2-4** Determine the boundary conditions in a prismatic solid with the cross section shown in Fig. P2-4. Note that the face $AB$ is fixed.

**2-5** Determine the boundary conditions in a disk shown in Fig. P2-5.

**2-6** Derive Eq. (2-31).

**2-7** The stress matrix $S$ at a point $M$ is given by

$$S = \begin{bmatrix} 2 & 1 & 2 \\ 1 & 3 & 0 \\ 2 & 0 & 2 \end{bmatrix}$$

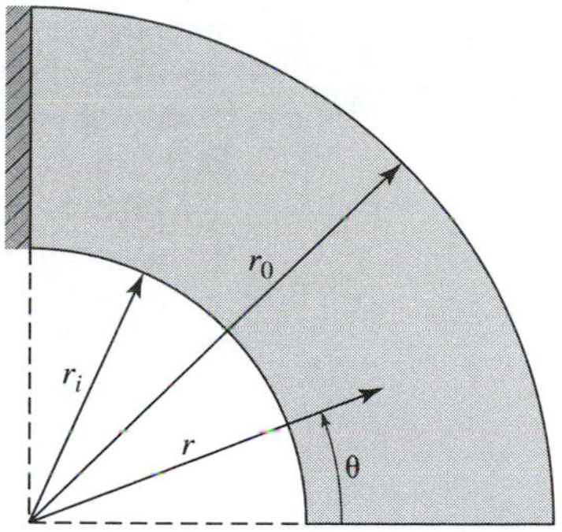

Figure P2-5

Draw a plane $T$ through $M$ with the direction cosines

$$n_x = \frac{1}{\sqrt{3}}, \qquad n_y = \frac{1}{\sqrt{3}}, \qquad n_z = \frac{1}{\sqrt{3}}$$

Find:

(a) The traction vector $t^{(n)}$ acting on $T$ at $M$.

(b) Its Cartesian components $t_x^{(n)}$, $t_y^{(n)}$, and $t_z^{(n)}$.

(c) Its normal component $\sigma_n$.

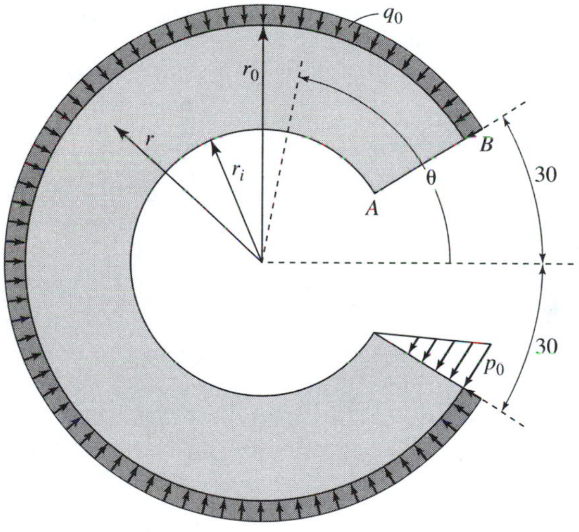

Figure P2-4

**2-8** A solid, rectangular bar is subject to uniform tension $q$ per unit area (Fig. P2-8). The state of stress inside the bar was found to be also uniform such that $\sigma_{xx} = q$, and all other stress components are equal to zero. Find the normal component $\sigma_n$ and the tangential component $\tau$ of the traction $\boldsymbol{t}^{(n)}$ acting at an inside point on the plane with the outer normal $\boldsymbol{N} = \boldsymbol{i} + 2\boldsymbol{j} + \boldsymbol{k}$. *Hint:* $\boldsymbol{n} = \boldsymbol{N}/|\boldsymbol{N}|$.

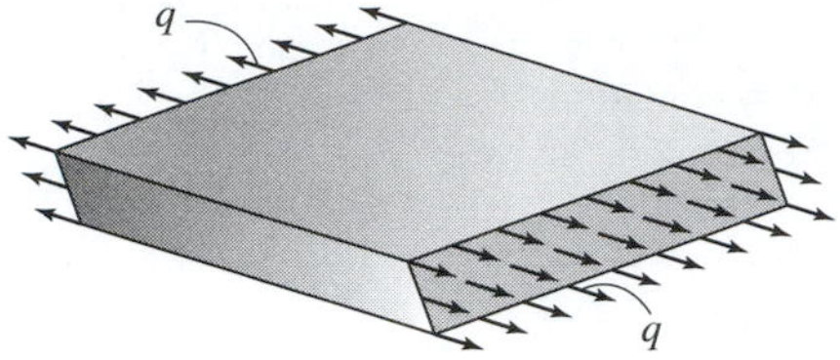


Figure P2-8

**2-9** Stress components at a point are known, except for the component $\sigma_{yy}$,

$$S = \begin{bmatrix} 1 & 0 & 1 \\ 0 & \sigma_{yy} & 2 \\ 1 & 2 & 0 \end{bmatrix}$$

Select $\sigma_{yy}$ so that there will be a traction-free plane through the point. Determine its unit normal.

**2-10** In the beam shown in Fig. P2-10 consider a plane perpendicular to the $Oxy$ plane, forming a 30° angle with the $Oy$ axis and passing through the point $M$ with the coordinates $x = L/2$, $y = 0$. Determine the stresses acting on this plane at $M$ using the local Cartesian coordinates $x'$, $y'$, $z'$ with the origin at $M$, where the plane $Mx'y'$ is identical to the plane $Oxy$.

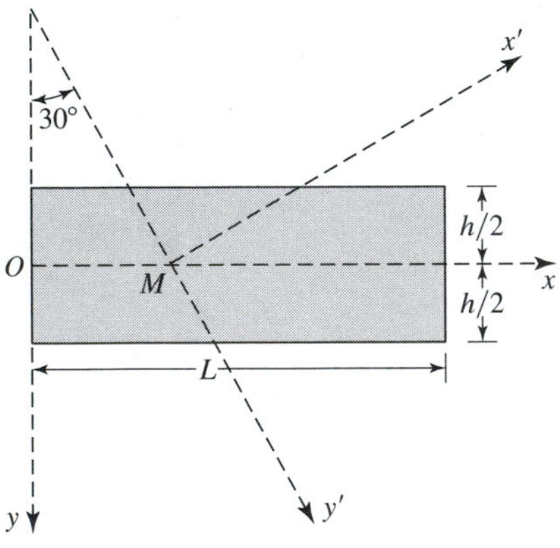


Figure P2-10

**2-11** The state of stress at a point $M$ of a solid is described in a Cartesian coordinate system $x$, $y$, $z$ by

$$S = \begin{bmatrix} 0 & 1 & 2 \\ 1 & 2 & 3 \\ 2 & 3 & 4 \end{bmatrix} \text{lb/in}^2$$

Determine the elements of the stress matrix at the same point $M$ relative to a Cartesian coordinate system $x'$, $y'$, $z'$ if the direction cosines are

$$n_{x'x} = 0, \qquad n_{x'y} = 0, \qquad n_{x'z} = 1,$$

$$n_{y'x} = n_{y'y} = \frac{\sqrt{2}}{2}, \qquad n_{y'z} = 0,$$

$$n_{z'x} = -n_{z'y} = \frac{\sqrt{2}}{2}, \qquad n_{z'z} = 0$$

**2-12** Derive Eqs. (2-37) and (2-38).

**2-13** Obtain Eqs. (2-42) from Eq. (2-44). Show all your steps.

**2-14** Use matrix multiplication to prove the validity of the expressions (2-52).

**2-15** Check whether the beam from Example 2-2 is in equilibrium.

**2-16** Given the stress distribution

$$\sigma_{xx} = -\frac{Pxy}{I_z}, \qquad \sigma_{yy} = 0,$$

$$\sigma_{xy} = -\frac{P}{2I_z}\left(\frac{h^2}{4} - y^2\right),$$

$$\sigma_{zz} = \sigma_{xz} = \sigma_{yz} = 0$$

in the narrow beam shown in Fig. P2-10. Plot the stress distribution along $x = 0$, $x = L/2$,

$y = 0$, $y = h/2$, and $x = L$. Analyze the equilibrium of the beam and discuss the results.

**2-17** Consider the beam of Fig. P2-10 assuming the stress distribution in the form

$$\sigma_{xx} = \frac{q}{5bh^3}\left[20y^3 - 30y\left(x^2 - \frac{L^2}{4}\right) - 3yh^2\right]$$

$$\sigma_{yy} = \frac{q}{2bh^3}(-4y^3 + 3yh^2 - h^3)$$

$$\sigma_{xy} = \frac{q}{2bh^3}(12xy^2 - 3xh^2)$$

$$\sigma_{zz} = \sigma_{xz} = \sigma_{yz} = 0$$

Determine the stress distribution along $x = 0$, $x = L/2$, $x = L$, $y = 0$, and $y = h/2$. Analyze the equilibrium of the beam and discuss the results.

**2-18** The stresses in the beam shown in Fig. P2-10 are

$$\sigma_{xx} = -\frac{M}{I_z}y,$$

$$\sigma_{yy} = \sigma_{zz} = \sigma_{xy} = \sigma_{xz} = \sigma_{yz} = 0$$

Discuss the equilibrium of the beam.

**2-19** Let the stresses in the beam in Fig. P2-10 be

$$\sigma_{xx} = \frac{2qy}{bh^3}\left[-3L^2\left(1 + \frac{h^2}{10L^2}\right) + 2y^2 + 3\left(2 - \frac{x}{L}\right)xL\right]$$

$$\sigma_{yy} = -\frac{q}{2bh^3}[h^3 - (3h^2 - 4y^2)y]$$

$$\sigma_{xy} = -\frac{6q}{bh^3}\left(\frac{h^2}{4} - y^2\right)(x - L)$$

$$\sigma_{zz} = \sigma_{xz} = \sigma_{yz} = 0$$

Determine the stress distribution along $x = 0$, $x = L$, $y = 0$, $y = h/2$, and $y = -h/2$. Discuss the overall equilibrium of the beam.

**2-20** Derive Eqs. (2-53) showing all the details. Write it also in matrix form.

**2-21** Assume that $b_x = b_y = b_z = 0$. Verify whether the stress system of Problem 2-16 satisfies the equilibrium equation at any point of the beam.

**2-22** Repeat Problem 2-21, for the stress system defined in Problem 2-17.

**2-23** Repeat Problem 2-21, for the stress system of Problem 2-18.

**2-24** Derive the last two of Eqs. (2-58). Show all your steps.

**2-25** Determine whether the following stress system satisfies the equilibrium Eqs. (2-60):

$$\sigma_{rr} = P\left(r + \frac{r_i^2 r_0^2}{r^3} - \frac{r_i^2 + r_0^2}{r}\right)\sin\theta$$

$$\sigma_{\theta\theta} = P\left(3r - \frac{r_i^2 r_0^2}{r^3} - \frac{r_i^2 + r_0^2}{r}\right)\sin\theta$$

$$\sigma_{r\theta} = -P\left(r + \frac{r_i^2 r_0^2}{r^3} - \frac{r_i^2 + r_0^2}{r}\right)\cos\theta$$

$$\sigma_{zz} = \sigma_{rz} = \sigma_{\theta z} = 0$$

*Note:* Assume that $b_r = b_\theta = b_z = 0$. What stress boundary conditions are satisfied?

**2-26** When the state of stress is axisymmetric, then $\sigma_{r\theta} = \sigma_{rz} = 0$ while the other stress components are independent of $\theta$. Show that if the body force is zero, then the equations of equilibrium (2-58) are identically satisfied by the stresses represented by

$$\sigma_{rr} = \frac{\partial^2 F}{\partial z^2} + \frac{1}{r}\frac{\partial G}{\partial r},$$

$$\sigma_{\theta\theta} = \frac{\partial^2 F}{\partial z^2} + \frac{\partial^2 G}{\partial r^2},$$

$$\sigma_{zz} = \frac{\partial^2 F}{\partial r^2} + \frac{1}{r}\frac{\partial F}{\partial r},$$

$$\sigma_{rz} = -\frac{\partial^2 F}{\partial r \partial z}$$

where $F(r, z)$ and $G(r, z)$ are arbitrary functions.

## Section 2-3

**2-27** The stress matrix at a point $M$ of the solid is

$$S = \begin{bmatrix} 1.0 & 0.5 & -1.0 \\ 0.5 & 2.0 & -1.5 \\ -1.0 & -1.5 & -1.0 \end{bmatrix} \times 10^3 \text{ psi}$$

Find the stress invariants, the principal stresses, and the corresponding principal directions.

**2-28** The stress matrix is

$$S = \begin{bmatrix} 1 & 0 & 2 \\ 0 & 3 & 1 \\ 2 & 1 & 4 \end{bmatrix}$$

(a) Determine the principal stresses.

(b) Find the direction cosines of the normal to the plane on which acts $\sigma_{max}$.

**2-29** By deriving Eq. (2-69) prove that at any point the sum of normal stresses is an invariant. Show all the details.

**2-30** Derive Eqs. (2-73) from Eqs. (2-72).

**2-31** Derive Eq. (2-80) relating maximum shearing stress to principal stresses.

**2-32** Derive Eq. (2-90).

**2-33** Derive Eq. (2-91).

**2-34** The stress matrix is

$$S = \begin{bmatrix} 2 & 1 & 0 \\ 1 & 4 & 2 \\ 0 & 2 & 3 \end{bmatrix}$$

Find:

(a) The principal stresses.

(b) The octahedral shear stress.

## Section 2-4

**2-35** Use Eqs. (2-109) to derive the inequalities (2-110).

## Section 2-5

**2-36** Use any symbolic manipulation package to derive the following equations:

(a) Eqs. (2-31), Section 2-2.

(b) Eqs. (2-37) and (2-38), Section 2-2.

(c) Eqs. (2-49), Section 2-2.

(d) Eqs. (2-52), Secton 2-2.

(e) Eq. (2-69), Section 2-3.

(f) Eqs. (2-72), Section 2-3.

(g) Eq. (2-102), Section 2-3.

## REFERENCES


[1] Dowling, N. E., *Mechanical Behavior of Materials,* 2nd ed. (Prentice-Hall, Upper Saddle River, NJ, 1999).

[2] Fung, Y. C., *A First Course in Continuum Mechanics,* 2nd ed. (Prentice-Hall, Englewood Cliffs, NJ, 1977).

[3] Kreidler, D. L., et al., *An Introduction to Linear Analysis* (Addison-Wesley, Reading, MA, 1966).

[4] *Macsyma,* version 13 (Symbolics, Inc., Burlington, MA, 1988).

[5] Malvern, L., *Introduction to the Mechanics of a Continuous Medium,* (Prentice-Hall, Englewood Cliffs, NJ, 1969).

[6] *Maple V* (Waterloo Maple Inc., Waterloo, ON, Canada, 1998).

[7] *Mathcad7,* professional ed. (MathSoft, Cambridge, MA, 1997).

# DISPLACEMENT AND STRAIN

As we indicated in Section 2-2-9, the equilibrium equations contain more unknowns than equations so that in order to complete the formulation, additional equations must be developed. These fall into two categories. One category is the strain-displacement equations, which will be developed in this chapter. These equations give relationships between the displacements that a body undergoes and the strains that result in the body. Both strain and displacement are geometric concepts and therefore have no obvious connection with the concept of stress. The remaining category of equations is the stress-strain equations or, more generally, constitutive equations. These equations provide the necessary link between stress and strain, and their development will be taken up in Chapter 4.

We begin by deriving relationships between the strains and the displacement components and then show that the strains are not all independent, but are related through the compatibility equations. We then show that the strain-displacement relationships completely describe the state of strain at a point in a body. The concept of principal strains is then developed, along with several other useful concepts concerning strain. We close the chapter by deriving the strain-displacement and compatibility equations in polar coordinates.

## 3-1 INTRODUCTION

When loads are applied to a body made up of a nonrigid material, the shape of the body changes and we say that the body *deforms*. What we mean is that the material points of the body change position relative to one another so that some points move closer together while others move apart. In Fig. 3-1 the position of a body prior to loading is indicated by $B$, and after loading its position is indicated by $B'$. The two points $P$ and $Q$ in the body before deformation are now located at $P'$ and $Q'$, and the distance between them has changed. Point $P$ has undergone a displacement $u_P$ while point $Q$ has undergone a displacement $u_Q$. Note that displacements are vector quantities since they have directional orientation, and that they are measured from the undeformed configuration

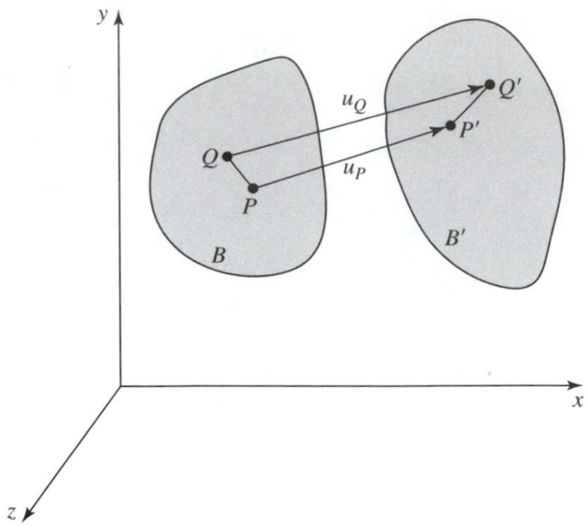


Figure 3-1 Change of position and deformation

of the body. The displacement that a material point undergoes is, in general, a function of the point's position within the body. Thus, for any point with coordinates $x$, $y$, $z$ we may write

$$\boldsymbol{u} = \boldsymbol{u}(x, y, z)$$

We will assume that the displacement varies continuously throughout the body, that it is a single-valued function of position, and also that it is differentiable. In Section 3-3 we will derive an equation whose satisfaction guarantees the validity of these assumptions.

The results obtained in this chapter involve only geometric considerations. Since there is no reference to material behavior, the results apply to all materials, including fluids. The only restriction on the results, as will be seen in Section 3-2, is that the strains be infinitesimal.

## 3-2 STRAIN-DISPLACEMENT EQUATIONS

We want to develop expressions relating strain and displacement. A body can undergo two types of strain—normal strain and shear strain. *Normal* strain, designated by $\varepsilon$, is the ratio of the change in length to the original length of a line segment. For example, in Fig. 3-1 the normal strain undergone by the line segment $PQ$ is

$$\varepsilon = \frac{P'Q' - PQ}{PQ} \tag{3-1}$$

*Shear* strain, designated by $\gamma$, is the change in angle between two originally perpendicular lines. In Fig. 3-2 the lines $PQ$ and $PR$ are originally perpendicular. In the deformed configuration, the angle between these lines is $\theta$. The change in angle is $\pi/2 - \theta$, or the sum of $\alpha$ and $\beta$, which are both positive, as shown. The shearing strain is therefore given by

$$\gamma = \alpha + \beta \tag{3-2}$$

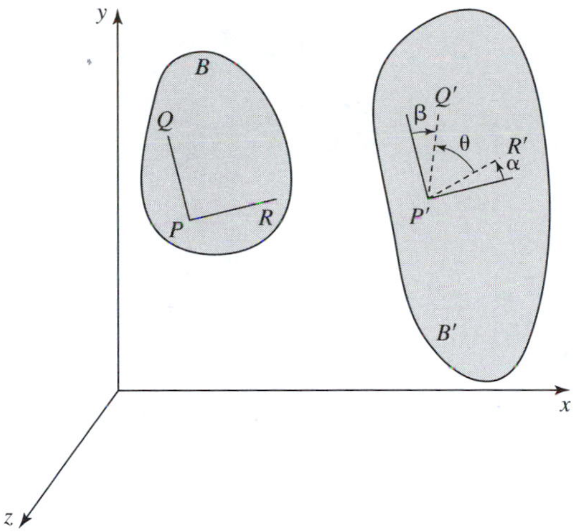


Figure 3-2

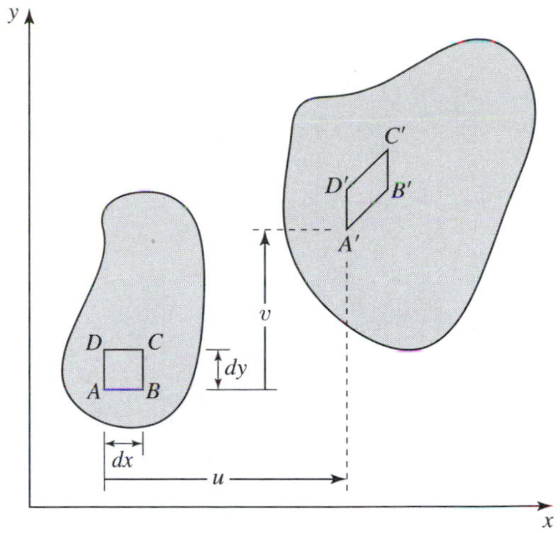


Figure 3-3

Now let us consider what happens to an infinitesimal element of material as the body within which it is located deforms, as shown in Fig. 3-3. Point $A$ on the element moves to a new location $A'$ and in so doing undergoes a displacement $u$ in the $x$ direction and $v$ in the $y$ direction. Point $B$ moves to $B'$, and since point $B$ is located a distance $dx$ from point $A$, its displacement will, in general, be different from that of point $A$. However, in the unloaded configuration points $A$ and $B$ are only a differential distance apart. Therefore we expect that the displacement of point $B$ will differ from that of point $A$ by only a differential amount. The displacement of point $B$ in the $x$ direction is then $u + du$, and in the $y$ direction it is $v + dv$. These can be written as $u + (\partial u/\partial x)\, dx$ and $v + (\partial v/\partial y)\, dx$, respectively.* Point $D$ can be treated similarly, resulting in the displacements shown in Fig. 3-4. From the figure we see that $A'B'$ is the hypotenuse of a right triangle, so that we can write

$$(A'B')^2 = \left(dx + \frac{\partial u}{\partial x}\, dx\right)^2 + \left(\frac{\partial v}{\partial x}\, dx\right)^2 \tag{3-3}$$

The normal strain in the $x$ direction is given by

$$\varepsilon_{xx} = \frac{A'B' - AB}{AB} = \frac{A'B' - dx}{dx} \tag{3-4}$$

---

*Mathematically we first express $u$ at point $B$ as a Taylor series about point $A$:

$$u_B = u_A + dx \left.\frac{\partial u}{\partial x}\right|_{x=x_A} + \frac{(dx)^2}{2!} \left.\frac{\partial^2 u}{\partial x^2}\right|_{x=x_A} + \cdots$$

Then since $dx$ is an infinitesimal quantity, higher order terms can be neglected by truncating the series after the second term. A similar approach is taken for $v$. Since $u$ is a function of both $x$ and $y$, can you explain why none of the $y$ terms appear in this expansion?

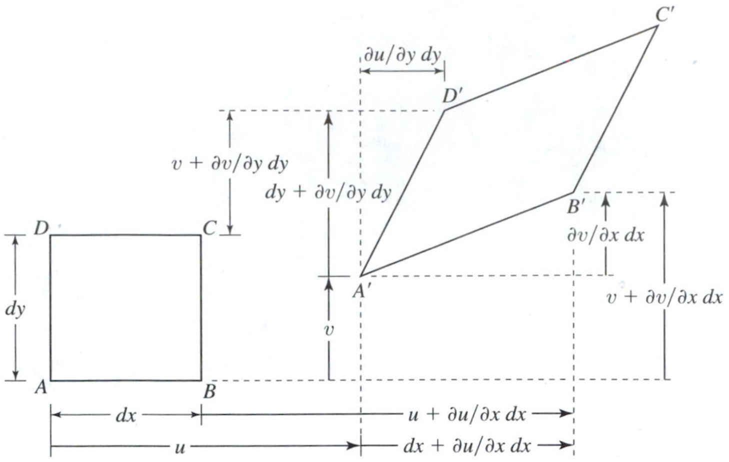


Figure 3-4 Deformation of an infinitesimal element

But

$$A'B' = dx\sqrt{\left(1+\frac{\partial u}{\partial x}\right)^2 + \left(\frac{\partial v}{\partial x}\right)^2} \tag{3-5}$$

We now make the assumption that derivatives of the displacements are small compared to 1, so that their squares are small compared to the displacement derivatives themselves and can be neglected. On expanding Eq. (3-5) and using this assumption we get

$$A'B' = dx\sqrt{1+2\frac{\partial u}{\partial x}}$$

The right-hand side of this expression is now expanded in a power series in the displacement derivative and, consistent with the preceding assumption, powers of this derivative greater than the first are neglected. This gives

$$A'B' = dx\left(1+\frac{\partial u}{\partial x}\right)$$

and substituting this expression into Eq. (3-4) and dividing by $dx$,

$$\varepsilon_{xx} = \frac{\partial u}{\partial x} \tag{3-6}$$

Proceeding in a similar manner with the line segments $AD$ and $A'D'$, we get the normal strain in the $y$ direction as

$$\varepsilon_{yy} = \frac{\partial v}{\partial y} \tag{3-7}$$

Recalling the definition of shearing strain from Eq. (3-2) and Fig. 3-2, we note that the lines $AB$ and $AD$ in Fig. 3-4 are perpendicular to each other, and after deformation the angle between these lines is the angle between $A'B'$ and $A'D'$. As shown in Fig. 3-2, $\alpha$ is the angle between the horizontal and line $A'B'$ while $\beta$ is the angle between the vertical and line $A'D'$. Therefore from Fig. 3-4,

$$\alpha = \tan^{-1}\left(\frac{\dfrac{\partial v}{\partial x}\,dx}{dx + \dfrac{\partial u}{\partial x}\,dx}\right) \tag{3-8}$$

Since displacement derivatives are assumed to be small compared to 1, $\partial u/\partial x$ can be neglected and Eq. (3-8) becomes

$$\alpha = \tan^{-1}\frac{\partial v}{\partial x}$$

and since $\partial v/\partial x$ is small, it follows that

$$\alpha = \frac{\partial v}{\partial x}$$

Similarly,

$$\beta = \frac{\partial u}{\partial y}$$

Thus as seen in Fig. 3-5, when the displacement derivatives are small, $\partial v/\partial x$ and $\partial u/\partial y$ are the angles $\alpha$ and $\beta$, respectively, through which sides $AB$ and $AD$ rotate. We then find that, using Eq. (3-2),

$$\gamma_{xy} = \frac{\partial u}{\partial y} + \frac{\partial v}{\partial x} \tag{3-9}$$

The subscripts indicate that this is the shearing strain for lines that were originally parallel to the $x$ and $y$ axes.

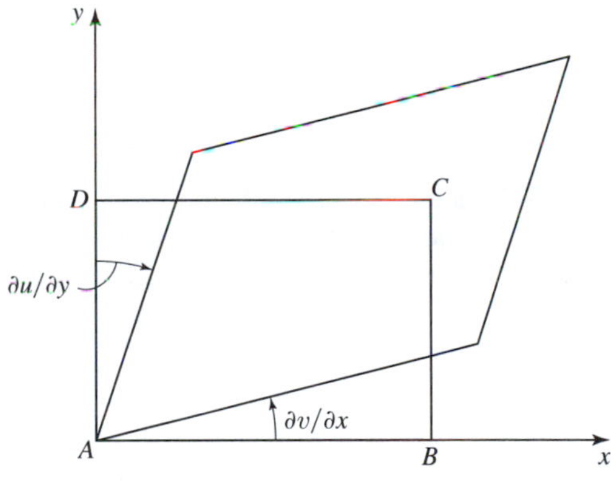


**Figure 3-5** Shearing strains

The strains in Eqs. (3-6), (3-7), and (3-9) are often referred to as *infinitesimal* strains because of our assumption concerning the displacement derivatives.

The shear strain given by Eq. (3-9) is often referred to as the *engineering* shear strain. In contrast, the *mathematical* shear strain, designated as $\varepsilon_{xy}$, is often used in the study of the theory of elasticity and continuum mechanics. It is related to the engineering shear strain as follows:

$$\varepsilon_{xy} = \tfrac{1}{2}\gamma_{xy} \tag{3-10}$$

The difference in definition is reconciled in the stress-strain relations so that a given displacement field produces the same stress field, regardless of which definition of shear strain is used.

At this point it would be worthwhile for the reader to verify that the strain $\varepsilon_{xx}$ computed using lines $DC$ and $D'C'$ in Fig. 3-4 is the same as that computed using lines $AB$ and $A'B'$ if products of displacement derivatives and powers greater than the first are neglected, as was done in our derivation of Eq. (3-6). Similarly, the result of calculating $\varepsilon_{yy}$ using $BC$ and $B'C'$ will be the same as given in Eq. (3-7). Consequently, when the strains are infinitesimal, the element $ABCD$ deforms into a parallelogram rather than a general quadrilateral.

For three-dimensional problems, in addition to the displacement components $u$ and $v$ there is a third component, $w$, directed along the $z$ axis. The infinitesimal strain-displacement relations for these problems are

$$\begin{aligned}
\varepsilon_{xx} &= \frac{\partial u}{\partial x}, & \gamma_{xy} &= \frac{\partial u}{\partial y} + \frac{\partial v}{\partial x} \\
\varepsilon_{yy} &= \frac{\partial v}{\partial y}, & \gamma_{yz} &= \frac{\partial v}{\partial z} + \frac{\partial w}{\partial y} \\
\varepsilon_{zz} &= \frac{\partial w}{\partial z}, & \gamma_{zx} &= \frac{\partial w}{\partial x} + \frac{\partial u}{\partial z}
\end{aligned} \tag{3-11}$$

We note from their definitions that the shear strains are symmetric with respect to their indices,

$$\gamma_{xy} = \gamma_{yx}, \qquad \gamma_{yz} = \gamma_{zy}, \qquad \gamma_{zx} = \gamma_{xz}$$

## 3-3 COMPATIBILITY

Equations (3-6), (3-7), and (3-9) give relationships between strains and displacements for two-dimensional problems. If the displacements are given, these equations provide three expressions from which to determine the three strains. On the other hand, if the three strains are given, these expressions provide three equations from which to determine only two unknown displacements, namely, $u$ and $v$. Thus if Eq. (3-7) is integrated to find $v$, the result can be used in Eq. (3-9) to find $u$. However, Eq. (3-6) can also be used to find $u$, and we have no guarantee that the two expressions for $u$ will be the same! That is, the displacement expression for $u$ need not be single-valued. For example, let us consider a hypothetical deformed body in which the strains are given as a function of position as follows:

$$\varepsilon_{xx} = \frac{\partial u}{\partial x} = y$$

$$\gamma_{xy} = \frac{\partial u}{\partial y} + \frac{\partial v}{\partial x} = xy$$

$$\varepsilon_{yy} = \frac{\partial v}{\partial y} = y$$

Then, on integrating the expressions for $\varepsilon_{xx}$ and $\varepsilon_{yy}$, we get

$$u = yx + f(y)$$

$$v = \tfrac{1}{2}y^2 + g(x)$$

where $f(y)$ and $g(x)$ are arbitrary functions resulting from integration. On the other hand, by substituting the expression for $v$ into Eq. (3-9) we get

$$\gamma_{xy} = \frac{\partial u}{\partial y} + g'(x)$$

But $\gamma_{xy} = xy$. Therefore, equating these two,

$$\frac{\partial u}{\partial y} = xy - g'(x)$$

so that, on integrating,

$$u = \tfrac{1}{2}xy^2 - yg'(x) + h(x)$$

where $h(x)$ is an arbitrary function resulting from integration. We now have two different expressions for calculating $u$, and they cannot be equal because the second contains $h(x)$, a function of $x$ alone, whereas the first cannot contain a function of $x$ only. Thus if two individuals were asked to find the displacements from a given set of strains and one chose to use Eqs. (3-6) and (3-7) while the other used Eqs. (3-7) and (3-9), their answers would be different even though each presumably solved the same problem.

Another problem with the strain-displacement equations is that continuous strains can lead to displacements that are not continuous. Thus, as shown in Fig. 3-6, while all the differential elements of the body may be connected in the unstrained state, it may happen that in the strained state the elements are not connected but can overlap or have gaps between them. As an example, consider the following continuous strain field, which is shown in Fig. 3-7:

$$\varepsilon_{xx} = \begin{cases} x, & x < 1 \\ 1, & x > 1 \end{cases}$$

This yields the following displacement field:

$$u = \tfrac{1}{2}x^2 + g(y), \qquad x < 1$$

$$u = x + f(y), \qquad x > 1$$

The displacement field, which is shown in Fig. 3-8, contains a jump of magnitude $\frac{1}{2} + f(y) - g(y)$ which is not, in general, zero.

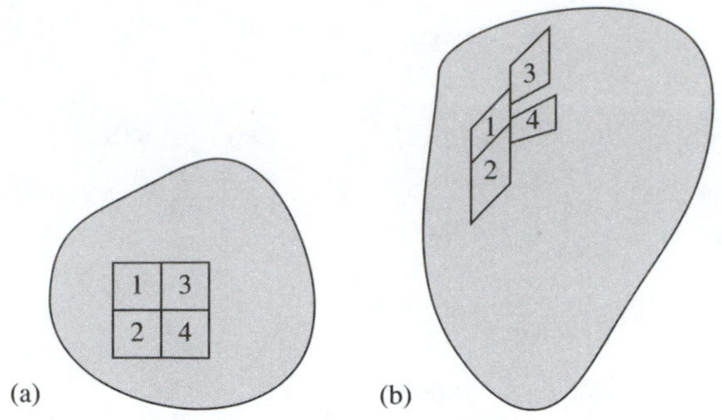


**Figure 3-6** Discontinuous displacements

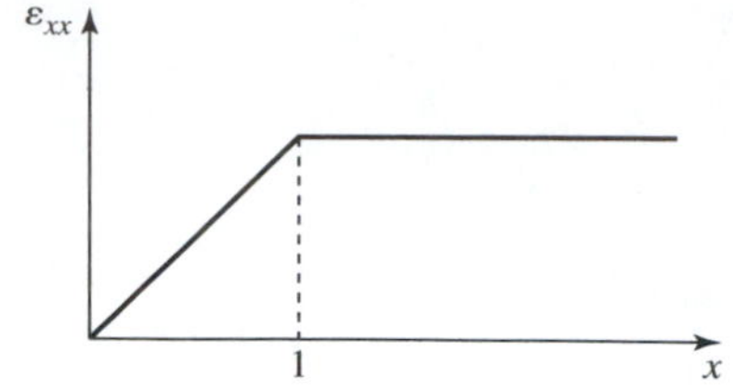


**Figure 3-7**

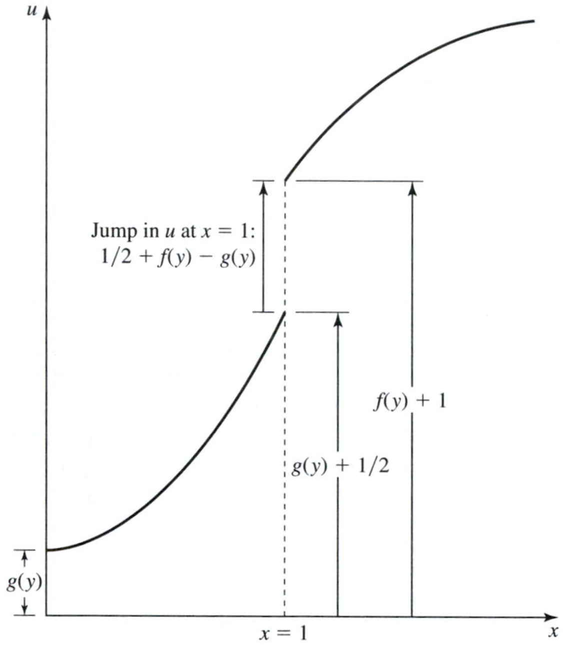


**Figure 3-8**

Thus it seems likely that all three of the strains cannot be specified independently, and that a relationship among them must exist that will ensure that the displacements are both continuous and single-valued. The appropriate relationship, referred to as *compatibility equation,* is easily found for two-dimensional problems.

First differentiate the expression for shear strain, Eq. (3-9), with respect to both $x$ and $y$ to give

$$\frac{\partial^2 \gamma_{xy}}{\partial x\,\partial y} = \frac{\partial^2}{\partial x\,\partial y}\left(\frac{\partial u}{\partial y}\right) + \frac{\partial^2}{\partial x\,\partial y}\left(\frac{\partial v}{\partial x}\right)$$

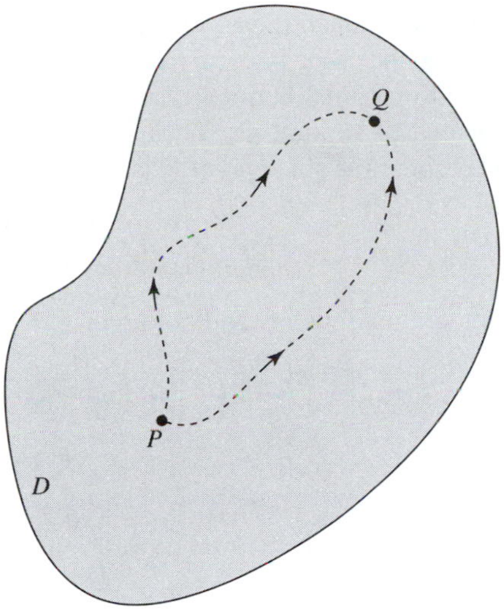


Figure 3-9

Now if the displacements along with their first and second mixed partial derivatives are continuous, the order of partial differentiation can be interchanged. Thus we can write

$$\frac{\partial^2 \gamma_{xy}}{\partial x\, \partial y} = \frac{\partial^2}{\partial y^2}\left(\frac{\partial u}{\partial x}\right) + \frac{\partial^2}{\partial x^2}\left(\frac{\partial v}{\partial y}\right)$$

The terms in parentheses are, respectively, $\varepsilon_{xx}$ and $\varepsilon_{yy}$. Therefore we have finally

$$\frac{\partial^2 \gamma_{xy}}{\partial x\, \partial y} = \frac{\partial^2 \varepsilon_{xx}}{\partial y^2} + \frac{\partial^2 \varepsilon_{yy}}{\partial x^2} \tag{3-12}$$

This provides the necessary and sufficient relationship among the strains to guarantee continuity of the displacements.

We will now show that Eq. (3-12) must be satisfied in order for the displacements to be single-valued. To this end, consider a two-dimensional domain $D$ within which there are two arbitrary points $P$ and $Q$, as shown in Fig. 3-9. If the displacement vector is continuous, then we can write for the component in the $x$ direction,

$$u_Q = u_P + \int_P^Q du \tag{3-13}$$

where the integral is evaluated along any continuous path between points $P$ and $Q$. Now if the displacement $u_Q$ is to be single-valued, its value must be the same no matter which path is taken between points $P$ and $Q$. Thus the condition that $u_Q$ be single-valued is that the integral in Eq. (3-13) be path independent. To derive conditions under which this is true, we note that, by the chain rule and through the use of Eqs. (3-6) and (3-9),

$$du = \frac{\partial u}{\partial x}\,dx + \frac{\partial u}{\partial y}\,dy = \varepsilon_{xx}\,dx + \left(\gamma_{xy} - \frac{\partial v}{\partial x}\right)dy$$

and therefore,

$$u_Q = u_P + \int_P^Q \left[\varepsilon_{xx}\,dx + \left(\gamma_{xy} - \frac{\partial v}{\partial x}\right)dy\right] \tag{3-14}$$

We would like all quantities in the integrand to be in terms of strains, and all quantities appearing outside of the integral to be referred to point $P$. If the last term of the integrand is integrated by parts, then appearing outside of the integral will be a term involving $\partial v/\partial x$ evaluated at point $Q$. We therefore replace $dy$ in that integral with $d(y - y_Q)$, which is permissible since point $Q$ is a fixed point. Then we have

$$\int_P^Q \frac{\partial v}{\partial x}\,dy = \int_P^Q \frac{\partial v}{\partial x}\,d(y - y_Q)$$

and on integrating by parts and noting that

$$d\left(\frac{\partial v}{\partial x}\right) = \frac{\partial^2 v}{\partial x^2}\,dx + \frac{\partial^2 v}{\partial x\,\partial y}\,dy$$

we get

$$\int_P^Q \frac{\partial v}{\partial x}\,dy = \left(\frac{\partial v}{\partial x}\right)_P (y_Q - y_P) - \int_P^Q \left[(y - y_Q)\frac{\partial^2 v}{\partial x^2}\,dx + (y - y_Q)\frac{\partial^2 v}{\partial x\,\partial y}\,dy\right]$$

But

$$\frac{\partial^2 v}{\partial x^2} = \frac{\partial}{\partial x}\left(\frac{\partial v}{\partial x} + \frac{\partial u}{\partial y}\right) - \frac{\partial^2 u}{\partial x\,\partial y} = \frac{\partial \gamma_{xy}}{\partial x} - \frac{\partial \varepsilon_{xx}}{\partial y}$$

and

$$\frac{\partial^2 v}{\partial x\,\partial y} = \frac{\partial \varepsilon_{yy}}{\partial x}$$

Thus,

$$\int_P^Q \frac{\partial v}{\partial x}\,dy = \left(\frac{\partial v}{\partial x}\right)_P (y_Q - y_P)$$
$$- \int_P^Q \left[(y - y_Q)\left(\frac{\partial \gamma_{xy}}{\partial x} - \frac{\partial \varepsilon_{xx}}{\partial y}\right)dx + (y - y_Q)\frac{\partial \varepsilon_{yy}}{\partial x}\,dy\right]$$

When this is substituted into Eq. (3-14), we get the following expression:

$$u_Q = u_P - \left(\frac{\partial v}{\partial x}\right)_P (y_Q - y_P) + \int_P^Q \left\{\left[\varepsilon_{xx} + (y - y_Q)\left(\frac{\partial \gamma_{xy}}{\partial x} - \frac{\partial \varepsilon_{xx}}{\partial y}\right)\right] dx + \left[\gamma_{xy} + (y - y_Q)\frac{\partial \varepsilon_{yy}}{\partial x}\right] dy\right\} \tag{3-15}$$

A necessary and sufficient condition for an integral of the form

$$\int_P^Q F\,dx + G\,dy \tag{3-16}$$

to be path independent is that (see, for example, [2, sec. 5.6, pp. 291–301])

$$\frac{\partial F}{\partial y} - \frac{\partial G}{\partial x} = 0 \tag{3-17}$$

On comparing Eqs. (3-15) and (3-16) we see that

$$F = \varepsilon_{xx} + (y - y_Q)\left(\frac{\partial \gamma_{xy}}{\partial x} - \frac{\partial \varepsilon_{xx}}{\partial y}\right)$$

$$G = \gamma_{xy} + (y - y_Q)\frac{\partial \varepsilon_{yy}}{\partial x}$$

These are now substituted into Eq. (3-17), resulting in an equation with certain terms multiplied by $(y - y_Q)$ and others that are not. The terms that are not multiplied by $(y - y_Q)$ cancel, leaving the following expression:

$$(y - y_Q)\left(\frac{\partial^2 \gamma_{xy}}{\partial x\,\partial y} - \frac{\partial^2 \varepsilon_{xx}}{\partial y^2} - \frac{\partial^2 \varepsilon_{yy}}{\partial x^2}\right) = 0$$

Since this must be true for all $y$ in $D$, Eq. (3-12) follows. Consequently Eq. (3-12) provides a necessary and sufficient condition to ensure that the displacement field is continuous and single-valued.

It is readily verified that if the $v$ component of displacement is required to be single-valued, then Eq. (3-12) again follows.

The strain-displacement equations in two dimensions contain five unknowns—two displacement components and three strains. If a particular problem is solved in such a way that the displacements are determined first and then the strains, the strain-displacement equations provide three independent equations. On the other hand, if the strains are determined first, then the strain-displacement equations provide only two independent equations. However, in that case compatibility gives an additional equation. In either case, then, kinematical considerations provide three additional equations but five additional unknowns. Coupled with the two-dimensional equilibrium equations given by Eq. (2-55), our problem now contains five equations but eight unknowns.

For three-dimensional problems there are six compatibility equations (see, for example, [5, sec. 2.3, pp. 21–23]),

$$\begin{aligned}
&\frac{\partial^2 \varepsilon_{xx}}{\partial y^2} + \frac{\partial^2 \varepsilon_{yy}}{\partial x^2} = \frac{\partial^2 \gamma_{xy}}{\partial x\,\partial y} \\
&\frac{\partial^2 \varepsilon_{yy}}{\partial z^2} + \frac{\partial^2 \varepsilon_{zz}}{\partial y^2} = \frac{\partial^2 \gamma_{yz}}{\partial y\,\partial z} \\
&\frac{\partial^2 \varepsilon_{zz}}{\partial x^2} + \frac{\partial^2 \varepsilon_{xx}}{\partial z^2} = \frac{\partial^2 \gamma_{zx}}{\partial z\,\partial x} \\
&2\frac{\partial^2 \varepsilon_{xx}}{\partial y\,\partial z} = \frac{\partial}{\partial x}\left(-\frac{\partial \gamma_{yz}}{\partial x} + \frac{\partial \gamma_{zx}}{\partial y} + \frac{\partial \gamma_{xy}}{\partial z}\right) \\
&2\frac{\partial^2 \varepsilon_{yy}}{\partial z\,\partial x} = \frac{\partial}{\partial y}\left(-\frac{\partial \gamma_{zx}}{\partial y} + \frac{\partial \gamma_{xy}}{\partial z} + \frac{\partial \gamma_{yz}}{\partial x}\right) \\
&2\frac{\partial^2 \varepsilon_{zz}}{\partial x\,\partial y} = \frac{\partial}{\partial z}\left(-\frac{\partial \gamma_{xy}}{\partial z} + \frac{\partial \gamma_{yz}}{\partial x} + \frac{\partial \gamma_{zx}}{\partial y}\right)
\end{aligned} \tag{3-18}$$

It happens that these six equations are not independent and that by suitable manipulation they can be reduced to three independent fourth-order partial differential equations. However, the equations given here are generally preferred because in practice these fourth-order equations are quite difficult to use.

## 3-4 SPECIFICATION OF THE STATE OF STRAIN AT A POINT

Now that we have developed expressions for normal and shear strains in terms of displacement derivatives, it is necessary to determine whether these strains are adequate to completely describe the state of strain at a point. We will do this by showing that with the three strain components $\varepsilon_{xx}$, $\varepsilon_{yy}$, $\gamma_{xy}$ the normal strain of a line element with any orientation in the $xy$ plane can be found, as can the change in angle between two perpendicular lines for any orientation.

We begin by considering an element $dx$ by $d\hat{y}$, as shown in Fig. 3-10. We wish to calculate the normal strain (that is, the unit elongation) of the line $AC$, which makes an angle $\theta$ with line $AB$. The axis along which line $AC$ lies will be referred to as the $x'$ axis, the length of the line $AC$ is $ds$, and the length of the line $BC$ will be designated as $dy$ to distinguish it from the element dimension $d\hat{y}$. Figure 3-11 shows the triangular element $ABC$ before and after deformation, when points $A$, $B$, and $C$ have moved to $A'$, $B'$, and $C'$, respectively. From the definition of normal strain, given by Eq. (3-1), we have

$$\varepsilon_{x'x'} = \frac{A'C' - AC}{AC} = \frac{A'C' - ds}{ds}$$

and therefore

$$ds(1 + \varepsilon_{x'x'}) = A'C' \tag{3-19}$$

From geometry,

$$(A'C')^2 = (dx + du)^2 + (dy + dv)^2 \tag{3-20}$$

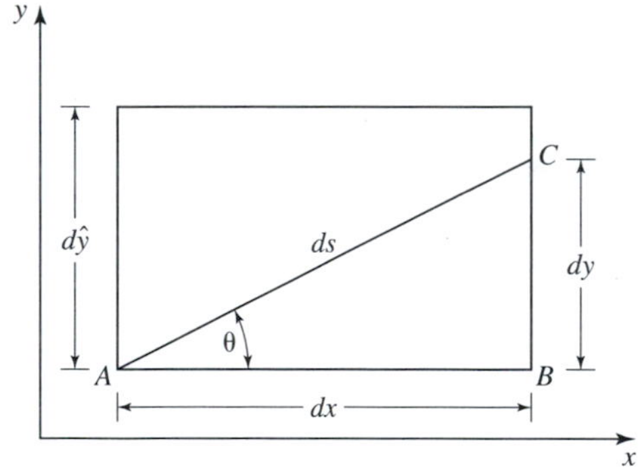


Figure 3-10

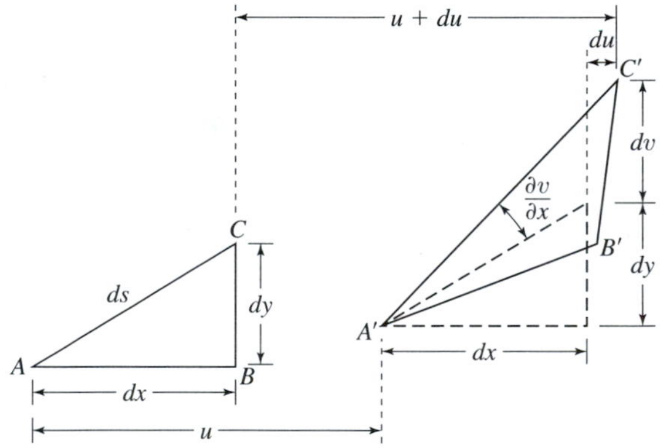


Figure 3-11 Deformation of a triangular element

With the chain rule, $du$ and $dv$ are given by

$$du = \frac{\partial u}{\partial x}dx + \frac{\partial u}{\partial y}dy$$
$$dv = \frac{\partial v}{\partial x}dx + \frac{\partial v}{\partial y}dy \tag{3-21}$$

When these are substituted into Eq. (3-20) we get, on expanding and collecting terms,

$$(A'C')^2 = ds^2 + \left[2\frac{\partial u}{\partial x} + \left(\frac{\partial u}{\partial x}\right)^2 + \left(\frac{\partial v}{\partial x}\right)^2\right](dx)^2$$
$$+ \left[2\frac{\partial v}{\partial y} + \left(\frac{\partial u}{\partial y}\right)^2 + \left(\frac{\partial v}{\partial y}\right)^2\right](dy)^2$$
$$+ 2\left(\frac{\partial u}{\partial y} + \frac{\partial v}{\partial x} + \frac{\partial u}{\partial x}\frac{\partial u}{\partial y} + \frac{\partial v}{\partial x}\frac{\partial v}{\partial y}\right)dx\,dy \tag{3-22}$$

Here we have used the fact that

$$dx^2 + dy^2 = ds^2$$

Equation (3-22) is now substituted into the square of Eq. (3-19). After dividing by $ds$ and noting that

$$\frac{dx}{ds} = \cos\theta, \qquad \frac{dy}{ds} = \sin\theta \tag{3-23}$$

we get

$$(1 + \varepsilon_{x'x'})^2 = 1 + \left[2\frac{\partial u}{\partial x} + \left(\frac{\partial u}{\partial x}\right)^2 + \left(\frac{\partial v}{\partial x}\right)^2\right]\cos^2\theta$$
$$+ \left[2\frac{\partial v}{\partial y} + \left(\frac{\partial u}{\partial y}\right)^2 + \left(\frac{\partial v}{\partial y}\right)^2\right]\sin^2\theta$$
$$+ 2\left(\frac{\partial u}{\partial y} + \frac{\partial v}{\partial x} + \frac{\partial u}{\partial x}\frac{\partial u}{\partial y} + \frac{\partial v}{\partial x}\frac{\partial v}{\partial y}\right)\cos\theta\sin\theta$$

By expanding the left-hand side, neglecting products of strains as well as products of displacement derivatives, we get

$$\varepsilon_{x'x'} = \varepsilon_{xx}\cos^2\theta + \varepsilon_{yy}\sin^2\theta + \gamma_{xy}\sin\theta\cos\theta \tag{3-24}$$

where Eqs. (3-6), (3-7), and (3-9) have been used. The normal strain for a line perpendicular to line $AC$ is obtained from this expression simply by replacing $\theta$ by $\theta + \pi/2$, yielding

$$\varepsilon_{y'y'} = \varepsilon_{xx}\sin^2\theta + \varepsilon_{yy}\cos^2\theta - \gamma_{xy}\sin\theta\cos\theta \tag{3-25}$$

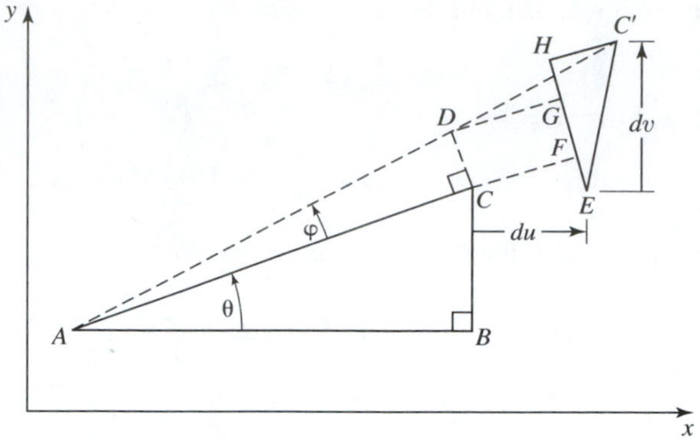


Figure 3-12 Change in orientation between lines $AC$ and $AC'$

In order to get an expression for $\gamma_{x'y'}$ we first calculate the change in orientation between lines $AC$ and $AC'$. As shown in Fig. 3-12, this change in orientation is given by the angle $\varphi$ where, since $\varphi$ is small,

$$\tan\varphi \approx \varphi = \frac{CD}{AC} = \frac{CD}{ds} \tag{3-26}$$

Now from the geometry

$$\begin{aligned} CD &= FG = EH - EF - GH \\ &= dv\cos\theta - du\sin\theta - DC'\tan\varphi \end{aligned} \tag{3-27}$$

where

$$DC' = \varepsilon_{x'x'}\,ds$$

Thus substituting Eqs. (3-21) for $du$ and $dv$ in Eq. (3-27), using the result in Eq. (3-26), rearranging, and noting that $\varepsilon_{x'x'}\varphi$ is small compared to 1, we get

$$\varphi = \frac{\partial v}{\partial x}\frac{dx}{ds}\cos\theta + \frac{\partial v}{\partial y}\frac{dy}{ds}\cos\theta - \frac{\partial u}{\partial x}\frac{dx}{ds}\sin\theta - \frac{\partial u}{\partial y}\frac{dy}{ds}\sin\theta$$

Finally, by using Eqs. (3-6), (3-7), (3-9), and (3-23) it follows that

$$\varphi = -(\varepsilon_{xx} - \varepsilon_{yy})\sin\theta\cos\theta + \frac{\partial v}{\partial x}\cos^2\theta - \frac{\partial u}{\partial y}\sin^2\theta \tag{3-28}$$

To get the change in angle for a line oriented in the $y$ direction, we substitute $\theta + \pi/2$ for $\theta$ to obtain

$$\varphi_{\theta+\pi/2} = (\varepsilon_{xx} - \varepsilon_{yy})\sin\theta\cos\theta + \frac{\partial v}{\partial x}\sin^2\theta - \frac{\partial u}{\partial y}\cos^2\theta$$

Thus since the shear strain is the change in angle between two originally perpendicular lines, we have

$$\gamma_{x'y'} = \varphi - \varphi_{\theta+\pi/2} = -2(\varepsilon_{xx} - \varepsilon_{yy}) \sin\theta \cos\theta + \gamma_{xy}(\cos^2\theta - \sin^2\theta) \quad (3\text{-}29)$$

Note that $\varphi_{\theta+\pi/2}$ is subtracted from $\varphi$ because both angles are positive in the counterclockwise direction. In the original derivation of shear strain given in Section 3-2, $\alpha$ was positive in the counterclockwise direction whereas $\beta$ was positive in the clockwise direction (see Fig. 3-2).

Equations (3-24), (3-25), and (3-29) are usually written in terms of $2\theta$, in which case they become

$$\begin{aligned}
\varepsilon_{x'x'} &= \frac{\varepsilon_{xx} + \varepsilon_{yy}}{2} + \frac{\varepsilon_{xx} - \varepsilon_{yy}}{2}\cos 2\theta + \frac{1}{2}\gamma_{xy}\sin 2\theta \\
\varepsilon_{y'y'} &= \frac{\varepsilon_{xx} + \varepsilon_{yy}}{2} - \frac{\varepsilon_{xx} - \varepsilon_{yy}}{2}\cos 2\theta - \frac{1}{2}\gamma_{xy}\sin 2\theta \qquad (3\text{-}30) \\
\gamma_{x'y'} &= -(\varepsilon_{xx} - \varepsilon_{yy})\sin 2\theta + \gamma_{xy}\cos 2\theta
\end{aligned}$$

These equations show that the two-dimensional strains on an element of material surrounding a point can be found if $\varepsilon_{xx}$, $\varepsilon_{yy}$, and $\gamma_{xy}$ are known, regardless of the orientation of the element. Thus knowledge of the shearing strain and normal strains in two perpendicular directions is sufficient to completely determine the two-dimensional state of strain at a point.

For three-dimensional problems, Eqs. (3-30) are replaced by the following equations (see [5, p. 20]):

$$\begin{aligned}
\varepsilon_{x'x'} &= n_{x'x}^2\varepsilon_{xx} + n_{x'y}^2\varepsilon_{yy} + n_{x'z}^2\varepsilon_{zz} + n_{x'x}n_{x'y}\gamma_{xy} + n_{x'y}n_{x'z}\gamma_{yz} + n_{x'z}n_{x'x}\gamma_{zx} \\
\varepsilon_{y'y'} &= n_{y'x}^2\varepsilon_{xx} + n_{y'y}^2\varepsilon_{yy} + n_{y'z}^2\varepsilon_{zz} + n_{y'x}n_{y'y}\gamma_{xy} + n_{y'y}n_{y'z}\gamma_{yz} + n_{y'z}n_{y'x}\gamma_{zx} \\
\varepsilon_{z'z'} &= n_{z'x}^2\varepsilon_{xx} + n_{z'y}^2\varepsilon_{yy} + n_{z'z}^2\varepsilon_{zz} + n_{z'x}n_{z'y}\gamma_{xy} + n_{z'y}n_{z'z}\gamma_{yz} + n_{z'z}n_{z'x}\gamma_{zx} \\
\gamma_{x'y'} &= 2(n_{x'x}n_{y'x}\varepsilon_{xx} + n_{x'y}n_{y'y}\varepsilon_{yy} + n_{x'z}n_{y'z}\varepsilon_{zz}) + (n_{x'x}n_{y'y} + n_{x'y}n_{y'x})\gamma_{xy} \\
&\quad + (n_{x'y}n_{y'z} + n_{x'z}n_{y'y})\gamma_{yz} + (n_{x'x}n_{y'z} + n_{x'z}n_{y'x})\gamma_{zx} \qquad (3\text{-}31) \\
\gamma_{y'z'} &= 2(n_{y'x}n_{z'x}\varepsilon_{xx} + n_{y'y}n_{z'y}\varepsilon_{yy} + n_{y'z}n_{z'z}\varepsilon_{zz}) + (n_{y'x}n_{z'y} + n_{y'y}n_{z'x})\gamma_{xy} \\
&\quad + (n_{y'y}n_{z'z} + n_{y'z}n_{z'y})\gamma_{yz} + (n_{y'x}n_{z'z} + n_{y'z}n_{z'x})\gamma_{zx} \\
\gamma_{z'x'} &= 2(n_{z'x}n_{x'x}\varepsilon_{xx} + n_{z'y}n_{x'y}\varepsilon_{yy} + n_{z'z}n_{x'z}\varepsilon_{zz}) + (n_{z'x}n_{x'y} + n_{z'y}n_{x'x})\gamma_{xy} \\
&\quad + (n_{z'y}n_{x'z} + n_{z'z}n_{x'y})\gamma_{yz} + (n_{z'x}n_{x'z} + n_{z'z}n_{x'x})\gamma_{zx}
\end{aligned}$$

where $n_{x'x}, \ldots$ are the direction cosines defined in Chapter 2.

### 3-4-1 Strain Gages

In experimental testing of structural parts and assemblies it is often necessary to determine the stress present in critical areas in order to confirm the analytical stress predictions used in design. Stress, however, cannot be measured directly. Rather, it is necessary to measure the strain at a particular location in the part or assembly and then compute the stress using an appropriate stress-strain law such as one of those to be discussed in the next chapter. The accurate experimental determination of strain is therefore important in determining stress. A device commonly used to measure the strain in a structure is the electrical resistance strain gage. (An interesting history of the strain gage can be found in [4].) Its operation is based on the 1856 observation of Lord Kelvin that when copper and iron wires are subjected to mechanical strain their electrical resistance changes. Thus by measuring the change in electrical resistance of the strain gage the strain can be determined. Current strain gages are often made of a metal foil that is photoetched to form a grid configuration. The strain gage is bonded to the surface of the structure. An example of these *bonded foil resistance strain gages* is shown in Fig. 3-13.

It is important to note that a strain gage measures the normal strain along the axis of the gage, as indicated by the arrows in Fig. 3-13. In order to completely determine the state of strain at a particular location it is necessary to use three strain gages arranged in a pattern referred to as a *strain rosette*.

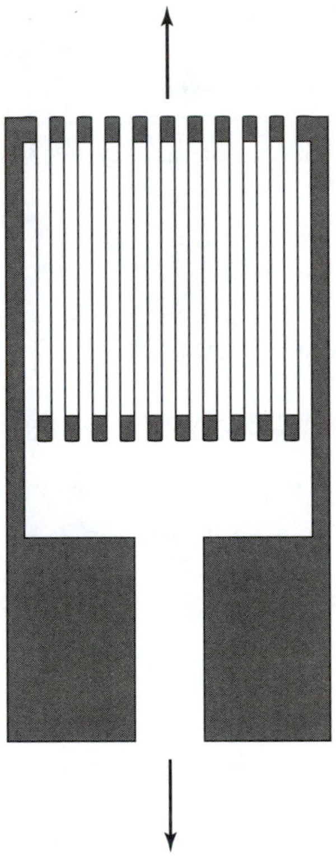

Figure 3-13 Bonded foil resistance strain gage

## EXAMPLE 3-1

Three strain gages are located at 0°, 45°, and 90°, as shown in Fig. 3-14, to form a pattern referred to as a *three-element rectangular rosette*. Strain gages *A*, *B*, and *C*, located at $\theta_A$, $\theta_B$, and $\theta_C$, respectively, read strains $\varepsilon_A$, $\varepsilon_B$, and $\varepsilon_C$. Now consider three different orientations of the $x'$ axis to coincide with the three strain gage axes. Then, when the first of Eqs. (3-30) is used with $\theta$ equal to, successively, $\theta_A$, $\theta_B$, and $\theta_C$, the following equations result:

$$\varepsilon_A = \varepsilon_{xx}$$

$$\varepsilon_B = \tfrac{1}{2}(\varepsilon_{xx} + \varepsilon_{yy} + \gamma_{xy})$$

$$\varepsilon_C = \varepsilon_{yy}$$

Alternatively, the equation for $\varepsilon_C$ can be obtained from the second of Eqs. (3-30) with $\theta = \theta_A$. These equations are now solved to give

$$\varepsilon_{xx} = \varepsilon_A$$

$$\varepsilon_{yy} = \varepsilon_C$$

$$\gamma_{xy} = 2\varepsilon_B - \varepsilon_A - \varepsilon_C$$

Thus with the strain gages oriented as shown in Fig. 3-14, and for the coordinate system shown, the two normal strains can be obtained directly from the strain gage readings and the shear strain found from a straightforward combination of the strain gage readings.

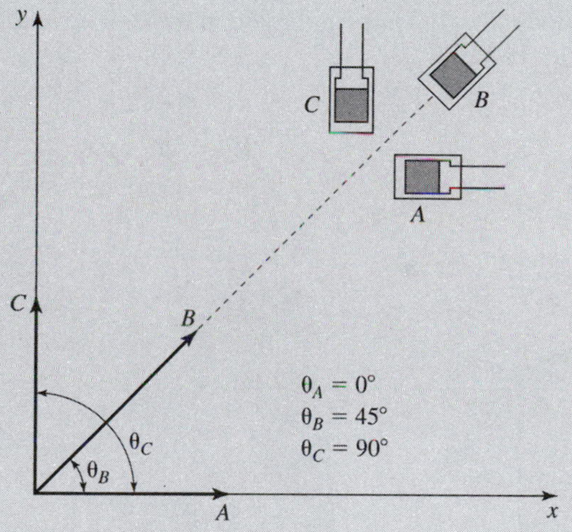


**Figure 3-14**

## 3-5 ROTATION

For two-dimensional problems the displacement components are functions of the two coordinates $x$ and $y$, that is,

$$u = u(x, y)$$
$$v = v(x, y)$$

If the displacement of a point $P$, whose coordinates are $(x, y)$, in a body is given by its components $u$ and $v$, then the displacement of a point located a differential distance from $P$, that is, located at $(x + dx, y + dy)$, is given by $u + du$ and $v + dv$. The differentials $du$ and $dv$ are given by Eqs. (3-21). However, these can be rewritten as

$$\begin{aligned} du &= \frac{\partial u}{\partial x} dx + \frac{1}{2}\left(\frac{\partial u}{\partial y} + \frac{\partial v}{\partial x}\right) dy + \frac{1}{2}\left(\frac{\partial u}{\partial y} - \frac{\partial v}{\partial x}\right) dy \\ dv &= \frac{\partial v}{\partial y} dy + \frac{1}{2}\left(\frac{\partial u}{\partial y} + \frac{\partial v}{\partial x}\right) dx + \frac{1}{2}\left(\frac{\partial v}{\partial x} - \frac{\partial u}{\partial y}\right) dx \end{aligned} \tag{3-32}$$

or

$$\begin{aligned} du &= \varepsilon_{xx}\, dx + \tfrac{1}{2}\gamma_{xy}\, dy - \omega_z\, dy \\ dv &= \varepsilon_{yy}\, dy + \tfrac{1}{2}\gamma_{xy}\, dx + \omega_z\, dx \end{aligned} \tag{3-33}$$

where the quantity

$$\omega_z = \frac{1}{2}\left(\frac{\partial v}{\partial x} - \frac{\partial u}{\partial y}\right) \tag{3-34}$$

is referred to as the *rotation*. Note that the rotation is in radians and, as indicated by the subscript, it is about the $z$ axis.

In order to better understand the concept of rotation, let us consider what happens when a body undergoes a rigid rotation through an angle $\varphi$ about point $A$, as shown in Fig. 3-15. We have

$$du = -\tan\varphi\, dy$$
$$dv = \tan\varphi\, dx$$

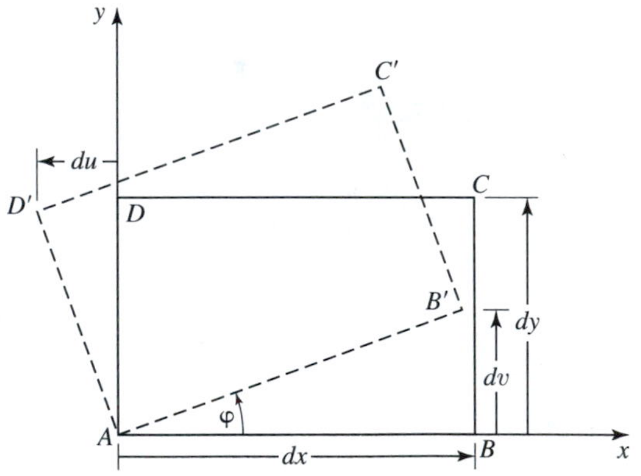


**Figure 3-15** Body undergoing rigid rotation through angle $\varphi$

where $\varphi$ is in radians to be consistent with the units for the rotation. If $\tan\varphi$ is small (implying that $\varphi$ is a small angle), then these become

$$du = -\varphi\, dy$$
$$dv = \varphi\, dx \tag{3-35}$$

Now under a rigid rotation the strains $\varepsilon_{xx}$, $\varepsilon_{yy}$, and $\gamma_{xy}$ are all zero so that from Eqs. (3-33) we have

$$du = -\omega_z\, dy$$
$$dv = \omega_z\, dx$$

and on comparing these with Eqs. (3-35) we conclude that

$$\omega_z = \varphi$$

We note from Fig. 3-5 that when the strains are infinitesimal, the displacement derivatives $\partial v/\partial x$ and $\partial u/\partial y$ represent the angles through which sides $AB$ and $AD$, respectively, rotate. Thus for problems in which the strains are infinitesimal our assertion that $\omega_z$ is a rotation is justified.

Since the shear strain is zero for rigid rotations, we have from Eqs. (3-9) and (3-34),

$$\omega_z = \frac{\partial v}{\partial x} = -\frac{\partial u}{\partial y}$$

In order to gain further understanding of the concept of rotation, let us first note that $\omega_z$ represents the average angle through which sides $AB$ and $AD$ rotate. This is perhaps easiest to see if we let $\partial v/\partial x$ be positive and $\partial u/\partial y$ be negative so that both sides rotate clockwise, as shown in Fig. 3-16. As an example, if $\partial v/\partial x = 0.1$ rad (5.730°) and $\partial u/\partial y = -0.06$ rad (−3.348°), then the configuration of the element is as shown in Fig. 3-17. The rotation is calculated from Eq. (3-34) to be

$$\omega_z = 0.08 \text{ rad } (4.584°)$$

From the figure it is clear that this represents the average of the rotation of sides $AB$ and $AD$.

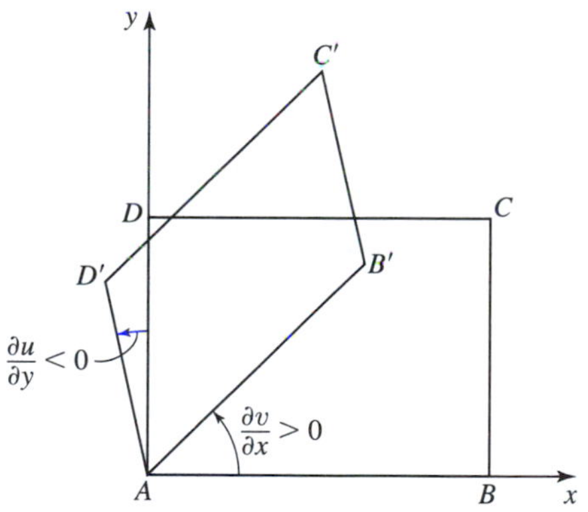


Figure 3-16

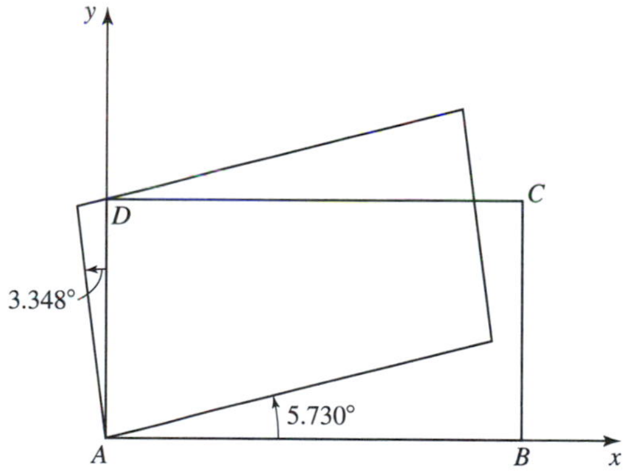


Figure 3-17

Now assume that $\varepsilon_{xx} = 0$ and $\varepsilon_{yy} = 0$. Then from Eqs. (3-33) we have

$$du = \tfrac{1}{2}\gamma_{xy}\,dy - \omega_z\,dy$$

$$dv = \tfrac{1}{2}\gamma_{xy}\,dx - \omega_z\,dx$$

Using Eq. (3-9) and the numbers from the previous example, we find that

$$\gamma_{xy}/2 = 0.02 \text{ rad } (1.146^\circ)$$

Consequently, the deformation of the element can be thought of as occurring in two steps:

1. A rigid rotation of $\omega_z = 0.08$ rad (4.584°) about point $A$
2. From this configuration, additional rotation of sides $AB$ and $AD$ toward each other through an amount $\gamma_{xy}/2 = 0.02$ rad (1.146°)*

Figure 3-18*a* shows the initial and final configurations of the element while Fig. 3-18*b* and *c* shows the two steps needed to attain this configuration. Note that only the second step results in deformation of the element. The first step is simply a rigid rotation.

By referring to Fig. 3-4 we see that the displacement of the element is given by $u + du$ and $v + dv$, where $u$ and $v$ represent a rigid translation of the element. Thus with Eqs. (3-33) and our previous discussion on rotations, we see that the displacement of the element is the superposition of four parts: a rigid translation $(u, v)$, a rigid rotation $\omega_z$, a stretching of the sides of the element due to the normal strains $\varepsilon_{xx}$ and $\varepsilon_{yy}$, and a change in the right angle between two adjacent sides due to the shearing strain $\gamma_{xy}$. This can be seen in Fig. 3-19, where Fig. 3-19*a* shows the initial and final configurations whereas Fig. 3-19*b–e* shows the rigid translation, rigid rotation, normal strain, and shear strain, respectively. If it happens that the element undergoes no rigid body motion, then the deformation is referred to as *pure* deformation. For pure deformation we have, from Eqs. (3-33),

$$\begin{aligned} du &= \varepsilon_{xx}\,dx + \tfrac{1}{2}\gamma_{xy}\,dy \\ dv &= \varepsilon_{yy}\,dy + \tfrac{1}{2}\gamma_{xy}\,dx \end{aligned} \tag{3-36}$$

and since $\omega_z = 0$, it follows that $\partial u/\partial y$ and $\partial v/\partial x$ are equal so that the shearing strain is given by

$$\gamma_{xy} = 2\frac{\partial v}{\partial x} = 2\frac{\partial u}{\partial y}$$

The deformed configuration of an element in a state of pure deformation is shown in Fig. 3-20.

In the next section we will show that $\omega_z$ gives the angular displacement of the principal strain axes.

---

*Note that because of our definition, given in Section 3-2 and shown in Fig. 3-2, of positive rotations, $AB$ rotates counterclockwise and $AD$ rotates clockwise.

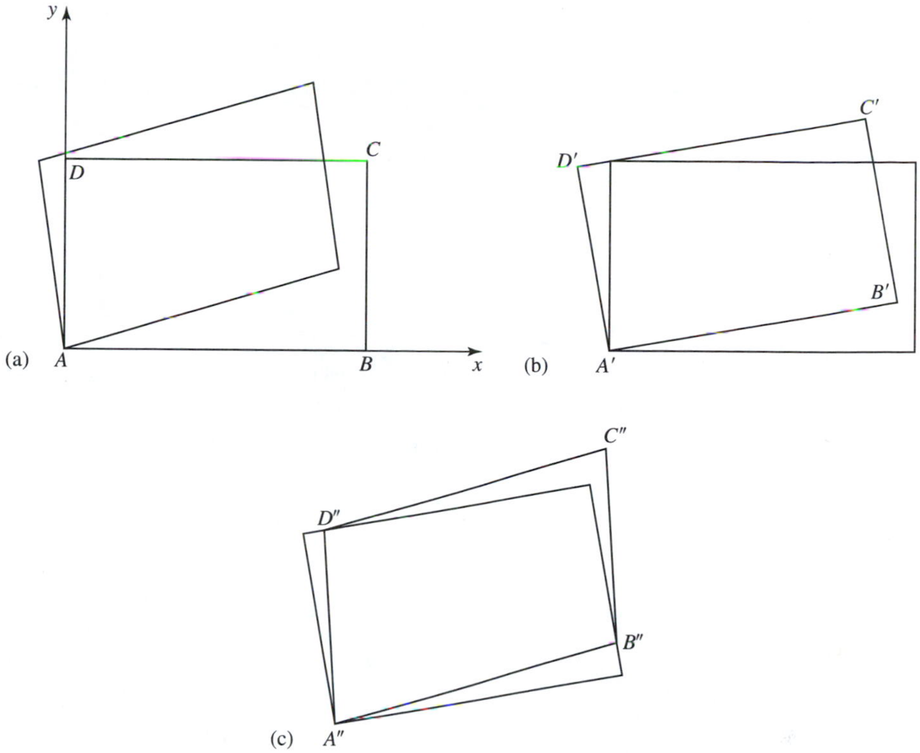


Figure 3-18

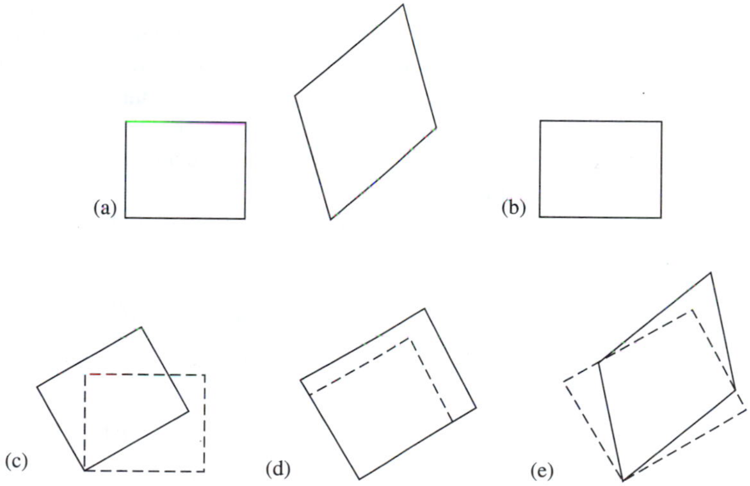


Figure 3-19 Displacement of element as the superposition of four parts

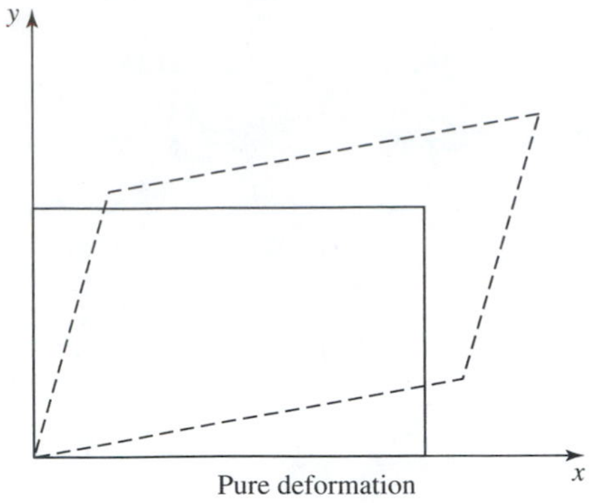


**Figure 3-20** Pure deformation

For three-dimensional problems the rotations are

$$\omega_x = \frac{1}{2}\left(\frac{\partial w}{\partial y} - \frac{\partial v}{\partial z}\right)$$

$$\omega_y = \frac{1}{2}\left(\frac{\partial u}{\partial z} - \frac{\partial w}{\partial x}\right) \qquad (3\text{-}37)$$

$$\omega_z = \frac{1}{2}\left(\frac{\partial v}{\partial x} - \frac{\partial u}{\partial y}\right)$$

## 3-6 PRINCIPAL STRAINS

As we saw in Section 3-4, a line segment $AC$ of length $ds$ oriented at an angle $\theta$ relative to the $x$ axis, as shown in Fig. 3-12, undergoes, in general, both an elongation and a change of orientation. An interesting question to pose is whether there is any orientation $\theta$ for line segment $AC$ in which its orientation is unaltered when the element shown in Fig. 3-12 is subjected to pure deformation. In other words, under conditions of pure deformation, is there an orientation $\theta$ for which the angle $\varphi$ is zero? If such an orientation exists, the element in its deformed state will appear as shown in Fig. 3-21, where $\boldsymbol{n}$ is a unit vector collinear with line segment $AC$. From the figure we can write

$$AC = ds\,\boldsymbol{n}$$

$$CC' = \varepsilon\, AC = \varepsilon\, ds\,\boldsymbol{n}$$

where $\varepsilon$, the proportionality constant, is the strain of line segment $AC$. However, $CC'$ can also be written as

$$CC' = du\,\boldsymbol{i} + dv\,\boldsymbol{j}$$

where $du$ and $dv$ are the horizontal and vertical components, respectively, of $CC'$. Now since

$$\boldsymbol{n} = n_x\,\boldsymbol{i} + n_y\,\boldsymbol{j}$$

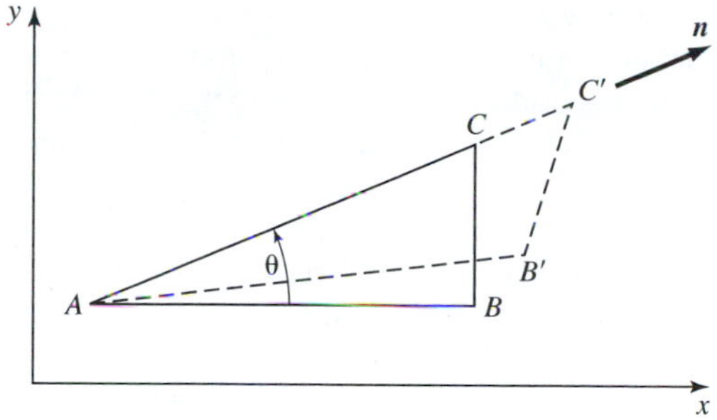


Figure 3-21 Orientation θ for which the angle φ is zero

we have

$$du\,\boldsymbol{i} + dv\,\boldsymbol{j} = \varepsilon\,ds\,n_x\,\boldsymbol{i} + \varepsilon\,ds\,n_y\,\boldsymbol{j}$$

or

$$du = \varepsilon n_x\,ds, \qquad dv = \varepsilon n_y\,ds \tag{3-38}$$

Since the strain results from pure deformation, the rotation must be zero and consequently $du$ and $dv$ are given by Eqs. (3-36). These can be substituted into Eqs. (3-38) and the resulting expressions divided by $ds$. After using Eqs. (3-23) and collecting terms we get the following equations:

$$\begin{aligned} (\varepsilon_{xx} - \varepsilon)n_x + \tfrac{1}{2}\gamma_{xy}n_y &= 0 \\ \tfrac{1}{2}\gamma_{xy}n_x + (\varepsilon_{yy} - \varepsilon)n_y &= 0 \end{aligned} \tag{3-39}$$

In matrix form these equations are

$$\begin{bmatrix} (\varepsilon_{xx} - \varepsilon) & \tfrac{1}{2}\gamma_{xy} \\ \tfrac{1}{2}\gamma_{xy} & (\varepsilon_{yy} - \varepsilon) \end{bmatrix} \begin{Bmatrix} n_x \\ n_y \end{Bmatrix} = \begin{Bmatrix} 0 \\ 0 \end{Bmatrix} \tag{3-40}$$

A nontrivial solution to this system of equations exists only if the determinant of the coefficient matrix is zero (see [1, p. 13]). Thus,

$$\begin{vmatrix} (\varepsilon_{xx} - \varepsilon) & \tfrac{1}{2}\gamma_{xy} \\ \tfrac{1}{2}\gamma_{xy} & (\varepsilon_{yy} - \varepsilon) \end{vmatrix} = (\varepsilon_{xx} - \varepsilon)(\varepsilon_{yy} - \varepsilon) - \frac{1}{4}\gamma_{xy}^2 = 0$$

or, on rearranging,

$$\varepsilon^2 - (\varepsilon_{xx} + \varepsilon_{yy})\varepsilon + \varepsilon_{xx}\varepsilon_{yy} - \tfrac{1}{4}\gamma_{xy}^2 = 0 \tag{3-41}$$

The two roots of this quadratic equation are given by

$$\varepsilon_1, \varepsilon_2 = \frac{\varepsilon_{xx} + \varepsilon_{yy}}{2} \pm \sqrt{\left(\frac{\varepsilon_{xx} - \varepsilon_{yy}}{2}\right)^2 + \left(\frac{1}{2}\gamma_{xy}\right)^2} \tag{3-42}$$

The ratio $n_y/n_x = \tan\theta$ can be found from either of Eqs. (3-39). From the first we have

$$\tan\theta = \frac{n_y}{n_x} = \frac{\varepsilon - \varepsilon_{xx}}{\tfrac{1}{2}\gamma_{xy}} \tag{3-43}$$

Thus since there are two values of $\varepsilon$ that satisfy Eqs. (3-39), there are also two directions for which, under conditions of pure deformation, line segment $AC$ undergoes no change in its orientation. One of these corresponds to $\varepsilon_1$, the other to $\varepsilon_2$. Later in this section we will show that these two directions are perpendicular to each other.

The strains $\varepsilon_1$ and $\varepsilon_2$ are referred to as *principal strains,* and the corresponding directions $\theta_1$ and $\theta_2$ as *principal directions*.

In introductory mechanics of materials courses it is customary to express Eq. (3-43) in terms of tan $2\theta$. To do this we note that

$$\tan 2\theta = \frac{2\tan\theta}{1-\tan^2\theta}$$

In this expression tan $\theta$ can be eliminated by substituting Eq. (3-43). After simplifying, this results in the following expression for tan $2\theta$:

$$\tan 2\theta = \gamma_{xy}\frac{\varepsilon - \varepsilon_{xx}}{\left(\frac{1}{2}\gamma_{xy}\right)^2 - (\varepsilon - \varepsilon_{xx})^2} \tag{3-44}$$

The denominator of this equation can be further simplified. To do so we write [see Eq. (3-41)]

$$\begin{aligned}\left(\tfrac{1}{2}\gamma_{xy}\right)^2 &= \varepsilon^2 - (\varepsilon_{xx} + \varepsilon_{yy})\varepsilon + \varepsilon_{xx}\varepsilon_{yy} \\ &= (\varepsilon - \varepsilon_{xx})(\varepsilon - \varepsilon_{yy})\end{aligned}$$

When this is substituted into the denominator, it becomes

$$D \equiv \left(\tfrac{1}{2}\gamma_{xy}\right)^2 - (\varepsilon - \varepsilon_{xx})^2 = (\varepsilon_{xx} - \varepsilon_{yy})(\varepsilon - \varepsilon_{xx})$$

and with this expression Eq. (3-44) becomes

$$\tan 2\theta = \frac{\gamma_{xy}}{\varepsilon_{xx} - \varepsilon_{yy}} \tag{3-45}$$

An examination of a plot of the tangent function between 0 and 360° indicates that there are two values of $2\theta$, differing by 180°, that satisfy Eq. (3-45). The corresponding values of $\theta$ are usually designated $\theta_1$ and $\theta_2$.

The shear strain acting on planes defined by $\theta_1$ and $\theta_2$ can be found from the last of Eqs. (3-30). If $\theta$ is a principal direction, then from Eq. (3-45),

$$\sin 2\theta = \frac{\gamma_{xy}}{\varepsilon_{xx} - \varepsilon_{yy}}\cos 2\theta$$

and therefore we have, from Eqs. (3-30),

$$\gamma_{x'y'} = 0$$

Thus *the shear strain acting on an element oriented in a principal direction is zero.*

Since we have two principal directions $\theta_1$ and $\theta_2$, it is natural to enquire about the relationship between these two directions. We begin by considering Eq. (3-43) for $\theta_2$,

$$\left(\frac{n_x}{n_y}\right)_2 = \tan\theta_2 = \frac{\varepsilon_2 - \varepsilon_{xx}}{\frac{1}{2}\gamma_{xy}}$$

However, from Eqs. (3-42) we have

$$\varepsilon_1 + \varepsilon_2 = \varepsilon_{xx} + \varepsilon_{yy}$$

and therefore we can write

$$\tan\theta_2 = \frac{\varepsilon_{yy} - \varepsilon_1}{\frac{1}{2}\gamma_{xy}} \tag{3-46}$$

Now from Eqs. (3-39) we have

$$\frac{\varepsilon_{yy} - \varepsilon_1}{\frac{1}{2}\gamma_{xy}} = -\left(\frac{n_x}{n_y}\right)_1$$

With Eq. (3-43) this becomes

$$\frac{\varepsilon_{yy} - \varepsilon_1}{\frac{1}{2}\gamma_{xy}} = -\cot\theta_1 = \tan\left(\theta_1 + \frac{\pi}{2}\right)$$

and when this is used in Eq. (3-46), we get

$$\tan\theta_2 = \tan\left(\theta_1 + \frac{\pi}{2}\right)$$

or

$$\theta_2 = \theta_1 + \frac{\pi}{2}$$

which shows that the principal directions are perpendicular. It is worth noting that $\theta_2$ is 90° *counterclockwise* from $\theta_1$.

We will now show that the rotation $\omega_z$ is the angular displacement of the principal axes. To do this we note that Eq. (3-28) gives the angular rotation $\varphi$ of a line element whose orientation is $\theta$. This equation can be written in terms of $2\theta$, $\omega_z$, and $\gamma_{xy}$ as

$$\varphi = -\tfrac{1}{2}(\varepsilon_{xx} - \varepsilon_{yy})\sin 2\theta + \tfrac{1}{2}\gamma_{xy}\cos 2\theta + \omega_z \tag{3-47}$$

But if the line element is oriented along a principal axis, then $\theta$ is given in terms of the strains by Eq. (3-45), which can be rewritten as

$$\sin 2\theta = \frac{\gamma_{xy}}{\varepsilon_{xx} - \varepsilon_{yy}}\cos 2\theta$$

Now when this is substituted into Eq. (3-47) and the resulting expression simplified, we get

$$\omega_z = \varphi$$

thereby proving our assertion that $\omega_z$ is the angular rotation of the principal axes of an element.

For three-dimensional problems Eqs. (3-40) become

$$\begin{bmatrix} \varepsilon_{xx}-\varepsilon & \frac{1}{2}\gamma_{xy} & \frac{1}{2}\gamma_{xz} \\ \frac{1}{2}\gamma_{xy} & \varepsilon_{yy}-\varepsilon & \frac{1}{2}\gamma_{yz} \\ \frac{1}{2}\gamma_{xz} & \frac{1}{2}\gamma_{yz} & \varepsilon_{zz}-\varepsilon \end{bmatrix} \begin{Bmatrix} n_x \\ n_y \\ n_z \end{Bmatrix} = \begin{Bmatrix} 0 \\ 0 \\ 0 \end{Bmatrix} \tag{3-48}$$

A nontrivial solution to this set of equations exists only if

$$\begin{vmatrix} \varepsilon_{xx}-\varepsilon & \frac{1}{2}\gamma_{xy} & \frac{1}{2}\gamma_{xz} \\ \frac{1}{2}\gamma_{xy} & \varepsilon_{yy}-\varepsilon & \frac{1}{2}\gamma_{yz} \\ \frac{1}{2}\gamma_{xz} & \frac{1}{2}\gamma_{yz} & \varepsilon_{zz}-\varepsilon \end{vmatrix} = 0 \tag{3-49}$$

A cubic equation in $\varepsilon$ results from the expansion of this determinant (see Chapter 2). The roots of this equation, $\varepsilon_1$, $\varepsilon_2$, and $\varepsilon_3$, give the principal strains. Once these are determined, the principal directions can be found by substituting each of the principal strains successively in Eqs. (3-48). Note, however, that because Eqs. (3-48) are homogeneous, $n_x$, $n_y$, and $n_z$ cannot be found explicitly without an additional equation. Since $\boldsymbol{n}$ is a unit vector, this additional equation is provided by Eq. (2-65). Without it, only the ratios $n_y/n_x$ and $n_z/n_x$ can be found. In fact, these ratios can be obtained from any two of Eqs. (3-48), and are sufficient to determine the principal directions. It can be shown that the three principal directions corresponding to the principal strains are mutually perpendicular.

## EXAMPLE 3-2

It will be shown in Section 4-13 that, in a multiaxial stress state, yielding begins when a certain combination of principal stresses reaches a critical value. For an isotropic material, the principal stress axes coincide with the principal strain axes (Section 4-2-1). Consequently when working with strain gages it is important to be able to determine, using the three-element rosette, for example, both the principal strains and the principal directions. Since, at the surface, there are two principal strains and one principal direction, it should be possible to determine these three quantities with three strain gage readings. To do this, the results of Example 3-1 are substituted in Eqs. (3-42) and (3-45). This yields

$$\varepsilon_1 = \frac{\varepsilon_A+\varepsilon_C}{2} + \frac{1}{2}\sqrt{(\varepsilon_A-\varepsilon_C)^2+(2\varepsilon_B-\varepsilon_A-\varepsilon_C)^2}$$

$$\varepsilon_2 = \frac{\varepsilon_A+\varepsilon_C}{2} - \frac{1}{2}\sqrt{(\varepsilon_A-\varepsilon_C)^2+(2\varepsilon_B-\varepsilon_A-\varepsilon_C)^2}$$

$$\tan 2\theta = \frac{2\varepsilon_B-\varepsilon_A-\varepsilon_C}{\varepsilon_A-\varepsilon_C}$$

## 3-7 STRAIN INVARIANTS

We will now show that certain combinations of the strains, called *strain invariants,* remain constant regardless of how the element is oriented. For the sake of simplicity we return to the two-dimensional formulation to show this. Recall that for a quadratic equation whose leading coefficient is unity, the sum of the roots is equal to the negative of the coefficient of the first power of the unknown whereas the product of the roots is equal to the free term in the quadratic. Thus if $\varepsilon_1$ and $\varepsilon_2$ are the roots of Eq. (3-41), then we can write

$$\begin{aligned} \varepsilon_1 + \varepsilon_2 &= \varepsilon_{xx} + \varepsilon_{yy} \\ \varepsilon_1 \varepsilon_2 &= \varepsilon_{xx}\varepsilon_{yy} - \tfrac{1}{4}\gamma_{xy}^2 \end{aligned} \tag{3-50}$$

Since these are independent of the element orientation $\theta$, it is also true that

$$\begin{aligned} \varepsilon_1 + \varepsilon_2 &= \varepsilon_{x'x'} + \varepsilon_{y'y'} \\ \varepsilon_1 \varepsilon_2 &= \varepsilon_{x'x'}\varepsilon_{y'y'} - \tfrac{1}{4}\gamma_{x'y'} \end{aligned} \tag{3-51}$$

By equating Eqs. (3-50) and (3-51) we see that

$$\begin{aligned} \varepsilon_{xx} + \varepsilon_{yy} &= \varepsilon_{x'x'} + \varepsilon_{y'y'} \\ \varepsilon_{xx}\varepsilon_{yy} - \tfrac{1}{4}\gamma_{xy} &= \varepsilon_{x'x'}\varepsilon_{y'y'} - \tfrac{1}{4}\gamma_{x'y'} \end{aligned} \tag{3-52}$$

for any element orientation. In other words, the quantities

$$\begin{aligned} J_1 &= \varepsilon_{xx} + \varepsilon_{yy} \\ J_2 &= \varepsilon_{xx}\varepsilon_{yy} - \tfrac{1}{4}\gamma_{xy}^2 \end{aligned} \tag{3-53}$$

are *invariant* with respect to element rotations. Accordingly, $J_1$ and $J_2$ are called the *first* and *second strain invariants,* respectively. In terms of the principal strains these invariants are given by

$$\begin{aligned} J_1 &= \varepsilon_1 + \varepsilon_2 \\ J_2 &= \varepsilon_1 \varepsilon_2 \end{aligned} \tag{3-54}$$

Note that if the strains are expressed in matrix form,

$$[\boldsymbol{\varepsilon}] = \begin{bmatrix} \varepsilon_{xx} & \tfrac{1}{2}\gamma_{xy} \\ \tfrac{1}{2}\gamma_{xy} & \varepsilon_{yy} \end{bmatrix} \tag{3-55}$$

then

$$\begin{aligned} J_1 &= \mathrm{tr}[\boldsymbol{\varepsilon}] \\ J_2 &= \det[\boldsymbol{\varepsilon}] \end{aligned} \tag{3-56}$$

With these, we see that Eq. (3-41) can be written as

$$\varepsilon^2 - J_1 \varepsilon + J_2 = 0 \tag{3-57}$$

For three-dimensional problems there are three strain invariants, given in terms of principal strains as (see [3, p. 18] and also Chapter 2)

$$\begin{aligned} J_1 &= \varepsilon_1 + \varepsilon_2 + \varepsilon_3 \\ J_2 &= -(\varepsilon_1\varepsilon_2 + \varepsilon_2\varepsilon_3 + \varepsilon_3\varepsilon_1) \\ J_3 &= \varepsilon_1\varepsilon_2\varepsilon_3 \end{aligned} \tag{3-58}$$

If the strains are written in matrix form as

$$[\boldsymbol{\varepsilon}] = \begin{bmatrix} \varepsilon_{xx} & \frac{1}{2}\gamma_{xy} & \frac{1}{2}\gamma_{zx} \\ \frac{1}{2}\gamma_{xy} & \varepsilon_{yy} & \frac{1}{2}\gamma_{yz} \\ \frac{1}{2}\gamma_{zx} & \frac{1}{2}\gamma_{yz} & \varepsilon_{zz} \end{bmatrix} \tag{3-59}$$

then the strain invariants are

$$\begin{aligned} J_1 &= \mathrm{tr}[\boldsymbol{\varepsilon}] = \varepsilon_{xx} + \varepsilon_{yy} + \varepsilon_{zz} \\ J_2 &= -\begin{vmatrix} \varepsilon_{yy} & \frac{1}{2}\gamma_{yz} \\ \frac{1}{2}\gamma_{yz} & \varepsilon_{zz} \end{vmatrix} - \begin{vmatrix} \varepsilon_{xx} & \frac{1}{2}\gamma_{zx} \\ \frac{1}{2}\gamma_{zx} & \varepsilon_{yy} \end{vmatrix} - \begin{vmatrix} \varepsilon_{xx} & \frac{1}{2}\gamma_{xy} \\ \frac{1}{2}\gamma_{xy} & \varepsilon_{yy} \end{vmatrix} \\ &= \tfrac{1}{4}\left(\gamma_{xy}^2 + \gamma_{yz}^2 + \gamma_{zx}^2\right) - (\varepsilon_{xx}\varepsilon_{yy} + \varepsilon_{yy}\varepsilon_{zz} + \varepsilon_{zz}\varepsilon_{xx}) \\ J_3 &= \det[\boldsymbol{\varepsilon}] = \varepsilon_{xx}\varepsilon_{yy}\varepsilon_{zz} + \tfrac{1}{4}\varepsilon_{xy}\varepsilon_{yz}\varepsilon_{zx} - \tfrac{1}{4}\left(\varepsilon_{xx}\varepsilon_{yz}^2 + \varepsilon_{yy}\varepsilon_{zx}^2 + \varepsilon_{zz}\varepsilon_{xy}^2\right) \end{aligned} \tag{3-60}$$

## 3-8 VOLUME CHANGES AND DILATATION

If a small volume of material of dimensions *dx*, *dy*, and *dz* is subjected to normal strains $\varepsilon_{xx}$, $\varepsilon_{yy}$, and $\varepsilon_{zz}$, the lengths of the sides parallel to the *x*, *y*, and *z* axes change by amounts $\varepsilon_{xx}\,dx$, $\varepsilon_{yy}\,dy$, and $\varepsilon_{zz}\,dz$, respectively, so that the new lengths of the sides are $(1+\varepsilon_{xx})\,dx$, $(1+\varepsilon_{yy})\,dy$, and $(1+\varepsilon_{zz})\,dz$. Therefore the volume of the element in its deformed configuration is given by

$$V_{\mathrm{def}} = (1+\varepsilon_{xx})(1+\varepsilon_{yy})(1+\varepsilon_{zz})\,dx\,dy\,dz$$

But in its undeformed configuration the element volume is $dx\,dy\,dz$. Thus the change in volume is given by

$$\Delta V = V_{\mathrm{def}} - dx\,dy\,dz = (\varepsilon_{xx} + \varepsilon_{yy} + \varepsilon_{zz})\,dx\,dy\,dz$$

where, for consistency with the infinitesimal strain theory, products of two or more strains have been neglected. The change in volume per unit volume is then given by

$$\Delta \equiv \frac{\Delta V}{V} = \varepsilon_{xx} + \varepsilon_{yy} + \varepsilon_{zz} \tag{3-61}$$

where $\Delta$ is referred to as the *dilatation*. Note that by neglecting the products of strains, what has been neglected are the small crosshatched areas shown in Fig. 3-22 and the volumes associated with them.

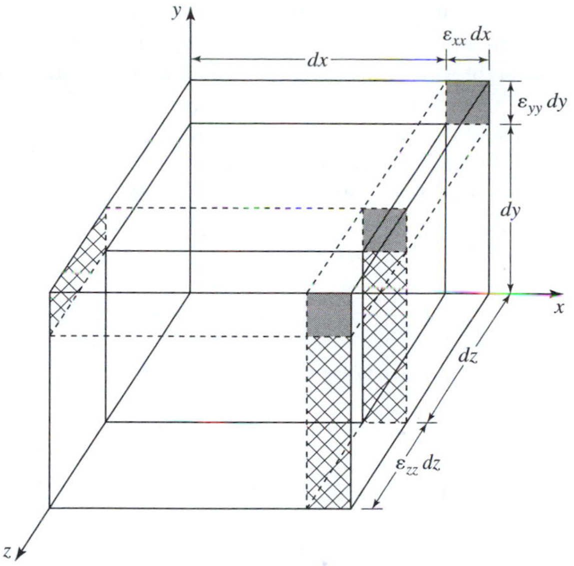


Figure 3-22

Now if the element is subjected to pure shear, say, $\gamma_{xy} = \gamma_{yz} = \gamma_{zx} = \gamma$, then since all normal strains are zero, it is easy to show that there is no change in the volume of the element. With some algebra it can be shown that, within the assumptions of the infinitesimal strain theory, if the element is subjected to both normal and shear strains, the change in volume per unit volume is still given by Eq. (3-61).

Thus the sum of the normal strains gives a measure of the change in volume of an element subjected to strain. In general, the dilatation varies from point to point within the body. However, if the dilatation is multiplied by $dx\,dy\,dz$ and integrated over the entire volume of the body, then the change in volume of the body can be calculated,

$$\Delta\overline{V} = \int_{\overline{V}} \Delta\,dx\,dy\,dz \tag{3-62}$$

Although the dilatation plays a limited role in problems involving only elastic deformation, it is an important quantity in problems involving plastic deformation.

## 3-9 STRAIN DEVIATOR

In Section 2-3-6 we introduced the deviatoric stress matrix and mentioned the role that these stresses play in determining the onset of yielding in a material. We can also define, in a similar way, the deviatoric strain matrix. We begin by noting that if the three normal strains acting on an element are equal and the shear strains are zero, then the element undergoes only a change in volume. With the normal strains equal there is no linear distortion because the lengths of all sides of the element change in the same proportion. There is no angular distortion of the element since the shear strains are all zero. Thus there is no change in the shape of the element, only a change in its volume.

When an element with an arbitrary spatial orientation is subjected to a general infinitesimal strain field consisting of both normal and shear strains, its deformation can be thought of as consisting of a change in volume accompanied by a change in shape (see, for example, Fig. 3-19*d* and *e*). Since the volume change is given by Eq. (3-61), the strain field that produces *only* that change in volume (and no change in shape) consists of equal normal strains of magnitude $\frac{1}{3}\Delta$ and zero shear strains. When this strain field is subtracted from the general strain field given, in matrix form, by Eq. (3-59), the resulting strain field produces only a change of shape in the element. That strain field, referred to as the *deviatoric strain field,* is given by

$$[D_\varepsilon] = \begin{bmatrix} \varepsilon_{xx} - \frac{1}{3}\Delta & \frac{1}{2}\gamma_{xy} & \frac{1}{2}\gamma_{zx} \\ \frac{1}{2}\gamma_{xy} & \varepsilon_{yy} - \frac{1}{3}\Delta & \frac{1}{2}\gamma_{yz} \\ \frac{1}{2}\gamma_{zx} & \frac{1}{2}\gamma_{yz} & \varepsilon_{zz} - \frac{1}{3}\Delta \end{bmatrix} \tag{3-63}$$

where $[D_\varepsilon]$ is the *strain deviator matrix.* When the element is oriented along principal axes, we have

$$[D_\varepsilon] = \begin{bmatrix} \varepsilon_1 - \frac{1}{3}\Delta & 0 & 0 \\ 0 & \varepsilon_2 - \frac{1}{3}\Delta & 0 \\ 0 & 0 & \varepsilon_3 - \frac{1}{3}\Delta \end{bmatrix}$$

$$= \begin{bmatrix} \dfrac{2\varepsilon_1 - \varepsilon_2 - \varepsilon_3}{3} & 0 & 0 \\ 0 & \dfrac{2\varepsilon_2 - \varepsilon_3 - \varepsilon_1}{3} & 0 \\ 0 & 0 & \dfrac{2\varepsilon_3 - \varepsilon_1 - \varepsilon_2}{3} \end{bmatrix} \tag{3-64}$$

The invariants of the strain deviator matrix are computed as follows:

$$\begin{aligned} J_{1D} &= \text{tr}[D_\varepsilon] = 0 \\ J_{2D} &= -\begin{vmatrix} D_{\varepsilon_{11}} & D_{\varepsilon_{12}} \\ D_{\varepsilon_{21}} & D_{\varepsilon_{22}} \end{vmatrix} - \begin{vmatrix} D_{\varepsilon_{22}} & D_{\varepsilon_{23}} \\ D_{\varepsilon_{32}} & D_{\varepsilon_{33}} \end{vmatrix} - \begin{vmatrix} D_{\varepsilon_{11}} & D_{\varepsilon_{13}} \\ D_{\varepsilon_{31}} & D_{\varepsilon_{33}} \end{vmatrix} \end{aligned} \tag{3-65}$$

where $D_{\varepsilon_{ij}}$ is the element in row $i$ and column $j$ in Eq. (3-63). In terms of the strains and the dilatation $J_2$ becomes

$$J_{2D} = -\varepsilon_{xx}\varepsilon_{yy} - \varepsilon_{yy}\varepsilon_{zz} - \varepsilon_{zz}\varepsilon_{xx} + \tfrac{1}{3}\Delta^2 + \tfrac{1}{4}\left(\gamma_{xy}^2 + \gamma_{yz}^2 + \gamma_{zx}^2\right) \tag{3-66}$$

or, in terms of the strains,

$$J_{2D} = \tfrac{1}{6}\left[(\varepsilon_{xx} - \varepsilon_{yy})^2 + (\varepsilon_{yy} - \varepsilon_{zz})^2 + (\varepsilon_{zz} - \varepsilon_{xx})^2 + \tfrac{3}{2}\left(\gamma_{xy}^2 + \gamma_{yz}^2 + \gamma_{zx}^2\right)\right] \tag{3-67}$$

By direct expansion we can show that

$$e_1^2 + e_2^2 + e_3^2 = \tfrac{1}{3}[(\varepsilon_{xx} - \varepsilon_{yy})^2 + (\varepsilon_{yy} - \varepsilon_{zz})^2 + (\varepsilon_{zz} - \varepsilon_{xx})^2]$$

where

$$e_{xx} = \varepsilon_{xx} - \tfrac{1}{3}\Delta, \ldots$$

and therefore we can also write $J_2$ as

$$J_{2D} = \tfrac{1}{2}\left[\varepsilon_{xx}^2 + \varepsilon_{yy}^2 + \varepsilon_{zz}^2 + \tfrac{1}{2}\left(\gamma_{xy}^2 + \gamma_{yz}^2 + \gamma_{zx}^2\right)\right] \tag{3-68}$$

In terms of principal strains we have, from Eq. (3-64),

$$J_{2D} = -(e_1 e_2 + e_2 e_3 + e_3 e_1) \tag{3-69}$$

where

$$\varepsilon_i = \varepsilon_i - \tfrac{1}{3}\Delta, \qquad i = 1, 2, 3 \tag{3-70}$$

or

$$J_{2D} = \tfrac{1}{6}[(\varepsilon_1 - \varepsilon_2)^2 + (\varepsilon_2 - \varepsilon_3)^2 + (\varepsilon_3 - \varepsilon_1)^2] \tag{3-71}$$

Finally,

$$J_{3D} = \det[D_\varepsilon] = e_1 e_2 e_3 \tag{3-72}$$

These invariants are related to the invariants of the strain matrix as follows:

$$\begin{aligned} J_{1D} &= 0 \\ J_{2D} &= \tfrac{1}{3}\left(J_1^2 + 3J_2\right) \\ J_{3D} &= \tfrac{1}{27}\left(2J_1^3 + 9J_1 J_2 + 27J_3\right) \end{aligned} \tag{3-73}$$

The strain deviator plays an important role in the theory of plasticity because it is known that there is no volume change during plastic deformation.

## 3-10 STRAIN-DISPLACEMENT EQUATIONS IN POLAR COORDINATES

Thus far we have considered the displacement components $u$ and $v$ associated with a Cartesian coordinate system. However, displacement is a vector quantity, and therefore its components can be expressed in any convenient coordinate system. In cylindrical coordinates one component, $u_r$, is in the radial direction and the other component, $u_\theta$, is in the tangential direction, as shown in Fig. 3-23. The strain components in the cylindrical coordinate system can be found in terms of the displacement components $u_r$ and $u_\theta$ by examining the deformation of an infinitesimal element in that coordinate system, as we did in Section 3-2 for the Cartesian coordinate system or by using the transformation equations developed in Section 3-4. We opt here for the latter approach.

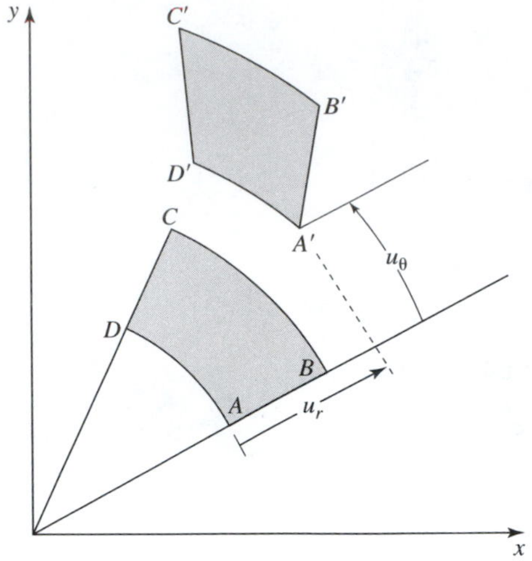


Figure 3-23 Radial and tangential components of displacements

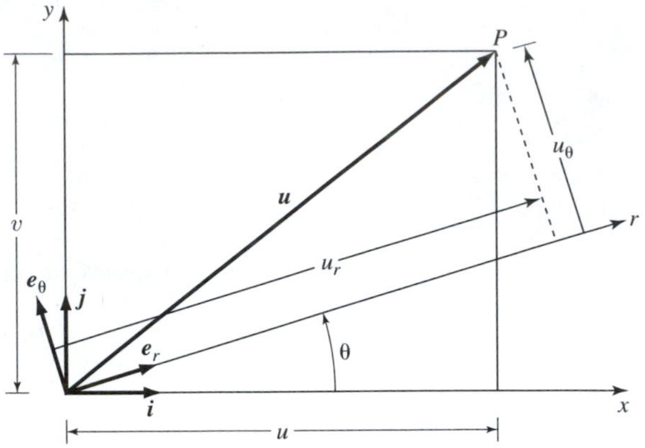


Figure 3-24 Cartesian and polar components of displacement

From Eqs. (3-24), (3-25), and (3-29) we have

$$\begin{aligned}
\varepsilon_{rr} &= \varepsilon_{xx}\cos^2\theta + \varepsilon_{yy}\sin^2\theta + \gamma_{xy}\sin\theta\cos\theta \\
\varepsilon_{\theta\theta} &= \varepsilon_{xx}\sin^2\theta + \varepsilon_{yy}\cos^2\theta - \gamma_{xy}\sin\theta\cos\theta \\
\gamma_{r\theta} &= 2(\varepsilon_{yy} - \varepsilon_{xx})\sin\theta\cos\theta + \gamma_{xy}(\cos^2\theta - \sin^2\theta)
\end{aligned} \tag{3-74}$$

In these equations the radial and tangential directions are along the $x'$ and $y'$ axes, respectively.

Now consider the vector $\boldsymbol{u}$ shown in Fig. 3-24 extending from the origin to point $P$. The components of $\boldsymbol{u}$ in the $r$, $\theta$ coordinate system are $u_r$ and $u_\theta$ in the $r$ and $\theta$ directions, respectively. If the unit vectors associated with the $r$ and $\theta$ directions are $\boldsymbol{e}_r$ and $\boldsymbol{e}_\theta$, respectively, then we can write

$$\boldsymbol{u} = u_r\,\boldsymbol{e}_r + u_\theta\,\boldsymbol{e}_\theta \tag{3-74a}$$

We would like to find the components of displacement in the $x$, $y$ coordinate system, with unit vectors $\boldsymbol{i}$ and $\boldsymbol{j}$ in terms of the components in the $r$, $\theta$ system. Since the dot product of $\boldsymbol{u}$ with $\boldsymbol{i}$ gives the projection of the vector $\boldsymbol{i}$ on the $x$ axis, we have that

$$\begin{aligned}
u = \boldsymbol{u}\cdot\boldsymbol{i} &= u_r(\boldsymbol{e}_r\cdot\boldsymbol{i}) + u_\theta(\boldsymbol{e}_\theta\cdot\boldsymbol{i}) \\
&= u_r\cos\theta - u_\theta\sin\theta
\end{aligned} \tag{3-75}$$

Similarly, in the $y$ direction,

$$\begin{aligned}
v = \boldsymbol{u}\cdot\boldsymbol{j} &= u_r(\boldsymbol{e}_r\cdot\boldsymbol{j}) + u_\theta(\boldsymbol{e}_\theta\cdot\boldsymbol{j}) \\
&= u_r\sin\theta + u_\theta\cos\theta
\end{aligned} \tag{3-76}$$

By use of the chain rule we find that

$$\begin{aligned}\frac{\partial}{\partial x} &= \frac{\partial}{\partial r}\frac{\partial r}{\partial x} + \frac{\partial}{\partial \theta}\frac{\partial \theta}{\partial x}\\ \frac{\partial}{\partial y} &= \frac{\partial}{\partial r}\frac{\partial r}{\partial y} + \frac{\partial}{\partial \theta}\frac{\partial \theta}{\partial y}\end{aligned} \tag{3-77}$$

and since

$$x = r\cos\theta, \qquad y = r\sin\theta$$

so that

$$\begin{aligned}r^2 &= x^2 + y^2\\ \theta &= \tan^{-1}\left(\frac{y}{x}\right)\end{aligned}$$

it follows that

$$\begin{aligned}\frac{\partial r}{\partial x} &= \frac{x}{r}, \qquad \frac{\partial \theta}{\partial x} = -\frac{y}{r^2}\\ \frac{\partial r}{\partial y} &= \frac{y}{r}, \qquad \frac{\partial \theta}{\partial y} = \frac{x}{r^2}\end{aligned} \tag{3-78}$$

Thus Eqs. (3-77) become

$$\begin{aligned}\frac{\partial}{\partial x} &= \cos\theta\,\frac{\partial}{\partial r} - \frac{1}{r}\sin\theta\,\frac{\partial}{\partial \theta}\\ \frac{\partial}{\partial y} &= \sin\theta\,\frac{\partial}{\partial r} + \frac{1}{r}\cos\theta\,\frac{\partial}{\partial \theta}\end{aligned} \tag{3-79}$$

Equations (3-79) show how derivatives with respect to $r$ and $\theta$ are transformed into those with respect to $x$ and $y$. With these and Eq. (3-75) the strain component $\varepsilon_{xx}$ in Cartesian coordinates, given by Eq. (3-6), becomes

$$\begin{aligned}\varepsilon_{xx} = \frac{\partial u}{\partial x} &= \cos\theta\,\frac{\partial u}{\partial r} - \sin\theta\,\frac{1}{r}\frac{\partial u}{\partial \theta}\\ &= \cos\theta\,\frac{\partial}{\partial r}(u_r\cos\theta - u_\theta\sin\theta) - \sin\theta\,\frac{1}{r}\frac{\partial}{\partial \theta}(u_r\cos\theta - u_\theta\sin\theta)\end{aligned}$$

or

$$\varepsilon_{xx} = \frac{1}{r}\left(u_r + \frac{\partial u_\theta}{\partial \theta}\right)\sin^2\theta - \left(\frac{1}{r}\frac{\partial u_r}{\partial \theta} + \frac{\partial u_\theta}{\partial r} - \frac{1}{r}u_\theta\right)\sin\theta\cos\theta + \frac{\partial u_r}{\partial r}\cos^2\theta \tag{3-80}$$

Similarly, from Eqs. (3-7) and (3-9),

$$
\begin{aligned}
\varepsilon_{yy} &= \frac{\partial v}{\partial y} = \sin\theta \frac{\partial v}{\partial r} + \cos\theta \frac{1}{r}\frac{\partial v}{\partial \theta} \\
&= \frac{\partial u_r}{\partial r}\sin^2\theta + \left(\frac{1}{r}\frac{\partial u_r}{\partial \theta} + \frac{\partial u_\theta}{\partial r} - \frac{1}{r}u_\theta\right)\sin\theta\cos\theta + \frac{1}{r}\left(u_r + \frac{\partial u_\theta}{\partial \theta}\right)\cos^2\theta \\
\gamma_{xy} &= \left(\frac{1}{r}\frac{\partial u_r}{\partial \theta} + \frac{\partial u_\theta}{\partial r} - \frac{1}{r}u_\theta\right)(\cos^2\theta - \sin^2\theta) \\
&\quad + 2\left(\frac{\partial u_r}{\partial r} - \frac{1}{r}u_r - \frac{1}{r}\frac{\partial u_\theta}{\partial \theta}\right)\sin\theta\cos\theta
\end{aligned} \tag{3.81}
$$

When these are substituted into Eqs. (3-74) and the resulting expressions simplified, we find that

$$
\begin{aligned}
\varepsilon_{rr} &= \frac{\partial u_r}{\partial r} \\
\varepsilon_{\theta\theta} &= \frac{u_r}{r} + \frac{1}{r}\frac{\partial u_\theta}{\partial \theta} \\
\gamma_{r\theta} &= \frac{1}{r}\frac{\partial u_r}{\partial \theta} + \frac{\partial u_\theta}{\partial r} - \frac{u_\theta}{r}
\end{aligned} \tag{3-82}
$$

Equations (3-82) give the strain components in the cylindrical coordinate system in terms of the displacement components for that system.

Note that the expression for $\varepsilon_{\theta\theta}$ contains two terms. The second of these is the contribution to the strain in the tangential direction that results from displacement in that direction, as shown in Fig. 3-25. The first term, however, shows that a displacement in the radial direction can cause strain in the tangential direction. One way of understanding this phenomenon is to think of a ring subjected to internal pressure. As shown in Fig. 3-26, due to the pressure the ring undergoes an increase in its radius by an amount $u_r$. The increase in circumference caused by the pressure is

$$\Delta C = 2\pi u_r$$

and since the original circumference is $2\pi r$, there is a strain in the tangential direction given by

$$\frac{\Delta C}{C} = \frac{u_r}{r}$$

The total strain in the tangential direction is simply the superposition of these two effects.

The shear strain $\gamma_{r\theta}$ consists of three terms, the first of which is the change in angular orientation of the curved side $AB$, as shown in Fig. 3-27. The second term gives the change in angular orientation of side $AD$. Not all of this change contributes to the

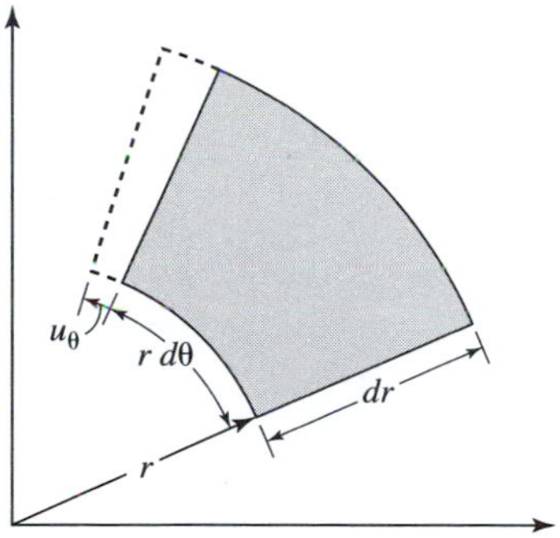


Figure 3-25

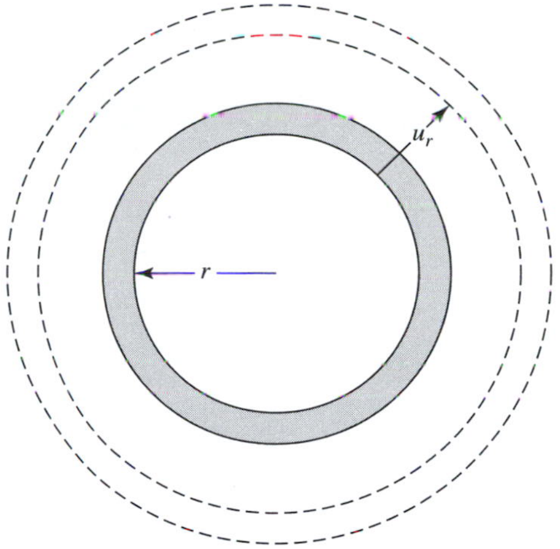


Figure 3-26

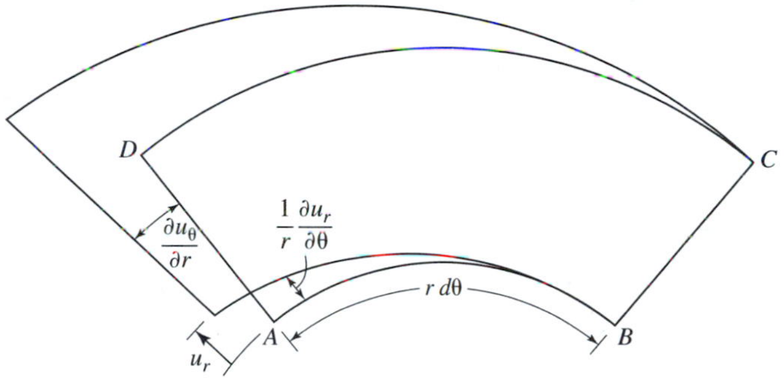


Figure 3-27 Change in angular orientation of curved side *AB*

shear strain, however. Figure 3-28 shows that the angular orientation of side *AD* can change even though the right angle $\angle DAB$ remains unchanged. Thus the last term in Eqs. (3-82) subtracts that portion of the change in angle of side *AD* that results from stretching the element rather than changing $\angle DAB$. Remember that shear strain is the change in angle between two originally perpendicular lines, and in Fig. 3-28 lines *AB* and *AD* are still perpendicular even though the element has deformed.

We now derive the compatibility equation for two-dimensional problems in cylindrical coordinates. One way to accomplish this is to transform Eq. (3-12) into cylindrical coordinates by first using Eqs. (3-80) and (3-81) to express $\varepsilon_{xx}$, $\varepsilon_{yy}$, and $\gamma_{xy}$ in terms of $\varepsilon_{rr}$, $\varepsilon_{\theta\theta}$, and $\gamma_{r\theta}$ (note that the angle is now $-\theta$ because we are going from the $r$, $\theta$ system to the $x$, $y$ system, as shown in Fig. 3-29) and then using Eqs. (3-79) to obtain expressions for the second derivatives, including the mixed derivative. This approach is very lengthy, however, and will not be pursued here.

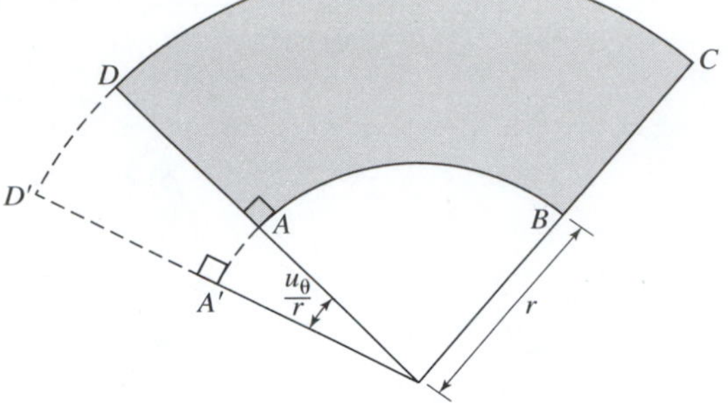


Figure 3-28

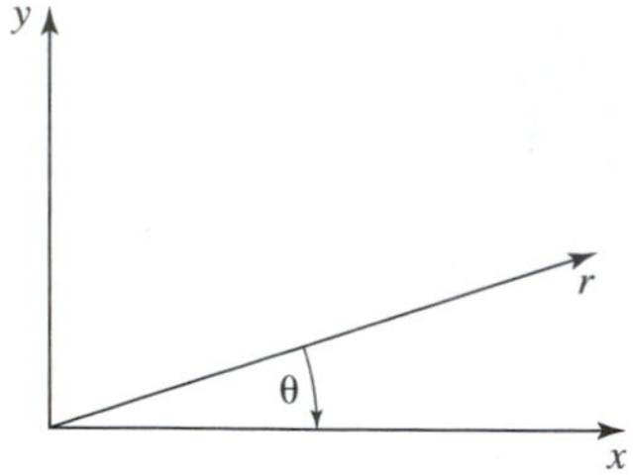


Figure 3-29

Instead, we note that each of the normal strains appearing in Eq. (3-12) is differentiated twice with respect to the coordinate direction opposite to the direction of the strain and that the shear strain is differentiated once with respect to each coordinate direction. It seems reasonable to expect that the same kinds of derivatives will appear in cylindrical coordinates. Thus we start by computing

$$\frac{1}{r}\frac{\partial^2 \gamma_{r\theta}}{\partial r \partial \theta} = -\frac{1}{r^3}\frac{\partial^2 u_r}{\partial \theta^2} + \frac{1}{r^2}\frac{\partial^3 u_r}{\partial r \partial \theta^2} + \frac{1}{r}\frac{\partial^3 u_\theta}{\partial r^2 \partial \theta} + \frac{1}{r^3}\frac{\partial u_\theta}{\partial \theta} - \frac{1}{r^2}\frac{\partial^2 u_\theta}{\partial r \partial \theta} \tag{3-83}$$

Then from Eqs. (3-82) we have

$$\frac{\partial^2 \varepsilon_{\theta\theta}}{\partial r^2} = \frac{2}{r^3}u_r - \frac{2}{r^2}\frac{\partial u_r}{\partial r} + \frac{1}{r}\frac{\partial^2 u_r}{\partial r^2} + \frac{2}{r^3}\frac{\partial u_\theta}{\partial \theta} - \frac{2}{r^2}\frac{\partial^2 u_\theta}{\partial r \partial \theta} + \frac{1}{r}\frac{\partial^3 u_\theta}{\partial r^2 \partial \theta}$$

This expression is used to replace the last three terms in Eq. (3-83), and Eqs. (3-82) are used to replace the second term. This leads to

$$\begin{aligned}\frac{1}{r}\frac{\partial^2 \gamma_{r\theta}}{\partial r \partial \theta} = &-\frac{1}{r^3}\frac{\partial^2 u_r}{\partial \theta^2} + \frac{1}{r^2}\frac{\partial^2 \varepsilon_{rr}}{\partial \theta^2} + \frac{\partial^2 \varepsilon_{\theta\theta}}{\partial r^2} - \frac{2}{r^3}u_r + \frac{2}{r^2}\frac{\partial u_r}{\partial r} \\ &- \frac{1}{r}\frac{\partial^2 u_r}{\partial r^2} - \frac{1}{r^3}\frac{\partial u_\theta}{\partial \theta} + \frac{1}{r^2}\frac{\partial u_\theta}{\partial r \partial \theta}\end{aligned} \tag{3-84}$$

We note that compatibility equations express a relationship among the strains which, as discussed in Section 3-3, cannot be specified independently. Therefore it is necessary to replace the remaining displacement terms in Eq. (3-84) with terms involving strains. In order to accomplish this we note that Eq. (3-84) can be rewritten as

$$\begin{aligned}\frac{1}{r}\frac{\partial^2 \gamma_{r\theta}}{\partial r \partial \theta} = &-\frac{1}{r^2}\frac{\partial}{\partial \theta}\left(\frac{1}{r}\frac{\partial u_r}{\partial \theta} + \frac{\partial u_\theta}{\partial r} - \frac{u_\theta}{r}\right) + \frac{1}{r^2}\frac{\partial^2 u_\theta}{\partial r \partial \theta} - \frac{1}{r^3}\frac{\partial u_\theta}{\partial \theta} \\ &+ \frac{1}{r^2}\frac{\partial^2 \varepsilon_{rr}}{\partial \theta^2} + \frac{\partial^2 \varepsilon_{\theta\theta}}{\partial r^2} - \frac{2}{r^3}u_r + \frac{2}{r^2}\frac{\partial u_r}{\partial r} - \frac{1}{r}\frac{\partial^2 u_r}{\partial r^2} - \frac{1}{r^3}\frac{\partial u_\theta}{\partial \theta} + \frac{1}{r^2}\frac{\partial^2 u_\theta}{\partial r \partial \theta}\end{aligned}$$

or, after collecting terms,

$$\frac{1}{r}\frac{\partial^2\gamma_{r\theta}}{\partial r\partial\theta} = -\frac{1}{r^2}\frac{\partial}{\partial\theta}\left(\frac{1}{r}\frac{\partial u_r}{\partial\theta} + \frac{\partial u_\theta}{\partial r} - \frac{u_\theta}{r}\right) + \frac{1}{r^2}\frac{\partial^2\varepsilon_{rr}}{\partial\theta^2} + \frac{\partial^2\varepsilon_{\theta\theta}}{\partial r^2}$$
$$-\frac{1}{r}\left(\frac{\partial^2 u_r}{\partial r^2}\right) + \frac{2}{r}\left(-\frac{1}{r^2}u_r + \frac{1}{r}\frac{\partial u_r}{\partial r} - \frac{1}{r^2}\frac{\partial u_\theta}{\partial\theta} + \frac{1}{r}\frac{\partial^2 u_\theta}{\partial r\partial\theta}\right)$$

Now the first term in parentheses is simply $\gamma_{r\theta}$, the second one is $\partial\varepsilon_{rr}/\partial r$, and the last one is $\partial\varepsilon_{\theta\theta}/\partial r$. Therefore we have

$$\frac{1}{r}\frac{\partial^2\gamma_{r\theta}}{\partial r\partial\theta} = -\frac{1}{r^2}\frac{\partial\gamma_{r\theta}}{\partial\theta} + \frac{1}{r^2}\frac{\partial^2\varepsilon_{rr}}{\partial\theta^2} + \frac{\partial^2\varepsilon_{\theta\theta}}{\partial r^2} - \frac{1}{r}\frac{\partial\varepsilon_{rr}}{\partial r} + \frac{2}{r}\frac{\partial\varepsilon_{\theta\theta}}{\partial r}$$

or, on rearranging, the desired compatibility equation becomes

$$\frac{1}{r}\frac{\partial^2\gamma_{r\theta}}{\partial r\partial\theta} + \frac{1}{r^2}\frac{\partial\gamma_{r\theta}}{\partial\theta} = \frac{1}{r^2}\frac{\partial^2\varepsilon_{rr}}{\partial\theta^2} + \frac{\partial^2\varepsilon_{\theta\theta}}{\partial r^2} - \frac{1}{r}\frac{\partial\varepsilon_{rr}}{\partial r} + \frac{2}{r}\frac{\partial\varepsilon_{\theta\theta}}{\partial r} \tag{3-85}$$

## PROBLEMS

**3-1** The definition of normal strain given at the beginning of Section 3-2, namely, the ratio of change in length to *original* length, is referred to as *engineering* strain. Normal strain can also be defined as the ratio of change in length to *current* length. This measure of normal strain is called *true* strain. The rod shown in Fig. P3-1 has an original length $l$. If its current and final lengths are $x$ and $l_f$, respectively, determine the true strain in terms of the engineering strain $\varepsilon$.

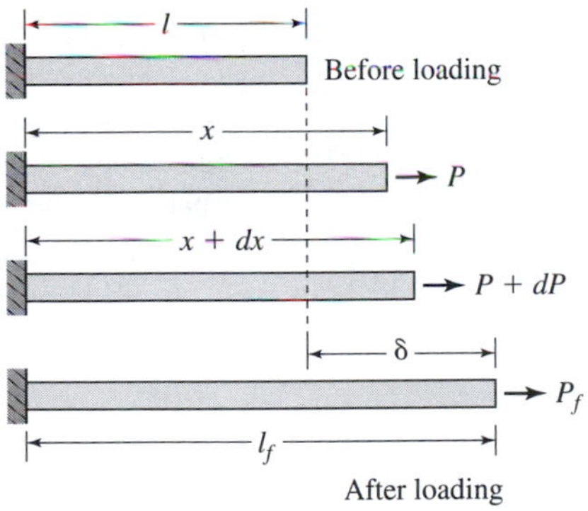


Figure P3-1

**3-2** One often encounters structural members bounded by two parallel plane surfaces. The dimension between the bounding planes is called the *thickness,* and if it is small compared with the in-plane dimensions, the structural member is called a *plate* (see Fig. P3-2). Examples of plates are a flat pane of glass and the flat ends of a pressure vessel.

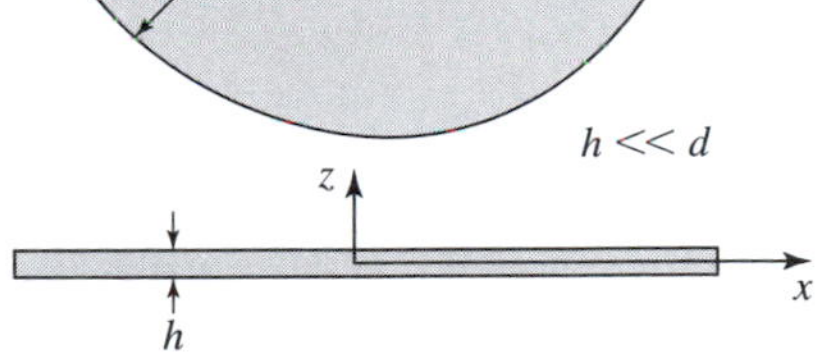

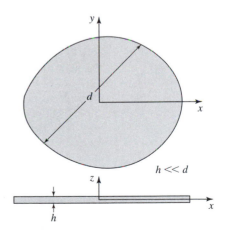

Figure P3-2

In developing a theory describing the bending behavior of plates, certain assumptions are made regarding the variation of the displacement components $u$, $v$, $w$ (in the $x$, $y$, $z$ directions, respectively) through the thickness of the plate, which is taken to be the $z$ direction. One such assumption is as follows:

$$u(x, y, z) = z\psi_x(x, y)$$

$$v(x, y, z) = z\psi_y(x, y)$$

$$w(x, y, z) = \overline{w}(x, y)$$

where $z$ is measured from a point midway between the bounding planes.

(a) Determine the six strain components that result from this assumed displacement field.

(b) Draw an infinitesimal element of material in the $xz$ plane and provide a physical interpretation of $\psi_x$. What is the physical interpretation of $\psi_y$?

---

**3-3** A bar of steel ($\nu = 0.285$) is subjected to a uniaxial stress state resulting in a strain state in which $\varepsilon_{xx}$ is nonzero, $\varepsilon_{yy} = -\nu\varepsilon_{xx}$, and $\gamma_{xy} = 0$. Compute values of $\varepsilon_{x'x'}$ for values of $\theta$ from 0° to 360°, in 1° increments. $\theta$ is the angle from the $x$ axis to the $x'$ axis. Display your results in a polar plot.

---

**3-4** Repeat Problem 3-3 for a general biaxial strain state in which $\varepsilon_{xx}$ and $\varepsilon_{yy}$ are both arbitrary and $\gamma_{xy} = 0$. Develop polar plots for the following specific cases:

(a) $\varepsilon_{yy} = \varepsilon_{xx}$

(b) $\varepsilon_{yy} = -\varepsilon_{xx}$

(c) $\varepsilon_{yy} = 0.285\varepsilon_{xx}$

---

**3-5** Three strain gages are arranged in a three-element rectangular rosette, as shown in Fig. 3-14. Determine the strains $\varepsilon_{xx}$, $\varepsilon_{yy}$, and $\gamma_{xy}$ for an angular orientation $\beta$ relative to the axis of strain gage $A$ in terms of $\varepsilon_A$, $\varepsilon_B$, $\varepsilon_C$, and $\beta$.

---

**3-6** In a delta rosette three strain gages are arranged as shown in the Fig. P3-6. Determine the strains $\varepsilon_{xx}$, $\varepsilon_{yy}$, and $\gamma_{xy}$ in terms of $\varepsilon_A$, $\varepsilon_B$, and $\varepsilon_C$.

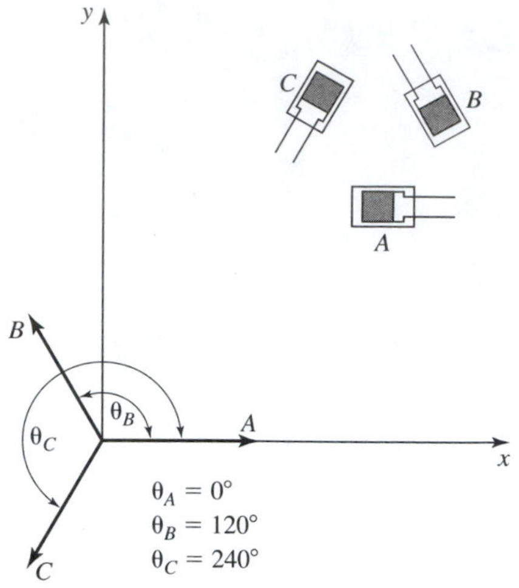


Figure P3-6

---

**3-7** For the strain gage configuration in Problem 3-6 determine the strains $\varepsilon_{xx}$, $\varepsilon_{yy}$, and $\gamma_{xy}$ for an angular orientation $\beta$ relative to the axis of strain gage $A$ in terms of $\varepsilon_A$, $\varepsilon_B$, $\varepsilon_C$, and $\beta$.

---

**3-8** For the strain gage configuration in Problem 3-6 determine the principal strains and the principal directions in terms of $\varepsilon_A$, $\varepsilon_B$, and $\varepsilon_C$.

---

**3-9** For the three-element rectangular rosette and the delta rosette, determine an expression for the maximum shear stress in terms of $\varepsilon_A$, $\varepsilon_B$, and $\varepsilon_C$.

---

**3-10** A plate with a hole of radius $a$ is subjected to uniaxial loading, as shown in the Fig. P3-10. If the dimensions of the plate are large compared to the radius of the hole, the displacement field is given by

$$u_r = \frac{A}{r}\left[\frac{1-\nu}{1+\nu}r^2 + a^2 + \left(\frac{4a^2}{1+\nu} + r^2 - \frac{a^4}{r^2}\right)\cos 2\theta\right]$$

$$u_\theta = -\frac{A}{r}\left(2\frac{1-\nu}{1+\nu}a^2 + r^2 + \frac{a^4}{r^2}\right)\sin 2\theta$$

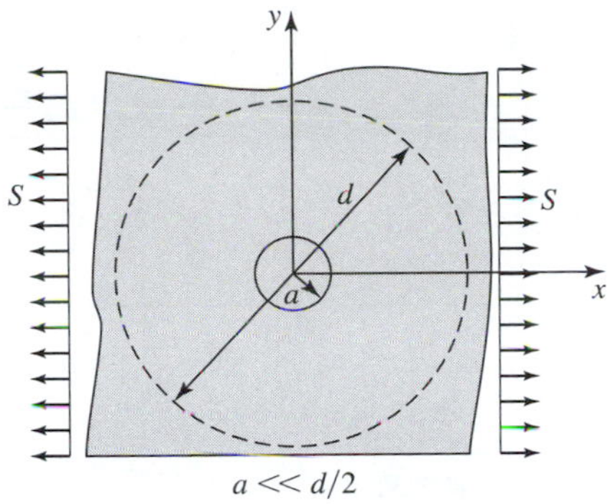


Figure P3-10

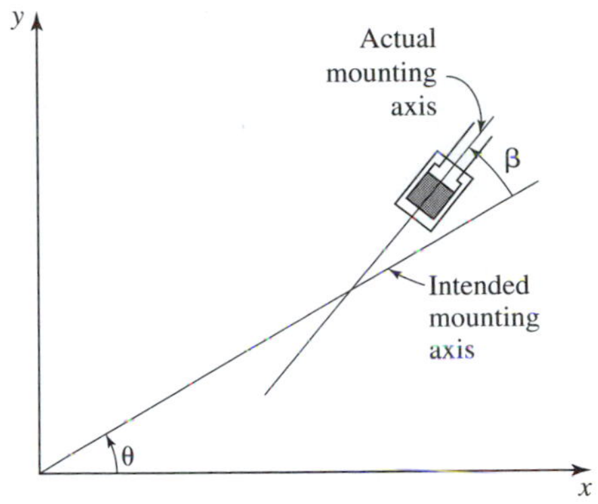


Figure P3-11

where $A = S(1+\nu)/2E$, $S$ is the applied stress, $E$ is Young's modulus, and $\nu$ is Poisson's ratio.

(a) Compute the strains $\varepsilon_{rr}/A$, $\varepsilon_{\theta\theta}/A$, and $\gamma_{xy}/A$ in terms of the dimensionless radius $R = r/a$.

(b) Show that the compatibility equation is satisfied with the given displacement field.

(c) Plot $\varepsilon_{rr}/A$ and $\varepsilon_{\theta\theta}/A$ versus $R$ for $\theta = 0°$ and $90°$ for $\nu = .30$.

(d) For the two angular locations in part (c), determine the value of $R$ for which $\varepsilon_{rr}$ and $\varepsilon_{\theta\theta}$ have their largest magnitudes, and the value of each strain at that point.

**3-11** One source of error in the use of strain gages is due to the misalignment of the gage. If the gage is attached to the surface with a small angular error relative to its intended mounting axis, the gage will not accurately give the strain along the intended mounting axis. Consider a biaxial strain field yielding principal strains $\varepsilon_1$ and $\varepsilon_2$.

(a) Determine an expression for the error and also the percent error in *gage* reading for a strain gage angular mounting error of $\pm\beta$ relative to an intended mounting axis whose angular location is $\theta$ relative to the principal strain axis, as shown in Fig. P3-11.

(b) Identify the three parameters that influence the percent error in gage reading.

(c) At what angular location of the strain gage does a given mounting error have a maximum effect on the gage error? a minimum effect?

(d) For a uniaxial stress state producing strains $\varepsilon_{xx}$ and $\varepsilon_{yy} = -\nu\varepsilon_{xx}$, develop a graph of percent error in strain indication versus intended mounting angle for mounting angles $\theta$ between $0°$ and $90°$, and mounting errors $\beta$ up to $\pm 10°$ in $1°$ increments. Use $\nu = 0.285$.

**3-12** The parallelogram *ABCD* is scribed on the surface of a plate, as shown in Fig. P3-12*a*. Loads are applied to the plate and the parallelogram deforms as shown in Fig. P3-12*b* (the magnitudes of deformation have been exaggerated for clarity). If the resulting strain state is uniform, determine strains $\varepsilon_{xx}$, $\varepsilon_{yy}$, $\gamma_{xy}$ as well as strains $\varepsilon_{x'x'}$, $\varepsilon_{y'y'}$, $\gamma_{x'y'}$. Verify that your solution satisfies Eqs. (3-52).

**3-13** Find the principal strains and principal directions for Problem 3-12.

**3-14** In the finite element method, to be discussed in Chapter 7, a structure with a complicated shape is modeled as a collection of small, but finite, elements with simple shapes such as triangles or quadrilaterals. A displacement distribution is then assumed within each element. One possible element is the rectangle shown in Fig. P3-14. The displacement field

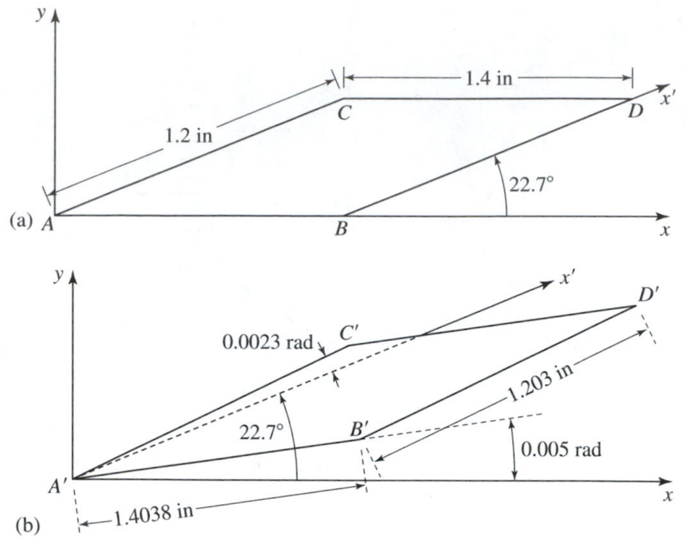


Figure P3-12

assumed for this element is

$$u(x, y) = a_0 + a_1 x + a_2 y + a_3 xy$$

$$v(x, y) = a_4 + a_5 x + a_6 y + a_7 xy$$

Determine the strains associated with this displacement distribution and put your results in the matrix form $\{\varepsilon\} = [\boldsymbol{x}]\{\boldsymbol{a}\}$, where $\{\varepsilon\}$ is a column matrix containing the strains $\varepsilon_{xx}$, $\varepsilon_{yy}$, and $\varepsilon_{xy}$, $[\boldsymbol{x}]$ contains combinations of the variables $x$ and $y$, and $\{\boldsymbol{a}\}$ is a column matrix containing the constants $a_1, \ldots, a_7$. Assume that the displacements at the corners $i, j, k$, and $l$ are known. Let $u_i = u(x_i, y_i)$, $v_i = v(x_i, y_i)$, and so on. Obtain the column matrix $\{\boldsymbol{a}\}$ in terms of $\{\boldsymbol{u}_{\text{node}}\}$, where $\{\boldsymbol{u}_{\text{node}}\} = [u_i v_i u_j v_j \cdots u_l v_l]^T$, and then obtain a matrix relationship between $\{\varepsilon\}$ and $\{\boldsymbol{u}_{\text{node}}\}$.

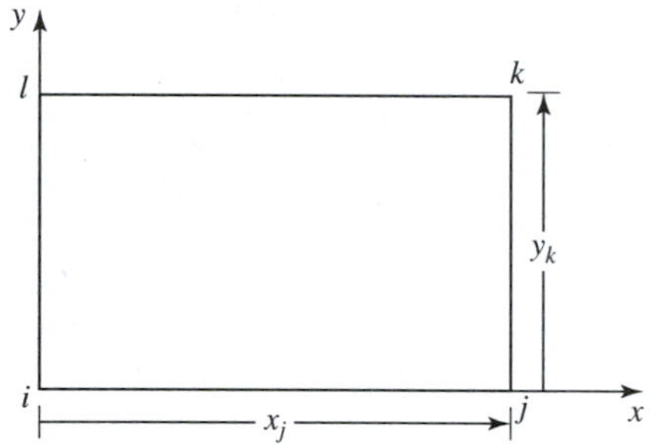


Figure P3-14

## REFERENCES

[1] Hildebrand, F. B., *Methods of Applied Mathematics,* 2nd ed. (Prentice-Hall, Englewood Cliffs, NJ, 1965).

[2] Kaplan, W., *Advanced Calculus,* 3rd ed. (Addison-Wesley, Reading, MA, 1984).

[3] Sokolnikoff, I. S., *Mathematical Theory of Elasticity,* 2nd ed. (McGraw-Hill, New York, 1956).

[4] Stein, P. K., "Strain/Stress," *Measurements & Control,* 24, pp.118–133 (1990).

[5] Wang, C. T., *Applied Elasticity* (McGraw-Hill, New York, 1953).

# 4 RELATIONSHIPS BETWEEN STRESS AND STRAIN

In this chapter we will develop relationships between stress and strain for several types of material behavior. We begin with a general discussion of material behavior and then proceed to develop stress-strain relationships for elastic materials that are isotropic. This is followed by a discussion of anisotropic materials and the development of stress-strain relationships for several classes of these materials. We then develop some simple stress-strain models for viscoelastic materials and perfectly plastic materials. Next we develop stress-strain relationships for problems in which the temperature is not uniform. Finally we discuss piezoelectric materials and develop their constitutive equations. The reader should be aware that in Sections 4-5 and 4-13-2 use is made of certain results that are obtained in Chapter 5.

## 4-1 INTRODUCTION

In Chapter 2 we derived relationships between the various stresses acting on an element of material by invoking Newton's second law, and in Chapter 3 we introduced the concept of strain and developed relationships between the strains and the displacement components. It is worthwhile pointing out that in deriving the equations of motion and the strain-displacement relations, the only restrictions placed on the quantities involved are that the stresses and displacements be continuous functions of position and that the strains be infinitesimal. Since these equations were developed without the need to consider any specific material or material behavior, they are valid for all materials and material behaviors.

We note, however, that we are still faced with a situation where the number of unknowns in our problem exceeds the number of equations that we have developed so far. Thus additional equations are needed before our problem is completely formulated. Furthermore, our everyday experience suggests that there is a relationship between stress and strain (or, force and deflection). If we hold one end of a rubber band in our hand and add a weight to the other end, the rubber band stretches. Add more weight and the amount of stretch increases; decrease the weight and it decreases. It is this relationship between stress and strain that provides the additional equations necessary to complete the formulation of our problem.

Before attempting to derive any equations relating stress and strain it is worth spending a few minutes discussing materials in a fairly general way. Materials are really quite complicated, and it would be futile to attempt to describe with mathematics everything that happens when a material is strained. Therefore in order to derive the necessary equations we develop an *idealization* of an actual material in which the primary behavior of the material is retained but secondary effects are ignored. For example, the stress-strain diagram for steel developed from experimental data shows that, up to a certain point, the relationship between stress and strain is approximately linear. Consequently it is customary to idealize the stress-strain relationship for steel as a straight line up to that point.

Real materials consist of many grains, each of which consists of many atoms. However, we will adopt a macroscopic point of view and assume that the material is a continuous substance and that its properties (for example, the elastic modulus) vary continuously throughout the material. We will ignore the effects of individual grains and their boundaries, arguing that the dimensions of the bodies we will be studying are large compared to those of individual grains. The idea here is that a small volume of material within the body consists of many grains and that the *average* properties of small volumes such as these will vary continuously throughout the body.

If the properties of a material, measured in a certain direction, are the same at each point of the material, we say the material is *homogeneous.* Otherwise it is *nonhomogeneous.* In this text we will consider all materials to be homogeneous. For a material such as wood the material properties along the direction of the fibers are different than those perpendicular to the fibers. Wood is an example of an *anisotropic* material, because its material properties vary with the directional orientation. If the material properties do not vary with the directional orientation, as is the commonly assumed situation for steel, then the material is said to be *isotropic*. Of course, no material is truly isotropic, but many are nearly so. For example, metals consist of many grains, each of which is anisotropic. But for many metals, from a macroscopic point of view, the average material properties for a small volume of material consisting of many grains are the same regardless of the direction in which they are measured. This is because of the random orientation of the grains. Certain manufacturing processes such as drawing or rolling tend to destroy this random orientation and yield a material that is anisotropic.

Figure 4-1 shows the stress-strain diagram for a length of thin steel wire when tensile loads, directed along the axis of the wire, are applied at its ends. Up to a certain point on the diagram, here point $A$, the curve is essentially a straight line, indicating a linear relationship between stress and strain. This point is referred to as the *proportional* limit. Beyond the proportional limit, the curve deviates from a straight line and slightly thereafter the *elastic* limit, point $B$, is reached. The elastic limit is significant because for stresses below this value, the material will return to its original configuration when the load is removed and no permanent deformation will occur. As a practical matter, the proportional limit and the elastic limit can be considered coincident. Until the elastic limit is reached, if the load on the wire is reduced, the material will unload along the same curve $OAB$ that was followed during the loading phase. However, beyond the elastic limit, unloading will occur along a line that is parallel to the original loading curve, as shown in Fig. 4-1 by curve $CD$. When the stress reaches zero, indicating that the load is

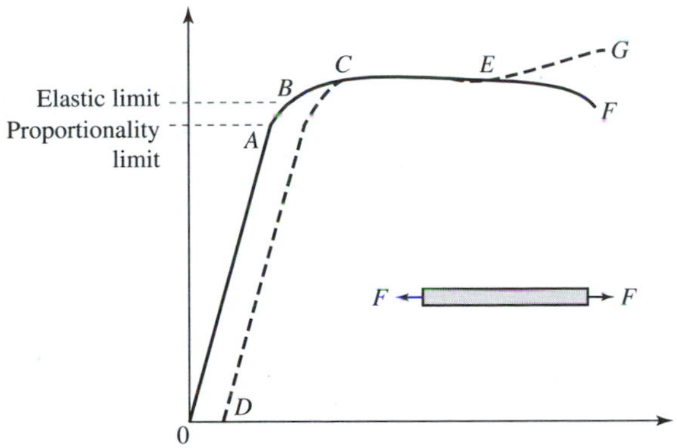


Figure 4-1

completely removed, the resulting strain is greater than zero, indicating that the material has experienced permanent deformation. If the material is subsequently reloaded, the material will follow the path *DC* until the original stress-strain curve is reached and then proceed in the direction of increasing strain along the original curve. Ultimately, as the load continues to increase, the material will fracture at point *F*. The fracture stress is not the highest stress on the curve because here we have plotted engineering stress* versus strain. If true stress had been plotted instead, the stress-strain curve would have continued to rise, as shown by the dashed line *EG*.

Virtually all engineering materials have some range within which they will return to their original configuration upon removal of the load. It is common to refer to this region as the *elastic region* and to materials loaded in this region as *elastic materials*.

For metals and other engineering materials it is common for the behavior in the elastic region to be linear, as in Fig. 4-1. Materials that exhibit this characteristic are called *linearly elastic* materials, or simply *linear* materials. On the other hand, there are elastic materials such as rubber for which no portion of the stress-strain curve is linear. Such materials are referred to as *nonlinear* materials, and a typical stress-strain curve is shown in Fig. 4-2. In this book we will restrict our attention to materials whose behavior within the elastic region is linear.

Our discussions up to this point have been essentially independent of time. Within broad limits, the relation between stress and strain is not influenced by the speed with which the loading is applied, and the final stress-strain state does not change if the load is held steady for an extended period of time. There are situations, however, in which the strain can change over time even though the load remains constant. This phenomenon is known as *creep,* and for metals it commonly occurs in environments in which the temperature is elevated, such as in gas turbine engines.

---

*The reader should recall from the study of strength of materials that *engineering* stress is defined as the force in the wire divided by the original cross-sectional area of the wire. *True* stress, on the other hand, is the force divided by the current cross-sectional area. For problems involving infinitesimal strains the difference between these stresses is negligible and the engineering stress is generally used because it is easier to compute. However, as the strain becomes larger the difference becomes significant (for example, when the wire experiences pronounced necking).

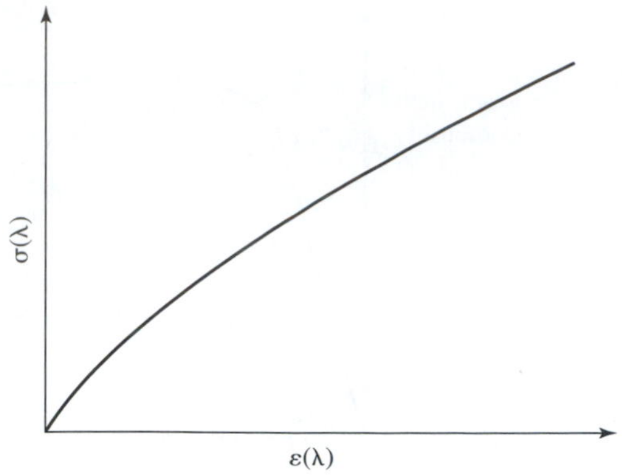


Figure 4-2

Conceivably it would be possible to develop a relationship between stress and strain that would be valid over the entire stress-strain curve. An approach such as this would not be very practical, however, because the resulting equations would be very complicated and quite difficult to use. In order to avoid the practical consequences associated with permanent deformation, most components are designed so that stresses do not exceed the elastic limit. Consequently there is no advantage to such complicated equations. Therefore stress-strain equations are generally developed to describe certain types of material behavior such as elastic behavior or plastic behavior.

In the remaining sections of this chapter we will develop relationships between stress and strain for several types of materials and material behavior.

## 4-2 ISOTROPIC MATERIALS—A PHYSICAL APPROACH

Although our emphasis thus far has been on two-dimensional problems, in order to develop adequate models describing material behavior for two-dimensional problems it is necessary to start with a three-dimensional stress-strain law and reduce it to the appropriate two-dimensional case. For even though the stresses acting on an element of material may be confined to act in a common plane, there will be a response to these stresses in the out-of-plane direction. For example, when tensile stresses directed along the $x$ axis are applied to the cube of material shown in Fig. 4-3, there is a corresponding elongation of the material in the $x$ direction, as shown. However, this elongation is accompanied by a contraction of the material in the $y$ direction and also in the $z$ direction. This contraction is known as *Poisson's effect,** and it has been observed experimentally that the contraction in the $y$ direction is related to the extension in the $x$ direction as follows:

$$\varepsilon_{yy} = -\nu\varepsilon_{xx} \tag{4-1}$$

*Siméon Denis Poisson (1781–1840), French mathematician.

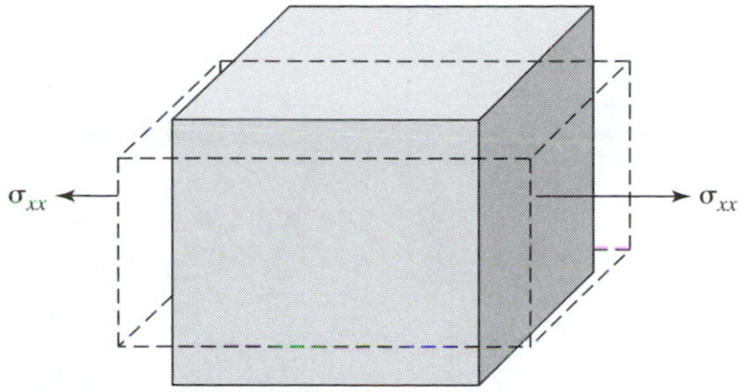


Figure 4-3 Poisson's effect

where $\nu$ is *Poisson's ratio,* a material property. For an isotropic material the contraction in the $z$ direction must be the same as that in the $y$ direction. We can therefore write

$$\varepsilon_{zz} = -\nu\varepsilon_{xx} \tag{4-2}$$

Now if a stress in the $y$ direction is superimposed on the element, the corresponding elongation in that direction will be accompanied by a contraction in both the $x$ and the $z$ directions. Thus, although the stresses are confined to the $xy$ plane, the material response has a component in the $z$ direction. On the other hand, if the material is constrained from contracting in the $z$ direction by some external mechanism, a greater stress in the $x$ direction will be required to produce the same strain in that direction. Consequently the in-plane behavior in two-dimensional problems is influenced by behavior in the out-of-plane direction. For this reason we will first develop the three-dimensional stress-strain law for isotropic materials and then discuss how these are reduced to the two-dimensional case.

In order to develop the stress-strain relations for isotropic materials we consider an element of material with stresses acting as shown in Fig. 4-4. This represents the final stress state of the element. To achieve this stress state we begin by applying the stress $\sigma_{xx}$, as shown in Fig. 4-3. This causes the element to elongate in the $x$ direction and to contract in the $y$ and $z$ directions. Since the material is linearly elastic, the stress-strain curve is as shown by line $OA$ in Fig. 4-1 and we have

$$\sigma_{xx} = E\varepsilon_{xx}^{(1)} \tag{4-3}$$

where $E$ is the slope of the stress-strain curve. It is a material property referred to as the *elastic modulus,* the *modulus of elasticity,* or *Young's modulus.** Note that it is the stress required to produce a strain of one unit, assuming the material remains elastic. Table 4-1 contains values of Young's modulus and Poisson's ratio for a few materials. The superscript (1) indicates that this is the first of a sequence of strains to be computed.

From Eq. (4-3) we find that

$$\varepsilon_{xx}^{(1)} = \frac{1}{E}\sigma_{xx} \tag{4-4}$$

---

*Thomas Young (1773–1829), English scientist.

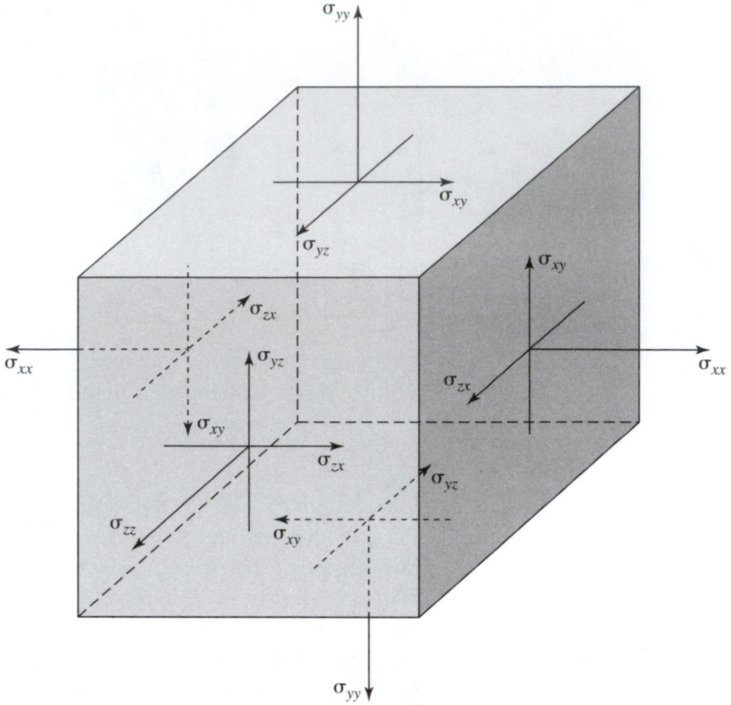


Figure 4-4

**TABLE 4-1 Typical Values of Young's Modulus and Poisson's Ratio**

| | Young's Modulus (psi) | Poisson's Ratio |
|---|---|---|
| Pure metals | | |
| Aluminum | $10.0 \times 10^6$ | 0.33 |
| Copper | $16.0 \times 10^6$ | 0.36 |
| Gold | $10.8 \times 10^6$ | 0.42 |
| Lead | $2.0 \times 10^6$ | 0.40–0.45 |
| Wrought alloys | | |
| Aluminum | $10.6 \times 10^6$–$11.4 \times 10^6$ | 0.33 |
| Low-carbon steel | $29.0 \times 10^6$ | 0.27–0.30 |
| 304 stainless steel | $27.4 \times 10^6$ | 0.283 |
| Magnesium | $6.5 \times 10^6$ | 0.35 |
| Monel 400 | $26.0 \times 10^6$ | 0.32 |
| Inconel X750 | $31.0 \times 10^6$ | 0.29 |
| Phosphor bronze | $15.0 \times 10^6$ | 0.20 |
| Titanium A55 | $16.0 \times 10^6$ | 0.361 |
| Yellow brass | $14.0 \times 10^6$ | 0.34 |
| Cast alloys | | |
| Aluminum | $10.6 \times 10^6$–$11.4 \times 10^6$ | |
| Iron | $13.4 \times 10^6$ | 0.27 |

As noted previously, the elongation in the $x$ direction is accompanied by a contraction in the $y$ and $z$ directions. Thus using Eqs. (4-1) and (4-2),

$$\varepsilon_{yy}^{(1)} = -\nu\varepsilon_{xx}^{(1)} \tag{4-5}$$

$$\varepsilon_{zz}^{(1)} = -\nu\varepsilon_{xx}^{(1)} \tag{4-6}$$

With Eq. (4-4), Eqs. (4-5) and (4-6) become

$$\begin{aligned} \varepsilon_{yy}^{(1)} &= -\nu\frac{\sigma_{xx}}{E} \\ \varepsilon_{zz}^{(1)} &= -\nu\frac{\sigma_{xx}}{E} \end{aligned} \tag{4-7}$$

We now add the stress $\sigma_{yy}$, as shown in Fig. 4-5. Note that the resulting strains are calculated based on the *original* element configuration, not its current configuration. Since the strains are small, this poses no problems. We also note that, because the material is isotropic, the stress-strain curve in the $y$ direction is the same as that in the $x$ direction. Thus the strains due to $\sigma_{yy}$ acting alone are

$$\begin{aligned} \varepsilon_{xx}^{(2)} &= -\nu\frac{\sigma_{yy}}{E} \\ \varepsilon_{yy}^{(2)} &= \frac{\sigma_{yy}}{E} \\ \varepsilon_{zz}^{(2)} &= -\nu\frac{\sigma_{yy}}{E} \end{aligned}$$

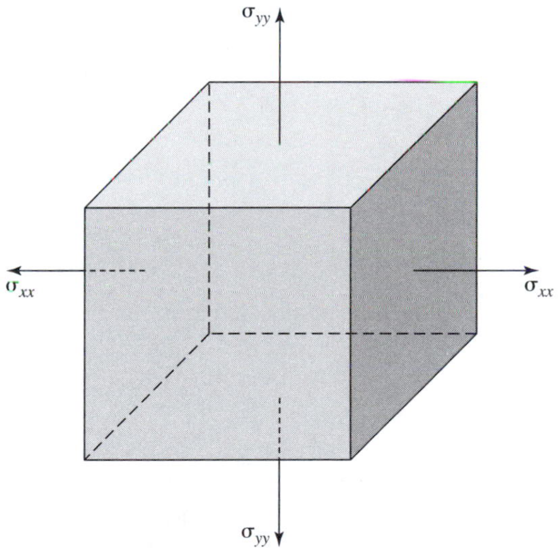


Figure 4-5

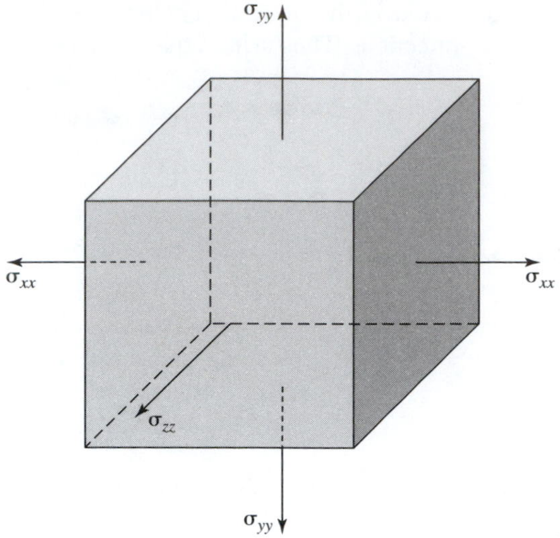


Figure 4-6

When added to the strains given by Eqs. (4-4) and (4-7), the total strain in each of the three directions is

$$\varepsilon_{xx} = \frac{1}{E}(\sigma_{xx} - \nu\sigma_{yy})$$

$$\varepsilon_{yy} = \frac{1}{E}(\sigma_{yy} - \nu\sigma_{xx})$$

$$\varepsilon_{zz} = -\frac{\nu}{E}(\sigma_{xx} + \sigma_{yy})$$

We now add the stress $\sigma_{zz}$, as shown in Fig. 4-6, and note that the stress-strain law in the $z$ direction is the same as that in the $x$ and $y$ directions for isotropic materials. The strains due to the stress $\sigma_{zz}$ acting alone are

$$\varepsilon_{xx}^{(3)} = -\nu\frac{\sigma_{zz}}{E}$$

$$\varepsilon_{yy}^{(3)} = -\nu\frac{\sigma_{zz}}{E}$$

$$\varepsilon_{zz}^{(3)} = \frac{\sigma_{zz}}{E}$$

and therefore the total strains due to $\sigma_{xx}$, $\sigma_{yy}$, and $\sigma_{zz}$ are

$$\varepsilon_{xx} = \frac{1}{E}[\sigma_{xx} - \nu(\sigma_{yy} + \sigma_{zz})] \tag{4-8a}$$

$$\varepsilon_{yy} = \frac{1}{E}[\sigma_{yy} - \nu(\sigma_{xx} + \sigma_{zz})] \tag{4-8b}$$

$$\varepsilon_{zz} = \frac{1}{E}[\sigma_{zz} - \nu(\sigma_{xx} + \sigma_{yy})] \tag{4-8c}$$

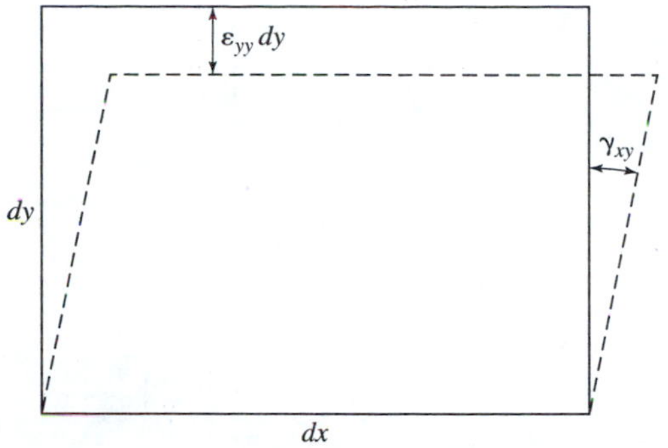


Figure 4-7

Next we consider the two-dimensional element shown in Fig. 4-7 to which a shear strain $\gamma_{xy}$ is applied. The deformed configuration of the element is shown by the dashed lines and we wish to determine the strain in the $y$ direction, $\varepsilon_{yy}$. We have

$$\varepsilon_{yy}\,dy = -(1 - \cos\gamma_{xy})\,dy$$

$$\varepsilon_{yy} = -(1 - \cos\gamma_{xy})$$

However, recalling that the shear strains are infinitesimal quantities, we have

$$1 - \cos\gamma_{xy} \approx 1 - \left(1 - \tfrac{1}{2}\gamma_{xy}^2\right) = \tfrac{1}{2}\gamma_{xy}^2 \approx 0$$

and therefore,

$$\varepsilon_{yy} \approx 0$$

That is, the normal strains are not influenced by shear strains. Thus the addition of shear strains will not alter Eqs. (4-8).

For a linearly elastic material the shear stresses as well as the normal stresses follow a linear stress-strain law. Thus when the shear stress $\sigma_{xy}$ is applied in the presence of normal stresses, as shown in Fig. 4-8, we have

$$\sigma_{xy} = G\gamma_{xy} \tag{4-9}$$

where $G$, the *shear modulus,* or *modulus of rigidity,* is the slope of the shear stress versus shear strain curve.

There is no Poisson effect for shear. The shear strain $\gamma_{xy}$ does not cause shear strains $\gamma_{yz}$ or $\gamma_{zx}$. Thus the shear strain $\gamma_{xy}$ is unaffected by the addition of shear stresses $\sigma_{yz}$ and $\sigma_{zx}$. Consequently we can write

$$\sigma_{yz} = G\gamma_{yz}, \qquad \sigma_{zx} = G\gamma_{zx} \tag{4-10}$$

Equations (4-9) and (4-10) can be readily solved for strains to give

$$\gamma_{xy} = \frac{1}{G}\sigma_{xy} \tag{4-11$a$}$$

$$\gamma_{yz} = \frac{1}{G}\sigma_{yz} \tag{4-11$b$}$$

$$\gamma_{zx} = \frac{1}{G}\sigma_{zx} \tag{4-11$c$}$$

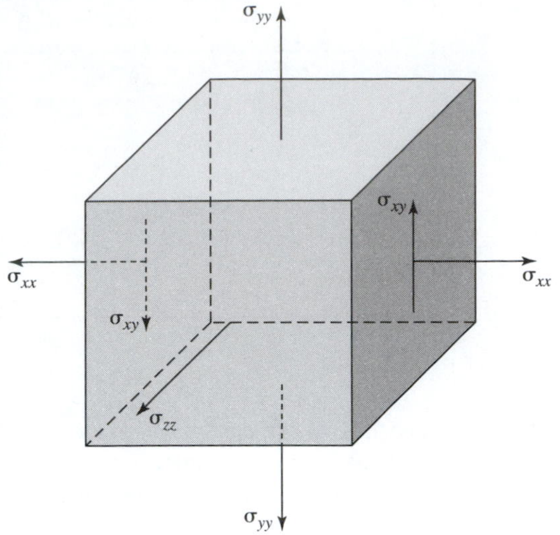


Figure 4-8

These equations, together with Eqs. (4-8), provide relationships for determining the six strain components in terms of the six stress components for an isotropic, homogeneous, linearly elastic material. Because the material is considered homogeneous as well as isotropic, the material properties contained in these equations are independent of position and directional orientation and can therefore be considered material constants. Although there appear to be three material constants in these relationships, we will show shortly that the shear modulus can be expressed in terms of Young's modulus and Poisson's ratio, thereby reducing the number of independent material constants to two.

It is often necessary to determine the six stress components from the strain components. This can be done by solving Eqs. (4-8) for the stresses, which results in the following expressions:

$$\begin{aligned}\sigma_{xx} &= \frac{E}{(1+\nu)(1-2\nu)}[(1-2\nu)\varepsilon_{xx}+\nu\Delta]\\ \sigma_{yy} &= \frac{E}{(1+\nu)(1-2\nu)}[(1-2\nu)\varepsilon_{yy}+\nu\Delta]\\ \sigma_{zz} &= \frac{E}{(1+\nu)(1-2\nu)}[(1-2\nu)\varepsilon_{zz}+\nu\Delta]\end{aligned} \tag{4-12}$$

where $\Delta$ is the dilatation, given by Eq. (3-61). Together with Eqs. (4-9) and (4-10) these provide the six stress-strain relationships.

## 4-2-1 Coincidence of Principal Stress and Principal Strain Axes

In Chapters 2 and 3 we established expressions defining the principal directions for stress and strain, respectively. We will now show that for an isotropic material the principal axes for stress and those for strain coincide. To do this we consider an element subjected to an arbitrary stress field and suppose that the orientation of the element is such that the element axes lie along the principal strain axes. Then we have

$$\gamma_{xy} = \gamma_{yz} = \gamma_{zx} = 0$$

It follows from Eqs. (4-9) and (4-10) that in this orientation

$$\sigma_{xy} = \sigma_{yz} = \sigma_{zx} = 0$$

Since the shear stresses are all zero, we conclude that these axes are also principal stress axes. Thus *for isotropic materials, the principal axes of stress are coincident with the principal strain axes.*

## 4-2-2 Relationship between *G* and *E*

With the coincidence of the principal stress and strain axes established, we will now show that the shear modulus $G$ is not an independent elastic constant, but can be determined from Young's modulus and Poisson's ratio. To do this, we consider a two-dimensional problem in the $xy$ plane. For this problem the principal directions of strain are given by Eq. (3-45) as

$$\tan 2\theta_\varepsilon = \frac{\gamma_{xy}}{\varepsilon_{xx} - \varepsilon_{yy}}$$

where the notation $\theta_\varepsilon$ indicates that these are the principal strain directions. Now from Eqs. (4-8*a*, *b*) we have

$$\varepsilon_{xx} - \varepsilon_{yy} = \frac{1}{E}(1 + \nu)(\sigma_{xx} - \sigma_{yy})$$

This together with Eq. (4-11*a*) is substituted into the preceding expression for principal strains to give

$$\tan 2\theta_\varepsilon = \frac{E/2(1+\nu)}{G} \frac{2\sigma_{xy}}{\sigma_{xx} - \sigma_{yy}}$$

With Eq. (2-76) this becomes

$$\tan 2\theta_\varepsilon = \frac{E}{2(1+\nu)G} \tan 2\theta_\sigma$$

However, since the principal directions for stress and strain are coincident for isotropic materials, we must have

$$G = \frac{E}{2(1+\nu)} \tag{4-13}$$

Thus the shear modulus is readily found from Young's modulus and Poisson's ratio.

## 4-2-3 Bulk Modulus

Another useful property for isotropic materials is the bulk modulus, which is a measure of the change of volume that a material specimen undergoes when subjected to a *hydrostatic stress state*. This is the stress state that a material would experience when submersed in a pressurized fluid. A small sample of material placed close to the ocean floor would experience a stress state that, for practical purposes, is hydrostatic. This stress state is characterized by

$$\sigma_{xx} = \sigma_{yy} = \sigma_{zz} = -p$$

$$\sigma_{xy} = \sigma_{yz} = \sigma_{zx} = 0$$

We now substitute these into Eqs. (4-12) and add the resulting equations together to get

$$-3p = \frac{E}{1+\nu}(\varepsilon_{xx} + \varepsilon_{yy} + \varepsilon_{zz}) + \frac{3E\nu}{(1+\nu)(1-2\nu)}\Delta$$

or with Eq. (3-61),

$$p = -K\Delta \tag{4-14}$$

where

$$K = \frac{E}{3(1-2\nu)} \tag{4-15}$$

is the *bulk modulus*. With Eq. (4-14) written as

$$\Delta = -\frac{1}{K}p \tag{4-16}$$

we see that the reciprocal of the bulk modulus gives the change of volume produced by a unit hydrostatic pressure. We note that, by virtue of Eqs. (4-11), a hydrostatic stress state produces no shear strains.

Now if we add Eqs. (4-12) we get

$$\sigma_{xx} + \sigma_{yy} + \sigma_{zz} = \frac{E}{1-2\nu}\Delta$$

or

$$3\sigma_M = \frac{E}{1-2\nu}\Delta \tag{4-17}$$

where $\sigma_M$, the mean stress, is defined by Eq. (2-93).

On dividing Eq. (4-17) by 3 we get

$$\sigma_M = K\Delta \tag{4-18}$$

Thus the bulk modulus is the material constant that relates the dilatation and the mean stress. Note that it is not an independent elastic constant, but can be determined from Young's modulus and Poisson's ratio, as shown in Eq. (4-15).

## 4-3 TWO-DIMENSIONAL STRESS-STRAIN LAWS—PLANE STRESS AND PLANE STRAIN

In certain instances it is possible to reduce the stress-strain law developed in the preceding section to its two-dimensional version.

### 4-3-1 Plane Stress

When the conditions given by Eqs. (2-54) apply, we say that a state of plane stress exists. Under these conditions the stress-strain equations given by Eqs. (4-8) and (4-11) become

$$\begin{aligned} \varepsilon_{xx} &= \frac{1}{E}(\sigma_{xx} - \nu\sigma_{yy}) \\ \varepsilon_{yy} &= \frac{1}{E}(\sigma_{yy} - \nu\sigma_{xx}) \\ \gamma_{xy} &= \frac{1}{G}\sigma_{xy} \end{aligned} \tag{4-19}$$

and

$$\varepsilon_{zz} = -\frac{\nu}{E}(\sigma_{xx} + \sigma_{yy}) \tag{4-20a}$$

$$\gamma_{yz} = 0 \tag{4-20b}$$

$$\gamma_{zx} = 0 \tag{4-20c}$$

We note that when conditions of plane stress exist, the normal strain in the $z$ direction is not zero even though the normal stress in that direction is zero. This is due to Poisson's effect. However, once the stresses $\sigma_{xx}$ and $\sigma_{yy}$ are found, $\varepsilon_{zz}$ can be readily determined from Eq. (4-20$a$).

Equations (4-19) can be solved to give stresses in terms of strains,

$$\begin{aligned} \sigma_{xx} &= \frac{E}{1-\nu^2}(\varepsilon_{xx} + \nu\varepsilon_{yy}) \\ \sigma_{yy} &= \frac{E}{1-\nu^2}(\varepsilon_{yy} + \nu\varepsilon_{xx}) \\ \sigma_{xy} &= G\gamma_{xy} \end{aligned} \tag{4-21}$$

With the assumptions of plane stress, we note that Eqs. (2-53), (3-11), and (3-18) reduce to Eqs. (2-55), (3-6), (3-7), (3-9), and (3-12).

## EXAMPLE 4-1

A rectangular material sample 2 in tall and 1 in wide is subjected to a stress of 50,000 psi on its upper edge. The bottom edge of the sample is constrained from displacing in the vertical, or $y$, direction but is free to expand or contract in the horizontal, or $x$, direction. The stresses throughout the sample are uniform. Determine the displacement in the $y$ direction at the top edge of the sample. Determine the displacement of the vertical edge in the $x$ direction. Assume that the material is low carbon steel, and that conditions of plane stress apply.

### SOLUTION

For the top edge of the sample the stress vector has components

$$t_x^{(n)} = 0, \qquad t_y^{(n)} = 50{,}000\,\text{psi}, \qquad t_z^{(n)} = 0$$

while the components of the outer normal are

$$n_x = 0, \qquad n_y = 1, \qquad n_z = 0$$

When these are used in Eqs. (2-14), the following stresses result along the top edge:

$$\sigma_{xy} = 0, \qquad \sigma_{yy} = 50{,}000\,\text{psi}, \qquad \sigma_{yz} = 0$$

On the right vertical edge, the components of the stress vector are all zero while the components of the outer normal are

$$n_x = 1, \qquad n_y = 0, \qquad n_z = 0$$

Equations (2-14) show that along that edge,

$$\sigma_{xx} = 0, \qquad \sigma_{xy} = 0, \qquad \sigma_{xz} = 0$$

Since the stress distribution is uniform, it follows that

$$\sigma_{xx} = 0, \qquad \sigma_{yy} = 50{,}000\,\text{psi}, \qquad \sigma_{xy} = 0$$

throughout the body and, with Eqs. (4-19) and (4-20), that the strains are also uniform. The reader can readily verify that a uniform stress distribution satisfies Eqs. (2-53) if the body forces are zero. Now, for low carbon steel, Table 4-1 gives $E = 29{,}000{,}000$ psi and $\nu = 0.27$–$0.30$. We will choose a value of $\nu = 0.28$. Equations (4-19) then give

$$\varepsilon_{xx} = -\frac{\nu}{E}\sigma_{yy} = -\frac{0.28}{29{,}000{,}000}50{,}000$$

$$\varepsilon_{xx} = -0.00048\ \text{in/in}$$

$$\varepsilon_{yy} = \frac{\sigma_{yy}}{E} = \frac{50{,}000}{29{,}000{,}000}$$

$$\varepsilon_{yy} = 0.001724\ \text{in/in}$$

Assume that the origin of the coordinate system is at the center of the bottom edge. With Eqs. (3-6) and (3-7), the displacements at the top and at the left and right edges are, respectively,

$$v(x, 2) = 0.0034 \text{ in}, \qquad u(-0.5, y) = 0.0002 \text{ in}, \qquad u(0.5, y) = -0.0002 \text{ in}$$

to four decimal places. Note that the vertical edges displace toward the vertical centerline.

## EXAMPLE 4-2

Re-solve the previous example with the vertical edges constrained from displacing in the $x$ direction, but free to displace vertically. Determine the vertical displacement of the top edge, and the stresses required to prevent the vertical edges from displacing. Compare these results with those of the previous example.

### SOLUTION

As in the previous example, the stresses, and hence the strains, are uniform. Consequently the displacements are, at most, linear functions of position. However, since the vertical edges are constrained from displacing in the $x$ direction, the displacement component $u$ is zero everywhere and therefore $\varepsilon_{xx}$ is also zero. Thus,

$$\varepsilon_{xx} = 0, \qquad \sigma_{yy} = 50{,}000 \text{ psi}$$

and Eqs. (4-19) reduce to

$$\sigma_{xx} = \frac{E\nu}{1-\nu^2}\varepsilon_{yy}, \qquad \sigma_{yy} = \frac{E}{1-\nu^2}\varepsilon_{yy}$$

The second equation is now solved for $\varepsilon_{yy}$ and the result used in the first equation to give

$$\sigma_{xx} = \nu\sigma_{yy}, \qquad \varepsilon_{yy} = \frac{1-\nu^2}{E}\sigma_{yy}$$

With the values of Young's modulus and Poisson's ratio from the previous example, these give

$$\sigma_{xx} = 14{,}000 \text{ psi}, \qquad \varepsilon_{yy} = 0.001589 \text{ in/in}$$

The resulting vertical displacement of the top edge is $v(x, 2) = 0.0032$ in, about 6 percent less than in the previous example. The stress $\sigma_{xx}$ is required to return the vertical edges in the previous example to their undeformed positions. In so doing, Poisson's effect comes into play and reduces the vertical displacement of the top edge.

## 4-3-2 Plane Strain

Alternatively, we can assume that the displacement component in the $z$ direction is zero and the remaining components are independent of $z$,

$$u = u(x, y), \qquad v = v(x, y), \qquad w = 0 \tag{4-22}$$

Bodies for which these assumptions are satisfied are in a state of *plane strain*. In an approximate sense, conditions of plane strain exist in long cylindrical bodies whose loading is independent of $z$, as shown in Fig. 4-9, and whose ends are fixed. Then $w$, the displacement in the $z$ direction, must be zero at the ends. Because the body is long and the loads and geometry do not vary in the $z$ direction, $w$ is therefore approximately zero everywhere. Hence the strain in the $z$ direction must be approximately zero.

With Eqs. (4-9), (4-10), and (4-12) and the assumptions of plane strain we have

$$\begin{aligned}
\sigma_{xx} &= \frac{E}{(1+\nu)(1-2\nu)}[(1-\nu)\varepsilon_{xx} + \nu\varepsilon_{yy}] \\
\sigma_{yy} &= \frac{E}{(1+\nu)(1-2\nu)}[(1-\nu)\varepsilon_{yy} + \nu\varepsilon_{xx}] \\
\sigma_{xy} &= G\gamma_{xy}
\end{aligned} \tag{4-23}$$

and

$$\begin{aligned}
\sigma_{zz} &= \frac{\nu E}{(1+\nu)(1-2\nu)}(\varepsilon_{xx} + \varepsilon_{yy}) \\
\sigma_{yz} &= 0 \\
\sigma_{zx} &= 0
\end{aligned} \tag{4-24}$$

Note that for plane strain, although $\varepsilon_{zz}$ is zero, $\sigma_{zz}$ is not. This again is due to Poisson's effect.

Equations (4-23) can be solved for the strains to give

$$\begin{aligned}
\varepsilon_{xx} &= \frac{1-\nu^2}{E}\left(\sigma_{xx} - \frac{\nu}{1-\nu}\sigma_{yy}\right) \\
\varepsilon_{yy} &= \frac{1-\nu^2}{E}\left(\sigma_{yy} - \frac{\nu}{1-\nu}\sigma_{xx}\right) \\
\gamma_{xy} &= \frac{1}{G}\sigma_{xy}
\end{aligned} \tag{4-25}$$

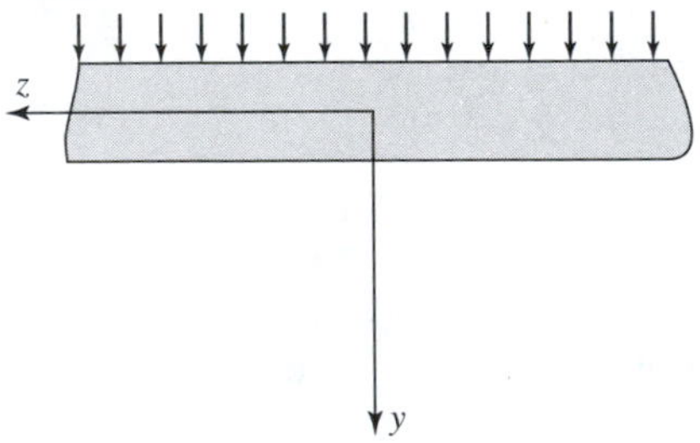


**Figure 4-9** Long cylindrical bodies whose loading is independent of $z$ and with fixed ends—plane strain

We note that with the assumptions of plane strain Eqs. (2-53), (3-11), and (3-18) again reduce to Eqs. (2-55), (3-6), (3-7), (3-9), and (3-12). The conditions of plane strain imply that $\sigma_{zz}$ is independent of $z$. Consequently, when Eqs. (4-24) are used in the last of Eqs. (2-49), it follows that the body force component $b_z$ must be zero if plane strain conditions exist.

At the end of Section 3-3 we noted that for two-dimensional problems we had developed six equations that contained nine unknowns and that three more equations were needed to complete the problem formulation. These three additional equations are provided by Eqs. (4-19) or (4-21) in the case of plane stress and Eqs. (4-23) or (4-25) in the case of plane strain. Thus with the addition of the stress-strain equations for either plane stress or plane strain the derivation of the governing equations for two-dimensional problems involving isotropic, homogeneous, linearly elastic materials is complete.

We close this section by noting that Eqs. (4-25) can be obtained from Eqs. (4-19) simply by replacing $E$ with $E/(1-\nu^2)$ and $\nu$ with $\nu/(1-\nu)$. Since the material constants appear only in the stress-strain equations, it follows that if we have a plane stress solution to a particular problem, the solution to the corresponding plane strain problem can be obtained simply by replacing $E$ with $E/(1-\nu^2)$ and $\nu$ with $\nu/(1-\nu)$ everywhere in the plane stress solution.

## 4-4 RESTRICTIONS ON ELASTIC CONSTANTS FOR ISOTROPIC MATERIALS

For a one-dimensional problem we know that a positive normal stress produces a positive normal strain, and that a positive shear stress produces a positive shear strain. Thus $E$ and $G$ are positive. Since they are also constant, we conclude that they are always positive. We also expect a finite value of stress to result from a finite strain. Consequently $E$ and $G$ cannot be infinite. Now since $E$ is finite and positive, we see from Eq. (4-13) that $\nu > -1$ in order for $G$ to be also finite and positive.

Similarly, when an isotropic body is subjected to hydrostatic pressure, we anticipate that the volume will decrease by a finite amount. Then we see from Eq. (4-16) that the bulk modulus $K$ must also be positive and finite. Since $E$ is positive, we conclude from Eq. (4-15) that $\nu < 1/2$. Thus we have the following constraints on the elastic constants:

$$0 < E < \infty, \qquad -1 < \nu < \tfrac{1}{2} \tag{4-26}$$

It is worth pointing out that if Poisson's ratio takes on a value of 1/2, then from Eqs. (4-15) and (4-16) there will be no decrease in volume regardless of the hydrostatic pressure applied. That is, the material is *incompressible*. There are no materials that are truly incompressible, although rubber is nearly so. Cork has a Poisson's ratio that is approximately zero. Although, by the inequality (4-26), it is physically possible to have negative Poisson's ratios, for all common engineering materials Poisson's ratio is positive. Recently, however, materials with a negative Poisson's ratio have been developed [10].

## 4-5 ANISOTROPIC MATERIALS

Up to this point our study of materials and their properties has been limited to isotropic materials. While many materials of interest to engineers are isotropic, at least in an approximate sense, many of the materials that will play important roles in future technological developments, such as composite materials and certain polymers, are anisotropic. Thus it is important for engineers to develop a knowledge of the behavior of these materials.

For the most general linearly elastic anisotropic material, a particular component of stress, say $\sigma_{xx}$, is assumed to depend on all six components of strain. So we have

$$\sigma_{xx} = C_{11}\varepsilon_{xx} + C_{12}\varepsilon_{yy} + C_{13}\varepsilon_{zz} + C_{14}\gamma_{yz} + C_{15}\gamma_{zx} + C_{16}\gamma_{xy}$$

where, for a homogeneous material, the coefficients $C_{11}, C_{12}, \ldots$ are constants. When the remaining five stress components are represented in this manner, we get

$$\begin{aligned} \sigma_{yy} &= C_{21}\varepsilon_{xx} + C_{22}\varepsilon_{yy} + \cdots + C_{26}\gamma_{xy} \\ &\vdots \\ \sigma_{xy} &= C_{61}\varepsilon_{xx} + C_{62}\varepsilon_{yy} + \cdots + C_{66}\gamma_{xy} \end{aligned}$$

These can be put in matrix form,

$$\begin{Bmatrix} \sigma_{xx} \\ \sigma_{yy} \\ \sigma_{zz} \\ \sigma_{yz} \\ \sigma_{zx} \\ \sigma_{xy} \end{Bmatrix} = \begin{bmatrix} C_{11} & C_{12} & C_{13} & C_{14} & C_{15} & C_{16} \\ C_{21} & C_{22} & C_{23} & C_{24} & C_{25} & C_{26} \\ C_{31} & C_{32} & C_{33} & C_{34} & C_{35} & C_{36} \\ C_{41} & C_{42} & C_{43} & C_{44} & C_{45} & C_{46} \\ C_{51} & C_{52} & C_{53} & C_{54} & C_{55} & C_{56} \\ C_{61} & C_{62} & C_{63} & C_{64} & C_{65} & C_{66} \end{bmatrix} \begin{Bmatrix} \varepsilon_{xx} \\ \varepsilon_{yy} \\ \varepsilon_{zz} \\ \gamma_{yz} \\ \gamma_{zx} \\ \gamma_{xy} \end{Bmatrix}$$

It is shown in Section 5-14, through energy considerations, that the coefficient matrix in the preceding expression is symmetric. Thus we can write

$$\begin{Bmatrix} \sigma_{xx} \\ \sigma_{yy} \\ \sigma_{zz} \\ \sigma_{yz} \\ \sigma_{zx} \\ \sigma_{xy} \end{Bmatrix} = \begin{bmatrix} C_{11} & C_{12} & C_{13} & C_{14} & C_{15} & C_{16} \\ C_{12} & C_{22} & C_{23} & C_{24} & C_{25} & C_{26} \\ C_{13} & C_{23} & C_{33} & C_{34} & C_{35} & C_{36} \\ C_{14} & C_{24} & C_{34} & C_{44} & C_{45} & C_{46} \\ C_{15} & C_{25} & C_{35} & C_{45} & C_{55} & C_{56} \\ C_{16} & C_{26} & C_{36} & C_{46} & C_{56} & C_{66} \end{bmatrix} \begin{Bmatrix} \varepsilon_{xx} \\ \varepsilon_{yy} \\ \varepsilon_{zz} \\ \gamma_{yz} \\ \gamma_{zx} \\ \gamma_{xy} \end{Bmatrix} \tag{4-27}$$

or

$$\{\boldsymbol{\sigma}\} = [\boldsymbol{C}]\{\boldsymbol{\varepsilon}\} \tag{4-28}$$

where

$$\{\boldsymbol{\sigma}\} = [\sigma_{xx} \quad \sigma_{yy} \quad \sigma_{zz} \quad \sigma_{yz} \quad \sigma_{zx} \quad \sigma_{xy}]^T$$
$$\{\boldsymbol{\varepsilon}\} = [\varepsilon_{xx} \quad \varepsilon_{yy} \quad \varepsilon_{zz} \quad \gamma_{yz} \quad \gamma_{zx} \quad \gamma_{xy}]^T \tag{4-29}$$

$$[\boldsymbol{C}] = \begin{bmatrix} C_{11} & C_{12} & C_{13} & C_{14} & C_{15} & C_{16} \\ C_{12} & C_{22} & C_{23} & C_{24} & C_{25} & C_{26} \\ C_{13} & C_{23} & C_{33} & C_{34} & C_{35} & C_{36} \\ C_{14} & C_{24} & C_{34} & C_{44} & C_{45} & C_{46} \\ C_{15} & C_{25} & C_{35} & C_{45} & C_{55} & C_{56} \\ C_{16} & C_{26} & C_{36} & C_{46} & C_{56} & C_{66} \end{bmatrix} \tag{4-30}$$

$[\boldsymbol{C}]$ is referred to as the *elasticity matrix* or *stiffness matrix*. It is worth mentioning that the order of the shear terms in the stress and strain matrices, while perhaps not the most obvious order, is the usual order in which these terms are written.

When using matrix notation for the stress-strain relations, it is customary to replace the two-subscript notation for stress and strain with a notation that uses only a single subscript. This is accomplished by first letting $x = 1$, $y = 2$, and $z = 3$. Thus, for example, $\sigma_{xx} = \sigma_{11}, \sigma_{xy} = \sigma_{12}, \ldots$. The stress matrix then reads

$$\{\boldsymbol{\sigma}\} = [\sigma_{11} \quad \sigma_{22} \quad \sigma_{33} \quad \sigma_{23} \quad \sigma_{31} \quad \sigma_{12}]^T \tag{4-31}$$

With the single-subscript notation this becomes

$$\{\boldsymbol{\sigma}\} = [\sigma_1 \quad \sigma_2 \quad \sigma_3 \quad \sigma_4 \quad \sigma_5 \quad \sigma_6]^T \tag{4-32}$$

Similarly for the strains,

$$\{\boldsymbol{\varepsilon}\} = [\varepsilon_1 \quad \varepsilon_2 \quad \varepsilon_3 \quad \varepsilon_4 \quad \varepsilon_5 \quad \varepsilon_6]^T \tag{4-33}$$

An easy way to remember the order of the shear terms is to note that in the transformation from double-subscript notation to single-subscript notation, $\sigma_{ij} = \sigma_k$, where $i, j = 1, 2, 3$ and $k = 1, 2, \ldots, 6$. For each of the shear terms, $i + j + k$ must equal 9. Thus, for example, if $i = 2$ and $j = 3$, then $k = 4$.

Since Eq. (4-27) contains 21 independent constants, we see that 21 elastic constants are required to describe the most general anisotropic material. This is in contrast to an isotropic material for which there are only two independent elastic constants (typically, Young's modulus and Poisson's ratio).

If a one-to-one relationship between stress and strain is assumed, Eq. (4-27) can be solved to obtain the strains in terms of the stresses,

$$\begin{Bmatrix} \varepsilon_{xx} \\ \varepsilon_{yy} \\ \varepsilon_{zz} \\ \gamma_{yz} \\ \gamma_{zx} \\ \gamma_{xy} \end{Bmatrix} = \begin{bmatrix} S_{11} & S_{12} & S_{13} & S_{14} & S_{15} & S_{16} \\ S_{12} & S_{22} & S_{23} & S_{24} & S_{25} & S_{26} \\ S_{13} & S_{23} & S_{33} & S_{34} & S_{35} & S_{36} \\ S_{14} & S_{24} & S_{34} & S_{44} & S_{45} & S_{46} \\ S_{15} & S_{25} & S_{35} & S_{45} & S_{55} & S_{56} \\ S_{16} & S_{26} & S_{36} & S_{46} & S_{56} & S_{66} \end{bmatrix} \begin{Bmatrix} \sigma_{xx} \\ \sigma_{yy} \\ \sigma_{zz} \\ \sigma_{yz} \\ \sigma_{zx} \\ \sigma_{xy} \end{Bmatrix} \tag{4-34}$$

or

$$\{\boldsymbol{\varepsilon}\} = [S]\{\boldsymbol{\sigma}\} \tag{4-35}$$

where

$$[S] = \begin{vmatrix} S_{11} & S_{12} & S_{13} & S_{14} & S_{15} & S_{16} \\ S_{12} & S_{22} & S_{23} & S_{24} & S_{25} & S_{26} \\ S_{13} & S_{23} & S_{33} & S_{34} & S_{35} & S_{36} \\ S_{14} & S_{24} & S_{34} & S_{44} & S_{45} & S_{46} \\ S_{15} & S_{25} & S_{35} & S_{45} & S_{55} & S_{56} \\ S_{16} & S_{26} & S_{36} & S_{46} & S_{56} & S_{66} \end{vmatrix} \tag{4-36}$$

is the *compliance matrix* and the elements of $[S]$ are the *compliances*.

## 4-6 MATERIAL SYMMETRIES

As we have seen in Section 4-5, in its most general form the stress-strain law for elastic materials contains 21 independent elastic constants. This represents a material that is fully anisotropic. However, many materials of practical interest contain certain material symmetries with respect to their elastic properties. These symmetries result in simplifications to the stress-strain law by reducing the number of independent elastic constants. Since the material symmetries we will be discussing pertain to the elastic properties of the material, we will refer to these symmetries as *elastic* symmetries. We note that other kinds of material symmetries are possible, such as symmetries with respect to optical, electrical, and thermal properties.

To understand better what is meant by elastic symmetries, consider the line $OA$ shown in Fig. 4-10. Its mirror image with respect to the plane $z = 0$ is $OA'$. A volume element $V$ of material is situated along line $OA$ as shown, and the mirror image of this

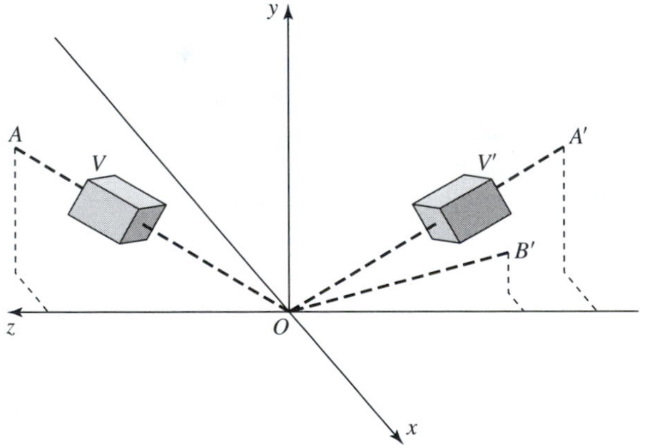


Figure 4-10 Elastic symmetry

element, $V'$, is shown along line $OA'$. If the elastic properties of element $V$ are the same as the corresponding properties of the mirror image element $V'$ for any orientation of line $OA$, the plane $z = 0$ is a plane of elastic symmetry. For example, when a tensile stress, directed along line $OA$, acts on an element $V$, a certain elongation of the element along line $OA$ will result. If the same stress directed along line $OA'$ is applied to element $V'$ and the same elongation along $OA'$ for element $V'$ results, then the elastic properties governing the two elongations are the same. Note that the properties of an element situated along line $OB'$, which is not a mirror image of line $OA$, in general are not the same as those of the element along line $OA$. Note also that for homogeneous materials it is the *orientation* of the element that is important, not its location. In fact, for homogeneous materials the properties do not vary along line $OA$ and its extension, nor do they vary along line $OA'$ and its extension. Thus if the plane $z = 0$ is a plane of symmetry, so is any parallel plane $z =$ constant. Finally we point out that if the volume element is rotated through an angle about line $OA$, its mirror image volume element is rotated through the same angle about line $OA'$. It is these two elements whose corresponding material properties must be examined to determine whether the plane $z = 0$ is a plane of elastic symmetry.

## 4-7 MATERIALS WITH A SINGLE PLANE OF ELASTIC SYMMETRY

Let us determine the structure of the elasticity matrix for a material with a single plane of elastic symmetry. Crystals whose crystalline structure is monoclinic are examples of materials possessing a single plane of elastic symmetry. Iron aluminide [21], gypsum [3], and talc [3] are examples of such materials as are, under certain conditions, ice [3], sulfur [21], selenium [21], and plutonium [21]. For such a material, the elastic coefficients in the stress-strain law must remain unchanged when subjected to a transformation that represents a reflection in the symmetry plane. If $z = 0$ is the symmetry plane, such a transformation is represented by

$$x' = x, \qquad y' = y, \qquad z' = -z$$

The direction cosines for this transformation are shown in Table 4-2.

In the primed coordinate system the stress-strain law is

$$\begin{Bmatrix} \sigma_{x'x'} \\ \sigma_{y'y'} \\ \sigma_{z'z'} \\ \sigma_{y'z'} \\ \sigma_{z'x'} \\ \sigma_{x'y'} \end{Bmatrix} = \begin{bmatrix} C'_{11} & C'_{12} & C'_{13} & C'_{14} & C'_{15} & C'_{16} \\ & C'_{22} & C'_{23} & C'_{24} & C'_{25} & C'_{26} \\ & & C'_{33} & C'_{34} & C'_{35} & C'_{36} \\ & & & C'_{44} & C'_{45} & C'_{46} \\ & \text{sym} & & & C'_{55} & C'_{56} \\ & & & & & C'_{66} \end{bmatrix} \begin{Bmatrix} \varepsilon_{x'x'} \\ \varepsilon_{y'y'} \\ \varepsilon_{z'z'} \\ \gamma_{y'z'} \\ \gamma_{z'x'} \\ \gamma_{x'y'} \end{Bmatrix} \tag{4-37}$$

while in the unprimed coordinate system it is given by Eq. (4-27).

TABLE 4-2

| | $x$ | $y$ | $z$ |
|---|---|---|---|
| $x'$ | $n_{x'x} = 1$ | $n_{x'y} = 0$ | $n_{x'z} = 0$ |
| $y'$ | $n_{y'x} = 0$ | $n_{y'y} = 1$ | $n_{y'z} = 0$ |
| $z'$ | $n_{z'x} = 0$ | $n_{z'y} = 0$ | $n_{z'z} = -1$ |

In order for $z = 0$ to be a plane of symmetry, the elements of $[\boldsymbol{C}']$ in Eq. (4-37) must be the same as those of $[\boldsymbol{C}]$ in Eq. (4-27). Therefore we have

$$\begin{Bmatrix} \sigma_{x'x'} \\ \sigma_{y'y'} \\ \sigma_{z'z'} \\ \sigma_{y'z'} \\ \sigma_{z'x'} \\ \sigma_{x'y'} \end{Bmatrix} = \begin{bmatrix} C_{11} & C_{12} & C_{13} & C_{14} & C_{15} & C_{16} \\ & C_{22} & C_{23} & C_{24} & C_{25} & C_{26} \\ & & C_{33} & C_{34} & C_{35} & C_{36} \\ & & & C_{44} & C_{45} & C_{46} \\ & \text{sym} & & & C_{55} & C_{56} \\ & & & & & C_{66} \end{bmatrix} \begin{Bmatrix} \varepsilon_{x'x'} \\ \varepsilon_{y'y'} \\ \varepsilon_{z'z'} \\ \gamma_{y'z'} \\ \gamma_{z'x'} \\ \gamma_{x'y'} \end{Bmatrix} \tag{4-38}$$

Now, with Eqs. (2-39), (2-40), (3-31), and the direction cosines, the relationship between the stresses and strains in the primed and unprimed coordinate systems is

$$\sigma_{x'x'} = \sigma_{xx}, \quad \sigma_{y'y'} = \sigma_{yy}, \quad \sigma_{z'z'} = \sigma_{zz}, \quad \sigma_{y'z'} = -\sigma_{yz}, \quad \sigma_{z'x'} = -\sigma_{zx}, \quad \sigma_{x'y'} = \sigma_{xy}$$
$$\varepsilon_{x'x'} = \varepsilon_{xx}, \quad \varepsilon_{y'y'} = \varepsilon_{yy}, \quad \varepsilon_{z'z'} = \varepsilon_{zz}, \quad \gamma_{y'z'} = -\gamma_{yz}, \quad \gamma_{z'x'} = -\gamma_{zx}, \quad \gamma_{x'y'} = \gamma_{xy}$$

Consequently Eq. (4-38) can be written as

$$\begin{Bmatrix} \sigma_{xx} \\ \sigma_{yy} \\ \sigma_{zz} \\ -\sigma_{yz} \\ -\sigma_{zx} \\ \sigma_{xy} \end{Bmatrix} = \begin{bmatrix} C_{11} & C_{12} & C_{13} & C_{14} & C_{15} & C_{16} \\ & C_{22} & C_{23} & C_{24} & C_{25} & C_{26} \\ & & C_{33} & C_{34} & C_{35} & C_{36} \\ & \text{sym} & & C_{44} & C_{45} & C_{46} \\ & & & & C_{55} & C_{56} \\ & & & & & C_{66} \end{bmatrix} \begin{Bmatrix} \varepsilon_{xx} \\ \varepsilon_{yy} \\ \varepsilon_{zz} \\ -\gamma_{yz} \\ -\gamma_{zx} \\ \gamma_{xy} \end{Bmatrix} \tag{4-39}$$

Now if we multiply the fourth and fifth rows in Eq. (4-39) by $-1$ and note that the minus signs in the strains alter the signs of the fourth and fifth columns of the elasticity matrix, Eq. (4-39) can be written as

$$\begin{Bmatrix} \sigma_{xx} \\ \sigma_{yy} \\ \sigma_{zz} \\ \sigma_{yz} \\ \sigma_{zx} \\ \sigma_{xy} \end{Bmatrix} = \begin{bmatrix} C_{11} & C_{12} & C_{13} & -C_{14} & -C_{15} & C_{16} \\ & C_{22} & C_{23} & -C_{24} & -C_{25} & C_{26} \\ & & C_{33} & -C_{34} & -C_{35} & C_{36} \\ & & & C_{44} & C_{45} & -C_{46} \\ & \text{sym} & & & C_{55} & -C_{56} \\ & & & & & C_{66} \end{bmatrix} \begin{Bmatrix} \varepsilon_{xx} \\ \varepsilon_{yy} \\ \varepsilon_{zz} \\ \gamma_{yz} \\ \gamma_{zx} \\ \gamma_{xy} \end{Bmatrix} \tag{4-40}$$

For $z = 0$ to be a plane of symmetry, the elasticity matrices in Eqs. (4-27) and (4-40) must be equal, so we must require, for example, that

$$-C_{14} = C_{14}$$

or

$$C_{14} = 0$$

In a similar manner we find that

$$C_{15} = C_{24} = C_{25} = C_{34} = C_{35} = C_{46} = C_{56} = 0$$

The resulting stress-strain equation is

$$\begin{Bmatrix} \sigma_{xx} \\ \sigma_{yy} \\ \sigma_{zz} \\ \sigma_{yz} \\ \sigma_{zx} \\ \sigma_{xy} \end{Bmatrix} = \begin{bmatrix} C_{11} & C_{12} & C_{13} & 0 & 0 & C_{16} \\ & C_{22} & C_{23} & 0 & 0 & C_{26} \\ & & C_{33} & 0 & 0 & C_{36} \\ & & & C_{44} & C_{45} & 0 \\ & \text{sym} & & & C_{55} & 0 \\ & & & & & C_{66} \end{bmatrix} \begin{Bmatrix} \varepsilon_{xx} \\ \varepsilon_{yy} \\ \varepsilon_{zz} \\ \gamma_{yz} \\ \gamma_{zx} \\ \gamma_{xy} \end{Bmatrix} \tag{4-41}$$

We see, then, that for monoclinic materials, the number of independent elastic constants is reduced from 21 to 13. We note that, in contrast to isotropic materials as defined in Eqs. (4-9), (4-10), and (4-12), each normal stress depends on the shear strain $\gamma_{xy}$, and the shear stress $\sigma_{xy}$ depends on all three normal strains. Only the shear stresses $\sigma_{yz}$ and $\sigma_{zx}$ are independent of the normal strains. However, each of these shear stresses depends on *both* shear strains $\gamma_{yz}$ and $\gamma_{zx}$.

## 4-8 ORTHOTROPIC MATERIALS

We now consider a material that has a second plane of elastic symmetry, $y = 0$, which is perpendicular to the single plane $z = 0$ discussed in the preceding section. For a material such as this, the elastic properties along line $OA$ are the same as those along lines $OA'$ and $OA''$, as shown in Fig. 4-11. The elements of the elastic matrix remain unchanged under a reflection in the $z = 0$ plane and also a reflection in the $y = 0$ plane. The transformation representing reflection in the $y = 0$ plane is given by

$$x' = x, \qquad y' = -y \qquad z' = z$$

and the direction cosines for the transformation are shown in Table 4-3.

For a material with a single plane of elastic symmetry the stress-strain relationship in the unprimed coordinate system is given by Eq. (4-41) while in the primed coordinate

**TABLE 4-3**

| | x | y | z |
|---|---|---|---|
| x′ | $n_{x'x} = 1$ | $n_{x'y} = 0$ | $n_{x'z} = 0$ |
| y′ | $n_{y'x} = 0$ | $n_{y'y} = -1$ | $n_{y'z} = 0$ |
| z′ | $n_{z'x} = 0$ | $n_{z'y} = 0$ | $n_{z'z} = 1$ |

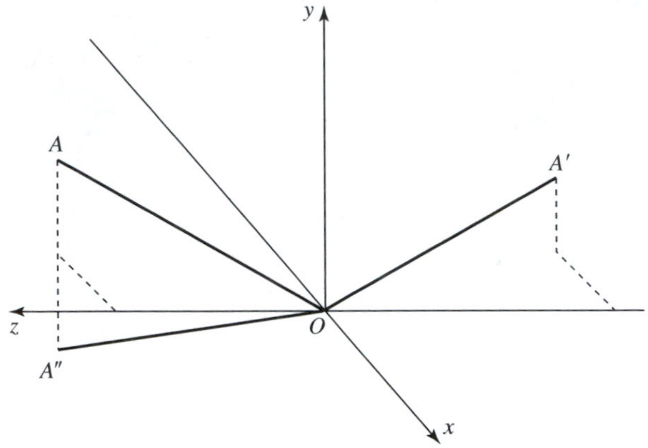


**Figure 4-11** Material with two planes of elastic symmetry

system we have

$$\begin{Bmatrix} \sigma_{x'x'} \\ \sigma_{y'y'} \\ \sigma_{z'z'} \\ \sigma_{y'z'} \\ \sigma_{z'x'} \\ \sigma_{x'y'} \end{Bmatrix} = \begin{bmatrix} C'_{11} & C'_{12} & C'_{13} & 0 & 0 & C'_{16} \\ & C'_{22} & C'_{23} & 0 & 0 & C'_{26} \\ & & C'_{33} & 0 & 0 & C'_{36} \\ & & & C'_{44} & C'_{45} & 0 \\ & \text{sym} & & & C'_{55} & 0 \\ & & & & & C'_{66} \end{bmatrix} \begin{Bmatrix} \varepsilon_{x'x'} \\ \varepsilon_{y'y'} \\ \varepsilon_{z'z'} \\ \gamma_{y'z'} \\ \gamma_{z'x'} \\ \gamma_{x'y'} \end{Bmatrix}$$

In order for $y = 0$ to be an additional plane of elastic symmetry, the elements of $[\boldsymbol{C}']$ in the preceding equation and those of $[\boldsymbol{C}]$ in Eq. (4-41) must be equal. Thus we can write

$$\begin{Bmatrix} \sigma_{x'x'} \\ \sigma_{y'y'} \\ \sigma_{z'z'} \\ \sigma_{y'z'} \\ \sigma_{z'x'} \\ \sigma_{x'y'} \end{Bmatrix} = \begin{bmatrix} C_{11} & C_{12} & C_{13} & 0 & 0 & C_{16} \\ & C_{22} & C_{23} & 0 & 0 & C_{26} \\ & & C_{33} & 0 & 0 & C_{36} \\ & & & C_{44} & C_{45} & 0 \\ & \text{sym} & & & C_{55} & 0 \\ & & & & & C_{66} \end{bmatrix} \begin{Bmatrix} \varepsilon_{x'x'} \\ \varepsilon_{y'y'} \\ \varepsilon_{z'z'} \\ \gamma_{y'z'} \\ \gamma_{z'x'} \\ \gamma_{x'y'} \end{Bmatrix} \tag{4-42}$$

Equations (2-39), (2-40), and (3-31) together with the direction cosines give us the following relationships between the stresses and strains in the primed and unprimed coordinate systems:

$$\sigma_{x'x'} = \sigma_{xx}, \quad \sigma_{y'y'} = \sigma_{yy}, \quad \sigma_{z'z'} = \sigma_{zz}, \quad \sigma_{y'z'} = -\sigma_{yz}, \quad \sigma_{z'x'} = \sigma_{zx}, \quad \sigma_{x'y'} = -\sigma_{xy}$$
$$\varepsilon_{x'x'} = \varepsilon_{xx}, \quad \varepsilon_{y'y'} = \varepsilon_{yy}, \quad \varepsilon_{z'z'} = \varepsilon_{zz}, \quad \gamma_{y'z'} = -\gamma_{yz}, \quad \gamma_{z'x'} = \gamma_{zx}, \quad \gamma_{x'y'} = -\gamma_{xy}$$

These are substituted into Eq. (4-42), giving the following stress-strain relationship:

$$\begin{Bmatrix} \sigma_{xx} \\ \sigma_{yy} \\ \sigma_{zz} \\ \sigma_{yz} \\ \sigma_{zx} \\ \sigma_{xy} \end{Bmatrix} = \begin{bmatrix} C_{11} & C_{12} & C_{13} & 0 & 0 & -C_{16} \\ & C_{22} & C_{23} & 0 & 0 & -C_{26} \\ & & C_{33} & 0 & 0 & -C_{36} \\ & & & C_{44} & -C_{45} & 0 \\ & \text{sym} & & & C_{55} & 0 \\ & & & & & C_{66} \end{bmatrix} \begin{Bmatrix} \varepsilon_{xx} \\ \varepsilon_{yy} \\ \varepsilon_{zz} \\ \gamma_{yz} \\ \gamma_{zx} \\ \gamma_{xy} \end{Bmatrix}$$

By comparing this equation with Eq. (4-41) we see that for the two to be equivalent we must impose the additional requirements

$$C_{16} = C_{26} = C_{36} = C_{45} = 0$$

so that the stress-strain equations for a material with two perpendicular planes of elastic symmetry become

$$\begin{Bmatrix} \sigma_{xx} \\ \sigma_{yy} \\ \sigma_{zz} \\ \sigma_{yz} \\ \sigma_{zx} \\ \sigma_{xy} \end{Bmatrix} = \begin{bmatrix} C_{11} & C_{12} & C_{13} & 0 & 0 & 0 \\ & C_{22} & C_{23} & 0 & 0 & 0 \\ & & C_{33} & 0 & 0 & 0 \\ & & & C_{44} & 0 & 0 \\ & \text{sym} & & & C_{55} & 0 \\ & & & & & C_{66} \end{bmatrix} \begin{Bmatrix} \varepsilon_{xx} \\ \varepsilon_{yy} \\ \varepsilon_{zz} \\ \gamma_{yz} \\ \gamma_{zx} \\ \gamma_{xy} \end{Bmatrix} \tag{4-43}$$

Note that Eq. (4-43) contains only nine independent material constants.

The stress-strain law for materials with three mutually perpendicular planes of elastic symmetry can be constructed in a similar manner by considering the additional transformation

$$x' = -x, \qquad y' = y, \qquad z' = z$$

which represents reflection in the plane $x = 0$.

The direction cosines and the resulting stress-strain law for this transformation can be obtained easily, and it can be shown that there are no additional restrictions placed on the elastic constants. Consequently the stress-strain law for a material with three mutually perpendicular planes of elastic symmetry is the same as for a material with only two perpendicular planes of elastic symmetry. More importantly, this result implies that a material possessing two perpendicular planes of elastic symmetry must also possess a third mutually perpendicular plane of elastic symmetry. This can also be seen from Fig. 4-12, in which $OA'''$ is the extension of $OA''$. Since, for homogeneous materials, the properties of an element situated anywhere along $OA''$ and its extension must be the same, it follows that an element situated along $OA'''$ must have the same properties as the corresponding element located along $OA'$. This is because an element located

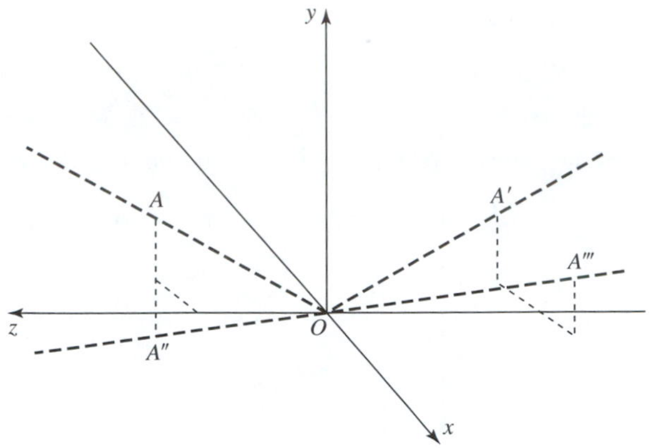


Figure 4-12

along $OA''$, for a material with two planes of elastic symmetry, has the same properties as the corresponding element along $OA$ which, in turn, has the same properties as an element located along $OA'$. Thus if $z = 0$ is a plane of symmetry, $x = 0$ must also be such a plane.

Materials having three mutually perpendicular planes of elastic symmetry are called *orthotropic* materials. The term orthotropic is a shortened form of orthogonally anisotropic. As shown in Eq. (4-43), orthotropic materials have only nine independent elastic constants.

Axes parallel to the intersections of the mutually perpendicular planes of symmetry define the *principal material directions*. It is along these axes that the stress-strain law takes on the relatively simple form of Eq. (4-43). For stresses calculated in coordinate systems whose axes are not parallel to the principal material directions, the elastic coefficient matrix may not contain any zero elements. This is shown in Section 4-8-3 for plane stress problems. However, there can still be only nine independent elastic constants. Thus regardless of the choice of coordinate system, all of the elements of the elastic coefficient matrix can be expressed in terms of the nine elastic constants contained in Eq. (4-43).

It is interesting to note that the structure of the elastic coefficient matrix for orthotropic materials with respect to the symmetry planes is similar to that for isotropic materials in that each normal stress depends on all three normal strains but not on the shear strains, and each shear stress depends only on the corresponding shear strain. However, each of the three shear moduli is different, and these apparently cannot be determined from the elastic constants associated with the normal stresses, as is done with isotropic materials. Further, the values of the elastic constants for each of the three normal stresses are different, in contrast to isotropic materials. For example, the elastic constant multiplying $\varepsilon_{xx}$ when calculating $\sigma_{xx}$ is different than the elastic constant multiplying $\varepsilon_{yy}$ when calculating $\sigma_{yy}$.

Equation (4-43) can be separated into three equations that give normal stresses in terms of normal strains and three that give shear stresses in terms of shear strains. From these it is a straightforward task to determine the normal strains in terms of normal stresses and the shear strains in terms of shear stresses. When put in matrix form, these equations are

$$
\begin{Bmatrix} \varepsilon_{xx} \\ \varepsilon_{yy} \\ \varepsilon_{zz} \\ \gamma_{yz} \\ \gamma_{zx} \\ \gamma_{xy} \end{Bmatrix} = \begin{bmatrix} S_{11} & S_{12} & S_{13} & 0 & 0 & 0 \\ & S_{22} & S_{23} & 0 & 0 & 0 \\ & & S_{33} & 0 & 0 & 0 \\ & & & S_{44} & 0 & 0 \\ & \text{sym} & & & S_{55} & 0 \\ & & & & & S_{66} \end{bmatrix} \begin{Bmatrix} \sigma_{xx} \\ \sigma_{yy} \\ \sigma_{zz} \\ \sigma_{yz} \\ \sigma_{zx} \\ \sigma_{xy} \end{Bmatrix} \tag{4-44}
$$

where, as noted previously, the elements of the coefficient matrix are the compliances.

It is important to point out that for a given material sample there may be no visual indication of the principal material directions. Several methods can be used to determine principal material directions in a material suspected to be orthotropic. One of these is to run experiments at various angles in an attempt to identify directions in which applied normal stresses do not induce shear strains.

The layers, or laminae, from which fiber-reinforced composites are constructed are good examples of orthotropic materials. These laminae consist of a layer of long, continuous fibers arranged unidirectionally, surrounded by a matrix that supports and protects the fibers, and also transmits loads between fibers. A typical configuration is shown in Fig. 4-13. The fibers, made from materials such as graphite or boron, are the primary load carriers in the laminae. The matrix is usually made from a material that is less stiff than the fibers, such as epoxy. The principal material directions can be visually identified in a fiber-reinforced lamina—the longitudinal direction is usually referred to as the $x$ direction, the $y$ direction is perpendicular to the $x$ direction and in the plane of the lamina, and the $z$ direction is perpendicular to the plane of the lamina.

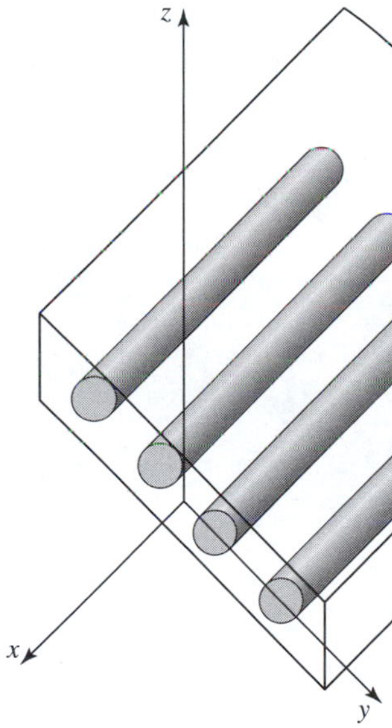

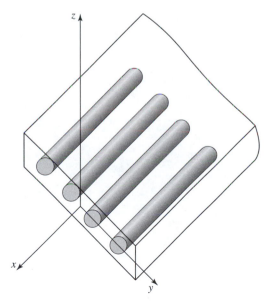

**Figure 4-13** Typical fiber-reinforced layer

Another example of an orthotropic material is the polymer from which beverage bottles are currently made. This polymer is polyethyleneterephthalate, or PET. In the cylindrical portion of the bottle the principal material directions are the longitudinal, tangential, and radial directions.

### 4-8-1 Engineering Material Constants for Orthotropic Materials

It is desirable to determine the material constants for orthotropic materials in terms of engineering constants such as Young's modulus and Poisson's ratio, as was done in Section 4-2 for isotropic materials. From an engineering point of view, these are much more readily understood than the quantities appearing in the coefficient matrix of Eq. (4-44). In order to determine a relationship between the elements of the compliance matrix and the engineering constants for orthotropic materials, we consider an element of material whose principal material directions are oriented along the $x$, $y$, $z$ coordinate axes, as shown in Fig. 4-14. Now if a tensile stress is applied to the $x$ faces of the element, as shown in Fig. 4-14, we can write

$$\sigma_{xx} = E_x \varepsilon_{xx} \tag{4-45}$$

If, instead, a tensile stress is applied to the $y$ faces, we get

$$\sigma_{yy} = E_y \varepsilon_{yy} \tag{4-46}$$

Similarly, for a tensile stress applied to the $z$ faces,

$$\sigma_{zz} = E_z \varepsilon_{zz} \tag{4-47}$$

In these equations, the quantities $E_x$, $E_y$, and $E_z$ are the Young's moduli in the $x$, $y$, and $z$ directions, respectively. Because the material is orthotropic, these quantities, which are called *generalized Young's moduli,* are not equal.

As in the case of an isotropic material, an extension of the element in the $x$ direction is accompanied by a contraction in both the $y$ and the $z$ directions. However, the

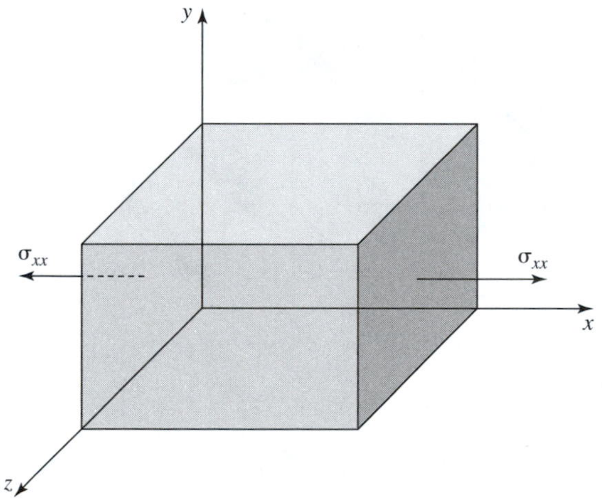


Figure 4-14

contraction in the $y$ direction will differ from that in the $z$ direction owing to the orthotropic nature of the material. Consequently we must write

$$\varepsilon_{yy} = -\nu_{xy}\varepsilon_{xx}, \qquad \varepsilon_{zz} = -\nu_{xz}\varepsilon_{xx} \tag{4-48}$$

where $\nu_{xy}$ and $\nu_{xz}$ are generalized Poisson's ratios. The second subscript refers to the direction of transverse strain when the element is stressed in the direction indicated by the first subscript. Thus, for example, $\nu_{xy}$ is the contraction in the $y$ direction due to a unit strain in the $x$ direction. Note that there are six generalized Poisson's ratios for an orthotropic material. These are

$$\nu_{xy} \qquad \nu_{xz} \qquad \nu_{yx} \qquad \nu_{yz} \qquad \nu_{zx} \qquad \nu_{zy}$$

When we solve Eq. (4-45) for $\varepsilon_{xx}$ and substitute the result in Eqs. (4-48) we get

$$\varepsilon_{xx} = \frac{1}{E_x}\sigma_{xx}, \qquad \varepsilon_{yy} = -\frac{\nu_{xy}}{E_x}\sigma_{xx}, \qquad \varepsilon_{zz} = -\frac{\nu_{xz}}{E_x}\sigma_{xx} \tag{4-49}$$

If we now apply a normal stress in the $y$ direction, we get

$$\varepsilon_{xx} = -\frac{\nu_{xy}}{E_y}\sigma_{yy}, \qquad \varepsilon_{yy} = \frac{1}{E_y}\sigma_{yy}, \qquad \varepsilon_{zz} = -\frac{\nu_{zy}}{E_y}\sigma_{yy} \tag{4-50}$$

Finally, a normal stress applied in the $z$ direction gives

$$\varepsilon_{xx} = -\frac{\nu_{zx}}{E_z}\sigma_{zz}, \qquad \varepsilon_{yy} = -\frac{\nu_{zy}}{E_z}\sigma_{zz}, \qquad \varepsilon_{zz} = \frac{1}{E_z}\sigma_{zz} \tag{4-51}$$

If all three normal stresses are applied simultaneously, then the total normal strain in the element in each direction is found by summing Eqs. (4-49)–(4-51),

$$\begin{aligned}
\varepsilon_{xx} &= \frac{1}{E_x}\sigma_{xx} - \frac{\nu_{yx}}{E_y}\sigma_{yy} - \frac{\nu_{zx}}{E_z}\sigma_{zz} \\
\varepsilon_{yy} &= -\frac{\nu_{xy}}{E_x}\sigma_{xx} + \frac{1}{E_y}\sigma_{yy} - \frac{\nu_{zy}}{E_z}\sigma_{zz} \\
\varepsilon_{zz} &= -\frac{\nu_{xz}}{E_x}\sigma_{xx} - \frac{\nu_{yz}}{E_y}\sigma_{yy} + \frac{1}{E_z}\sigma_{zz}
\end{aligned} \tag{4-52}$$

On comparing these with the first three of Eqs. (4-44) we find that

$$\begin{aligned}
&S_{11} = \frac{1}{E_x}, \qquad S_{22} = \frac{1}{E_y}, \qquad S_{33} = \frac{1}{E_z} \\
&S_{12} = -\frac{\nu_{yx}}{E_y} = -\frac{\nu_{xy}}{E_x}, \qquad S_{13} = -\frac{\nu_{zx}}{E_z} = -\frac{\nu_{xz}}{E_x}, \qquad S_{23} = -\frac{\nu_{zy}}{E_z} = -\frac{\nu_{yz}}{E_y}
\end{aligned} \tag{4-53}$$

The double equality in the last three of Eqs. (4-53) arises because of the symmetry of the compliance matrix. We note specifically that it is *not* the generalized Poisson's ratios that are symmetric. This difference is illustrated in Fig. 4-15.

Whereas with isotropic materials the relationship between shear stress and shear strain is the same in any of the coordinate planes, for orthotropic materials these

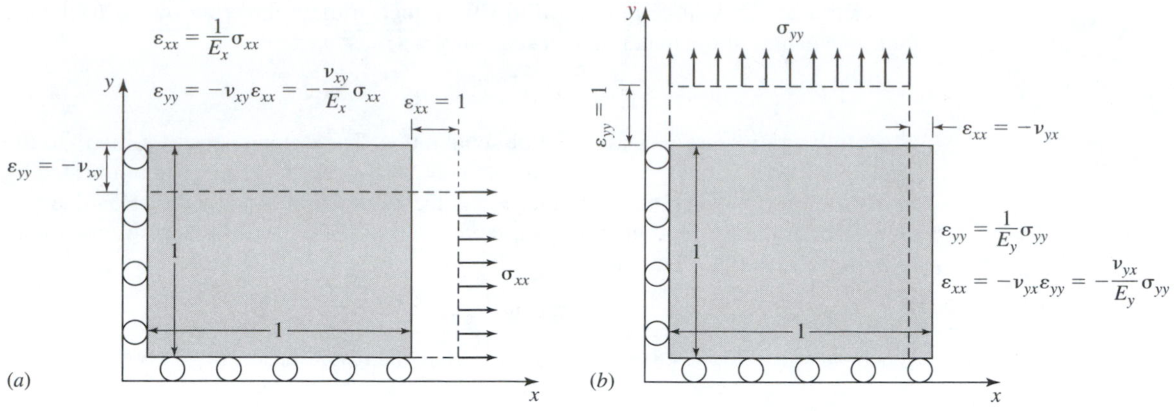


Figure 4-15

relationships are not the same. Thus we have

$$\gamma_{xy} = \frac{1}{G_{xy}}\sigma_{xy}, \qquad \gamma_{yz} = \frac{1}{G_{yz}}\sigma_{yz}, \qquad \gamma_{zx} = \frac{1}{G_{zx}}\sigma_{zx} \tag{4-54}$$

where the subscripts on the shear moduli indicate the plane of the shear moduli.

Although it appears that we have 12 physical constants (three generalized Young's moduli, six generalized Poisson's ratios, and three generalized shear moduli), only nine of these are independent. From the last three of Eqs. (4-53) we can eliminate, for example, three of the Poisson's ratios by writing

$$\nu_{xy} = \frac{E_x}{E_y}\nu_{yx}, \qquad \nu_{yz} = \frac{E_y}{E_z}\nu_{zy}, \qquad \nu_{zx} = \frac{E_z}{E_x}\nu_{xz} \tag{4-55}$$

Thus we have only nine independent physical constants for an orthotropic material, as Eqs. (4-43) or (4-44) require.

As with isotropic materials, it is possible to place restrictions on the values of the elastic constants for orthotropic materials. The interested reader is referred to Jones [9, pp. 41–45].

## 4-8-2 Orthotropic Materials under Conditions of Plane Stress

Often applications involving orthotropic materials make use of thin layers and the loading is confined to the plane of the layer. For example, in fiber-reinforced composites each lamina in the laminate is a thin layer. Thus there are many situations in which orthotropic materials are subjected to conditions of plane stress.

In order to derive the stress-strain relations for plane stress in orthotropic materials we consider a thin layer of material that lies in the $xy$ plane and recall that all stresses in the $z$ direction are zero. Then, from Eqs. (4-44) and (4-53) we have

$$\begin{Bmatrix} \varepsilon_{xx} \\ \varepsilon_{yy} \\ \gamma_{xy} \end{Bmatrix} = \begin{bmatrix} \dfrac{1}{E_x} & -\dfrac{\nu_{yx}}{E_y} & 0 \\ -\dfrac{\nu_{yx}}{E_y} & \dfrac{1}{E_y} & 0 \\ 0 & 0 & \dfrac{1}{G_{xy}} \end{bmatrix} \begin{Bmatrix} \sigma_{xx} \\ \sigma_{yy} \\ \sigma_{xy} \end{Bmatrix} \tag{4-56}$$

In addition we have

$$\varepsilon_{zz} = -\frac{\nu_{zx}}{E_z}\sigma_{xx} - \frac{\nu_{zy}}{E_z}\sigma_{yy}, \qquad \gamma_{yz} = 0, \qquad \gamma_{zx} = 0 \tag{4-57}$$

Equation (4-56) can be written in compact form as

$$\{\boldsymbol{\varepsilon}\} = [\overline{S}]\{\boldsymbol{\sigma}\} \tag{4-58}$$

where

$$\{\boldsymbol{\sigma}\} = [\,\sigma_{xx} \quad \sigma_{yy} \quad \sigma_{xy}]^T$$

$$\{\boldsymbol{\varepsilon}\} = [\,\varepsilon_{xx} \quad \varepsilon_{yy} \quad \gamma_{xy}]^T$$

$$[\overline{S}] = \begin{bmatrix} \dfrac{1}{E_x} & -\dfrac{\nu_{yx}}{E_y} & 0 \\ -\dfrac{\nu_{yx}}{E_y} & \dfrac{1}{E_y} & 0 \\ 0 & 0 & \dfrac{1}{G_{xy}} \end{bmatrix} \tag{4-59}$$

Since the stiffness matrix is the inverse of the compliance, we find, after some calculations, that

$$\{\boldsymbol{\sigma}\} = [\overline{C}]\{\boldsymbol{\varepsilon}\} \tag{4-60}$$

where

$$[\overline{C}] = [\overline{S}]^{-1} = \begin{bmatrix} \overline{C}_{11} & \overline{C}_{12} & 0 \\ \overline{C}_{12} & \overline{C}_{22} & 0 \\ 0 & 0 & \overline{C}_{33} \end{bmatrix} \tag{4-61}$$

and

$$\begin{aligned} \overline{C}_{11} &= \frac{E_x}{1 - \nu_{xy}\nu_{yx}}, \qquad \overline{C}_{22} = \frac{E_y}{1 - \nu_{xy}\nu_{yx}}, \qquad \overline{C}_{33} = G_{xy} \\ \overline{C}_{12} &= \frac{\nu_{xy}E_y}{1 - \nu_{xy}\nu_{yx}} = \frac{\nu_{yx}E_x}{1 - \nu_{xy}\nu_{yx}} \end{aligned} \tag{4-62}$$

### 4-8-3 Stress-Strain Relations in Coordinates Other than the Principal Material Coordinates

One of the significant advantages of materials such as fiber-reinforced composites is that the laminate can be constructed in such a way that it provides maximum strength for the anticipated loading conditions. This is accomplished by careful orientation of the various laminae that make up the laminate. The principal material directions of individual laminae may therefore vary from lamina to lamina. In studying the properties of the laminate, however, one must ultimately express the stress-strain relationships for each lamina in a common coordinate system. For example, Fig. 4-16 shows a lamina with principal material coordinates $x'$ and $y'$, but the obvious choice of coordinates based on the geometry and loading is $x$ and $y$. Thus some means by which the stress-strain relations can be transformed from one coordinate system to another is needed.

We begin by defining the *body coordinate system* as the coordinate system used to define the geometry and the loading for the problem (for example, the $x$, $y$ coordinate system in Fig. 4-16). We consider the case of plane stress, where the transformation law for stress is given by Eqs. (2-39) and (2-40) and for strain by Eqs. (3-31). In matrix form these are

$$\begin{Bmatrix} \sigma_{x'x'} \\ \sigma_{y'y'} \\ \sigma_{x'y'} \end{Bmatrix} = \begin{bmatrix} n_{x'x}^2 & n_{x'y}^2 & 2n_{x'x}n_{x'y} \\ n_{y'x}^2 & n_{y'y}^2 & 2n_{y'x}n_{y'y} \\ n_{x'x}n_{y'x} & n_{x'y}n_{y'y} & n_{x'x}n_{y'y} + n_{x'y}n_{y'x} \end{bmatrix} \begin{Bmatrix} \sigma_{xx} \\ \sigma_{yy} \\ \sigma_{xy} \end{Bmatrix} \tag{4-63a}$$

$$\begin{Bmatrix} \varepsilon_{x'x'} \\ \varepsilon_{y'y'} \\ \gamma_{x'y'} \end{Bmatrix} = \begin{bmatrix} n_{x'x}^2 & n_{x'y}^2 & n_{x'x}n_{x'y} \\ n_{y'x}^2 & n_{y'y}^2 & n_{y'x}n_{y'y} \\ 2n_{x'x}n_{y'x} & 2n_{x'y}n_{y'y} & n_{x'x}n_{y'y} + n_{x'y}n_{y'x} \end{bmatrix} \begin{Bmatrix} \varepsilon_{xx} \\ \varepsilon_{yy} \\ \gamma_{xy} \end{Bmatrix} \tag{4-63b}$$

We note that the transformation matrices for stress and strain are slightly different. This can be rectified, however, by replacing $\gamma_{xy}$ and $\gamma_{x'y'}$ with $\gamma_{xy}/2$ and $\gamma_{x'y'}/2$ in Eq. (4-63*b*). In order to maintain the equality, we must divide the last row of the transformation matrix by 2 and multiply the last column by 2. This gives

$$\begin{Bmatrix} \varepsilon_{x'x'} \\ \varepsilon_{y'y'} \\ \gamma_{x'y'}/2 \end{Bmatrix} = \begin{bmatrix} n_{x'x}^2 & n_{x'y}^2 & 2n_{x'x}n_{x'y} \\ n_{y'x}^2 & n_{y'y}^2 & 2n_{y'x}n_{y'y} \\ n_{x'x}n_{y'x} & n_{x'y}n_{y'y} & n_{x'x}n_{y'y} + n_{x'y}n_{y'x} \end{bmatrix} \begin{Bmatrix} \varepsilon_{xx} \\ \varepsilon_{yy} \\ \gamma_{xy}/2 \end{Bmatrix} \tag{4-64}$$

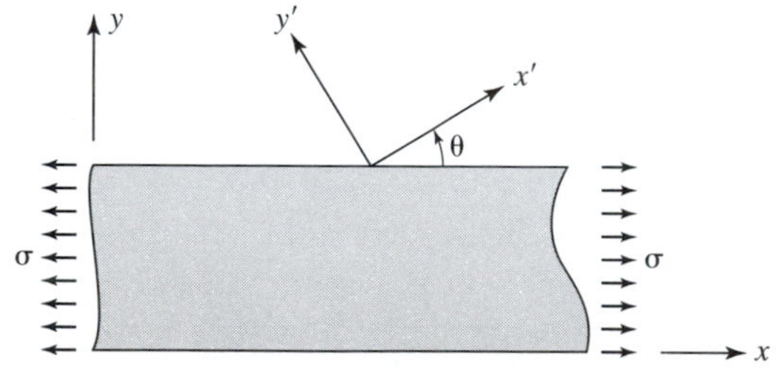


Figure 4-16

In compact form Eqs. (4-63*a*) and (4-64) become

$$\{\boldsymbol{\sigma}'\} = [\boldsymbol{R}]\{\boldsymbol{\sigma}\} \tag{4-65a}$$

$$\{\bar{\boldsymbol{\varepsilon}}'\} = [\boldsymbol{R}]\{\bar{\boldsymbol{\varepsilon}}\} \tag{4-65b}$$

where

$$\{\bar{\boldsymbol{\varepsilon}}\} = [\varepsilon_{xx} \quad \varepsilon_{yy} \quad \gamma_{xy}/2]^T \tag{4-66a}$$

$$[\boldsymbol{R}] = \begin{bmatrix} n_{x'x}^2 & n_{x'y}^2 & 2n_{x'x}n_{x'y} \\ n_{y'x}^2 & n_{y'y}^2 & 2n_{y'x}n_{y'y} \\ n_{x'x}n_{y'x} & n_{x'y}n_{y'y} & n_{x'x}n_{y'y} + n_{x'y}n_{y'x} \end{bmatrix}$$
$$= \begin{bmatrix} \cos^2\theta & \sin^2\theta & 2\sin\theta\cos\theta \\ \sin^2\theta & \cos^2\theta & -2\sin\theta\cos\theta \\ -\sin\theta\cos\theta & \sin\theta\cos\theta & \cos^2\theta - \sin^2\theta \end{bmatrix} \tag{4-66b}$$

and $\{\boldsymbol{\sigma}\}$ is defined by the first of Eqs. (4-59). $\theta$ is the angle between the axes of the body coordinate system and the principal axes of the lamina, as shown in Fig. 4-16. Definitions of the primed quantities follow from these by replacing the unprimed subscripts with primed quantities.

In a manner similar to Reuter [18] we note that $\{\bar{\boldsymbol{\varepsilon}}'\}$ can be obtained from $\{\boldsymbol{\varepsilon}\}$ as follows:

$$\begin{Bmatrix} \bar{\varepsilon}_{x'x'} \\ \bar{\varepsilon}_{y'y'} \\ \bar{\gamma}_{x'y'}/2 \end{Bmatrix} = \begin{bmatrix} 1 & 0 & 0 \\ 0 & 1 & 0 \\ 0 & 0 & \frac{1}{2} \end{bmatrix} \begin{Bmatrix} \varepsilon_{x'x'} \\ \varepsilon_{y'y'} \\ \gamma_{x'y'} \end{Bmatrix} \tag{4-67}$$

or, in compact form,

$$\{\bar{\boldsymbol{\varepsilon}}'\} = [A]\{\boldsymbol{\varepsilon}'\} \tag{4-68}$$

where

$$[A] = \begin{bmatrix} 1 & 0 & 0 \\ 0 & 1 & 0 \\ 0 & 0 & \frac{1}{2} \end{bmatrix} \tag{4-69}$$

Equation (4-68) must hold for any orientation of the primed coordinate system relative to the unprimed system. Therefore it also applies to the strains in the unprimed coordinate system, that is,

$$\{\bar{\boldsymbol{\varepsilon}}\} = [A]\{\boldsymbol{\varepsilon}\} \tag{4-70}$$

Now let us write Eq. (4-60) in the primed coordinate system,

$$\{\boldsymbol{\sigma}'\} = [\overline{C}]\{\boldsymbol{\varepsilon}'\} \tag{4-71}$$

Equation (4-68) can be readily solved for $\{\boldsymbol{\varepsilon}'\}$ to give

$$\{\boldsymbol{\varepsilon}'\} = [A]^{-1}\{\overline{\boldsymbol{\varepsilon}}'\}$$

When this and Eqs. (4-65*a*) are substituted into Eq. (4-71) and (4-65*b*) is used in the result, we obtain

$$[\boldsymbol{R}]\{\boldsymbol{\sigma}\} = [\overline{C}][A]^{-1}[\boldsymbol{R}]\{\overline{\boldsymbol{\varepsilon}}\}$$

Equation (4-70) is now substituted for $\{\overline{\boldsymbol{\varepsilon}}\}$ and both sides are premultiplied by $[\boldsymbol{R}]^{-1}$ to give

$$\{\boldsymbol{\sigma}\} = [\boldsymbol{R}]^{-1}[\overline{C}][A]^{-1}[\boldsymbol{R}][A]\{\boldsymbol{\varepsilon}\} \tag{4-72}$$

From Eq. (4-66*b*) the inverse of $[\boldsymbol{R}]$ can be readily computed. It is

$$[\boldsymbol{R}]^{-1} = \begin{bmatrix} \cos^2\theta & \sin^2\theta & -2\sin\theta\cos\theta \\ \sin^2\theta & \cos^2\theta & 2\sin\theta\cos\theta \\ \sin\theta\cos\theta & -\sin\theta\cos\theta & \cos^2\theta - \sin^2\theta \end{bmatrix}$$

which can be written as

$$[\boldsymbol{R}]^{-1} = [\hat{\mathbf{1}}][\boldsymbol{R}][\hat{\mathbf{1}}]$$

where

$$[\hat{\mathbf{1}}] = \begin{bmatrix} 1 & 0 & 0 \\ 0 & 1 & 0 \\ 0 & 0 & -1 \end{bmatrix} \tag{4-73}$$

Now computing $[A]^{-1}[\boldsymbol{R}][A]$ and comparing the result with $[\boldsymbol{R}]^{-1}$ we see that

$$[A]^{-1}[\boldsymbol{R}][A] = ([\boldsymbol{R}]^{-1})^T = ([\hat{\mathbf{1}}][\boldsymbol{R}][\hat{\mathbf{1}}])^T = [\hat{\mathbf{1}}][\boldsymbol{R}]^T[\hat{\mathbf{1}}]$$

With this Eq. (4-72) becomes

$$\{\boldsymbol{\sigma}\} = [\hat{\mathbf{1}}][\boldsymbol{R}][\hat{\mathbf{1}}][\overline{C}][\hat{\mathbf{1}}][\boldsymbol{R}]^T[\hat{\mathbf{1}}]\{\boldsymbol{\varepsilon}\} \tag{4-74}$$

But with Eqs. (4-61) and (4-73) we have

$$[\hat{\mathbf{1}}][\overline{C}][\hat{\mathbf{1}}] = [\overline{C}]$$

so that we can write Eq. (4-74) as

$$\{\boldsymbol{\sigma}\} = [\hat{\boldsymbol{R}}][\overline{C}][\hat{\boldsymbol{R}}]^T\{\boldsymbol{\varepsilon}\} \tag{4-75}$$

where

$$[\hat{\boldsymbol{R}}] = [\hat{\mathbf{1}}][\boldsymbol{R}] = \begin{bmatrix} \cos^2\theta & \sin^2\theta & 2\sin\theta\cos\theta \\ \sin^2\theta & \cos^2\theta & -2\sin\theta\cos\theta \\ \sin\theta\cos\theta & -\sin\theta\cos\theta & -(\cos^2\theta - \sin^2\theta) \end{bmatrix} \tag{4-76}$$

It should be noted that the operation

$$[\hat{\boldsymbol{R}}][\overline{\boldsymbol{C}}][\hat{\boldsymbol{R}}]^T$$

simply transforms the elasticity matrix into the body coordinate system. When the indicated matrix multiplication is carried out, we get

$$\{\boldsymbol{\sigma}\} = [\hat{\boldsymbol{C}}]\{\boldsymbol{\varepsilon}\} \tag{4-77}$$

where the symmetric matrix $[\hat{\boldsymbol{C}}]$ is given by

$$[\hat{\boldsymbol{C}}] \equiv [\hat{\boldsymbol{R}}][\overline{\boldsymbol{C}}][\hat{\boldsymbol{R}}]^T$$

and its elements are

$$\begin{aligned}
\hat{C}_{11} &= \overline{C}_{11}\cos^4\theta + \overline{C}_{22}\sin^4\theta + 2(\overline{C}_{12} + 2\overline{C}_{33})\sin^2\theta\cos^2\theta \\
\hat{C}_{12} &= \hat{C}_{21} = (\overline{C}_{11} + \overline{C}_{12} - 4\overline{C}_{33})\sin^2\theta\cos^2\theta + \overline{C}_{12}(\sin^4\theta + \cos^4\theta) \\
\hat{C}_{13} &= \hat{C}_{31} = (\overline{C}_{11} - \overline{C}_{12} - 2\overline{C}_{33})\sin\theta\cos^3\theta + (\overline{C}_{12} - \overline{C}_{22} + 2\overline{C}_{33})\sin^3\theta\cos\theta \\
\hat{C}_{22} &= \overline{C}_{11}\sin^4\theta + \overline{C}_{22}\cos^4\theta + 2(\overline{C}_{12} + 2\overline{C}_{33})\sin^2\theta\cos^2\theta \\
\hat{C}_{23} &= \hat{C}_{32} = (\overline{C}_{11} - \overline{C}_{12} - 2\overline{C}_{33})\sin^3\theta\cos\theta + (\overline{C}_{12} - \overline{C}_{22} + 2\overline{C}_{33})\sin\theta\cos^3\theta \\
\hat{C}_{33} &= (\overline{C}_{11} + \overline{C}_{22} - 2\overline{C}_{12} - 2\overline{C}_{33})\sin^2\theta\cos^2\theta + \overline{C}_{33}(\sin^4\theta + \cos^4\theta)
\end{aligned} \tag{4-78}$$

With Eq. (4-77) and (4-78) the stress-strain relations for each lamina in the laminate can be written in the body coordinate system. Note that, in general, $[\hat{\boldsymbol{C}}]$ contains no zero elements, but that each of its six independent elements can be obtained from the four elements of $[\overline{\boldsymbol{C}}]$, given in Eqs. (4-61) and (4-62).

It is left to the excercises to show that

$$\begin{aligned}
\hat{S}_{11} &= \overline{S}_{11}\cos^4\theta + \overline{S}_{22}\sin^4\theta + (2\overline{S}_{12} + \overline{S}_{33})\sin^2\theta\cos^2\theta \\
\hat{S}_{12} &= \overline{S}_{21} = (\overline{S}_{11} + \overline{S}_{22} - \overline{S}_{33})\sin^2\theta\cos^2\theta + \overline{S}_{12}(\sin^4\theta + \cos^4\theta) \\
\hat{S}_{13} &= \hat{S}_{31} = (2\overline{S}_{11} - 2\overline{S}_{12} - \overline{S}_{33})\sin\theta\cos^3\theta - (2\overline{S}_{22} - 2\overline{S}_{12} - \overline{S}_{33})\sin^3\theta\cos\theta \\
\hat{S}_{22} &= \overline{S}_{11}\sin^4\theta + \overline{S}_{22}\cos^4\theta + (2\overline{S}_{12} + \overline{S}_{33})\sin^2\theta\cos^2\theta \\
\hat{S}_{23} &= \overline{S}_{32} = (2\overline{S}_{11} - 2\overline{S}_{12} - \overline{S}_{33})\sin^3\theta\cos\theta - (2\overline{S}_{22} - 2\overline{S}_{12} - \overline{S}_{33})\sin\theta\cos^3\theta \\
\hat{S}_{33} &= 2(2\overline{S}_{11} + 2\overline{S}_{22} - 4\overline{S}_{12} - \overline{S}_{33})\sin^2\theta\cos^2\theta + \overline{S}_{33}(\sin^4\theta + \cos^4\theta)
\end{aligned} \tag{4-79}$$

and therefore

$$
\begin{aligned}
\frac{1}{E_x} &= \frac{1}{E_{x'}}\cos^4\theta + \left(\frac{1}{G_{x'y'}} - \frac{2\nu_{x'y'}}{E_{x'}}\right)\sin^2\theta\cos^2\theta + \frac{1}{E_{y'}}\sin^4\theta \\
\frac{1}{E_y} &= \frac{1}{E_{x'}}\sin^4\theta + \left(\frac{1}{G_{x'y'}} - \frac{2\nu_{x'y'}}{E_{x'}}\right)\sin^2\theta\cos^2\theta + \frac{1}{E_{y'}}\cos^4\theta \\
\frac{1}{G_{xy}} &= 2\left(\frac{2}{E_{x'}} + \frac{2}{E_{y'}} + \frac{4\nu_{x'y'}}{E_{x'}} - \frac{1}{G_{x'y'}}\right)\sin^2\theta\cos^2\theta + \frac{1}{G_{x'y'}}(\sin^4\theta + \cos^4\theta) \\
\nu_{xy} &= E_x\left[\frac{\nu_{x'y'}}{E_{x'}}(\sin^4\theta + \cos^4\theta) - \left(\frac{1}{E_{x'}} + \frac{1}{E_{y'}} - \frac{1}{G_{x'y'}}\right)\sin^2\theta\cos^2\theta\right] \qquad (4.80) \\
\hat{S}_{13} &= \frac{\eta_{xy,x}}{E_y} = \left(2\frac{1}{E_{x'}} + 2\frac{\nu_{x'y'}}{E_{x'}} - \frac{1}{G_{x'y'}}\right)\sin\theta\cos^3\theta \\
&\quad - \left(2\frac{1}{E_{y'}} + 2\frac{\nu_{x'y'}}{E_{x'}} - \frac{1}{G_{x'y'}}\right)\cos\theta\sin^3\theta \\
\hat{S}_{23} &= \frac{\eta_{xy,y}}{E_y} = -\left(2\frac{1}{E_{y'}} + 2\frac{\nu_{x'y'}}{E_{x'}} - \frac{1}{G_{x'y'}}\right)\sin\theta\cos^3\theta \\
&\quad + \left(2\frac{1}{E_{x'}} + 2\frac{\nu_{x'y'}}{E_{x'}} - \frac{1}{G_{x'y'}}\right)\cos\theta\sin^3\theta
\end{aligned}
$$

where the primed quantities are in the material coordinate system. $\eta_{xy,x}$ and $\eta_{xy,y}$ are the *coefficients of mutual influence*. The presence of these coupling terms shows that for an orthotropic material, in general, a shear strain causes normal stresses and, conversely, normal strains cause shear stress. Note that this coupling exists unless the $x$ and $y$ axes coincide with the material axes.

## 4-9 TRANSVERSELY ISOTROPIC MATERIALS

If, at every point in a material, there exist parallel planes in which the elastic properties are isotropic, the material is called *transversely isotropic*. Wood is an example of such a material. The elastic properties measured in plane perpendicular to the direction of the fibers are approximately equal for any directional orientation within the plane. However, they are not the same as the properties measured along the fibers.

Ice found in bodies of water that are relatively unaffected by currents, such as lakes or the ocean, is another example of a material that is approximately transversely isotropic. Ice crystals in these bodies of water form in such a way that one of the crystal axes is oriented vertically, whereas the remaining two axes lie in the horizontal plane. Individual crystals are highly anisotropic, and the orientation of their axes in the horizontal plane varies randomly from crystal to crystal. However, a sample of ice consisting of many crystals has elastic properties that are approximately isotropic in the horizontal plane.

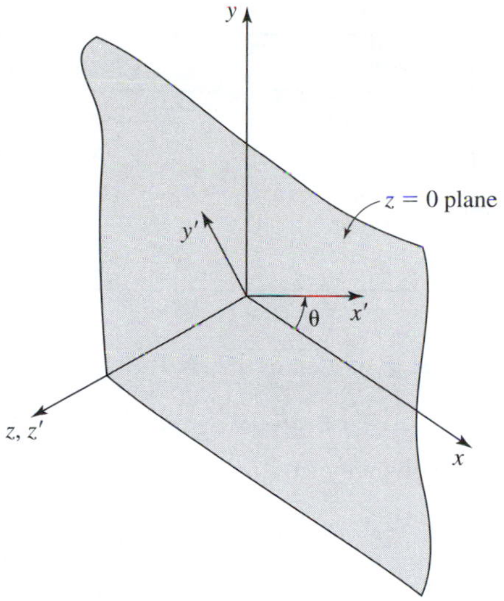


**Figure 4-17** Transverse isotropy

In order to develop the appropriate stress-strain relationship for transversely isotropic materials let us assume that the material is isotropic in planes parallel to the $z = 0$ plane. For the material to be transversely isotropic, the stress-strain law for the $x'$, $y'$, $z'$ coordinate system, where the $x'$ axis is rotated through an angle $\theta$ about the $z$ axis, as shown in Fig. 4-17, must be the same as the stress-strain law for the $x$, $y$, $z$ coordinate system. Recalling the definition of the direction cosines as given by Eqs. (B-13), we have the relationships in Table 4-4. Then, with Eqs. (2-39), (2-40), and (3-31) the stresses and strains in the primed coordinate system are given by

$$
\begin{aligned}
\sigma_{x'x'} &= \sigma_{xx}\cos^2\theta + \sigma_{yy}\sin^2\theta + \sigma_{xy}\sin 2\theta \\
\sigma_{y'y'} &= \sigma_{xx}\sin^2\theta + \sigma_{yy}\cos^2\theta - \sigma_{xy}\sin 2\theta \\
\sigma_{z'z'} &= \sigma_{zz} \\
\sigma_{y'z'} &= \sigma_{yz}\cos\theta - \sigma_{zx}\sin\theta \\
\sigma_{z'x'} &= \sigma_{yz}\sin\theta + \sigma_{zx}\cos\theta \\
\sigma_{x'y'} &= -\tfrac{1}{2}\sigma_{xx}\sin 2\theta + \tfrac{1}{2}\sigma_{yy}\sin 2\theta + \sigma_{xy}(\cos^2\theta - \sin^2\theta) \\
\varepsilon_{x'x'} &= \varepsilon_{xx}\cos^2\theta + \varepsilon_{yy}\sin^2\theta + \tfrac{1}{2}\gamma_{xy}\sin 2\theta \\
\varepsilon_{y'y'} &= \varepsilon_{xx}\sin^2\theta + \varepsilon_{yy}\cos^2\theta - \tfrac{1}{2}\gamma_{xy}\sin 2\theta \\
\varepsilon_{z'z'} &= \varepsilon_{zz} \\
\gamma_{y'z'} &= \gamma_{yz}\cos\theta - \gamma_{zx}\sin\theta \\
\gamma_{z'x'} &= \gamma_{yz}\sin\theta + \gamma_{zx}\cos\theta \\
\gamma_{x'y'} &= -\varepsilon_{xx}\sin 2\theta + \varepsilon_{yy}\sin 2\theta + \gamma_{xy}(\cos^2\theta - \sin^2\theta)
\end{aligned}
$$

TABLE 4-4

| | x | y | z |
|---|---|---|---|
| x′ | $n_{x'x} = \cos\theta$ | $n_{x'y} = \cos(270^\circ + \theta) = \sin\theta$ | $n_{x'z} = 0$ |
| y′ | $n_{y'x} = \cos(90^\circ + \theta) = -\sin\theta$ | $n_{y'y} = \cos\theta$ | $n_{y'z} = 0$ |
| z′ | $n_{z'x} = 0$ | $n_{z'y} = 0$ | $n_{z'z} = 1$ |

For convenience, these are written in matrix form as

$$\{\boldsymbol{\sigma}'\} = [\boldsymbol{R}]\{\boldsymbol{\sigma}\}, \qquad \{\boldsymbol{\varepsilon}'\} = [\hat{\boldsymbol{R}}]\{\boldsymbol{\varepsilon}\} \tag{4-81}$$

where $\{\boldsymbol{\sigma}\}$ and $\{\boldsymbol{\varepsilon}\}$ are given by Eqs. (4-29) and $[\boldsymbol{R}]$, the stress transformation matrix, is given by

$$[\boldsymbol{R}] = \begin{bmatrix} \cos^2\theta & \sin^2\theta & 0 & 0 & 0 & \sin 2\theta \\ \sin^2\theta & \cos^2\theta & 0 & 0 & 0 & -\sin 2\theta \\ 0 & 0 & 1 & 0 & 0 & 0 \\ 0 & 0 & 0 & \cos\theta & -\sin\theta & 0 \\ 0 & 0 & 0 & \sin\theta & \cos\theta & 0 \\ -\frac{1}{2}\sin 2\theta & \frac{1}{2}\sin 2\theta & 0 & 0 & 0 & \cos 2\theta \end{bmatrix} \tag{4-82}$$

whereas $[\hat{\boldsymbol{R}}]$, the strain transformation matrix, is given by

$$[\hat{\boldsymbol{R}}] = \begin{bmatrix} \cos^2\theta & \sin^2\theta & 0 & 0 & 0 & \frac{1}{2}\sin 2\theta \\ \sin^2\theta & \cos^2\theta & 0 & 0 & 0 & -\frac{1}{2}\sin 2\theta \\ 0 & 0 & 1 & 0 & 0 & 0 \\ 0 & 0 & 0 & \cos\theta & -\sin\theta & 0 \\ 0 & 0 & 0 & \sin\theta & \cos\theta & 0 \\ -\sin 2\theta & \sin 2\theta & 0 & 0 & 0 & \cos 2\theta \end{bmatrix} \tag{4-83}$$

For the primed coordinate system we have, on requiring the elasticity matrix for the two coordinate systems to be the same,

$$\{\boldsymbol{\sigma}'\} = [\boldsymbol{C}]\{\boldsymbol{\varepsilon}'\}$$

and by substituting Eqs. (4-81*a*, *b*) for $\{\boldsymbol{\sigma}'\}$ and $\{\boldsymbol{\varepsilon}'\}$, respectively, we get

$$[\boldsymbol{R}]\{\boldsymbol{\sigma}\} = [\boldsymbol{C}]\{\hat{\boldsymbol{R}}\}\{\boldsymbol{\varepsilon}\} \tag{4-84}$$

But in the unprimed coordinate system Eq. (4-28) must hold. When this is premultiplied by $[\boldsymbol{R}]$, we get

$$[\boldsymbol{R}]\{\boldsymbol{\sigma}\} = [\boldsymbol{R}][\boldsymbol{C}]\{\boldsymbol{\varepsilon}\} \tag{4-85}$$

In order for the material to be transversely isotropic, Eqs. (4-84) and (4-85) must be equal. Consequently,

$$[\boldsymbol{C}][\hat{\boldsymbol{R}}]\{\boldsymbol{\varepsilon}\} = [\boldsymbol{R}][\boldsymbol{C}]\{\boldsymbol{\varepsilon}\}$$

or

$$[\boldsymbol{A}]\{\boldsymbol{\varepsilon}\} = \{\boldsymbol{0}\} \tag{4-86}$$

where

$$[\boldsymbol{A}] = [\boldsymbol{B}] - [\hat{\boldsymbol{B}}] \tag{4-87}$$

and

$$[\boldsymbol{B}] = [\boldsymbol{C}][\hat{\boldsymbol{R}}], \qquad [\hat{\boldsymbol{B}}] = [\boldsymbol{R}][\boldsymbol{C}] \tag{4-88}$$

The first row of Eq. (4-86) is

$$A_{11}\varepsilon_{xx} + A_{12}\varepsilon_{yy} + \cdots + A_{16}\gamma_{xy} = 0$$

This must be true for arbitrary values of the strains, which are linearly independent quantities. Therefore we must have

$$A_{11} = A_{12} = \cdots = A_{16} = 0$$

Similar results are obtained from the remaining rows of Eq. (4-86), so that we have

$$A_{ij} = 0, \qquad i = 1, 2, \ldots, 6, \quad j = 1, 2, \ldots, 6 \tag{4-89}$$

or

$$[\boldsymbol{A}] = [\boldsymbol{0}] \tag{4-90}$$

for a material that is transversely isotropic. With Eqs. (4-30) and (4-83) we find that the elements of $[\boldsymbol{B}]$ are

$$\begin{aligned}
B_{i1} &= C_{i1}\cos^2\theta + C_{i2}\sin^2\theta - C_{i6}\sin 2\theta \\
B_{i2} &= C_{i1}\sin^2\theta + C_{i2}\cos^2\theta + C_{i6}\sin 2\theta \\
B_{i3} &= C_{i3} \\
B_{i4} &= C_{i4}\cos\theta + C_{i5}\sin\theta \\
B_{i5} &= -C_{i4}\sin\theta + C_{i5}\cos\theta \\
B_{i6} &= \tfrac{1}{2}C_{i1}\sin 2\theta - \tfrac{1}{2}C_{i2}\sin 2\theta + C_{i6}\cos 2\theta
\end{aligned} \tag{4-91}$$

where $i = 1, 2, \ldots, 6$, whereas with Eqs. (4-30) and (4-82) the elements of $[\hat{\boldsymbol{B}}]$ are

$$\begin{aligned}
\hat{B}_{1j} &= C_{1j}\cos^2\theta + C_{2j}\sin^2\theta + C_{6j}\sin 2\theta \\
\hat{B}_{2j} &= C_{1j}\sin^2\theta + C_{2j}\cos^2\theta - C_{6j}\sin 2\theta \\
\hat{B}_{3j} &= C_{3j} \\
\hat{B}_{4j} &= C_{4j}\cos\theta - C_{5j}\sin\theta \\
\hat{B}_{5j} &= C_{4j}\sin\theta + C_{5j}\cos\theta \\
\hat{B}_{6j} &= -\tfrac{1}{2}C_{1j}\sin 2\theta + \tfrac{1}{2}C_{2j}\sin 2\theta + C_{6j}\cos 2\theta
\end{aligned} \tag{4-92}$$

where $j = 1, 2, \ldots, 6$.

Substitution of these into Eq. (4-89) yields the elements of $[\boldsymbol{A}]$. Equation (4-90) represents 36 equations, two of which are identically satisfied (for example, $A_{33} = C_{33} - C_{33} = 0$). The remaining equations are somewhat unwieldy, a situation that can be improved by noting that the transpose of Eq. (4-90) gives

$$[\boldsymbol{A}]^T = [\mathbf{0}]$$

so that

$$[\boldsymbol{a}] \equiv [\boldsymbol{A}] + [\boldsymbol{A}]^T = [\mathbf{0}] \tag{4-93}$$

Here $[\boldsymbol{a}]$ is symmetric, and therefore Eq. (4-93) has the advantage of containing, at most, 21 independent equations, considerably fewer than Eq. (4-90). Coincidentally, the equations are also simpler than those of Eq. (4-90). Of Eq. (4-93), two equations are identically satisfied; the remaining equations are

$$\begin{aligned}
a_{11} &= -2C_{16}\sin 2\theta = 0\\
a_{12} &= (C_{16} - C_{26})\sin 2\theta = 0\\
a_{13} &= -C_{36}\sin 2\theta = 0\\
a_{14} &= C_{15}\sin\theta - C_{46}\sin 2\theta = 0\\
a_{15} &= -C_{14}\sin\theta - C_{56}\sin 2\theta = 0\\
a_{16} &= \left(\tfrac{1}{2}C_{11} - \tfrac{1}{2}C_{12} - C_{66}\right)\sin 2\theta = 0\\
a_{22} &= 2C_{26}\sin 2\theta = 0\\
a_{23} &= C_{36}\sin 2\theta = 0\\
a_{24} &= C_{25}\sin\theta + C_{46}\sin 2\theta = 0\\
a_{25} &= -C_{24}\sin\theta + C_{56}\sin 2\theta = 0\\
a_{26} &= \left(\tfrac{1}{2}C_{12} - \tfrac{1}{2}C_{22} + C_{66}\right)\sin 2\theta = 0\\
a_{34} &= C_{35}\sin\theta = 0\\
a_{35} &= -C_{34}\sin\theta = 0\\
a_{36} &= \tfrac{1}{2}(C_{13} - C_{23})\sin 2\theta = 0\\
a_{45} &= (-C_{44} + C_{55})\sin\theta = 0\\
a_{46} &= \tfrac{1}{2}(C_{14} - C_{24})\sin 2\theta + C_{56}\sin\theta = 0\\
a_{55} &= -2C_{45}\sin\theta = 0\\
a_{56} &= \tfrac{1}{2}(C_{15} - C_{25})\sin 2\theta + C_{46}\sin\theta = 0\\
a_{66} &= (C_{16} - C_{26})\sin 2\theta = 0
\end{aligned} \tag{4-94}$$

For a transversely isotropic material these must be satisfied for arbitrary values of $\theta$. From those members of Eq. (4-94) that contain only one trigonometric function we conclude that

$$
\begin{aligned}
C_{16} &= 0 \\
C_{26} &= C_{16} = 0 \\
C_{36} &= 0 \\
C_{66} &= \tfrac{1}{2}(C_{11} - C_{12}) \\
C_{66} &= \tfrac{1}{2}(C_{22} - C_{12}) \\
C_{35} &= 0 \\
C_{34} &= 0 \\
C_{23} &= C_{13} \\
C_{55} &= C_{44} \\
C_{45} &= 0
\end{aligned}
\tag{4-95}
$$

The remaining equations contain both $\sin\theta$ and $\sin 2\theta$. These are linearly independent, so for arbitrary values of $\theta$ the equations can be zero only if each coefficient is zero. Thus we find that

$$
\begin{aligned}
C_{14} &= 0, \qquad C_{15} = 0, \qquad C_{24} = 0 \\
C_{25} &= 0, \qquad C_{46} = 0, \qquad C_{56} = 0
\end{aligned}
\tag{4-96}
$$

With these results, the 21 equations resulting from Eq. (4-93) are satisfied. The material coefficient matrix for a transversly istoropic material is then given as

$$
\begin{Bmatrix} \sigma_{xx} \\ \sigma_{yy} \\ \sigma_{zz} \\ \sigma_{yz} \\ \sigma_{zx} \\ \sigma_{xy} \end{Bmatrix}
=
\begin{bmatrix}
C_{11} & C_{12} & C_{13} & 0 & 0 & 0 \\
 & C_{11} & C_{13} & 0 & 0 & 0 \\
 & & C_{33} & 0 & 0 & 0 \\
 & \text{sym} & & C_{44} & 0 & 0 \\
 & & & & C_{44} & 0 \\
 & & & & & \frac{1}{2}(C_{11} - C_{12})
\end{bmatrix}
\begin{Bmatrix} \varepsilon_{xx} \\ \varepsilon_{yy} \\ \varepsilon_{zz} \\ \gamma_{yz} \\ \gamma_{zx} \\ \gamma_{xy} \end{Bmatrix}
\tag{4-97}
$$

We note that for transversely isotropic materials there are only five independent elastic constants.

## 4-10 ISOTROPIC MATERIALS—A MATHEMATICAL APPROACH

In Section 4-2 we obtained relationships between stress and strain for an isotropic material based on physical observations concerning the behavior of an element of material. In this section we show that these same relationships can be obtained using a mathematical approach in which the independence of directional orientation of the elastic properties is used to reduce the number of independent elastic constants in Eq. (4-27).

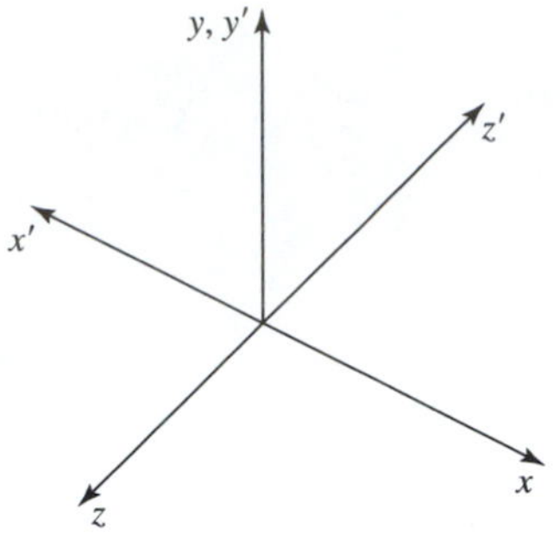


Figure 4-18

To begin we show that for an isotropic medium the principal stress axes are coincident with the principal strain axes.* Let us assume that the $x$, $y$, $z$ coordinate axes line up with the principal strain axes. Then we have

$$\gamma_{yz} = \gamma_{zx} = \gamma_{xy} = 0$$

and from Eq. (4-27),

$$\sigma_{yz} = C_{14}\varepsilon_{xx} + C_{24}\varepsilon_{yy} + C_{34}\varepsilon_{zz} \tag{4-98}$$

We now consider an $x'$, $y'$, $z'$ coordinate system constructed by rotating 180° about the $y$ axis, as shown in Fig. 4-18. This system also lines up with the principal axes. Thus we can write

$$\sigma_{y'z'} = C_{14}\varepsilon_{x'x'} + C_{24}\varepsilon_{y'y'} + C_{34}\varepsilon_{z'z'} \tag{4-99}$$

The transformation is given as

$$x' = -x, \qquad y' = y, \qquad z' = -z$$

and the direction cosines for the transformation are shown in Table 4-5.

From Eqs. (2-37) we find that

$$\sigma_{y'z'} = -\sigma_{yz}$$

and from Eqs. (3-31),

$$\varepsilon_{x'x'} = \varepsilon_{xx}, \qquad \varepsilon_{y'y'} = \varepsilon_{yy}, \qquad \varepsilon_{z'z'} = \varepsilon_{zz}$$

so that Eq. (4-99) can be written as

$$-\sigma_{yz} = C_{14}\varepsilon_{xx} + C_{24}\varepsilon_{yy} + C_{34}\varepsilon_{zz}$$

By comparing this equation with Eq. (4-98) we conclude that

$$\sigma_{yz} = -\sigma_{yz}$$

or

$$\sigma_{yz} = 0$$

*Note that in Section 4-2 the approach taken to show this required that the stress-strain law be known *in advance*.

**TABLE 4-5**

| | x | y | z |
|---|---|---|---|
| x′ | $n_{x'x} = -1$ | $n_{x'y} = 0$ | $n_{x'z} = 0$ |
| y′ | $n_{y'x} = 0$ | $n_{y'y} = 1$ | $n_{y'z} = 0$ |
| z′ | $n_{z'x} = 0$ | $n_{z'y} = 0$ | $n_{z'z} = -1$ |

Similarly, we can show that

$$\sigma_{zx} = 0, \qquad \sigma_{xy} = 0$$

Thus we conclude that for an isotropic medium the principal stress axes and the principal strain axes are coincident, and we will refer to them simply as the principal axes.

In order to derive the stress-strain relationships for an isotropic material we begin by assuming a coordinate system aligned along the principal axes so that

$$\gamma_{yz} = \gamma_{zx} = \gamma_{xy} = 0$$

$$\sigma_{yz} = \sigma_{zx} = \sigma_{xy} = 0$$

Equation (4-27) then becomes

$$\begin{Bmatrix} \sigma_1 \\ \sigma_2 \\ \sigma_3 \\ 0 \\ 0 \\ 0 \end{Bmatrix} = \begin{bmatrix} C_{11} & C_{12} & C_{13} & C_{14} & C_{15} & C_{16} \\ C_{12} & C_{22} & C_{23} & C_{24} & C_{25} & C_{26} \\ C_{13} & C_{23} & C_{33} & C_{34} & C_{35} & C_{36} \\ C_{14} & C_{24} & C_{34} & C_{44} & C_{45} & C_{46} \\ C_{15} & C_{25} & C_{35} & C_{45} & C_{55} & C_{56} \\ C_{16} & C_{26} & C_{36} & C_{46} & C_{56} & C_{66} \end{bmatrix} \begin{Bmatrix} \varepsilon_1 \\ \varepsilon_2 \\ \varepsilon_3 \\ 0 \\ 0 \\ 0 \end{Bmatrix} \tag{4-100}$$

The last three equations of this set are

$$\begin{bmatrix} C_{14} & C_{24} & C_{34} \\ C_{15} & C_{25} & C_{35} \\ C_{16} & C_{26} & C_{36} \end{bmatrix} \begin{Bmatrix} \varepsilon_1 \\ \varepsilon_2 \\ \varepsilon_3 \end{Bmatrix} = \begin{Bmatrix} 0 \\ 0 \\ 0 \end{Bmatrix}$$

and these must be satisfied for arbitrary values of the strains $\varepsilon_1$, $\varepsilon_2$, and $\varepsilon_3$. Therefore we must require that

$$C_{14} = C_{24} = C_{34} = C_{15} = C_{25} = C_{35} = C_{16} = C_{26} = C_{36} = 0$$

The first three equations of Eq. (4-100) are

$$\begin{Bmatrix} \sigma_1 \\ \sigma_2 \\ \sigma_3 \end{Bmatrix} = \begin{bmatrix} C_{11} & C_{12} & C_{13} \\ C_{12} & C_{22} & C_{23} \\ C_{13} & C_{23} & C_{33} \end{bmatrix} \begin{Bmatrix} \varepsilon_1 \\ \varepsilon_2 \\ \varepsilon_3 \end{Bmatrix} \tag{4-101}$$

Now apply the following strain field:

$$\varepsilon_1 = \varepsilon, \qquad \varepsilon_2 = \varepsilon_3 = 0 \tag{4-102}$$

From Eq. (4-101) this gives

$$\sigma_1^{(1)} = C_{11}\varepsilon, \qquad \sigma_2^{(1)} = C_{12}\varepsilon, \qquad \sigma_3^{(1)} = C_{13}\varepsilon$$

Next we apply the strain field

$$\varepsilon_2 = \varepsilon, \qquad \varepsilon_1 = \varepsilon_3 = 0$$

with the following stresses resulting:

$$\sigma_1^{(2)} = C_{12}\varepsilon, \qquad \sigma_2^{(2)} = C_{22}\varepsilon, \qquad \sigma_3^{(2)} = C_{23}\varepsilon$$

Note that this amounts to simply rotating the strain field given by Eqs. (4-102) through an angle of 90°. Thus if the material is isotropic, we must have $\sigma_1^{(1)} = \sigma_2^{(2)}$, $\sigma_2^{(1)} = \sigma_3^{(2)}$, $\sigma_3^{(1)} = \sigma_1^{(2)}$. It follows therefore that

$$C_{11} = C_{22}, \qquad C_{12} = C_{13} = C_{23} \tag{4-103}$$

Proceeding in a similar manner, from the strain field

$$\varepsilon_3 = \varepsilon, \qquad \varepsilon_1 = \varepsilon_2 = 0$$

we get the following stresses:

$$\sigma_1^{(3)} = C_{13}\varepsilon, \qquad \sigma_2^{(3)} = C_{23}\varepsilon, \qquad \sigma_3^{(3)} = C_{33}\varepsilon$$

If the material is isotropic, we have $\sigma_3^{(3)} = \sigma_2^{(2)}$, $\sigma_1^{(3)} = \sigma_3^{(2)}$, and $\sigma_2^{(3)} = \sigma_1^{(3)}$, and therefore,

$$C_{33} = C_{22}, \qquad C_{12} = C_{13} = C_{23} \tag{4-104}$$

For an isotropic material the elastic constants are independent of directional orientation. Eqs. (4-103) and (4-104) are therefore valid for any orientation, and Eq. (4-27) becomes

$$\begin{Bmatrix} \sigma_{xx} \\ \sigma_{yy} \\ \sigma_{zz} \\ \sigma_{yz} \\ \sigma_{zx} \\ \sigma_{xy} \end{Bmatrix} = \begin{bmatrix} C_{11} & C_{12} & C_{12} & 0 & 0 & 0 \\ C_{12} & C_{11} & C_{12} & 0 & 0 & 0 \\ C_{12} & C_{12} & C_{11} & 0 & 0 & 0 \\ 0 & 0 & 0 & C_{44} & C_{45} & C_{46} \\ 0 & 0 & 0 & C_{45} & C_{55} & C_{56} \\ 0 & 0 & 0 & C_{46} & C_{56} & C_{66} \end{bmatrix} \begin{Bmatrix} \varepsilon_{xx} \\ \varepsilon_{yy} \\ \varepsilon_{zz} \\ \varepsilon_{yz} \\ \varepsilon_{zx} \\ \varepsilon_{xy} \end{Bmatrix} \tag{4-105}$$

From Eq. (4-105) we have for the shear stresses,

$$\begin{Bmatrix} \sigma_{yz} \\ \sigma_{zx} \\ \sigma_{xy} \end{Bmatrix} = \begin{bmatrix} C_{44} & C_{45} & C_{46} \\ C_{45} & C_{55} & C_{56} \\ C_{46} & C_{56} & C_{66} \end{bmatrix} \begin{Bmatrix} \gamma_{yz} \\ \gamma_{zx} \\ \gamma_{xy} \end{Bmatrix}$$

We proceed in the same manner as we did for the normal stresses. First we apply the shear strains

$$\gamma_{yz} = \gamma, \qquad \gamma_{zx} = \gamma_{xy} = 0$$

then the shear strains

$$\gamma_{yz} = 0, \qquad \gamma_{zx} = \gamma, \qquad \gamma_{xy} = 0$$

and finally

$$\gamma_{yz} = 0, \qquad \gamma_{zx} = 0, \qquad \gamma_{xy} = \gamma$$

If the material is isotropic, the corresponding stresses for each strain field must be equal so that the following relationships among the elastic constants must hold:

$$C_{44} = C_{55} = C_{66}, \qquad C_{45} = C_{46} = C_{56} \tag{4-106}$$

Therefore Eq. (4-105) becomes

$$\begin{Bmatrix} \sigma_{xx} \\ \sigma_{yy} \\ \sigma_{zz} \\ \sigma_{yz} \\ \sigma_{zx} \\ \sigma_{xy} \end{Bmatrix} = \begin{bmatrix} C_{11} & C_{12} & C_{12} & 0 & 0 & 0 \\ C_{12} & C_{11} & C_{12} & 0 & 0 & 0 \\ C_{12} & C_{12} & C_{11} & 0 & 0 & 0 \\ 0 & 0 & 0 & C_{44} & C_{45} & C_{45} \\ 0 & 0 & 0 & C_{45} & C_{44} & C_{45} \\ 0 & 0 & 0 & C_{45} & C_{45} & C_{44} \end{bmatrix} \begin{Bmatrix} \varepsilon_{xx} \\ \varepsilon_{yy} \\ \varepsilon_{zz} \\ \gamma_{yz} \\ \gamma_{zx} \\ \gamma_{xy} \end{Bmatrix} \tag{4-107}$$

At this point all but four of the elastic constants appearing in Eq. (4-27) have been eliminated. We will now proceed to show that one of the constants appearing in Eq. (4-107) is zero while the other is related to $C_{11}$ and $C_{12}$.

From Eq. (4-107) we can write

$$\sigma_{yz} = C_{44}\gamma_{yz} + C_{45}\gamma_{zx} + C_{45}\gamma_{xy} \tag{4-108}$$

The stress $\sigma_{yz}$ can also be related to the principal stresses through the transformation laws given in Eqs. (2-37). To do so, first construct an equation for $\sigma_{y'z'}$ as explained below Eqs. (2-37):

$$\begin{aligned}\sigma_{y'z'} &= \sigma_{xx} n_{y'x} n_{xz'} + \sigma_{yy} n_{y'y} n_{yz'} + \sigma_{zz} n_{y'z} n_{zz'} + \sigma_{xy}(n_{y'y} n_{xz'} + n_{y'x} n_{yz'}) \\ &\quad + \sigma_{xz}(n_{y'z} n_{xz'} + n_{zz'} n_{y'x}) + \sigma_{yz}(n_{y'z} n_{yz'} + n_{y'y} n_{zz'})\end{aligned}$$

For the principal stress state this becomes,

$$\sigma_{y'z'} = \sigma_1 n_{y'x} n_{xz'} + \sigma_2 n_{y'y} n_{yz'} + \sigma_3 n_{y'z} n_{zz'}$$

Finally, we interchange the primed and unprimed subscripts to get:

$$\sigma_{yz} = \sigma_1 n_{yx'} n_{x'z} + \sigma_2 n_{yy'} n_{y'z} + \sigma_3 n_{yz'} n_{z'z} \tag{4-109}$$

The principal stresses are related to the principal strains through Eq. (4-101). With these and Eqs. (4-103) and (4-104), Eq. (4-109) becomes

$$\begin{aligned}\sigma_{yz} = {} & n_{yx'} n_{x'z}(C_{11}\varepsilon_1 + C_{12}\varepsilon_2 + C_{12}\varepsilon_3) + n_{yy'} n_{y'z}(C_{12}\varepsilon_1 + C_{11}\varepsilon_2 + C_{12}\varepsilon_3) \\ & + n_{yz'} n_{z'z}(C_{12}\varepsilon_1 + C_{12}\varepsilon_2 + C_{11}\varepsilon_3)\end{aligned}$$

After collecting terms we get

$$\begin{aligned}\sigma_{yz} &= (n_{yx'} n_{x'z} + n_{yy'} n_{y'z} + n_{yz'} n_{z'z}) C_{12}\Delta \\ &\quad + (C_{11} - C_{12})(n_{yx'} n_{x'z}\varepsilon_1 + n_{yy'} n_{y'z}\varepsilon_2 + n_{yz'} n_{z'z}\varepsilon_3)\end{aligned}$$

where $\Delta$ is the dilatation.

Noting, from Eq. (B-18) that the first term to the right of the equal sign is zero, we get

$$\sigma_{yz} = (C_{11} - C_{12})(n_{yx'}n_{x'z}\varepsilon_1 + n_{yy'}n_{y'z}\varepsilon_2 + n_{yz'}n_{z'z}\varepsilon_3) \tag{4-110}$$

According to the strain transformation laws given by Eqs. (3-31), for principal strains, after interchanging primed and unprimed subscripts as was done for obtaining $\sigma_{yz}$, we can write

$$\gamma_{yz} = 2(n_{yx'}n_{x'z}\varepsilon_1 + n_{yy'}n_{y'z}\varepsilon_2 + n_{yz'}n_{z'z}\varepsilon_3)$$

Consequently Eq. (4-110) becomes

$$\sigma_{yz} = \tfrac{1}{2}(C_{11} - C_{12})\gamma_{yz}$$

By comparing this with Eq. (4-108) we conclude that

$$C_{44} = \frac{C_{11} - C_{12}}{2}, \qquad C_{45} = 0$$

No further reduction in the number of elastic constants is possible. Thus for an isotropic material we get

$$\begin{Bmatrix} \sigma_{xx} \\ \sigma_{yy} \\ \sigma_{zz} \\ \sigma_{yz} \\ \sigma_{zx} \\ \sigma_{xy} \end{Bmatrix} = \begin{bmatrix} C_{11} & C_{12} & C_{12} & 0 & 0 & 0 \\ C_{12} & C_{11} & C_{12} & 0 & 0 & 0 \\ C_{12} & C_{12} & C_{11} & 0 & 0 & 0 \\ 0 & 0 & 0 & \dfrac{C_{11} - C_{12}}{2} & 0 & 0 \\ 0 & 0 & 0 & 0 & \dfrac{C_{11} - C_{12}}{2} & 0 \\ 0 & 0 & 0 & 0 & 0 & \dfrac{C_{11} - C_{12}}{2} \end{bmatrix} \begin{Bmatrix} \varepsilon_{xx} \\ \varepsilon_{yy} \\ \varepsilon_{zz} \\ \gamma_{yz} \\ \gamma_{zx} \\ \gamma_{xy} \end{Bmatrix} \tag{4-111}$$

By comparing these equations with Eqs. (4-12) we see that

$$C_{11} = \frac{E(1-\nu)}{(1+\nu)(1-2\nu)}, \qquad C_{12} = \frac{E\nu}{(1+\nu)(1-2\nu)} \tag{4-112}$$

Equation (4-111) can also be written in terms of the *Lamé constants* $\lambda$ and $\mu$. These are the elastic constants that are often used in the theory of elasticity (see, for example, [11, pp. 13–14]).

To introduce the Lamé constants we first add and subtract $C_{12}\varepsilon_{xx}$ from the first equation of Eq. (4-111), $C_{12}\varepsilon_{yy}$ from the second, and $C_{12}\varepsilon_{zz}$ from the third. This gives

$$\begin{aligned} \sigma_{xx} &= (C_{11} - C_{12})\varepsilon_{xx} + C_{12}\Delta \\ \sigma_{yy} &= (C_{11} - C_{12})\varepsilon_{yy} + C_{12}\Delta \\ \sigma_{zz} &= (C_{11} - C_{12})\varepsilon_{zz} + C_{12}\Delta \end{aligned} \tag{4-113}$$

where $\Delta$ is the dilatation given by Eq. (3-61). Now let

$$\frac{C_{11} - C_{12}}{2} = \mu, \qquad C_{12} = \lambda \tag{4-114}$$

Then Eqs. (4-113) become

$$\sigma_{xx} = \lambda\Delta + 2\mu\varepsilon_{xx}, \qquad \sigma_{yy} = \lambda\Delta + 2\mu\varepsilon_{yy}, \qquad \sigma_{zz} = \lambda\Delta + 2\mu\varepsilon_{zz} \tag{4-115}$$

and the last three equations of system (4-111) become

$$\sigma_{yz} = \mu\gamma_{yz}, \qquad \sigma_{zx} = \mu\gamma_{zx}, \qquad \sigma_{xy} = \mu\gamma_{xy} \tag{4-116}$$

By comparing Eqs. (4-115) with Eqs. (4-12) we see that the Lamé constants can be expressed in terms of the physical constants as

$$\lambda = \frac{E\nu}{(1+\nu)(1-2\nu)}, \qquad \mu = \frac{E}{2(1+\nu)} \tag{4-117}$$

and from these it follows that

$$E = \frac{\mu(3\lambda + 2\mu)}{\lambda + \mu}, \qquad \nu = \frac{\lambda}{2(\lambda + \mu)} \tag{4-118}$$

From Eq. (4-13) we note that $\mu = G$.

Finally, with Eqs. (4-118) the expression for the bulk modulus given in Eq. (4-15) becomes

$$K = \lambda + \tfrac{2}{3}\mu \tag{4-119}$$

## 4-11 STRESS-STRAIN RELATIONS FOR VISCOELASTIC MATERIALS

When metals are subjected to high stresses at elevated temperatures, they exhibit a phenomenon known as *creep* where, over time, the material continues to deform even though the stress remains constant, as shown in Fig. 4-19. Creep is a common consideration in the design of gas turbine engines, where metal temperatures can exceed 2000°F. Without designing for creep, rotating parts could eventually interfere with nonrotating parts, causing a great deal of damage.

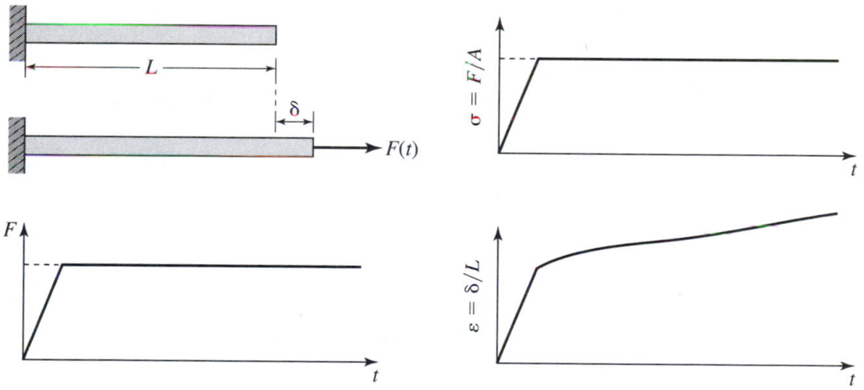


Figure 4-19 Creep

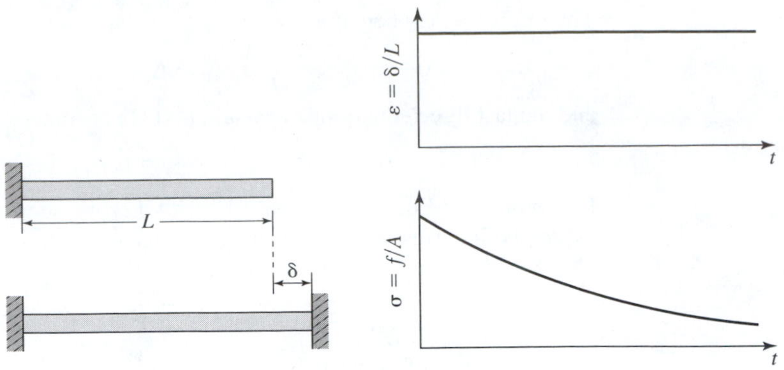


Figure 4-20 Stress relaxation

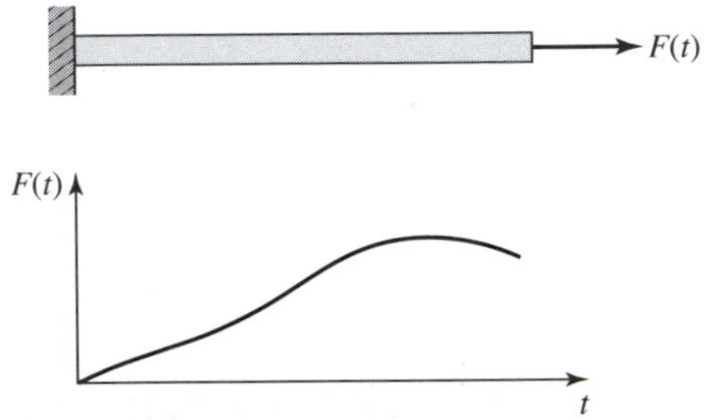


Figure 4-21

Some materials, including concrete and certain polymers, exhibit creep even at room temperature. Thus an understanding of creep is important in modern engineering applications.

Creep is a phenomenon that occurs at constant stress. On the other hand if, under the same conditions, the strain is held constant, then, over time, the stress will decrease, as shown in Fig. 4-20. This phenomenon is known as *stress relaxation*.

For a material undergoing creep, the current strain depends on the entire stress history. In order to see this history dependence let us consider a rod fixed at one end and subjected to a force $F(t)$ at the other end, as shown in Fig. 4-21. We wish to compute the displacement at the free end of the rod at time $t$, $u(t)$. Let us assume initially that a constant force $F_0$ is applied at $t = 0$, as shown in Fig. 4-22*a*. Now if the material is elastic, the displacement will change when the load is applied and remain constant thereafter, as shown in Fig. 4-22*b*, and we can write a relationship between force and displacement as

$$u(t) = \begin{cases} 0, & t < 0 \\ \hat{c}F_0, & t > 0 \end{cases}$$

where $\hat{c}$ is a proportionality constant.

If the load is applied at some time $\tau$, as shown in Fig. 4-23*a*, we can write

$$u(t) = \begin{cases} 0, & t < \tau \\ \hat{c}F_0, & t > \tau \end{cases}$$

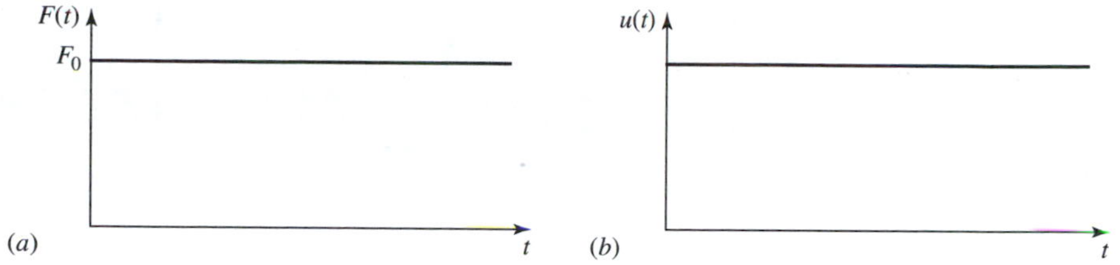


Figure 4-22

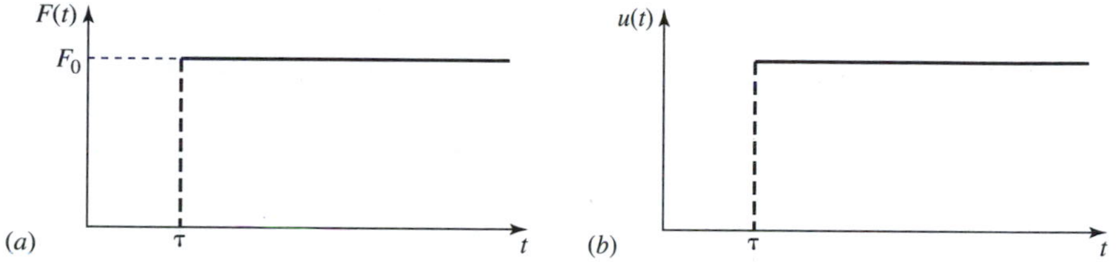


Figure 4-23

Note that this merely shifts the displacement curve shown in Fig. 4-22*b* to the right by $\tau$ units, as shown in Fig. 4-23*b*.

However, since the material undergoes creep, the displacement changes with time from the purely elastic displacement even though the load remains constant, as shown in Fig. 4-24*b*. Thus since we have assumed a linear relationship between force and deflection, we can write

$$u(t) = \begin{cases} 0, & t < 0 \\ \hat{c}(t)F_0, & t > 0 \end{cases}$$

where $\hat{c}$ now depends on time.

If the load is applied at some time $\tau$, as shown in Fig. 4-25*a*, then we have

$$u(t) = \begin{cases} 0, & t < \tau \\ \hat{c}(t-\tau)F_0, & t > \tau \end{cases} \tag{4-120}$$

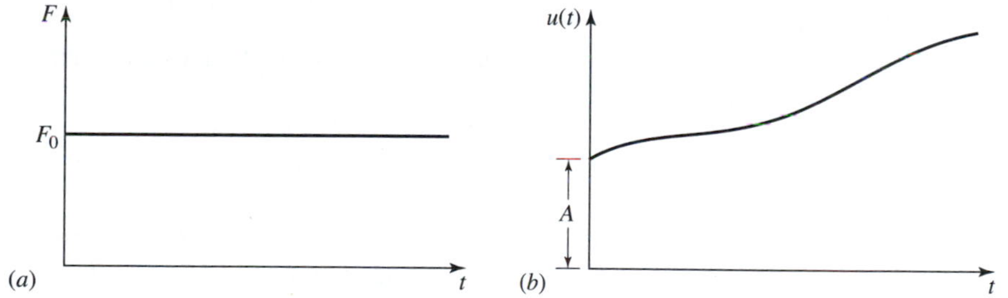


Figure 4-24

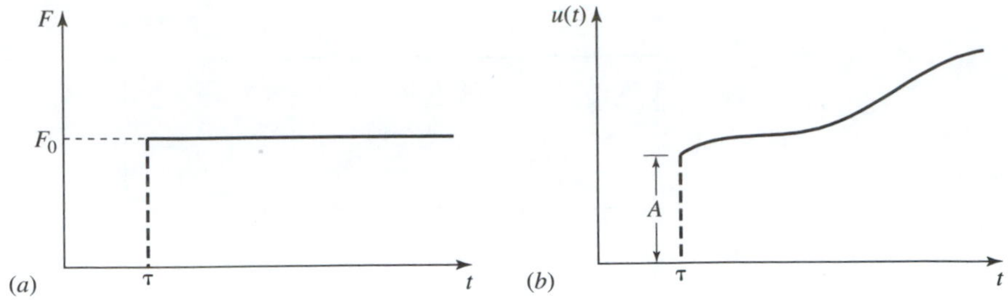


Figure 4-25

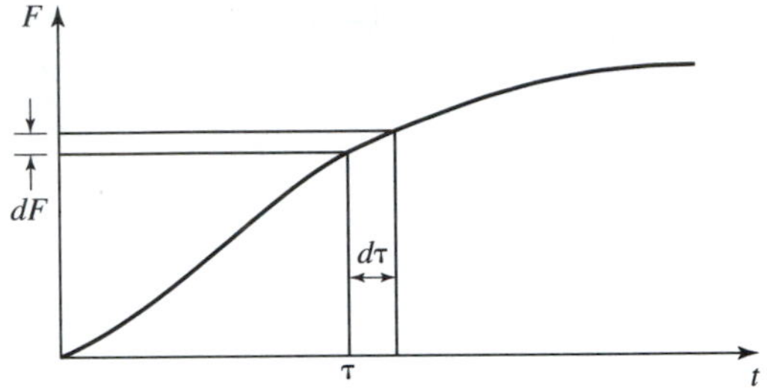


Figure 4-26

Again we see that the effect is to shift the displacement curve shown in Fig. 4-24*a* to the right by $\tau$ units, as shown in Fig. 4-25*b*. Thus $\hat{c}$ depends on the *elapsed* time $t - \tau$, that is, how long the load has been applied.

Now let us assume that the load changes with time, as shown in Fig. 4-26. In a small increment of time $d\tau$ occurring at time $\tau$, the load increment is

$$dF = \left.\frac{\partial F}{\partial t}\right|_{t=\tau} d\tau = \frac{\partial F}{\partial \tau}\, d\tau$$

and this increment produces a displacement increment of $du(t)$ at time $t$. This load increment remains constant as time increases (even though the load itself is changing), and therefore the elapsed time between the application of this load increment and the determination of the displacement increment is $t - \tau$. Thus by Eq. (4-120) we can write

$$du(t) = \begin{cases} 0, & \tau > t \\ \hat{c}(t-\tau)\dfrac{\partial F}{\partial \tau}\, d\tau, & \tau < t \end{cases}$$

The actual displacement at time $t$ is simply the superposition of all of the displacement increments from $\tau = 0$,

$$u(t) = \int_0^t \hat{c}(t-\tau)\frac{\partial F}{\partial \tau}\, d\tau = \int_0^t \hat{c}(t-\tau)F'(\tau)\, d\tau \tag{4-121}$$

Thus if the coefficient $\hat{c}$, called the *creep function,* can be determined, the displacement can be obtained. It is worth mentioning that the upper limit of integration in Eq. (4-121) is $t$ because $du(t) = 0$ for $\tau > t$, which simply states the obvious fact that the displacement at time $t$ is unaffected by loads applied at later times.

By following a similar argument we can obtain the force in terms of the displacement and the *relaxation function* $\hat{r}$ as

$$F(t) = \int_0^t \hat{r}(t - \tau)u'(\tau)\, d\tau \tag{4-122}$$

Note from Eq. (4-120) that the creep function is the *displacement resulting at time* t *from a unit load applied at time* $\tau$ *and then held constant*. Conversely, the relaxation function is the *force resulting at time* t *from a unit displacement applied at time* $\tau$ *and then held constant*.

If the load in Eq. (4-121) is replaced by $bF(t)$, where $b$ is a constant, then the displacement is changed by the same factor $b$. Thus Eq. (4-121) provides a linear relationship between force and deflection. A similar argument shows that Eq. (4-122) is also a linear relationship.

By replacing $u$ with $u/L$ and $F$ with $F/A$ in Eq. (4-121) and incorporating the constant $A/L$ in $\hat{c}$, we can write

$$\varepsilon(t) = \int_0^t c(t - \tau)\sigma(\tau)\, d\tau \tag{4-123}$$

An analogous expression holds for Eq. (4-122).

Several physical models, consisting of linear springs and dashpots, have been proposed to represent the creep phenomenon. Three of these are shown in Fig. 4-27.

A linear spring produces an instantaneous displacement that is proportional to the load, whereas a dashpot produces an instantaneous velocity that is proportional to the load. Thus for the spring we can write

$$F_s = Gu_s \tag{4-124}$$

and for the dashpot,

$$F_d = \eta\dot{u}_d \tag{4-125}$$

where $G$ is the spring constant, $\eta$ is the coefficient of viscosity, and the superposed dot indicates differentiation with respect to time.

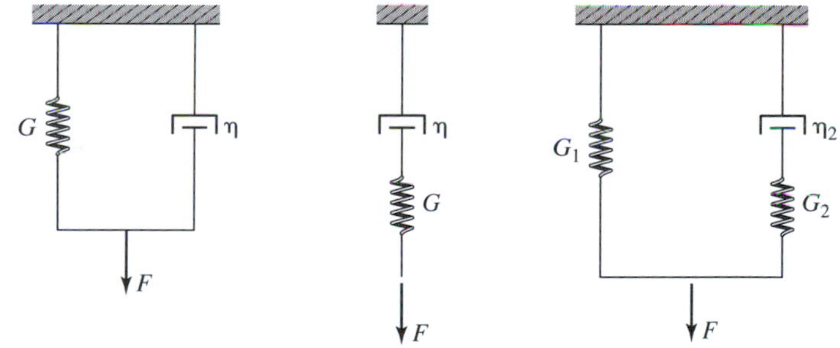


(*a*) Kelvin-Voigt model (*b*) Maxwell model (*c*) Standard linear model

**Figure 4-27** Examples of physical models proposed to represent creep phenomenon

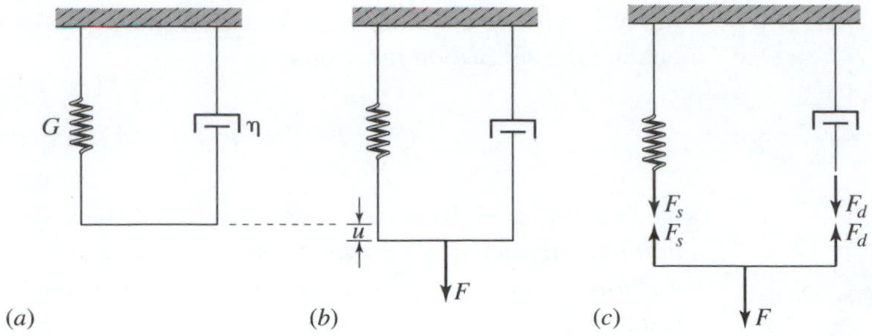


Figure 4-28 Kelvin–Voigt model

We will now develop a relationship between force and deflection for each of these models, and from these determine the creep functions.

For the Kelvin–Voigt model we have the free-body diagram shown in Fig. 4-28*c*, from which we can write

$$F = F_s + F_d$$

By using Eqs. (4-124) and (4-125) and noting from Fig. 4-28*b* that the spring and dashpot deflections $u_s$ and $u_d$, respectively, are equal, this yields

$$F = Gu + \eta\dot{u} \tag{4-126}$$

This is a first-order differential equation defining $u$ in terms of $F$. Its solution requires one initial condition. Since creep is a time-dependent phenomenon, following the initial elastic deflection there will be an additional deflection that depends on time. Thus for our models the initial condition is the elastic deflection, determined from the models by assuming that the dashpot is "locked." Then from Fig. 4-27*a* the initial condition associated with Eq. (4-126) is

$$u(0) = 0 \tag{4-127}$$

For the Maxwell model we obtain the following relationship from Fig. 4-29*b*:

$$u = u_s + u_d$$

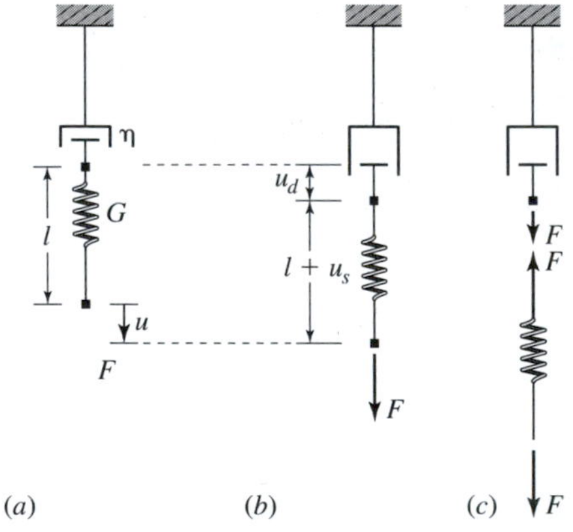


Figure 4-29 Maxwell model

or

$$\dot{u} = \dot{u}_s + \dot{u}_d$$

With Eqs. (4-124) and (4-125) and the free-body diagram in Fig. 4-29*c* this gives

$$\dot{u} = \frac{\dot{F}}{G} + \frac{F}{\eta} \tag{4-128}$$

From Fig. 4-29*a* we see that for this model,

$$u(0) = \frac{F(0)}{G} \tag{4-129}$$

By using a similar approach we have for the standard linear model,

$$\dot{F} + \frac{G_2}{\eta_2}F = \frac{G_1 G_2}{\eta_2}u + (G_1 + G_2)\dot{u} \tag{4-130}$$

and

$$u(0) = \frac{F(0)}{G_1 + G_2} \tag{4-131}$$

In terms of stress and strain, Eqs. (4-126), (4-128), and (4-130) become

$$\sigma = G\varepsilon + \eta\dot{\varepsilon} \tag{4-132}$$

$$\dot{\varepsilon} = \frac{\dot{\sigma}}{G} + \frac{\sigma}{\eta} \tag{4-133}$$

$$\dot{\sigma} + \frac{G_2}{\eta_2}\sigma = \frac{G_1 G_2}{\eta_2}\varepsilon + (G_1 + G_2)\dot{\varepsilon} \tag{4-134}$$

Equations (4-126), (4-128), and (4-130) provide three stress-strain relations for creep, whereas Eq. (4-121) gives another stress-strain relation provided the creep function is known. It should be noted that the standard linear model shown in Fig. 4-27*c* is a combination of the Kelvin–Voigt and Maxwell models—if $G_2$ is infinite, the standard linear model reduces to the Kelvin–Voigt model, whereas if $G_1$ is zero, it reduces to the Maxwell model.

If we recall that the creep function is the displacement at time $t$ resulting from a constant unit force applied at time $\tau$, then we can find the creep function for the models shown in Fig. 4-27 by solving Eqs. (4-126)–(4-131). In order to do this we consider the load shown in Fig. 4-30 and note that $F(t) = 1$. Thus Eqs. (4-126) and (4-127) become

$$\dot{u} + \frac{G}{\eta}u = \frac{1}{\eta}, \qquad u(0) = 0$$

The complete solution to this differential equation is the sum of the complementary solution and the particular solution, and is readily found to be

$$u = Ae^{-Gt/\eta} + \frac{1}{G}$$

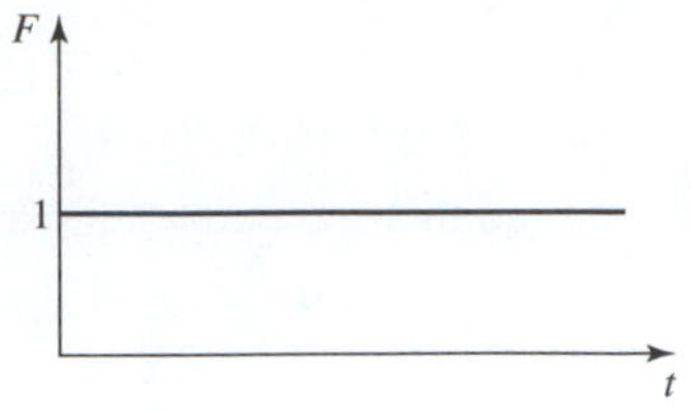


Figure 4-30

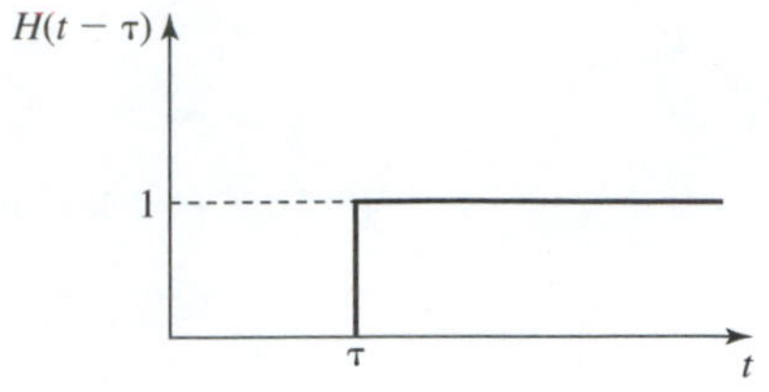


Figure 4-31 Heaviside's function

The constant $A$ is determined from the initial condition to be

$$A = -\frac{1}{G}$$

so that we have

$$u(t) = \frac{1}{G}(1 - e^{-Gt/\eta}) \tag{4-135}$$

Thus when the load $F_0 = 1$, shown in Fig. 4-24*a*, is applied to the Kelvin–Voigt model, the displacement shown in Fig. 4-24*b* is given by Eq. (4-135).

We now introduce *Heaviside's function* (also known as *unit step function*), as shown in Fig. 4-31. Mathematically, Heaviside's function is defined as

$$H(t - \tau) = \begin{cases} 0, & t < \tau \\ 1, & t > \tau \end{cases} \tag{4-136}$$

Note that Heaviside's function has a value of zero when its argument is negative and a value of unity when it is positive. It is undefined when its argument is zero. With this definition we find that

$$H(t) = \begin{cases} 0, & t < 0 \\ 1, & t > 0 \end{cases}$$

Therefore Eq. (4-135) can be written as

$$u(t) = \frac{1}{G}(1 - e^{-Gt/\eta})H(t) \tag{4-137}$$

Now if the load is applied at time $\tau > 0$, as shown in Fig. 4-25*a*, then there will be no displacement until time $\tau$, as shown in Fig. 4-25*b*. This solution can be obtained from Eq. (4-137) simply by replacing $t$ with $t - \tau$,

$$u(t) = \frac{1}{G}\left[1 - e^{-G(t-\tau)/\eta}\right]H(t - \tau)$$

Note, on comparing this with Eq. (4-135), that the displacement curves are the same except for a shift of $\tau$ units to the right. This gives the displacement at time $t$ resulting from a unit constant load applied at time $\tau$ and is therefore the desired creep function. That is, for the Kelvin–Voigt model,

$$c(t - \tau) = \frac{1}{G}\left[1 - e^{-G(t-\tau)/\eta}\right]H(t - \tau) \tag{4-138}$$

For the Maxwell model, Eq. (4-128) is easily integrated to give

$$u(t) = \frac{F(t)}{G} + \frac{1}{\eta}\int F(t)\,dt + A$$

where $A$ is a constant of integration. Thus for the load shown in Fig. 4-28 and the initial condition given by Eq. (4-129), we get

$$u(t) = \left(\frac{1}{G} + \frac{t}{\eta}\right) H(t)$$

and therefore the creep function for the Maxwell model is

$$\hat{c}(t) = \left(\frac{1}{G} + \frac{t}{\eta}\right) H(t) \tag{4-139}$$

For the standard linear model the creep function is given by

$$\hat{c}(t) = \frac{1}{G_1}\left[1 - \frac{G_2}{G_1 + G_2}\exp\left(-\frac{G_1}{\eta_2}\frac{G_2}{G_1 + G_2}t\right)\right] H(t) \tag{4-140}$$

Figure 4-32 shows the response of each of the models to a constant load applied at time $t_1$ and removed at time $t_2$. For the Kelvin–Voigt model shown in Fig. 4-32*a* there is no initial elastic deflection because the dashpot and the spring are in parallel. Rather, the displacement increases gradually until the load is removed. This increase occurs at a decreasing rate because the spring carries more of the load as the displacement increases. Note that if the load is applied for a long enough time, the displacement will cease to increase because the spring force will exactly balance the load. On removal of the load the displacement will decrease to zero because of the tension in the spring. This decrease will be gradual because the force in the spring is balanced by the dashpot.

For the Maxwell model there is an initial elastic displacement followed by a linear increase of the displacement with time, as shown in Fig. 4-32*b*. This linear increase

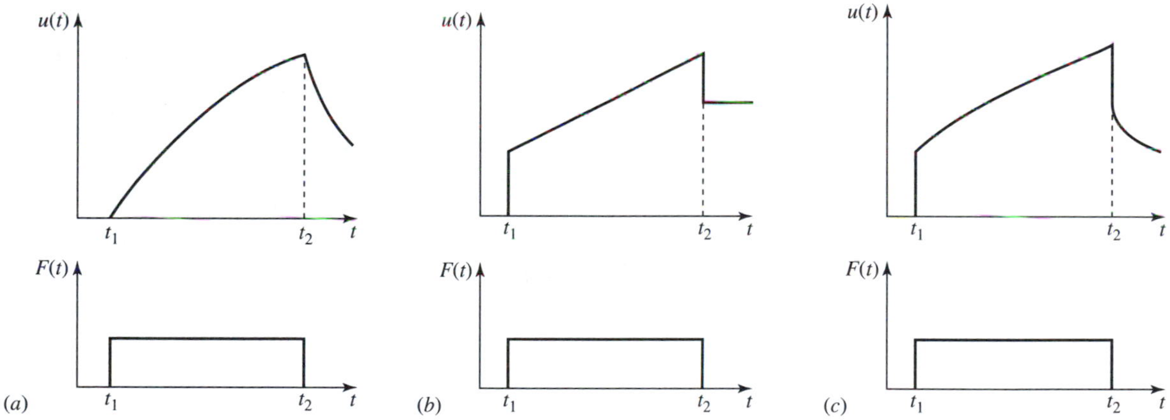


**Figure 4-32** Responses of three models to constant load applied at time $t_1$ and removed at time $t_2$

continues until the load is removed, at which point there is a sudden decrease in displacement as the tension in the spring is removed. There is no further change in displacement because, in contrast to the Kelvin–Voigt model, there is no force acting on the dashpot once the load is removed.

The standard linear model combines the attributes of the Kelvin–Voigt and Maxwell models. On application of the load there is a sudden increase in displacement followed by a gradual increase because of the dashpot. This is followed by a sudden decrease when the load is removed as the tension in spring $G_2$ is totally removed and the tension in spring $G_1$ is partially removed. A more gradual decrease in displacement follows as the remaining tension in spring $G_1$, which is balanced by the dashpot, is removed. This behavior is shown in Fig. 4-32*c*.

Because of the viscous effects of the dashpot in these models, materials exhibiting this type of behavior are referred to as *viscoelastic materials*.

The stress-strain relations presented here for one-dimensional viscoelastic solids can be generalized to two- and three-dimensional solids. The books by Fung [5, chap. 15] and Shames and Cozzarelli [19, chap. 6, pt. B] are recommended to those readers interested in such generalizations.

## 4-12 MATERIAL BEHAVIOR BEYOND THE ELASTIC LIMIT

In this section we will develop relationships between stress and strain for materials that are loaded beyond their elastic limits. As we noted at the beginning of this chapter, in order to arrive at a manageable set of equations we often idealize the behavior of materials by introducing certain simplifying assumptions. Although materials that are anisotropic may undergo plastic deformation, in keeping with the introductory nature of this section, we will assume that the materials we are considering are both isotropic and homogeneous.

As we noted earlier in this chapter, when a thin metal wire is stressed beyond the yield point, plastic deformation occurs—on unloading to a point of zero stress, a permanent deformation exists. Let us now examine this behavior in more detail, with the intention of developing a mathematical description of the material behavior beyond the elastic limit.

Figure 4-33 shows a typical stress-strain curve. Let us assume that our thin metal wire has been loaded beyond the yield point, say, to point 2. At this point the load is reduced and unloading follows curve 2–3, which is nearly parallel to the elastic loading curve, 0–1. On reaching point 3, permanent deformation has occurred, although the stress is zero. The permanent strain that exists in the wire is the plastic strain $\varepsilon^p$, which is shown in the figure. On reloading, the material behavior follows curve 3–2 and intersects the original stress-strain curve at point 2.

The narrow loop formed by curve 2–3–2 is referred to as a *hysteresis loop*. As a practical matter, the hysteresis loop is extremely narrow, and therefore it is usually ignored.

If, at any point along curve 3–2, the loading is reversed, the unloading will proceed along the same curve until, when the load is totally removed, point 3 is reached. Thus curve 3–2 is another elastic curve that is parallel to the original elastic curve 0–1. Along

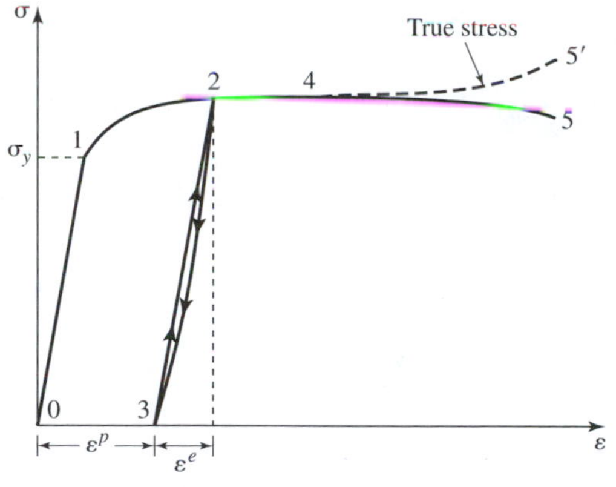


**Figure 4-33** Typical stress–strain curve for materials that are loaded beyond their elastic limits

this new elastic curve, yielding does not begin until the original stress-strain curve is intersected at point 2. Thus on unloading and subsequent reloading a new yield point is established whose value is greater than, or at least equal to, the original yield stress. This effect is known as *work hardening*.

As seen in Fig. 4-33, for loading in the plastic region, that is, beyond the initial yield point (point 1), the *total strain* can be expressed as the sum of the elastic strain and the plastic strain. The *elastic strain* $\varepsilon^e$ is the decrease in strain that results from unloading to a state of zero stress, while the *plastic strain* is the strain that remains in the specimen after the load is removed.

Loading beyond point 2 continues along the original stress-strain curve. Until point 4, the highest point on the curve, is reached, the process proceeds as described in the previous paragraphs. Beyond this point the specimen "necks down," and in the region of the neck the stress state is very complicated because it is three-dimensional and the material becomes highly anisotropic. Beyond point 4 the engineering stress (load divided by original area) decreases, although the true stress (load divided by the actual neck area) continues to increase. At point 5 the specimen fractures. The stress at point 4 is the *ultimate strength,* or *tensile strength,* of the material. Point 4 marks the upper limit of the useful portion of the stress-strain curve.

In practice, plasticity problems can be classified as those involving large plastic strains and those involving small plastic strains. Metal forming processes usually involve large plastic strains. In these problems, because the plastic behavior of the material is so dominant, the elastic behavior of the material is often ignored. On the other hand, in machine design and structural design the plastic strains are on the same order of magnitude as the elastic strains, so both the elastic and the plastic behavior of the material must be considered.

Up to this point our discussion has been concerned only with one-dimensional problems. For these, the stress-strain curve derived from a standard tensile test has served our needs because the onset of yielding as well as the material behavior beyond yield can be readily determined from such a curve. However, most engineering problems of interest involve stress states that are multidimensional, so the questions of when yielding begins and what the stress-strain relations for the material are after the onset of

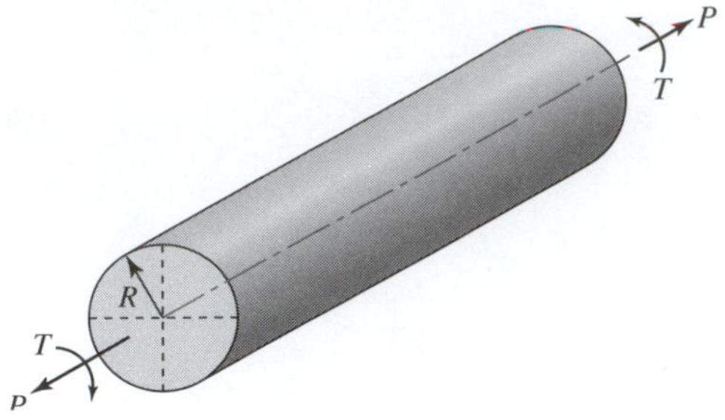


Figure 4-34

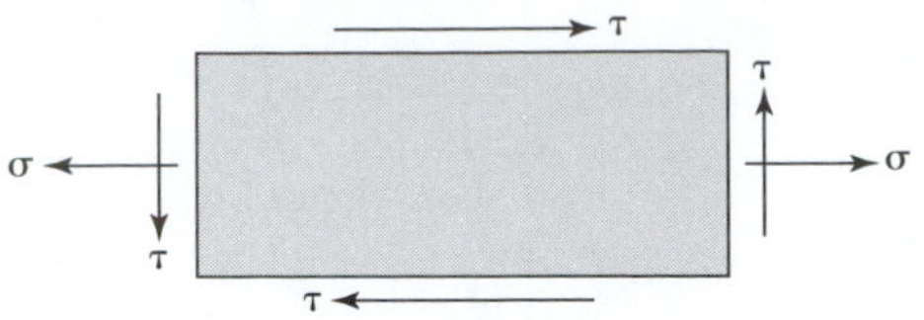


Figure 4-35

yielding naturally arise. *The fundamental tasks of plasticity theory, then, are to establish a yield criterion and to develop stress-strain relations valid in the plastic region.*

When a body is subjected to a multiaxial stress state that results in yielding, it is important to know not only the final stress state, but also the order in which the loads were applied to produce it. This dependence on the loading history can easily be seen in the following example.

Consider a bar with circular cross section of radius $R$ and cross-sectional area $A$ subjected to a tensile load $P$ and a torque $T$, as shown in Fig. 4-34. The tensile load will produce a tensile stress $\sigma = P/A$, acting along the axis of the bar, while the torque will produce a shear stress at the outer surface of $\tau = TR/J$, where $J$ is the polar moment of inertia of the cross section. These are shown in Fig. 4-35. We assume that some combination of $\sigma$ and $\tau$ produces yielding, and that this combination is represented by the line $AB$ in Fig. 4-36. Thus a stress state that lies below line $AB$ is elastic, one that lies above $AB$ is plastic, while one that lies on $AB$ represents the onset of yielding.

With the bar unloaded, apply $P$ until $\sigma$ is just past yield, point 1 in Fig. 4-36*a*. Now unload to some point 2 below yield. Finally, while maintaining this value of $P$, apply $T$ so that the shear stress $\tau$ is just below yield, as indicated by point 3. Note that because yielding occurred when $P$ was applied, there is a plastic strain $\varepsilon_p$ in the bar. However, since yield was not exceeded when $T$ was applied, there is no plastic shear strain. Thus we have the following plastic strains:

$$\varepsilon^p \neq 0, \qquad \gamma^p = 0$$

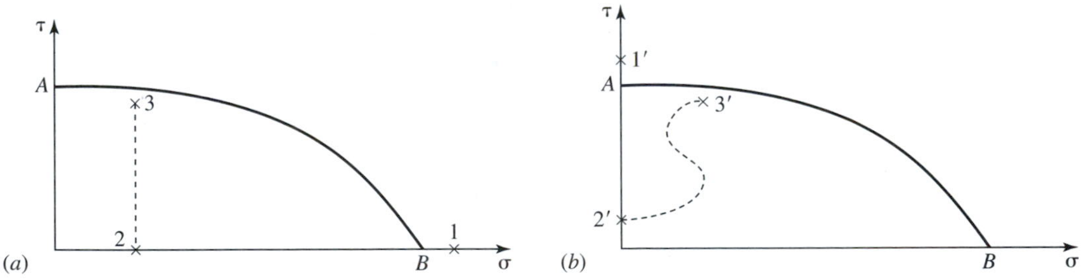


Figure 4-36

Now consider the same unloaded bar and apply the torque $T$ so that the shear stress $\tau$ is just past yield, as shown by point $1'$ in Fig. 4-36*b*, then unload to some point $2'$ that is below yield. Now, without going above line $AB$, use any combination of $P$ and $T$ to reach point $3'$, which is the same as point 3 in Fig. 4-36*a*. The resulting plastic strains are

$$\varepsilon^p = 0, \qquad \gamma^p \neq 0$$

Note that in the two examples the final stress state is the same. Therefore the stresses are the same. By following the process on a stress-strain diagram for normal stresses and one for shear stresses, it is readily seen that the elastic strains are also the same. However, the plastic strains are different. Thus we conclude that *the final strain state cannot be determined solely from the final stress state*. How the final stress state was achieved, that is, the history of the loading, must also enter into our considerations.

It is important to note that, in the preceding examples, time is not a factor. The final state is the same whether the loading is carried out over several minutes or several days.

The history dependence associated with plastic deformation can be introduced by determining the *plastic strain increment* $d\varepsilon^p$ that occurs at each stage of loading and summing over all of the strain increments to determine the actual plastic strain,

$$\varepsilon^p = \varepsilon^p(0) + \int d\varepsilon^p$$

where $\varepsilon^p(0)$ is the initial plastic strain. The strain increment can be written as

$$d\varepsilon^p = \dot{\varepsilon}^p dt$$

where $t$ is a rate parameter and $\dot{\varepsilon}^p$ is the strain rate. Then we have

$$\varepsilon^p = \varepsilon^p(0) + \int_0^t \dot{\varepsilon}^p dt \tag{4-141}$$

The rate parameter is not necessarily time. It is some parameter, such as number of cycles.

Once the plastic strain is determined, the total strain is found by adding the elastic strain given by the generalized Hooke's law, as shown in Fig. 4-33. Therefore once we have a way to specify $\dot{\varepsilon}^p$, our formulation is complete.

Many models of the uniaxial stress-strain curve have been proposed, some of which are shown in Fig. 4-37. The first curve is the most general and is valid for the largest range of strain. The second and third curves are good representations when the plastic strains are on the same order as the elastic strains. The last three curves are appropriate for problems where the plastic strains are much larger than the elastic strains. For these models the unloading curve is shown by the dashed line in Fig. 4-37*d*. The curves for perfectly plastic materials, curves 3 and 6, represent materials, such as mild steel, that have a flat portion immediately following the point of initial yield, as shown in Fig. 4-38.

We will formulate a plasticity theory for *linear elastic–perfectly plastic* materials (also called *linear elastic–ideally plastic* materials), that is, materials for which there is no work hardening. For such a material, the stress-strain curve is given in Fig. 4-37*c*. It is characterized by the fact that the stress can never exceed the initial yield strength of the material even though the strain continues to increase. A theory such as this will be valid for materials, such as mild steel, when the plastic strains are of comparable magnitude to the elastic strains.

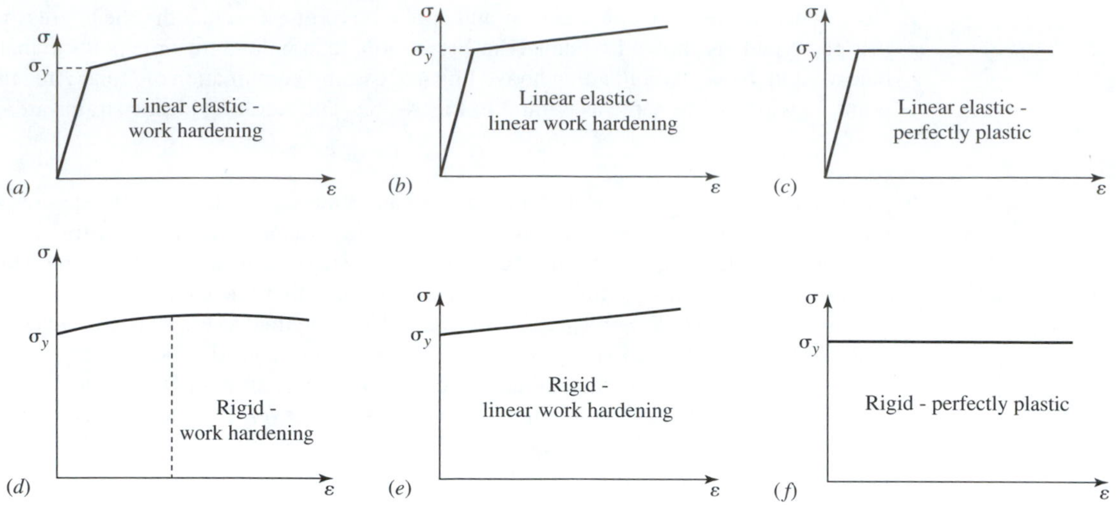


**Figure 4-37** Models of the uniaxial stress–strain curve

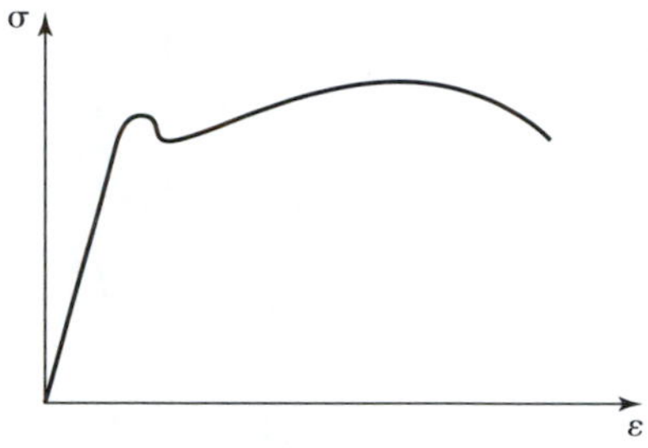


**Figure 4-38** Stress–strain curve for mild steel

## 4-12-1 Additional Experimental Observations

There are two experimental observations regarding the behavior of materials loaded beyond the yield point that will prove very useful in developing a theory of plasticity. The first is the observation that when a body is subjected to hydrostatic pressure, even to values well above the yield strength of the material, the effect on the yield point is negligible. Thus if a body is subjected to a stress state that causes permanent deformation and a hydrostatic stress state is superimposed on the body, *the permanent deformation remains unchanged.*

The second experimental observation is that there is no volume change associated with plastic deformation. That is, if a body is loaded to the yield point and its volume determined, as the loading is increased beyond the yield point there is no further change in volume. Thus in the plastic region the body behaves essentially as an incompressible material. If we view the total deformation of a body as the sum of a change in volume and a change in shape (i.e., a *distortion*), we conclude that *plastic deformation is associated with changes in shape.*

## 4-13 CRITERIA FOR YIELDING

As was noted earlier, when considering a multiaxial stress state, the question of when yielding begins is not a trivial one. In this section we will present two theories that attempt to answer this question for ductile materials and compare these theories with experimental results.

### 4-13-1 Maximum Shear Theory

From the observations that yielding is not affected by hydrostatic stress and that plastic deformation is associated with distortion, it is not unreasonable to assume that yielding is governed by shear. The maximum shear theory, or Tresca theory, assumes that yielding begins when the maximum shear stress reaches a critical value $k$, which is constant for perfectly plastic materials,

$$\tau_{\max} = k \tag{4-142}$$

Now consider an element of material oriented so that the sides are parallel to the principal planes, as shown in Fig. 4-39*a*. Then the three Mohr's circles for the element are shown in Fig. 4-39*b*, and

$$\tau_{\max} = \text{largest of} \begin{cases} \dfrac{|\sigma_1 - \sigma_2|}{2} \\ \dfrac{|\sigma_1 - \sigma_3|}{2} \\ \dfrac{|\sigma_2 - \sigma_3|}{2} \end{cases}$$

If the stress state is biaxial with $\sigma_3 = 0$, the situation simplifies to that shown in Fig. 4-40, and we have

$$\tau_{\max} = \text{largest of} \begin{cases} \dfrac{|\sigma_1 - \sigma_2|}{2} \\ \dfrac{|\sigma_1|}{2} \\ \dfrac{|\sigma_2|}{2} \end{cases} \tag{4-143}$$

In order to determine $k$ we consider the one-dimensional problem of a slender rod subjected to a tensile stress. For this problem,

$$\sigma_1 = \sigma_{xx}, \qquad \sigma_2 = \sigma_3 = 0 \tag{4-144}$$

and yielding begins when $\sigma_{xx} = S_y$, where $S_y$ is the yield strength of the material. From Eqs. (4-142) and (4-143) we conclude that for this problem yielding begins when

$$\tau_{\max} = k = \frac{S_y}{2}$$

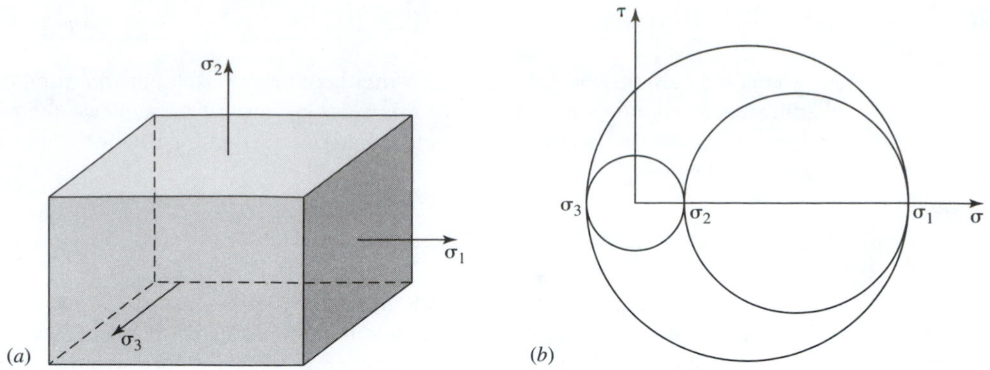


Figure 4-39

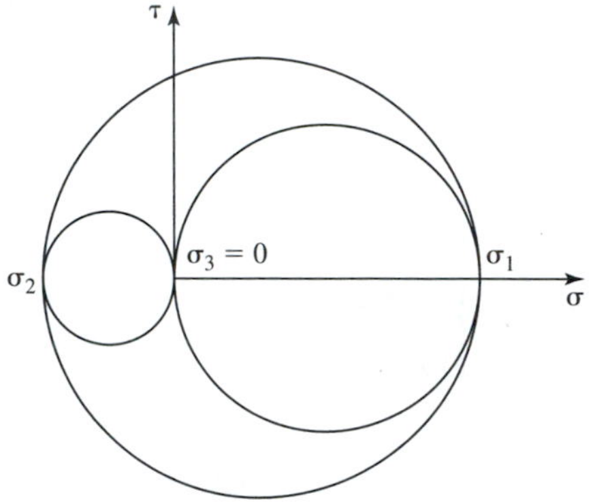


Figure 4-40

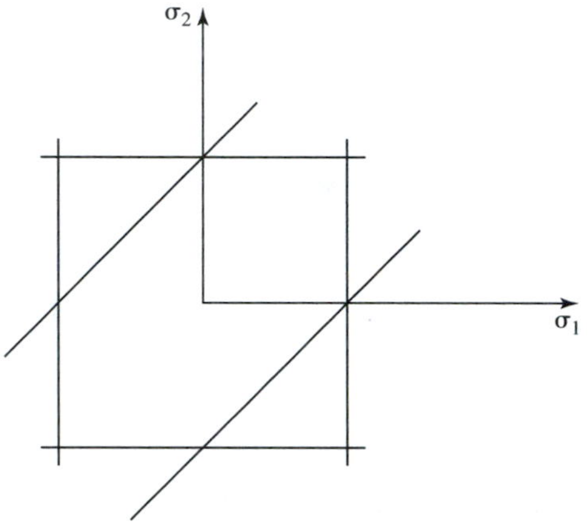


Figure 4-41 The lines indicating the onset of yielding

Since $k$ is a constant, we have, from Eqs. (4-142) and (4-143), that, in general, yielding begins when

$$\text{largest of} \begin{Bmatrix} |\sigma_1 - \sigma_2| \\ |\sigma_1| \\ |\sigma_2| \end{Bmatrix} = S_y \tag{4-145}$$

and in the three-dimensional case yielding begins when

$$\text{largest of} \begin{Bmatrix} |\sigma_1 - \sigma_2| \\ |\sigma_1 - \sigma_3| \\ |\sigma_2 - \sigma_3| \end{Bmatrix} = S_y \tag{4-146}$$

Graphically the first equation of Eq. (4-145) is represented by the 45° lines in Fig. 4-41 while the second and third equations are represented by the vertical and horizontal lines, respectively. The maximum shear theory predicts the onset of yielding in a biaxial stress

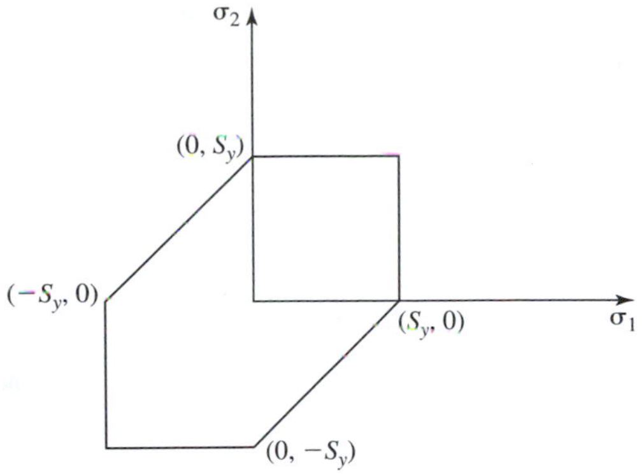


Figure 4-42 Points inside the hexagon represent elastic stress states

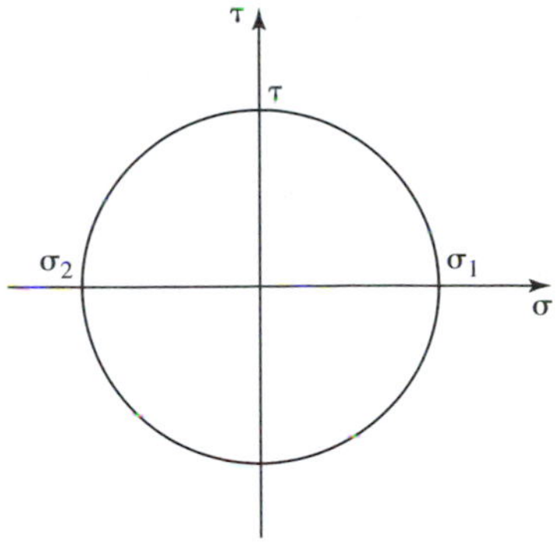


Figure 4-43

state whenever a point in principal stress space falls on the hexagon shown in Fig. 4-42. Points inside the hexagon represent elastic stress states. For elastic–perfectly plastic behavior, points outside the hexagon are unreachable because the stress state can never increase beyond the state that causes the initial yielding.

From Eq. (4-145) the maximum shear theory predicts that in situations where pure shear exists, so that $\sigma_1 = -\sigma_2 = \tau$, as shown in Fig. 4-43, yielding begins when the shear stress reaches a value of one-half the yield stress in simple tension,

$$\tau = 0.5S_y \tag{4-147}$$

## 4-13-2 Distortion Energy Theory

This theory, an alternative to the maximum shear theory, makes use of the observation that yielding is associated with distortion, or shape change, rather than volume change. It states that yielding begins when the strain energy caused by distortion, or the *distortion energy* $\tilde{u}_d$ reaches a critical value $k$, which is again constant for perfectly plastic materials,

$$\tilde{u}_d = k \tag{4-148}$$

To compute the distortion energy we subtract from the strain energy that portion resulting from a volume change only. Denoting by $\tilde{u}_V$ that portion of the strain energy resulting from a volume change, we can write

$$\tilde{u}_d = \tilde{u} - \tilde{u}_V \tag{4-149}$$

In terms of principal stresses the strain energy is found from Eq. (5-72*a*) by setting the shear stresses to zero,

$$\tilde{u} = \frac{1}{2E}\left[\sigma_1^2 + \sigma_2^2 + \sigma_3^2 - 2\nu(\sigma_1\sigma_2 + \sigma_2\sigma_3 + \sigma_3\sigma_1)\right] \tag{4-150}$$

In the special case where all of the principal stresses have a common value of $\sigma$ this becomes

$$\tilde{u} = \frac{3(1-2\nu)}{2E}\sigma^2 \tag{4-151}$$

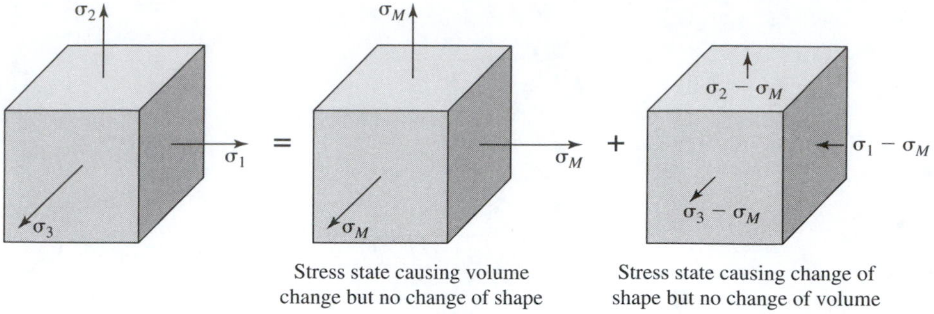


Figure 4-44

Now for a general stress state consisting of three normal and three shear stresses it is the mean stress, defined by Eq. (2-93), that produces a change in volume, as shown by Eq. (4-18) and discussed at the beginning of Section 3-9. Since the sum of the normal stresses is invariant, we can write

$$\sigma_M = \tfrac{1}{3}(\sigma_{xx} + \sigma_{yy} + \sigma_{zz}) = \tfrac{1}{3}(\sigma_1 + \sigma_2 + \sigma_3) \tag{4-152}$$

The principal stress field can be decomposed into one that produces only a volume change and another that produces only a change in shape, as shown in Fig. 4-44. The strain energy associated with the change in volume is found by substituting the mean stress $\sigma_M$ for $\sigma$ in Eq. (4-151). This is then expressed in terms of principal stresses using Eq. (4-152) to give

$$\tilde{u}_V = \frac{3(1-2\nu)}{2E}\left(\frac{\sigma_1 + \sigma_2 + \sigma_3}{3}\right)^2$$

On expansion this yields

$$\tilde{u}_V = \frac{1-2\nu}{6E}\left[\sigma_1^2 + \sigma_2^2 + \sigma_3^2 + 2(\sigma_1\sigma_2 + \sigma_2\sigma_3 + \sigma_3\sigma_1)\right]$$

When this and Eq. (4-150) are substituted into Eq. (4-149), the following expression for distortion energy results:

$$\tilde{u}_d = \frac{1+\nu}{3E}\left(\sigma_1^2 + \sigma_2^2 + \sigma_3^2 - \sigma_1\sigma_2 - \sigma_2\sigma_3 - \sigma_3\sigma_1\right) \tag{4-153}$$

We now consider again the uniaxial tension test discussed in conjunction with the maximum shear theory, and expressed by Eqs. (4-144). With that information in Eq. (4-153) and the subsequent use of Eq. (4-148) we find that

$$\tilde{u}_d = k = \frac{1+\nu}{3E}S_y^2 \tag{4-154}$$

and that, since $k$ is a constant, yielding begins under general loading when

$$S_y = \sqrt{\sigma_1^2 + \sigma_2^2 + \sigma_3^2 - \sigma_1\sigma_2 - \sigma_2\sigma_3 - \sigma_3\sigma_1} \tag{4-155}$$

For the biaxial stress state with $\sigma_3 = 0$ this becomes

$$S_y = \sqrt{\sigma_1^2 + \sigma_2^2 - \sigma_1\sigma_2} \tag{4-156}$$

In order to obtain a graphical interpretation we note that Eq. (4-156) can be written as

$$S_y^2 = \frac{(\sigma_1 + \sigma_2)^2}{4} + \frac{(\sigma_1 - \sigma_2)^2}{4/3}$$

or, equivalently,

$$\frac{\left[\sqrt{2}(\sigma_1 + \sigma_2)/2\right]^2}{\left(\sqrt{2}S_y\right)^2} + \frac{\left[\sqrt{2}(\sigma_1 - \sigma_2)/2\right]^2}{\left(\sqrt{2}S_Y/\sqrt{3}\right)^2} = 1 \tag{4-157}$$

We now let

$$x = \frac{\sqrt{2}}{2}(\sigma_1 + \sigma_2) \tag{4-158$a$}$$

$$y = \frac{\sqrt{2}}{2}(\sigma_1 - \sigma_2) \tag{4-158$b$}$$

$$a = \sqrt{2}S_Y \tag{4-158$c$}$$

$$b = \frac{\sqrt{2}S_Y}{\sqrt{3}} \tag{4-158$d$}$$

in which case Eq. (4-157) becomes

$$\frac{x^2}{a^2} + \frac{y^2}{b^2} = 1$$

This is clearly the equation of an ellipse with major and minor axes given by Eqs. (4-158*c*) and (4-158*d*), respectively. From Eqs. (4-158*a*, *b*) we see that the line $y = 0$ is the line $\sigma_2 = \sigma_1$ whereas the line $x = 0$ is the line $\sigma_2 = -\sigma_1$. Thus the ellipse described by Eq. (4-156) is relative to a coordinate system rotated 45° counterclockwise from the $\sigma_1$, $\sigma_2$ coordinate system, as shown in Fig. 4-45. Stress states falling on

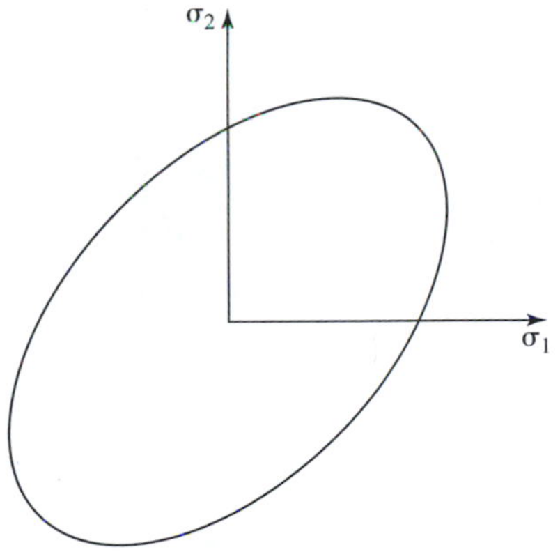


Figure 4-45

the interior of the ellipse represent elastic states. Yielding is represented by all stress states lying on the ellipse. For perfectly plastic materials, stress states lying outside of the ellipse are unreachable.

For a state of pure shear the Mohr's circle is again as shown in Fig. 4-43, and we have

$$\sigma_1 = -\sigma_2 = \tau$$

When this is substituted into Eq. (4-156), we find that yielding begins when the shear stress reaches a value of

$$\tau = \frac{\sqrt{3}}{3} S_Y \approx 0.577 S_y \tag{4-159}$$

The distortion energy theory was first proposed, based on purely mathematical considerations, by von Mises [14] and is therefore referred to as the *von Mises yield criterion*. The physical interpretation of the Mises criterion in terms of the strain energy of distortion is due to Hencky [6]. Accordingly, the theory is sometimes referred to as the *von Mises–Hencky criterion.*

### 4-13-3 Comparison of the Two Theories

By comparing Eqs. (4-147) and (4-159) we see that the distortion energy theory predicts that yielding in pure shear will begin at a slightly higher stress than predicted by the maximum shear theory. In fact, if the hexagon of the maximum shear theory and the ellipse of the distortion energy theory are drawn on a common set of axes, as in Fig. 4-46, we see that the maximum shear theory is the more conservative of the two because it predicts yielding for certain stress states that are considered elastic in the distortion energy theory.

The question arises, of course, as to which theory shows better agreement with experimental results. This question was addressed in a set of experiments by Taylor and Quinney [20], in which a thin-walled tube was subjected to combined tension and torsion. The geometry is shown in Fig. 4-47. Although the stress field near the ends of the

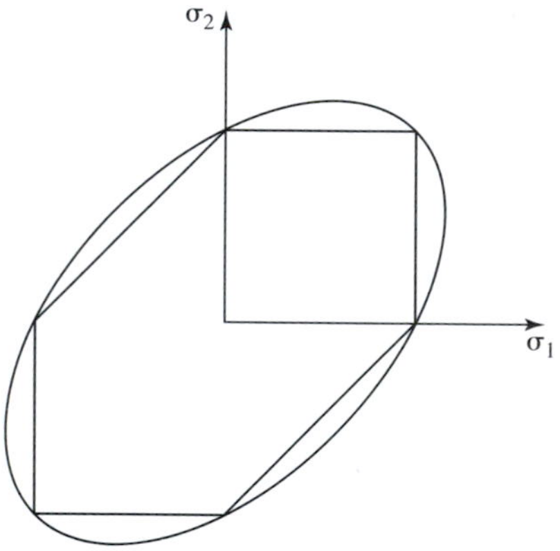


Figure 4-46

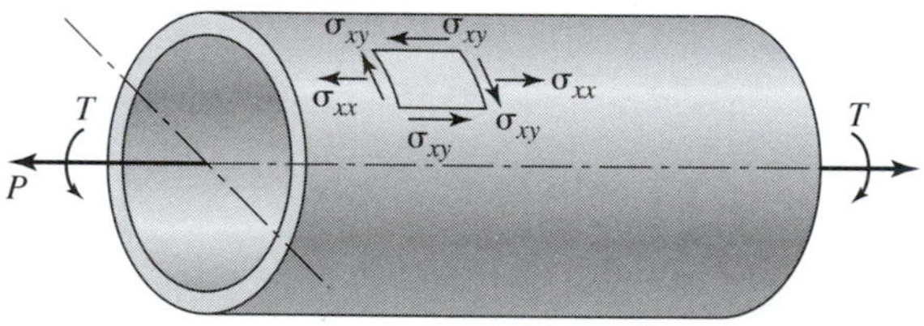


Figure 4-47

tube is quite complicated because of the constraints on the tube ends, away from the ends the stress field reduces to one in which the only nonzero stress components are $\sigma_{xx}$ and $\sigma_{xy}$, as shown in the figure. The utility of this configuration lies in the fact that the normal stress and the shear stress can be controlled independently by varying the tensile load and the torque, respectively.

From Eq. (2-80) we have the maximum shear stress, with $\sigma_{yy} = 0$, given by

$$\tau_{\max} = \sqrt{\left(\frac{\sigma_{xx}}{2}\right)^2 + \sigma_{xy}^2}$$

According to the maximum shear theory, yielding begins when

$$\tau_{\max} = \tfrac{1}{2} S_y$$

Therefore we have

$$S_y^2 = \sigma_{xx}^2 + 4\sigma_{xy}^2 \tag{4-160}$$

Now from Eqs. (2-75) and (2-80) we have, with $\sigma_{yy} = 0$,

$$\sigma_{1,2} = \frac{\sigma_{xx}}{2} \pm \tau_{\max}$$

As noted above Eq. (4-156), we have chosen the principal stress $\sigma_3 = 0$ for the biaxial stress state. This accounts for the notational difference between the above equation and Eq. (2-75). When these expressions for $\sigma_1$ and $\sigma_2$ are substituted into Eq. (4-156) and Eq. (2-80) is used for $\tau_{\max}$, we get

$$S_y^2 = \sigma_{xx}^2 + 3\sigma_{xy}^2 \tag{4-161}$$

for the distortion energy theory. By dividing both sides by $S_y^2$, Eqs. (4-160) and (4-161) can be put in the following form:

$$\left(\frac{\sigma_{xx}}{S_y}\right)^2 + \frac{(\sigma_{xy}/S_y)^2}{(1/2)^2} = 1$$

$$\left(\frac{\sigma_{xx}}{S_y}\right)^2 + \frac{(\sigma_{xy}/S_y)^2}{(1/\sqrt{3})^2} = 1$$

Both of these are ellipses in the $\sigma_{xx}$, $\sigma_{xy}$ plane. They are shown, along with the results of Taylor and Quinney, in Fig. 4-48, reprinted by permission of Prentice-Hall, Upper Saddle River, NJ, from [13, p. 90]. It is seen that the experimental data show better agreement with the distortion energy theory, but that the maximum shear theory is conservative in its predictions. It is generally agreed that the distortion energy theory is the more accurate of the two theories, and thus it is the theory most often used.

From a computational point of view the distortion energy theory is the easier of the two to use because once the principal stresses are found, yielding can be determined directly from Eqs. (4-155) or (4-156). On the other hand, the hexagon of the maximum shear theory can easily be drawn, so that as a graphical tool, it is the easier of the two theories to use.

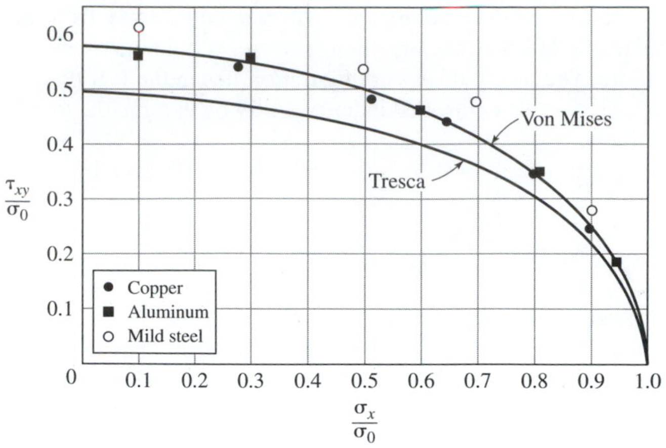


Figure 4-48 Experimental results of Taylor and Quinney

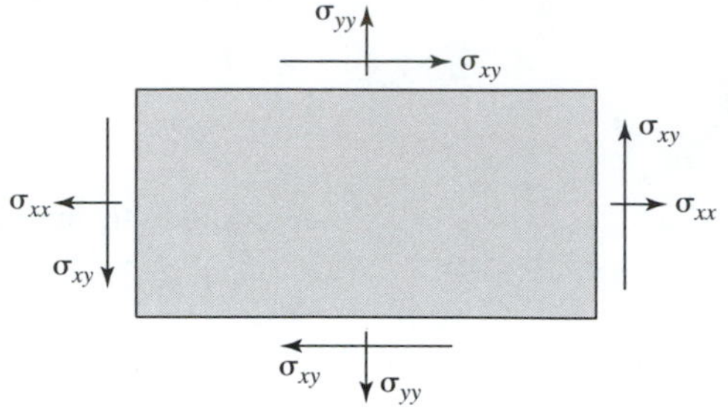


Figure 4-49

It is worth mentioning that a general biaxial stress state in which each of the stresses can be controlled independently can be achieved by simply adding internal pressure to the thin-walled tube described. The tangential stress $\sigma_{yy}$ is then controlled independently of the other two stresses by varying the internal pressure. The resulting stress state is shown in Fig. 4-49.

Finally we note that Eq. (4-153) can be written as follows:

$$\tilde{u}_d = \frac{1+\nu}{E}\left\{\frac{1}{6}[(\sigma_1-\sigma_2)^2+(\sigma_2-\sigma_3)^2+(\sigma_3-\sigma_1)^2]\right\}$$

Now from Eq. (2-101) the term in square brackets is simply the second invariant of the deviatoric stress $I_{2D}$. This is because, with the shear stresses set to zero, the stresses appearing in the second of Eqs. (2-100) are principal stresses. Thus we can write

$$\tilde{u}_d = \frac{1+\nu}{E} I_{2D}$$

or

$$\tilde{u}_d = \frac{1}{2G} I_{2D}$$

where $G$ is the shear modulus given by Eq. (4-13). During plastic deformation of perfectly plastic materials the value of the strain energy is given by Eq. (4-154). Thus the second invariant of the strain deviator is constant with a value of

$$I_{2D} = \tfrac{1}{3} S_y^2 \tag{4-162}$$

when a perfectly plastic material is undergoing plastic deformation.

When von Mises developed his yield criterion, he noted that yielding should involve some combination of the deviatoric stress components because yielding is unaffected by hydrostatic stress. For isotropic materials, the particular combination of deviatoric stresses involved should be invariant with regard to coordinate transformations. Thus, he reasoned, a yield criterion should depend on some combination of the second and third invariants of the stress deviator, $I_{2D}$ and $I_{3D}$. He chose $I_{2D} =$ constant because of its inherent simplicity.

## 4-14 STRESS-STRAIN RELATIONS FOR ELASTIC–PERFECTLY PLASTIC MATERIALS

The stress-strain relations that will be introduced in this section are the Prandtl–Reuss equations. They were first proposed by Prandtl [16] in 1924 and, in a more general form, by Reuss [17] in 1930. We note that plastic behavior is independent of hydrostatic stress. Therefore it seems reasonable that the stress components in our plastic stress-strain law should be the deviatoric stress components. Further, the key variable during plastic flow is the strain *rate* and not the strain itself. Therefore we simply assume that at any instant the plastic strain deviation rate is proportional to the stress deviation at that instant,

$$\frac{\dot{e}^p_{xx}}{\sigma'_{xx}} = \frac{\dot{e}^p_{yy}}{\sigma'_{yy}} = \frac{\dot{e}^p_{zz}}{\sigma'_{zz}} = \frac{\frac{1}{2}\dot{\gamma}^p_{xy}}{\sigma_{xy}} = \frac{\frac{1}{2}\dot{\gamma}^p_{yz}}{\sigma_{yz}} = \frac{\frac{1}{2}\dot{\gamma}^p_{zx}}{\sigma_{zx}} = \lambda \tag{4-163}$$

where $\lambda$ is a nonnegative constant. Equations (4-163), in terms of the actual stress components given in Eqs. (2-95), are

$$\begin{aligned}
\dot{e}^p_{xx} &= \lambda\sigma'_{xx} = \tfrac{2}{3}\lambda\left[\sigma_{xx} - \tfrac{1}{2}(\sigma_{yy} + \sigma_{zz})\right] \\
\dot{e}^p_{yy} &= \lambda\sigma'_{yy} = \tfrac{2}{3}\lambda\left[\sigma_{yy} - \tfrac{1}{2}(\sigma_{zz} + \sigma_{xx})\right] \\
\dot{e}^p_{zz} &= \lambda\sigma'_{zz} = \tfrac{2}{3}\lambda\left[\sigma_{zz} - \tfrac{1}{2}(\sigma_{xx} + \sigma_{yy})\right] \\
\tfrac{1}{2}\dot{\gamma}^p_{xy} &= \lambda\sigma_{xy} \\
\tfrac{1}{2}\dot{\gamma}^p_{yz} &= \lambda\sigma_{yz} \\
\tfrac{1}{2}\dot{\gamma}^p_{zx} &= \lambda\sigma_{zx}
\end{aligned} \tag{4-164}$$

Thus the plastic stress-strain equations are known if $\lambda$ can be determined. This can be done with the aid of the von Mises yield criterion. We begin by squaring each of Eqs. (4-164), adding the first three of the squared equations, dividing the result by 2, and then adding to it the last three of the squared equations. This gives

$$\begin{aligned}
&\tfrac{1}{2}\left(\dot{e}^{p^2}_{xx} + \dot{e}^{p^2}_{yy} + \dot{e}^{p^2}_{zz}\right) + \tfrac{1}{4}\dot{\gamma}^{p^2}_{xy} + \tfrac{1}{4}\dot{\gamma}^{p^2}_{yz} + \tfrac{1}{4}\dot{\gamma}^{p^2}_{zx} \\
&\qquad = \lambda^2\left[\tfrac{1}{2}\left(\sigma'^2_{xx} + \sigma'^2_{yy} + \sigma'^2_{zz}\right) + \sigma^2_{xy} + \sigma^2_{yz} + \sigma^2_{zx}\right]
\end{aligned} \tag{4-165}$$

The bracketed term in this equation is $I_{2D}$, as shown by Eq. (2-101). The term on the left is the second invariant of the strain rate deviator $J_{2D}$, as seen from Eq. (3-68). Thus with

Eq. (3-67) we have

$$\tfrac{1}{2}\left(\dot{e}_{xx}^{p^2} + \dot{e}_{yy}^{p^2} + \dot{e}_{zz}^{p^2}\right) + \tfrac{1}{4}\left(\dot{\gamma}_{xy}^{p^2} + \dot{\gamma}_{yz}^{p^2} + \dot{\gamma}_{zx}^{p^2}\right)$$
$$= \tfrac{1}{6}\left[\left(\dot{\varepsilon}_{xx}^{p} - \dot{\varepsilon}_{yy}^{p}\right)^2 + \left(\dot{\varepsilon}_{yy}^{p} - \dot{\varepsilon}_{zz}^{p}\right)^2 + \left(\dot{\varepsilon}_{zz}^{p} - \dot{\varepsilon}_{xx}^{p}\right)^2 + \tfrac{3}{2}\left(\dot{\gamma}_{xy}^{p^2} + \dot{\gamma}_{yz}^{p^2} + \dot{\gamma}_{zx}^{p^2}\right)\right] \quad (4\text{-}166)$$

Analogous to Mendelson [13, p. 102] we define the *effective plastic strain* rate as follows:

$$\dot{\varepsilon}_p = \frac{\sqrt{2}}{3}\left[\left(\dot{\varepsilon}_{xx}^{p} - \dot{\varepsilon}_{yy}^{p}\right)^2 + \left(\dot{\varepsilon}_{yy}^{p} - \dot{\varepsilon}_{zz}^{p}\right)^2 + \left(\dot{\varepsilon}_{zz}^{p} - \dot{\varepsilon}_{xx}^{p}\right)^2 + \frac{3}{2}\left(\dot{\gamma}_{xy}^{p^2} + \dot{\gamma}_{yz}^{p^2} + \dot{\gamma}_{zx}^{p^2}\right)\right]^{1/2} \quad (4\text{-}167)$$

Thus Eq. (4-165) becomes, with Eqs. (4-166) and (4-167),

$$\tfrac{3}{4}\dot{\varepsilon}_p^2 = \lambda^2 I_{2D}$$

From this it follows that

$$\lambda = \frac{\sqrt{3}}{2}\frac{\dot{\varepsilon}_p}{\sqrt{I_{2D}}} \quad (4\text{-}168)$$

However, during plastic deformation of perfectly plastic materials the value of $I_{2D}$, as determined by the distortion energy theory, is given by Eq. (4-162). Therefore Eq. (4-168) becomes

$$\lambda = \frac{3}{2}\frac{\dot{\varepsilon}_p}{S_y}$$

With this value of $\lambda$ Eqs. (4-164) become

$$\begin{aligned}
\dot{\varepsilon}_{xx}^{p} &= \frac{\dot{\varepsilon}_p}{S_y}\left[\sigma_{xx} - \frac{1}{2}(\sigma_{yy} + \sigma_{zz})\right] \\
\dot{\varepsilon}_{yy}^{p} &= \frac{\dot{\varepsilon}_p}{S_y}\left[\sigma_{yy} - \frac{1}{2}(\sigma_{zz} + \sigma_{xx})\right] \\
\dot{\varepsilon}_{zz}^{p} &= \frac{\dot{\varepsilon}_p}{S_y}\left[\sigma_{zz} - \frac{1}{2}(\sigma_{xx} + \sigma_{yy})\right] \\
\dot{\gamma}_{xy}^{p} &= \frac{3}{2}\frac{\dot{\varepsilon}_p}{S_y}\sigma_{xy} \\
\dot{\gamma}_{yz}^{p} &= \frac{3}{2}\frac{\dot{\varepsilon}_p}{S_y}\sigma_{yz} \\
\dot{\gamma}_{zx}^{p} &= \frac{3}{2}\frac{\dot{\varepsilon}_p}{S_y}\sigma_{zx}
\end{aligned} \quad (4\text{-}169)$$

In the first three of these equations we have made use of the fact that the deviatoric plastic strain rate components and the actual plastic strain rate components are the same because the material undergoes no volume change during plastic deformation.

## 4-15 STRESS-STRAIN RELATIONS WHEN THE TEMPERATURE FIELD IS NONUNIFORM

Up to this point our discussion of stress-strain relations has been restricted to situations in which the temperature distribution in the body is uniform. However, there are many practical problems in which the temperature varies from point to point in the body. Such problems can be found in the design of devices like heat engines and nuclear reactors.

Temperature-induced stresses arise, in the case of a uniform temperature distribution, if the body is constrained. On the other hand such stresses will arise if the temperature field is nonuniform, even though the body may be unrestrained. To see this, consider an unrestrained body subjected to a uniform temperature field. The body is composed of many differential elements that are "connected" together because of the continuity of the material. Each of these elements undergoes thermal expansion. However, since the temperature field is uniform, the thermal expansion of each element is the same and no stresses are required to keep the elements connected. If the temperature field is nonuniform, then two adjacent elements have different temperatures and therefore different amounts of thermal expansion. Because material continuity requires that the elements remain connected, stresses must arise in order to maintain this continuity.

We define the *reference temperature* $T_0$ as that uniform temperature associated with a state in which the body is free from both stress and strain. Let the actual temperature of the body, which need not be uniform, be denoted by $\hat{T}$ and the deviation of the actual temperature from the reference temperature by $T$,

$$T = \hat{T} - T_0 \tag{4-170}$$

Let us consider a body composed of an isotropic material and remove from that body a differential element of material at temperature $T$. This temperature will bring about a change of volume, but not a change of shape, in an isotropic material. Thus associated with this temperature will be normal strains but no shear strains. These strains are given by

$$\begin{gathered} \varepsilon_{xx}^T = \alpha T, \qquad \varepsilon_{yy}^T = \alpha T, \qquad \varepsilon_{zz}^T = \alpha T \\ \gamma_{xy}^T = \gamma_{yz}^T = \gamma_{zx}^T = 0 \end{gathered} \tag{4-171}$$

where the superscript $T$ indicates that these are thermal strains, and $\alpha$ is the *coefficient of linear thermal expansion*. This coefficient is defined as the change in length per unit length per degree change in temperature. If, in addition, there are stresses acting on the element, the *total* strain on the element, given by Eqs. (3-11), in terms of displacement gradients will be

$$\begin{aligned} \varepsilon_{xx} &= \varepsilon_{xx}^m + \varepsilon_{xx}^T \\ \varepsilon_{yy} &= \varepsilon_{yy}^m + \varepsilon_{yy}^T \\ \varepsilon_{zz} &= \varepsilon_{zz}^m + \varepsilon_{zz}^T \\ \gamma_{xy} &= \gamma_{xy}^m \\ \gamma_{yz} &= \gamma_{yz}^m \\ \gamma_{zx} &= \gamma_{zx}^m \end{aligned}$$

where quantities with the superscript $m$ are *mechanical* strains due to the stresses. The mechanical strains are given in terms of stresses by Eqs. (4-8)–(4-10). The total strains are therefore given by

$$\begin{aligned}
\varepsilon_{xx} &= \frac{1}{E}[\sigma_{xx} - \nu(\sigma_{yy} + \sigma_{zz})] + \alpha T \\
\varepsilon_{yy} &= \frac{1}{E}[\sigma_{yy} - \nu(\sigma_{zz} + \sigma_{xx})] + \alpha T \\
\varepsilon_{zz} &= \frac{1}{E}[\sigma_{zz} - \nu(\sigma_{xx} + \sigma_{yy})] + \alpha T \\
\gamma_{xy} &= \frac{1}{G}\sigma_{xy} \\
\gamma_{yz} &= \frac{1}{G}\sigma_{yz} \\
\gamma_{zx} &= \frac{1}{G}\sigma_{zx}
\end{aligned} \tag{4-172}$$

In the presence of a temperature field the dilatation, defined by Eq. (3-62), is given by

$$\Delta = \varepsilon_{xx} + \varepsilon_{yy} + \varepsilon_{zz} = \varepsilon_{xx}^m + \varepsilon_{yy}^m + \varepsilon_{zz}^m + 3\alpha T = \Delta^m + 3\alpha T \tag{4-173}$$

or, with Eq. (4-18),

$$\Delta = \frac{\sigma_M}{K} + 3\alpha T \tag{4-174}$$

where $\sigma_M$ is the mean stress defined by Eq. (2-93).

Equations (4-172) can be solved readily to give the stresses in terms of the total strains and the temperature,

$$\begin{aligned}
\sigma_{xx} &= \frac{E}{(1+\nu)(1-2\nu)}[(1-2\nu)\varepsilon_{xx} + \nu\Delta] - \frac{E}{1-2\nu}\alpha T \\
\sigma_{yy} &= \frac{E}{(1+\nu)(1-2\nu)}[(1-2\nu)\varepsilon_{yy} + \nu\Delta] - \frac{E}{1-2\nu}\alpha T \\
\sigma_{zz} &= \frac{E}{(1+\nu)(1-2\nu)}[(1-2\nu)\varepsilon_{zz} + \nu\Delta] - \frac{E}{1-2\nu}\alpha T \\
\sigma_{xy} &= G\gamma_{xy} \\
\sigma_{yz} &= G\gamma_{yz} \\
\sigma_{zx} &= G\gamma_{zx}
\end{aligned} \tag{4-175}$$

It should be pointed out that the theory presented here yields equations that are linear in temperature. As such, it is valid for relatively small temperature changes compared to the reference temperature,

$$\frac{|\hat{T} - T_0|}{T_0} \ll 1 \tag{4-176}$$

where $\hat{T}$ and $T_0$ are absolute temperatures. Practically speaking, the theory can be considered valid for a range of temperatures over which Young's modulus, Poisson's ratio, and the coefficient of linear thermal expansion can be considered constant.

## 4-16 STRESS-STRAIN RELATIONS FOR PIEZOELECTRIC MATERIALS

Certain materials exhibit interesting behavior whereby an electric field is generated when the material is strained. This is always accompanied by the converse behavior in which strain occurs when the material is placed in an electric field. These are referred to as the *direct piezoelectric effect* and the *converse piezoelectric effect,* respectively, and materials exhibiting these effects are called *piezoelectric materials*. Both effects are linear so that with the converse piezoelectric effect, for example, the strain is proportional to the electric field intensity and changes sign when the sign of the electric field changes.

Interesting historical accounts of early developments in piezoelectricity and its applications can be found in the references [4, chap. 1], [12, sec. 1.2]. The direct piezoelectric effect was first observed in quartz crystals in 1880, and the converse piezoelectric effect in 1881, by Pierre and Jacques Curie. However, it was not utilized until late in World War I, when Langevin used quartz crystals to generate and receive sound waves underwater for the purpose of detecting submarines. Although Langevin's development came too late in the war to be of use in combating submarines, it was the forerunner of modern sonar, and it also led to the development of ultrasonics for nondestructive material testing. Recent application areas that make use of piezoelectric materials are *active vibration control,* where a sensor (which could be piezoelectric) is used to sense vibration and send an electric signal to a piezoelectric actuator which provides corrective antivibration, and *smart structures,* where embedded piezoelectric sensors can detect damage such as delamination in a composite, or can detect excessive deflection in a flexible structure and signal a piezoelectric actuator to provide stiffening.

Piezoelectric behavior is tied to the crystal structure of a material, a detailed discussion of which is beyond the scope of the present text. However, some discussion of crystal structure will help in understanding this behavior and in explaining why certain materials are piezoelectric and others are not. Additional information can be found in the references [2], [4], [12], [15].

*Crystalline materials* are those in which the atoms are arranged in a pattern that repeats itself periodically, the particular arrangement of the atoms being referred to as the *crystal structure*. The basic arrangement of atoms in a crystal is shown by the crystal's *unit cell,* and the periodic repetition forms the *lattice structure* of the crystal, as shown in Fig. 4-50, where $a$, $b$, and $c$ are the axes of the unit cell, which are not necessarily orthogonal. Crystals fall into one of seven *crystal groups,* depending on the lengths of the sides of the unit cell, $a_0$, $b_0$, $c_0$, and the angles between the unit cell's axes, $\alpha, \beta, \gamma$, as shown in Fig. 4-50*a*. *Cubic crystals* are those with unit cells for which $a_0 = b_0 = c_0$ and $\alpha = \beta = \gamma = \pi/2$; *tetragonal crystals* are those for which

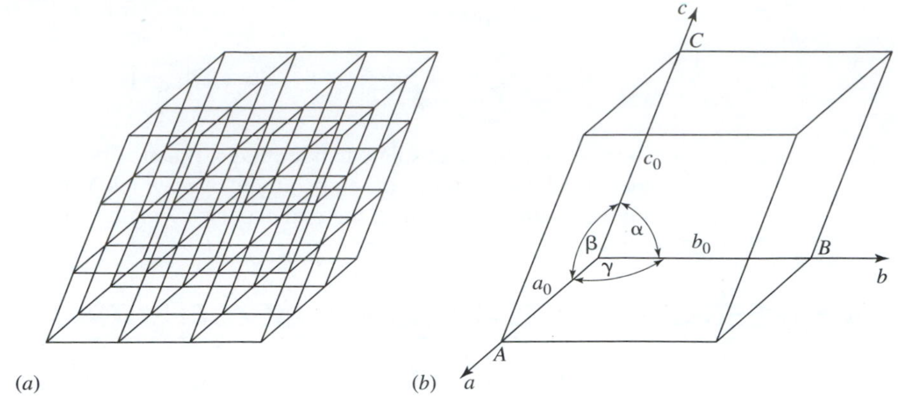


**Figure 4-50** The lattice structure of the crystal

$a_0 = b_0 \neq c_0$ and $\alpha = \beta = \gamma = \pi/2$; *triclinic* (or *rhombohedral*) *crystals* are those for which $a_0 = b_0 = c_0$ and $\alpha = \beta = \gamma \neq \pi/2$; and *orthorhombic crystals* are those for which $a_0 \neq b_0 \neq c_0$ and $\alpha = \beta = \gamma = \pi/2$. The remaining crystal groups are *monoclinic, triclinic,* and *hexagonal.*

Crystal are, in general, anisotropic. However, as noted in Section 4-1, a macroscopic sample of polycrystalline material can be isotropic if it contains many crystals that are oriented randomly.

Unit cells often possess symmetries in their structures, which determine the number of independent material properties needed to describe certain kinds of behavior (elastic behavior, for example) of the crystal. As we saw in Sections 4-6 through 4-10, the number of elastic constants needed to describe a material can range anywhere from 2 to 21, depending on the symmetry of the material.

Based on these symmetries, 32 crystal classes can be identified. The particular crystal class to which a given crystalline material belongs is determined experimentally through the use of X-ray diffraction or neutron diffraction.

Some crystal structures have unit cells that possess *point symmetry*. This occurs if there is a point from which a ray extended in an *arbitrary* direction encounters features that are identical to features at the same distance from the point when the ray is extended in the *opposite* direction. A unit cell may exhibit point symmetry with respect to certain of its features but not others. In Fig. 4-51, for example, point *O* is a point of symmetry with respect to the square dots because a ray extended from point *O* to *any* square dot also encounters a square dot exactly the same distance from *O* when extended in the opposite direction. Conversely, there is no point symmetry with respect to the triangular dots. *Centrosymmetric* crystals have unit cells that exhibit point symmetry with respect to *all* of their features. The point of symmetry is referred to as the *center of symmetry* and necessarily lies at the geometric center of the unit cell. This is because the geometry of a unit cell is one of its features. Of the 32 crystal classes, 11 are centrosymmetric. Examples are face-centered cubic and body-centered cubic, which are common crystal classes for metals. Isotropic materials also possess a center of symmetry. The importance of centrosymmetry is that *centrosymmetric crystals cannot be piezoelectric*.

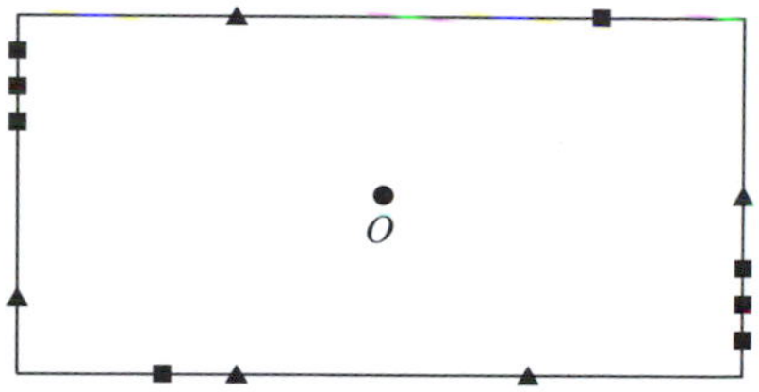


Figure 4-51

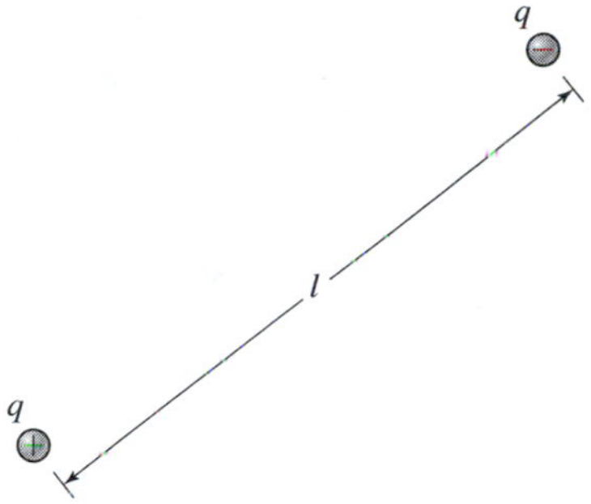


Figure 4-52 An electric dipole

Of the 21 crystal classes that lack a center of symmetry, only one is not piezoelectric. Hence *20 crystal classes exhibit piezoelectric behavior.*

Piezoelectric behavior is directly linked to the motion of point charges in the crystal during deformation. Recall from elementary physics that when two point charges of equal magnitude $q$ but of opposite sign are separated by a distance $l$, as shown in Fig. 4-52, the combination is referred to as an *electric dipole*. An *electric moment,* or *dipole moment* $\boldsymbol{p}$ exists and is given by

$$\boldsymbol{p} = q\boldsymbol{l}$$

By convention $\boldsymbol{l}$ is directed from the negative charge to the positive charge. In a crystal, where several electric dipoles exist, the *net* dipole moment is the vector sum of the individual dipole moments. It depends on the distance between the centroid of the positive charges and the centroid of the negative charges, as indicated in Fig. 4-53. A crystal possessing a net dipole moment is said to be *polar*. The *polarization* $\boldsymbol{P}$ is defined as the net dipole moment per unit volume, averaged over the volume of the unit cell. If the centroids are coincident, there is no net dipole moment and the crystal is *nonpolar* (i.e., $\boldsymbol{P} = \boldsymbol{0}$). When the crystal is deformed, the distance between the centroids of the positive and negative charges may change so that a nonpolar crystal in the undeformed state may become polar in the deformed state. If there is a net dipole moment, then an electric field exists in the crystal that is parallel to the vector connecting the two centroids.

Figure 4-54 shows the charge distribution for sodium chloride, which has a face-centered cubic crystal structure. Because this is a centrosymmetric crystal, the centroid

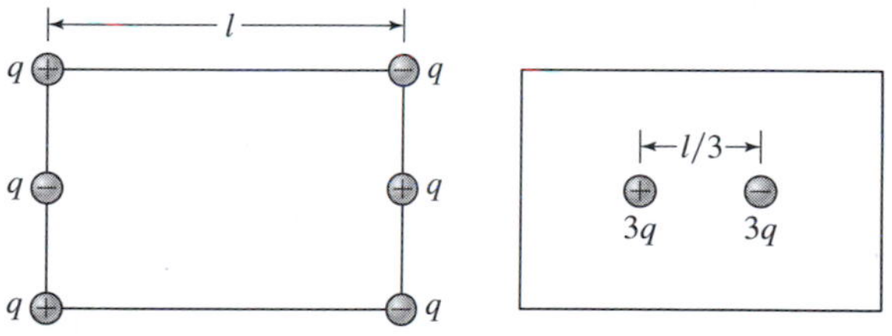


Actual charge distribution    Equivalent charge distribution

Figure 4-53

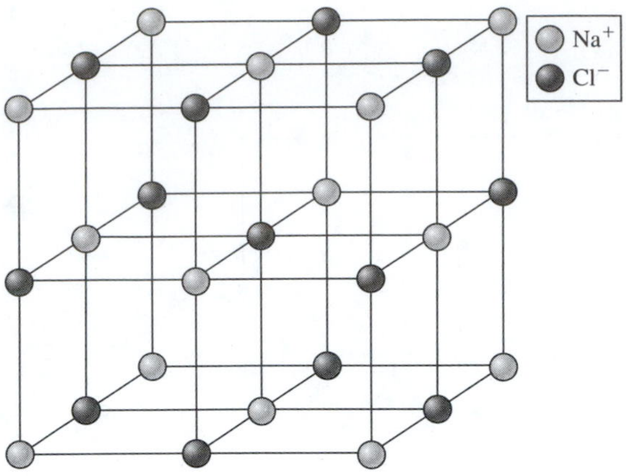


Figure 4-54 The charge distribution for sodium chloride

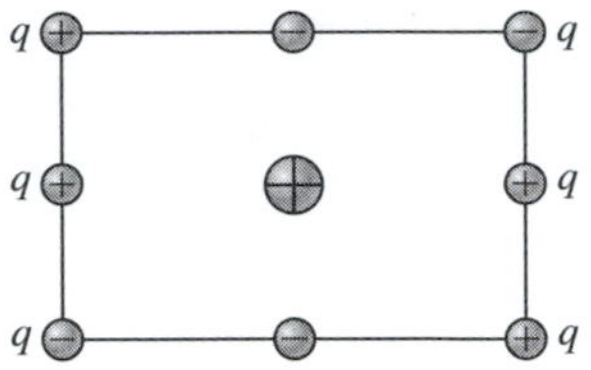


Figure 4-55

of the positive charges is coincident with that of the negative charges and the crystal is nonpolar (i.e., $\boldsymbol{P} = \mathbf{0}$). Now let us see what happens when a centrosymmetric crystal is strained. For simplicity, consider a fictitious crystal with the arrangement of charges shown in Fig. 4-55. Certainly for the charges the geometric center of the crystal is a point of symmetry. Also, the centroids of the positive and negative charges are coincident and are located at the geometric center of the crystal, as shown. Suppose that the crystal is compressed, as shown in Fig. 4-56. Although the faces of the crystal change position, because of symmetry the location of the centroids of both positive and negative charges remains the same and the distance between them is still zero. Consequently the polarization of the crystal remains zero after straining. If the crystal is subjected to shear, as shown in Fig. 4-57, the centroid of the positive charges moves the same amount as that of the negative charges, so the distance between them remains zero and, again, the polarization remains zero. With all other combinations of strain, the distance between the centroids of positive and negative charges remains zero, so the crystal remains nonpolar. We see, then, that this arrangement of charges cannot result in piezoelectric behavior. By using similar arguments, the distance between the centroids of positive and negative charges for *all* centrosymmetric crystals, when strained, can be shown to be zero. Hence, as asserted previously, no centrosymmetric crystal can be piezoelectric.

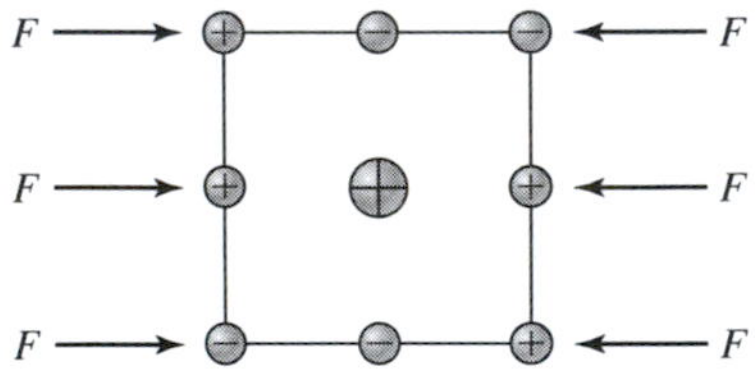


Figure 4-56

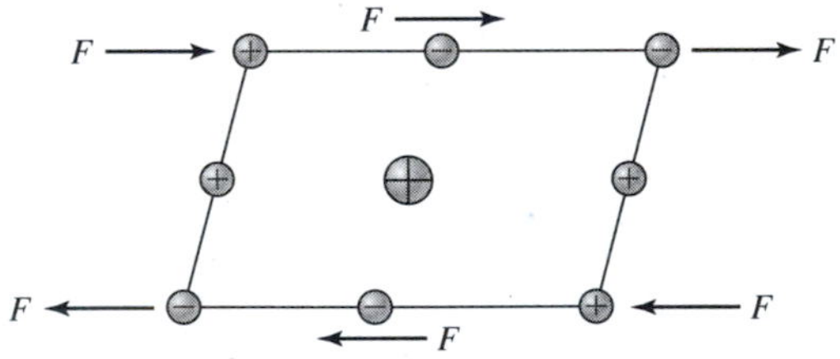


Figure 4-57

It is important to note that in noncentrosymmetric crystals the centroids of positive and negative charges may or may not be coincident in the unstrained state. If they are coincident, however, when the crystal is strained, the centroids of the positive and negative charges move by different amounts, resulting in a net dipole moment. If the centroids are not coincident, in the unstrained state the crystal, in addition to being piezoelectric, is also *pyroelectric,* which means that a net dipole moment is also produced by a temperature change in the crystal. Of the 20 crystal classes that are piezoelectric, 10 are also pyroelectric (see, for example, [2, p. 408]).

In general, the crystal structure is temperature dependent, and above a critical temperature known as the *Curie temperature,* crystals assume a centrosymmetric form. Hence above the Curie temperature, which is a crystal property, the piezoelectric effect, if present, is lost.

In order to make use of the piezoelectric effect in a material, electrodes must be applied to opposite surfaces and electrical leads attached. Silver is a common electrode material and can be applied by a variety of methods, two of which are sputtering and painting on silver ink. Figure 4-58 shows a typical configuration. Note that the electrodes form a parallel-plate capacitor with the piezoelectric material acting as the dielectric. When the material is strained, the resulting electric field manifests itself as a voltage difference across the electrodes. The relationship between electric field intensity and voltage is

$$\boldsymbol{E} = -\nabla V \tag{4-177}$$

where the del operator $\nabla$ is defined in Cartesian coordinates as

$$\nabla = \boldsymbol{i}\frac{\partial}{\partial x} + \boldsymbol{j}\frac{\partial}{\partial y} + \boldsymbol{k}\frac{\partial}{\partial z} \tag{4-178}$$

and $\boldsymbol{i}$, $\boldsymbol{j}$, and $\boldsymbol{k}$ are mutually orthogonal unit vectors.

*Polycrystalline materials* contain many crystals. Examples are ceramics and certain polymers. Polymers are either *amorphous,* in which case they contain no crystals, or *semicrystalline,* in which case crystals are embedded in an amorphous matrix. Without special processing, crystals in polycrystalline materials are oriented randomly. Hence the dipole moments of individual crystals, upon straining, tend to cancel each other so that the net dipole moment of a macroscopic sample is zero. The sample is therefore not piezoelectric, even though its individual crystals may be.

Polycrystalline materials composed of piezoelectric crystals can be made piezoelectric by proper processing of the material. For ceramics, the material is heated to just below its Curie temperature and a strong electric field applied across the electrodes. At

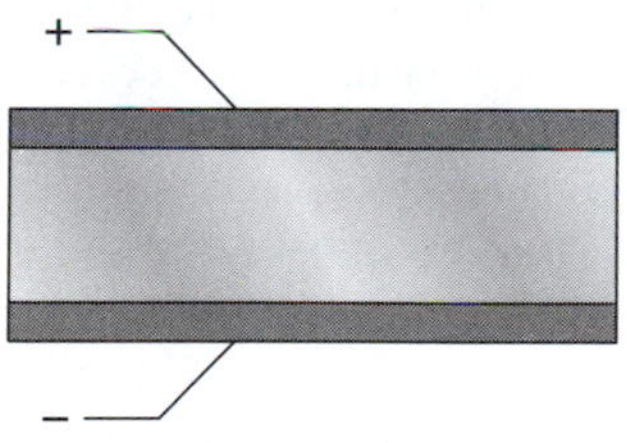


Figure 4-58

this elevated temperature the electric dipoles can change orientation, and in the presence of the electric field they become more or less aligned* with the direction of the field, commonly referred to as the *poling direction*. The material is then cooled, locking the electric dipoles in their new orientations, and the electric field removed, leaving the material sample with a net dipole moment when strained. This process is referred to as *poling*, and the direction of the electric field is the *poling direction*. It should be noted that if the temperature of the ceramic is subsequently raised to the vicinity of its Curie temperature, the electric dipoles will again assume a random orientation and the piezoelectric effect will be lost. As a result of the poling process, a common axis of each crystal (for ceramics, the $c$ axis) becomes aligned with the poling direction. The remaining two crystal axes, however, are still oriented randomly about the aligned axis. Therefore poled ceramics are *transversely isotropic* materials.

In semicrystalline polymers, molecules are linked to each other in a chain. In some polymers the chain configuration is such that a net dipole moment exists in individual crystals, and therefore individual crystals are piezoelectric. In others, PVDF for example, individual crystals are not piezoelectric. However, by stretching the polymer at elevated temperatures a favorable chain configuration is obtained in which individual crystals have a net dipole moment. This stretching can be accomplished in several ways, one of which is to pass a strip of the polymer through two pairs of rollers with one pair rotating faster than the other. Once the polymer is stretched, the net dipole moments of individual crystals in a macroscopic sample are still randomly oriented about the chain axis, resulting in no net piezoelectric effect. To overcome this, the stretched polymer must be poled in much the same manner as was described for ceramics.

All materials commonly thought of as piezoelectric are electrical insulators. However, since they are not perfect insulators, when a piezoelectric material is deformed, the resulting electric field decays with time, even though the deformation remains constant. For most piezoelectric materials this decay occurs over a few microseconds. Therefore the material must be strained repetitively in order to maintain an electric field which, since the piezoelectric effect is linear, will also be repetitive. On the other hand, if a constant electric field is applied, the resulting strain is sustained as long as the field is active.

Metals are not piezoelectric. Many metals have a centrosymmetric crystal structure, such as face-centered cubic, body-centered cubic, or close-packed hexagonal. Others may lack a center of symmetry, but because metals are good conductors of electricity, the electric field developed as a result of deformation decays instantaneously so that it is of no practical value.

One might expect, based on our discussions thus far, that pyroelectric materials produce an electric field in their unstrained state since such materials possess a net dipole moment in that state. This is not the case, however, because the molecules of air surrounding the material become ionized and therefore the air acts as a conductor. Thus it is really the *change* in distance between the centroids of the positive and negative charges that produces an electric field. For example, if the force system shown in Fig. 4-56 is applied to the crystal shown in Fig. 4-53, the distance $l$ changes, and therefore the distance between the centroids of the positive and negative charges changes,

---

*The alignment is not perfect. However, the axes of individual crystals will generally be aligned to within a few degrees of the poling direction.

thereby producing an electric field. Hence the noncentrosymmetric crystal shown in Fig. 4-53 is piezoelectric. Note, however, that if a shear force directed to the right is applied to the top surface in Fig. 4-53 and one directed to the left is applied to the bottom surface, the centroids both move the same amount. Consequently, with this charge arrangement no electric field is produced by shear forces applied to the top and bottom surfaces. However, with other charge arrangements shear forces can produce an electric field. What happens if shear forces are applied vertically to the right and left surfaces in Fig. 4-53?

For some materials the piezoelectric effect, although present, is too small to be of practical value. If, for example, a strain is applied to the material, the resulting electrical signal may be so small that it is indistinguishable from the electrical noise present in the system. Nylon is an example of a material whose piezoelectric effect is, for practical purposes, negligible.

In order to provide a mathematical description of piezoelectric behavior, three mutually orthogonal axes that form a right-handed system, identified and labeled 1, 2, and 3, are customarily employed.* For crystals these are assigned according to certain conventions involving the crystallographic axes. The positive sense of certain specified axes is chosen such that the corresponding piezoelectric constant is positive. Details can be found in [7].

In ceramics and polymers the 3-axis is in the poling direction with positive taken as the direction of the positive electric field used in poling. As noted earlier, poled ceramics are transversely isotropic so that perpendicular to the poling direction they are nondirectional. Therefore the orientation of the 1- and 2-axes is arbitrary except that the resulting coordinate system should be right-handed.

In piezoelectric polymers that are stretched to obtain a favorable chain orientation, the 1-axis is in the stretch direction and the 3-axis is in the poling direction. The direction of the 2-axis is chosen to form a right-handed coordinate system with the 1- and 3-axes. The positive sense of the 1-axis is arbitrary and the positive sense of the 3-axis is determined as it is for ceramics.

Because piezoelectric materials are electromechanical materials, there is an overlap of notation since certain symbols have one meaning in mechanics and another in electromagnetics. For example, in mechanics the strain is usually designated by the symbol $\varepsilon$, but in electromagnetics $\varepsilon$ is the dielectric permittivity. In order to avoid ambiguity, we will therefore adopt the notation contained in the IEEE Standard on Piezoelectricity [7] in which the pertinent symbols and their meanings are shown in Table 4-6. Also, since the electric field is represented by $E$, where necessary we will refer to Young's modulus as $Y$.

Let us now consider the piezoelectric material shown in Fig. 4-59. For definiteness we will assume that the material is a ceramic electroded and poled as shown. For the sake of simplicity, assume the material is one-dimensional, that is, we assume that stresses, strains, and electric fields all act in the poling direction, and behavior normal to the poling direction can be ignored. If a stress is applied, a strain results and, as we have discussed, the material becomes polarized in proportion to the applied stress. We can therefore write

$$P_T = dT$$

*These axes are sometimes referred to as $X$, $Y$, and $Z$.

TABLE 4-6

| Symbol | Meaning |
|---|---|
| $D$ | Electric displacement |
| $E$ | Electric field |
| $S$ | Strain |
| $T$ | Stress |
| $\varepsilon$ | Dielectric permittivity |
| $s$ | Compliance |
| $c$ | Stiffness |
| $d$ | Piezoelectric strain coefficient |

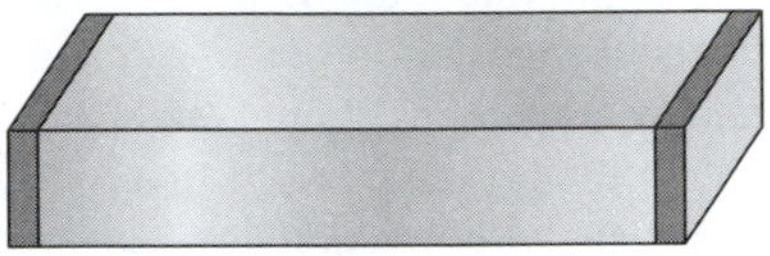


Figure 4-59

where $P_T$ is the polarization resulting from stress and $d$ is the *piezoelectric coefficient,* or *piezoelectric modulus,* a constant. $d$ is the amount of polarization resulting from a unit stress. From its definition given in the preceding, the dimensions of polarization are *charge per area*. It follows then that the dimensions of $d$ are *charge per force*. Typical units for $d$ are *coulombs per newton* (C/N) in the SI system and *coulombs per pound force* (C/lbf) in the U.S. customary system.

In the absence of stress it is known from elementary physics that the material will become polarized if it is placed in an electric field. The polarization is given by

$$P_E = \chi \varepsilon_0 E$$

where $P_E$ is the polarization resulting from the electric field, the constant $\chi$ is the dielectric susceptibility, and $\varepsilon_0$ is the permittivity of a vacuum, a universal constant. Its value is, in *farads per meter* (F/m),

$$\varepsilon_0 = 8.854 \times 10^{-12} \tag{4-179}$$

By superposition, if the material is in an electric field and subjected to a stress, the polarization is the sum of $P_T$ and $P_E$. Therefore,

$$P = dT + \chi \varepsilon_0 E \tag{4-180}$$

In the study of piezoelectricity it is customary to replace the polarization with the *electric displacement D,* defined as

$$D = \varepsilon_0 E + P$$

Therefore Eq. (4-180) becomes

$$D = dT + \varepsilon E \tag{4-181}$$

where $\varepsilon = \varepsilon_0(1 + \chi)$ is the *permittivity*. Note that the dimensions of electric displacement are the same as those of polarization.

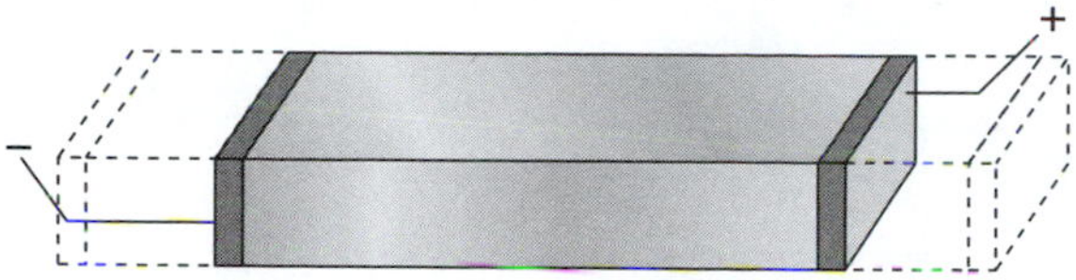


Figure 4-60

Now suppose the material is unconstrained and electrical leads are attached to the electrodes. A voltage differential is applied, as indicated in Fig. 4-60. The voltage differential is in the same direction as the poling, and therefore the material expands as shown. We can write the strain in terms of the electric field intensity as

$$S_E = aE \tag{4-182}$$

where $S_E$ is the strain due to the applied electric field and $a$ is a constant. Based on considerations of thermodynamics, it is possible to show that $a = d$. Therefore Eq. (4-182) can be written as

$$S_E = dE \tag{4-183}$$

Note that for Eq. (4-183) to be dimensionally correct the dimensions for $d$ are *length per volt*. Typical units for $d$ are (in/in)/(V/in) and (mm/mm)/(V/m). The reader should verify that coulomb/newton and (mm/mm)/(V/m) are equivalent units.

If there is no electric field but the material is subjected to a stress, the resulting strain $S_T$ is

$$S_T = sT$$

where $s$ is the compliance. For a one-dimensional formulation, $s = 1/Y$. The strain due to both the stress and the electric field is therefore given by

$$S = sT + dE \tag{4-184}$$

From Eqs. (4-181) and (4-184) note that

$$S = S(T, E), \qquad D = D(T, E)$$

that is, both the strain and the electric displacement are functions of the stress and the electric field. From Eq. (4-184) note that

$$\frac{\partial S}{\partial T} = s$$

that is, $s$ is the ratio of strain increment to stress increment *at constant electric field,* a fact that we signify by placing a superscript $E$ on $s$. Then Eq. (4-184) becomes

$$S = s^E T + dE \tag{4-185}$$

Similar reasoning shows that the permittivity in Eq. (4-181) is taken at constant stress so that Eq. (4-181) should be written as

$$D = dT + \varepsilon^T E \tag{4-186}$$

For coupled phenomena such as piezoelectricity it is important to recognize the conditions under which material constants are measured. For example, suppose a piezoelectric

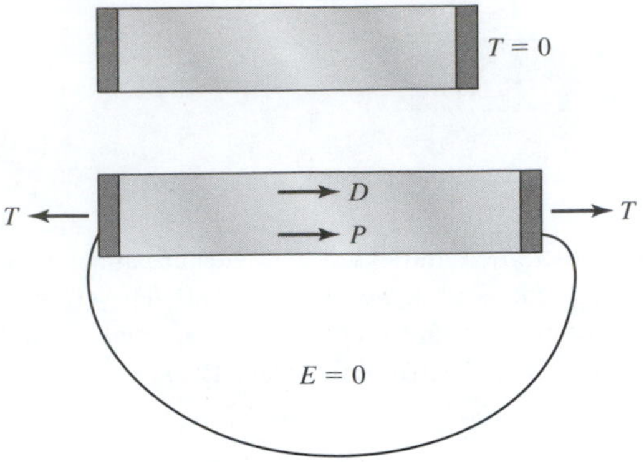


Figure 4-61 A short circuited ($E = 0$) piezoelectric sample

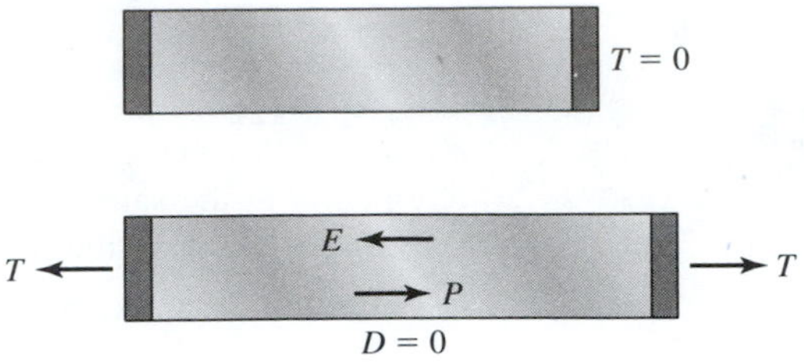


Figure 4-62

sample is short-circuited ($E = 0$), as shown in Fig. 4-61, a force is applied and incremented, and the resulting strains are measured. Then the data are used in Eq. (4-185), which reduces to

$$S = s^E T \tag{4-187}$$

The value of compliance obtained from the experimental data is therefore *at constant electric field,* as indicated by Eq. (4-187). On the other hand, suppose the same experiment is run, but this time with an open circuit ($D = 0$), as shown in Fig. 4-62. Then from Eqs. (4-185) and (4-186),

$$S = s^E T + dE, \qquad 0 = dT + \varepsilon^T E$$

The resulting electric field is no longer constant, but depends on the stress, as shown in the second equation. With it the first equation becomes

$$S = s^D T \tag{4-188}$$

where $s^D$, the compliance measured *at constant electric displacement,* is given by

$$s^D = s^E - \frac{d^2}{\varepsilon^T} \tag{4-189}$$

Thus the value of compliance measured under open-circuit conditions is not the same as that measured under short-circuit conditions. Using a similar approach, it will be seen that one can measure either the permittivity at constant stress, $\varepsilon^T$, or the permittivity at constant strain, $\varepsilon^S$. When using the equations of piezoelectricity, therefore, it is important to understand the *conditions under which the material constants have been measured.*

To extend these ideas to three-dimensional problems, consider again the situation where electrical leads are attached to the electrodes and a voltage differential is applied. The strain in the 3-direction will be accompanied, in general, by strains in the 1- and 2-directions, as well as by shear strains. We therefore write

$$\begin{aligned} S_1 &= d_{31} E_3, & S_2 &= d_{32} E_3, & S_3 &= d_{33} E_3 \\ S_4 &= d_{34} E_3, & S_5 &= d_{35} E_3, & S_6 &= d_{36} E_3 \end{aligned} \tag{4-190}$$

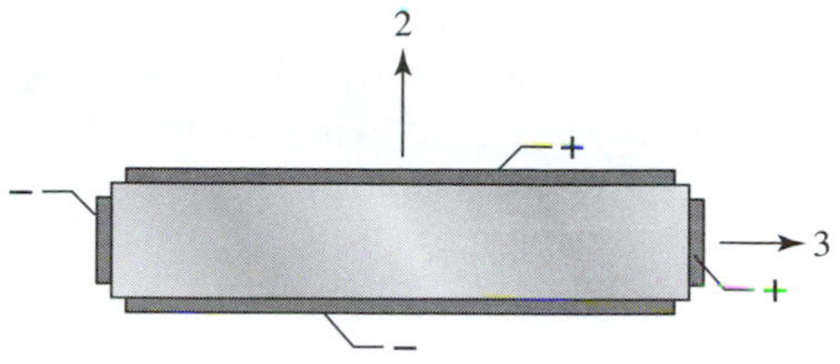


Figure 4-63

where the first subscript in $d_{ij}$ gives the direction of the electric field and the second the direction of the strain. Assume now that electrodes have been attached to the 2-faces and a field of intensity $E_2$ applied, as shown in Fig. 4-63. This causes a normal strain in the 2-direction, which is accompanied by normal strains in the 1- and 3-directions as well as shear strains in all three directions. A similar situation results if electrodes are applied to the 1-faces and a field $E_1$ is applied. The strain in the 3-direction, for example, can therefore be written as

$$S_3 = d_{13}E_1 + d_{23}E_2 + d_{33}E_3 \tag{4-191}$$

Thus there is a contribution to the strain in the 3-direction from the applied electric field in each of the directions. In general, all six strain components are influenced by the applied electric field in each of the directions 1, 2, and 3. Therefore equations similar to Eq. (4-191) can be constructed for each strain component. In matrix form these are written as

$$\{\boldsymbol{S}\} = [\boldsymbol{d}]^T\{\boldsymbol{E}\}$$

where the strain matrix and the electric field intensity matrix are, respectively,

$$\{\boldsymbol{S}\}^T = [S_1 \quad S_2 \quad S_3 \quad S_4 \quad S_5 \quad S_6] \tag{4-192}$$

$$\{\boldsymbol{E}\}^T = [E_1 \quad E_2 \quad E_3] \tag{4-193}$$

and the piezoelectric strain matrix is

$$[\boldsymbol{d}] = \begin{bmatrix} d_{11} & d_{12} & d_{13} & d_{14} & d_{15} & d_{16} \\ d_{21} & d_{22} & d_{23} & d_{24} & d_{25} & d_{26} \\ d_{31} & d_{32} & d_{33} & d_{34} & d_{35} & d_{36} \end{bmatrix} \tag{4-194}$$

Note that the superscript $T$ outside the braces or brackets signifies matrix transposition. Equation (4-194) shows that there are 18 piezoelectric strain constants, in general. In practice, however, some of these are zero owing to crystal symmetry and the remaining nonzero constants are not all independent.

If the material is subjected to mechanical loads, additional strains will result and are given by Eq. (4-34). The strain due to both the applied electric field and applied mechanical loads is the sum of the two effects,

$$\{\boldsymbol{S}\} = [\boldsymbol{s}^E]\{\boldsymbol{T}\} + [\boldsymbol{d}]^T\{\boldsymbol{E}\} \tag{4-195}$$

where, again, the superscript $E$ indicates that the elements of the compliance matrix are obtained at constant electric field. The order of elements in the stress and strain matrices are given by Eqs. (4-29*a*) or (4-32) and Eqs. (4-29*b*) or (4-33), respectively. On the other hand, if stresses are applied to a piezoelectric material, the resulting electric displacement is given by

$$\{\boldsymbol{D}\} = [\boldsymbol{d}]\{\boldsymbol{T}\}$$

where the electric displacement matrix is

$$\{\boldsymbol{D}\}^T = [D_1 \quad D_2 \quad D_3] \tag{4-196}$$

In the absence of stress the electric displacement produced by an applied electric field is

$$\{\boldsymbol{D}\} = [\boldsymbol{\varepsilon}^T]\{\boldsymbol{E}\}$$

where $[\boldsymbol{\varepsilon}]$ is the $3 \times 3$ permittivity matrix. The electric displacement due to both an applied stress and an electric field is therefore given by

$$\{\boldsymbol{D}\} = [\boldsymbol{d}]\{\boldsymbol{T}\} + [\boldsymbol{\varepsilon}^T]\{\boldsymbol{E}\} \tag{4-197}$$

Equations (4-195) and (4-197) give the general constitutive equations for piezoelectric materials. The specific forms of the $[\boldsymbol{s}^E]$ and $[\boldsymbol{d}]$ matrices for all 32 crystal classes can be found in numerous references (for example, [4], [1], [7]). We give here the form of the matrices and the numerical values of the elements for three piezoelectric materials.

### Quartz ($SiO_2$)

This is the material in which the piezoelectric effect was first observed, but its use as a piezoelectric material has largely been superseded by piezoceramics, which have a much greater piezoelectric effect. Although there are many crystalline forms of quartz, we refer here specifically to α-quartz, which crystallizes at temperatures below 573°C. The crystal goes through an α→β transformation at that temperature and the resulting β-quartz has a different crystal structure. Crystals of α-quartz belong to the trigonal system with 32 symmetry.* For this crystal class the 3-axis is coincident with the crystal's $c$ axis and the 1-axis is coincident with one of the crystal's $a$ axes. The positive sense of the 1-axis is chosen so that $d_{11}$ is positive while the positive sense of the 3-axis is arbitrary. The 2-axis forms a right-handed system with the 1- and 3-axes. Because of the crystal symmetry of this crystal class, the forms of the compliance, piezoelectric, and

---

*Briefly, with this type of symmetry, pronounced "three-two" symmetry, the first digit, 3, means that one axis (for quartz, the $c$ axis) is a 3-fold axis of symmetry. That is, an observer located in a plane normal to the $c$ axis, and at a fixed distance from the axis, encounters identical features every 1/3 revolution about the $c$ axis. In this plane, therefore, three axes, each radiating from the $c$ axis and lying 120° apart, bound regions of identical features. The second digit, 2, indicates that each of these axes is a 2-fold axis of symmetry. In general, an observer located in a plane normal to an $n$-fold axis of symmetry encounters identical features every $1/n$ revolution about the axis.

dielectric matrices are [1, app. 2]

$$[s^E] = \begin{bmatrix} s_{11} & s_{12} & s_{13} & s_{14} & 0 & 0 \\ s_{12} & s_{11} & s_{13} & -s_{14} & 0 & 0 \\ s_{13} & s_{13} & s_{33} & 0 & 0 & 0 \\ s_{14} & -s_{14} & 0 & s_{44} & 0 & 0 \\ 0 & 0 & 0 & 0 & s_{44} & 2s_{14} \\ 0 & 0 & 0 & 0 & 2s_{14} & 2(s_{11} - s_{12}) \end{bmatrix} \tag{4-198}$$

$$[\boldsymbol{d}] = \begin{bmatrix} d_{11} & -d_{11} & 0 & d_{14} & 0 & 0 \\ 0 & 0 & 0 & 0 & -d_{14} & -2d_{11} \\ 0 & 0 & 0 & 0 & 0 & 0 \end{bmatrix} \tag{4-199}$$

$$[\boldsymbol{\varepsilon}^T] = \begin{bmatrix} \varepsilon_{11} & 0 & 0 \\ 0 & \varepsilon_{11} & 0 \\ 0 & 0 & \varepsilon_{33} \end{bmatrix} \tag{4-200}$$

where [1, app. 2]

$$\begin{aligned} s_{11} &= 12.77 \times 10^{-12}\ \text{m}^2/\text{N} \\ s_{33} &= 9.60 \times 10^{-12}\ \text{m}^2/\text{N} \\ s_{44} &= 20.04 \times 10^{-12}\ \text{m}^2/\text{N} \\ s_{12} &= -1.79 \times 10^{-12}\ \text{m}^2/\text{N} \\ s_{13} &= -1.22 \times 10^{-12}\ \text{m}^2/\text{N} \\ s_{14} &= -4.50 \times 10^{-12}\ \text{m}^2/\text{N} \end{aligned} \tag{4-201}$$

$$\begin{aligned} d_{11} &= -2.3\ \text{pC/N} \\ d_{14} &= -0.67\ \text{pC/N} \end{aligned} \tag{4-202}$$

$$\begin{aligned} \varepsilon_{11}/\varepsilon_0 &= 4.52 \\ \varepsilon_{33}/\varepsilon_0 &= 4.68 \end{aligned} \tag{4-203}$$

From these forms it is seen that only six independent elastic constants, two independent piezoelectric constants, and two independent values of permittivity are needed to describe the piezoelectric behavior of quartz.

### Lead Zirconate-titanate (PZT)

This is a piezoelectric ceramic that is commonly used as the active element in ultrasonic transducers because it shows a much greater piezoelectric effect than quartz. Piezoelectric ceramics can also be readily fabricated into a variety of shapes and sizes and can therefore be tailored to a particular application. Properties given here are for the specific form, PZT-5H, which has a Curie temperature of 340°C. Since poled ceramics are transversely isotropic, the forms of the [$\mathbf{s}$], [$\mathbf{d}$], and [$\boldsymbol{\varepsilon}$] matrices are [1, app. 2]

$$[\mathbf{s}^E] = \begin{bmatrix} s_{11} & s_{12} & s_{13} & 0 & 0 & 0 \\ s_{12} & s_{11} & s_{13} & 0 & 0 & 0 \\ s_{13} & s_{13} & s_{33} & 0 & 0 & 0 \\ 0 & 0 & 0 & s_{44} & 0 & 0 \\ 0 & 0 & 0 & 0 & s_{44} & 0 \\ 0 & 0 & 0 & 0 & 0 & 2(s_{11} - s_{12}) \end{bmatrix} \tag{4-204}$$

$$[\mathbf{d}] = \begin{bmatrix} 0 & 0 & 0 & 0 & d_{15} & 0 \\ 0 & 0 & 0 & d_{15} & 0 & 0 \\ d_{31} & d_{31} & d_{33} & 0 & 0 & 0 \end{bmatrix} \tag{4-205}$$

$$[\boldsymbol{\varepsilon}^T] = \begin{bmatrix} \varepsilon_{11} & 0 & 0 \\ 0 & \varepsilon_{11} & 0 \\ 0 & 0 & \varepsilon_{33} \end{bmatrix} \tag{4-206}$$

where [1, app. 2]

$$\begin{aligned} s_{11} &= 16.50 \times 10^{-12} \text{ m}^2/\text{N} \\ s_{33} &= 20.70 \times 10^{-12} \text{ m}^2/\text{N} \\ s_{44} &= 43.5 \times 10^{-12} \text{ m}^2/\text{N} \\ s_{12} &= -4.78 \times 10^{-12} \text{ m}^2/\text{N} \\ s_{13} &= -8.45 \times 10^{-12} \text{ m}^2/\text{N} \end{aligned} \tag{4-207}$$

$$\begin{aligned} d_{15} &= 741 \text{ pC/N} \\ d_{31} &= -274 \text{ pC/N} \\ d_{33} &= 593 \text{ pC/N} \end{aligned} \tag{4-208}$$

$$\begin{aligned} \varepsilon_{11}/\varepsilon_0 &= 3130 \\ \varepsilon_{33}/\varepsilon_0 &= 3400 \end{aligned} \tag{4-209}$$

### Polyvinylidene fluoride (PVDF or $PVF_2$)

This is a semicrystalline polymer $(-CH_2-CF_2-)_n$ with a crystallinity of 40%–50%. For reasons to be noted, this material is available only in thin films, with thicknesses ranging up to about 100 μm. As a polymer it is very flexible and can easily be applied to curved surfaces. In contrast to ceramics and many crystals, PVDF is not brittle. This polymer does not have a Curie temperature, per se, but has a rather low melting temperature in the range of 165°C to 180°C, which limits its use as a piezoelectric material. A very high-voltage gradient of about 100 V/μm is required to pole PVDF. Thus a 25-μm film of this material, about 0.001 in, requires 2500 V for poling. At these voltages, equipment and safety considerations impose practical limitations on the thickness of PVDF. Crystals in PVDF belong to the orthorhombic system with *mm2* symmetry.* The stretching process causes the *a* axis of each crystal to become aligned with the stretch direction and poling causes the *c* axis to become aligned with the poling direction. It follows that the *b* axis of each crystal must also be aligned with a common direction. Each crystal must therefore have the same orientation and, as a consequence, the symmetry of stretched, poled PVDF is the same as that of its individual crystals. The $[\boldsymbol{s}]$, $[\boldsymbol{d}]$, and $[\boldsymbol{\varepsilon}]$ matrices thus have the following forms [1, app. 2]:

$$[\boldsymbol{s}^E] = \begin{bmatrix} s_{11} & s_{12} & s_{13} & 0 & 0 & 0 \\ s_{12} & s_{22} & s_{23} & 0 & 0 & 0 \\ s_{13} & s_{23} & s_{33} & 0 & 0 & 0 \\ 0 & 0 & 0 & s_{44} & 0 & 0 \\ 0 & 0 & 0 & 0 & s_{55} & 0 \\ 0 & 0 & 0 & 0 & 0 & s_{66} \end{bmatrix} \tag{4-210}$$

$$[\boldsymbol{d}] = \begin{bmatrix} 0 & 0 & 0 & 0 & d_{15} & 0 \\ 0 & 0 & 0 & d_{24} & 0 & 0 \\ d_{31} & d_{32} & d_{33} & 0 & 0 & 0 \end{bmatrix} \tag{4-211}$$

$$[\boldsymbol{\varepsilon}^T] = \begin{bmatrix} \varepsilon_{11} & 0 & 0 \\ 0 & \varepsilon_{22} & 0 \\ 0 & 0 & \varepsilon_{33} \end{bmatrix} \tag{4-212}$$

where [22]

$$\begin{aligned} s_{11} &= 365 \times 10^{-12}\ \text{m}^2/\text{N} \\ s_{22} &= 424 \times 10^{-12}\ \text{m}^2/\text{N} \\ s_{33} &= 472 \times 10^{-12}\ \text{m}^2/\text{N} \\ s_{12} &= -110 \times 10^{-12}\ \text{m}^2/\text{N} \\ s_{13} &= -209 \times 10^{-12}\ \text{m}^2/\text{N} \\ s_{23} &= -192 \times 10^{-12}\ \text{m}^2/\text{N} \end{aligned} \tag{4-213}$$

$$\begin{aligned} d_{15} &= -27\ \text{pC/N} \\ d_{24} &= 23\ \text{pC/N} \\ d_{31} &= 21\ \text{pC/N} \\ d_{32} &= 2.3\ \text{pC/N} \\ d_{33} &= -26\ \text{pC/N} \end{aligned} \tag{4-214}$$

$$\begin{aligned} \varepsilon_{11}/\varepsilon_0 &= 6.9 \\ \varepsilon_{22}/\varepsilon_0 &= 8.6 \\ \varepsilon_{33}/\varepsilon_0 &= 7.6 \end{aligned} \tag{4-215}$$

Thus nine elastic constants, five piezoelectric constants, and three dielectric constants are needed to describe the piezoelectric behavior of PVDF.

---

*Briefly, the crystallographic *c* axis is a 2-fold axis of symmetry, and two perpendicular mirror planes, indicated by the symbol *m*, intersect along the *c* axis.

## EXAMPLE 4-3

Let us consider an actuator made from each of these materials. An electric field is applied and we wish to determine the resulting strains. We assume that there is no applied stress. We wish to generate normal strains in each case in order to change the thickness of the piezoelectric material. Therefore we must chose the direction of the electric field accordingly. For example, from the form of the $[\boldsymbol{d}]$ matrix for quartz, only a field in the 1-direction will produce a normal strain, whereas for PZT and PVDF only a field in the 3-direction will produce normal strains. Let us assume a field of 1000 V/mm. The resulting strains for each material, electroded as shown in Fig. 4-64, are therefore

$$\{\boldsymbol{S}_{\text{quartz}}\} = [-2.30 \times 10^{-6} \quad 2.30 \times 10^{-6} \quad 0 \quad -0.67 \times 10^{-6} \quad 0 \quad 0]^T$$

$$\{\boldsymbol{S}_{\text{PZT}}\} = [-274 \times 10^{-6} \quad -274 \times 10^{-6} \quad 593 \times 10^{-6} \quad 0 \quad 0 \quad 0]^T$$

$$\{\boldsymbol{S}_{\text{PVDF}}\} = [21 \times 10^{-6} \quad 2.30 \times 10^{-6} \quad -26 \times 10^{-6} \quad 0 \quad 0 \quad 0]^T$$

The behaviors of these three materials differ markedly. Each undergoes the maximum strain in the direction of the electric field, although for quartz and PVDF that strain is a contraction whereas it is an expansion for PZT. For quartz one of the lateral strains is of the same magnitude but opposite in sign, but the other lateral strain is zero; only quartz shows a shear strain, although it is quite small compared to the normal strains.

PZT shows the same strain in each lateral direction, which is to be expected since, as transversely isotropic materials, perpendicular to the poling direction ceramics are isotropic. It was noted earlier that the piezoelectric effect in PZT is much greater than that in quartz. The equations show this quite clearly since the magnitude of the strains in PZT are two orders of magnitude greater than those in quartz. They are also an order of magnitude greater than the largest strains in PVDF. This makes PZT an excellent piezoelectric material for ultrasonic transducers since, for a given electric field, it will produce the strongest ultrasonic signal of the three materials.

Although the strain in the electric field direction in PVDF is as large as the strain in the stretch direction, the resulting displacement in the field direction is very small because, as a thin film material, the thickness in that direction is extremely small, as noted before. Since the strain in the 2-direction is an order of magnitude less than that in the other directions, in practice it is often neglected. Consequently in PVDF the most significant strain is that in the 1-direction, which results from an electric field in the 3-direction.

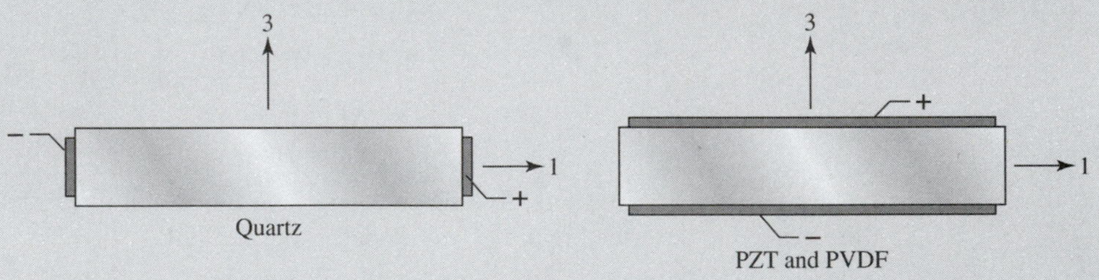


**Figure 4-64**

For quartz and PZT it is possible to apply an electric field in the 2-direction to produce shear strains without normal strains because the appropriate faces can be electroded. A similar situation holds in the remaining direction (the 1-direction) for PZT but not for quartz, where an electric field in the remaining direction (the 3-direction) produces no piezoelectric strain. With PVDF, however, the thinness of the film makes it impossible to apply electrodes to the film edges. Thus an electric field can be applied only through the thickness of the PVDF film.

## PROBLEMS

**4-1** Derive Eqs. (4-12) from Eqs. (4-8).

**4-2** Show that, in the cylindrical coordinates $r$, $\theta$, $z$,

$$\varepsilon_{rr} = \frac{1}{E}[\sigma_{rr} - \nu(\sigma_{\theta\theta} + \sigma_{zz})]$$

$$\varepsilon_{\theta\theta} = \frac{1}{E}[\sigma_{\theta\theta} - \nu(\sigma_{rr} + \sigma_{zz})]$$

$$\varepsilon_{zz} = \frac{1}{E}[\sigma_{zz} - \nu(\sigma_{rr} + \sigma_{\theta\theta})]$$

$$\varepsilon_{r\theta} = \frac{1}{G}\sigma_{r\theta}$$

$$\varepsilon_{\theta z} = \frac{1}{G}\sigma_{\theta z}$$

$$\varepsilon_{zr} = \frac{1}{G}\sigma_{zr}$$

*Hint:* Use Eqs. (2-49), (3-31), (4-8), and (4-11).

**4-3** Solve the equations in Problem 4-2 for stresses in terms of the strains and then use Eqs. (3-82) to show that, in the case of plane stress,

$$\sigma_{rr} = \frac{E}{1-\nu^2}(\varepsilon_{rr} + \nu\varepsilon_{\theta\theta})$$

$$\sigma_{\theta\theta} = \frac{E}{1-\nu^2}(\varepsilon_{\theta\theta} + \nu\varepsilon_{rr})$$

$$\sigma_{r\theta} = G\gamma_{r\theta}$$

$$\varepsilon_{zz} = -\frac{\nu}{1-\nu}(\varepsilon_{rr} + \varepsilon_{\theta\theta})$$

**4-4** An isotropic, elastic material fills a cup whose side surface is circular and whose bottom surface is flat. The cup is perfectly rigid and there is no friction between the elastic material and the cup. A pressure $p$ is applied uniformly to the unconstrained (top) surface of the material. Determine the resulting stresses when Poisson's ratio is (a) 0.0; (b) 0.5. Find the dilatation for each of these cases. Provide a physical explanation of your mathematical results.

**4-5** For an orthotropic material under conditions of plane stress, derive expressions for the matrix elements in Eq. (4-61) in terms of $\overline{S}_{11}, \overline{S}_{12}, \ldots$. Then use Eq. (4-59) to obtain Eqs. (4-62).

**4-6** Derive an expression for $[\hat{S}]$ in terms of $[\overline{S}]$ and $[\hat{\boldsymbol{R}}]$. Determine the elements of $[\hat{S}]$ and from these deduce Eqs. (4-80).

**4-7** A certain glass-fiber and polyester resin composite lamina has properties in the principal material coordinates of $E_{x'} = 5.8 \times 10^6$ psi, $E_{y'} = 1.2 \times 10^6$ psi, $G_{x'y'} = 5.7 \times 10^5$ psi, and $\nu_{x'y'} = 0.26$. Plot the variation of $E_x/E_{x'}$, $G_{xy}/G_{x'y'}$, $\eta_{xy,y}$, and $\nu_{x'y'}$ as a function of $\theta$ for $0 \leq \theta \leq \pi/2$.

**4-8** Figure P4-8 shows the basic structure of a hexagonal crystal, referred to as *primitive hexagonal*. Atom positions are indicated by

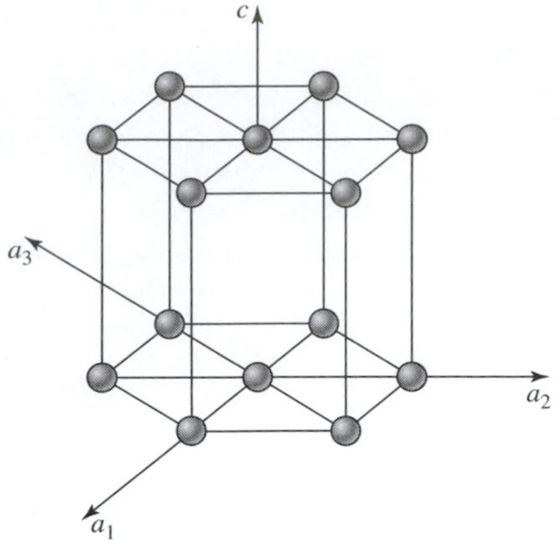


Figure P4-8

the spheres. The plane formed by axes $a_1$, $a_2$, and $a_3$ is the basal plane. The $c$ axis is normal to the basal plane. The positive $a_1$, $a_2$, and $a_3$ axes are 120° apart. Taken together with their negative portions, these axes divide the basal plane into 60° segments. The elastic properties of the crystal at a point in one segment of the basal plane, or any plane parallel to it, are the same at the corresponding point in the other 5 segments. Consequently, the elastic properties repeat themselves for every 60° of rotation about the $c$ axis.

Obtain the stiffness matrix for a hexagonal crystal. (*Hint:* Follow the basic steps of Section 4-9 for a transversely isotropic material. Let the crystal's $c$ axis coincide with the $z$ axis in an $x$, $y$, $z$ coordinate system.)

**4-9** Show graphically that a material with two planes of elastic symmetry must possess a third plane of elastic symmetry.

**4-10** In order to gain an understanding of the meaning of the various Poisson's ratios for orthotropic materials, consider a two-dimensional problem for an element of dimensions $a \times a$. Apply a unit stress in the $x$ direction and compute the displacements. Then compute the strains when the same stress is applied in the $y$ direction. What conclusions can you draw?

**4-11** A cylindrical plug of length $L$ and circular cross-section of radius $a$ has its curved surface compressed uniformly an amount $\delta$. The radial displacement is given by $\delta r/a$ where $r$ is the radius from the axis of the plug, which is free to expand and contract longitudinally because its lateral surface is frictionless and its end surfaces are unconstrained. Find the stresses on the surfaces of the plug.

**4-12** Show that by proper choice of constants the standard linear model reduces to the Kelvin–Voigt model or the Maxwell model.

**4-13** Show how to compute the curves in Fig. 4-32 mathematically. *Hint:* Let $F_0 = 1$. Subtract load $F_0$ at $t = \tau_2$ using Heaviside's function. Assume $t_1 = 10$ sec, $t_2 = 20$ sec, $G = G_2 = 15\ \text{in}^{-1}$, $\eta = \eta_2 = 3000$, $G_1 = 30\ \text{in}^{-1}$.

**4-14** An alternative to the yield theories presented in Section 4-13-2 for ductile materials is the *octahedral shear theory*. The arguments for the theory are as follows. It is known that ductile materials fail by shear. The octahedral normal stress is the *mean* or *hydrostatic* stress, which we know does not influence yielding. Depending on the stress state, the octahedral shear stress lies somewhere in the range $\frac{1}{2}\sqrt{3}\tau_{max} \le \tau_{oct} \le \tau_{max}$, and so is never too much smaller than the maximum shear stress. However, there are eight octahedral shear planes but only four maximum shear planes. $\tau_{oct}$ therefore has a greater statistical likelihood of lying on or near a crystal plane favorably oriented for slip, and this may overcome the disadvantage of $\tau_{oct}$ being slightly smaller than $\tau_{max}$.

The octahedral shear theory states that yielding begins when the actual octahedral shear stress reaches a critical value $k$, which is found from the one-dimensional tensile test, as was done for the maximum shear theory and the von Mises theory. Develop the yield condition for the octahedral shear theory and compare it with the yield conditions for the maximum shear and von Mises theories.

## REFERENCES


[1] Auld, B. A., *Acoustic Fields and Waves in Solids,* vol. 1, 2nd ed. (Krieger Publishing, Malabar, FL, 1990).

[2] Bloss, F. D., *Crystallography and Crystal Chemistry* (Holt, Rinehart and Winston, New York, 1971).

[3] Bolz, R. E., and G. L. Tuve, Eds., *Handbook of Tables for Applied Engineering Science,* 2nd ed. (CRC Press, Cleveland, OH, 1973).

[4] Cady, W. G., *Piezoelectricity* (McGraw-Hill, New York, 1946).

[5] Fung, Y. C., *Foundations of Solid Mechanics* (Prentice Hall, Englewood Cliffs, NJ, 1965).

[6] Hencky, H., "Zur Theorie plastischer Deformationen und der hierdurch im Material hervorgerufenen Nebenspannungen," *Z. angew. Math. Mech.,* **4**, pp. 323–334 (1924).

[7] IEEE Standard on Piezoelectricity, *IEEE Trans. Sonics and Ultrasonics,* **31**, no. 2 (1984).

[8] Ikeda, T., *Fundamentals of Piezoelectricity* (Oxford University Press, Oxford, UK, 1996).

[9] Jones, R. M., *Mechanics of Composite Materials* (McGraw-Hill, New York, 1975).

[10] Lakes, R., "Foam Structures with a Negative Poisson's Ratio," *Science,* **235**, pp. 1038–1140 (1987).

[11] Mal, A. K., and S. J., Singh, *Deformation of Elastic Solids* (Prentice-Hall, Englewood Cliffs, NJ, 1991).

[12] Mason, W. P., *Piezoelectric Crystals and Their Application to Ultrasonics* (Van Norstrand, New York, 1950).

[13] Mendelson, A., *Plasticity: Theory and Applicaton,* reprint ed. (Krieger Publishing, Malabar, FL, 1983, © 1968).

[14] Mises, R. von, "Mechanik der festen Körper im plastisch deformablen Zustand," *Goettinger Nachr., Math.-Phys. Kl.,* pp. 582–592 (1913).

[15] Nye, J. F., *Physical Properties of Crystals* (Oxford University Press, London, UK, 1957).

[16] Prandtl, L., "Spannungsverteilung in plastischen Körpern," in *Proc. 1st Congr. Appl. Mech.* (Delft), pp. 43–54 (1924).

[17] Reuss, E., "Berücksichtigung der elastischen Formänderungen in der Plastizitätstheorie," *Z. angew. Math. Mech.,* **10**, pp. 266–274 (1930).

[18] Reuter, R. C., Jr., "Concise Property Tranformation Relations for an Anisotropic Lamina," *J. Composite Materials,* **5**, pp. 270–272 (1971).

[19] Shames, I. H., and F. A., Cozzarelli, *Elastic and Inelastic Stress Analysis* (Prentice Hall, Englewood Cliffs, NJ, 1992).

[20] Taylor, G. I., and H., Quinney, "The Plastic Distortion of Metals," *Phil. Trans. Roy. Soc. (London),* **A230**, pp. 323–362 (1931).

[21] Touloukian, Y. S., R. K. Kirby, R. E. Taylor, and P. D. Desai, *Thermophysical Properties of Matter,* vol. 12, *Thermal Expansion: Metallic Elements and Alloys* (Plenum Publishing, New York, 1975).

[22] Wang, H., Q. M. Zhang, and L. E., Cross, "Piezoelectric, Dielectric, and Elastic Properties of Polyvinylidene Fluoride/Trifluoroethylene," *J. Appl. Phys.,* **74**, pp. 3394–3398 (1993).

# ENERGY CONCEPTS

Many problems in the mechanics of solids can be solved in a very efficient way through the use of certain theorems involving the concept of energy. In some cases, use of these theorems leads to an exact solution requiring much less effort on the part of the analyst than the Newtonian methods, which involve the use of Newton's first and second laws, learned thus far. In those cases where an exact solution is not feasible, these energy theorems may be used to develop approximate methods that are extremely powerful. It is these approximate methods that enable engineers to solve the complex problems typically encountered in designing the many technological devices used by our society.

In this chapter we develop the first and second laws of thermodynamics for continuous media and then specialize these for problems in the mechanics of solids. Once the laws of thermodynamics are developed, we introduce the concept of strain energy and then proceed to develop several energy theorems useful for solving problems in solid mechanics.

## 5-1 FUNDAMENTAL CONCEPTS AND DEFINITIONS

In the nomenclature of thermodynamics a *system* is a particular collection of matter being studied. It may consist of a part of a material body, an entire body, or perhaps several bodies. We envision a system as being separated from all remaining matter by a clearly defined *system boundary*. All matter not contained in the system constitutes the *surroundings*. The influence of the surroundings on the system is accounted for by allowing certain quantities (heat, for example) to cross the system boundary. In this regard, a system is similar to the free-body diagram used in the study of mechanics. In the free-body diagram some material is isolated from the remaining material for the purpose of focused study. The action of the remaining material on the material contained in the free-body diagram is represented by forces applied to the boundary of the free-body diagram.

An *open* system is capable of exchanging mass with its surroundings while a *closed* system is not. Consideration in this book is restricted to closed systems. A system that is not capable of interacting with its surroundings is referred to as an *isolated* system. As a practical matter, many systems are treated as isolated systems because the influence of the surroundings on the behavior of the system under study is negligible. For example, if the gas contained in an insulated cylinder is pressurized by means of a piston, the system may be considered to be the gas itself. For the purpose of computing the temperature of the gas, if it is assumed that the cylinder is perfectly insulated so that no heat is transferred, the system is an isolated system.

A system can be described in terms of certain measurable characteristics called *properties*. Examples of properties are pressure, temperature, and density. It is often desirable to refer to properties and other quantities on a unit mass basis. In that case the quantity is referred to as a *specific* quantity. For example, specific energy is energy per unit mass. We will identify specific quantities by the use of lowercase letters so that, for example, if $E$ is the energy of the system, then its specific energy is $e$.

A *uniform* system is one in which the properties are constant throughout the system, whereas in a *continuous* system the properties vary from point to point.

The *state* of a system is known when all its properties have known values. A system may change from one state to another, in which case it is said to have performed a *process*. For example, a rod made of a deformable material and fixed at one end will, upon application of a load at its free end, deform. Its state changes from an unstressed state to a stressed state. The particular sequence of states that a system goes through during a process is the *path*. There is no unique path between two states. For example, the load on our one-dimensional rod may be increased in such a way that the stress changes gradually and monotonically from zero to 2000 psi, or it may be varied such that the stress increases to 1000 psi, then decreases to 500 psi, then increases to 3000 psi, and finally decreases to 2000 psi. Although in each case the final state, as characterized by the stress of 2000 psi, is the same, the sequence of states through which the rod passed, and hence the path, is different.

In order for a characteristic of the system to be a property, it must be *independent* of the path. In other words, the difference in properties between two states does not depend on the path taken to get from one to the other.

If, at the end of a process, the final state is the same as the initial state (meaning that there is no net change in the properties of the system), the system has undergone a *cycle*.

When a system interacts with its surroundings, its properties change with time. However, if the properties of the surroundings do not change, then, after a certain length of time (which depends on the nature of the interaction) no further change in the system properties will be observed. The system has then reached a state of *thermodynamic equilibrium*. For a continuous system to be in thermodynamic equilibrium, the distribution of properties (pressure distribution, for example) must not change with time. For such systems we say that the properties have reached an *equilibrium distribution* [2, pp. 215–216]. The properties of a system and hence its state can only be determined if the system is in thermodynamic equilibrium. Thus for any real system undergoing a process, although its initial and final states can be determined, intermediate states and hence the path cannot be determined. Fortunately it is possible to idealize many real processes as ones in which, at any point in the process, the deviation from equilibrium is infinitesimal. These idealizations are referred to as *quasiequilibrium* or *quasistatic* processes, and during such a process, all states that the system passes through may be considered equilibrium states. With this idealization, the path of the process may be determined. If a system is undergoing a quasistatic process, it can be made to follow the original path in the opposite direction. For this reason, quasistatic processes are *reversible*. Since a quasistatic process is an idealization, all real processes are *irreversible*.

## 5-2 WORK

In mechanics we define *work* by considering a path $C$, as shown in Fig. 5-1, along which a point $Q$ moves. Attached to $Q$ is a force $\boldsymbol{F}$, which, as it moves from point 1 to point 2 along $C$, may change in both magnitude and spatial orientation. The location of $\boldsymbol{F}$ along the path is given by a position vector $\boldsymbol{r}$ from a fixed point to $Q$. As $Q$ moves an infinitesimal amount along $C$ to $Q'$, the position vector changes from $\boldsymbol{r}$ to $\boldsymbol{r} + d\boldsymbol{r}$, where $d\boldsymbol{r}$ is the vector from $Q$ to $Q'$. The infinitesimal amount of work done by the force $\boldsymbol{F}$ in moving an amount $d\boldsymbol{r}$ along $C$ is defined as

$$dW = \boldsymbol{F} \cdot d\boldsymbol{r} \tag{5-1}$$

Note that only the component of $\boldsymbol{F}$ that is parallel to $d\boldsymbol{r}$ (or tangent to the path at $Q$) contributes to $W$. The component of $\boldsymbol{F}$ that is perpendicular to the path at $Q$ does no work. It follows from Eq. (5-1) that the work done by the force $\boldsymbol{F}$ as it moves along $C$ from point 1 to point 2 is

$$W = \int_1^2 \boldsymbol{F} \cdot d\boldsymbol{r} \tag{5-2}$$

From the definition of the dot product we see that

$$\boldsymbol{F} \cdot d\boldsymbol{r} = F \cos\theta \, ds \tag{5-3}$$

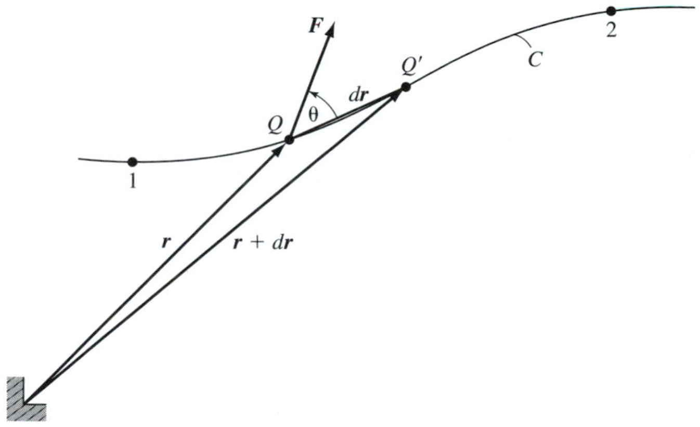


Figure 5-1

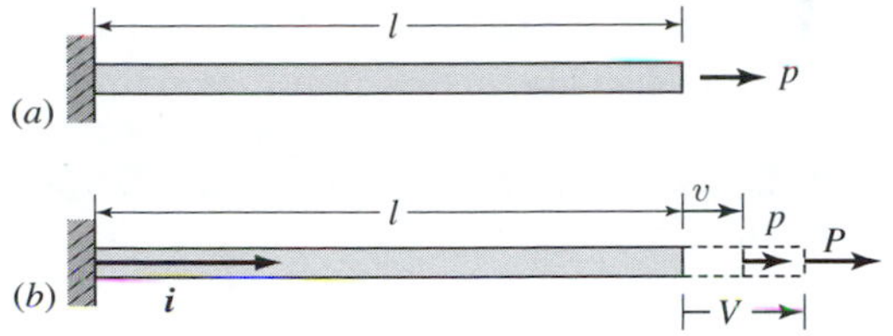


Figure 5-2

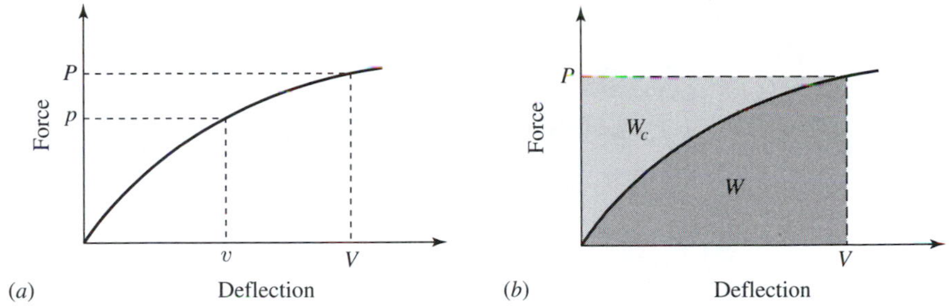


Figure 5-3

where θ is the angle between $\boldsymbol{F}$ and $d\boldsymbol{r}$ at $Q$, $F$ is the magnitude of $\boldsymbol{F}$, and $ds$ is the length of arc associated with $d\boldsymbol{r}$. Here we have assumed that for infinitesimal quantities, $ds \approx |d\boldsymbol{r}|$.

To apply the definition of work to a deformable body, we consider the one-dimensional problem of a rod of initial length $l$, fixed at one end and subjected to an applied force of magnitude $p$, directed along the axis of the rod, at the other end, as shown in Fig. 5-2*a*. As the load is gradually increased to $P$, the rod elongates an amount $V$, as shown in Fig. 5-2*b* and in the force-displacement curve for the material (Fig. 5-3*a*). At some intermediate point in the loading process the load has a value of $p$ while the displacement is $v$. We establish a coordinate axis along the axis of the rod, and associate with this axis the unit vector $\boldsymbol{i}$, as shown in Fig. 5-2*b*. The force vector in Eq. (5-2) then becomes

$$\boldsymbol{F} = p\boldsymbol{i} \tag{5-4}$$

Attached to the fixed end of the rod and extending to the loaded end is a position vector given by

$$\boldsymbol{r} = (l + v)\boldsymbol{i} \tag{5-5}$$

From this it follows that

$$d\boldsymbol{r} = dv\,\boldsymbol{i} \tag{5-6}$$

The variable of integration here is $v$, which ranges from zero in the unloaded state to $V$. Thus with Eqs. (5-4) and (5-6), and with $\boldsymbol{i} \cdot \boldsymbol{i} = 1$, Eq. (5-2) takes the form

$$W = \int_0^V p\,dv \tag{5-7}$$

If we compare this equation with Fig. 5-3*b*, we see that the work done by a force increasing in magnitude from zero to $P$ is simply the area under the force-displacement curve.

In the special case where a linear relationship exists between force and displacement so that

$$p = Cv \tag{5-8}$$

where $C$ is a constant, Eq. (5-7) becomes

$$W = C\int_0^V v\,dv = \frac{1}{2}CV^2 \tag{5-9}$$

Alternatively, using Eq. (5-8), Eq. (5-9) can be written as either

$$W = \tfrac{1}{2}PV \tag{5-10}$$

or

$$W = \frac{P^2}{2C} \tag{5-11}$$

It is customary in mechanics to refer to the area above the force-displacement curve as the *complementary work*. Thus from Fig. 5-3*b* we have

$$W_c = \int_0^P v\,dp = PV - W \tag{5-12}$$

Substitution of Eq. (5-10) into Eq. (5-12) shows that work and complementary work are equal for materials with a linear relationship between force and displacement.

It appears to be a point of confusion among readers first encountering the notion of work as it applies to deformable bodies with a linear force-displacement relationship as to when the factor of $1/2$ should be included in the calculation of work and when it should not. In the foregoing example, note that there is a change in the displacement that occurs *simultaneously* with the change in force. It is this simultaneous change in force and displacement, along with the linear relationship between those quantities, that causes the factor of $1/2$ to appear in the expressions for work given by Eqs. (5-9)–(5-11). In order to further clarify this point, let us consider the following example.

## EXAMPLE 5-1

Calculate the work done by the force $P$ in the following two-step process. First a one-dimensional rod of length $l$ with a linear force-displacement relationship is subjected to a force whose magnitude increases from zero to some value $P$, causing the rod to elongate by an amount $V$. Then, while the load is held fixed at $P$, the rod is heated in such a way that an additional elongation $V$ takes place due to thermal expansion.

## SOLUTION

The force-displacement diagram for this process is shown in Fig. 5-4. It consists of two straight-line segments—one while the load is increasing and the other while the rod is undergoing thermal expansion at fixed load. The work done by $\boldsymbol{P}$ during this process is given by

$$W = \int_0^V p\,dv + P\int_V^{2V} dv \tag{a}$$

which, with Eq. (5-8), gives

$$W = \tfrac{1}{2}PV + PV = \tfrac{3}{2}PV \tag{b}$$

Note that, again, the work done during the process is the area under the force-displacement curve. During the first part of the process, the force causes the displacement, whereas during the second part of the process, the displacement is caused by the heating of the rod, not by the force. Nevertheless, since the force moves in the direction of the displacement during the second part of the process, it still does work. It is always helpful, in visualizing what is happening in these problems, to draw the force-displacement diagram for the process.

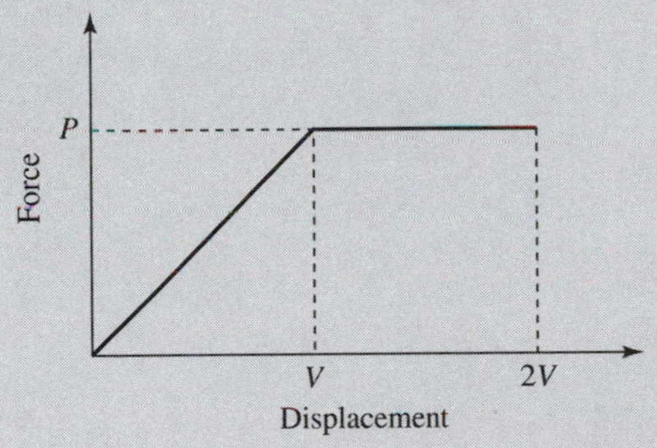


**Figure 5-4**

The following problem provides another example.

## EXAMPLE 5-2

A concentrated load is slowly applied at the midpoint of a simply supported beam, as shown in Fig. 5-5*a*. Once the beam touches the stop, the load continues to increase until its value is twice that required to reach the stop. With this load applied, the beam is then heated in such a way that its midpoint returns to its original location. A linear relationship

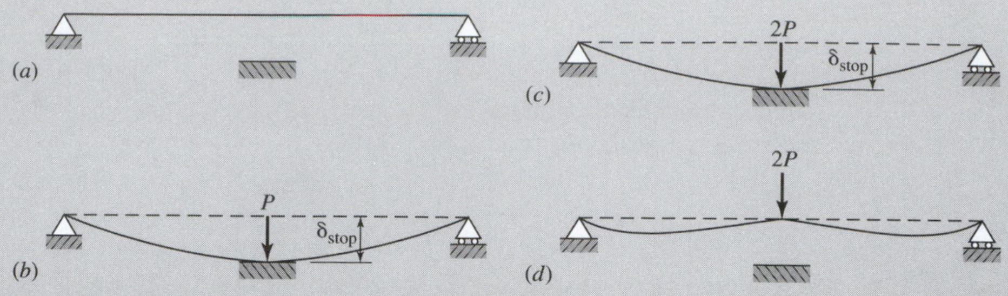


Figure 5-5

exists between force and deflection. Draw a single force-deflection diagram for the entire process. Compute the net work done by the force assuming that the force–deflection relationship is linear.

## SOLUTION

Let $P$ be the load required to reach the stop. As the load increases from 0 to $P$, the midpoint deflection increases from 0 to $\delta_{stop}$. If the undeformed state of the beam is state 1 and the deformed state when the beam just contacts the stop is state 2, as shown in Fig. 5-5*b*, then the force-deflection diagram is as shown in Fig. 5-6*a*. Note that in this

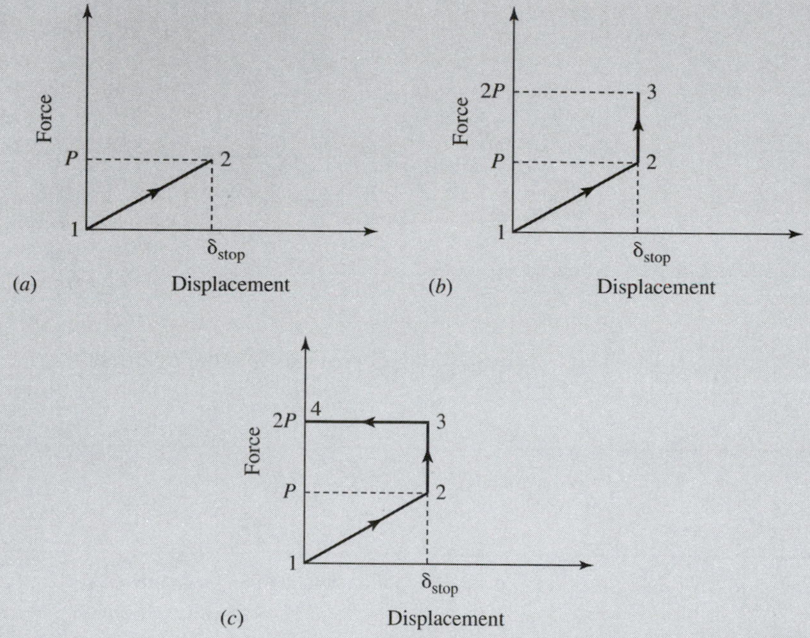


Figure 5-6

part of the loading process the deflection increases as the load is increased. The work done in going from state 1 to state 2 is

$$W_{1/2} = \tfrac{1}{2} P \delta_{\text{stop}}$$

The beam having reached the stop, the load continues to increase to a value of $2P$. However, since the beam is in contact with the stop, no further increase in displacement occurs. The situation with the beam contacting the stop and the load at a value of $2P$ is referred to as state 3, and is shown in Fig. 5-5*c*. The force-deflection diagram in going from state 1 to state 3 is shown in Fig. 5-6*b*. Note that there is no work done in going from state 2 to state 3. Thus,

$$W_{1/3} = W_{1/2} + W_{2/3} = \tfrac{1}{2} P \delta_{\text{stop}}$$

Heat is now added to the beam in such a way that the midpoint returns to its original location, as shown in Fig. 5-5*d*. This is state 4. During this part of the process the midpoint displacement changes from $\delta_{\text{stop}}$ to 0, but the load remains constant at $2P$. This is shown in the force-deflection diagram of Fig. 5-6*c*. The work done in going from state 3 to state 4 is*

$$W_{3/4} = 2P(0 - \delta_{\text{stop}}) = -2P\delta_{\text{stop}}$$

and the net work done in going from state 1 to state 4 is

$$W_{\text{net}} = W_{1/4} = W_{1/2} + W_{2/3} + W_{3/4} = -\tfrac{3}{2} P \delta_{\text{stop}}$$

It is worth stating that through an appropriate (and probably somewhat artificial) manipulation of load and heat input, we have managed to apply a load of $2P$ to a beam while causing no net displacement at the load application point. Further, contrary to what our intuition may have told us, the net work done by the load is not zero.

*If Eq. (5-2) is used, we have $W_{3/4} = \int_3^4 \boldsymbol{F} \cdot d\boldsymbol{r} = 2P \int_{\delta_{\text{stop}}}^0 d\delta = -2P\delta_{\text{stop}}$.

When several forces are applied to a deformable body, the work done by these forces is simply the sum of the work done by each individual force,

$$W = \sum_{i=1}^{n} W_i \tag{5-13}$$

## EXAMPLE 5-3

Calculate the work done by two forces of magnitude $P$ and $Q$ applied at points 1 and 2, respectively, on the cantilever beam shown in Fig. 5-7*a*. We will assume that the force $\boldsymbol{P}$ is applied first, followed by $\boldsymbol{Q}$, and that the force–deflection relationship is linear.

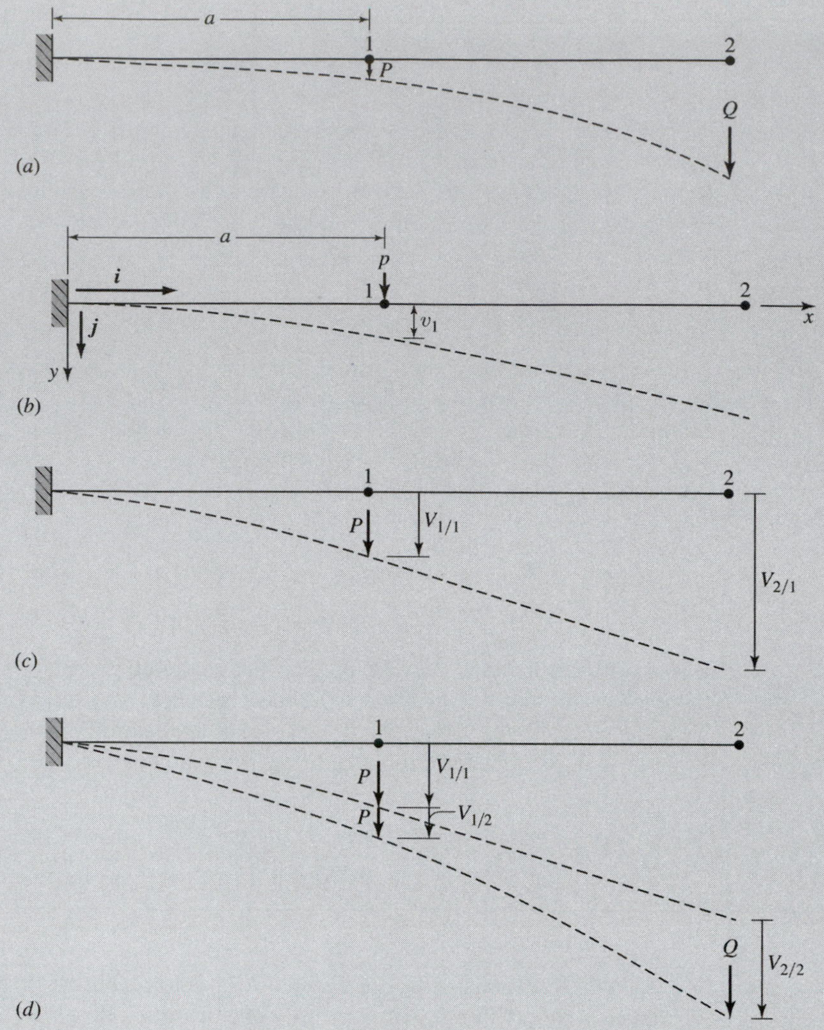


**Figure 5-7**

## SOLUTION

At point 1 the load is increased from zero to some intermediate value $p$, causing the beam to deform as shown in Fig. 5-7$b$. The force-displacement diagram for point 1 is shown in Fig. 5-8$a$. By establishing coordinate axes as shown in Fig. 5-7$b$ and drawing a position vector from the fixed end of the beam to point 1, we have

$$\boldsymbol{F} = p\boldsymbol{j} \tag{a}$$

$$\boldsymbol{r} = a\boldsymbol{i} + v\boldsymbol{j} \tag{b}$$

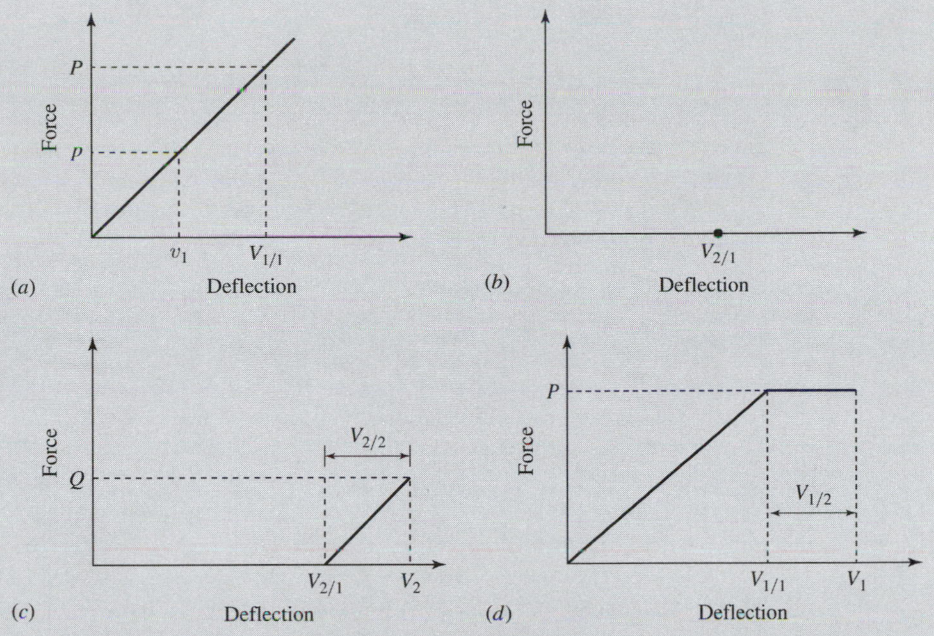


**Figure 5-8**

and therefore,

$$d\boldsymbol{r} = dv\,\boldsymbol{j} \tag{c}$$

Loading continues at point 1 until $P$ is reached. The resulting displacement is $V_{1/1}$, where the first subscript refers to the point at which the displacement is measured and the second to the point at which the force is applied. The work done by the force $\boldsymbol{P}$ is then given by Eq. (5-2) as

$$W = \int_0^{V_{1/1}} p\,dv = \frac{1}{2} P V_{1/1} \tag{d}$$

The configuration of the beam at this point is shown in Fig. 5-7*c*. The force-displacement diagram for point 2 during this part of the process is shown in Fig. 5-8*b*. Although the displacement at point 2 is $V_{2/1}$, since there is no force applied at point 2, there is no work done by the displacement.

Now, with the load at point 1 held at $P$, the force $\boldsymbol{Q}$ is applied at point 2. As shown in Fig. 5-7*d*, this causes an additional displacement of $V_{2/2}$ at point 2 so that the total displacement $V_2$ is the sum of $V_{2/1}$ and $V_{2/2}$. The force-displacement diagram for point 2 at this point in the process is shown in Fig. 5-8*c*. There is also an additional displacement at point 1 of $V_{1/2}$. Since the load at point 1 stays at $P$, the force-displacement diagram for point 1 at this point in the process is as shown in Fig. 5-8*d*. The amount of work done by $\boldsymbol{Q}$ in moving through the distance $V_2$ is

$$W = \int_{V_{2/1}}^{V_{2/1}+V_{2/2}} q\, dv = \frac{1}{2} Q V_{2/2} \tag{e}$$

Note that this is the area under the force-displacement diagram for point 2. As the load is applied at point 2, additional work is done by the force $\boldsymbol{P}$ at point 1 as it moves through the displacement $V_{1/2}$. Since the force $\boldsymbol{P}$ is constant during this part of the process, the additional work is given by

$$W = P V_{1/2} \tag{f}$$

The total amount of work done by the force $\boldsymbol{P}$ is found by adding Eqs. (*d*) and (*f*),

$$W = \tfrac{1}{2} P V_{1/1} + P V_{1/2} \tag{g}$$

Finally the total amount of work done by the applied forces $\boldsymbol{P}$ and $\boldsymbol{Q}$ is

$$W = \tfrac{1}{2} P V_{1/1} + \tfrac{1}{2} Q V_{2/2} + P V_{1/2} \tag{h}$$

In general, the work done by forces applied to a deformable body depends on the order in which the forces are applied. However, as we will show in Section 5-5, if the loading is such that the material remains elastic, the work done by the applied forces is independent of the order in which these forces are applied. That is, the elastic loading process is one that is *path independent* because the final state of the system is independent of the path (in this case, the order of loading) taken to get from the initial to the final state. In contrast, more general loading processes are *path dependent* because the final state depends on the order in which the loads are applied. A loading process that extends beyond the elastic limit is an example of a process that is path dependent, as explained in Section 4-12 when discussing Figs. 4-34–4-36.

With the definition of work given in Eq. (5-2) and with Eq. (5-13), the work done by a distributed load applied to a deformable body can be found. This calculation is left to the exercises.

By considering the force in Eq. (5-2) to be tangent to and traveling along a circular path, as shown in Fig. 5-9, using Eq. (5-3), and recalling the definition of a moment, the reader should be able to verify that the work done by a moment in rotating through an angle $\varphi$ is given by

$$W = \int_1^2 M\, d\varphi \tag{5-14}$$

## 5-2-1 Work Done by Stresses Acting on an Infinitesimal Element

We now calculate the work done by the stresses acting on an infinitesimal element as it deforms, assuming that the stress resultants act at the midpoints of the sides and that the body forces act at the centroid of the element. The relationship between the element in

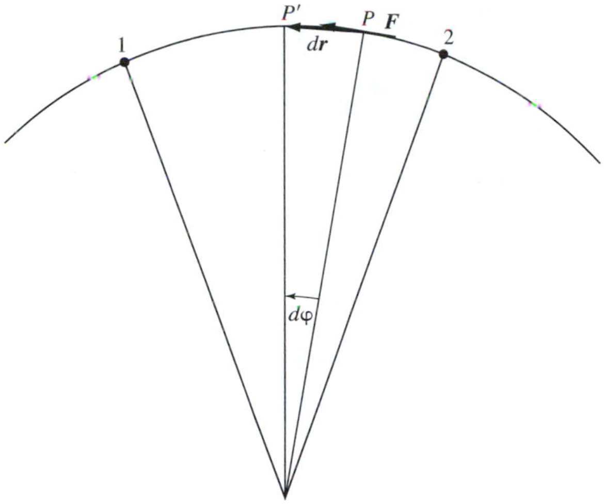


Figure 5-9

its deformed configuration and in its undeformed configuration is shown in Fig. 5-10. For simplicity, we consider first the work done by the stresses acting in the $x$ direction, as shown in Fig. 5-11. Now the force due to the stress $\sigma_{xx}$ acting on the left of the element is $\sigma_{xx} h\, dy$, where $h$ is the thickness of the element in the $z$ direction. The displacement of the midpoint of side $A'D'$, where the stress resultant is assumed to act, is the displacement of point $A'$ plus the additional displacement of the midpoint of side $A'D'$ relative to $A'$, that is, $u + (\partial u/\partial y)\, dy/2$. Note that $dy/2$ is used here because of our assumption that the stress resultants act at the midpoints of the sides. The work done by the stress $\sigma_{xx}$ is therefore

$$dW_{\sigma_{xx}} = -\sigma_{xx} d\left(u + \frac{1}{2}\frac{\partial u}{\partial y}\, dy\right) h\, dy \tag{5-15}$$

The work done by the remaining stresses acting in the $x$ direction is calculated in a similar way, so that we have

$$\begin{aligned}
dW = &-\sigma_{xx} d\left(u + \frac{1}{2}\frac{\partial u}{\partial y}\, dy\right) h\, dy \\
&+ \left(\sigma_{xx} + \frac{\partial \sigma_{xx}}{\partial x}\, dx\right) d\left(u + \frac{\partial u}{\partial x}\, dx + \frac{1}{2}\frac{\partial u}{\partial y}\, dy\right) h\, dy \\
&- \sigma_{xy} d\left[\frac{1}{2}\left(u + u + \frac{\partial u}{\partial x}\right)\right] h\, dx + \left(\sigma_{xy} + \frac{\partial \sigma_{xy}}{\partial y}\, dy\right) \\
&\times d\left[\frac{1}{2}\left(u + \frac{\partial u}{\partial y}\, dy + u + \frac{\partial u}{\partial x}\, dx + \frac{\partial u}{\partial y}\, dy\right)\right] h\, dx \\
&+ F_x d\left(u + \frac{1}{2}\frac{\partial u}{\partial y}\, dy + \frac{1}{2}\frac{\partial u}{\partial x}\, dx\right) h\, dx\, dy
\end{aligned}$$

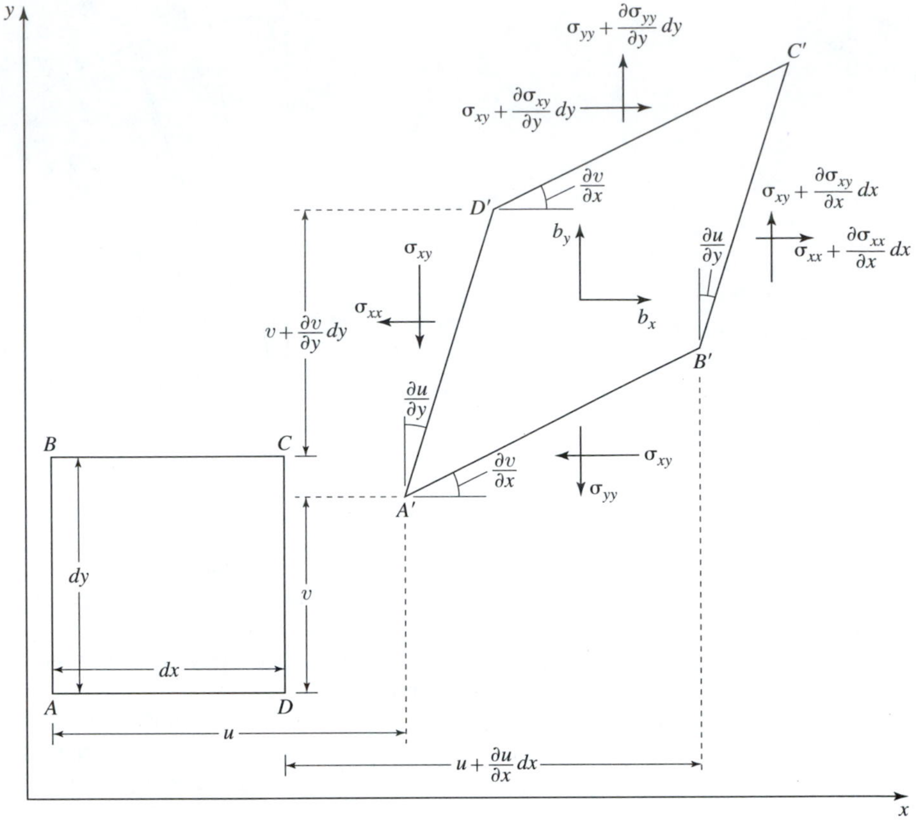


**Figure 5-10** The element in its deformed and in its undeformed configuration

After expanding and simplifying we get

$$
\begin{aligned}
dW = {} & \sigma_{xx} d\left(\frac{\partial u}{\partial x} dx\right) h\, dy + \frac{\partial \sigma_{xx}}{\partial x} d\left(u + \frac{\partial u}{\partial x} dx + \frac{1}{2}\frac{\partial u}{\partial y} dy\right) h\, dx\, dy \\
& + \sigma_{xy} d\left(\frac{\partial u}{\partial y} dy\right) h\, dx + \frac{\partial \sigma_{xy}}{\partial y} d\left(u + \frac{\partial u}{\partial y} dy + \frac{1}{2}\frac{\partial u}{\partial x} dx\right) h\, dx\, dy \\
& + b_x d\left(u + \frac{1}{2}\frac{\partial u}{\partial y} dy + \frac{1}{2}\frac{\partial u}{\partial x} dx\right) h\, dx\, dy
\end{aligned} \tag{5-16}
$$

Now since $dx$ is the original length of the element in the $x$ direction and is therefore fixed, and since, for infinitesimal strains, $\varepsilon_{xx} = \partial u / \partial x$, we have

$$
d\left(\frac{\partial u}{\partial x} dx\right) = d\varepsilon_{xx}\, dx
$$

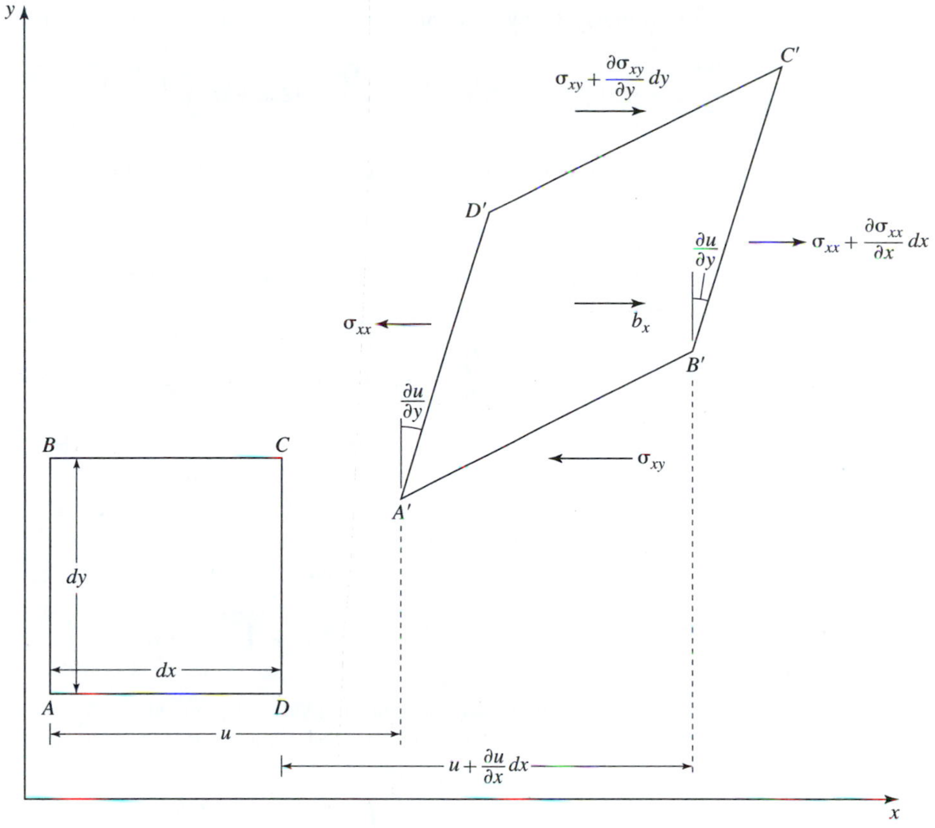


Figure 5-11

Similarly,

$$d\left(\frac{\partial u}{\partial y} dy\right) = d\varepsilon_{yy}\, dy$$

By using these results and collecting terms, we can write Eq. (5-16) as

$$dW = \left[\left(\frac{\partial \sigma_{xx}}{\partial x} + \frac{\partial \sigma_{xy}}{\partial y} + b_x\right) du + \left(\frac{\partial \sigma_{xx}}{\partial x} + \frac{1}{2}\frac{\partial \sigma_{xy}}{\partial y} + \frac{1}{2} b_x\right) d\varepsilon_{xx}\, dx \right.$$
$$\left. + \left(\frac{\partial \sigma_{xy}}{\partial y} + \frac{1}{2}\frac{\partial \sigma_{xx}}{\partial x} + \frac{1}{2} b_x\right) d\left(\frac{\partial u}{\partial y}\right) dy + \sigma_{xx}\, d\varepsilon_{xx} + \sigma_{xy} d\left(\frac{\partial u}{\partial y}\right)\right] h\, dx\, dy \tag{5-17}$$

The first parenthesis is zero because equilibrium must hold. The second and third parentheses are multiplied by $dx$ and $dy$, respectively, and therefore are negligibly small compared with the remaining terms. Therefore Eq. (5-17) becomes

$$dW = \left[\sigma_{xx}\, d\varepsilon_{xx} + \sigma_{xy} d\left(\frac{\partial u}{\partial y}\right)\right] h\, dx\, dy$$

Similarly, the work done by forces acting in the $y$ direction is

$$dW = \left[\sigma_{yy}\, d\varepsilon_{yy} + \sigma_{xy} d\left(\frac{\partial v}{\partial x}\right)\right] h\, dx\, dy$$

so that the work done by all forces acting on the element is

$$dW = (\sigma_{xx}\, d\varepsilon_{xx} + \sigma_{yy}\, d\varepsilon_{yy} + \sigma_{xy}\, d\gamma_{xy}) h\, dx\, dy \tag{5-18}$$

where Eq. (3-9) has been used. Integration from the initial unstrained state to the final strained state yields

$$W = \left(\int_0^{\varepsilon_{xx}} \sigma_{xx}\, d\varepsilon_{xx} + \int_0^{\varepsilon_{yy}} \sigma_{yy}\, d\varepsilon_{yy} + \int_0^{\gamma_{xy}} \sigma_{xy}\, d\gamma_{xy}\right) h\, dx\, dy \tag{5-19}$$

Because we have used the strain-displacement relationships for infinitesimal strains, Eq. (5-19) is valid for any problem in which the strains are small. Note that there is no reference to material behavior in Eq. (5-19), and therefore it is valid for any material.

Since $h\, dx\, dy$ is simply the volume of the element, we can divide by that quantity to get the work $\tilde{w}$ per unit volume,

$$\tilde{w} = \int_0^{\varepsilon_{xx}} \sigma_{xx}\, d\varepsilon_{xx} + \int_0^{\varepsilon_{yy}} \sigma_{yy}\, d\varepsilon_{yy} + \int_0^{\gamma_{xy}} \sigma_{xy}\, d\gamma_{xy} \tag{5-20}$$

For a linearly elastic material in a state of plane stress, the relations between stress and strain are given by Eqs. (4-21). Substitution of these into Eq. (5-20) gives

$$\tilde{w} = \frac{E}{1-\nu^2}\int_0^{\varepsilon_{xx}} (\varepsilon_{xx} + \nu\varepsilon_{yy})\, d\varepsilon_{xx}$$

$$+ \frac{E}{1-\nu^2}\int_0^{\varepsilon_{yy}} (\varepsilon_{yy} + \nu\varepsilon_{xx})\, d\varepsilon_{yy} + G\int_0^{\gamma_{xy}} \gamma_{xy}\, d\gamma_{xy}$$

We will show in Section 5-5 that the work done on an elastic body is independent of the order in which the loads are applied. Let us therefore envision the strains to be applied in the following order:

1. Apply $\varepsilon_{xx}$ while holding $\varepsilon_{yy}$ and $\gamma_{xy}$ at zero.
2. Apply $\varepsilon_{yy}$ while holding $\varepsilon_{xx}$ fixed and $\gamma_{xy}$ at zero.
3. Apply $\gamma_{xy}$ while holding $\varepsilon_{xx}$ and $\varepsilon_{yy}$ fixed.

From step 1 we have

$$\tilde{w}_1 = \frac{E}{1-\nu^2}\int_0^{\varepsilon_{xx}} \varepsilon_{xx}\, d\varepsilon_{xx} = \frac{1}{2}\frac{E}{1-\nu^2}\varepsilon_{xx}^2$$

and from step 2 we have

$$\tilde{w}_2 = \frac{E}{1-\nu^2}\int_0^{\varepsilon_{yy}} (\varepsilon_{yy} + \nu\varepsilon_{xx})\, d\varepsilon_{yy} = \frac{E}{1-\nu^2}\left(\frac{1}{2}\varepsilon_{yy}^2 + \nu\varepsilon_{xx}\varepsilon_{yy}\right)$$

while from step 3 we have

$$\tilde{w}_3 = G \int_0^{\gamma_{xy}} \gamma_{xy}\, d\gamma_{xy} = \frac{1}{2} G \gamma_{xy}^2$$

Combining these we get

$$\tilde{w} = \frac{1}{2} \frac{E}{1 - \nu^2} \varepsilon_{xx}^2 + \frac{E}{1 - \nu^2} \left( \frac{1}{2} \varepsilon_{yy}^2 + \nu \varepsilon_{xx} \varepsilon_{yy} \right) + \frac{1}{2} G \gamma_{xy}^2$$

Rearranging terms and using Eqs. (4-21) we get

$$\tilde{w} = \tfrac{1}{2}(\sigma_{xx}\varepsilon_{xx} + \sigma_{yy}\varepsilon_{yy} + \sigma_{xy}\gamma_{xy}) \tag{5-21}$$

It would be worthwhile for the reader to verify that if we had assumed the body to be in a state of plane strain, Eq. (5-21) would still result. In fact, Eq. (5-21) is valid for any two-dimensional problem involving the use of a linearly elastic material.

The corresponding result for three-dimensional problems is

$$\tilde{w} = \tfrac{1}{2}(\sigma_{xx}\varepsilon_{xx} + \sigma_{yy}\varepsilon_{yy} + \sigma_{zz}\varepsilon_{zz} + \sigma_{xy}\gamma_{xy} + \sigma_{xz}\gamma_{xz} + \sigma_{yz}\gamma_{yz}) \tag{5-22}$$

This result states that for a linearly elastic material, the work done on an infinitesimal element by stresses acting on that element is simply one-half of the sum of the product of stress (a measure of force) and strain (a measure of displacement). Note the similarity between Eqs. (5-22) and (5-10).

For three-dimensional problems, Eq. (5-18) becomes

$$dW = (\sigma_{xx}\, d\varepsilon_{xx} + \sigma_{yy}\, d\varepsilon_{yy} + \sigma_{zz}\, d\varepsilon_{zz} + \sigma_{xy}\, d\gamma_{xy} + \sigma_{xz}\, d\gamma_{xz} + \sigma_{yz}\, d\gamma_{yz})h\, dx\, dy\, dz \tag{5-23}$$

and the work per unit volume is

$$\begin{aligned} \tilde{w} = &\int_0^{\varepsilon_{xx}} \sigma_{xx}\, d\varepsilon_{yy} + \int_0^{\varepsilon_{yy}} \sigma_{yy}\, d\varepsilon_{xx} + \int_0^{\varepsilon_{zz}} \sigma_{zz}\, d\varepsilon_{zz} \\ &+ \int_0^{\gamma_{xy}} \sigma_{xy}\, d\gamma_{xy} + \int_0^{\gamma_{xz}} \sigma_{xz}\, d\gamma_{xz} + \int_0^{\gamma_{yz}} \sigma_{yz}\, d\gamma_{yz} \end{aligned} \tag{5-24}$$

## 5-3 FIRST LAW OF THERMODYNAMICS

We give here a brief discussion of the first law as it applies to uniform systems, and then extend the concepts to continuous systems. For readers unfamiliar with the elementary concepts of thermodynamics, it is suggested that texts such as Van Wylan and Sonntag [8] or Kestin [3] be consulted.

For a closed system undergoing a cycle, the first law of thermodynamics is given as follows:

$$\oint \tilde{d}H = -\oint \tilde{d}W \tag{5-25}$$

where $H$ is the heat transferred to the system by the surroundings and $W$ is the work done on the system by the surroundings. The tildes over the differential symbols for heat and work indicate that these quantities are path dependent, and hence their differentials are not exact differentials. Note that this form of the first law is slightly different than that commonly used by thermodynamicists, who generally take positive work to be work done on the surroundings by the system. This results in a difference of sign in the work term. However, in mechanical systems work is almost always done on the system by forces external to the system, so the form of the first law given here reduces considerably the possibility of inadvertently dropping a minus sign. It also results in a positive work term in the energy equation when forces and displacements are in the same direction and in a negative work term when they are in opposite directions.

If, rather than a cycle, the system undergoes a process that takes it from state 1 to state 2, the first law of thermodynamics may be written in incremental form as

$$dE = \tilde{d}H + \tilde{d}W \tag{5-26}$$

where $E$ is the energy of the system. Energy is path independent and therefore it is a property of the system, and its differential is an exact differential. Equation (5-26) states that at any point in a process there is an imbalance between work and heat, which is the energy of the system.

It is customary to write the energy of the system as the sum of the internal energy $I$ and the kinetic energy $K$,

$$E = I + K \tag{5-27}$$

where

$$K = \tfrac{1}{2}M\boldsymbol{v} \cdot \boldsymbol{v} \tag{5-28}$$

$M$ being the mass of the system and $\boldsymbol{v}$ the velocity vector. Then Eq. (5-26) becomes

$$dI + dK = \tilde{d}H + \tilde{d}W \tag{5-29}$$

If the center of mass of a rigid body undergoes a change in elevation, then it is often convenient to write the energy of the system as follows:

$$E = I + K + T \tag{5-30}$$

where $T$ is the potential energy of the system, given by

$$T = Mgz \tag{5-31}$$

in which $g$ is the gravitational acceleration and $z$ is the height of the center of mass above an arbitrary reference plane. Then Eq. (5-26) becomes

$$dI + dK + dT = \tilde{d}H + \tilde{d}W \tag{5-32}$$

It is important to note that the differential quantities in Eq. (5-32) refer to changes in the system quantities (i.e., the kinetic energy of the entire system) as the *process* proceeds from some initial state to a final state.

The first law can also be written as a rate equation simply by dividing by $dt$,

$$\frac{dI}{dt} + \frac{dK}{dt} = \dot{H} + \dot{W} \tag{5-33}$$

where

$$\dot{H} = \frac{\tilde{d}H}{dt} \tag{5-34}$$

$$\dot{W} = \frac{\tilde{d}W}{dt} \tag{5-35}$$

are the heat transfer rate and the power, respectively.

The first law as given by Eqs. (5-29), (5-32), or (5-33) applies to uniform systems, that is, systems in which there is no spatial variation of the properties describing the system. On the other hand, our experience suggests that in most systems these properties do indeed vary spatially within the system. For example, we know from elementary beam theory that stress varies from point to point in a beam that is subjected to bending loads. These systems are referred to as continuous systems because their properties generally vary in a continuous manner. Whereas for a uniform system the properties are variables, in a continuous system the properties are functions whose value at any instant of time at a particular point within the system depends on the $x$, $y$, $z$ coordinates of that point.

Since this text deals with continuous systems, it is desirable to develop the first law of thermodynamics for continuous systems. Let us examine an arbitrary volume $\overline{V}$ of material with surface $A$ and unit outer normal $\boldsymbol{n}$, contained within a region $R$. This material is removed from $R$, as shown in Fig. 5-12. Acting in $\overline{V}$ are the body forces $\boldsymbol{b}$, while the surface tractions $\boldsymbol{t}$ act on $A$. In addition, a heat flux vector $\boldsymbol{q}$, with typical units of BTU/in$^2$ · min, acts at the surface. A particular point $P$ on the surface moves with velocity $\partial \boldsymbol{u}/\partial t$, while the force acting is $\boldsymbol{t}\, dA$. Consequently the power associated with $P$ is $\boldsymbol{t} \cdot (\partial \boldsymbol{u}/\partial t)\, dA$. Over the entire surface $A$ the power due to the surface tractions is given by

$$\dot{W}_t = \int_A \boldsymbol{t} \cdot \frac{\partial \boldsymbol{u}}{\partial t}\, dA$$

In the study of continuous systems it is customary to include in the work term the work done by the body forces as well as that done by the surface tractions. The total power is

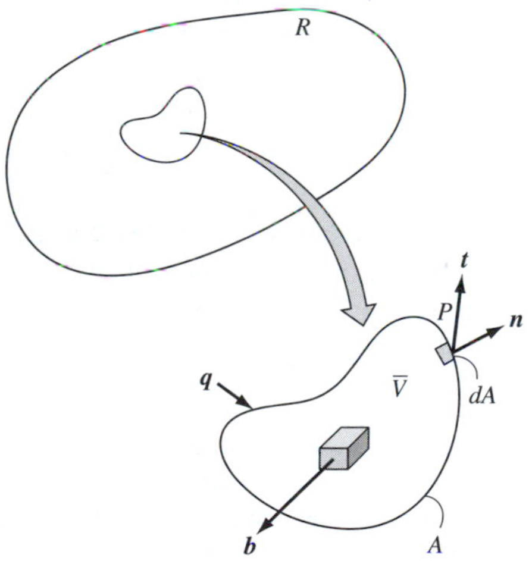


Figure 5-12

then given by

$$\dot{W} = \int_A \boldsymbol{t} \cdot \frac{\partial \boldsymbol{u}}{\partial t}\, dA + \int_{\overline{V}} \boldsymbol{b} \cdot \frac{\partial \boldsymbol{u}}{\partial t}\, d\overline{V} \tag{5-36}$$

The heat flux vector can be decomposed into components normal and tangent to the surface. The heat crossing the system boundary at $P$ is then given by $-\boldsymbol{q} \cdot \boldsymbol{n}\, dA$, with the negative sign arising because heat transferred to the system is considered positive. The total heat transfer rate is then

$$\dot{H} = -\int_A \boldsymbol{q} \cdot \boldsymbol{n}\, dA \tag{5-37}$$

We write the internal energy as the volume integral of the specific internal energy $i$,

$$I = \int_{\overline{V}} \rho i\, d\overline{V}$$

where $\rho$ is the mass density (mass per unit volume). Then,

$$\frac{dI}{dt} = \frac{d}{dt} \int_{\overline{V}} \rho i\, d\overline{V} \tag{5-38}$$

But this can be written as

$$\frac{dI}{dt} = \frac{d}{dt} \int_{\overline{V}} \rho i\, d\overline{V} = \frac{d}{dt} \int_m i\, dm = \int_m \frac{di}{dt}\, dm = \int_{\overline{V}} \frac{di}{dt} \rho\, d\overline{V} \tag{5-39}$$

where $m$ is the mass of the system. It is permissible to change the order of differentiation and integration when integrating over the mass because mass is conserved.

The specific internal energy $i$ depends on quantities such as strain, which vary from point to point within the body, and also with time. Thus we can write

$$i = i(x, y, z, t)$$

where $x$, $y$, $z$ give the location of the particle under observation. It is this particle in which we are observing the property $i$ change with time. Now with the chain rule we have

$$di = \frac{\partial i}{\partial x} dx + \frac{\partial i}{\partial y} dy + \frac{\partial i}{\partial z} dz + \frac{\partial i}{\partial t} dt$$

The first three terms on the right-hand side of this equation represent the change in $i$ as we change our point of observation from a particle located at $(x, y, z)$ to one located at $(x + dx, y + dy, z + dz)$. Since we are interested only in what happens to the property $i$ associated with the particle at $(x, y, z)$ as time increases, these first three terms are zero.*

*In the study of fluid motion these terms are not zero because an observer viewing a particular point in space observes different particles passing that point as the fluid flows. Thus the observer sees a change in some characteristic of the fluid at the point of observation because as time changes, different particles occupy the point of observation.

Thus, dividing by $dt$ and using the result in Eq. (5-39) we have

$$\frac{dI}{dt} = \int_{\overline{V}} \rho \frac{\partial i}{\partial t} \, d\overline{V} \tag{5-40}$$

The kinetic energy is found by calculating the kinetic energy of a volume element $d\overline{V}$ and integrating over the volume:

$$K = \frac{1}{2} \int_{\overline{V}} \rho \frac{\partial \boldsymbol{u}}{\partial t} \cdot \frac{\partial \boldsymbol{u}}{\partial t} \, d\overline{V} \tag{5-41}$$

By following the arguments used in arriving at Eq. (5-40) we get

$$\frac{dK}{dt} = \frac{1}{2} \int_{\overline{V}} \rho \frac{\partial}{\partial t} \left( \frac{\partial \boldsymbol{u}}{\partial t} \cdot \frac{\partial \boldsymbol{u}}{\partial t} \right) d\overline{V}$$

or, after expanding the integrand,

$$\frac{dK}{dt} = \int_{\overline{V}} \rho \left( \frac{\partial^2 u}{\partial t^2} \frac{\partial u}{\partial t} + \frac{\partial^2 v}{\partial t^2} \frac{\partial v}{\partial t} + \frac{\partial^2 w}{\partial t^2} \frac{\partial w}{\partial t} \right) d\overline{V} \tag{5-42}$$

The quantities required in the first law, namely, Eqs. (5-36), (5-37), (5-40), and (5-42), now contain both surface and volume integrals. In order to proceed further we first need to replace the surface integrals with volume integrals, or vice versa. This can be accomplished through the use of Gauss's theorem (see, for example, [1, p. 116] for a derivation of Gauss's theorem), which reads

$$\int_A \boldsymbol{F} \cdot \boldsymbol{n} \, dA = \int_{\overline{V}} \nabla \cdot \boldsymbol{F} d\overline{V} \tag{5-43}$$

where $\boldsymbol{n}$ is the unit outer normal to $A$, $\boldsymbol{F}$ is any vector that is continuous in $\overline{V}$ and on $A$, and the del operator is given by Eq. (4-178).

The dot product $\nabla \cdot \boldsymbol{F}$ is therefore given by

$$\nabla \cdot \boldsymbol{F} = \boldsymbol{i} \cdot \frac{\partial \boldsymbol{F}}{\partial x} + \boldsymbol{j} \cdot \frac{\partial \boldsymbol{F}}{\partial y} + \boldsymbol{k} \cdot \frac{\partial \boldsymbol{F}}{\partial z} \tag{5-44}$$

and Gauss's theorem can be written in an expanded form as

$$\int_S (F_x n_x + F_y n_y + F_z n_z) \, dS = \int_{\overline{V}} \left( \frac{\partial F_x}{\partial x} + \frac{\partial F_y}{\partial y} + \frac{\partial F_z}{\partial z} \right) d\overline{V} \tag{5-45}$$

With Eq. (5-43) the surface integral in Eq. (5-37) can be written as

$$\int_A \boldsymbol{q} \cdot \boldsymbol{n} \, dA = \int_{\overline{V}} \nabla \cdot \boldsymbol{q} \, d\overline{V} \tag{5-46}$$

A similar approach can be taken with the work term. We first carry out the indicated dot multiplication in the integrand to give

$$\int_A \boldsymbol{t} \cdot \frac{\partial \boldsymbol{u}}{\partial t} \, dA = \int_A \left( t_x \frac{\partial u}{\partial t} + t_y \frac{\partial v}{\partial t} + t_z \frac{\partial w}{\partial t} \right) dA \tag{5-47}$$

But from Eqs. (2-14),

$$t_x = \sigma_{xx}n_x + \sigma_{yx}n_y + \sigma_{zx}n_z$$
$$t_y = \sigma_{xy}n_x + \sigma_{yy}n_y + \sigma_{zy}n_z$$
$$t_z = \sigma_{xz}n_x + \sigma_{yz}n_y + \sigma_{zz}n_z$$

so that

$$\int_A \boldsymbol{t} \cdot \frac{\partial \boldsymbol{u}}{\partial t}\, dS = \int_A \left[ \left( \sigma_{xx}\frac{\partial u}{\partial t} + \sigma_{xy}\frac{\partial v}{\partial t} + \sigma_{xz}\frac{\partial w}{\partial t} \right) n_x + \left( \sigma_{yx}\frac{\partial u}{\partial t} + \sigma_{yy}\frac{\partial v}{\partial t} + \sigma_{yz}\frac{\partial w}{\partial t} \right) n_y + \left( \sigma_{zx}\frac{\partial u}{\partial t} + \sigma_{zy}\frac{\partial v}{\partial t} + \sigma_{zz}\frac{\partial w}{\partial t} \right) n_z \right] dA \tag{5-48}$$

Now if the first expression in parentheses is identified as $F_x$, the second as $F_y$, and the third as $F_z$, then Eq. (5-48) becomes, with the use of Eq. (5-45),

$$\int_A \boldsymbol{t} \cdot \frac{\partial \boldsymbol{u}}{\partial t}\, dA = \int_{\overline{V}} \left[ \frac{\partial}{\partial x}\left( \sigma_{xx}\frac{\partial u}{\partial t} + \sigma_{xy}\frac{\partial v}{\partial t} + \sigma_{xz}\frac{\partial w}{\partial t} \right) + \frac{\partial}{\partial y}\left( \sigma_{yx}\frac{\partial u}{\partial t} + \sigma_{yy}\frac{\partial v}{\partial t} + \sigma_{yz}\frac{\partial w}{\partial t} \right) + \frac{\partial}{\partial z}\left( \sigma_{zx}\frac{\partial u}{\partial t} + \sigma_{zy}\frac{\partial v}{\partial t} + \sigma_{zz}\frac{\partial w}{\partial t} \right) \right] d\overline{V}$$

or, on expanding, collecting terms, and using Eqs. (3-11),

$$\int_A \boldsymbol{t} \cdot \frac{\partial \boldsymbol{u}}{\partial t}\, dA = \int_{\overline{V}} \left[ \left( \frac{\partial \sigma_{xx}}{\partial x} + \frac{\partial \sigma_{yx}}{\partial y} + \frac{\partial \sigma_{zx}}{\partial z} \right)\frac{\partial u}{\partial t} + \left( \frac{\partial \sigma_{xy}}{\partial x} + \frac{\partial \sigma_{yy}}{\partial y} + \frac{\partial \sigma_{zy}}{\partial z} \right)\frac{\partial v}{\partial t} + \left( \frac{\partial \sigma_{xz}}{\partial x} + \frac{\partial \sigma_{yz}}{\partial y} + \frac{\partial \sigma_{zz}}{\partial z} \right)\frac{\partial w}{\partial t} + \sigma_{xx}\dot{\varepsilon}_{xx} + \sigma_{yy}\dot{\varepsilon}_{yy} + \sigma_{zz}\dot{\varepsilon}_{zz} + \sigma_{xy}\dot{\gamma}_{xy} + \sigma_{xz}\dot{\gamma}_{xz} + \sigma_{yz}\dot{\gamma}_{yz} \right] d\overline{V}$$

Therefore Eq. (5-36) becomes,

$$\dot{W} = \int_{\overline{V}} \left[ \left( \frac{\partial \sigma_{xx}}{\partial x} + \frac{\partial \sigma_{yx}}{\partial y} + \frac{\partial \sigma_{zx}}{\partial z} + b_x \right)\frac{\partial u}{\partial t} + \left( \frac{\partial \sigma_{xy}}{\partial x} + \frac{\partial \sigma_{yy}}{\partial y} + \frac{\partial \sigma_{zy}}{\partial z} + b_y \right)\frac{\partial v}{\partial t} + \left( \frac{\partial \sigma_{xz}}{\partial x} + \frac{\partial \sigma_{yz}}{\partial y} + \frac{\partial \sigma_{zz}}{\partial z} + b_z \right)\frac{\partial w}{\partial t} + \sigma_{xx}\dot{\varepsilon}_{xx} + \sigma_{yy}\dot{\varepsilon}_{yy} + \sigma_{zz}\dot{\varepsilon}_{zz} + \sigma_{xy}\dot{\gamma}_{xy} + \sigma_{xz}\dot{\gamma}_{xz} + \sigma_{yz}\dot{\gamma}_{yz} \right] d\overline{V} \tag{5-49}$$

With Eqs. (5-40), (5-42), (5-46), and (5-49) the first law, Eq. (5-33), becomes

$$\begin{aligned}\int_{\overline{V}} \Bigg[ & \left( \frac{\partial \sigma_{xx}}{\partial x} + \frac{\partial \sigma_{yx}}{\partial y} + \frac{\partial \sigma_{zx}}{\partial z} + b_x - \rho \frac{\partial^2 u}{\partial t^2} \right) \frac{\partial u}{\partial t} \\ & + \left( \frac{\partial \sigma_{xy}}{\partial x} + \frac{\partial \sigma_{yy}}{\partial y} + \frac{\partial \sigma_{zy}}{\partial z} + b_y - \rho \frac{\partial^2 v}{\partial t^2} \right) \frac{\partial v}{\partial t} \\ & + \left( \frac{\partial \sigma_{xz}}{\partial x} + \frac{\partial \sigma_{yz}}{\partial y} + \frac{\partial \sigma_{zz}}{\partial z} + b_z - \rho \frac{\partial^2 w}{\partial t^2} \right) \frac{\partial w}{\partial t} + \sigma_{xx}\dot{\varepsilon}_{xx} \\ & + \sigma_{yy}\dot{\varepsilon}_{yy} + \sigma_{zz}\dot{\varepsilon}_{zz} + \sigma_{xy}\dot{\gamma}_{xy} + \sigma_{xz}\dot{\gamma}_{xz} + \sigma_{yz}\dot{\gamma}_{yz} - \nabla \cdot \boldsymbol{q} - \rho \frac{\partial i}{\partial t} \Bigg] d\overline{V} = 0\end{aligned}$$

Each of the terms in parentheses is zero by virtue of the momentum equations. We note that Eqs. (2-53), the equilibrium equations, were derived by requiring that the sum of the forces acting on a differential element of material be zero, assuming that the element is at rest. If the element is not at rest then, by Newton's second law, the sum of the forces acting on the element must equal the product of the mass of the element and its acceleration. Thus the right-hand side of the first of Eqs. (2-53) becomes $\rho\, d^2u/dt^2$, where $u$ is the displacement component in the $x$ direction. Similar terms appear in the last two of Eqs. (2-53) with the displacement components $v$ and $w$, respectively. The resulting equations are the *momentum equations*. Consequently,

$$\int_{\overline{V}} \left( \sigma_{xx}\dot{\varepsilon}_{xx} + \sigma_{yy}\dot{\varepsilon}_{yy} + \sigma_{zz}\dot{\varepsilon}_{zz} + \sigma_{xy}\dot{\gamma}_{xy} + \sigma_{xz}\dot{\gamma}_{xz} + \sigma_{yz}\dot{\gamma}_{yz} - \nabla \cdot \boldsymbol{q} - \rho \frac{\partial i}{\partial t} \right) d\overline{V} = 0$$

Since the volume $\overline{V}$ is *arbitrary,* the foregoing integral can be zero only if the integrand itself is zero. We have then

$$\rho \frac{\partial i}{\partial t} = \sigma_{xx}\dot{\varepsilon}_{xx} + \sigma_{yy}\dot{\varepsilon}_{yy} + \sigma_{zz}\dot{\varepsilon}_{zz} + \sigma_{xy}\dot{\gamma}_{xy} + \sigma_{zz}\dot{\gamma}_{zz} + \sigma_{yz}\dot{\gamma}_{yz} - \nabla \cdot \boldsymbol{q} \tag{5-50}$$

This is the local form of the first law. Note that it is only valid for infinitesimal strains because in converting the surface integral for work to a volume integral, the strain-displacement relations for infinitesimal strains [Eqs. (3-11)] were used.

## 5-4 SECOND LAW OF THERMODYNAMICS

All real processes are irreversible, and therefore a real system, on reaching its final state, cannot be returned to its initial state simply by reversing the direction of the process. The second law of thermodynamics is a statement of this experimental observation. Although the second law can be stated in several different forms, we will use a form referred to as the *Clausius inequality,* which is given as

$$dS \geq \frac{\tilde{d}H}{T} \tag{5-51}$$

where $S$ is the *entropy* of the system and $T$ is the *absolute temperature* at which heat is received by the system. Since entropy is a property of the system, its differential is an exact differential.

The equality in Eq. (5-51) holds only for a reversible process,

$$dS = \left(\frac{\tilde{d}H}{T}\right)_{\text{rev}} \tag{5-52}$$

It is interesting to note that although $\tilde{d}H$ is an inexact differential, $\tilde{d}H/T$ is an exact differential if the process is reversible.

Of all conceivable processes, the only ones that are physically possible are those that satisfy the inequality (5-51).

Dividing by $dt$, the second law can be written as a rate inequality,

$$\frac{dS}{dt} \geq \frac{\dot{H}}{T} \tag{5-53}$$

It is often desirable to write the second law as an equality, in which case inequality (5-53) becomes

$$\frac{dS}{dt} = \frac{\dot{H}}{T} + \dot{\Psi} \tag{5-54}$$

where $\Psi$ is designated the *entropy produced by the irreversibilities*. It is not a property of the system, and hence its differential is inexact. This equation may be viewed as a statement of the balance of entropy. It states that the change of entropy of a system results from the flow of entropy crossing the system boundary due to heat crossing the boundary, and from the entropy produced by the irreversible process. We note that $\dot{\Psi}$ must be nonnegative.

In developing a local form of the second law it is important to note that both inequality (5-51) and Eq. (5-53) apply to a uniform system whose instantaneous temperature $T$ is the same throughout the volume of the system, and which exchanges heat with surroundings of uniform temperature. Thus the inequality (5-51) cannot be applied to a continuous medium of finite size merely by writing $dS/dt$ as a volume integral and $\dot{H}/T$ as a surface integral, as was done with the first law. Rather, since the temperature varies throughout a continuous medium, $\dot{H}$ must first be evaluated for a vanishingly small subsystem and then divided by $T$. Thus we have

$$\dot{H} = -\int_{\Delta A} \boldsymbol{q} \cdot \boldsymbol{n}\, dA$$

$$\frac{\dot{H}}{T} = -\frac{1}{T}\int_{\Delta A} \boldsymbol{q} \cdot \boldsymbol{n}\, dA$$

where $\Delta A$ is the surface area of the subsystem. When Gauss's theorem is applied to the surface integral, we get

$$\frac{\dot{H}}{T} = -\frac{1}{T}\int_{\Delta \overline{V}} \nabla \cdot \boldsymbol{q}\, d\overline{V} \tag{5-55}$$

where $\Delta\overline{V}$ is the volume of the subsystem. The entropy can be written as the volume integral of the specific entropy $s$,

$$S = \int_{\Delta \overline{V}} \rho s\, d\overline{V}$$

By proceeding in the same manner as used to arrive at Eq. (5-40), we have

$$\frac{dS}{dt} = \int_{\Delta\overline{V}} \rho \frac{\partial s}{\partial t}\, d\overline{V} \tag{5-56}$$

Thus with Eqs. (5-55) and (5-56) the Clausius inequality becomes

$$\int_{\Delta\overline{V}} \left( \rho \frac{\partial s}{\partial t} + \frac{1}{T} \nabla \cdot \boldsymbol{q} \right) d\overline{V} \geq 0$$

Since this inequality must be true for *any* subsystem, it follows that

$$\rho \frac{\partial s}{\partial t} + \frac{1}{T} \nabla \cdot \boldsymbol{q} \geq 0 \tag{5-57}$$

The local form of Eq. (5-54) is

$$\rho \frac{\partial s}{\partial t} + \frac{1}{T} \nabla \cdot \boldsymbol{q} = \sum_{i=1}^{n} \dot{\psi}_i \tag{5-58}$$

where

$$\dot{\Psi} = \sum_{i=1}^{n} \int_{\Delta\overline{V}} \dot{\psi}_i\, d\overline{V} \tag{5-59}$$

and $\dot{\psi}_i$ is the entropy production per unit volume per unit time due to the $i$th source of irreversibility. Note, however, that entropy production due to the transfer of heat resulting from the temperature gradient is not contained in Eq. (5-58) because $T$ is assumed uniform throughout the subsystem. This equation can be written in the equivalent form

$$\rho \frac{\partial s}{\partial t} + \nabla \cdot \left( \frac{\boldsymbol{q}}{T} \right) = \boldsymbol{q} \cdot \nabla \left( \frac{1}{T} \right) + \sum_{i=1}^{n} \dot{\psi}_i \tag{5-60}$$

That this equation is indeed a statement of local entropy balance can be seen by noting that the term immediately preceding the equal sign can be interpreted, by appealing to Gauss's theorem, as flow of entropy across the boundary of the infinitesimal volume $dV$. Entropy production from heat transfer caused by the temperature gradient is given by the term immediately following the equal sign. This term must be nonnegative because the entropy of the system cannot be decreased. The nonnegative nature of this term can readily be seen by expanding,

$$\boldsymbol{q} \cdot \nabla \left( \frac{1}{T} \right) = -\frac{1}{T^2} (\boldsymbol{q} \cdot \nabla T) = -\frac{1}{T^2} \left( q_x \frac{\partial T}{\partial x} + q_y \frac{\partial T}{\partial y} + q_z \frac{\partial T}{\partial z} \right)$$

Now a positive temperature gradient produces a negative flow of heat, that is, if $\partial T / \partial x$ is positive, then $q_x$ must be negative, and so on. Note that the term can be zero only if the temperature is constant or if there is no flow of heat. Consequently the term $\boldsymbol{q} \cdot \nabla(1/T)$ is nonnegative, and Eq. (5-60) may be written as an inequality,

$$\rho \frac{\partial s}{\partial t} + \nabla \cdot \left( \frac{\boldsymbol{q}}{T} \right) \geq 0 \tag{5-61}$$

If the process is reversible (which implies that there is no spatial temperature gradient), this becomes

$$\rho \frac{\partial s}{\partial t} + \frac{\nabla \cdot \boldsymbol{q}}{T} = 0 \tag{5-62}$$

## 5-5 SOME SIMPLE APPLICATIONS INVOLVING THE FIRST LAW

In this section we present two results that follow from the first law. To begin, we note that:

> In the absence of thermal effects, the work done by the forces acting on an elastic body is independent of the order in which the forces are applied.

To see this, we refer to Eq. (5-25) which, in the absence of thermal effects, states that the cyclic integral of the work done by the forces acting on the body must be zero. Assume that the work done by the forces acting on an elastic body depends on the order in which the loads are applied. Then if loads are applied to the body in a certain order and removed in a different order, the work done in loading will be different than that done in unloading, and the resulting cyclic work integral will be different than zero, in violation of Eq. (5-25). Hence our assumption must be incorrect.

### EXAMPLE 5-4

The rod shown in Fig. 5-13 is made from a linearly elastic material and undergoes small displacements due to the application of forces of magnitude $P$ and $Q$ at points 1 and 2, respectively, along the axis of the rod. Find the work done if $P$ is applied first, followed by $Q$. Then recalculate the work with $Q$ applied first, followed by $P$.

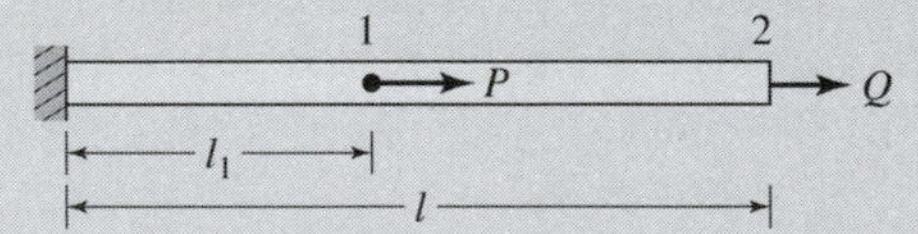


**Figure 5-13**

### SOLUTION

If the force $\boldsymbol{P}$ is applied at point 1, the work done is

$$W = \frac{1}{2} \frac{P^2 l_1}{AE}$$

We now apply the force $\boldsymbol{Q}$ at point 2. In addition to the work done by $\boldsymbol{Q}$ of $\frac{1}{2} Q^2 l / AE$, there is additional work done by $\boldsymbol{P}$ because point 1 moves through an additional displacement when $\boldsymbol{Q}$ is applied. This additional work is

$$P\left(\frac{Ql}{AE}\right)\left(\frac{l_1}{l}\right)$$

Hence the total work done is

$$W_1 = \frac{1}{2}\frac{P^2 l_1}{AE} + \frac{1}{2}\frac{Q^2 l}{AE} + P\frac{Q l_1}{AE}$$

Now if the force $\boldsymbol{Q}$ is applied first, the work done by $\boldsymbol{Q}$ is

$$W = \frac{1}{2}\frac{Q^2 l}{AE}$$

The force $\boldsymbol{P}$ is then applied at point 1 and the work done by $\boldsymbol{P}$ is $\frac{1}{2}P^2 l_1/AE$. Additional work is done by $\boldsymbol{Q}$ because the force $\boldsymbol{P}$ causes the end of the rod to move an additional amount. When $\boldsymbol{P}$ is applied at point 1, the material between points 1 and 2 moves as a rigid body. Therefore the displacement at point 2 due to the application of force $\boldsymbol{P}$ is the same as the displacement at point 1. Thus the additional work done by $\boldsymbol{Q}$ is $Q(P l_1/AE)$. Consequently the total work done with this order of loading is

$$W_2 = \frac{1}{2}\frac{Q^2 l}{AE} + \frac{1}{2}\frac{P^2 l_1}{AE} + Q\frac{P l_1}{AE}$$

Note that the total work done is the same regardless of whether $\boldsymbol{P}$ or $\boldsymbol{Q}$ is applied first.

## 5-5-1 Maxwell's Reciprocity Theorem

A very useful reciprocity theorem can be derived from the idea that for elastic bodies the work done is independent of the order in which the loads are applied. We restrict our attention to linearly elastic bodies subjected to infinitesimal strains, in which thermal effects are absent. For simplicity, we start by considering the beam shown in Fig. 5-14*a*, for which the displacements are assumed to be small. Two arbitrary points 1 and 2 are identified on the beam as shown, and the forces $\boldsymbol{F}_1$ and $\boldsymbol{F}_2$ are applied at points 1 and 2, respectively, in the directions shown. We wish to compute the work done by these two forces as the beam deforms. In order to do so, we use the following notation for displacements.

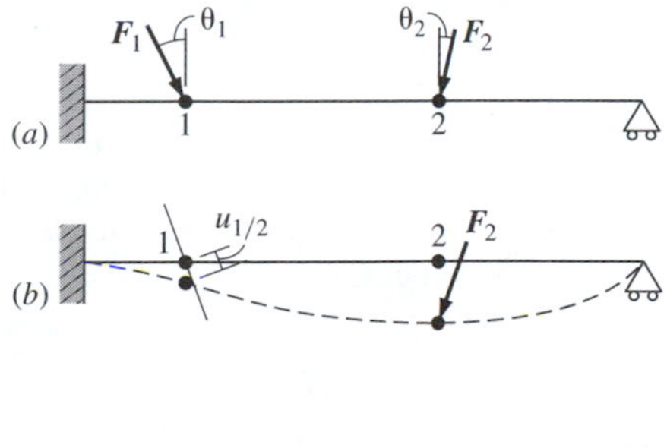


Figure 5-14

The displacement $u_{i/j}$ is the component of the displacement vector $\boldsymbol{u}$ at point $i$ and at an angle $\theta_i$ due to the force $\boldsymbol{F}_j$ acting at point $j$ at an angle $\theta_j$. Thus with the force $\boldsymbol{F}_2$ applied to the beam as shown in Fig. 5-14*b*, the displacement $u_{1/2}$ is as shown. Note that this is not the transverse displacement, nor is it the magnitude of the displacement vector from the undeformed to the deformed position of point 1. Rather, it is the component of the displacement vector in the direction of $\boldsymbol{F}_1$.

We now calculate the work done when $\boldsymbol{F}_1$ is applied first, followed by $\boldsymbol{F}_2$. When $\boldsymbol{F}_1$ is applied, the work done is given by

$$W = \tfrac{1}{2} F_1 u_{1/1}$$

Application of the force $\boldsymbol{F}_2$ results in additional work in the amount of $\frac{1}{2} F_2 u_{2/2} + F_1 u_{1/2}$, and a total work $W_1$ of

$$W_1 = \tfrac{1}{2} F_1 u_{1/1} + \tfrac{1}{2} F_2 u_{2/2} + F_1 u_{1/2}$$

We now apply the force $\boldsymbol{F}_2$ first. Since the displacements are small, the geometry changes in the beam are insignificant and the displacement of point 2 is the same as when $\boldsymbol{F}_2$ was added in the previous case. Thus we have

$$W = \tfrac{1}{2} F_2 u_{2/2}$$

The force $\boldsymbol{F}_1$ is applied next, resulting in additional work given by $\frac{1}{2} F_1 u_{1/1} + F_2 u_{2/1}$. In this case the total work $W_2$ is

$$W_2 = \tfrac{1}{2} F_2 u_{2/2} + \tfrac{1}{2} F_1 u_{1/1} + F_2 u_{2/1}$$

Because the work done does not depend on the order in which the loads are applied, $W_1$ and $W_2$ must be equal. Therefore we have

$$F_1 u_{1/2} = F_2 u_{2/1}$$

If both forces are equal in magnitude, this reduces to

$$u_{1/2} = u_{2/1} \tag{5-63}$$

This expression states that:

> In the absence of thermal effects, the displacement at point 1 in the direction defined by $\theta_1$ due to a force acting at point 2 at an angle $\theta_2$ is the same as the displacement at point 2 in the direction defined by $\theta_2$ that results when the same force is applied at point 1 at an angle $\theta_1$.

This result is referred to as *Maxwell's reciprocity theorem*.

## EXAMPLE 5-5

Consider the cantilever beam shown in Fig. 5-15*a*. Equation (5-63) states that if a force $\boldsymbol{F}$ is applied perpendicular to the beam at point 2 and the transverse displacement at point 1 determined, that displacement is the same as the transverse displacement at point 2 when

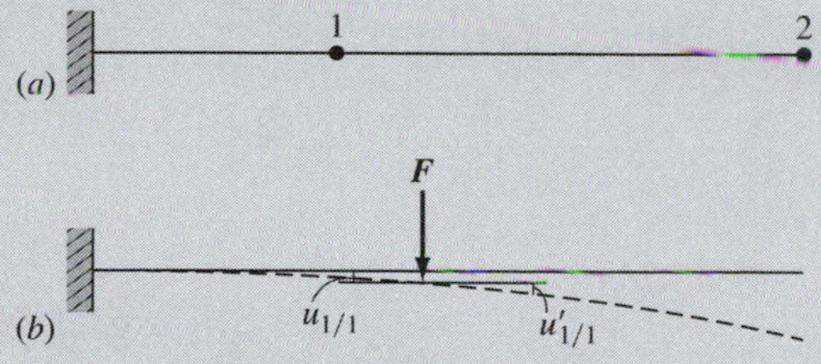


**Figure 5-15**

a force $\boldsymbol{F}$ perpendicular to the beam is applied at point 1. Show that, although this result is not at all obvious, it is true.

## SOLUTION

From elementary mechanics of materials textbooks (see [2, p. 737]) for a force $\boldsymbol{F}$ applied to the free end of a cantilever beam of length $l$ we have

$$u_{1/2} = \frac{Fa^3}{6EI}(3l - a)$$

where $a$ is the distance from the fixed end of the beam to point 1, at which $u$ is measured. Now when a force $F$ is applied at point 1, as shown in Fig. 5-15*b*, because the displacements are small, we find that

$$u_{2/1} = u_{1/1} + u'_{1/1}(l - a)$$

where $u'_{1/1}$ is the slope of the beam at point 1. But we have [2, p. 737]

$$u_{1/1} = \frac{Fa^3}{3EI}, \qquad u'_{1/1} = \frac{Fa^2}{2EI}$$

Thus $u_{2/1}$ is given by

$$u_{2/1} = \frac{Fa^2}{6EI}(3l - a)$$

which is the same as $u_{1/2}$.

Maxwell's reciprocity theorem can also be derived for the case when one of the loads, say at point 2, is a moment, as shown in Fig. 5-16*a*. In that case the displacement "in the direction of the force" is a rotational displacement, that is, a slope. Thus if $F_2$ is a moment, the displacement $u_{2/2}$ is the slope at point 2 due to the moment at that point, while $u_{2/1}$ is the slope at point 2 resulting from a force at point 1, as shown in Fig. 5-16*b* and *c*. Similarly, $u_{1/2}$ is the displacement at point 1 due to a moment at point 2. Thus if we think of the force $F$ as a generalized force that includes both linear forces and rotational forces (i.e., moments), and the displacement $u$ as a generalized displacement that includes both linear and rotational displacements (i.e., slopes), then Eq. (5-63) is a general statement of Maxwell's reciprocity theorem.

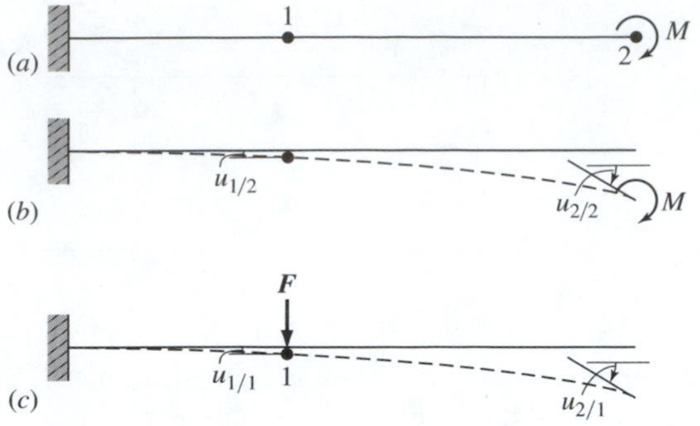


Figure 5-16

Note that in using Maxwell's reciprocity theorem with both forces and moments, the *magnitudes* of the force and of the moment must be equal, and the units must be consistent. Thus if the moment is given in inch-pounds, the force must be in pounds and the displacement in inches. As usual, the slope is measured in radians.

Although we have derived Maxwell's reciprocity theorem for the specific case of a beam, it applies to any linear elastic body for which the strains are infinitesimal. For example, in Fig. 5-17 the forces $\boldsymbol{F}_1$ and $\boldsymbol{F}_2$ are in directions $\boldsymbol{d}_1$ and $\boldsymbol{d}_2$, which need not lie in a plane. Thus Maxwell's reciprocity theorem applies to a large number of structures encountered by practicing engineers.

In Chapter 7 we will develop a powerful numerical method, called the finite element method, for analyzing complex structures. This method leads to a large system of linear algebraic equations which are solved on a computer. With Maxwell's reciprocity theorem we can prove that the coefficient matrix associated with this system of equations is a symmetric matrix. This important observation makes it possible to reduce the amount of computer storage space required for the matrix by almost 50 percent.

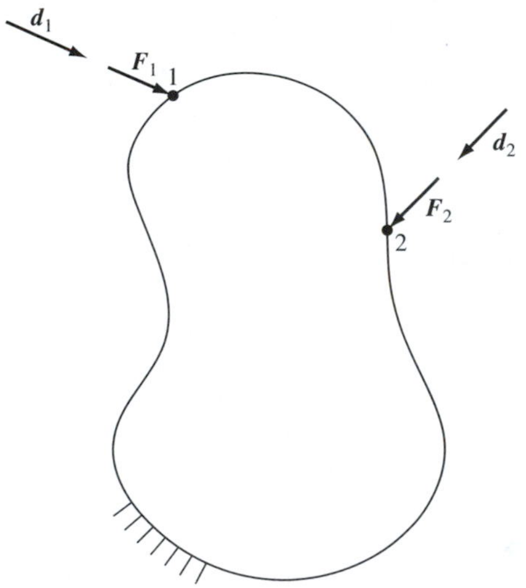


Figure 5-17

## 5-6 STRAIN ENERGY

Let us now consider a reversible process. In that case we have from Eq. (5-62)

$$\nabla \cdot \boldsymbol{q} = -T\rho \frac{\partial s}{\partial t} \tag{5-64}$$

and therefore the first law, given by Eq. (5-50), can be written as

$$\rho \frac{\partial i}{\partial t} = \sigma_{xx}\dot{\varepsilon}_{xx} + \sigma_{yy}\dot{\varepsilon}_{yy} + \sigma_{zz}\dot{\varepsilon}_{zz} + \sigma_{xy}\dot{\gamma}_{xy} + \sigma_{xz}\dot{\gamma}_{xz} + \sigma_{yz}\dot{\gamma}_{yz} + T\rho \frac{\partial s}{\partial t}$$

On multiplying by $dt$ we get

$$\rho\, di = \sigma_{xx}\, d\varepsilon_{xx} + \sigma_{yy}\, d\varepsilon_{yy} + \sigma_{zz}\, d\varepsilon_{zz} + \sigma_{xy}\, d\gamma_{xy} + \sigma_{xz}\, d\gamma_{xz} + \sigma_{yz}\, d\gamma_{yz} + T\rho\, ds \tag{5-65}$$

We now consider an *isentropic* process, that is, one in which the entropy $s$ is constant. Note that an *adiabatic* process (one in which no heat is transferred so that $q$ is zero) is a special case of the more general isentropic process. For this process Eq. (5-65) becomes

$$\rho\, di = \sigma_{xx}\, d\varepsilon_{xx} + \sigma_{yy}\, d\varepsilon_{yy} + \sigma_{zz}\, d\varepsilon_{zz} + \sigma_{xy}\, d\gamma_{xy} + \sigma_{xz}\, d\gamma_{xz} + \sigma_{yz}\, d\gamma_{yz} \tag{5-66}$$

We note that the right-hand side of this expression contains no explicit reference to the temperature. Thus $i$ does not depend explicitly on temperature. Now stresses and strains are related through some form of constitutive law, and therefore $i$ can be viewed as a function of either stress or strain (which, in turn, depend on position and time). Let us assume that it is a function of the strains, as suggested by the structure of Eq. (5-66),

$$i = i(\varepsilon_{xx}, \varepsilon_{yy}, \varepsilon_{zz}, \gamma_{xy}, \gamma_{xz}, \gamma_{yz})$$

Thus the specific internal energy results only from the work done by the stresses acting on an element as it undergoes strain. Consequently, under these conditions the specific internal energy is referred to as the specific strain energy.

For infinitesimal deformations the mass density $\rho$ can be considered constant throughout the process, so that we may write

$$\rho\, di = d(\rho i) = d\tilde{u} \tag{5-67}$$

where $\tilde{u}$ is the strain energy per unit volume, or the strain energy density. Thus Eq. (5-66) becomes

$$d\tilde{u} = \sigma_{xx}\, d\varepsilon_{xx} + \sigma_{yy}\, d\varepsilon_{yy} + \sigma_{zz}\, d\varepsilon_{zz} + \sigma_{xy}\, d\gamma_{xy} + \sigma_{xz}\, d\gamma_{xz} + \sigma_{yz}\, d\gamma_{yz} \tag{5-68}$$

Since $\tilde{u}$ is a function of the strains, the chain rule gives

$$d\tilde{u} = \frac{\partial \tilde{u}}{\partial \varepsilon_{xx}}\, d\varepsilon_{xx} + \frac{\partial \tilde{u}}{\partial \varepsilon_{yy}}\, d\varepsilon_{yy} + \cdots + \frac{\partial \tilde{u}}{\partial \gamma_{yz}}\, d\gamma_{yz} \tag{5-69}$$

and therefore, when Eq. (5-69) is used and all terms are moved to the left of the equal sign, Eq. (5-68) becomes,

$$\left(\sigma_{xx} - \frac{\partial \tilde{u}}{\partial \varepsilon_{xx}}\right) d\varepsilon_{xx} + \left(\sigma_{yy} - \frac{\partial \tilde{u}}{\partial \varepsilon_{yy}}\right) d\varepsilon_{yy} + \left(\sigma_{zz} - \frac{\partial \tilde{u}}{\partial \varepsilon_{zz}}\right) d\varepsilon_{zz}$$
$$+ \left(\sigma_{xy} - \frac{\partial \tilde{u}}{\partial \gamma_{xy}}\right) d\gamma_{xy} + \left(\sigma_{xz} - \frac{\partial \tilde{u}}{\partial \gamma_{xz}}\right) d\gamma_{xz} + \left(\sigma_{yz} - \frac{\partial \tilde{u}}{\partial \gamma_{yz}}\right) d\gamma_{yz} = 0$$

Now since the strains are linearly independent, we have

$$\begin{aligned}
\sigma_{xx} &= \frac{\partial \tilde{u}}{\partial \varepsilon_{xx}}, & \sigma_{xy} &= \frac{\partial \tilde{u}}{\partial \gamma_{xy}} \\
\sigma_{yy} &= \frac{\partial \tilde{u}}{\partial \varepsilon_{yy}}, & \sigma_{xz} &= \frac{\partial \tilde{u}}{\partial \gamma_{xz}} \\
\sigma_{zz} &= \frac{\partial \tilde{u}}{\partial \varepsilon_{zz}}, & \sigma_{yz} &= \frac{\partial \tilde{u}}{\partial \gamma_{yz}}
\end{aligned} \tag{5-70}$$

Thus if the strain energy density function is known, the stresses can be computed by simple differentiation.

It is worth noting the similarity between Eqs. (5-68) and (5-24). In fact, upon integrating Eq. (5-68) over the strains we see that

$$\tilde{u} = w \tag{5-71}$$

That is, the work done by the stresses acting on an element is stored in the element as strain energy. If the material is linearly elastic, so the relationship between stress and strain is described by the generalized Hooke's law. Then by using Eq. (5-22) we obtain

$$\tilde{u} = \tfrac{1}{2}(\sigma_{xx}\varepsilon_{xx} + \sigma_{yy}\varepsilon_{yy} + \sigma_{zz}\varepsilon_{zz} + \sigma_{xy}\gamma_{xy} + \sigma_{xz}\gamma_{xz} + \sigma_{yz}\gamma_{yz}) \tag{5-72}$$

A form involving only stresses and elastic constants is obtained by using the generalized Hooke's law, Eqs. (4-8) and (4-11):

$$\tilde{u} = \frac{1}{2E}\left[\sigma_{xx}^2 + \sigma_{yy}^2 + \sigma_{zz}^2 - 2\nu(\sigma_{xx}\sigma_{yy} + \sigma_{yy}\sigma_{zz} + \sigma_{xx}\sigma_{zz})\right] + \frac{1}{2G}\left(\sigma_{xy}^2 + \sigma_{yz}^2 + \sigma_{xz}^2\right)$$

If, instead, Eqs. (4-9), (4-10), and (4-12) are used to eliminate stresses in Eq. (5-72) we get,

$$\tilde{u} = \frac{E}{2(1-\nu^2)}\left[\varepsilon_{xx}^2 + \varepsilon_{yy}^2 + \varepsilon_{zz}^2 + 2\nu(\varepsilon_{xx}\varepsilon_{yy} + \varepsilon_{yy}\varepsilon_{zz} + \varepsilon_{xx}\varepsilon_{zz})\right] + \frac{G}{2}\left(\gamma_{xy}^2 + \gamma_{yz}^2 + \gamma_{xz}^2\right)$$

This gives the general expression for the strain energy stored in an element whose material is linearly elastic, and which undergoes infinitesimal strains.

The strain energy density can be shown to be a nonnegative quantity. To do this we eliminate the stresses in Eq. (5-72), then introduce the bulk modulus $K$ and the shear modulus $G$ as material properties because these are always positive. Finally we show that only squares of the strains appear in the resulting equations. We begin by substituting Eqs. (4-9), (4-10), and (4-12) for the stresses in Eq. (5-72). This gives

$$2\tilde{u} = \frac{E}{1+\nu}\left(\varepsilon_{xx}^2 + \varepsilon_{yy}^2 + \varepsilon_{zz}^2\right) + \frac{E\nu}{(1+\nu)(1-2\nu)}\Delta^2 + G\left(\gamma_{xy}^2 + \gamma_{yz}^2 + \gamma_{zx}^2\right)$$

With Eqs. (4-13) and (4-117) this becomes

$$2\tilde{u} = 2G\left(\varepsilon_{xx}^2 + \varepsilon_{yy}^2 + \varepsilon_{zz}^2\right) + \lambda\Delta^2 + G\left(\gamma_{xy}^2 + \gamma_{yz}^2 + \gamma_{zx}^2\right) \tag{5-72$a$}$$

The necessary terms are now added to (and then subtracted from) the first parentheses to make it a perfect square, and then Eq. (3-61) is used, yielding

$$2\tilde{u} = (\lambda + 2G)\Delta^2 + G\left(\gamma_{xy}^2 + \gamma_{yz}^2 + \gamma_{zx}^2\right) - 4G(\varepsilon_{xx}\varepsilon_{yy} + \varepsilon_{yy}\varepsilon_{zz} + \varepsilon_{zz}\varepsilon_{xx})$$

This can be written as

$$2\tilde{u} = \left(\lambda + \tfrac{2}{3}G\right)\Delta^2 + G\left(\gamma_{xy}^2 + \gamma_{yz}^2 + \gamma_{zx}^2\right) + 4G\left(\tfrac{1}{3}\Delta^2 - \varepsilon_{xx}\varepsilon_{yy} - \varepsilon_{yy}\varepsilon_{zz} - \varepsilon_{zz}\varepsilon_{xx}\right)$$

From Eq. (4-119), $\lambda + \frac{2}{3}G$ is recognized as the bulk modulus. With this, and expanding $\Delta^2$ in the last parentheses, we get, after simplifying,

$$2\tilde{u} = K\Delta^2 + G\left(\gamma_{xy}^2 + \gamma_{yz}^2 + \gamma_{zx}^2\right) + \tfrac{2}{3}G\left[(\varepsilon_{xx} - \varepsilon_{yy})^2 + (\varepsilon_{yy} - \varepsilon_{zz})^2 + (\varepsilon_{zz} - \varepsilon_{xx})^2\right]$$

Since $G$ and $K$ are always posivite, we see from this expression that the strain energy density is always positive or zero, and that it can only be zero when all of the strains are zero. Functions possessing this property are said to be *positive definite*. Although the proof has been carried out for isotropic materials, it can be shown that the strain energy density for anisotropic materials is also positive definite.

Integrating over the volume of the body, the total strain energy $U$ stored in the body is given,

$$U = \int_{\overline{V}} \tilde{u}\, d\overline{V} = \frac{1}{2}\int_{\overline{V}} (\sigma_{xx}\varepsilon_{xx} + \sigma_{yy}\varepsilon_{yy} + \sigma_{zz}\varepsilon_{zz} + \sigma_{xy}\gamma_{xy} + \sigma_{xz}\gamma_{xz} + \sigma_{yz}\gamma_{yz})\, d\overline{V} \tag{5-73}$$

Since it is the *change* in strain energy density that is of interest in solving problems, the state of zero strain energy can be assigned somewhat arbitrarily. It is customary to assume that the strain energy density is zero when all of the strains are zero. Thus for some process which proceeds from a state of zero strain to a final state characterized by nonzero strains, the change of strain energy density during the process is found from Eq. (5-68) to be

$$\begin{aligned}\tilde{u} = &\int_0^{\varepsilon_{xx}} \sigma_{xx}\, d\varepsilon_{xx} + \int_0^{\varepsilon_{yy}} \sigma_{yy}\, d\varepsilon_{yy} + \int_0^{\varepsilon_{zz}} \sigma_{zz}\, d\varepsilon_{zz} + \int_0^{\gamma_{xy}} \sigma_{xy}\, d\gamma_{xy} \\ &+ \int_0^{\gamma_{xz}} \sigma_{xz}\, d\gamma_{xz} + \int_0^{\gamma_{yz}} \sigma_{yz}\, d\gamma_{yz}\end{aligned} \tag{5-74}$$

where the upper limits in the integrals represent the strain in the final state.

## 5-6-1 Complementary Energy

In Eq. (5-12) we have defined complementary work as the area *above* the force-deflection curve. In an analogous way we can define the complementary energy density. When this is generalized to include all six stress components, we arrive at the following:

$$\begin{aligned}\tilde{u}_c = &\int_0^{\sigma_{xx}} \varepsilon_{xx}\, d\sigma_{xx} + \int_0^{\sigma_{yy}} \varepsilon_{yy}\, d\sigma_{yy} + \int_0^{\sigma_{zz}} \varepsilon_{zz}\, d\sigma_{zz} \\ &+ \int_0^{\sigma_{xy}} \gamma_{xy}\, d\sigma_{xy} + \int_0^{\sigma_{xz}} \gamma_{xz}\, d\sigma_{xz} + \int_0^{\sigma_{yz}} \gamma_{yz}\, d\sigma_{yz}\end{aligned} \tag{5-75}$$

where $\tilde{u}_c$ is the complementary energy. From this definition it follows that

$$d\tilde{u}_c = \varepsilon_{xx}\,d\sigma_{xx} + \varepsilon_{yy}\,d\sigma_{yy} + \varepsilon_{zz}\,d\sigma_{zz} + \gamma_{xy}\,d\sigma_{xy} + \gamma_{xz}\,d\sigma_{xz} + \gamma_{yz}\,d\sigma_{yz} \tag{5-76}$$

From Eq. (5-76) we see that the complementary energy depends on the stresses,

$$\tilde{u}_c = \tilde{u}_c(\sigma_{xx}, \sigma_{yy}, \sigma_{zz}, \sigma_{xy}, \sigma_{xz}, \sigma_{yz})$$

Then

$$d\tilde{u}_c = \frac{\partial \tilde{u}_c}{\partial \sigma_{xx}}\,d\sigma_{xx} + \frac{\partial \tilde{u}_c}{\partial \sigma_{yy}}\,d\sigma_{yy} + \frac{\partial \tilde{u}_c}{\partial \sigma_{zz}}\,d\sigma_{zz} + \frac{\partial \tilde{u}_c}{\partial \sigma_{xy}}\,d\sigma_{xy} + \frac{\partial \tilde{u}_c}{\partial \sigma_{xz}}\,d\sigma_{xz} + \frac{\partial \tilde{u}_c}{\partial \sigma_{yz}}\,d\sigma_{yz} \tag{5-77}$$

From Eqs. (5-76) and (5-77) we have the following relationships:

$$\begin{aligned} \frac{\partial \tilde{u}_c}{\partial \sigma_{xx}} &= \varepsilon_{xx}, & \frac{\partial \tilde{u}_c}{\partial \sigma_{xy}} &= \gamma_{xy} \\ \frac{\partial \tilde{u}_c}{\partial \sigma_{yy}} &= \varepsilon_{yy}, & \frac{\partial \tilde{u}_c}{\partial \sigma_{xz}} &= \gamma_{xz} \\ \frac{\partial \tilde{u}_c}{\partial \sigma_{zz}} &= \varepsilon_{zz}, & \frac{\partial \tilde{u}_c}{\partial \sigma_{yz}} &= \gamma_{yz} \end{aligned} \tag{5-78}$$

Finally we note that if the material is linear, then from Eqs. (4-9), (4-10), (4-12), and (5-75) we have that

$$\tilde{u}_c = \tfrac{1}{2}(\sigma_{xx}\varepsilon_{xx} + \sigma_{yy}\varepsilon_{yy} + \sigma_{zz}\varepsilon_{zz} + \sigma_{xy}\gamma_{xy} + \sigma_{xz}\gamma_{xz} + \sigma_{yz}\gamma_{yz}) \tag{5-79}$$

and therefore for materials with a linear stress-strain relationship,

$$\tilde{u} = \tilde{u}_c \tag{5-80}$$

## 5-6-2 Strain Energy in Beams

In the case of beams, the expression for strain energy given by Eq. (5-73) can be specialized further so that it is expressed in terms of the usual beam quantities, namely, the internal longitudinal force $F$, the internal transverse shear force $V$, and the internal bending moment $M$, as shown in Fig. 5-18*a*. For simplicity we restrict our attention to problems in which the neutral axis of the beam is coincident with the $x$ axis and the loads are in the $xz$ plane. Then $\sigma_{yy} = 0$, $\sigma_{yz} = 0$, and $\sigma_{xy} = 0$. On recalling that in elementary beam theory $\sigma_{zz}$, the normal stress perpendicular to the beam axis, is considered negligibly small, Eq. (5-73) becomes

$$U = \frac{1}{2}\int_{\overline{V}} (\sigma_{xx}\varepsilon_{xx} + \sigma_{xz}\gamma_{xz})\,d\overline{V} \tag{5-81}$$

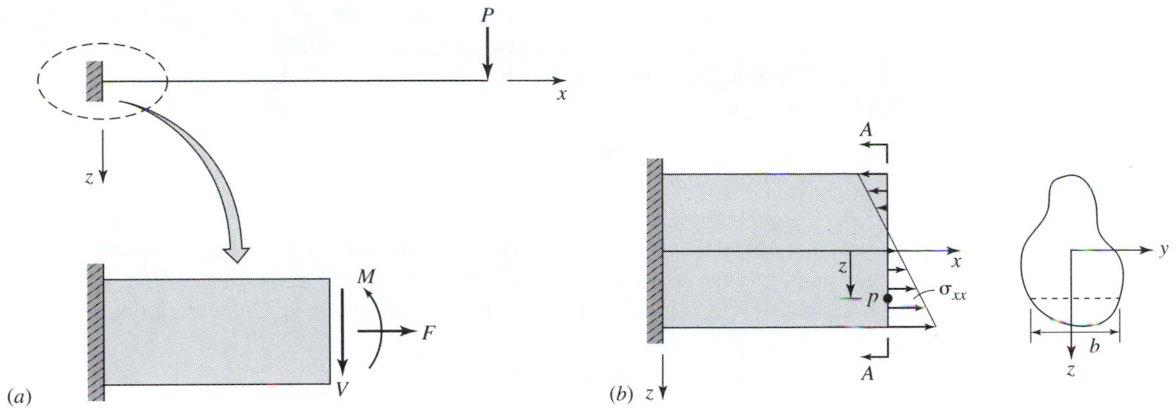


Figure 5-18

or

$$U = \frac{1}{2}\int_{\overline{V}}\left(\frac{\sigma_{xx}^2}{E} + \frac{\sigma_{xz}^2}{G}\right) d\overline{V} \tag{5-82}$$

For a beam the volume integral can be obtained by integrating over the cross-sectional area and then over the length,

$$U = \frac{1}{2}\int_0^l \int_A \left(\frac{\sigma_{xx}^2}{E} + \frac{\sigma_{xz}^2}{G}\right) dA\, dx \tag{5-83}$$

From this expression we have the strain energy expressed in terms of the stresses $\sigma_{xx}$, $\sigma_{xz}$,

$$U = U(\sigma_{xx}, \sigma_{xz})$$

However, for a beam we would like the strain energy to be expressed in terms of the beam quantities shown in Fig. 5-18*a*,

$$U = U(F, M, V)$$

Now in a beam the stress $\sigma_{xx}$ at $p$ results from the bending moment $M$ and the longitudinal force $F$, as shown in Fig. 5-18*b*, and is given by

$$\sigma_{xx} = \frac{Mz}{I} + \frac{F}{A} \tag{5-84}$$

where $z$ is the distance from the neutral axis to $p$, $I$ is the moment of inertia of the cross section about the $y$ axis, and $A$ is the area of the cross section. The shearing stress results from the shearing force $V$ and is given by

$$\sigma_{xz} = \frac{VQ}{Ib} \tag{5-85}$$

where $Q$ is the first moment of the cross section outside of point $p$ and $b$ is the width (in the $y$ direction) of the cross section at $p$, as shown in Fig. 5-18$b$. Note that both $Q$ and $b$ are functions of $z$. On squaring Eq. (5-84) and substituting the result in the first integral in Eq. (5-83), we get

$$\int_A \frac{\sigma_{xx}^2}{E}\,dA = \frac{M^2}{EI^2}\int_A z^2\,dA + \frac{2MF}{IA}\int_A z\,dA + \frac{F^2}{A^2}\int_A dA$$

The first integral on the right is the moment of inertia about the $y$ axis $I$, whereas the last integral is simply the cross-sectional area $A$. The second integral is the first moment of the area, and since $z$ is measured from the centroid, it is zero. Consequently,

$$\int_A \frac{\sigma_{xx}^2}{E}\,dA = \frac{M^2}{EI} + \frac{F^2}{EA} \tag{5-86}$$

Substitution of Eq. (5-85) in the second integral on the right-hand side of Eq. (5-83) gives

$$\int_A \frac{\sigma_{xz}^2}{G}\,dA = \frac{V^2}{GI^2}\int_A \frac{Q^2}{b^2}\,dA$$

This can be written as

$$\int_A \frac{\sigma_{xz}^2}{G}\,dA = \frac{\kappa V^2}{GA} \tag{5-87}$$

where $\kappa$ is a form factor, given by

$$\kappa = \frac{A}{I^2}\int_A \frac{Q^2}{b^2}\,dA \tag{5-88}$$

Since $dA = b\,dz$, the form factor can be written as

$$\kappa = \frac{A}{I^2}\int_z \frac{Q^2}{b}\,dz \tag{5-89}$$

Note that the form factor depends only on the geometry of the cross section. For a solid rectangular cross section its value is 1.20, whereas for a solid circular cross section it is 10/9.

With Eqs. (5-86) and (5-87), Eq. (5-83) becomes

$$U = \int_0^l \left(\frac{M^2}{2EI} + \frac{F^2}{2EA} + \frac{\kappa V^2}{2GA}\right)dx \tag{5-90}$$

If the beam is subjected to a torque $T$ that varies along its length, Eq. (5-90) becomes

$$U = \int_0^l \left(\frac{M^2}{2EI} + \frac{F^2}{2EA} + \frac{\kappa V^2}{2GA} + \frac{T^2}{2GK}\right)dx$$

where $K$ is a shape constant for the cross section. For circular or annular cross sections it is the polar moment of inertia about the centroid. $K$ is less than the polar moment of inertia for other cross sections. Formulas for $K$ for various cross sections can be found in [9, table 20, pp. 348–359].

## EXAMPLE 5-6

Determine the strain energy stored in the column shown in Fig. 5-19*a*, with a load $P$ applied to its free end.

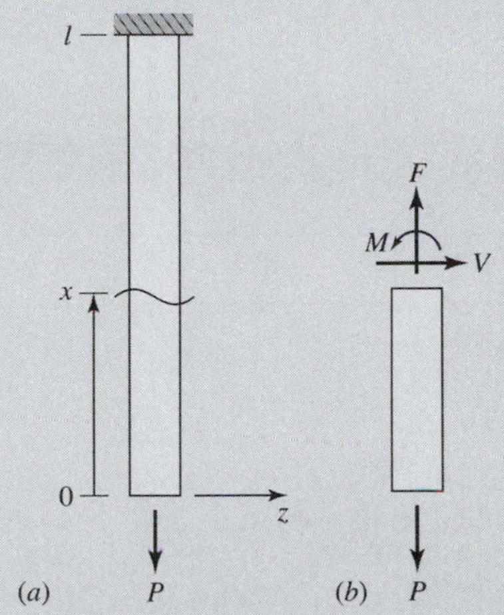


**Figure 5-19**

## SOLUTION

A cut is made $x$ units from the free end, and Fig. 5-19*b* shows the free-body diagram of the material below the cut. Considerations of equilibrium for the free-body diagram give the following values for the internal forces: $F = P$, $V = 0$, and $M = 0$. Therefore Eq. (5-88) becomes

$$U = \int_0^l \frac{P^2}{2AE}\,dx$$

and since $P$ is constant,

$$U = \frac{P^2 l}{2AE} \tag{a}$$

Since there are no thermal effects present, the strain energy should equal the work done by the force $P$. To verify that this is the case, note that for a column the longitudinal displacement $u_0$ at the free end is given by

$$u_0 = \frac{Pl}{AE} \tag{b}$$

and therefore,

$$W = \frac{1}{2} P u_0 = \frac{P^2 l}{2AE}$$

Thus,

$$U = W$$

Note that with Eqs. (*a*) and (*b*) we can write

$$U = \frac{u_0^2 AE}{2l} \qquad (c)$$

or

$$U = \tfrac{1}{2} P u_0 \qquad (d)$$

## EXAMPLE 5-7

Now let us determine the strain energy stored in a cantilever beam that is acted upon by a uniformly distributed load of magnitude $q$, as shown in Fig. 5-20*a*.

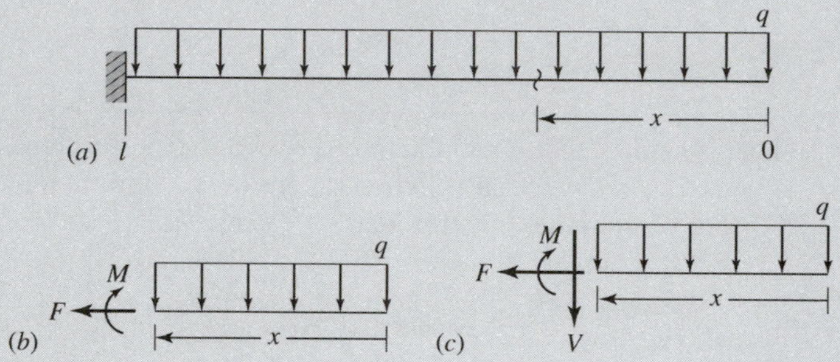


**Figure 5-20**

### SOLUTION

For many of the problems encountered in practice the strain energy caused by shear is small compared to that caused by bending. Assume that this is the case, so that $V = 0$. Then Eq. (5-90) reduces to

$$U = \int_0^l \left( \frac{M^2}{2EI} + \frac{F^2}{2EA} \right) dx \qquad (a)$$

From the free-body diagram in Fig. 5-20*b* it follows that

$$F = 0, \qquad M = -\tfrac{1}{2} q x^2$$

Therefore Eq. (*a*) gives

$$U = \int_0^l \frac{q^2 x^4}{8EI}\, dx$$

or

$$U = \frac{q^2 l^5}{40EI} \tag{b}$$

Thus when the effects of only the bending moment are considered, the strain energy stored in a cantilever beam subjected to a uniformly distributed load is given by Eq. (*b*).

Now, let us re-solve the beam problem, but this time we will include the effects of transverse shear. From the free-body diagram shown in Fig. 5-20*c* we find that

$$F = 0, \qquad M = -\tfrac{1}{2} q x^2, \qquad V = qx$$

and when these are substituted into Eq. (*a*) in Example 5-7, we get

$$U = \int_0^l \left( \frac{q^2 x^4}{8EI} + \frac{\kappa q^2 x^2}{2GA} \right) dx$$

or, carrying out the integration,

$$U = \frac{q^2 l^5}{40EI} + \frac{\kappa q^2 l^3}{6GA}$$

where the first term gives the bending contribution and the second the shear contribution. By factoring out the bending term, the expression for strain energy can be written as

$$U = \frac{q^2 l^5}{40EI} \left[ 1 + \frac{40}{3}(1+\nu)\frac{I}{A}\frac{\kappa}{l^2} \right] \tag{5-91}$$

where we have used Eq. (4-13), which states that

$$G = \frac{E}{2(1+\nu)}$$

Now for the common engineering materials the value of Poisson's ratio is between 0.25 and 0.33. Therefore we will assume a representative value of 0.3. With this, Eq. (5-91) becomes

$$U = \frac{q^2 l^5}{40EI} \left( 1 + 17.33 \frac{\kappa I}{A l^2} \right) \tag{5-92}$$

The importance of transverse shear to the strain energy is determined by the size of the second term in the parentheses when compared to 1. If this term is small compared to 1, then shear effects are negligible, whereas if it is not small compared to 1, shear effects cannot be neglected. For the sake of definiteness we will assume that the beam cross section is rectangular. Then

$$I = \frac{bh^3}{12}, \qquad A = bh, \qquad \kappa = 1.2$$

and Eq. (5-92) becomes

$$U = \frac{q^2 l^5}{40EI}\left(1 + 1.73\frac{h^2}{l^2}\right) \tag{5-93}$$

Thus we see that the importance of transverse shear to the strain energy is determined by the length-to-thickness ratio of the beam. If the beam is short and stubby, like a gear tooth, then $l/h$ is close to 1 and shear makes a significant contribution to the strain energy. On the other hand, if the beam is long and slender, as is often the case, then the contribution of transverse shear to the overall strain energy of the beam is minimal. For example, if $l/h = 10$, less than 2 percent of the beam's strain energy is due to shear. In this case it is reasonable to neglect the effects of transverse shear when computing the strain energy, as was done in the previous example. In fact, a common rule of thumb is to neglect the contribution of transverse shear to the strain energy stored in a beam when the beam's length-to-thickness ratio is greater than 10.

We return now to Eq. (5-32). With this equation some interesting problems can be solved. For instance, we know that if a block of weight $B$ rests on a column of length $l$ and cross-sectional area $A$, then the static deflection of the column end along its axis is given by

$$u_{\text{ST}} = \frac{Bl}{AE} \tag{5-94}$$

However, if the block is dropped from a height $h$ above the column, the problem is considerably more complicated. In that case the concept of strain energy plays a central role in determining the column deflection, as seen in the following example.

## EXAMPLE 5-8

A block is dropped from a height $h$ above a column, as shown in Fig. 5-21$a$. Find the deflection of the column.

### SOLUTION

In this problem the block falls a distance $h$ prior to coming into contact with the column. After it contacts the column, the block travels an additional distance $u_0$ as the column deforms. At the instant the column reaches its maximum displacement $u_0$, the downward velocity of the block is zero and it is about to start moving upward as it rebounds. We are interested in calculating the maximum displacement $u_0$ of the end of the column.

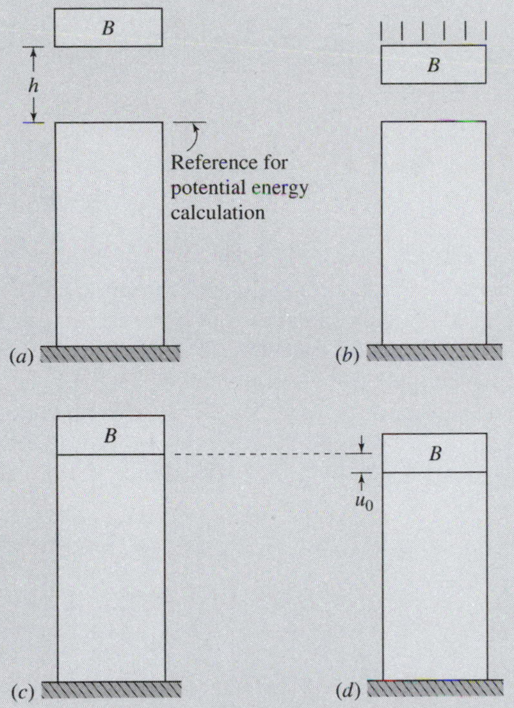


**Figure 5-21**

From the standpoint of thermodynamics this problem can be viewed as consisting of three processes. The first process is the free fall of the block through a height $h$; the second process is the additional displacement $u_0$ of the block while it is in contact with the column; and the third process is the displacement of the column by an amount $u_0$. Let us assume that potential energy is calculated relative to the free end of the column in its undeformed location, as shown in Fig. 5-21*a*. For the first process, state 1 is when the block is located a distance $h$ above the column, as shown in Fig. 5-21*a*, whereas state 2 is when the block has dropped an amount $h$ and is about to contact the column, as shown in Fig. 5-21*c*. The first law then gives

$$I_{1/2} + T_2 - T_1 + K_2 - K_1 = W_{1/2}$$

No heat is transferred during the process, and the internal energy is zero because the block has not deformed. The work is zero because no forces cross the system boundary, in this case, the surface of the block. In state 1 the potential and kinetic energies are given by

$$T_1 = Bh, \qquad K_1 = 0$$

whereas in state 2 the potential energy is zero and the kinetic energy is unknown. Thus from the first law we find that

$$K_2 = Bh \tag{a}$$

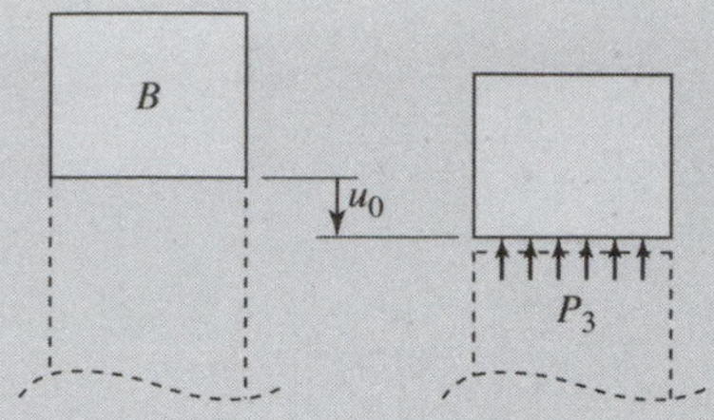


Figure 5-22

Now during the second process the block goes from state 2 to state 3, in which the block has displaced an additional amount $u_0$ and is at rest. During the second process, since the block is in contact with the column, a force acts on the block, as shown in Fig. 5-22. However, the magnitude of this force is unknown (note that, because of the dynamic effects, it is not simply the weight of the block). The first law for this process is

$$T_3 - T_2 + K_3 - K_2 = W_{2/3}$$

The potential energy in state 3 is $-Bu_0$, where the minus sign arises because the displacement is below the reference line, and the kinetic energy of the block is zero. Thus with Eq. (*a*) we have

$$W_{2/3} = -B(h + u_0) \tag{b}$$

The negative sign indicates that the force $P_3$ in Fig. 5-22 and the displacement $u_0$ through which it moves are in opposite directions.

Now the third process, shown in Fig. 5-23, is that of the column deforming. We know that the strain energy stored in the column must equal the work done by the external forces acting on it,

$$U = W$$

Now equilibrium across the interface dictates that the force acting on the column is the opposite of the force acting on the block. Consequently the work done by this force acting on the column is $-W_{2/3}$. Thus,

$$U = B(h + u_0) \tag{c}$$

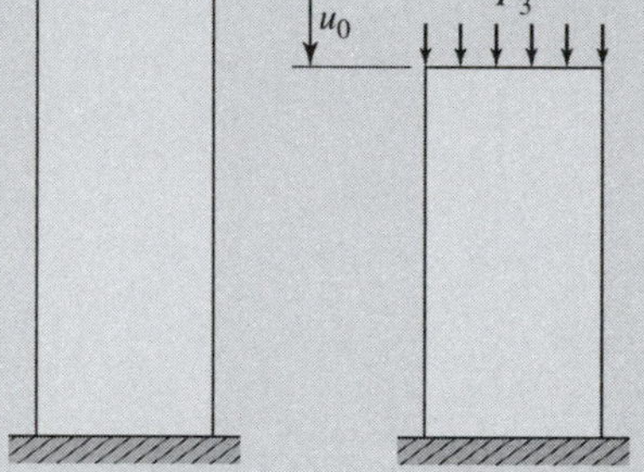


Figure 5-23

But with Eq. ($c$) of Example 5-6 this becomes

$$\frac{AEu_0^2}{2l} = B(h + u_0)$$

or, after rearranging,

$$u_0^2 - \frac{2lB}{AE}u_0 - \frac{2lBh}{AE} = 0$$

This is a quadratic in $u_0$, whose roots are

$$u_0 = \frac{Bl}{AE} \pm \sqrt{\left(\frac{Bl}{AE}\right)^2 + 2h\frac{Bl}{AE}}$$

or with Eq. (5-94),

$$u_0 = u_{ST}\left(1 + \sqrt{1 + \frac{2h}{u_{ST}}}\right) \tag{d}$$

Note that the radical gives a correction to the static displacement when the block is dropped from a height $h$. From this, two limiting cases can be examined. First, let $h$ be large so that $2h/u_{ST} \gg 1$. Then Eq. ($d$) gives

$$u_0 = \sqrt{2hu_{ST}}$$

This applies when the block falls a great distance before coming into contact with the column. On the other hand, when $h = 0$, Eq. ($d$) gives

$$u_0 = 2u_{ST}$$

Thus if the block is held so that it is barely contacting the column and then suddenly released, the displacement of the column is twice what it is when the weight of the block is applied slowly.

It is interesting to note that the net result of the three processes described in this example, and as evidenced in Eq. ($c$), is that the net change in the potential energy of the block as it moves through a change in elevation $h + u$ is stored in the column as strain energy.

## 5-7 CASTIGLIANO'S THEOREM

Let us now consider a beam, as shown in Fig. 5-24$a$, to which two loads $P_1$ and $P_2$ are applied. Although the beam shown is simply supported, we could use other boundary conditions so long as rigid body motion of the beam is prevented. The displacement $\delta_2$, resulting from the combination of loads $P_1$ and $P_2$, is that component of the displacement $\boldsymbol{u}_2$

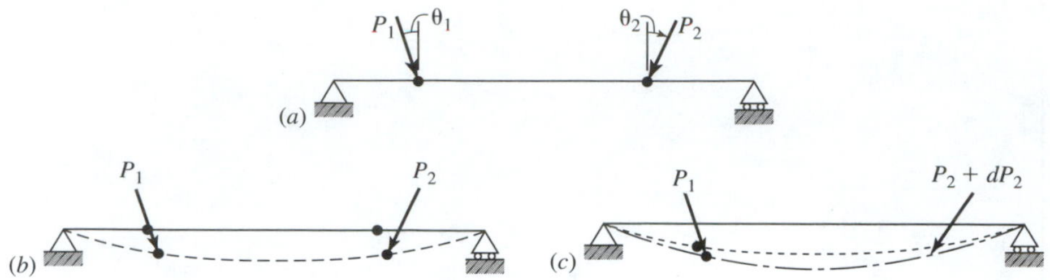


**Figure 5-24**

that is in the direction defined by $\theta_2$, that is, in the direction of the load $P_2$. Note that $\delta_2$ will change if either $P_1$ or $P_2$ changes. The strain energy stored in the beam is $U(P_1, P_2)$, while the work done by the applied loads is $W(P_1, P_2)$. Further, from the first law of thermodynamics,

$$U_1(P_1, P_2) = W_1(P_1, P_2) \tag{5-95}$$

With the beam in the configuration shown in Fig. 5-24*b*, we now increase $P_2$ by an infinitesimal amount $dP_2$. This is accomplished simply by adding a load $dP_2$ whose point of application and direction are the same as for $P_2$. This causes an additional displacement of the beam, as shown in Fig. 5-24*c*, and increases the strain energy by a differential amount. Thus,

$$U_2(P_1, P_2) = U_1(P_1, P_2) + dU$$

Now with the chain rule we have

$$dU = \frac{\partial U}{\partial P_1} dP_1 + \frac{\partial U}{\partial P_2} dP_2$$

but in our case only $P_2$ was changed, so that $dP_1 = 0$. Consequently we have

$$U_2(P_1, P_2) = U_1(P_1, P_2) + \frac{\partial U}{\partial P_2} dP_2 \tag{5-96}$$

The work done by the applied loads also changes so that for the configuration shown in Fig. 5-24*c* the total work done by the applied loads is $W_2(P_1, P_2)$ and the first law requires that

$$U_2(P_1, P_2) = W_2(P_1, P_2) \tag{5-97}$$

We now need to calculate $W_2$ in terms of $W_1$. However, this is easier to do if we apply the load $dP_2$ first, as shown in Fig. 5-25*a*. The work done is therefore

$$\tilde{W} = \tfrac{1}{2}\, dP_2\, d\delta_2$$

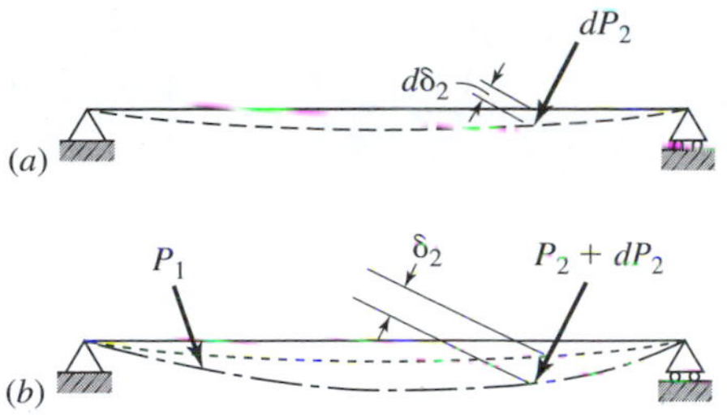


Figure 5-25

Now we add the loads $P_1$ and $P_2$, which cause the beam to deform to the configuration shown in Fig. 5-25*b*. The work done by the addition of loads $P_1$ and $P_2$ is

$$\hat{W} = W(P_1, P_2) + \delta_2\, dP_2$$

where the first term on the right is the work done by loads $P_1$, $P_2$ and the second term is the additional work done as the load $dP_2$ moves through the displacement $\delta_2$. The final configuration, shown in Fig. 5-25*b*, is the same as the final configuration shown in Fig. 5-24*c*. Therefore,

$$W_2(P_1, P_2) = W(P_1, P_2) + \delta_2\, dP_2 + \tfrac{1}{2}\, dP_2\, d\delta_2$$

Further, for small deflections the work $W_1(P_1, P_2)$ done in moving from the undeformed configuration to that shown in Fig. 5-24*b* is the same as the work $W(P_1, P_2)$ done by the loads $P_1$, $P_2$ in moving from the configuration shown in Fig. 5-25*a* to that shown in Fig. 5-25*b*. Thus $W(P_1, P_2) = W_1(P_1, P_2)$ and we can write

$$W_2(P_1, P_2) = W_1(P_1, P_2) + \delta_2\, dP_2 + \tfrac{1}{2}\, dP_2\, d\delta_2$$

Substituting this expression into Eq. (5-97), using Eqs. (5-96) and (5-95), then dividing by $dP_2$ we get

$$\frac{\partial U}{\partial P_2} = \delta_2 + \frac{1}{2}\, d\delta_2$$

Note, however, that the second term on the right is small compared to the other two terms and may therefore be neglected. This gives

$$\frac{\partial U}{\partial P_2} = \delta_2$$

This is a mathematical statement of Castigliano's second theorem. In words, it states the following:

- Calculate the strain energy in the beam due to $P_1$ and $P_2$.
- Differentiate the expression for strain energy with respect to $P_2$.
- What results is the deflection $\delta_2$, at the point of application of the load $P_2$ and in the direction of that load, due to both loads $P_1$ and $P_2$.

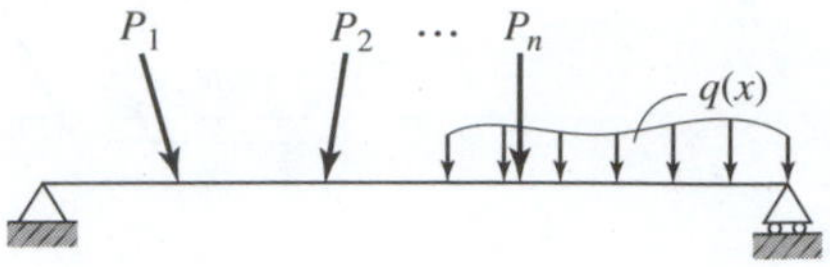


Figure 5-26

The derivation of Castigliano's second theorem is easily generalized to give

$$\frac{\partial U}{\partial P_i} = \delta_i, \qquad i = 1, 2, \ldots, n \tag{5-98}$$

where

$$U = U(P_1, P_2, \ldots, P_n, q) \tag{5-99}$$

In the expression for strain energy $P_i$ are concentrated forces or moments and $q_i$ are distributed loads, as shown in Fig. 5-26. In words, the generalized form of *Castigliano's second theorem* states the following:

- Calculate the strain energy due to *all* loads acting on the beam.
- Differentiate the expression for strain energy with respect to a particular *concentrated* load.
- What results is the deflection, *at the point of application of the load and in the direction of that load,* due to *all* applied loads, both concentrated and distributed.

Castigliano's second theorem provides a means of determining deflections at the points where concentrated loads are applied. In contrast with the Newtonian methods generally learned in strength of materials courses, it does not give the entire deflection curve.

Note that if $P_i$ is a moment, then the corresponding deflection is an angular deflection, or slope. Thus if the expression for strain energy is differentiated with respect to a moment, the resulting deflection is the slope at that point.

It should be emphasized that in using Castigliano's second theorem the differentiation *must* be with respect to a concentrated force or moment. No meaningful result is obtained if differentiation is done with respect to a distributed load.

If the strain energy is determined in terms of deflections rather than forces, a derivation analogous to that which led to Castigliano's second theorem leads to the following result:

$$\frac{\partial U}{\partial \delta_i} = P_i, \qquad i = 1, 2, \ldots, n \tag{5-100}$$

where

$$U = U(\delta_1, \delta_2, \ldots, \delta_n) \tag{5-101}$$

This is referred to as *Castigliano's first theorem.*

## EXAMPLE 5-9

As an example of the use of Castigliano's second theorem, let us find the deflection of the free end of the column shown in Fig. 5-27.

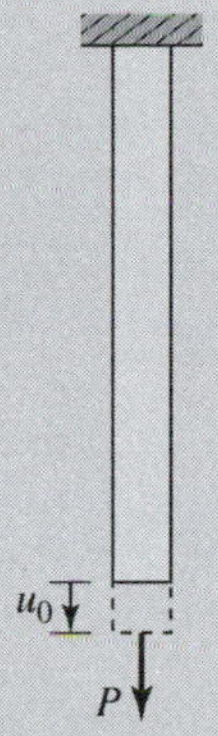


**Figure 5-27**

### SOLUTION

Since this deflection occurs at the point of application of the load $P$ and in the direction of that load, we simply differentiate the expression for strain energy of the column, Eq. ($a$) in Example 5-6, with respect to $P$:

$$u_0 = \frac{\partial U}{\partial P} = \frac{\partial}{\partial P}\left(\frac{P^2 l}{AE}\right)$$

Thus,

$$u_0 = \frac{Pl}{AE}$$

which is the familiar result obtained in courses in strength of materials.

## EXAMPLE 5-10

As a second example, let us look at a cantilever beam with a concentrated load applied at its free end, as shown in Fig. 5-28$a$. We wish to calculate the deflection at the free end of this beam, neglecting transverse shear.

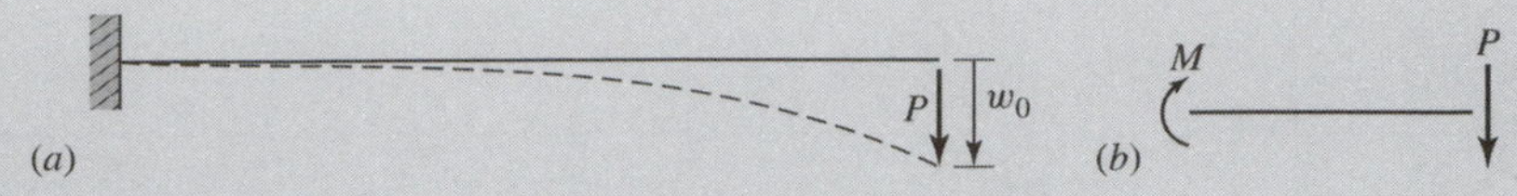


**Figure 5-28**

## SOLUTION

With the aid of the free-body diagram shown in Fig. 5-28*b* the strain energy for this problem is found to be

$$U = \frac{P^3 l^3}{6EI}$$

The deflection $w_0$ is in the same direction as the load $P$, and therefore to find it we simply differentiate the expression for strain energy with respect to the load $P$:

$$w_0 = \frac{\partial U}{\partial P} = \frac{\partial}{\partial P}\left(\frac{P^3 l^3}{6EI}\right)$$

Then,

$$w_0 = \frac{Pl^2}{3EI}$$

This result should be familiar from the study of strength of materials. It should be recalled that in strength of materials the calculated beam deflections typically do not include the effects of transverse shear. The importance of this effect on the deflection can be obtained with Castigliano's theorem by including transverse shear effects in the strain energy, as was done in arriving at Eq. (5-93). When Poisson's ratio is 0.30 and the cross section is rectangular, the strain energy for the problem shown in Fig. 28*a* then becomes

$$U = \frac{P^2 l^3}{6EI}\left[1 + 0.78\left(\frac{h}{l}\right)^2\right] \tag{5-102}$$

Thus by differentiating Eq. (5-102) with respect to $P$ we get the deflection, including that due to transverse shear, at the free end of the cantilever beam,

$$w_0 = \frac{\partial U}{\partial P} = \frac{Pl^3}{3EI}\left[1 + 0.78\left(\frac{h}{l}\right)^2\right]$$

Note that the error in neglecting transverse shear is less than 1% if the length-to-thickness ratio of the beam exceeds 10.

## EXAMPLE 5-11

Consider the cantilever beam shown in Fig. 5-29. In this figure both a moment and a bending force are applied to the beam. Find the slope at the free end of the beam, assuming that the length-to-thickness ratio exceeds 10.

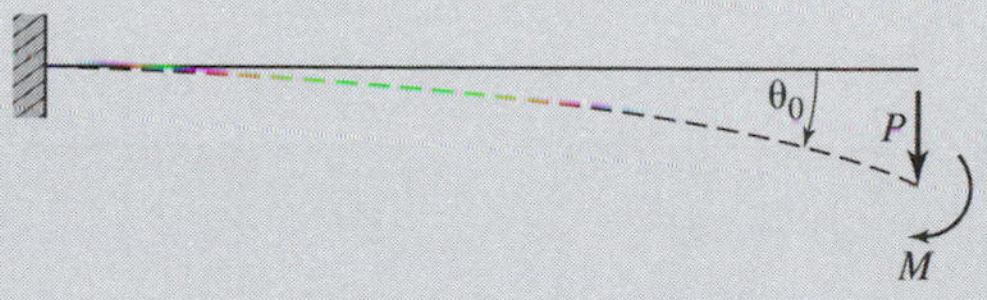


**Figure 5-29**

## SOLUTION

Since the length-to-thickness ratio of the beam is to exceed 10, the effects of transverse shear can be neglected and, on drawing a free-body diagram and summing forces and moments, we find that the internal longitudinal force is zero while the internal moment is given by

$$M = -Px - M_0$$

Thus the strain energy is

$$U = \int_0^l \frac{(Px + M_0)^2}{2EI}\,dx \tag{a}$$

where $U$ depends on both $P$ and $M$. Of course we could integrate Eq. ($a$) and obtain an algebraic expression for the strain energy, but let us wait. To calculate the slope at the free end of the cantilever beam, we differentiate the expression for strain energy with respect to $M$ to obtain

$$\theta_0 = \frac{\partial U}{\partial M_0} = \int_0^l \frac{Px + M_0}{EI}\,dx$$

and carrying out the integration, we get

$$\theta_0 = \frac{Pl^2}{2EI} + \frac{M_0 l}{EI} \tag{b}$$

Note that by waiting to integrate until after the differentiation was carried out the task of integrating was made a little easier. It happens frequently in the use of Castigliano's theorem that carrying out the differentiation prior to integrating reduces the amount of work necessary to perform the integration. This is because the internal loads appear as second-order quantities in the strain energy. After differentiation they are reduced to first order. Additional applications of this approach can be found in Chapter 8.

The result in Eq. ($b$) gives the slope at the free end of the cantilever beam due to both the bending load $P$ and the moment $M$ acting simultaneously. Note that had we chosen to differentiate Eq. ($a$) with respect to $P$ we would have obtained the bending deflection at the free end of the beam.

Equation (*b*) in Example 5-11 is valid for any value of $P$ and $M_0$, including zero. Thus by choosing $M_0$ to be zero in Eq. (*b*) we obtain the slope at the free end of a cantilever beam with a concentrated load at its free end, as shown in Fig. 5-28. By applying a moment to the end of the beam and ultimately setting it to zero, a means has been found by which a displacement in a direction other than that of the applied load can be determined. When $M_0$ is used in this manner, it is referred to as a *dummy load.*

In order to reinforce the concept of a dummy load, consider the following example.

## EXAMPLE 5-12

For the cantilever beam shown in Fig. 5-30*a*, and subjected to a uniformly distributed load of intensity $q$, find the displacement at the free end. Assume that the length-to-thickness ratio of the beam exceeds 10 so that the effects of transverse shear can be neglected.

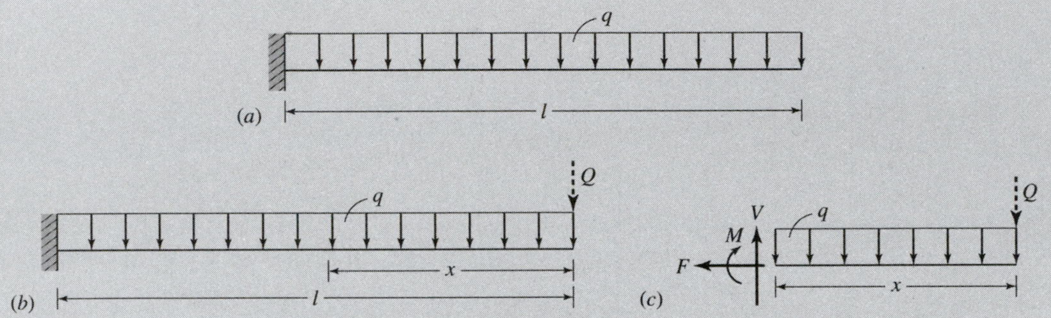


**Figure 5-30**

### SOLUTION

Note that Castigliano's theorem gives a displacement *at a point,* and it is therefore necessary, as indicated earlier in this section, to differentiate the strain energy with respect to a *concentrated* load. However, in this example the load on the beam is a distributed one. Therefore in order to determine the displacement at the free end of the beam, we need to introduce a concentrated load at that point. This load is a dummy load, that is, the concentrated load $Q$ indicated by the dashed arrow in Fig. 5-30*b*. Although $Q$ is a dummy load, it must be included in our calculation of the strain energy. Then, at an appropriate point in the solution process, it is set to zero.

With $x$ measured from the free end of the beam, as shown in Fig. 5-30*b*, the free-body diagram for a beam segment is shown in Fig. 5-30*c*. From that free-body diagram the internal moment is found to be

$$M = -\tfrac{1}{2}qx^2 - Qx$$

The strain energy is then given by

$$U(q, Q) = \int_0^l \frac{\left(\frac{1}{2}qx^2 + Qx\right)^2}{2EI} dx$$

Castigliano's theorem then states that the displacement at the free end is found by differentiating the strain energy with respect to $Q$,

$$\delta_0 = \frac{\partial U(q, Q)}{\partial Q}$$

As in Example 5-11, it is convenient to differentiate first and then integrate. This gives

$$\delta_0 = \int_0^l \frac{\left(\frac{1}{2}qx^2 + Qx\right)x}{EI} dx$$

Recall that the sole purpose for introducing the dummy load $Q$ was to provide a concentrated load at the free end with respect to which differentiation would be carried out. Since the differentiation is now complete, $Q$ serves no further purpose and may be set to zero in order to simplify the integration. This gives

$$\delta_0 = \int_0^l \frac{\frac{1}{2}qx^3}{EI} dx$$

By carrying out the indicated integration the displacement at the free end is found to be

$$\delta_0 = \frac{ql^4}{8EI}$$

Thus the dummy load concept has enabled us to use Castigliano's theorem to find the free-end deflection due to a distributed load. Note that if the dummy load had been applied at some other point on the beam, then we could have found the displacement at that point. Finally, if a dummy moment had been applied at some point on the beam, we could have found the slope at that point due to the distributed load.

Further examples involving the use of the dummy load concept will be presented in Chapter 8.

Castigliano's second theorem provides a valuable tool for calculating deflections and slopes in beams when our interest is in determining these quantities at certain points, rather than determining the deflection or slope distribution throughout the entire beam. The following three points are worth emphasizing again:

- In order to use Castigliano's second theorem the differentiation must be with respect to *concentrated* forces or moments.
- In general, differentiation of the strain energy can be carried out before the strain energy expression is integrated, if it is convenient to do so.
- Displacements at points where no concentrated loads are applied can be found by introducing a dummy load.

In Sections 5-10 and 5-12 we will generalize Castigliano's first and second theorems and show that they may be applied to elastic bodies of *arbitrary* shape. In the applications chapters we will solve additional problems using this powerful tool.

## 5-8 PRINCIPLE OF VIRTUAL WORK

We turn our attention now to the development of some very powerful methods for solving problems in the mechanics of deformable bodies. These methods build on the energy concepts we studied thus far in this chapter, and their development hinges on understanding the idea of a virtual, or imagined, change.

To start with, let us picture a body that is acted upon by a set of equilibrium forces. Now imagine that the body is in a slightly different position, even though the forces remain unchanged. This is not an actual displacement of the body but only one that we have *imagined* to occur. Note that there is a certain degree of arbitrariness in the new position of the body because we only required the new position to be a "slightly different" one, and therefore there are many options. There is no cause and effect relationship at work here because no forces, not even gravitational forces, were used to put the body in its new position—we simply imagined that it happened.

Displacements of the type imagined in the preceding paragraph are, of course, somewhat abstract. However, displacements similar to these, called virtual displacements, are very useful in developing certain energy theorems in mechanics, and lead to some very useful problem-solving techniques.

We define a *virtual displacement* as one that

- Is infinitesimal
- Occurs at an instant of time
- Is kinematically admissible
- Is otherwise arbitrary.

By "infinitesimal" we mean that a virtual displacement is on the order of those that are allowed in linear elasticity. A virtual displacement "occurs at an instant of time," in contrast to an actual displacement, which occurs over an interval of time $dt$ during which both forces and system constraints may change. By "kinematically admissible" we mean that a virtual displacement satisfies all of the kinematic constraints of the problem. Thus for a deformable body a virtual displacement must satisfy the strain-displacement equations [Eqs. (3-11)] and the compatibility equations [Eqs. (3-18)]. In addition a virtual displacement must be consistent with the kinematic constraints imposed on the body. For example, if the displacement at the free end of a cantilever beam is specified as 0.10 cm, then the virtual displacement at that point must be zero because if it is not, then the kinematic constraint that the free end displacement be 0.10 cm is violated. On the other hand, if the force at the free end of a cantilever beam is specified as 100 N, then the virtual displacement at that point need not be zero because there is no kinematic constraint at the free end. Rather, the constraint is on the force. By requiring the virtual displacement to be "otherwise arbitrary" we mean that no further restrictions can be placed on a virtual displacement.

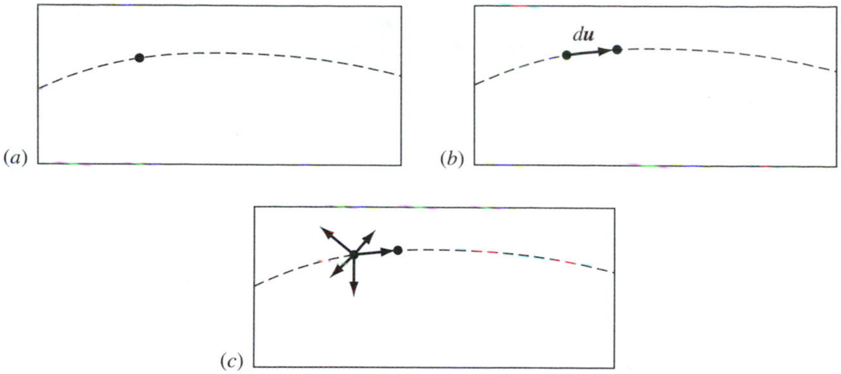


Figure 5-31 Actual and virtual displacements of a particle

Virtual displacements are indicated by the notation $\delta \boldsymbol{u}$. This notation serves to distinguish virtual displacements from actual infinitesimal displacements, $d\boldsymbol{u}$. In order to gain a better understanding of virtual displacements let us look at a few examples.

Figure 5-31*a* shows the location at time $t$ of a particle moving from left to right along the trajectory indicated by the dashed line. At time $t + dt$ the particle is in a new location, as indicated in Fig. 5-31*b*, and the actual displacement of the particle along the trajectory in the time increment $dt$ is $d\boldsymbol{u}$. However, as shown in Fig. 5-31*c*, infinitely many virtual displacements are possible for the particle because there are no kinematic constraints on the particle. One of the possible virtual displacements is $d\boldsymbol{u}$.

In Fig. 5-32*a* we have a simple pendulum. Because of the kinematic constraint that the rod cannot change its length, the particle at the end of the pendulum is constrained to move in a circular path. Thus the choice of virtual displacements is restricted to those that are along the circular arc, as shown in Fig. 5-32*b*.

Another kind of pendulum is the sliding pendulum shown in Fig. 5-33*a* at time $t$. Let us assume that the sliding motion of point $a$ is some prescribed function of time $x(t)$. Figure 5-33*b* shows the actual location of the pendulum at time $t + dt$, and the incremental displacement $d\boldsymbol{u}$. However, because virtual displacements happen at an instant of time and must be consistent with the kinematic constraints of the system, the only possible virtual displacements are those along the circular path shown in Fig. 5-33*c*. This is because at time $t$ the slider is constrained to be in a location determined by the prescribed function $x(t)$. Thus the prescribed motion of the slider imposes a kinematic constraint on the system, which limits the possible choices of virtual displacements because the location of the slider is not arbitrary. Whereas in the previous

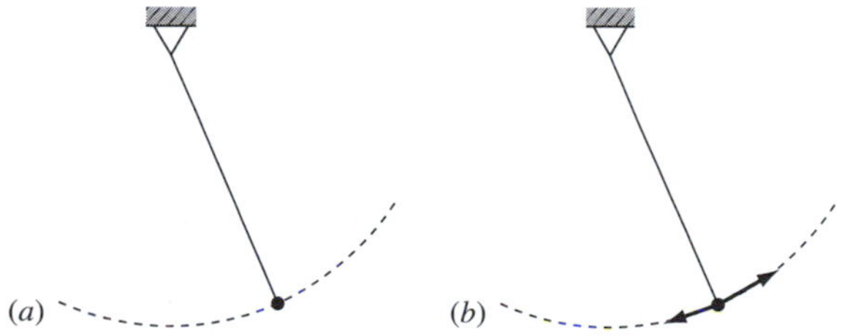


Figure 5-32 A simple pendulum

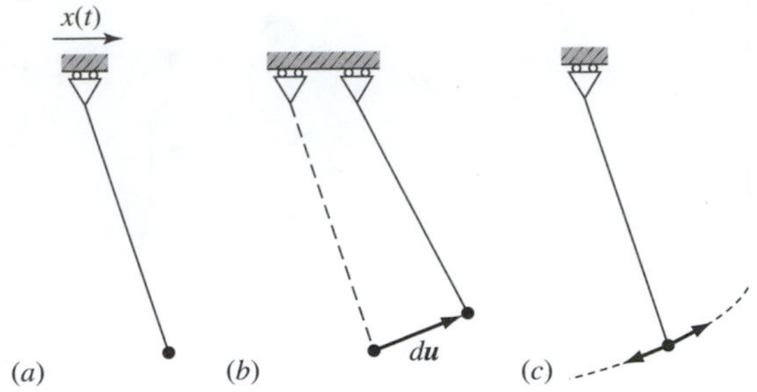


**Figure 5-33** A sliding pendulum

example the actual displacement $\boldsymbol{du}$ and the virtual displacement $\delta\boldsymbol{u}$ *could* be the same, in this example they cannot.

The quantity $\delta\boldsymbol{u}$ is often referred to as the *variation of* $\boldsymbol{u}$, and we will use that phrase from time to time. Since the symbol $\delta$ implies quantities of infinitesimal size, we might expect that the rules for calculating the variation of products, quotients, and so on, are analogous to the corresponding rules for calculating differentials. This is indeed the case, and we list here some of the more common results:

$$\delta(uv) = u\delta v + v\delta u \tag{5-103a}$$

$$\delta\left(\frac{u}{v}\right) = \frac{v\delta u - u\delta v}{v^2} \tag{5-103b}$$

$$\delta\left(\frac{du}{dx}\right) = \frac{d}{dx}(\delta u) \tag{5-103c}$$

$$\delta\int f\,dx = \int(\delta f)\,dx \tag{5-103d}$$

The one important difference is that

$$dF(u, v, t) = \frac{\partial f}{\partial u}\,du + \frac{\partial f}{\partial v}\,dv + \frac{\partial f}{\partial t}\,dt$$

whereas

$$\delta f(u, v, t) = \frac{\partial f}{\partial u}\,\delta u + \frac{\partial f}{\partial v}\,\delta v \tag{5-104}$$

because a variation occurs at an instant of time.

## 5-8-1 Principle of Virtual Work for Particles and Rigid Bodies

To develop the principle of virtual work we start by considering a particle that is acted upon by several forces $\boldsymbol{F}_1, \boldsymbol{F}_2, \ldots, \boldsymbol{F}_n$. If the particle undergoes a virtual displacement, as shown in Fig. 5-34, then the forces acting on the particle do a certain amount of work. Since the particle undergoes a virtual displacement, the work done is referred to as the

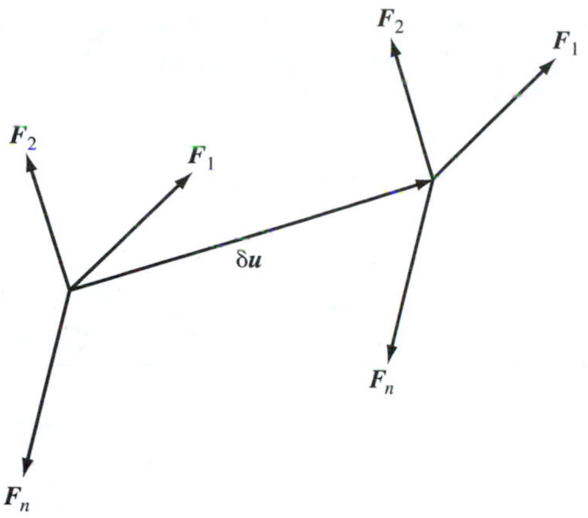


Figure 5-34

virtual work and is designated $\delta W$. The virtual work is given by

$$\delta W = \boldsymbol{F}_1 \cdot \delta\boldsymbol{u} + \boldsymbol{F}_2 \cdot \delta\boldsymbol{u} + \cdots + \boldsymbol{F}_n \cdot \delta\boldsymbol{u}$$

or, after collecting terms,

$$\delta W = (\boldsymbol{F}_1 + \boldsymbol{F}_2 + \cdots + \boldsymbol{F}_n) \cdot \delta\boldsymbol{u} = \sum_{i=1}^{n} \boldsymbol{F}_i \cdot \delta\boldsymbol{u}$$

If the particle is in static equilibrium, its virtual work must be zero since

$$\sum_{i=1}^{n} \boldsymbol{F}_i = \boldsymbol{0}$$

On the other hand if the virtual work done on the particle is zero, then the particle must be in static equilibrium because in that case

$$\left(\sum_{i=1}^{n} \boldsymbol{F}_i\right) \cdot \delta\boldsymbol{u} = 0$$

Since the virtual displacement $\delta\boldsymbol{u}$ must be arbitrary (in both magnitude and direction), it follows that

$$\sum_{i=1}^{n} \boldsymbol{F}_i = \boldsymbol{0}$$

Consequently:

> A necessary and sufficient condition for a particle to be in static equilibrium is that the virtual work done on the particle by the forces acting on it be zero.

This statement is known as the *principle of virtual work* for a particle. This principle is easily extended to rigid bodies since we may think of these as made up of rigidly connected particles. If the body is in equilibrium, it follows that each particle is also in equilibrium, and therefore, since the virtual work of each particle must be zero, the virtual work done on the rigid body must be zero. Further, since the internal forces between any two adjacent particles are equal in magnitude but opposite in direction, the net virtual work done by the internal forces is zero. The principle of virtual work for a rigid body then states that:

> A necessary and sufficient condition for a rigid body to be in static equilibrium is that the virtual work done on the body by the *external* forces acting on it be zero.

As it applies to particles and rigid bodies, the principle of virtual work offers no particular advantage over other methods for problems involving static equilibrium because these problems are relatively simple. However, for problems involving a system of interconnected rigid bodies, the principle of virtual work has a couple of advantages. We restrict our attention to problems in which the connections are ideal, that is, connections that are rigid and frictionless. For these problems the virtual work done by the internal forces at the connections is zero since these forces occur in equal and opposite pairs. Further, the reaction forces at the fixed supports do no virtual work because the virtual displacement at a fixed support must be zero to be consistent with the kinematic constraints of the system. Thus the only forces that have the potential to do virtual work are the applied forces. The principle of virtual work then states that:

> A necessary and sufficient condition for a system of interconnected rigid bodies with ideal connections to be in static equilibrium is that the virtual work done by the applied forces acting on it be zero.

To see how the principle of virtual work can be applied, consider the following example.

## EXAMPLE 5-13

Determine the force $P$ required to balance the load $F$ for the structure shown in Fig. 5-35*a*, assuming that the connection at $C$ is both rigid and frictionless and point $B$ is constrained to remain in contact with the frictionless surface.

### SOLUTION

Figure 5-35*b* shows the system after point $B$ has undergone a virtual displacement $\delta x_B$. Because of the kinematic constraint provided by the hinge, point $C$ undergoes a virtual displacement whose $x$ and $y$ components are $\delta x_C$ and $\delta y_C$. With the coordinate system shown, both $P$ and $F$ are negative. The principle of virtual work then gives

$$\delta W = (-P)\,\delta y_C + (-F)\,\delta x_B = 0 \qquad (a)$$

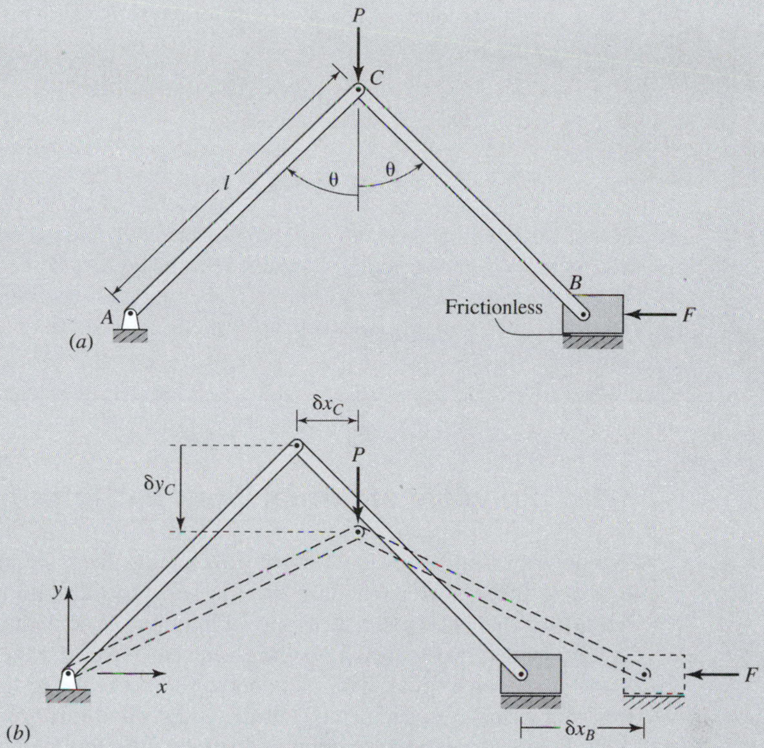


**Figure 5-35**

Note that the virtual displacements are positive in positive coordinate directions. Because of the kinematic constraint provided by the hinge at $C$, $\delta x_B$ and $\delta y_C$ are not independent. In fact, from the geometry of the problem we can write

$$x_B = 2l \sin\theta$$

$$y_C = l \cos\theta$$

and recalling that the symbol $\delta$ is analogous to the differential symbol, we have

$$\delta x_B = 2l(\cos\theta)\,\delta\theta$$
$$\delta y_C = -l(\sin\theta)\,\delta\theta \qquad (b)$$

So we see that $\delta x_B$ and $\delta y_C$ are related through the common quantity $\delta\theta$. These expressions also show that a positive value for $\delta x_B$ results in a negative value for $\delta y_C$, thus bringing the sign of the virtual work done by $P$ into agreement with the physical observation that $P$ does positive virtual work if $\delta x_B$ is positive. Substitution of Eqs. ($b$) into the expression for virtual work given by Eq. ($a$) results in an expression that depends on the virtual displacement $\delta\theta$,

$$\delta W = (P\sin\theta - 2F\cos\theta)l\,\delta\theta = 0$$

Since the virtual displacement $\delta\theta$ must be arbitrary, the term in parentheses must be zero if the virtual work is to be zero. This gives

$$P \sin\theta - 2F \cos\theta = 0$$

or

$$F = \tfrac{1}{2} P \tan\theta$$

Note that to arrive at this result by summing forces and moments, the free-body diagrams for each member of the structure would have to be constructed, and this would involve the reaction forces and the internal forces at the hinge as well as the applied forces. The principle of virtual work provides a simpler method for solving this problem.

## 5-8-2 Principle of Virtual Work for Deformable Bodies

We now turn our attention toward developing the principle of virtual work for deformable bodies. We envision a body to be in equilibrium under the action of surface tractions $\boldsymbol{T}$ and body forces $\boldsymbol{b}$, as shown in Fig. 5-36, and imagine each point of the body to have undergone a virtual displacement $\delta\boldsymbol{u}$. Note that the virtual displacement may vary from point to point within the body, subject to the restrictions stated in our definition of a virtual displacement. Thus the body is deformed owing to the virtual displacement varying from point to point in the body. At point $P$ on the surface an infinitesimal amount of virtual work is done because the force $\boldsymbol{T}\,dA$ moves through a virtual dis-

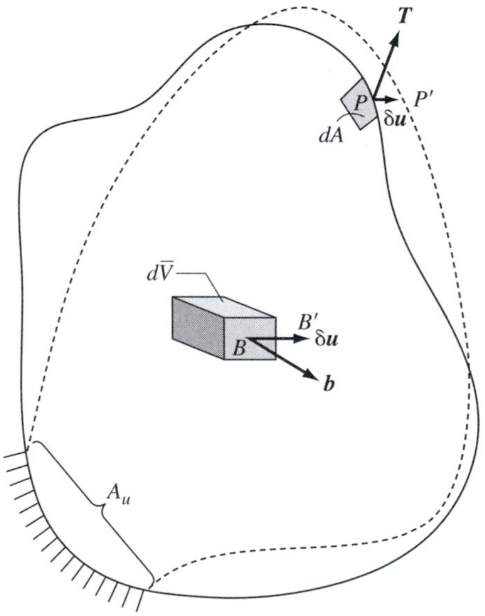


Figure 5-36 Virtual displacements of a body in equilibrium

placement $\delta \boldsymbol{u}$. This is given by

$$d(\delta W_T) = (\boldsymbol{T} \cdot \delta \boldsymbol{u})\, dA \tag{5-105}$$

where the subscript $T$ indicates that the virtual work is due to surface tractions. Similarly, at point $B$ the body force does virtual work in the amount of

$$d(\delta W_b) = (\boldsymbol{b} \cdot \delta \boldsymbol{u})\, d\overline{V} \tag{5-106}$$

Consequently the virtual work done by the surface tractions and body forces acting over the entire body is

$$\delta W = \int_{\overline{V}} d(\delta W_b) + \int_A d(\delta W_T)$$

or, using Eqs. (5-105) and (5-106),

$$\delta W = \int_{\overline{V}} \boldsymbol{b} \cdot \delta \boldsymbol{u}\, d\overline{V} + \int_A \boldsymbol{T} \cdot \delta \boldsymbol{u}\, dA \tag{5-107}$$

The first integral can be expanded to give

$$\int_{\overline{V}} \boldsymbol{b} \cdot \delta \boldsymbol{u}\, d\overline{V} = \int_{\overline{V}} (b_x \delta u + b_y \delta v)\, d\overline{V} \tag{5-108}$$

The surface integral can also be expanded to give

$$\int_A \boldsymbol{T} \cdot \delta \boldsymbol{u}\, dA = \int_A (T_x \delta u + T_y \delta v)\, dA$$

With the use of Eqs. (2-14) this becomes, after rearranging,

$$\int_A \boldsymbol{T} \cdot \delta \boldsymbol{u}\, dA = \int_A [(\sigma_{xx} \delta u + \sigma_{xy} \delta v) n_x + (\sigma_{yx} \delta u + \sigma_{yy} \delta v) n_y]\, dA$$

The surface integrals can be converted to volume integrals using Gauss's theorem [Eq. (5-45)]. Thus we have

$$\int_A \boldsymbol{T} \cdot \delta \boldsymbol{u}\, dA = \int_{\overline{V}} \left[ \frac{\partial}{\partial x}(\sigma_{xx} \delta u + \sigma_{xy} \delta v) + \frac{\partial}{\partial y}(\sigma_{yx} \delta u + \sigma_{yy} \delta v) \right] d\overline{V} \tag{5-109}$$

Next we expand the right-hand side of Eq. (5-109), collect terms, and use Eq. (5-103*c*) to interchange the order of differentiation and variation. This gives

$$\int_A \boldsymbol{T} \cdot \delta \boldsymbol{u}\, dA = \int_{\overline{V}} \left[ \left( \frac{\partial \sigma_{xx}}{\partial x} + \frac{\partial \sigma_{yx}}{\partial y} \right) \delta u + \left( \frac{\partial \sigma_{xy}}{\partial x} + \frac{\partial \sigma_{yy}}{\partial y} \right) \delta v \right.$$
$$\left. + \sigma_{xx} \delta \left( \frac{\partial u}{\partial x} \right) + \sigma_{yy} \delta \left( \frac{\partial v}{\partial y} \right) + \sigma_{xy} \delta \left( \frac{\partial v}{\partial x} \right) + \sigma_{yx} \delta \left( \frac{\partial u}{\partial y} \right) \right] d\overline{V}$$

Since the virtual displacement must be consistent with the kinematic constraints of the problem, the strain-displacement equations must be satisfied. Thus if the strains are infinitesimal, using Eqs. (3-11), noting that $\sigma_{xy} = \sigma_{yx}$, and collecting terms, we have

$$\int_A \boldsymbol{T} \cdot \delta \boldsymbol{u}\, dA = \int_{\overline{V}} \left[ \left( \frac{\partial \sigma_{xx}}{\partial x} + \frac{\partial \sigma_{xy}}{\partial y} \right) \delta u + \left( \frac{\partial \sigma_{xy}}{\partial x} + \frac{\partial \sigma_{yy}}{\partial y} \right) \delta v \right.$$
$$\left. + \sigma_{xx}\delta\varepsilon_{xx} + \sigma_{yy}\delta\varepsilon_{yy} + \sigma_{xy}\delta\gamma_{xy} \right] d\overline{V}$$

Now when this, together with Eq. (5-108), is substituted into Eq. (5-107), the virtual work becomes

$$\delta W = \int_{\overline{V}} \left[ \left( \frac{\partial \sigma_{xx}}{\partial x} + \frac{\partial \sigma_{xy}}{\partial y} + b_x \right) \delta u + \left( \frac{\partial \sigma_{xy}}{\partial x} + \frac{\partial \sigma_{yy}}{\partial y} + b_y \right) \delta v \right.$$
$$\left. + \sigma_{xx}\delta\varepsilon_{xx} + \sigma_{yy}\delta\varepsilon_{yy} + \sigma_{xy}\delta\gamma_{xy} \right] d\overline{V} \tag{5-110}$$

With Eqs. (2-55) the two terms in parentheses are zero because we stated that the body was in equilibrium under the action of the surface tractions and body forces, so that Eq. (5-110) reduces to

$$\delta W = \int_{\overline{V}} (\sigma_{xx}\delta\varepsilon_{xx} + \sigma_{yy}\delta\varepsilon_{yy} + \sigma_{xy}\delta\gamma_{xy})\, d\overline{V} \tag{5-111}$$

This equation gives the principle of virtual work for deformable bodies. Note that, in contrast to problems involving particles and rigid bodies where the virtual work of the applied forces must be zero, for a deformable body subjected to infinitesimal strains the virtual work of the applied forces is equal to the virtual work done by the stresses in acting through the virtual strains. Of course, for rigid bodies the virtual strains must be zero. Substituting for $\delta W$ in Eq. (5-107), we get

$$\int_{\overline{V}} \boldsymbol{b} \cdot \delta \boldsymbol{u}\, d\overline{V} + \int_A \boldsymbol{T} \cdot \delta \boldsymbol{u}\, dA = \int_{\overline{V}} (\sigma_{xx}\delta\varepsilon_{xx} + \sigma_{yy}\delta\varepsilon_{yy} + \sigma_{xy}\delta\gamma_{xy})\, d\overline{V} \tag{5-112}$$

Notice that the boundary of the body consists of two parts. On one part, $A_\sigma$, the tractions are specified, whereas on the other part, $A_u$, displacements are specified. The surface integral can be separated into two parts, so that we have

$$\int_A \boldsymbol{T} \cdot \delta \boldsymbol{u}\, dA = \int_{A_\sigma} \boldsymbol{T} \cdot \delta \boldsymbol{u}\, dA + \int_{A_u} \boldsymbol{T} \cdot \delta \boldsymbol{u}\, dA$$

The integral on $A_u$ must be zero because the virtual displacements on $A_u$ must be zero if they are to be kinematically admissible. Equation (5-112) therefore reduces to

$$\int_{\overline{V}} \boldsymbol{b} \cdot \delta \boldsymbol{u}\, d\overline{V} + \int_{A_\sigma} \boldsymbol{T} \cdot \delta \boldsymbol{u}\, dA = \int_{\overline{V}} (\sigma_{xx}\delta\varepsilon_{xx} + \sigma_{yy}\delta\varepsilon_{yy} + \sigma_{xy}\delta\gamma_{xy})\, d\overline{V} \tag{5-113}$$

No material behavior has entered into the derivation of Eq. (5-113), so the principle of virtual work is valid for any material so long as the strains are infinitesimal.

## 5-9 THEOREM OF MINIMUM TOTAL POTENTIAL ENERGY

Since the surface tractions and body forces remain unchanged as the body undergoes a virtual displacement, we can write

$$\int_{\overline{V}} \delta(\boldsymbol{b} \cdot \boldsymbol{u})\, d\overline{V} + \int_{A_\sigma} \delta(\boldsymbol{T} \cdot \boldsymbol{u})\, dA = \int_{\overline{V}} (\sigma_{xx}\delta\varepsilon_{xx} + \sigma_{yy}\delta\varepsilon_{yy} + \sigma_{xy}\delta\gamma_{xy})\, d\overline{V} \qquad (5\text{-}114)$$

If the material is elastic (with, possibly, a nonlinear stress-strain law), then Eqs. (5-70) can be substituted in the right-hand side of Eq. (5-114). This gives

$$\int_{\overline{V}} \delta(\boldsymbol{b} \cdot \boldsymbol{u})\, d\overline{V} + \int_{A_\sigma} \delta(\boldsymbol{T} \cdot \boldsymbol{u})\, dA = \int_{\overline{V}} \left( \frac{\partial \tilde{u}}{\partial \varepsilon_{xx}} \delta\varepsilon_{xx} + \frac{\partial \tilde{u}}{\partial \varepsilon_{yy}} \delta\varepsilon_{yy} + \frac{\partial \tilde{u}}{\partial \gamma_{xy}} \delta\gamma_{xy} \right) d\overline{V}$$

Interchanging the order of variation and integration we get the following expression:

$$\delta\left( U - \int_{\overline{V}} \boldsymbol{b} \cdot \boldsymbol{u}\, d\overline{V} - \int_{A_\sigma} \boldsymbol{T} \cdot \boldsymbol{u}\, dA \right) = 0$$

The quantity in parentheses is referred to as the total potential energy of the body. Thus we may write

$$\delta\Pi = 0 \qquad (5\text{-}115)$$

where $\Pi$, the total potential energy, is given by

$$\Pi = U - \int_{\overline{V}} \boldsymbol{b} \cdot \boldsymbol{u}\, d\overline{V} - \int_{A_\sigma} \boldsymbol{T} \cdot \boldsymbol{u}\, dA \qquad (5\text{-}116)$$

Equation (5-115) states that for admissible virtual displacements $\delta\boldsymbol{u}$ the variation of the total potential energy of the body is zero when the body is in its equilibrium configuration.

In other words, of all the displacement fields that satisfy the boundary conditions of the problem, the displacement field that satisfies the equilibrium equations makes the variation of the total potential energy of the body zero.

It is possible to prove a stronger statement than this if we consider the change in potential energy that occurs when the body undergoes a virtual displacement $\delta\boldsymbol{u}$ from its equilibrium configuration, that is, when the equilibrium displacements $\boldsymbol{u}$ are replaced by $\boldsymbol{u} + \delta\boldsymbol{u}$. In order to simplify the mathematics we consider a two-dimensional problem so that the strains are reduced to $\varepsilon_{xx}$, $\varepsilon_{yy}$, and $\gamma_{xy}$. If we think of $\Pi$ as a function of the displacements, then we can define

$$\overline{\Pi} = \Pi(\boldsymbol{u} + \delta\boldsymbol{u})$$

Thus we have

$$\overline{\Pi} = \int_{\overline{V}} \tilde{u}(\varepsilon_{xx} + \delta\varepsilon_{xx}, \varepsilon_{yy} + \delta\varepsilon_{yy}, \gamma_{xy} + \delta\gamma_{xy})\, d\overline{V}$$
$$- \int_{A_\sigma} \boldsymbol{T} \cdot (\boldsymbol{u} + \delta\boldsymbol{u})\, dA - \int_{\overline{V}} \boldsymbol{b} \cdot (\boldsymbol{u} + \delta\boldsymbol{u})\, d\overline{V}$$

and therefore the difference $\overline{\Pi} - \Pi$ is given by

$$\begin{aligned}\overline{\Pi} - \Pi = &\int_{\overline{V}} \tilde{u}(\varepsilon_{xx} + \delta\varepsilon_{xx}, \varepsilon_{yy} + \delta\varepsilon_{yy}, \gamma_{xy} + \delta\gamma_{xy})\, d\overline{V} \\ &- \int_{A_\sigma} \boldsymbol{T} \cdot (\boldsymbol{u} + \delta\boldsymbol{u})\, dA - \int_{\overline{V}} \boldsymbol{b} \cdot (\boldsymbol{u} + \delta\boldsymbol{u})\, d\overline{V} \\ &- \int_{\overline{V}} \tilde{u}(\varepsilon_{xx}, \varepsilon_{yy}, \gamma_{xy})\, d\overline{V} + \int_{A_\sigma} \boldsymbol{T} \cdot \boldsymbol{u}\, dA + \int_{\overline{V}} \boldsymbol{b} \cdot \boldsymbol{u}\, d\overline{V}\end{aligned}$$

or

$$\begin{aligned}\overline{\Pi} - \Pi = &\int_{\overline{V}} \tilde{u}(\varepsilon_{xx} + \delta\varepsilon_{xx}, \varepsilon_{yy} + \delta\varepsilon_{yy}, \gamma_{xy} + \delta\gamma_{xy})\, d\overline{V} \\ &- \int_{\overline{V}} \tilde{u}(\varepsilon_{xx}, \varepsilon_{yy}, \gamma_{xy})\, d\overline{V} - \int_{A_\sigma} \boldsymbol{T} \cdot \delta\boldsymbol{u}\, dA - \int_{\overline{V}} \boldsymbol{b} \cdot \delta\boldsymbol{u}\, d\overline{V}\end{aligned} \tag{5-117}$$

To examine the term $\tilde{u}(\varepsilon_{xx} + \delta\varepsilon_{xx}, \varepsilon_{yy} + \delta\varepsilon_{yy}, \gamma_{xy} + \delta\gamma_{xy})$ we expand it in a power series about the point $\varepsilon_{xx}, \varepsilon_{yy}, \gamma_{xy}$. This gives

$$\begin{aligned}\tilde{u}(\varepsilon_{xx} + \delta\varepsilon_{xx}, \varepsilon_{yy} + \delta\varepsilon_{yy}, \gamma_{xy} + \delta\gamma_{xy}) = &\tilde{u}(\varepsilon_{xx}, \varepsilon_{yy}, \gamma_{xy}) + \sigma_{xx}\delta\varepsilon_{xx} \\ &+ \sigma_{yy}\delta\varepsilon_{yy} + \sigma_{xy}\delta\gamma_{xy} + a\end{aligned} \tag{5-118}$$

where

$$\begin{aligned}a = &\frac{1}{2}\frac{\partial^2\tilde{u}}{\partial\varepsilon_{xx}^2}(\delta\varepsilon_{xx})^2 + \frac{1}{2}\frac{\partial^2\tilde{u}}{\partial\varepsilon_{yy}^2}(\delta\varepsilon_{yy})^2 + \frac{1}{2}\frac{\partial^2\tilde{u}}{\partial\gamma_{xy}^2}(\delta\gamma_{xy})^2 \\ &+ \frac{1}{2}\frac{\partial^2\tilde{u}}{\partial\varepsilon_{xx}\partial\varepsilon_{yy}}(\delta\varepsilon_{xx}\delta\varepsilon_{yy}) + \frac{1}{2}\frac{\partial^2\tilde{u}}{\partial\varepsilon_{xx}\partial\gamma_{xy}}(\delta\varepsilon_{xx}\delta\gamma_{xy}) \\ &+ \frac{1}{2}\frac{\partial^2\tilde{u}}{\partial\varepsilon_{yy}\partial\gamma_{xy}}(\delta\varepsilon_{yy}\delta\gamma_{xy}) + \cdots\end{aligned}$$

and we have used Eqs. (5-70) in obtaining the coefficients of $\delta\varepsilon_{xx}, \delta\varepsilon_{yy}, \delta\gamma_{xy}$. With this expression Eq. (5-117) becomes

$$\begin{aligned}\overline{\Pi} - \Pi = &\int_{\overline{V}} (\sigma_{xx}\delta\varepsilon_{xx} + \sigma_{yy}\delta\varepsilon_{yy} + \sigma_{xy}\delta\gamma_{xy})\, d\overline{V} - \int_{A_\sigma} \boldsymbol{T} \cdot \delta\boldsymbol{u}\, dA \\ &- \int_{\overline{V}} \boldsymbol{b} \cdot \delta\boldsymbol{u}\, d\overline{V} + \int_{\overline{V}} a\, d\overline{V}\end{aligned}$$

Now the first three integrals sum to zero because of the principle of virtual work [Eq. (5-113)], leaving

$$\overline{\Pi} - \Pi = \int_{\overline{V}} a\, d\overline{V} \tag{5-119}$$

Let us now assume that the body has not been preloaded. Then when the body is in a state of zero strain it is also in a state of zero stress. We also assume that the reference state for the strain energy density is a state of zero strain, that is, the strain energy density is zero when a state of zero strain exists. Then if we examine the limiting form of Eq. (5-118) as the strains tend to zero, we find

$$\tilde{u}(\delta\varepsilon_{xx}, \delta\varepsilon_{yy}, \delta\gamma_{xy}) = a$$

Finally we recall that the strain energy density is positive definite and therefore

$$\tilde{u}(\delta\varepsilon_{xx}, \delta\varepsilon_{yy}, \delta\gamma_{xy}) \geq 0$$

That being the case it follows that $a \geq 0$ and hence, from Eq. (5-119),

$$\overline{\Pi} - \Pi \geq 0$$

That is, for any virtual displacement of the body relative to its equilibrium configuration there is an increase in the potential energy of the body. This enables us to state the *theorem of minimum total potential energy:*

> Of all the displacement configurations that satisfy the boundary conditions the configuration that is the true equilibrium configuration is the one that renders the total potential energy of the body an absolute minimum.

We note that if concentrated loads are allowed to act on the surface and within the volume of the body at points 1, 2, . . . , $n$, as shown in Fig. 5-37, then the total potential

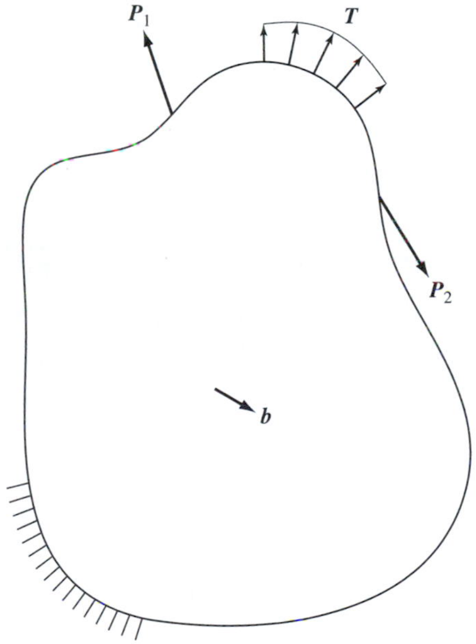


Figure 5-37

energy given in Eq. (5-116) becomes*

$$\Pi = \int_{\bar{V}} \tilde{u}\, d\bar{V} - \int_{A_\sigma} \boldsymbol{T} \cdot \boldsymbol{u}\, dA - \int_{\bar{V}} \boldsymbol{b} \cdot \boldsymbol{u}\, d\bar{V} - \sum_{i=1}^{n} \boldsymbol{P}_i \cdot \boldsymbol{u}_i \qquad (5\text{-}120)$$

or, with Eq. (5-73)

$$\Pi = U - \int_{A_\sigma} \boldsymbol{T} \cdot \boldsymbol{u}\, dA - \int_{\bar{V}} \boldsymbol{b} \cdot \boldsymbol{u}\, d\bar{V} - \sum_{i=1}^{n} \boldsymbol{P}_i \cdot \boldsymbol{u}_i \qquad (5\text{-}121)$$

## 5-10 APPLICATIONS OF THE THEOREM OF MINIMUM TOTAL POTENTIAL ENERGY

The theorem of minimum total potential energy plays an important role in the study of the mechanics of solid bodies, and applications of this important theorem will be found in Chapters 8 and 10. Here and in the next section we give but a few examples.

### EXAMPLE 5-14

First let us determine the displacement at the end of a column of length $l$, as shown in Fig. 5-27.

#### SOLUTION

Since there are no distributed loads or body forces, Eq. (5-121) reduces to

$$\Pi = U - \sum_{i=1}^{n} \boldsymbol{P}_i \cdot \boldsymbol{u}_i$$

Now there is only one load applied to the column. Thus,

$$\sum_{i=1}^{n} \boldsymbol{P}_i \cdot \boldsymbol{u}_i = P u_0$$

The strain energy is given in terms of displacements by Eq. (*c*) of Example 5-6. Thus the total potential energy is

$$\Pi = \frac{AE}{2l} u_0^2 - P u_0$$

*It should be noted that Eq. (5-120) follows directly from Eq. (5-116) when the concentrated loads acting on the body are represented by the Dirac delta function, introduced in Section 8-1-2.

Note that $\Pi$ depends only on the displacement $u_0$. The variation of potential energy is therefore given by

$$\delta\Pi = \left(\frac{AE}{l}u_0 - P\right)\delta u_0$$

and since $\delta u_0$ must be arbitrary, Eq. (5-115) can only be satisfied if

$$u_0 = \frac{Pl}{AE} \tag{a}$$

Is the total potential energy a minimum? In order to find out, let us calculate $\overline{\Pi}$ at some displacement $u_0 + \Delta u_0$,

$$\overline{\Pi} = \frac{AE}{2l}(u_0 + \Delta u_0)^2 - P(u_0 + \Delta u_0)$$

Therefore

$$\begin{aligned}\overline{\Pi} - \Pi &= \frac{AE}{2l}(u_0 + \Delta u_0)^2 - P(u_0 + \Delta u_0) - \frac{AE}{2l}u_0^2 + P\Delta u_0 \\ &= \left(\frac{AE}{l}u_0 - P\right)\Delta u_0 + \frac{AE}{2l}(\Delta u_0)^2\end{aligned}$$

Now the term in the first parentheses is zero by Eq. (*a*). Consequently,

$$\overline{\Pi} - \Pi = \frac{AE}{2l}(\Delta u_0)^2 \geq 0$$

The theorem of minimum total potential energy can also be used to show that Castigliano's first theorem has much more general applicability than was evident in the derivation of Section 5-7. In order to see this, we consider a body of arbitrary shape that is fixed against rigid body motions and subjected to concentrated loads $P_1, P_2, \ldots, P_n$, as shown in Fig. 5-38. Then the strain energy can be determined in terms of the displacements $u_1, u_2, \ldots, u_n$, at the points of load application and in the direction of the load. We also note that

$$\boldsymbol{u}_i \cdot \boldsymbol{P}_i = u_i P_i$$

That is, the quantity $\boldsymbol{u}_i \cdot \boldsymbol{P}_i$ is simply the product of a load in a given direction and the displacement component at that point in the direction of the load. Note that this interpretation is the same as was used in deriving Castigliano's theorem in Section 5-7. Then since there are no surface tractions or body forces acting we have, from Eq. (5-121),

$$\Pi = U - \sum_{i=1}^{n} u_i P_i$$

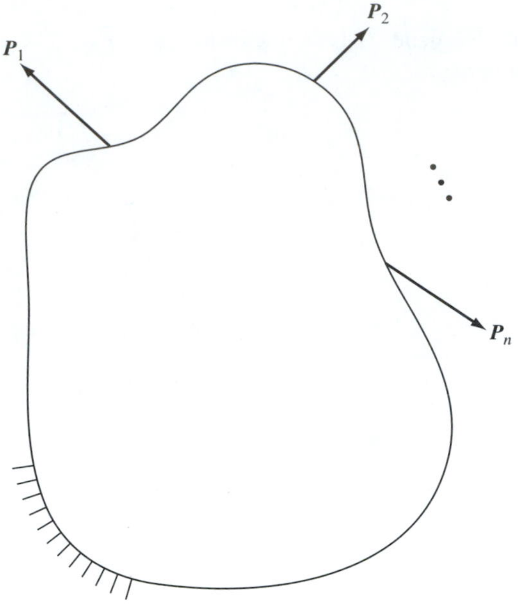


Figure 5-38

Now when points $1, 2, \ldots, n$ are subjected to the virtual displacements $\delta u_1, \delta u_2, \ldots, \delta u_n$, we have

$$\delta \Pi = \frac{\partial U}{\partial u_1}\delta u_1 + \frac{\partial U}{\partial u_2}\delta u_2 + \cdots + \frac{\partial U}{\partial u_n}\delta u_n - P_1\delta u_1 - P_2\delta u_2 - \cdots - P_n\delta u_n = 0$$

or, after collecting terms,

$$\left(\frac{\partial U}{\partial u_1} - P_1\right)\delta u_1 + \left(\frac{\partial U}{\partial u_2} - P_2\right)\delta u_2 + \cdots + \left(\frac{\partial U}{\partial u_n} - P_n\right)\delta u_n = 0$$

But each of the virtual displacements $\delta u_1, \delta u_2, \ldots, \delta u_n$ must be independent of all of the other virtual displacements in order for the virtual displacements to be arbitrary. Therefore for the preceding expression to be true for arbitrary virtual displacements we must have

$$\frac{\partial U}{\partial u_i} = P_i, \qquad i = 1, 2, \ldots, n$$

This, of course, is the same as Eq. (5-100). However, since we have just derived Castigliano's first theorem from the principle of minimum total potential energy, it is valid under the same restrictions as this principle, namely, that the shape of the body can be arbitrary, that the material may be nonlinear but must be elastic, and that the strains must be infinitesimal.

## 5-11 RAYLEIGH–RITZ METHOD

The theorem of minimum total potential energy finds some of its greatest applications in the development of approximate methods for solving problems concerning the mechanics of deformable bodies. As an example of these we will discuss the Rayleigh–Ritz method, in which the idea is to assume a certain functional form for the displacement field that satisfies the kinematic boundary conditions of the problem and as many of the remaining boundary conditions as possible. Satisfaction of the *kinematic boundary conditions* (that is, boundary conditions pertaining to kinematic quantities such as displacement) by the assumed displacement field is essential since the virtual displacement calculated from this displacement field must not violate the kinematic constraints of the problem. The total potential energy is then calculated based on the assumed displacement field. Minimization of the total potential energy with respect to this displacement field then allows the determination of the unknown constants contained in the assumed displacement field.

Equation (5-121) can be put in a form that is more convenient for beam problems. Since there are no body forces present, the expression for total potential energy reduces to

$$\Pi = U - \int_{A_\sigma} \boldsymbol{T} \cdot \boldsymbol{u}\, dA - \sum_{i=1}^{n} \boldsymbol{P}_i \cdot \boldsymbol{u}_i$$

Now for beams the tractions and concentrated forces are normal to the beam axis as is the nonzero displacement component. Taking the beam axis as the $x$ axis and the bending plane as the $xz$ plane, we have

$$\Pi = U - \int_{A_\sigma} T_z w\, dA - \sum_{i=1}^{n} P_i w_i$$

Furthermore the surface traction integral for a beam can be expressed in the following manner:

$$\int_{A_\sigma} T_z w\, dA = \int_0^l \left[ \int_p T_z w\, dp \right] dx$$

where the inner integral is taken over $p$, the perimeter of the cross section. The deflection $w$ is assumed to be uniform over the cross section and therefore can be taken outside the integral around $p$. The surface traction acting on a beam is also normal to the beam axis and, when integrated over the perimeter of the cross section, yields the distributed load $q$. Thus,

$$\int_{A_\sigma} T_z w\, dA = \int_0^l q w\, dx$$

Consequently, for beams the total potential energy is given by

$$\Pi = U - \int_0^l q w\, dx - \sum_{i=1}^{n} w_i P_i \tag{5-122}$$

We recall that for beams,

$$M = -EI \frac{d^2 w}{dx^2}$$

and therefore by using this in Eq. (5-90) and the result in Eq. (5-122) we get, in the absence of longitudinal forces and transverse shear,

$$\Pi = \int_0^l \frac{EI}{2}\left(\frac{d^2w}{dx^2}\right)^2 dx - \int_0^l qw\,dx - \sum_{i=1}^n w_i P_i \tag{5-123}$$

## EXAMPLE 5-15

Use the Rayleigh–Ritz method to determine an approximate expression for the displacement distribution for the simply supported beam shown in Fig. 5-39.

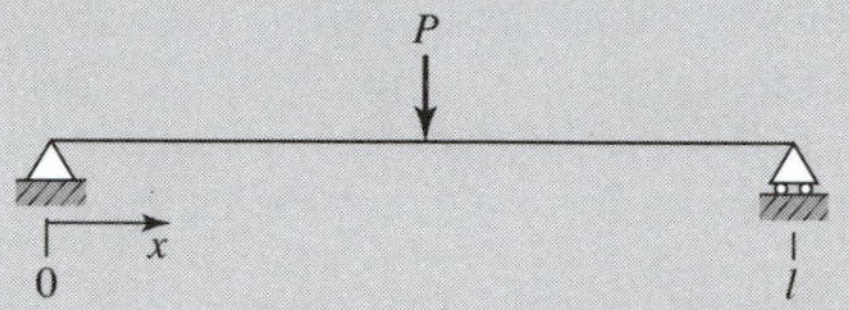


**Figure 5-39**

## SOLUTION

For the Rayleigh–Ritz method we assume a functional form for the displacement $w$. Let us approximate the displacement $w$ by

$$w \approx a \sin\frac{\pi x}{l} \tag{a}$$

where $a$ is a constant to be determined. Note that the assumed displacement field satisfies the kinematic boundary conditions of the problem, namely, that the displacements are zero at the ends of the beam. In this case, although not required by the method, the assumed displacement field also satisfies the remaining two boundary conditions, which require the moment, and hence the curvature, to be zero at the ends of the beam. At midspan the assumed displacement field gives

$$w_{l/2} \approx a$$

Consequently the expression for total potential energy becomes

$$\Pi = \frac{1}{2}EI\int_0^l \left[-\left(\frac{\pi}{l}\right)^2 a\sin\frac{\pi x}{l}\right]^2 dx - Pa$$

or

$$\Pi = \frac{1}{4}EI\left(\frac{\pi}{l}\right)^4 a^2 l - Pa$$

From Eq. $(a)$ we note that the displacement field is completely determined except for the constant $a$. Thus the virtual displacement $\delta w$ is given by

$$\delta w = \delta a \sin\frac{\pi x}{l} \tag{b}$$

Note that in computing the virtual displacement $\delta w$ there is no variation of the variable $x$, which gives the location of a particle that is undergoing a virtual displacement. If $x$ were allowed to undergo a variation, it would indicate that a virtual displacement could come about not only by a variation of the coefficient $a$ but also by changing the particle under observation.

We note further that the total potential energy is a function of the single unknown $a$. Thus we can write

$$\Pi = \Pi(a)$$

and

$$\delta\Pi = \frac{\partial \Pi}{\partial a}\,\delta a$$

In order for this expression to be zero, as required by the theorem of minimum potential energy, we must have

$$\frac{\partial \Pi}{\partial a} = 0$$

since the virtual displacement, as characterized by $\delta a$, must be arbitrary. Consequently from Eq. ($b$) we have

$$\frac{\partial \Pi}{\partial a} = \frac{1}{2} EI \frac{\pi^4}{l^3} a - P = 0$$

or

$$a = \frac{2Pl^3}{EI\pi^4}$$

The approximate expression for the displacement is then found from Eq. ($a$) as

$$w \approx \frac{2Pl^3}{EI\pi^4} \sin \frac{\pi x}{l} \qquad (c)$$

At the midpoint of the beam the approximate displacement is

$$w_{l/2} = \frac{2Pl^3}{EI\pi^4} = \frac{Pl^3}{48.7EI}$$

whereas the actual displacement, as found by direct integration of the beam flexure equation, is

$$(w_{l/2})_{\text{exact}} = \frac{Pl^3}{48EI}$$

Hence the error is 1.4%. Considering the simplicity of the assumed displacement field, the results are quite good. The error can be reduced by using more terms in the displacement field, as is done in the next example.

## EXAMPLE 5-16

Use the Rayleigh–Ritz method to determine an approximate expression for the displacement distribution for the beam shown in Fig. 5-40.

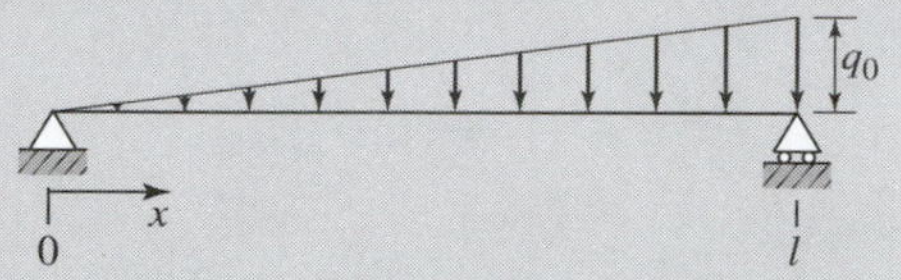


**Figure 5-40**

## SOLUTION

We assume a displacement field of the form

$$w \approx a \sin \frac{\pi x}{l} + b \sin \frac{2\pi x}{l} \tag{a}$$

where $a$ and $b$ are constants to be determined. We note that the displacement is completely determined except for $a$ and $b$, so that the virtual displacement is

$$\delta w \approx \delta a \sin \frac{\pi x}{l} + \delta b \sin \frac{2\pi x}{l}$$

and since this must be arbitrary, it follows that both $\delta a$ and $\delta b$ must be arbitrary. With Eq. ($a$) the total potential energy [Eq. (5-123)] becomes

$$\Pi = \frac{1}{2} EI \int_0^l \left[ \left(\frac{\pi}{l}\right)^2 a \sin \frac{\pi x}{l} + \left(\frac{2\pi}{l}\right)^2 b \sin \frac{2\pi x}{l} \right]^2 dx$$
$$- q_0 \int_0^l \frac{x}{l} \left[ a \sin \frac{\pi c}{l} + b \sin \frac{2\pi x}{l} \right] dx$$

Thus $\Pi = \Pi(a, b)$ and therefore,

$$\delta \Pi = \frac{\partial \Pi}{\partial a} \delta a + \frac{\partial \Pi}{\partial b} \delta b \tag{b}$$

Since both $\delta a$ and $\delta b$ must be arbitrary, this can only be zero if their coefficients are each zero,

$$\frac{\partial \Pi}{\partial a} = 0, \qquad \frac{\partial \Pi}{\partial b} = 0$$

Thus from Eq. (*a*) we have

$$\frac{\partial \Pi}{\partial a} = EI \int_0^l \left[ \left(\frac{\pi}{l}\right)^4 a \sin^2 \frac{\pi x}{l} + \left(\frac{2\pi}{l}\right)^2 \left(\frac{\pi}{l}\right)^2 b \sin \frac{2\pi x}{l} \sin \frac{\pi x}{l} \right] dx - q_0 \int_0^l \frac{x}{l} \sin \frac{\pi x}{l}\, dx = 0$$

$$\frac{\partial \Pi}{\partial b} = EI \int_0^l \left[ \left(\frac{\pi}{l}\right)^2 \left(\frac{2\pi}{l}\right) a \sin \frac{\pi x}{l} \sin \frac{2\pi x}{l} + \left(\frac{2\pi}{l}\right)^4 b \sin^2 \frac{2\pi x}{l} \right] dx - q_0 \int_0^l \frac{x}{l} \sin \frac{2\pi x}{l}\, dx = 0$$

or, after carrying out the indicated integration and solving for $a$ and $b$,

$$a = \frac{2q_0 l^4}{EI\pi^5}$$

$$b = -\frac{q_0 l^4}{16EI\pi^5}$$

The approximate displacement is then given by

$$w \approx \frac{q_0 l^4}{EI\pi^5} \left( 2 \sin \frac{\pi x}{l} - \frac{1}{16} \sin \frac{2\pi x}{l} \right) \qquad (c)$$

The maximum displacement occurs at $0.519l$, and its actual value, to three significant digits, is given by [2, p. 741],

$$(w_{\max})_{\text{exact}} = 0.00654 \frac{q_0 l^4}{EI}$$

From Eq. (*c*) the approximate displacement at $0.519l$ is

$$w_{\max} = 0.0065482 \frac{q_0 l^4}{EI}$$

giving an error of about 0.13%. With only one term the error is approximately doubled.

It is important to note that the stresses in the beam are proportional to the moment that is related to the second derivative of the displacement. Thus, for example, from Eq. (*c*) of Example 5-15,

$$M = -EI \frac{d^2 w}{dx^2} = \frac{2Pl}{\pi^2} \sin \frac{\pi x}{l}$$

The maximum moment occurs at the midpoint of the beam and is given by

$$M_{l/2} = \frac{2Pl}{\pi^2} = \frac{1}{4.93}Pl$$

whereas the actual moment at that point is

$$(M_{l/2})_{\text{exact}} = \tfrac{1}{4}Pl$$

Thus the error in the moment is 18.86%.

The moment, and therefore the stress in the beam, is not as accurate as the displacement. This illustrates the difficulty of accurately determining moments and stresses in problems where a displacement field is assumed. It may be necessary to use many terms in the assumed displacement field in order to determine the moment distribution in the beam accurately.

In general, with the Rayleigh–Ritz method we assume a displacement field that is a linear combination of approximating functions of the following form:

$$w(x) = \varphi_0(x) + a_1\varphi_1(x) + a_2\varphi_2(x) + \cdots + a_n\varphi_n(x)$$

The approximating functions $\varphi_j$, $j = 1, 2, \ldots, n$, are chosen such that the resulting displacement field is *admissible,* which means that it must satisfy the kinematic conditions of the problem. These kinematic conditions are the displacement boundary conditions and also the compatibility conditions. Thus $\varphi_0$ is chosen to satisfy the prescribed displacement boundary conditions whereas the $\varphi_j$, $j = 1, 2, \ldots, n$, are chosen to satisfy the homogeneous counterparts to these conditions. In addition, all of the $\varphi_j$ must be continuous. As an example, for the cantilever beam with prescribed displacement at its free end, shown in Fig. 5-41, the following conditions must be satisfied:

$$\varphi_0(0) = 0, \qquad \varphi_0(l) = u_l, \qquad \varphi_0'(0) = 0$$
$$\varphi_j(0) = 0, \qquad \varphi_j(l) = 0, \qquad \varphi_j'(0) = 0, \qquad j = 1, 2, \ldots, n$$

Although the $\varphi_j$ are not required to satisfy the remaining conditions (boundary conditions on moment and shear, for example), an accurate solution will be obtained with fewer terms if these remaining conditions are satisfied.

While the choice of approximating functions is arbitrary, except for satisfying the kinematic constraints of a problem, a few practical guidelines can be established.

- The approximating functions should be simple to use. Functions such as powers of $x$, sines, or cosines are usually preferred over more complicated choices.
- The approximating functions should be linearly independent.
- As more functions are used in the approximation, it is reasonable to expect the approximate solution to approach the exact solution. In order for this to occur, the

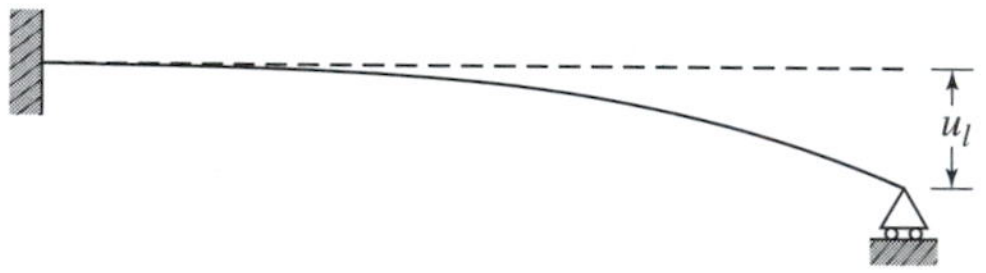


Figure 5-41

set of approximating functions must be *complete*. This means that as more terms are added to the approximation, none can be omitted. It is especially important that all low-order terms be present. Neither of the approximations

$$u(x) = a_0 + a_1 x + a_2 x^3 + a_2 x^4$$
$$u(x) = b_0 + b_1 \cos(2\pi x/l) + b_2 \cos(3\pi x/l)$$

is complete because in the first, the $x^2$ term is missing, whereas in the second, the $\cos(px/l)$ term is missing.

It should be noted that whenever the displacements of a structure are constrained, the stiffness of the structure is increased. Thus a beam that is fixed at one end and pinned at the other is stiffer than a beam that is pinned at both ends. This is because by constraining its slope at one end, the stiffness of the simply supported beam is increased. With the Rayleigh–Ritz method, a structure can displace only into configurations that are permitted by the choice of approximating functions. It is unlikely that our assumed displacement field contains the exact displacement field. Thus the resulting structure is stiffer than the actual one, and the resulting displacements are, on average, underestimated. This does not necessarily mean that the displacements are underestimated at each point in the structure. However, it does mean that the strain energy is underestimated. Since the strain energy equals the work done by the external loads, it follows that if a structure is subjected to a single concentrated external load, the displacement at the load is smaller than the actual displacement at that point.

Further applications of the Rayleigh–Ritz method will be found in Chapters 8 and 10.

## 5-12 PRINCIPLE OF MINIMUM COMPLEMENTARY ENERGY

Instead of examining the behavior of a deformable body subjected to a virtual displacement field, we could have examined the behavior of the body under the action of virtual tractions. Virtual surface tractions are defined in the same way as virtual displacements, except that instead of being kinematically admissible they must by statically admissible. This means that virtual surface tractions must satisfy the traction boundary conditions, they must be related to the stresses through Eqs. (2-14), and the stresses arising from them must satisfy the equations of equilibrium, Eqs. (2-53).

We note that during a virtual change in the surface tractions and body forces the displacements as well as the volume of the body remain unchanged. We envision a body that is subjected to virtual surface tractions in equilibrium under the action of surface tractions $\boldsymbol{t}$ and body forces $\boldsymbol{b}$. The complementary virtual work is then defined as

$$\delta W_c = \int_V \boldsymbol{u} \cdot \delta\boldsymbol{b}\, d\overline{V} + \int_A \boldsymbol{u} \cdot \delta\boldsymbol{t}\, dA \tag{5-124}$$

where $W_c$ is the complementary work. Equation (5-124) is referred to as the complementary virtual work because of its similarity to the expression for complementary work given by Eq. (5-12). The second integral on the right of Eq. (5-124) can be expanded to give

$$\int_A u \cdot \delta t\, dA = \int_A (u\delta t_x + v\delta t_y)\, dA$$

The use of Eqs. (2-14) in this expression yields

$$\int_A \boldsymbol{u} \cdot \delta \boldsymbol{t}\, dA = \int_A [(u\delta\sigma_{xx} + v\delta\sigma_{xy})n_x + (u\delta\sigma_{xy} + v\delta\sigma_{yy})n_y]\, dA$$

and the use of Gauss's theorem gives

$$\int_A \boldsymbol{u} \cdot \delta \boldsymbol{t}\, dA = \int_{\overline{V}} \left[\frac{\partial}{\partial x}(u\delta\sigma_{xx} + v\delta\sigma_{xy}) + \frac{\partial}{\partial y}(u\delta\sigma_{xy} + v\delta\sigma_{yy})\right] d\overline{V}$$

Expanding this and using the strain-displacement relations for infinitesimal strains [Eqs. (3-11)], yields

$$\int_A \boldsymbol{u} \cdot \delta \boldsymbol{t}\, dA = \int_{\overline{V}} \left\{\varepsilon_{xx}\delta\sigma_{xx} + \varepsilon_{yy}\delta\sigma_{yy} + \gamma_{xy}\delta\sigma_{xy} + u\left[\delta\left(\frac{\partial\sigma_{xx}}{\partial x}\right) + \delta\left(\frac{\partial\sigma_{xy}}{\partial y}\right)\right]\right.$$
$$\left. + v\left[\delta\left(\frac{\partial\sigma_{yy}}{\partial y}\right) + \delta\left(\frac{\partial\sigma_{xy}}{\partial x}\right)\right]\right\} d\overline{V} \qquad (5\text{-}125)$$

When the first integral on the right-hand side of Eq. (5-124) is expanded, and that result plus Eq. (5-125) are substituted into Eq. (5-124), we get

$$\int_{\overline{V}} \boldsymbol{u} \cdot \delta \boldsymbol{b}\, d\overline{V} + \int_A \boldsymbol{u} \cdot \delta \boldsymbol{t}\, dA = \int_{\overline{V}} \left\{\varepsilon_{xx}\delta\sigma_{xx} + \varepsilon_{yy}\delta\sigma_{yy} + \gamma_{xy}\delta\sigma_{xy}\right.$$
$$+ u\left[\frac{\partial(\delta\sigma_{xx})}{\partial x} + \frac{\partial(\delta\sigma_{xy})}{\partial_y} + \delta b_x\right]$$
$$\left. + v\left[\frac{\partial(\delta\sigma_{yy})}{\partial y} + \frac{\partial(\delta\sigma_{yx})}{\partial x} + \delta b_y\right]\right\} d\overline{V}$$

where Eq. (5-103*c*) has been used. But the terms in brackets must be zero because the virtual stresses must satisfy Eqs. (2-55), so that

$$\int_{\overline{V}} \boldsymbol{u} \cdot \delta \boldsymbol{F}\, d\overline{V} + \int_A \boldsymbol{u} \cdot \delta \boldsymbol{t}\, dA = \int_{\overline{V}} (\varepsilon_{xx}\delta\sigma_{xx} + \varepsilon_{yy}\delta\sigma_{yy} + \gamma_{xy}\delta\sigma_{xy})\, d\overline{V} \qquad (5\text{-}126)$$

Finally we note that the virtual surface tractions must be zero on the portion of the boundary where tractions are specified. Consequently Eq. (5-126) can be written as

$$\int_{\overline{V}} \boldsymbol{u} \cdot \delta \boldsymbol{b}\, d\overline{V} + \int_{A_u} \boldsymbol{u} \cdot \delta \boldsymbol{t}\, dA = \int_{\overline{V}} (\varepsilon_{xx}\delta\sigma_{xx} + \varepsilon_{yy}\delta\sigma_{yy} + \gamma_{xy}\delta\sigma_{xy})\, d\overline{V} \qquad (5\text{-}127)$$

Equation (5-127) is a mathematical statement of the principle of virtual complementary work. The only restrictions on this equation are that the strains be infinitesimal. This equation places no restrictions on the nature of the stress-strain relationship for the material.

If the material is elastic the strain terms in Eq. (5-127) can be replaced by Eq. (5-78) so that, by analogy with Eq. (5-77),

$$\int_{\overline{V}} \boldsymbol{u} \cdot \delta \boldsymbol{b}\, d\overline{V} + \int_{A_u} \boldsymbol{u} \cdot \delta \boldsymbol{t}\, dA = \int_{\overline{V}} \left(\frac{\partial \tilde{u}_c}{\partial\sigma_{xx}}\delta\sigma_{xx} + \frac{\partial \tilde{u}_c}{\partial\sigma_{yy}}\delta\sigma_{yy} + \frac{\partial \tilde{u}_c}{\partial\sigma_{xy}}\delta\sigma_{xy}\right) d\overline{V}$$
$$= \int_{\overline{V}} \delta\tilde{u}_c\, d\overline{V}$$

By noting that neither the displacements nor the volume change during the application of virtual surface tractions, this can be written as

$$\delta\left(\int_{\overline{V}} \boldsymbol{u}\cdot\boldsymbol{b}\,d\overline{V} + \int_{A_u} \boldsymbol{u}\cdot\boldsymbol{t}\,dA - \int_{\overline{V}} \tilde{u}_c\,d\overline{V}\right) = 0$$

or

$$\delta\Pi_c = 0 \tag{5-128}$$

where $\Pi_c$, the *complementary energy,* is given by

$$\Pi_c = \int_{\overline{V}} u_c\,d\overline{V} - \int_{\overline{V}} \boldsymbol{F}\cdot\boldsymbol{u}\,d\overline{V} - \int_{A_u} \boldsymbol{t}\cdot\boldsymbol{u}\,dA \tag{5-129}$$

Equation (5-128) states that of all the stress fields that satisfy the equations of equilibrium and the traction boundary conditions, the true stress field is the one for which the variation in total complementary energy is zero.

It can further be shown that if the body is in stable equilibrium, the total complementary energy is a minimum. The proof is similar to that used for the theorem of minimum total potential energy. We then have the *theorem of minimum total complementary energy:*

> Of all the stress fields that satisfy the equations of equilibrium and the traction boundary conditions the true stress field is the one for which the total complementary energy is a minimum.

Note that the only material restriction imposed in deriving Eq. (5-128) is that the material be elastic. Thus the theorem of minimum complementary energy applies to all elastic materials, either linear or nonlinear, for which the strains are infinitesimal.

It is interesting to note that in deriving Eq. (5-128), equilibrium equations, traction boundary conditions, and strain-displacement equations were used. In fact, the only equations that were not explicitly employed were the compatibility equations. However, the theorem of minimum complementary energy may be stated in the following alternative way:

> Of all the stress fields that satisfy the equations of equilibrium and the traction boundary conditions, the stress field that also satisfies the compatibility equations is the one for which the complementary energy is a minimum.

The proof of this alternative form can be found in Fung [1, pp. 295–296].

If concentrated loads are included, the complementary energy is given by (see [1], p. 11)

$$\Pi_c = \int_{\overline{V}} \tilde{u}_c\,d\overline{V} - \int_{\overline{V}} \boldsymbol{b}\cdot\boldsymbol{u}\,d\overline{V} - \int_{A_u} \boldsymbol{t}\cdot\boldsymbol{u}\,dA - \sum_{i=1}^{n} \boldsymbol{P}_i\cdot\boldsymbol{u}_i \tag{5-130a}$$

or, with Eq. (5-73),

$$\Pi_c = U - \int_{\overline{V}} \boldsymbol{b}\cdot\boldsymbol{u}\,d\overline{V} - \int_{A_u} \boldsymbol{t}\cdot\boldsymbol{u}\,dA - \sum_{i=1}^{n} \boldsymbol{P}_i\cdot\boldsymbol{u}_i \tag{5-130b}$$

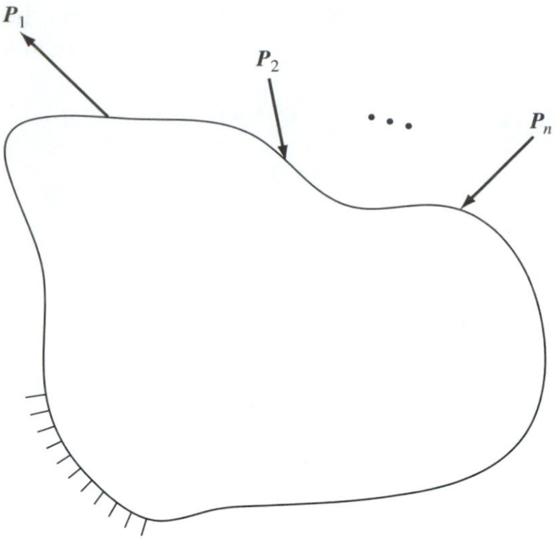


Figure 5-42

As an example of the application of the theorem of minimum total complementary energy, consider a body constrained against rigid body motion and subjected to concentrated loads, as shown in Fig. 5-42. The complementary energy is then given by

$$\Pi_c = U_c - \sum_{i=1}^{n} \boldsymbol{P}_i \cdot \boldsymbol{u}_i$$

Thus,

$$U_c = U_c(\boldsymbol{P}_1, \boldsymbol{P}_2, \ldots, \boldsymbol{P}_n)$$

The variation of the total complementary energy must be zero,

$$\delta\Pi_c = \frac{\partial U_c}{\partial \boldsymbol{P}_1} \cdot \delta\boldsymbol{P}_1 + \frac{\partial U_c}{\partial \boldsymbol{P}_2} \cdot \delta\boldsymbol{P}_2 + \cdots + \frac{\partial U_c}{\partial \boldsymbol{P}_n} \cdot \delta\boldsymbol{P}_n - \boldsymbol{u}_1 \cdot \delta\boldsymbol{P}_1 - \boldsymbol{u}_2 \cdot \delta\boldsymbol{P}_2 - \cdots - \boldsymbol{u}_n \cdot \delta\boldsymbol{P}_n = 0$$

Thus, because the virtual surface tractions (here the virtual concentrated loads) must be arbitrary we have

$$\frac{\partial U_c}{\partial P_i} = u_i, \qquad i = 1, 2, \ldots, n \tag{5-131}$$

This expression is referred to as the generalized Castigliano's second theorem. If the material is linear, then $U_c = U$, and this becomes Castigliano's second theorem,

$$\frac{\partial U}{\partial P_i} = u_i, \qquad i = 1, 2, \ldots, n$$

as derived in Section 5-7. We note from this that Castigliano's second theorem can be applied to bodies of arbitrary shape and that if we use the generalized second theorem, the material need not be linear.

## 5-13 BETTI–RAYLEIGH RECIPROCAL THEOREM

We now derive an important generalization of Maxwell's reciprocity theorem and demonstrate its use. We begin by considering a body with volume $\overline{V}$ and surface $A$ in equilibrium with displacements $\boldsymbol{u}^a$ due to surface tractions $\boldsymbol{t}^a$ and body forces $\boldsymbol{b}^a$. Now consider a second state of equilibrium of the same body due to surface tractions $\boldsymbol{t}^b$ and body forces $\boldsymbol{b}^b$, resulting in displacements $\boldsymbol{u}^b$. Let us calculate the work done by the surface tractions $\boldsymbol{t}^a$ and body forces $\boldsymbol{b}^a$ in acting through the displacements $\boldsymbol{u}^b$. This gives

$$W_1 = \int_{\overline{V}} \boldsymbol{b}^a \cdot \boldsymbol{u}^b \, d\overline{V} = \int_A \boldsymbol{t}^a \cdot \boldsymbol{u}^b \, dA \tag{5-132}$$

We will show that this is the same as the work done by the surfaces tractions $\boldsymbol{t}^b$ and body forces $\boldsymbol{b}^b$ in acting through the displacements $\boldsymbol{u}^a$. To do this we expand the surface integral in Eq. (5-132) to give

$$\int_A \boldsymbol{t}^a \cdot \boldsymbol{u}^b \, dA = \int_A \left(t_x^a u^b + t_y^a v^b\right) dA$$

or, using Eqs. (2-14) and collecting terms,

$$\int_A \boldsymbol{t}^a \cdot \boldsymbol{u}^b \, dA = \int_A \left[\left(\sigma_{xx}^a u^b + \sigma_{xy}^a v^b\right) n_x + \left(\sigma_{xy}^a u^b + \sigma_{yy}^a v^b\right) n_y\right] dA$$

With the aid of Gauss's theorem this becomes

$$\int_A \boldsymbol{t}^a \cdot \boldsymbol{u}^b \, dA = \int_{\overline{V}} \left[\frac{\partial}{\partial x}\left(\sigma_{xx}^a u^b + \sigma_{xy}^a v^b\right) + \frac{\partial}{\partial y}\left(\sigma_{xy}^a u^b + \sigma_{yy}^a v^b\right)\right] d\overline{V}$$

Expanding, collecting terms, and using the strain-displacement relations for infinitesimal strains, we have

$$\begin{aligned}\int_A \boldsymbol{t}^a \cdot \boldsymbol{u}^b \, dA = &\int_{\overline{V}} \left[\left(\frac{\partial \sigma_{xx}^a}{\partial x} + \frac{\partial \sigma_{xy}^a}{\partial y}\right) u^b + \left(\frac{\partial \sigma_{xy}^a}{\partial x} + \frac{\partial \sigma_{yy}^a}{\partial y}\right) v^b\right] d\overline{V} \\ &+ \int_{\overline{V}} \left(\sigma_{xx}^a \varepsilon_{xx}^b + \sigma_{yy}^a \varepsilon_{yy}^b + \sigma_{xy}^a \gamma_{xy}^b\right) dV\end{aligned}$$

Thus Eq. (5-132) becomes

$$\begin{aligned}W_1 = &\int_{\overline{V}} \left[\left(\frac{\partial \sigma_{xx}^a}{\partial x} + \frac{\partial \sigma_{xy}^a}{\partial y} + b_x^a\right) u^b + \left(\frac{\partial \sigma_{xy}^a}{\partial x} + \frac{\partial \sigma_{yy}^a}{\partial y} + b_y^a\right) v^b\right] d\overline{V} \\ &+ \int_{\overline{V}} \left(\sigma_{xx}^a \varepsilon_{xx}^b + \sigma_{yy}^a \varepsilon_{yy}^b + \sigma_{xy}^a \gamma_{xy}^b\right) d\overline{V}\end{aligned}$$

Now the first two integrals on the right are zero because the body is in equilibrium. Therefore,

$$W_1 = \int_{\overline{V}} \left(\sigma_{xx}^a \varepsilon_{xx}^b + \sigma_{yy}^a \varepsilon_{yy}^b + \sigma_{xy}^a \gamma_{xy}^b\right) d\overline{V}$$

For a linear elastic material under a state of plane stress we have, from Eqs. (4-21),

$$\sigma_{xx}^a = \frac{E}{1-\nu^2}\left(\varepsilon_{xx}^a + \nu\varepsilon_{yy}^a\right)$$
$$\sigma_{yy}^a = \frac{E}{1-\nu^2}\left(\varepsilon_{yy}^a + \nu\varepsilon_{xx}^a\right)$$
$$\sigma_{xy}^a = G\gamma_{xy}^a$$

Consequently,

$$W_1 = \int_{\overline{V}} \left\{ \frac{E}{1-\nu^2}\left[\varepsilon_{xx}^a\varepsilon_{xx}^b + \nu\left(\varepsilon_{xx}^a\varepsilon_{yy}^b + \varepsilon_{yy}^a\varepsilon_{xx}^b\right) + \varepsilon_{yy}^a\varepsilon_{yy}^b\right] + G\gamma_{xy}^a\gamma_{xy}^b \right\} d\overline{V} \quad (5\text{-}133)$$

Similarly, if we calculate the work done by the surface tractions $\boldsymbol{t}^b$ and body forces $\boldsymbol{b}^b$ in acting through the displacements $\boldsymbol{u}^a$ we get

$$W_2 = \int_{\overline{V}} \boldsymbol{b}^b \cdot \boldsymbol{u}^a \, d\overline{V} + \int_A \boldsymbol{t}^b \cdot \boldsymbol{u}^a \, dA$$

This expression is the same as Eq. (5-132) except that the superscripts $a$ and $b$ are interchanged. Therefore interchanging the superscripts $a$ and $b$ in Eq. (5-133) we get

$$W_2 = \int_{\overline{V}} \left\{ \frac{E}{1-\nu^2}\left[\varepsilon_{xx}^b\varepsilon_{xx}^a + \nu\left(\varepsilon_{xx}^b\varepsilon_{yy}^a + \varepsilon_{yy}^b\varepsilon_{xx}^a\right) + \varepsilon_{yy}^b\varepsilon_{xx}^a\right] + G\gamma_{xy}^b\gamma_{xy}^a \right\} d\overline{V}$$

We note from this that

$$W_1 = W_2$$

or

$$\int_{\overline{V}} \boldsymbol{b}^a \cdot \boldsymbol{u}^b \, d\overline{V} + \int_A \boldsymbol{t}^a \cdot \boldsymbol{u}^b \, dA = \int_{\overline{V}} \boldsymbol{b}^b \cdot \boldsymbol{u}^a \, d\overline{V} + \int_A \boldsymbol{t}^b \cdot \boldsymbol{u}^a \, dA \quad (5\text{-}134)$$

This same equation would have resulted if we had assumed conditions of plane strain or if the problem had been three-dimensional, so long as the material was linearly elastic. This expression is known as the *Betti–Rayleigh reciprocal theorem*. In words, it states that:

> For a body composed of linearly elastic material, the work done by a body in equilibrium under the action of surface tractions $\boldsymbol{t}^a$ and body forces $\boldsymbol{b}^a$ in moving through displacements $\boldsymbol{u}^b$ is the same as the work done by a body in equilibrium under the action of surface tractions $\boldsymbol{t}^b$ and body forces $\boldsymbol{b}^b$ moving through displacements $\boldsymbol{u}^a$.

If the system of loads designated $a$ contains concentrated loads at points $1, 2, \ldots, n$ throughout the body, as shown in Fig. 5-43, and the $b$ system of loads contains concentrated loads at points $1, 2, \ldots, m$ (not necessarily at the same locations as the $a$ system),

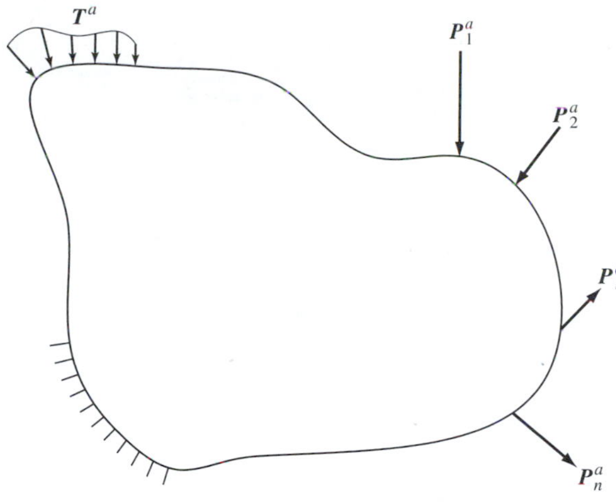


Figure 5-43

then Eq. (5-134) becomes (see [2], pp. 5–6)

$$\int_{\overline{V}} \boldsymbol{b}^a \cdot \boldsymbol{u}^b \, d\overline{V} + \int_A \boldsymbol{t}^a \cdot \boldsymbol{u}^b \, dA + \sum_{i=1}^{n} \boldsymbol{P}_i^a \cdot \boldsymbol{u}_i^b$$
$$= \int_{\overline{V}} \boldsymbol{b}^b \cdot \boldsymbol{u}^a \, d\overline{V} + \int_A \boldsymbol{t}^b \cdot \boldsymbol{u}^a \, dA + \sum_{i=1}^{m} \boldsymbol{P}_i^b \cdot \boldsymbol{u}_i^a \qquad (5\text{-}135)$$

where $\boldsymbol{u}_i^a$ is the displacement at the point of application of the concentrated load $\boldsymbol{P}^b$ but arising from the combined effect of the loads of system $a$.

Now consider the case where there are no surface tractions or body forces and only a single concentrated load in each load system, say at point 1 for load system $a$ and at point 2 for system $b$, as shown in Fig. 5-44. Then Eq. (5-135) reduces to

$$\boldsymbol{P}_1^a \cdot \boldsymbol{u}_1^b = \boldsymbol{P}_2^b \cdot \boldsymbol{u}_2^a \qquad (5\text{-}136)$$

If the loads are both of unit magnitude, the left-hand side of Eq. (5-136) is simply the component of the displacement vector $\boldsymbol{u}_1^b$ in the direction of the unit load acting at point 1, but caused by the unit load at point 2. The right-hand side of Eq. (5-136) bears a similar interpretation. Thus with loads of unit magnitude, Eq. (5-136) is simply Maxwell's reciprocity theorem. Consequently, Maxwell's reciprocity theorem is simply a special case of the more general Betti–Rayleigh reciprocal theorem.

As an application of the Betti–Rayleigh reciprocal theorem consider a beam bending problem where the beam cross section is uniform. Then, in the absence of body forces, Eq. (5-135) becomes

$$\int_0^l \boldsymbol{t}^a \cdot \boldsymbol{u}^b \, dx + \sum_{i=1}^{n} \boldsymbol{P}_i^a \cdot \boldsymbol{u}_i^b = \int_0^l \boldsymbol{t}^b \cdot \boldsymbol{u}^a \, dx + \sum_{i=1}^{m} \boldsymbol{P}_i^b \cdot \boldsymbol{u}_i^a$$

But for a beam the traction integrals in Eq. (5-135) can be written as integrals taken around the perimeter of the cross section and then along the length, as was done in arriving at Eq. (5-122). For a beam subjected to bending loads the traction integral taken around the perimeter of the cross section is simply the distributed load per unit length $q$,

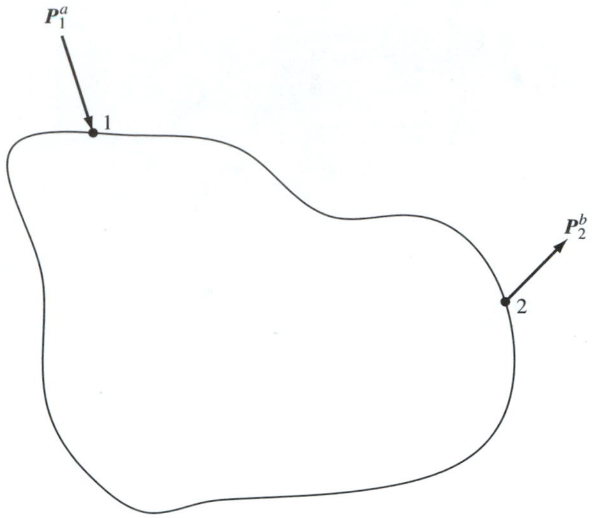


Figure 5-44

and the nonzero displacement component is the bending displacement $w$. Thus we have

$$\int_0^l q^a w^b \, dx + \sum_{i=1}^{n} P_i^a w_i^b = \int_0^l q^b w^a \, dx + \sum_{i=1}^{m} P_i^b w_i^a \tag{5-137}$$

where the loads $P_i$ are the components of $\boldsymbol{P}_i$ in the direction of the bending displacement $w$.

We now restrict our attention to cases in which there is only a single concentrated load acting on the beam. Thus Eq. (5-137) reduces to

$$\int_0^l q^a w^b \, dx + P_1^a w_1^b = \int_0^l q^b w^a \, dx + P_2^b w_2^a \tag{5-138}$$

Now let the $a$ system of loads consist only of a concentrated load of unit magnitude acting at a point $\xi$, as shown in Fig. 5-45, that is,

$$P_1^a = 1, \qquad q^a = 0$$

We will designate the resulting displacement as $g(x, \xi)$, that is, $w^a = g(x, \xi)$ is the displacement at point $x$ that results from a unit concentrated load acting at point $\xi$, as shown in Fig. 5-45. The $b$ system of loads consists of the distributed load $q(x)$, as shown in Fig. 5-46, so that we have

$$q^b = q(x), \qquad P_2^b = 0$$

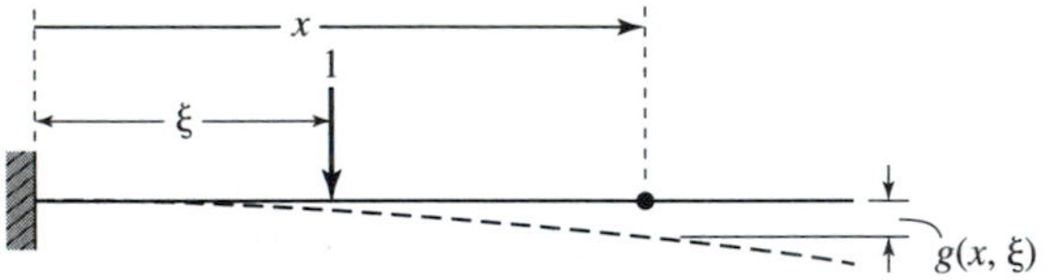


Figure 5-45

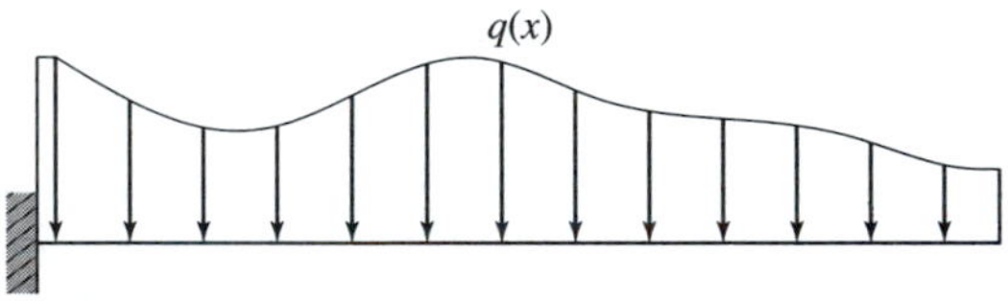


Figure 5-46

The displacement $w_1^b$ is the displacement $w$ at point $\xi$ due to $q^b = q(x)$. Thus Eq. (5-138) becomes

$$w(\xi) = \int_0^l q(x) g(x, \xi)\, dx$$

As a matter of convenience let us interchange $x$ and $\xi$,

$$w(x) = \int_0^l q(\xi) g(x, \xi)\, d\xi \tag{5-139}$$

where we have made use of the fact that, by virtue of Maxwell's reciprocity theorem, $g(\xi, x) = g(x, \xi)$. That is, $g$ is symmetric in its arguments.

Equation (5-139) states that, for a certain set of boundary conditions, if we know the displacement distribution due to a concentrated unit load acting at point $\xi$, then we can calculate the displacement $w(x)$ due to any distributed load simply by integrating as shown.

## EXAMPLE 5-17

Determine the displacement distribution for the simply supported beam shown in Fig. 5-47.

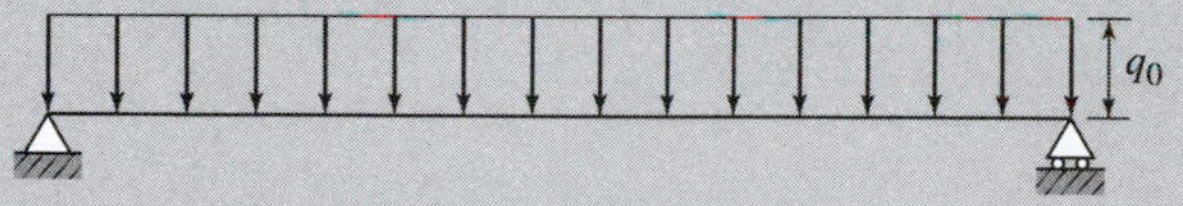


**Figure 5-47**

## SOLUTION

For a simply supported beam, if a concentrated load of unit magnitude acts at $\xi$, as shown in Fig. 5-48, the resulting displacement at point $x$ is given by (see [6, p. 106]*)

$$g(x, \xi) = \begin{cases} \dfrac{(l-\xi)x}{6EIl}[2l(l-x) - (l-\xi)^2 - (l-x)^2], & \xi \geq x \\ \dfrac{(l-x)\xi}{6EIl}[2l(l-\xi) - (l-x)^2 - (l-\xi)^2], & \xi \leq x \end{cases} \tag{a}$$

As an aside, the reader should note that interchanging $x$ and $\xi$ on both sides of Eqs. (*a*), including the inequalities, yields expressions for $g(\xi, x)$ that are identical to the original expressions, demonstrating the symmetry of $g(x, \xi)$.

---

*Although this is not the latest edition of Roark's book, it has the advantage of giving the desired formula directly. Readers using the 6th edition will find it necessary to do some manipulation to obtain Eq. (*a*).

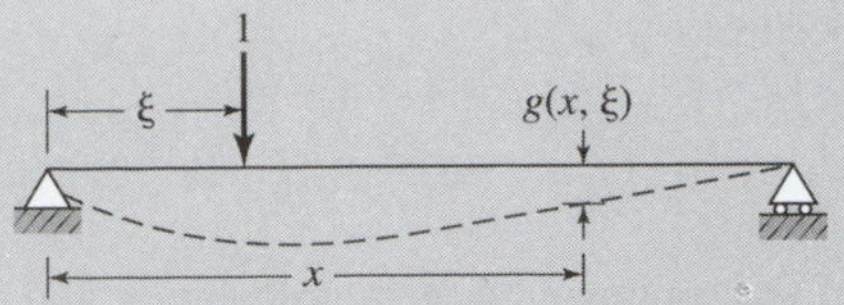


**Figure 5-48**

Now if the simply supported beam is subjected to a uniformly distributed load of magnitude $q_0$, as shown in Fig. 5-47, the displacement $w(x)$ is given by Eqs. (5-139) and ($a$),

$$w(x) = \frac{q_0}{6EIl}\left\{\int_0^x \xi(l-x)[2l(l-\xi)-(l-x)^2-(l-\xi)^2]\,d\xi \right.$$
$$\left. + \int_x^l x(l-\xi)[2l(l-x)-(l-\xi)^2-(l-x)^2]\,d\xi\right\}$$

Carrying out the indicated integration and collecting terms, the displacement $w(x)$ is found to be

$$w(x) = \frac{q_0 x}{24EI}(l^3 - 2lx^2 + x^3)$$

The unit concentrated load solution $g(x, \xi)$ is known as a Green's function, and the method of finding displacements for arbitrarily distributed loads using the concentrated load solution is called the Green's function method. Further applications of this method will be considered in Chapters 8 and 10.

We note here that the concentrated load solutions for all common beam boundary conditions are readily available. Thus by storing these on a computer, specifying the boundary conditions desired and the load distribution, virtually all beam problems can be solved simply by writing a computer program to perform the integration indicated in Eq. (5-139).

## 5-14 GENERAL STRESS-STRAIN RELATIONSHIPS FOR ELASTIC MATERIALS

We will close this chapter by showing the general form of the stress-strain equations for homogeneous materials that are anisotropic. In order to do this we note that no particular stress-strain relation has been used to derive Eqs. (5-70). The only assumptions are that the strains are infinitesimal, that thermal effects are negligible, and that the material remains elastic. If we think of the strain energy density function as a function of the strains and recall that the strains are infinitesimal quantities, then the strain energy density function can be expanded in a power series of the strains,

$$\begin{aligned}\tilde{u} = C_0 + A_1\varepsilon_{xx} + A_2\varepsilon_{yy} + \cdots + A_6\gamma_{yz} + B_1\varepsilon_{xx}^2 + B_2\varepsilon_{yy}^2 + \cdots + B_6\gamma_{yz}^2 \\ + B_7\varepsilon_{xx}\varepsilon_{yy} + B_8\varepsilon_{xx}\varepsilon_{zz} + \cdots + B_{11}\varepsilon_{xx}\gamma_{yz} + B_{12}\varepsilon_{yy}\varepsilon_{zz} + B_{13}\varepsilon_{yy}\gamma_{xy} \\ + \cdots + B_{15}\varepsilon_{yy}\gamma_{yz} + B_{16}\varepsilon_{zz}\gamma_{xy} + B_{17}\varepsilon_{zz}\gamma_{xz} + B_{18}\varepsilon_{zz}\gamma_{yz} + B_{19}\gamma_{xy}\gamma_{xz} \\ + B_{20}\gamma_{xy}\gamma_{yz} + B_{21}\gamma_{xz}\gamma_{yz} + \text{higher order terms}\end{aligned} \tag{5-140}$$

Since we assumed the material to be homogeneous (but not isotropic), then $C_0$, $A_1, \ldots, B_1, \ldots$ are constants. Now $\tilde{u}$ is really the change in strain energy density relative to some reference state, so we can consider the state of zero strain to be a state of zero strain energy density. Then by setting the strains equal to zero in Eq. (5-140) and requiring $\tilde{u}$ to be zero it follows that

$$C_0 = 0$$

Equations (5-70) together with Eq. (5-140) give

$$\begin{aligned}\sigma_{xx} &= \frac{\partial \tilde{u}}{\partial \varepsilon_{xx}} = A_1 + 2B_1\varepsilon_{xx} + B_7\varepsilon_{yy} + B_8\varepsilon_{zz} + B_9\gamma_{xy} + B_{10}\gamma_{xz} + B_{11}\gamma_{yz} \\ \sigma_{yy} &= \frac{\partial \tilde{u}}{\partial \varepsilon_{yy}} = A_2 + 2B_2\varepsilon_{yy} + B_7\varepsilon_{xx} + B_{12}\varepsilon_{zz} + B_{13}\gamma_{xy} + B_{14}\gamma_{xz} + B_{15}\gamma_{yz} \\ \sigma_{zz} &= \frac{\partial \tilde{u}}{\partial \varepsilon_{zz}} = A_3 + 2B_3\varepsilon_{zz} + B_8\varepsilon_{xx} + B_{12}\varepsilon_{yy} + B_{16}\gamma_{xy} + B_{17}\gamma_{xz} + B_{18}\gamma_{yz} \\ \sigma_{xy} &= \frac{\partial \tilde{u}}{\partial \gamma_{xy}} = A_4 + 2B_4\gamma_{xy} + B_9\varepsilon_{xx} + B_{13}\varepsilon_{yy} + B_{16}\varepsilon_{zz} + B_{19}\gamma_{xz} + B_{20}\gamma_{yz} \\ \sigma_{xz} &= \frac{\partial \tilde{u}}{\partial \gamma_{xz}} = A_5 + 2B_5\gamma_{xz} + B_{10}\varepsilon_{xx} + B_{14}\varepsilon_{yy} + B_{17}\varepsilon_{zz} + B_{19}\gamma_{xy} + B_{21}\gamma_{yz} \\ \sigma_{yz} &= \frac{\partial \tilde{u}}{\partial \gamma_{yz}} = A_6 + 2B_6\gamma_{yz} + B_{11}\varepsilon_{xx} + B_{15}\varepsilon_{yy} + B_{18}\varepsilon_{zz} + B_{20}\gamma_{xy} + B_{21}\gamma_{xz}\end{aligned} \tag{5-141}$$

We will assume that the body is not prestressed so that it is in a stress-free state when all the strains are zero. Then

$$A_1 = 0,\ A_2 = 0, \ldots, A_6 = 0$$

Consequently Eqs. (5-141) become, in matrix form,

$$\begin{Bmatrix}\sigma_{xx} \\ \sigma_{yy} \\ \sigma_{zz} \\ \sigma_{xy} \\ \sigma_{xz} \\ \sigma_{yz}\end{Bmatrix} = \begin{bmatrix}\hat{B}_1 & B_7 & B_8 & B_9 & B_{10} & B_{11} \\ B_7 & \hat{B}_2 & B_{12} & B_{13} & B_{14} & B_{15} \\ B_8 & B_{12} & \hat{B}_3 & B_{16} & B_{17} & B_{18} \\ B_9 & B_{13} & B_{16} & \hat{B}_4 & B_{19} & B_{20} \\ B_{10} & B_{14} & B_{17} & B_{19} & \hat{B}_5 & B_{21} \\ B_{11} & B_{15} & B_{18} & B_{20} & B_{21} & \hat{B}_6\end{bmatrix} \begin{Bmatrix}\varepsilon_{xx} \\ \varepsilon_{yy} \\ \varepsilon_{zz} \\ \gamma_{xy} \\ \gamma_{xz} \\ \gamma_{yz}\end{Bmatrix} \tag{5-142}$$

where $\hat{B}_i = 2B_i$, $i = 1, 2, \ldots, 6$.

Since the coefficient matrix is symmetric, Eq. (5-142) shows that for a general anisotropic, homogeneous elastic material there are 21 elastic constants in the constitutive law. In the case of an isotropic material these 21 constants reduce to two and the constitutive law becomes the familiar generalized Hooke's law given in Chapter 4.

Often an anisotropic material possesses certain axes of material symmetry, in which case the number of elastic constants lies somewhere between 2 and 21, depending on the nature of these symmetries. In any case the values of these constants must be determined experimentally.

## PROBLEMS

### Section 5-2

**5-1** A particle of weight $W$ is hung from a rigid, weightless rod of length $l$. The particle is pulled slowly, so that equilibrium exists at all times, by a variable horizontal force $P$. This force starts at zero and gradually increases until the particle makes an angle $\theta_0$ with the vertical. Draw a free-body diagram of the particle and find the work done by each of the forces acting on it.

**5-2** Each of the beams shown in Fig. P5-2 is of length $l$. In each case, determine the total work done by the applied loads. Assume the force-displacement relationship to be linear in each case.

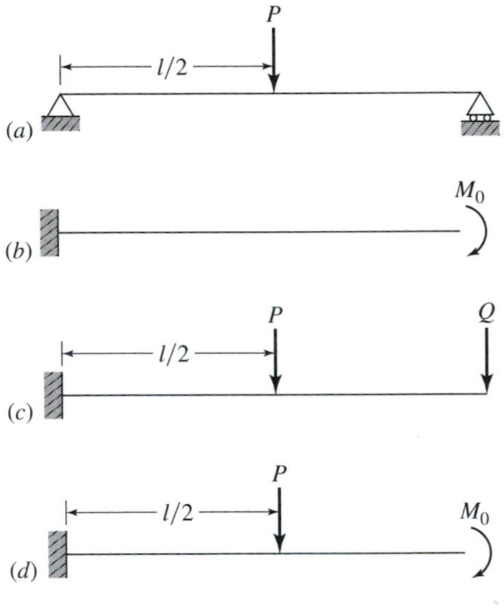


Figure P5-2

**5-3** For each of the cases in Problem 5-2, use the appropriate formulas for the displacements to determine the total work done in terms of applied loads, geometry, and material properties.

**5-4** For cases ($c$) and ($d$) of Problem 5-2 show that the work done is the same if the loads are applied in a different order.

**5-5** A load $P$ is applied to the simple truss shown in Fig. P5-5. The material is linear elastic. Determine the vertical displacement at point 2, and the work done by $P$ in terms of load, geometry, and material properties. Assume that the deflections are small compared to the member lengths.

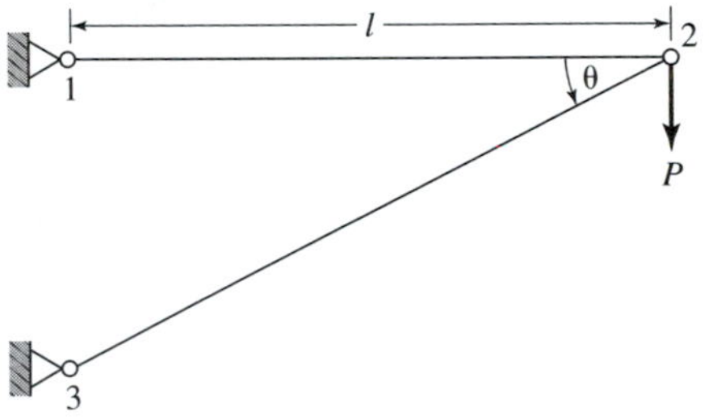


Figure P5-5

**5-6** The simple truss shown in Fig. P5-6 is constructed of a linear elastic material. A load $P$ is applied at point 2, and then the temperature of the truss members is raised by an amount $\Delta T$. Determine a formula for the work done by $P$ in terms of the load, the geometry, and the material properties. Assume

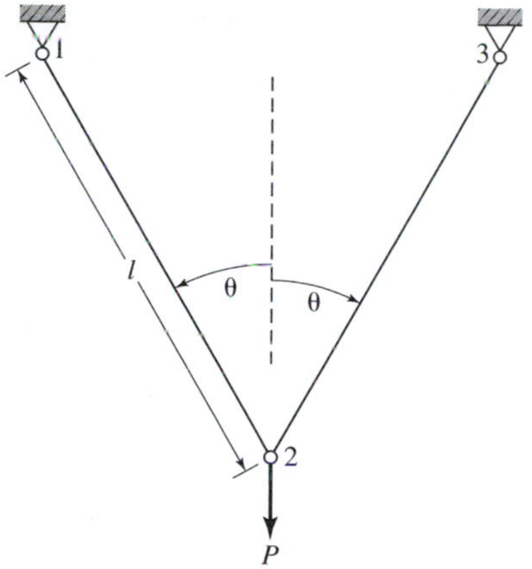


Figure P5-6

that the deflections are small compared to the member lengths.

The two truss members are steel wire, 0.0625 in diameter and 36 in long, for which $E = 30 \times 10^6$ psi and $\alpha = 8 \times 10^{-6}$ in/in-°F. If $\theta = 30^\circ$, $P = 400$ lbf, and $\Delta T = 300$ °F, determine the work done.

**5-7** The column shown in Fig. P5-7 has a cross-sectional area $A$. It is constructed from a material with the following stress-strain law:

$$\sigma = k\varepsilon^n$$

where $k$ and $n$ are constant properties of the material. Plot stress versus strain and determine a formula for the work done by $P$ in terms of the load, the geometry, and the material properties when $k = 10^6$ psi and $n = 1/2$. Assume that the displacement is small compared to the column length, so that the resulting strains are small. Compute the work done if the column is a wire with a diameter of 0.0625 in, 36 in long, and $P = 200$ lbf.

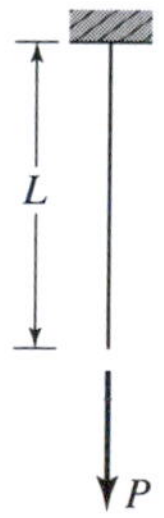


Figure P5-7

**5-8** Repeat Problem 5-7 if the stress-strain law is

$$\sigma = \sigma_0 \tanh \frac{\varepsilon}{\varepsilon_0}$$

where $\sigma_0$ and $\varepsilon_0$ are material constants. Compute the work done using the conditions in Problem 5-7 with $\sigma_0 = 90{,}000$ psi and $\varepsilon_0 = 0.004$.

**5-9** Repeat Problem 5-7 if the stress-strain law is

$$\sigma = c(a + \varepsilon)^n$$

where $c$, $a$, and $n$ are material constants. Compute the work done using the conditions in Problem 5-7 with $P = 350$ lbf, $c = 225{,}000$ psi, $a = 0.001$, and $n = 1/4$.

## Section 5-5

**5-10** A beam of length $L$, with a linear relationship between force and deflection, is subjected to a distributed bending load $q(x)$, as shown in the Fig. P5-10. If the displacement distribution for the beam is $w(x)$, show that the work done by the distributed load is

$$W = \frac{1}{2} \int_0^L q(x)w(x)\,dx$$

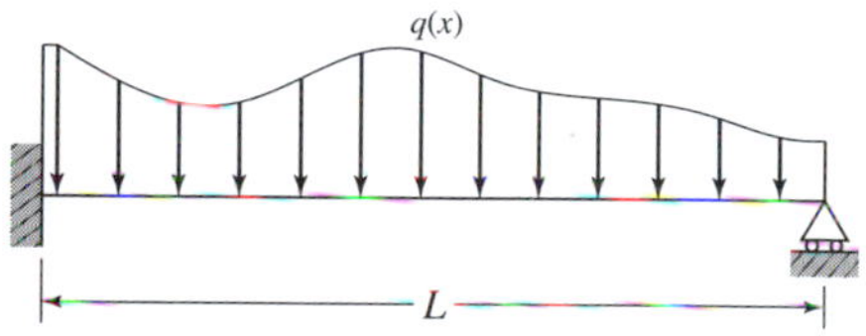


Figure P5-10

**5-11** Use the result of Problem 5-10 to determine the work done by a uniformly distributed load acting on a cantilever beam. Compare your result with the result obtained in Example 5-7. Assume that the beam material is linear elastic.

**5-12** Using the result obtained in Problem 5-10, find the work done by the distributed

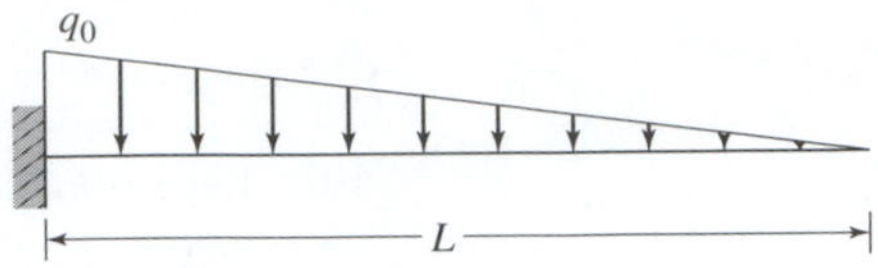


Figure P5-12

load acting on the beam shown in Fig. P5-12. Assume that the beam material is linear elastic.

## Section 5-6

**5-13** Use Eq. (5-89) to derive the form factor for a rectangular cross section.

**5-14** Use Eq. (5-89) to derive the form factor for a solid circular cross section.

**5-15** Compute the strain energy stored in each of the beams in Problem 5-2. Verify in each case that the strain energy equals the work done as obtained in Problem 5-3. Neglect shear.

**5-16** The column shown in Fig. P5-16 is loaded by its own weight. If the weight per unit length is $\gamma$, determine the strain energy stored in the column.

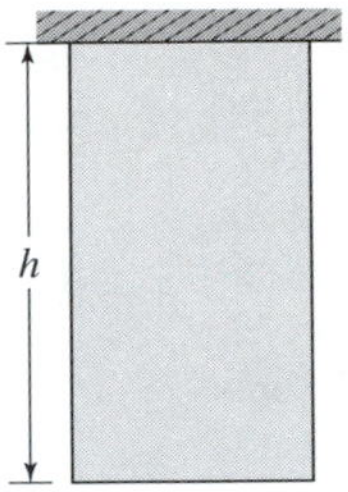


Figure P5-16

**5-17** For the cantilever beam shown in Fig. P5-17 calculate the strain energy stored in the beam and the work done by the load at the free end. Then equate these two quantities and obtain an expression for the deflection at the free end of the beam. Neglect shear in the strain energy calculation.

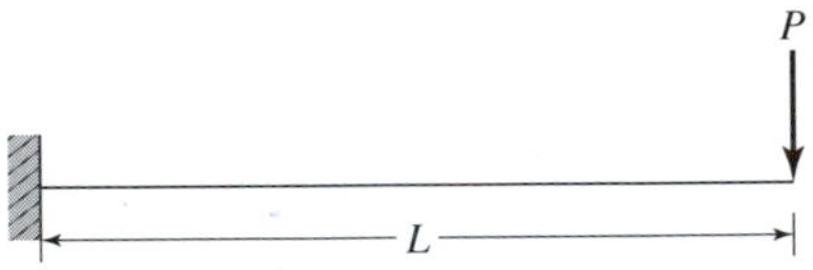


Figure P5-17

**5-18** For case (*a*) of Problem 5-2 equate the work obtained in that problem to the strain energy computed in Problem 5-15 and deduce a formula for the midspan displacement. Verify that this agrees with the formula obtained from mechanics of materials. Repeat for the slope at the end of the beam in case (*b*) of Problem 5-2.

**5-19** Find the strain energy stored in each of the beams shown in Fig. P5-19 assuming that the length-to-thickness ratio exceeds 10. For case (*b*) show that the strain energy equals the work calculated in Problem 5-12.

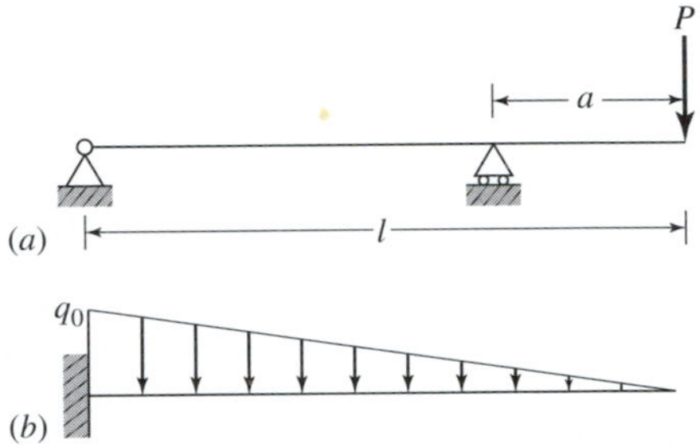


Figure P5-19

**5-20** The cantilever beam shown in Fig. P5-20 makes an angle $\alpha$ with the horizontal and is loaded by its own weight. Determine the strain energy stored in the beam, assuming shear can be neglected. With your solution, examine the limiting cases of $\alpha = 0°$ and $\alpha = 90°$. Neglect shear.

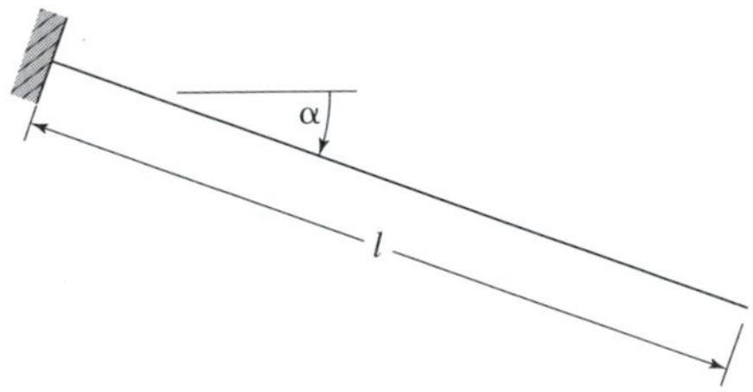


Figure P5-20

## Section 5-7

**5-21** Use Castigliano's theorem to determine the transverse displacement at the location of the load in Problem 5-2, case (*a*), and in Problem 5-19, case (*a*), and the slope at the free end of the beam in Problem 5-2, case (*b*).

**5-22** For the cantilever beam in Problem 5-17, use Castigliano's theorem to determine the transverse displacement at the free end if the effects of shear are included. What is the increase in displacement caused by the inclusion of shear? When is the inclusion of shear important? Assume a circular cross section.

**5-23** Use Castigliano's theorem to determine the vertical displacement at the point where the load is applied for the structure shown in Problem 5-5.

**5-24** Repeat Problem 5-23 for the structure shown in Problem 5-6.

**5-25** Use Castigliano's theorem to determine the vertical displacement at the load for the cases shown in Fig. P5-25.

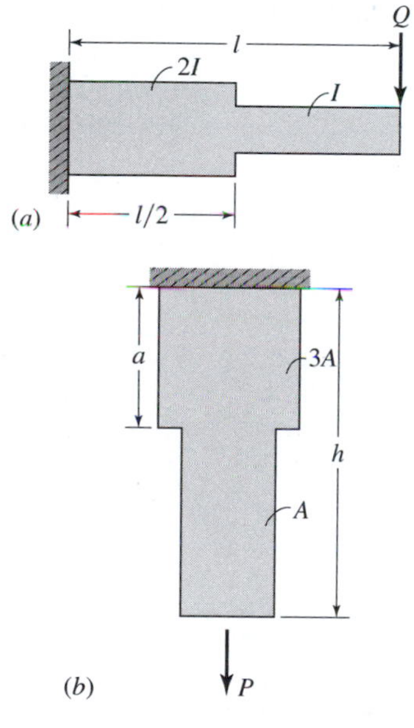


Figure P5-25

**5-26** Use Castigliano's theorem to determine the displacement at the midspan and at the free end of the beam shown in Problem 5-2, case (*c*).

**5-27** Use Castigliano's theorem to determine the displacement at the midspan and the slope at the free end of the beam shown in Problem 5-2, case (*d*).

**5-28** Use Castigliano's theorem to find the vertical displacement at the free end of the beam shown in Problem 5-20.

**5-29** Use Castigliano's theorem to find the vertical displacement at the free end of the beam shown in Problem 5-19, case (*b*).

**5-30** Use Castigliano's theorem to find the slope at the free end of the beam shown in Problem 5-19, case (*b*).

**5-31** Use Castigliano's theorem to find the displacement at the free end of the column shown in Problem 5-16.

**5-32** When Eq. (5-71) is integrated over the volume of the body, it follows that

$$U = W_{\text{ext}}$$

where $U$ is the strain energy stored in the body and $W_{\text{ext}}$ is the work done by the external forces acting on the body. For a linear elastic material we know that work and complementary work are equal, and therefore it follows that for a body made of such a material,

$$dU = dW^c_{\text{ext}} \tag{a}$$

where the superscript indicates complementary work. If a single load is applied to a beam, then the force-deflection diagram shows that

$$dW^c_{\text{ext}} = \delta\, dP$$

If several loads are applied to the beam, it follows that

$$dW^c_{\text{ext}} = \delta_1\, dP_1 + \delta_2\, dP_2 + \cdots + \delta_n\, dP_n \tag{b}$$

Use Eqs. (*a*) and (*b*) to derive Castigliano's second theorem.

*Hint:* Equations $(a)$ and $(b)$, taken together, imply that $U = U(P_1, P_2, \ldots, P_n)$. Use the chain rule to express $dU$ in terms of the differentials of the loads.

## Section 5-11

**5-33** Use the Rayleigh–Ritz method to find an approximate deflection distribution for a simply supported beam loaded as shown in Fig. P5-33*a* and *b*.

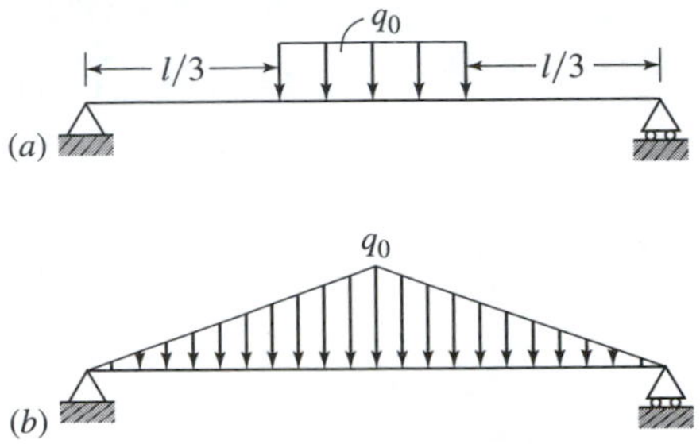


Figure P5-33

**5-34** Use the Rayleigh–Ritz method to find an approximate deflection distribution for a cantilever beam loaded as shown in Fig. P5-37. A displacement field that satisfies the kinematic boundary conditions of the problem is $w(x) = ax^2 + bx^3 + cx^4$.

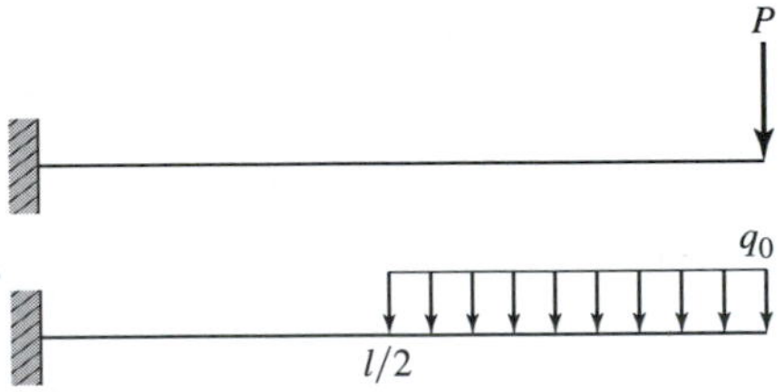


Figure P5-34

## Section 5-13

**5-35** (a) Use the Green's function approach to determine the deflection distribution in the cantilever beam shown in Fig. P5-35.

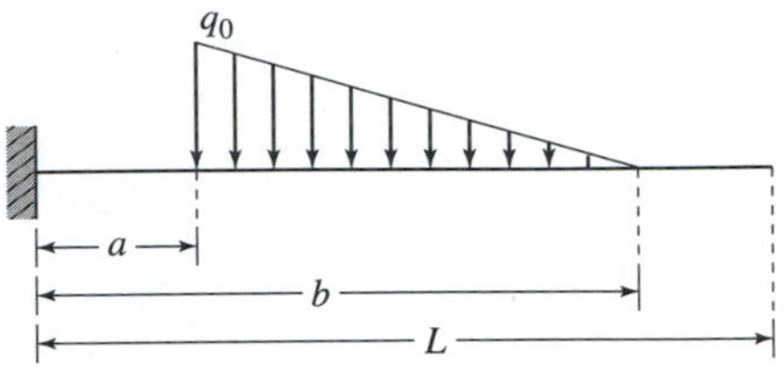


Figure P5-35

(b) Specialize your result for the case when $a = 0$ and $b = L$.

(c) Specialize your result for the case when $a = \frac{3}{4}L$ and $b = L$.

**5-36** Show that the Green's function used in Problem 5-38 is symmetric with respect to its two arguments.

**5-37** Use the Green's function approach to determine the deflection distribution in a simply supported beam for the two loading cases shown in Problem 5-33.

## REFERENCES


[1] Fung, Y. C., *Foundations of Solid Mechanics* (Prentice-Hall, Englewood Cliffs, NJ, 1965).

[2] Gere, J. M., and S. P. Timoshenko, *Mechanics of Materials,* 2nd ed. (PWS Publishers, Boston, MA, 1984).

[3] Kestin, J., *A Course in Thermodynamics,* vol. I (Blaisdell, Waltham, MA, 1966).
[4] Langhaar, H. L., *Energy Methods in Applied Mechanics* (Wiley, New York, 1962).
[5] Reddy, J. N., *Energy and Variational Methods in Applied Mechanics* (Wiley, New York, 1984).
[6] Roark, R. J., *Formulas for Stress and Strain,* 4th ed. (McGraw-Hill, New York, 1965).
[7] Shames, I. H., and C. L. Dym, *Energy and Finite Element Methods in Structural Mechanics* (McGraw-Hill, New York, 1985).
[8] Van Wylan, G. J., and R. E. Sonntag, *Fundamentals of Classical Thermodynamics,* 3rd ed. (Wiley, New York, 1985).
[9] Young, W. C., *Roark's Formulas for Stress and Strain,* 6th ed. (McGraw-Hill, New York, 1989).

# NUMERICAL METHODS I

Most problems in stress analysis are described by differential equations. Sometimes these can be solved analytically with little effort. More often, however, the geometry of the part to be analyzed, its material properties, and so on, introduce considerable complications, thus making an analytical solution impossible to obtain. In such situations numerical methods play a very important role. In this chapter we develop basic ideas for the methods of finite differences, iteration, and collocation. The numerical method of finite elements is treated separately in Chapter 7. In this chapter, for comparison purposes, we solve one simple example using various methods. Applications to stress analysis can be found in other chapters of this text.

## 6-1 METHOD OF FINITE DIFFERENCES

The goal of the finite differences method is the replacement of the differential equation by a system of algebraic equations. This is achieved by replacing the derivatives by the so-called difference quotients.

### 6-1-1 Application to Ordinary Differential Equations

We begin by solving an ordinary differential equation for the unknown $y(x)$, where $a \leq x \leq b$. First we divide the interval $a$, $b$ into $n$ equal* parts, as shown in Fig. 6-1,

$$h = \frac{b-a}{n} \tag{6-1}$$

*In general the parts do not have to be equal.

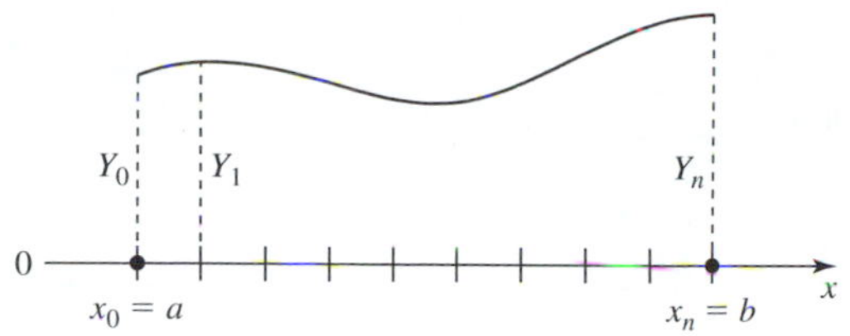


Figure 6-1

Next the endpoints of the subintervals thus obtained are labeled

$$x_0 = a, \quad x_1 = a + h, \quad x_2 = a + 2h, \ldots, \quad x_i = a + ih, \ldots, \quad x_n = a + nh = b \tag{6-2}$$

Points $x_i$ are called *pivotal points* (*pivots* for short), and $h$ is called a *pivotal interval*. Finally the approximate values of the unknown function $y(x)$ at $x = x_i$ are denoted by $Y_i$. Now these can be related to the derivatives of $y(x)$, which appear in the differential equation. There is a variety of ways of doing this. We choose the representation known as *central differences* (see, for example, [1]). The first step in this development will be to expand $y(x)$ into a Taylor series in the neighborhood of $x = x_0$. This series has the form

$$y(x_0 + h) = y(x_0) + hy'(x_0) + \frac{h^2}{2}y''(x_0) + \cdots \tag{6-3}$$

Next, replacing $h$ in Eq. (6-3) by $-h$, we obtain

$$y(x_0 - h) = y(x_0) - hy'(x_0) + \frac{h^2}{2}y''(x_0) + \cdots \tag{6-4}$$

Subtracting Eq. (6-4) from Eq. (6-3), neglecting terms with higher powers of $h$, and solving the resulting equation with respect to $y'(x)$ yields

$$y'(x_0) = \frac{1}{2h}[y(x_0 + h) - y(x_0 - h)] \tag{6-5}$$

Now we replace this by

$$\frac{1}{2h}\Delta Y_0 = \frac{1}{2h}(Y_1 - Y_{-1}) \tag{6-6}$$

where $\Delta Y_0$ is called the first difference and $\Delta Y_0/2h$ the first difference quotient of $y(x)$ at $x = x_0$. Here

$$Y_1 = y(x_0 + h), \qquad Y_{-1} = y(x_0 - h)$$

Thus we use the following expression to replace the first derivative at an arbitrary point $x_i$:

$$\left(\frac{\partial y}{\partial x}\right)_i = \frac{1}{2h}\Delta Y_i = \frac{1}{2h}(Y_{i+1} - Y_{i-1}) \tag{6-7}$$

where

$$Y_{i+1} = y(x_i + h), \qquad Y_{i-1} = y(x_i - h)$$

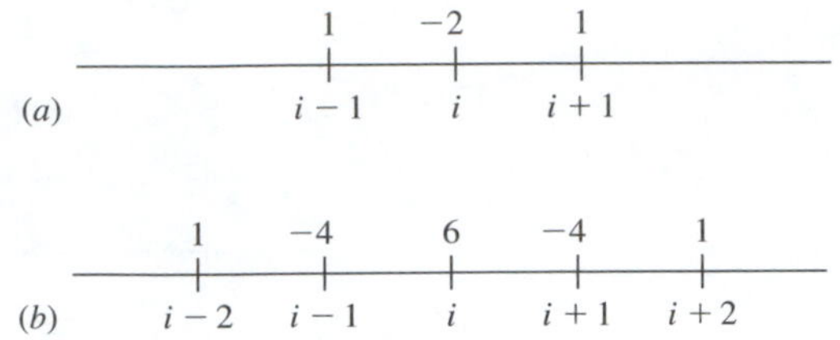


**Figure 6-2** Stencils corresponding to the second and fourth differences

The next step is to add Eqs. (6-3) and (6-4) and to solve the result for $y''(x)$,

$$y''(x_0) = \frac{1}{h^2}[y(x_0 + h) - 2y(x_0) + y(x_0 - h)] \tag{6-8}$$

In analogy to Eq. (6-6), this is written in the form

$$\frac{1}{h^2}\Delta^2 Y_0 = \frac{1}{h^2}(Y_1 - 2Y_0 + Y_{-1}) \tag{6-9}$$

and again at an arbitrary point,

$$\left(\frac{\partial^2 y}{\partial x^2}\right)_i = \frac{1}{h^2}\Delta^2 Y_i = \frac{1}{h^2}(Y_{i+1} - 2Y_i + Y_{i-1}) \tag{6-10}$$

where $\Delta^2 Y_i / h^2$ is the *second difference quotient* at $x = x_i$. Higher order difference quotients are readily evaluated by taking into account higher order derivatives in Eqs. (6-3) and (6-4). Thus, for example, the third-order difference quotient, which replaces $y'''(x)$, is given by

$$\left(\frac{\partial^3 y}{\partial x^3}\right)_i = \frac{1}{2h^3}\Delta^3 Y_i = \frac{1}{2h^3}(Y_{i+2} - 2Y_{i+1} + 2Y_{i-1} - Y_{i-2}) \tag{6-11}$$

and the fourth-order difference quotient used to replace $y^{\text{iv}}(x)$ becomes

$$\left(\frac{\partial^4 y}{\partial x^4}\right)_i = \frac{1}{h^4}\Delta^4 Y_i = \frac{1}{h^4}(Y_{i+2} - 4Y_{i+1} + 6Y_i - 4Y_{i-1} + Y_{i-2}) \tag{6-12}$$

and so on. To facilitate the application of finite difference equations, Milne [2] (see also [1]) introduced the concept of a *stencil*. This term is applied to an array of coefficients set out in a pattern corresponding to the points they are to multiply. Thus stencils corresponding to the second and fourth differences evaluated at a point *i*, Eqs. (6-10) and (6-12), are shown in Fig. 6-2. The stencils are particularly useful when solving partial difference equations, as will be seen in Section 6-1-2.

Now that we have developed expressions for finite difference quotients up to the fourth order we can easily solve any ordinary differential equation up to the fourth order. We begin by replacing the differential equation by a system of difference equations. Next these are evaluated at each of the pivots.* Finally, the boundary conditions are also replaced by difference equations.

---

*Usually they are evaluated at internal pivots. It is, however, sometimes necessary to use boundary pivots as well (see [1]).

## EXAMPLE 6-1

Use finite differences to solve the following second-order differential equation:

$$y''(x) + xy = x \tag{a}$$

for $0 \le x \le 1$. The boundary conditions are

$$y(0) = 0, \qquad y(1) = 0 \tag{b}$$

## SOLUTION

We begin by replacing Eq. ($a$) by a corresponding finite difference equation,

$$\frac{1}{h^2}(Y_{i+1} - 2Y_i + Y_{i-1}) + x_i Y_i = x_i, \qquad i = 0, 1, \ldots, n \tag{c}$$

In arriving at Eq. ($c$), Eq. (6-10) has been used. The boundary conditions ($b$) become

$$Y_0 = Y_n = 0 \tag{d}$$

At this point $n$ must be chosen, and we assume that $n = 4$, and thus $h = 0.25$. The pivotal points are now shown in Fig. 6-3. We thus have five unknowns: $Y_0$, $Y_1$, $Y_2$, $Y_3$, and $Y_4$. From Eq. ($d$), $Y_0 = Y_4 = 0$ and therefore the number of unknowns is reduced to three: $Y_1$, $Y_2$, and $Y_3$. Next Eq. ($c$) is applied to each of the pivots: $i = 1$, $i = 2$, and $i = 3$. Placing the pivot $i$ of the stencil at points 1, 2, and 3, respectively (see corresponding Fig. 6-3$b$, $c$, and $d$) we obtain the first part of Eq. ($c$). Consequently the difference equations are

$$(i = 1) \qquad Y_2 - 2Y_1 + Y_0 + 0.25^3 Y_1 = 0.25 \times 0.25^2$$

$$(i = 2) \qquad Y_3 - 2Y_2 + Y_1 + 0.25^2 \times 0.5Y_2 = 0.50 \times 0.25^2$$

$$(i = 3) \qquad Y_4 - 2Y_3 + Y_2 + 0.25^2 \times 0.75Y_3 = 0.75 \times 0.25^2$$

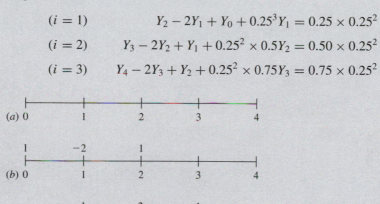

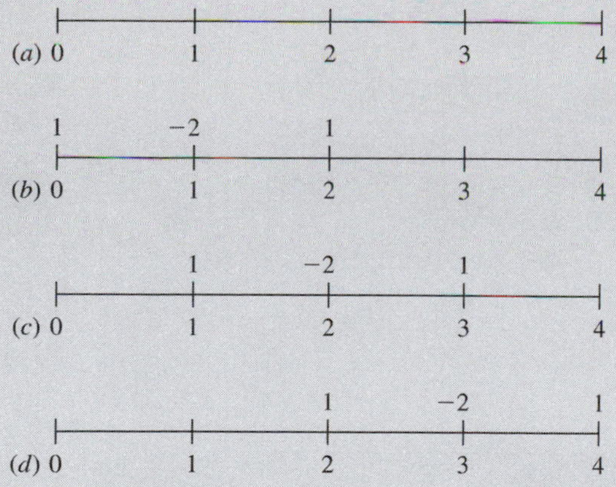


**Figure 6-3** Solving a second order differential equation: the pivotal points

and substituting Eq. ($d$) into these expressions,

$$\begin{aligned} -1.9375Y_1 + Y_2 &= 0.015625 \\ Y_1 - 1.96875Y_2 + Y_3 &= 0.03125 \\ Y_2 - 1.953125Y_3 &= 0.046875 \end{aligned} \tag{e}$$

The solution to Eqs. ($e$) is

$$Y_1 = -0.0428058, \qquad Y_2 = -0.0673114, \qquad Y_3 = -0.0584634$$

## 6-1-2 Application to Partial Differential Equations

We assume now that the unknown function $w(x, y)$ is the solution to a partial differential equation, and that $x$ and $y$ are Cartesian coordinates. We further assume, for simplicity, that the function $w(x, y)$ is defined in a rectangle $a \times b$ (Fig. 6-4). The development that follows is a generalization of the procedure described in Section 6-1-1. The first step in this development will be to cover the area in which the function $w(x, y)$ is defined by a rectangular mesh, as shown in Fig. 6-4. We assume that each rectangle of this mesh has a dimension $h$ by $l$, such that

$$h = \frac{a}{m}, \qquad l = \frac{b}{n} \tag{6-13}$$

and that

$$\beta = \frac{h}{l} \tag{6-14}$$

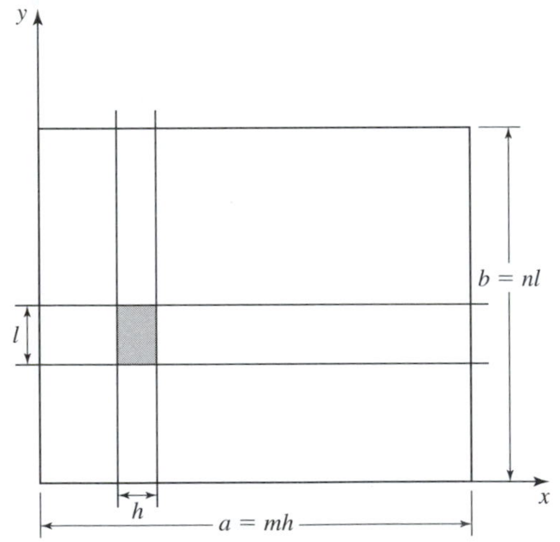


Figure 6-4 The area of definition of the function $w(x, y)$ and the rectangular mesh

where $m$ and $n$ are the total numbers of rectangles in the $x$ and $y$ directions, respectively. Here the *pivots* are the points of intersection of the mesh lines. Now assign the numbers $i$ and $k$ to an arbitrary pivot. These numbers specify its $x$, $y$ position. Both $i$ and $k$ run from 0 to $m$ and $n$, respectively, and they take only integer values,

$$0 \le i \le m, \qquad 0 \le k \le n$$

Since the goal is to replace partial derivatives of $w(x, y)$ in any differential equation by the corresponding difference quotients, let us examine $\partial w/\partial x$. To this end we fix $k$ and change $i$. This represents an increment in the $x$ direction. Then Eq. (6-7) becomes

$$\left(\frac{\partial w}{\partial x}\right)_{i,k} = \frac{1}{2h}\Delta_x W_{i,k} = \frac{1}{2h}(W_{i+1,k} - W_{i-1,k}) \tag{6-15}$$

where $\Delta_x W_{i,k}$ is the first difference quotient in the $x$ direction.* Similarly, to derive an expression corresponding to $\partial w/\partial y$, $i$ is fixed and $k$ is changed to give

$$\left(\frac{\partial w}{\partial y}\right)_{i,k} = \frac{1}{2l}\Delta_y W_{i,k} = \frac{1}{2l}(W_{i,k+1} - W_{i,k-1}) \tag{6-16}$$

The second-order partial derivatives $\partial^2 w/\partial x^2$ and $\partial^2 w/\partial y^2$ are readily evaluated using Eq. (6-10),

$$\begin{aligned}\left(\frac{\partial^2 w}{\partial x^2}\right)_{i,k} &= \frac{1}{h^2}(W_{i+1,k} - 2W_{i,k} + W_{i-1,k})\\ \left(\frac{\partial^2 w}{\partial y^2}\right)_{i,k} &= \frac{1}{l^2}(W_{i,k+1} - 2W_{i,k} + W_{i,k-1})\end{aligned} \tag{6-17}$$

which can also be written in terms of second difference quotients,

$$\left(\frac{\partial^2 w}{\partial x^2}\right)_{i,k} = \frac{\Delta_x^2 W_{i-1,k}}{h^2}, \qquad \left(\frac{\partial^2 w}{\partial y^2}\right)_{i,k} = \frac{\Delta_y^2 W_{i,k-1}}{l^2} \tag{6-18}$$

In general, an even-order partial derivative can be replaced by the following expression (see [1, p. 264]):

$$\left(\frac{\partial^{m+n} w}{\partial x^m \partial y^n}\right)_{i,k} = \frac{\Delta_x^m \Delta_y^n W_{i-\frac{m}{2},k-\frac{n}{2}}}{h^m l^n} \tag{6-19}$$

We now use Eq. (6-19) to determine the replacement for $\partial^4 w/\partial x^4$. Since $m = 4$ and $n = 0$ we get

$$\left(\frac{\partial^4 w}{\partial x^4}\right)_{i,k} = \frac{\Delta_x^4 W_{i-2,k}}{h^4} \tag{6-20}$$

This can be written as

$$\left(\frac{\partial^4 w}{\partial x^4}\right)_{i,k} = \frac{\Delta_x^2 \left(\Delta_x^2 W_{i-2,k}\right)}{h^4}$$

*This is not a unique definition of the first difference quotient.

and the first of Eqs. (6-18) is then used in the result,

$$\left(\frac{\partial^4 w}{\partial x^4}\right)_{i,k} = \frac{\Delta_x^2\left(\Delta_x^2 W_{i-2,k}\right)}{h^4} = \frac{1}{h^4}\Delta_x^2(W_{i,k} - 2W_{i-1,k} + W_{i-2,k})$$

$$= \frac{1}{h^4}[(W_{i+2,k} - 2W_{i+1,k} + W_{i,k}) - 2(W_{i+1,k} - 2W_{i,k} + W_{i-1,k})$$

$$+ (W_{i,k} - 2W_{i-1,k} + W_{i-2,k})]$$

$$= \frac{1}{h^4}(W_{i+2,k} - 4W_{i+1,k} + 6W_{i,k} - 4W_{i-1,k} + W_{i-2,k}) \tag{6-21}$$

The values of the other derivatives at *i*, *k* are readily obtained,

$$\left(\frac{\partial^4 w}{\partial y^4}\right)_{i,k} = \frac{1}{l^4}(W_{i,k+2} - 4W_{i,k+1} + 6W_{i,k} - 4W_{i,k-1} + W_{i,k-2}) \tag{6-22}$$

$$\left(\frac{\partial^4 w}{\partial x^2 \partial y^2}\right)_{i,k} = \frac{1}{h^2 l^2}(W_{i+1,k+1} - 2W_{i+1,k} + W_{i+1,k-1} - 2W_{i,k+1}$$

$$+ 4W_{i,k} - 2W_{i,k-1} + W_{i-1,k+1} - 2W_{i-1,k} + W_{i-1,k-1}) \tag{6-23}$$

With Eq. (6-7) and the first of Eqs. (6-17) the value of the third-order partial derivative $\partial^3 w/\partial x^3$ at *i*, *k* can be written as

$$\left(\frac{\partial^3 w}{\partial x^3}\right)_{i,k} = \frac{\partial}{\partial x}\left(\frac{\partial^2 w}{\partial x^2}\right)_{i,k} = \frac{1}{2h}\left[\left(\frac{\partial^2 w}{\partial x^2}\right)_{i+1,k} - \left(\frac{\partial^2 w}{\partial x^2}\right)_{i-1,k}\right] \tag{6-24}$$

In order to evaluate the terms appearing on the right-hand sides of Eqs. (6-24), Eq. $(6\text{-}17)_1$ has to be used. To do so, we replace *i* by $i + 1$ or by $i - 1$ in Eq. $(6\text{-}17)_1$ and obtain, respectively, the first or the second term in Eqs. (6-24). Consequently, Eqs. (6-24) become

$$\left(\frac{\partial^3 w}{\partial x^3}\right)_{i,k} = \frac{1}{2h^3}(W_{i+2,k} - 2W_{i+1,k} + 2W_{i-1,k} - W_{i-2,k}) \tag{6-25}$$

There are few alternative expressions for the second-order mixed derivative (see [1, p. 264] and [3, p. 146]). Following Mitchell we choose

$$\left(\frac{\partial^2 w}{\partial x \partial y}\right)_{i,k} = \frac{1}{4hl}(W_{i+1,k+1} - W_{i+1,k-1} - W_{i-1,k+1} + W_{i-1,k-1}) \tag{6-26}$$

It is now convenient to introduce stencils (see Section 6-1-1). For $\nabla^2 w$ at the pivot *i*, *k* we have

$$(\nabla^2 w)_{i,k} = \frac{1}{h^2}\left(\frac{\partial^2 w}{\partial x^2}\right)_{i,k} + \frac{1}{l^2}\left(\frac{\partial^2 w}{\partial y^2}\right)_{i,k}$$

$$= \frac{1}{h^2}[-2(1+\beta^2)W_{i,k} + W_{i+1,k} + W_{i-1,k} + \beta^2(W_{i,k+1} + W_{i,k-1})] \tag{6-27}$$

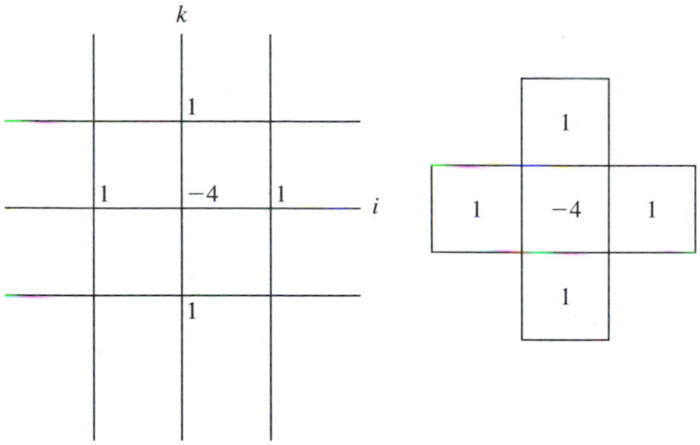


Figure 6-5

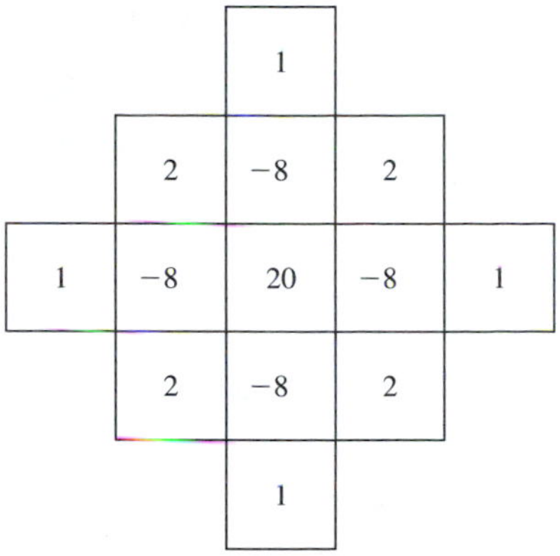


Figure 6-6 Stencil for a square mesh

In arriving at Eq. (6-27), Eqs. (6-17) have been used. A stencil corresponding to this expression for the case of a square mesh ($\beta = 1$) is shown in Fig. 6-5. To calculate $\nabla^2\nabla^2 w$ at $i$, $k$,

$$(\nabla^2\nabla^2 w)_{i,k} = \left(\frac{\partial^4 w}{\partial x^4}\right)_{i,k} + 2\left(\frac{\partial^4 w}{\partial x^2 \partial y^2}\right)_{i,k} + \left(\frac{\partial^4 w}{\partial y^4}\right)_{i,k} \tag{6-28}$$

we substitute Eqs. (6-21), (6-22), and (6-23) into Eq. (6-28). This yields the following expression:

$$\begin{aligned}(\nabla^2\nabla^2 w)_{i,k} = \frac{1}{h^2}[&2(3 + 4\beta^2 + 3\beta^4)W_{i,k} - 4(1 + \beta^2)(W_{i+1,k} + W_{i-1,k}) \\ &- 4\beta^2(1 + \beta^2)(W_{i,k+1} + W_{i,k-1}) + 2\beta^2(W_{i+1,k+1} + W_{i+1,k-1} \\ &+ W_{i-1,k+1} + W_{i-1,k-1}) + W_{i+2,k} + W_{i-2,k} + \beta^4(W_{i,k+2} + W_{i,k-2})]\end{aligned} \tag{6-29}$$

For $h = l$ (a square mesh) this becomes

$$\begin{aligned}(\nabla^2\nabla^2 w)_{i,k} = 20W_{i,k} &- 8(W_{i+1,k} + W_{i,k+1} + W_{i-1,k} + W_{i,k-1}) \\ &+ 2(W_{i+1,k+1} + W_{i-1,k+1} + W_{i+1,k-1} + W_{i-1,k-1}) \\ &+ W_{i+2,k} + W_{i-2,k} + W_{i,k+2} + W_{i,k-2}\end{aligned} \tag{6-30}$$

and the stencil is shown in Fig. 6-6. The pivotal deflection $W_{i,k}$ appearing in Eq. (6-30), and in previous equations given in this section, has two subscripts. They identify the position of the pivot but make it difficult to use matrix notation. However, we can readily remove this obstacle by assigning to the pivots arbitrary, preferably sequential numbers. Now $W_{i,k}$ is replaced by $W_i$, where $i$ stands for the number associated with the pivot. Thus the right-hand side of an expression such as Eq. (6-30) can be put in the form of a product of two matrices,

$$\begin{bmatrix} c_{11} & c_{12} & \cdot & \cdot & \cdot & c_{1n} \\ \cdot & \cdot & \cdot & \cdot & \cdot & \cdot \\ c_{n1} & c_{n2} & \cdot & \cdot & \cdot & c_{nn} \end{bmatrix} \cdot \begin{Bmatrix} W_1 \\ W_2 \\ \cdot \\ W_n \end{Bmatrix} \tag{6-31}$$

where $n$ is the number of internal pivots. We now can easily use the stencils developed before. To do so we place the center of the stencil [such as box 20 for the stencil of Eq. (6-30)] at the pivot under consideration and develop an appropriate finite difference equation. Then this procedure is repeated for all internal pivots. This will be explained in detail in Example 6-2.

## EXAMPLE 6-2

Solve the partial differential equation

$$\nabla^2\nabla^2 w(x, y) = q/D \tag{a}$$

defined over a square region $a \times a$ and subject to the boundary conditions

$$w = 0 \text{ at } x = 0, \quad w = 0 \text{ at } x = a, \quad w = 0 \text{ at } y = 0, \quad w = 0 \text{ at } y = a \tag{b}$$

$$\frac{\partial^2 w}{\partial x^2} = 0 \text{ at } x = 0, \quad \frac{\partial^2 w}{\partial x^2} = 0 \text{ at } x = a,$$
$$\frac{\partial^2 w}{\partial y^2} = 0 \text{ at } y = 0, \quad \frac{\partial^2 w}{\partial y^2} = 0 \text{ at } y = a \tag{c}$$

where $q$ and $D$ are constants. Equations ($a$), ($b$), and ($c$) describe the bending of a thin rectangular plate subject to a uniform load $q$, as will be shown in Chapter 10.

### SOLUTION

We now introduce a square mesh and replace the derivatives appearing in Eq. ($a$) by finite differences. Thus using Eq. (6-30), Eq. ($a$) becomes

$$\begin{aligned} 20W_{i,k} - 8(W_{i+1,k} + W_{i,k+1} + W_{i-1,k} + W_{i,k-1}) + 2(W_{i+1,k+1} + W_{i-1,k+1} \\ + W_{i+1,k-1} + W_{i-1,k-1}) + W_{i+2,k} + W_{i-2,k} + W_{i,k+2} + W_{i,k-2} = Q \end{aligned} \tag{d}$$

in which $Q = qh^4/D$. Equation ($d$) is applicable at all pivots. Because of the symmetry of the plate and load, the displacement $w(x, y)$ is symmetric with respect to the lines $x = a/2$ and $y = a/2$ (Fig. 6-7). Therefore it is sufficient to consider only the shaded portion of the plate (Fig. 6-7). At this point we choose a grid with $h = l = a/4$, as shown in Fig. 6-8. Next we label the pivots: the internal pivots are $\alpha, \ldots, \mu$, the boundary pivots are $A, \ldots, I$. Note that the labeling of the pivots is affected by the symmetry of the deflection with respect to the diagonal E-$\mu$. That means that the presence of the

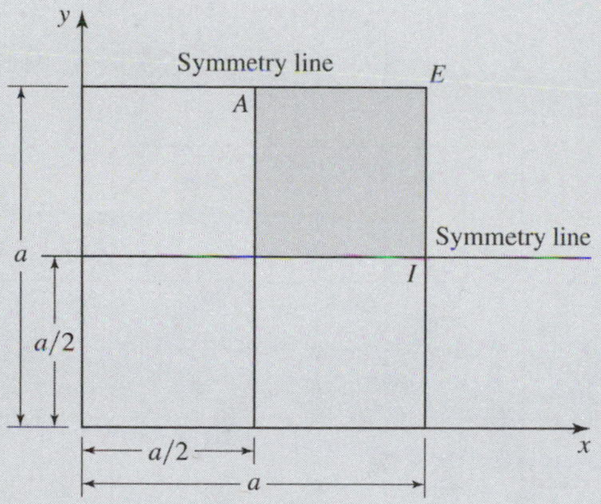


**Figure 6-7**

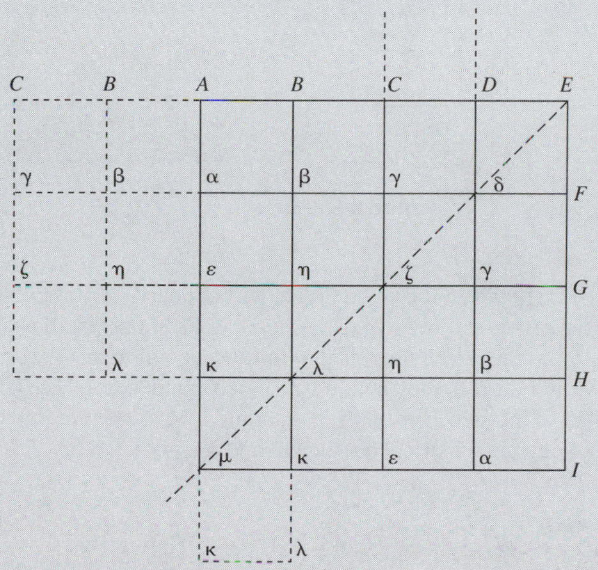


**Figure 6-8** The grid and the pivots

same label at different pivots indicates equality of deflections at those pivots. Finally we replace the boundary conditions (*b*) by suitable difference conditions,

$$W_A = W_B = W_C = W_D = W_E = W_F = W_G = W_H = W_I = 0 \tag{e}$$

while, using Eqs. (6-17), boundary conditions (*c*) become

$$\begin{aligned} W_\alpha - 2W_A + W_{\bar{\alpha}} &= 0, \qquad W_\beta - 2W_B + W_{\bar{\beta}} = 0 \\ W_\gamma - 2W_C + W_{\bar{\gamma}} &= 0, \qquad W_\delta - 2W_D + W_{\bar{\delta}} = 0 \end{aligned} \tag{f}$$

We note that the pivots with overlined indices, $\bar{\alpha}, \ldots, \bar{\delta}$, are located outside the area of

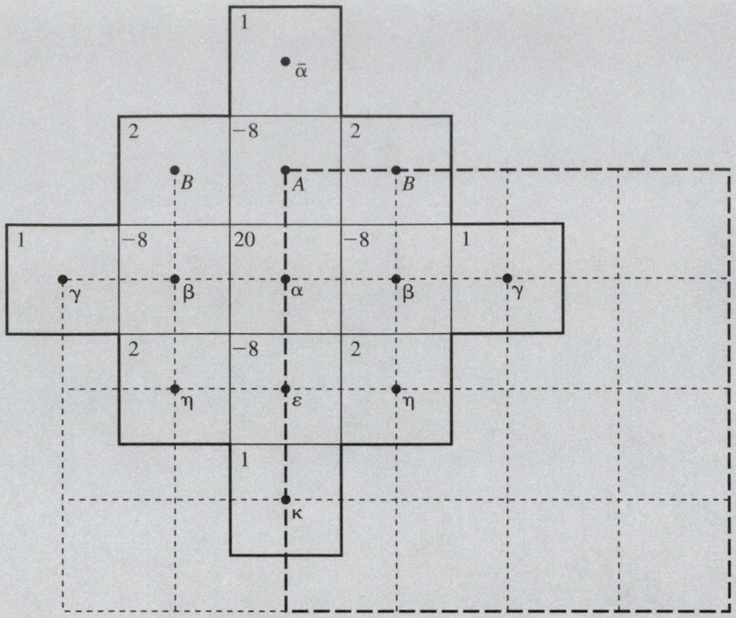


**Figure 6-9** Central box

definition of $w(x, y)$. However, the same pivots will appear as the result of placing the stencil at the internal pivots $\alpha$ through $\delta$. Thus deflections at these will be eliminated by using conditions $(f)$. In the developments that follow, we will place the stencil at all internal pivots shown in Fig. 6-8, thus obtaining the left-hand side of Eq. $(d)$. First let us place the central box of the stencil from Fig. 6-6 at pivot $\alpha$, as seen in Fig. 6-9. We show in each stencil box both the numerical coefficient and the pivot's label. We then find that in this case Eq. $(d)$ becomes

$$\begin{aligned} W_{\bar{\alpha}} + 2W_B - 8W_A + 2W_B + W_\gamma - 8W_\beta + 20W_\alpha \\ - 8W_\beta + W_\gamma + 2W_\eta - 8W_\varepsilon + 2W_\eta + W_\kappa = Q \end{aligned} \tag{g}$$

Using conditions $(e)$ and $(f)$ to set $W_A = W_B = 0$ and to eliminate from Eq. $(g)$ $W_{\bar{\alpha}}$ in favor of $W_\alpha$, we get

$$19W_\alpha - 16W_\beta + 2W_\gamma + 4W_\eta - 8W_\varepsilon + W_\kappa = Q \tag{h}$$

The next step is to place the center of the stencil at pivot $\beta$, as shown in Fig. 6-10. This gives

$$\begin{aligned} W_{\bar{\beta}} + 2W_A - 8W_B + 2W_C + W_\beta - 8W_\alpha + 20W_\beta \\ - 8W_\gamma + W_\delta + 2W_\varepsilon - 8W_\eta + 2W_\zeta + W_\lambda = Q \end{aligned}$$

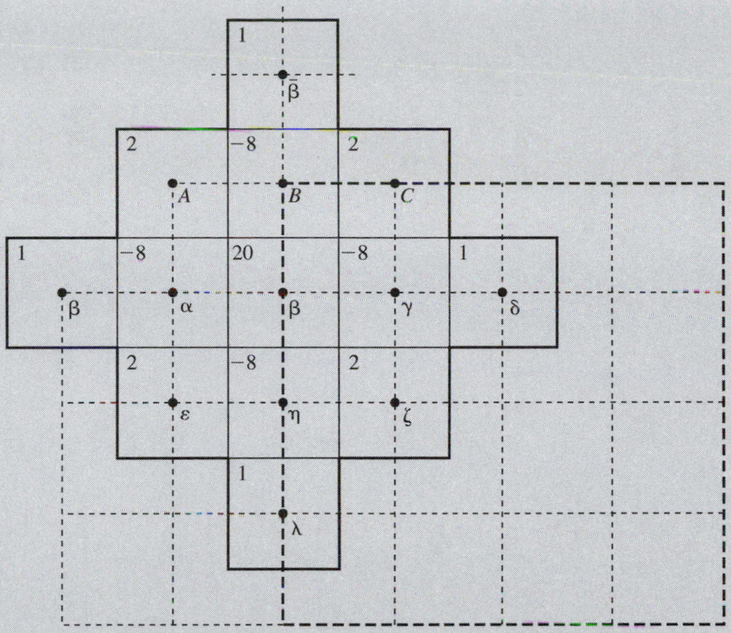

**Figure 6-10** Center of the stencil at the pivot $b$

With Eqs. ($e$) and ($f$),

$$-8W_{\alpha} + 20W_{\beta} - 8W_{\gamma} + W_{\delta} + 2W_{\varepsilon} - 8W_{\eta} + 2W_{\zeta} + W_{\lambda} = Q \tag{i}$$

Proceeding in a similar manner with the other pivots we get

$$\begin{aligned}
W_{\alpha} - 8W_{\beta} + 21W_{\gamma} - 8W_{\delta} + 3W_{\eta} - 8W_{\zeta} &= Q \\
2W_{\beta} - 16W_{\gamma} + 18W_{\delta} + 2W_{\zeta} &= Q \\
-8W_{\alpha} + 4W_{\beta} + 20W_{\varepsilon} - 16W_{\eta} + 2W_{\zeta} - 8W_{\kappa} + 4W_{\lambda} + W_{\mu} &= Q \\
2W_{\alpha} - 8W_{\beta} + 3W_{\gamma} - 8W_{\varepsilon} + 23W_{\eta} - 8W_{\zeta} + 3W_{\kappa} - 8W_{\lambda} &= Q \\
4W_{\beta} - 16W_{\gamma} + 2W_{\delta} + 2W_{\varepsilon} - 16W_{\eta} + 20W_{\zeta} + 2W_{\lambda} &= Q \\
W_{\alpha} - 8W_{\varepsilon} + 6W_{\eta} + 25W_{\kappa} - 16W_{\lambda} - 8W_{\mu} &= Q \\
2W_{\beta} + 4W_{\varepsilon} - 16W_{\eta} + 2W_{\zeta} - 16W_{\kappa} + 22W_{\lambda} + 2W_{\mu} &= Q \\
4W_{\varepsilon} - 32W_{\kappa} + 8W_{\lambda} + 20W_{\mu} &= Q
\end{aligned} \tag{j}$$

Thus we have 10 equations [($h$), ($i$), and ($j$)] in 10 unknown deflections, $W_{\alpha}, \ldots, W_{\mu}$. When written in matrix form, these equations become

$$[\boldsymbol{C}]\{\boldsymbol{W}\} = \{\boldsymbol{Q}\} \tag{k}$$

where

$$[C] = \begin{bmatrix} 19 & -16 & 2 & 0 & -8 & 4 & 0 & 1 & 0 & 0 \\ -8 & 20 & -8 & 1 & 2 & -8 & 2 & 0 & 1 & 0 \\ 1 & -8 & 21 & -8 & 0 & 3 & -8 & 0 & 0 & 0 \\ 0 & 2 & -16 & 18 & 0 & 0 & 2 & 0 & 0 & 0 \\ -8 & 4 & 0 & 0 & 20 & -16 & 2 & -8 & 4 & 1 \\ 2 & -8 & 3 & 0 & -8 & 23 & -8 & 3 & -8 & 0 \\ 0 & 4 & -16 & 2 & 2 & -16 & 20 & 0 & 2 & 0 \\ 1 & 0 & 0 & 0 & -8 & 6 & 0 & 25 & -16 & -8 \\ 0 & 2 & 0 & 0 & 4 & -16 & 2 & -16 & 22 & 2 \\ 0 & 0 & 0 & 0 & 4 & 0 & 0 & -32 & 8 & 20 \end{bmatrix} \qquad (l)$$

and

$$\{W\} = \begin{Bmatrix} W_\alpha \\ W_\beta \\ W_\gamma \\ W_\delta \\ W_\varepsilon \\ W_\eta \\ W_\zeta \\ W_\kappa \\ W_\lambda \\ W_\mu \end{Bmatrix}, \qquad \{Q\} = \begin{Bmatrix} Q \\ Q \\ Q \\ Q \\ Q \\ Q \\ Q \\ Q \\ Q \\ Q \end{Bmatrix} \qquad (m)$$

where $Q = qh^4/D = q(a/8)^4/D = qa^4/4096$. Solving this system for the unknown deflections gives

$$\begin{aligned} &W_\alpha = 6.663Q, \quad W_\beta = 6.207Q, \quad W_\gamma = 4.860Q, \quad W_\delta = 2.714Q \\ &W_\varepsilon = 12.03Q, \quad W_\eta = 11.20Q, \quad W_\zeta = 8.743Q, \quad W_\kappa = 15.44Q \\ &W_\lambda = 14.36Q, \quad W_\mu = 16.61Q \end{aligned} \qquad (n)$$

Maximum deflection appears, as expected, at the center pivot $\mu$,

$$\max w(x, y) = W_\mu = 16.61Q = 0.004055\frac{qa^4}{D}$$

This compares very well with the exact result $\max w(x, y) = 0.00406qa^4/D$ obtained in Section 10-3.

## 6-2 METHOD OF ITERATION

The advantage of the method of iteration is its simplicity. We begin by writing the differential equation to be solved in the form

$$y^{(n)}(x) = f\left(x, y, y', \ldots, y^{(n-1)}\right) \tag{6-32}$$

where $y^{(n)}$ is the highest order derivative and $f(x, y, y', \ldots, y^{(n-1)})$ represents all remaining terms, and

$$0 \leq x \leq 1$$

It is assumed that $n$ boundary conditions are given at the ends of the interval [0, 1]. Next an initial guess of the solution (the *zeroth iteration*) is made,

$$y(x) = Ay_0(x) \tag{6-33}$$

where $A$ is a constant and $y_0(x)$ satisfies the boundary conditions. Substituting Eq. (6-33) into Eq. (6-32) and integrating we get the *first iteration,*

$$y(x) = y_1(x) \tag{6-34}$$

Here $y_1(x)$ includes constants of integration evaluated using the boundary conditions. Now the constant $A$ is determined by equating the values of $y_0(x)$ and $y_1(x)$ at any* point $x$ inside [0, 1]. We now repeat this procedure several times, comparing successive approximations at a number of selected points. We stop this process when the comparison is satisfactory.

Note that (1) the process is not necessarily convergent and (2) if it is convergent, the selection of the zeroth iteration $y_0(x)$ may affect the rate of convergence.

### EXAMPLE 6-3

Use the iteration method to solve the differential equation given in Example 6-1,

$$y''(x) + xy = x, \qquad \text{with } y = 0 \text{ at } x = 0, \text{ and } y = 0 \text{ at } x = 1 \tag{a}$$

### SOLUTION

First we represent the differential equation in the form

$$y'' = -xy + x \tag{b}$$

The next step is to select $y_0(x)$. Assume, for instance,

$$y_0(x) = Ax(1 - x) \tag{c}$$

---

*Preferably a point where $y(x)$ is expected to reach maximum.

which satisfies the boundary conditions. Equation (*c*) is now substituted into the right-hand side of Eq. (*b*) to give

$$y_1'' = -Ax^2(1-x) + x = x - Ax^2 + Ax^3 \qquad (d)$$

Integrating this expression twice we get

$$y_1(x) = \frac{1}{6}x^3 - \frac{1}{12}Ax^4 + \frac{1}{20}Ax^5 + C_1x + C_2 \qquad (e)$$

Then, substituting Eq. (*e*) into the boundary conditions (*a*), we get the constants $C_1$ and $C_2$,

$$C_1 = -\frac{1}{6} + \frac{1}{30}A, \qquad C_2 = 0 \qquad (f)$$

Next Eqs. (*f*) are substituted into Eq. (*e*), and the expected maximum values of $y_0(x)$ and $y_1(x)$ are then compared,

$$y_0\left(\tfrac{1}{2}\right) = y_1\left(\tfrac{1}{2}\right)$$

to give

$$A = -0.2637 \qquad (g)$$

Finally, using Eqs. (*f*) and (*g*), Eq. (*e*) becomes

$$y_1(x) = -0.01318x^5 + 0.02197x^4 + 0.16667x^3 - 0.17546x \qquad (h)$$

Proceeding in a similar manner with $y_1(x)$, we get

$$y_2''(x) = -x(-0.01318x^5 + 0.02197x^4 + 0.16667x^3 - 0.17546x) + x$$

and therefore, integrating and using boundary conditions (*a*),

$$y_2(x) = 0.00235x^8 - 0.00523x^7 - 0.005556x^6 + 0.0146x^4 + 0.1667x^3 - 0.1776x \qquad (i)$$

The values of consecutive iterations at pivots are compiled in Table 6-1.

Inspection of these results reveals that although the initial guess was wrong, the subsequent iterations show a tendency to converge. This opinion is reinforced by the results obtained in Section 6-1 (see also Example 6-4, which follows). Normally we would not have the luxury of knowing the results obtained by using the other method. We would then continue the iteration process until a satisfactory agreement between two consecutive iterations has been reached.

**TABLE 6-1**

| $x$ | $y_0(x)$ | $y_1(x)$ | $y_2(x)$ |
|---|---|---|---|
| 0.25 | 0.1875 | −0.0412 | −0.0417 |
| 0.50 | 0.2500 | −0.0659 | −0.0671 |
| 0.75 | 0.1875 | −0.0575 | −0.0593 |

## 6-3 METHOD OF COLLOCATION

In the method of collocation it is first assumed that the solution to the differential equation can be approximated by a judiciously selected function of the variable $x$ and of a number of parameters $a_1, a_2, \ldots, a_n$,

$$y(x) \simeq y(x, a_1, a_2, \ldots, a_n) \tag{6-35}$$

The next step is to choose $n$ points, called *collocation points*. These points must be located within the range of definition of $x$ and usually distributed uniformly within this range.* Finally, substituting Eq. (6-35) into the differential equation to be solved gives an equation depending on $n$ parameters $a_1, a_2, \ldots, a_n$,

$$f(x, a_1, a_2, \ldots, a_n) = 0 \tag{6-36}$$

We then find the unknown parameters requiring that Eq. (6-36) be satisfied at the collocation points. Therefore we have finally

$$f(x_1, a_1, a_2, \ldots, a_n) = 0, \quad f(x_2, a_1, a_2, \ldots, a_n) = 0, \ldots, \quad f(x_n, a_1, a_2, \ldots, a_n) = 0 \tag{6-37}$$

which is a system of $n$ linear algebraic equations in $n$ unknown parameters.

### EXAMPLE 6-4

The problem solved in Example 6-1 is repeated again,

$$y'' + xy = x, \qquad y(0) = y(1) = 0 \tag{a}$$

### SOLUTION

We assume that the solution is a polynomial,

$$y(x) = \sum_{n=1}^{3} a_n x^n (1 - x) \tag{b}$$

Substituting this expression into Eq. ($a$) we get

$$a_1(-2 + x^2 - x^3) + a_2(2 - 6x + x^3 - x^4) + a_3(6x - 12x^2 + x^4 - x^5) = x \tag{c}$$

Now let us choose three collocation points,

$$x_1 = 0.25, \qquad x_2 = 0.50, \qquad x_3 = 0.75 \tag{d}$$

*There are other, more sophisticated varieties of the collocation method. The reader is advised to consult the references.

TABLE 6-2

| $x$ | Iteration $y_0(x)$ | $y_1(x)$ | $y_2(x)$ | Finite Differences | Collocation | Polymath |
|---|---|---|---|---|---|---|
| 0.00 | 0.0000 | 0.0000 | 0.0000 | 0.0000 | 0.0000 | 0.0000 |
| 0.25 | 0.1875 | −0.0412 | −0.0417 | −0.0428 | −0.0412 | −0.0412 |
| 0.50 | 0.2500 | −0.0659 | −0.0671 | −0.0673 | −0.0661 | −0.0661 |
| 0.75 | 0.1875 | −0.0575 | −0.0593 | −0.0585 | −0.0578 | −0.0571 |
| 1.00 | 0.0000 | 0.0000 | 0.0000 | 0.0000 | 0.0000 | 0.0000 |

With these substituted, in sequence, into Equation ($c$), we obtain three algebraic equations,

$$\begin{aligned} -1.953125a_1 + 0.511719a_2 + 0.752930a_3 &= 0.25 \\ -1.875000a_1 - 0.937500a_2 + 0.031250a_3 &= 0.50 \\ -1.859375a_1 - 2.394531a_2 - 2.170898a_3 &= 0.75 \end{aligned} \tag{e}$$

We then find

$$a_1 = -0.173529, \qquad a_2 = -0.185998, \qquad a_3 = 0.008307 \tag{f}$$

Thus using Eq. ($b$), the approximate solution becomes

$$y(x) = -0.0173529x - 0.012469x^2 + 0.194305x^3 - 0.008307x^4 \tag{g}$$

The values of $y(x)$ obtained here and in Examples 6-1 and 6-2 are compiled in Table 6-2, where they are also compared with the results calculated using the software Polymath* (see also [4]).

*Developed by Professors M. Shacham of Ben Gurion University, Beer-Sheva, Israel, and M. B. Cutlip of the University of Connecticut, Storrs. Available from the CACHE Corporation, P. O. Box 7939, Austin, TX 78713-7939.

## PROBLEMS

**6-1** Use finite differences to solve the following differential equations with associated boundary conditions. Compare the results with exact solutions.

(a) $y'' + y = x, \quad y = y(x),$
$0 \le x \le 1, \quad y(0) = 0, \quad y(1) = 0$

(b) $y'' + y = x, \quad y = y(x),$
$0 \le x \le 1, \quad y'(0) = 0, \quad y(1) = 0$

(c) $y'' + y = x, \quad y = y(x),$
$0 \le x \le 1, \quad y(0) = 0, \quad y'(1) = 0$

**6-2** Use finite differences to solve the following differential equation:

$$\frac{d^4 y}{dx^4} = q, \qquad y = y(x), \qquad q = \text{constant}$$

with the boundary conditions $y(0) = 0$, $y''(0) = 0$, $y(1) = 0$, and $y''(1) = 0$.

**6-3** Find the finite difference expression corresponding to

$$\nabla^2 \nabla^2 \nabla^2 w(x, y)$$

**6-4** Draw a stencil for Problem 6-3. Assume $h = 1$.

**6-5** Use the iteration method to solve the differential equations given in Problems 6-1(a)–(c).

**6-6** Use the collocation method to solve the differential equations given in Problems 6-1(a)–(c).

## REFERENCES


[1] Collatz, L., *The Numerical Treatment of Differential Equations* (Springer, Berlin, 1960).

[2] Milne, W. E., *Numerical Solution of Differential Equations* (Wiley, New York and London, 1953).

[3] Mitchell, A. R., *Computational Methods in Partial Differential Equations* (Wiley, London, 1969).

[4] Shacham, M., and M. B. Cutlip, *Chem. Eng. Education,* 22, pp. 18ff. (1988).

# NUMERICAL METHODS II: FINITE ELEMENTS

In this chapter we will develop a method of numerical analysis referred to as the *finite element method*. It will provide us with a general method for analyzing bodies of complex shapes such as those commonly encountered by engineers.

## 7-1 INTRODUCTION

Most of the remaining chapters of this book deal with various analytical techniques for analyzing problems concerning the stress and deformation of solid bodies. For the most part the goal in these chapters is to obtain an exact solution to the governing equations for a particular problem. This is possible as long as the geometry—that is, the shape of the boundary—is a simple shape, which can be described using a standard coordinate system. As we discussed in Chapter 6, it is the shape of the boundary that determines the coordinate system to be used. It might be preferable to solve the differential equations for each of the problems in the Cartesian coordinate system if for no other reason than for the great amount of experience we would gain working in this coordinate system. However, application of the boundary conditions can only be accomplished with a great deal of difficulty unless the body happens to be rectangular with sides parallel to the Cartesian axes. For example, to specify a constant pressure $p$ on the surface of a cylinder of radius $b$, we simply write $\sigma_{rr} = -p$ at $r = b$ in cylindrical coordinates. The same condition is quite complicated if written in Cartesian coordinates.

Now in the practice of engineering it is a rare situation in which the shape of the body we wish to analyze conforms to one of the familiar coordinate systems. Usually we are asked to design and analyze such diverse items as automobile crankshafts, gears, turbine engine blades, pump casings, centrifugal impellers, and numerous other devices that cannot be conveniently described using standard coordinate systems. In analyzing complex shapes we are usually forced to abandon the idea of obtaining an exact solution, and we use an approximate method of solution that has the ability to handle complex shapes. In so doing we usually replace the actual problem with a model of the actual problem and then analyze the model. How well the solution for the model represents the actual solution depends on the quality of the assumptions used in developing the model. For example, in the Ritz method the accuracy of the solution when compared with the exact solution depends on the choice of functions used to represent the displacement.

## 7-2 TWO-DIMENSIONAL FRAMES

We will define a *frame* as a structure that consists of several *members,* each of which can be considered, for analysis purposes, to be a beam. Generally speaking, at least some of the members have their ends welded or bolted in such a way that there is no relative motion—either translation or rotation—between the members at the point of attachment. At such points of attachment we assume that there is both displacement compatibility and slope compatibility among all members that are joined at that point.

In a *two-dimensional frame,* or *plane frame,* all members of the structure and all applied loads lie in a common plane both before and after deformation. (Structures in which all members lie in a plane, and have loads applied normal to that plane, are called *grids*.) This implies that the loads are applied at the shear center of each member, and that one of the principal axes of inertia of each member lies in the common plane.

In order to facilitate the developments that follow, we will assume that, at all joints of the structure, compatibility of displacements and slopes exists among *all* members forming the joint, that the weight of the members is negligible, and that loads are applied to the structure only at its joints. While these assumptions may seem to seriously impair our ability to solve problems of practical interest, it turns out that once the basic method of solution is formulated, the modifications required to overcome these imposed constraints are easily accomplished.

Let us agree that, for our purposes, the term "force" includes both linear forces and rotational forces (moments), and that the term "displacement" includes both linear displacements and rotational displacements (slope). Thus at each joint of the structure, we can apply three forces: two perpendicular linear forces and one rotational force. Likewise, there can occur at each joint of the structure three displacements: two linear displacements in perpendicular directions and one rotational displacement.

Figure 7-1 shows a typical frame structure. Note that at each joint in the structure, either a displacement is known and the corresponding force is to be determined, or vice versa. We cannot specify both a displacement and the corresponding force at a joint, nor can we leave both unspecified. For example, in Fig. 7-1, at joint 1 the displacements in the $X$ and $Y$ directions as well as the slope are zero. The corresponding reaction forces are unknown. At joint 3 the applied moment is 1200 in-lb while the forces applied in the $X$ and $Y$ directions are zero. Here the displacements and slope must be determined. Likewise, at joint 5 the displacement in the $Y$ direction is zero, and the corresponding reaction force is unknown, but because of the rollers, the force in the $X$ direction is zero, and the corresponding displacement is to be determined. Since joint 5 is prevented from rotating, we know its slope is zero, but the resulting moment is not known. Our goal is to develop a method of analysis whereby we can *systematically* determine the unknown displacements and reaction forces, as well as the internal forces in each member.

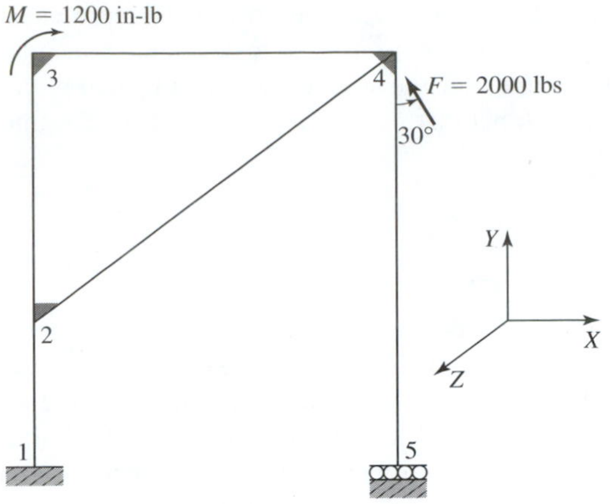


Figure 7-1

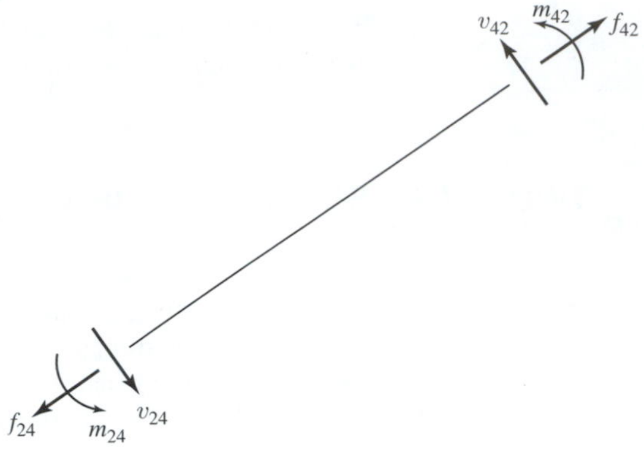


Figure 7-2

Figure 7-2 shows the diagonal member of the structure in Fig. 7-1, along with the internal forces acting at the ends of the member. If we knew the values of these end forces then, from equilibrium considerations, we could determine the internal forces at any other point in the member. On the other hand, if we knew the displacements of both ends of the member, then we would have a set of boundary conditions to use in the differential equations governing bending and extension of the beam. The solution of these equations would yield internal forces at other points in the member. Thus if we could determine either forces or displacements at the ends of each member, we would be able to obtain, by methods already known to us, the displacements, forces, and, consequently, stresses at any point in the structure.

We will use the $X$, $Y$ coordinate system shown to describe the structure in Fig. 7-1. We will assume that the orientation of the $Z$ axis is such that the resulting $X$, $Y$, $Z$ coordinate system is right-handed. The location and orientation of the coordinate system is otherwise arbitrary. Since this coordinate system is used to describe the entire structure, it will be referred to as the *global coordinate system,* and will be designated by uppercase letters.

For identification purposes we will number the joints of the structure. Members can be conveniently referred to using the joint numbers associated with the ends of the member. For example, the horizontal member in Fig. 7-1 would be referred to as member 3-4 or as member 4-3. Numbering of the joints can be random, as long as all integers between 1 and $N$ are used for a structure with $N$ joints. Later we will see that by numbering the joints judiciously we can minimize the amount of computer storage and time required to solve a problem.

## 7-3 OVERALL APPROACH

At this point we will count the number of unknown quantities, and compare this to the number of equations available. Referring again to Fig. 7-1, let us disassemble the structure and draw the free-body diagram for each member, as shown in Fig. 7-3. From this

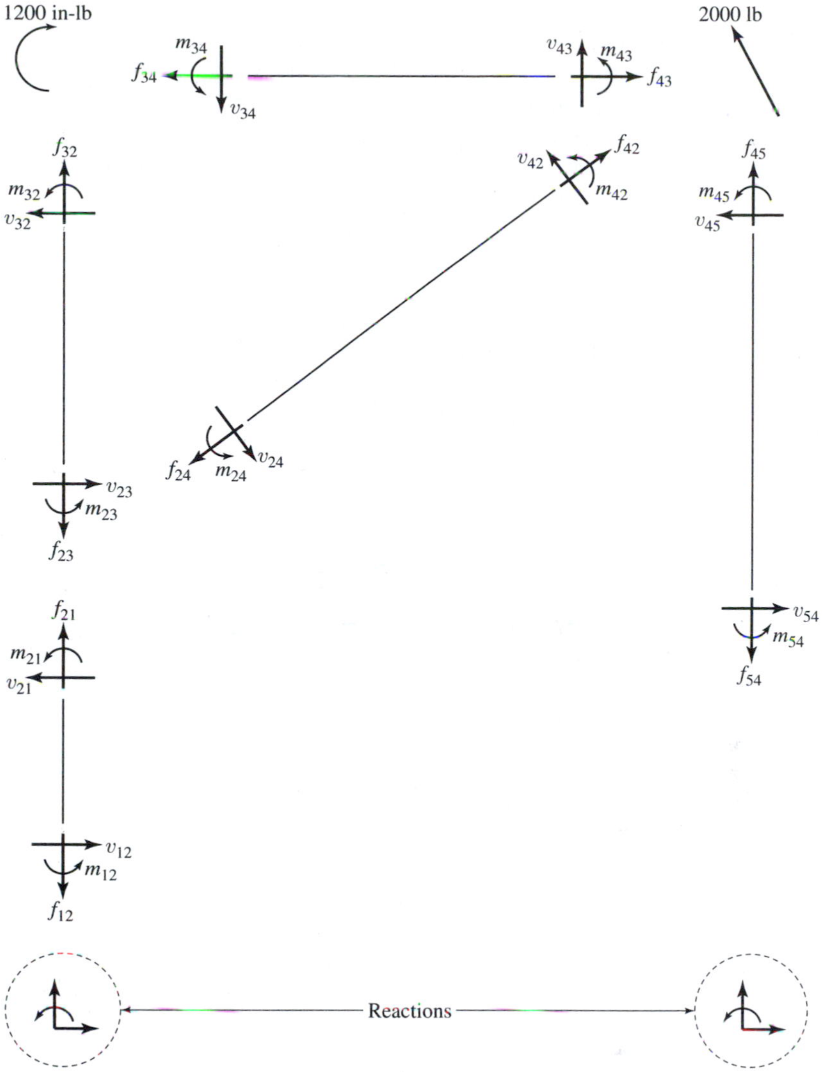


**Figure 7-3** Free-body diagram for disassembled structure

figure we see that for each member we need to determine the three force components and three displacement components at each end of the member. In addition, we need to find, in general, three reaction forces at each support. Consequently, for a structure consisting of $M$ members and $S$ supports, there are $12M + 3S$ unknown quantities to be determined. For the structure shown in Fig. 7-1, there are five members and two supports, giving a total of 66 unknowns.

However, at each joint in the structure, both force and moment equilibrium must hold—at the supports the member end forces must be in equilibrium with the reaction forces, and at the remaining joints, the member end forces must be in equilibrium with the applied loads. Force and moment equilibrium must also hold across each member: The forces and moments at one end of a member must be in equilibrium with those at the other end. Thus for a structure with $N$ joints and $M$ members there are $3N + 3M$ equilibrium equations.

For each member there are, of course, two ends so that for a structure with $M$ members there are $2M$ ends. At each joint of the structure where several member ends meet, compatibility of both displacements and slope must exist. For a joint consisting of two member ends we have one set of compatibility conditions (three compatibility equations), for a joint with three member ends we have two sets of compatibility conditions (between the ends of the first and second members, and the second and third members), and so on. That is, at each joint of the structure there is one less compatibility condition than the number of member ends forming the joint. Since the total number of member ends must equal $2M$, distributed over $N$ joints, we must have $2M - N$ compatibility conditions, or $3(2M - N)$ compatibility equations for the structure.

At each support of the structure three quantities are known. For example, at a fixed support, such as at joint 1 in Fig. 7-1, all three displacements are zero. At a pinned joint, the two linear displacements are zero, as is the moment. If the structure contains an elastic support, then although neither the displacement nor the corresponding force is known, the relationship between the two is known. For example, if the slope at a pinned joint is controlled by a torsional spring, the slope and moment are related by the expression $M = k\theta$, where the constant $k$ is the torsional stiffness of the spring. It follows then that there are $3S$ additional equations for a structure with $S$ supports.

At this point there are $3N + 3M + 3(2M - N) + 3S = 9M + 3S$ equations, but $12M + 3S$ unknowns.

However, we can write equations relating the member end forces and displacements. For example, if we impose longitudinal and transverse displacements as well as a slope to the free end of a cantilever beam, we can calculate the resulting forces at the free end using Castigliano's theorem. These member force-displacement relationships provide three additional equations for each member of the structure, relating three of the end forces of the member to the six end displacements. With these, the total number of equations available is $12M + 3S$, the same as the number of unknowns.

For the structure shown in Fig. 7-1 we have 15 equilibrium equations, 15 compatibility equations (six at joints 2 and 4, three at joint 3), 6 support or boundary conditions (at joint 1 all displacements are zero, whereas at joint 5 the vertical displacement and slope are zero, and the horizontal force is zero), and 30 force-displacement relationships for a total of 66 equations. Since the number of unknowns is also 66, the problem can be solved.

## 7-4 MEMBER FORCE-DISPLACEMENT RELATIONSHIPS

In order to develop relationships between the forces and displacements at the end of a member, let us consider member $i$-$j$, as shown in Fig. 7-4. In an actual structure the letters $i$ and $j$ will be replaced by the joint numbers associated with the member. As a matter of convenience, since we are used to thinking of forces acting along the axis of the member and perpendicular to the axis, we will establish a *local coordinate system* for the

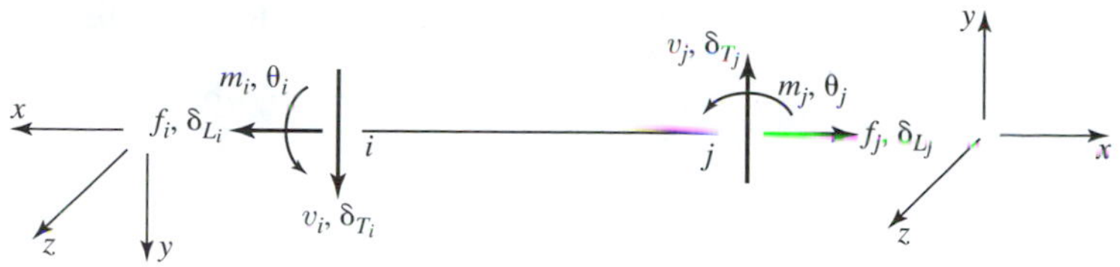


Figure 7-4

ends of the member, and designate this system by lowercase letters. Positive forces and displacements will be directed in positive coordinate directions and relationships will be developed between the member end forces and displacements in the local coordinate system. Later it will be necessary to transform these quantities into the global coordinate system so that the relationships for all members are in a common coordinate system. The actual transformation will depend on the orientation of a particular member relative to the global coordinate system.

We will choose the local coordinate system as follows. The local $x$ axis at each end of the member is directed away from the ends of the member; the local $z$ axis is directed perpendicular to and out of the plane of the paper; the local $y$ axis forms a right-handed system with $x$ and $z$. Notice that this definition of the local coordinate system results in a different orientation of the local coordinate system at each end of the member.

Now we fix end $i$ and apply positive end forces at $j$, as shown in Fig. 7-5. We will use Castigliano's theorem to determine the resulting displacements at $j$. From the figure and Eq. (5-90) the expression for the strain energy of the beam is

$$U(f_j, v_j, m_j) = \int_0^l \left[ \frac{(m_j + v_j x)^2}{2EI} + \frac{f_j^2}{2EA} \right] dx \tag{7-1}$$

where $x$ is measured from the free end, and transverse shear effects have been neglected. Quantities in lowercase letters are in the local coordinate system of the member. Differentiation of this expression with respect to $f_j$, $v_j$, and $m_j$ gives $\delta_{L_j}$, $\delta_{T_j}$, and $\theta_j$,

$$\begin{aligned} \delta_{L_j} &= \frac{f_j l}{EA} \\ \delta_{T_j} &= \frac{v_j l^3}{3EI} + \frac{m_j l^2}{2EI} \\ \theta_j &= \frac{v_j l^2}{2EI} + \frac{m_j l}{EI} \end{aligned} \tag{7-2}$$

Figure 7-5

These can be solved to give forces in terms of displacements,

$$
\begin{aligned}
f_j &= \frac{EA\delta_{L_j}}{l} \\
v_j &= \frac{12EI\delta_{T_j}}{l^3} - \frac{6EI\theta_j}{l^2} \\
m_j &= -\frac{6EI\delta_{T_j}}{l^2} + \frac{4EI\theta_j}{l}
\end{aligned}
\tag{7-3}
$$

Written in matrix form, these expressions become

$$
\begin{Bmatrix} f_j \\ v_j \\ m_j \end{Bmatrix} = \begin{bmatrix} \frac{EA}{l} & 0 & 0 \\ 0 & \frac{12EI}{l^3} & -\frac{6EI}{l^2} \\ 0 & -\frac{6EI}{l^2} & \frac{4EI}{l} \end{bmatrix} \begin{Bmatrix} \delta_{L_j} \\ \delta_{T_j} \\ \theta_j \end{Bmatrix}
\tag{7-4}
$$

To develop an understanding of the various quantities in the square matrix in Eq. (7-4), it will be beneficial to look at a few simple problems and interpret the results. First, let us consider a case in which

$$
\delta_{L_j} = 1, \qquad \delta_{T_j} = 0, \qquad \theta_j = 0
\tag{7-5}
$$

The displaced member is shown in Fig. 7-6. From Eq. (7-4) we see that the forces developed are

$$
f_j = \frac{AE}{l}, \qquad v_j = 0, \qquad m_j = 0
\tag{7-6}
$$

Thus $AE/l$ is the force required to produce a unit longitudinal displacement, all other displacements being zero. Since we are dealing with linear relationships between forces and displacements, it takes three times as much force to produce a longitudinal displacement of three units. $AE/l$ is a measure of the resistance of the member to the displacement, or a measure of its *stiffness*.

Now consider the following displacement configuration:

$$
\delta_{L_j} = 0, \qquad \delta_{T_j} = 1, \qquad \theta_j = 0
\tag{7-7}
$$

Figure 7-6

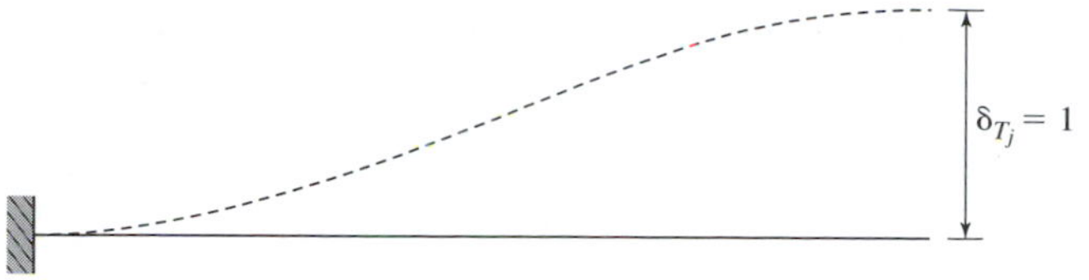


Figure 7-7

Here we have imposed on the beam a lateral displacement of one unit, whereas all other displacements are held at zero, as shown in Fig. 7-7. From Eq. (7-4) we see that the forces required to produce the given displacements are

$$f_j = 0, \qquad v_j = \frac{12EI}{l^3}, \qquad m_j = -\frac{6EI}{l^2} \tag{7-8}$$

In other words, to produce a lateral displacement of one unit, with all other displacement components zero, a lateral force of $12EI/l^3$ is required, as well as a bending moment of $6EI/l^2$ acting in the clockwise direction. Again, since we are dealing with linear relationships, if the lateral displacement is doubled, the forces and moment required to produce it are also doubled. The quantities $12EI/l^3$ and $-6EI/l^2$ are also seen to be measures of the stiffness of the member.

If we were to apply a unit slope to the end of the beam while holding the longitudinal and lateral displacements at zero, as shown in Fig. 7-8, and determine the forces and moment required to produce these displacements, we would see that the quantities $-6EI/l^2$ and $4EI/l$ appearing in the third row of the coefficient matrix in Eq. (7-4) also measure the stiffness of the beam. Consequently all the elements of the coefficient matrix in Eq. (7-4) represent stiffnesses of the member. The matrix itself is referred to as a *stiffness matrix.*

Now the forces at end $i$ of the beam must be in equilibrium with those at end $j$, so we must have, referring to Fig. 7-4 and using Eqs. (7-3),

$$\begin{aligned} f_i &= f_j = \frac{EA\delta_{L_j}}{l} \\ v_i &= v_j = \frac{12EI\delta_{T_j}}{l^3} - \frac{6EI\theta_j}{l^2} \\ m_i &= -m_j - v_j l = -\frac{6EI\delta_{T_j}}{l^2} + \frac{2EI\theta_j}{l} \end{aligned} \tag{7-9}$$

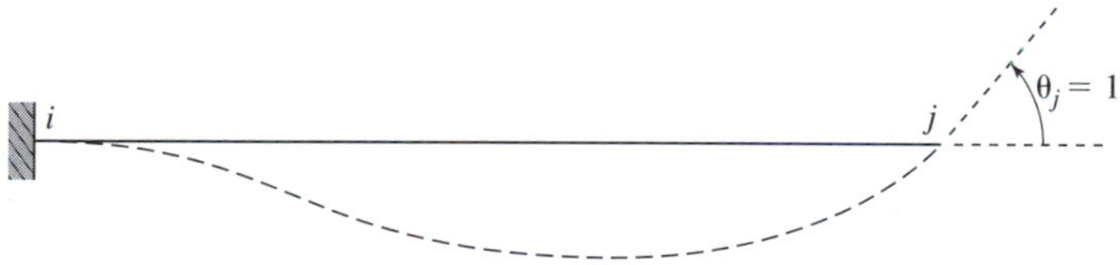


Figure 7-8

In matrix form these become

$$\begin{Bmatrix} f_i \\ v_i \\ m_i \end{Bmatrix} = \begin{bmatrix} \frac{EA}{l} & 0 & 0 \\ 0 & \frac{12EI}{l^3} & -\frac{6EI}{l^2} \\ 0 & -\frac{6EI}{l^2} & \frac{2EI}{l} \end{bmatrix} \begin{Bmatrix} \delta_{L_j} \\ \delta_{T_j} \\ \theta_j \end{Bmatrix} \tag{7-10}$$

When applying the displacement configurations used before to Eq. (7-10), it will become clear that the coefficient matrix in Eq. (7-10) is also a stiffness matrix.

If we now fix end $j$ of member $i$-$j$ and apply positive forces at end $i$, we will find that the resulting relationships between forces and displacements are those given in Eqs. (7-4) and (7-10), but with the subscripts $i$ and $j$ interchanged,

$$\begin{Bmatrix} f_i \\ v_i \\ m_i \end{Bmatrix} = \begin{bmatrix} \frac{EA}{l} & 0 & 0 \\ 0 & \frac{12EI}{l^3} & -\frac{6EI}{l^2} \\ 0 & -\frac{6EI}{l^2} & \frac{4EI}{l} \end{bmatrix} \begin{Bmatrix} \delta_{L_i} \\ \delta_{T_i} \\ \theta_i \end{Bmatrix} \tag{7-11}$$

$$\begin{Bmatrix} f_j \\ v_j \\ m_j \end{Bmatrix} = \begin{bmatrix} \frac{EA}{l} & 0 & 0 \\ 0 & \frac{12EI}{l^3} & -\frac{6EI}{l^2} \\ 0 & -\frac{6EI}{l^2} & \frac{2EI}{l} \end{bmatrix} \begin{Bmatrix} \delta_{L_i} \\ \delta_{T_i} \\ \theta_i \end{Bmatrix} \tag{7-12}$$

This should not be a surprising result since Eqs. (7-4) and (7-10) are valid for any orientation of the beam shown in Fig. 7-5, including the orientation obtained by rotating the beam 180 degrees about end $i$.

Since we are dealing here with linear problems, the principle of superposition holds. Therefore if we first impose displacements on end $j$ while holding end $i$ fixed, and then, while preventing end $j$ from displacing further, subject end $i$ to displacements, the forces and moment developed at end $i$ will be the sum of those given by Eqs. (7-10) and (7-11). Similarly, the forces and moment at end $j$ will be given by the sum of Eqs. (7-4) and (7-12). That is, by combining Eqs. (7-10) and (7-11) the forces at $i$ are given as

$$\begin{Bmatrix} f_i \\ v_i \\ m_i \end{Bmatrix} = \begin{bmatrix} \frac{EA}{l} & 0 & 0 \\ 0 & \frac{12EI}{l^3} & -\frac{6EI}{l^2} \\ 0 & -\frac{6EI}{l^2} & \frac{4EI}{l} \end{bmatrix} \begin{Bmatrix} \delta_{L_i} \\ \delta_{T_i} \\ \theta_i \end{Bmatrix} + \begin{bmatrix} \frac{EA}{l} & 0 & 0 \\ 0 & \frac{12EI}{l^3} & -\frac{6EI}{l^2} \\ 0 & -\frac{6EI}{l^2} & \frac{2EI}{l} \end{bmatrix} \begin{Bmatrix} \delta_{L_j} \\ \delta_{T_j} \\ \theta_j \end{Bmatrix} \tag{7-13}$$

or

$$\begin{Bmatrix} f_i \\ v_i \\ m_i \end{Bmatrix} = \begin{bmatrix} \dfrac{EA}{l} & 0 & 0 & \dfrac{EA}{l} & 0 & 0 \\ 0 & \dfrac{12EI}{l^3} & -\dfrac{6EI}{l^2} & 0 & \dfrac{12EI}{l^3} & -\dfrac{6EI}{l^2} \\ 0 & -\dfrac{6EI}{l^2} & \dfrac{4EI}{l} & 0 & -\dfrac{6EI}{l^2} & \dfrac{2EI}{l} \end{bmatrix} \begin{Bmatrix} \delta_{L_i} \\ \delta_{T_i} \\ \theta_i \\ \delta_{L_j} \\ \delta_{T_j} \\ \theta_j \end{Bmatrix} \tag{7-14}$$

Similarly, the forces at end $j$ are given by

$$\begin{Bmatrix} f_j \\ v_j \\ m_j \end{Bmatrix} = \begin{bmatrix} \dfrac{EA}{l} & 0 & 0 & \dfrac{EA}{l} & 0 & 0 \\ 0 & \dfrac{12EI}{l^3} & -\dfrac{6EI}{l^2} & 0 & \dfrac{12EI}{l^3} & -\dfrac{6EI}{l^2} \\ 0 & -\dfrac{6EI}{l^2} & \dfrac{2EI}{l} & 0 & -\dfrac{6EI}{l^2} & \dfrac{4EI}{l} \end{bmatrix} \begin{Bmatrix} \delta_{L_i} \\ \delta_{T_i} \\ \theta_i \\ \delta_{L_j} \\ \delta_{T_j} \\ \theta_j \end{Bmatrix} \tag{7-15}$$

To establish our confidence in Eqs. (7-14) and (7-15), let us work a few examples. First assume that end $j$ is given a longitudinal displacement of 1 unit, end $i$ is given a longitudinal displacement of $-1$ unit, and all other displacement components are zero. If we recall the definition of positive displacements at ends $i$ and $j$, we see that the displacement is simply a rigid longitudinal translation, and therefore we would expect the resulting forces at $i$ to be zero. From Eq. (7-14) we find that

$$f_i = \frac{AE}{l}(-1+1) = 0, \qquad v_i = 0, \qquad m_i = 0 \tag{7-16}$$

From Eq. (7-15) we see that the forces at $j$ are also zero.

Now suppose that end $j$ is given a longitudinal displacement of 1 unit and end $i$ is also given a longitudinal displacement of 1 unit. Again, all other displacements are zero. Then Eq. (7-14) gives

$$f_i = \frac{AE}{l}(1+1) = \frac{2AE}{l}, \qquad v_i = 0, \qquad m_i = 0 \tag{7-17}$$

What we have done is equivalent to moving the member rigidly 1 unit to the left and then, while holding end $i$ fixed, displacing end $j$ 2 units to the right. Thus we would intuitively expect the longitudinal reaction force at $i$ to be $2AE/l$, as predicted by Eq. (7-14). A similar check at end $j$ via Eq. (7-15) shows that the longitudinal force there is also $2AE/l$, and the other forces are zero.

Finally let us assume that end $i$ is given a counterclockwise rotation of $\varphi$ units, as is end $j$, and then end $j$ is given a lateral displacement of $\varphi l$. This results in a rigid counterclockwise rotation about $i$ of $\varphi$ units, and should produce no forces at $i$ and $j$. That this is indeed the case is readily seen from Eqs. (7-14) and (7-15).

Equations (7-14) and (7-15) can be combined to give a single matrix equation relating the forces and displacements at ends $i$ and $j$,

$$\begin{Bmatrix} f_i \\ v_i \\ m_i \\ f_j \\ v_j \\ m_j \end{Bmatrix} = \begin{bmatrix} \frac{EA}{l} & 0 & 0 & \frac{EA}{l} & 0 & 0 \\ 0 & \frac{12EI}{l^3} & -\frac{6EI}{l^2} & 0 & \frac{12EI}{l^3} & -\frac{6EI}{l^2} \\ 0 & -\frac{6EI}{l^2} & \frac{4EI}{l} & 0 & -\frac{6EI}{l^2} & \frac{2EI}{l} \\ \frac{EA}{l} & 0 & 0 & \frac{EA}{l} & 0 & 0 \\ 0 & \frac{12EI}{l^3} & -\frac{6EI}{l^2} & 0 & \frac{12EI}{l^3} & -\frac{6EI}{l^2} \\ 0 & -\frac{6EI}{l^2} & \frac{2EI}{l} & 0 & -\frac{6EI}{l^2} & \frac{4EI}{l} \end{bmatrix} \begin{Bmatrix} \delta_{L_i} \\ \delta_{T_i} \\ \theta_i \\ \delta_{L_j} \\ \delta_{T_j} \\ \theta_j \end{Bmatrix} \tag{7-18}$$

The complete stiffness matrix for the member is therefore a $6 \times 6$ matrix. Note that it is a symmetric matrix, as we might have anticipated from Maxwell's reciprocity theorem. For example, if a unit rotation is imposed on end $j$ of a beam, the resulting shear force at $i$, from Eq. (7-18), is $-6EI/l^2$. Now if a unit lateral displacement is imposed at $i$, Maxwell's reciprocity theorem tells us that the moment at $j$ is $-6EI/l^2$. Consequently we expect elements 2, 6 and 6, 2 of the coefficient matrix in Eq. (7-18) to be the same.

To write Eq. (7-18) in a more compact form, we will introduce the following notation:

$$\begin{matrix} f = p_1, & v = p_2, & m = p_3 \\ \delta_L = \delta_1, & \delta_T = \delta_2, & \theta = \delta_3 \end{matrix} \tag{7-19}$$

With these, we can write

$$\begin{Bmatrix} f_i \\ v_i \\ m_i \end{Bmatrix} = \begin{Bmatrix} p_1 \\ p_2 \\ p_3 \end{Bmatrix}_{ij} = \boldsymbol{p}_{ij} \tag{7-20}$$

and

$$\begin{Bmatrix} \delta_{L_i} \\ \delta_{T_i} \\ \theta_i \end{Bmatrix} = \begin{Bmatrix} \delta_1 \\ \delta_2 \\ \delta_3 \end{Bmatrix}_{ij} = \boldsymbol{\delta}_{ij} \tag{7-21}$$

and so on. In the notation $ij$ on the right of these expressions, the first letter refers to the end of the member $i$-$j$ for which these expressions apply. We now use these definitions in Eq. (7-18) and the following results:

$$\begin{Bmatrix} \boldsymbol{p}_{ij} \\ \boldsymbol{p}_{ji} \end{Bmatrix} = \begin{bmatrix} \boldsymbol{k}_{ii}^{j} & \boldsymbol{k}_{ij} \\ \boldsymbol{k}_{ji} & \boldsymbol{k}_{jj}^{i} \end{bmatrix} \begin{Bmatrix} \boldsymbol{\delta}_{ij} \\ \boldsymbol{\delta}_{ji} \end{Bmatrix} \tag{7-22}$$

where

$$\boldsymbol{k}_{ij} = \boldsymbol{k}_{ji} = \begin{bmatrix} \frac{EA}{l} & 0 & 0 \\ 0 & \frac{12EI}{l^3} & -\frac{6EI}{l^2} \\ 0 & -\frac{6EI}{l^2} & \frac{2EI}{l} \end{bmatrix} \tag{7-23}$$

$$\boldsymbol{k}_{ii}^{j} = \boldsymbol{k}_{jj}^{i} = \begin{bmatrix} \frac{EA}{l} & 0 & 0 \\ 0 & \frac{12EI}{l^3} & -\frac{6EI}{l^2} \\ 0 & -\frac{6EI}{l^2} & \frac{2EI}{l} \end{bmatrix} \tag{7-24}$$

Equation (7-23) relates the forces at one end of the member to the displacements at the other end of the member. It is referred to as a *cross stiffness matrix*. Equation (7-24) relates the forces at one end of a member to the displacements at the same end. It is referred to as a *direct stiffness matrix*. In both sets of expressions, the first subscript refers to the end of the member at which forces are applied, whereas the second refers to the end at which displacements are imposed. In Eq. (7-24) both subscripts are the same, so a superscript is used to completely identify the member. For example, referring to Fig. 7-1, each of the members that meet at joint 2 would have a direct stiffness matrix identified as $\boldsymbol{k}_{22}$. Since the lengths, elastic moduli, moments of inertia, and cross-sectional areas of these members are not necessarily the same, the direct stiffness matrix $\boldsymbol{k}_{22}$ for member 2-3 will be different from that for member 2-1. The direct stiffness matrix for member 2-3 therefore is identified as $\boldsymbol{k}_{22}^{3}$, whereas that for member 2-1 is identified as $\boldsymbol{k}_{22}^{1}$, so that the member to which the direct stiffness matrix applies is uniquely identified.

If we expand Eq. (7-22), we will have two expressions, one giving the forces at end $i$ of member $i$-$j$, the other giving the forces at end $j$. Note that the second expression can be obtained from the first simply by interchanging indices $i$ and $j$ on both sides of the expression. Consequently if we remember the first expression, we can always get the second.

We now have expressions relating the end forces and displacements of each member in the structure. However, the forces and displacements are expressed in the local coordinate system associated with the ends of the member. Our next task is to transform the forces and displacements so that these quantities for each member are expressed in a common coordinate system. The forces $p_1$ and $p_2$ are components of a vector, as are the displacements $\delta_1$ and $\delta_2$. Since the local $z$ axis has the same orientation as the global $Z$ axis (see Fig. 7-1), the rotational force (moment) $p_3$ is the same in both the local coordinate system and the global system. Similarly, the rotational displacement (slope) $\delta_3$ is the same in both coordinate systems. (Remember that the slope measures the *change* in angular orientation of the member end between the deformed configuration and the undeformed configuration.)

Referring to Fig. 7-9 and Appendix B (Eq. B-21), the force components $p_1$ and $p_2$ at end $i$ of the member can be expressed in terms of those in the global coordinate

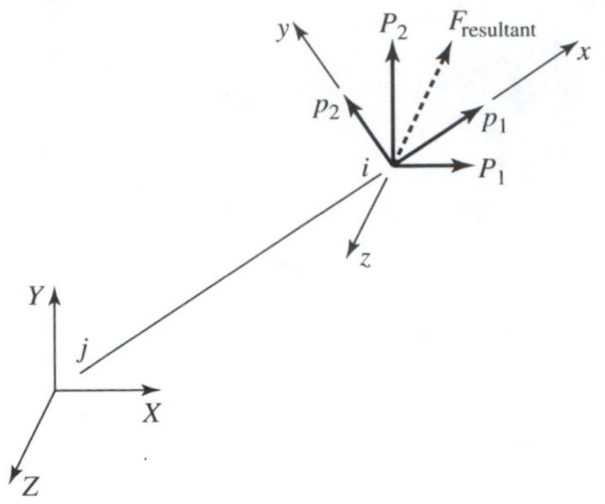


Figure 7-9 Global and local components of applied forces

system as follows:

$$\begin{Bmatrix} p_1 \\ p_2 \end{Bmatrix} = \begin{bmatrix} \cos\alpha & \sin\alpha \\ -\sin\alpha & \cos\alpha \end{bmatrix} \begin{Bmatrix} P_1 \\ P_2 \end{Bmatrix} \tag{7-25}$$

where $P_1$ and $P_2$ are components of force in the global directions $X$ and $Y$, respectively, and $\alpha$ is measured counterclockwise from the global $X$ axis to the local $x$ axis at end $i$. Since the moment is the same in both the global and the local coordinate systems, $p_3 = P_3$, and we can write

$$\boldsymbol{p}_{ij} = \boldsymbol{R}_{ij}\boldsymbol{P}_{ij} \tag{7-26}$$

where

$$\boldsymbol{P}_{ij}^T = [P_1 \quad P_2 \quad P_3]_{ij} \tag{7-27}$$

and

$$\boldsymbol{R}_{ij} = \begin{bmatrix} \cos\alpha & \sin\alpha & 0 \\ -\sin\alpha & \cos\alpha & 0 \\ 0 & 0 & 1 \end{bmatrix} \tag{7-28}$$

Similarly, for the displacements,

$$\boldsymbol{\delta}_{ij} = \boldsymbol{R}_{ij}\boldsymbol{\Delta}_{ij} \tag{7-29}$$

where

$$\boldsymbol{\Delta}_{ij}^T = [\Delta_1 \quad \Delta_2 \quad \Delta_3]_{ij} \tag{7-30}$$

Here $\Delta_1$ and $\Delta_2$ are the displacement components in the $X$ and $Y$ directions, respectively, and $\Delta_3$ is the rotation about the $Z$ axis.

We want to express forces and displacements in the global coordinate system in terms of those in the local coordinate system. Thus since $\boldsymbol{R}_{ij}$ is an orthogonal matrix,

$$\boldsymbol{P}_{ij} = \boldsymbol{R}_{ij}^T\boldsymbol{p}_{ij}, \qquad \boldsymbol{\Delta}_{ij} = \boldsymbol{R}_{ij}^T\boldsymbol{\delta}_{ij} \tag{7-31}$$

If we interchange $i$ and $j$ in the preceding expressions, then we get the corresponding relationships for end $j$ of member $i$-$j$. $\boldsymbol{R}_{ji}$ is the rotation matrix associated with end $j$ of member $i$-$j$. It can be formed simply by replacing $\alpha$ with $\alpha + \pi$ in $\boldsymbol{R}_{ij}$.

The first equation in Eqs. (7-22), when expanded, gives

$$p_{ij} = k_{ii}^j \delta_{ij} + k_{ij} \delta_{ji} \tag{7-32}$$

Using Eq. (7-29), we can write this as

$$p_{ij} = k_{ii}^j R_{ij} \Delta_{ij} + k_{ij} R_{ji} \Delta_{ji} \tag{7-33}$$

We now premultiply this by $R_{ij}^T$ and substitute the result into the first of Eqs. (7-31) to get the following expression:

$$P_{ij} = K_{ii}^j \Delta_{ij} + K_{ij} \Delta_{ji} \tag{7-34}$$

where

$$K_{ii}^j = R_{ij}^T k_{ii}^j R_{ij}, \qquad K_{ij} = R_{ij}^T k_{ij} R_{ji} \tag{7-35}$$

By applying the same procedure to the second of Eqs. (7-22), we will find that the expression for $P_{ji}$ in terms of displacements in the global coordinate system can be obtained from the expression for $P_{ij}$ by interchanging $i$ and $j$. Therefore in the global coordinate system, the force-displacement relationships are

$$\begin{Bmatrix} P_{ij} \\ P_{ji} \end{Bmatrix} = \begin{bmatrix} K_{ii}^j & K_{ij} \\ K_{ji} & K_{jj}^i \end{bmatrix} \begin{Bmatrix} \Delta_{ij} \\ \Delta_{ji} \end{Bmatrix} \tag{7-36}$$

or

$$P = K\Delta \tag{7-37}$$

where $K$ is the *member stiffness matrix in the global coordinate system.*

We recall from Appendix A that the transpose of a product of matrices is equal to the product of transposes taken in reverse order, that is,

$$(ABC)^T = C^T B^T A^T \tag{7-38}$$

Then, from the first of Eqs. (7-35) and the symmetry of $k_{ii}^j$ it follows that

$$K_{ii}^{j^T} = K_{ii}^j \tag{7-39}$$

and from the second of Eqs. (7-35) and Eq. (7-23),

$$K_{ij}^T = K_{ji} \tag{7-40}$$

Now since

$$K^T = \begin{bmatrix} K_{ii}^{j^T} & K_{ji}^T \\ K_{ij}^T & K_{jj}^{i^T} \end{bmatrix} \tag{7-41}$$

we see, from the preceding expressions, that

$$K^T = K \tag{7-42}$$

That is, the symmetry of the member stiffness matrix is preserved in the transformation from local to global coordinates.

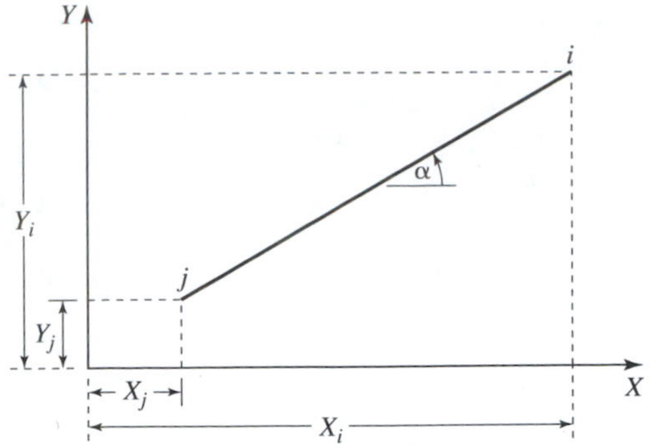


Figure 7-10

The elements of the rotation matrix can be readily found from the member coordinates. From Fig. 7-10

$$\cos\alpha = \frac{X_i - X_j}{L}, \qquad \sin\alpha = \frac{Y_i - Y_j}{L} \tag{7-43}$$

where

$$L = \sqrt{(X_i - X_j)^2 + (Y_i - Y_j)^2} \tag{7-44}$$

is the length of member *i-j*. Therefore the rotation matrix given by Eq. (7-28) becomes

$$\boldsymbol{R}_{ij} = \begin{bmatrix} \dfrac{X_i - X_j}{L} & \dfrac{Y_i - Y_j}{L} & 0 \\ -\dfrac{Y_i - Y_j}{L} & \dfrac{X_i - X_j}{L} & 0 \\ 0 & 0 & 1 \end{bmatrix} \tag{7-45}$$

The rotation matrix $\boldsymbol{R}_{ji}$ is formed by interchanging *i* and *j* in $\boldsymbol{R}_{ij}$. Note that this amounts to simply changing the sign of the first two rows of $\boldsymbol{R}_{ij}$. A few moments of reflection should convince the reader that the preceding form of the rotation matrix is valid regardless of the orientation of the member, and regardless of which end is labeled *i*.

## 7-5 ASSEMBLING THE PIECES

Now that we have developed the relationships between forces and displacements at the ends of the members, the available number of equations is, in fact, equal to the number of unknown quantities. However, we need to develop an algorithm that can be readily converted into a working computer program. Above all, our algorithm must be sufficiently general so that any problem falling within the constraints given in Section 7-2 can be handled routinely by our program.

Remember that we have three sets of equations available—equilibrium of forces at the joints, compatibility of displacements at the joints, and boundary conditions. The

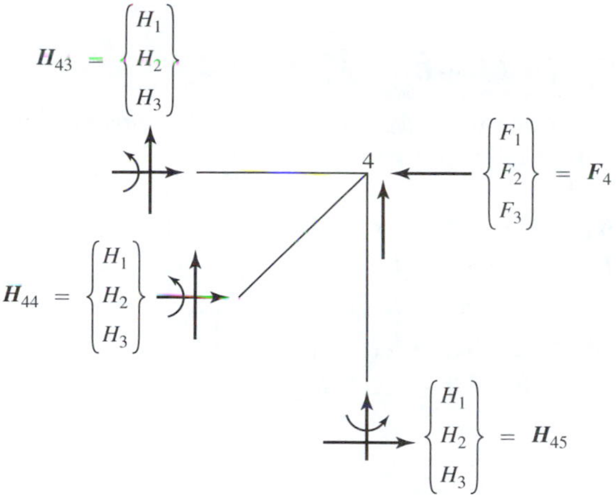


Figure 7-11 Free-body diagram of joint 4

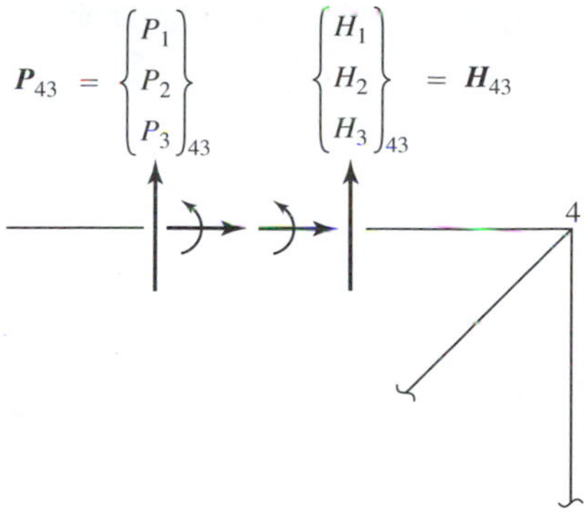


Figure 7-12

remaining sets of equations, the member force deflection relationships and the member equilibrium equations, have already been used in obtaining Eq. (7-18).

Let us take a look now at joint 4 of the structure shown in Fig. 7-1. If we cut each member forming this joint infinitesimally close to it and draw the free-body diagram of the joint, as shown in Fig. 7-11, static equilibrium requires that the applied forces at the joint be balanced by the internal forces of the members, that is,

$$\boldsymbol{F}_4 + \boldsymbol{H}_{43} + \boldsymbol{H}_{42} + \boldsymbol{H}_{45} = \mathbf{0} \tag{7-46}$$

The quantities appearing in Eq. (7-46) are referred to the global coordinate system. But since equilibrium must hold across the cut (see Fig. 7-12), $\boldsymbol{H}_{43}$ is equal in magnitude and opposite in direction to the member end forces $\boldsymbol{P}_{43}$. Similar results hold for the remaining members at joint 4, so that the equilibrium condition becomes

$$\boldsymbol{F}_4 = \boldsymbol{P}_{43} + \boldsymbol{P}_{42} + \boldsymbol{P}_{45} \tag{7-47}$$

We now use the member force-displacement relationships [Eq. (7-36)] and get

$$\boldsymbol{F}_4 = \boldsymbol{K}^3_{44}\boldsymbol{\Delta}_{43} + \boldsymbol{K}_{43}\boldsymbol{\Delta}_{34} + \boldsymbol{K}^2_{44}\boldsymbol{\Delta}_{42} + \boldsymbol{K}_{42}\boldsymbol{\Delta}_{24} + \boldsymbol{K}^5_{44}\boldsymbol{\Delta}_{45} + \boldsymbol{K}_{45}\boldsymbol{\Delta}_{54} \tag{7-48}$$

This expression can be reduced by recognizing that the displacements at end 4 of each member forming the joint must be compatible. Consequently,

$$\boldsymbol{\Delta}_{43} = \boldsymbol{\Delta}_{42} = \boldsymbol{\Delta}_{45} \equiv \boldsymbol{\Delta}_4 \tag{7-49}$$

Since the displacements at joint 4 are identical for each member, we will drop the double subscript notation and refer to it simply as $\boldsymbol{\Delta}_4$. Likewise we will refer to the displacement at joint 3 as $\boldsymbol{\Delta}_3$, that of joint 2 as $\boldsymbol{\Delta}_2$, and so on. With these, Eq. (7-48) becomes

$$\boldsymbol{F}_4 = \boldsymbol{K}_{42}\boldsymbol{\Delta}_2 + \boldsymbol{K}_{43}\boldsymbol{\Delta}_3 + \left(\boldsymbol{K}^2_{44} + \boldsymbol{K}^3_{44} + \boldsymbol{K}^5_{44}\right)\boldsymbol{\Delta}_4 + \boldsymbol{K}_{45}\boldsymbol{\Delta}_5 \tag{7-50}$$

or

$$\boldsymbol{F}_4 = \boldsymbol{K}_{42}\boldsymbol{\Delta}_2 + \boldsymbol{K}_{43}\boldsymbol{\Delta}_3 + \boldsymbol{K}_{44}\boldsymbol{\Delta}_4 + \boldsymbol{K}_{45}\boldsymbol{\Delta}_5 \tag{7-51}$$

where

$$\boldsymbol{K}_{44} = \boldsymbol{K}_{44}^2 + \boldsymbol{K}_{44}^3 + \boldsymbol{K}_{44}^5 \tag{7-52}$$

We now apply the conditions of equilibrium and compatibility to the remaining joints in the structure and finally arrive at the following set of equations:

$$\begin{Bmatrix} \boldsymbol{F}_1 \\ \boldsymbol{F}_2 \\ \boldsymbol{F}_3 \\ \boldsymbol{F}_4 \\ \boldsymbol{F}_5 \end{Bmatrix} = \begin{bmatrix} \boldsymbol{K}_{11} & \boldsymbol{K}_{12} & \boldsymbol{0} & \boldsymbol{0} & \boldsymbol{0} \\ \boldsymbol{K}_{21} & \boldsymbol{K}_{22} & \boldsymbol{K}_{23} & \boldsymbol{K}_{24} & \boldsymbol{0} \\ \boldsymbol{0} & \boldsymbol{K}_{32} & \boldsymbol{K}_{33} & \boldsymbol{K}_{34} & \boldsymbol{0} \\ \boldsymbol{0} & \boldsymbol{K}_{42} & \boldsymbol{K}_{43} & \boldsymbol{K}_{44} & \boldsymbol{K}_{45} \\ \boldsymbol{0} & \boldsymbol{0} & \boldsymbol{0} & \boldsymbol{K}_{54} & \boldsymbol{K}_{55} \end{bmatrix} \begin{Bmatrix} \boldsymbol{\Delta}_1 \\ \boldsymbol{\Delta}_2 \\ \boldsymbol{\Delta}_3 \\ \boldsymbol{\Delta}_4 \\ \boldsymbol{\Delta}_5 \end{Bmatrix} \tag{7-53}$$

where

$$\boldsymbol{K}_{ii} = \sum_n \boldsymbol{K}_{ii}^n \tag{7-54}$$

and the summation is over all members $i$–$n$ forming joint $i$. Equation (7-53) can be abbreviated as

$$\boldsymbol{F} = \boldsymbol{K\Delta} \tag{7-55}$$

where $\boldsymbol{K}$ is the stiffness matrix for the entire structure.

In Eq. (7-53) the column matrix of loads represents the external loads acting on the structure. Each of its elements $\boldsymbol{F}_i$ is itself a $3 \times 1$ column matrix with components $F_1$, $F_2$, $F_3$ representing forces in the global $X$ and $Y$ directions and a moment about the global $Z$ axis. The elements $\boldsymbol{F}_2$, $\boldsymbol{F}_3$, and $\boldsymbol{F}_4$ are the applied loads and are therefore known. For example, from Fig. 7-1, $\boldsymbol{F}_2^T = [0\ \ 0\ \ 0]$, $\boldsymbol{F}_3^T = [0\ \ 0\ \ 1200]$, and $\boldsymbol{F}_4^T = [-1000\ \ 1732\ \ 0]$. The first component of $\boldsymbol{F}_5$ is also 0. The element $\boldsymbol{F}_1$ and the last two components of $\boldsymbol{F}_5$ are reaction forces and must be determined. Each element of the column matrix representing the joint displacements is a $3 \times 1$ column matrix whose components are the displacements in the $X$ and $Y$ directions and the slope about the $Z$ axis. We know that all components of $\boldsymbol{\Delta}_1$ are zero and that the last two components of $\boldsymbol{\Delta}_5$ are zero. The first component of $\boldsymbol{\Delta}_5$ and all components of $\boldsymbol{\Delta}_2$, $\boldsymbol{\Delta}_3$, and $\boldsymbol{\Delta}_4$ must be determined. Each of the elements $\boldsymbol{K}_{ij}$ of the stiffness matrix is a $3 \times 3$ matrix. The fifteen equations in Eq. (7-53) contain five unknown reaction forces and ten unknown displacements.

Several observations regarding the structure stiffness matrix can be made by referring to Fig. 7-1. First the size of the matrix is determined by the number of joints, not the number of members, in the structure. Second, the row and column locations of nonzero elements of the matrix correspond to joint pairs in the structure that are connected by a single member. For example, joints 2 and 4 are connected by member 2-4, and the structure stiffness matrix contains nonzero entries for both elements 2,4 and 4,2, but joints 2 and 5 are not connected by a single member (there is no member 2-5) and correspondingly both elements 2,5 and 5,2 in the structure stiffness matrix are zero. The location of the nonzero elements is determined by the numbering scheme used for the joints so that if, for example, joint numbers 1 and 5 were interchanged, the structure stiffness matrix would be as shown in Fig. 7-13, although the structure itself has not changed.

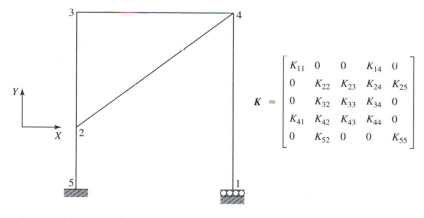

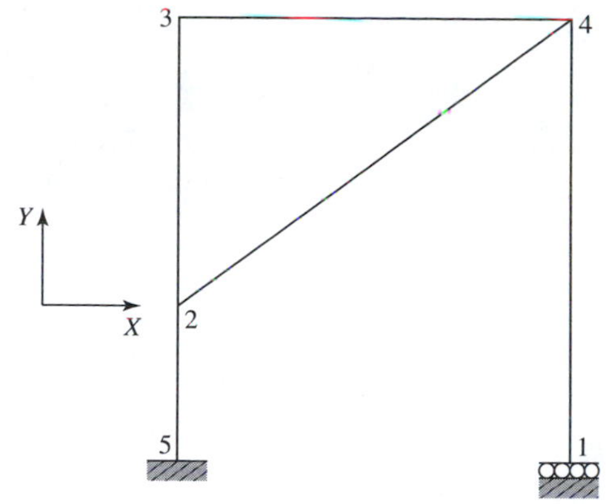


$$\boldsymbol{K} = \begin{bmatrix} K_{11} & 0 & 0 & K_{14} & 0 \\ 0 & K_{22} & K_{23} & K_{24} & K_{25} \\ 0 & K_{32} & K_{33} & K_{34} & 0 \\ K_{41} & K_{42} & K_{43} & K_{44} & 0 \\ 0 & K_{52} & 0 & 0 & K_{55} \end{bmatrix}$$

Figure 7-13 Changing the joint numbers changes the structure stiffness matrix

Returning to Fig. 7-1, we note that none of the diagonal elements of the matrix are zero, and each is formed by adding the direct stiffness contribution of each member at that joint. That is, element $\boldsymbol{K}_{22}$ is formed by adding the direct stiffness matrices $\boldsymbol{K}_{22}^n$ for each member at joint 2.

From Eqs. (7-39) and (7-54) we see that the matrices $\boldsymbol{K}_{ii}$ are symmetric. This, together with Eq. (7-40), leads us to conclude that the structure stiffness matrix is symmetric.

Finally we note that the elements of the structure stiffness matrix depend only on the geometry and material properties of the members, not on the applied loads or constraints.

We will generalize some of these observations in what follows. For any structure subjected to statical loading we know that the work done by the external loads $W_{\text{EXT}}$ and the strain energy $U$ stored in the structure are equal.

$$U = W_{\text{EXT}} \tag{7-56}$$

Now a typical joint of the structure can have applied to it three forces (two linear and one rotational) and will undergo, in general, a displacement consisting of two linear components and one rotational component. Therefore at each joint in the structure the work done by the external forces is

$$W_{\text{EXT}_i} = \tfrac{1}{2}(F_1\Delta_1 + F_2\Delta_2 + F_3\Delta_3)_i \tag{7-57}$$

This can be written in matrix notation as

$$W_{\text{EXT}_i} = \frac{1}{2}[F_1 \quad F_2 \quad F_3]_i \begin{Bmatrix} \Delta_1 \\ \Delta_2 \\ \Delta_3 \end{Bmatrix}_i \tag{7-58}$$

or

$$W_{\text{EXT}_i} = \tfrac{1}{2}\boldsymbol{F}_i^T \boldsymbol{\Delta}_i \tag{7-59}$$

For an entire structure consisting of $n$ joints we then have

$$U = \tfrac{1}{2}\left(\boldsymbol{F}_1^T\boldsymbol{\Delta}_1 + \boldsymbol{F}_2^T\boldsymbol{\Delta}_2 + \cdots + \boldsymbol{F}_n^T\boldsymbol{\Delta}_n\right) \tag{7-60}$$

In an actual structure some of these terms may be zero either because no external loads are applied or because the joint is constrained. Now the forces appearing in Eq. (7-60) are functions of the displacements by virtue of the force-displacement relationships so that if we differentiate the foregoing expression for strain energy with respect to, say, $\boldsymbol{\Delta}_1$, we get the following:

$$\frac{\partial U}{\partial \boldsymbol{\Delta}_1} = \frac{1}{2}\left(\frac{\partial \boldsymbol{F}_1^T}{\partial \boldsymbol{\Delta}_1}\boldsymbol{\Delta}_1 + \frac{\partial \boldsymbol{\Delta}_1}{\partial \boldsymbol{\Delta}_1}\boldsymbol{F}_1 + \frac{\partial \boldsymbol{F}_2^T}{\partial \boldsymbol{\Delta}_1}\boldsymbol{\Delta}_2 + \frac{\partial \boldsymbol{\Delta}_2}{\partial \boldsymbol{\Delta}_1}\boldsymbol{F}_2 + \cdots + \frac{\partial \boldsymbol{F}_n^T}{\partial \boldsymbol{\Delta}_1}\boldsymbol{\Delta}_n + \frac{\partial \boldsymbol{\Delta}_n}{\partial \boldsymbol{\Delta}_1}\boldsymbol{F}_n\right) \tag{7-61}$$

From the definition of partial differentiation we see that

$$\frac{\partial \boldsymbol{\Delta}_m}{\partial \boldsymbol{\Delta}_1} = \begin{cases}\boldsymbol{I}, & m = 1\\ \boldsymbol{0}, & m \neq 1\end{cases} \tag{7-62}$$

so that Eq. (7-61) reduces to

$$\frac{\partial U}{\partial \boldsymbol{\Delta}_1} = \frac{1}{2}\left(\frac{\partial \boldsymbol{F}_1^T}{\partial \boldsymbol{\Delta}_1}\boldsymbol{\Delta}_1 + \boldsymbol{F}_1 + \frac{\partial \boldsymbol{F}_2^T}{\partial \boldsymbol{\Delta}_1}\boldsymbol{\Delta}_2 + \cdots + \frac{\partial \boldsymbol{F}_n^T}{\partial \boldsymbol{\Delta}_1}\boldsymbol{\Delta}_n\right) \tag{7-63}$$

But Castigliano's first theorem gives (see Chapter 5)

$$\frac{\partial U}{\partial \Delta_1} = F_1, \qquad \frac{\partial U}{\partial \Delta_2} = F_2, \qquad \frac{\partial U}{\partial \Delta_3} = F_3 \tag{7-64}$$

Hence,

$$\frac{\partial U}{\partial \boldsymbol{\Delta}_i} = \boldsymbol{F}_i \tag{7-65}$$

Substituting into Eq. (7-63) gives

$$\boldsymbol{F}_1 = \frac{\partial \boldsymbol{F}_1^T}{\partial \boldsymbol{\Delta}_1}\boldsymbol{\Delta}_1 + \frac{\partial \boldsymbol{F}_2^T}{\partial \boldsymbol{\Delta}_1}\boldsymbol{\Delta}_2 + \cdots + \frac{\partial \boldsymbol{F}_n^T}{\partial \boldsymbol{\Delta}_1}\boldsymbol{\Delta}_n \tag{7-66}$$

If we now differentiate Eq. (7-60) successively with respect to $\boldsymbol{\Delta}_2, \boldsymbol{\Delta}_3, \ldots, \boldsymbol{\Delta}_n$ we arrive at the following set of equations:

$$\begin{Bmatrix}\boldsymbol{F}_1\\ \boldsymbol{F}_2\\ \boldsymbol{F}_3\\ \vdots\\ \boldsymbol{F}_n\end{Bmatrix} = \begin{bmatrix}\frac{\partial \boldsymbol{F}_1^T}{\partial \boldsymbol{\Delta}_1} & \frac{\partial \boldsymbol{F}_2^T}{\partial \boldsymbol{\Delta}_1} & \frac{\partial \boldsymbol{F}_3^T}{\partial \boldsymbol{\Delta}_1} & \cdots & \frac{\partial \boldsymbol{F}_n^T}{\partial \boldsymbol{\Delta}_1}\\ \frac{\partial \boldsymbol{F}_1^T}{\partial \boldsymbol{\Delta}_2} & \frac{\partial \boldsymbol{F}_2^T}{\partial \boldsymbol{\Delta}_2} & \frac{\partial \boldsymbol{F}_3^T}{\partial \boldsymbol{\Delta}_2} & \cdots & \frac{\partial \boldsymbol{F}_n^T}{\partial \boldsymbol{\Delta}_2}\\ \frac{\partial \boldsymbol{F}_1^T}{\partial \boldsymbol{\Delta}_3} & \frac{\partial \boldsymbol{F}_2^T}{\partial \boldsymbol{\Delta}_3} & \frac{\partial \boldsymbol{F}_3^T}{\partial \boldsymbol{\Delta}_3} & \cdots & \frac{\partial \boldsymbol{F}_n^T}{\partial \boldsymbol{\Delta}_3}\\ \vdots & \vdots & \vdots & & \vdots\\ \frac{\partial \boldsymbol{F}_1^T}{\partial \boldsymbol{\Delta}_n} & \frac{\partial \boldsymbol{F}_2^T}{\partial \boldsymbol{\Delta}_n} & \frac{\partial \boldsymbol{F}_3^T}{\partial \boldsymbol{\Delta}_n} & \cdots & \frac{\partial \boldsymbol{F}_n^T}{\partial \boldsymbol{\Delta}_n}\end{bmatrix}\begin{Bmatrix}\boldsymbol{\Delta}_1\\ \boldsymbol{\Delta}_2\\ \boldsymbol{\Delta}_3\\ \vdots\\ \boldsymbol{\Delta}_n\end{Bmatrix} \tag{7-67}$$

The column matrix of forces contains all external forces acting on the structure, including reaction forces, and the column matrix of displacements contains all displacements of the structure, including those at the supports. Typically some elements of $\boldsymbol{F}$ are known whereas others are unknown. Corresponding to each known $\boldsymbol{F}_i$ is an unknown displacement, while corresponding to each unknown force we have a known displacement.

That the coefficient matrix of Eq. (7-67) is symmetric can be shown as follows. From Castigliano's first theorem we have

$$\boldsymbol{F}_i^T = \left(\frac{\partial U}{\partial \boldsymbol{\Delta}_i}\right)^T \tag{7-68}$$

If we differentiate this expression with respect to $\boldsymbol{\Delta}_j$, we get

$$\frac{\partial \boldsymbol{F}_i^T}{\partial \boldsymbol{\Delta}_j} = \frac{\partial}{\partial \boldsymbol{\Delta}_j}\left(\frac{\partial U}{\partial \boldsymbol{\Delta}_i}\right)^T \tag{7-69}$$

Now interchanging the order of differentiation of $U$ and referring to Appendix A (Eq. A-16), it follows that

$$\frac{\partial \boldsymbol{F}_i^T}{\partial \boldsymbol{\Delta}_j} = \left[\frac{\partial}{\partial \boldsymbol{\Delta}_i}\left(\frac{\partial U}{\partial \boldsymbol{\Delta}_j}\right)^T\right]^T = \left(\frac{\partial \boldsymbol{F}_j^T}{\partial \boldsymbol{\Delta}_i}\right)^T \tag{7-70}$$

In other words, the coefficient matrix is symmetric. By taking advantage of the symmetry we can write

$$\begin{Bmatrix} \boldsymbol{F}_1 \\ \boldsymbol{F}_2 \\ \boldsymbol{F}_3 \\ \vdots \\ \boldsymbol{F}_n \end{Bmatrix} = \begin{bmatrix} \dfrac{\partial \boldsymbol{F}_1^T}{\partial \boldsymbol{\Delta}_1} & \dfrac{\partial \boldsymbol{F}_1^T}{\partial \boldsymbol{\Delta}_2} & \dfrac{\partial \boldsymbol{F}_1^T}{\partial \boldsymbol{\Delta}_3} & \cdots & \dfrac{\partial \boldsymbol{F}_1^T}{\partial \boldsymbol{\Delta}_n} \\ \dfrac{\partial \boldsymbol{F}_2^T}{\partial \boldsymbol{\Delta}_1} & \dfrac{\partial \boldsymbol{F}_2^T}{\partial \boldsymbol{\Delta}_2} & \dfrac{\partial \boldsymbol{F}_2^T}{\partial \boldsymbol{\Delta}_3} & \cdots & \dfrac{\partial \boldsymbol{F}_2^T}{\partial \boldsymbol{\Delta}_n} \\ \dfrac{\partial \boldsymbol{F}_3^T}{\partial \boldsymbol{\Delta}_1} & \dfrac{\partial \boldsymbol{F}_3^T}{\partial \boldsymbol{\Delta}_2} & \dfrac{\partial \boldsymbol{F}_3^T}{\partial \boldsymbol{\Delta}_3} & \cdots & \dfrac{\partial \boldsymbol{F}_3^T}{\partial \boldsymbol{\Delta}_n} \\ \vdots & \vdots & \vdots & & \vdots \\ \dfrac{\partial \boldsymbol{F}_n^T}{\partial \boldsymbol{\Delta}_1} & \dfrac{\partial \boldsymbol{F}_n^T}{\partial \boldsymbol{\Delta}_2} & \dfrac{\partial \boldsymbol{F}_n^T}{\partial \boldsymbol{\Delta}_3} & \cdots & \dfrac{\partial \boldsymbol{F}_n^T}{\partial \boldsymbol{\Delta}_n} \end{bmatrix} \begin{Bmatrix} \boldsymbol{\Delta}_1 \\ \boldsymbol{\Delta}_2 \\ \boldsymbol{\Delta}_3 \\ \vdots \\ \boldsymbol{\Delta}_n \end{Bmatrix} \tag{7-71}$$

A typical row of this matrix, the second, for example, reads as follows:

$$\boldsymbol{F}_2 = \frac{\partial \boldsymbol{F}_2^T}{\partial \boldsymbol{\Delta}_1}\boldsymbol{\Delta}_1 + \frac{\partial \boldsymbol{F}_2^T}{\partial \boldsymbol{\Delta}_2}\boldsymbol{\Delta}_2 + \frac{\partial \boldsymbol{F}_2^T}{\partial \boldsymbol{\Delta}_3}\boldsymbol{\Delta}_3 + \cdots + \frac{\partial \boldsymbol{F}_2^T}{\partial \boldsymbol{\Delta}_n}\boldsymbol{\Delta}_n \tag{7-72}$$

The coefficient $\partial \boldsymbol{F}_2^T/\partial \boldsymbol{\Delta}_3$ gives the change in $\boldsymbol{F}_2$ resulting from a unit change in the displacement $\boldsymbol{\Delta}_3$ with all other displacements being held fixed. Therefore the term $(\partial \boldsymbol{F}_2^T/\partial \boldsymbol{\Delta}_3)\boldsymbol{\Delta}_3$ gives the change in $\boldsymbol{F}_2$ due to the actual change in the displacement at joint 3. The other terms are interpreted in a similar manner. Since the structure is originally in an unloaded, undeformed state, Eq. (7-72) tells us that the load developed at joint 2 is the sum of the individual contributions due to the displacements of each joint in the structure.

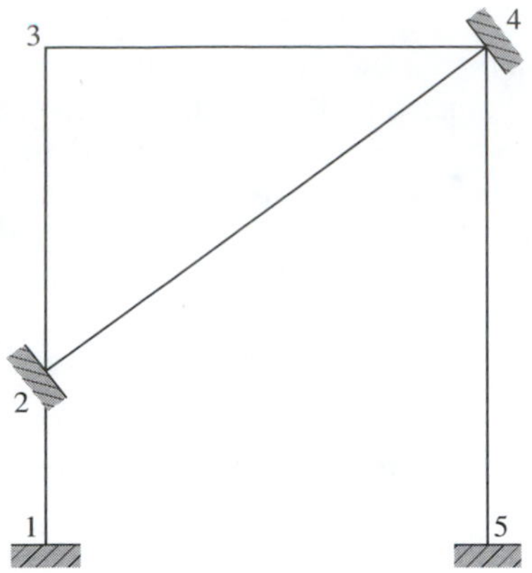


Figure 7-14

If we write Eq. (7-71) for the structure shown in Fig. 7-1 and compare it with Eq. (7-53), we see that

$$\frac{\partial \boldsymbol{F}_i^T}{\partial \boldsymbol{\Delta}_j} = \boldsymbol{K}_{ij} \tag{7-73}$$

That is, the elements of the coefficient matrix in Eq. (7-71) are the elements of the structure stiffness matrix.

Returning now to Fig. 7-1, let us examine the term $\partial \boldsymbol{F}_1^T/\partial \boldsymbol{\Delta}_3$. We recall again the definition of a partial derivative, which, applied to the preceding expression, requires us to hold all displacements except $\boldsymbol{\Delta}_3$ fixed, make an infinitesimal change in $\boldsymbol{\Delta}_3$, and observe the resulting change in $\boldsymbol{F}_1^T$.

With reference to Fig. 7-14, we see that since there is no member connecting directly joints 1 and 3, $\partial \boldsymbol{F}_1^T/\partial \boldsymbol{\Delta}_3 = \boldsymbol{0}$. Since there is a member that directly connects joints 3 and 4, we see that an infinitesimal change in $\boldsymbol{\Delta}_3$ will result in a change in $\boldsymbol{F}_4$. Therefore $\partial \boldsymbol{F}_4^T/\partial \boldsymbol{\Delta}_3 \neq \boldsymbol{0}$. This enables us to conclude that if two joints $i$ and $j$ are not directly connected by a member, the corresponding element $ij$ in the stiffness matrix is $\boldsymbol{0}$.

In order to assemble the structure stiffness matrix, we can proceed as follows:

1. Set all $3N \times 3N$ elements of the structure stiffness matrix equal to zero.
2. Calculate the elements of the member stiffness matrix for member 1 in local coordinates using Eqs. (7-23 and 7-24).
3. Construct the rotation matrices associated with member 1 from Eq. (7-45).
4. Calculate the elements of the member stiffness matrix for member 1 in global coordinates using Eqs. (7-35).
5. Add the elements of the $3 \times 3$ submatrices $\boldsymbol{K}_{ii}^j$, $\boldsymbol{K}_{jj}^i$, $\boldsymbol{K}_{ij}$, and $\boldsymbol{K}_{ji}$ to the $3 \times 3$ locations $i$-$i$, $j$-$j$, $i$-$j$, and $j$-$i$, respectively.
6. Repeat steps 2 through 5 for each of the remaining members of the structure.

Note that since we are *adding* the member elements to the elements of the structure stiffness matrix, no special consideration has to be given to the $3 \times 3$ submatrices situated along the diagonal of the structure stiffness matrix, which are formed by adding the direct stiffness contributions of each member making up a particular joint. If member 2-3 is processed after member 2-5, for example, the elements of the direct stiffness matrix $\boldsymbol{K}_{22}^{3}$ are simply added to $\boldsymbol{K}_{22}$, which already contains the elements of $\boldsymbol{K}_{22}^{5}$.

## 7-6 SOLVING THE PROBLEM

Before we decide on our approach to solving Eq. (7-55), let us examine the stiffness matrix for the structure. We are going to show that the matrix is singular if rigid body motion of the structure is possible. To do this, let the structure shown in Fig. 7-15 undergo an arbitrary rigid body motion, as shown. Then, since no forces are involved in displacing the body rigidly, Eq. (7-55) gives

$$\boldsymbol{K\Delta} = \boldsymbol{0} \tag{7-74}$$

But if $\boldsymbol{K}$ is nonsingular, then the only solution to the equations is

$$\boldsymbol{\Delta} = \boldsymbol{0} \tag{7-75}$$

which contradicts our initial assumption that $\boldsymbol{\Delta}$ is nonzero. Equation (7-74) possesses a nontrivial solution if and only if det $\boldsymbol{K} = 0$. We thus conclude that $\boldsymbol{K}$ is a singular matrix if the structure is capable of undergoing rigid body motion.

If we provide a sufficient number of constraints to suppress rigid body motion, then, as will be seen shortly, we will effectively reduce the size of the stiffness matrix, and the reduced stiffness matrix will be nonsingular.

To visualize the solution of the set of Eqs. (7-55), we will assume that a suitable number of constraints have been imposed to prevent rigid body motion—there may be more, but there must be at least that many. The displacement vector contains, for each

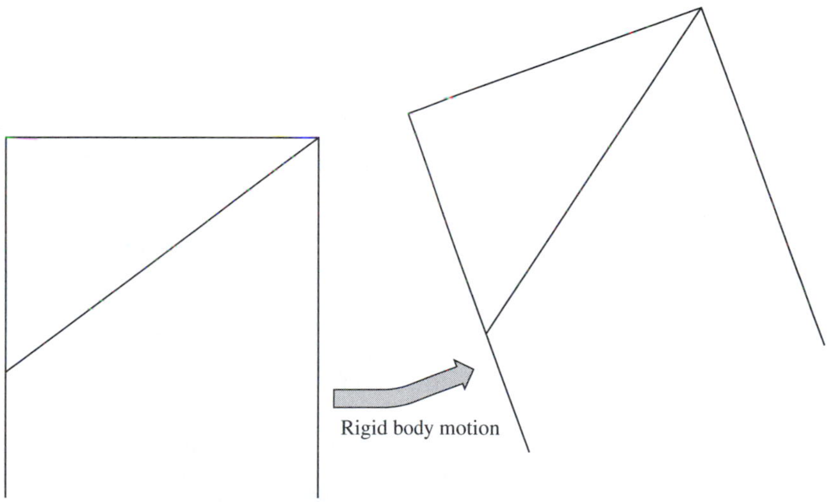


Figure 7-15

joint in the structure, two translational components and one rotational component—$3N$ components for a structure with $N$ joints. Some of these are known, whereas the remainder are to be determined. The load vector also contains three components for each joint in the structure. Here also certain elements are known, the remainder being unknown. But as we have discussed previously, for each component of displacement that is known, the corresponding component of the force vector is unknown, and vice versa. Thus for each row of the force vector that contains a known quantity, the corresponding row of the displacement vector contains an unknown quantity.

Now, let us rearrange the set of equations so that the known elements of the force vector appear first. Of course, this involves also rearranging the rows of the stiffness matrix. The result looks like this,

$$\begin{Bmatrix} \boldsymbol{F}_k \\ -- \\ \boldsymbol{F}_u \end{Bmatrix} = \begin{bmatrix} \text{-----}\boldsymbol{K}\text{-----} \end{bmatrix} \{\boldsymbol{\Delta}\} \tag{7-76}$$

Next we will rearrange the elements of the displacement vector so that the unknown components of displacement appear first. A few moments of reflection should convince the reader that this will require a corresponding rearrangement of the columns of the stiffness matrix, which gives

$$\begin{Bmatrix} \boldsymbol{F}_k \\ \boldsymbol{F}_u \end{Bmatrix} = \begin{bmatrix} \boldsymbol{K}_{I,I} & \boldsymbol{K}_{I,II} \\ \boldsymbol{K}_{II,I} & \boldsymbol{K}_{II,II} \end{bmatrix} \begin{Bmatrix} \boldsymbol{\Delta}_u \\ \boldsymbol{\Delta}_k \end{Bmatrix} \tag{7-77}$$

Expanding these equations, we get

$$\boldsymbol{F}_k = \boldsymbol{K}_{I,I}\,\boldsymbol{\Delta}_u + \boldsymbol{K}_{I,II}\,\boldsymbol{\Delta}_k \tag{7-78$a$}$$

$$\boldsymbol{F}_u = \boldsymbol{K}_{II,I}\,\boldsymbol{\Delta}_u + \boldsymbol{K}_{II,II}\,\boldsymbol{\Delta}_k \tag{7-78$b$}$$

Equation (7-78*a*) can be rearranged to give

$$\boldsymbol{K}_{I,I}\,\boldsymbol{\Delta}_u = \boldsymbol{A} \tag{7-79}$$

where

$$\boldsymbol{A} \equiv \boldsymbol{F}_k - \boldsymbol{K}_{I,II}\boldsymbol{\Delta}_k \tag{7-80}$$

Since the number of unknown components of displacement equals the number of known force components, the matrix $\boldsymbol{K}_{I,I}$ is a square matrix, and we can write

$$\boldsymbol{\Delta}_u = \boldsymbol{K}_{I,I}^{-1}\boldsymbol{A} \tag{7-81}$$

At this point all of the displacement components are known, and the unknown force components can be found from Eq. (7-78*b*). Typically, these are the reaction forces at the supports.

Once the displacements have been determined, the member end forces in local coordinates can be found from Eqs. (7-29) and (7-22). From these, shear and bending moment diagrams for each member can be constructed.

Let us return now to Eq. (7-77). In many problems that we will encounter, all the displacement components that are specified are zero, that is, displacement constraints are imposed on the structure. In that case the submatrices $\boldsymbol{K}_{I,II}$ and $\boldsymbol{K}_{II,II}$ are never used—they are simply multiplied by a zero vector. Now if we were not interested in the reaction forces, the submatrix $\boldsymbol{K}_{II,I}$ would also not be used. In this case it would not be necessary to actually rearrange the rows and columns of the stiffness matrix as we did in Eq. (7-77). Rather, we could simply cross out or eliminate those rows and the corresponding columns of the stiffness matrix in Eq. (7-55) that are associated with displacement components known to be zero, and "compress" the stiffness matrix. The resulting stiffness matrix would, of course, be the same as $\boldsymbol{K}_{I,I}$.

We could take this line of thought one step further and ask the question, "If we don't use all elements of the stiffness matrix, why do we need to calculate all of them?" The answer is that we do not need to calculate those elements of the stiffness matrix that are never used. In fact, many computer programs employing the direct stiffness method are written in such a way that only those elements of the stiffness matrix associated with "active," that is, unconstrained, degrees of freedom are calculated. The benefit of this approach is that the amount of computer memory required to solve a problem is minimized. It is not at all unusual in actual engineering structures to have several hundred degrees of freedom, and the resulting stiffness matrix would require copious amounts of memory. Consequently anything that can be done to reduce the amount of required memory is considered very worthwhile.

While we are on the subject of reducing the required amount of computer memory, recall that previously we showed that the stiffness matrix contains many zero elements, and that the location of the nonzero elements is governed by the numbering scheme used for the joints. It is possible, by judicious numbering of the joints, to obtain a stiffness matrix in which all of the nonzero elements lie within a band centered about the diagonal of the matrix. The width of the band is determined by the member with the largest difference in joint numbers. We can minimize the bandwidth by numbering the joints in such a way that the difference in joint numbers associated with each member is a minimum. For example, compare the two stiffness matrices for the structures shown in Fig. 7-16. The only difference in the structures is the numbering of the joints.

Obtaining a banded stiffness matrix is important because we know that all elements outside a certain region are zero, and therefore there is no need to store these elements in memory. If we use Gauss elimination to solve the system of equations, then it turns out that at each step of the forward reduction process, only a small portion of the coefficient matrix is affected. It is possible, then, to develop a Gauss elimination algorithm that retains in memory only that portion of the stiffness matrix that is affected by the

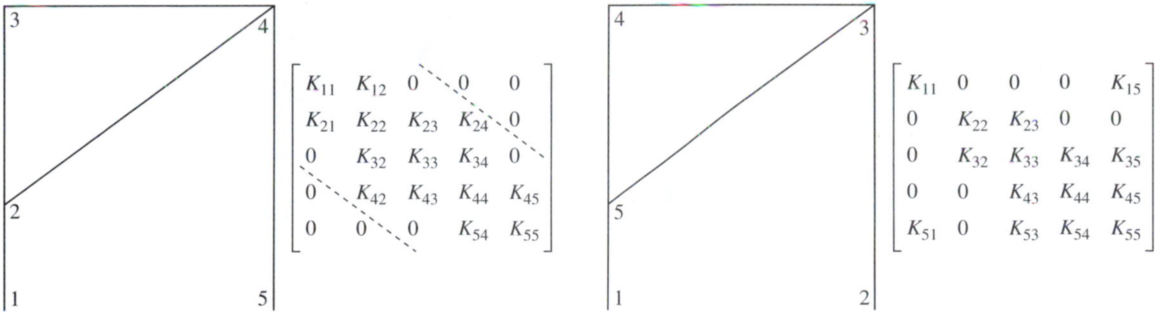


$$\begin{bmatrix} K_{11} & K_{12} & 0 & 0 & 0 \\ K_{21} & K_{22} & K_{23} & K_{24} & 0 \\ 0 & K_{32} & K_{33} & K_{34} & 0 \\ 0 & K_{42} & K_{43} & K_{44} & K_{45} \\ 0 & 0 & 0 & K_{54} & K_{55} \end{bmatrix} \qquad \begin{bmatrix} K_{11} & 0 & 0 & 0 & K_{15} \\ 0 & K_{22} & K_{23} & 0 & 0 \\ 0 & K_{32} & K_{33} & K_{34} & K_{35} \\ 0 & 0 & K_{43} & K_{44} & K_{45} \\ K_{51} & 0 & K_{53} & K_{54} & K_{55} \end{bmatrix}$$

**Figure 7-16** Only difference in structures is the numbering of the joints, but stiffness matrices are different

current reduction operations. The remainder of the stiffness matrix—those portions that have already been processed and those that have yet to be processed—can be stored on disk or tape. With this kind of solution algorithm, very large problems can be solved with computers that have only a modest amount of memory.

A more detailed discussion of the benefits of a banded stiffness matrix, and of the actual development of solution algorithms that make advantageous use of this bandedness, is beyond the scope of the present text. The interested reader should refer to texts such as [1, sec. 2.7], [2, secs. 6.2.3, 7.2.3], and [7, sec. 20.5], which are devoted to the subject of finite element analysis.

## 7-7 EXAMPLE

Analyze the structure shown in Fig. 7-17 to determine the joint displacements and the reactions. The complete set of equations for the structure is seen to be

$$\begin{Bmatrix} \boldsymbol{F}_1 \\ \boldsymbol{F}_2 \\ \boldsymbol{F}_3 \end{Bmatrix} = \begin{bmatrix} \boldsymbol{K}_{11} & \boldsymbol{K}_{12} & \boldsymbol{0} \\ \boldsymbol{K}_{21} & \boldsymbol{K}_{22} & \boldsymbol{K}_{23} \\ \boldsymbol{0} & \boldsymbol{K}_{32} & \boldsymbol{K}_{33} \end{bmatrix} \begin{Bmatrix} \boldsymbol{\Delta}_1 \\ \boldsymbol{\Delta}_2 \\ \boldsymbol{\Delta}_3 \end{Bmatrix}$$

where

$$\begin{aligned} \boldsymbol{K}_{11} &= \boldsymbol{K}_{11}^2 \\ \boldsymbol{K}_{22} &= \boldsymbol{K}_{22}^1 + \boldsymbol{K}_{22}^3 \\ \boldsymbol{K}_{33} &= \boldsymbol{K}_{33}^2 \end{aligned} \tag{a}$$

The rotation matrices, given by Eq. (7-45), are as follows:

$$\boldsymbol{R}_{12} = \begin{bmatrix} 0.832 & -0.555 & 0.0 \\ 0.555 & 0.832 & 0.0 \\ 0.0 & 0.0 & 1.0 \end{bmatrix}, \qquad \boldsymbol{R}_{23} = \begin{bmatrix} -1.0 & 0.0 & 0.0 \\ 0.0 & -1.0 & 0.0 \\ 0.0 & 0.0 & 1.0 \end{bmatrix} \tag{b}$$

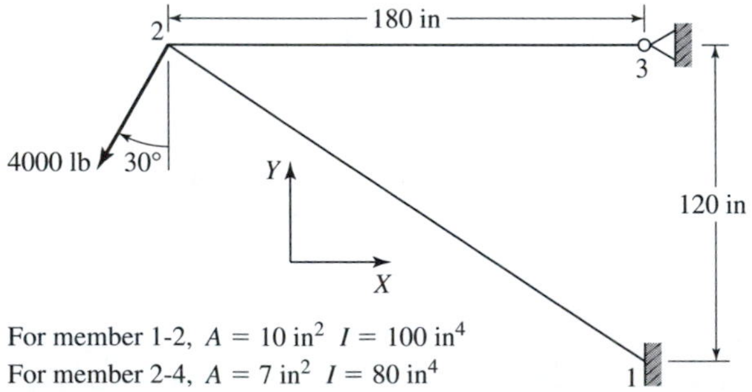


Figure 7-17

The rotation matrices for the opposite ends of the members can be found from these by simply changing the signs of the first two rows.

The direct member stiffness matrices in local coordinates are readily calculated from Eq. (7-24),

$$k_{11}^2 = \begin{bmatrix} 1.387 & 0.0 & 0.0 \\ 0.0 & 0.003556 & -0.3846 \\ 0.0 & -0.3846 & 55.47 \end{bmatrix} \times 10^6$$

$$k_{22}^3 = \begin{bmatrix} 1.167 & 0.0 & 0.0 \\ 0.0 & 0.004938 & -0.4444 \\ 0.0 & -0.4444 & 53.33 \end{bmatrix} \times 10^6 \qquad (c)$$

The cross stiffness matrices given by Eq. (7-23) are found from these by dividing element 3,3 by 2.

The member stiffness matrices in global coordinates can now be found using Eqs. (7-35), (*b*), and (*c*),

$$K_{11}^2 = \begin{bmatrix} 0.9612 & -0.6384 & -0.2133 \\ -0.6384 & 0.4292 & -0.3200 \\ -0.2133 & -0.3200 & 55.47 \end{bmatrix} \times 10^6, \quad K_{12} = \begin{bmatrix} -0.9612 & 0.6384 & -0.2133 \\ 0.6384 & -0.4292 & -0.3200 \\ 0.2133 & 0.3200 & 27.74 \end{bmatrix} \times 10^6$$

$$K_{21} = \begin{bmatrix} -0.9612 & 0.6384 & 0.2133 \\ 0.6384 & -0.4292 & 0.3200 \\ -0.2133 & -0.3200 & 27.74 \end{bmatrix} \times 10^6, \quad K_{22}^1 = \begin{bmatrix} 0.9612 & -0.6384 & 0.2133 \\ -0.6384 & 0.4292 & 0.3200 \\ 0.2133 & 0.3200 & 55.47 \end{bmatrix} \times 10^6$$

$$K_{22}^3 = \begin{bmatrix} 1.167 & 0.0 & 0.0 \\ 0.0 & 0.004938 & 0.4444 \\ 0.0 & 0.4444 & 53.33 \end{bmatrix} \times 10^6, \quad K_{23} = \begin{bmatrix} -1.167 & 0.0 & 0.0 \\ 0.0 & -0.004938 & 0.4444 \\ 0.0 & -0.4444 & 26.67 \end{bmatrix} \times 10^6$$

$$K_{32} = \begin{bmatrix} -1.167 & 0.0 & 0.0 \\ 0.0 & -0.004938 & -0.4444 \\ 0.0 & 0.4444 & 26.67 \end{bmatrix} \times 10^6, \quad K_{33}^2 = \begin{bmatrix} 1.167 & 0.0 & 0.0 \\ 0.0 & 0.004938 & -0.4444 \\ 0.0 & -0.4444 & 53.33 \end{bmatrix} \times 10^6$$

From these and Eq. (*a*) we see that

$$K_{22} = \begin{bmatrix} 2.128 & -0.6384 & 0.2133 \\ -0.6384 & 0.4341 & 0.7645 \\ 0.2133 & 0.7645 & 108.8 \end{bmatrix} \times 10^6$$

The global stiffness matrix for the structure can now be assembled according to the steps outlined in Section 7-6. This leads to the following set of equations for the structure shown in Fig. 7-17:

$$\begin{Bmatrix} P_{1X} \\ P_{1Y} \\ M_1 \\ P_{2X} \\ P_{2Y} \\ M_2 \\ P_{3X} \\ P_{3Y} \\ M_3 \end{Bmatrix} = \begin{bmatrix} 0.9612 & -0.6384 & -0.2133 & -0.9612 & 0.6384 & -0.2133 & 0.000 & 0.000 & 0.000 \\ -0.6384 & 0.4292 & -0.3200 & 0.6384 & -0.4292 & -0.3200 & 0.000 & 0.000 & 0.000 \\ -0.2133 & -0.3200 & 55.47 & 0.2133 & 0.3200 & 27.74 & 0.000 & 0.000 & 0.000 \\ -0.9612 & 0.6384 & 0.2133 & 2.128 & -0.6384 & 0.2133 & -1.167 & 0.000 & 0.000 \\ 0.6384 & -0.4292 & 0.3200 & -0.6384 & 0.4341 & 0.7645 & 0.000 & -0.004938 & 0.4444 \\ -0.2133 & -0.3200 & 27.74 & 0.2133 & 0.7645 & 108.8 & 0.000 & -0.4444 & 26.67 \\ 0.000 & 0.000 & 0.000 & -1.167 & 0.000 & 0.000 & 1.167 & 0.000 & 0.000 \\ 0.000 & 0.000 & 0.000 & 0.000 & -0.004938 & -0.4444 & 0.000 & 0.004938 & -0.4444 \\ 0.000 & 0.000 & 0.000 & 0.000 & 0.4444 & 26.67 & 0.000 & -0.4444 & 53.33 \end{bmatrix} \begin{Bmatrix} \Delta_{1X} \\ \Delta_{1Y} \\ \theta_1 \\ \Delta_{2X} \\ \Delta_{2Y} \\ \theta_2 \\ \Delta_{3X} \\ \Delta_{3Y} \\ \theta_3 \end{Bmatrix} \times 10^6$$

The boundary conditions have yet to be applied. Consequently, as shown earlier, the coefficient matrix is singular. (Note, for example, that the sum of rows 4 and 7 is the negative of row 1. Hence the determinant of the coefficient matrix is zero.) At node 1 all three degrees of freedom are suppressed, whereas at node 3 the translational degrees of freedom in the $X$ and $Y$ directions have been suppressed. Thus in the force-displacement equations for the structure in Fig. 7-17 we should cross out rows 1, 2, 3, 7, and 8 along with the corresponding columns. Substituting the loads shown in Fig. 7-17, the reduced system of equations reads

$$\begin{Bmatrix} -2000 \\ -3464 \\ 0.0 \\ 0.0 \end{Bmatrix} = \begin{bmatrix} 2.128 & -0.6384 & 0.2133 & 0.0 \\ -0.6384 & 0.4341 & 0.7645 & 0.4444 \\ 0.2133 & 0.7645 & 108.8 & 26.67 \\ 0.0 & 0.4444 & 26.67 & 53.33 \end{bmatrix} \begin{Bmatrix} \Delta_{2X} \\ \Delta_{2Y} \\ \theta_2 \\ \theta_3 \end{Bmatrix} \times 10^6$$

The solution of this set of equations gives the unknown displacements,

$$\begin{aligned} \Delta_{2_X} &= -6.141 \times 10^{-3} \text{ in} \\ \Delta_{2_Y} &= -17.32 \times 10^{-3} \text{ in} \\ \theta_2 &= 0.1120 \times 10^{-3} \text{ rad} \\ \theta_3 &= 0.08817 \times 10^{-3} \text{ rad} \end{aligned} \qquad (d)$$

The unknown reaction forces can now be found from Eq. (7-78*b*). Since in this example $\boldsymbol{\Delta}_k = \mathbf{0}$, Eq. (7-78*b*) is simply

$$\boldsymbol{F}_u = \boldsymbol{K}_{II,I}\boldsymbol{\Delta}_u$$

where

$$\boldsymbol{K}_{II,I} = \begin{bmatrix} -0.9612 & 0.6384 & -0.2133 & 0.000 \\ 0.6384 & -0.4292 & -0.3200 & 0.000 \\ 0.2133 & 0.3200 & 27.74 & 0.000 \\ -1.167 & 0.000 & 0.000 & 0.000 \\ 0.000 & -0.004938 & -0.4444 & -0.4444 \end{bmatrix} \times 10^6$$

This, together with Eqs. (*d*), yields, at joint 1,

$$F_X = -5.165 \times 10^3 \text{ lb}$$
$$F_Y = 3.468 \times 10^3 \text{ lb}$$
$$M = -3.741 \times 10^3 \text{ in-lb}$$

and at joint 3,

$$F_X = 7.165 \times 10^3 \text{ lb}$$
$$F_Y = -3.527 \text{ lb}$$

As a check on the accuracy of the solution, note that the sum of the forces in the $X$ direction gives $-4.883 \times 10^{-4}$; in the $Y$ direction, $-4.029 \times 10^{-4}$; and the sum of the moments about joint 1 gives $1.211 \times 10^{-1}$. These are small compared to the applied and reaction forces, indicating that overall equilibrium is maintained.

It is now a simple matter to determine the member end forces in local coordinates using Eqs. (7-29) and (7-22). From these, shear and bending moment diagrams can be drawn for each member and the integrity of the structure assessed.

## 7-8 NOTES CONCERNING THE STRUCTURE STIFFNESS MATRIX

We have already shown that the structure stiffness matrix is symmetric and that until constraints preventing rigid body motion are imposed on the structure it is also singular. We will now proceed to show that once rigid body motion is prevented, the stiffness matrix is nonsingular and all of its diagonal elements are positive. In order to accomplish this we will first define a positive definite matrix, then state some of its properties, and finally show that the structure stiffness matrix is positive definite.

A symmetric square matrix $\boldsymbol{A}$ is said to be positive definite if and only if

$$\boldsymbol{xAx} \geq \boldsymbol{0} \tag{7-82}$$

for any real $\boldsymbol{x} > \boldsymbol{0}$, and zero only if $\boldsymbol{x} = \boldsymbol{0}$.

The principal minors of a square matrix $\boldsymbol{A}$ of order $n$ are the determinants of square arrays of order $i \leq n$ whose main diagonals lie along the main diagonal of $\boldsymbol{A}$. If $\boldsymbol{A}$ is positive definite, all of its principal minors are positive. The proof of this is given in texts on linear algebra (see, for example, Hildebrand [4, sec. 1.20]).

If $\boldsymbol{A}$ is a square matrix of order $n$, its principal minors are

$$\begin{aligned}
i = 1:&\quad A_{11}, A_{22}, \ldots, A_{nn} \\
i = 2:&\quad \begin{vmatrix} A_{11} & A_{12} \\ A_{21} & A_{22} \end{vmatrix}, \quad \begin{vmatrix} A_{33} & A_{35} \\ A_{53} & A_{55} \end{vmatrix}, \quad \begin{vmatrix} A_{44} & A_{45} \\ A_{54} & A_{55} \end{vmatrix}, \ldots \\
&\quad \vdots \\
i = n:&\quad \begin{vmatrix} A_{11} & A_{12} & \cdots & A_{1n} \\ A_{21} & A_{22} & \cdots & A_{2n} \\ \vdots & \vdots & & \vdots \\ A_{n1} & A_{n2} & \cdots & A_{nn} \end{vmatrix}
\end{aligned} \tag{7-83}$$

The strain energy for the structure given by Eq. (7-60) can be written as

$$U = \tfrac{1}{2}\boldsymbol{F}^T\boldsymbol{\Delta} \tag{7-84}$$

but from Eq. (7-55),

$$\boldsymbol{F}^T = \boldsymbol{\Delta}^T\boldsymbol{K}^T \tag{7-85}$$

If we substitute this into the expression for $U$ and recognize that $\boldsymbol{K}$ is symmetric, then

$$U = \tfrac{1}{2}\boldsymbol{\Delta}^T\boldsymbol{K}\boldsymbol{\Delta} \tag{7-86}$$

We have shown in Section 5-6 that $U$ is positive definite. Now from the definition of a positive definite quantity and from Eq. (5-72*c*), $U$ can only be zero if the strains are zero throughout the structure. This, of course, can occur only if $\boldsymbol{\Delta} = \boldsymbol{0}$ or if the structure undergoes rigid body motion. If the structure has been constrained against rigid body motion, then

$$\boldsymbol{\Delta}^T\boldsymbol{K}\boldsymbol{\Delta} > 0 \tag{7-87}$$

It is 0 only if $\boldsymbol{\Delta} = \boldsymbol{0}$. Hence $\boldsymbol{K}$ is positive definite. Since, to prevent rigid body motion, three displacement components must be zero, the stiffness matrix $\boldsymbol{K}$ is not the full structure stiffness matrix. Rather, it would result from the imposition of rigid body constraints. Note also that the imposition of additional displacement constraints does not alter the positive definite nature of $\boldsymbol{K}$ since not all components of $\boldsymbol{\Delta}$ would be zero.

From the properties of positive definite matrices we conclude that after imposing rigid body constraints the structure stiffness matrix is nonsingular since its determinant is positive. Therefore Eq. (7-79) does, in fact, possess a solution.

We note also that all diagonal elements of $\boldsymbol{K}$ are positive.

## 7-9 FINITE ELEMENT ANALYSIS

The stiffness method of analysis developed in the preceding sections has provided us with a straightforward method of analyzing frame structures. However, many of the structures that we will be required to analyze cannot be represented by a collection of beams. Most machine parts have shapes that cannot be approximated as beams, yet accurate stress analysis of these parts is important if we are to determine the structural integrity of a machine. In the remaining sections of this chapter we will extend the ideas developed thus far to the analysis of complex shapes. The method that we will develop is referred to as *finite element analysis*. The ideas used in this method are a direct extension of those developed in the preceding sections.

In finite element analysis, we envision the actual solid as being replaced with a collection of small, but finite elements of relatively simple shapes. For example, we may replace the actual solid with a similar one made up of a collection of triangles, as shown in Fig. 7-18. The triangles enable us to approximate the curved boundary of the actual solid. The quality of the approximation is dependent, to some extent, on the number of triangles used to model or approximate the curved boundary. We then make some assumptions regarding the behavior of the individual elements. For example, we could

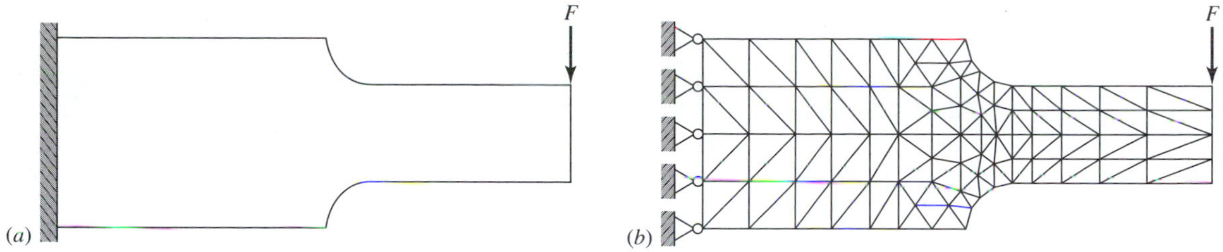


Figure 7-18 (*a*) Actual machine part, (*b*) finite element model using triangular elements

assume that within each triangular element the displacement varies linearly in each coordinate direction. Now, except for special load distributions, the displacement of the actual solid will not be a linear function of position but rather, some sort of smooth curve. If we require compatibility of the displacements at the triangle vertices in the model, then the effect of our assumption of linear displacement variation in each triangular element will be to replace the smooth curve representing the actual displacement with one that is piecewise linear. The resulting displacement distribution in the model will then be an approximation of the actual displacement distribution. It seems reasonable to expect that as we use more elements to model the same problem, the accuracy of the approximation should become better. This is, in fact, the case up to a certain point, beyond which computer round-off error becomes significant, and accuracy begins to decrease.

Development of the finite element method of analysis parallels the development of the stiffness method for frame analysis presented in the first part of this chapter. Once we have decided on an element shape, we develop the element stiffness matrix in the local coordinate system of the element, then transform the stiffness matrix into global coordinates, and invoke the conditions of equilibrium and compatibility at the joints (called *nodes* in finite element analysis) to develop the stiffness matrix for the entire assemblage of elements. In fact, a beam is simply one kind of finite element, which is used in the analysis of frame structures.

There are a couple of differences between the analysis of frame structures and the more general finite element analysis. In the analysis of frames the location of the elements is obvious—each beam in the structure constitutes one element. But in applying finite element analysis to a structural component, we must first decide on the number and location of elements, that is, we must *discretize* the component and actually develop the finite element model. In so doing, we must exercise our engineering judgment in order to ensure that the behavior of the model is a reasonably accurate representation of the behavior of the actual component. For example, in areas where we would anticipate a large stress gradient, such as in the vicinity of holes or other stress concentrators, we would need to use a sufficient number of elements in the model to depict the rapid variation in stress accurately.

The stiffness matrix that we developed for a beam is exact to within the limitations of elementary beam theory. However, for more general finite element models it will be necessary to make some assumptions regarding the behavior of the element. Except under special loading situations, the actual behavior of the element—that predicted by theory—will differ from our assumed behavior. That is to say, the assumed behavior will satisfy the theory on which the model is based only in an approximate sense.

Nevertheless, the finite element method has proven to be an accurate and reliable method of analysis, and it is undoubtedly one of the most significant developments in the history of engineering. Since its introduction in 1956 [6] this method has seen rapid growth, and its applications range from problems in structural analysis, both static and dynamic, to those in heat transfer, fluid mechanics, and electrostatics. It is used extensively by industry, in product design and in research and development, and has led to design savings in both weight and cost, as well as improved reliability.

Our goal in this chapter is to introduce the reader to the basics of finite element analysis. In so doing we will choose an approach that emphasizes the physical aspects of finite element analysis as it applies to structural analysis. Our approach is by no means the most advanced nor the most general approach to finite element analysis. However, by following the developments outlined in the following sections, interested readers should be able to expand their knowledge of the finite element method through any of the excellent references on the subject ([1], [2], and [7] represent a sampling of these).

For our purposes we will assume that the body we are analyzing is two-dimensional, in a state of plane stress, and that its behavior is described by the linear theory of elasticity for isotropic, homogeneous materials. We will develop the stiffness matrix for a triangular element, assuming that the displacement field within the element is a linear function of position, the so-called *constant strain triangle*.

## 7-10 CONSTANT STRAIN TRIANGLE

Let us consider the triangular element shown in Fig. 7-19. The vertices, or nodes, of the triangle are labeled $i$, $j$, and $k$. We establish a local coordinate system for the element in which the $x$ axis extends from $i$ to $j$, the $z$ axis is out of the plane, and the $y$ axis forms a right-handed coordinate system with $x$ and $z$. We assume that at each node of the triangle, forces in the $x$ and $y$ directions can be applied. These cause displacements $u$ and $v$ in the $x$ and $y$ directions, respectively, throughout the element. We would like to establish relationships between the displacements at the nodes and the nodal forces. Since each node can displace in both the $x$ and $y$ directions, the element has six degrees of freedom.

Since it is not possible to determine an exact relationship between nodal forces and displacements for this element, we will make the assumption that the displacements $u$ and $v$ vary linearly within the element. Thus, we can write

$$\begin{aligned} u &= A_0 + A_1 x + A_2 y \\ v &= B_0 + B_1 x + B_2 y \end{aligned} \tag{7-88}$$

The following conditions hold at the nodes:

$$\text{At } i, \qquad x = y = 0; \quad u = u_i, \quad v = v_i \tag{7-89a}$$

$$\text{At } j, \qquad x = x_j, \quad y = 0; \quad u = u_j, \quad v = v_j \tag{7-89b}$$

$$\text{At } k, \qquad x = x_k, \quad y = y_k; \quad u = u_k, \quad v = v_k \tag{7-89c}$$

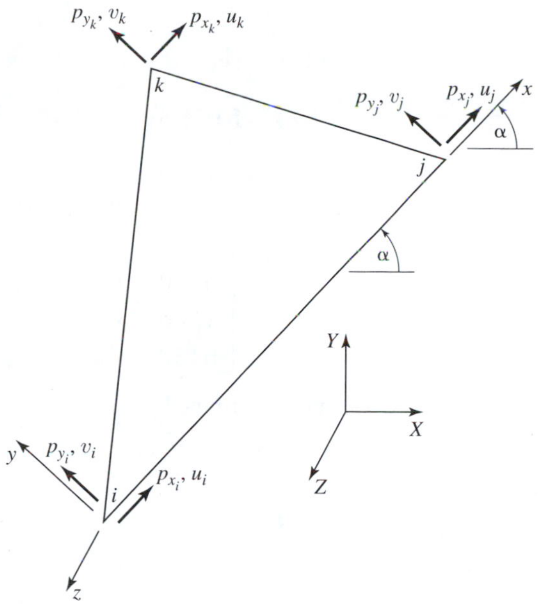


Figure 7-19

In these expressions subscripted quantities represent nodal values. For example, $u_j$ is the displacement component in the local $x$ direction at node $j$. We can determine the constants $A_m$, $B_m$, $m = 0, 1, 2$, in Eqs. (7-88) in terms of the six nodal displacements. Substituting $x = 0$ and $y = 0$ in Eqs. (7-88) and using Eq. (7-89$a$), we have

$$u_i = A_0, \qquad v_i = B_0 \tag{7-90}$$

Similarly, from Eqs. (7-89$b$, $c$) and (7-88),

$$\begin{aligned} u_j &= A_0 + A_1 x_j, & v_j &= B_0 + B_1 x_j \\ u_k &= A_0 + A_1 x_k + A_2 y_k, & v_k &= B_0 + B_1 x_k + B_2 y_k \end{aligned} \tag{7-91}$$

In matrix form these become

$$\begin{Bmatrix} u_i \\ v_i \\ u_j \\ v_j \\ u_k \\ v_k \end{Bmatrix} = \begin{bmatrix} 1 & 0 & 0 & 0 & 0 & 0 \\ 0 & 0 & 0 & 1 & 0 & 0 \\ 1 & x_j & 0 & 0 & 0 & 0 \\ 0 & 0 & 0 & 1 & x_j & 0 \\ 1 & x_k & y_k & 0 & 0 & 0 \\ 0 & 0 & 0 & 1 & x_k & y_k \end{bmatrix} \begin{Bmatrix} A_0 \\ A_1 \\ A_2 \\ B_0 \\ B_1 \\ B_2 \end{Bmatrix} \tag{7-92}$$

These can be written more compactly as

$$\boldsymbol{u}_{\text{node}} = \boldsymbol{cA} \tag{7-93}$$

where

$$\boldsymbol{u}_{\text{node}}^T = [u_i \quad v_i \quad u_j \quad v_j \quad u_k \quad v_k] \tag{7-94}$$

$$\boldsymbol{A}^T = [A_0 \quad A_1 \quad A_2 \quad B_0 \quad B_1 \quad B_2] \tag{7-95}$$

$$\boldsymbol{c} = \begin{bmatrix} 1 & 0 & 0 & 0 & 0 & 0 \\ 0 & 0 & 0 & 1 & 0 & 0 \\ 1 & x_j & 0 & 0 & 0 & 0 \\ 0 & 0 & 0 & 1 & x_j & 0 \\ 1 & x_k & y_k & 0 & 0 & 0 \\ 0 & 0 & 0 & 1 & x_k & y_k \end{bmatrix} \tag{7-96}$$

From Eq. (7-93), $\boldsymbol{A}$ can be determined,

$$\boldsymbol{A} = \boldsymbol{c}^{-1}\boldsymbol{u}_{\text{node}} \tag{7-97}$$

The inverse of $\boldsymbol{c}$ is*

$$\boldsymbol{c}^{-1} = \frac{1}{x_j y_k} \begin{bmatrix} x_j y_k & 0 & 0 & 0 & 0 & 0 \\ -y_k & 0 & y_k & 0 & 0 & 0 \\ x_k - x_j & 0 & -x_k & 0 & -x_j & 0 \\ 0 & x_j y_k & 0 & 0 & 0 & 0 \\ 0 & -y_k & 0 & y_k & 0 & 0 \\ 0 & x_k - x_j & 0 & -x_k & 0 & x_j \end{bmatrix} \tag{7-98}$$

*This can readily be seen by interchanging the rows of Eq. (7-92) to give

$$\begin{Bmatrix} u_i \\ u_j \\ u_k \\ v_i \\ v_j \\ v_k \end{Bmatrix} = \begin{bmatrix} 1 & 0 & 0 & 0 & 0 & 0 \\ 1 & x_j & 0 & 0 & 0 & 0 \\ 1 & x_k & y_k & 0 & 0 & 0 \\ 0 & 0 & 0 & 1 & 0 & 0 \\ 0 & 0 & 0 & 1 & x_j & 0 \\ 0 & 0 & 0 & 1 & x_k & y_k \end{bmatrix} \begin{Bmatrix} A_0 \\ A_1 \\ A_2 \\ B_0 \\ B_1 \\ B_2 \end{Bmatrix}$$

Since the coefficient matrix is a lower triangular matrix, its determinant is the product of the diagonal elements. This being nonzero, the inverse of the matrix exists, and since the first three equations are uncoupled from the last three, the inverse can be readily found through inversion by partitioning. This yields

$$\begin{Bmatrix} A_0 \\ A_1 \\ A_2 \\ B_0 \\ B_1 \\ B_2 \end{Bmatrix} = \frac{1}{x_j y_k} \begin{bmatrix} x_j y_k & 0 & 0 & 0 & 0 & 0 \\ -y_k & y_k & 0 & 0 & 0 & 0 \\ x_k - x_j & -x_k & x_j & 0 & 0 & 0 \\ 0 & 0 & 0 & x_j y_k & 0 & 0 \\ 0 & 0 & 0 & -y_k & y_k & 0 \\ 0 & 0 & 0 & x_k - x_j & -x_k & x_j \end{bmatrix} \begin{Bmatrix} u_i \\ u_j \\ u_k \\ v_i \\ v_j \\ v_k \end{Bmatrix}$$

When the nodal displacement vector is rearranged to correspond with Eq. (7-94), the resulting coefficient matrix is given by Eq. (7-98).

Note that all nodal displacements are contained in $\boldsymbol{u}_{\text{node}}$, and that if we define the $2 \times 1$ column matrix $\boldsymbol{u}_i$ as

$$\boldsymbol{u}_i^T = [u \quad v]_i \tag{7-99}$$

then $\boldsymbol{u}_{\text{node}}$ can be written as

$$\boldsymbol{u}_{\text{node}}^T = [\boldsymbol{u}_i \quad \boldsymbol{u}_j \quad \boldsymbol{u}_k] \tag{7-100}$$

Now for two-dimensional problems the strains are given in terms of the displacements by Eqs. (3-6), (3-7), and (3-9). With the displacements given by Eqs. (7-88), these become

$$\begin{aligned} \varepsilon_{xx} &= \frac{\partial u}{\partial x} = A_1 \\ \varepsilon_{yy} &= \frac{\partial v}{\partial y} = B_1 \\ \gamma_{xy} &= \frac{\partial u}{\partial y} + \frac{\partial v}{\partial x} = A_2 + B_1 \end{aligned} \tag{7-101}$$

From these expressions the origin of the phrase "constant strain triangle" becomes apparent. In matrix notation,

$$\boldsymbol{\varepsilon} = \boldsymbol{dA} \tag{7-102}$$

where

$$\boldsymbol{\varepsilon}^T = [\varepsilon_{xx} \quad \varepsilon_{yy} \quad \gamma_{xy}] \tag{7-103}$$

$$\boldsymbol{d} = \begin{bmatrix} 0 & 1 & 0 & 0 & 0 & 0 \\ 0 & 0 & 0 & 0 & 0 & 1 \\ 0 & 0 & 1 & 0 & 1 & 0 \end{bmatrix} \tag{7-104}$$

Substituting Eq. (7-97) into Eq. (7-102) we get

$$\boldsymbol{\varepsilon} = \boldsymbol{b}\boldsymbol{u}_{\text{node}} \tag{7-105}$$

where

$$\boldsymbol{b} = \boldsymbol{d}\boldsymbol{c}^{-1} \tag{7-106}$$

The constitutive equations for isotropic, elastic materials in a state of plane stress are given by Eqs. (4-21). In matrix form these are

$$\boldsymbol{\sigma} = \boldsymbol{D}\boldsymbol{\varepsilon} \tag{7-107}$$

where

$$\boldsymbol{\sigma}^T = [\sigma_{xx} \quad \sigma_{yy} \quad \sigma_{xy}] \tag{7-108}$$

$$\boldsymbol{D} = \frac{E}{1 - \nu^2} \begin{bmatrix} 1 & \nu & 0 \\ \nu & 1 & 0 \\ 0 & 0 & \dfrac{1-\nu}{2} \end{bmatrix} \tag{7-109}$$

Using Eq. (7-105) this becomes

$$\boldsymbol{\sigma} = \boldsymbol{D}\boldsymbol{b}\boldsymbol{u}_{\text{node}} \tag{7-110}$$

We now substitute the matrix expressions for stress and strain into our expression for strain energy [Eq. (5-73)], which in matrix form is given by

$$U = \frac{1}{2}\int_{\overline{V}} \boldsymbol{\sigma}^T \boldsymbol{\varepsilon}\, d\overline{V} \tag{7-111}$$

This yields

$$U = \frac{h}{2}\int_{A} \boldsymbol{u}^T_{\text{node}}\boldsymbol{b}^T\boldsymbol{D}\boldsymbol{b}\boldsymbol{u}_{\text{node}}\, d\mathcal{A} \tag{7-112}$$

where $h$ is the element thickness, $\mathcal{A}$ is its area, and we have recognized that $\boldsymbol{D}$ is symmetric.

The nodal displacements do not depend on the element coordinates nor, because of the assumed material homogeneity, do the elements of matrix $\boldsymbol{D}$. The simple nature of the assumed displacement field led to a matrix $\boldsymbol{b}$ whose elements are also independent of position. Therefore all matrix quantities can be taken outside the integral, resulting in the following expression for strain energy of the element,

$$U = \frac{h}{2}\boldsymbol{u}^T_{\text{node}}\boldsymbol{b}^T\boldsymbol{D}\boldsymbol{b}\boldsymbol{u}_{\text{node}}\mathcal{A} \tag{7-113}$$

The nodal forces $\boldsymbol{p}_{\text{node}}$ are given by

$$\boldsymbol{p}^T_{\text{node}} = [\boldsymbol{p}_i \quad \boldsymbol{p}_j \quad \boldsymbol{p}_k] \tag{7-114}$$

where

$$\boldsymbol{p}^T_i = [p_x \quad p_y]_i \tag{7-115}$$

They are determined from the strain energy by employing Castigliano's first theorem,

$$\boldsymbol{p}_{\text{node}} = \frac{\partial U}{\partial \boldsymbol{u}_{\text{node}}} = \mathcal{A}h\boldsymbol{b}^T\boldsymbol{D}\boldsymbol{b}\boldsymbol{u}_{\text{node}} \tag{7-116}$$

Details of the differentiation can be found in Appendix A.

Equation (7-116) relates nodal forces and displacements and can be written in the familiar form

$$\boldsymbol{p}_{\text{node}} = \boldsymbol{k}_e\boldsymbol{u}_{\text{node}} \tag{7-117}$$

Thus the element stiffness matrix is given by

$$\boldsymbol{k}_e = \mathcal{A}h\boldsymbol{b}^T\boldsymbol{D}\boldsymbol{b} \tag{7-118}$$

But for the triangular element shown in Fig. 7-19 the area is

$$\mathcal{A} = \tfrac{1}{2}x_j y_k \tag{7-119}$$

Therefore the element stiffness matrix becomes

$$\boldsymbol{k}_e = \tfrac{1}{2} h \boldsymbol{b}^T \boldsymbol{D} \boldsymbol{b} x_j y_k \tag{7-120}$$

Since $\boldsymbol{b}$ is a $3 \times 6$ matrix and $\boldsymbol{D}$ is a $3 \times 3$ matrix, the size of the element stiffness matrix is $(6 \times 3)(3 \times 3)(3 \times 6) = 6 \times 6$. That $\boldsymbol{k}_e$ is symmetric follows directly from Eq. (7-120).

Note that the simple form of the assumed displacement field enabled us to evaluate the integral in Eq. (7-112) exactly. Had our assumed displacement field been more complicated, it would have been necessary to resort to numerical integration to evaluate the right-hand side of Eq. (7-112). This is commonly done in finite element analysis and poses no real difficulty.

The local displacement components for each node are related to the global components $U$ and $V$ through the transformation equations given in Appendix B. Referring to Fig. 7-19, these are

$$\begin{Bmatrix} u \\ v \end{Bmatrix}_i = \begin{bmatrix} \cos\alpha & \sin\alpha \\ -\sin\alpha & \cos\alpha \end{bmatrix} \begin{Bmatrix} U \\ V \end{Bmatrix}_i \tag{7-121}$$

or

$$\boldsymbol{u}_i = \boldsymbol{r} \boldsymbol{U}_i \tag{7-122}$$

where

$$\boldsymbol{r} = \begin{bmatrix} \cos\alpha & \sin\alpha \\ -\sin\alpha & \cos\alpha \end{bmatrix} \tag{7-123}$$

Since the local coordinate system is the same for each node, we can write

$$\boldsymbol{u}_{\text{node}} = \boldsymbol{R}_e \boldsymbol{U}_{\text{node}} \tag{7-124}$$

where

$$\boldsymbol{R}_e = \begin{bmatrix} \boldsymbol{r} & \boldsymbol{0} & \boldsymbol{0} \\ \boldsymbol{0} & \boldsymbol{r} & \boldsymbol{0} \\ \boldsymbol{0} & \boldsymbol{0} & \boldsymbol{r} \end{bmatrix} \tag{7-125}$$

From Fig. 7-19 we see that

$$\begin{aligned} \cos\alpha &= \frac{X_j - X_i}{L_{ij}} \\ \sin\alpha &= \frac{Y_j - Y_i}{L_{ij}} \\ L_{ij} &= \sqrt{(X_i - X_j)^2 + (Y_i - Y_j)^2} \end{aligned} \tag{7-126}$$

In a similar manner we can calculate the local force components $\boldsymbol{p}_{\text{node}}$ in terms of the global components $\boldsymbol{P}_{\text{node}}$ as

$$\boldsymbol{p}_{\text{node}} = \boldsymbol{R}_e \boldsymbol{P}_{\text{node}} \tag{7-127}$$

The rotation matrix $\boldsymbol{R}_e$ is an orthogonal matrix, so we can write

$$\boldsymbol{U}_{\text{node}} = \boldsymbol{R}_e^T \boldsymbol{u}_{\text{node}} \tag{7-128a}$$

$$\boldsymbol{P}_{\text{node}} = \boldsymbol{R}_e^T \boldsymbol{p}_{\text{node}} \tag{7-128b}$$

Therefore if we premultiply Eq. (7-117) by $\boldsymbol{R}_e^T$ and use Eqs. (7-128*b*) and (7-124), we get

$$\boldsymbol{P}_{\text{node}} = \boldsymbol{R}_e^T \boldsymbol{k}_e \boldsymbol{R}_e \boldsymbol{U}_{\text{node}} \tag{7-129}$$

or

$$\boldsymbol{P}_{\text{node}} = \boldsymbol{K}_e \boldsymbol{U}_{\text{node}} \tag{7-130}$$

where

$$\boldsymbol{K}_e = \boldsymbol{R}_e^T \boldsymbol{k}_e \boldsymbol{R}_e \tag{7-131}$$

Since both $\boldsymbol{R}_e$ and $\boldsymbol{k}_e$ are $6 \times 6$ matrices, the global stiffness matrix $\boldsymbol{K}_e$ is also $6 \times 6$. The symmetry of $\boldsymbol{K}_e$ follows directly from Eq. (7-131).

Equation (7-130) can be expanded to give

$$\begin{Bmatrix} \boldsymbol{P}_i \\ \boldsymbol{P}_j \\ \boldsymbol{P}_k \end{Bmatrix} = \begin{bmatrix} \boldsymbol{K}_{ii} & \boldsymbol{K}_{ij} & \boldsymbol{K}_{ik} \\ \boldsymbol{K}_{ji} & \boldsymbol{K}_{jj} & \boldsymbol{K}_{jk} \\ \boldsymbol{K}_{ki} & \boldsymbol{K}_{kj} & \boldsymbol{K}_{kk} \end{bmatrix} \begin{Bmatrix} \boldsymbol{U}_i \\ \boldsymbol{U}_j \\ \boldsymbol{U}_k \end{Bmatrix} \tag{7-132}$$

By examining Eq. (7-132) we see that the $2 \times 2$ submatrices situated along the main diagonal of $\boldsymbol{K}_e$ give the forces at a node resulting from unit displacements at the same node. These submatrices are the direct stiffness matrices. The remaining $2 \times 2$ submatrices of $\boldsymbol{K}_e$ give the forces at one node resulting from unit displacements at another node, and are therefore the cross stiffness matrices.

## 7-11 ELEMENT ASSEMBLY

Now that we have developed the force-displacement relationships for the constant strain triangle, we can proceed to develop the assembly procedure used to construct the stiffness matrix for our structural model from the element stiffness matrices. Referring to the model shown in Fig. 7-20, let us look at the six elements enclosed by the dashed line. These are shown in Fig. 7-21. The nodes are identified by lowercase letters, the elements

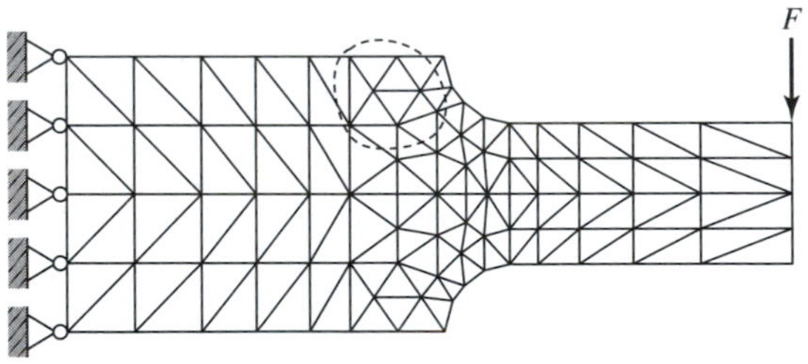


Figure 7-20

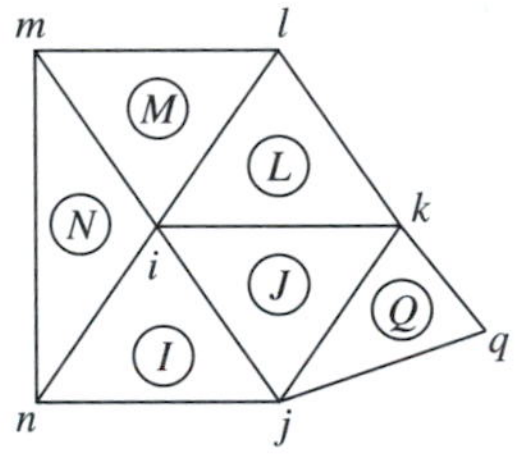


Figure 7-21

by encircled uppercase letters. Let us assume, for generality, that an external load acts at node $i$. This load has components in the global $X$ and $Y$ directions and, in matrix form, is given by

$$\boldsymbol{F}_i^T = [F_X \quad F_Y]_i \tag{7-133}$$

Here $\boldsymbol{F}_i$ must be in equilibrium with the nodal forces at $i$ of each element that connects at node $i$. Therefore,

$$\boldsymbol{F}_i = \boldsymbol{P}_i^I + \boldsymbol{P}_i^J + \boldsymbol{P}_i^L + \boldsymbol{P}_i^M + \boldsymbol{P}_i^N \tag{7-134}$$

where the subscripts refer to the node in question and the superscripts to the element. From Eq. (7-132) we have, for element $J$,

$$\boldsymbol{P}_i^J = \boldsymbol{K}_{ii}^J \boldsymbol{U}_i^J + \boldsymbol{K}_{ij}^J \boldsymbol{U}_j^J + \boldsymbol{K}_{ik}^J \boldsymbol{U}_k^J \tag{7-135}$$

where, again, the subscripts refer to nodes and the superscripts to elements. The $\boldsymbol{K}_{ij}^J$ here are $2 \times 2$ submatrices of the stiffness matrix for element $J$. Similar expressions can be written for the remaining elements appearing in Eq. (7-134), so that the equation becomes

$$\begin{aligned} \boldsymbol{F}_i = {} & (\boldsymbol{K}_{ii}^I \boldsymbol{U}_i^I + \boldsymbol{K}_{ij}^I \boldsymbol{U}_j^I + \boldsymbol{K}_{ip}^I \boldsymbol{U}_p^I) + (\boldsymbol{K}_{ii}^J \boldsymbol{U}_i^J + \boldsymbol{K}_{ij}^J \boldsymbol{U}_j^J + \boldsymbol{K}_{ik}^J \boldsymbol{U}_k^J) \\ & + (\boldsymbol{K}_{ii}^L \boldsymbol{U}_i^L + \boldsymbol{K}_{ik}^L \boldsymbol{U}_k^L + \boldsymbol{K}_{il}^L \boldsymbol{U}_l^L) + (\boldsymbol{K}_{ii}^M \boldsymbol{U}_i^M + \boldsymbol{K}_{il}^M \boldsymbol{U}_l^M + \boldsymbol{K}_{im}^M \boldsymbol{U}_m^M) \\ & + (\boldsymbol{K}_{ii}^N \boldsymbol{U}_i^N + \boldsymbol{K}_{im}^N \boldsymbol{U}_m^N + \boldsymbol{K}_{in}^N \boldsymbol{U}_n^N) \end{aligned} \tag{7-136}$$

Compatibility of displacements at nodes $i$ through $n$ requires that

$$\begin{aligned} \boldsymbol{U}_i^I &= \boldsymbol{U}_i^J = \boldsymbol{U}_i^L = \boldsymbol{U}_i^M = \boldsymbol{U}_i^N \equiv \boldsymbol{U}_i \\ \boldsymbol{U}_k^J &= \boldsymbol{U}_k^L = \boldsymbol{U}_k^Q \equiv \boldsymbol{U}_k \\ \boldsymbol{U}_j^I &= \boldsymbol{U}_j^J = \boldsymbol{U}_j^Q \equiv \boldsymbol{U}_j \\ \boldsymbol{U}_l^L &= \boldsymbol{U}_l^M \equiv \boldsymbol{U}_l \\ \boldsymbol{U}_m^M &= \boldsymbol{U}_m^N \equiv \boldsymbol{U}_m \\ \boldsymbol{U}_n^N &= \boldsymbol{U}_n^I \equiv \boldsymbol{U}_n \end{aligned} \tag{7-137}$$

With these, Eq. (7-136) becomes

$$\begin{aligned} \boldsymbol{F}_i = {} & (\boldsymbol{K}_{ii}^I + \boldsymbol{K}_{ii}^J + \boldsymbol{K}_{ii}^L + \boldsymbol{K}_{ii}^M + \boldsymbol{K}_{ii}^N)\boldsymbol{U}_i + (\boldsymbol{K}_{ij}^I + \boldsymbol{K}_{ij}^J)\boldsymbol{U}_j + (\boldsymbol{K}_{ik}^J + \boldsymbol{K}_{ik}^L)\boldsymbol{U}_k \\ & + (\boldsymbol{K}_{il}^L + \boldsymbol{K}_{il}^M)\boldsymbol{U}_l + (\boldsymbol{K}_{im}^M + \boldsymbol{K}_{im}^N)\boldsymbol{U}_m + (\boldsymbol{K}_{in}^N + \boldsymbol{K}_{in}^I)\boldsymbol{U}_n \end{aligned} \tag{7-138}$$

This equation gives the force at node $i$ in terms of displacements at nodes $i$, $j$, $k$, $l$, $m$, and $n$. Note, in particular, that the only nodes whose displacements contribute to the force at node $i$ are those that are directly connected to node $i$ by an element. For example, the displacement at node $q$ does not contribute to the force at node $i$ because there is no direct connection between $q$ and $i$. Similarly, displacements of nodes outside the dashed line in Fig. 7-20 will not contribute to the force at $i$ since they are not directly connected to $i$.

The coefficient of $\boldsymbol{U}_i$ is the sum of the direct stiffness contributions at node $i$, whereas the coefficients of the remaining displacements are sums of the appropriate cross stiffness contributions with respect to node $i$. The coefficient for a particular displacement, say, $\boldsymbol{U}_l$, contains contributions only from those elements that contain both the displacement node (in this case, node $l$) and the load node (node $i$).

Note that we required the displacements to be compatible at the nodes, but not the displacement gradients. Therefore the nodes are pinned.

Equations similar to Eq. (7-138) can be written for each node in the structure, so that we can write

$$\boldsymbol{F} = \boldsymbol{K}_s \boldsymbol{U} \tag{7-139}$$

where the subscript $s$ indicates that this equation is for the entire structure.

It is interesting to note that although the structure shown in Fig. 7-20 contains a large number of nodes, only a small number of nodes appears in Eq. (7-138). This indicates that a large percentage of the elements of the structure stiffness matrix are zero. This is typical of the stiffness matrices encountered in finite element analysis, which is why it is particularly important to number the nodes in such a way that the nonzero elements are contained within a relatively narrow band, and that a solution routine which takes advantage of the banded nature of the coefficient matrix be used. In this way, very large problems can be solved on computers with relatively small amounts of memory.

The actual assembly of the structure stiffness matrix follows the same procedure used for planar rigid frames. Since each node has two degrees of freedom, we start with a null matrix containing twice as many rows and columns as there are nodes in the structure. Then we calculate the element stiffness matrix for the first element of the structure. The location of each of the nine $2 \times 2$ submatrices of the element stiffness matrix in the structure stiffness matrix is determined by the actual node numbers associated with the element. Each of the $2 \times 2$ submatrices is *added* to the appropriate location in the structure stiffness matrix. This process is repeated for each element in the structure. Since the $2 \times 2$ submatrices for each element are added to whatever exists in the corresponding location of the structure stiffness matrix, summing of the appropriate submatrices as shown in Eq. (7-138) takes place automatically.

The vector of external loads $\boldsymbol{F}$ can be constructed from the external load information for the problem. As for our planar rigid frame problem, corresponding to each known load component there must be an unknown displacement component and vice versa.

## 7-12 EXAMPLE

As an example, consider the problem of a 5-in square plate with a 1-in-diameter hole located at the center. Two opposing edges of the plate are subjected to a normal traction of 1000 psi, as shown in Fig. 7-22*a*. We would like to find the stress concentration factor due to the hole.

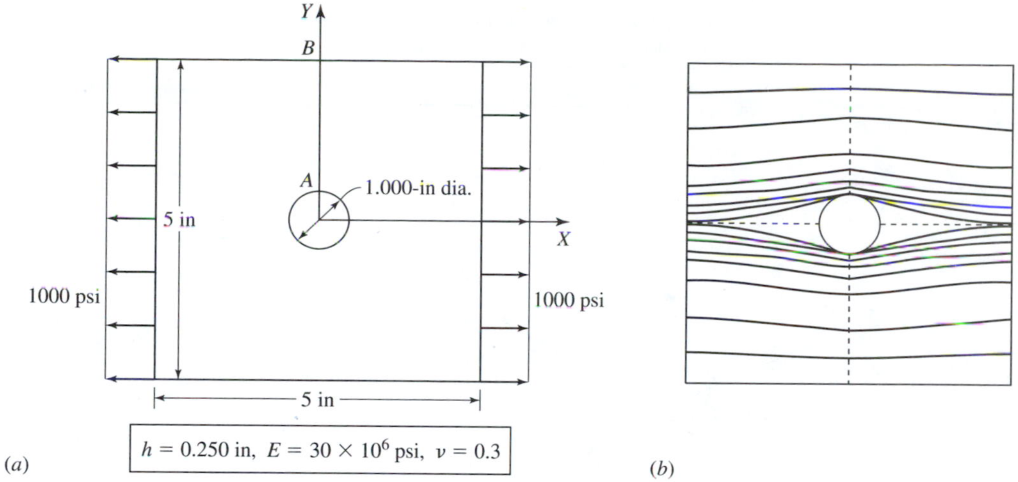


Figure 7-22

Because of the symmetry of the problem, it is necessary to analyze only one quadrant of the plate, thereby reducing the amount of computational effort required to solve the problem considerably. Based on the fluid flow analogy for the problem, shown in Fig. 7-22*b*, we anticipate high stresses near the hole at point *A* and also a large stress gradient in the vicinity of the hole. Therefore our finite element model should have many relatively small elements in the neighborhood of the hole to predict the rapid change in stress anticipated in that area accurately.

The finite element model used in this problem is shown in Fig. 7-23. The two edges of symmetry are placed on rollers so that there is no displacement in a direction normal to the edge, and no traction parallel to the edge. Rigid body motion is also constrained by the rollers. The distributed load is replaced with concentrated loads at the nodes under the assumption that, for each element with a side on the loaded edge, the two nodes on the loaded side carry an equal share of the resultant load. For example, on element 111, nodes 67 and 68 each carry one-half of the resultant load, 1000 psi × 0.32 in, or 160 lb/in (of thickness). Note, however, that node 67 also carries one-half of the resultant load associated with element 95, so the total concentrated load applied to node 67 is 500 lb/in. The particular finite element program used to perform the analysis gives stresses at the element centroid and also at the midpoint of the element sides. The maximum value of the normal stress $\sigma_{xx}$ as predicted by the computer program, is 3305 psi at the midpoint of the side of element 1 bounded by nodes 1 and 9.

According to the usual definition of stress concentration factor,

$$k_T = \frac{\sigma_{\text{max}}}{\sigma_{\text{nom}}}$$

where $\sigma_{\text{nom}}$ is calculated using the net area. For this problem the net area is the cross-sectional area of the plate at the hole center. The nominal stress is thus 1250 psi. This

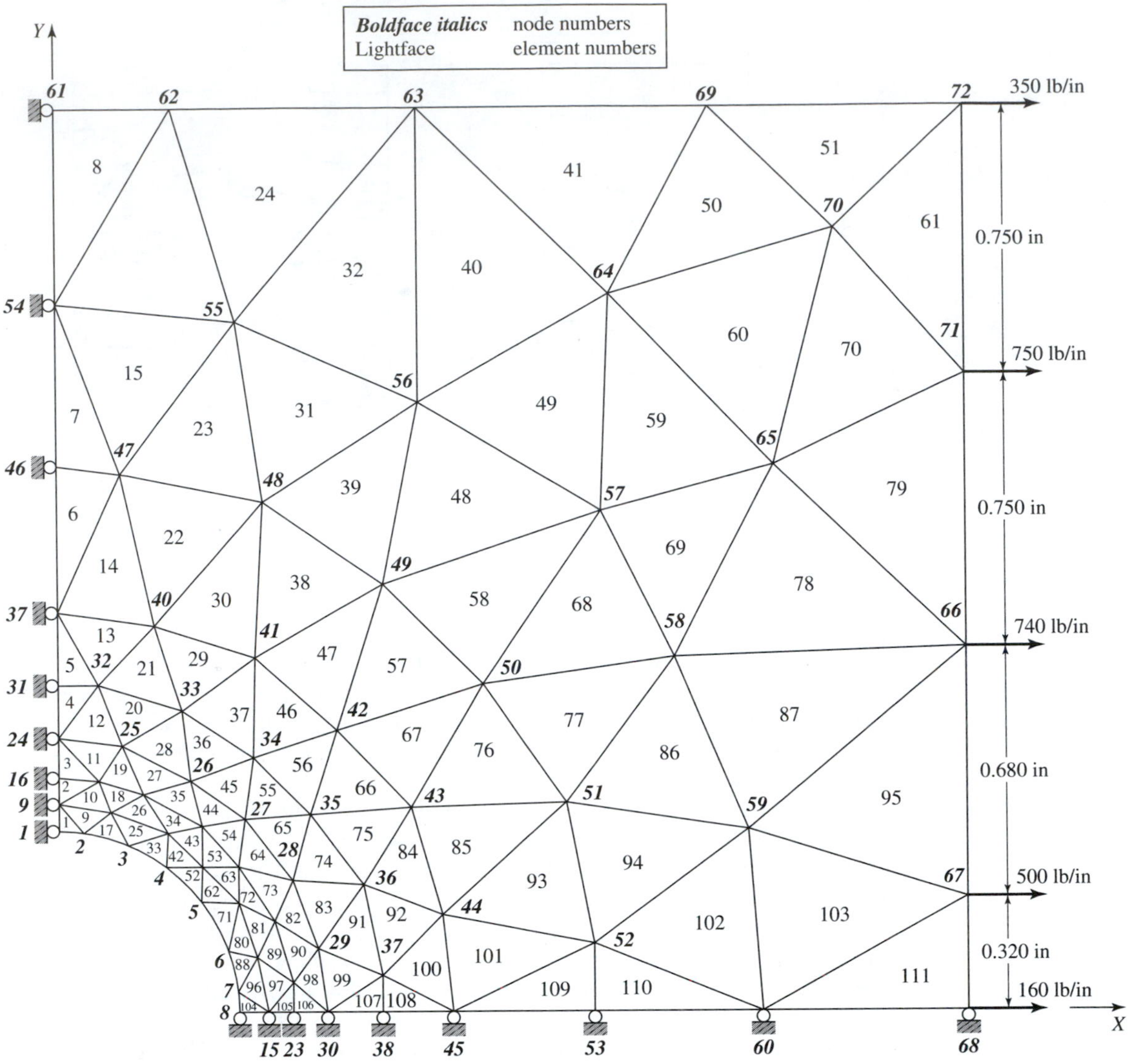


**Figure 7-23** Finite element model for plate with hole

yields a stress concentration factor of 2.64, slightly higher than the actual stress concentration factor for this problem of 2.51, as given in Peterson [5, p. 84]. A finer breakup in the vicinity of the hole would result in a better estimation of the stress concentration factor. The normal stress distribution along line *AB* of Fig. 7-22*a* is shown in Fig. 7-24.

This problem contains 126 active degrees of freedom. Hence 126 simultaneous equations must be solved. However, the maximum difference in node numbers for any element is 8, yielding a bandwidth of 31, that is, for a given row, all the nonzero elements are contained in the 31 columns centered on the main diagonal.

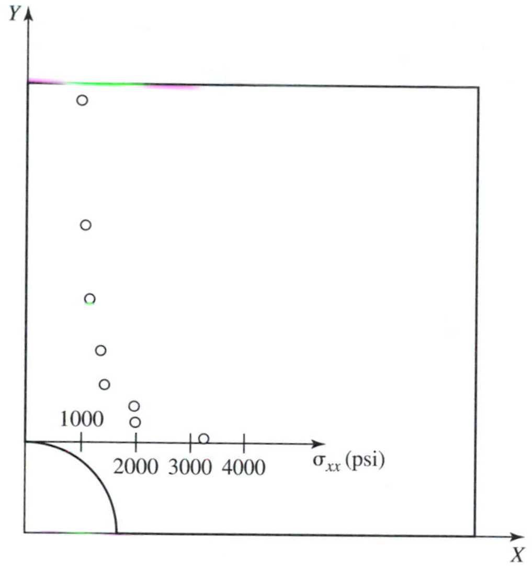


Figure 7-24

## 7-13 NOTES ON USING FINITE ELEMENT PROGRAMS

### 7-13-1 Interelement Compatibility

In our development of the finite element method, we assumed that the elements were connected to each other only at their nodes. We then required compatibility of displacements at the nodes, but did not concern ourselves with what happened along the boundary between two elements. Now, if we simply draw an assemblage of finite elements on an actual structural part and then load the part, of course we would see that across the boundaries of the elements compatibility is maintained as the part deforms. However, our mathematical model is only that—a model of the actual structure. As such, its behavior will not be the same as that of the actual structure, and therefore compatibility at the element boundaries cannot simply be presumed.

For our constant strain triangle, the assumed displacement field is linear, implying that the curvature along its boundaries is zero. This means that the element can deform only in such a way that its sides remain straight lines. Referring to Fig. 7-25, we see that

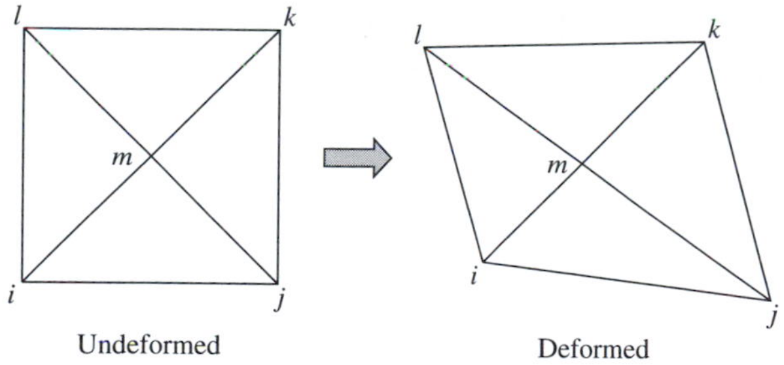


Figure 7-25 Interelement compatibility preserved

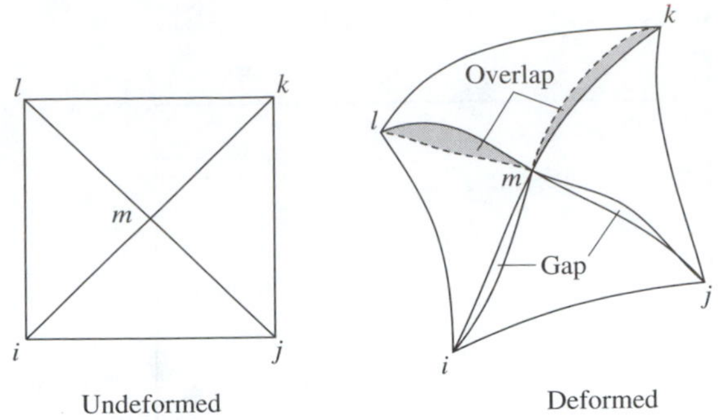


Figure 7-26 Interelement compatibility violated

for this element, compatibility across the element boundaries, or *interelement compatibility,* is preserved. Hence with constant strain elements compatibility throughout the model is maintained.

For other types of elements this is not necessarily the case. If the assumed displacement is not linear, then the elements might deform, as shown in Fig. 7-26, with gaps and/or overlaps along the element boundaries. Here interelement compatibility is clearly not satisfied.

In plate bending elements and thin-shell elements, which contain rotational as well as translational degrees of freedom, it is difficult to maintain slope compatibility accross element boundaries. It is also difficult to maintain interelement compatibility in problems where more than one type of element is used.

At first, lack of interelement compatibility may seem to be a serious problem. Indeed for many problems it is. Surprisingly, however, in many other problems some incompatibility along element boundaries is desirable, since it tends to compensate for the inherent overstiffness of finite elements discussed next.

## 7-13-2 Inherent Overstiffness in a Finite Element

In developing the stiffness matrix for the constant-strain triangle, we assumed a certain form for the displacement field. In essence, we decided how we would allow the element to deform, and thereby prevented the element from deforming into the correct shape except under special loading situations. The element, and therefore the structure, is constrained from deforming the way it wants to. This has the effect of stiffening the structure. As an example, think of a cantilever beam with a lateral load applied at the free end. If the center of the beam is constrained from deflecting laterally, a larger load is required to produce a specified end displacement than if the center were not constrained. Therefore the constraint has the effect of stiffening the beam.

Since the model is stiffer than the actual structure, if loads are specified throughout the structure, the calculated displacements will be less than the actual displacements. The strain energy, which must equal the work done by the external forces, will then be less than the actual strain energy since the actual loads are moving through distances that are underestimated. Conversely, if displacements are specified, the resulting loads will be greater than the actual loads since more force is required to deform a stiff structure. The strain energy will be greater than the actual value. This does not mean that all loads in the structure will be overestimated, but on the average, the calculated loads will be higher than the actual ones. Similarly, if loads are specified, on average, the calculated displacements will be low. In either case, stresses, which are determined from the displacements, will be underestimated.

Overstiffness is present in any finite element based on an assumed displacement field, since the actual displacement field will coincide with the assumed field only under special loading situations. In certain elements this effect is compensated for, to some degree, by the incompatibility along interelement boundaries. For those elements that employ numerical integration, overstiffness can be reduced by appropriately choosing the order of the numerical integration scheme (see, for example, [2, sec. 4.7]).

## 7-13-3 Bending and the Constant Strain Triangle

With its linear displacement field, the constant strain triangle is not well suited to problems in which significant bending is present. Bending, such as in a beam, implies curvature. But the curvature in the constant strain triangle is zero—its sides always remain straight. In beam problems the strain varies through the thickness, and the element edges parallel to the neutral axis must curve, something the constant strain triangle cannot do. Consequently in problems where bending may be significant, many constant strain triangles must be used in order to obtain accurate results.

As an example we will analyze the cantilever beam shown in Fig. 7-27 using the finite element model shown in Fig. 7-28. The model contains 128 elements, 85 nodes, and 160 active degrees of freedom. The analytical solution to this problem can be found in Gere and Timoshenko [3, p. 413]. For the parameters given in Fig. 7-27, the lateral displacement at point $A$ is 0.022 in. On the other hand, the displacement of the finite element model is 0.019 in, or about 86% of the actual displacement. The actual bending stress at point $B$ is 4500 psi, while the finite element model gives 3787 psi, low by about 16%.

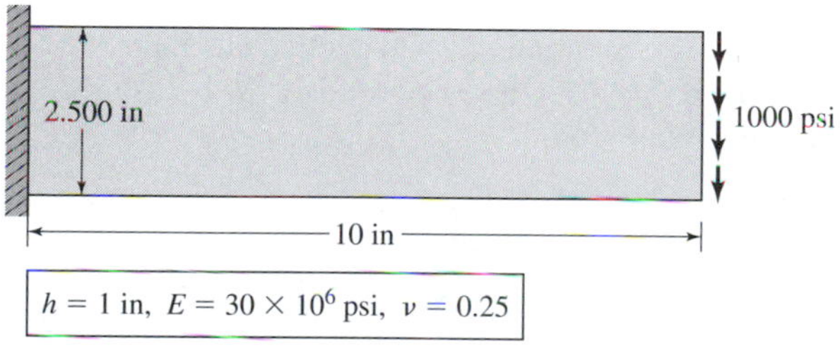


Figure 7-27

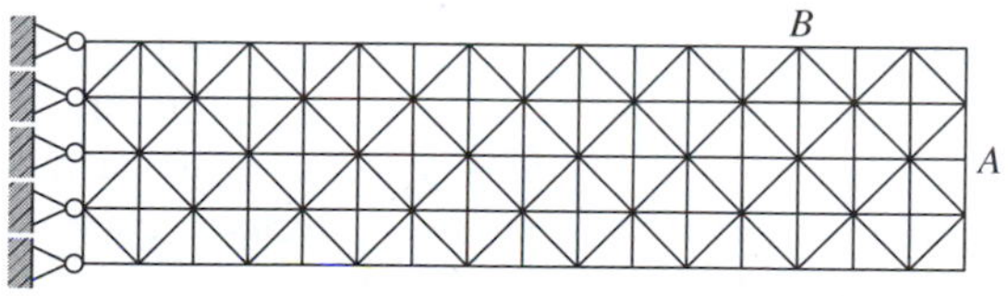


Figure 7-28

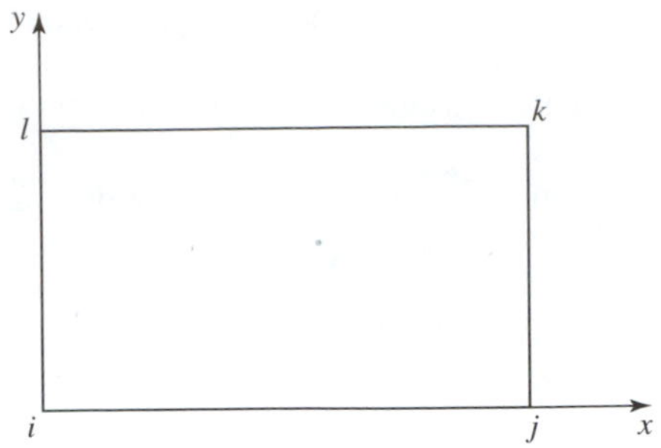


Figure 7-29

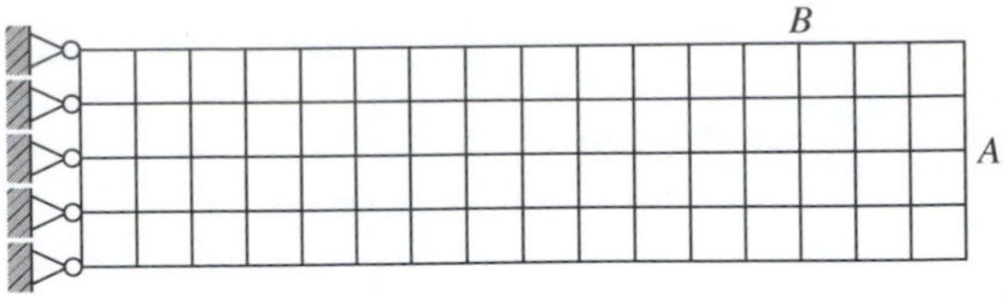


Figure 7-30

These results can be improved considerably with the use of the rectangular element shown in Fig. 7-29. This element is a special case of the more general four-node quadrilateral element. It has 8 degrees of freedom, and the assumed displacement field is

$$\begin{aligned} u &= A_0 + A_1 x + A_2 y + A_3 xy \\ v &= B_0 + B_1 x + B_2 y + B_3 xy \end{aligned} \tag{7-140}$$

This can also be written as

$$\begin{aligned} u &= (\overline{A}_0 + \overline{A}_1 x)(\overline{A}_2 + \overline{A}_3 y) \\ v &= (\overline{B}_0 + \overline{B}_1 x)(\overline{B}_2 + \overline{B}_3 y) \end{aligned} \tag{7-141}$$

Thus we see that the displacement varies linearly along the lines $y =$ constant and also along the lines $x =$ constant. For this reason the element is often referred to as *bilinear element*.

The stiffness matrix for this element can be developed using the same method that was employed for the constant strain triangle. Although more complicated than the constant strain triangle, the integral analogous to Eq. (7-112) can be evaluated analytically.

A finite element model for the problem shown in Fig. 7-27 using the rectangular element is shown in Fig. 7-30. Notice that the nodes are identical to those of Fig. 7-28, but that only 64 elements are needed. With this model the displacement at $A$ is 0.021 in, low by about 5%.

## 7-14 CLOSURE

In this chapter we have introduced the reader to the fundamental concepts of the finite element method. In order to stress the physical concepts involved, we started our discussion with the analysis of frame structures, involving only a knowledge of the behavior of beams. Virtually all of the ideas underlying finite element analysis are contained in Sections 7-2 through 7-8. Thus the reader with a firm grasp of this material already has a pretty good idea of what finite element analysis is all about. The additional details required for the finite element analysis of continuous bodies are addressed in Sections 7-9 through 7-13. This chapter is, of necessity, introductory in nature. As such, many important topics have not been addressed. For instance, we might enquire as to the conditions under which the finite element solution converges to the theoretical solution as the mesh is refined. Isoparametric elements, a subject that could easily occupy several chapters, have not been considered, nor have the numerical integration techniques usually employed with these elements. Other important topics include programming strategies, numerical difficulties and their avoidance, finite element analysis of plates and shells, finite element solutions of dynamics and vibration problems, and solution of nonlinear problems such as problems in plasticity.

The reader has no doubt already realized that the development of the finite element model for a particular problem and the accurate generation of the data for that model are not trivial tasks. This involves a certain amount of engineering judgment, gained through experience, on the part of the analyst.

In order to emphasize to the reader the physical concepts underlying the finite element method in structural analysis we have chosen to keep the mathematical approach to finite element analysis simple. However, many of the more advanced concepts of finite element analysis can only be understood with the aid of advanced mathematical tools. For readers with a background in variational calculus, it is possible to develop the finite element method for linear elasticity problems in a very elegant and satisfying manner by means of the theorem of minimum total potential energy. With this approach, the essential features of stiffness matrices for very complex elements become apparent, as does the method of structure stiffness matrix assembly.

It is significant that in linear structural analysis, the finite element method can be developed from basic physical concepts. However, in many fields of science (conduction heat transfer, for example) such an approach is not possible. In these fields the development of the finite element method must proceed along more mathematical lines.

Developments by the mathematics community in the area of finite element analysis have shown that the method is, generally speaking, simply an approximate method for solving partial differential equations. Consequently it is possible to study the finite element method from a purely mathematical viewpoint, devoid of any constraints imposed by physical considerations. This approach makes it possible to apply the method to physical problems for which a finite element formulation was previously unknown.

In closing this chapter we hope that the reader has gained a basic understanding of the finite element method as it applies to problems in linear elasticity, and that we have perhaps aroused the curiosity of some readers to the point where they will want to learn more of the finite element method, either through independent study or through a formal course on the subject.

## PROBLEMS

**7-1** Obtain the counterpart to Eq. (7-4) when the effects of transverse shear are included.

**7-2** The following local coordinate system is defined for member $ij$ shown in Fig. P7-2: the $x$ axis extends along the axis of the member from joint $i$ to joint $j$; the $z$ axis extends out of

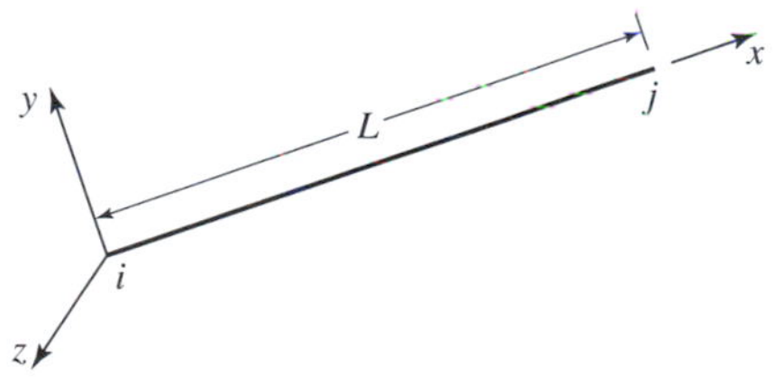


Figure P7-2

the plane; and the local $y$ axis forms a right-handed coordinate system with $x$ and $z$. With this definition, the orientation of the local coordinate system is the same for each end of the member. Derive the force-deflection relations for the member with this local coordinate system.

**7-3** For the member shown in Fig. P7-3 derive a general expression for the stiffness matrices $[K_{ii}^j]$ and $[K_{ij}]$ in the global coordinate system in terms of $A$, $E$, $I$, $L$, and $\alpha$. Verify your result by applying it to member 4-7 in Problem 7-8.

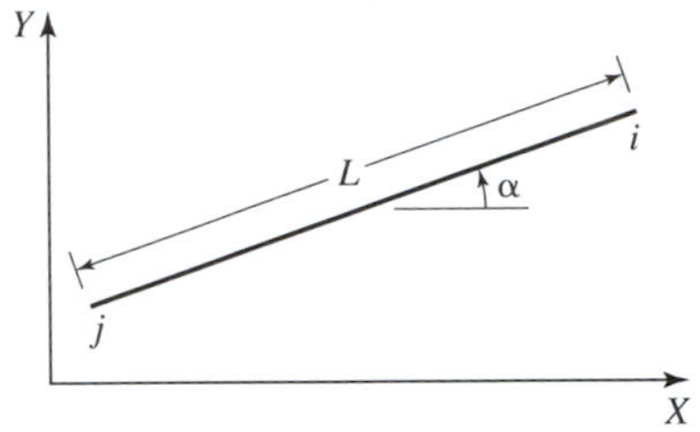


Figure P7-3

**7-4** In the absence of loads at points other than the ends, the beam bending equation reduces to $w^{iv} = 0$ and therefore has the solution $w(x) = a_0 + a_1 x + a_2 x^2 + a_3 x^3$, where $w$ is the transverse displacement. The longitudinal displacement $u$ is linear: $u(x) = b_0 + b_1 x$.

(a) Determine the constants $a_0, a_1, a_2, a_3, b_0, b_1$ in terms of the member end displacements $u_i, w_i, \theta_i, u_j, w_j, \theta_j$.

(b) Write the displacements in the following form:

$$w(x) = f_1(x)w_i + f_2(x)\theta_i + f_3(x)w_j + f_4(x)\theta_j$$

$$u(x) = f_5(x)u_i + f_6(x)u_j$$

and determine the functions $f_i(x)$, $i = 1, 2, \ldots, 6$.

(c) Use the expressions in Eq. (5-123) and then employ the theorem of minimum total potential energy to determine the stiffness matrix for the beam shown in Fig. P7-4. Note that the first integral in Eq. (5-123) is the strain energy due only to bending, and therefore the strain energy due to extension,

$$\int_0^l \frac{EA}{2}\left(\frac{du}{dx}\right)^2 dx$$

must be added.

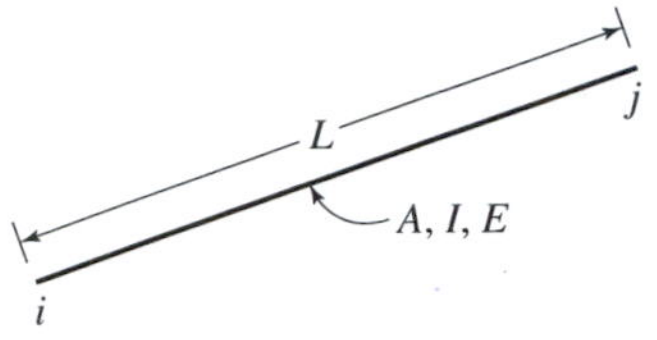


Figure P7-4

**7-5** Suppose one member of a frame is connected to the other members by frictionless pins, as shown in Fig. P7-5. How can the stiffness matrix given in Eq. (7-18) be used to represent this truss member?

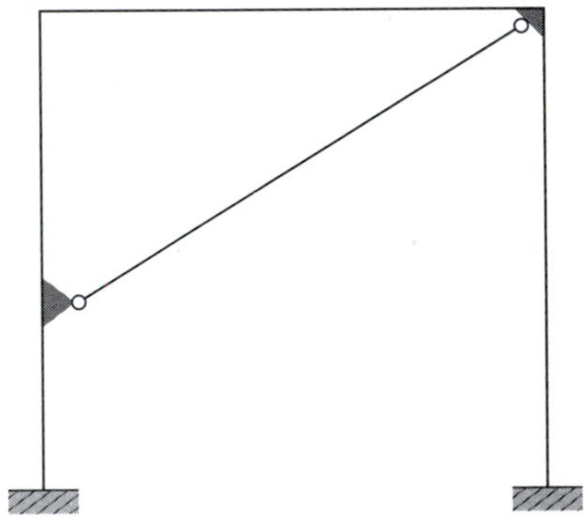

Figure P7-5

**7-6** The ring shown in Fig. P7-6$a$ is 70 ft in diameter and made from structural steel shapes (for example, I-beams). The ring, standing alone, cannot support its own weight, so cables have been added that radiate outward from the center, similar to bicycle spokes. These cables can carry only tensile loads, they can carry

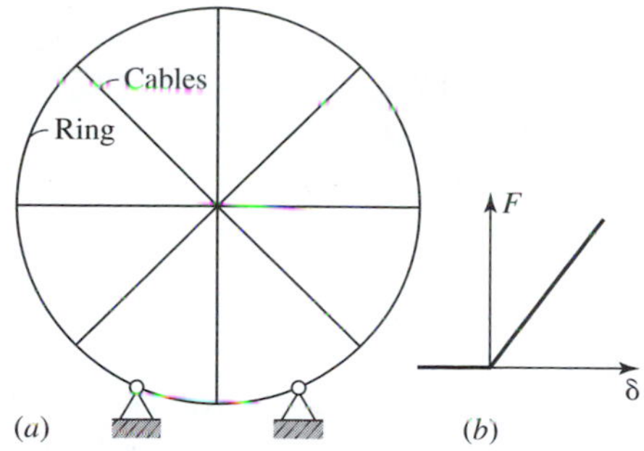


Figure P7-6

neither compression nor bending. Because of their inability to carry compression, the longtudinal force-deflection curve is nonlinear, as shown in Fig. P7-6*b*.

(a) Explain how the stiffness method of analysis developed in this chapter can be used in an iterative scheme to analyze the structure.

(b) What *two* checks must be made to determine whether the correct behavior of the cables has been captured?

**7-7** It is sometimes desirable to write Eq. (7-22) in the following form:

$$\begin{Bmatrix} p_{ij} \\ \delta_{ij} \end{Bmatrix} = [T] \begin{Bmatrix} p_{ji} \\ \delta_{ji} \end{Bmatrix}$$

Here $[T]$ is referred to as the *transfer matrix* because the behavior at end $j$ is transferred to end $i$ through this matrix. Obtain the transfer matrix.

*Problems 7-8 to 7-11*

The bicycle frame shown in Fig. P7-8 has been analyzed. The $X$ and $Y$ coordinates, in inches, are given by the numbers in parentheses. The following joint deflections were found:

| Joint | $\Delta_1$ (in) | $\Delta_2$ (in) | $\Delta_3$ (rad) |
|---|---|---|---|
| 1 | 0.00 | 0.00 | −0.1225E-02 |
| 2 | 0.2520E-01 | −0.1162E-01 | −0.6351E-03 |
| 3 | 0.2296E-01 | −0.1234E-01 | −0.9434E-03 |
| 4 | −0.2003E-02 | −0.2130E-01 | −0.1828E-02 |
| 5 | 0.2342E-01 | −0.4335E-01 | −0.2121E-03 |
| 6 | 0.2264E-01 | −0.4371E-01 | −0.3479E-03 |
| 7 | 0.2337E-01 | −0.4328E-01 | 0.9090E-03 |
| 8 | 0.1093 | 0.00 | 0.7614E-02 |

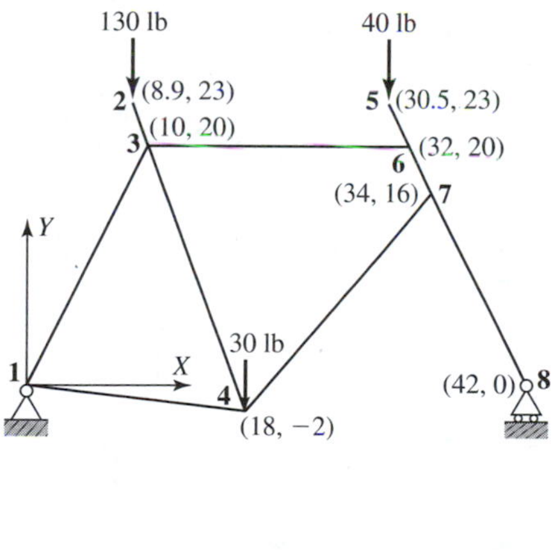


Figure P7-8

The structure has the following properties:

- For all members, $E = 30.0 \times 10^6$ psi
- For member 1-4, $A = 0.1022$ in$^2$, $I = 0.002780$ in$^4$
- For all other members, $A = 0.1340$ in$^2$, $I = 0.02470$ in$^4$

**7-8** Determine the member end forces in the local coordinate system for members 4-7, 6-7, and 7-8. Sketch each member in its undeformed and deformed configurations. Draw the shear and moment diagrams for each member.

**7-9** Determine the column vector of the member end forces of each member forming joint 7 in the global coordinate system, and verify that the joint is in equilibrium.

**7-10** Determine the column vector of the member end forces of each member forming joint 4 in the global coordinate system, and verify that the joint is in equilibrium.

**7-11** Determine the strain energy stored in the bicycle frame.

*Problems 7-12 to 7-15*

For each member in the structure shown in Fig. P7-12, $A = 7.45$ in$^2$, $I = 123.39$ in$^4$, and $E = 10.5 \times 10^6$ psi. The structure has been analyzed and the following displacements were obtained:

| Joint | *X* Displacement (in) | *Y* Displacement (in) | Slope (rad) |
|---|---|---|---|
| 1 | −0.3619 | −0.0196 | 0.0654 |
| 2 | 0.00 | 0.00 | −0.0322 |
| 3 | 0.00 | 0.00 | 0.00 |
| 4 | 0.00 | 0.00 | −0.0319 |
| 5 | 0.00 | 0.00 | 0.00 |

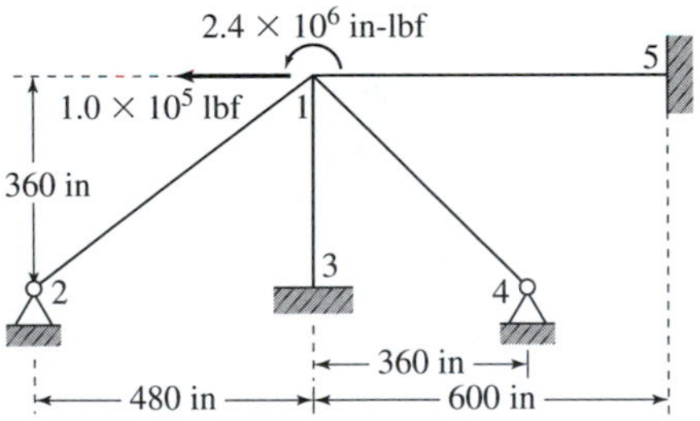


Figure P7-12

**7-12** Construct the rotation matrices for each member of the structure and use these to determine the member displacements in the local coordinate system. Make a sketch of each member showing its undeformed and deformed configurations.

**7-13** Construct the stiffness matrix for each member and, using the deflections in the local coordinate system, find the member end forces for each member. Make a sketch of each member showing the member end forces that act on the member. Draw the shear and bending moment diagrams for each member.

**7-14** Determine the strain energy stored in the structure.

**7-15** Obtain the member end forces in the global coordinate system for each member. With these, verify that joint 1 is in equilibrium. Draw the free-body diagram of the structure, indicating the reaction forces. Verify that the structure is in equilibrium.

**7-16** For the structure shown in Fig. P7-16 determine the following:

(a) The stiffness matrices in local coordinates.

(b) The rotation matrices.

(c) The stiffness matrices in global coordinates.

(d) The global stiffness matrix for the structure.

(e) The reduced stiffness matrix $K_{I,I}$.

(f) The unknown displacements.

(g) Draw a free-body diagram of the structure showing numerical values for the reactions and applied loads.

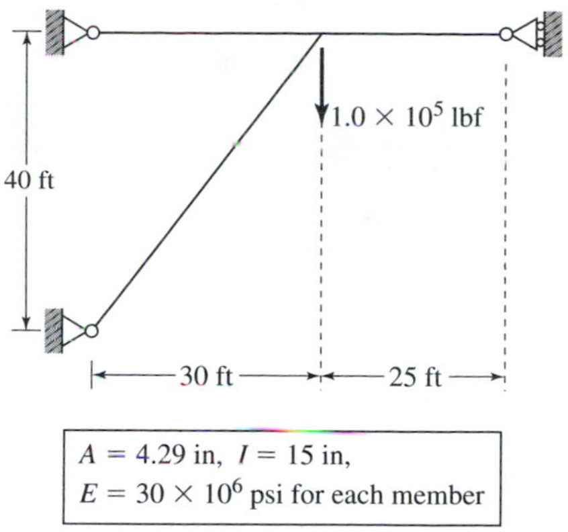


Figure P7-16

**7-17** Equation (7-88) can be written as

$$\boldsymbol{u} = H(x, y)\boldsymbol{A}$$

where $\boldsymbol{u} = [u \ v]^T$ and $\boldsymbol{A}$ is given by Eq. (7-95). With Eq. (7-97) this becomes

$$\boldsymbol{u} = N(x, y)\boldsymbol{u}_{\text{node}}$$

where $N(x, y) = H(x, y)\boldsymbol{c}^{-1}$ is the matrix of shape functions. Obtain the matrix of shape functions for a constant strain triangle.

**7-18** Obtain the matrix of shape functions for a bilinear rectangular element whose displacement field is given by either Eqs. (7-140) or Eqs. (7-141).

**7-19** Use the principle of minimum total potential energy to obtain Eq. (7-118).

## REFERENCES

[1] Cook, R. D., *Concepts and Applications of Finite Element Analysis,* 2nd ed. (Wiley, New York, 1981).

[2] Bathe, K. J., and E. L. Wilson, *Numerical Methods in Finite Element Analysis* (Prentice-Hall, Englewood Cliffs, NJ, 1976).

[3] Gere, J. M., and S. P. Timoshenko, *Mechanics of Materials,* 2nd ed. (PWS Engineering, Boston, MA, 1984).

[4] Hildebrand, F. B., *Methods of Applied Mathematics,* 2nd ed. (Prentice-Hall, Englewood Cliffs, NJ, 1965).

[5] Peterson, R. E., *Stress Concentration Design Factors* (Wiley, New York, 1953).

[6] Turner, M. J., R. W. Clough, H. C. Martin, and L. J. Topp, "Stiffness and Deflection Analysis of Complex Structures," *J. Aero. Sci.,* 23, pp. 805–823 (1956).

[7] Zienkiewicz, O. C., *The Finite Element Method in Engineering Science* (McGraw-Hill, London, 1971).

# BEAMS

In this chapter various advanced topics from simple beam theory are discussed. We begin with several methods of solution for bending of continuous beams. Next, unsymmetric bending is analyzed, followed by the analysis of curved beams (including rings and arches). The section on beams on elastic foundations constitutes a large portion of the chapter and also includes, because of mathematical analogy, storage tanks. The emphasis throughout the chapter is on the derivation and usage of influence functions (Green's functions). Symbolic manipulation packages are often used, and some outputs of sessions are provided. Deflections are calculated using mainly Castigliano's theorem. Limit analysis is treated very briefly, as are thermal stresses. This is followed by the application of approximate methods, which are illustrated by examples. Several Fortran programs are included on the disk. The chapter ends with an extensive discussion of piezoelectric and also composite beams. Several examples are solved.

## 8-1 BENDING OF CONTINUOUS BEAMS

### 8-1-1 Introduction

In this section we consider the bending of thin elastic beams with an arbitrary number of spans. Each span may have a different but constant stiffness and may carry arbitrarily distributed or concentrated loads. The equation for bending of an elastic beam is known to the reader from an introductory course of mechanics of materials (see [1], [21, p. 273]). It has the form

$$EIw^{\text{iv}}(x) = q(x) \tag{8-1}$$

where $E$ is Young's modulus, $I$ the moment of inertia of the beam cross section, $w(x)$ the deflection, and $q(x)$ the applied load. The derivation of Eq. (8-1) is based on an analysis of the equilibrium of a beam element, and it is repeated, with some modifications, in

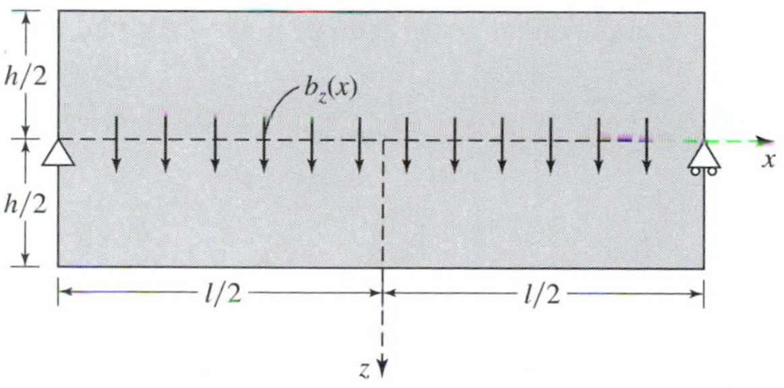


Figure 8-1

Section 8-4-1. Here we will show that under certain conditions this equation can also be derived using the equilibrium equation (2-55) (where the variable $y$ was replaced by $z$) obtained for an arbitrary two-dimensional solid,

$$\frac{\partial \sigma_{xx}}{\partial x} + \frac{\partial \sigma_{xz}}{\partial z} + b_x = 0$$
$$\frac{\partial \sigma_{xz}}{\partial x} + \frac{\partial \sigma_{zz}}{\partial z} + b_z = 0 \tag{8-2}$$

where $b_x$ and $b_z$ are components of the body forces. The beam to be analyzed, shown in Fig. 8-1, is assumed to have a rectangular cross section. We now make the assumptions, known from simple beam theory, that (1) the plane cross sections remain plane after deformation, (2) the displacement $v$ in the $y$ direction is zero, and (3) the deflection $w$ depends only on $x$,

$$u = u(x) = -z\frac{\partial w}{\partial x}, \qquad v = 0, \qquad w = w(x) \tag{8-3}$$

It is further assumed that the normal stress component in the $z$ direction is negligible,* that is, $\sigma_{zz} = 0$, and that

$$b_x = 0, \qquad b_z = b_z(x) \tag{8-4}$$

Using Eqs. (3-11) and Hooke's law (see Chapter 4) we have

$$\varepsilon_{xx} = \frac{\partial u}{\partial x} = -z\frac{\partial^2 w}{\partial x^2}, \qquad \sigma_{xx} = E\varepsilon_{xx} = -Ez\frac{\partial^2 w}{\partial x^2} \tag{8-5}$$

Substituting Eqs. (8-4) and (8-5) into the first of Eqs. (8-2) we obtain

$$\frac{\partial \sigma_{xz}}{\partial z} = -\frac{\partial \sigma_{xx}}{\partial x} = Ez\frac{\partial^3 w}{\partial x^3}$$

Integrating both sides with respect to $z$ gives

$$\sigma_{xz} = \frac{1}{2}Ez^2\frac{\partial^3 w}{\partial x^3} + C_1(x) \tag{8-6}$$

*This is an approximation, but since $\sigma_{zz}$ is assumed to vanish at $z = h/2$ and at $z = -h/2$ and since $h \ll l$, the stress $\sigma_{zz}$ is much smaller than the stress $\sigma_{xx}$, which is the basis for the design.

Now we assume that no tangential stress is applied to either of the horizontal faces of the beam,

$$\sigma_{xz}|_{z=-h/2} = \sigma_{xz}|_{z=h/2} = 0$$

This yields

$$0 = \frac{1}{8}Eh^2\frac{\partial^3 w}{\partial x^3} + C_1(x)$$

and substituting this expression into Eq. (8-6),

$$\sigma_{xz} = \frac{E}{2}\left(z^2 - \frac{h^2}{4}\right)\frac{\partial^3 w}{\partial x^3} \tag{8-7}$$

We note that Eqs. (3-11) and (8-3) imply that

$$\gamma_{xz} = \frac{\partial u}{\partial z} + \frac{\partial w}{\partial x} = -\frac{\partial w}{\partial x} + \frac{\partial w}{\partial x} = 0$$

Using Hooke's law we thus find that, contrary to the result (8-6), $\sigma_{xz} = 0$. This contradiction is the price we pay for making simplifying assumptions regarding the deformation.

Now that we found the shear stress [Eq. (8-7)], we use the second of Eqs. (8-2),

$$\frac{\partial \sigma_{xz}}{\partial x} + b_z = 0$$

obtaining

$$\frac{E}{2}\left(z^2 - \frac{h^2}{4}\right)\frac{\partial^4 w}{\partial x^4} + b_z = 0 \tag{8-8}$$

Integrating Eq. (8-8) with respect to $z$ along the height of the beam, we get

$$\frac{E}{2}\left(\frac{z^3}{3} - \frac{h^2}{4}z\right)\Bigg|_{-h/2}^{h/2}\frac{d^4 w}{dx^4} + \int_{-h/2}^{h/2} b_z(x)\,dz = 0 \tag{8-9}$$

or

$$-\frac{Eh^3}{12}\frac{d^4 w}{dx^4} + \int_{-h/2}^{h/2} b_z(x)\,dz = 0 \tag{8-10}$$

Now we multiply this by the width $b$ of the cross section. This gives again, as expected, Eq. (8-1),

$$EI\frac{d^4 w}{dx^4} = q(x) \tag{8-11}$$

Here $I = bh^3/12$ is the moment of inertia of the rectangular cross section of the beam and $q(x) = b_z bh$ is the resultant of the body force per unit length of the beam.

A continuous beam is a multispan structure such as seen in Fig. 8-2. Since the number of unknown reactions for each of the beams shown exceeds 3, which is the number of equations of equilibrium, they are *statically indeterminate*. The beam shown in Fig. 8-2*a* is $5 - 3 = 2$ times statically indeterminate whereas the beam of Fig. 8-2*b* is

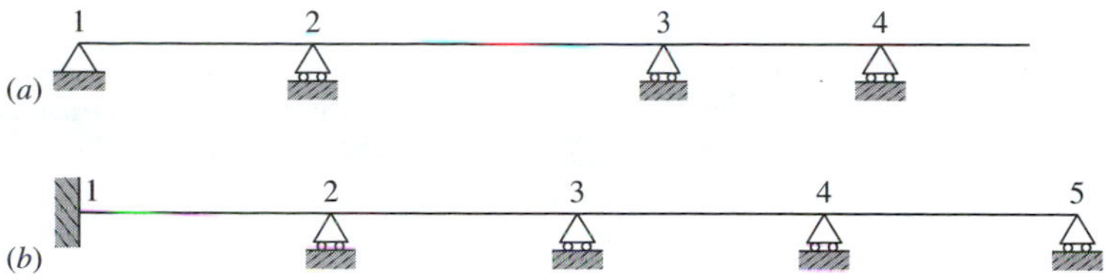


Figure 8-2

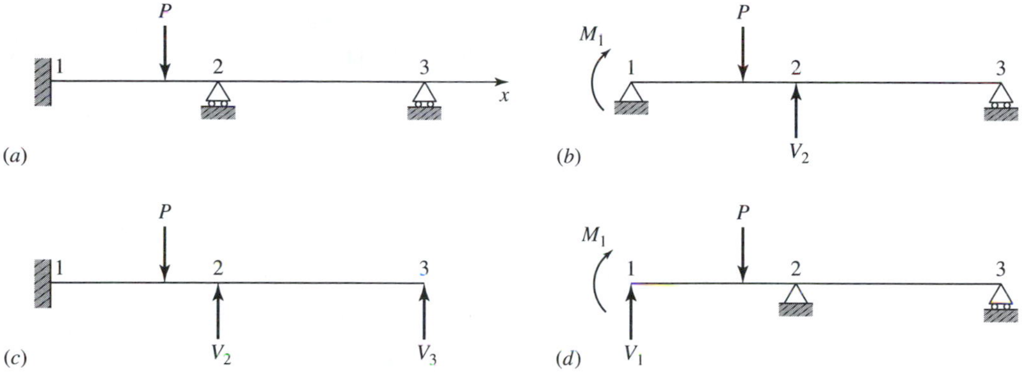


Figure 8-3 Replacement of statically indeterminate beams by equivalent statically determinate beams

$7 - 3 = 4$ times statically indeterminate. The first objective of the analysis, as always in statically indeterminate problems, is to replace the given beam by an equivalent statically determinate one. This can be achieved in a variety of ways, as shown, for instance, in Fig. 8-3. The beams in Fig. 8-3*b*, *c*, and *d* become equivalent to Fig. 8-3*a* when one pair of the following conditions is met

$$\varphi_1 = 0, \qquad w_2 = 0 \tag{8-12$a$}$$

$$w_2 = 0, \qquad w_3 = 0 \tag{8-12$b$}$$

$$\varphi_1 = 0, \qquad w_1 = 0 \tag{8-12$c$}$$

Here $w$ is the vertical deflection and $\varphi = dw/dx$ is the slope. Hence in any of the cases shown in Fig. 8-3 we have two unknown quantities and two corresponding conditions [Eqs. (8-12)]. Whatever method is chosen to calculate the unknown quantities, it always leads in this case to two algebraic equations in two unknowns, that is, the choice of an equivalent beam does not affect the number of algebraic equations to be ultimately solved. However, the structure of these equations depends on this choice. Once the unknown quantities have been found, the final objective—determining deflections and stresses—is easily achieved using simple beam theory.

From an introductory course the reader is familiar with a procedure based on the *three moments* theorem. As a consequence of this procedure the matrix of the coefficients of the unknowns becomes banded with the bandwidth of no more than three elements. This was of course very important, particularly when using an iteration method or Gaussian elimination. Since continuous beams have usually a rather small number of spans, the importance of bandedness faded with the advent of the modern computer (although it is still very important when dealing with large systems of equations).

In the remainder of this section we will show other approaches, one consisting in integrating the differential equation for bending, the other in applying Castigliano's theorem to an arbitrary equivalent beam.

## 8-1-2 Method of Initial Parameters

One method for analyzing continuous beams is the method of *initial parameters*. It is closely related to the method of transfer functions, popular in studying vibrations of beams, and it deserves our attention. Its advantage in the application to static problems lies in the convenience of having formulas for all static quantities (displacements, slopes, bending moments, and shear forces) once the superfluous quantities are determined. Consider a beam of constant stiffness $EI$ with completely arbitrary boundary conditions at the supports (Fig. 8-4*a*). We begin by removing all the intermediate supports, rendering the system free of constraints (Fig. 8-4*b*). Next we recall the differential equation for the bending of a beam,

$$EIy^{\text{iv}}(x) = q(x)$$

Finally we incorporate into $q(x)$, appearing on the right-hand side, the unknown intermediate reactions shown in Fig. 8-4*b* and treat them as if they were known concentrated forces. To do this we introduce some new mathematical concepts. First we represent a concentrated force, say, $P = 1$, as if it were a continuous load of the intensity $\delta = 1/\varepsilon$ distributed over a very short segment $\varepsilon$ (Fig. 8-5). The resultant of this load is $\delta \cdot \varepsilon = 1$.

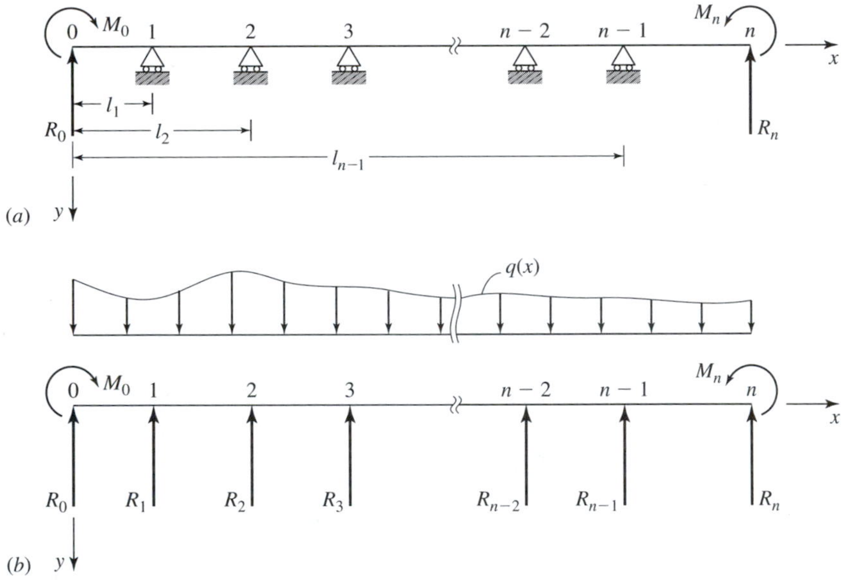


Figure 8-4

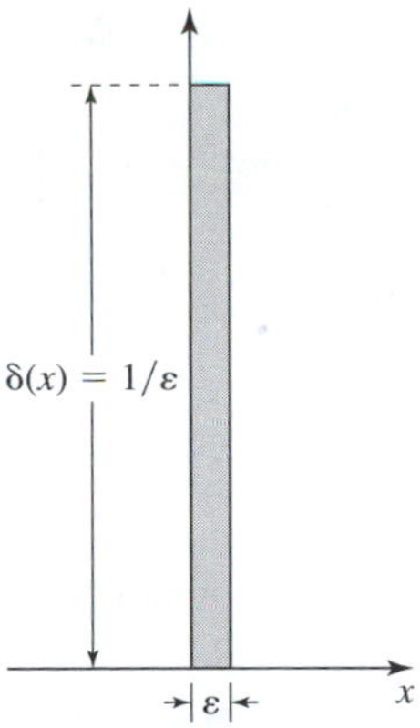


Figure 8-5 Derivation of Dirac's delta

Next we decrease $\varepsilon$. The intensity $\delta(x)$ increases automatically, holding the resultant equal to 1. In the limit, when $\varepsilon \to 0$, the intensity $\delta \to \infty$ while the product $\delta \cdot \varepsilon = 1$. We can represent this analytically as

$$\int_0^\infty \delta(x)\, dx = 1 \tag{8-13}$$

which says that the resultant of the load of intensity $\delta(x)$ equals 1. The quantity $\delta(x)$, which has the property (8-13) and which behaves according to the law

$$\delta(x) = \begin{cases} \infty, & x = 0 \\ 0, & \text{otherwise} \end{cases} \tag{8-14}$$

is called *Dirac's delta** or *impulse function* (see, for example, [24, pp. 79–80]). Moving the origin by $a$ we change the definitions (8-13) and (8-14) to

$$\int_0^\infty \delta(x-a)\, dx = 1, \qquad \delta(x-a) = \begin{cases} \infty, & x = a \\ 0, & \text{otherwise} \end{cases} \tag{8-15}$$

From Eq. (8-15) we have

$$\int_0^\infty f(x)\delta(x-a)\, dx = f(a) \tag{8-16}$$

Now we see that a concentrated force $P$ located at $x = a$ can be represented analytically as a continuous load,

$$P\delta(x-a) \tag{8-17}$$

Integrating Dirac's delta $\delta(x-a)$ with respect to $x$ we note that the value of the integral is zero as long as $x < a$, whereas by virtue of Eq. (8-15) the value becomes 1 when $x$ exceeds $a$. This can be written in the form

$$\int_0^x \delta(x-a)\, dx = H(x-a) \tag{8-18}$$

*Paul Adrien M. Dirac (1902–1984), English physicist.

where $H(x-a)$, known as the *Heaviside** or *step function,* has the following properties:

$$H(x-a) = \begin{cases} 1, & x > a \\ 0, & x < a \end{cases} \tag{8-19}$$

Using these definitions we have

$$\begin{aligned} EIy^{\text{iv}}(x) = q(x) - R_1\delta(x-L_1) - R_2\delta(x-L_2) - \cdots \\ - R_{n-1}\delta(x-L_{n-1}), \qquad 0 < x < L \end{aligned} \tag{8-20}$$

Then, integrating Eq. (8-20), we obtain

$$EIy''' = \int_0^x q(x_1)\,dx_1 - R_1H(x-L_1) - \cdots - R_{n-1}H(x-L_{n-1}) + C_1 \tag{8-21}$$

At this point we change $x_1$ to $x_2$ and $x$ to $x_1$ in Eq. (8-21). Finally, multiplying the entire equation by $dx_1$ and integrating with respect to $x_1$ in the range from 0 to $x$, we get

$$\begin{aligned} EIy''(x) = \int_0^x \int_0^{x_1} q(x_2)\,dx_1\,dx_2 - R_1(x-L_1)H(x-L_1) - \cdots \\ - R_{n-1}(x-L_{n-1})H(x-L_{n-1}) + C_1x + C_2 \end{aligned} \tag{8-22}$$

A similar procedure is repeated, giving

$$\begin{aligned} EIy'(x) = \int_0^x \int_0^{x_1} \int_0^{x_2} q(x_3)\,dx_1\,dx_2\,dx_3 - R_1\frac{(x-L_1)^2}{2}H(x-L_1) - \cdots \\ - R_{n-1}\frac{(x-L_{n-1})^2}{2}H(x-L_{n-1}) + C_1\frac{x^2}{2} + C_2x + C_3 \end{aligned} \tag{8-23}$$

and similarly,

$$\begin{aligned} EIy(x) = \int_0^x \int_0^{x_1} \int_0^{x_2} \int_0^{x_3} q(x_4)\,dx_1\,dx_2\,dx_3\,dx_4 - R_1\frac{(x-L_1)^3}{6}H(x-L_1) - \cdots \\ - R_{n-1}\frac{(x-L_{n-1})^3}{6}H(x-L_{n-1}) + C_1\frac{x^3}{6} + C_2\frac{x^2}{2} + C_3x + C_4 \end{aligned} \tag{8-24}$$

We notice that four unknown constants ($C_1$, $C_2$, $C_3$, $C_4$) and $n-1$ unknown reactions appear in Eq. (8-24). Next we replace the quadruple integral, depending on $q(x)$, by a single integral. To do so we use the formula [13, pp. 222–225]

$$\int_0^x \int_0^{x_1} \int_0^{x_2} \int_0^{x_3} q(x_4)\,dx_4 = \frac{1}{3!}\int_0^x (x-x_1)^3 q(x_1)\,dx_1 \tag{8-25}$$

---

*Oliver Heaviside (1850–1925), English electrical engineer.

The constants $C$ are in turn eliminated in favor of the values of $y(x)$ and its derivatives at $x = 0$. By denoting

$$y_0 = y(x), \qquad y_0' = y'(x), \qquad M_0 = -EIy''(x), \qquad R_0 = -EIy'''(x), \qquad x = 0$$

we obtain from Eqs. (8-21)–(8-24),

$$-R_0 = C_1, \qquad -M_0 = C_2, \qquad EIy_0' = C_3, \qquad EIy_0 = C_4 \tag{8-26}$$

Now Eqs. (8-21)–(8-24) take the form

$$\begin{aligned}
-V(x) = EIy'''(x) &= \int_0^x q(x_1)\,dx_1 - R_1 H(x - L_1) - \cdots \\
&\quad - R_{n-1} H(x - L_{n-1}) - R_0 \\
-M(x) = EIy''(x) &= \int_0^x (x - x_1) q(x_1)\,dx_1 - R_1 (x - L_1) H(x - L_l) - \cdots \\
&\quad - R_{n-1}(x - L_{n-1}) H(x - L_{n-1}) - R_0 x - M_0 \\
EIy'(x) &= \frac{1}{2}\int_0^x (x - x_1)^2 q(x_1)\,dx_1 - R_1 \frac{(x - L_1)^2}{2} H(x - L_1) - \cdots \\
&\quad - R_{n-1}\frac{(x - L_{n-1})^2}{2} H(x - L_{n-1}) - R_0 \frac{x^2}{2} - M_0 x + EIy_0' \\
EIy(x) &= \frac{1}{3!} q_0 \int_0^x (x - x_1)^3\,dx_1 - R_1 \frac{(x - L)^3}{6} H(x - L) - R_0 \frac{x^3}{6} - M_0 \frac{x^2}{2}
\end{aligned} \tag{8-27}$$

Although there are still $n + 3$ unknowns in these expressions, two are now clearly associated with the boundary conditions at $x = 0$. Thus for a simply supported end,

$$y_0 = M_0 = 0 \tag{8-28}$$

for a clamped end,

$$y_0 = y_0' = 0 \tag{8-29}$$

and for a free end,

$$M_0 = R_0 = 0 \tag{8-30}$$

The remaining unknowns are calculated using two boundary conditions at the end $x = L$ and conditions at the intermediate supports,

$$y(x) = 0, \qquad \text{at } x = L_1,\ x = L_2, \ldots, x = L_{n-1} \tag{8-31}$$

With all the unknowns determined, we calculate the static quantities using Eqs. (8-27).

## EXAMPLE 8-1

Use the method of initial parameters to determine the deflections and the bending moments for the beam shown in Fig. 8-6.

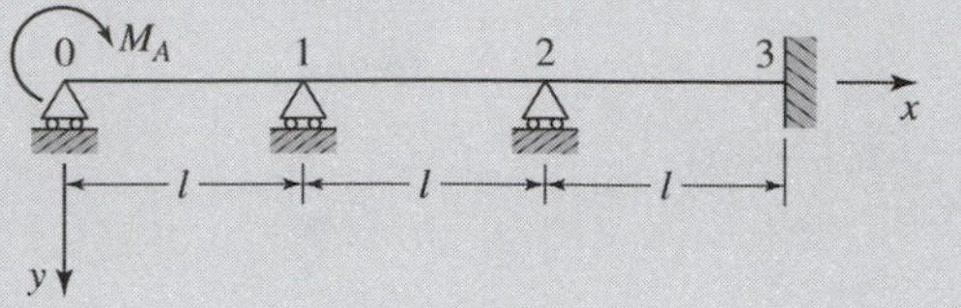


**Figure 8-6**

## SOLUTION

We have now $n = 3$, $q(x) = 0$, $L_1 = l$, $L_2 = 2l$, and $L_3 = 3l$. The boundary conditions at the end $x = 0$ are

$$y_0 = 0, \qquad M_0 = M_A$$

so that

$$EIy(x) = -R_1\frac{(x-l)^3}{6}H(x-l) - R_2\frac{(x-2l)^3}{6}H(x-2l) - R_0\frac{x^3}{6}$$
$$- M_A\frac{x^2}{2} + EIy'_0 \qquad (a)$$

$$EIy'(x) = -R_1\frac{(x-l)^2}{2}H(x-l) - R_2\frac{(x-2l)^2}{2}H(x-2l) - R_0\frac{x^2}{2}$$
$$- M_Ax + EIy'_0 \qquad (b)$$

$$-M(x) = EIy''(x) = -R_1(x-l)H(x-l) - R_2(x-2l)H(x-2l) - R_0x - M_A \qquad (c)$$

$$-V(x) = EIy'''(x) = -R_1H(x-l) - R_2H(x-2l) - R_0 \qquad (d)$$

Two boundary conditions at $x = 3l$ are

$$M(x) = -EIy''(x) = 0, \qquad y'(x) = 0, \qquad \text{at } x = 3l \qquad (e)$$

Substituting Eqs. ($c$) and ($b$) into Eq. ($e$) results in

$$-R_1 \cdot 2l - R_2 \cdot l - R_0 \cdot 3l - M_A = 0 \qquad (f)$$

$$-R_1\frac{(2l)^2}{2} - R_2\frac{l^2}{2} - R_0\frac{(3l)^2}{2} - M_A \cdot 3l + EIy'_0 = 0 \qquad (g)$$

We also have the conditions $n - 1 = 2$ [Eq. (8-31)] at the intermediate supports, and they yield the following.

(a) At support 1 ($x = l$),

$$EIy(l) = -R_0\frac{l^3}{6} - M_A\frac{l^2}{2} + EIy_0'l = 0 \tag{h}$$

(b) At support 2 ($x = 2l$),

$$EIy(2l) = -R_1\frac{l^3}{6} - R_0\frac{(2l)^3}{6} - M_A\frac{(2l)^2}{2} + EIy_0'(2l) = 0 \tag{i}$$

Solving the system of Eqs. ($f$), ($g$), ($h$), and ($i$) we obtain

$$R_0 = -\frac{24}{19}\frac{M_A}{l}, \qquad R_1 = \frac{30}{19}\frac{M_A}{l}, \qquad R_2 = -\frac{7}{19}\frac{M_A}{l}, \qquad y_0' = \frac{15}{38}\frac{M_A l}{EI}$$

## 8-1-3 Application of Castigliano's Theorem

When applying Castigliano theorem we can readily determine redundant quantities in a statically indeterminate beam. The procedure is straightforward and easily programmable. Consider first a beam with $N$ redundant constraints, as shown in Fig. 8-7$a$. Let us use as the equivalent system the statically determinate beam of Fig. 8-7$b$. To make both systems identical we require the displacements $y_i$ at all intermediate supports in the direction of the reactions to be zero. This leads to the following system of conditional equations:

$$\begin{aligned} y_1 &= w_{11} + w_{12} + \cdots + w_{1N} + w_{1q} = 0 \\ y_2 &= w_{21} + w_{22} + \cdots + w_{2N} + w_{2q} = 0 \\ &\vdots \\ y_N &= w_{N} + w_{N2} + \cdots + w_{NN} + w_{Nq} = 0 \end{aligned} \tag{8-32}$$

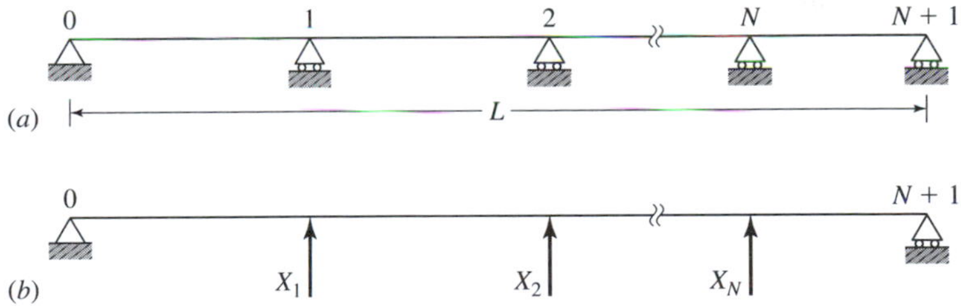


Figure 8-7 Statically indeterminate beam with $N$ redundant constraints

where $w_{ij}$ denotes the displacement at support $i$ due to the reaction $X_j$, whereas $w_{iq}$ is the displacement at $i$ due to the external load. We now represent the displacements $w_{ij}$ in the following form:

$$w_{ij} = X_j \delta_{ij} \tag{8-33}$$

where $\delta_{ij}$ is the displacement at $i$ in the direction of $X_i$ due to the unit force applied at $j$ in the direction of $X_j$. The system of conditional equations (8-32) thus becomes

$$\begin{aligned} X_1\delta_{11} + X_2\delta_{12} + \cdots + X_n\delta_{1N} + w_{1q} &= 0 \\ X_1\delta_{21} + X_2\delta_{22} + \cdots + X_n\delta_{2N} + w_{2q} &= 0 \\ &\vdots \\ X_1\delta_{N1} + X_2\delta_{N2} + \cdots + X_N\delta_{NN} + w_{Nq} &= 0 \end{aligned} \tag{8-34}$$

To calculate the coefficients $\delta_{ij}$, known as the *influence coefficients,* we employ Castigliano's theorem (although we could use another method as well). First we note that the coefficients $\delta_{ij}$ are symmetric by virtue of the *reciprocity theorem,*

$$\delta_{ji} = \delta_{ij} \tag{8-35}$$

Next we observe that each of these coefficients can be obtained using the same formula. Indeed, placing a unit force at $j$ we attach also a dummy load at $i$ (Fig. 8-8). Taking into account only the strain energy due to bending by the forces $\Pi$ and $P = 1$, we find

$$\delta_{ij} = \frac{\partial U}{\partial \Pi} = \frac{1}{2EI}\frac{\partial}{\partial \Pi}\int_0^L \overline{M}_{ij}^2\, dx = \frac{1}{EI}\int_0^L \overline{M}_{ij}\frac{\partial \overline{M}_{ij}}{\partial \Pi}\, dx, \qquad \text{when } \Pi = 0 \tag{8-36}$$

where $\overline{M}_{ij}$ is the bending moment due to the force $P = 1$ located at $x = x_j$ and $\Pi$ located at $x_i$. The reaction at 0 due to the action of both forces is

$$R_0 = \Pi\frac{L - x_i}{L} + \frac{L - x_j}{L}$$

so that the bending moment becomes

$$\begin{aligned} \overline{M}_{ij} &= \Pi x\left(1 - \frac{x_i}{L}\right) + x\left(1 - \frac{x_j}{L}\right), && 0 < x < x_i \\ \overline{M}_{ij} &= \Pi x_i\left(1 - \frac{x}{L}\right) + x\left(1 - \frac{x_j}{L}\right), && x_i < x < x_j \\ \overline{M}_{ij} &= \Pi x_i\left(1 - \frac{x}{L}\right) + x_j\left(1 - \frac{x}{L}\right), && x_j < x \end{aligned} \tag{8-37}$$

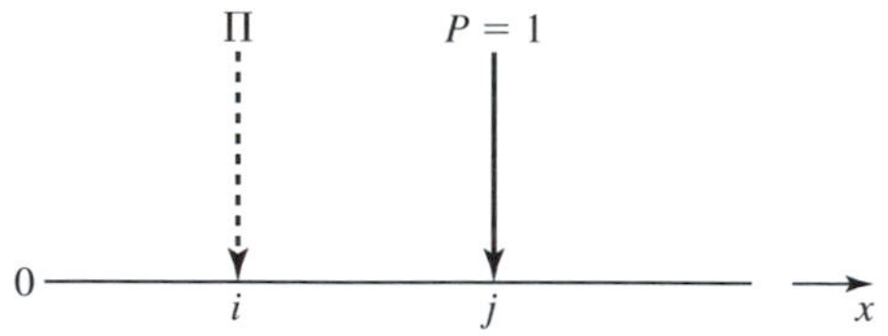

Figure 8-8

Therefore,

$$\frac{\partial \overline{M}_{ij}}{\partial \Pi} = x\left(1 - \frac{x_i}{L}\right), \qquad 0 < x < x_i$$
$$\frac{\partial \overline{M}_{ij}}{\partial \Pi} = x_i\left(1 - \frac{x}{L}\right), \qquad x_i < x < x_j \qquad (8\text{-}38)$$
$$\frac{\partial \overline{M}_{ij}}{\partial \Pi} = x_i\left(1 - \frac{x}{L}\right), \qquad x_j < x$$

Next Eqs. (8-37) (with $\Pi = 0$) are substituted into Eq. (8-36), and Eqs. (8-38) are then used in the result,

$$\delta_{ij} = \frac{1}{EI}\left[\int_0^{x_i}\left(1 - \frac{x_j}{L}\right)\left(1 - \frac{x_i}{L}\right)x^2\,dx + \int_{x_i}^{x_j} x_i x\left(1 - \frac{x_j}{L}\right)\left(1 - \frac{x}{L}\right)dx + \int_{x_j}^{L} x_i x_j\left(1 - \frac{x}{L}\right)\left(1 - \frac{x}{L}\right)dx\right]$$
$$= \frac{x_i}{6EI}\left(1 - \frac{x_j}{L}\right)(2Lx_j - x_i^2 - x_j^2), \qquad x_j > x_i, \quad i, j = 1, \ldots, N \qquad (8\text{-}39)$$

The terms $w_{iq}$ appearing in Eqs. (8-34) are obtained from

$$w_{iq} = \frac{\partial U_q}{\partial \Pi} = \frac{1}{2EI}\frac{\partial}{\partial \Pi}\int_0^L M_{iq}^2\,dx = \frac{1}{EI}\int_0^L M_{iq}\frac{\partial M_{iq}}{\partial \Pi}\,dx, \qquad \Pi = 0 \qquad (8\text{-}40)$$

Here $U_q$ is the strain energy due to bending by the external load and by the concentrated force $\Pi$ applied at $x = x_i$. $M_{iq}$ is the bending moment caused by the same loads.

## EXAMPLE 8-2

Determine the reactions and the bending moments for the three-span beam shown in Fig. 8-9*a*.

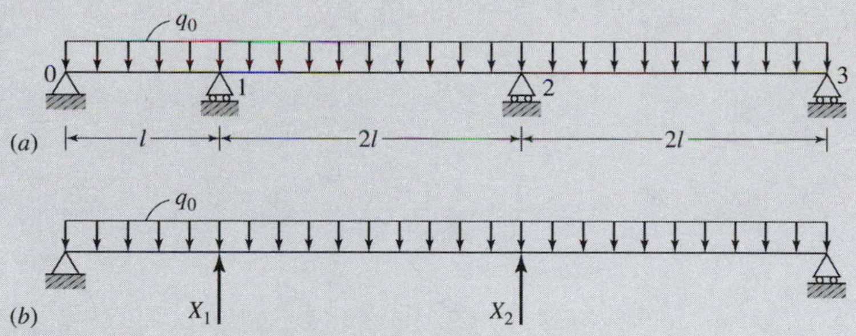


**Figure 8-9**

## SOLUTION

The beam shown in Fig. 8-9*b* is chosen as a statically determinate equivalent of the original beam so that the reactions $X_1$ and $X_2$ are the unknowns. The conditional equations (8-34) become now

$$X_1\delta_{11} + X_2\delta_{12} + w_{1q} = 0$$
$$X_1\delta_{21} + X_2\delta_{22} + w_{2q} = 0 \qquad (a)$$

where $\delta_{ij}$ is calculated from Eq. (8-39). Since $x_1 = l$, $x = 3l$, and $L = 5l$, we obtain

$$\delta_{11} = \frac{l^3(5-1)}{30}(10-1-1) = \frac{16l^3}{15EI} = 1.0667\frac{l^3}{EI}$$

Similarly,

$$\delta_{22} = \frac{36l^3}{15EI} = 2.4\frac{l^3}{EI}$$
$$\delta_{12} = \delta_{21} = \frac{20l^3}{15EI} = 1.3333\frac{l^3}{EI} \qquad (b)$$

To obtain $w_{iq}$ we consider the beam shown in Fig. 8-10. The reaction at 0 is given by

$$R_0 = \frac{5q_0 l}{2} - \Pi\left(1 - \frac{x_i}{5l}\right)$$

Thus the bending moment becomes

$$M_{iq} = \left[\frac{5q_0 l}{2} - \Pi\left(1 - \frac{x_i}{5l}\right)\right]x - \frac{q_0 x^2}{2}, \qquad 0 < x < x_i$$

$$M_{iq} = \left[\frac{5q_0 l}{2} - \Pi\left(1 - \frac{x_i}{5l}\right)\right]x + \Pi(x - x_i) - \frac{q_0 x^2}{2}, \qquad x_i < x$$

Substituting these and the derivatives $\partial M_{iq}/\partial\Pi$ into Eq. (8-40) we obtain

$$w_{iq} = -\frac{q_0}{2EI}\left[\int_0^{x_i}(5lx - x^2)\left(1 - \frac{x_i}{5l}\right)x\,dx + \int_{x_i}^{5l}(5lx - x^2)x_i\left(1 - \frac{x}{5l}\right)dx\right]$$
$$= -\frac{q_0 x_i(125l^3 - 10lx_i^2 + x_i^3)}{24EI}$$

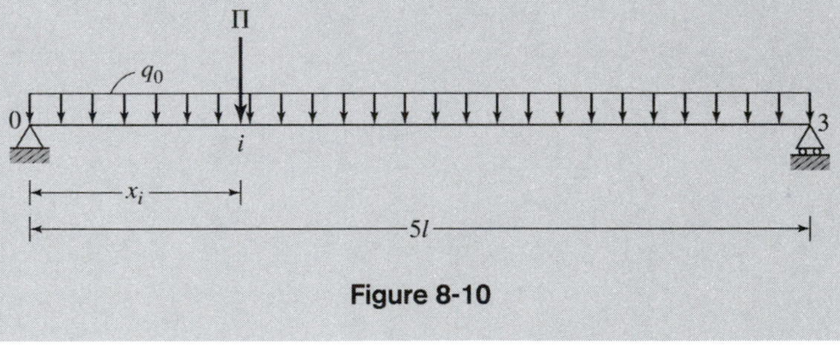


**Figure 8-10**

Since $x_i = x_1 = l$ when $i = 1$, and $x_i = x_2 = 3l$ when $i = 2$, we have

$$w_{1q} = -\frac{q_0 l^4}{24EI}(116) = -4.833\frac{q_0 l^4}{EI}$$
$$w_{2q} = -\frac{186 q_0 l^4}{24EI} = -7.75\frac{q_0 l^4}{EI} \tag{c}$$

With Eqs. ($b$) and ($c$), Eqs. ($a$) become

$$1.0667X_1 + 1.3333X_2 - 4.8333q_0 l = 0$$
$$1.3333X_1 + 2.4X_2 - 7.75q_0 l = 0$$

with the solution

$$X_1 = 1.6183q_0 l, \qquad X_2 = 2.3301q_0 l$$

The diagram of the bending moments is shown in Fig. 8-11.

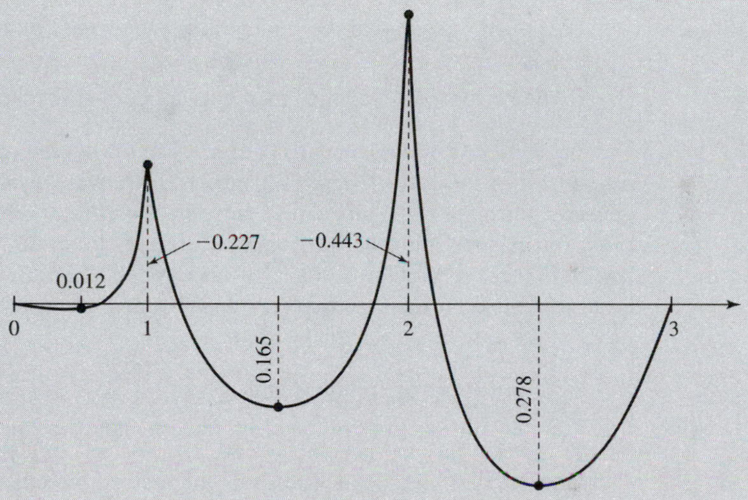


**Figure 8-11**

The procedure is very similar when one end of the beam is clamped or free (Fig. 8-12$a$, $c$). In the former case we may remove all the supports, except for support 0, treating the reactions $X_1, \ldots, X_{N+1}$ as unknowns; in the latter we may leave two outside supports, removing all the remaining ones, and treat $X_2, \ldots, X_N$ as unknowns. We can obviously use many other equivalent statically determinate systems instead of the systems shown in Fig. 8-12$b$ and $d$.

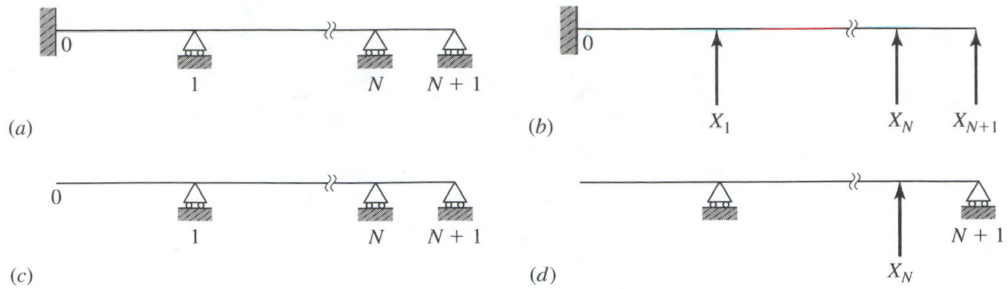


Figure 8-12

## 8-2 UNSYMMETRIC BENDING OF STRAIGHT BEAMS

Symmetric bending of a beam occurs when:

- Its cross section has an axis of symmetry
- Applied forces lie in the symmetry plane formed by this axis and by the neutral axis
- The vectors of the applied moments are perpendicular to this plane

This case of bending, occurring entirely in the symmetry plane, was covered in an introductory course of strength of materials. Here we consider the bending of a homogeneous, linearly elastic prismatic beam with an arbitrary cross section (Fig. 8-13) carrying forces and moments that do not necessarily lie in the same plane. It is assumed, however, that every plane of loads passes through a centroidal axis.* We introduce the arbitrary, but centroidal, coordinates $\xi$ and $\eta$ and make the same assumptions as are made in the case of symmetric bending:

- Deformations are infinitesimal
- Plane cross sections of the beam remain plane after deformation

The latter implies that the normal strain in the axial direction is a linear function of the cross-sectional coordinates $\xi$, $\eta$. Since for a linearly elastic solid stress is a linear function of strain, it depends linearly on the coordinates $\xi$ and $\eta$ or on any other coordinates obtained from $\xi$ and $\eta$ by rotation and translation. The expression for the normal stress $\sigma_{xx}$ takes on a particularly simple form when referred to the principal axes of inertia. Thus our first objective is to find the principal axes. These are shown in Fig. 8-14 as $y$ and $z$, whereas $\xi$ and $\eta$ are arbitrary centroidal axes. The angle $\alpha$ is to be found. Using

*As can be found, for instance, in [8] for many solid cross sections the centroid almost coincides with the so-called *shear center*. If the plane of loads passes through the shear center twisting is avoided.

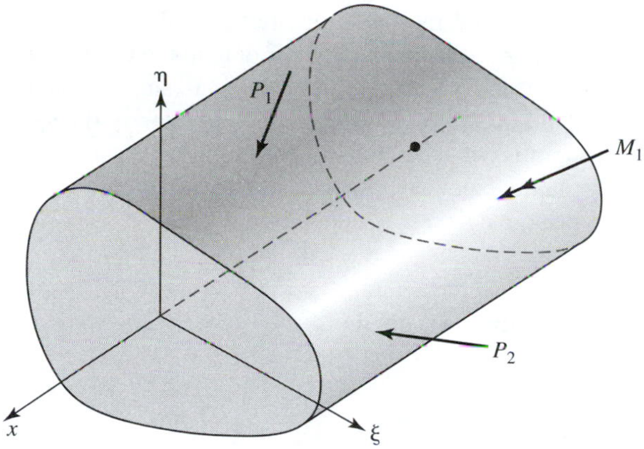


Figure 8-13 Unsymmetric bending of prismatic beam

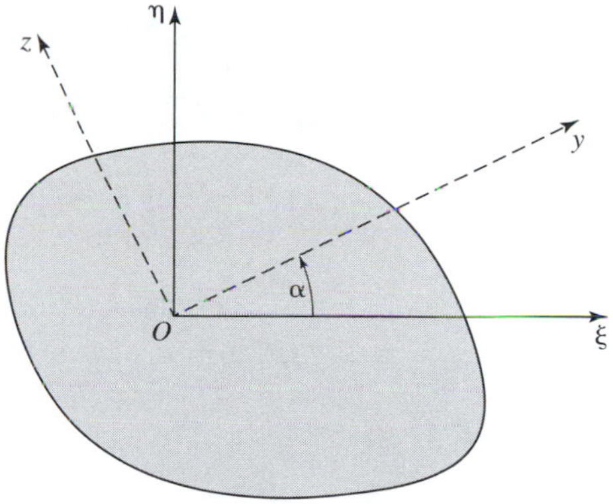


Figure 8-14 Principal axes of inertia

the definition of moment of inertia,

$$I_{yy} = \int_A z^2\, dA, \qquad I_{zz} = \int_A y^2\, dA, \qquad I_{yz} = \int_A yz\, dA$$
$$I_{\xi\xi} = \int_A \eta^2\, dA, \qquad I_{\eta\eta} = \int_A \xi^2\, dA, \qquad I_{\xi\eta} = \int_A \xi\eta\, dA \tag{8-41}$$

and substituting the coordinate transformation

$$y = \xi\cos\alpha + \eta\sin\alpha, \qquad z = -\xi\sin\alpha + \eta\cos\alpha$$

into the last three of the Eqs. (8-41), these become

$$\begin{aligned} I_{yz} &= \int_A (\xi\cos\alpha + \eta\sin\alpha)(-\xi\sin\alpha + \eta\cos\alpha)\, dA \\ &= -I_{\eta\eta}\sin\alpha\cos\alpha + I_{\xi\xi}\sin\alpha\cos\alpha + I_{\xi\eta}(\cos^2\alpha - \sin^2\alpha) \\ I_{yy} &= I_{\xi\xi}\cos^2\alpha + I_{\eta\eta}\sin^2\alpha - I_{\xi\eta}\sin 2\alpha \\ I_{zz} &= I_{\xi\xi}\sin^2\alpha + I_{\eta\eta}\cos^2\alpha + I_{\xi\eta}\sin 2\alpha \end{aligned} \tag{8-42}$$

When trigonometric formulas are used in Eqs. (8-42), these can be written as

$$I_{yy} = \frac{I_{\xi\xi} + I_{\eta\eta}}{2} + \frac{I_{\xi\xi} - I_{\eta\eta}}{2}\cos 2\alpha - I_{\xi\eta}\sin 2\alpha \tag{8-43a}$$

$$I_{zz} = \frac{I_{\xi\xi} + I_{\eta\eta}}{2} - \frac{I_{\xi\xi} - I_{\eta\eta}}{2}\cos 2\alpha + I_{\xi\eta}\sin 2\alpha \tag{8-43b}$$

$$I_{yz} = \frac{I_{\xi\xi} - I_{\eta\eta}}{2}\sin 2\alpha + I_{\xi\eta}\cos 2\alpha \tag{8-43c}$$

We note the similarity of the relations for the moments of inertia in the $y, z$ and in $\xi, \eta$ coordinate systems [Eqs. (8-43)] and the stress or strain components in the $x$, $y$ and in $\xi, \eta$ coordinate systems [Eqs. (2-73) and (3-30), respectively]. It shows that the moments of inertia and the stress and strain components transform in exactly the same way when a Cartesian coordinate system is rotated. This indicates that all these quantities belong mathematically to the same category, known as Cartesian tensors of rank 2 [9, pp. 43–55]. For $y, z$ to be principal axes we must have

$$I_{yz} = 0$$

Substituting here Eq. (8-43$c$) and solving for $\alpha$ we obtain

$$\tan 2\alpha = \frac{2I_{\xi\eta}}{I_{\eta\eta} - I_{\xi\xi}} \tag{8-44}$$

Thus we determined the angle $\alpha$. When Eq. (8-44) is substituted into Eqs. (8-43$a$) and (8-43$b$), we get the principal moments of inertia $I_{yy}$ and $I_{zz}$ in terms of the arbitrary centroidal moments of inertia $I_{\xi\xi}$, $I_{\eta\eta}$, and $I_{\xi\eta}$,

$$\begin{aligned} I_{yy} &= \frac{1}{2}\left[I_{\xi\xi} + I_{\eta\eta} + \sqrt{(I_{\xi\xi} - I_{\eta\eta})^2 + 4I_{\xi\eta}^2}\right] \\ I_{zz} &= \frac{1}{2}\left[I_{\xi\xi} + I_{\eta\eta} - \sqrt{(I_{\xi\xi} - I_{\eta\eta})^2 + 4I_{\xi\eta}^2}\right] \end{aligned} \tag{8-45}$$

In the developments that follow we assume that the principal axes of inertia $y$ and $z$ have already been found. Now let us decompose the bending moment $M$ into two components, $M_y$ and $M_z$ (Fig. 8-15). According to our previous comments regarding the linearity of stress we can write

$$\sigma_{xx} = a + by + cz \tag{8-46}$$

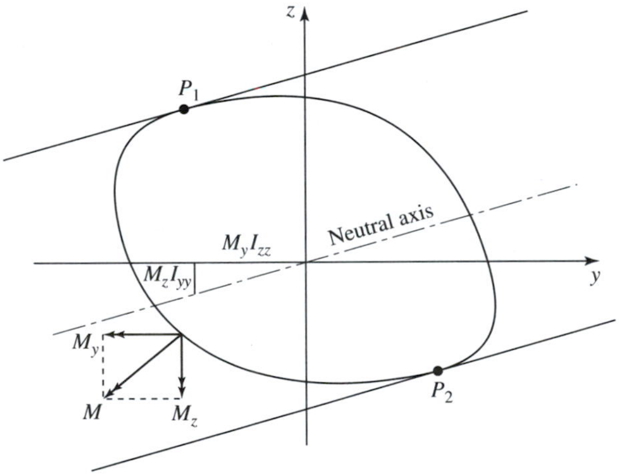


**Figure 8-15** Decomposition of bending moment vector

where $a$, $b$, $c$ are constants to be determined using equilibrium equations,

$$\int_A \sigma_{xx}\, dA = 0, \qquad \int_A \sigma_{xx} z\, dA - M_y = 0, \qquad -\int_A \sigma_{xx} y\, dA - M_z = 0 \tag{8-47}$$

The first of Eqs. (8-47) indicates that the axial force acting on the cross section is zero. The other two Eqs. (8-47) define the bending moments $M_y$ and $M_z$ in terms of the sums of the moments of the normal stress with respect to the principal axes $y$ and $z$. Substituting Eq. (8-46) into Eqs. (8-47) and using the fact that $y$ and $z$ are principal centroidal axes of inertia,

$$\int_A y\, dA = \int_A z\, dA = \int_A yz\, dA = 0$$

we obtain

$$a = 0, \qquad cI_{yy} - M_y = 0, \qquad -bI_{zz} - M_z = 0$$

With these,

$$a = 0, \qquad b = -\frac{M_z}{I_{zz}}, \qquad c = \frac{M_y}{I_{yy}}$$

Thus Eq. (8-46) becomes

$$\sigma_{xx} = -\frac{M_z y}{I_{zz}} + \frac{M_y z}{I_{yy}} \tag{8-48}$$

It is evident that when either $M_y$ or $M_z$ equals zero, the neutral axis is $y = 0$ or $z = 0$, respectively. In the general case the condition that $\sigma_{xx} = 0$ at the neutral axis gives, with Eq. (8-48),

$$z = \frac{M_z I_{yy}}{M_y I_{zz}} y \tag{8-49}$$

Maximum $\sigma_{xx}$ is found at those points on the contour that are farthest away from the neutral axis (points $P_1$ and $P_2$ in Fig. 8-15). Recalling that the problem is linear, we can use the superposition principle to calculate the deflections. Each of the component moments $M_y$ and $M_z$ produces deflection in the principal planes $Oxz$ and $Oxy$, respectively. These deflections, denoted by $w_z$ and $w_y$, are calculated using the differential equations for bending,

$$EI_{yy}\frac{d^2 w_z}{dx^2} = M_y, \qquad EI_{zz}\frac{d^2 w_y}{dx^2} = M_z \tag{8-50}$$

Substituting $w_z$ and $w_y$ into Pythagoras' formula gives the expression for the total deflection $w$,

$$w(x) = \sqrt{w_y^2(x) + w_z^2(x)} \tag{8-51}$$

We note that in general the inclination of the vector $\boldsymbol{M}$, and thus the ratio $M_y/M_z$, is not constant.

## EXAMPLE 8-3

A simply supported beam of the cross section shown in Fig. 8-16*a* carries loads in the vertical and horizontal planes and an inclined bending moment $M_C$ (Fig. 8-16*b*). Find the maximum normal stress and maximum deflection. Assume that $M_C = Pa$.

In order to determine the maximum stress $\sigma_{xx}$ we must (a) find the bending moments at the points of application of $P_1$, $P_2$, and $M_C$ (that is where the maximum bending moment may occur), (b) find the centroid of the cross section, (c) find the principal axes and the principal moments of inertia $I_{yy}$ and $I_{zz}$, (d) determine the position of the neutral axis in each of the critical cross sections using Eq. (8-49), and (e)–(g) calculate the stress in each of the critical cross sections using Eq. (8-48).

### SOLUTION

(a) First we determine, for convenience, the reaction components $R_{A\xi_1}$ and $R_{A\eta_1}$

$$R_{A\eta_1} = \frac{P}{4} + M_C\frac{\sqrt{2}}{2a}, \qquad R_{A\xi_1} = \frac{3P}{4} - M_C\frac{\sqrt{2}}{2a}$$

at support $A$ in the directions $\eta_1$ and $\xi_1$ (see Fig. 8-16*c*). Next we find the bending moments $M_{\xi_1}(x)$ and $M_{\eta_1}(x)$ at an arbitrary location $x$,

$$M_{\xi_1}(x) = \left(\frac{P}{4} + M_C\frac{\sqrt{2}}{2a}\right)x - M_C\frac{\sqrt{2}}{2}H\left(x - \frac{a}{2}\right) - P\left(x - \frac{3a}{4}\right)H\left(x - \frac{3a}{4}\right)$$

$$M_{\eta_1}(x) = \left(\frac{3P}{4} - M_C\frac{\sqrt{2}}{2a}\right)x - P\left(x - \frac{a}{4}\right)H\left(x - \frac{a}{4}\right) + M_C\frac{\sqrt{2}}{2}H\left(x - \frac{a}{2}\right)$$

where $H(x)$ is the step function (Heaviside function). From these we find

$$M_{\xi_1} = \frac{Pa}{8} + M_C\frac{\sqrt{2}}{4}, \qquad M_{\eta_1} = \frac{Pa}{8} - M_C\frac{\sqrt{2}}{4}, \qquad \text{at } x = \frac{a}{2}$$

$$M_{\xi_1} = \left(\frac{P}{4} + M_C\frac{\sqrt{2}}{2a}\right)\frac{a}{4}, \qquad M_{\eta_1} = \left(\frac{3P}{4} - M_C\frac{\sqrt{2}}{2a}\right)\frac{a}{4}, \qquad \text{at } x = \frac{a}{4}$$

$$M_{\xi_1} = \frac{3Pa}{16} + M_C\frac{3\sqrt{2}}{8}, \qquad M_{\eta_1} = \frac{Pa}{16} + M_C\frac{\sqrt{2}}{2}, \qquad \text{at } x = \frac{3a}{4}$$

(b) The centroid is found from the conditions

$$S_\xi = \int_A \eta\, dA = 0, \qquad S_\eta = \int_A \xi\, dA = 0 \tag{a}$$

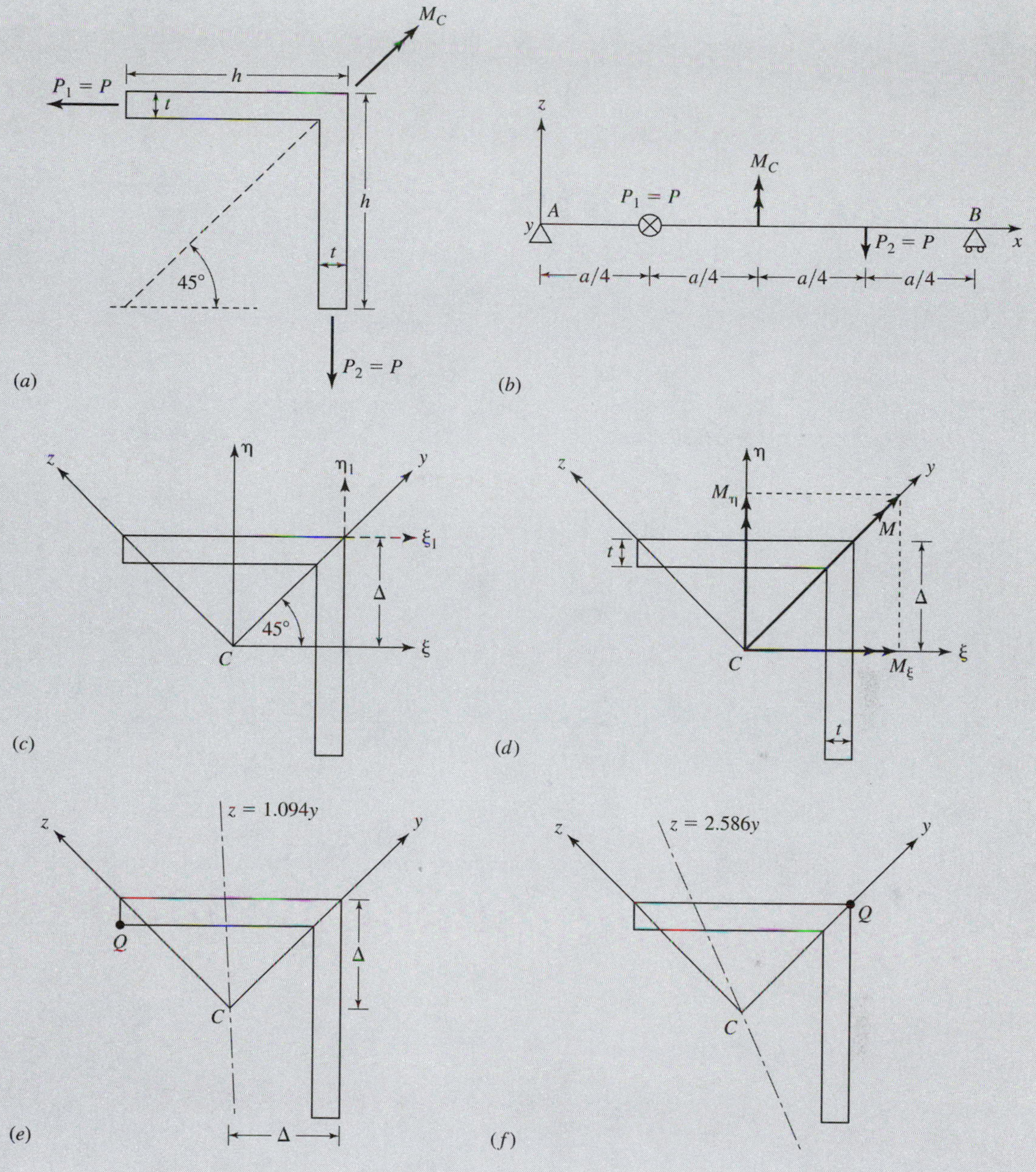


**Figure 8-16** Simply supported beam under loads acting in different planes

where $\xi$ and $\eta$ are still unknown centroidal axes. Furthermore, because of symmetry, the centroid lies on the symmetry line. Thus using

$$\xi = \xi_1 + \Delta, \qquad \eta = \eta_1 + \Delta$$

we obtain from either of the conditions (*a*),

$$\Delta = \frac{t^2 - th - h^2}{2(t - 2h)}$$

(c) The symmetry of the cross section accounts for $I_{\xi\xi} = I_{\eta\eta}$. Therefore with Eq. (8-44),

$$\tan 2\alpha = \infty, \qquad 2\alpha = \frac{\pi}{2}, \qquad \alpha = \frac{\pi}{4}$$

and the principal axes $y$ and $z$ are located as shown in Fig. 8-16*c*. In order to find the principal moments of inertia $I_{yy}$ and $I_{zz}$, we first determine $I_{\xi\xi}$, $I_{\eta\eta}$, and $I_{\xi\eta}$,

$$I_{\xi\xi} = I_{\eta\eta} = \frac{t}{3}(h^3 + ht^2 - t^3) - t\Delta(h^2 + ht - t^2) + t\Delta^2(2h - t)$$

$$I_{\xi\eta} = \tfrac{1}{4}t^4 - t^3\Delta + t^2\Delta^2$$

Substituting here previously calculated $\Delta$* we obtain

$$I_{\xi\xi} = I_{\eta\eta} = \frac{t(h^2 - ht + t^2)(5h^2 - 5ht + t^2)}{12(2h - t)}$$

$$I_{\xi\eta} = -\frac{h^2 t(h - t)^2}{4(2h - t)}$$

Then the values of the principal moments of inertia follow from Eqs. (8-45),

$$I_{yy} = \frac{t(2h^4 - 4h^3 t + 8h^2 t^2 - 6ht^3 + t^4)}{12(2h - t)}$$

$$I_{zz} = \frac{t(4h^3 - 6h^2 t + 4ht^2 - t^3)}{12}$$

(d) The position of the neutral axis depends on the magnitudes of the components $M_y$ and $M_z$ of the bending moment. We see from Fig. 8-16*d* that

$$M_y = \frac{\sqrt{2}}{2}(M_\xi - M_\eta), \qquad M_z = \frac{\sqrt{2}}{2}(M_\xi + M_\eta)$$

*The software package Mathcad7, professional edition, was used here.

where $M_\xi = M_{\xi_1}$ and $M_\eta = M_{\eta_1}$. With Eq. (8-49) we find the expression for the neutral axis,

$$z = \frac{M_z}{M_y} y = \frac{M_\xi + M_\eta}{M_\xi - M_\eta} y$$

Substituting for $M_\xi$ and $M_\eta$ the values calculated before, we obtain with $M_C = Pa$,

$$z = \frac{P}{-\dfrac{P}{2} + P\sqrt{2}} y = 1.094y, \qquad \text{at } x = \frac{a}{4}$$

$$z = \frac{P\sqrt{2}}{4P} y = 0.3536y, \qquad \text{at } x = \frac{a}{2}$$

$$z = \frac{\dfrac{P\sqrt{2}}{8} + \dfrac{3}{8}P}{\dfrac{P\sqrt{2}}{16} + \dfrac{1}{8}P} y = 2.586y, \qquad \text{at } x = \frac{3a}{4}$$

(e) When calculating the maximum stresses it is assumed that $t = 0.1h$. At $x = a/4$ the neutral axis is located as shown in Fig. 8-16*e*. On account of the linearity of the stress $\sigma_{xx}$, its maximum occurs at the point $Q$, which is farthest from the neutral axis. First we find the coordinates of point $Q$,

$$y_Q = \frac{h\sqrt{2}}{2} - \Delta\sqrt{2}, \qquad z_Q = \frac{h\sqrt{2}}{2}$$

Next we use Eq. (8-48) substituting $y_Q$ and $z_Q$ for $y$ and $z$ and applying the formulas obtained for the principal moments of inertia $I_{yy}$ and $I_{zz}$. This yields the following result:

$$\max \sigma_{xx} = \frac{1}{h^3}(60.6583M_\xi - 75.5341M_\eta)$$

Recalling our assumption that $M_C = Pa$, we have at $x = a/4$, $M_\xi = 0.2268Pa$, $M_\eta = 0.0107Pa$. Finally,

$$\max \sigma_{xx} = 12.95\frac{Pa}{h^3}$$

(f) Similarly, after determining the location of the neutral axis corresponding to $x = a/2$, we find that at $x = a/2$ the maximum stress occurs at point $P$ with the coordinates

$$y_P = \frac{(h+t)\sqrt{2}}{2} - \Delta\sqrt{2}, \qquad z_P = \frac{(h-t)\sqrt{2}}{2}$$

and equals

$$\max \sigma_{xx} = 41.0462 \frac{Pa}{h^3}$$

(g) Finally, at $x = 3a/4$, point $Q$ (Fig. 8-16*f*) is again farthest away from the neutral axis. We find the maximum stress to be

$$\max \sigma_{xx} = -14.4520 \frac{Pa}{h^3}$$

## 8-3 CURVED BEAMS

Curved beams can be subdivided into two major categories—beams loaded out of plane and beams loaded in their planes. In the first category, discussed in Sections 8-3-1 and 8-3-2, the application of the load does not change the natural curvature of the beam (within the assumption of small deformation). In the second category a change occurs affecting the moment-deflection relation (see also Chapter 5) and thus the differential equation for deflection. On the other hand, while torsion always occurs in a beam bent out of plane, even though no external torque is applied, it is never present in a beam bent in its own plane unless a torque is applied.

### 8-3-1 Out-of-Plane Loaded Beams and Rings

A thin curved beam loaded out of its plane is assumed to behave similarly to a straight beam to the extent that any plane cross section remains plane after the deformation. In order to determine the deflection of such a beam, we could derive a differential equation for the deflection by analyzing the equilibrium of an infinitesimal element, or we might apply Castigliano's theorem. Using the second approach, which is easier, we disregard the effect of shear on the strain energy and assume that there are no axial forces. Thus we take into account only the strain energy due to the bending and to the torsion since both will obviously be generated by any load. Considering for simplicity a curved cantilever with a circular axis (Fig. 8-17), let us determine the vertical deflection $w$ at the point of application of the concentrated force $P$. The expression for the strain energy due to bending and torsion is

$$U = \int_0^S \frac{M^2\,ds}{2EI} + \int_0^S \frac{T^2\,ds}{2GJ}$$

Recalling Castigliano's principle, we get

$$w_B = \frac{\partial U}{\partial P} = \frac{1}{EI}\int_0^S M \frac{\partial M}{\partial P}\,ds + \frac{1}{GJ}\int_0^S T \frac{\partial T}{\partial P}\,ds$$

where $J$ represents an integral equal to the polar moment of inertia when the cross section is circular [1, p. 166]. Substituting here $s = R\varphi$, $ds = R\,d\varphi$, replacing the limits of

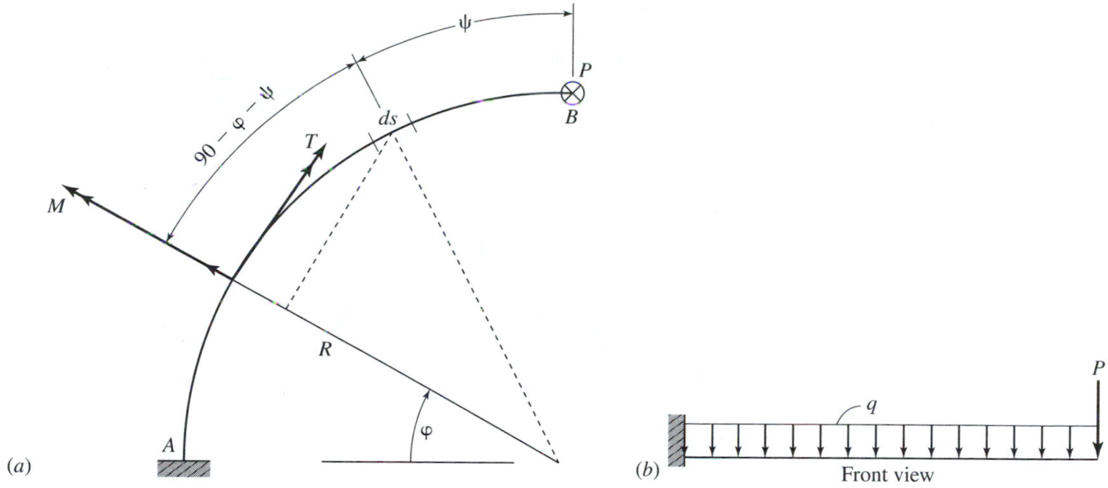


**Figure 8-17** Curved cantilever with circular axis

integration $s = 0$ and $s = S$ by $\varphi = 0$ and $\varphi = \pi/2$, and calculating the following expressions for $M$, $T$, and for their derivatives with respect to $P$,

$$M = PR\sin(90^\circ - \varphi) + qR^2 \int_0^{90^\circ - \varphi} \sin(90^\circ - \varphi - \psi)\, d\psi$$

$$= PR\cos\varphi + qR^2(1 - \sin\varphi)$$

$$T = PR(1 - \sin\varphi) + qR^2 \int_0^{90^\circ - \varphi} [1 - \cos(90^\circ - \varphi - \psi)]\, d\psi$$

$$= PR(1 - \sin\varphi) + qR^2\left(\frac{\pi}{2} - \varphi - \cos\varphi\right)$$

$$\frac{\partial M}{\partial P} = R\cos\varphi, \qquad \frac{\partial T}{\partial P} = R(1 - \sin\varphi)$$

we obtain for the deflection at point $B$,

$$w_B = \frac{R^3}{EI}\int_0^{\frac{\pi}{2}} [P\cos\varphi + qR(1 - \sin\varphi)]\cos\varphi\, d\varphi$$

$$+ \frac{R^3}{GJ}\int_0^{\frac{\pi}{2}} \left[P(1 - \sin\varphi) + qR\left(\frac{\pi}{2} - \varphi - \cos\varphi\right)\right](1 - \sin\varphi)\, d\varphi$$

$$= \frac{R^3}{EI}\left(qR + \frac{P\pi}{4}\right) + \frac{R^3}{8GJ}[-2P(8 + 3\pi) + qR(\pi - 2)^2]$$

The evaluation of the deflection of a similar cantilever but with an arbitrarily curved axis $y = f(x)$ (Fig. 8-18$a$) is more complicated. We first determine the bending moment

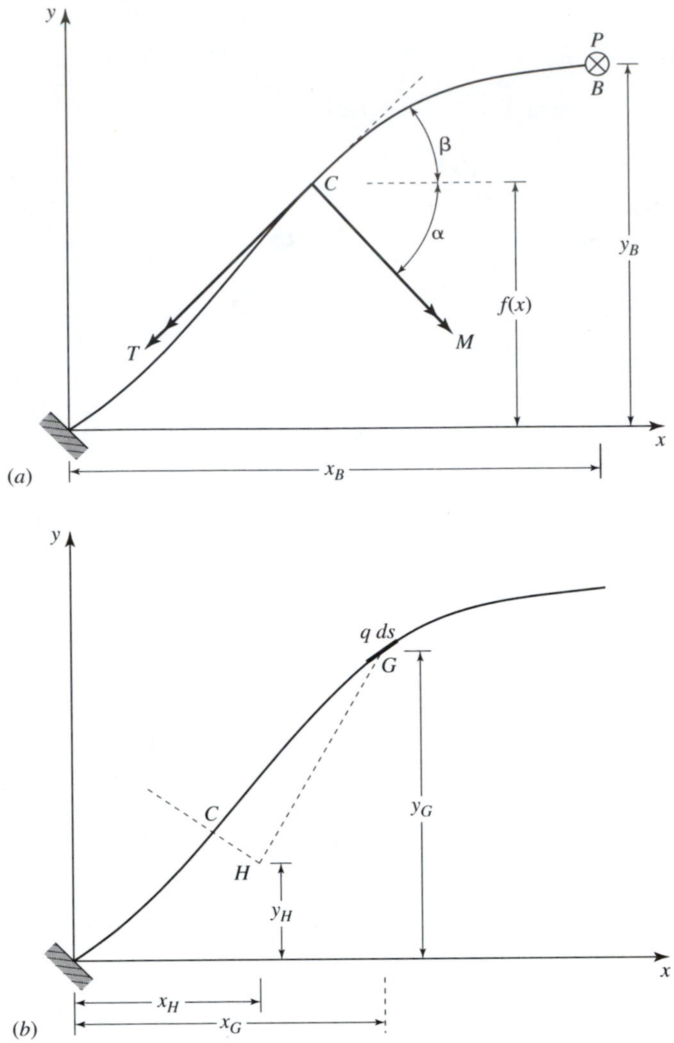


Figure 8-18

and the torque at an arbitrary cross section (point $C$). The $x$ and $y$ components of the moment at $C$ are

$$M_y(x) = P(x_B - x), \qquad M_x(x) = -P[y_B - f(x)] \tag{8-52}$$

Next we find the bending moment $M^P(x)$ and the torque $T^P(x)$ due to the concentrated force $P$,

$$\begin{aligned} M^P(x) &= M_x(x)\cos\alpha - M_y(x)\sin\alpha \\ T^P(x) &= -M_x(x)\cos\beta - M_y(x)\sin\beta \end{aligned} \tag{8-53}$$

However,

$$\alpha = \frac{\pi}{2} - \beta$$

Hence,

$$M^P(x) = M_x(x)\sin\beta - M_y(x)\cos\beta \tag{8-54}$$

where $\beta$ is the angle between the tangent to the curve $y = f(x)$ and the $x$ axis: $\tan\beta = f'(x)$. Furthermore, using trigonometric identities, we find that

$$\sin\beta = \frac{f'(x)}{\sqrt{1+[f'(x)]^2}}, \qquad \cos\beta = \frac{1}{\sqrt{1+[f'(x)]^2}}$$

Next these are substituted into Eqs. (8-53) and (8-54)

$$\begin{aligned} M^P(x) &= \frac{P}{\sqrt{1+[f'(x)]^2}}[f(x)f'(x) - y_B f'(x) - x_B + x] \\ T^P(x) &= \frac{P}{\sqrt{1+[f'(x)]^2}}[y_B - f(x) - x_B f'(x) + x f'(x)] \end{aligned} \tag{8-55}$$

To determine the moments due to a continuously distributed load we have to apply integration. We begin by deriving the infinitesimal bending moment at the cross section $C$ due to an infinitesimal load $q\,ds$, located at $G$. It equals $(q\,ds)\cdot GH$ (see Fig. 8-18*b*), where

$$\begin{aligned} x_H &= \frac{f'}{1+(f')^2}\left(x_G f' - y_G + \frac{1}{f'}x_C + y_C\right) \\ y_H &= \frac{f'}{1+(f')^2}\left(-x_G + \frac{1}{f'}y_G + x_C + y_C f'\right) \end{aligned} \tag{8-56}$$

and where $f'(x)$ is evaluated at $C$.* The bending moment $M_C^q$ becomes now

$$M_C^q = \int_C^B (q\,ds)\sqrt{(x_G - x_H)^2 + (y_G - y_H)^2}$$

Since $C$ is arbitrary, set

$$x_C = x, \qquad y_C = f(x), \qquad x_G = \xi, \qquad y_G = f(\xi)$$

$$ds = \sqrt{dx^2 + dy^2} = \sqrt{1+[f'(\xi)]^2}\,d\xi$$

thus transforming the expression $M_C^q$ into

$$M^q(x) = q\int_x^{x_B} \sqrt{(\xi - x_H)^2 + [f(\xi) - y_H]^2}\sqrt{1+[f'(\xi)]^2}\,d\xi \tag{8-57}$$

*The results (8-56) are obtained by finding the coordinates of point $H$ of the intersection of a line through $G$ parallel to the tangent to $y(x)$ at $C$ with a normal to $y(x)$ through $C$.

A similar, rather unwieldy, expression can be readily obtained for the torque $T^q(x)$ at $C$. This is left as an exercise for the reader. Despite the complicated form of these expressions, the further procedure is quite simple, as is evident from the following outline. First, to evaluate the integral of $M(\partial M/\partial P)$ appearing in this example the transformation is used,

$$\int_0^S M\frac{\partial M}{\partial P}\,ds = \int_0^R M(x)\frac{\partial M(x)}{\partial P}\sqrt{1+[f'(x)]^2}\,dx$$

$$= \int_0^R [M^P(x)+M^q(x)]\frac{\partial M^P(x)}{\partial P}\sqrt{1+[f'(x)]^2}\,dx \qquad (8\text{-}58)$$

Next the expressions (8-55) and (8-57) are substituted for $M^P$ and $M^q$ followed by numerical integration. When a curved beam is statically indeterminate, then the removal of redundant constraints and their replacement by the unknown reactions makes the system statically determinate. The unknown quantities are found by imposing the conditions making the system geometrically equivalent to the original one. For an application see Section 8-3-2 on Biezeno's theorem.

## 8-3-2 Transversely Loaded Circular Ring Supported by Three or More Supports (Biezeno's Theorem) [2]

We now consider a circular ring carrying a load normal to its plane and supported by three or more supports (Fig. 8-19). We begin with an auxiliary problem of an externally statically determinate ring supported at three points (Fig. 8-20*a*). Biezeno's theorem to be derived relates in a simple manner the statically indeterminate internal forces in the ring (bending moments, torques, and shear forces) to its geometry and to given external forces and their locations. First we remove the supports and assume that the reactions $R_1$, $R_2$, $R_3$ have already been found from equilibrium equations (the free-body diagram is shown in Fig. 8-20*b*). Next we choose an arbitrary point $A$ and, to simplify notation, we assign new symbols $F_i$ to all external forces (Fig. 8-20*c*), denoting by $\alpha_i$ their angular locations with respect to $OA$. It is assumed that the downward direction of the forces $F_i$ is positive. Let us now make a cut at $A$, attaching unknown internal forces—the bending moment $M_A$, the torque $T_A$, and the vertical shear force $V_A$—to both sides of

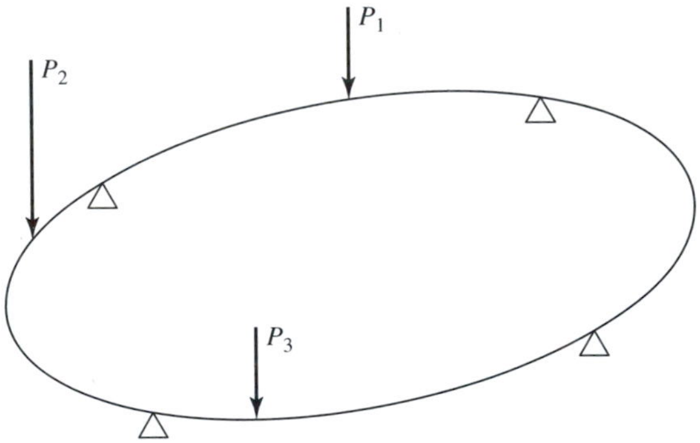


Figure 8-19 Circular ring on three or more supports

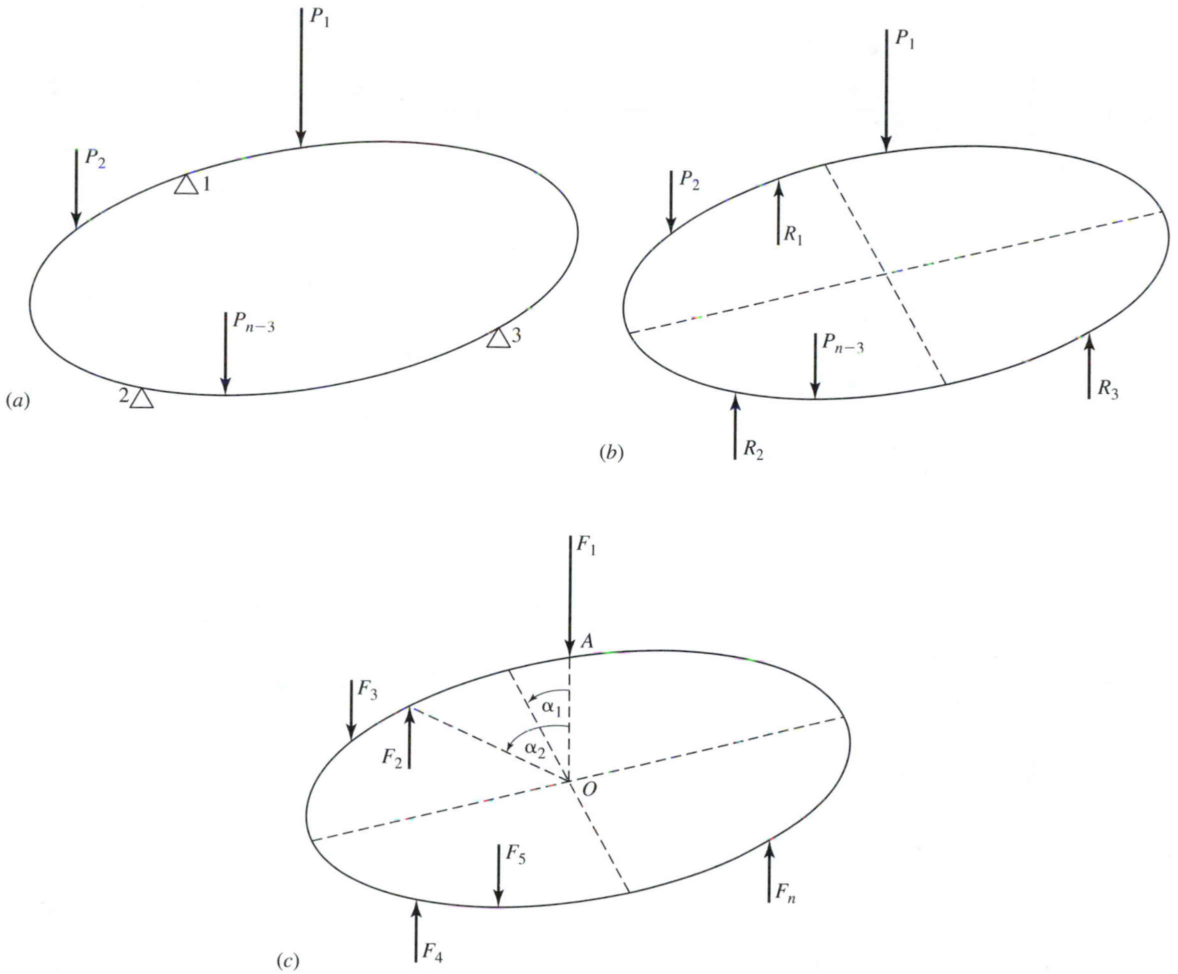


**Figure 8-20** Externally statically determinate ring

the cut. Figure 8-21 shows positive directions of these quantities. They will be determined by requiring that both sides of the cut do not deform relative to each other. Thus, by Castigliano's theorem,

$$\frac{\partial U}{\partial M_A} = 0, \qquad \frac{\partial U}{\partial T_A} = 0, \qquad \frac{\partial U}{\partial V_A} = 0 \tag{8-59}$$

where $U$ is the total strain energy. It is further assumed that the strain energy due to shear is negligible. Therefore,

$$U = \frac{R}{2EI}\int_0^{2\pi} M^2\, d\theta + \frac{R}{2GI_0}\int_0^{2\pi} T^2\, d\theta \tag{8-60}$$

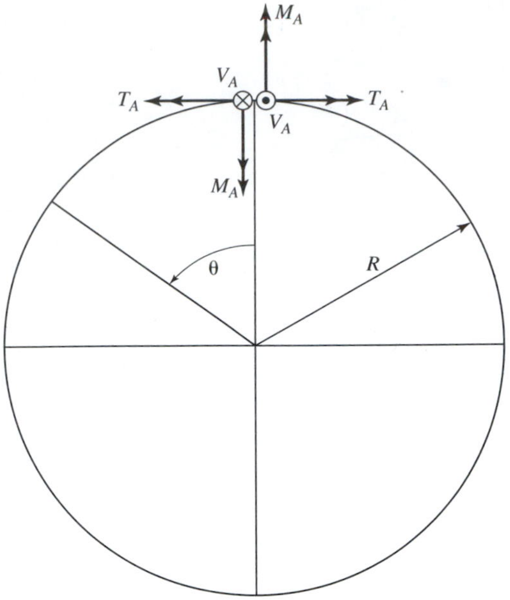


Figure 8-21

Now we calculate the bending moment $M$ and the torque $T$ at an arbitrary cross section (see also Section 8-3-1),

$$M = M_A \cos\theta - T_A \sin\theta - V_A R \sin\theta - \sum_{m=1}^{k} F_m R H(\theta - \alpha_m) \sin(\theta - \alpha_m) \tag{8-61}$$

$$T = M_A \sin\theta + T_A \cos\theta - V_A R(1 - \cos\theta) - \sum_{m=1}^{k} F_m R H(\theta - \alpha_m)[1 - \cos(\theta - \alpha_m)] \tag{8-62}$$

where $k$ is the number of forces $F_i$ between the cross section $A(\theta = 0)$ and the arbitrary cross section $\theta$, and $H$ is the Heaviside function. Substituting Eq. (8-60) into Eqs. (8-59), we obtain

$$\frac{1}{EI}\int_0^{2\pi} M \cos\theta \, d\theta + \frac{1}{GI_0}\int_0^{2\pi} T \sin\theta \, d\theta = 0 \tag{8-63}$$

$$-\frac{1}{EI}\int_0^{2\pi} M \sin\theta \, d\theta + \frac{1}{GI_0}\int_0^{2\pi} T \cos\theta \, d\theta = 0 \tag{8-64}$$

$$\frac{1}{EI}\int_0^{2\pi} M \sin\theta \, d\theta + \frac{1}{GI_0}\int_0^{2\pi} T(1 - \cos\theta) \, d\theta = 0 \tag{8-65}$$

Adding Eqs. (8-64) and (8-65) gives

$$\int_0^{2\pi} T \, d\theta = 0 \tag{8-66}$$

We note that the forces $F_m$ are multiplied by the Heaviside function, and therefore the integrals in Eqs. (8-63)–(8-65) range from $\alpha_m$ to $2\pi$ for the terms including these forces. We also note that $k$ becomes now $n$ (all forces are between $A$ and $\theta$ when $\theta = 2\pi$). Hence substituting Eq. (8-62) into Eq. (8-66) we obtain

$$M_A \int_0^{2\pi} \sin\theta\, d\theta + T_A \int_0^{2\pi} \cos\theta\, d\theta - V_A R \int_0^{2\pi} (1 - \cos\theta)\, d\theta$$
$$+ \sum_{m=1}^{n} F_m R \int_{\alpha_m}^{2\pi} \cos(\theta - \alpha_m)\, d\theta - \sum_{m=1}^{n} F_m R \int_{\alpha_m}^{2\pi} d\theta = 0$$

so that evaluating the integrals,

$$-2V_A R\pi + \sum_{m=1}^{n} F_m R \sin(2\pi - \alpha_m) - \sum_{m=1}^{n} F_m R(2\pi - \alpha_m) = 0 \qquad \text{(8-67)}$$

and dividing by $2\pi R$ we obtain

$$V_A + \sum_{m=1}^{n} F_m - \frac{1}{2\pi} \sum_{m=1}^{n} F_m \alpha_m - \frac{1}{2\pi} \sum_{m=1}^{n} F_m \sin(2\pi - \alpha_m) = 0 \qquad \text{(8-68)}$$

While the first sum in Eq. (8-68) vanishes by virtue of the equilibrium of the system in the vertical direction, the last one is zero because it represents the sum of the moments of all forces acting on the original system about the $OA$ diameter. Finally then

$$V_A = \frac{1}{2\pi} \sum_{m=1}^{n} F_m \alpha_m \qquad \text{(8-69)}$$

Similarly the substitution of Eqs. (8-61) and (8-62) into Eq. (8-63) yields

$$\frac{1}{EI}\left[M_A \pi - \sum_{m-1}^{n} F_m R \int_{\alpha_m}^{2\pi} \cos\theta \sin(\theta - \alpha_m)\, d\theta\right]$$
$$+ \frac{1}{GI_0}\left[M_A \pi - \sum_{m=1}^{n} F_m R \int_{\alpha_m}^{2\pi} [1 - \cos(\theta - \alpha_m)] \sin\theta\, d\theta\right] = 0 \qquad \text{(8-70)}$$

Integrating,

$$\int_{\alpha_m}^{2\pi} \cos\theta \sin(\theta - \alpha_m)\, d\theta = \frac{1}{2}\alpha_m \sin\alpha_m - \pi \sin\alpha_m$$

$$\int_{\alpha_m}^{2\pi} [1 - \cos(\theta - \alpha_m)] \sin\theta\, d\theta = \frac{1}{2}\alpha_m \sin\alpha_m - \pi \sin\alpha_m - 1 + \cos\alpha_m$$

so that Eq. (8-70) becomes

$$\frac{1}{EI}\left[M_A\pi-\sum_{m=1}^{n}F_mR\left(\frac{1}{2}\alpha_m-\pi\right)\sin\alpha_m\right]$$
$$+\frac{1}{GI_0}\left[M_A\pi-\sum_{m=1}^{n}F_mR\left(\frac{\alpha_m}{2}-\pi\right)\sin\alpha_m+\sum_{m=1}^{n}F_mR(1-\cos\alpha_m)\right]=0 \tag{8-71}$$

The equilibrium equations of external forces acting on the original system are

$$\sum_{m=1}^{n}F_m=0,\qquad \sum_{m=1}^{n}F_mR\sin\alpha_m=0,\qquad \sum_{m=1}^{n}F_mR\cos\alpha_m=0 \tag{8-72}$$

Using these equations, Eq. (8-71) reduces to

$$M_A=\frac{1}{2\pi}\sum_{m=1}^{n}F_mR\alpha_m\sin\alpha_m \tag{8-73}$$

Similarly, uncomplicated transformations of Eq. (8-64) yield

$$T_A=-\frac{1}{2\pi}\sum_{m=1}^{n}F_mR\alpha_m(1-\cos\alpha_m) \tag{8-74}$$

Expressions (8-69), (8-73), and (8-74) constitute Biezeno's theorem. The presence of the distributed load only slightly complicates the calculations, as is evident from the following. Let us assume, for example, that in addition to concentrated forces, the ring carries also a uniformly distributed load (positive downward) over its entire length. Now the expressions for the bending moment $M$ [Eq. (8-61)] and for the torque $T$ [Eq. (8-62)] become

$$M=M_A\cos\theta-T_A\sin\theta-V_AR\sin\theta$$
$$-\sum_{m=1}^{k}F_mR\sin(\theta-\alpha_m)H(\theta-\alpha_m)-q_0R^2\int_0^{\theta}\sin(\theta-\psi)\,d\psi \tag{8-75}$$

$$T=M_A\sin\theta+T_A\cos\theta-V_AR(1-\cos\theta)$$
$$-\sum_{m=1}^{k}F_mR[1-\cos(\theta-\alpha_m)]H(\theta-\alpha_m)-q_0R^2\int_0^{\theta}[1-\cos(\theta-\psi)]\,d\psi \tag{8-76}$$

Equations (8-63)–(8-66) are still valid, but the substitution of Eq. (8-76) brings Eq. (8-68), after some manipulations, to the following form,

$$V_A+\sum_{m=1}^{n}F_m-\frac{1}{2\pi}\sum_{m=1}^{n}F_m\alpha_m+\frac{1}{2\pi}\sum_{m=1}^{n}F_m\sin\alpha_m+q_0R\pi=0 \tag{8-77}$$

The conditions of overall equilibrium [Eqs. (8-72)] now become

$$\sum_{m=1}^{n} F_m + 2q_0 R\pi = 0, \qquad \sum_{m=1}^{n} F_m R \sin\alpha_m = 0, \qquad \sum_{m=1}^{n} F_m R \cos\alpha_m = 0 \tag{8-78}$$

We then find that with Eq. (8-74),

$$V_A = \frac{1}{2\pi}\sum_{m=1}^{n} F_m\alpha_m + q_0 R\pi \tag{8-79}$$

The expressions for $M_A$ and $T_A$ are modified to

$$M_A = \frac{1}{2\pi}\sum_{m=1}^{n} F_m R\alpha_m \sin\alpha_m - q_0 R^2 \tag{8-80}$$

$$T_A = -\frac{1}{2\pi}\sum_{m=1}^{n} F_m R\alpha_m(1-\cos\alpha_m) - q_0 R^2\pi \tag{8-81}$$

The application of this theorem is presented in the following example.

## EXAMPLE 8-4

Determine the bending moments, torques, and shear forces in a ring supported by three equally spaced supports and carrying a uniformly distributed load $q_0$ (Fig. 8-22*a*).

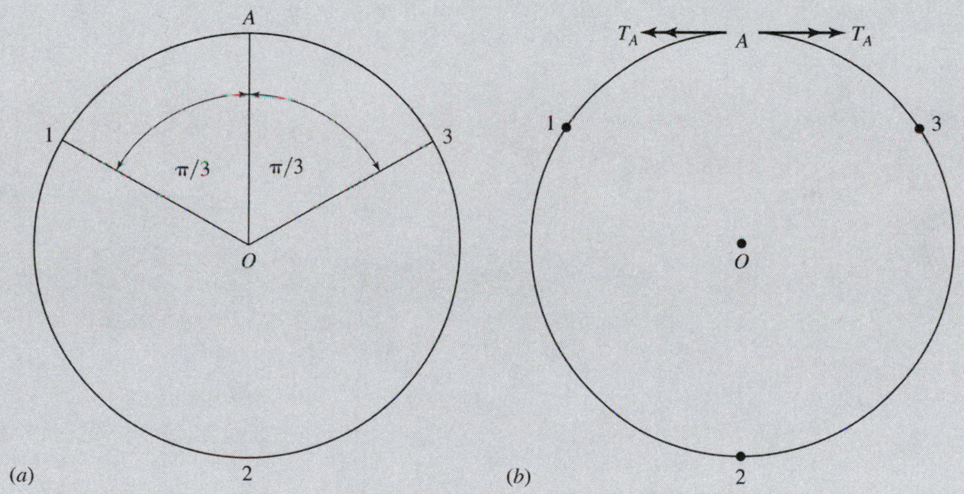


Figure 8-22

## SOLUTION

Using previously introduced notation, we have

$$F_1 = -R_1, \qquad F_2 = -R_2, \qquad F_3 = -R_3$$

$$\alpha_1 = \frac{\pi}{3}, \qquad \alpha_2 = \pi, \qquad \alpha_3 = \frac{5\pi}{3}$$

We now make a cut through an arbitrary point $A$. Next we clamp one side of this cut and calculate the internal forces at this cross section,

$$V_A = \frac{1}{2\pi}\frac{\pi}{3}(F_1 + 3F_2 + 5F_3) + q_0 R\pi$$

$$M_A = \frac{R}{2\pi}\frac{\pi}{3}\left(F_1 \sin\frac{\pi}{3} + 3F_2 \sin\pi + 5F_3 \sin\frac{5\pi}{3}\right) - q_0 R^2$$

$$T_A = -\frac{R}{2\pi}\frac{\pi}{3}\left[F_1\left(1 - \cos\frac{\pi}{3}\right) + 3F_2(1 - \cos\pi) + 5F_3\left(1 - \cos\frac{5\pi}{3}\right)\right] - q_0 R^2\pi$$

Equilibrium equations (8-78) become

$$F_1 + F_2 + F_3 + 2q_0 R\pi = 0$$

$$F_1 \sin\frac{\pi}{3} + F_2 \sin\pi + F_3 \sin\frac{5\pi}{3} = 0$$

$$F_1 \cos\frac{\pi}{3} + F_2 \cos\pi + F_3 \cos\frac{5\pi}{3} = 0$$

from which it follows that

$$F_1 = F_2 = F_3 = -\tfrac{2}{3}q_0 R\pi$$

Therefore,

$$V_A = 0, \qquad M_A = \left(-1 + \frac{2\sqrt{3}}{9}\pi\right) q_0 R^2 = 0.2092 q_0 R^2, \qquad T_A = 0$$

This result confirms what we could have guessed on account of the symmetry of the system with respect to line $OA$: both the shear force $V_A$ and the torque $T_A$ must be zero. The latter conclusion becomes evident when we attach equal and opposite torques $T_A$ at both faces of the cut made at $A$ (Fig. 8-22*b*). For if $T_A$ were different than zero, the segment $A1$ would be twisted outward, and the segment $A3$, inward. Thus the deformation would not be symmetric with respect to $OA$, contrary to the assumed symmetry of the system and of the load. With Eqs. (8-61), (8-62), (8-75), and (8-76) we obtain

$$M = \left(-1 + \frac{2\sqrt{3}}{\pi}\right) q_0 R^2 \cos\theta - q_0 R^2(\theta - \cos\theta) = q_0 R^2 \left(-1 + \frac{2\sqrt{3}}{9}\pi \cos 0\right)$$

$$T = \left(-1 + \frac{2\sqrt{3}}{9}\pi\right) q_0 R^2 \sin\theta - q_0 R^2(\theta - \sin\theta) = \left(-\theta + \frac{2\sqrt{3}}{9}\pi \sin\theta\right) q_0 R^2$$

$$V = q_0 R\theta, \qquad 0 < \theta < \frac{\pi}{3}$$

$$M = \left(-1 + \frac{2\sqrt{3}}{9}\pi \cos\theta\right) q_0 R^2 + \frac{2}{3} q_0 R^2 \pi \sin\left(\theta - \frac{\pi}{3}\right)$$

$$T = \left(-\theta + \frac{2\sqrt{3}}{9}\pi \sin\theta\right) q_0 R^2 + \frac{2}{3} q_0 R^2 \pi \left[1 - \cos\left(\theta - \frac{\pi}{3}\right)\right]$$

$$V = q_0 R\theta - \frac{2}{3} q_0 R\pi, \qquad \frac{\pi}{3} < \theta < \pi$$

$$M = \left(-1 + \frac{2\sqrt{3}}{9}\pi \cos\theta\right) q_0 R^2 + \frac{2}{3} q_0 R^2 \pi \left[\sin\left(\theta - \frac{\pi}{3}\right) + \sin(\theta - \pi)\right]$$

$$T = \left(-\theta + \frac{2\sqrt{3}}{9}\pi \sin\theta\right) q_0 R^2 + \frac{2}{3} q_0 R^2 \pi \left[1 - \cos\left(\theta - \frac{\pi}{3}\right) + 1 - \cos(\theta - \pi)\right]$$

$$V = q_0 R\theta - \frac{4}{3} q_0 R\pi, \qquad \pi < \theta < \frac{5\pi}{3}$$

$$M = \left(-1 + \frac{2\sqrt{3}}{9}\pi \cos\theta\right) q_0 R^2$$
$$+ \frac{2}{3} q_0 R^2 \pi \left[\sin\left(\theta - \frac{\pi}{3}\right) + \sin(\theta - \pi) + \sin\left(\theta - \frac{5\pi}{3}\right)\right]$$

$$T = \left(-\theta + \frac{2\sqrt{3}}{9}\pi \sin\theta\right) q_0 R^2$$
$$+ \frac{2}{3} q_0 R^2 \pi \left[1 - \cos\left(\theta - \frac{\pi}{3}\right) + 1 - \cos(\theta - \pi) + 1 - \cos\left(\theta - \frac{5\pi}{3}\right)\right]$$

$$V = q_0 R\theta - 2 q_0 R\pi, \qquad \frac{5\pi}{3} < \theta < 2\pi$$

Because of the symmetry of the system, the diagrams of $M$, $T$, and $V$ are, as expected, the same for each span. Typical diagrams are shown in Fig. 8-23.

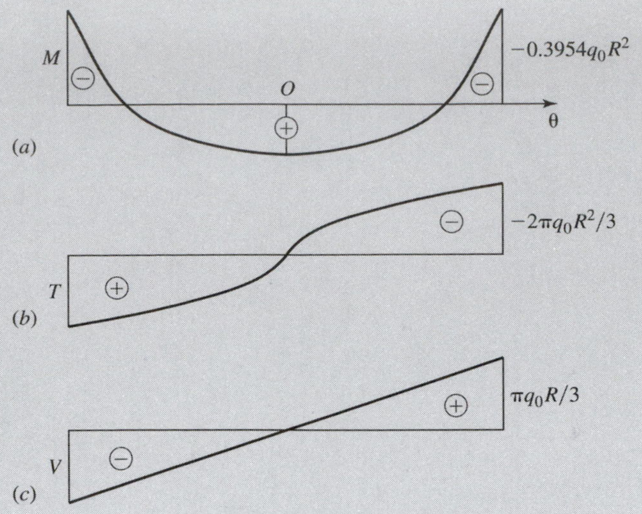


**Figure 8-23**

When the number of supports is greater than three, the system is both externally and internally statically indeterminate. However, if the supports are equally spaced and if each span carries an identical load symmetric with respect to the midspan cross section, all reactions can be easily calculated and the Biezeno formulas are again applicable. In the case of asymmetry we have to use, in addition to the equilibrium equations, the condition that the vertical displacement be zero at the redundant supports. Then if there are $N$ supports ($N > 3$), any $N - 3$ of them may be selected as redundant and the conditions set

$$\frac{\partial U}{\partial X_1} = \frac{\partial U}{\partial X_2} = \cdots = \frac{\partial U}{\partial X_{N-3}} = 0 \tag{8-82}$$

Here $X_1, \ldots, X_{N-3}$ are the reactions at redundant supports and $U$ is the strain energy expressed again by Eq. (8-60). One has to remember, however, that $M$ and $T$ depend also on $X_1, \ldots, X_{N-3}$ temporarily treated as known. Equations (8-82) together with three equations of overall equilibrium provide us with means to calculate all $N$ reactions.

## EXAMPLE 8-5

Consider a ring supported by four supports, as shown in Fig. 8-24, and carrying a uniform load $q_0$.

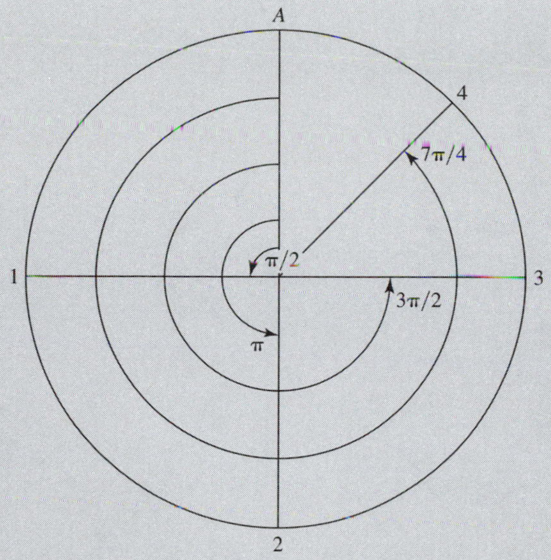


Figure 8-24

## SOLUTION

Assigning the reference cross section $A$ arbitrarily (see Fig. 8-24), we select the reaction at support 4 as the redundant quantity. By denoting

$$F_1 = -R_1, \qquad F_2 = -R_2, \qquad F_3 = -R_3, \qquad F_4 = -X$$

we have using Eqs. (8-79)–(8-81),

$$V_A = \tfrac{1}{2}\left(F_1 \cdot \tfrac{1}{2} + F_2 \cdot 1 + F_3 \cdot \tfrac{3}{2} + F_4 \cdot \tfrac{7}{4}\right) + q_0 R\pi$$

$$M_A = \frac{R}{2}\left(F_1 \cdot \frac{1}{2}\sin\frac{\pi}{2} + F_2 \cdot 1 \sin\pi + F_3 \cdot \frac{3}{2}\sin\frac{3\pi}{2} + F_4 \cdot \frac{7}{4}\sin\frac{7\pi}{4}\right) - q_0 R^2$$

$$= \frac{R}{2}\left(\frac{1}{2}F_1 - \frac{3}{2}F_3 - \frac{7\sqrt{2}}{8}F_4\right) - q_0 R^2 \qquad (a)$$

$$T_A = -\frac{R}{2}\left[F_1 \cdot \frac{1}{2}\left(1 - \cos\frac{\pi}{2}\right) + F_2 \cdot 1(1 - \cos\pi)\right.$$

$$\left. + F_3 \cdot \frac{3}{2}\left(1 - \cos\frac{3\pi}{2}\right) + F_4 \cdot \frac{7}{4}\left(1 - \cos\frac{7\pi}{4}\right)\right] - q_0 R^2 \pi$$

$$= -\frac{R}{2}\left[\frac{1}{2}F_1 + 2F_2 + \frac{3}{2}F_3 + \frac{7}{4}F_4\left(1 - \frac{\sqrt{2}}{2}\right)\right] - q_0 R^2 \pi$$

Equilibrium equations (8-78) now become

$$F_1 + F_2 + F_3 + F_4 + 2q_0 R\pi = 0$$

$$F_1 - F_3 - \frac{\sqrt{2}}{2} F_4 = 0$$

$$-F_2 + \frac{\sqrt{2}}{2} F_4 = 0$$

These give $F_1$, $F_2$, and $F_3$ in terms of the redundant force $F_4 = -X$,

$$F_1 = -\frac{1}{2} F_4 - q_0 R\pi, \qquad F_2 = \frac{\sqrt{2}}{2} F_4, \qquad F_3 = -\frac{1+\sqrt{2}}{2} F_4 - q_0 R\pi \qquad (b)$$

Substituting Eqs. (*b*) into Eqs. (*a*) we get

$$V_A = \frac{3-\sqrt{2}}{8} F_4, \qquad M_A = \frac{4-\sqrt{2}}{16} F_4 R + \left(-1 + \frac{\pi}{2}\right) q_0 R^2$$

$$T_A = -\frac{6-5\sqrt{2}}{16} F_4 R$$

Finally, with Eqs. (8-75) and (8-76), the expressions for the bending moment $M$ and the torque $T$ are obtained. They depend on $\theta$ and on the unknown $F_4$. To calculate $F_4$ we apply the condition (8-82), which now takes the following form:

$$\frac{\partial U}{\partial F_4} = 0$$

or

$$\frac{1}{EI} \int_0^{2\pi} M \frac{\partial M}{\partial F_4}\, d\theta + \frac{1}{GI_0} \int_0^{2\pi} T \frac{\partial T}{\partial F_4}\, d\theta = 0$$

At this point we substituted here the expressions for $M$ and $T$, and in doing so we used the software package Mathcad7, professional edition [18]. The conditional equation to determine $F_4$ becomes

$$\frac{i}{EI} + \frac{j}{GI_0} = 0$$

where

$$i = \frac{R^2}{128}\left[\left(78\pi - 32 + 4\pi\sqrt{2}\right) F_4 + \left(-32\sqrt{2} + 16\pi + 4\pi\sqrt{2}\right) q_0 \pi R\right]$$

$$j = \frac{1}{128\sqrt{2}}\left[\left(82\sqrt{2}\pi - 8\pi - 192 - 96\sqrt{2}\right) F_4 R^2 + \left(-192\pi + 28\sqrt{2}\pi^2 + 24\pi^2\right) q_0 R^3\right]$$

Simplifying the results we obtain

$$F_4 = -\frac{0.17798 + 0.043084\alpha}{1.8033 + 0.063083\alpha} q_0 R$$

where

$$\alpha = \frac{EI}{GJ_0} = \frac{EI}{EJ_0} \cdot 2(1+\nu) = 2(1+\nu)\frac{I}{J_0}$$

From Eq. (*b*) we find

$$F_1 = -\left(\pi - \frac{0.08899 + 0.021542\alpha}{1.8033 + 0.063083\alpha}\right) q_0 R$$

$$F_2 = -\frac{0.12585 + 0.030465\alpha}{1.8033 + 0.063083\alpha} q_0 R$$

$$F_3 = -\left(\pi - \frac{0.21484 + 0.052007\alpha}{1.8033 + 0.063083\alpha}\right) q_0 R$$

We note that in the absence of support 4 the reactions of the remaining three supports would be

$$F_1 = F_3 = -3.1416 q_0 R$$

$$F_2 = 0$$

as can be seen easily by using the equilibrium equation $\Sigma M = 0$ with respect to line 1-3. In this case the presence of support 4 only slightly changes the load distribution. Numerical values of the reactions for various cross sections are given in Table 8-1.

Now all the reactions are known and it is a simple matter to obtain all the internal forces from expressions (*a*). The diagram of the bending moment $M$ is shown in Fig. 8-25 for the case of a circular cross section.

**TABLE 8-1**

| Cross Section | $\alpha$ | $F_1$ | $F_2$ | $F_3$ | $F_4$ |
|---|---|---|---|---|---|
| Circular | 1.30 | −3.0796 | −0.0877 | −2.9919 | −0.1240 |
| Rectangular ($a = b$) | 1.51 | −3.0776 | −0.0905 | −2.9871 | −0.1280 |
| Rectangular ($a = 30b$) | 0.65 | −3.0857 | −0.0790 | −3.0068 | −0.1117 |
| Rectangular ($a = 30b$) | 585 | −2.8137 | −0.4637 | −2.3500 | −0.6558 |
| Statically determinate case ($F_4 = 0$) | — | −3.1416 | 0 | −3.1416 | 0 |
| Limit | 0 | −3.0922 | −0.0698 | −3.0225 | −0.0987 |
| Limit | $\infty$ | −2.8001 | −0.4829 | −2.3172 | −0.6830 |

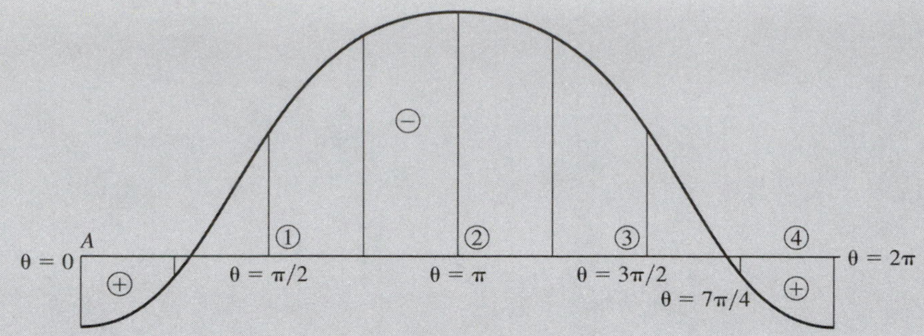


Figure 8-25

## 8-3-3 In-Plane Loaded Curved Beams (Arches) and Rings

### 8-3-3-1 Thin Curved Beams

The assumption that the height of the cross section of a curved bar is small compared to the radius of curvature of its centerline leads, as will be shown later, to the same expression for the strain energy as for a straight beam. Also the hypothesis that plane cross sections remain plane leads for an elastic beam to a linear stress distribution. The derivation of the differential equation for the deflection curve is more complicated than for a straight beam [16, pp. 447–450] and will be replaced here by the application of Castigliano's theorem. The first step in this development will be to take into account the strain energy due to bending and compression and to neglect the influence of shear,

$$U = \int_0^S \frac{M^2\,ds}{2\,EI} + \int_0^S \frac{N^2\,ds}{2\,EA} \tag{8-83}$$

Let us analyze, as an example, a circular arch with built-in ends carrying a concentrated force applied at its axis of symmetry (Fig. 8-26). Because of the symmetry we may

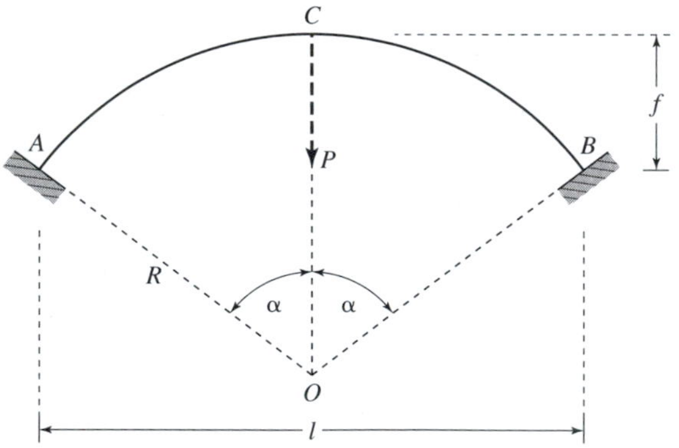


Figure 8-26

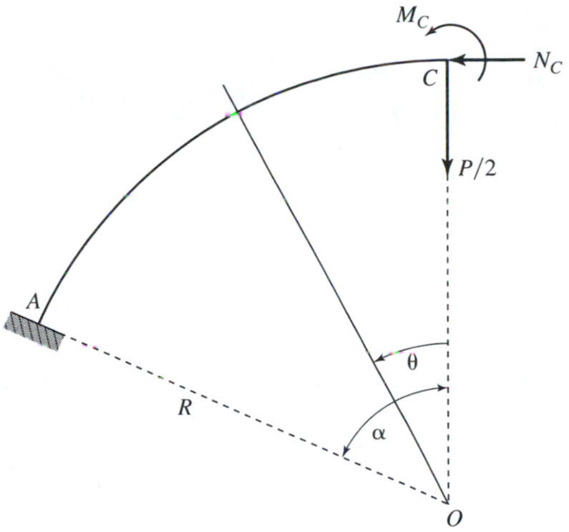


Figure 8-27

cut the arch at $C$ and analyze its part $AC$ subject to the action of the concentrated force $P/2$ of the unknown bending moment $M^C$ and the unknown axial force $N^C$, as seen in Fig. 8-27. The shear force is obviously equal to zero at $C$. The symmetry of the original arch provides us with two conditions needed to determine $M^C$ and $N^C$,

$$\varphi_C = \frac{\partial w}{\partial s} = 0, \qquad u_C = 0 \tag{8-84}$$

where $u_C$ is the horizontal displacement at $C$. By applying Castigliano's theorem with Eq. (8-84) we obtain

$$\begin{aligned} \varphi_C &= \frac{\partial U}{\partial M_C} = \frac{1}{EI}\int_0^S M\frac{\partial M}{\partial M_C}\,ds + \frac{1}{EA}\int_0^S N\frac{\partial N}{\partial M_C}\,ds = 0 \\ u_C &= \frac{\partial U}{\partial N_C} = \frac{1}{EI}\int_0^S M\frac{\partial M}{\partial N_C}\,ds + \frac{1}{EA}\int_0^S N\frac{\partial N}{\partial N_C}\,ds = 0 \end{aligned} \tag{8-85}$$

By using the polar coordinate $\theta$ we replace $ds$ by $R\,d\theta$ and find that the bending moment $M$ and the axial force $N$ are given by

$$\begin{aligned} M &= M_C + N_C R(1 - \cos\theta) \\ N &= -\frac{P}{2}\sin\theta - N_C \end{aligned} \tag{8-86}$$

Therefore,

$$\frac{\partial M}{\partial M_C} = 1, \qquad \frac{\partial N}{\partial M_C} = 0, \qquad \frac{\partial M}{\partial N_C} = R(1 - \cos\theta), \qquad \frac{\partial N}{\partial N_C} = -1$$

so that Eqs. (8-85) become

$$\frac{\partial U}{\partial M_C} = \frac{R}{EI}\int_0^{\alpha}\left[M_C + N_C R(1-\cos\theta) - \frac{P}{2}R\sin\theta\right]d\theta = 0$$

$$\frac{\partial U}{\partial N_C} = \frac{R^2}{EI}\int_0^{\alpha}\left[M_C + N_C R(1-\cos\theta) - \frac{PR}{2}\sin\theta\right](1-\cos\theta)\,d\theta + \frac{R}{EA}\int_0^{\alpha}\left(\frac{P}{2}\sin\theta + N_C\right)d\theta = 0 \tag{8-87}$$

Then, integrating expressions (8-87) and simplifying the results, we obtain

$$N_C = \frac{\dfrac{P}{2}\left[\dfrac{1}{4}(\cos^2\alpha - 1) + \dfrac{\sin\alpha}{\alpha}(1-\cos\alpha) - \beta(1-\cos\alpha)\right]}{\dfrac{1}{2}\alpha - \dfrac{\sin^2\alpha}{\alpha} + \dfrac{\sin 2\alpha}{4} + \alpha\beta}$$

$$M_C = -\frac{1}{\alpha}N_C R(\alpha - \sin\alpha) + \frac{PR}{2\alpha}(1-\cos\alpha) \tag{8-88}$$

where the quantity

$$\beta = \frac{I}{AR^2} \tag{8-89}$$

indicates the influence of the axial forces. It is often negligibly small, particularly for flat arches, that is, for which $f/l \ll 1$ (see Fig. 8-26), because then $I \ll AR^2$. Using expressions (8-88) we may determine the solutions for various angles $\alpha$. Thus, for example, when $\alpha = \pi/2$ (a semicircular arch), we obtain

$$N_C = \frac{P}{2}\left(-\frac{1}{2} + \frac{2}{\pi} - \beta\right)\bigg/\left(\frac{\pi^2 - 8}{4\pi} + \frac{\beta\pi}{2}\right)$$

$$M_C = \frac{PR}{\pi}\,\frac{\pi^2 - 2\pi + 8 - 4(\pi - 2)\beta}{\pi^2 - 8 + 2\pi^2\beta} \tag{8-90}$$

Similarly, when $\alpha = \pi$ (a full ring, as shown in Fig. 8-28), we get

$$N_C = -\frac{2P\beta}{\pi(1+2\beta)}, \qquad M_C = \frac{PR}{\pi}\left(1 + \frac{2\beta}{1+2\beta}\right) \tag{8-91}$$

We now find from Eqs. (8-86) the bending moment and the axial force at an arbitrary cross section $\theta$ of a full ring,

$$M(\theta) = \frac{PR}{2}\left[\frac{2}{\pi} + \frac{4\beta}{\pi(1+2\beta)}\cos\theta\right]$$

$$N(\theta) = -\frac{P}{2}\left[\sin\theta - \frac{4\beta}{\pi(1+2\beta)}\right] \tag{8-92}$$

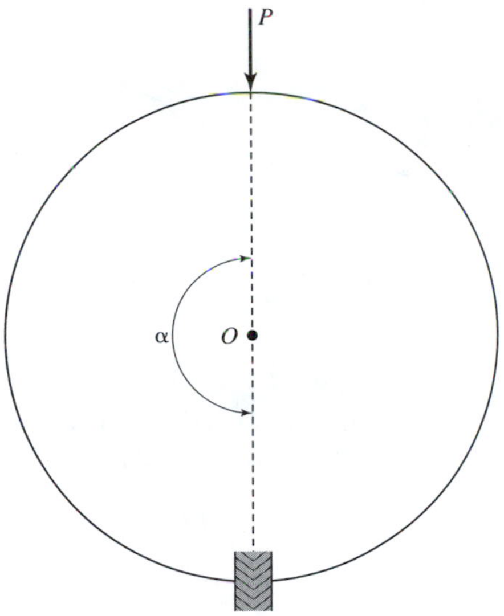


Figure 8-28 Full ring under the action of a concentrated force

When $\beta \ll 1$, the effect of the axial forces on the strain energy can be neglected and we obtain from Eqs. (8-92)

$$M(\theta) = \frac{PR}{\pi}, \qquad N(\theta) = -\frac{P}{2}\sin\theta \tag{8-93}$$

Finally we calculate the stresses in the ring from the standard expression

$$\sigma(\theta) = \frac{M(\theta)z}{I} + \frac{N(\theta)}{A} \tag{8-94}$$

In order to find max $\sigma$ we set $z = -h/2$ and determine the corresponding location $\theta$ from the condition

$$\frac{d\sigma(\theta)}{d\theta} = 0 \tag{8-95}$$

Substituting Eqs. (8-92) for $M(\theta)$ and $N(\theta)$, this becomes

$$-\frac{h}{R}\left[-\cos\theta - \frac{4\beta}{\pi(1+2\beta)}\sin\theta\right] - 2\beta\cos\theta = 0 \tag{8-96}$$

For small $\beta$ this yields

$$\cos\theta = 0 \tag{8-97}$$

indicating that the maximum stress occurs at $\theta = \pi/2$. It equals

$$\max\sigma(\theta) = -\frac{PRh}{4I}\left(\frac{2}{\pi} - 1\right) - \frac{P}{2A} = -\frac{PR^2}{4I}\left[\frac{h}{R}\left(\frac{2}{\pi} - 1\right) - 2\beta\right]$$

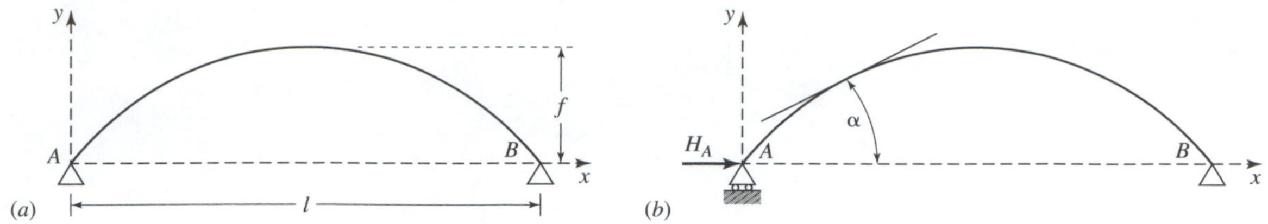


Figure 8-29

In arriving at this expression Eqs. (8-92) and (8-95) have been used. Recalling that β was already neglected previously, we finally obtain

$$\max \sigma(\theta) = -\frac{PRh}{4I}\left(\frac{2}{\pi} - 1\right) \tag{8-98}$$

Next consider an arch on hinged supports (Fig. 8-29*a*) with an arbitrarily curved axis $y(x) = f(x)$ and carrying an arbitrary load. This statically indeterminate structure is transformed into a statically determinate one by removing a horizontal constraint from one of the supports. We select support $A$ and replace it by the unknown reaction $H_A$ (Fig. 8-29*b*). To determine this unknown quantity, we require that the horizontal displacement at $A$ be the same as in the original structure, that is,

$$u_A = 0 \tag{8-99}$$

With Castigliano's theorem, Eq. (8-99) becomes

$$\frac{\partial U}{\partial H_A} = 0 \tag{8-100}$$

where

$$U = \int_0^S \frac{M^2}{2EI_z}\, ds + \int_0^S \frac{N^2}{2EA}\, ds \tag{8-101}$$

The bending moment $M$ and the axial force $N$ at an arbitrary cross section are given by

$$M = M_0 - H_A y, \qquad N = N_0 - H_A \cos\alpha \tag{8-102}$$

in which $M_0$ and $N_0$ are the bending moment and the axial force in the equivalent statically determinate arch whereas α is the angle between the $x$ axis and the tangent to the neutral axis. Substituting these into Eq. (8-101) and differentiating the resulting expression with respect to $H_A$ we obtain

$$-\int_0^S \frac{(M_0 - H_A y)y}{EI_z}\, ds - \int_0^S \frac{(N_0 - H_A \cos\alpha)\cos\alpha}{EA}\, ds = 0$$

Solving this equation with respect to $H_A$ yields

$$H_A = \frac{\displaystyle\int_0^S \frac{M_0 y\, ds}{EI_z} + \int_0^S \frac{N_0 \cos\alpha}{EA}\, ds}{\displaystyle\int_0^S \frac{y^2\, ds}{EI_z} + \int_0^S \frac{\cos^2\alpha}{EA}\, ds} \tag{8-103}$$

Since $y = y(x)$, $ds = \sqrt{1 + y'^2}\, dx$, and $\cos\alpha = 1/\sqrt{1 + y'^2}$, Eq. (8-103) can also be put in the form

$$H_A = \frac{\displaystyle\int_0^l \frac{M_0(x) y \sqrt{1 + y'^2}\, dx}{EI_z} + \int_0^l \frac{N_0(x)\, dx}{EA}}{\displaystyle\int_0^l \frac{y^2 \sqrt{1 + y'^2}\, dx}{EI_z} + \int_0^l \frac{dx}{EA\sqrt{1 + y'^2}}} \tag{8-104}$$

For flat arches $y'(x) \ll 1$ and the second integrals in both the numerator and the denominator of Eq. (8-104) are negligible in comparison with the first ones. Thus Eq. (8-104) becomes

$$H_A = \frac{\displaystyle\int_0^l M_0(x) y\, dx}{\displaystyle\int_0^l y^2\, dx} \tag{8-105}$$

Now with Eq. (8-105) the bending moment $M$ and the axial force $N$ are readily obtained from Eqs. (8-102).

The results obtained in this section involve only thin beams. Another problem occurs for a thick curved beam. Then the expressions for the stress and for the strain energy change. This is discussed in the next section.

#### 8-3-3-2 Thick Curved Beams

A structural or machine element can often be modeled as a curved bar whose thickness is not small in comparison to the radius of curvature. Specifically this is true when analyzing hooks, links, eye-shaped ends of bars, and so on (Fig. 8-30). Let us derive in this section more accurate expressions for the strain energy and for the normal stress distribution in a cross section. We will assume initially that the arch is subject to pure bending. An element of arch is shown in Fig. 8-31 before and after the deformation (dashed line). The segment $n$-$n$ represents a portion of the neutral axis. Thus its length does not change as a result of the deformation. Since we have assumed that originally plane cross sections remain plane during the deformation, the straight segment $d$-$b$ changes with the deformation into the straight segment $d'$-$b'$. In order to determine the normal strain in the direction tangent to the axis of the arch, we analyze the extension of a fiber $\alpha$-$\alpha$ at a distance $y$ from the neutral axis. Its original length is $(R - y)\, d\varphi$ whereas its elongation

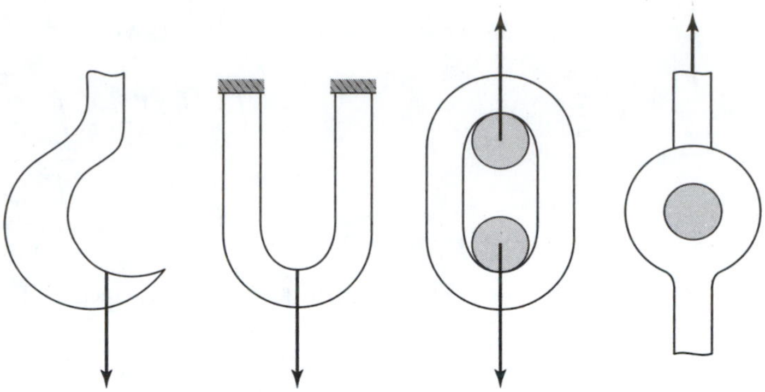

Figure 8-30 Examples of thick curved beams

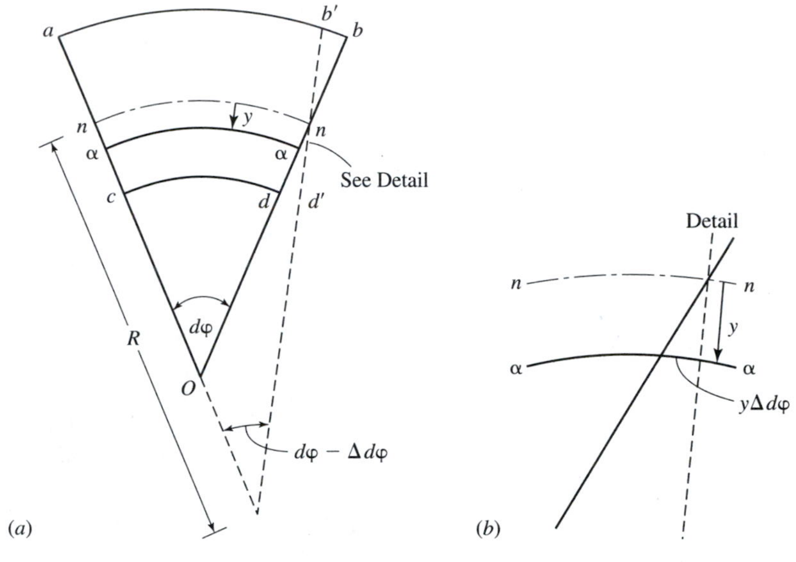


Figure 8-31 Element of arch before and after deformation

is $y\Delta\, d\varphi$, where $R$ is the radius of curvature of the neutral axis. Recalling the definition of the normal strain, we obtain

$$\varepsilon_x = \frac{y\Delta\, d\varphi}{(R-y)\, d\varphi} \tag{8-106}$$

Thus using Hooke's law and disregarding the radial strain, the normal stress is obtained,

$$\sigma_x = E\varepsilon_x = \frac{Ey\Delta\, d\varphi}{(R-y)\, d\varphi} \tag{8-107}$$

Two unknown quantities appear in expression (8-107)—the radius $R$ of the curvature of the neutral axis and the change $\Delta\, d\varphi$ of the angle $d\varphi$. In order to determine these we employ the equations of equilibrium. They require the resultant of normal stresses to be zero and the resultant moment of these stresses with respect to the neutral axis to equal the bending moment $M$ at this cross section,

$$\int_A \sigma_x \, dA \equiv \frac{E\Delta\, d\varphi}{d\varphi} \int_A \frac{y\, dA}{R - y} = 0 \tag{8-108a}$$

$$\int_A \sigma_x y \, dA \equiv \frac{E\Delta\, d\varphi}{d\varphi} \int_A \frac{y^2\, dA}{R - y} \tag{8-108b}$$

Now the second integral in Eqs. (8-108) can be transformed. With the help of Eq. (8-108$a$) we find

$$\int_A \frac{y^2\, dA}{R - y} = \int_A \frac{y(y - R + R)\, dA}{R - y} = -\int_A y\, dA + R \int_A \frac{y\, dA}{R - y} = -\int_A y\, dA$$

where the resulting expression is the first moment of the cross section with respect to the neutral axis. Denoting by $e$ the distance between the neutral axis and the centroid and recalling that the static moment with respect to the centroid is zero, we find readily that (Fig. 8-32)

$$\int_A y\, dA = \int_A (y_1 - e)\, dA = -\int_A e\, dA = -eA$$

Therefore

$$\int_A \frac{y^2\, dA}{R - y} = eA$$

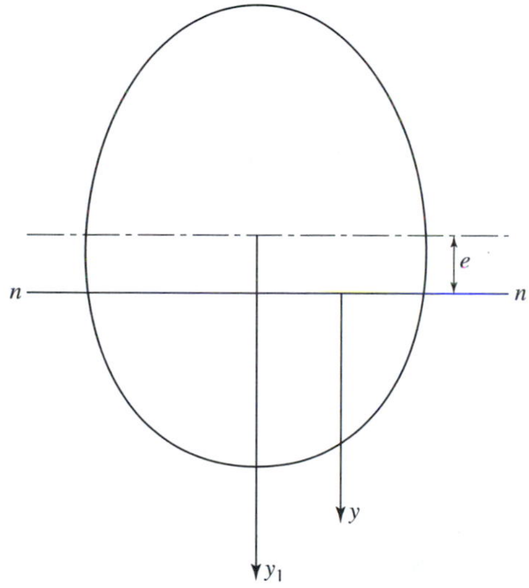


Figure 8-32

Substituting this into Eq. (8-108*b*) we have finally

$$\Delta\, d\varphi = \frac{M}{EAe}\, d\varphi \tag{8-109}$$

Consequently

$$\sigma_x = \frac{My}{Ae(R-y)} \tag{8-110}$$

In order to calculate $R$ from Eq. (8-108*a*) we make the substitution $v = R - y$ (Fig. 8-33) so that

$$\int_A \frac{y\, dA}{R-y} \equiv \int_A \frac{(R-v)\, dA}{v} = 0 \tag{8-111}$$

from which we find

$$R = \frac{A}{K} \tag{8-112}$$

where

$$K \equiv \int_{R_1}^{R_2} \frac{b(v)\, dv}{v} \tag{8-113}$$

In a few cases closed-form expressions for $K$ can be found [25, p. 70]:

- For a rectangle (Fig. 8-34*a*), $K = b \ln\left(\frac{c}{a}\right)$
- For a trapezoid (Fig. 8-34*b*), $K = \left(\frac{b_1 c - b_2 a}{h}\right) \ln\left(\frac{c}{a}\right) - (b_1 - b_2)$

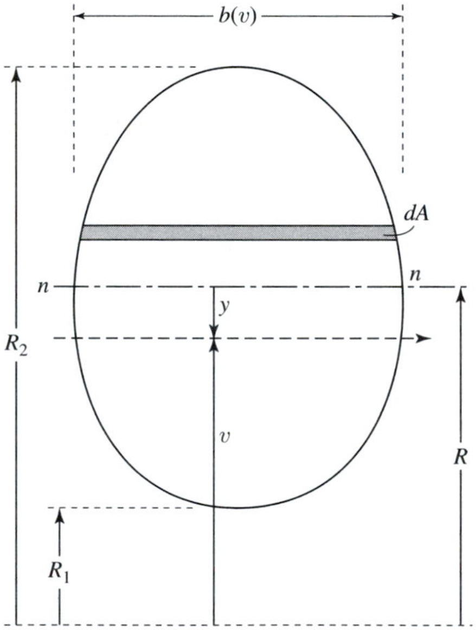


Figure 8-33

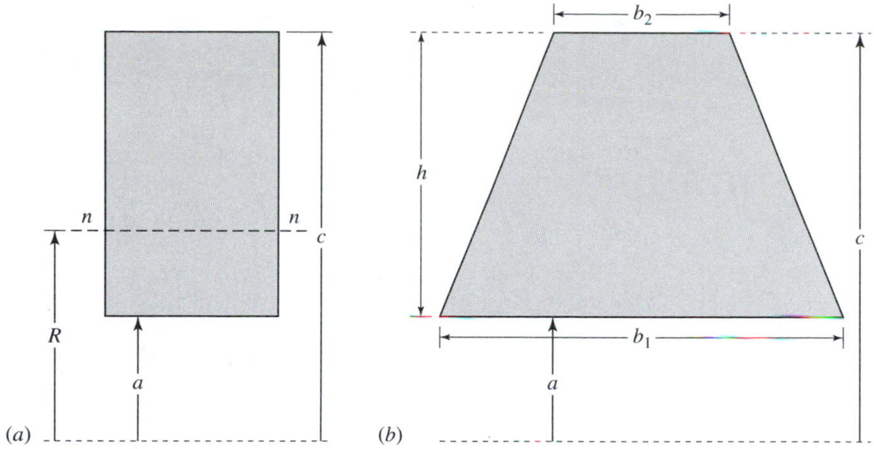


Figure 8-34

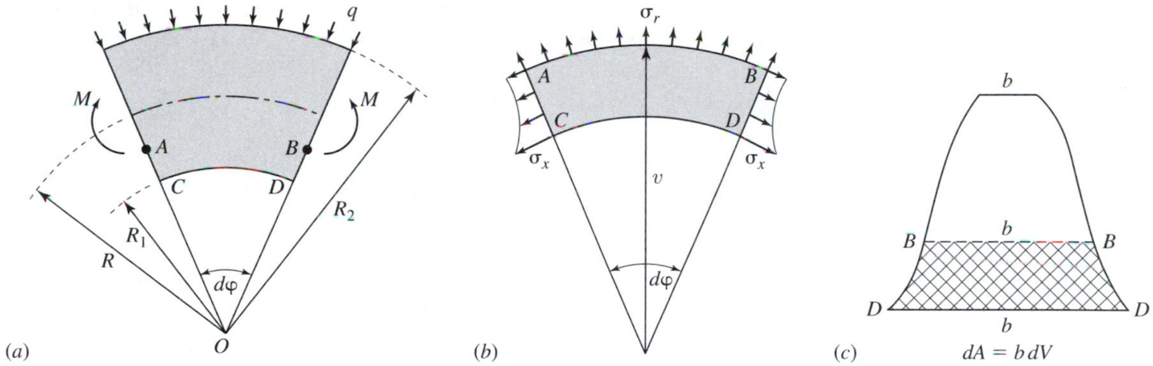


Figure 8-35 Element of thick curved beam

Closed-form expressions for other shapes can be obtained as well, but they are often very involved. It is simpler to evaluate $K$ using numerical integration.*

In deriving Eq. (8-107) we ignored the radial strain and thus, in consequence, the radial stress. The calculations show, however, that for thick curved beams the radial stress $\sigma_r$ may be of the same order of magnitude as the normal stress $\sigma_x$. By assuming again pure bending we will derive a relatively simple expression for $\sigma_r$. We begin by taking a slice of the beam with two very close radial cuts. This element carries only the bending moments $M$ and the external load $q$ (Fig. 8-35*a*). Now consider a portion *ABCD* of this element (Fig. 8-35*b*). In order to derive the equation of equilibrium we have to examine also the cross section of the element (Fig. 8-35*c*). The radial resultant of all

*A Fortran program 8-1 is provided on the diskette for this purpose.

forces acting on *ABCD* is

$$\sigma_r v b \, d\varphi - 2 \int_{R_1}^{v} \sigma_x \, dA \sin \frac{d\varphi}{2} = 0$$

By using $\sin(d\varphi/2) = d\varphi/2$ and substituting Eq. (8-110) we obtain

$$\sigma_r = \frac{1}{bv} \int_{R_1}^{v} \sigma_x \, dA = \frac{1}{bv} \int_{R_1}^{v} \frac{My}{Ae(R-y)} \, dA = \frac{M}{Aebv} \int_{R_1}^{v} \frac{R-v}{v} \, dA$$

where $v = R - y$. Therefore we have finally

$$\sigma_r = \frac{M}{Aebv} \left( \int_{R_1}^{v} \frac{R}{v} \, dA - A_v \right)$$

where $A_v$ is the cross-sectional area cross-hatched in Fig. 8-35*c*,

$$A_v \equiv \int_{R_1}^{v} dA$$

In order to calculate the displacements of a thick ring (Fig. 8-36) we take into account the bending moment $M$, the axial force $N$, and the shear force $V$. Next we use Castigliano's theorem, and this requires adjustments in the strain energy. With Eq. (8-109) the elementary energy due to bending equals (see also Chapter 5)

$$dU_1 = \frac{1}{2} M \Delta_1 \, d\varphi = \frac{M^2}{2EAe} \, d\varphi = \frac{M^2 \, ds}{2EARe}$$

while the strain energy due to elongation by the axial forces $N$ is given by

$$dU_2 = \frac{N^2 \, ds}{2EA}$$

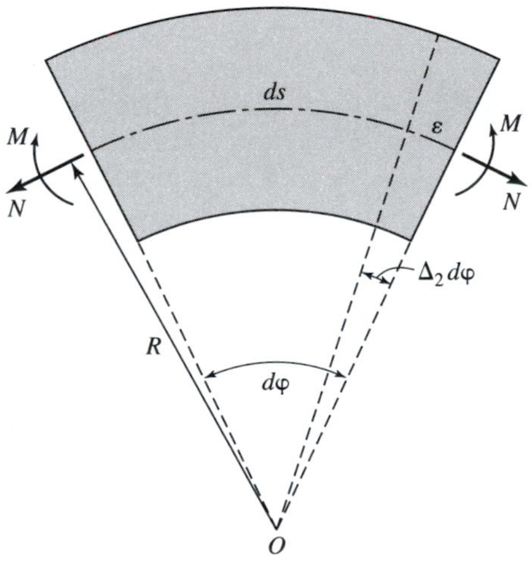


Figure 8-36

However, an elongation $N\,ds/EA$ generates a change $\Delta_2\,d\varphi$ of the angle $d\varphi$ between neighboring cross sections,

$$\Delta_2\,d\varphi = \frac{N\,ds}{EAR}$$

which contributes to the additional work done by the bending moments $M$. The change $\Delta_2 d\varphi$ makes the angle $d\varphi$ larger. Thus the work done by the bending moments $M$ is negative. The additional accumulation of strain energy is therefore

$$dU_3 = -M\Delta_2\varphi = -\frac{MN\,ds}{EAR}$$

Finally, in view of the relative thickness of the ring, the strain energy due to shear is more important now than in the other cases. Using the standard expression we get

$$dU_4 = \frac{\alpha V^2\,ds}{2GA}$$

Integrating the sum of the elementary strain energies along the centerline* of the bar, we find the total strain energy for a thick curved bar,

$$U = \int_0^S \left(\frac{M^2}{2EAeR} + \frac{N^2}{2EA} - \frac{MN}{EAR} + \frac{\alpha V^2}{2GA}\right) ds \qquad \text{(8-114)}$$

## EXAMPLE 8-6

Calculate the radial displacement of the end $B$ of the curved bar shown in Fig. 8-37, assuming that the cross-sectional area is constant and the centerline is an arc of a circle.

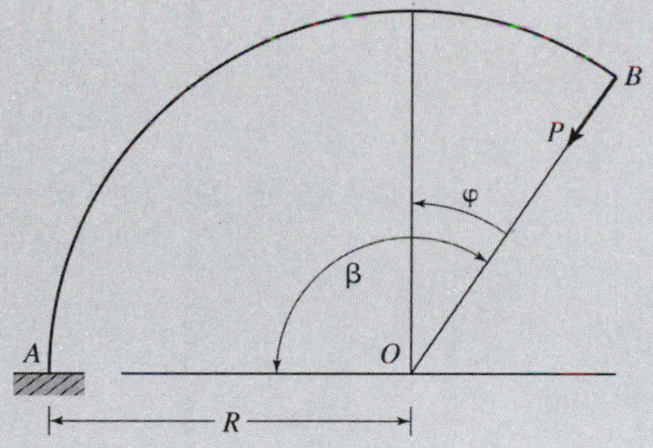


Figure 8-37

*A centerline is a line joining the centroids of the cross sections of the arch.

**SOLUTION**

At an arbitrary cross section $\varphi$ we have

$$M = -PR\sin\varphi, \qquad N = -P\sin\varphi, \qquad V = -P\cos\varphi$$

Substituting these into Eq. (8-114) and integrating gives

$$U = \frac{P^2R}{EA}\left[\left(\frac{R}{2e} - \frac{1}{2}\right)\frac{\beta - \sin\beta\cos\beta}{2} + \alpha(1+\nu)\frac{\beta + \sin\beta\cos\beta}{2}\right] \tag{8-115}$$

The radial displacement at $B$ is therefore

$$\delta_B = \frac{\partial U}{\partial P} = \frac{PR}{EA}\left[\left(\frac{R}{e} - 1\right)\frac{\beta - \sin\beta\cos\beta}{2} + \alpha(1+\nu)(\beta + \sin\beta\cos\beta\right]$$

Let us analyze the behavior of $\delta_B$ with varying angle $\beta$. In order to find a local extremum (if any) we differentiate $\delta_B$ with respect to $\beta$, obtaining the following equation:

$$\frac{1}{2}\left(\frac{R}{e} - 1\right)(1 - \cos 2\beta) + \alpha(1+\nu)(1+\cos 2\beta) = 0$$

This yields the expression for the extremal value of $\beta$,

$$\cos 2\beta = \frac{\dfrac{R}{e} - 1 + 2\alpha(1+\nu)}{\dfrac{R}{e} - 1 - 2\alpha(1+\nu)}$$

To interpret this result assume a rectangular cross section for which $e = h^2/12R$. By using the same data as Timoshenko [25]: ($\nu = 0.3$, $\alpha = 1.2$), we obtain

$$\cos 2\beta = \frac{12\dfrac{R^2}{h^2} + 2.12}{12\dfrac{R^2}{h^2} - 4.12}$$

Since $R/h > 0.5$ it is easy to see that the above equation does not have any real roots. Thus $\delta_B$ is increasing monotonically with $\beta$. Using the same data, we find

$$\delta_B = \frac{PR}{EA}\left[\frac{6R^2}{h^2}(\beta - \sin\beta\cos\beta) + 1.06\beta + 2.06\sin\beta\cos\beta\right] \tag{8-116}$$

With $\beta = \pi/2$ this gives [25]

$$\delta_B = \frac{PR\pi}{4EA}\left(\frac{12R^2}{h^2} + 2.12\right) \tag{8-117}$$

## 8-3-4 Bending, Stretching, and Twisting of Springs

In this section we examine helical and conical springs of cross-sectional dimensions much smaller than both the helix radius $R$ (centroidal radius) and the average radius $R_{av}$ of the conical curve.

### 8-3-4-1 Helical Springs [6, pp. 548–550]

Let us consider a helical spring fixed at one end (point $C$ at the end of the horizontal segment $B$–$C$) and subject to a force $P$ and moments $M_1$, $M_2$, and $M_3$ applied at the other end (Fig. 8-38). We will use Castigliano's theorem to find the maximum deflection, which obviously occurs at end $B$ carrying the load,

$$\max v = v_B = \frac{\partial U}{\partial P} \tag{8-118}$$

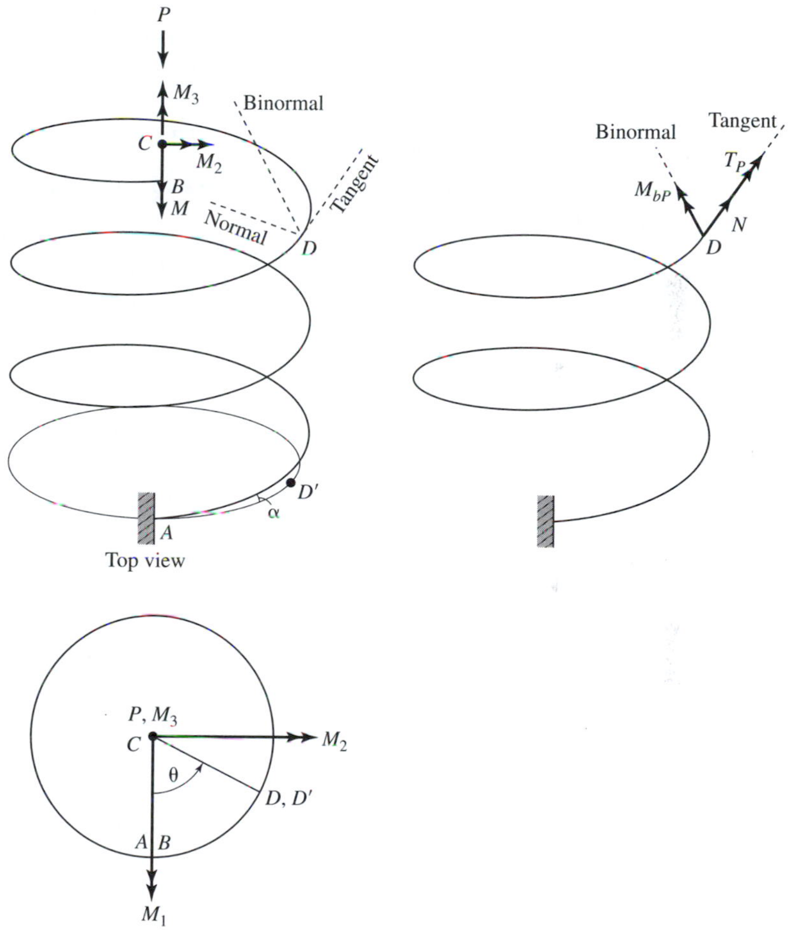


Figure 8-38 Helical spring fixed at one end

Taking into account the strain energy from bending, torsion, and compression due to the axial force but ignoring the effects of shear, we obtain

$$U = U_b + U_t + U_a \tag{8-119}$$

where

$$\begin{aligned} U_b &= \frac{1}{2EI_n}\int_0^S M_n^2\,ds + \frac{1}{2EI_b}\int_0^S M_b^2\,ds \\ U_t &= \frac{1+\nu}{EJ}\int_0^S T^2\,ds \\ U_a &= \frac{1}{2EA}\int_0^S N^2\,ds \end{aligned} \tag{8-120}$$

Here $M_n$ and $M_b$ denote bending moments with the vectors directed along the normal and the binormal, respectively. Next we determine the internal forces. In a polar coordinate system the equation of the helix representing the neutral axis of the beam is

$$x = R\cos\theta, \qquad y = R\sin\theta, \qquad z = k\theta \tag{8-121}$$

where $k$ is a constant known as the *pitch* of the helix. With $n$ denoting the number of coils and $\alpha$ the *pitch angle,* we have

$$\begin{aligned} \tan\alpha &= \frac{2\pi k}{2\pi R} = \frac{k}{R} \\ k &= R\tan\alpha \end{aligned} \tag{8-122}$$

The axial force $N$ at an arbitrary cross section $D$ is generated by the applied load $P$ alone,

$$\begin{aligned} N &= -P\sin\alpha = -P\frac{\tan\alpha}{\sqrt{1+\tan^2\alpha}} \\ &= -P\frac{k/R}{\sqrt{1+k^2/R^2}} = -P\frac{k}{\sqrt{k^2+R^2}} \end{aligned} \tag{8-123}$$

The force $P$ also creates the torque $T_P$ and the bending moment $M_{bP}$,

$$\begin{aligned} T_P &= PR\cos\alpha \\ M_{bP} &= PR\sin\alpha \end{aligned} \tag{8-124}$$

Each of the applied moments $M_1$, $M_2$, and $M_3$ contributes to $T$, $M_b$, and $M_n$. We have

$$\begin{aligned} T_M &= M_1\sin\theta\cos\alpha - M_2\cos\theta\cos\alpha - M_3\sin\alpha \\ M_{bM} &= M_1\sin\theta\sin\alpha - M_2\cos\theta\sin\alpha + M_3\cos\alpha \\ M_{nM} &= M_1\cos\theta + M_2\sin\theta \end{aligned} \tag{8-125}$$

Substituting $T = T_P + T_M$, and so on, and putting $ds = R\,d\theta/\cos\alpha$, $S = 2\pi Rn$, we find the following expressions for the strain energy:

$$U_a = \frac{P^2k^2R^2\pi n}{EA(k^2+R^2)\cos\alpha} \tag{8-126a}$$

$$\begin{aligned}
U_b &= \frac{R}{2EI_n\cos\alpha}\int_0^{2\pi Rn}(M_1\cos\theta + M_2\sin\theta)^2\,d\theta \\
&\quad + \frac{R}{2EI_b\cos\alpha}\int_0^{2\pi Rn}(PR\sin\alpha + M_1\sin\theta\sin\alpha \\
&\quad - M_2\cos\theta\sin\alpha + M_3\cos\alpha)^2\,d\theta \\
&= \frac{R}{2EI_n\cos\alpha}\left[(M_1^2+M_2^2)R\pi n + \frac{1}{4}(M_1^2 - M_2^2)\sin 4\pi Rn\right. \\
&\quad \left. - \frac{1}{2}M_1M_2(\cos 4\pi Rn - 1)\right] + \frac{R}{2EI_b\cos\alpha}\left[P^2R^2\,2\pi Rn\sin^2\alpha\right. \\
&\quad + M_1^2\frac{1}{2}\sin^2\alpha\left(2\pi Rn - \frac{1}{2}\sin 4\pi Rn\right) + M_2^2\frac{1}{2}\sin^2\alpha\left(2\pi Rn + \frac{1}{2}\sin 4\pi Rn\right) \\
&\quad + M_3^2\,2\pi Rn\cos^2\alpha + 2PM_1R\sin^2\alpha(1 - \cos 2\pi Rn) \\
&\quad - 2PM_2R\sin^2\alpha\sin 2\pi Rn + 2PM_3R\sin\alpha\,2\pi Rn\cos\alpha \\
&\quad + \frac{1}{2}M_1M_2\sin^2\alpha(\cos 4\pi Rn - 1) + 2M_1M_3\sin\alpha\cos\alpha(1 - \cos 2\pi Rn) \\
&\quad \left. - M_1M_3\sin\alpha\cos\alpha\sin 2\pi Rn\right]
\end{aligned} \tag{8-126b}$$

In order to determine the displacement components we used the software package Mathcad7 [18]. After simplification the following expression is obtained for the deflection:

$$\begin{aligned}
w &= \frac{PRL}{Ec}\left[\frac{s^2}{A} + \frac{2R^2(1+\nu)c^2}{J} + \frac{R^2s^2}{I_b}\right] \\
&\quad + \frac{M_3R^2Ls}{E}\left[-\frac{2(1+\nu)}{J} + \frac{1}{I_b}\right] \\
&\quad + M_1\frac{R^2}{Ec}\left[\frac{2c^2(1+\nu)}{J} + \frac{s^2}{I_b}\right](1 - \cos L) \\
&\quad + M_2\frac{R^2\sin L}{Ec}\left[-\frac{2c^2(1+\nu)}{J} - \frac{s^2}{I_b}\right]
\end{aligned} \tag{8-127}$$

where $s \equiv \sin\alpha$ and $c \equiv \cos\alpha$. When the cross section is circular then $J = 2I_b$. Denoting $I_b = \pi d^4/64$ and $L = 2\pi n$, we obtain from here (see also the result in [8, p. 263])

$$w = \frac{2PR^3\pi n}{E\cos\alpha}\left[\frac{\sin^2\alpha}{AR^2} + \frac{64(1+N\cos^2\alpha)}{\pi d^4}\right] - \frac{128\nu M_3 R^2 n\sin\alpha}{Ed^4}$$
$$+ \frac{64M_1R^2(1+\nu\cos^2\alpha)(1-\cos 2\pi n)}{E\pi d^4\cos\alpha} - \frac{64M_2R^2(1+\nu\cos^2\alpha)\sin 2\pi n}{E\pi d^4\cos\alpha} \tag{8-128}$$

This indicates that with $n$ an integer the moments $M_1$ and $M_2$ have no influence on the extension or the compression $w$ of the spring. Proceeding in a similar manner we obtain the maximum angle of twist,

$$\gamma = -\frac{128\nu PR^2 n\sin\alpha}{Ed^4} + \frac{128M_3Rn}{Ed^4\cos\alpha}(1+\nu\sin^2\alpha)$$
$$+ \frac{M_1Rs(1-\cos L)}{E}\left[\frac{1}{I_b} - \frac{2(1+\nu)}{J}\right]$$
$$- \frac{M_2Rs\sin L}{E}\left[\frac{1}{I_b} - \frac{2(1+\nu)}{J}\right] \tag{8-129}$$

For a circular cross section $J = 2I_b = \pi d^4/32$, and this result reduces to the formula (see also [8, pp. 262–264])

$$\gamma = -\frac{128\nu PR^2 n\sin\alpha}{Ed^4} + \frac{128M_3Rn}{Ed^4\cos\alpha}(1+\nu\sin^2\alpha)$$
$$- \nu[M_1(1-\cos 2\pi n) - M_2\sin 2\pi n]\frac{64R\sin\alpha}{E\pi d^4} \tag{8-130}$$

Now the maximum slope in the direction of bending by $M_1$ is found to be

$$\varphi_1 = \frac{PR^2}{Ec}(1-\cos L)\left[\frac{2(1+\nu)c^2}{J} + \frac{s^2}{I_b}\right]$$
$$+ \frac{M_3R(1-\cos L)}{E}\left[\frac{1}{I_b} - \frac{2(1+\nu)}{J}\right]$$
$$+ \frac{M_1RL}{2Ec}\left[\frac{1}{I_n} + \frac{s^2}{I_b} - \frac{s^2\sin 2L}{2I_bL} + \frac{\sin 2L}{2I_nL} + \frac{2(1+\nu)c^2}{J} - \frac{c^2\sin 2L(1+\nu)}{JL}\right]$$
$$+ \frac{M_2R}{4Ec}\left[-\frac{s^2(1-\cos 2L)}{I_b} + \frac{1-\cos 2L}{I_n} - \frac{2c^2(1+\nu)(1-\cos 2L)}{J}\right] \tag{8-131}$$

which for a circular cross section ($J = 2I_b = 2I_n = \pi d^4/32$) becomes (see also [8, p. 265, eq. 7.14.12])

$$\varphi_1 = \frac{32PR^2(1-\cos 2n\pi)}{E\pi d^4 \cos\alpha}(1+\nu\cos 2\alpha) - \frac{32\nu M_3 R(1-\cos 4n\pi)\sin\alpha}{E\pi d^4} + \frac{32M_1 Rn}{Ed^4\cos\alpha}$$
$$\times\left[1+\sin^2\alpha + (1-\sin^2\alpha)\frac{\sin 4n\pi}{4\pi n} + (1+\nu)\cos^2\alpha\left(1-\frac{\sin 4\pi n}{4\pi n}\right)\right]$$
$$-\frac{16M_2R(1-\cos 4\pi n)}{E\pi d^4\cos\alpha}\nu\cos^2\alpha \qquad (8\text{-}132)$$

The maximum slope $\varphi_2$ in the direction of bending by $M_2$ is also calculated,

$$\varphi_2 = \frac{PR^2\sin L}{Ec}\left[-\frac{2(1+\nu)c^2}{J} - \frac{s^2}{I_b}\right] + \frac{M_3Rs\sin L}{E}\left[-\frac{1}{I_b}+\frac{2(1+\nu)}{J}\right]$$
$$+\frac{M_1R(1-\cos 2L)}{4Ec}\left[-\frac{s^2}{I_b}+\frac{1}{I_n}-\frac{2(1+\nu)c^2}{J}\right]$$
$$+\frac{M_2RL}{2Ec}\left[\frac{1}{I_n}+\frac{s^2}{I_b}+\frac{s^2\sin 2L}{2I_bL}-\frac{\sin 2L}{2I_nL}+\frac{2(1+\nu)c^2}{J}+\frac{c^2(1+\nu)\sin 2L}{JL}\right]$$
$$(8\text{-}133)$$

which for a circular cross section becomes

$$\varphi_2 = -\frac{32PR^2\sin 2\pi n}{E\pi d^4\cos\alpha}(1+\nu\cos^2\alpha)$$
$$+\frac{32\nu M_3R\sin 2\pi n\sin\alpha}{E\pi d^4} - \frac{8M_1R(1-\cos 4\pi n)\cos\alpha}{E\pi d^4}\nu$$
$$+\frac{32M_2Rn}{Ed^4\cos\alpha}\left[1+\sin^2\alpha - \frac{\sin 4\pi n\cos^2\alpha}{4\pi n}\right.$$
$$\left.+(1+\nu)\cos^2\alpha + 2(1+\nu)\frac{\sin 4\pi n\cos^2\alpha}{4\pi n}\right] \qquad (8\text{-}134)$$

### 8-3-4-2 Conical Springs [6, p. 552]

The deflections and stresses in a conical spring fixed at one end, as seen in Fig. 8-39, are calculated in a manner similar to the previous case. The only difference consists in that

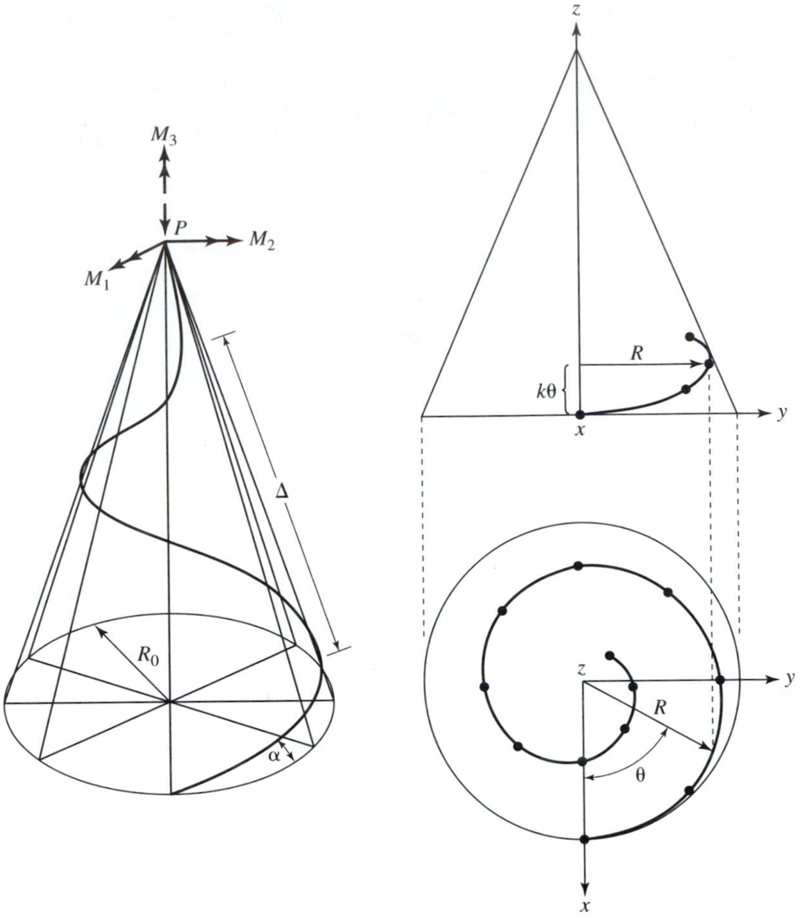


Figure 8-39 Conical spring fixed at one end

$R$ is no longer a constant but can be represented as

$$R = R_0 - \Delta\theta \tag{8-135}$$

where $\Delta$ is a given constant. The apex angle of the cone is $\beta$ and the length of the spring is $L$. The coordinates of an arbitrary point on the spring are given by

$$\begin{aligned} x &= R\cos\theta \\ y &= R\sin\theta \\ z &= k\theta \end{aligned} \tag{8-136}$$

The expressions for $N$, $T$, $M_n$, and $M_b$ as well as the general expressions for the strain energies are the same as for a helical spring (see Section 8-3-4-1). When integrating one must keep in mind that $R$ is now variable. The results of integration, obtained using the software package Mathcad7, professional edition [18], are given next. The angle of twist becomes

$$\gamma = \frac{PL\sin\alpha}{3E}\left[\frac{(GR_0\Delta L - 6R_0^2 - 2\Delta^2 L^2)(1+\nu)}{J} + \frac{3R_0^2 - 3R_0\Delta L + \Delta^2 L^2}{I_b}\right]$$
$$+ \frac{M_3 L}{E}\left[\frac{\sin^2\alpha(2R_0 - \Delta L)(1+\nu)}{J\cos\alpha} + \frac{(2R_0 - \Delta L)\cos\alpha}{2I_b}\right]$$
$$+ \frac{M_1\sin\alpha}{E}\left\{\frac{[-2R_0(1-\cos L) + 2\Delta(\sin L - L\cos L)](1+\nu)}{J}\right.$$
$$\left. + \frac{R_0(1-\cos L) + \Delta(L\cos L - \sin L)}{I_b}\right\}$$
$$+ \frac{M_2\sin\alpha}{E}\left\{\frac{[2R_0\sin L + 2\Delta(1 - L\sin L - \cos L)](1+\nu)}{J}\right.$$
$$\left. + \frac{-R_0\sin L + \Delta(L\sin L - 1 + \cos L)}{I_b}\right\} \tag{8-137}$$

For a circular cross section this becomes

$$\gamma = \frac{PL\sin\alpha}{3E}\left[-\frac{\nu(3R_0^2 - 3R_0\Delta L + \Delta^2 L^2)}{I_b}\right]$$
$$+ \frac{M_3 L}{2E\cos\alpha}\left[\frac{2R_0(1+\nu\sin^2\alpha) - \Delta L(1+\nu\sin^2\alpha)}{I_b}\right]$$
$$+ \frac{M_1\nu\sin\alpha}{E}\left[\frac{-R_0(1-\cos L) + \Delta(\sin L - L\cos L)}{I_b}\right]$$
$$+ \frac{M_2\nu\sin\alpha}{E}\left[\frac{R_0\sin L + \Delta(1 - L\sin L - \cos L)}{I_b}\right] \tag{8-138}$$

When $\Delta = 0$ this reduces to the previously obtained expression for the angle of twist in the helical spring. When $n$ in $L = 2\pi n$ is an integer, we obtain the following expression for the angle of twist of a conical spring with circular cross section and an integer number of coils:

$$\gamma = -\frac{PL\nu\sin\alpha}{3EI_b}(3R_0^2 - 3R_0\Delta L + \Delta^2 L^2)$$
$$+ \frac{M_3 L}{2EI_b\cos\alpha}[2R_0(1+\nu\sin^2\alpha) - \Delta L(1+\nu\sin^2\alpha)]$$
$$- \frac{M_1 L\nu\sin\alpha}{EI_b} \tag{8-139}$$

Note that in this case $M_2$ has no effect on $\gamma$. The deflection $w$ for $n$ integer (integer number of coils) is given by

$$w = \frac{PL}{E\cos\alpha}\left[\frac{\left(4R_0^3 - 6R_0^2 L\Delta + 4R_0 L^2\Delta^2 - L^3\Delta^3\right)(1+\nu)\cos^2\alpha}{2J}\right.$$

$$\left. + \frac{\left(R_0 - \frac{1}{2}L\Delta\right)\sin^2\alpha}{A} + \frac{\left(4R_0^3 - 6R_0^2 L\Delta + 4R_0 L^2\Delta^2 - L^3\Delta^3\right)\sin^2\alpha}{4I_b}\right]$$

$$+ \frac{M_3 L \sin\alpha}{E}\left[\frac{2\left(-3R_0^2 + 3R_0 L\Delta - L^2\Delta^2\right)(1+\nu)}{3J} + \frac{3R_0^2 - 3R_0 L\Delta + L^2\Delta^2}{3I_b}\right]$$

$$+ \frac{M_1}{E\cos\alpha}\left[\frac{2L\Delta(2R_0 - L\Delta)(1+\nu)\cos^2\alpha}{J} + \frac{L\Delta(2R_0 - L\Delta)\sin^2\alpha}{I_b}\right]$$

$$+ \frac{M_2}{E\cos\alpha}\left[\frac{-4L\Delta^2(1+\nu)\cos^2\alpha}{J} - \frac{2L\Delta^2\sin^2\alpha}{I_b}\right] \tag{8-140}$$

Finally, when $n$ is an integer and $J = 2I_b$ (circular cross section), this becomes

$$w = \frac{PL}{E\cos\alpha}\left[\frac{\left(4R_0^3 - 6R_0^2 L\Delta + 4R_0 L^2\Delta^2 - L^3\Delta^3\right)(1+\nu)\cos^2\alpha}{4I_b}\right.$$

$$\left. + \frac{(2R_0 - L\Delta)\sin^2\alpha}{2A}\right] - \frac{M_3 L\nu\sin\alpha}{3EI_b}\left(3R_0^2 - 3R_0 L\Delta + L^2\Delta^2\right)$$

$$+ \frac{M_1 L\Delta}{EI_b\cos\alpha}(2R_0 - L\Delta)(1+\nu)\cos^2\alpha - \frac{2M_2 L\Delta^2}{EI_b\cos\alpha}(1+\nu)\cos^2\alpha \tag{8-141}$$

When $\alpha \ll 1$ and when the only load is the concentrated force $P$, this becomes*

$$w = \frac{Pn\pi}{GJ}\left[R_0^2 + (R_0 - 2n\pi\Delta)^2\right](R_0 - n\pi\Delta) \tag{8-142}$$

## 8-4 BEAMS ON ELASTIC FOUNDATIONS

We consider a beam resting on an elastic solid (a "foundation") and carrying a load, as shown in Fig. 8-40*a*. To explain by simple examples the differences between two basic models describing a beam's behavior we assume that it is so rigid that it can be treated as undeformable. On account of symmetry the beam will move downward by a constant

*This result agrees with [6, p. 552, eq. 12.20b] (when $\alpha \ll 1$ and when other notation is used).

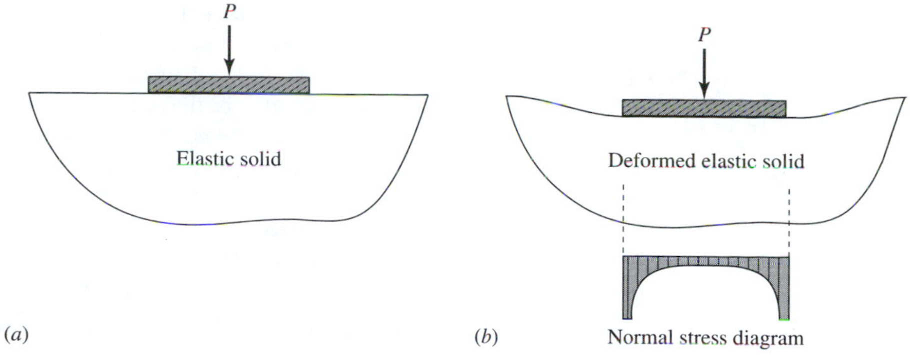


**Figure 8-40** Stamp problem

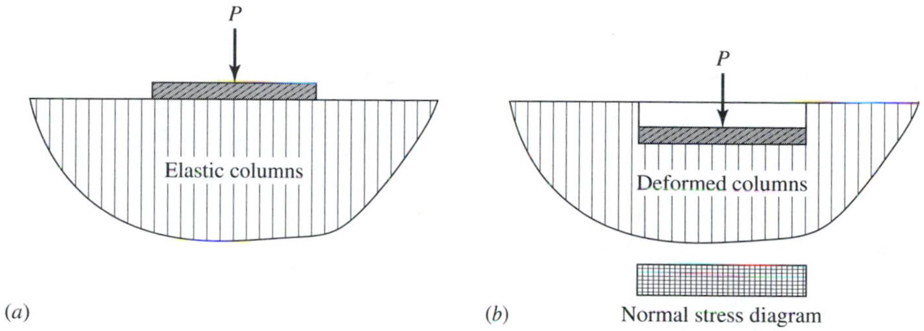


**Figure 8-41** Beam on spring foundation

amount, as seen in Fig. 8-40*b*. Such a problem, known as a stamp problem, has been solved using equations of the theory of elasticity (see, for example [10, pp. 45–60]). It was found that the diagram of the normal stress under the rigid stamp has the form shown in Fig. 8-40*b*. It is seen that the stress is not only nonuniform, but that it even grows infinitely near the ends of the stamp. Now we consider another case, shown in Fig. 8-41*a*. This time the "foundation" is composed of many identical independent elastic columns (with no friction between adjacent columns). It is evident that although the rigid beam will be displaced in exactly the same way as before, the response of the foundation will be quite different. Since the ends of all the columns are displaced by the same amount, their responses—the end stresses—will also be identical, as shown in Fig. 8-41*b*. In the second model the stress distribution along the beam is proportional to the deformation, in the first it is not.

Materials such as soil (the foundation) do not behave like either of these models. In fact the behavior of soil is a complicated problem, which depends on many factors (porosity, humidity, compaction, and others). The behavior of soil is often described as that of a *viscoelastic* material (see Chapter 5). No wonder that while choosing between two hypotheses, each with its own deficiencies, we select the one allowing for a relatively simple mathematical description, and that is the model shown in Fig. 8-41*b*, known as *Winkler's foundation* [28, p. 182]. Hence simply stated: the intensity of the reaction of the foundation at a point is proportional to the displacement at the same point. We also assume that the foundation is bilateral, that is, the beam cannot "lift off." The coefficient of proportionality is often denoted by $k$ and known as the *foundation modulus*. Let us add that a beam on an elastic foundation does not model exclusively a beam resting on the ground, but also, and more accurately, a beam on a series of closely and regularly spaced elastic supports (ties of a railroad track). This model is also used to derive the equation of equilibrium of a thin-walled storage tank (see Section 8-4-4).

## 8-4-1 Equilibrium Equation for a Straight Beam

In order to derive the equation for deflection we cut from a beam on an elastic foundation (Fig. 8-42*a*) an infinititesimal slice, shown magnified in Fig. 8-42*b*. It is subject to a force $ky\,dx$, representing the resultant of the reaction of the foundation to the bending moment $M$ and to the shear force $V$. The modulus $k$ is measured in units of force per $y\,dx$, that is, in pounds per square inch (psi). All the assumptions of the theory of thin beams are still valid. The equation of equilibrium $\Sigma F_y = 0$ becomes

$$-V + V + \frac{dV}{dx}\,dx + q(x)\,dx - ky\,dx = 0$$

or

$$\frac{dV}{dx} = -q(x) + ky \tag{8-143}$$

The equilibrium equation $\Sigma M = 0$ yields

$$M - M - \frac{dM}{dx}\,dx + V\,dx - q(x)\frac{dx^2}{2} + ky\frac{dx^2}{2} = 0$$

which, when small quantities of higher order are neglected, becomes

$$\frac{dM}{dx} = V \tag{8-144}$$

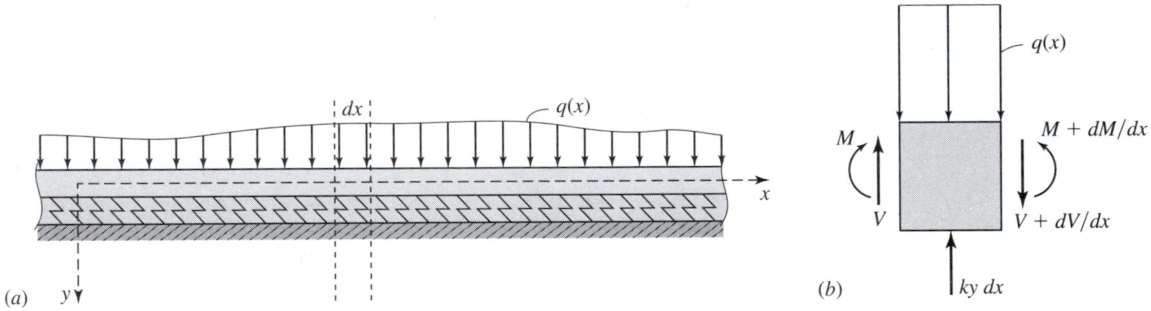


Figure 8-42 Beam on Winkler's foundation

Combining Eqs. (8-143) and (8-144) we arrive at the following relation:

$$\frac{d^2M}{dx^2} = -q(x) + ky \tag{8-145}$$

Note that the foundation acts as a distributed load of magnitude $ky$. Finally it is known from the deformation theory of simple beams [21, p. 272] that

$$M = -EI\frac{d^2y}{dx^2} \tag{8-146}$$

Substituting Eq. (8-146) into Eq. (8-145) we obtain

$$EI\frac{d^4y}{dx^4} + ky = q(x) \tag{8-147}$$

which is transformed to

$$y^{\text{iv}} + \alpha^4 y = \frac{q(x)}{EI} \tag{8-148}$$

where

$$\alpha = \sqrt[4]{\frac{k}{EI}} \tag{8-149}$$

Proceeding in the standard way with the reduced equation (8-148), $y^{\text{iv}} + \alpha^4 y = 0$, we assume a solution of the form

$$y = e^{rx}$$

This leads to the auxiliary equation

$$r^4 + \alpha^4 = 0 \tag{8-150}$$

from which it follows that

$$r = \sqrt[4]{-\alpha^4} \equiv \alpha\sqrt[4]{-i} \tag{8-151}$$

where $i = \sqrt{-1}$. As can be verified by substitution, expression (8-151) takes the following four values [12, pp. 232–233]:

$$\begin{aligned} r_1 &= \frac{\alpha}{\sqrt{2}}(1+i), & r_2 &= \frac{\alpha}{\sqrt{2}}(1-i) \\ r_3 &= \frac{\alpha}{\sqrt{2}}(-1+i), & r_4 &= -\frac{\alpha}{\sqrt{2}}(1+i) \end{aligned} \tag{8-152}$$

Introducing for convenience the symbol

$$\beta = \sqrt[4]{\frac{\alpha^4}{4}} = \sqrt[4]{\frac{k}{4EI}} \tag{8-153}$$

we write the expressions (8-152) as

$$r_1 = \beta(1+i), \qquad r_2 = \beta(1-i), \qquad r_3 = \beta(-1+i), \qquad r_4 = -\beta(1+i)$$

Thus the general solution to the reduced equation becomes

$$\begin{aligned} y(x) &= C_1 e^{\beta x + i\beta x} + C_2 e^{\beta x - i\beta x} + C_3 e^{-\beta x + i\beta x} + C_4 e^{-\beta x - i\beta x} \\ &\equiv e^{\beta x}\left(C_1 e^{i\beta x} + C_2 e^{-i\beta x}\right) + e^{-\beta x}\left(C_3 e^{i\beta x} + C_4 e^{-i\beta x}\right) \end{aligned} \tag{8-154}$$

By using the following representations of trigonometric functions [12, p. 232],

$$\sin \beta x = \frac{e^{i\beta x} - e^{-i\beta x}}{2i}, \qquad \cos \beta x = \frac{e^{i\beta x} + e^{-i\beta x}}{2}$$

we can easily cast Eq. (8-154) in the form

$$y(x) = e^{\beta x}(D_1 \cos \beta x + D_2 \sin \beta x) + e^{-\beta x}(D_3 \cos \beta x + D_4 \sin \beta x) \tag{8-155}$$

Four boundary conditions are needed to determine the constants $D_1$, $D_2$, $D_3$, and $D_4$, and this is exactly the number available—two boundary conditions for each support.

## 8-4-2 Infinite Beams

Occasionally it is convenient to analyze an infinite beam instead of a finite one. Since no infinite beam exists in reality, it is merely used as a model for a beam sufficiently long and a load appropriately located for the effects of the boundaries to be disregarded. In return we get a simpler solution than that for a finite beam.

We begin by analyzing an infinite beam carrying a concentrated force $P$ (Fig. 8-43). Since the beam extends to infinity on both sides of the force $P$, the deflection will obviously be symmetric with respect to the point $O$. Therefore we introduce the $x$ axis with origin under the force $P$ and consider the part of the beam shown in Fig. 8-44. Two boundary conditions are provided by the "end" at infinity. It is intuitively obvious that the deflection will decrease when we move away from the point of application of the concentrated force. By inspection of Eq. (8-155) we see that to satisfy this assertion, both $D_1$ and $D_2$ must be zero,

$$D_1 = D_2 = 0$$

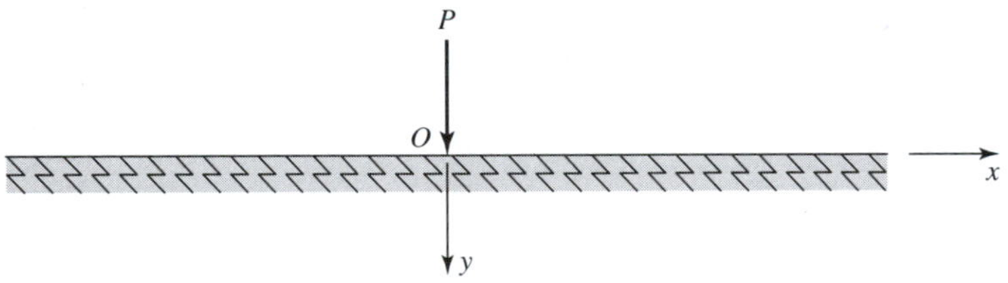


Figure 8-43

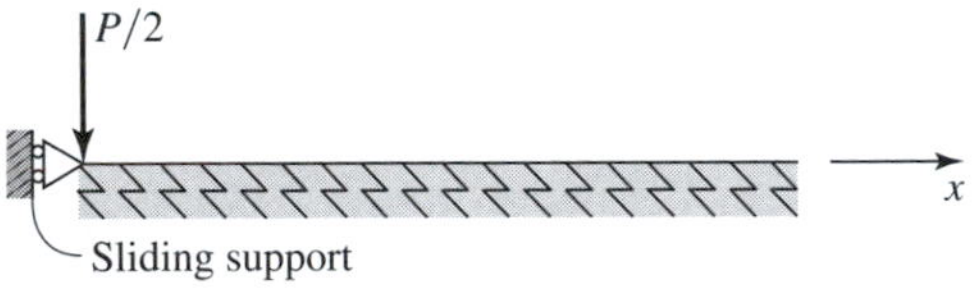


Figure 8-44

Therefore Eq. (8-155) becomes

$$y(x) = e^{-\beta x}(D_3 \cos \beta x + D_4 \sin \beta x) \tag{8-156}$$

It is convenient to introduce the following functions:

$$F_2(\beta x) = e^{-\beta x} \sin \beta x$$

$$F_4(\beta x) = e^{-\beta x} \cos \beta x$$

$$F_1(\beta x) = -\frac{1}{\beta}\frac{dF_4(\beta x)}{dx} = e^{-\beta x}(\cos \beta x + \sin \beta x) \equiv F_2(\beta x) + F_4(\beta x) \tag{8-157}$$

$$F_3(\beta x) = \frac{1}{\beta}\frac{dF_2(\beta x)}{dx} = e^{-\beta x}(\cos \beta x - \sin \beta x) \equiv F_4(\beta x) - F_2(\beta x)$$

We note that

$$\frac{dF_1(\beta x)}{dx} = -2\beta F_2(\beta x), \qquad \frac{dF_3(\beta x)}{dx} = -2\beta F_4(\beta x) \tag{8-158}$$

Equation (8-156) is now rewritten in the form

$$y(x) = D_3 F_4(\beta x) + D_4 F_2(\beta x) \tag{8-159}$$

Hence,

$$y'(x) = -D_3 \beta F_1(\beta x) + D_4 \beta F_3(\beta x) \tag{8-160}$$

$$M(x) = EIy''(x) = 2EI\beta^2[D_3 F_2(\beta x) - D_4 F_4(\beta x)] \tag{8-161}$$

$$Q(x) = EIy'''(x) = 2EI\beta^3[D_3 F_3(\beta x) + D_4 F_1(\beta x)] \tag{8-162}$$

Equations (8-160)–(8-162) are applicable to any infinite beam on an elastic foundation carrying a symmetrically distributed load. Two constants, $D_3$ and $D_4$, appearing in these expressions are determined by using two boundary conditions at $x = 0$. In the present case, because of the symmetry, the slope is zero at $x = 0$ and the shear force equals $P/2$. Thus,

$$y' = 0, \qquad EIy''' = \frac{P}{2}, \qquad \text{at } x = 0 \tag{8-163}$$

Applying the boundary conditions (8-163) we note that $F_1(0) = F_3(0) = 1$, obtaining with Eqs. (8-160) and (8-162),

$$-D_3 + D_4 = 0, \qquad 2\beta^3(D_3 + D_4) = \frac{P}{2EI} \tag{8-164}$$

so that

$$D_3 = D_4 = \frac{P}{8\beta^3 EI} \tag{8-165}$$

Substituting this and expressions (8-157) into Eq. (8-159) we obtain the following expression for the deflection:

$$y(x) = \frac{P}{8\beta^3 EI}[F_2(\beta x) + F_4(\beta x)] \equiv \frac{P}{8\beta^3 EI}F_1(\beta x) \equiv \frac{P\beta}{2k}F_1(\beta x) \qquad \text{(8-166)}$$

From the definition of $F_1(\beta x)$ [Eq. (8-157)] it is seen that $y(x)$ is decaying at a rate depending on the magnitude of $\beta$. The remaining quantities are readily evaluated by differentiation:

$$y'(x) = \frac{P}{8\beta^2 EI}F_1'(\beta x) = -\frac{P}{4\beta^2 EI}F_2(\beta x) = -\frac{P\beta^2}{k}F_2(\beta x)$$

$$M(x) = EIy''(x) = -\frac{P}{4\beta^2}F_2'(\beta x) = -\frac{P}{4\beta}F_3(\beta x) \qquad \text{(8-167)}$$

$$V(x) = EIy'''(x) = -\frac{P}{4\beta}F_3'(\beta x) = \frac{P}{2}F_4(\beta x)$$

These results have wider applications than it seems at first sight. Let us consider, for example, an infinite beam carrying two concentrated forces $P_1$ and $P_2$, placing the origin of the $x$ axis under $P_1$ (Fig. 8-45*a*). We now use the principle of superposition, evaluating the effect of each force separately and then adding the results (Fig. 8-45*b* and *c*). The deflection $y_1(x)$ due to the force $P_1$ is, according to Eq. (8-166),

$$y_1(x) = \frac{P_1\beta}{2k}F_1(\beta|x|)$$

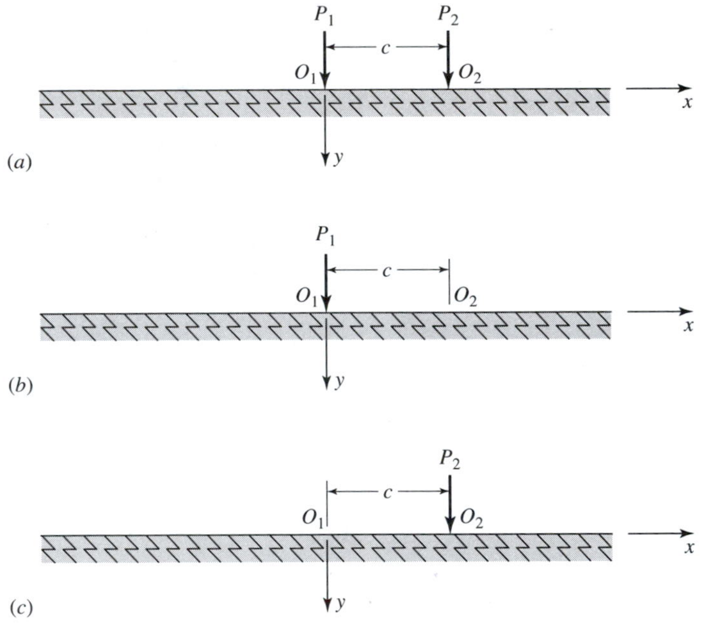


Figure 8-45

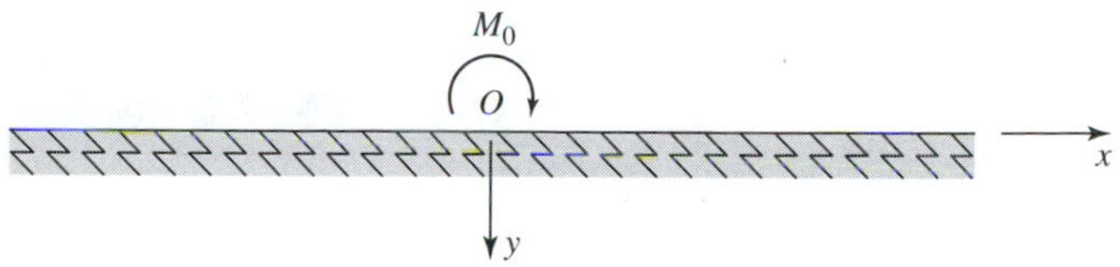


Figure 8-46

Similarly, the deflection $y_2(x_2)$ due to the force $P_2$ is

$$y_2(x_2) = \frac{P_2\beta}{2k} F_1(\beta|x_2|)$$

but $x_2 = x - c$ (Fig. 8-45*a*). Hence,

$$y_2(x) = \frac{P_2\beta}{2k} F_1(\beta|x - c|)$$

Therefore the total deflection is

$$y(x) = \frac{\beta}{2k}[P_1 F_1(\beta|x|) + P_2 F_1(\beta|x - c|)]$$

The slope, the bending moment, and the shear force are calculated in the same way. Another application involves using the deflection (8-166) as the influence function (Green's function*) to obtain the deflection and other quantities due to an arbitrary load. This is shown in the section on Green's functions (Section 8-5). As a word of caution: if a concentrated load were the only force applied to the beam, a negative deflection would occur causing the "lifting" of the beam from the foundation. Since we assumed in our analysis a continuous contact between the beam and the foundation, the solution would no longer be valid, and the analysis would have to be repeated. In this case we would use the first solution to assume that the part of the beam detached from the foundation must obey the differential equation for the bending of simple beams. The remaining portion would still be treated as a beam on an elastic foundation. The results obtained in this way would be likely to show a discrepancy between the assumed and the calculated lengths of the detached regions. A reanalysis would thus be needed, and this iterative process would have to be continued until satisfactory results are obtained. Normally, however, continuous loads are present instead of, or in addition to, concentrated forces and usually compensate for the "lifting" phenomenon. For now we shall consider as another example the case of an infinite beam carrying a concentrated moment $M_0$ (Fig. 8-46). Locating the origin of the $x$ axis at the point of application of $M_0$, we notice that the deflection of the beam is antisymmetric with respect to this point. Thus,

$$y(-x) = -y(x) \tag{8-168}$$

This implies also that the bending moment is antisymmetric,

$$EIy''(-x) = -EIy''(x) \tag{8-169}$$

*George Green (1793–1841), English mathematician.

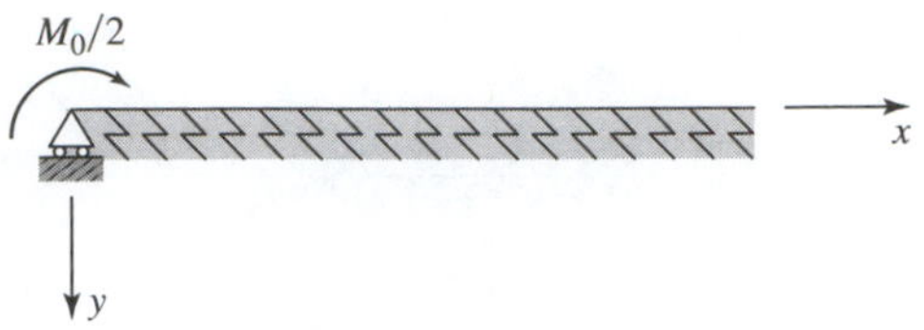


Figure 8-47

and that consequently it is sufficient to analyze only the portion $x > 0$ of the beam with one-half of the applied moment attached to its left end (Fig. 8-47). Here again $D_1 = D_2 = 0$ so that Eqs. (8-159)–(8-162) are still valid and the two boundary conditions needed are

$$y = 0, \qquad EIy'' = -\frac{M_0}{2}, \qquad \text{at } x = 0 \tag{8-170}$$

Substituting here Eqs. (8-159) and (8-161) we obtain

$$D_3 = 0, \qquad -2D_4\beta^2 = -\frac{M_0}{2EI}$$

or

$$D_4 = \frac{M_0}{4EI\beta^2} = \frac{M_0\beta^2}{k}$$

so that

$$y(x) = \frac{M_0\beta^2}{k}F_2(\beta x) \tag{8-171}$$

$$y'(x) = \frac{M_0\beta^3}{k}F_3(\beta x) \tag{8-172}$$

$$M(x) = -\frac{M_0}{2}F_4(\beta x) \tag{8-173}$$

$$V(x) = \frac{M_0\beta}{2}F_1(\beta x) \tag{8-174}$$

The result (8-171) could also have been obtained using the reciprocity theorem (see Chapter 5). In order to prove this, note that the deflection at point $A$ due to the moment equal to 1 applied at $x = 0$ equals the slope at $x = 0$ due to the force equal to 1 applied at $A$. On the other hand, the deflection of the beam is symmetric with respect to the point of application of the force $P = 1$. Thus the slope at $A$ due to the force $P = 1$ at $x = 0$ equals the slope at $O$ with a minus sign due to the force $P = 1$ at $x = A$ (see Fig. 8-48). From the results in the previous example we find that

$$y'(x) = \frac{\beta^2}{k}F_2(\beta x)$$

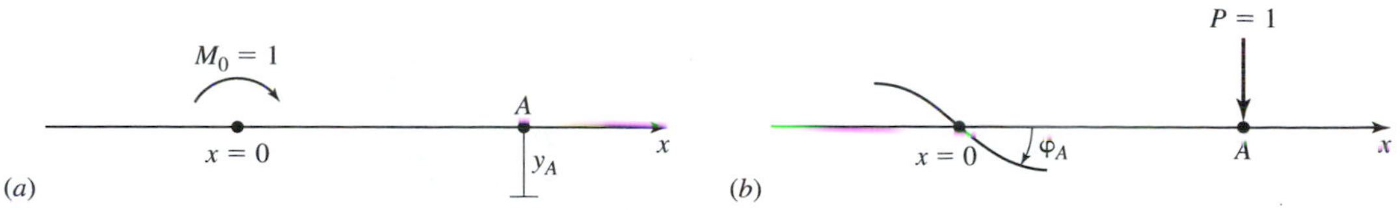


Figure 8-48

is therefore equal to the deflection at $x$ due to the moment equal to 1 at $x = 0$, so that we obtain again the result (8-171),

$$y(x) = \frac{M_0\beta^2}{k} F_2(\beta x)$$

Thanks to the symmetry or antisymmetry of the load we were able to replace an infinite beam by a semi-infinite one. Sometimes when the effect of one boundary must be taken into account, we model a finite beam by a semi-infinite beam. To satisfy the boundary conditions at infinity we must require again that

$$D_1 = D_2 = 0$$

Therefore Eqs. (8-159)–(8-162) are still valid, except that the particular solutions and their derivatives must be added. This yields the following expressions:

$$\begin{aligned} y(x) &= D_3 F_4(\beta x) + D_4 F_2(\beta x) + y_p(x) \\ y'(x) &= -D_3\beta F_1(\beta x) + D_4\beta F_3(\beta x) + y_p'(x) \\ M(x) &= 2EI\beta^2[D_3 F_2(\beta x) - D_4 F_4(\beta x)] + EIy_p''(x) \\ Q(x) &= 2EI\beta^3[D_3 F_3(\beta x) + D_4 F_1(\beta x)] + EIy_p'''(x) \end{aligned} \tag{8-175}$$

Let us consider, for example, a semi-infinite beam carrying a uniform load $q_0$ (Fig. 8-49). Here the boundary conditions at $x = 0$ are

$$y = 0, \qquad y' = 0$$

and since $y_p(x) = q_0/4\beta^4 EI \equiv q_0/k$, the following system of algebraic equations results:

$$D_3 + \frac{q_0}{k} = 0, \qquad -D_3\beta + D_4\beta = 0$$

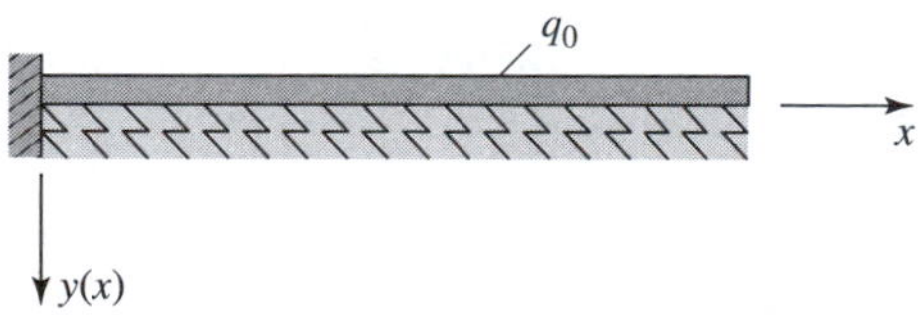


Figure 8-49

Solving this we obtain

$$D_3 = D_4 = -\frac{q_0}{k} \tag{8-176}$$

Substituting expression (8-176) into Eqs. (8-175) we find the following expressions for the deflection $y(x)$ and the bending moment $M(x)$:

$$\begin{aligned} y(x) &= \frac{q_0}{k}[1 - F_1(\beta x)] \\ M(x) &= \frac{q_0}{2\beta^2} F_3(\beta x) \equiv \frac{q_0}{2\beta^2} e^{-\beta x}(\cos\beta x - \sin\beta x) \end{aligned} \tag{8-177}$$

Inspection of Eqs. (8-177) reveals that

$$\lim_{x=\infty} y(x) = \frac{q_0}{k}$$

In order to find the maximum bending moment we differentiate $M(x)$,

$$M'(x) = -\frac{q_0}{\beta} F_4(\beta x) \equiv -\frac{q_0}{\beta} e^{-\beta x} \cos\beta x$$

and note that the equation $M'(x) = 0$ has roots at $x = (2n - 1)\pi/2\beta$, where $n$ is a positive integer. The magnitude of $M(x)$ at these points is

$$\begin{aligned} M(x) &= \frac{q_0}{2\beta^2} e^{-(2n-1)\pi/2} \left[ \cos\frac{(2n-1)\pi}{2} - \sin\frac{(2n-1)\pi}{2} \right] \\ &= \frac{q_0}{2\beta^2} e^{-(2n-1)\pi/2} \cos n\pi \end{aligned}$$

Next we note that $\cos n\pi = \pm 1$, and that $e^{-(2n-1)\pi/2}$ is rapidly decreasing. Therefore the largest local maximum of the bending moment $M(x)$ occurs for $n = 1$ at $x = \pi/2\beta$, where it equals

$$\max M = -\frac{q_0}{2\beta^2} e^{-\pi/2}$$

On the other hand, however, at $x = 0$,

$$M = \frac{q_0}{2\beta^2}$$

which obviously is the absolute maximum. The graphs of $y(x)$ and $M(x)$ are shown in Fig. 8-50.

## 8-4-3 Finite Beams

To find a solution for a finite beam we must use the complete expression (8-155) with four unknown constants. First we represent this expression in a more convenient form,

$$y(x) = D_1 F_8(\beta x) + D_2 F_6(\beta x) + D_3 F_4(\beta x) + D_4 F_2(\beta x) \tag{8-178}$$

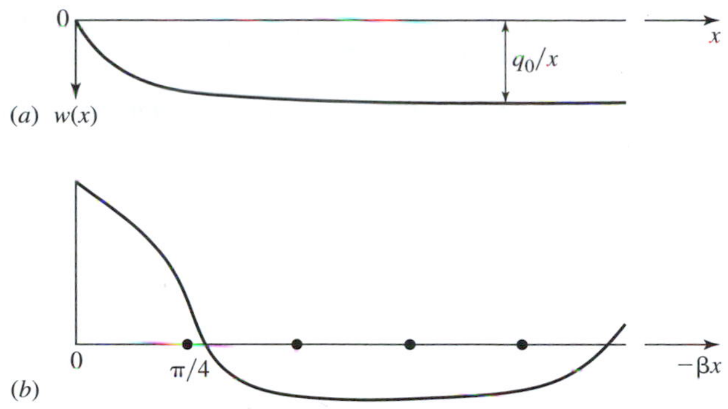


Figure 8-50 Graphs of $y(x)$ and $M(x)$

where

$$F_8(\beta x) = e^{\beta x}\cos\beta x, \qquad F_6(\beta x) = e^{\beta x}\sin\beta x \tag{8-179}$$

Next we introduce the following functions:

$$\begin{aligned} F_5(\beta x) &= -\frac{1}{\beta}\frac{dF_8(\beta x)}{dx} = e^{\beta x}(\sin\beta x - \cos\beta x) \equiv F_6(\beta x) - F_8(\beta x) \\ F_7(\beta x) &= \frac{1}{\beta}\frac{dF_6(\beta x)}{dx} = e^{\beta x}(\sin\beta x + \cos\beta x) \equiv F_6(\beta x) + F_8(\beta x) \end{aligned} \tag{8-180}$$

We note that

$$\frac{dF_5(\beta x)}{dx} = 2\beta F_6(\beta x), \qquad \frac{dF_7(\beta x)}{dx} = 2\beta F_8(\beta x)$$

Then we find the following expressions:

$$y'(x) = \beta[-D_1F_5(\beta x) + D_2F_7(\beta x) - D_3F_1(\beta x) + D_4F_3(\beta x)] \tag{8-181}$$

$$M(x) = EIy''(x) = 2EI\beta^2[-D_1F_6(\beta x) + D_2F_8(\beta x) + D_3F_2(\beta x) - D_4F_4(\beta x)] \tag{8-182}$$

$$V(x) = EIy'''(x) = 2EI\beta^3[-D_1F_7(\beta x) - D_2F_5(\beta x) + D_3F_3(\beta x) + D_4F_1(\beta x)] \tag{8-183}$$

We have to add to these expressions the particular solutions or their derivatives. Since two boundary conditions are available at each end of the beam, we may determine the constants $D_1$, $D_2$, $D_3$, and $D_4$ for each admissible combination of boundary conditions and for each type of load. Now the particular solutions will be derived for two cases: (1) a uniform load and (2) a concentrated force, which is particularly important for determining the Green's function (see Section 8-5).

1. *Particular solution for a uniform load* $q(x) = q_0$ We begin by writing Eq. (8-148) in the form

$$y^{\text{iv}}(x) + 4\beta^4 y = \frac{q_0}{EI} \tag{8-184}$$

It is evident that $y_p = A =$ constant is a solution if

$$A = \frac{q_0}{4\beta^4 EI} \equiv \frac{q_0}{k}$$

Thus,

$$y_p = \frac{q_0}{k}, \qquad y_p' = y_p'' = y_p''' = 0 \tag{8-185}$$

2. *Particular solution for a concentrated force P applied at* $x = \xi$ We have seen in Section 8-4-2 that a force $P$ at $x = 0$ generates a displacement $P\beta F_1(\beta|x|)/2k$ at an arbitrary point $x$ of an infinite beam. Consequently a force $P$ at $x = \xi$ generates a displacement $P\beta F_1(\beta|x-\xi|)/2k$ at an arbitrary point $x$. Since a particular solution depends on the differential equation but not on the boundary conditions, this result is also applicable in the present case,

$$\begin{aligned} y_p &= \frac{P\beta}{2k} F_1(\beta|x-\xi|) \\ y_p' &= -\frac{P\beta^2}{k} F_2(\beta|x-\xi|) \\ y_p'' &= -\frac{P\beta^3}{k} F_3(\beta|x-\xi|) \\ y_p''' &= \frac{2P\beta^4}{k} F_4(\beta|x-\xi|) \end{aligned} \tag{8-186}$$

It can be verified by substitution that for $x \neq \xi$ this satisfies the reduced equation

$$y^{\text{iv}} + 4\beta^4 y = 0$$

The beam shown in row 1, Table 8-2, will be analyzed as an example. The general solution to the equilibrium equation is now obtained by adding expressions (8-178) and (8-185),

$$y(x) = D_1 F_8(\beta x) + D_2 F_6(\beta x) + D_3 F_4(\beta x) + D_4 F_2(\beta x) + \frac{q_0}{k} \tag{8-187}$$

By using the boundary conditions

$$y = 0, \qquad y'' = 0, \qquad \text{at } x = \frac{l}{2} \tag{8-188}$$

and the symmetry conditions

$$y' = 0, \qquad y''' = 0, \qquad \text{at } x = 0 \tag{8-189}$$

**TABLE 8-2**

| Schematics | Constants of Integration and Maximum Deflection |
|---|---|
| 1. 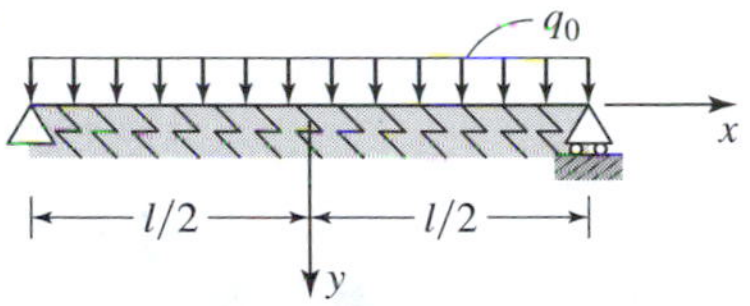  | $D_1 = -\frac{q_0}{k}\cosh\frac{\beta l}{2}\cos\frac{\beta l}{2}\Big/(\cosh\beta l + \cos\beta l)$ <br> $D_2 = -\frac{q_0}{k}\sinh\frac{\beta l}{2}\sin\frac{\beta l}{2}\Big/(\cosh\beta l + \cos\beta l)$ <br> $y(0) = \frac{q_0}{k}\left[1 - 2\cosh\frac{\beta l}{2}\cos\frac{\beta l}{2}\Big/(\cosh\beta l + \cos\beta l)\right]$ |
| 2. 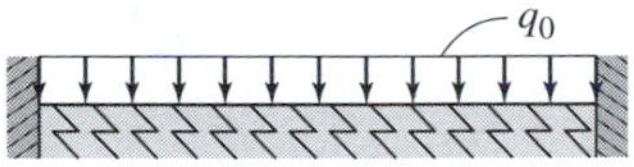  | $D_1 = -\frac{q_0}{k}\left(\cosh\frac{\beta l}{2}\sin\frac{\beta l}{2} + \sinh\frac{\beta l}{2}\cos\frac{\beta l}{2}\right)\Big/(\sinh\beta l + \sin\beta l)$ <br> $D_2 = -\frac{q_0}{k}\left(\cosh\frac{\beta l}{2}\sin\frac{\beta l}{2} - \sinh\frac{\beta l}{2}\cos\frac{\beta l}{2}\right)\Big/(\sinh\beta l + \sin\beta l)$ <br> $y(0) = \frac{q_0}{k}\left[1 - \left(\cosh\frac{\beta l}{2}\sin\frac{\beta l}{2} + \sinh\frac{\beta l}{2}\cos\frac{\beta l}{2}\right)\Big/(\sinh\beta l + \sin\beta l)\right]$ |
| 3. General expression | $y(x) = D_1F_8(\beta x) + D_2F_6(\beta x) + D_3F_4(\beta x) + D_4F_2(\beta x) + \frac{q_0}{k}$ |

we get

$$D_1F_8\left(\frac{\beta l}{2}\right) + D_2F_6\left(\frac{\beta l}{2}\right) + D_3F_4\left(\frac{\beta l}{2}\right) + D_4F_2\left(\frac{\beta l}{2}\right) + \frac{q_0}{k} = 0$$

$$-D_1F_6\left(\frac{\beta l}{2}\right) + D_2F_8\left(\frac{\beta l}{2}\right) + D_3F_2\left(\frac{\beta l}{2}\right) - D_4F_4\left(\frac{\beta l}{2}\right) = 0$$

$$D_1 + D_2 - D_3 + D_4 = 0$$

$$-D_1 + D_2 + D_3 + D_4 = 0$$

Therefore,

$$D_3 = D_1, \qquad D_4 = -D_2$$

We note that these results are the consequence of the symmetry conditions, Eqs. (8-189). They are therefore valid for any symmetric case, independent of the boundary conditions at the ends. The other two equations become then

$$D_1\left[F_8\left(\frac{\beta l}{2}\right) + F_4\left(\frac{\beta l}{2}\right)\right] + D_2\left[F_6\left(\frac{\beta l}{2}\right) - F_2\left(\frac{\beta l}{2}\right)\right] + \frac{q_0}{k} = 0$$

$$-D_1\left[F_6\left(\frac{\beta l}{2}\right) - F_2\left(\frac{\beta l}{2}\right)\right] + D_2\left[F_8\left(\frac{\beta l}{2}\right) + F_4\left(\frac{\beta l}{2}\right)\right] = 0$$

This system of algebraic equations has been solved using the Macsyma [17] symbolic manipulation package. The resulting values of the constants $D_1$ and $D_2$ are shown in Table 8-2, as are the constants corresponding to a beam with built-in ends, and carrying a uniform load.

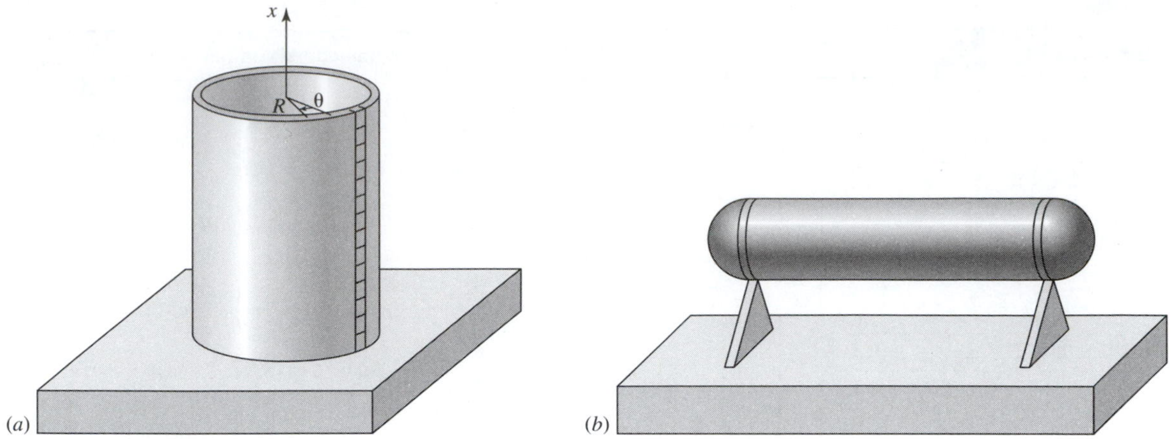


Figure 8-51 Thin-walled circular storage tank

The maximum deflection is in agreement with the result given by Timoshenko [25, p. 22]. Once the constants $D_1$, $D_2$, $D_3$, and $D_4$ are known we find $y(x)$, $y'(x)$, $M(x)$, and $V(x)$ from Eqs. (8-178)–(8-183). The entries in Table 8-2 are convenient when closed-form analytical solutions are needed. Otherwise the interactive Fortran program 8-2 may be useful.* It calculates all the necessary quantities for various values of $x$, giving the option of choosing either a uniform load or a concentrated force in the middle of the span. In addition several combinations of the boundary conditions are available.

## 8-4-4 Stresses in Storage Tanks†

Equations derived in Section 8-4-1 have an unexpected application. Consider thin-walled circular tanks, as shown in Fig. 8-51. They both belong to a larger category of structures called *shells,* which is not discussed in this text. While Fig. 8-51*a* shows a complete cylinder attached to a rigid foundation and serving as a storage tank for liquids or granular materials, the tank in Fig. 8-51*b* has end caps in addition to a cylindrical central part. The caps are shown here as curved surfaces, but they could be flat as well. They are attached to the cylinder with a rigid ring reinforcing the junction. Tanks of this type are used for the storage of pressurized gases. Let us derive an approximate differential equation for the bending of such tanks subject to an axisymmetric load. It will be shown that this equation is analogous to the equation for beams on elastic foundations. This makes all previously obtained solutions applicable for comparable boundary conditions.

We begin by considering the cross-hatched strip in Fig. 8-51*a*, bent in the radial direction both by the inside pressure and by the "hoop stresses" $\sigma_{\theta\theta}$ (Fig. 8-52). Let us

---

*Program 8-2 is provided on the diskette.

†Note that in this section there is some reference to material from Chapter 10.

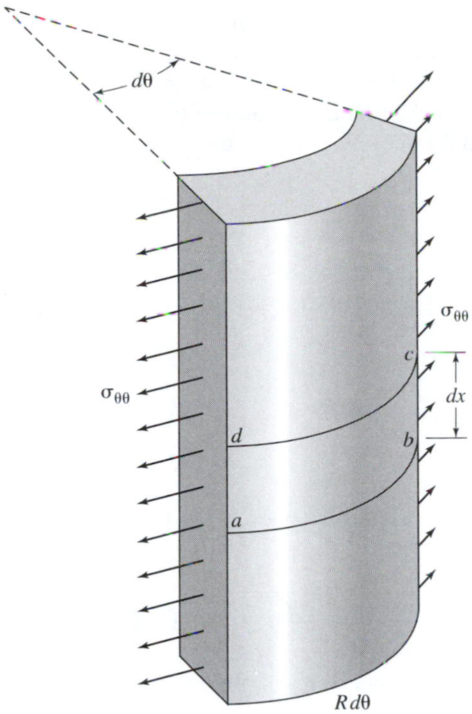


Figure 8-52 Strip element of storage tank wall

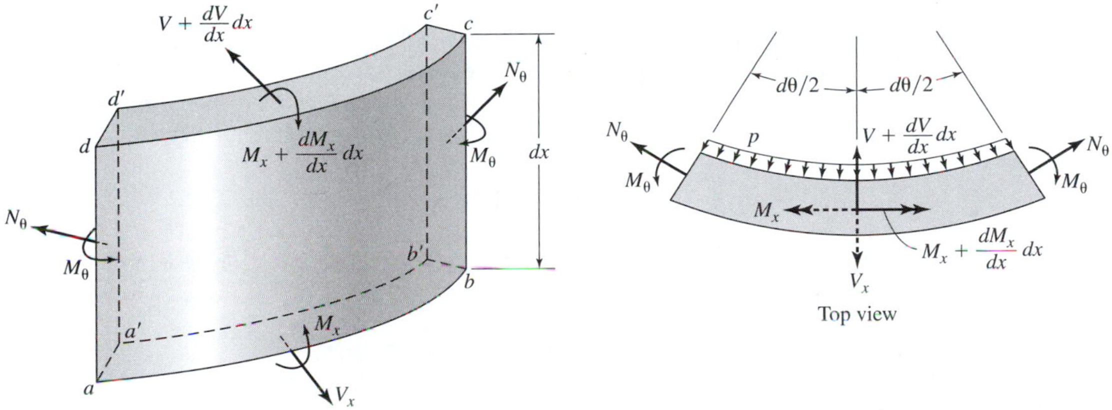


Figure 8-53 Infinitesimal element of strip

now analyze the equilibrium of an infinitesimal element *abcd* (Fig. 8-52) cut from this strip. This element is shown magnified in Fig. 8-53. Because of the symmetry of the structure and of the load, no shear stresses appear on the faces $aa'dd'$ and $bb'cc'$. The normal stresses are assumed to be uniform throughout the thickness—an approximation valid for thin-walled tanks. Their resultants are circumferential forces $N_\theta$. Also the bending moments $M_\theta$, caused by the change in curvature (see also Chapter 10), are present on these faces. On faces $abb'a'$ and $dcc'd'$ of the element there appear the bending moments $M_x$ and $M_x + (dM_x/dx)\,dx$ and the shear forces $V_x$ and $V_x + (dV_x/dx)\,dx$. No axial forces exist on $dcc'd'$ and $abb'a'$ if the weight of the tank is neglected. The circumferential forces $N_\theta$ generate the radial component

$$2N_\theta \sin\frac{d\theta}{2}\,dx$$

where $dx$ is the height of the element. The radial components of the shear forces $V_x$ and $V_x + (dV_x/dx)\,dx$ and of the pressure $p$ are

$$\frac{dV_x}{dx}\,dx\,R\,d\theta$$

and

$$-p(x)R\,d\theta\,dx$$

The term $dx$ is present here because all forces are defined per unit height. Adding all three terms we obtain

$$2N_\theta \sin\frac{d\theta}{2}\,dx + \frac{dV_x}{dx}\,dx\,R\,d\theta - p(x)R\,d\theta\,dx = 0$$

By using $\sin(d\theta/2) = d\theta/2$ and dividing by $R\,d\theta\,dx$ we get

$$\frac{N_\theta}{R} + \frac{dV_x}{dx} - p(x) = 0 \tag{8-190}$$

Next we calculate the sum of the moments of all forces about a line, say the line tangent to $ab$, which is parallel to the direction of the vector $M_x$, noting that it must be zero. To do so, first observe that the vectors $M_\theta$ are normal to $M_x$, and therefore their projection on $M_x$ is zero. Next the contribution of $M_x$ and $M_x + (dM_x/dx)\,dx$ is readily evaluated as

$$M_xR\,d\theta - \left(M_x + \frac{dM_x}{dx}\,dx\right)R\,d\theta$$

where $R\,d\theta$ is the length of the arc. The moment of the shear forces about the tangent to $ab$ is

$$\left(V_x + \frac{dV_x}{dx}\right)dx\,R\,d\theta$$

To obtain the moment produced by the circumferential forces $N_\theta$ we expand these into two components each, one parallel to $ab$,

$$N_\theta \cos\frac{d\theta}{2}$$

and the other one normal to it,

$$N_\theta \sin\frac{d\theta}{2}$$

Only the latter components produce a moment about *ab* given by

$$2N_\theta \sin\frac{d\theta}{2}\, dx\, \frac{dx}{2}$$

Finally the moment of the applied pressure $p$ becomes

$$p\, dx\, R\, d\theta \frac{dx}{2}$$

Combining all terms, we get

$$M_x R\, d\theta - \left(M_x + \frac{dM_x}{dx} dx\right) R\, d\theta + \left(V_x + \frac{dV_x}{dx} dx\right) dx\, R\, d\theta$$
$$+ 2N_\theta \sin\frac{d\theta}{2}\, dx\, \frac{dx}{2} + p\, dx\, R\, d\theta \frac{dx}{2} = 0$$

Simplifying, neglecting higher order terms, and dividing by $R\, dx\, d\theta$, we get

$$-\frac{dM_x}{dx} + V_x = 0 \qquad (8\text{-}191)$$

The next step is to calculate $V_x$ from Eq. (8-191) and to substitute it into Eq. (8-190),

$$\frac{d^2 M_x}{dx^2} = -\frac{N_\theta}{R} + p(x) \qquad (8\text{-}192)$$

In order to relate $M_x$ and $N_\theta$ to the radial deflection $w(x)$ we note that the latter causes the circumferential extension (or contraction) $2\pi(R + w) - 2\pi R = 2\pi w$. Thus the corresponding strain is (see also Chapter 3)

$$e_{\theta\theta} = \frac{2\pi(R + w) - 2\pi R}{2\pi R} = \frac{w}{R} \qquad (8\text{-}193)$$

Therefore,

$$N_\theta = \sigma_{\theta\theta} t = E e_{\theta\theta} t = \frac{Et}{R} w \qquad (8\text{-}194)$$

We also note that the element *abcd* is in a situation similar to that of the infinitesimal element of a plate in bending (Chapter 10). Since here $w$ does not depend on $\theta$, we have

$$M_x = D\frac{d^2 w}{dx^2} \qquad (8\text{-}195)$$

where $D = Et^3/12(1 - \nu^2)$. Substituting expressions (8-194) and (8-195) into Eq. (8-192) we get

$$D\frac{d^4 w}{dx^4} + \frac{Et}{R^2}\, w = p \qquad (8\text{-}196)$$

Dividing Eq. (8-196) by $D$ and introducing the quantities

$$k = \frac{Et}{R^2}, \qquad \beta^4 = \frac{Et}{4DR^2} = \frac{3(1-\nu^2)}{t^2R^2} \tag{8-197}$$

we make it analogous to Eq. (8-148),

$$\frac{d^4w}{dx^4} + 4\beta^4 w = \frac{p}{D} \tag{8-198}$$

## EXAMPLE 8-7

Determine the maximum stress in a tank completely filled with water. Assume that the bottom plate is perfectly rigid.

### SOLUTION

The pressure distribution is

$$p(x) = p_0\left(1 - \frac{x}{h}\right) \tag{a}$$

The particular solution corresponding to $p(x)$ given by Eq. ($a$) has the form

$$w(x) = A + Bx \tag{b}$$

where $A$ and $B$ are constants. With Eqs. ($b$) and ($a$) substituted into Eq. (8-198) we get

$$4\beta^4(A + Bx) = \frac{p_0}{D}\left(1 - \frac{x}{h}\right) \tag{c}$$

Then, equating the coefficients of the same powers of $x$, we find

$$A = \frac{p_0}{4\beta^4 D}, \qquad B = -\frac{p_0}{4\beta^4 Dh} \tag{d}$$

Consequently the general solution of Eq. (8-198) is [see also Eq. (8-178)]

$$w(x) = D_1F_8(\beta x) + D_2F_6(\beta x) + D_3F_4(\beta x) + D_4F_2(\beta x) + \frac{p_0}{4\beta^4 D}\left(1 - \frac{x}{h}\right) \tag{e}$$

The constants of integration are determined using the boundary conditions

$$\begin{aligned} w(x) = 0, \quad w'(x) = 0, &\qquad \text{at } x = 0 \\ w''(x) = 0, \quad w'''(x) = 0, &\qquad \text{at } x = h \end{aligned} \tag{f}$$

Next expression ($e$) is substituted here. The definitions of the funtion $F_2(\beta x)$, and so on, and the expressions (8-181)–(8-190) are used for the derivatives of $w(x)$, yielding

$$D_1 + D_3 + \frac{p_0}{4\beta^4 D} = 0$$

$$\beta[-D_1 + D_2 - D_3 + D_4] - \frac{p_0}{4\beta^4 Dh} = 0$$

$$-D_1 F_6(\beta h) + D_2 F_8(\beta h) + D_3 F_2(\beta h) - D_4 F_4(\beta h) = 0$$

$$-D_1 F_7(\beta h) - D_2 F_5(\beta h) + D_3 F_3(\beta h) + D_4 F_1(\beta h) = 0$$

This system of equations has been solved using the software package Mathcad7, professional edition [18]. Then,

$$\begin{aligned}
D_1 &= -\frac{P}{\Delta}\{\beta h[F_1(F_2 + F_8) + F_2 F_5 - F_3(F_4 + F_8) - F_4 F_5] - F_1 F_8 + F_4 F_5\} \\
D_2 &= -\frac{P}{\Delta}\{\beta h[F_1(F_2 + F_6) - F_2 F_7 + F_3(F_4 + F_6) + F_4 F_7] \\
&\qquad - F_1(F_2 + F_6) - F_4(F_3 + F_7)\} \\
D_3 &= -\frac{P}{\Delta}\{\beta h[F_1(F_6 - F_8) + F_4(F_5 + F_7) + F_5 F_6 + F_7 F_8] + F_1 F_8 - F_4 F_5\} \\
D_4 &= -\frac{P}{\Delta}\{\beta h[F_2(F_5 + F_7) + F_3(F_8 - F_6) + F_5 F_6 + F_7 F_8] \\
&\qquad - F_2 F_5 - F_3 F_8 - F_5 F_6 - F_7 F_8\}
\end{aligned} \qquad (g)$$

where

$$P = \frac{p_0}{4\beta^4 D}$$

$$\Delta = \beta h[F_1(F_2 + F_6) + F_2 F_5 + F_3(F_4 + F_8) + F_4 F_7 + F_5 F_6 + F_7 F_8]$$

and where $F_i \equiv F_i(\beta h)$. In terms of the original functions, expressions ($g$) become

$$D_1 = -\frac{P}{\Delta_1}[\beta h(e^{-2\beta h} + 2\cos^2\beta h - 2\sin\beta h\cos\beta h + 1) - 2\cos^2\beta h]$$

$$D_2 = -\frac{P}{\Delta_1}[\beta h(e^{-2\beta h} + 2\sin\beta h\cos\beta h - 2\sin^2\beta h - 1) - 1 - 2\sin\beta h\cos\beta h - e^{-2\beta h}]$$

$$D_3 = -\frac{P}{\Delta_1}[\beta h(e^{2\beta h} + 2\sin\beta h\cos\beta h + 2\sin^2\beta h - 1) + 2 - 2\sin^2\beta h]$$

$$D_4 = -\frac{P}{\Delta_1}[\beta h(e^{2\beta h} - 2\sin\beta h\cos\beta h + 2\sin^2\beta h + 1) - e^{2\beta h} + 2\sin\beta h\cos\beta h - 1]$$

where

$$\Delta_1 = 2\beta h(1 + \cos 2\beta h)$$

With the deflection $w(x)$ now known, we determine the bending moment from expression (8-195),

$$M_x \equiv D\frac{d^2w}{dx^2} = 2\beta^2 D[-D_1F_6(\beta x) + D_2F_8(\beta x) + D_3F_2(\beta x) - D_4F_4(\beta x)]$$

The resulting expressions would include fewer terms if we could treat the strip, cut out from the tank, as a semi-infinite, rather than finite, beam. The validity of such an approximation will be determined by comparing the exact results to the approximate results for various values of $\beta h$. For a semi-infinite beam we assume

$$w(x) = D_3F_4(\beta x) + D_4F_2(\beta x) + \frac{p_0}{4\beta^4 D}\left(1 - \frac{x}{h}\right)$$

and we find the integration constants using two boundary conditions at $x = 0$,

$$w(x) = 0, \qquad w'(x) = 0$$

This yields

$$D_3 + \frac{p_0}{4\beta^4 D} = 0$$

$$\beta(-D_3 + D_4) - \frac{p_0}{4\beta^4 Dh} = 0$$

Therefore,

$$D_3 = -\frac{p_0}{4\beta^4 D}$$

$$D_4 = \frac{p_0}{4\beta^5 Dh}(1 - \beta h)$$

so that

$$w(x) = \frac{p_0}{4\beta^4 D}\left[-F_4(\beta x) + \frac{1-\beta h}{\beta h}F_2(\beta x) + 1 - \frac{x}{h}\right]$$

and

$$M_x = \frac{p_0}{2\beta^2}\left[-F_2(\beta x) - \frac{1-\beta h}{\beta h}F_4(\beta x)\right]$$

The results obtained using Fortran program 8-3 show that using approximate expressions has no appreciable effect on the deflection (except for the very tip) when $\beta h \geq 4.80$ and on the bending moment $M_x$ when $\beta h \geq 6.00$.

## 8-5 INFLUENCE FUNCTIONS (GREEN'S FUNCTIONS) FOR BEAMS

### 8-5-1 Straight Beams

The displacement of a beam carrying a concentrated force $P = 1$ is known as the *Green's function* or *influence function* for the displacement. When the Green's function for the displacement is given, one can find, by means of integration, the displacement of the same beam due to an arbitrary load. Let us consider, for example, a single-span beam with still unspecified boundary conditions, carrying the load $P = 1$ at $x = a$ (Fig. 8-54). We shall derive the expression for the Green's function by integrating the differential equation for deflection [21, p. 273],

$$EI\frac{d^4y}{dx^4} = q(x) \tag{8-199}$$

We begin our analysis by representing the concentrated load in the form (see Section 8-1-3)

$$q(x) = \delta(x - a) \tag{8-200}$$

Substituting Eq. (8-200) into Eq. (8-199), we get

$$EIy^{\text{iv}} = \delta(x - a) \tag{8-201}$$

By integrating this and using Eqs. (8-15) and (8-21) we obtain

$$EIy''' = H(x - a) + C_1 \tag{8-202}$$

Consecutive integrations of Eq. (8-202) give

$$EIy'' = (x - a)H(x - a) + C_1x + C_2 \tag{8-203}$$

$$EIy' = \frac{(x-a)^2}{2}H(x - a) + C_1\frac{x^2}{2} + C_2x + C_3 \tag{8-204}$$

$$EIy = \frac{(x-a)^3}{6}H(x - a) + C_1\frac{x^3}{6} + C_2\frac{x^2}{2} + C_3x + C_4 \tag{8-205}$$

Next the constants $C_1$, $C_2$, $C_3$, and $C_4$ are determined for various cases of support.

**1.** *Simply supported beam* The boundary conditions are

$$y = 0, \quad EIy'' = 0, \qquad \text{at } x = 0$$
$$y = 0, \quad EIy'' = 0, \qquad \text{at } x = l$$

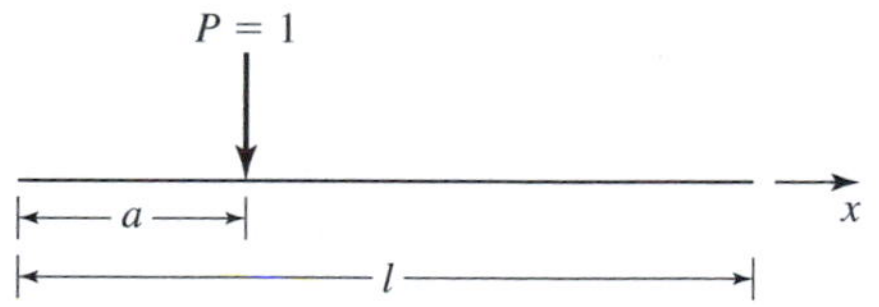


Figure 8-54

Therefore,

$$C_4 = 0, \qquad C_2 = 0$$

$$\frac{(l-a)^3}{6} + C_1\frac{l^3}{6} + C_3 l = 0$$

$$l - a + C_1 l = 0$$

so that

$$C_1 = -\frac{l-a}{l}, \qquad C_3 = \frac{(l-a)(2l-a)}{6l}a$$

Consequently the Green's function for a simply supported beam is

$$G(x,a) \equiv y(x,a)$$

$$= \frac{1}{EI}\left[\frac{(x-a)^3}{6}H(x-a) + \frac{l-a}{6l}(-x^2 + 2al - a^2)x\right] \tag{8-206}$$

**2.** *Beam clamped on both ends* Now with the boundary conditions

$$y = 0, \quad y' = 0, \qquad \text{at } x = 0$$

$$y = 0, \quad y' = 0, \qquad \text{at } x = l$$

we obtain

$$C_4 = 0, \qquad C_3 = 0$$

$$\frac{(l-a)^3}{6} + C_1\frac{l^3}{6} + C_2\frac{l^2}{2} = 0$$

$$\frac{(l-a)^2}{2} + C_1\frac{l^2}{2} + C_2 l = 0$$

Solving these equations we get

$$C_1 = -\frac{(l+2a)(l-a)^2}{l^3}, \qquad C_2 = \frac{(l-a)^2 a}{l^2}$$

so that the Green's function becomes

$$G(x,a) = \frac{1}{EI}\left[\frac{(x-a)^3}{6}H(x-a) + \frac{(l-a)^2}{2l^2}\left(-\frac{l+2a}{3l}x + a\right)x^2\right] \tag{8-207}$$

This and other results obtained in a similar way are listed in the Table 8-3. It can be shown, and this is left for the reader, that $G(x,a) = G(a,x)$. This symmetry of the Green's function is the consequence of the reciprocity theorem (see Chapter 5).

As we already mentioned, once the Green's function for a given set of boundary conditions is known, it is very easy to calculate the displacement of a beam with the same boundary conditions under any load. Let us consider, for example, a beam with unspecified boundary conditions, carrying a variable load $q(x)$ (Fig. 8-55*a*), assuming that the Green's function $G(x,\xi)$ for this beam is known. $G(x,\xi)$ gives the deflection at a

TABLE 8-3

| Schematics | Green's Function for Displacement $G(x, a)$ |
|---|---|
| 1.  | $\frac{1}{EI}\left[\frac{(x-a)^3}{6}H(x-a) + \frac{l-a}{6}l(-x^2+2al-a^2)x\right]$ |
| 2.  | $\frac{1}{EI}\left[\frac{(x-a)^3}{6}H(x-a) + \frac{(l-a)^2}{2l^2}\left(-\frac{l+2a}{3l}x+a\right)x^2\right]$ |
| 3.  | $\frac{1}{EI}\left[\frac{(x-a)^3}{6}H(x-a) + \frac{x^2}{6}(3a-x)\right]$ |
| 4.  | $\frac{1}{EI}\left\{\frac{(x-a)^3}{6}H(x-a) - \frac{l-a}{12}\left[(2l^2+2al-a^2)\left(\frac{x}{l}\right)^3 - 3a(2l-a)\left(\frac{x}{l}\right)^2\right]\right\}$ |

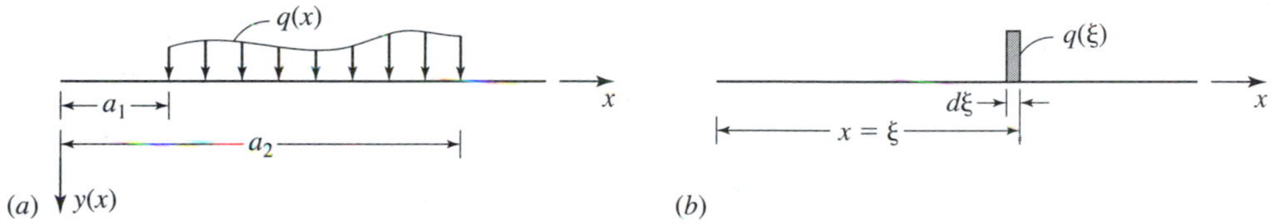


Figure 8-55

point $x$ due to a unit force applied at a point $\xi$. We envision the load $q(x)$ as being composed of infinitesimal load elements, one of which, located at $x = \xi$, is shown in Fig. 8-55*b*. The resultant force equivalent to this load is $q(\xi)\,d\xi$, thus the deflection generated by it equals $G(x, \xi)q(\xi)\,d\xi$. By adding the effects of all these infinitesimal loads over the length of the beam, and using the principle of superposition, we obtain the following formula:

$$y(x) = \int_{a_1}^{a_2} G(x, \xi)q(\xi)\,d\xi \tag{8-208}$$

## EXAMPLE 8-8

Apply the Green's function to find the displacement of the beam shown in Fig. 8-56.

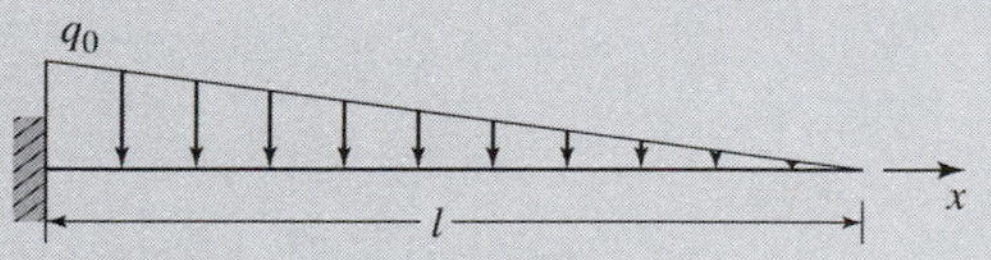


**Figure 8-56** Finding displacement of a beam when Green's function is known

**SOLUTION**

First use entry 3 from Table 8-3,

$$EIy(x) = \int_0^l q(a)\frac{x^2}{6}(3a - x)\,da + \int_0^x q(a)\frac{(x-a)^3}{6}\,da$$

But $q(x) = q_0(l - x)/l$. Therefore,

$$EIy(x) = \frac{q_0 x^2}{6l}\int_0^l (l - a)(3a - x)\,da = \frac{q_0}{6}l\left[\frac{xl^2(l-x)}{2} + \frac{x^4}{20}(5l - x)\right]$$

Proceeding in a similar manner we determine the Green's function for a two-span beam. Consider, for example, the beam shown in Fig. 8-57. The first step will be to remove the middle support and replace it by the unknown reaction $R$. Now the differential equation for deflection [Eq. (8-201)] becomes

$$EIy^{\text{iv}} = \delta(x - a) - R\delta(x - l_1) \tag{8-209}$$

Thus, integrating,

$$EIy''' = H(x - a) - RH(x - l_1) + C_1 \tag{8-210}$$

$$EIy'' = (x - a)H(x - a) - R(x - l_1)H(x - l_1) + C_1 x + C_2 \tag{8-211}$$

$$EIy' = \frac{(x-a)^2}{2}H(x - a) - R\frac{(x-l_1)^2}{2}H(x - l_1) + C_1\frac{x^2}{2} + C_2 x + C_3 \tag{8-212}$$

$$EIy = \frac{(x-a)^3}{6}H(x - a) - R\frac{(x-l_1)^3}{6}H(x - l_1) + C_1\frac{x^3}{6} + C_2\frac{x^2}{2} + C_3 x + C_4 \tag{8-213}$$

The boundary conditions and the condition at the intermediate support are

$$\begin{aligned} y &= 0, \quad y'' = 0, \qquad \text{at } x = 0 \\ y &= 0, \quad y'' = 0, \qquad \text{at } x = l \\ y &= 0, \qquad\qquad\quad\; \text{at } x = l_1 \end{aligned}$$

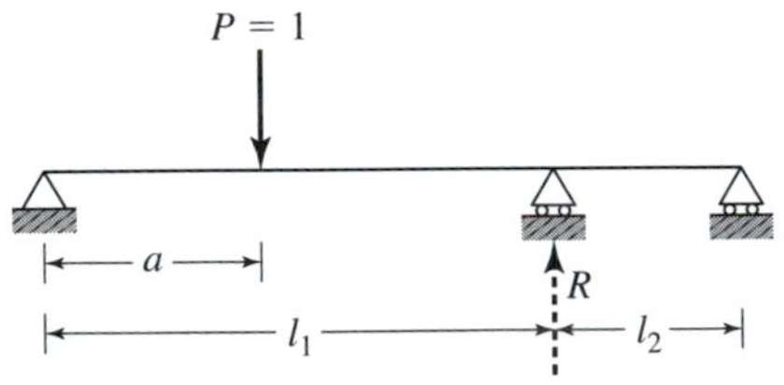


Figure 8-57

With Eqs. (8-213) and (8-211) we get

$$C_2 = C_4 = 0$$

$$\frac{(l-a)^3}{6} - R\frac{(l-l_1)^3}{6} + C_1\frac{l^3}{6} + C_3 l = 0$$

$$l - a - R(l - l_1) + C_1 l = 0$$

with

$$\frac{(l_1-a)^3}{6} + C_1\frac{l_1^3}{6} + C_3 l_1 = 0, \qquad a \le l_1$$

$$C_1\frac{l_1^3}{6} + C_3 l_1 = 0, \qquad a \ge l_1$$

Solving this system of equations we have, using the software package Mathcad7, professional edition [18],

$$C_1 = \frac{1}{2l_1 l}[3all_1 - a^3 - ll_1(l + l_1) + l_1 l_2(l - a)]$$

$$C_3 = \frac{1}{12 l_1 l_2 l}\left[-5al_1^3 l - a^3 l_1 l + 4al_1^2 l^2 - a^3 l_1^2 + al_1^4 + 2a^3 l^2 - 6a^2 l_1^2 l^2 + 6a^2 l_1^2 l\right]$$

$$R = \frac{a}{2l_1^2 l_2 l}\left(-l_1^2 + 2l_1 l - a^2\right)$$

when $a \le l_1$. Note that for $a = l_1$, $C_1 = C_3 = 0$, $R = 1$.

When $a \ge l_1$ we find that

$$C_1 = \frac{(a-l_1)\left[(l-a)^2 - l_2^2\right]}{2l_1 l^2}$$

$$C_3 = \frac{(a-l_1)\left[(l-a)^2 - l_2^2\right] l_1}{12 l^2}$$

$$R = \frac{1}{2l_1 l_2 l}\left\{2l_1 l(l-a) + (a - l_1)\left[(l-a)^2 - l_2^2\right]\right\}$$

Therefore, the Green's function becomes

$$y(x) = \frac{1}{6EI}[(x-a)^3 H(x-a) - R(x-l_1)^3 H(x-l_1) + C_1 x^3 + 6C_3 x] \qquad \text{(8-214)}$$

This and other results obtained in a similar way are listed in the Table 8-4.

We may also generate the Green's function due to a concentrated unit moment applied at $x = \xi$ (Fig. 8-58$a$). This can be done directly or by using the Green's function $G(x, \xi)$ due to a concentrated unit force. In the latter case we represent the moment as a pair of equal but opposite forces $1/\Delta\xi$ located in proximity of the point $x = \xi$ at a

TABLE 8-4

| Schematic | Constants $C_1, \ldots, C_4$ and middle support reaction $R$ |
|---|---|
| 1. 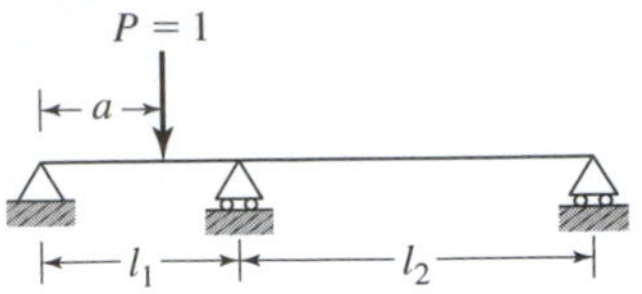 | $C_2 = C_4 = 0$ |
| | $R = a\left(l_1^2 + 2l_1 l - a^2\right)/\left(2l_1^2 l_2\right)$ |
| | $C_1 = (a - l_1)\left(2l_1 l_2 - al_1 - a^2 + 2l_1^2\right)/\left(2l_1^2 l\right)$ |
| | $C_3 = a(a - l_1)\left(3al_1 + 2al_2 - 3l_1^2 - 4l_1 l_2\right)/(12 l l_1)$ when $a \le l_1$ |
| | $C_2 = C_4 = 0$ |
| | $R = (a - l)\left(a^2 - 2al + l_1^2\right)/\left(2l_1 l_2^2\right)$ |
| | $C_1 = (a - l_1)(a - l_1 - 2l_2)(a - l)/(2 l l_1 l_2)$ |
| | $C_3 = -l_1(a - l_1)(a - l_1 - 2l_2)(a - l)/(12 l l_2)$ when $a \ge l_1$ |
| 2. 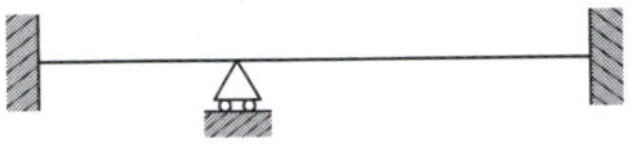 | $C_3 = C_4 = 0$ |
| | $R = a\left[al_1^2(3l - 2a) - ll_1(3l^2 - 3al + 2a^2) - l^2(3l^2 - 6al + 2a^2)\right]/\left[l_1 l_2\left(2l_1^3 + l_1^2 l - 3l^3\right)\right]$ |
| | $C_1 = -\left[2l_1^3 + 3(l - 2a)l_1^2 + 3(l - a)(l - 2a)l_1 - a(3l^2 - 6al + a^2)\right]/\left[l_1\left(2l_1^2 + 3l_1 l + 3l^2\right)\right]$ |
| | $C_2 = a\left[l_1^2(2l - a) + a(a - 3l)l_1 + a^2 l\right]/\left[l\left(2l_1^2 + 3l_1 l + 3l^2\right)\right]$ when $a \le l_1$ |
| | $C_3 = C_4 = 0$ |
| | $R = -(l - a)^2\left[(l + 2a)l_1^3 - 3al^3\right]/\left[l_1 l_2^3\left(2l_1^2 + 3l_1 l + 3l^2\right)\right]$ |
| | $C_1 = 3l(l - a)^2(l_1 - l)/\left[l_1(l_1 - l)\left(2l_1^2 + 3l_1 l + 3l^2\right)\right]$ |
| | $C_2 = l_1^2(l - a)^2(l_1 - a)/\left[ll_2(2l_1^2 + 3l_1 l + 3l^2)\right]$ when $a \ge l_1$ |
| 3. 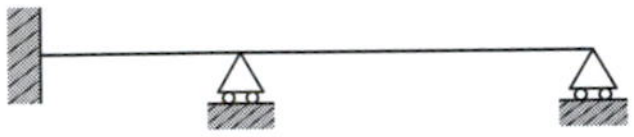 | $C_3 = C_4 = 0$ |
| | $R = -a\left[(3l - a^2)al_1^2 + (-6l^2 + 3al - a^2)ll_1 - 6l^4 + 9al^3 - a^2 l^2\right]/\left[(l - l_1)\left(l_1^3 - ll_1^2 - 3l^2 l_1 - 6l^3\right)\right]$ |
| | $C_1 = -\left[l_1^4 - (1 + 3a)l_1^3 + 3(-l^2 + al + a^2)l_1^2 - (6l^3 - 12al^2 + 3a^2 l + a^3)l_1 + 6al^3 - 9a^2 l^2 + a^3 l\right]/\left[l_1\left(l_1^3 - ll_1^2 - 3l^2 l_1 - 6l^3\right)\right]$ |
| | $C_2 = a\left[l_1^3 - 4ll_1^2 + (6l - a)al_1 - 2a^2 l\right]/\left(l_1^3 - ll_1^2 - 3l^2 l_1 - 6l^3\right)$ when $a \le l_1$ |
| | $C_3 = C_4 = 0$ |
| | $R = (l - a)\left[(2l^2 - 2al - a^2)l_1^3 - 3al^3(2l - a)\right]/\left[l_1(l_1 - l)^2\left(l_1^3 - l_1^2 l - 3l_1 l^2 - 6l^3\right)\right]$ |
| | $C_1 = (l - a)l^2\left[3l_1^2 - 2ll_1 + 3a(2l - a)\right]/\left[l_1 l_2\left(l_1^3 - l_1^2 l - 3l_1 l^2 - 6l^3\right)\right]$ |
| | $C_2 = -(l - a)l_1^2\left[l_1^2 - 2l_1 l - a(a - 2l)\right]/\left[l_1 l_2\left(l_1^3 - l_1^2 l - 3l_1 l^2 - 6l^3\right)\right]$ when $a \ge l_1$ |
| 4. Green's function | $y(x) = \dfrac{1}{EJ}\left[(x - a)^3 H(x - a) - R(x - l_1)^3 H(x - l_1) + C_1 x^3 + 3C_2 x^2 + 6C_3 x + 6C_4\right]$ |

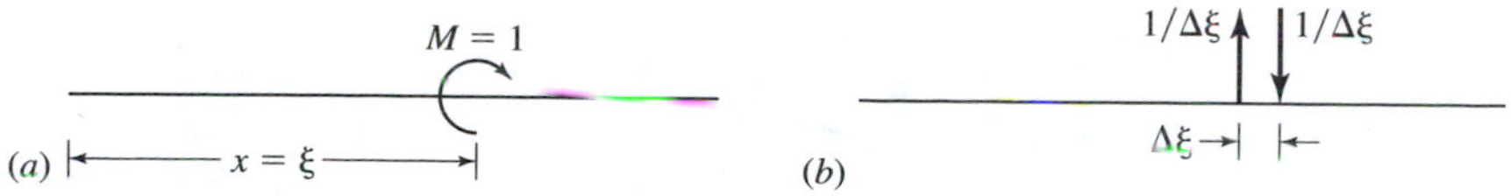


Figure 8-58

distance $\Delta\xi$ from each other (Fig. 8-58*b*). By adding the effects of their action we find that the deflection at $\xi$ is

$$y(x, \xi) = \frac{1}{\Delta\xi}[-G(x, \xi) + G(x, \xi + \Delta\xi)]$$

We obtain the Green's function $G_M(x, \xi)$ due to the moment $M = 1$ as the limit of $y(x, \xi)$ as $\Delta\xi \to 0$,

$$G_M(x, \xi) = \lim_{\Delta\xi \to 0} y(x, \xi) = \lim_{\Delta\xi \to 0} \frac{G(x, \xi + \Delta\xi) - G(x, \xi)}{\Delta\xi} = \frac{dG(x, \xi)}{d\xi} \quad (8\text{-}215)$$

The Green's function $G_M$ may be useful in determining the deflection due to continuously distributed external moments.

## 8-5-2 Straight Beams on Elastic Foundations

Now let us derive the expressions for the Green's function for infinite, semi-infinite, and finite beams on elastic foundations with various boundary conditions. The Green's function for an infinite beam has already been derived in Section 8-4-2, where we considered the effects of a concentrated force $P$. Thus using Eq. (8-166) we find that for $P = 1$ we get

$$y(x) = \frac{\beta}{2k} F_1(\beta|x|) \quad (8\text{-}216)$$

This expression tells us what is the deflection at an arbitrary point of the beam due to a concentrated force $P = 1$ applied at $x = 0$ (Fig. 8-59*a*). On the other hand, the Green's function represents the deflection at a point $N$ due to a concentrated force $P = 1$ located at another point $Q$ (Fig. 8-59*b*). By introducing a coordinate axis $x$ with the origin $O_1$ (Fig. 8-59*b*) we assume the coordinates of points $Q$ and $N$ to be $\xi$ and $x$, respectively. To make the beams in Fig. 8-59*a* and *b* identical, we change the coordinate $x$ by substituting (see Fig. 8-59*c*)

$$z = x - \xi \quad (8\text{-}217)$$

Now we see that the deflection at $N$ due to the force $P = 1$ at $Q$ equals [see Eq. (8-216)]

$$G(z, \xi) = \frac{\beta}{2k} F_1(\beta z)$$

Substituting Eq. (8-217) we finally obtain

$$G(x, \xi) = \frac{\beta}{2k} F_1(\beta|x - \xi|) \quad (8\text{-}218)$$

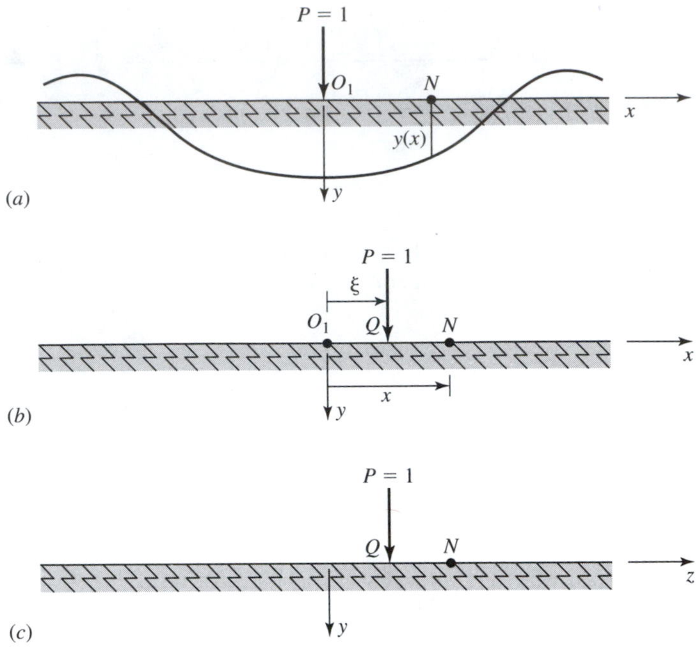


Figure 8-59

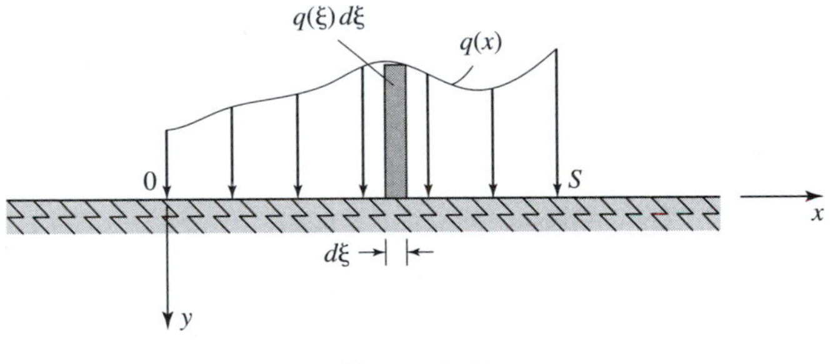


Figure 8-60

This is the Green's function for an infinite beam on an elastic foundation. In order to obtain the deflection $y(x)$ due to an arbitrary load $q(x)$ the function $G(x, \xi)$ can be utilized in the following way (see also Section 8-5-1). First we introduce a coordinate $x$ with the origin located, for example, at the left-hand side of the load (Fig. 8-60). Next we take an infinitesimal element of $q(x)$, $q(\xi)\, d\xi$. Its resultant behaves like a concentrated force $P = q(\xi)\, d\xi$ applied at $x = \xi$, and therefore the deflection at $x$ equals

$$y_q = \frac{\beta q(\xi)\, d\xi}{2k} F_1(\beta|x - \xi|)$$

Thus using the principle of superposition, we "add" the effects of similar infinitesimal forces by integrating and obtain

$$y(x) = \frac{\beta}{2k} \int_0^S q(\xi) F_1(\beta|x - \xi|)\, d\xi \tag{8-219}$$

The Green's function for a semi-infinite beam depends on the boundary conditions and can be determined by proceeding as follows. We begin by writing the general solution for the beam shown in Fig. 8-61. It contains the expression (8-159) obtained for an infinite beam,

$$y(x) = D_3 F_4(\beta x) + D_4 F_2(\beta x) + \frac{\beta}{2k} F_1(\beta|x - \xi|) \tag{8-220}$$

The boundary conditions at infinity are already satisfied and the conditions at $x = 0$ are

$$y = 0, \qquad y'' = 0 \tag{8-221}$$

Substituting here Eqs. (8-220), (8-157), and (8-161), and putting $x = 0$, we obtain

$$D_3 + \frac{\beta}{2k} F_1(\beta\xi) = 0, \qquad -2D_4\beta^2 - \frac{1}{4EI\beta} F_3(\beta\xi) = 0 \tag{8-222}$$

Therefore,

$$D_3 = -\frac{\beta}{2k} F_1(\beta\xi), \qquad D_4 = -\frac{\beta}{2k} F_3(\beta\xi)$$

Substituting these back into Eq. (8-220) we get the Green's function,

$$G(x, \xi) = \frac{\beta}{2k}[-F_4(\beta x)F_1(\beta\xi) - F_2(\beta x)F_3(\beta\xi) + F_1(\beta|x - \xi|)] \tag{8-223}$$

Proceeding in a similar manner we derive the Green's function for the semi-infinite beam shown in Fig. 8-62. The boundary conditions

$$y = 0, \quad y' = 0, \qquad \text{at } x = 0 \tag{8-224}$$

and the general solution, Eq. (8-220), yield

$$D_3 + \frac{\beta}{2k} F_1(\beta\xi) = 0, \qquad -D_3\beta + D_4\beta = 0$$

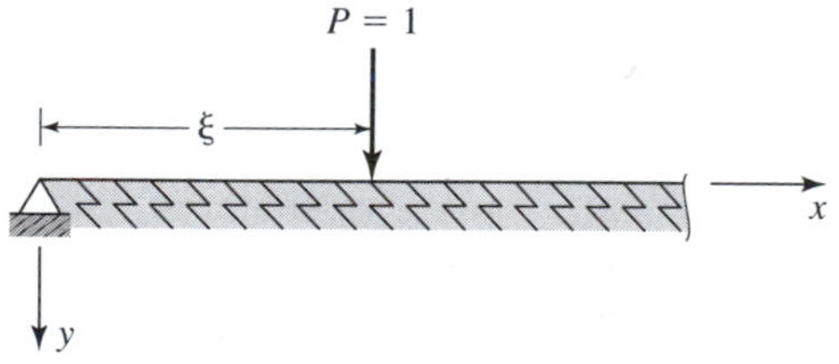


Figure 8-61

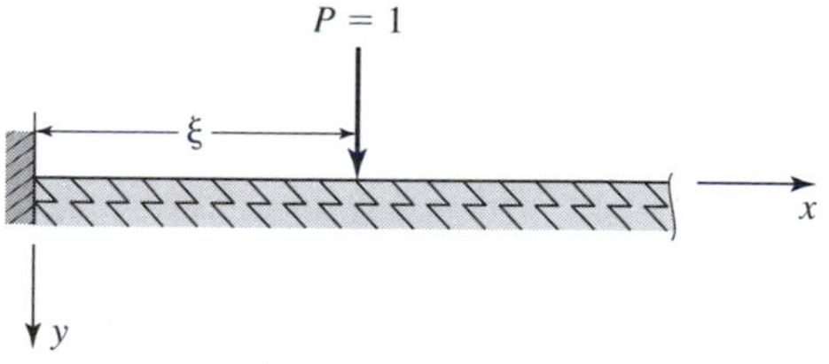


Figure 8-62

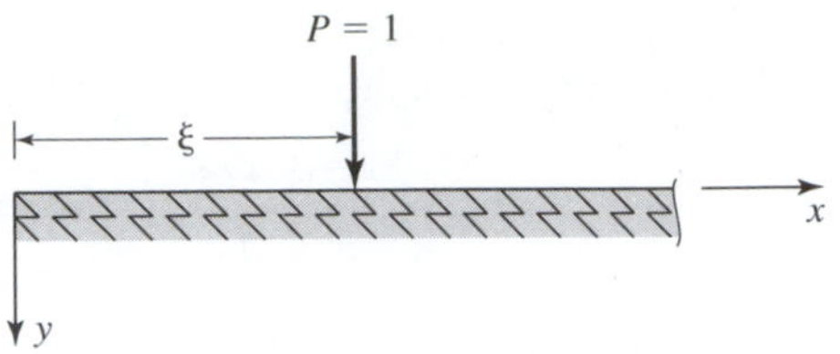


Figure 8-63

Thus,

$$D_3 = D_4 = -\frac{\beta}{2k} F_1(\beta\xi)$$

and the Green's function for this case becomes

$$G(x, \xi) = \frac{\beta}{2k}\{-[F_4(\beta x) + F_2(\beta x)]F_1(\beta\xi) + F_1(\beta|x - \xi|)\} \tag{8-225}$$

Finally, for the beam in Fig. 8-63 we have the boundary conditions

$$y'' = 0, \quad y''' = 0, \qquad \text{at } x = 0 \tag{8-226}$$

which in conjunction with Eq. (8-220) yield

$$-2D_4\beta^2 - \frac{1}{4EI\beta}F_3(\beta\xi) = 0, \qquad 2D_3\beta^3 + 2D_4\beta^3 + \frac{1}{2EI}F_4(\beta\xi) = 0$$

Therefore,

$$D_3 = \frac{\beta}{2k}[F_3(\beta\xi) - 2F_4(\beta\xi)], \qquad D_4 = -\frac{\beta}{2k}F_3(\beta\xi)$$

and the Green's function is obtained in the form

$$G(x, \xi) = \frac{\beta}{2k}\{F_4(\beta x)[F_3(\beta\xi) - 2F_4(\beta\xi)] - F_1(\beta|x - \xi|)F_3(\beta\xi)\} \tag{8-227}$$

Now for a finite beam with the span length $l$ we use the general solution (8-178) and the particular solution corresponding to $P = 1$ at $x = \xi$ [Eq. $(8\text{-}186)_1$],

$$y(x) = D_1F_8(\beta x) + D_2F_6(\beta x) + D_3F_4(\beta x) + D_4F_2(\beta x) + \frac{\beta}{2k}F_1(\beta|x - \xi|) \tag{8-228}$$

This expression, combined with the appropriate boundary conditions, generates the Green's function $G(x, \xi)$. A considerable amount of manipulations is needed to obtain the results listed below for which we used the symbolic manipulation package Macsyma [17].

A list of the constants of integration $D_1$, $D_2$, $D_3$, and $D_4$ appearing in expression (8-228) for the Green's functions follows.

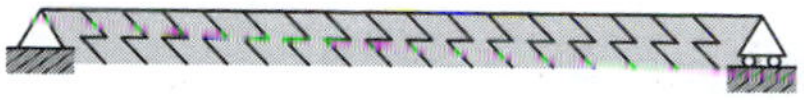

Figure 8-64

Figure 8-65

Figure 8-66

**Case 1: Simply Supported Ends (Fig. 8-64)**

$$
\begin{aligned}
D_1 &= \{\beta e^{\beta\xi}[H_1 \sin\beta(l-\xi) - H_2 \cos\beta(l-\xi)] \\
&\quad + \beta e^{-\beta\xi}[-H_3 \sin\beta\xi + H_4 \cos\beta\xi]\}/\Delta_1 \\
D_2 &= \{\beta e^{\beta\xi}[-H_1 \cos\beta(l-\xi) - H_2 \sin\beta(l-\xi)] \\
&\quad + \beta e^{-\beta\xi}[H_3 \cos\beta\xi + H_4 \sin\beta\xi]\}/\Delta_1 \\
D_3 &= \{\beta e^{\beta\xi}[-H_1 \sin\beta(l-\xi) + H_2 \cos\beta(l-\xi)] \\
&\quad + \beta e^{-\beta\xi}[-H_{17} \sin\beta\xi - H_{18} \cos\beta\xi]\}/\Delta_1 \\
D_4 &= \{\beta e^{\beta\xi}[-H_1 \cos\beta(l-\xi) - H_2 \sin\beta(l-\xi)] \\
&\quad + \beta e^{-\beta\xi}[-H_{17} \cos\beta\xi + H_{18} \sin\beta\xi]\}/\Delta_1
\end{aligned}
$$

**Case 2: One End Simply Supported, the Other End Clamped (Fig. 8-65)**

$$
\begin{aligned}
D_1 &= \{\beta e^{\beta\xi}[-H_5 \sin\beta(l-\xi) + H_6 \cos\beta(l-\xi)] \\
&\quad + \beta e^{-\beta\xi}[H_8 \sin\beta\xi + H_7 \cos\beta\xi]\}/\Delta_2 \\
D_2 &= \{\beta e^{\beta\xi}[H_{10} \cos\beta(l-\xi) + H_{14} \sin\beta(l-\xi)] \\
&\quad + \beta e^{-\beta\xi}[-H_8 \cos\beta\xi + H_{11} \sin\beta\xi]\}/\Delta_2 \\
D_3 &= \{\beta e^{\beta\xi}[H_5 \sin\beta(l-\xi) - H_6 \cos\beta(l-\xi)] \\
&\quad + \beta e^{-\beta\xi}[H_{12} \sin\beta\xi - H_{48} \cos\beta\xi]\}/\Delta_2 \\
D_4 &= \{\beta e^{\beta\xi}[H_{10} \cos\beta(l-\xi) + H_{14} \sin\beta(l-\xi)] \\
&\quad + \beta e^{-\beta\xi}[H_{12} \cos\beta\xi - H_{16} \sin\beta\xi]\}/\Delta_2
\end{aligned}
$$

**Case 3: One End Simply Supported, the Other End Free (Fig. 8-66)**

$$
\begin{aligned}
D_1 &= \{\beta e^{\beta\xi}[-H_{10} \sin\beta(l-\xi) + H_{14} \cos\beta(l-\xi)] \\
&\quad + \beta e^{-\beta\xi}[H_8 \sin\beta\xi + H_{11} \cos\beta\xi]\}/\Delta_3 \\
D_2 &= \{\beta e^{\beta\xi}[H_5 \cos\beta(l-\xi) + H_6 \sin\beta(l-\xi)] \\
&\quad + \beta e^{-\beta\xi}[-H_8 \cos\beta\xi + H_7 \sin\beta\xi]\}/\Delta_3 \\
D_3 &= \{\beta e^{\beta\xi}[H_{10} \sin\beta(l-\xi) - H_{14} \cos\beta(l-\xi)] \\
&\quad + \beta e^{-\beta\xi}[H_{12} \sin\beta\xi + H_{16} \cos\beta\xi]\}/\Delta_3 \\
D_4 &= \{\beta e^{\beta\xi}[H_5 \cos\beta(l-\xi) + H_6 \sin\beta(l-\xi)] \\
&\quad + \beta e^{-\beta\xi}[H_{12} \cos\beta\xi + H_{48} \sin\beta\xi]\}/\Delta_3
\end{aligned}
$$

Figure 8-67

Figure 8-68

Figure 8-69

**Case 4: Clamped Ends (Fig. 8-67)**

$$D_1 = \{\beta e^{\beta\xi}[H_{22}\sin\beta(l-\xi) - H_2\cos\beta(l-\xi)] + \beta e^{-\beta\xi}[-H_{49}\sin\beta\xi - H_{24}\cos\beta\xi]\}/\Delta_4$$

$$D_2 = \{\beta e^{\beta\xi}[H_{26}\cos\beta(l-\xi) - H_{25}\sin\beta(l-\xi)] + \beta e^{-\beta\xi}[-H_4\cos\beta\xi - H_{51}\sin\beta\xi]\}/\Delta_4$$

$$D_3 = \{\beta e^{\beta\xi}[-H_{22}\sin\beta(l-\xi) + H_2\cos\beta(l-\xi)] + \beta e^{-\beta\xi}[-H_{18}\sin\beta\xi + H_{29}\cos\beta\xi]\}/\Delta_4$$

$$D_4 = \{\beta e^{\beta\xi}[-H_{22}\cos\beta(l-\xi) - H_{30}\sin\beta(l-\xi)] + \beta e^{-\beta\xi}[-H_{18}\cos\beta\xi - H_{52}\sin\beta\xi]\}/\Delta_4$$

**Case 5: One End Clamped, the Other End Free (Fig. 8-68)**

$$D_1 = \{\beta e^{\beta\xi}[H_1\sin\beta(l-\xi) - H_{38}\cos\beta(l-\xi)] + \beta e^{-\beta\xi}[-H_{53}\sin\beta\xi + H_{34}\cos\beta\xi]\}/\Delta_5$$

$$D_2 = \{\beta e^{\beta\xi}[-H_{36}\cos\beta(l-\xi) - H_{35}\sin\beta(l-\xi)] + \beta e^{-\beta\xi}[-H_4\cos\beta\xi - H_{54}\sin\beta\xi]\}/\Delta_5$$

$$D_3 = \{\beta e^{\beta\xi}[-H_1\sin\beta(l-\xi) + H_{38}\cos\beta(l-\xi)] + \beta e^{-\beta\xi}[-H_{18}\sin\beta\xi - H_{40}\cos\beta\xi]\}/\Delta_5$$

$$D_4 = \{\beta e^{\beta\xi}[-H_{41}\cos\beta(l-\xi) - H_{38}\sin\beta(l-\xi)] + \beta e^{-\beta\xi}[-H_{18}\cos\beta\xi - H_{55}\sin\beta\xi]\}/\Delta_5$$

**Case 6: Free Ends (Fig. 8-69)**

$$D_1 = \{\beta e^{\beta\xi}[-H_{26}\sin\beta(l-\xi) - H_{25}\cos\beta(l-\xi)] + \beta e^{-\beta\xi}[H_4\sin\beta\xi - H_{51}\cos\beta\xi]\}/\Delta_6$$

$$D_2 = \{\beta e^{\beta\xi}[-H_{22}\cos\beta(l-\xi) - H_2\sin\beta(l-\xi)] + \beta e^{-\beta\xi}[H_{49}\cos\beta\xi - H_{24}\sin\beta\xi]\}/\Delta_6$$

$$D_3 = \{\beta e^{\beta\xi}[-H_{22}\sin\beta(l-\xi) + H_{30}\cos\beta(l-\xi)] + \beta e^{-\beta\xi}[-H_{18}\sin\beta\xi + H_{52}\cos\beta\xi]\}/\Delta_6$$

$$D_4 = \{\beta e^{\beta\xi}[-H_{22}\cos\beta(l-\xi) - H_2\sin\beta(l-\xi)] + \beta e^{-\beta\xi}[-H_{18}\cos\beta\xi - H_{29}\sin\beta\xi]\}/\Delta_6$$

The following symbols appear in Cases 1–6:

$$\Delta_1 = 2k[(1 + e^{2\beta l})^2 - 4e^{2\beta l}\cos^2\beta l]$$
$$\Delta_2 = \Delta_3 = 2k(-1 + e^{4\beta l} - 2e^{2\beta l}\sin 2\beta l)$$
$$\Delta_4 = \Delta_6 = 2k[1 + e^{4\beta l} - 2e^{2\beta l}(3 - 2\cos^2\beta l)]$$
$$\Delta_5 = 2k[(1 + e^{2\beta l})^2 + 4e^{2\beta l}\cos^2\beta l]$$

Other symbols include

$$H_1 = G_1 \sin\beta l + G_2 \cos\beta l$$
$$H_2 = G_1 \sin\beta l - G_2 \cos\beta l$$
$$H_3 = G_1 \sin^2\beta l + e^{2\beta l} + G_2 \cos^2\beta l$$
$$H_4 = -G_1 \sin^2\beta l + e^{2\beta l}\sin 2\beta l - G_2 \cos^2\beta l$$
$$H_5 = G_3 \sin\beta l + G_1 \cos\beta l$$
$$H_6 = G_2 \sin\beta l - G_1 \cos\beta l$$
$$H_7 = G_3 \sin^2\beta l + e^{2\beta l}\sin 2\beta l + G_1 \cos^2\beta l$$
$$H_{11} = G_2 \sin^2\beta l + e^{2\beta l}\sin 2\beta l + G_4 \cos^2\beta l$$
$$H_{12} = (G_2 \sin^2\beta l + \sin 2\beta l - G_1 \cos^2\beta l)e^{2\beta l}$$
$$H_{14} = G_2 \sin\beta l - G_4 \cos\beta l$$
$$H_{16} = (G_2 \sin^2\beta l + \sin 2\beta l + G_5 \cos^2\beta l)e^{2\beta l}$$
$$H_{17} = (G_1 \sin^2\beta l - \sin 2\beta l - G_2 \cos^2\beta l)e^{2\beta l}$$
$$H_{18} = (G_1 \sin^2\beta l + \sin 2\beta l - G_2 \cos^2\beta l)e^{2\beta l}$$
$$H_{22} = G_2 \cos\beta l + G_4 \sin\beta l$$
$$H_{24} = G_2 \cos^2\beta l - e^{2\beta l}\sin 2\beta l + G_4 \cos^2\beta l$$
$$H_{25} = G_1 \sin\beta l + 3G_2 \cos\beta l$$
$$H_{26} = G_5 \sin\beta l - G_2 \cos\beta l$$
$$H_{29} = (G_5 \sin^2\beta l - \sin 2\beta l + G_2 \cos^2\beta l)e^{2\beta l}$$
$$H_{30} = G_8 \sin\beta l - G_2 \cos\beta l$$
$$H_{32} = H_{18}$$
$$H_{34} = -G_1 \sin^2\beta l + e^{2\beta l}\sin 2\beta l - G_3 \cos^2\beta l$$
$$H_{35} = G_1 \sin\beta l + G_6 \cos\beta l$$
$$H_{36} = -3G_1 \sin\beta l + G_2 \cos\beta l$$
$$H_{38} = G_1 \sin\beta l - G_3 \cos\beta l$$
$$H_{40} = (G_1 \sin^2\beta l + \sin 2\beta l + G_6 \cos^2\beta l)e^{2\beta l}$$
$$H_{41} = G_1 \sin\beta l + G_9 \cos\beta l$$
$$H_{48} = (G_6 \sin^2\beta l - \sin 2\beta l + G_1 \cos^2\beta l)e^{2\beta l}$$
$$H_{51} = G_4 \sin^2\beta l + 3e^{2\beta l}\sin 2\beta l + G_2 \cos^2\beta l$$
$$H_{52} = (-G_4 \sin^2\beta l - \sin 2\beta l - 3G_2 \cos^2\beta l)e^{2\beta l}$$
$$H_{53} = G_1 \sin^2\beta l - e^{2\beta l}\sin 2\beta l + G_9 \cos^2\beta l$$
$$H_{54} = G_1 \sin^2\beta l + 3e^{2\beta l}\sin 2\beta l + G_3 \cos^2\beta l$$

where

$$G_1 = 1 + e^{2\beta l}, \quad G_2 = 1 - e^{2\beta l}, \quad G_3 = 1 + 3e^{2\beta l}, \quad G_4 = 1 - 3e^{2\beta l}$$
$$G_5 = 3 - e^{2\beta l}, \quad G_6 = 3 + e^{2\beta l}, \quad G_8 = 1 - 7e^{2\beta l}, \quad G_9 = 1 + 7e^{2\beta l}$$

## 8-6 THERMAL EFFECTS

We examine in this section the asymmetric bending of a beam subject to temperature changes. Symmetric bending is then obtained as a special case. (For a more extensive treatment of this subject see [4, p. 307].) It is assumed that no mechanical loads are applied. Since we consider only small deformations and a linear stress-strain relationship, the principle of superposition is valid. Thus simultaneous mechanical and thermal effects can be analyzed separately and the resulting stresses and deformations added. The basis for our analysis is the stress-strain relation for a thermoelastic solid (see Chapter 4),

$$\sigma_{xx} = E\varepsilon_{xx} - E\alpha\Delta T \tag{8-229}$$

Following the procedure adopted in Section 8-2, the expression for normal stress due to temperature changes alone is now assumed to be

$$\sigma_{xx} = a_T + b_T y + c_T z - E\alpha\Delta T \tag{8-230}$$

The unknown constants $a_T$, $b_T$, and $c_T$ will be determined using the equilibrium equations, which now are

$$\int_A \sigma_{xx}\, dA = 0, \qquad \int_A \sigma_{xx} z\, dA = 0, \qquad \int_A \sigma_{xx} y\, dA = 0 \tag{8-231}$$

It is further assumed that as in Section 8-2, $y$ and $z$ are the principal centroidal axes of inertia. Thus with Eq. (8-231) we obtain the following results:

$$a_T = \frac{P_T}{A}, \qquad b_T = \frac{M_{Tz}}{I_{zz}}, \qquad c_T = \frac{M_{Ty}}{I_{yy}} \tag{8-232}$$

where

$$P_T = E\alpha \int_A \Delta T\, dA, \qquad M_{Ty} = E\alpha \int_A \Delta T z\, dA, \qquad M_{Tz} = E\alpha \int_A \Delta T y\, dA \tag{8-233}$$

Consequently the thermal stress equals

$$\sigma_{xx} = \frac{P_T}{A} - E\alpha\Delta T + \frac{M_{Tz}}{I_{zz}} y + \frac{M_{Ty}}{I_{yy}} z \tag{8-234}$$

and the corresponding strain follows from Eq. (8-229),

$$\varepsilon_{xx} = \frac{1}{E}\sigma_{xx} + \alpha\Delta T \tag{8-235}$$

When the cross section and the change of the temperature are symmetric with respect to one of the principal axes, say $Oz$,

$$\Delta T(x, -y, z) = \Delta T(x, y, z)$$

then $M_{Tz} = 0$, and the expression for the stress reduces to

$$\sigma_{xx} = \frac{P_T}{A} - E\alpha\Delta T + \frac{M_{Ty}}{I_{yy}} z \tag{8-236}$$

When the temperature varies only along the $x$ axis then, regardless of the geometry of the cross section,

$$M_{Ty} = M_{Tz} = 0, \qquad P_T = EA\alpha\Delta T$$

Hence,

$$\sigma_{xx} = 0, \qquad \varepsilon_{xx} = \alpha\Delta T$$

## EXAMPLE 8-9

Determine the stresses in an elastic rod of a symmetric cross section fitted between two rigid walls and then heated by $\Delta T(x, y, z) = \Delta T(x)$ (Fig. 8-70).

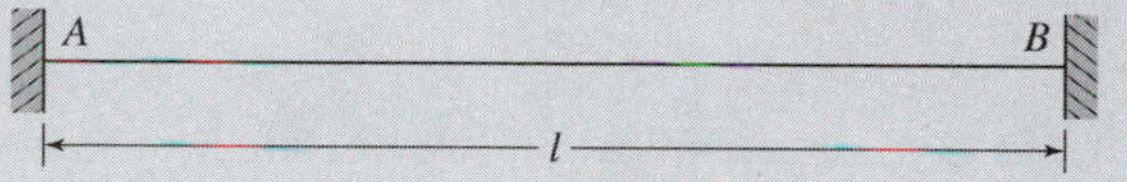


**Figure 8-70** Statically indeterminate heated rod

## SOLUTION

A statically determinate rod, obtained by removing the constraint at $B$, is subject to the temperature change $\Delta T$ and to the unknown reaction $R_B$ (Fig. 8-71). The net effect of these two factors is a zero displacement at $B$,

$$u_B = 0$$

First we find the displacement at $B$ due to the *force* $R_B$ by using the principle of superposition,

$$u_B^{(1)} = -\frac{R_B l}{EA}$$

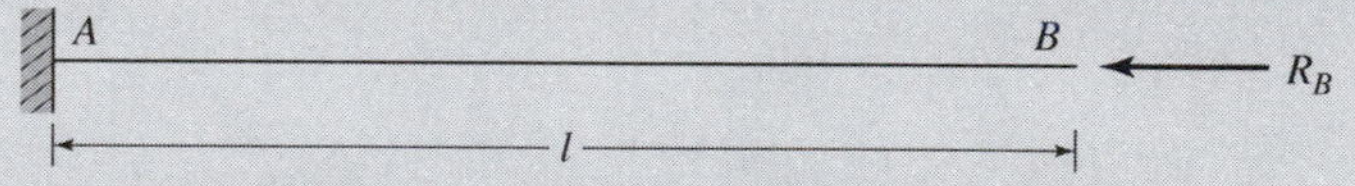


**Figure 8-71** Equivalent statically determinate rod

Next we find that the stress due to $\Delta T$ alone is zero while the corresponding strain equals

$$\varepsilon_{xx} = \alpha \Delta T$$

and the resulting displacement at $B$ is

$$u_B^{(2)} = \int_0^l \alpha \Delta T \, dx = \alpha \int_0^l \Delta T \, dx$$

From the condition

$$u_B = u_B^{(1)} + u_B^{(2)} = 0$$

we obtain the reaction $R_B$,

$$R_B = \frac{EA\alpha}{l} \int_0^l \Delta T \, dx$$

and the stress equals

$$\sigma_{xx} = -\frac{R_B}{A} = -\frac{E\alpha}{l} \int_0^l \Delta T \, dx$$

## 8-7 COMPOSITE BEAMS

We consider in this section the bending of thin straight beams composed of several layers made from different orthotropic materials. We derive a differential equation analogous to that of homogeneous beams and solve some problems.

### 8-7-1 Stresses, Bending Moments, and Bending Stiffness of a Laminated Beam

A beam composed of $N$ perfectly bonded layers of various orthotropic materials is shown in Fig. 8-72. Such a beam is known as a *laminated beam*. The usual assumptions of the thin beam theory are also valid here:

1. The plane cross sections, initially perpendicular to the axis of the beam, remain plane and perpendicular after deformation.

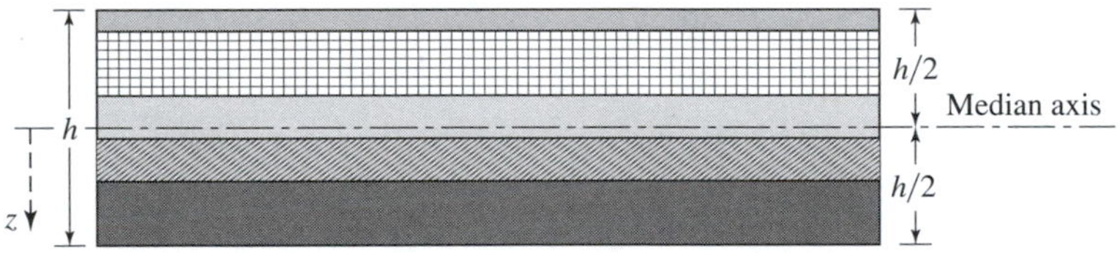


Figure 8-72 Laminated beam

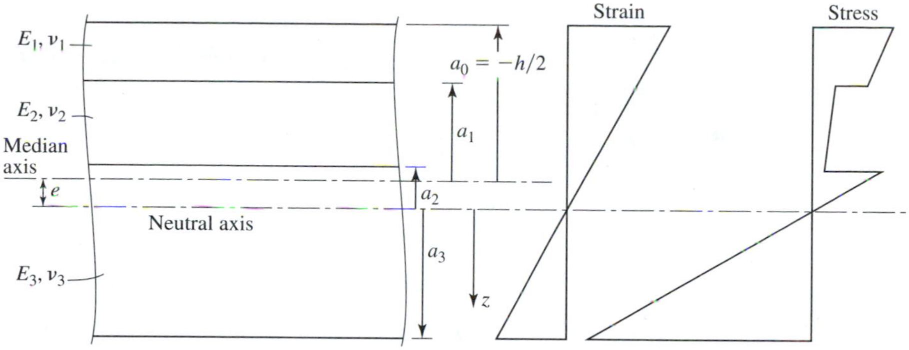


**Figure 8-73** Three-layer laminated beam

2. All stresses are functions only of $x$ and $z$.
3. The total thickness $h$ of the beam is small compared to the span length.
4. The normal stress $\sigma_{zz}$ is zero.
5. The strains $\gamma_{xz}$ and $\gamma_{yz}$ are zero.

As a result of assumption 1 the strains are linear functions of the distance $z$ from the still unknown neutral axis,

$$e_{xx} = -z\frac{d^2y}{dx^2} \tag{8-237}$$

where $y = y(x)$ is the deflection. Let us consider a three-layer beam (Fig. 8-73) as an example. The resulting expressions will be generalized easily to be valid for a beam with an arbitrary number of layers. Although the strain is continuous across the interfaces, the stress is not. This is due to the assumed validity of Hooke's law and to the differences in the Young's moduli for adjacent layers. The following relation between stress and deformation exists for each layer:

$$\sigma_{xx} = E_k e_{xx} = -E_k z\frac{d^2y}{dx^2}, \qquad k = 1, 2, 3 \tag{8-238}$$

where $E_k$ is Young's modulus in the direction $x$ for the $k$th layer. This relation is obviously valid in the general case for any of the $N$ layers,

$$\sigma_{xx} = -E_k z\frac{d^2y}{dx^2}, \qquad k = 1, 2, \ldots, N \tag{8-239}$$

In order to determine the position of the neutral axis we assume that it is located at a distance $e$ from the median axis (Fig. 8-73). The condition that the resultant of $\sigma_{xx}$ is zero,

$$\int_{a_0+e}^{a_1+e} \sigma_{xx}\,dz + \int_{a_1+e}^{a_2+e} \sigma_{xx}\,dz + \int_{a_2+e}^{a_3+e} \sigma_{xx}\,dz = 0$$

takes the form

$$E_1\left(\overline{a}_1^2 - \overline{a}_0^2\right) + E_2\left(\overline{a}_2^2 - \overline{a}_1^2\right) + E_3\left(\overline{a}_3^2 - \overline{a}_2^2\right) = 0$$

Substituting,

$$\overline{a}_0 = a_0 + e, \qquad \overline{a}_1 = a_1 + e, \qquad \overline{a}_2 = a_2 + e, \qquad \overline{a}_3 = a_3 + e$$

we obtain the magnitude of $e$,

$$e = \frac{E_1\left(a_0^2 - a_1^2\right) + E_2\left(a_1^2 - a_2^2\right) + E_3\left(a_2^2 - a_3^2\right)}{2[E_1(a_1 - a_0) + E_2(a_2 - a_1) + E_3(a_3 - a_2)]}$$

Once the position of the neutral axis is known, we apply the usual definition of the bending moment,

$$M = b\int_{\overline{a}_0}^{\overline{a}_3} \sigma_{xx} z\, dz = -b\frac{d^2y}{dx^2}\left[\int_{\overline{a}_0}^{\overline{a}_1} E_1 z^2\, dz + \int_{\overline{a}_1}^{\overline{a}_2} E_2 z^2\, dz + \int_{\overline{a}_2}^{\overline{a}_3} E_3 z^2\, dz\right] \tag{8-240}$$

This can be put in the form

$$M = -D\frac{d^2y}{dx^2} \tag{8-241}$$

where the *bending stiffness* of a three-layer composite beam is given by

$$D \equiv -\frac{b}{3}\sum_{k=1}^{3} E_k\left(\overline{a}_{k-1}^3 - \overline{a}_k^3\right) \tag{8-242}$$

It is evident that for $N$ layers this becomes

$$D \equiv -\frac{b}{3}\sum_{k=1}^{N} E_k\left(\overline{a}_{k-1}^3 - \overline{a}_k^3\right) \tag{8-243}$$

The shear stresses $\sigma_{xz}$ at a distance $z_1$ from the neutral axis are related to the shear force $V$ in the same way as for a simple beam, that is [21, p. 137],

$$\sigma_{xz} = \frac{VQ}{Ib} \tag{8-244}$$

where $Q$ is the first moment about the neutral axis of the portion of the cross section above $z = z_1$,

$$Q = \int_{z_1}^{h/2} z\, dA \tag{8-245}$$

$I$ is the moment of inertia of the cross section about the neutral axis, and $b$ is the width of the cross section at $z = z_1$. For a beam of rectangular cross section Eq. (8-244) becomes [21, p. 137]

$$\sigma_{xz} = \frac{V}{2I}\left[\left(\frac{h}{2}\right)^2 - z^2\right] \tag{8-246}$$

## 8-7-2 Differential Equation for Deflection of a Laminated Beam

Our goal is now to determine the relation between the shear force $V$ and the applied load $q(x)$. First consider the infinitesimal beam element shown in Fig. 8-74. Adding the vertical projections of all forces we find

$$V - q(x)\,dx - \left(V + \frac{dV}{dx}\,dx\right) = 0$$

Therefore,

$$\frac{dV}{dx} = -q(x) \qquad \text{(8-247)}$$

The relation between the bending moment $M$ and the shear force $V$ is obtained by taking the sum of the moments about a line through point $A$, perpendicular to the figure. This sum must also be zero,

$$M - \left(M + \frac{dM}{dx}\,dx\right) + \left(V + \frac{dV}{dx}dx\right)dx - q(x)\,dx\frac{dx}{2} = 0$$

By ignoring higher order infinitesimals we reduce this to

$$V = \frac{dM}{dx} \qquad \text{(8-248)}$$

Combining Eqs. (8-247) and (8-248) we obtain

$$\frac{d^2M}{dx^2} = -q(x)$$

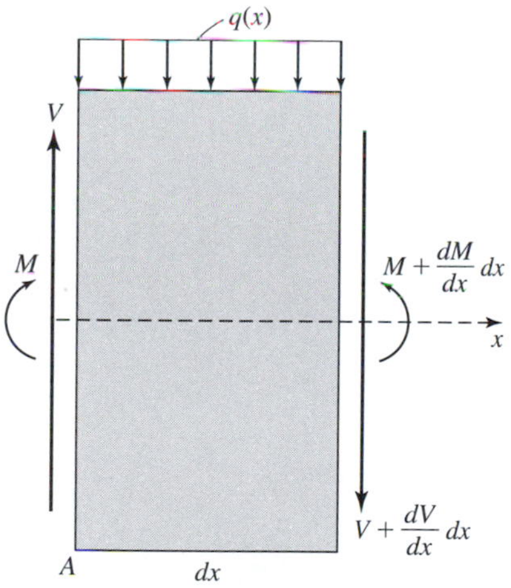


**Figure 8-74** Infinitesimal element of laminated beam

and then, substituting here Eq. (8-241), we get the deflection equation for a laminated beam,

$$D\frac{d^4y}{dx^4} = q(x) \tag{8-249}$$

It is identical with the deflection equation for a simple beam [see Eq. (8-1)] except that the stiffness $EI$ must be replaced by $D$ defined in Eq. (8-242). Thus the analysis of the bending of laminated beams is the same as that for simple beams.

## 8-8 LIMIT ANALYSIS [19, pp. 300–326]

So far we only analyzed elastic beams. When a beam, or another structure composed of thin rods, reaches yield stress at some cross section, then this section behaves like a hinge. Thus the beam transforms into a mechanism whose parts can rotate around this hinge, causing the collapse of the system (Fig. 8-75). However, when a statically indeterminate beam, made of a ductile material, reaches yield stress at a certain cross section, it may still carry loads. This is studied by the theory of *limit analysis* based on the theory of plasticity. A few examples will help in understanding this theory. Let us consider the three-bar symmetric truss shown in Fig. 8-76. It is assumed that each bar has the same cross-sectional area $A$ and is made of the same material with Young's modulus $E$ in the elastic range. First we calculate the forces and the deformations of each bar. We note that the truss is statically indeterminate and we remove bar 2-4, replacing its action by the unknown force $X$ (Fig. 8-77). From the equilibrium equation of node 4 we

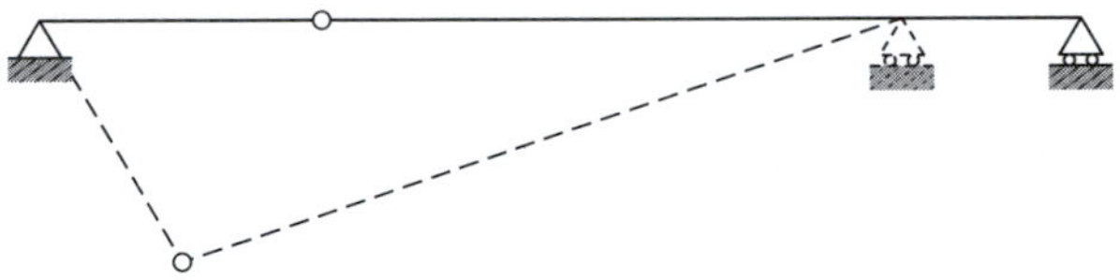

Figure 8-75 Collapse of beam with plastic hinge

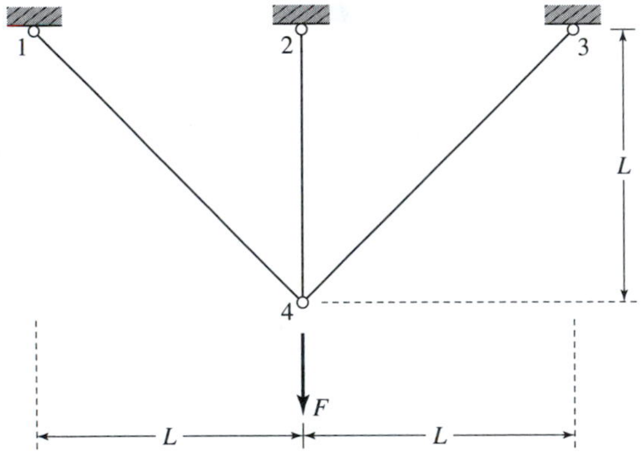


Figure 8-76

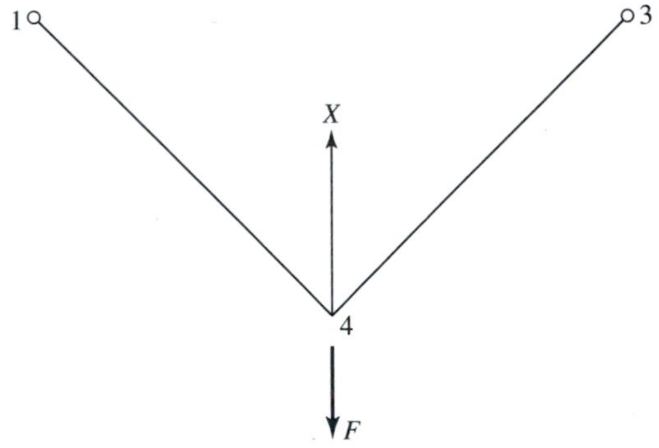


Figure 8-77

find that the force $N$ in each remaining bar is

$$N = \frac{F - X}{\sqrt{2}} \tag{8-250}$$

In order to determine $X$ we use the conditional equation

$$v_4 = \Delta_{2-4} \tag{8-251}$$

which states that the vertical displacement of the system 1-4-3 at node 4 equals the extension of the removed bar 2-4. However,

$$\Delta_{2-4} = \frac{XL}{EA} \tag{8-252}$$

whereas

$$v_4 = \Delta_{1-4}\sqrt{2} = \frac{NL\sqrt{2}}{EA}\sqrt{2} = \frac{2NL}{EA} \tag{8-253}$$

where the first of Eq. (8-253) results from geometric considerations. Substituting Eqs. (8-252) and (8-253) into Eq. (8-251) we find that

$$X = 2N$$

or, using Eq. (8-250), we get

$$\begin{aligned} X &= \frac{2F}{2+\sqrt{2}}, \qquad & N &= \frac{F}{2+\sqrt{2}} \\ \Delta_{2-4} &= \frac{2FL}{(2+\sqrt{2})EA}, \qquad & \Delta_{1-4} &= \frac{FL\sqrt{2}}{(2+\sqrt{2})EA} \end{aligned} \tag{8-254}$$

This implies that the maximum stress will occur in bar 2-4. We will assume that $F_e = kF$ causes the maximum stress to reach the yield stress $Y$. Here $F_e$ is the collapse load (also called the *limit load*), and $k$ is the coefficient of proportionality known as the *safety factor*. We find

$$Y = \frac{X}{A} = \frac{2F_e}{(2+\sqrt{2})A} = \frac{2kF}{(2+\sqrt{2})A} = 0.586k\frac{F}{A}$$

or, if $F$ is to be determined,

$$k = \frac{YA}{0.586F} = 1.705\frac{YA}{F}$$

so that the allowable working load equals

$$F = \frac{YA}{0.586k} = 1.705\frac{YA}{k} \tag{8-255}$$

Because the structure is statically indeterminate, it would not collapse even if it were, as assumed, elastic and brittle. The consequence would be the destruction of bar 2-4, and then the remaining two bars, 1-4 and 3-4, would carry the load $F$. It is readily seen from Eq. (8-250) that the force in each of these bars would now equal

$$N = \frac{F}{\sqrt{2}} = 0.707F$$

with the safety factor reduced to

$$k = \frac{YA}{0.707F} = 1.414\frac{YA}{F}$$

and the allowable load for a given $k$ becoming

$$F = \frac{YA}{0.707k} = 1.414\frac{YA}{k} \tag{8-256}$$

At this point we assume that the material of the bar is elastic–perfectly plastic (see Chapter 4). Consequently the bar will not fracture when the stress reaches the yield level, but it will start to flow. With $F$ increasing beyond its present level, the strain in bar 2-4 continues to increase, but the stress remains at $Y$. On the other hand, the stresses in bars 1-4 and 3-4 increase until they also reach the yield level, $N = YA$. The ultimate level of $F$, called the *carrying capacity* of the truss, is then

$$F_e = YA + YA\sqrt{2} = 2.414YA \tag{8-257}$$

The analysis for beams in bending is similar to that shown for the truss, although more complicated. Again, if we assume that the material is elastic, then we allow the working load to reach a level for which the maximum stress in the beam reaches the yield stress divided by the safety factor. In the development that follows we neglect the influence of the shear forces, axial forces, and torques, if any, on the carrying capacity of the beam. We examine a beam under the action of various transverse loads, assuming that the location of the maximum bending moment due to these loads is known. The bending moment is defined, as usual, as the resultant moment of the normal stress in the cross section with respect to the neutral axis,

$$M = \int_{-h/2}^{h/2} \sigma_x(y) by\, dy \tag{8-258}$$

Let us also assume that the plane cross section remains plane for the deformed beam. Therefore the strain $\varepsilon_x$ varies linearly with the distance from the neutral axis. Denoting by $K = K(x)$ the curvature of the neutral axis, we have then

$$\varepsilon_x = Ky$$

As long as the entire beam remains linearly elastic we find from Hooke's law that

$$\sigma_x = E\varepsilon_x = EKy \tag{8-259}$$

Substituting Eq. (8-259) into Eq. (8-258) gives the following relation between the bending moment and the curvature in the elastic range:

$$M = EK \int_{-h/2}^{h/2} by^2\, dy = EK\frac{bh^3}{12} = EKI \tag{8-260}$$

where $I$ is the moment of inertia of the cross section. Now the increase of the load generates a proportional increase of the bending moment, which reaches its absolute elastic maximum when the maximum elastic stress $\sigma_x$ reaches the yield point,

$$|\sigma_x| = Y, \qquad \text{at } y = \pm h/2$$

that is, when $EKh = 2Y$. Substituting this into expression (8-260) for the bending moment we obtain

$$M_{\max} = \frac{1}{12}EKbh^3 = \frac{bh^2}{12}EKh = \frac{bh^2}{6}Y \tag{8-261}$$

A further increase of the load will not change the stress in the outer fibers, which are at yield. This excess load will, however, be transferred to the neighboring fibers until the normal stress there also reaches the yield. This results in the intermediate stress distribution shown in Fig. 8-78. Recalculating the relation between the bending moment and the stress we get

$$\begin{aligned} M &= 2\int_{\zeta h/2}^{(1-\zeta)h/2} Yby\,dy + b\int_{-\zeta h/2}^{\zeta h/2} EKy^2\,dy \\ &= 2b\left[\int_{\zeta h/2}^{(1-\zeta)h/2} Yy\,dy + \int_0^{\zeta h/2} EKy^2\,dy\right] \\ &= 2b\left[Y\frac{1-\zeta}{2}h\frac{1-\zeta}{4}h + EK\frac{\zeta^3h^3}{24}\right] \\ &= \frac{bh^2}{12}[3(1-\zeta^2)Y + EK\zeta^3h] \end{aligned} \tag{8-262}$$

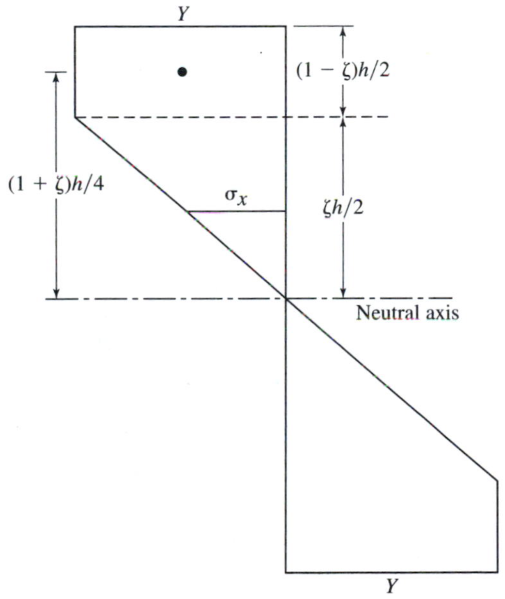


**Figure 8-78** Intermediate stress distribution

But now

$$|\sigma_x| = Y, \qquad \text{at } y = \pm\zeta h/2$$

or from Eq. (8-259),

$$2Y = EK\zeta h$$

so that

$$K = \frac{2Y}{E\zeta h}$$

Substituting this into the expression (8-262) for $M$ we get

$$M = \frac{bh^2}{12} Y[3(1-\zeta^2) + 2\zeta^2] = \frac{1}{12} Ybh^2(3-\zeta^2)$$

We keep increasing the load until no elastic region is left (Fig. 8-79). Now $\zeta = 0$, and the limiting bending moment corresponding to this fully plastic range is

$$M_0 = \tfrac{1}{4} Ybh^2$$

Thus a bending moment of this magnitude will cause collapse of the beam. It is now necessary to find when the collapse will occur. This is easy to determine for a statically determinate beam.

Consider, for example, a cantilever carrying at its tip a concentrated force $F$ (Fig. 8-80). Collapse will occur when the maximum bending moment

$$M_{\max} = -Fl$$

attains the yield value $M_0$, namely, $-Fl = M_0$. Therefore the collapse load is $F_0 = M_0/l$. More analytical effort is needed to analyze a statically indeterminate beam such

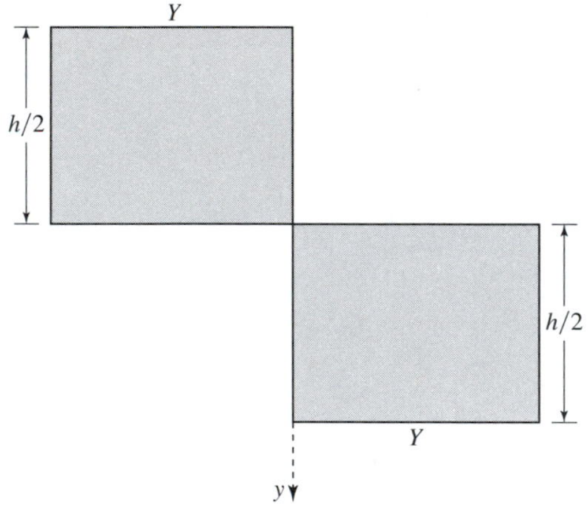


Figure 8-79 Fully plastic range

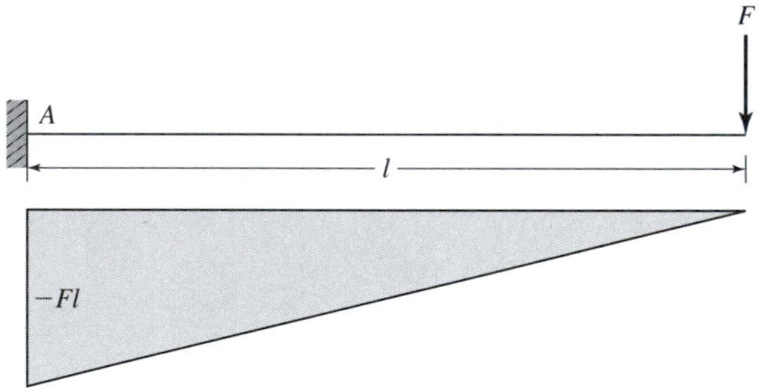


Figure 8-80 Collapse of cantilever carrying a concentrated force

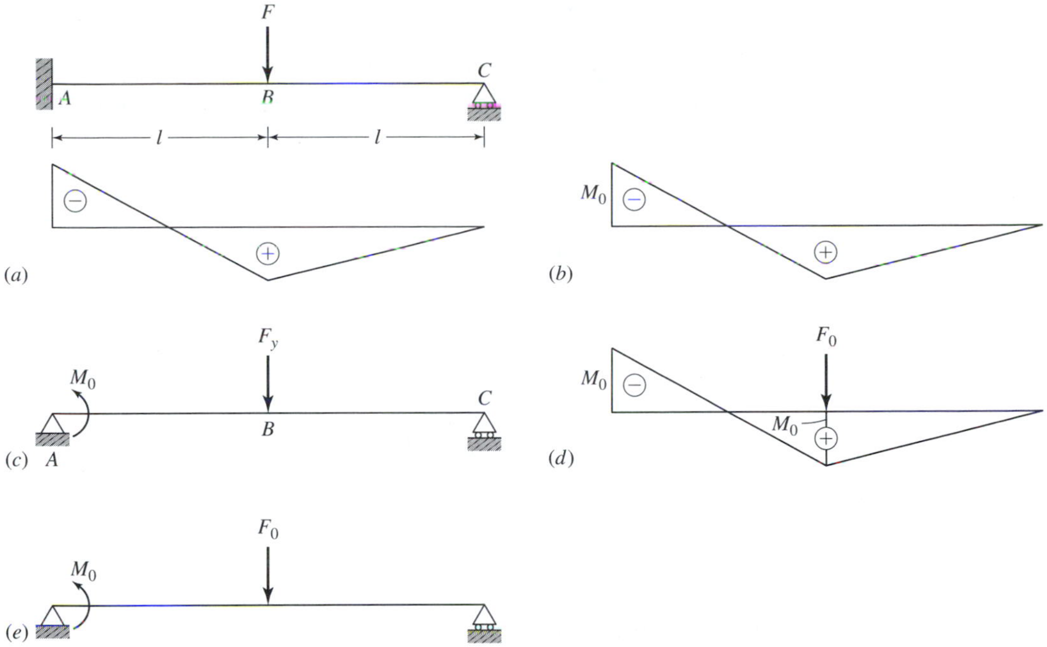


**Figure 8-81** Statically indeterminate beam, various stages

as the one shown, for example, in Fig. 8-81. We first determine the moment distribution in the fully elastic state (Fig. 8-81*a*). Since the maximum bending moment appears at $A$, that is where $M_0$ will occur first (Fig. 8-81*b*). Let the corresponding value of $F$ be $F_Y$. Hence the load of magnitude $F_Y$ will cause the formation of a plastic hinge at $A$. Unlike in the previous example, this hinge does not transform the beam into a mechanism, but merely replaces the existing structure by a simply supported beam (Fig. 8-81*c*). A further increase of the force $F$ beyond its value $F_Y$ causes an increase of the bending moments outside hinge $A$ until a second hinge forms, which occurs at $B$. Now we have a mechanism by which the collapse of the beam will take place. The moment diagram just preceding the collapse is shown in Fig. 8-81*d*. Considering the equilibrium of the beam in Fig. 8-81*e* and the expression for the bending moment at $B$, we get

$$2R_Al - M_0 - F_0 l = 0, \qquad R_A l - M_0 = M_0$$

Solving this system of equations we find that

$$R_A = \frac{2M_A}{l}, \qquad 4M_0 - M_0 - F_0 l = 0, \qquad F_0 = \frac{3M_0}{l}$$

Finally, let us analyze a simple frame carrying two concentrated forces (Fig. 8-82*a*). This frame is twice statically indeterminate, so that three plastic hinges are needed to turn it into a mechanism, thus causing collapse. The plastic hinges may appear at any of the points in which the bending moment reaches a local maximum. A hypothetical bending moment distribution is shown in Fig. 8-82*b*. Therefore each of the points 1, 2, 3, and 4 is

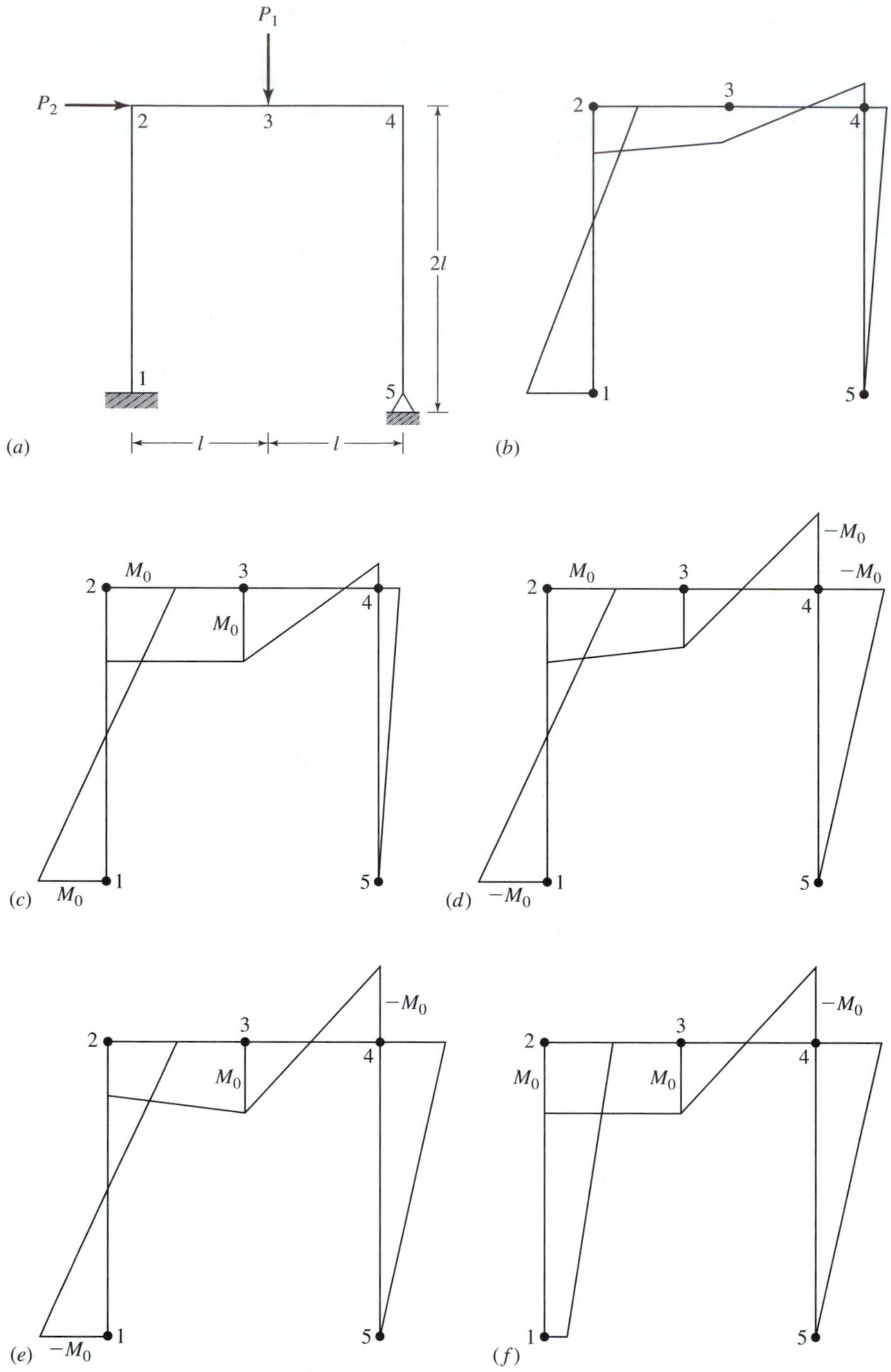


Figure 8-82

a candidate for the plastic hinge. There are four possible collapse mechanisms with plastic hinges formed at 1-2-3, 1-2-4, 1-3-4, and 2-3-4, but only one of these is correct. To find which one, we will consider each of the collapse modes separately. The bending moments at 1, 2, 3, and 4 are not independent, and the relations between them are common to all collapse modes under consideration. Denoting by $V$ and $H$ the vertical and horizontal reactions at node 5 we calculate the bending moments at all critical points. Thus,

$$M_4 = -2Hl, \qquad M_3 = -2Hl + Vl$$
$$M_2 = -2Hl + 2Vl - P_1 l, \qquad M_1 = 2Vl - P_1 l - 2P_2 l$$

By eliminating $H$ and $V$ we obtain

$$M_2 - 2M_3 + M_4 = -P_1 l, \qquad M_1 - 2M_3 + 2M_4 = -P_1 l - 2P_2 l \tag{8-263}$$

Let us first consider the collapse mechanism shown in Fig. 8-82*c*. We find that the moments at the plastic hinge should equal

$$M_1 = -M_0, \qquad M_2 = M_0, \qquad M_3 = M_0$$

Now from Eqs. (8-263) we determine $M_4$, $P_1$, and $P_2$,

$$\begin{aligned} M_4 &= 2M_3 - M_2 - P_1 l = M_0 - P_1 l \\ 2P_2 l &= -M_1 + 2M_3 - 2M_4 - P_1 l = M_0 + 2M_0 - 2M_0 + 2P_1 l - P_1 l \\ &= M_0 + P_1 l \end{aligned} \tag{8-264}$$

If $P_2 = \alpha P_1$, where $\alpha$ is known, then from the equations

$$M_4 = M_0 - P_1 l, \qquad 2P_1 \alpha l = M_0 + P_1 l \tag{8-265}$$

we find $P_1(2\alpha - 1)l = M_0$. Hence,

$$P_1 = \frac{M_0}{(2\alpha - 1)l}, \qquad M_4 = M_0 - \frac{M_0}{2\alpha - 1} = \frac{2(\alpha - 1)M_0}{2\alpha - 1} \tag{8-266}$$

Since $2(\alpha - 1) < 2\alpha - 1$, $M_4$ is less than $M_0$. Thus this mechanism is possible. For the mechanism shown in Fig. 8-82*d* we get

$$M_1 = -M_0, \qquad M_2 = M_0, \qquad M_4 = -M_0$$
$$2M_3 = M_4 + M_2 + P_1 l = -M_0 + M_0 + P_1 l = P_1 l$$
$$M_1 = -M_0 = 2M_3 - 2M_4 - P_1 l - 2P_1 l = P_1 l + 2M_0 - P_1 l - 2P_2 l = 2M_0 - 2P_2 l$$

Therefore, $3M_0 = 2P_2 l$, so that

$$P_2 = \frac{3M_0}{2l}, \qquad M_3 = \frac{P_1 l}{2} \tag{8-267}$$

If $P_2 = \alpha P_1$, then

$$P_1 \equiv \frac{P_2}{\alpha} = \frac{3M_0}{2l\alpha}, \qquad M_3 = \frac{P_1 l}{2} = \frac{3M_0}{4\alpha} \tag{8-268}$$

This mechanism is possible for $\alpha > 3/4$ because otherwise $M_3 > M_0$, which contradicts our assumptions. For the mechanism shown in Fig. 8-82*e* we have

$$M_1 = -M_0, \qquad M_3 = M_0, \qquad M_4 = -M_0$$

$$M_2 = 2M_3 - M_4 - P_1 l = 2M_0 + M_0 - P_1 l = 3M_0 - P_1 l$$

$$2P_2 l = -P_1 l - M_1 + 2M_3 - 2M_4 = -P_1 l + M_0 + 2M_0 + 2M_0 = -P_1 l + 5M_0$$

Hence, $2P_1 \alpha l = -P_1 l + 5M_0$ so that

$$P_1 = \frac{5M_0}{(2\alpha + 1)l}, \qquad M_2 = 3M_0 - \frac{5M_0}{2\alpha + 1} = \frac{6\alpha + 1}{2\alpha + 1} M_0 > M_0 \tag{8-269}$$

Since here $M_2 > M_0$, which contradicts our assumptions, the mechanism in Fig. 8-82*e* is not admissible.

Finally for the collapse mechanism shown in Fig. 8-82*f* we obtain

$$32M_2 = M_3 = M_0, \qquad M_4 = -M_0$$

$$P_1 l = -M_2 + 2M_3 - M_4 = -M_0 + 2M_0 + M_0 = 2M_0$$

so that

$$P_1 = \frac{2M_0}{l}, \qquad M_1 = 2M_3 - 2M_4 - P_1 l - 2P_2 l = 2M_0 + 2M_0 - 2M_0 - 2P_2 l$$
$$= 2M_0 - 2P_2 l$$

But $P_2 = \alpha P_1$. Therefore,

$$M_1 = 2M_0 - 2\alpha P_1 l = 2M_0 - 4M_0 \alpha = 2(1 - 2\alpha) M_0 \tag{8-270}$$

This is admissible as long as $|M_1| \leq M_0$, that is, as long as

$$\tfrac{1}{4} \leq \alpha \leq \tfrac{3}{4}$$

Limit analysis of more complicated structures requires advanced techniques (see, for example, [14]).

## 8-9 FOURIER SERIES AND APPLICATIONS

Fourier series not only constitute an important class of trial functions for application with the Rayleigh–Ritz method (see Section 8-10), but it can also be used directly to solve the differential equation for the deflection of a beam. Consider an arbitrary function $y(x)$ (Fig. 8-83). Assuming that it does not tend to infinity for $0 \leq x \leq l$, we

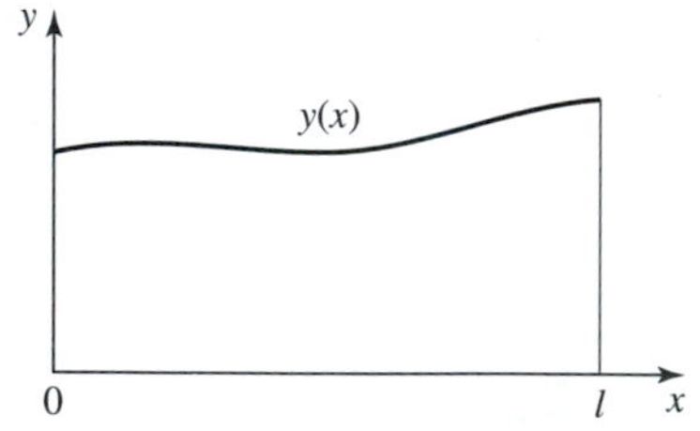


Figure 8-83

represent it as an infinite series,

$$y(x) = a_1 \sin \frac{\pi x}{l} + a_2 \sin \frac{2\pi x}{l} + \cdots = \sum_{n=1}^{\infty} a_n \sin \frac{n\pi x}{l} \tag{8-271}$$

or, alternatively, as

$$y(x) = b_0 + b_1 \cos \frac{\pi x}{l} + b_2 \cos \frac{2\pi x}{l} + \cdots = b_0 + \sum_{n=1}^{\infty} b_n \cos \frac{n\pi x}{l} \tag{8-272}$$

called *Fourier sine series* and *Fourier cosine series,* respectively. The coefficients

$$\begin{aligned} a_n &\equiv \frac{2}{l} \int_0^l y(x) \sin \frac{n\pi x}{l}\, dx, \qquad n = 1, 2, \ldots, \infty \\ b_0 &\equiv \frac{1}{l} \int_0^l y(x)\, dx \\ b_n &\equiv \frac{2}{l} \int_0^l y(x) \cos \frac{n\pi x}{l}\, dx, \qquad n = 1, \ldots, \infty \end{aligned} \tag{8-273}$$

are called *Fourier coefficients* of the function $y(x)$. It can be shown [26, pp. 75ff.] that the Fourier series of $y(x)$ converges to $y(x)$. This means that if we were able to represent the sum of the infinite series in closed form, then the result would equal *exactly* $y(x)$. When such a representation is impossible, then the convergence guarantees that increasing the number of terms taken into account increases the accuracy of the result. It is also crucial to note that sine and cosine series can be obtained for the same function. One of the important features of the sets of functions $\sin n\pi x/l$ and $\cos n\pi x/l$ is their *orthogonality,* which means that

$$\begin{aligned} \int_0^l \sin \frac{n\pi x}{l} \sin \frac{m\pi x}{l}\, dx = 0, &\qquad m \neq n (m, n = 1, 2, \ldots, \infty) \\ \int_0^l \cos \frac{n\pi x}{l} \cos \frac{m\pi x}{l}\, dx = 0, &\qquad m \neq n (m, n = 0, 1, \ldots, \infty) \end{aligned} \tag{8-274}$$

We begin with an example concerning the bending of a simply supported beam.

## EXAMPLE 8-10

Find the deflection $y(x)$ of the beam shown in Fig. 8-84.

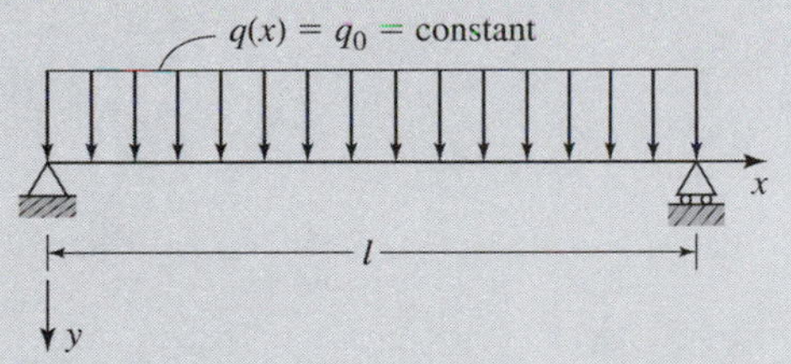


**Figure 8-84**

**SOLUTION**

The equation for bending is

$$EIy^{\text{iv}}(x) = q(x) \tag{a}$$

and the unknown deflection $y(x)$ must also satisfy the boundary conditions

$$y(0) = y(l) = 0, \qquad M(0) = EIy''(x)|_{x=0} = 0, \qquad M(l) = EIy''(x)|_{x=l} = 0 \tag{b}$$

First we expand $y(x)$ into a Fourier sine series,

$$y(x) = \sum_{n=1}^{\infty} y_n \sin\frac{n\pi x}{l} \tag{c}$$

where $y_n$ are Fourier coefficients to be determined,

$$y_n = \frac{2}{l}\int_0^l y(x)\sin\frac{n\pi x}{l}\,dx \tag{d}$$

The expansion into a sine series rather than a cosine series has been chosen because representation ($c$) identically satisfies all the boundary conditions ($b$). Next we also expand the load $q(x) = q_0$ into a Fourier sine series,

$$q(x) = \frac{2}{l}\sum_{n=1}^{\infty} q_n \sin\frac{n\pi x}{l} \tag{e}$$

and we calculate $q_n$ using the definition of Fourier coefficients,

$$\begin{aligned} q_n &= \frac{2}{l}\int_0^l q(x)\sin\frac{n\pi x}{l}\,dx = \frac{2q_0}{l}\int_0^l \sin\frac{n\pi x}{l}\,dx \\ &= -\frac{2q_0}{n\pi}\cos\frac{n\pi x}{l}\Big|_0^l = \frac{2q_0}{n\pi}(1-\cos n\pi) = \frac{2q_0}{n\pi}[1-(-1)^n] \end{aligned} \tag{f}$$

We now replace in Eq. ($a$) $y(x)$ and $q(x)$ by the series ($c$) and ($e$). Series ($c$) must be differentiated four times, which can be done "term by term" in view of the boundary conditions ($b$).* Therefore,

$$y^{\text{iv}} = \sum_{n=1}^{\infty}\left(\frac{n\pi}{l}\right)^4 \sin\frac{n\pi x}{l}$$

and Eq. ($a$) becomes

$$EI\sum_{n=1}^{\infty}\left(\frac{n\pi}{l}\right)^4 y_n \sin\frac{n\pi x}{l} = \frac{2q_0}{ln\pi}\sum_{n=1}^{\infty}[1-(-1)^n]\sin\frac{n\pi x}{l}$$

which can be written as

$$\sum_{n=1}^{\infty}\left\{EI\left(\frac{n\pi}{l}\right)^4 y_n - \frac{2q_0}{ln\pi}[1-(-1)^n]\right\}\sin\frac{n\pi x}{l} = 0 \tag{g}$$

---

*As will be explained later, the differentiation "term by term" is not always correct.

Expression ($g$) is the Fourier series of the function on the right-hand side. Since this function is identically equal to zero, its Fourier coefficient is also zero. This implies that the term in braces in Eq. ($g$) is zero for any value of $n$. Hence,

$$EI\left(\frac{n\pi}{l}\right)^4 y_n = \frac{2q_0}{ln\pi}[1-(-1)^n], \qquad n = 1, 2, \ldots, \infty$$

From here

$$y_n = \frac{4q_0 l^3}{EIn^5\pi^5}, \qquad n = 1, 3, \ldots, \infty \tag{h}$$

and the Fourier series for the deflection, obtained from Eq. ($c$), is

$$y(x) = \frac{4q_0 l^3}{EI\pi^5} \sum_{n=1,3,\ldots}^{\infty} \frac{\sin(n\pi x/l)}{n^5} \tag{i}$$

This expression looks differently than the closed-form solution to the differential equation ($a$). However, after some manipulations we can put Eq. ($i$) in the same form. Using tables (see [11, position 1.443.7]) we find that

$$\sum_{n=1}^{\infty} \frac{\sin(n\pi x/l)}{n^5} = \frac{\pi^4}{90}\left(\frac{\pi x}{l}\right) - \frac{\pi^2}{36}\left(\frac{\pi x}{l}\right)^3 + \frac{\pi}{48}\left(\frac{\pi x}{l}\right)^4 - \frac{1}{240}\left(\frac{\pi x}{l}\right)^5 \tag{j}$$

Next we write

$$\sum_{n=1,3,\ldots}^{\infty} \frac{\sin(n\pi x/l)}{n^5} = \sum_{n=1}^{\infty} \frac{\sin(n\pi x/l)}{n^5} - \sum_{n=2,4,\ldots}^{\infty} \frac{\sin(n\pi x/l)}{n^5} \tag{k}$$

Moreover, replacing $n$ by $2k$ on the right-hand side of the last series, we obtain

$$\sum_{n=2,4,\ldots}^{\infty} \frac{\sin(n\pi x/l)}{n^5} = \sum_{2k=2,4,\ldots}^{\infty} \frac{\sin(2k\pi x/l)}{(2k)^5} = \frac{1}{32}\sum_{k=1}^{\infty} \frac{\sin(2k\pi x/l)}{k^5} \tag{l}$$

By using the result ($j$) we get

$$\sum_{k=1}^{\infty} \frac{\sin(2k\pi x/l)}{k^5} = \frac{\pi^4}{90}\left(\frac{2\pi x}{l}\right) - \frac{\pi^2}{36}\left(\frac{2\pi x}{l}\right)^3 + \frac{\pi}{48}\left(\frac{2\pi x}{l}\right)^4 - \frac{1}{240}\left(\frac{2\pi x}{l}\right)^5 \tag{m}$$

Substituting expressions ($j$), ($l$), and ($m$) into Eq. ($k$), we obtain the same expression as we would by integrating the differential equation ($a$),

$$y(x) = \frac{\pi^4}{36}\frac{\pi x}{l} - \frac{\pi^2}{48}\left(\frac{\pi x}{l}\right)^3 + \frac{\pi}{96}\left(\frac{\pi x}{l}\right)^4$$

We may use this to obtain the well-known result

$$y_{\max} = y|_{x=l/2} = \frac{q_0 l^3}{EI}\left\{\frac{1}{96}\left(\frac{1}{2}\right) - \frac{1}{48}\left(\frac{1}{2}\right)^3 + \frac{1}{96}\left(\frac{1}{2}\right)^4\right\} = \frac{5q_0 l^3}{384EI}$$

Now we consider a general case of bending a straight beam with arbitrary boundary conditions under an arbitrary load $q(x)$. The equation for deflection is again

$$EIy^{\text{iv}}(x) = q(x) \tag{8-275}$$

Employing the expansion into sine series (we could choose a cosine series as well), we assume

$$y(x) = \sum_{n=1}^{\infty} y_n \sin \frac{n\pi x}{l} \tag{8-276}$$

where the coefficients $y_n$ are unknown. In an attempt to represent both sides of the differential equation in the same form, we also expand $q(x)$ into a sine series,

$$q(x) = \sum_{n=1}^{\infty} q_n \sin \frac{n\pi x}{l} \tag{8-277}$$

where the Fourier coefficients $q_n$ are found from

$$q_n = \frac{2}{l} \int_0^l q(x) \sin \frac{n\pi x}{l}\, dx, \qquad n = 1, 2, \ldots, \infty \tag{8-278}$$

Next $y(x)$, given now by Eq. (8-276), must be differentiated, and this must be done with particular care. It seems quite obvious and natural to calculate the first derivative of

$$f(x) = \sum_{n=1}^{\infty} f_n \sin \frac{n\pi x}{l} \tag{8-279}$$

as

$$f'(x) = \sum_{n=1}^{\infty} \frac{n\pi}{l} f_n \cos \frac{n\pi x}{l} \tag{8-280}$$

that is, differentiating "term by term." But this is often incorrect. It can be shown [26, p. 138] that the proper expression for the derivative of the series (8-279) is

$$f'(x) = \frac{1}{l}[f(l) - f(0)] + \sum_{n=1}^{\infty} \left[ \frac{n\pi}{l} f_n - \frac{2}{l} f(0) + \frac{2}{l}(-1)^n f(l) \right] \cos \frac{n\pi x}{l} \tag{8-281}$$

This reduces to the simpler expression, Eq. (8-280), only when

$$f(0) = f(l) = 0 \tag{8-282}$$

The cosine series, on the other hand, can always be differentiated "term by term" that is, when

$$z(x) = z_0 + \sum_{n=1}^{\infty} z_n \cos \frac{n\pi x}{l} \tag{8-283}$$

then

$$z'(x) = -\sum_{n=1}^{\infty} \frac{n\pi}{l} z_n \sin \frac{n\pi x}{l} \tag{8-284}$$

These expressions will now be employed to solve the differential equation (8-275). At this point we do not have to specify the boundary conditions, and we obtain

$$y''(x) = \sum_{n=1}^{\infty} A_n \sin \frac{n\pi x}{l} \tag{8-285}$$

where

$$A_n = -\frac{n\pi}{l}\left[\frac{n\pi}{l} y_n - \frac{2}{l} y(0) + \frac{2}{l}(1-)^n y(l)\right] \tag{8-286}$$

Differentiating Eq. (8-285) again and making appropriate modifications in Eq. (8-281), we find that

$$y'''(x) = \frac{1}{l}[y''(l) - y''(0)] + \sum_{n=1}^{\infty}\left[\frac{n\pi}{l} A_n - \frac{2}{l} y''(0) + \frac{2}{l}(-1)^n y''(l)\right] \cos \frac{n\pi x}{l} \tag{8-287}$$

Term-by-term differentiation of the expression (8-287) yields

$$y^{\text{iv}}(x) = -\sum_{n=1}^{\infty} \frac{n\pi}{l}\left[\frac{n\pi}{l} A_n - \frac{2}{l} y''(0) + \frac{2}{l}(-1)^n y''(l)\right] \sin \frac{n\pi x}{l} \tag{8-288}$$

Substituting this into the differential equation (8-275), replacing $A_n$ by Eq. (8-286), and relating the second derivatives of $y(x)$ to the bending moments,

$$M(0) = -EIy''(0), \qquad M(l) = -EIy''(l) \tag{8-289}$$

we finally get

$$EI \sum_{n=1}^{\infty}\left\{\frac{n^4\pi^4}{l^4} y_n + \frac{2}{l}\frac{n^3\pi^3}{l^3}[(-1)^n y(l) - y(0)] + \frac{2}{l}\frac{1}{EI}\frac{n\pi}{l}[(-1)^n M(l) - M(0)\right\} \times \sin \frac{n\pi x}{l} = \sum_{n=1}^{\infty} q_n \sin \frac{n\pi x}{l} \tag{8-290}$$

When the series on the right-hand side is moved to the left, the following expression is obtained:

$$\sum_{n=1}^{\infty} B_n \sin \frac{n\pi x}{l} = 0 \tag{8-291}$$

implying that $B_n$ is a Fourier coefficient of a function identically equal to 0. Therefore $B_n = 0$, and replacing $B_n$ by the terms appearing in Eq. (8-290) we find the following expression for the Fourier coefficient $y_n$ of the unknown deflection $y(x)$:

$$y_n = \left\{\frac{q_n}{EI} - \frac{2}{l}\frac{n\pi}{l}\frac{(-1)^n M(l) - M(0)}{EI} - \frac{2}{l}\frac{n^3\pi^3}{l^3}[(-1)^n y(l) - y(0)]\right\} \Bigg/ \frac{n^4\pi^4}{l^4} \tag{8-292}$$

Consequently the Fourier series for the deflection becomes

$$y(x) = \sum_{n=1}^{\infty} y_n \sin \frac{n\pi x}{l} \tag{8-293}$$

It is evident from Eq. (8-292) that the expression (8-293) represents the solution only when both the displacements and the bending moments are known at both ends. This occurs, for example, for a simply supported beam with the boundary conditions

$$y(0) = y(l) = M(0) = M(l) = 0 \tag{8-294}$$

so that

$$y_n = \frac{q_n}{EI(n^4\pi^4/l^4)} \sin \frac{n\pi x}{l} \tag{8-295}$$

Therefore the deflection equals

$$y(x) = \frac{1}{EI} \sum_{n=1}^{\infty} \frac{q_n}{n^4\pi^4/l^4} \sin \frac{n\pi x}{l} \tag{8-296}$$

Once $q_n$ is known, this infinite series can sometimes be replaced by a closed-form expression (see [11], [27, pp. 42–45], or [22]) as in Example 8-10. This procedure becomes more involved with one or more of the boundary values of $y(x)$ and $M(x)$, appearing in Eq. (8-292), unknown. Consider, for example, the beam shown in Fig. 8-85. Here the boundary conditions are

$$y(0) = y(l) = 0, \qquad y'(0) = 0, \qquad M(l) = 0 \tag{8-297}$$

With Eqs. (8-292) and (8-293) the deflection becomes

$$y(x) = \sum_{n=1}^{\infty} \frac{q_n + \dfrac{2n\pi}{l^2} M(0)}{EI \dfrac{n^4\pi^4}{l^4}} \sin \frac{n\pi x}{l} \tag{8-298}$$

In order to determine the unknown moment $M(0)$, we satisfy the boundary condition $y'(0) = 0$. To do so, we must differentiate the series (8-298). Fortunately, since $y(x)$ vanishes at both ends, term-by-term differentiation is correct and we find that

$$y'(x) = \frac{1}{EI} \sum_{n=1}^{\infty} \frac{q_n + \dfrac{2n\pi}{l^2} M(0)}{\dfrac{n^3\pi^3}{l^3}} \cos \frac{n\pi x}{l} \tag{8-299}$$

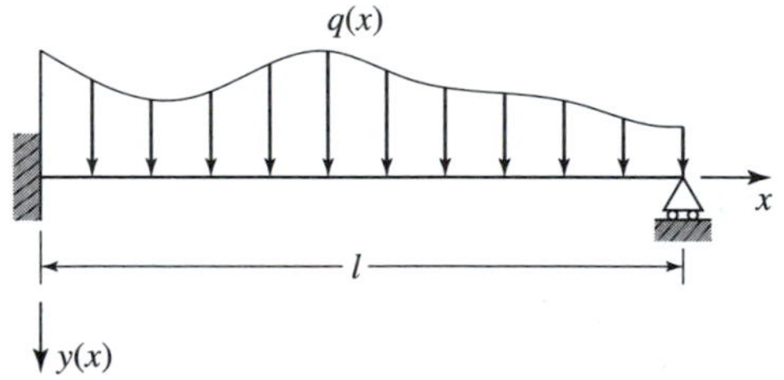


Figure 8-85

It follows from this that

$$y'(0) = \frac{1}{EI} \sum_{n=1}^{\infty} \left[ \frac{q_n}{n^3\pi^3/l^3} + \frac{2M(0)}{n^2\pi^2/l^2} \right] = 0$$

and consequently,

$$M(0) = -\frac{l^2}{\pi} \frac{\sum_{n=1}^{\infty} \frac{q_n}{n^3}}{\sum_{n=1}^{\infty} \frac{1}{n^2}} \tag{8-300}$$

Furthermore,

$$\sum_{n=1}^{\infty} \frac{1}{n^2} = \frac{\pi^2}{6} \tag{8-301}$$

so that finally

$$M(0) = -\frac{6l^2}{\pi^3} \sum_{n=1}^{\infty} \frac{q_n}{n^3} \equiv -\frac{6l^2}{\pi^3} \sum_{m=1}^{\infty} \frac{q_m}{m^3} \tag{8-302}$$

where the summation index $n$ has been replaced by $m$ to avoid confusion when substituting this expression into Eq. (8-298),

$$y(x) = \frac{l^4}{EI\pi^4} \sum_{n=1}^{\infty} \frac{q_n}{n^4} \sin\frac{n\pi x}{l} - \frac{12l^4}{EI\pi^6} \sum_{m=1}^{\infty} \frac{q_m}{m^3} \sum_{n=1}^{\infty} \frac{\sin(n\pi x/l)}{n^3} \tag{8-303}$$

The infinite series appearing here can sometimes be replaced by closed-form expressions, depending on the function $q(x)$. We note that the same procedure is applicable for other, more complicated boundary conditions and for other beam structures, such as beams on elastic foundations and curved beams.

## 8-10 APPROXIMATE METHODS IN THE ANALYSIS OF BEAMS

### 8-10-1 Finite Differences—Examples

Let us consider a simply supported beam of constant stiffness $EI$ carrying a uniform load $q$ = constant. (See [21] for a different way of solving this problem.) The differential equation for the deflection of this beam [21, p. 273]

$$EI\frac{d^4y}{dx^4} = q \tag{8-304}$$

is replaced by a difference equation (see Chapter 6),

$$Y_{i-2} - 4Y_{i-1} + 6Y_i - 4Y_{i+1} + Y_{i+2} = \frac{q_i h^4}{EI}, \qquad i = 0, 1, \ldots, n \tag{8-305}$$

where $q_i = q_0$ = constant. Because of the symmetry of the problem we put the origin of the coordinate system at the midspan (Fig. 8-86$a$) and consider only the portion of the

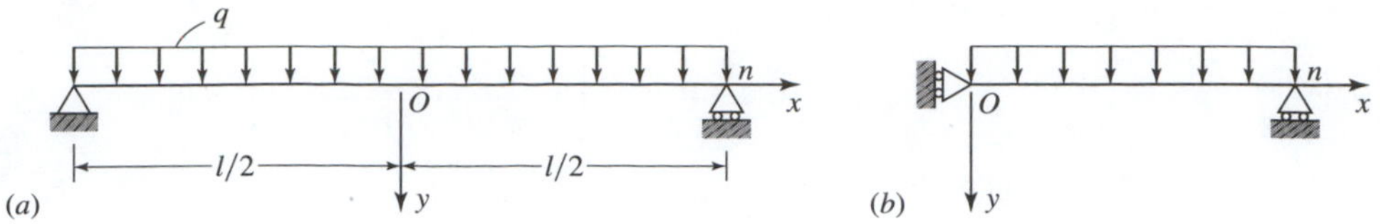


Figure 8-86

beam shown in Fig. 8-86*b*. The symmetry implies that

$$y'(0) = 0, \qquad Q(0) = EIy'''(0) = 0 \tag{8-306}$$

In addition to the boundary conditions (8-306) we also have the boundary conditions at $x = l/2$,

$$y(l/2) = 0, \qquad M(l/2) = EIy''(l/2) = 0 \tag{8-307}$$

Conditions (8-306) when applied at 0 as the pivotal point result in

$$Y_{-1} = Y_1, \qquad Y_{-2} - 2Y_{-1} = Y_2 - 2Y_1$$

Therefore,

$$Y_{-1} = Y_1, \qquad Y_{-2} = Y_2 \tag{8-308}$$

Boundary conditions (8-307) imply that at the pivotal point $x_n = l/2$ we have

$$Y_n = 0, \qquad Y_{n-1} - 2Y_n + Y_{n+1} = 0 \tag{8-309}$$

Together we have $n+5$ equations: $n+1$ Eq. (8-305), four Eqs. (8-308), and Eqs. (8-309). The number of unknowns also equals $n+5$. The smallest index of the unknown in Eq. (8-305) is $-2$ for $i = 0$ whereas the largest occurs for $i = n$ and equals $n+2$. Ultimately our equations become

$$\begin{aligned}
Y_{-2} - 4Y_{-1} + 6Y_0 - 4Y_1 + Y_2 &= \frac{\alpha}{(2n)^4} \\
Y_{-1} - 4Y_0 + 6Y_1 - 4Y_2 + Y_3 &= \frac{\alpha}{(2n)^4} \\
\cdots\cdots\cdots\cdots\cdots\cdots\cdots\cdots \\
Y_{n-3} - 4Y_{n-2} + 6Y_{n-1} - 4Y_n + Y_{n+1} &= \frac{\alpha}{(2n)^4} \\
Y_{n-2} - 4Y_{n-1} + 6Y_n - 4Y_{n+1} + Y_{n+2} &= \frac{\alpha}{(2n)^4} \\
Y_{-1} - Y_1 &= 0 \\
Y_{-2} - 2Y_{-1} + 2Y_1 - Y_2 &= 0 \\
Y_n &= 0 \\
Y_{n-1} - 2Y_n + Y_{n+1} &= 0
\end{aligned} \tag{8-310}$$

where

$$\alpha = \frac{ql^4}{EI} \tag{8-311}$$

The solution to this system gives the deflection $Y_i$ at all points of division. Specifically, we find that

$$\max Y = Y_0$$

We now easily determine the bending moment $M_2 = EIY_i''$ and the shear force $Q_i = EIY_1'''$ at any point $x_i$. This is achieved by using the expressions for $Y''$ and for $Y'''$ derived in Chapter 6. Thus,

$$\begin{aligned} M_i &= -\frac{EI}{h^2}(Y_{i-1} - 2Y_i + Y_{i+1}) = -\frac{4EIn^2}{l^2}(Y_{i-1} - 2Y_i + Y_{i+1}) \\ Q_i &= -\frac{EI}{2h^3}(-Y_{i-2} + 2Y_{i-1} - 2Y_{i+1} + Y_{i+2}) \\ &= -\frac{4EIn^3}{l^3}(-Y_{i-2} + 2Y_{i-1} - 2Y_{i+1} + Y_{i+2}) \end{aligned} \tag{8-312}$$

Equations (8-310) have been solved for $n \leq 20$. Table 8-5 compares the exact results with the results obtained by using the finite difference method for various $n$.

Next consider the bending of a finite beam with variable stiffness on an elastic foundation carrying a distributed, nonuniform load, symmetric about $x = 0$ (Fig. 8-87). In

TABLE 8-5

| $n$ | $\frac{EI}{ql^4} y_{max}$ | $\frac{M_{max}}{ql^2}$ | $\frac{Q_{max}}{ql}$ |
|---|---|---|---|
| 1 | 0.01563 | 0.1250 | −0.5000 |
| 2 | 0.01367 | 0.1250 | −0.5000 |
| 3 | 0.01331 | 0.1250 | −0.5000 |
| 5 | 0.01313 | 0.1250 | −0.5000 |
| 10 | 0.01305 | 0.1250 | −0.5000 |
| 20 | 0.01303 | 0.1250 | −0.5000 |
| Exact | 0.01302 | 0.1250 | −0.5000 |

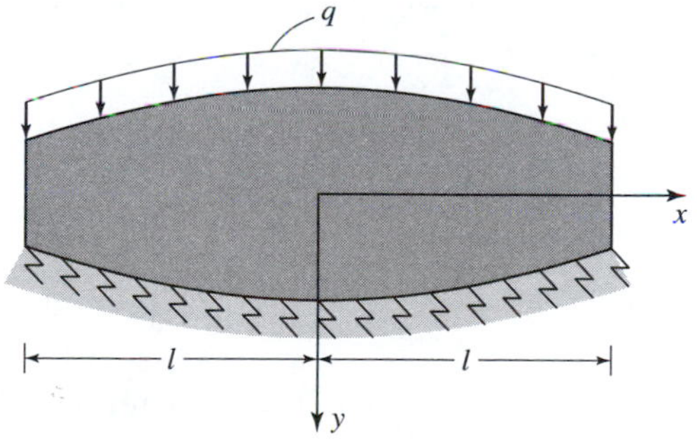


Figure 8-87

the following we use the data given by Collatz [7]. The differential equation for bending a beam of variable cross section on an elastic foundation takes the form [7, p. 152]

$$\frac{d^2}{dx^2}\left[EI(x)\frac{d^2y}{dx^2}\right] + ky = q(x) \tag{8-313}$$

The symmetry of the problem allows us to consider only one-half of the beam ($x > 0$) and implies again, like in the previous example, that

$$y'(0) = 0, \qquad y'''(0) = 0 \tag{8-314}$$

The boundary conditions at the end $x = 1$ are

$$y''(l) = 0, \qquad y'''(l) = 0 \tag{8-315}$$

Expanding Eq. (8-313) and assuming that, for example [7, p. 152],

$$EI(x) = \left(2 - \frac{x^2}{l^2}\right)EI_0, \qquad q(x) = \left(2 - \frac{x^2}{l^2}\right)q_0 \tag{8-316}$$

where $I_0$ and $q_0$ are constant, we obtain

$$EI_0\left(2 - \frac{x^2}{l^2}\right)\frac{d^4y}{dx^2} - \frac{4EI_0}{l^2}x\frac{d^3y}{dx^3} - \frac{2EI_0}{l^2}\frac{d^2y}{dx^2} + ky = \left(2 - \frac{x^2}{l^2}\right)q_0 \tag{8-317}$$

By putting now

$$\xi = \frac{x}{l}, \qquad \eta = \frac{EI_0}{q_0 l^4}y, \qquad K = \frac{kl^4}{EI_0} \tag{8-318}$$

we nondimensionalize Eq. (8-317), which then becomes

$$(2 - \xi^2)\frac{d^4\eta}{d\xi^4} - 4\xi\frac{d^3\eta}{d\xi^3} - 2\frac{d^2\eta}{d\xi^2} + K\eta = 2 - \xi^2 \tag{8-319}$$

By using the expressions from Chapter 6 and noting that the nondimensional quotient $h$ now becomes

$$h = \frac{1}{n}$$

and that the coordinate $\xi_i$ of a point can be expressed in the form

$$\xi_i = ih$$

we write the difference equation corresponding to Eq. (8-319) as

$$\begin{aligned}&\frac{1}{h^4}(2 - i^2h^2)(\eta_{i-2} - 4\eta_{i-1} + 6\eta_i - 4\eta_{i+1} + \eta_{i+2})\\ &\quad - \frac{2}{h^3}ih(-\eta_{i-2} + 2\eta_{i-1} - 2\eta_{i+1} + \eta_{i+2})\\ &\quad - \frac{2}{h^2}(\eta_{i-1} - 2\eta_i + \eta_{i+1}) + K\eta_i = 2 - i^2h^2\end{aligned} \tag{8-320}$$

TABLE 8-6*

| Deflection | $N = 2$ | $N = 3$ | $N = 4$ |
|---|---|---|---|
| $y_0$ | 0.04518 | 0.04524 | 0.04525 |
| $y_1$ | 0.04095 | 0.04354 | 0.04436 |
| $y_2$ | 0.03021 | 0.03884 | 0.04180 |
| $y_3$ | — | 0.03223 | 0.03798 |
| $y_4$ | — | — | 0.03349 |

*The first and last entries in each column represent the deflection at the symmetry line and at the end cross section, respectively.

This equation can be simplified to

$$[2 - ih^2(i-2)]\eta_{i-2} + [-8 + 2(2i^2 - 2i - 1)h^2]\eta_{i-1} + [12 - 2(3i^2 - 2)h^2 + Kh^4]\eta_i + [-8 + 2(2i^2 + 2i - 1)h^2]\eta_{i+1} + [2 - ih^2(i+2)]\eta_{i+2} = (2 - i^2h^2)h^4 \tag{8-321}$$

The symmetry conditions (8-314) now have the same form as in the previous example [see Eq. (8-313)],

$$\begin{aligned} \eta_{-1} - \eta_1 &= 0 \\ \eta_{-2} - 2\eta_{-1} - \eta_2 &= 0 \end{aligned} \tag{8-322}$$

whereas the boundary conditions (8-315) become

$$\begin{aligned} \eta_{n-1} - 2\eta_n + \eta_{n+1} &= 0 \\ -\eta_{n-2} + 2\eta_{n-1} - 2\eta_{n+1} + \eta_{n+2} &= 0 \end{aligned} \tag{8-323}$$

The results were obtained using an interactive Fortran program 8-4. The deflections for various $n$ (the number of subdivisions) are given in Table 8-6.

## 8-10-2 Rayleigh–Ritz Method—Examples

As we have learned from Chapter 5, the Rayleigh–Ritz method is one of the most widely used methods in stress analysis. It can be applied directly to the differential equation, or it can be used as a tool in a numerical method (such as the finite element method). Once the expression for the potential energy of the beam is determined, the only difficulty in implementing the method is a judicious selection of the admissible displacement functions (trial functions). Basically we can choose any set of functions satisfying the kinematic boundary conditions, but then there is no guarantee of the convergence of the results. It is therefore advisable to choose as admissible functions any set of orthogonal functions such as sines, cosines, or orthogonal polynomials. This selection would virtually guarantee an improvement in the results with the increase of the number of the terms in the assumed function.

## EXAMPLE 8-11

Let us apply the Rayleigh–Ritz method to find the deflection of a finite beam on an elastic foundation (Fig. 8-88).

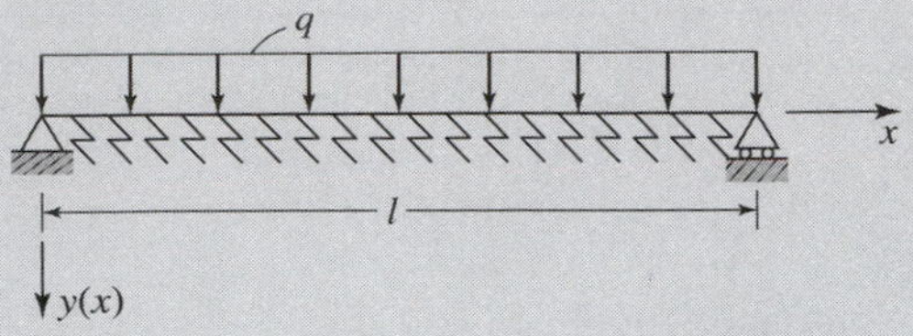


**Figure 8-88**

## SOLUTION

The potential energy is now (see Chapter 5)

$$\Pi = \frac{EI}{2}\int_0^l (w'')^2\,dx + \frac{1}{2}k\int_0^l w^2\,dx - q\int_0^l w\,dx \tag{8-324}$$

where the second integral represents the work done by the reaction forces of the foundation. The diplacement is assumed in the form of the following trigonometric series, which is a finite approximation to a Fourier series (see Section 8-8),

$$w(x) = a_1 \sin\frac{\pi x}{l} + a_3 \sin\frac{3\pi x}{l} + \cdots = \sum_{n=1,3,\ldots}^{N} a_n \sin\frac{n\pi x}{l} \tag{8-325}$$

Each term in this expression satisfies the kinematic boundary conditions $w(0) = w(l) = 0$ as well as the condition of the symmetry of the deflection implied by the symmetry of the system and of the load with respect to $x = l/2$. Substituting the expression (8-325) for $w(x)$ into Eq. (8-324), we obtain

$$\Pi = \frac{EI}{2}\int_0^l \left(\sum_{n=1,3}^{N} \frac{n^2\pi^2}{l^2} a_n \sin\frac{n\pi x}{l}\right)^2 dx + \frac{1}{2}k\int_0^l \left(\sum_{n=1,3}^{N} a_n \sin\frac{n\pi x}{l}\right)^2 dx$$
$$- q\int_0^l \sum_{n=1,3}^{N} a_n \sin\frac{n\pi x}{l}\,dx \tag{8-326}$$

The conditions for the minimum of potential energy become

$$\frac{\partial \Pi}{\partial a_n} = 0, \qquad n = 1, 3, \ldots, N \tag{8-327}$$

Note that although differentiation and integration are here interchangeable, the computational effort is always reduced when the differentiation is performed first. To avoid misunderstandings, we represent $\Pi$ in the following form:

$$\Pi = \frac{EI}{2}\int_0^l \left(\sum_{p=1,3}^{N} \frac{p^2\pi^2}{l^2} a_p \sin\frac{p\pi x}{l}\right)\left(\sum_{m=1,3}^{N} \frac{m^2\pi^2}{l^2} a_m \sin\frac{m\pi x}{l}\right) dx$$
$$+ \frac{1}{2}k\int_0^l \left(\sum_{p=1,3}^{N} a_p \sin\frac{p\pi x}{l}\right)\left(\sum_{m=1,3}^{N} a_m \sin\frac{m\pi x}{l}\right) dx$$
$$- q\int_0^l \sum_{p=1,3}^{N} a_p \sin\frac{p\pi x}{l}\, dx \tag{8-328}$$

Differentiating $\Pi$ with respect to $a_n$ we now notice that $a_n$ appears once in each of the series, and therefore,

$$\frac{\partial \Pi}{\partial a_n} = \frac{EI}{2}\int_0^l \left[\frac{n^2\pi^2}{l^2}\sin\frac{n\pi x}{l}\left(\sum_{m=1,3}^{N} \frac{m^2\pi^2}{l^2} a_m \sin\frac{m\pi x}{l}\right)\right.$$
$$\left. + \frac{n^2\pi^2}{l^2}\sin\frac{n\pi x}{l}\left(\sum_{p=1,3}^{N} \frac{p^2\pi^2}{l^2} a_p \sin\frac{p\pi x}{l}\right)\right] dx$$
$$+ \frac{1}{2}k\int_0^l \left[\sin\frac{n\pi x}{l}\left(\sum_{m=1,3}^{N} a_m \sin\frac{m\pi x}{l}\right) + \sin\frac{n\pi x}{l}\left(\sum_{p=1,3}^{N} a_p \sin\frac{p\pi x}{l}\right)\right] dx$$
$$- q\int_0^l \sin\frac{n\pi x}{l}\, dx = 0, \qquad n = 1, 3, \ldots, N$$

By using the fact that there are two sums under each integral, consisting of identical terms, and that

$$\int_0^l \sin\frac{n\pi x}{l}\sin\frac{m\pi x}{l}\, dx = \begin{cases} 0, & m \neq n \\ \dfrac{l}{2}, & m = n \end{cases} \tag{8-329}$$

we see that after the integration all sums disappear. This leads to the following expression:

$$\frac{\partial \Pi}{\partial a_n} = \frac{EIl}{2}\frac{n^4\pi^4}{l^4} a_n + \frac{kl}{2} a_n - \frac{ql}{n\pi}[1 - (-1)^n] = 0 \tag{8-330}$$

Therefore,

$$a_n = \frac{4q}{n\pi\left(EI\dfrac{n^4\pi^4}{l^4} + k\right)}, \qquad n = 1, 3, \ldots, N \tag{8-331}$$

and consequently the deflection curve is given by

$$w(x) = \frac{4q}{\pi}\sum_{n=1,3}^{N}\frac{\sin\dfrac{n\pi x}{l}}{n\left(EI\dfrac{n^4\pi^4}{l^4}+k\right)} \equiv \frac{4ql^4}{EI\pi^5}\sum_{n=1,3}^{N}\frac{\sin\dfrac{n\pi x}{l}}{n\left(n^4+\dfrac{kl^4}{EI\pi^4}\right)} \tag{8-332}$$

This series rapidly converges to the exact result.

## EXAMPLE 8-12

In the previous example we selected trigonometric series as a representation of the unknown deflection. Let us apply here a polynomial trial function to determine the deflection of the beam shown in Fig. 8-89,

$$w(x) = a_0 + a_1x + a_2x^2 + a_3x^3 + a_4x^4 \tag{8-333}$$

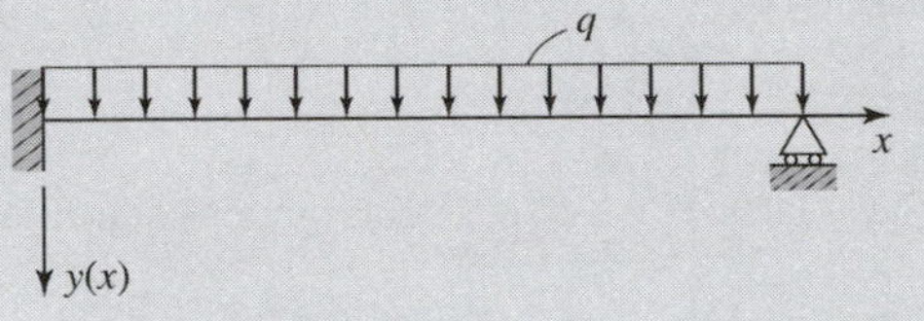


Figure 8-89

## SOLUTION

We require Eq. (8-333) to satisfy all the kinematic boundary conditions, which are

$$w(0) = 0, \qquad w'(0) = 0, \qquad w(l) = 0$$

This results in

$$a_0 = 0, \qquad a_1 = 0, \qquad a_2l^2 + a_3l^3 + a_4l^4 = 0$$

or

$$a_0 = 0, \qquad a_1 = 0, \qquad a_2 = -a_3l - a_4l^2$$

and consequently,

$$w(x) = a_3(x^3 - lx^2) + a_4(x^4 - l^2x^2) \tag{8-334}$$

The potential energy is

$$\Pi = \frac{EI}{2}\int_0^l (w'')^2\,dx - q\int_0^l w\,dx$$
$$= \frac{EI}{2}\int_0^l [a_3(6x-2l)+a_4(12x^2-2l^2)]^2\,dx$$
$$-q\int_0^l [a_3(x^3-lx^2)+a_4(x^4-l^2x^2)]\,dx \tag{8-335}$$

Thus the conditions that $\Pi$ take on the minimum value become

$$\frac{\partial\Pi}{\partial a_3} = EI\int_0^l (6x-2l)[a_3(6x-2l)+a_4(12x^2-2l^2)]\,dx$$
$$-q\int_0^l (x^3-lx^2)\,dx = 0$$
$$\frac{\partial\Pi}{\partial a_4} = EI\int_0^l (12x^2-2l^2)[a_3(6x-2l)+a_4(12x^2-2l^2)]\,dx$$
$$-q\int_0^l (x^4-l^2x^2)\,dx = 0$$

resulting in the following system of algebraic equations:

$$4EIa_3l^3 + 8EIa_4l^4 + \frac{ql^4}{12} = 0, \qquad 8EIa_3l^4 + \frac{84EIl^5}{5}a_4 + \frac{2ql^5}{15} = 0 \tag{8-336}$$

Solving these we get

$$a_3 = -\frac{5ql}{48EI} = -0.104166\frac{ql}{EI}, \qquad a_4 = \frac{q}{24EI} = 0.0416666\frac{q}{EI} \tag{8-337}$$

and the deflection becomes

$$w(x) = -\frac{ql}{24EI}\left[2.5(x^3-lx^2) - \frac{1}{l}(x^4-l^2x^2)\right]$$
$$\equiv \frac{q}{48EI}(2x^4-5x^3l+3x^2l^2) \tag{8-338}$$

which is identical to the exact result [5, p. 457]. This equivalence of the results is not a coincidence. We have chosen as our trial function a fourth-degree polynomial, which happens to be the form of the exact solution. Had we chosen a polynomial of a higher degree instead, we would have found that all the coefficients beyond $a_4$ were zero. On the other hand, a choice of a third-degree polynomial would give an approximate result.

## 8-11 PIEZOELECTRIC BEAMS

In Chapter 4 it was seen that when a piezoelectric material is placed in an electric field, the material undergoes expansion or contraction as well as shear. Which of these modes of deformation actually occurs in a particular piezoelectric material depends on the details of the piezoelectric strain matrix $\boldsymbol{d}$ given by Eq. (4-194), and the electric field vector $\boldsymbol{E}$. However, as was seen in Example 4-3, the strains are quite small, even for a field of 1000 V/mm. Consequently the resulting displacements will also be small, unless the dimensions of the material are quite large. When piezoelectric beams are used as actuators, it is generally desirable to maximize the displacement for a given voltage. One way to increase the displacement in a piezoelectric material is to induce bending. To accomplish this, a beam is constructed of two or more layers of piezoelectric material, as shown for the two-layer case in Fig. 8-90. The layers are arranged so that the poling direction is the same for each layer, and bonded so that no slip occurs at the interface between the layers. An electric circuit is constructed so that the electric field in each layer is different. If the electric fields in the two layers have the same magnitude but are directed oppositely, one layer will expand while the other contracts. Because slip between layers is prevented, the beam will bend. Alternatively, the layers can be arranged so that the poling directions are opposite but the electric fields are in the same direction. The arrangement shown, consisting of two piezoelectric layers, is referred to as a *bimorph,* and an arrangement consisting of more than two layers is called a *multimorph.*

In order to develop the equations describing the behavior of piezoelectric beams, we assume that the thickness and width (out-of-plane dimension) are small compared to the length of the beam. Further, we assume that the top and bottom surfaces are unconstrained, as are the edges of the beam parallel to the plane. Under these assumptions all elements of the stress vector, Eq. (4-32), are zero except $\sigma_1$, the component in the $x$ direction.

For definiteness consider the piezoelectric material to be PVDF. As noted at the end of Example 4-3, it is not possible to attach electrodes to the edges of PVDF because of the thinness of the material. Since it is only possible to attach electrodes to the material faces, the only nonzero component of the electric field vector given by Eq. (4-193) is $E_3$.

With the reduced stress and electric field vectors, and with the compliance and piezoelectric strain matrices for PVDF given by Eqs. (4-210) and (4-211), Eq. (4-195) yields the following constitutive equations:

$$\begin{aligned} S_1 &= s_{11}T_1 + d_{31}E_3 \\ S_2 &= s_{12}T_1 + d_{32}E_3 \\ S_3 &= s_{13}T_1 + d_{33}E_3 \end{aligned} \tag{8-339}$$

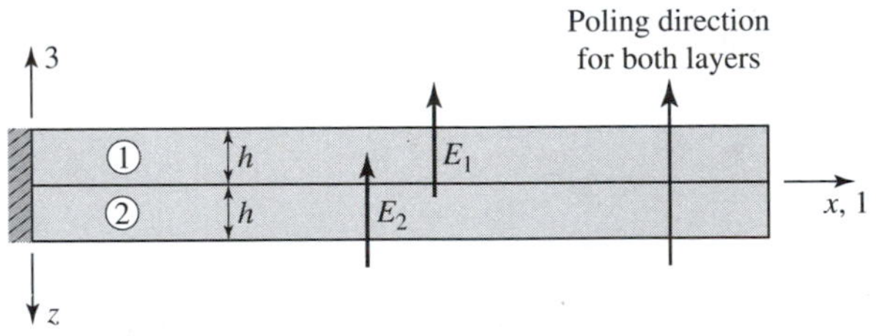


Figure 8-90 Two-layer piezoelectric beam

Equations (8-339) use the notation described in Chapter 4, where $S$ is the strain, $s$ is the compliance, $T$ is the stress, $d$ is the piezoelectric strain coefficient, and $E$ is the electric field. As explained in Chapter 4 and at the end of Section 2-2-5, subscripts 1, 2, and 3 correspond to the $X$, $Y$, and $Z$ directions for the material, respectively. Note that $d_{31}$ is the strain resulting in the 1-direction from a unit electric field in the 3-direction.

Only the first of Eqs. (8-339) is needed to formulate the piezoelectric beam problem. Once the problem is solved, the remaining equations can be used to obtain the strain in the $Y$ and $Z$ directions, if desired. The required elements of the compliance matrix are given by Eqs. (4-53). From these it is seen that $s_{12}$ and $s_{13}$ represent Poisson's effects. As noted in Example 4-3, for PVDF the piezoelectric strain in the 2-direction is an order of magnitude smaller than that in the 1-direction, and although the strain in the 3-direction is of the same order of magnitude as that in the 1-direction, the resulting displacement is quite small owing to the thinness of the PVDF material. Consequently the strains in the 2- and 3-directions are generally not of interest. Also, note from Fig. 8-90 that the beam coordinates differ from the material coordinates in that the orientation of the $z$ direction is not the same. This causes no problem because the $x$ directions are the same and the $y$ directions do not enter into the problem. With the compliance $s_{11}$ as given in Eq. (4-53), the first of Eqs. (8-339) becomes

$$S_x = \frac{1}{Y}T_x + d_{31}E \tag{8-340}$$

where $Y$ is used for the elastic modulus in the $x$ direction in order to avoid the confusion resulting from using the symbol $E$ for the electric field, as explained in Chapter 4, and it is understood that

$$E = E_3 \tag{8-341}$$

It should be mentioned that if Eqs. (4-204) and (4-205) are used in Eq. (4-195), it will be seen that Eq. (8-340) applies for ceramic PZT as well as for PVDF.

## 8-11-1 Piezoelectric Bimorph

In this section we consider a piezoelectric bimorph in which both layers are of equal thickness. Because of bending, the bimorph takes on a curvature, as shown in Fig. 8-91, which causes the upper layer to stretch and the lower layer to contract. Consequently a longitudinal force $N$ develops in each layer. In addition, a bending moment develops owing to the curvature of the layer. These are shown on the left in the figure. If the bimorph undergoes bending, but not extension, a net bending moment $M$ will act on the cross section of the bimorph, but the net longitudinal force must be zero, as shown on

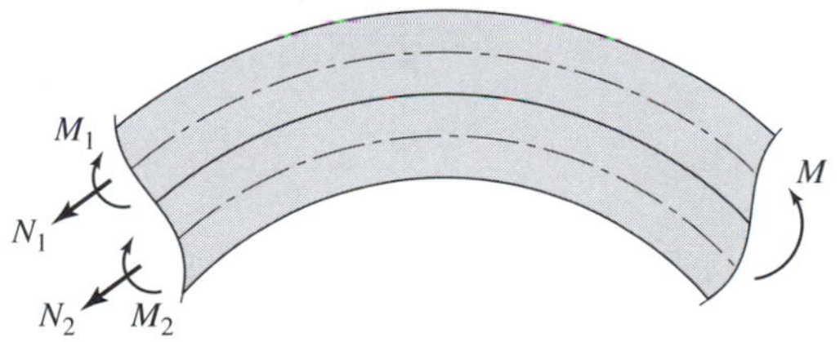


**Figure 8-91** Bent piezoelectric bimorph

the right. As the element becomes vanishingly small, it reduces to a thin slice and we find, by summing longitudinal forces and moments across the resulting slice, that

$$N_2 = -N_1 \tag{8-342a}$$

$$M = M_1 + M_2 + \frac{(N_2 - N_1)h}{2} = M_1 + M_2 - N_1 h \tag{8-342b}$$

The total strain $S_x$ in each layer consists of a longitudinal strain $S_{xL}$ and a bending strain $S_{xb}$,

$$S_x = S_{xL} + S_{xb} \tag{8-343}$$

If $T_x$ is the longitudinal stress $N/A$, where $A$ is the cross-sectional area of a layer, then since the electric field is uniform through the thickness of the layer, the longitudinal strain is given by Eq. (8-340),

$$S_{xL} = \frac{N}{AY} + d_{31}E \tag{8-344}$$

The bending strain is, from elementary strength of materials,

$$S_{xb} = -\frac{z}{r} \tag{8-345}$$

where $z$ is the distance of a layer from the neutral axis to the point where the bending strain is calculated, and $r$ is the radius of curvature of the layer, as shown in Fig. 8-92. Thus the total strain in each layer is

$$S_x(z) = \frac{N}{AY} + d_{31}E - \frac{z}{r} \tag{8-346}$$

With the assumption that each piezoelectric layer has the same dimensions and material properties, the strain in each layer is

$$\begin{aligned} S_{x1}(z_1) &= \frac{N_1}{AY} + d_{31}E_1 - \frac{z_1}{r_1} \\ S_{x2}(z_2) &= \frac{N_2}{AY} + d_{31}E_2 - \frac{z_2}{r_2} \end{aligned} \tag{8-347}$$

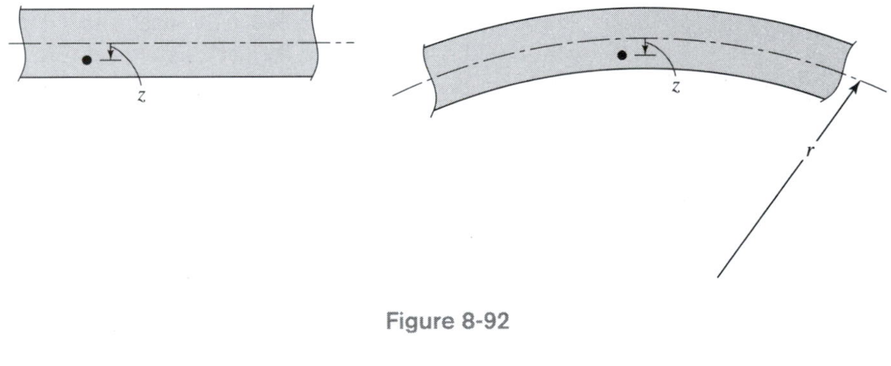


Figure 8-92

where $z_1$ and $z_2$ are the distances from the neutral axes of layers 1 and 2, respectively. At the interface between layers the strains must be equal,

$$S_{x1}\left(\frac{h}{2}\right) = S_{x2}\left(-\frac{h}{2}\right) \tag{8-348}$$

Equations (8-347) are now substituted into Eq. (8-348) together with Eq. (8-342$a$), and the resulting expression is solved for $N_1$,

$$N_1 = \frac{YAh}{4}\left(\frac{1}{r_2} + \frac{1}{r_1}\right) + \frac{YAd_{31}}{2}(E_2 - E_1) \tag{8-349}$$

Recall that, from elementary strength of materials,

$$M = -YI\frac{d^2w}{dx^2} = -YI\frac{1}{r} \tag{8-350}$$

where $I$ is the moment of inertia of the cross section of a layer and $w$ is the lateral, or bending, deflection. Thus for each layer,

$$\begin{aligned} M_1 &= -YI\frac{1}{r_1} \\ M_2 &= -YI\frac{1}{r_2} \end{aligned} \tag{8-351}$$

Equations (8-349) and (8-351) are now substituted into Eq. (8-342$b$) to yield

$$M = -Y\left(I + A\frac{h^2}{4}\right)\left(\frac{1}{r_1} + \frac{1}{r_2}\right) - \frac{YAhd_{31}}{2}(E_2 - E_1) \tag{8-352}$$

Since, for each layer, $h \ll r$, it follows that

$$r_1 \approx r_2 = r$$

where $r$ is the radius of curvature of the interface, in which case, after rearranging Eq. (8-352) and noting, from Eq. (8-350), that $1/r = d^2w/dx^2$,

$$YI_{\text{eq}}\frac{d^2w}{dx^2} = -M - AYd_{31}(E_2 - E_1)\frac{h}{2} \tag{8-353}$$

where

$$I_{\text{eq}} \equiv 2\left[I + A\left(\frac{h}{2}\right)^2\right] \tag{8-354}$$

Note that $I_{\text{eq}}$ is the equivalent moment of inertia, obtained via the parallel axis theorem.

Remember that $d_{31}E$ is the piezoelectric strain and $Yd_{31}E$ is therefore a stress. Thus $AYd_{31}E$ is a force and $AYd_{31}Eh/2$ is a moment resulting from the electric field. Consequently Eq. (8-353) can be written as

$$\frac{d^2w}{dx^2} = -\frac{M + M_{\text{piezo}}}{YI_{\text{eq}}} \tag{8-355}$$

where the piezoelectric moment $M_{\text{piezo}}$ is

$$M_{\text{piezo}} = AYd_{31}(E_2 - E_1)\frac{h}{2} \tag{8-356}$$

Note that, with electrodes applied along the entire piezoelectric layer, $M_{\text{piezo}}$ is uniform along the piezoelectric beam.

## EXAMPLE 8-13

Obtain an expression for the vertical deflection at the free end of a piezoelectric bimorph in the form of a cantilever beam, as shown in Fig. 8-93*a*.

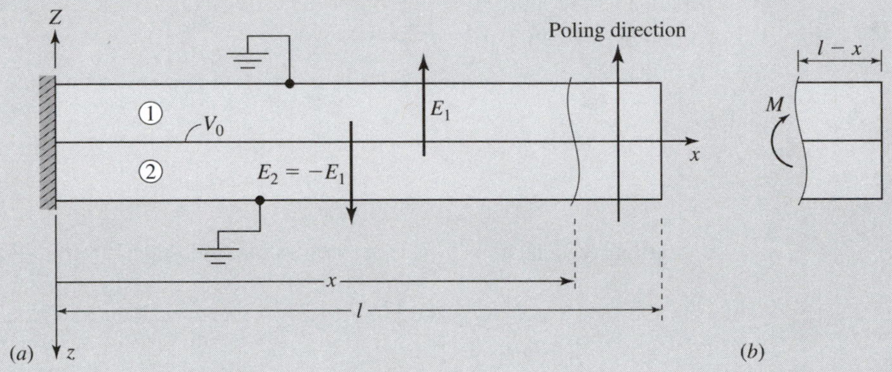


**Figure 8-93**

## SOLUTION

From the free-body diagram (Fig. 8-93*b*), it is seen that $M = 0$. Therefore Eq. (8-355) becomes

$$\frac{d^2w}{dx^2} = -\frac{M_{\text{piezo}}}{YI_{\text{eq}}} \tag{a}$$

When this is integrated twice and the boundary conditions $w(0) = 0$ and $dw/dx|_{x=0} = 0$ are used, the following expression for $w$ results:

$$w(x) = -\frac{M_{\text{piezo}}}{2YI_{\text{eq}}}x^2 \tag{b}$$

The maximum deflection $\delta$ occurs at $x = l$,

$$\delta = w(l) = -\frac{M_{\text{piezo}}}{2YI_{\text{eq}}}l^2 \tag{c}$$

The electric field imposed is such that $E_2 = -E_1$. Therefore the piezoelectric moment is, from Eq. (8-356),

$$M_{\text{piezo}} = -AYd_{31}E_1h \tag{d}$$

If the width of the piezoelectric layer is $b$ and its thickness is $h$, then

$$M_{\text{piezo}} = -Yd_{31}E_1bh^2 \tag{e}$$

and from Eq. (8-354),

$$I_{\text{eq}} = \tfrac{2}{3}bh^3 \tag{f}$$

so that

$$\delta = -\frac{-Yd_{31}E_1bh^2}{2Y\frac{2}{3}bh^3}l^2 = \frac{3}{4}d_{31}E_1\frac{l^2}{h} \tag{g}$$

Recall that $E_1$ is the electric field of layer 1 in the 3, or $Z$, direction so that, from Eq. (4-177),

$$E_1 = -\frac{\partial V}{\partial Z} \tag{h}$$

where $V$ is the voltage. It was noted in Chapter 4 that in the piezoelectric layer the 3, or $Z$, direction is the poling direction, and it follows that $E_1$ is positive in the poling direction. Therefore, from Fig. 8-93, since $Z = 0$ is the interface between layers,

$$E_1 = -\frac{\partial V}{\partial Z} = -\frac{V(h) - V(0)}{h - 0} = -\frac{0 - V_0}{h - 0} = \frac{V_0}{h} \tag{i}$$

and therefore,

$$\delta = \frac{3}{4}d_{31}V_0\left(\frac{l}{h}\right)^2 \tag{j}$$

Since a positive electric field is in the positive 3-direction established by the poling of the PVDF film, $E_1$, as shown in Fig. 8-90, is positive. Therefore since $d_{31}$ is, by Eqs. (4-214), positive, it follows from Eq. (8-340), with $T_x = 0$, that the sign of the strain in the upper layer is positive, indicating extension, whereas that in the bottom layer is negative, indicating contraction. This causes the bimorph to bend downward, that is, in the positive $z$ direction, as indicated by the positive sign in Eq. ($j$).

Now consider a standard film thickness for PVDF of 28 μm. The electrodes at the top and bottom surfaces of the piezoelectric film each add about 7.5 μm, resulting in a total film thickness of about 43 μm. From Eq. (4-214), $d_{31} = 21$ pC/N = 21 (mm/mm)/(V/m). If we assume a length of 2.54 cm and $V_0 = 100$ V, then from Eq. ($j$),

$$\delta = 0.55 \text{ mm}$$

It is worth emphasizing that we have a cantilever beam 1 in in length whose total thickness is about 86 μm (about 0.0034 in, roughly the thickness of a piece of paper), and a voltage of 100 V causes a deflection of slightly more than 0.5 mm. Thus the voltages required to produce modest deflections in piezoelectric actuators such as this are quite high. For this reason, as a safety measure, the outer surfaces of the bimorph are typically grounded, as shown in Fig. 8-93.

## 8-11-2 Piezoelectric Multimorph

Next let us look at a multimorph formulation where not all layers need be piezoelectric. This formulation is quite general and allows for many different combinations of piezoelectric and nonpiezoelectric layers in order to accomplish a specific design goal. We will adopt a fairly structured approach to the formulation, which has the advantage of yielding the desired equations quite readily.

First let us address the kinematics of deformation. We make the usual kinematic assumptions of elementary beam theory, namely, that straight lines perpendicular to the neutral axis before bending remain straight and perpendicular after bending, and the length of these lines is not changed during bending. Now refer to Fig. 8-94, which shows a point $P$ in a single-layer beam in its undeformed configuration and the same point $P'$ in the deformed configuration. A fixed coordinate system is established at some point along the beam such that the unit vector $\boldsymbol{i}$ lies along the neutral axis, as shown. Here $x$ and $z$ are the coordinates of $P$ in the $\boldsymbol{i}$ and $\boldsymbol{k}$ directions, respectively, and $u_0$ and $w_0$ are the $x$ and $z$ displacement components, respectively, of the *neutral axis* at a point $O$, where it is intersected by a perpendicular line passing through $P$.* If $\boldsymbol{X}_i$ and $\boldsymbol{X}_f$ are position vectors from the origin of the fixed coordinate system to points $P$ and $P'$, respectively, and $\boldsymbol{u}$ is the displacement vector from $P$ to $P'$, it follows from Fig. 8-94 that

$$\boldsymbol{u} = \boldsymbol{X}_f - \boldsymbol{X}_i \tag{8-357}$$

where

$$\boldsymbol{X}_i = x\boldsymbol{i} + z\boldsymbol{k} \tag{8-358}$$

$$\boldsymbol{X}_f = (x + u_0)\boldsymbol{i} + w_0\boldsymbol{k} + z\boldsymbol{k} - z\beta\boldsymbol{i} \tag{8-359}$$

The first two terms give the location of point $O'$, and the last two terms give the location of point $P'$ relative to $O'$. Note that for small angles $\sin\beta \approx \beta$ and $\cos\beta \approx 1$, and if lines perpendicular to the neutral axis remain unchanged in length, then $\overline{O'P'} = z$.

Equations (8-359) and (8-358) are now substituted into Eq. (8-357) to give

$$\boldsymbol{u} = (u_0 - z\beta)\boldsymbol{i} + w_0\boldsymbol{k} \tag{8-360}$$

*Note that since they measure the displacement of the *neutral axis* only,

$$\frac{\partial u_0}{\partial z} = 0, \qquad \frac{\partial w_0}{\partial z} = 0$$

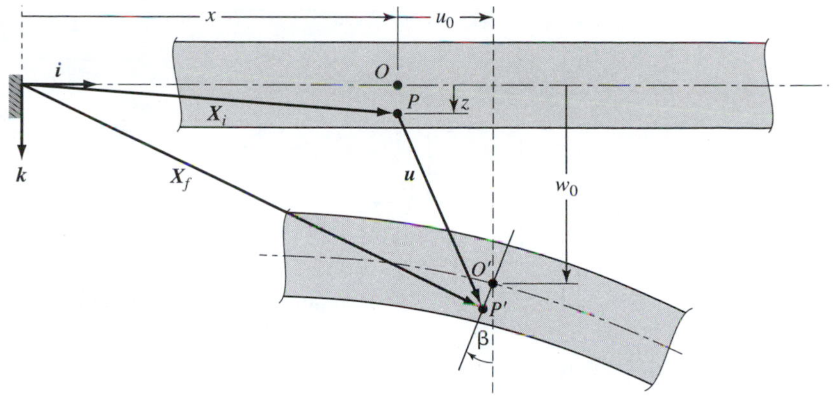


Figure 8-94 Undeformed and deformed configurations of single layer

Thus since $\boldsymbol{u} = u\boldsymbol{i} + v\boldsymbol{j} + w\boldsymbol{k}$, where $u$, $v$, and $w$ are the $x$, $y$, and $z$ components, respectively, of the displacement vector $\boldsymbol{u}$, then the displacement components of point $P$ are

$$u = u_0 - z\beta$$
$$v = 0 \qquad (8\text{-}361)$$
$$w = w_0$$

The strains are calculated using Eqs. (3-11) and yield

$$\varepsilon_x = \frac{\partial u_0}{\partial x} - z\frac{\partial \beta}{\partial x} \qquad (8\text{-}362a)$$

$$\varepsilon_z = 0 \qquad (8\text{-}362b)$$

$$\gamma_{xz} = -\beta + \frac{\partial w_0}{\partial x} \qquad (8\text{-}362c)$$

The three remaining strains are out-of-plane strains and are identically zero. Since, by assumption, straight lines perpendicular to the neutral axis before bending remain so after bending, it follows that $\gamma_{xz} = 0$ so that, from Eq. (8-362*c*),

$$\beta = \frac{\partial w_0}{\partial x} \qquad (8\text{-}363)$$

and therefore Eq. (8-362*a*) becomes

$$\varepsilon_x = S_x = \frac{\partial u_0}{\partial x} - z\frac{\partial^2 w_0}{\partial x^2} \qquad (8\text{-}364)$$

or

$$S_x = \varepsilon^0 - z\kappa^0 \qquad (8\text{-}365)$$

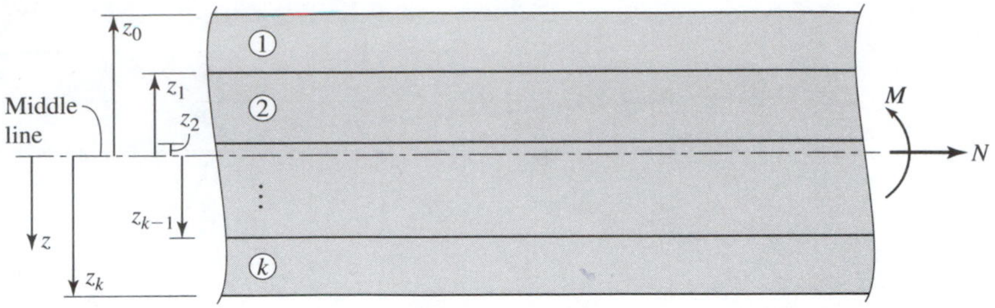


**Figure 8-95** Geometry of multimorph

where

$$\varepsilon^0 = \frac{du_0}{dx} \tag{8-366a}$$

$$\kappa^0 = \frac{d^2 w_0}{dx^2} \tag{8-366b}$$

Here $\varepsilon^0$ and $\kappa^0$ are the stretch and the curvature of the neutral axis, respectively, and the partial derivatives have been replaced with total derivatives because the quantities vary *only* with $x$.

Next we consider the multimorph consisting of $k$ layers shown in Fig. 8-95. The line midway between the upper and lower surfaces is referred to as the *middle line* and serves as a reference line for specifying dimensions in the $z$ direction. Note that it is not necessarily the neutral axis because the layers may have different properties, in which case the bending stress along the middle line may not be zero.

Mating surfaces between adjacent layers of the multimorph are assumed to be perfectly bonded so that no slip occurs between layers. Because of this, the displacements across layer boundaries are continuous. The previous kinematic assumptions, namely, that straight lines perpendicular to the middle line before bending remain straight and perpendicular after bending, and are unchanged in length, are assumed to hold for the multimorph. This is valid so long as the multimorph is thin. In the terminology of composite materials, the foregoing assumptions are those of *classical lamination theory*. Note that with these assumptions, strain compatibility at each interface, which was explicitly imposed in the bimorph formulation of Section 8-11-1 through Eq. (8-348), is automatically satisfied.

It is worth pointing out that a consequence of the foregoing assumptions is that the strain varies linearly across the multimorph but the stress may not because of possible differences in properties between layers.

Next we obtain the resulting force and moment for the multimorph. Note that, from Fig. 8-96, which shows the stress distribution $T_x$ and its resultants $M$ and $N$,

$$N = \int_{\mathcal{A}} T_x \, d\mathcal{A} = \int_z T_x b \, dz = \sum_{i=1}^{k} b_i \int_{z_{i-1}}^{z_i} T_{xi} \, dz \tag{8-367a}$$

$$M = \int_{\mathcal{A}} T_x z \, d\mathcal{A} = \int_z T_x b z \, dz = \sum_{i=1}^{k} b_i \int_{z_{i-1}}^{z_i} T_{x_i} z \, dz \tag{8-367b}$$

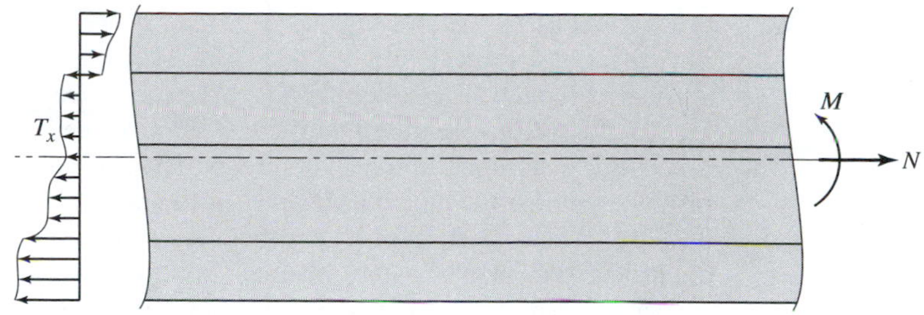


Figure 8-96 Stress distribution in multimorph

where $\mathcal{A}$ is the area of the entire cross section, $b$ is the width at point $z$, and $b_i$ is the width of layer $i$. Equation (8-340) is now solved for $T_x$, Eq. (8-365) is used to replace $S_x$, and the result is applied to layer $i$,

$$T_{x_i} = Y_i(\varepsilon^0 - z\kappa^0 - d_{31_i} E_i) \tag{8-368}$$

where $\varepsilon^0$ and $\kappa^0$ are, respectively, the stretch and the curvature of the *middle line*. Equation (8-368) is then used in Eqs. (8-367) to give

$$N = \sum_{i=1}^{k} \int_{z_{i-1}}^{z_i} b_i Y_i(\varepsilon^0 - z\kappa^0 - d_{31_i} E_i)\, dz \tag{8-369$a$}$$

$$M = \sum_{i=1}^{k} \int_{z_{i-1}}^{z_i} b_i Y_i(\varepsilon^0 - z\kappa^0 - d_{31_i} E_i) z\, dz \tag{8-369$b$}$$

When the indicated integration is carried out, Eqs. (8-369) become

$$N = A\varepsilon^0 - B\kappa^0 - N_{\text{piezo}} \tag{8-370$a$}$$

$$M = B\varepsilon^0 - D\kappa^0 - M_{\text{piezo}} \tag{8-370$b$}$$

where

$$A \equiv \sum_{i=1}^{k} b_i Y_i \int_{z_{i-1}}^{z_i} dz = \sum_{i=1}^{k} b_i Y_i (z_i - z_{i-1}) \tag{8-371$a$}$$

$$B \equiv \sum_{i=1}^{k} b_i Y_i \int_{z_{i-1}}^{z_i} z\, dz = \frac{1}{2} \sum_{i=1}^{k} b_i Y_i \left(z_i^2 - z_{i-1}^2\right) \tag{8-371$b$}$$

$$D \equiv \sum_{i=1}^{k} b_i Y_i \int_{z_{i-1}}^{z_i} z^2\, dz = \frac{1}{3} \sum_{i=1}^{k} b_i Y_i \left(z_i^3 - z_{i-1}^3\right) \tag{8-371$c$}$$

$$N_{\text{piezo}} \equiv \sum_{i=1}^{k} b_i Y_i d_{31_i} E_i \int_{z_{i-1}}^{z_i} dz = \sum_{i=1}^{k} b_i Y_i d_{31_i} E_i (z_i - z_{i-1}) \tag{8-371$d$}$$

$$M_{\text{piezo}} \equiv \sum_{i=1}^{k} b_i Y_i d_{31_i} E_i \int_{z_{i-1}}^{z_i} z\, dz = \frac{1}{2} \sum_{i=1}^{k} b_i Y_i d_{31_i} E_i \left(z_i^2 - z_{i-1}^2\right) \tag{8-371$e$}$$

From Eqs. (8-370) we see that $A$, $B$, and $D$ are the extensional, coupling, and bending stiffnesses, respectively. Note that the stiffnesses of individual layers combine as springs in parallel. The coupling stiffness couples extensional and bending behavior. In Eq. (8-370*a*), for example, a curvature $\kappa^0$ produces an internal extensional force $N$, even if the extensional strain $\varepsilon^0$ is zero, whereas in Eq. (8-370*b*) an extensional strain $\varepsilon^0$ produces an internal moment $M$, even if the curvature $\kappa^0$ is zero. This behavior arises because, in general, the layers have different geometries and elastic stiffnesses, and they are arranged in an asymmetric manner about the middle line.

$N_{\text{piezo}}$ and $M_{\text{piezo}}$ are, respectively, an extensional force and a bending moment induced by the piezoelectric behavior of the multimorph.

Three special cases can be obtained from Eqs. (8-371):

1. If the geometries (width and thickness) and the elastic moduli of the layers are *symmetric about the middle line,* then $B = 0$, a result that simplifies calculations considerably.
2. If the geometries and the elastic moduli are symmetric about the middle line, there are an *even* number of layers, and the electric field of a layer on one side of the middle line is the negative of that of the corresponding layer on the opposite side of the middle line, then $N_{\text{piezo}} = 0$. This will also occur with an *odd* number of layers if the layer through which the middle line passes is not piezoelectric.
3. If the geometries and the elastic moduli are symmetric about the middle line and the electric field of a layer on one side of the middle line is the same as the electric field of the corresponding layer on the opposite side of the middle line, then $M_{\text{piezo}} = 0$.

Proof of these statements is left to the exercises.

If $N = 0$, as is the case when the middle line is unconstrained in the $x$ direction, a situation that is often encountered in practice, then from Eq. (8-370*a*),

$$\varepsilon^0 = \frac{B}{A}\kappa^0 + \frac{N_{\text{piezo}}}{A} \tag{8-372}$$

and Eq. (8-370*b*) becomes

$$M = -D^*\kappa^0 - M^*_{\text{piezo}} \tag{8-373}$$

where

$$D^* = D - B^2/A \tag{8-374a}$$

$$M^*_{\text{piezo}} = M_{\text{piezo}} - N_{\text{piezo}}B/A \tag{8-374b}$$

Therefore Eq. (8-373) becomes, after rearranging and using Eq. (8-366*b*),

$$\frac{d^2w}{dx^2} = -\frac{M + M^*_{\text{piezo}}}{D^*} \tag{8-375}$$

In Eq. (8-375) the subscript has been dropped from $w$, but it is nevertheless understood that $w$ is the displacement of the middle line.

## EXAMPLE 8-14

Consider a cantilever beam with a load $P$ at the free end, which produces an unacceptably large displacement at the free end. In order to reduce the displacement, a piezoelectric layer is bonded to the beam, as shown in Fig. 8-97. If the thickness of the piezoelectric layer is $h_1$ and that of the beam is $h_2$, what voltage is needed to reduce the free-end deflection to zero?

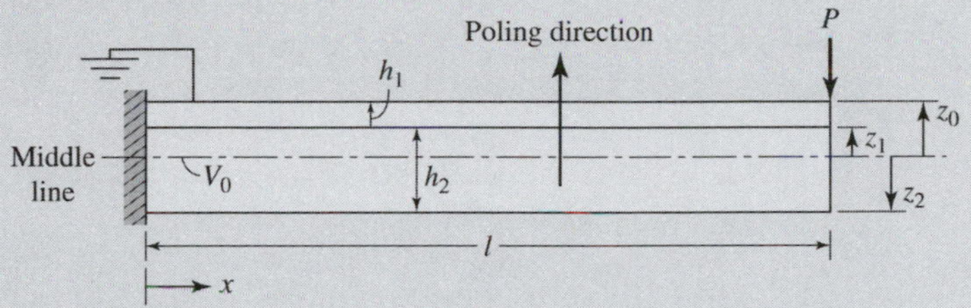


**Figure 8-97**

## SOLUTION

We begin by obtaining $M$ from Fig. 8-98 as

$$M = -P(l - x) \tag{a}$$

so that

$$\frac{d^2w}{dx^2} = \frac{P(l-x)}{D^*} - \frac{M^*_{\text{piezo}}}{D^*} \tag{b}$$

Integrating twice and using the boundary conditions $w(0) = 0$ and $dw/dx|_{x=0} = 0$, we get

$$w(x) = \frac{Px^2}{6D^*}(3l - x) - \frac{M^*_{\text{piezo}}}{2D^*}x^2 \tag{c}$$

and therefore,

$$\delta = w(l) = \frac{Pl^3}{3D^*} - \frac{M^*_{\text{piezo}}l^2}{2D^*} \tag{d}$$

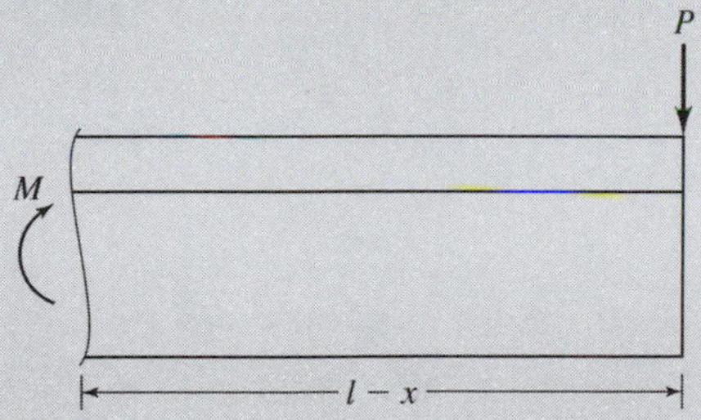


**Figure 8-98** Equilibrium of cut-off piece of a cantilever

If a voltage is applied to the piezoelectric layer with the intention of making the tip deflection zero, then

$$M^*_{\text{piezo}} = \tfrac{2}{3} Pl \tag{e}$$

From Fig. 8-97 we find that

$$\begin{aligned} z_0 &= -\frac{h_1 + h_2}{2} \\ z_1 &= -\frac{h_1 - h_2}{2} \\ z_2 &= \frac{h_1 + h_2}{2} \end{aligned} \tag{f}$$

and therefore from Eqs. (8-371),

$$\begin{aligned} A &= b(Y_1 h_1 + Y_2 h_2) \\ B &= \frac{b}{2} h_1 h_2 (Y_2 - Y_1) \\ N_{\text{piezo}} &= b Y_1 d_{31_1} E_1 h_1 \\ M_{\text{piezo}} &= -\frac{b}{2} h_1 h_2 Y_1 d_{31} E_1 \end{aligned} \tag{g}$$

and with Eq. (8-374*b*),

$$M^*_{\text{piezo}} = -\frac{b}{2} Y_1 Y_2 h_1 h_2 d_{31} E_1 \frac{h_1 + h_2}{h_1 Y_1 + h_2 Y_2} \tag{h}$$

For a thin piezoelectric layer,

$$E_1 = -\frac{dV}{dZ} = -\frac{V(h_1) - V(0)}{h_1 - 0} = -\frac{0 - V_0}{h_1 - 0} = \frac{V_0}{h_1} \tag{i}$$

Therefore,

$$M^*_{\text{piezo}} = -\frac{b}{2} Y_1 Y_2 h_2 d_{31} \frac{h_1 + h_2}{h_1 Y_1 + h_2 Y_2} V_0 \tag{j}$$

so that, from Eq. (*e*),

$$V_0 = -\frac{4}{3} \frac{Pl}{b Y_1 Y_2 h_2 d_{31} (h_1 + h_2)/(h_1 Y_1 + h_2 Y_2)} \tag{k}$$

or, after rearranging,

$$V_0 = -\frac{4}{3} \frac{1 + h_1 Y_1 / h_2 Y_2}{1 + h_1 / h_2} \frac{Pl}{b Y_1 h_2 d_{31}} \tag{l}$$

Now consider a steel beam of length 5 in, width 0.5 in, and thickness 0.125 in with a load of 2 lbf applied at the free end. From elementary strength of materials, the free-end deflection is

$$\delta = 0.034 \text{ in}$$

Suppose a 0.063-in-thick layer of the piezoelectric ceramic PZT is bonded to the beam. Then, with the properties of PZT given in Chapter 4, the preceding formula for $V_0$ shows that a voltage of

$$V_0 = -1720 \text{ V}$$

is needed to reduce the free-end deflection to zero. Thus a voltage of about $-50.5$ V is required for each 0.001-in reduction in the deflection of a steel beam.

## EXAMPLE 8-15

In constructing bimorphs and other piezoelectric devices the bond layer is often thin compared to the thickness of the other materials and is therefore ignored in the analysis. However, in the case of thin-film PVDF bimorphs the thickness of the bond layer, which typically is epoxy, is on the same order as the film thickness. Therefore it is reasonable to expect that the behavior of the bimorph is influenced by the epoxy layer. In order to assess the effect of the epoxy glue layer on the bimorph, we examine the problem in Example 8-13, but consider a glue layer of thickness $h_g$ that is inserted between the two piezoelectric films, as shown in Fig. 8-99. As in Example 8-13 the moment $M$ is zero; both the geometries and the elastic moduli of the layers are symmetric about the middle line so that $B$ is also zero. Thus, with Eqs. (8-374), Eq. (8-375) reduces to

$$\frac{d^2w}{dx^2} = -\frac{M_{\text{piezo}}}{D} \tag{a}$$

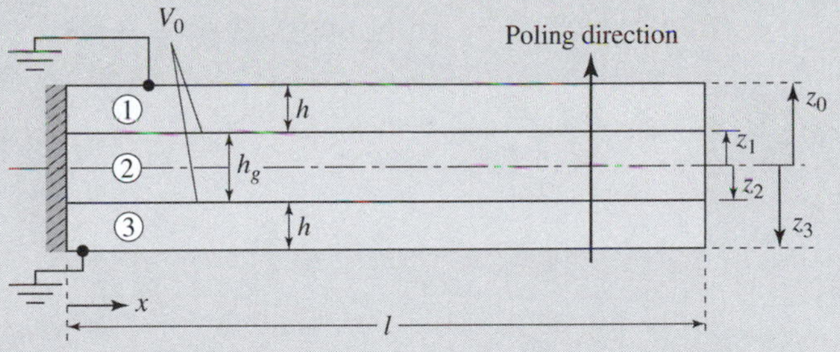


**Figure 8-99** Glue layer inserted between two piezoelectric films

The solution to this equation, subject to cantilever boundary conditions, is

$$w(x) = -\frac{M_{\text{piezo}}}{2D}x^2 \tag{b}$$

so that the free-end deflection δ is

$$\delta = w(l) = -\frac{M_{\text{piezo}}}{2D}l^2 \tag{c}$$

From Fig. 8-99 we obtain

$$\begin{aligned} z_0 &= -h - h_g/2 \\ z_1 &= -h_g/2 \\ z_2 &= h_g/2 \\ z_3 &= h + h_g/2 \end{aligned} \tag{d}$$

and therefore, from Eqs. (8-371), with $Y_3 = Y_1 = Y$, $Y_2 = Y_g$, $d_{31_3} = d_{31_1} = d_{31}$, $d_{31_2} = 0$,

$$\begin{aligned} D &= \frac{b}{12}(8Yh^3 + 12Yh^2h_g + 6Yhh_g^2 + Y_gh_g^3) \\ M_{\text{piezo}} &= \frac{b}{2}Yd_{31}h(h+h_g)(E_1 - E_2) \end{aligned} \tag{e}$$

The equation for $D$ can be manipulated to give a slightly different form,

$$D = \left\{2Y\left[\frac{bh^3}{12} + bh\left(\frac{h_g+h}{2}\right)^2\right] + Y_g\frac{bh_g^3}{12}\right\} \tag{f}$$

The term in square brackets gives the moment of inertia of the piezoelectric layer about the middle line which, in this case, is also the neutral axis. The first term in the braces is therefore the bending stiffness of the piezoelectric layers, whereas the second is the bending stiffness of the glue layer. Note that the layers behave like springs in parallel.

As in Example 8-13, we assume $E_2$ to be the negative of $E_1$ so that

$$M_{\text{piezo}} = bYd_{31}h(h+h_g)E_1 \tag{g}$$

Then the expression for δ becomes

$$\delta = -\frac{d_{31}h\left(1+\frac{h_g}{h}\right)E_1}{2\left\{2\left[\frac{1}{12} + \left(\frac{1+h_g/h}{2}\right)^2\right] + \frac{1}{12}\frac{Y_g}{Y}\left(\frac{h_g}{h}\right)^3\right\}}\left(\frac{l}{h}\right)^2 \tag{h}$$

and since, from Example 8-13, $E_1 = V_0/h$, then

$$\delta = -\frac{6\left(1+\dfrac{h_g}{h}\right)\left(\dfrac{l}{h}\right)^2}{\left\{2\left[1+3\left(1+\dfrac{h_g}{h}\right)^2\right]+\dfrac{Y_g}{Y}\left(\dfrac{h_g}{h}\right)^3\right\}} d_{31} V_0 \tag{i}$$

A typical thickness for the glue layer is about 25 μm [3, sec. 2.3]. The value of the elastic modulus for a particular epoxy, *Z*-poxy, with a 30-min curing time can be found through ultrasonic testing to be 4.24 GPa (see [3, sec. 2.2] for a description of the process). The elastic modulus for the PVDF film is obtained from the first of Eqs. (4-213) and the first of Eqs. (4-53) to be 2.74 GPa. With this information and the data for the piezoelectric layers from Example 8-13, a plot of tip displacement versus voltage for various $h_g/h$ ratios is shown in Fig. 8-100. Note that as the thickness of the glue layer increases, a greater voltage is needed to yield a desired tip displacement of the cantilever beam. Devices such as this piezoelectric beam are often used as actuators in smart structures and in vibration isolation of devices operating in a microgravity environment. Therefore it is generally desirable to use the lowest voltage possible to obtain a desired displacement. Thus the glue layer should be as thin as possible. Current efforts have produced glue layers as thin as 2 μm [23].

From the expression for the bending stiffness $D$, given by Eq. ($f$), and the numerical data (using $h_g = 25$ μm and $b = 5$ mm) we find that the bending stiffness of the piezoelectric film is

$$D_{\text{piezo}} = 2Y\left[\frac{bh^3}{12} + bh\left(\frac{h_g + h}{2}\right)^2\right] = 1.33\ \text{N}\cdot\text{mm}^2 \tag{j}$$

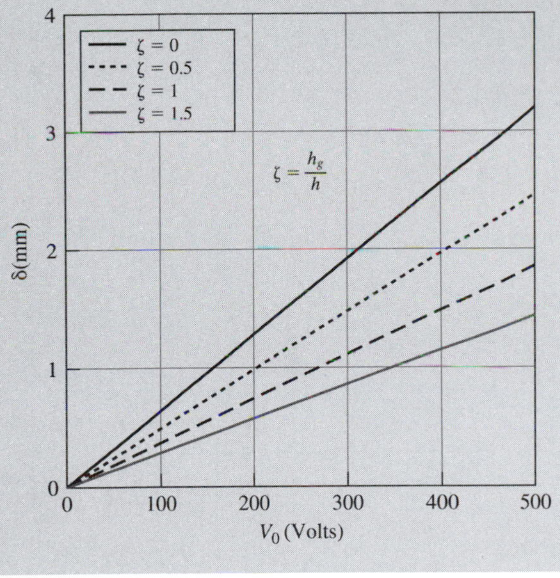


**Figure 8-100** Displacement versus voltage

whereas the bending stiffness of the glue layer is

$$D_g = Y_g \frac{bh_g^3}{12} = 0.028\ \mathrm{N \cdot mm^2} \tag{k}$$

The bending stiffness of the glue layer is seen to be quite insignificant compared to that of the piezoelectric film. The significant effect of the glue layer thickness on the overall bending stiffness is to move the piezoelectric film away from the neutral axis, thereby creating an I-beam effect.

We can define the *electrical stiffness* $k_e$ as the voltage needed to produce a unit tip displacement,

$$k_e = V_0/\delta \tag{l}$$

With the expression for $V_0$ obtained from Eq. ($i$), this becomes

$$k_e = \frac{2\left[1 + 3\left(1 + \frac{h_g}{h}\right)^2\right] + \frac{Y_g}{Y}\left(\frac{h_g}{h}\right)^3}{6\left(1 + \frac{h_g}{h}\right)\left(\frac{l}{h}\right)^2 d_{31}} \tag{m}$$

A plot of electrical stiffness versus $h_g/h$ is shown in Fig. 8-101. Note from expression ($m$) that the electrical stiffness increases with increasing $h_g/h$ but decreases with increasing $l/h$. Thus in a design situation, if the minimum attainable glue layer thickness does not yield a low enough electrical stiffness, additional reduction can be obtained by increasing $l/h$.

Finally, note that if $h_g = 0$, the expression for $\delta$ agrees with that obtained in Example 8-13.

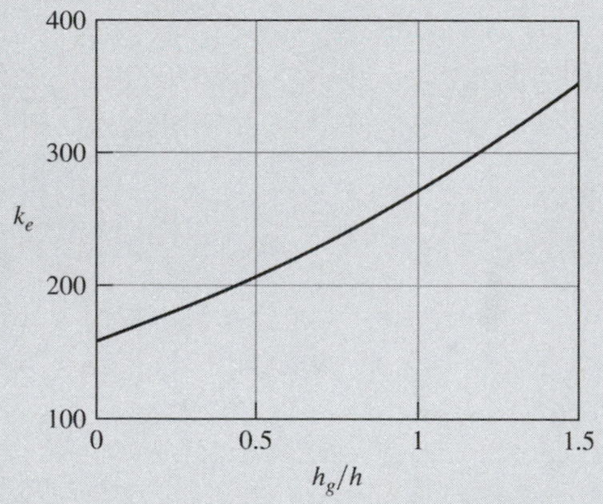


**Figure 8-101** Electrical stiffness versus $h_g/h$

## 8-11-3 Castigliano's Theorem for Piezoelectric Beams

As we saw in Chapter 5, Castigliano's first and second theorems are very useful for determining deflections in beam problems where the loading may be quite complex but the displacement is needed only at specific locations, rather than throughout the beam. It seems natural, then, to see whether Castigliano's theorem, in particular his second theorem, can be applied to piezoelectric beams. We begin by considering a one-dimensional problem where an element of material in a piezoelectric layer, whose behavior is governed by Eq. (8-340), is subjected first to a stress $T_x$ and then to an electric field $E$. We assume, as in Section 8-11, that the electric field is in the 3-direction and therefore the pertinent piezoelectric strain coefficient is $d_{31}$. The resulting stress-strain diagram is shown in Fig. 8-102.* The stress-strain diagram is not linear so that, from the discussion at the end of Section 5-12, we must use the generalized Castigliano's second theorem, given by Eq. (5-131), which requires use of the complementary energy.

From Eq. (5-72), for a one-dimensional problem the strain energy per unit volume $\tilde{u}$ is

$$\tilde{u} = \tfrac{1}{2} S_{x_m} T_x \tag{8-376}$$

where we have used the piezoelectric notation for stress and strain, and the subscript $m$ on the strain indicates that this is the *mechanical* strain found by subtracting the piezoelectric strain $d_{31}E$ from the total strain $S_x$,

$$S_{x_m} = S_x - d_{31}E$$

Equation (8-376) then becomes

$$\tilde{u} = \tfrac{1}{2}(S_x - d_{31}E)T_x \tag{8-377}$$

The complementary energy per unit volume $\tilde{u}_c$ is

$$\tilde{u}_c = S_x T_x - \tilde{u}$$

With Eq. (8-377) this becomes

$$\tilde{u}_c = \tfrac{1}{2} S_x T_x + \tfrac{1}{2} d_{31} E T_x$$

or, after rearranging,

$$\tilde{u}_c = \tfrac{1}{2}(S_x - d_{31}E)T_x + d_{31}ET_x \tag{8-378}$$

and, using Eq. (8-377),

$$\tilde{u}_c = \tilde{u} + d_{31}ET_x \tag{8-379}$$

*It is shown in elementary thermodynamics texts that electrical effects are a form of work. In Section 5-5 it was shown that the work done is independent of the order in which the forces (including, in the present case, the electrical "forces") are applied. Therefore there is no loss of generality in assuming that the mechanical force is applied first, followed by the electric field. It is left to the exercises for the reader to show that, if the electric field is applied first, followed by the mechanical force, the resulting expression for strain energy is given by Eq. (8-377).

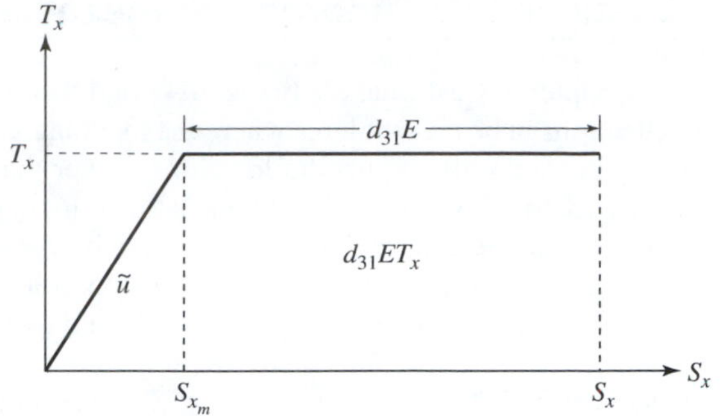


Figure 8-102 Stress-strain diagram

Equation (8-340) is now used to replace $T_x$ in Eq. (8-378),

$$\tilde{u}_c = \tfrac{1}{2}Y(S_x - d_{31}E)^2 + Yd_{31}E(S_x - d_{31}E)$$

When Eq. (8-364) is used for the total strain and the resulting expression expanded and simplified, we get

$$\tilde{u}_c = \tfrac{1}{2}Y\left[(\varepsilon^0)^2 + z^2(\kappa^0)^2 - 2z\varepsilon^0\kappa^0 - d_{31}^2E^2\right] \tag{8-380}$$

The total complementary energy for a volume $V$ is

$$U_c = \int_V \tilde{u}_c\,dV = \int_0^l \int_A \tilde{u}_c\,dA\,dx = \int_0^l \sum_{i=1}^{k} \int_z \tilde{u}_{c_i} b_i\,dz\,dx \tag{8-381}$$

where, for a beam, $l$ is the length and the remaining terms are defined following Eqs. (8-367). When Eq. (8-380) is used in Eq. (8-381) and the resulting integrals are evaluated, the following expression for complementary energy results:

$$U_c = \frac{1}{2}\int_0^l [A(\varepsilon^0)^2 + D(\kappa^0)^2 - 2B\varepsilon^0\kappa^0 - H]\,dx \tag{8-382}$$

where

$$H \equiv \sum_{i=1}^{k} b_i Y_i d_{31_i}^2 E_i^2 \int_{z_{i-1}}^{z_i} dz = \sum_{i=1}^{k} b_i Y_i d_{31_i}^2 E_i^2 (z_i - z_{i-1}) \tag{8-383}$$

and $A$, $B$, and $D$ are given by Eqs. (8-371*a–c*). Equations (8-370) can be readily solved for $\varepsilon^0$ and $\kappa^0$ to give

$$\begin{aligned} \varepsilon^0 &= \frac{D}{D^*}\frac{N_T}{A} - \frac{B}{A}\frac{M_T}{D^*} \\ \kappa^0 &= \frac{B}{A}\frac{N_T}{D^*} - \frac{M_T}{D^*} \end{aligned} \tag{8-384}$$

where

$$\begin{aligned} N_T &= N + N_{\text{piezo}} \\ M_T &= M + M_{\text{piezo}} \end{aligned} \tag{8-385}$$

Castigliano's second theorem [Eq. (5-131)] requires differentiation of the complementary energy with respect to either a concentrated force or a concentrated moment. Since $H$ contains only electrical terms, its derivative will necessarily be zero. Therefore we can define a reduced complementary energy $U_{cR}$ as

$$U_{cR} = \frac{1}{2}\int_0^l [A(\varepsilon^0)^2 + F(\kappa^0)^2 - 2B\varepsilon^0\kappa^0]\,dx \tag{8-386}$$

When this expression is used in Eq. (5-131), it is not necessary to calculate $H$.

Equations (8-384) can now be substituted into Eq. (8-386). This gives, after some algebra,

$$U_{cR} = \int_0^l \left(\frac{D}{D^*}\frac{N_T^2}{2A} - B\frac{N_T M_T}{AD^*} + 2\frac{M_T^2}{D^*}\right) dx \tag{8-387$a$}$$

Note that there is, in general, a coupling between extension and bending through the middle term in parentheses. This coupling disappears if the multimorph layers are symmetric about the middle line in both geometry and elastic modulus.

If $N = 0$, then Eq. (8-387$a$) becomes, after rearranging and dropping purely electrical terms because their derivatives with respect to concentrated forces and moments are zero,

$$U_{cR} = \int_0^l \frac{(M + M^*_{\text{piezo}})^2}{2D^*}\,dx \tag{8-387$b$}$$

Similarly, when $M = 0$ and similar steps are followed,

$$U_{cR} = \int_0^l \frac{D}{D^*}\frac{(N + N^*_{\text{piezo}})^2}{2A}\,dx \tag{8-388}$$

where

$$N^*_{\text{piezo}} = N_{\text{piezo}} - \frac{B}{D}M_{\text{piezo}} \tag{8-389}$$

## EXAMPLE 8-16

Obtain the free-end deflection of the cantilever beam in Example 8-15 using Castigliano's theorem.

### SOLUTION

We begin by applying a concentrated load, directed downward, at the free end, as shown in Fig. 8-103. As a matter of convenience, we measure $x$ from the free end. From the free-body diagram in Fig. 8-103 we get

$$M = -Px \tag{$a$}$$

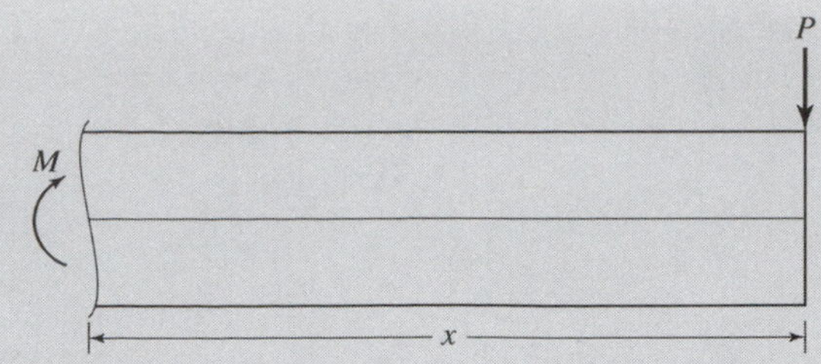


**Figure 8-103**

Since there is no longitudinal force, $N = 0$, so that Eq. (8-387*b*) applies. With Eq. (*a*) it becomes

$$U_{cR} = \int_0^l \frac{(-Px + M^*_{\text{piezo}})^2}{2D^*}\,dx \qquad (b)$$

Furthermore, because of the symmetry of the geometry and the elastic modulus, $B = 0$, and therefore with Eqs. (8-374), Eq. (*b*) reduces to

$$U_{cR} = \int_0^l \frac{(-Px + M_{\text{piezo}})^2}{2D}\,dx \qquad (c)$$

Equation (*c*) is differentiated with respect to $P$ to get, from Eq. (5-131),

$$\delta = \frac{\partial U_{cR}}{\partial P} = \int_0^l \frac{Px^2 - M_{\text{piezo}}x}{D}\,dx \qquad (d)$$

where $\delta$ is the free-end deflection. Now set $P = 0$ and carry out the indicated integration. This gives

$$\delta = -\frac{M_{\text{piezo}}}{2D}l^2 \qquad (e)$$

Note that this agrees with Eq. (*c*) in Example 8-15.

## EXAMPLE 8-17

Resolve the problem in Example 8-14 using Castigliano's theorem.

### SOLUTION

Since a load $P$ is applied at the free end of the cantilever, Fig. 8-103 can be used. The moment $M$ is then given by Eq. (*a*) in Example 8-16. Since, in this problem, $B \neq 0$, the complementary energy is given by Eq. (8-387*b*) which, when the moment expression is substituted, gives

$$U_{cR} = \int_0^l \frac{(-Px + M^*_{\text{piezo}})^2}{2D^*}\,dx \qquad (a)$$

The free-end displacement is found by differentiating Eq. (*a*) with respect to $P$,

$$\delta = \frac{\partial U_{cR}}{\partial P} = \int_0^l \frac{Px^2 - M^*_{\text{piezo}}x}{D^*}\,dx \qquad (b)$$

When the integration is carried out and the displacement set equal to zero, we obtain

$$M^*_{\text{piezo}} = \tfrac{2}{3}Pl \qquad (c)$$

which is in agreement with Eq. (*e*) of Example 8-14.

## 8-11-4 Thin Curved Piezoelectric Beams

For certain smart structure problems, such as active position control of satellite dishes or mirrors, the use of curved piezoelectric elements is necessary, or at least desirable. Furthermore, the use of curved actuators opens up additional possibilities in the design of piezoelectric actuators. For example, in particular applications there may be advantages to actuator configurations in the form of bellows or rings (see, for example, [3], [23]). As with straight piezoelectric beams, layers of material are bonded together to form a bimorph or, more generally, a multimorph. We restrict our attention to thin beams so that the assumptions of classical lamination theory can be assumed to hold, as was done in Section 8-11-2 for straight piezoelectric beams. We assume that the beams are arcs of circles so that their undeformed radius of curvature is constant over the beam.

As in Section 8-11-2, we begin by considering the kinematics of deformation of a single layer whose middle line has radius $R$. We assume, as with straight beams, that straight lines perpendicular to the middle line before bending remain straight and perpendicular after bending, and are unchanged in length. As shown in Fig. 8-104, the point $P$ in the undeformed configuration becomes point $P'$ in the deformed configuration, in which the radius of curvature of the middle line is $r$. $\boldsymbol{X}_i$ is a position vector from the origin of a fixed coordinate system to $P$ and $\boldsymbol{x}_f$ is the corresponding position vector for $P'$. It follows from the figure that the displacement vector $\boldsymbol{u}$, is

$$\boldsymbol{u} = \boldsymbol{x}_f - \boldsymbol{X}_i \qquad (8\text{-}390)$$

Figure 8-105 shows the curved beam in its undeformed configuration, where $\tilde{R}$ is the radius of a point $P$ located at a distance $z$ (assumed positive in the positive radial direction) from the middle line, $\boldsymbol{e}_r$ is a unit vector in the radial direction passing through $P$, and $\boldsymbol{X}_O$ is a position vector of point $O$, the point on the middle line where it is intersected by $\boldsymbol{e}_r$. $\boldsymbol{e}_\theta$ is a unit vector tangent to the middle line at $O$. It follows from the figure that

$$\boldsymbol{X}_i = \boldsymbol{X}_O + z\boldsymbol{e}_r \qquad (8\text{-}391)$$

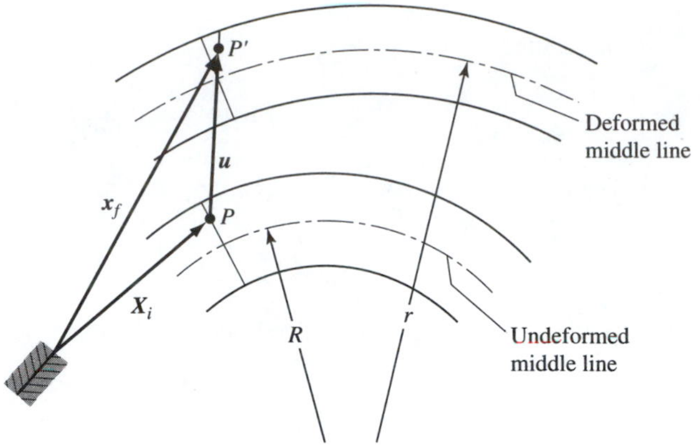


**Figure 8-104** Deformation of single-layer curved beam

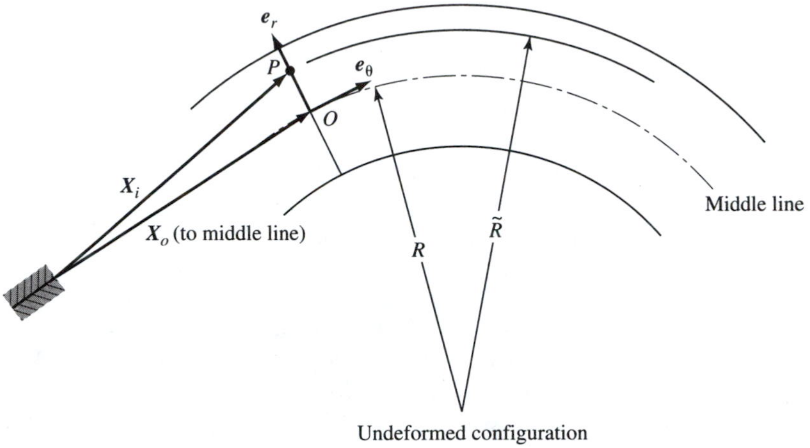


**Figure 8-105**

We wish to write the position vector $\boldsymbol{x}_f$ to point $P'$ in terms of $\boldsymbol{e}_r$ and $\boldsymbol{e}_\theta$, the unit vectors associated with the undeformed configuration. Figure 8-106 shows the geometry. Point $O$ moves to $O'$ during deformation, just as point $P$ moves to $P'$. The line $\overline{O'a}$ is parallel to $\boldsymbol{e}_r$, and $\beta$ is the small angle between $\overline{O'a}$ and $\overline{O'P'}$. $\boldsymbol{x}_O$ is the position vector of $O'$. By invoking the small angle approximations stated in Section 8-11-2 we find that

$$\boldsymbol{x}_f = \boldsymbol{x}_O + z\boldsymbol{e}_r + z\beta\boldsymbol{e}_\theta \tag{8-392}$$

With reference to Fig. 8-107, $\boldsymbol{x}_O$ can be expressed in terms of $w_O$ and $v_O$, the displacement components of point $O$ parallel to $\boldsymbol{e}_r$ and $\boldsymbol{e}_\theta$, respectively, as

$$\boldsymbol{x}_O = \boldsymbol{X}_O + v_O\boldsymbol{e}_\theta + w_O\boldsymbol{e}_r \tag{8-393}$$

Note that $w_O$ and $v_O$ are the radial and tangential displacement components, respectively, of the middle line. Equations (8-391)–(8-393) are now substituted into Eq. (8-390). After

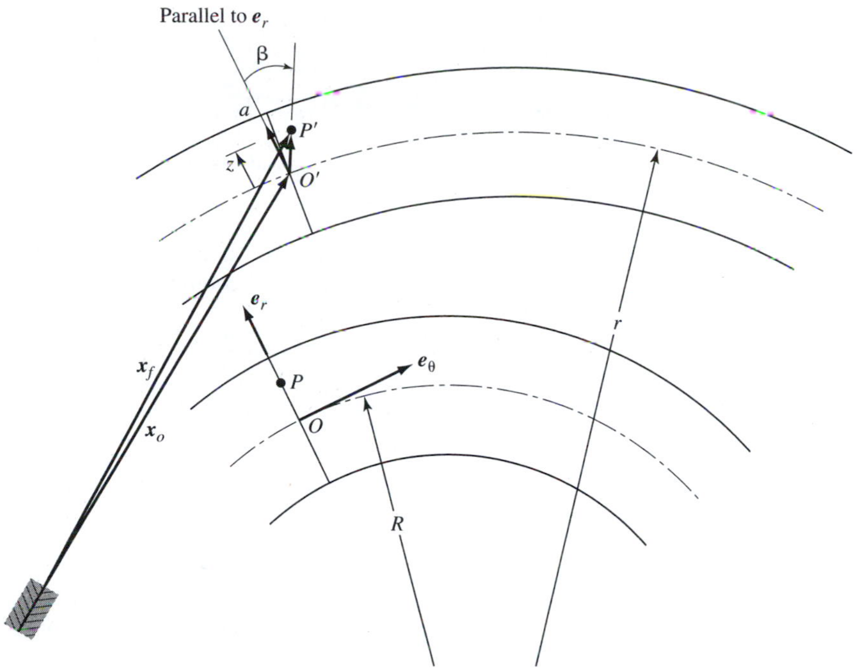


Figure 8-106 Undeformed and deformed configurations

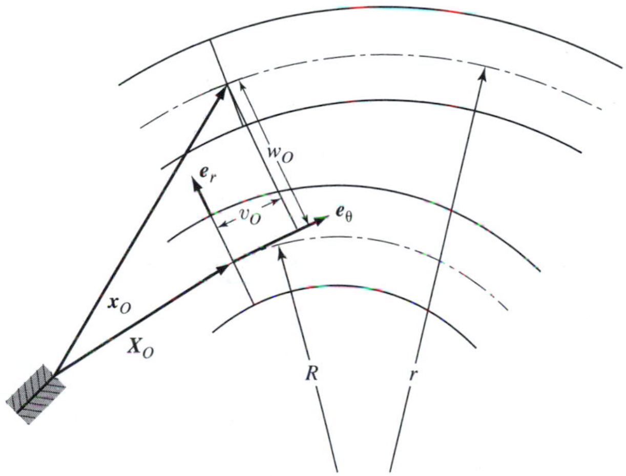


Figure 8-107 Position vector in terms of displacement components

simplifying and collecting terms we get

$$\boldsymbol{u} = w_O\boldsymbol{e}_r + (v_O + \beta z)\boldsymbol{e}_\theta \tag{8-394}$$

Comparing this with Eq. (3-74*a*) we find that

$$\begin{aligned} u_r &= w_O \\ u_\theta &= v_O + \beta z \end{aligned} \tag{8-395}$$

The strains in polar coordinates are calculated according to Eqs. (3-82) which, in the present notation where $\tilde{R}$ is the radial coordinate in the undeformed configuration, read

$$\begin{aligned} \varepsilon_r &= \frac{\partial u_r}{\partial \tilde{R}} \\ \varepsilon_\theta &= \frac{u_r}{\tilde{R}} + \frac{1}{\tilde{R}}\frac{\partial u_\theta}{\partial \theta} \\ \gamma_{r\theta} &= \frac{1}{\tilde{R}}\frac{\partial u_r}{\partial \theta} + \frac{\partial u_\theta}{\partial \tilde{R}} - \frac{u_\theta}{\tilde{R}} \end{aligned} \tag{8-396}$$

Note that

$$\tilde{R} = R + z = R\left(1 + \frac{z}{R}\right) \tag{8-397}$$

and therefore, with the chain rule and noting that $R$ is constant,

$$\frac{\partial}{\partial \tilde{R}} = \frac{\partial z}{\partial \tilde{R}}\frac{\partial}{\partial z} = \frac{\partial}{\partial z} \tag{8-398}$$

Since $w_O$ and $v_O$ are displacements of the middle line, and since $\overline{O'P'}$ is a straight line, it follows that

$$v_O = v_O(\theta), \qquad w_O = w_O(\theta), \qquad \beta = \beta(\theta) \tag{8-399}$$

With Eqs. (8-395) and (8-397)–(8-399) substituted into Eqs. (8-396) we find that $\varepsilon_r = 0$ and the remaining strains become

$$\varepsilon_\theta = \frac{1}{R\{1 + (z/R)\}}\left(w_O + \frac{dv_O}{d\theta} + z\frac{d\beta}{d\theta}\right) \tag{8-400a}$$

$$\gamma_{r\theta} = \frac{1}{R\{1 + (z/R)\}}\left(\frac{dw_O}{d\theta} + R\beta - v_O\right) \tag{8-400b}$$

Note that partial differentiation has been replaced with total differentiation because the relevant quantities depend only on the single variable $\theta$. Since we have assumed that straight lines perpendicular to the middle line before bending remain so after bending, it follows that $\gamma_{r\theta} = 0$, and therefore from Eq. (8-400*b*),

$$\beta = \frac{v}{R} - \frac{1}{R}\frac{dw}{d\theta} \tag{8-401}$$

where we have dropped the subscript $O$ on the displacement components but it is understood that $v$ and $w$ are middle line displacements. Note that $|z| \leq h/2$, and since, for a thin beam, $h \ll R$, it follows that $|z|/R \ll 1$. Thus Eq. (8-400$a$) can be written as

$$\varepsilon_\theta = \varepsilon^0 + \frac{z}{R}\frac{d\beta}{d\theta} \tag{8-402}$$

where

$$\varepsilon^0 = \frac{w}{R} + \frac{1}{R}\frac{dv}{d\theta} \tag{8-403}$$

is the stretch of the middle line.

We now examine the term $(1/R)d\beta/d\theta$. With reference to Fig. 8-108, which shows a segment of beam in its undeformed and deformed states, let $ds$ be the length of the arc $AB$,

$$ds = R\,d\theta \tag{8-404}$$

In the deformed configuration, starting with the line parallel to $\overline{OA}$ and working through the angles to the line parallel to $\overline{OB}$, we obtain the angle between $\overline{OA}$ and $\overline{OB}$ as

$$\beta + d\theta' - \left(\beta + \frac{d\beta}{ds}\,ds\right)$$

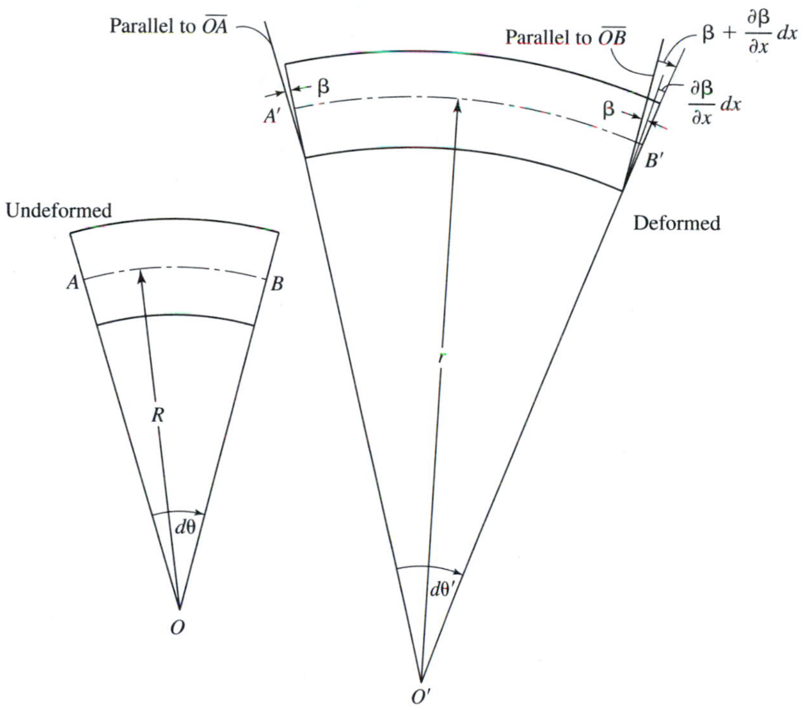


**Figure 8-108** Segment of beam in undeformed and deformed states

However, with reference to the undeformed configuration the angle between $\overline{OA}$ and $\overline{OB}$ is $d\theta$ so that, after simplifying,

$$d\theta = d\theta' - \frac{d\beta}{ds}\,ds \tag{8-405}$$

Now the arc length $A'B'$ is

$$ds' = r\,d\theta' = (1 + \varepsilon^0)\,ds \tag{8-406}$$

Equations (8-404) and (8-406) can be rearranged to give $d\theta$ and $d\theta'$, respectively, in terms of $ds$, and the results used in Eq. (8-405). After some simplification,

$$\frac{d\beta}{ds} = \frac{1 + \varepsilon^0}{r} - \frac{1}{R}$$

or, since $\varepsilon^0 \ll 1$,

$$\frac{d\beta}{ds} = \frac{1}{R}\frac{d\beta}{d\theta} = \frac{1}{r} - \frac{1}{R} \tag{8-407}$$

In other words, the quantity $(1/R)d\beta/d\theta$ gives the *change in curvature of the middle line* between the deformed and the undeformed configurations. Note that for a straight beam, $s = x$ and $1/R = 0$.

We now define

$$\kappa^0 = \frac{1}{R}\frac{d\beta}{d\theta} = \frac{1}{R^2}\frac{dv}{d\theta} - \frac{1}{R^2}\frac{d^2w}{d\theta^2} \tag{8-408}$$

where we have used Eq. (8-401). With this, Eq. (8-402) becomes

$$\varepsilon_\theta = S'_\theta = \varepsilon^0 + z\kappa^0 \tag{8-409}$$

Note that Eq. (8-408) reduces to Eq. (8-366*b*) if $R\,d\theta$ is replaced by $dx$, $w$ is replaced by $w_0$, and the sign difference is reconciled. Now we return briefly to Eqs. (8-401). As seen in Fig. 8-109, in the absence of shear strain, $\beta$ is the slope of the deformed middle

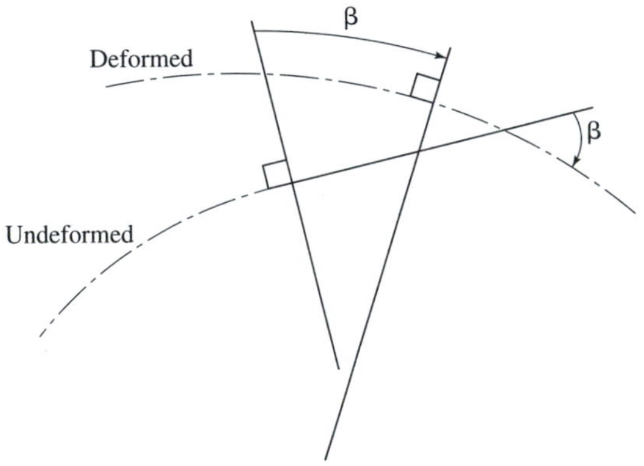


**Figure 8-109** Geometric interpretation of $\beta$

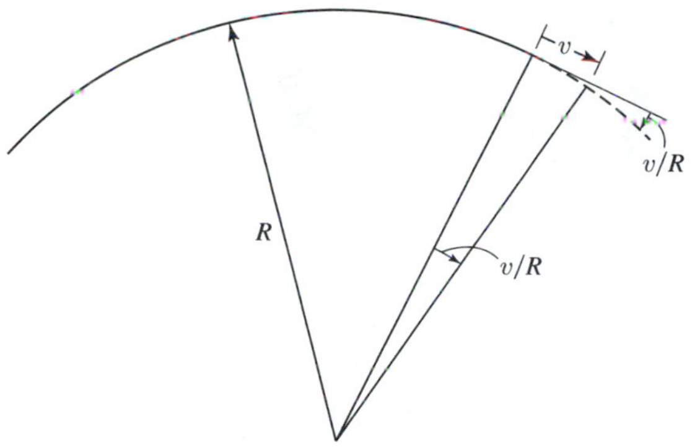


Figure 8-110 Physical significance of $v/R$

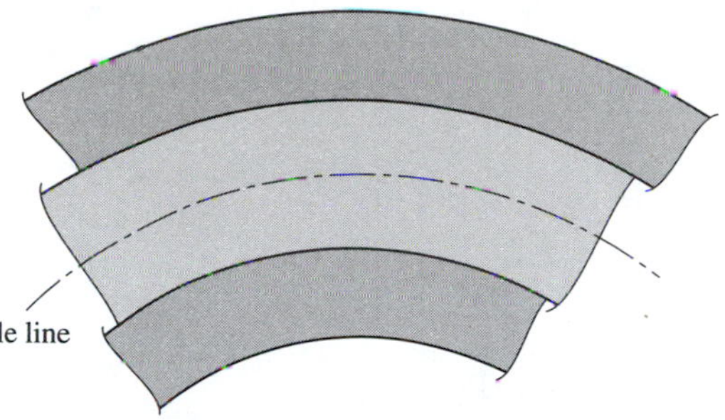


Figure 8-111 Middle line of multimorph

surface relative to the undeformed middle surface. With Eq. (8-404), Eq. (8-401) can be written as

$$\beta = \frac{v}{R} - \frac{dw}{ds} \tag{8-410}$$

For a straight beam, $R \to \infty$ and $s = x$ so that the slope is $-dw/dx$, as expected. However, for the curved beam the slope contains an additional term, $v/R$. The physical significance of this additional term is shown in Fig. 8-110. Note that even if there is no transverse displacement $w$, there is still a change of slope due to the tangential displacement $v$. Furthermore, a boundary condition of zero slope at a particular point gives, for a curved beam,

$$\frac{1}{R}\frac{dw}{d\theta} = \frac{v}{R} \tag{8-411}$$

at that point, whereas for a straight beam the condition of zero slope gives

$$\frac{dw}{dx} = 0$$

For a multimorph we define the middle line as being midway between the inner and outer surfaces, as shown in Fig. 8-111. As was the case with straight beams, the middle line is not necessarily the neutral axis because the layers may have different properties. We invoke the assumptions of classical lamination theory discussed in Section 8-11-2, in which case Eq. (8-409) holds for the multimorph.

Next we turn our attention to obtaining the force and moment resultants for the multimorph. Figure 8-112 shows the distribution of the tangential stress $T_\theta$ acting over a typical cross section of the multimorph along with the force resultant $N$ and moment resultant $M$. These are given by

$$\begin{aligned} N &= \int_{\mathcal{A}} T_\theta \, d\mathcal{A} = \int_z T_\theta b \, dz = \sum_{i=1}^{k} b_i \int_{z_{i-1}}^{z_i} T_{\theta_i} \, dz \\ M &= \int_{\mathcal{A}} T_\theta z \, d\mathcal{A} = \int_z T_\theta b z \, dz = \sum_{i=1}^{k} b_i \int_{z_{i-1}}^{z_i} T_{\theta_i} z \, dz \end{aligned} \tag{8-412}$$

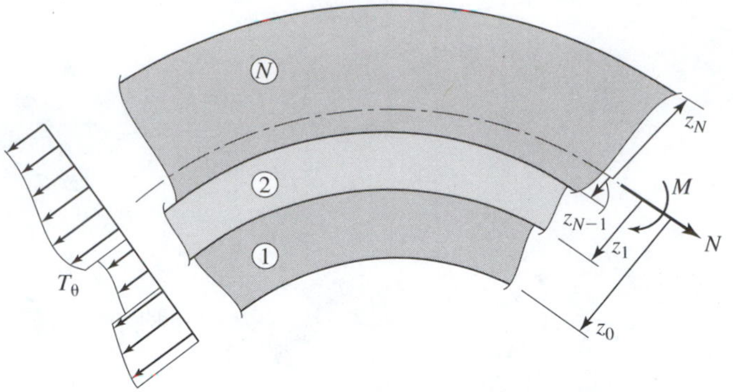


**Figure 8-112** Distribution of tangential stress over typical multimorph cross section

where $\mathcal{A}$ is the area of the multimorph cross section, $b$ is its width at point $z$, and $b_i$ is the width of layer $i$. The relationship between stress, strain, and electric field for layer $i$ is given by Eq. (8-368) which, in polar coordinates, reads

$$T_{\theta_i} = Y_i(S_\theta - d_{31_i}E_i) = Y_i(\varepsilon^0 + z\kappa^0 - d_{31_i}E_i) \tag{8-413}$$

where we have used Eq. (8-409), and $\varepsilon^0$, $\kappa^0$ are given by Eqs. (8-403) and (8-408), respectively. Equation (8-413) is used to replace $T_{\theta_i}$ in Eqs. (8-412) and yields, upon carrying out the indicated integration,

$$\begin{aligned} N &= A\varepsilon^0 + B\kappa^0 - N_{\text{piezo}} \\ M &= B\varepsilon^0 + D\kappa^0 - M_{\text{piezo}} \end{aligned} \tag{8-414}$$

The coefficients $A$, $B$, and $D$, as well as $N_{\text{piezo}}$ and $M_{\text{piezo}}$, are given by Eqs. (8-371). Equations (8-414) can be written in terms of displacements by substituting Eqs. (8-403) and (8-408),

$$N = A\left(\frac{w}{R} + \frac{1}{R}\frac{dv}{d\theta}\right) + B\left(\frac{1}{R^2}\frac{dv}{d\theta} - \frac{1}{R^2}\frac{d^2w}{d\theta^2}\right) - N_{\text{piezo}} \tag{8-415$a$}$$

$$M = B\left(\frac{w}{R} + \frac{1}{R}\frac{dv}{d\theta}\right) + D\left(\frac{1}{R^2}\frac{dv}{d\theta} - \frac{1}{R^2}\frac{d^2w}{d\theta^2}\right) - M_{\text{piezo}} \tag{8-415$b$}$$

If $M$ and $N$ can be found from statics alone, then Eqs. (8-415), together with the appropriate boundary conditions for $v$ and $w$, are sufficient to determine the displacement components $v$ and $w$.

If the geometry of the cross section and the elastic modulus are symmetric with respect to the middle line, then $B = 0$ and Eqs. (8-415) reduce to

$$N = A\left(\frac{w}{R} + \frac{1}{R}\frac{dv}{d\theta}\right) - N_{\text{piezo}} \tag{8-416$a$}$$

$$M = \frac{D}{R^2}\left(\frac{dv}{d\theta} - \frac{d^2w}{d\theta^2}\right) - M_{\text{piezo}} \tag{8-416$b$}$$

Here $v$ can be eliminated from Eq. (8-416$b$) by substituting Eq. (8-416$a$) with the following result:

$$\frac{d^2w}{d\theta^2} + w = R\frac{N + N_{\text{piezo}}}{A} - R^2\frac{M + M_{\text{piezo}}}{D} \tag{8-417}$$

Once $w$ is obtained from Eq. (8-417), $v$ can be found from Eq. (8-416$a$).

## EXAMPLE 8-18

Consider a piezoelectric ring resting on a frictionless, undeformable surface so that it is free to deform, as shown in Fig. 8-113. The ring consists of layers that are arranged so that the geometries and the elastic moduli are symmetric about the middle line, and the electric field produces a piezoelectric force $N_{\text{piezo}}$, but no piezoelectric moment. We wish to find the ring deflections $v$ and $w$.

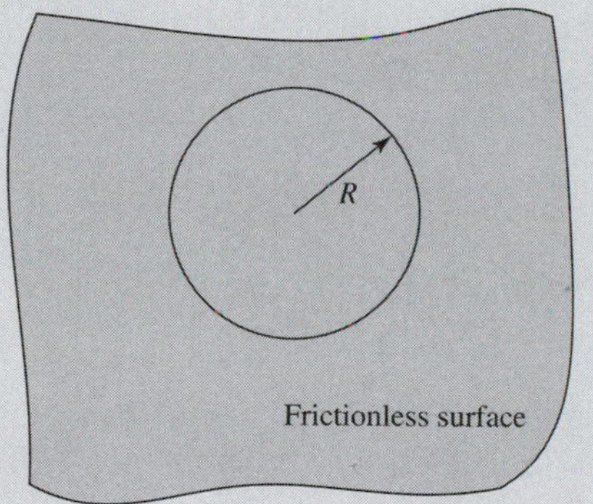


**Figure 8-113**

### SOLUTION

Because of symmetry about the vertical and horizontal lines passing through the center of the ring, we need only consider a quarter ring, as shown in Fig. 8-114$a$. The kinematic

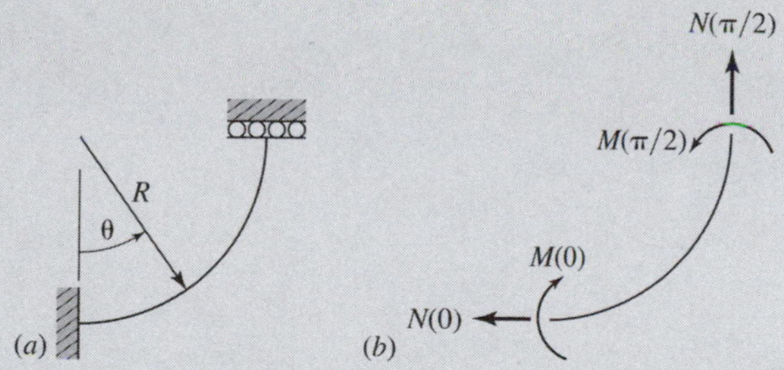


**Figure 8-114**

boundary conditions are

$$v(0) = 0, \qquad v(\pi/2) = 0 \tag{a}$$

$$\beta(0) = 0, \qquad \beta(\pi/2) = 0 \tag{b}$$

A free-body diagram of the quarter segment is shown in Fig. 8-114*b*. Since the shear on a plane of symmetry is zero, by summing forces in the horizontal and vertical directions we get

$$N(0) = 0, \qquad N(\pi/2) = 0 \tag{c}$$

Summation of the moments gives

$$M(0) = M(\pi/2) = M_0 \tag{d}$$

where $M_0$ is to be determined.

Next we consider the free-body diagram of a segment of angular length $\theta$, where $\theta$ is arbitrary, as shown in Fig. 8-115. The forces are decomposed into horizontal and vertical components and summed to give

$$\begin{aligned} Q \sin\theta + N \cos\theta &= 0 \\ -Q \cos\theta + N \sin\theta &= 0 \end{aligned} \tag{e}$$

Since the determinant of the coefficient matrix of $N$ and $Q$ is not zero, it follows that

$$N = 0, \qquad Q = 0, \qquad 0 < \theta < \pi/2 \tag{f}$$

By summing moments we find that

$$M = M(0) = M_0, \qquad 0 < \theta < \pi/2 \tag{g}$$

Thus we see that there can be no shear force or normal force in the ring, and the moment must be constant. When these results are used in Eqs. (8-416*a*) and (8-417), we get the following:

$$\frac{dv}{d\theta} = -w + \frac{R}{A} N_{\text{piezo}} \tag{h}$$

$$\frac{d^2w}{d\theta^2} + w = \frac{R}{A} N_{\text{piezo}} - R^2 \frac{M_0}{D} \tag{i}$$

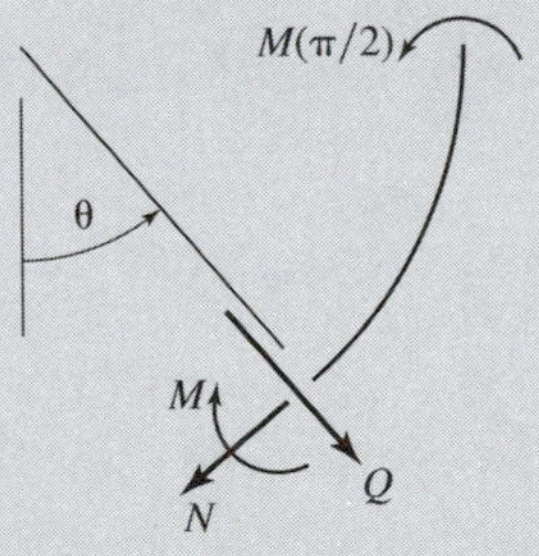


**Figure 8-115**

The solution to Eq. (*i*) is

$$w = a_1 \cos\theta + a_2 \sin\theta - \frac{R^2}{D} M_0 + \frac{R}{A} N_{\text{piezo}} \tag{j}$$

When this is substituted into Eq. (*h*) and the resulting equation integrated, we get $v$,

$$v = -a_1 \sin\theta + a_2 \cos\theta + \frac{R^2}{D} M_0 \theta + a_3 \tag{k}$$

where $a_3$ is a constant of integration. The expressions for $v$ and $w$ are now substituted into Eq. (8-401) to get the slope,

$$\beta = \frac{R}{D} M_0 \theta + \frac{1}{R} a_3 \tag{l}$$

When the slope is used in Eqs. (*b*), we find that

$$a_3 = 0, \qquad M_0 = 0 \tag{m}$$

Thus from Eq. (*g*) we see that the moment is zero throughout the ring.

When Eqs. (*m*) are substituted into Eq. (*k*) and Eqs. (*a*) are used, there results

$$a_1 = 0, \qquad a_2 = 0 \tag{n}$$

These are now used in Eqs. (*j*) and (*k*), from which the displacements follow,

$$v = 0 \tag{o}$$

$$w = \frac{R}{A} N_{\text{piezo}} \tag{p}$$

Thus there is no tangential displacement $v$ in the ring and the radial displacement $w$ is constant.

Now consider a piezoelectric bimorph consisting of layers of equal thickness $h$, width $b$, elastic modulus $Y$, and piezoelectric strain coefficient $d_{31}$. Then

$$z_0 = -h, \qquad z_1 = 0, \qquad z_2 = h \tag{q}$$

and from Eqs. (8-371*a*) and (8-371*d*),

$$\begin{aligned} A &= 2bhY \\ N_{\text{piezo}} &= bhYd_{31}(E_1 + E_2) \end{aligned} \tag{r}$$

With these, Eq. (*p*) gives

$$w = \tfrac{1}{2} R d_{31}(E_1 + E_2) \tag{s}$$

If the electric field in each layer has a value $E$, then

$$w = R d_{31} E \tag{t}$$

Thus, as we might have anticipated, the radial displacement $w$ is simply the radius of the ring multiplied by the piezoelectric strain.

Next consider a ring formed by two piezoelectric layers, each of thickness $h$, separated by a glue layer of thickness $h_g$. The width of each layer is $b$. The elastic modulus and the piezoelectric strain coefficient of each piezoelectric layer are $Y$ and $d_{31}$, respectively. The elastic modulus of the glue layer, which is not piezoelectric, is $Y_g$. In that case,

$$z_0 = -\left(h + \tfrac{1}{2}h_g\right), \qquad z_1 = -\tfrac{1}{2}h_g, \qquad z_2 = \tfrac{1}{2}h_g, \qquad z_3 = h + \tfrac{1}{2}h_g \tag{u}$$

and

$$A = 2bhY\left(1 + \frac{h_g}{2h}\frac{Y_g}{Y}\right)$$
$$N_{\text{piezo}} = bhYd_{31}(E_1 + E_2) \tag{v}$$

so that

$$w = \frac{1}{1 + \dfrac{h_g}{2h}\dfrac{Y_g}{Y}}\left[\frac{1}{2}Rd_{31}(E_1 + E_2)\right] \tag{w}$$

Thus, just as we found with the straight beam, a glue layer reduces the displacement for a given electric field, as compared with a ring without glue layer.

## EXAMPLE 8-19

Let us examine the half-ring shown in Fig. 8-116. It consists of layers that are arranged such that $B = 0$, and the electric field produces a piezoelectric moment $M_{\text{piezo}}$, but no piezoelectric force. We wish to find the ring deflection $w$ at the top of the ring, $\theta = \pi$.

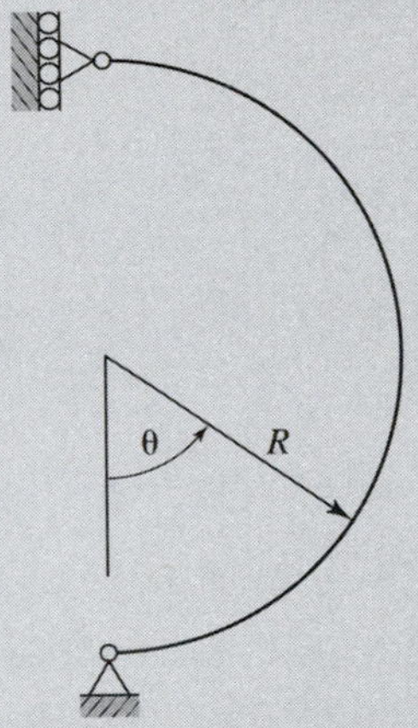


Figure 8-116

**SOLUTION**

Since the half-ring is weightless, it follows that $Q(0) = 0$. By considering a quarter-ring and proceeding as in Example 8-18, we find that

$$N = 0, \qquad Q = 0, \qquad M = 0, \qquad 0 < \theta < \pi \tag{a}$$

Thus all internal forces and moments in the half-ring are zero. In addition, we have the following displacement boundary conditions:

$$v(0) = 0, \qquad v(\pi) = 0 \tag{b}$$

$$w(0) = 0 \tag{c}$$

When Eqs. (*a*), together with $B = 0$ and $N_{\text{piezo}} = 0$, are used in Eqs. (8-416*a*) and (8-417), we get

$$\frac{dv}{d\theta} = -w \tag{d}$$

$$\frac{d^2w}{d\theta^2} + w = -R^2\frac{M_{\text{piezo}}}{D} \tag{e}$$

The solution to Eq. (*e*) is

$$w = a_1 \cos\theta + a_2 \sin\theta - \frac{R^2}{D}M_{\text{piezo}} \tag{f}$$

Equation (*c*) is used to obtain the value of $a_1$,

$$a_1 = R^2\frac{M_{\text{piezo}}}{D} \tag{g}$$

and therefore Eq. (*f*) becomes

$$w = -\frac{R^2}{D}M_{\text{piezo}}(1 - \cos\theta) + a_2 \sin\theta \tag{h}$$

Equation (*h*) is now substituted into Eq. (*d*) which, upon integration, gives

$$v = \frac{R^2}{D}M_{\text{piezo}}(\theta - \sin\theta) + a_2 \cos\theta + a_3 \tag{i}$$

where $a_3$ is a constant of integration. The remaining boundary conditions [Eqs. (*b*)], give

$$a_3 = -a_2$$

$$a_2 = \frac{\pi}{2}\frac{R^2}{D}M_{\text{piezo}} \tag{j}$$

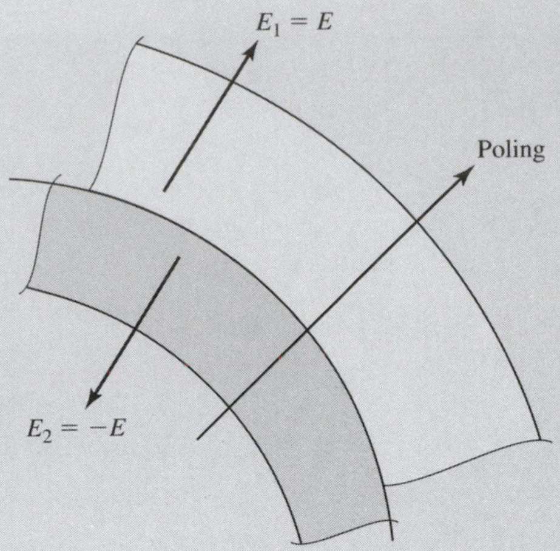


**Figure 8-117** Curved poled bimorph

and therefore the displacements $v$ and $w$ become

$$
\begin{aligned}
v &= \frac{R^2}{D} M_{\text{piezo}} \left[ \theta - \sin\theta - \frac{\pi}{2}(1 - \cos\theta) \right] \\
w &= \frac{R^2}{D} M_{\text{piezo}} \left[ -(1 - \cos\theta) + \frac{\pi}{2} \sin\theta \right]
\end{aligned} \qquad (k)
$$

The vertical deflection $\delta$ at $\theta = \pi$ is therefore

$$
\delta = w(\pi) = -2\frac{R^2}{D} M_{\text{piezo}} \qquad (l)
$$

If we envision a bimorph poled as shown in Fig. 8-117 with fields $E_1$ and $E_2$ equal in magnitude but oppositely directed, but with $E_1$ positive (that is, in the poling direction), then the outer layer will expand while the inner layer contracts. This causes a downward displacement at $\theta = \pi$, which is the reason for the negative sign in Eq. ($l$).

## EXAMPLE 8-20

Piezoelectric actuators in various configurations can be used as a means of actively controlling vibration, as explained in Section 4-16. For example, certain experiments conducted in space, in a zero-gravity environment, are sensitive to vibrations of the space vehicle that are transferred to the experimental apparatus. It is therefore desirable to use some form of active vibration control to reduce the vibrational accelerations reaching the experimental apparatus to the micro-$g$ range.

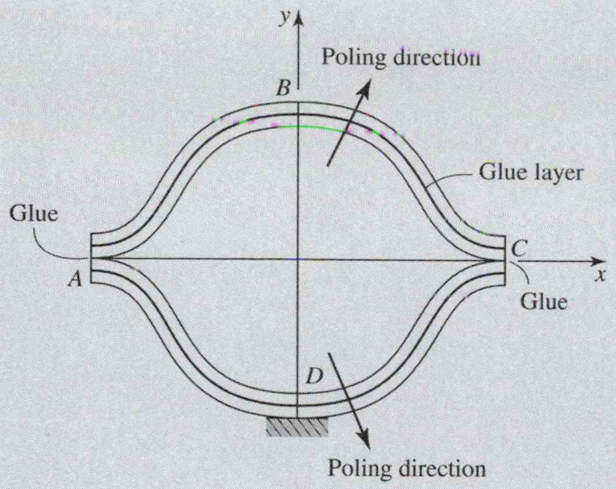


**Figure 8-118** Bellows actuator

Consider the bellows actuator shown in Fig. 8-118 consisting of PVDF bimorphs for the upper and lower halves. Each bimorph has the same cross-sectional dimensions and elastic properties, and the two PVDF films making up a bimorph are separated by a glue layer having different thicknesses and elastic moduli, and possibly different widths, than the PVDF films. The poling direction shown applies along the entire upper bimorph and along the entire lower bimorph. The upper and lower bimorphs are joined by gluing at their ends, points $A$ and $C$. An electric field is applied, which causes the bellows to expand or contract in the $y$ direction at point $B$, where typically a load or mass is attached to the actuator, by changing the curvature of portions of the bellows. The left and right ends of the bellows, points $A$ and $C$, are unconstrained and therefore free to move in the $x$ direction as the bellows displaces vertically. The bottom half of the bellows is a mirror image of the upper half, both mechanically and electrically, so only the upper half need be modeled. Furthermore, a vertical plane through the center of the bellows is a plane of symmetry so that only the left portion of the upper half is modeled. The glue surface attaching the upper and lower halves at points $A$ and $C$ is considered, for modeling purposes, to be large enough to maintain zero slope at those points as the bellows deforms. The quarter-actuator, shown schematically in Fig. 8-119, consists of two segments, 1 and 2, with opposite curvatures. The circled numbers serve to identify the piezoelectric films in each segment according to the scheme shown in Fig. 8-112.

## SOLUTION

The bellows is actuated by an electric field producing a piezoelectric moment $M_{\text{piezo}}$, but no piezoelectric force. Thus $N_{\text{piezo}}$ is zero. In order to maximize the displacement for a given electric field, it is necessary for both segments to simultaneously increase (or decrease) their curvatures. Thus the electric field in the two segments must be directed oppositely, as shown in Fig. 8-120. In order to accomplish this there must be an insulating gap in the electrode surfaces adjacent to the glue layer, at the inflection point, as noted in the figure. Otherwise a short circuit will result. This gap is put in place at the time the electrodes are applied to the PVDF film by the manufacturer. The electroded

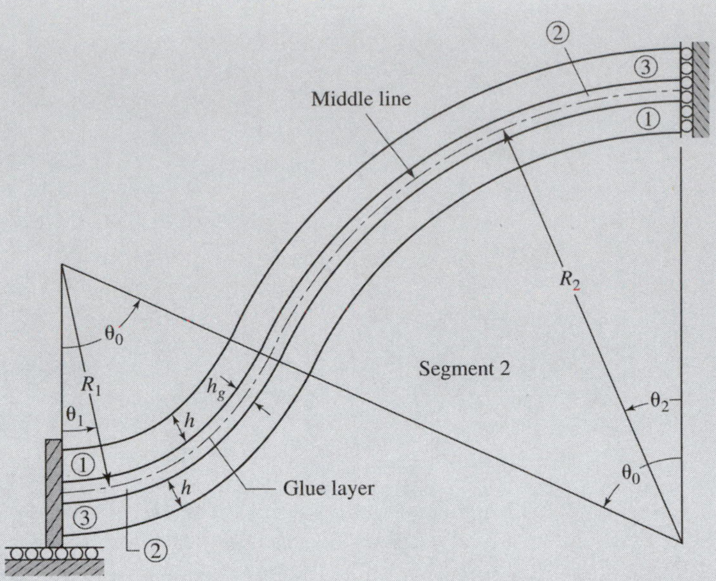


**Figure 8-119**

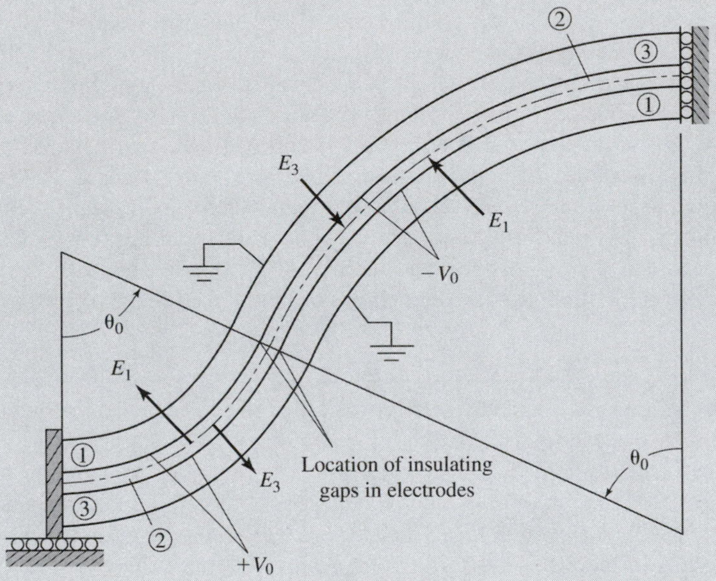


**Figure 8-120** Directions of electric fields in two segments

surfaces of both layers 1 and 3 away from the glue layer are continuous (no insulating gap) and are grounded. The electroded surfaces for layers 1 and 3 adjacent to the glue layer in segment 1 carry a voltage of $+V_0$ whereas those for segment 2 carry a voltage of $-V_0$.

As shown in Fig. 8-119, the radii of curvature of the middle lines of the two segments are $R_1$ and $R_2$. Since the two segments are required to form a smooth curve, the angular location of the interface is $\theta_0$ measured from either the left end of segment 1 or the right end of segment 2. Distances $s_1 = R_1\theta_1$ along segment 1 are measured from its left end, and $s_2 = R_2\theta_2$ along segment 2 from its right end. Hence the origins of the coordinate systems for the two segments are at the left end of segment 1 and the right end of segment 2. As a matter of convenience we will use $s_1$ and $s_2$ as measures of distance. The arc length of segment 1 is $S_1 = R_1\theta_0$ and of segment 2 it is $S_2 = R_2\theta_0$. From Fig. 8-119 the following kinematic constraints are seen to hold:

$$w_1(0) = 0, \qquad \beta_1(0) = 0 \tag{a}$$

$$v_2(0) = 0, \qquad \beta_2(0) = 0 \tag{b}$$

where subscripts 1 and 2 refer to segments 1 and 2, respectively. Furthermore,

$$N_1(0) = 0 \tag{c}$$

$$Q_2(0) = 0 \tag{d}$$

Overall equilibrium of the quarter-actuator gives

$$Q_1(0) = 0 \tag{e}$$

$$N_2(0) = 0 \tag{f}$$

A cut is now made at the inflection point, and the two segments are considered separately, as shown in Fig. 8-121, which also shows the positive internal forces and moments acting on the cut end of each segment. Since equilibrium must hold across the cut, we have

$$N_1(S_1) = N_2(S_2) \tag{g}$$

$$Q_1(S_1) = Q_2(S_2) \tag{h}$$

$$M_1(S_1) = -M_2(S_2) \tag{i}$$

Figure 8-122 shows the kinematic behavior of the two segments at the cut. The labeling of the tangential displacement $v$ has been omitted for clarity. The figure shows the radial displacement $w$, positive in both segments, and $\beta$, negative in both segments [Eq. (8-401)]. Therefore if the radial displacement in both segments is positive, as shown, a gap develops and is given by

$$g = w_1(S_1) + w_2(S_2)$$

Compatibility of displacements at the cut requires that the gap be zero so that

$$w_1(S_1) + w_2(S_2) = 0 \tag{j}$$

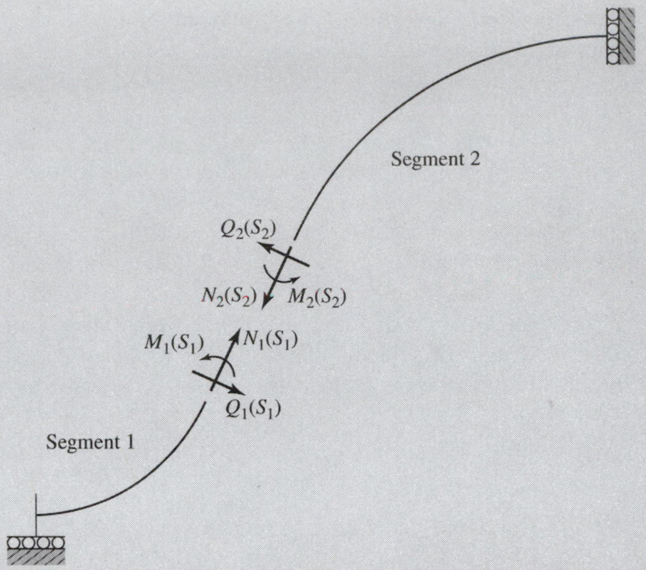


**Figure 8-121** Internal forces and moments acting on end cut of each segment

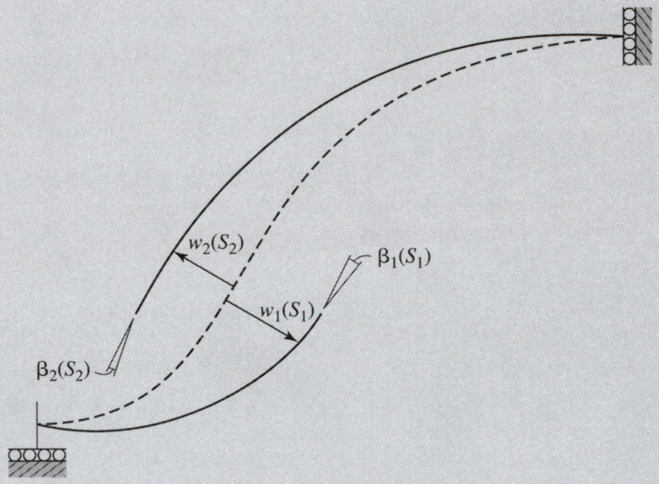


**Figure 8-122**

The tangential displacements $v$, when both are positive as shown in the figure, cause an overlap which, by virtue of displacement compatibility, must also be zero. Thus,

$$v_1(S_1) + v_2(S_2) = 0 \tag{k}$$

As seen in Fig. 8-122, a negative slope in segment 2 has the same sense as a negative slope in segment 1. Slope compatibility therefore requires that both slopes have the same sense and the same magnitude. Consequently,

$$\beta_1(S_1) = \beta_2(S_2) \tag{$l$}$$

In light of Eqs. ($c$)–($f$), equilibrium considerations for segments 1 and 2 require that

$$N_i(S_i) = 0, \qquad Q_i(S_i) = 0 \tag{$m$}$$

where it is understood here and in what follows that $i = 1, 2$ and also that

$$N_i(s_i) = 0, \qquad Q_i(s_i) = 0 \tag{$n$}$$

that is, there are no internal forces acting in either segment. Since the longitudinal force is zero in both segments, it follows from moment equilibrium that

$$M_i(S_i) = M_i(0) = M_{0_i} \tag{$o$}$$

where $M_0$ is a constant and

$$M_i(s_i) = M_i(0) = M_{0_i} \tag{$p$}$$

Equation ($p$) shows that the moment is constant throughout each segment. Equation ($i$), in conjunction with Eq. ($o$), gives

$$M_{0_2}(S_2) = -M_{0_1}(S_1) \tag{$q$}$$

It should be pointed out that, contrary to what Eqs. ($p$) and ($q$) seem to imply, there is no moment discontinuity across the interface between segments. The change in sign arises because the coordinate systems for the two segments are different, and thus the positive sense for moments changes. Therefore Eqs. ($p$) and ($q$) show that the moment is constant throughout the bellows. In fact, a few moments of reflection regarding Fig. 8-122 and Eq. ($l$) should convince the reader that if the piezoelectric moments applied to each segment are the same and if $R_1 = R_2$, then the moment throughout the bellows is zero. This is because, under the conditions stated, if the segments are disconnected at the interface, each will deform through the same slope, in both magnitude and sense, so no moment is required at the interface to maintain slope compatibility. Thus since it is constant, the moment throughout the actuator is necessarily zero.

The solution to Eq. (8-417) for segments 1 and 2 with $N$ and $N_{\text{piezo}}$ both zero, and $M = M_0$, is

$$w_i = a_i \cos\left(\frac{s_i}{R_i}\right) + b_i \sin\left(\frac{s_i}{R_i}\right) - \frac{M_{0_i} + M_{p_i}}{D} R_i^2 \tag{$r$}$$

where the transformation

$$s_i = R_i \theta_i \tag{$s$}$$

has been used. Also, to simplify notation, $M_{\text{piezo}}$ has been replaced with $M_p$. $v$ is given by Eq. (8-416$a$) as

$$\frac{dv_i}{ds_i} = -\frac{w_i}{R_i}$$

and therefore,

$$v_i = -a_i \sin\left(\frac{s_i}{R_i}\right) + b_i \cos\left(\frac{s_i}{R_i}\right) + \frac{M_{0_i} + M_{p_i}}{D} R_i s_i + c_i \qquad (t)$$

where $c_i$ is a constant of integration. The slope β is obtained from Eq. (8-401) and is given by

$$\beta_i = \frac{M_{0_i} + M_{p_i}}{D} s_i + \frac{c_i}{R_i} \qquad (u)$$

With the second of Eqs. (*a*) and (*b*) we find that

$$c_i = 0 \qquad (v)$$

Equation (*l*), in conjunction with Eqs. (*u*) and (*v*), gives

$$\frac{M_{0_1} + M_{p_1}}{D} S_1 = \frac{-M_{0_1} + M_{p_2}}{D} S_2$$

and therefore,

$$M_{0_1} = \frac{M_{p_2} R_2 - M_{p_1} R_1}{R_1 + R_2} \qquad (w)$$

Equation (*w*) gives the moment throughout the bellows in terms of the piezoelectric moments $M_{p_1}$ and $M_{p_2}$.

From the first of Eqs. (*a*) and Eq. (*r*) we find that

$$a_1 = \frac{M_{0_1} + M_{p_1}}{D} R_1^2 \qquad (x)$$

and with the first of Eqs. (*b*), along with Eqs. (*t*) and (*v*),

$$b_2 = 0 \qquad (y)$$

With the constants thus far determined, along with Eqs. (*r*) and (*t*), Eqs. (*j*) and (*k*) provide two equations for determining the remaining constants, $a_2$ and $b_1$,

$$\begin{aligned} a_2 \cos\theta_0 + b_1 \sin\theta_0 &= K_1(1 - \cos\theta_0) + K_2 \\ -a_2 \sin\theta_0 + b_1 \cos\theta_0 &= -K_1(\theta_0 - \sin\theta_0) - K_2\theta_0 \end{aligned} \qquad (z)$$

where

$$\begin{aligned} K_1 = a_1 &= \frac{M_{0_1} + M_{p_1}}{D} R_1^2 \\ K_2 &= \frac{-M_{0_1} + M_{p_2}}{D} R_2^2 \end{aligned} \qquad (aa)$$

Equations (*z*), when solved, give

$$\begin{aligned} a_2 &= (K_1 + K_2)(\cos\theta_0 + \theta_0 \sin\theta_0) - K_1 \\ b_1 &= (K_1 + K_2)(\sin\theta_0 - \theta_0 \cos\theta_0) \end{aligned} \qquad (bb)$$

But with Eqs. ($w$) and ($aa$),

$$K_1 + K_2 = \frac{M_{p_1} + M_{p_2}}{D} R_1 R_2 \tag{cc}$$

and therefore,

$$\begin{aligned} a_2 &= \frac{M_{p_1} + M_{p_2}}{D} R_1 R_2 (\cos\theta_0 + \theta_0 \sin\theta_0) - \frac{M_{p_2}}{D} \frac{R_1^2 R_2}{R_1 + R_2} \\ b_1 &= \frac{M_{p_1} + M_{p_2}}{D} R_1 R_2 (\sin\theta_0 - \theta_0 \cos\theta_0) \end{aligned} \tag{dd}$$

Finally we have

$$\delta = w_2(0) = a_2 - K_2 = (K_1 + K_2)(\cos\theta_0 + \theta_0 \sin\theta_0 - 1) \tag{ee}$$

or, with Eq. ($cc$),

$$\delta = \frac{M_{p_1} + M_{p_2}}{D} R_1 R_2 (\cos\theta_0 + \theta_0 \sin\theta_0 - 1) \tag{ff}$$

The electric fields $E_1$ and $E_3$ are taken to be the same for both segments 1 and 2 so that

$$M_{p_1} = M_{p_2} = M_{\text{piezo}}$$

and

$$\delta = 2 \frac{M_{\text{piezo}}}{D} R_1 R_2 (\cos\theta_0 + \theta_0 \sin\theta_0 - 1) \tag{gg}$$

Since both segments 1 and 2 consist of two piezoelectric layers, of equal thickness and elastic modulus, separated by a glue layer of different thickness and elastic modulus, $D$ is given by Eq. ($f$) of Example 8-15. The piezoelectric properties of each layer are also equal, and we impose electric fields such that the field in layer 2 is the negative of that in layer 1 for each segment. Therefore the piezoelectric moment is given by Eq. ($g$) of Example 8-15. Then Eq. ($gg$) becomes

$$\delta = 4 \frac{A_{\text{ef}} h_{\text{eq}} E_1}{I_{\text{eq}}} R_1 R_2 d_{31} (\cos\theta_0 + \theta_0 \sin\theta_0 - 1)$$

where $A_{\text{ef}}$, the cross-sectional area of the electroded film, is

$$A_{\text{ef}} = bh \tag{hh}$$

$$h_{\text{eq}} = \frac{h + h_g}{2} \tag{ii}$$

$$I_{\text{eq}} = 2\left(I + h_{\text{eq}}^2 A_{\text{ef}}\right) + \frac{Y_g}{Y} I_g \tag{jj}$$

with

$$I = \frac{bh^3}{12} \qquad (kk)$$

$$I_g = \frac{b_g h_g^3}{12} \qquad (ll)$$

In Eq. (*ll*) $b_g$ is the width of the glue layer.

The electric field through the thickness of the piezoelectric film is given by

$$E = -\frac{\partial V}{\partial x_3}$$

where, for the purpose of obtaining the electric field, $x_3$ is normal to the plane of the piezoelectric film, and positive in the poling direction. For the arrangement shown in Fig. 8-120, the electric field $E_1$ for both segments 1 and 2 is therefore given by

$$E_1 = \frac{V_0}{h_f} \qquad (mm)$$

where $h_f$ is the thickness of the piezoelectric film, *without* the electrodes. The deflection of the bellows actuator at point $B$, by virtue of the symmetry invoked at the beginning of the example, is

$$\Delta = 2\delta \qquad (nn)$$

and therefore the deflection of the bellows is

$$\Delta = 8R_1R_2\frac{A_{\text{ef}}}{I_{\text{eq}}}\frac{h_{\text{eq}}}{h_f}d_{31}V_0(\cos\theta_0 + \theta_0\sin\theta_0 - 1) \qquad (oo)$$

The length of the bimorph for the quarter-actuator, that is, the total arc length between points $A$ and $B$, is

$$L = S_1 + S_2 = (R_1 + R_2)\theta_0 \qquad (pp)$$

Therefore Eq. (*oo*) can be written in terms of $L$ as

$$\Delta = 8R_1R_2\frac{A_{\text{ef}}}{I_{\text{eq}}}\frac{h_{\text{eq}}}{h_f}d_{31}V_0\left[\cos\left(\frac{L}{R_1+R_2}\right) + \frac{L}{R_1+R_2}\sin\left(\frac{L}{R_1+R_2}\right) - 1\right] \qquad (qq)$$

If we consider $L$ to be constant (that is, the length of the bimorph is fixed) and allow $R_1$ and $R_2$ to increase, we obtain the large-radius counterpart to Eq. (*qq*),

$$\Delta = 4\frac{R_1R_2}{(R_1+R_2)^2}\frac{A_{\text{ef}}}{I_{\text{eq}}}\frac{h_{\text{eq}}}{h_f}d_{31}V_0L^2 \qquad (rr)$$

In arriving at this result it is necessary to retain two terms in the small-angle expansion for cosine, that is,

$$\cos(x) \approx 1 - \frac{x^2}{2}$$

If the second term is not included, then the approximation of $\cos(x)$ is inconsistent with the approximation of $x\sin(x)$.

For computational purposes we assume that $R_1 = R_2 = R$. Therefore Eqs. ($qq$) and ($rr$) become

$$\Delta = 8\frac{A_{\text{ef}}R^2}{I_{\text{eq}}}\frac{h_{\text{eq}}}{h_f}V_0 d_{31}(\cos\theta_0 + \theta_0\sin\theta_0 - 1) \tag{ss}$$

and

$$\Delta = \frac{A_{\text{ef}}}{I_{\text{eq}}}\frac{h_{\text{eq}}}{h_f}d_{31}V_0L^2 \tag{tt}$$

where

$$\theta_0 = \frac{L}{2R} \tag{uu}$$

The bimorphs for the bellows are constructed of PVDF film, with $b = 47.24$ mm, $L = 18.42$ mm, and $R = 31.5$ mm. The average thickness of the electroded film is $h = 40$ μm, the thickness of the unelectroded film is $h_f = 28$ μm, and the thickness of the glue layer is $h_g = 15$ μm.

The material properties for PVDF are found from the first of Eqs. (4-213), in conjunction with the first of Eqs. (4-53) and the third of Eqs. (4-214), while the elastic modulus for the epoxy glue is obtained as described in Example 8-15. The values are

$$Y = 2.74 \text{ Gpa} \qquad Y_g = 4.24 \text{ Gpa} \qquad d_{31} = 21 \text{ pC/N}$$

The resulting curve of displacement versus voltage, obtained from Eq. ($ss$), is shown in Fig. 8-123 for glue layer thicknesses of 0 μm and 15 μm. As was noted in an earlier

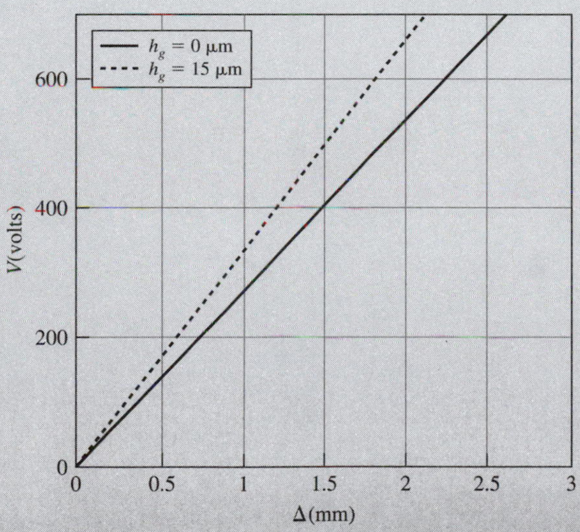


**Figure 8-123** Displacement versus voltage

example for the straight bimorph, the presence of the glue layer tends to reduce the displacement for a given voltage.

Figure 8-124 shows a comparison of the results predicted by Eq. (*ss*) and experimental values for the displacement obtained by testing three actuators with the dimensions and properties given in the foregoing. The figure shows that the agreement between predicted and experimental values, overall, is quite good. Up to about 400 volts, the

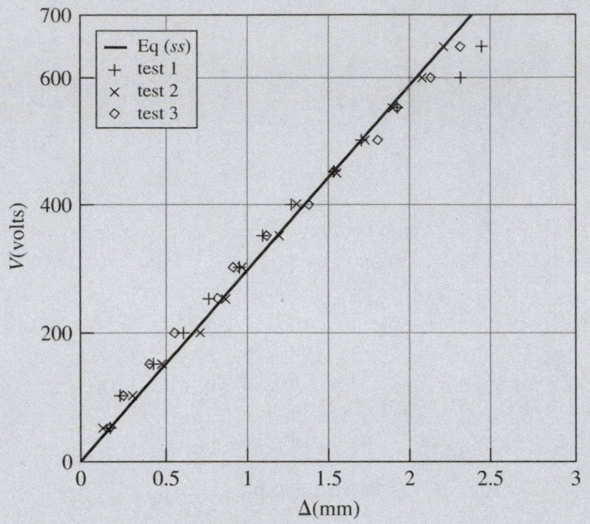


**Figure 8-124** Comparison of predicted and experimental values

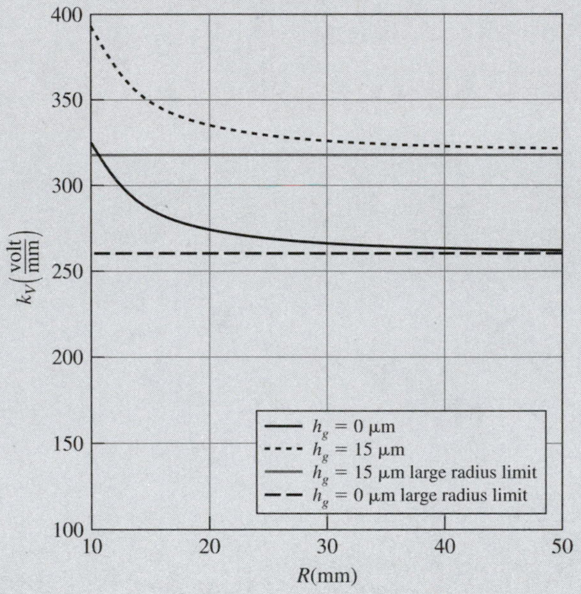


**Figure 8-125** Electrical stiffness $k_v$ as a function of $R$

agreement is excellent; beyond that Eq. (*ss*) predicts that a higher voltage is required to obtain a certain displacement that is found experimentally.

Figure 8-125 shows the electrical stiffness $k_v$ obtained from Eq. (*ss*) as

$$k_v = \frac{V_0}{\Delta} = \left[8\frac{AR^2}{I_{eq}}\frac{h_{eq}}{h_f}d_{31}(\cos\theta_0 + \theta_0\sin\theta_0 - 1)\right]^{-1} \qquad (vv)$$

as a function of $R$ for glue layer thicknesses of 0 μm and 15 μm. Also shown is the large radius limit obtained by Eq. (*tt*) as

$$k_v = \frac{h_f}{h_{eq}}\frac{I_{eq}}{AL^2d_{31}} \qquad (ww)$$

The figure shows that the stiffness is quite sensitive to small changes in $R$ when $R$ is small, and becomes less sensitive as $R$ increases. Thus from the point of view of actuator design, for actuators with small values of $R$ the stiffness will be sensitive to manufacturing tolerances on the radius.

If $M$ and $N$ cannot be obtained by considering the statical equilibrium of the external forces (applied forces and reactions at the supports), then additional equations are required. These are obtained by considering the equilibrium of internal forces.

Figure 8-126 shows the middle line of an element of a curved beam with arc length $ds = R\,d\theta$. An externally distributed load $q$ is applied to the beam, as shown in the figure. As a result of the external load and the piezoelectric effects, internal stress resultants $M$ and $N$, as well as a shear $Q$, develop within the element. In general these change from point to point in the ring. We follow the convention that positive internal forces and moments are directed in positive coordinate directions on positive element faces and in

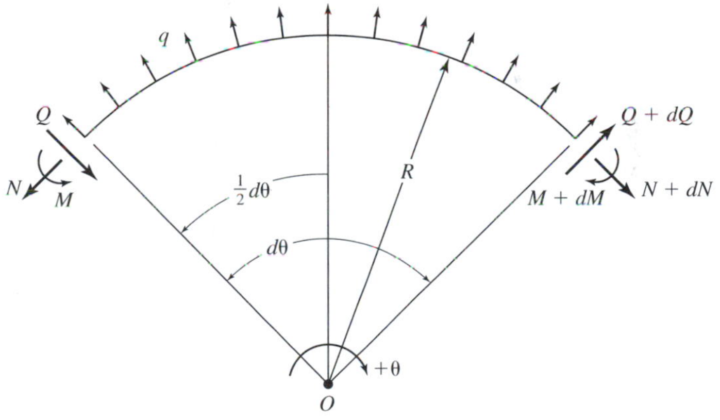


Figure 8-126 Middle line of an element of a curved beam

negative coordinate directions on negative element faces. External loads are taken to be positive in positive coordinate directions. We take the radial coordinate to be positive when directed radially outward and $\theta$ to be positive in the clockwise direction. Then, for a right-handed coordinate system the positive $z$ axis must be directed into the plane. Thus all forces and moments are positive, as shown in Fig. 8-126. After decomposing the forces into horizontal and vertical components, and taking advantage of the small-angle approximations for sin $\theta$ and cos $\theta$, we sum forces in the horizontal and vertical directions. This yields

$$\frac{dN}{d\theta} + Q = 0 \tag{8-418a}$$

$$\frac{dQ}{d\theta} - N = -Rq \tag{8-418b}$$

In obtaining these equations we have neglected terms containing products of differential quantities. Next we sum moments about point $O$. This gives

$$\frac{dM}{d\theta} + R\frac{dN}{d\theta} = 0 \tag{8-419}$$

Equation (8-419) can be integrated, in which case we find that

$$M + RN = \text{constant} \tag{8-420}$$

In other words, both $M$ and $N$ can change as we move around the curved beam, but they must do so in such a way that the quantity $M + RN$ does not change in value.

Equations (8-418) and (8-419) can be put in various useful forms. For example, Eq. (8-418) can be uncoupled by differentiating one and substituting the other. We then obtain

$$\begin{aligned} \frac{d^2N}{d\theta^2} + N &= Rq \\ \frac{d^2Q}{d\theta^2} + Q &= -R\frac{dq}{d\theta} \end{aligned} \tag{8-421}$$

If Eq. (8-418$a$) is used in Eq. (8-419), we get

$$\frac{dM}{d\theta} = RQ \tag{8-422}$$

When this is differentiated, it becomes

$$\frac{d^2M}{d\theta^2} = R\frac{dQ}{d\theta} \tag{8-423}$$

and when Eq. (8-418$b$) is used, the resulting equation is

$$\frac{d^2M}{d\theta^2} = RN - R^2q \tag{8-424}$$

Equations (8-419) and (8-424) can be rewritten in terms of displacements by substituting Eqs. (8-415). This gives

$$(AR^2 + 2BR + D)\frac{d^2v}{d\theta^2} - (BR + D)\frac{d^3w}{d\theta^3} + (AR^2 + BR)\frac{dw}{d\theta} = 0 \tag{8-425a}$$

$$D\frac{d^4w}{d\theta^4} - 2BR\frac{d^2w}{d\theta^2} + AR^2w - (BR + D)\frac{d^3v}{d\theta^3} + (AR^2 + BR)\frac{dv}{d\theta} = R^4q + R^3N_{\text{piezo}} \tag{8-425b}$$

These equations can be uncoupled to obtain a single equation for $w$ [15]. This is accomplished by rearranging Eq. (8-425$a$) to give

$$\frac{d^2v}{d\theta^2} = \frac{BR + D}{AR^2 + 2BR + D}\frac{d^3w}{d\theta^3} - \frac{AR^2 + BR}{AR^2 + 2BR + D}\frac{dw}{d\theta} \tag{8-426}$$

This equation is now integrated and differentiated and the results are substituted into Eq. (8-425$b$). After some algebra, the following equation is obtained for $w$:

$$\frac{d^4w}{d\theta^4} + 2\frac{d^2w}{d\theta^2} + w = \frac{R^4}{k_3}q + \frac{R^3}{k_3}N_{\text{piezo}} + a_5 \tag{8-427}$$

where

$$k_3 = \frac{(AD - B^2)R^2}{AR^2 + 2BR + D} \tag{8-428}$$

and $a_5$ is a constant of integration arising from the integration of Eq. (8-426).

Other pertinent quantities can be expressed in terms of $w$. For example, Eq. (8-415$a$) can be solved for $dv/d\theta$ and the result used in Eq. (8-415$b$). This gives

$$\frac{dv}{d\theta} = \frac{R^2}{AR + B}\left(\frac{B}{R^2}\frac{d^2w}{d\theta^2} - \frac{A}{R}w + N + N_{\text{piezo}}\right) \tag{8-429}$$

$$M = -\frac{AD - B^2}{AR + B}\frac{1}{R}\left(\frac{d^2w}{d\theta^2} + w\right) + \frac{BR + D}{AR + B}(N + N_{\text{piezo}}) - M_{\text{piezo}} \tag{8-430}$$

From Eq. (8-420),

$$N = -\frac{M}{R} + \frac{a_6}{R} \tag{8-431}$$

where $a_6$ is a constant. Equation (8-431) is used to eliminate $N$ from Eq. (8-430), and the resulting equation is solved for $M$. This yields, after some algebra,

$$M = -\frac{k_3}{R^2}\left(\frac{d^2w}{d\theta^2} + w\right) + k_1RN_{\text{piezo}} - k_2M_{\text{piezo}} + k_1a_6 \tag{8-432}$$

where

$$k_1 = \frac{BR + D}{AR^2 + 2BR + D} \tag{8-433}$$

$$k_2 = \frac{(AR + B)R}{AR^2 + 2BR + D} \tag{8-434}$$

Equation (8-431) then gives $N$ as

$$N = \frac{k_3}{R^3}\left(\frac{d^2w}{d\theta^2} + w\right) - k_1 N_{\text{piezo}} + \frac{k_2}{R} M_{\text{piezo}} + \frac{1}{R}(1 - k_1)a_6 \tag{8-435}$$

$Q$ is obtained from Eqs. (8-422) and (8-432) as

$$Q = \frac{k_3}{R^3}\left(\frac{d^2w}{d\theta^3} + \frac{dw}{d\theta}\right) \tag{8-436}$$

Finally we substitute Eq. (8-435) into Eq. (8-429),

$$\frac{dv}{d\theta} = k_1 \frac{d^2w}{d\theta^2} - k_2 w + k_4 R^2 M_{\text{piezo}} + k_4 R^3 N_{\text{piezo}} + k_4 R^2 a_6 \tag{8-437}$$

where

$$k_4 = \frac{1}{AR^2 + 2BR + D} \tag{8-438}$$

Integration of Eq. (8-437) gives $v$.

In problems where $q$ is a uniformly distributed load $q_0$, the complete solution to Eq. (8-427) is readily obtained as

$$w = a_1 \cos\theta + a_2 \sin\theta + a_3\theta \cos\theta + a_4\theta \sin\theta + \frac{R^4}{k_3} q_0 + \frac{R^3}{k_3} N_{\text{piezo}} + a_5 \tag{8-439}$$

The constants $a_i$ $(i = 1, 2, \ldots, 6)$ plus the constant that results from integrating Eq. (8-437) are found from the boundary conditions. It is worth mentioning that if the cross section is symmetric in its geometry and elastic modulus so that $B = 0$, the forms of Eq. (8-427) and of its solution, Eq. (8-439), remain unchanged. Only the constants $k_i$ $(i = 1, 2, 3, 4)$ change.

## EXAMPLE 8-21

The semicircular beam shown in Fig. 8-127 is pinned at both ends. A piezoelectric moment, but no piezoelectric force, is applied to the beam. Find the vertical deflection at $\theta = \pi/2$.

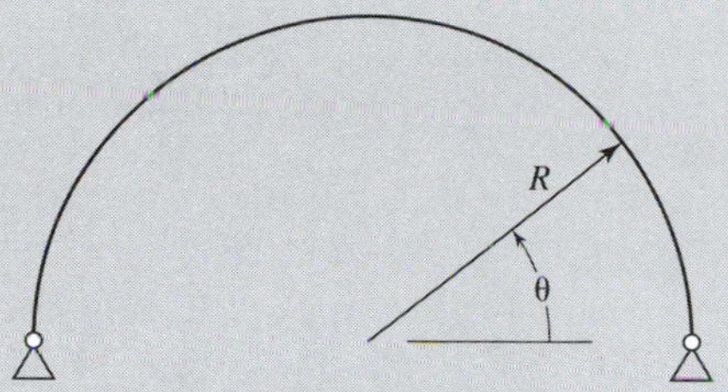


Figure 8-127

## SOLUTION

Owing to symmetry about a vertical plane through the center of curvature, we need consider only the quarter-ring shown in Fig. 8-128, for which the boundary conditions are

$$v(0) = 0, \qquad w(0) = 0 \tag{a}$$

$$v\left(\frac{\pi}{2}\right) = 0, \qquad \beta\left(\frac{\pi}{2}\right) = 0 \tag{b}$$

$$M(0) = 0, \qquad Q\left(\frac{\pi}{2}\right) = 0 \tag{c}$$

The free-body diagram of the quarter-ring shown in Fig. 8-129 gives the additional conditions

$$N(0) = 0 \tag{d}$$

With $q_0 = 0$ and $N_{\text{piezo}} = 0$ the expression for $w$ given by Eq. (8-439) results in

$$w = a_1 \cos\theta + a_2 \sin\theta + a_3\theta\cos\theta + a_4\theta\sin\theta + a_5 \tag{e}$$

and therefore $M$, $N$, and $Q$, from Eqs. (8-432), (8-435), and (8-436), are

$$M = -\frac{k_3}{R^2}(-2a_3\sin\theta + 2a_4\cos\theta + a_5) - k_2 M_{\text{piezo}} + k_1 a_6 \tag{f}$$

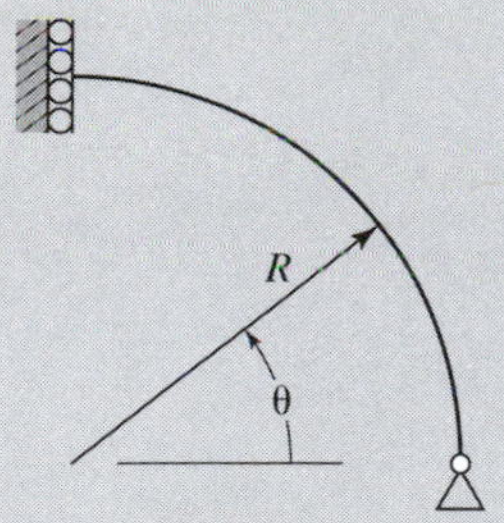


Figure 8-128

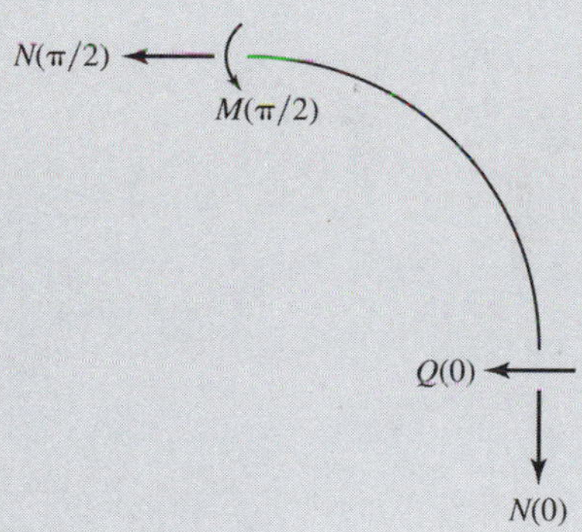


**Figure 8-129** Free-body diagram of quarter ring

$$N = \frac{k_3}{R^3}(-2a_3 \sin\theta + 2a_4 \cos\theta + a_5) + \frac{k_2}{R} M_{\text{piezo}} + \frac{k_2}{R} a_6 \qquad (g)$$

$$Q = \frac{k_3}{R^3}(2a_3 \cos\theta + 2a_4 \sin\theta) \qquad (h)$$

In obtaining Eq. ($g$) we have used the result that, from Eqs. (8-433) and (8-434),

$$k_1 + k_2 = 1$$

The second of Eqs. ($c$), in conjunction with Eq. ($h$), gives

$$a_4 = 0 \qquad (i)$$

The first of Eqs. ($c$) and Eqs. ($d$), ($f$), ($g$), in conjunction with Eq. ($i$), give

$$-\frac{k_3}{R^2} a_5 - k_2 M_{\text{piezo}} + k_1 a_6 = 0 \qquad (j)$$

$$\frac{k_3}{R^3} a_5 + \frac{k_2}{R} M_{\text{piezo}} + \frac{k_2}{R} a_6 = 0 \qquad (k)$$

When Eq. ($k$) is multiplied by $R$ and added to Eq. ($j$) we find that

$$a_6 = 0 \qquad (l)$$

Equation ($j$) then gives

$$a_5 = -\frac{k_2}{k_3} R^2 M_{\text{piezo}} \qquad (m)$$

Equation (8-437), together with the constants obtained thus far, gives $v$ as

$$v = -a_1 \sin\theta + a_2 \cos\theta - a_3[(1 - 2k_1)\cos\theta + \theta \sin\theta] + \left(\frac{k_2^2}{k_3} + k_4\right) R^2 M_{\text{piezo}}\theta + a_7 \qquad (n)$$

The slope, given by Eq. (8-401), with Eqs. ($e$), ($n$), and the constants that have been determined, is

$$R\beta = -2a_2 k_3 \cos\theta + \left(\frac{k_2^2}{k_3} + k_4\right) R^2 M_{\text{piezo}}\theta + a_7 \qquad (o)$$

Equation ($e$) and the second of Eqs. ($a$) give $a_1$ as

$$a_1 = \frac{k_2}{k_3} R^2 M_{\text{piezo}} \qquad (p)$$

Equation ($o$) and the second of Eqs. ($b$) give $a_7$ as

$$a_7 = -\frac{\pi}{2}\left(\frac{k_2^2}{k_3} + k_4\right) R^2 M_{\text{piezo}} \qquad (q)$$

With $a_1$ and $a_7$ determined, $a_3$ is obtained from Eq. ($n$) and the first of Eqs. ($b$),

$$a_3 = -\frac{2}{\pi}\frac{k_2}{k_3}R^2 M_{\text{piezo}} \tag{r}$$

Finally the remaining constant, $a_2$, is determined from Eq. ($n$) and the first of Eqs. ($a$),

$$a_2 = -\left[\frac{2}{\pi}\frac{k_2(1-2k_1)}{k_3} - \frac{\pi}{2}\left(\frac{k_2^2}{k_3} + k_4\right)\right] R^2 M_{\text{piezo}} \tag{s}$$

With the constants determined, Eq. ($e$) gives $w$ as

$$w = \left\{\frac{k_2}{k_3}\left[\left(1 - \frac{2}{\pi}\theta\right)\cos\theta - 1\right] - \left[\frac{2}{\pi}\frac{k_2(1-2k_1)}{k_3} - \frac{\pi}{2}\left(\frac{k_2^2}{k_3} + k_4\right)\sin\theta\right]\right\} R^2 M_{\text{piezo}} \tag{t}$$

and therefore,

$$\delta = w\left(\frac{\pi}{2}\right) = -\frac{k_2}{k_3}\left\{1 + \left[\frac{2}{\pi}(1-2k_1) - \frac{\pi}{2}\left(k_2 + \frac{k_3 k_4}{k_2}\right)\right]\right\} R^2 M_{\text{piezo}} \tag{u}$$

With Eqs. (8-428), (8-433), and (8-434) we find that

$$\frac{k_2}{k_3} = \frac{AR + B}{(AD - B^2)R} \tag{v}$$

$$1 - 2k_1 = \frac{AR^2 - D}{AR^2 + 2BR + D} \tag{w}$$

$$k_2 + \frac{k_3 k_4}{k_2} = \frac{AR}{AR + B} \tag{x}$$

and therefore Eq. ($u$) becomes

$$\delta = -\frac{AR + B}{AD - B^2}\left(1 - \frac{\pi}{2}\frac{AR}{AR + B} + \frac{2}{\pi}\frac{AR^2 - D}{AR^2 + D}\right) R M_{\text{piezo}} \tag{y}$$

Finally if $B = 0$, then

$$\delta = \left(\frac{\pi}{2} - 1 - \frac{2}{\pi}\frac{AR^2 - D}{AR^2 + D}\right) R^2 \frac{M_{\text{piezo}}}{D} \tag{z}$$

### 8-11-5 Castigliano's Theorem for Thin Curved Piezoelectric Beams

If the undeformed radius of curvature for a curved beam is small compared to the thickness of the beam, then the expression for the reduced complementary energy given by Eq. (8-387*a*) can be adapted to curved beams simply by replacing $x$ with $R\theta$ and integrating over an appropriate range of $\theta$. Thus,

$$U_{cR} = R\int_0^{\theta_0}\left(\frac{D}{D^*}\frac{N_T^2}{2A} - B\frac{N_T M_T}{AD^*} + 2\frac{M_T^2}{D^*}\right)d\theta \qquad (8\text{-}440)$$

Equation (8-440) is valid for $R/t \geq 10$, where $t$ is the beam thickness.

## EXAMPLE 8-22

Use Castigliano's theorem to solve the problem in Example 8-21 with $B = 0$.

### SOLUTION

Since we wish to obtain the displacement at $\theta = \pi/2$ and there is no concentrated load at that location, we apply a load $P$. In order to make the problem statically determinate we release the constraint preventing a displacement $w$ at $\theta = \pi$ and apply a reaction force $F$. This arrangement is shown in Fig. 8-130*a* and the corresponding free-body diagram in Fig. 8-130*b*. Ultimately $F$ is chosen so that $w(\pi) = 0$.

To get the internal forces and moments, we cut the beam somewhere in the range $0 < \theta < \pi/2$, as shown in Fig. 8-131*a*. The forces $P/2$ and $F$ are decomposed into components parallel and perpendicular to $N$. Then summing forces and moments, we

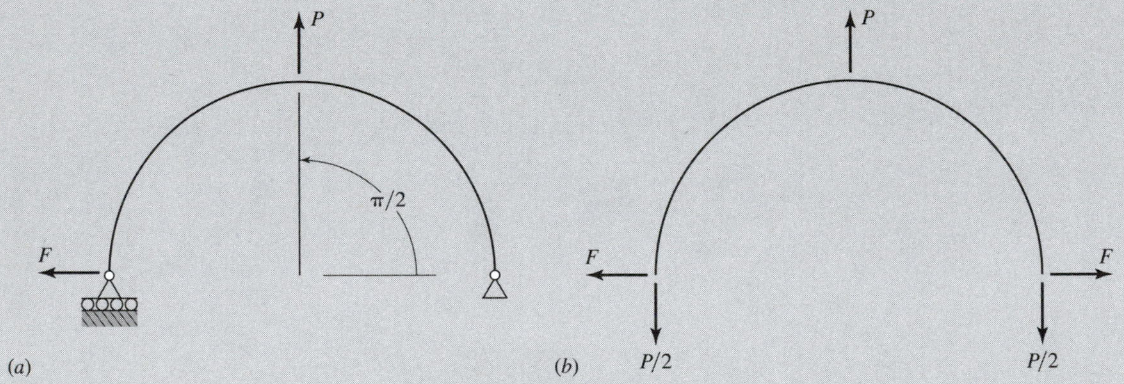


**Figure 8-130** Statically determinate equivalent and corresponding free-body diagram

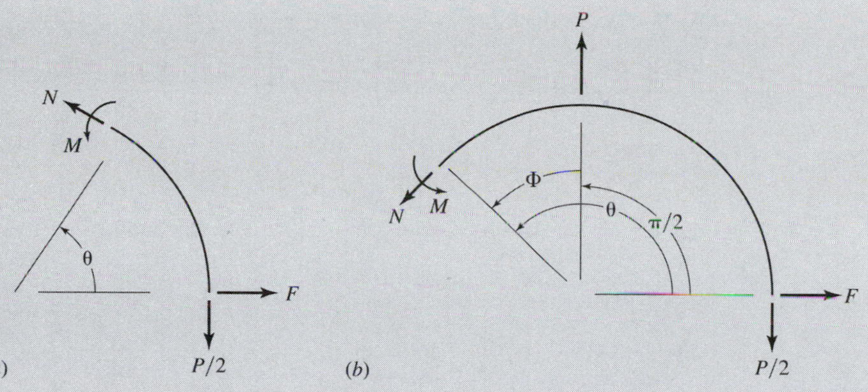


**Figure 8-131**

obtain for $0 < \theta < \pi/2$,

$$N = F \sin\theta$$

$$M = -NR + \frac{P}{2}R = -R\left[F\sin\theta - \frac{P}{2}(1-\cos\theta)\right] \tag{a}$$

For the remainder of the beam the free-body diagram in Fig. 8-131*b* applies, from where, for $\pi/2 < \theta < \pi$,

$$N = F\sin\theta - \frac{P}{2}\cos\theta, \qquad M = -R\left[F\sin\theta - \frac{P}{2}(1+\cos\theta)\right] \tag{b}$$

The reduced complementary energy, given by Eq. (8-440), is then

$$U_{cR} = R\int_0^{\pi/2}\left\{\frac{\left(F\sin\theta + \frac{P}{2}\cos\theta\right)^2}{2A} + \frac{\left\{-R\left[F\sin\theta - \frac{P}{2}(1-\cos\theta)\right] + M_{\text{piezo}}\right\}^2}{2D}\right\}d\theta$$

$$+ R\int_{\pi/2}^{\pi}\left\{\frac{\left(F\sin\theta - \frac{P}{2}\cos\theta\right)^2}{2A} + \frac{\left\{-R\left[F\sin\theta - \frac{P}{2}(1+\cos\theta)\right] + M_{\text{piezo}}\right\}^2}{2D}\right\}d\theta$$

Carrying out the integration, we get

$$U_{cR} = \frac{\pi}{4}\frac{D+AR^2}{AD}RF^2 + \left(\frac{1}{2}\frac{D-AR^2}{AD}RP - 2R^2\frac{M_{\text{piezo}}}{D}\right)F$$

$$+\frac{1}{16}\frac{\pi D + AR^2(3\pi-8)}{AD}RP^2 + \frac{1}{2}\frac{\pi-2}{D}R^2M_{\text{piezo}}P + \frac{1}{2}\frac{\pi}{D}RM_{\text{piezo}}^2 \tag{c}$$

This is now differentiated with respect to $P$ and then $F$. $P$ is then set equal to zero to give

$$\delta = w\left(\frac{\pi}{2}\right) = \frac{dU_{cR}}{dP} = \frac{1}{2}\frac{D - AR^2}{AD}RF + \frac{1}{2}\frac{\pi - 2}{D}R^2 M_{\text{piezo}} \qquad (d)$$

$$\Delta = w(\pi) = \frac{dU_{cR}}{dF} = \frac{\pi}{2}\frac{D + AR^2}{AD}RF - 2R^2\frac{M_{\text{piezo}}}{D} \qquad (e)$$

Equation ($e$) can now be solved for the reaction $F$ needed to maintain $\Delta = 0$. This gives

$$F = \frac{4}{\pi}\frac{A}{D + AR^2}RM_{\text{piezo}} \qquad (f)$$

With this value for the reaction, Eq. ($d$) gives the displacement $\delta$ as

$$\delta = \left(\frac{\pi}{2} - 1 - \frac{2}{\pi}\frac{AR^2 - D}{AR^2 + D}\right)R^2\frac{M_{\text{piezo}}}{D} \qquad (g)$$

which is in agreement with Eq. ($z$) of Example 8-21.

## PROBLEMS

### Section 8-1-2

**8-1** Apply the method of initial parameters to analyze the beam shown in Fig. P8-1.

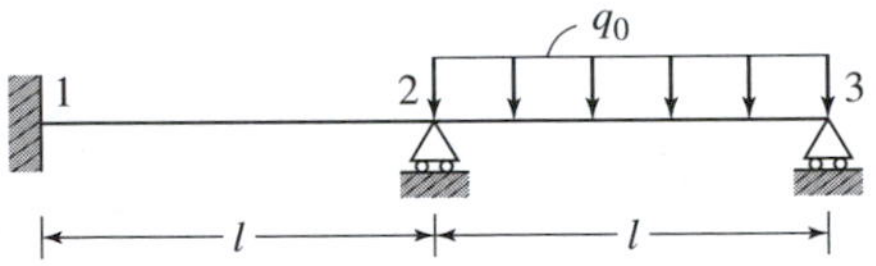


Figure P8-1

**8-2** Select from all two-span beams of total length $L$ and with the loads and supports shown in Fig. P8-2 such that the integral of the absolute values of the bending moments is minimum. *Note:* Assume that $J_2 = J_1$.

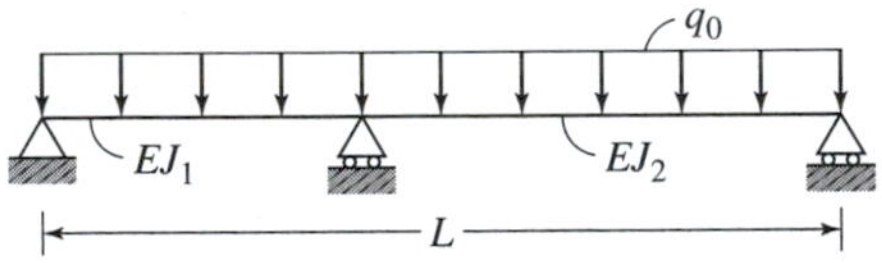


Figure P8-2

### Section 8-1-3

**8-3** Derive Eqs. (8-37).

**8-4** Use Castigliano's theorem to determine the deflection of a bent cantilever of a square cross section (Fig. P8-4) at a point $D$ in the direction normal to $CD$.

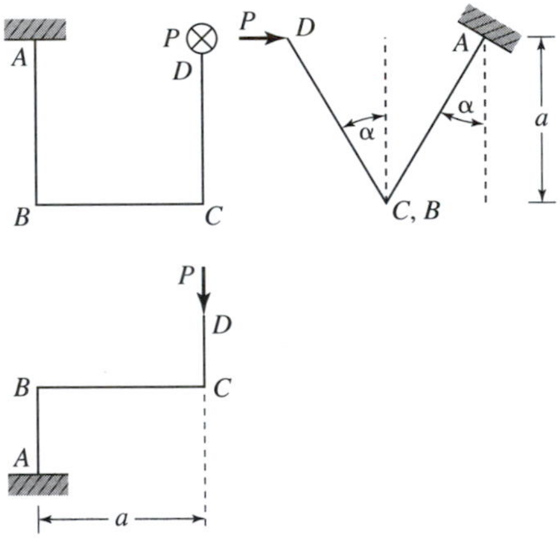


Figure P8-4

**8-5** Determine the bending moments in the beam shown in Fig. P8-5.

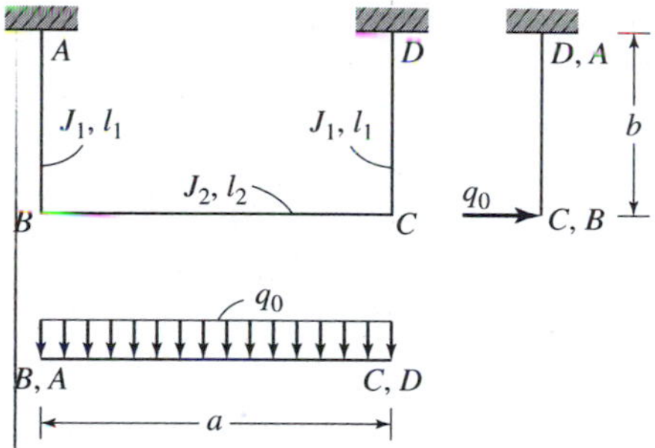


Figure P8-5

## Section 8-2

**8-6** Derive Eqs. (8-43) using Eqs. (8-42).

**8-7** Derive Eq. (8-44).

**8-8** Derive Eqs. (8-45).

**8-9** A cantilever beam of length $l$ and the cross section shown in Fig. P8-9 carries a load $P$ at the tip. Determine the maximum normal stress and the maximum deflection.

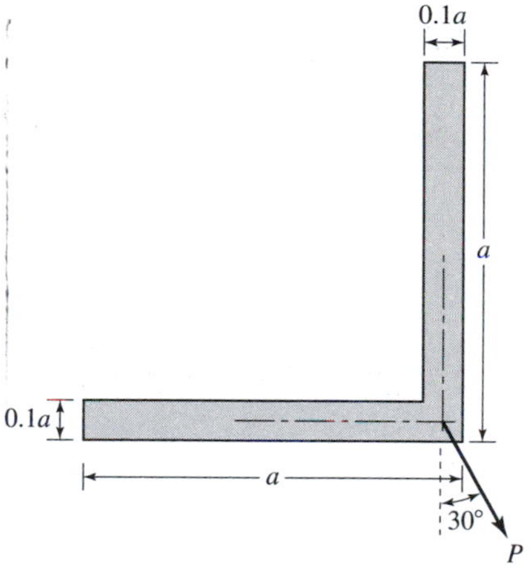


Figure P8-9

## Sections 8-3-1, 8-3-2

**8-10** Apply Castigliano's theorem to determine the bending moments in the statically indeterminate beam shown in Fig. P8-10.

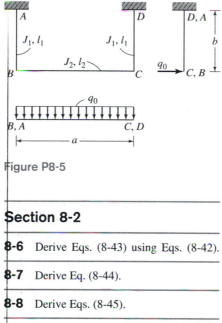

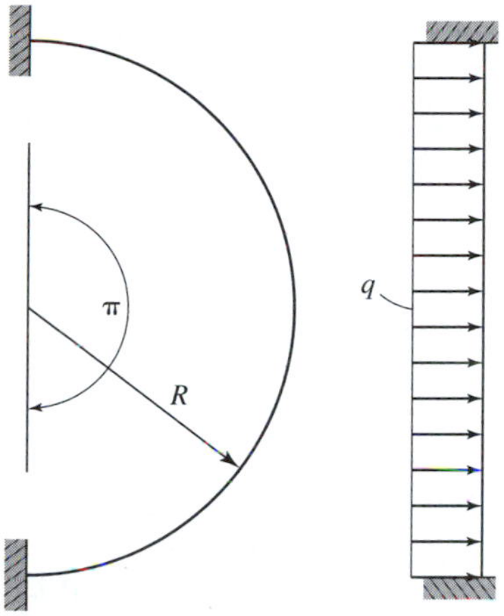


Figure P8-10

**8-11** Give a detailed derivation of Eqs. (8-61) and (8-62).

**8-12** Give a detailed derivation of Eqs. (8-63), (8-64), and (8-65).

**8-13** Give a detailed derivation of Eqs. (8-73) and (8-74).

**8-14** Derive Eq. (8-77).

**8-15** Modify Eqs. (8-73) and (8-74) to derive Eqs. (8-80) and (8-81).

**8-16** Determine the bending moments, the torques, and the shear forces in a ring supported at $A$, $B$, and $C$ and carrying a concentrated force at $D$ (Fig. P8-16).

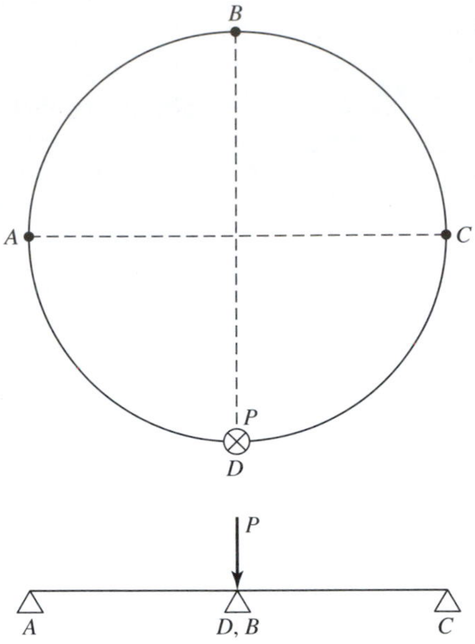


Figure P8-16

## Section 8-3-3

**8-17** Derive Eqs. (8-88) from Eqs. (8-87).

**8-18** Determine the axial forces and the bending moments for an arch with the parabolic axis shown in Fig. P8-18.

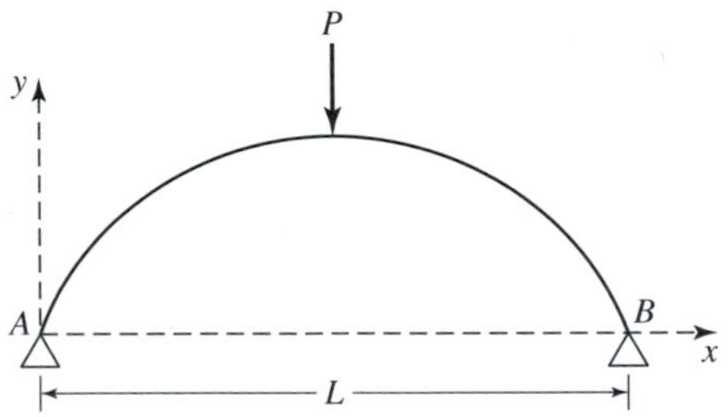


Figure P8-18

**8-19** Determine the axial forces and the bending moments for the arch in Fig. P8-18 but carrying a uniform load $q$ perpendicular to the $x$ axis.

**8-20** Derive Eq. (8-105) from Eq. (8-104).

**8-21** Find the maximum stresses in a thick ring of an arbitrary cross section (Fig. P8-21). *Note:* Use Castigliano's theorem.

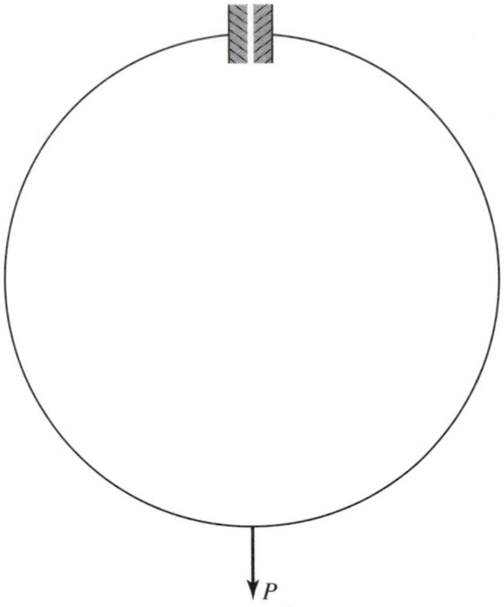


Figure P8-21

## Section 8-4

**8-22** Use the principle of superposition to determine the deflection $y(x)$ of an infinite beam on an elastic foundation carrying loads as shown in Fig. P8-22.

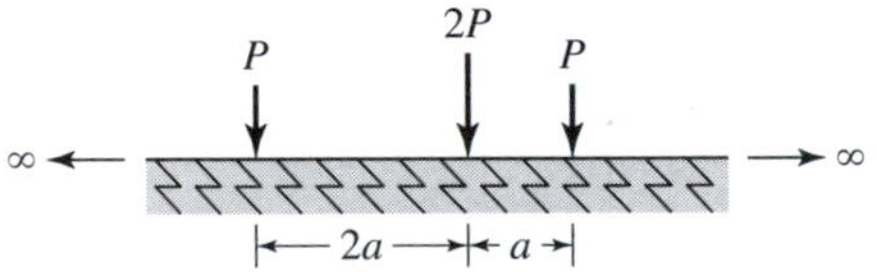


Figure P8-22

**8-23** Design a circular tank of minimum weight to be completely filled with water of a given volume $V$. Assume that the material of the tank is preassigned but that its height, its radius, and the constant thickness of its wall are design variables. The material of the foundation is not to be taken into account. The wall must withstand the maximum stress.

## Section 8-5

**8-24** Prove the symmetry of Green's functions from Table 8-3:

(a) Position 1.
(b) Position 2.
(c) Position 3.
(d) Position 4.

**8-25** Use the Green's function from Table 8-3, position 4, to find the deflection of a beam carrying a uniform load $q$.

**8-26** Use Green's function from Table 8-3, position 1, to find the deflection of a beam carrying a concentrated force $P$ at the middle.

**8-27** Verify position 2 in Table 8-3.

**8-28** Verify position 3 in Table 8-3.

**8-29** Use the Green's function from Table 8-4, position 1, to find the deflection of a two-span beam with $l_1 = 2l_2$ and carrying a uniform load $q$.

**8-30** Use the Green's function [Eq. (8-225)] to determine the deflection of a semi-infinite beam on an elastic foundation carrying a uniform load $q$ (Fig. P8-30).

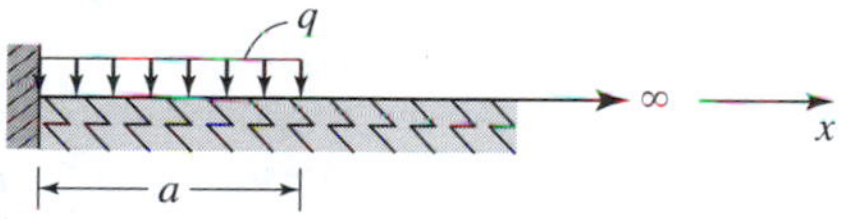


Figure P8-30

**8-31** Write a Fortran program to calculate the deflection of a finite beam (case 2) carrying a uniform load.

## Sections 8-6, 8-7

**8-32** Determine the bending moment due to the temperature change $\Delta T$, symmetric with respect to the $z$ axis, in a simply supported beam of rectangular cross section.

**8-33** Find the thermally induced stresses in a two-span beam (Fig. P8-33) whose one span is subjected to the temperature change

$$\Delta T(x, y, z) = T_0\left(z - \frac{h}{2}\right)\frac{2}{h}$$

whereas the other is kept at constant temperature $T_0$.

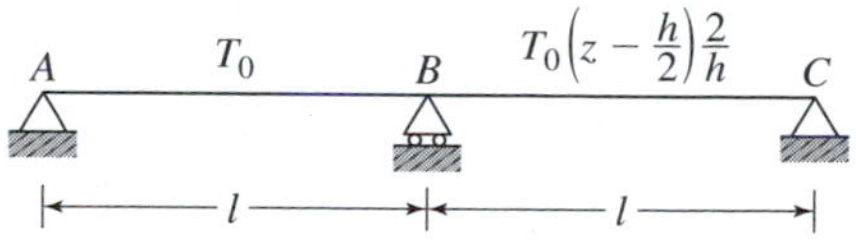


Figure P8-33

**8-34** Determine the maximum stress in a simply supported three-layer beam (Fig. P8-34) with the outer layers made of steel and the inner layer (the core) out of aluminum.

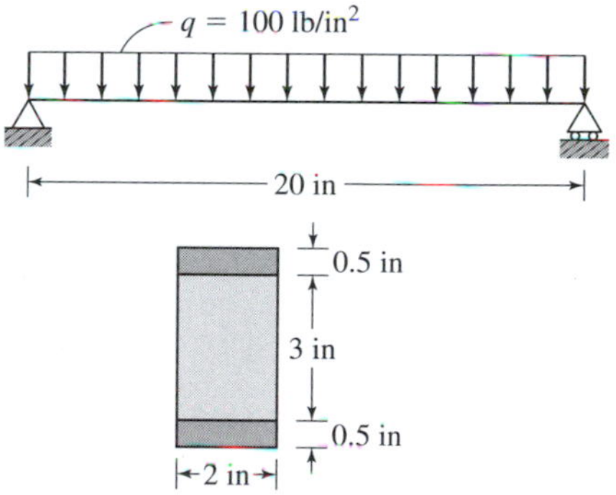


Figure P8-34

## Sections 8-9, 8-10

**8-35** Use Eq. (8-296) to determine the maximum deflection of a simply supported beam carrying a uniform load $q$. Compare the results with a closed-form solution.

**8-36** Use Fourier sine series to find the deflection of a clamped beam carrying a uniform load.

**8-37** Derive finite difference equations for a beam with variable height clamped on both ends. *Assume:* $J(x) = J_0(1 + x/l)$.

**8-38** Apply the Rayleigh–Ritz method to determine the deflection of a cantilever beam carrying a concentrated force $P$ at the tip. *Note:* Use a polynomial trial function.

**8-39** Apply the Rayleigh–Ritz method to determine the deflection of a two-span beam (Fig. P8-39). *Note*: Use a polynomial trial function.

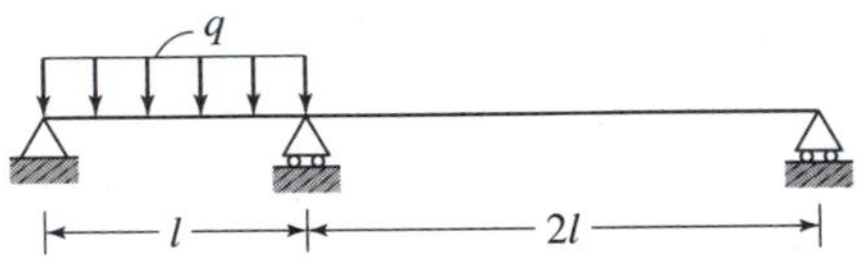


Figure P8-39

**8-40** Prove the statements in the three special cases obtained from Eqs. (8-371) in Section 8-11-2.

**8-41** A piezoelectric beam of length $L$ is in the form of a bimorph and is simply supported at its ends. Each layer of the bimorph has width $b$ and thickness $h$. Electrical and mechanical properties of each layer are identical. The poling direction and the orientation of the electric fields are as shown in Fig. 8-93. Obtain the displacement distribution in the beam as a function of the voltage. What is the deflection at the center of the beam?

**8-42** A simply supported beam made of low carbon steel is 5 in long, 1 in wide, and has a thickness of $1/16$ in. With a uniformly distributed load of 1.5 lbf/in, the beam has excessive deflection at its mid-point, and it is desired to add a piezoelectric layer to the top of the beam and another to the bottom in order to reduce the deflection. Both layers are PZT and are 0.020 in thick. The configuration of the layers, as well as the poling and electric fields, are as shown in Fig. 8-99, except that layer 2 is now low-carbon steel. If it is required to reduce the mid-point deflection by 65%, what voltage, $V_0$, is required?

The deflection reduction achieved by adding two piezoelectric layers comes about because of

1. stiffening of the low carbon steel beam by the PZT, even when no voltage is applied and
2. additional reduction in mid-point deflection due to the activation of the piezoelectric layers with a voltage $V_0$.

Determine the reduction in deflection due to each effect.

**8-43** Show that if the electric field is applied first, followed by a mechanical force, the resulting expression for strain energy is given by Eq. (8-377).

**8-44** Solve Problem 8-41 using Castigliano's theorem.

**8-45** Use Castigliano's theorem to derive an expression for the mid-span deflection in Problem 8-42.

**8-46** Use Castigliano's theorem to obtain Eq. ($gg$) in Example 8-20.

## REFERENCES

[1] Beer, F. P., and E. R. Johnston, Jr., *Mechanics of Materials* (McGraw-Hill, New York, 1992).
[2] Biezeno, C. B., and R. Grammel, transl., *Engineering Dynamics,* vol. II (Blackie, Glasgow, 1956).
[3] Bohannan, G. W., V. H. Schmidt, R. J. Conant, et al., "Piezoelectric Polymer Actuators in a Vibration Isolation Application," in *Proc. SPIE,* no. 3987, pp. 331–342 (2000).

[4] Boley, B. E., and J. H. Weiner, *Theory of Thermal Stresses* (Wiley, New York, 1960).

[5] Budynas, R. G., *Advanced Strength and Applied Stress Analysis* (McGraw-Hill, New York, 1977).

[6] Burr, A. H., *Mechanical Analysis and Design* (Elsevier, New York, 1982).

[7] Collatz, L., *The Numerical Treatment of Differential Equations,* 3rd ed., P. G. Williams, transl. (Springer, Berlin, 1960).

[8] Cook, R. D., and W. C. Young, *Advanced Mechanics of Materials* (Macmillan, New York, 1985).

[9] Fung, Y. C., *Foundations of Solid Mechanics* (Prentice-Hall, Englewood Cliffs, NJ, 1965).

[10] Gladwell, G. M. L., *Contact Problems in the Classical Theory of Elasticity* (Sijthoff & Noordhoff, Germantown, MD, 1980).

[11] Gradshtein, I. S., and I. M. Ryzhik, *Table of Integrals, Series, and Products,* A. Jeffrey, transl. (Academic Press, Orlando, FL, 1980).

[12] Greenberg, M. D., *Foundations of Applied Mathematics* (Prentice-Hall, Englewood Cliffs, NJ, 1978).

[13] Hildebrand, F. B., *Methods of Applied Mathematics,* 2nd ed. (Prentice-Hall, Englewood Cliffs, NJ, 1952).

[14] Hodge, P. G., Jr., *Plastic Analysis of Structures* (Robert E. Krieger, Malabar, FL, 1981).

[15] Larson, P. H., and J. R. Vinson, "The Use of Piezoelectric Materials in Curved Beams and Rings," in *Adaptive Structures and Material Systems,* vol. 35 (ASME, 1993), pp. 277–285.

[16] Love, A. E. H., *A Treatise on the Mathematical Theory of Elasticity,* 4th ed. (Dover, New York, 1944).

[17] *Macsyma,* version 13 (Symbolics, Burlington, MA, 1988).

[18] *Mathcad7,* professional ed. (MathSoft, Cambridge, MA, 1997).

[19] Mendelson, A., *Plasticity: Theory and Application* (Macmillan, New York, 1968).

[20] Nadai, A., *Die elastischen Platten* (Julius Springer, Berlin, 1925).

[21] Popov, E. P., *Mechanics of Materials* (Prentice-Hall, Englewood Cliffs, NJ, 1952).

[22] Solecki, R., and G. Zhao, Closed-Form Expressions for Some Trigonometric and Related Infinite Series Occurring in Solid Mechanics, *Journal of the Industrial Mathematic Society,* 38, pp. 115–129 (1988).

[23] Stampleman, D. S., and A. H. von Flotow, "Microgravity Isolation Mounts Based on Piezoelectric Film," in *Active Noise and Vibration Control* (ASME WAM, 1990), pp. 57–65.

[24] Thompson, W. T., *Theory of Vibrations with Applications* (Prentice-Hall, Englewood Cliffs, NJ, 1972).

[25] Timoshenko, S., *Strength of Materials,* pt. II, 2nd ed. (van Nostrand, New York, 1954).

[26] Tolstov, G. P., transl., *Fourier Series* (Dover, New York, 1960).

[27] Wheelon, A. D., *Tables of Summable Series and Integrals Involving Bessel Functions* (Holden-Day, San Francisco, 1968).

[28] Winkler, E., *Die Lehre von der Elastizität und Festigkeit* (Prague, 1867).

# 9 ELEMENTARY PROBLEMS IN TWO- AND THREE-DIMENSIONAL SOLID MECHANICS

Solid mechanics is based on general principles that lead to differential equations of stress equilibrium and, through stress-strain relationships, to differential equations for displacement components. Solving these is in general quite complicated and beyond the scope of this text. Some idealized problems of fundamental character which can be relatively simple are discussed in this chapter. These are axisymmetric compression or expansion and rotation of thick-walled cylinders and torsion of noncircular prisms. We also solve one more complicated problem, namely, bending of a high beam, using the finite differences method (see also Chapter 6).

## 9-1 PROBLEM FORMULATION—BOUNDARY CONDITIONS

The equilibrium equations derived in Chapter 2 for a three-dimensional solid have the form

$$\begin{aligned} \frac{\partial \sigma_{xx}}{\partial x} + \frac{\partial \sigma_{xy}}{\partial y} + \frac{\partial \sigma_{xz}}{\partial z} + X &= 0 \\ \frac{\partial \sigma_{xy}}{\partial x} + \frac{\partial \sigma_{yy}}{\partial y} + \frac{\partial \sigma_{yz}}{\partial z} + Y &= 0 \\ \frac{\partial \sigma_{xz}}{\partial x} + \frac{\partial \sigma_{yz}}{\partial y} + \frac{\partial \sigma_{zz}}{\partial z} + Z &= 0 \end{aligned} \tag{9-1}$$

Since only three equations in six unknowns are available, we evidently need additional equations relating stress components to displacement components. Replacing the unknown stress components $\sigma_{xx}, \ldots$ in Eqs. (9-1) by the unknown displacement components $u$, $v$, $w$, we make the number of unknowns in the resulting system equal to the number of equations. The nature of the resulting system of equations depends on the relationship between stress and strain, that is, on the constitutive equations (see Chapter 4).

The simplest example is a linearly elastic isotropic solid, which we shall consider here. Substituting the relations between strain and displacement components [see Eq. (3-11)]

$$
\begin{aligned}
&\varepsilon_{xx} = \frac{\partial u}{\partial x}, \qquad \varepsilon_{yy} = \frac{\partial v}{\partial y}, \qquad \varepsilon_{zz} = \frac{\partial w}{\partial z} \\
&\gamma_{xy} = \frac{\partial u}{\partial y} + \frac{\partial v}{\partial x}, \qquad \gamma_{xz} = \frac{\partial u}{\partial z} + \frac{\partial w}{\partial x}, \qquad \gamma_{yz} = \frac{\partial v}{\partial z} + \frac{\partial w}{\partial y}
\end{aligned}
\tag{9-2}
$$

into Hooke's law [Eq. (4-111)],

$$
\begin{aligned}
\sigma_{xx} &= \frac{E}{(1+\nu)(1-2\nu)}[(1-\nu)\varepsilon_{xx} + \nu(\varepsilon_{yy} + \varepsilon_{zz})] \\
\sigma_{yy} &= \frac{E}{(1-\nu)(1-2\nu)}[(1-\nu)\varepsilon_{yy} + \nu(\varepsilon_{xx} + \varepsilon_{zz})] \\
\sigma_{zz} &= \frac{E}{(1-\nu)(1-2\nu)}[(1-\nu)\varepsilon_{zz} + \nu(\varepsilon_{xx} + \varepsilon_{yy})] \\
\sigma_{xy} &= \frac{E}{2(1+\nu)}\gamma_{xy} \\
\sigma_{xz} &= \frac{E}{2(1+\nu)}\gamma_{xz} \\
\sigma_{yz} &= \frac{E}{2(1+\nu)}\gamma_{yz}
\end{aligned}
\tag{9-3}
$$

we find the following relations between stress components and displacement components:

$$
\begin{aligned}
\sigma_{xx} &= \frac{E}{(1+\nu)(1-2\nu)}\left[(1-\nu)\frac{\partial u}{\partial x} + \nu\left(\frac{\partial v}{\partial y} + \frac{\partial w}{\partial z}\right)\right] \\
\sigma_{yy} &= \frac{E}{(1-\nu)(1-2\nu)}\left[(1-\nu)\frac{\partial v}{\partial y} + \nu\left(\frac{\partial w}{\partial z} + \frac{\partial u}{\partial x}\right)\right] \\
\sigma_{zz} &= \frac{E}{(1-\nu)(1-2\nu)}\left[(1-\nu)\frac{\partial w}{\partial z} + \nu\left(\frac{\partial u}{\partial x} + \frac{\partial v}{\partial y}\right)\right] \\
\sigma_{xy} &= \frac{E}{2(1-\nu)}\left(\frac{\partial u}{\partial y} + \frac{\partial v}{\partial x}\right) \\
\sigma_{xz} &= \frac{E}{2(1-\nu)}\left(\frac{\partial u}{\partial z} + \frac{\partial w}{\partial x}\right) \\
\sigma_{yz} &= \frac{E}{2(1-\nu)}\left(\frac{\partial v}{\partial z} + \frac{\partial w}{\partial y}\right)
\end{aligned}
\tag{9-4}
$$

Substituting Eqs. (9-4) into the equations of equilibrium (9-1) and simplifying, we have

$$
\begin{aligned}
(1-2\nu)\nabla^2 u + \frac{\partial}{\partial x}\left(\frac{\partial u}{\partial x} + \frac{\partial v}{\partial y} + \frac{\partial w}{\partial z}\right) + \frac{2(1+\nu)(1-2\nu)}{E}X = 0 \\
(1-2\nu)\nabla^2 v + \frac{\partial}{\partial y}\left(\frac{\partial u}{\partial x} + \frac{\partial v}{\partial y} + \frac{\partial w}{\partial z}\right) + \frac{2(1+\nu)(1-2\nu)}{E}Y = 0 \\
(1-2\nu)\nabla^2 w + \frac{\partial}{\partial z}\left(\frac{\partial u}{\partial x} + \frac{\partial v}{\partial y} + \frac{\partial w}{\partial z}\right) + \frac{2(1+\nu)(1-2\nu)}{E}Z = 0
\end{aligned}
\tag{9-5}
$$

Equations (9-5) are known as *Navier's equations.** With $u = u(x, y)$, $v = v(x, y)$, and $w = 0$ they reduce to

$$
\begin{aligned}
(1-2\nu)\nabla^2 u + \frac{\partial}{\partial x}\left(\frac{\partial u}{\partial x} + \frac{\partial v}{\partial y}\right) + \frac{2(1+\nu)(1-2\nu)}{E}X = 0 \\
(1-2\nu)\nabla^2 v + \frac{\partial}{\partial y}\left(\frac{\partial u}{\partial x} + \frac{\partial v}{\partial y}\right) + \frac{2(1+\nu)(1-2\nu)}{E}Y = 0
\end{aligned}
\tag{9-6}
$$

In order to obtain a complete solution to the system of equations (9-5), we need the *boundary conditions*. An inspection of Eqs. (9-5) reveals that it is a system of second-order partial differential equations in $u$, $v$, and $w$.† By analogy with the ordinary differential equations (such as the equation for the deflection of a beam) we conclude that since each of the three unknown functions appears as a second-order derivative, we need three boundary conditions at each point of the boundary. Similarly, we see from Eqs. (9-6) that here two boundary conditions are needed for each boundary point. If we seem to have a different number of boundary conditions available than suggested here, it indicates that our logic in deriving those conditions is faulty. Recall that the bending of a beam is described by a fourth-order ordinary differential equation, and that the boundary conditions can be imposed on the derivatives of the unknowns up to the third order. Since here we have a system of second-order differential equations, we can impose the conditions on the derivatives of the unknown functions up to the first order. Physically it means that we can prescribe the displacements $u$, $v$, and $w$ and also the stresses, because by Hooke's law they are combinations of various first-order derivatives of $u$, $v$, and $w$. From all possible combinations of the boundary conditions only some are acceptable. Others violate certain minimum principles, as discussed in conjunction with the calculus of variations [5]. Without resorting to advanced mathematics, we can use some physical insight to decide with confidence which combinations of boundary conditions are acceptable. Consider, for simplicity, a beam. We know, for example, that we cannot require that both the deflection $y(x)$ and its third derivative $y'''(x)$ vanish at the same point $x = x_1$. Such a pair of boundary conditions would be contradictory because to force a deflection to be zero at a point, we would have to place a support at this point. Therefore the reaction present there would generate a shear force at the adjacent cross section so that at this point $y'''(x) \neq 0$. Similar reasoning may help us in deciding the acceptability of various combinations of boundary conditions for two- or three-dimensional solids.

---

*Claude L. M. H. Navier (1785–1836), French physicist.

†The solution to this system of equations is, in general, *unique*. For details see [4, p. 160].

## EXAMPLE 9-1

Let us examine a two-dimensional solid, as shown in Fig. 9-1. The boundary is composed of five distinct parts ($AB$, $BC$, $CD$, $DE$, $EA$) on each of which two boundary conditions are needed.

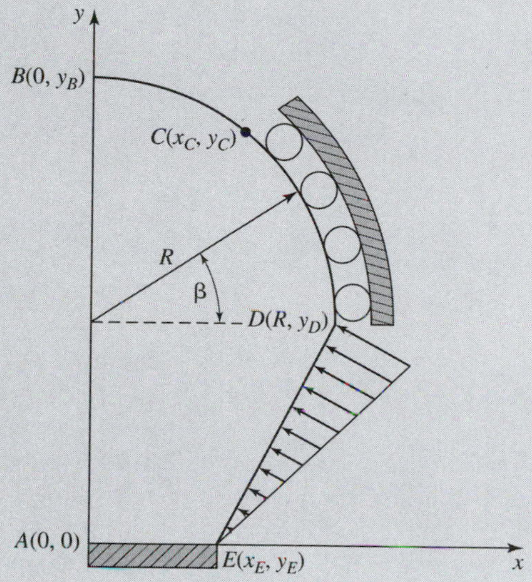


**Figure 9-1** Two-dimensional solid with various boundary conditions

## SOLUTION

We note that the section $AE$ is rigidly attached to some external structure so that the displacement components vanish there. We also note that the section $CD$ can slide tangentially, but its displacement is prevented in the normal direction. Moreover, it is also evident from Fig. 9-1 that sections $AB$ and $BC$ are stress-free, whereas section $ED$ is carrying a linearly varying normal load. Let us finish the description of the problem by emphasizing the arbitrariness of the coordinate system $x$, $y$ introduced here. We begin the analysis with the side $AB$, say, by asking ourselves what two quantities, among the available five—$u$, $v$, $\sigma_{xx}$, $\sigma_{xy}$, and $\sigma_{yy}$—are known. Since we have no information about the displacements, these will be known only after the solution of the problem. Thus our choice is limited to three stress components. Obviously one of the stress components is not applicable here. In order to decide which one, we examine a small element next to $AB$ (Fig. 9-2). Since the normal to it has the direction of the $x$ axis, only the components $\sigma_{xx}$ and $\sigma_{xy}$ are present here. The third component, $\sigma_{yy}$, appears on the side perpendicular to $AB$. By using the fact that side $AB$ carries no load, we find from the equilibrium that

$$\sigma_{xx} = 0, \qquad \sigma_{xy} = 0 \tag{a}$$

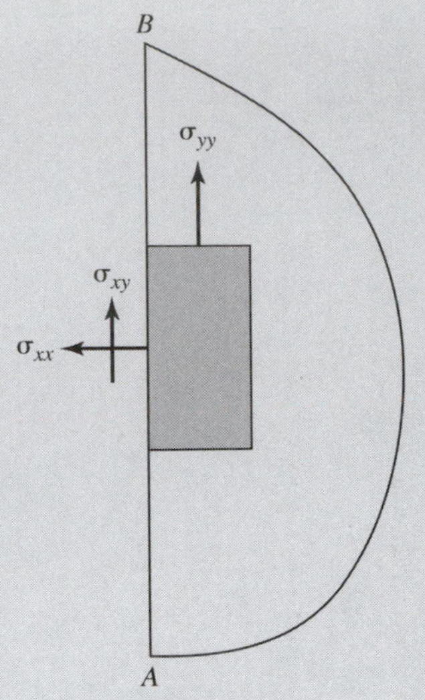


Figure 9-2

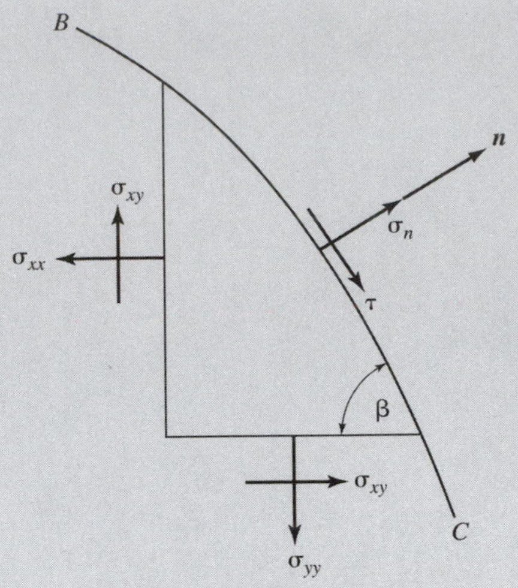


Figure 9-3

These are the boundary conditions to be satisfied on $AB$. Next comes the section $BC$ of the circular side $BD$. Let us consider a very small triangular element next to it (Fig. 9-3). We know from Chapter 2 that the stress vector acting on the inclined side of an element, such as the side forming a portion of $BC$, can be decomposed into the stress components $\sigma_n$ and $\tau$. These are normal and tangent to this side, respectively. From the equilibrium we find that

$$\sigma_n = 0, \qquad \tau = 0 \tag{b}$$

and since with $\sigma_{zz} = \sigma_{xz} = \sigma_{yz} = 0$, the expressions for $\sigma_n$ [Eq. (2-17)] and for $\tau$ [Eq. (2-27)] become

$$\begin{aligned} \sigma_n &= \sigma_{xx}n_x^2 + \sigma_{yy}n_y^2 + 2\sigma_{xy}n_x n_y \\ \tau &= \sqrt{(\sigma_{xx}n_x + \sigma_{xy}n_y)^2 + (\sigma_{xy}n_x + \sigma_{yy}n_y)^2 - \sigma_n^2} \end{aligned} \tag{c}$$

Then, substituting

$$n_x = \sin\theta, \qquad n_y = \cos\theta \tag{d}$$

we find, with Eqs. ($b$), that the boundary conditions on $BC$ are

$$\begin{aligned} \sigma_{xx}\sin^2\beta + \sigma_{yy}\cos^2\beta + 2\sigma_{xy}\sin\beta\cos\beta &= 0 \\ (\sigma_{yy} - \sigma_{xx})\sin\beta\cos\beta + \sigma_{xy}(\cos^2\beta - \sin^2\beta) &= 0 \end{aligned} \tag{e}$$

These results can be obtained directly from the equilibrium of the two-dimensional element in Fig. 9-3 while recalling that the "arrows" representing the stresses only show the

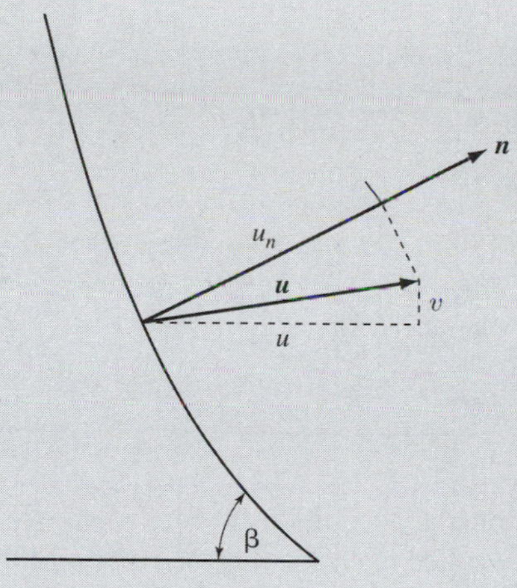


**Figure 9-4** Boundary condition at the free edge

stress intensities, not their resultants. Next we look on the portion *CD* of the circular part of the boundary and use again the element in Fig. 9-3. We see that still

$$\tau = 0 \tag{f}$$

but as $\sigma_n$ is now unknown, this becomes

$$(\sigma_{xx}n_x + \sigma_{xy}n_y)^2 + (\sigma_{xy}n_x + \sigma_{yy}n_y)^2 - \left(\sigma_{xx}n_x^2 + \sigma_{yy}n_y^2 + 2\sigma_{xy}n_x n_y\right)^2 = 0 \tag{g}$$

Also note that the displacement in the direction of the normal $\boldsymbol{n}$ is zero,

$$u_n = 0 \tag{h}$$

But (see Fig. 9-4)

$$u_n = \boldsymbol{u} \cdot \boldsymbol{n} = (u\boldsymbol{i} + v\boldsymbol{j}) \cdot (n_x\boldsymbol{i} + n_y\boldsymbol{j}) = un_x + vn_y$$

so that finally the boundary condition (*h*) becomes

$$u \sin\beta + v \cos\beta = 0 \tag{i}$$

where (see Fig. 9-1)

$$\cos\beta = \frac{x}{R}$$

The straight section *DE* of the boundary is subject to compressive load of linearly varying intensity $p(x)$. Thus (Fig. 9-3),

$$\sigma_n = -p(x), \qquad \tau = 0 \tag{j}$$

But from geometry we find that

$$p(x) = \frac{p_0}{R - x_E}(x - x_E) \tag{k}$$

Thus [see Eqs. (*e*)],

$$\sigma_{xx}\sin^2\alpha + \sigma_{yy}\cos^2\alpha + 2\sigma_{xy}\sin\alpha\cos\alpha = -\frac{p_0}{R - x_E}(x - x_E)$$
$$(\sigma_{yy} - \sigma_{xx})\sin\alpha\cos\alpha + \sigma_{xy}(\cos^2\alpha - \sin^2\alpha) = 0 \qquad (l)$$

Finally, since segment *AE* of the contour is fixed, the displacement vector equals zero so that its components are also zero. Thus the boundary conditions are

$$u = 0, \qquad v = 0 \qquad (m)$$

Another type of boundary condition appears when two or more solids share a common boundary, usually called *interface* (Fig. 9-5). Solid I in Fig. 9-5*a* (the *inclusion*) is inserted into solid II (the *matrix*), and the interface is the curved line separating them and forming a common boundary. Solids I and II in Fig. 9-5*b* are bonded along the interface. In both cases the differential equations to be solved for either of the solids are subject to the boundary conditions along the boundary that is not shared by the solids (*ABCD* for solid II in Fig. 9-5*a*; *BCDE* for solid I and *EFAB* for solid II in Fig. 9-5*b*). The boundary conditions at the interface, known as the *interface conditions* (but occasionally also as *continuity conditions*) usually relate the stresses to the displacements at the boundaries of adjacent solids. Assuming, for example, that inclusion I in Fig. 9-5*a* can rotate freely in its cavity, we find that the interface conditions require the interface displacements in the direction of the inclusion's center to be the same for both solids,

$$u_n^{(\mathrm{I})} = u_n^{(\mathrm{II})} \quad \text{at the interface} \qquad (9\text{-}7)$$

where the subscript $n$ indicates the "normal." The displacement components in the direction tangent to the interface need not be the same for solids I and II. On the other hand, the requirement for the inclusion to rotate freely implies the absence of friction, and therefore the tangential stresses must vanish on both sides of the interface,

$$\sigma_{\theta\theta}^{(\mathrm{I})} = 0, \qquad \sigma_{\theta\theta}^{(\mathrm{II})} = 0 \quad \text{at the interface} \qquad (9\text{-}8)$$

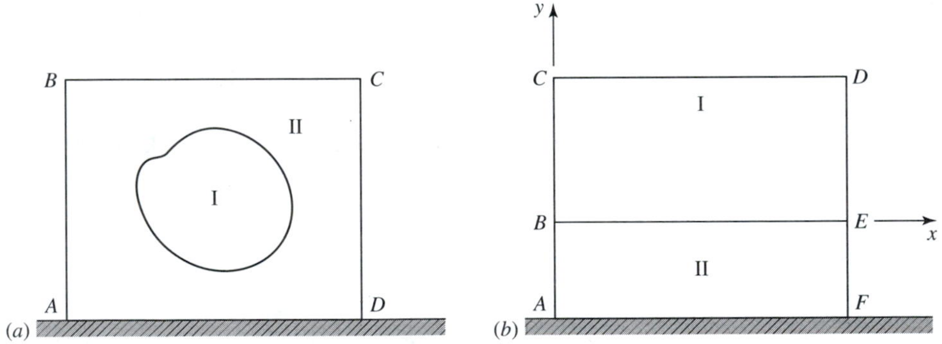


**Figure 9-5** The inclusion, the matrix, and the interface

Here the subscript $\theta$ is used because it seems natural to apply locally a polar coordinate system $r$, $\theta$ with the origin at the center of the inclusion. Moreover, from the equilibrium of small elements on both sides of the interface we deduce the equality of the radial stresses,

$$\sigma_{rr}^{(\mathrm{I})} = \sigma_{rr}^{(\mathrm{II})} \qquad \text{at the interface} \tag{9-9}$$

Note that a geometrically similar problem of an inclusion II press-fitted into a slightly smaller hole in solid I would change the interface condition. This time the normal displacements would differ by the amount of the radial interference (see Section 9-4) whereas the tangential stresses would no longer be zero. We would have the continuity of both the tangential displacements and the tangential stresses at the interface instead. Further complications may, however, occur should the completed stress and deformation analysis reveal a separation of the two solids at some segment of the interface. Changed interface conditions would, in this case, require repeated analysis. This is also true for the simpler geometry shown in Fig. 9-5*b*. Normally it would be assumed that solids I and II are bonded at the interface, the bond being achieved by, say, a thin layer of an adhesive. The interface conditions along *BE* ($y = 0$) would then be

$$u^{(\mathrm{I})} = u^{(\mathrm{II})}, \qquad v^{(\mathrm{I})} = v^{(\mathrm{II})}, \qquad \sigma_{yy}^{(\mathrm{I})} = \sigma_{yy}^{(\mathrm{II})}, \qquad \sigma_{xy}^{(\mathrm{I})} = \sigma_{xy}^{(\mathrm{II})} \tag{9-10}$$

If the strength of the bonding agent is smaller than that of solids I and II, then the stress analysis may uncover failure of the bond along a certain portion of *BE*. The remedy for this would be either using a stronger adhesive or reanalysis with changed interface conditions along the debonded region.

## 9-2 COMPATIBILITY OF ELASTIC STRESS COMPONENTS

In Section 3-3 we have shown that when the strain components are preassigned, the continuity of the displacements in a two-dimensional solid is ensured only if Eq. (3-12), the compatibility condition, is satisfied,

$$\frac{\partial^2 \gamma_{xy}}{\partial x\,\partial y} = \frac{\partial^2 \varepsilon_{xx}}{\partial y^2} + \frac{\partial^2 \varepsilon_{yy}}{\partial x^2} \tag{9-11}$$

Since the strains depend on the stresses (see Chapter 4), Eq. (9-11) implies that stress components are also subject to the limitations of compatibility. Let us consider a linearly elastic, isotropic solid. We begin with the case of plane strain, and substituting the strain-stress relations [Eqs. (4-25)]

$$\varepsilon_{xx} = \frac{1+\nu}{E}[(1-\nu)\sigma_{xx} - \nu\sigma_{yy}]$$

$$\varepsilon_{yy} = \frac{1+\nu}{E}[(1-\nu)\sigma_{yy} - \nu\sigma_{xx}]$$

$$\gamma_{xy} = 2\frac{1+\nu}{E}\sigma_{xy}$$

into Eq. (9-11), we obtain

$$2\frac{1+\nu}{E}\frac{\partial^2\sigma_{xy}}{\partial x\,\partial y} = \frac{1+\nu}{E}\left[(1-\nu)\frac{\partial^2\sigma_{xx}}{\partial y^2} - \nu\frac{\partial^2\sigma_{yy}}{\partial y^2} + (1-\nu)\frac{\partial^2\sigma_{yy}}{\partial x^2} - \nu\frac{\partial^2\sigma_{xx}}{\partial x^2}\right]$$

or

$$2\frac{\partial^2\sigma_{xy}}{\partial x\,\partial y} = -\nu\nabla^2\sigma_{xx} - \nu\nabla^2\sigma_{yy} + \frac{\partial^2\sigma_{xx}}{\partial y^2} + \frac{\partial^2\sigma_{yy}}{\partial x^2} \tag{9-12}$$

where

$$\nabla^2 \equiv \frac{\partial^2}{\partial x^2} + \frac{\partial^2}{\partial y^2} \tag{9-13}$$

Now, from the equilibrium equations (9-1) we find that in the two-dimensional case,

$$\begin{aligned}\frac{\partial\sigma_{xy}}{\partial y} &= -\frac{\partial\sigma_{xx}}{\partial x} - X \\ \frac{\partial\sigma_{xy}}{\partial x} &= -\frac{\partial\sigma_{yy}}{\partial y} - Y\end{aligned} \tag{9-14}$$

By differentiating the first of Eqs. (9-14) with respect to $x$ and the second with respect to $y$ we obtain

$$\begin{aligned}\frac{\partial^2\sigma_{xy}}{\partial x\,\partial y} &= -\frac{\partial^2\sigma_{xx}}{\partial x^2} - \frac{\partial X}{\partial x} \\ \frac{\partial^2\sigma_{xy}}{\partial x\,\partial y} &= -\frac{\partial\sigma_{yy}}{\partial y^2} - \frac{\partial Y}{\partial y}\end{aligned} \tag{9-15}$$

Substituting Eqs. (9-15) into the left-hand side of Eq. (9-12) we get

$$-\frac{\partial^2\sigma_{xx}}{\partial x^2} - \frac{\partial^2\sigma_{yy}}{\partial y^2} - \frac{\partial X}{\partial x} - \frac{\partial Y}{\partial y} = -\nu\nabla^2\sigma_{xx} - \nu\nabla^2\sigma_{yy} + \frac{\partial^2\sigma_{xx}}{\partial y^2} + \frac{\partial\sigma_{yy}}{\partial x^2}$$

which simplifies to

$$(1-\nu)\nabla^2\sigma_{xx} + (1-\nu)\nabla^2\sigma_{yy} = \frac{\partial X}{\partial x} + \frac{\partial Y}{\partial y}$$

or

$$\nabla^2(\sigma_{xx}+\sigma_{yy}) = \frac{1}{1-\nu}\left(\frac{\partial X}{\partial x} + \frac{\partial Y}{\partial y}\right) \tag{9-16}$$

This shows that in a plane strain problem the stress components cannot be assigned arbitrarily but, in addition to equations of equilibrium, must also satisfy Eq. (9-16), known as the *Beltrami–Mitchell compatibility condition*. In the case of plane stress (Section 4-3-1) we replace Poisson's ratio $\nu$ in Eq. (9-16) by $\nu/(1+\nu)$, thus obtaining the compatibility equation in the form

$$\nabla^2(\sigma_{xx}+\sigma_{yy}) = (1+\nu)\left(\frac{\partial X}{\partial x} + \frac{\partial Y}{\partial y}\right) \tag{9-17}$$

## EXAMPLE 9-2

Let the stresses in a two-dimensional solid equal

$$\sigma_{xx} = Ay, \qquad \sigma_{yy} = Axy, \qquad \sigma_{xy} = -\tfrac{1}{2}Ax^2 + Bx^2y \tag{a}$$

while the body forces are

$$X = -Bx^2, \qquad Y = -2Bxy \tag{b}$$

Are Eqs. (*a*) a possible state of stress?

## SOLUTION

First verify whether the equations of equilibrium are satisfied. To this end we substitute Eqs. (*a*) and (*b*) into Eqs. (9-14),

$$Bx^2 = 0 + Bx^2, \qquad -Ax + 2Bxy = -Ax + 2Bxy$$

Thus the equations of equilibrium are identically satisfied. Next we substitute Eqs. (*a*) and (*b*) into the compatibility equation (9-17). This time we get

$$0 = -2(1+\nu)Bx$$

which is impossible unless $B = 0$. Thus Eqs. (*a*) are not a possible state of stress when the body forces are preassigned according to Eqs. (*b*).

# 9-3 THICK-WALLED CYLINDERS AND CIRCULAR DISKS

In this section thick-walled circular cylinders or disks are investigated. Homogeneous and composite elastic and plastic problems are analyzed. The influence of temperature and the problem of shrink-fit are discussed. Examples of symbolic manipulation are supplied. Nonaxisymmetric problems are also discussed.

Circular disks and cylinders have many practical applications. Rotating shafts with press-fitted circular disks appear in many machines. Wheels of a wagon or of a car, a clutch, disk brakes, certain parts of turbines, and other elements all fall into this category. The stress analysis of listed problems is often very complicated, mainly because the loads are not axisymmetric. The geometry also becomes more complicated by the presence of holes, notches, and so on. In the following we will only be concerned with stress and strain analysis of axisymmetric problems.

## 9-3-1 Equilibrium Equation and Strains

Considering a cross section of the circular thick-walled tube shown in Fig. 9-6, we cut a slice of thickness $t$ of this tube. It is noted that such a slice differs from the disk only by the presence of stress acting on it in the axial direction (perpendicular to the figure). In

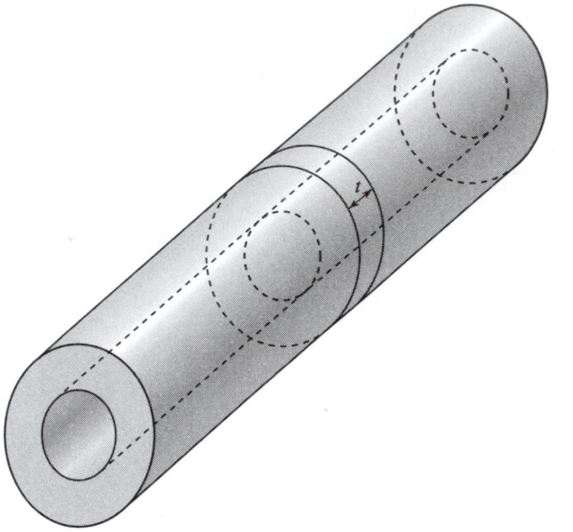


Figure 9-6

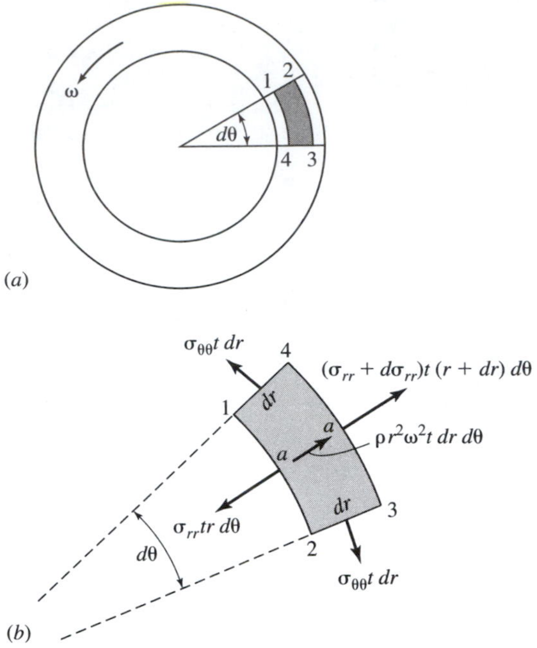


Figure 9-7

order to derive the equilibrium equation, let us examine an infinitesimal element (Figs. 9-6 and 9-7). As a first step we make the following assumptions:

1. The slice is subject to axisymmetric loads only. This assumption requires the loads to be uniform on either of the circular edges. Although it is dictated by the need of simplification of a complex problem, it covers a large segment of practically important problems. Among these is the effect of rotation with constant speed on stresses and displacements. On the other hand, the cases of nonuniform pressure (action of a belt on a pulley, of shoes on the brake disk) or variable speed (transient regiment of rotation) are excluded.
2. The effects of axisymmetric distribution of the temperature will be included at a later stage.
3. All deformations are assumed to be small compared to the characteristic dimension of the slice.

Figure 9-7 shows the free-body diagram of the slice shown in Fig. 9-6. This element is subject to uniform normal stresses $\sigma_{rr}$ on the curved edges (both the uniformity of the normal stresses and the absence of shear stresses are due to assumed axial symmetry of the loads). Also the straight edges are subject to normal stresses $\sigma_{\theta\theta}$ only. These are not uniform, but their variation along the infinitesimal interval $dr$ is negligible. We note that the magnitudes of the stresses on both straight edges are identical. Again, this is the consequence of axial symmetry. Finally we assume that the slice rotates with constant speed $\omega$. This generates an inertia force $r\omega^2$ per unit mass. Therefore the resultant inertia force per volume of the element is $\rho r^2\omega^2 t\, dr\, d\theta$, where $\rho$ is the mass density per unit volume of the material in question. In order to derive the equilibrium equations for the element in Fig. 9-7 we project all forces acting on it on the line $a$-$a$ bisecting the central angle $d\theta$, obtaining

$$(\sigma_{rr} + d\sigma_{rr})t(r + dr)\, d\theta - \sigma_{rr} tr\, d\theta - 2\sigma_{\theta\theta} t\, dr \sin\frac{d\theta}{2} + \rho r^2\omega^2 t\, dr\, d\theta = 0$$

Performing the indicated operations and noting that $d\theta \ll 1$ results in $\sin(d\theta/2) \approx d\theta/2$, we obtain an equation with two types of terms: (1) terms with two differentials, and (2) terms with three differentials. The latter are infinitesimal compared to the former and thus can be disregarded, which leads to the following equation:

$$\frac{d\sigma_{rr}}{dr} + \frac{\sigma_{rr} - \sigma_{\theta\theta}}{r} + \rho\omega^2 r = 0 \qquad (9\text{-}18)$$

Note that the same equation was obtained as a special case of the general equilibrium equation in Chapter 2, Eq. (2-58). Now we have a single differential equation with two unknown functions, $\sigma_{rr}$ and $\sigma_{\theta\theta}$. Recall that we have also the second equilibrium equation at our disposal which was obtained by projecting all forces on the direction perpendicular to $a$-$a$. This equation is, however, identically satisfied due to assumed symmetry. The next step is to calculate the strains in terms of the displacements and assume material properties so that the stresses can be related to the strains. To this end we consider again the element of Fig. 9-7, now shown in Fig. 9-8, in both the deformed (1′-2′-3′-4′) and the undeformed (1-2-3-4) configurations. We note that because deformation occurs only in the axial direction, there is no change of angles between the lines $r$ = constant and $\theta$ = constant. However, the lengths of the fibers located on the element of the arc (such as 1-2) or on the element of the radius (such as 1-4) change. Denoting by $u$ the radial displacement of point 1 and by $u + du$ that of point 4, and using the definition of strain, we calculate first the radial

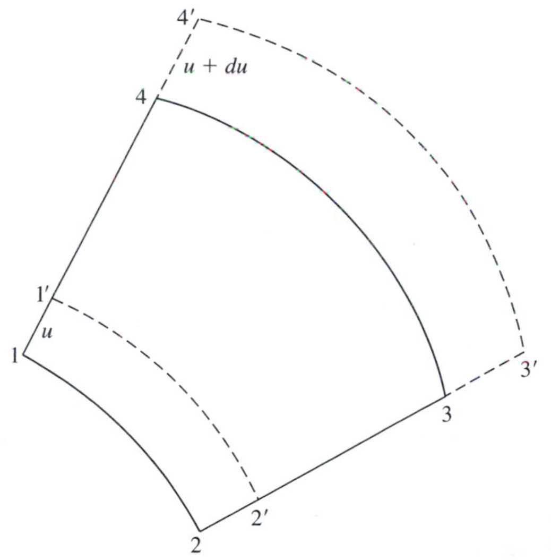


Figure 9-8 Deformed and undeformed configuration of the infinitesimal element.

strain $\varepsilon_{rr}$ as a relative change in length of the segment 1-4,

$$\varepsilon_{rr} = \frac{(dr + du) - dr}{dr} = \frac{du}{dr} \tag{9-19}$$

Similarly, the tangential strain $\varepsilon_{\theta\theta}$, which is the relative change in length of segment 1-2, becomes

$$\varepsilon_{\theta\theta} = \frac{(r + u)\, d\theta - r\, d\theta}{r\, d\theta} = \frac{u}{r} \tag{9-20}$$

Finally we can represent the equilibrium equation in terms of strains and, consequently, in terms of displacements.

## 9-3-2 Elastic, Homogeneous Disks and Cylinders

In order to associate Eq. (9-18) with the displacement $u$ we have to use the stress-strain relation. Thus for an elastic material we use Hooke's law, which for plane stress becomes

$$\begin{aligned} \sigma_{rr} &= \frac{E}{1 - \nu^2}(\varepsilon_{rr} + \nu\varepsilon_{\theta\theta}) \\ \sigma_{\theta\theta} &= \frac{E}{1 - \nu^2}(\varepsilon_{\theta\theta} + \nu\varepsilon_{rr}) \end{aligned} \tag{9-21}$$

If instead of a disk we analyze a long cylinder then, assuming uniform pressure, the cylinder approximately behaves as in the state of plane strain. Thus in the case of a cylinder (see comments in Section 4-3-2) we should replace $E$ with $E/(1 - \nu^2)$ and $\nu$ with $\nu/(1 - \nu)$ in Eqs. (9-21) and in all the following equations. Substituting Eqs. (9-12) and (9-20) into Eqs. (9-21), this becomes

$$\begin{aligned} \sigma_{rr} &= \frac{E}{1 - \nu^2}\left(\frac{du}{dr} + \nu\frac{u}{r}\right) \\ \sigma_{\theta\theta} &= \frac{E}{1 - \nu^2}\left(\frac{u}{r} + \nu\frac{du}{dr}\right) \end{aligned} \tag{9-22}$$

Next, expression (9-22) is substituted into Eq. (9-18), yielding

$$\frac{d^2u}{dr^2} + \frac{1}{r}\frac{du}{dr} - \frac{u}{r^2} = -\frac{1 - \nu^2}{E}\rho\omega^2 r \tag{9-23}$$

This is the equilibrium equation for an elastic disk (or cylinder) in terms of displacements.

Because the differential equation (9-23), called *Euler's equation*,* appears also in other problems (see bending of circular plates, Chapter 10), we will briefly summarize the method of its solution. First we note its characteristic feature. It is an equation with variable coefficients such that the exponent of power of the independent variable appearing in the coefficient equals the order of the derivative by which it is multiplied. Indeed, when we multiply the left-hand side of Eq. (9-23) by $r^2$ it becomes

*Leonhard Euler (1707–1783), Swiss mathematician.

$r^2u'' + ru' - u$,* exactly as required. The procedure used for solving this type of equation is independent of its order. It is assumed that

$$u = r^n \tag{9-24}$$

where $n$ is unknown. Substituting Eq. (9-24) into the reduced equation (9-23),

$$u'' + \frac{1}{r}u' - \frac{u}{r^2} = 0 \tag{9-25}$$

yields

$$n(n-1)r^{n-2} + nr^{n-2} - r^{n-2} = 0$$

Dividing this by $r^{n-2}$ we obtain the so-called *indicial* equation,

$$n(n-1) + n - 1 = 0 \tag{9-26}$$

with the roots

$$n_1 = +1, \qquad n_2 = -1 \tag{9-27}$$

Thus the general solution to the reduced equation (9-25) is

$$u = C_1 r^{n_1} + C_2 r^{n_2} = C_1 r + \frac{C_2}{r} \tag{9-28}$$

Next we assume a particular solution to the nonhomogeneous equation (9-23) in the form of a cubic polynomial. This choice is dictated by the form of the right-hand side of Eq. (9-23) after it has been multiplied by $r^2$. Therefore,

$$u_p = A + Br + Cr^2 + Dr^3 \tag{9-29}$$

When this is substituted into Eq. (9-23) multiplied by $r^2$, we obtain

$$2Cr^2 + 6Dr^3 + Br + 2Cr^2 + 3Dr^3 - A - Br - Cr^2 - Dr^3 = -\frac{1-\nu^2}{E}\rho\omega^2 r^3 \tag{9-30}$$

Comparing the coefficients of equal powers of $r$ on both sides of Eq. (9-30) we get

$$A = 0, \qquad C = 0, \qquad D = -\frac{1-\nu^2}{8E}\rho\omega^2 \tag{9-31}$$

Finally, with Eqs. (9-29) and (9-31), Eq. (9-28) for the radial displacement becomes

$$u(r) = C_1 r + \frac{C_2}{r} - \frac{1-\nu^2}{8E}\rho\omega^2 r^3 \tag{9-32}$$

We now substitute this into Eqs. (9-22) and find the following expressions for the stresses:

$$\sigma_{rr} = \frac{E}{1-\nu^2}\left[(1+\nu)C_1 - \frac{1-\nu}{r^2}C_2\right] - \frac{3+\nu}{8}\rho\omega^2 r^2 \tag{9-33a}$$

$$\sigma_{\theta\theta} = \frac{E}{1-\nu^2}\left[(1+\nu)C_1 + \frac{1-\nu}{r^2}C_2\right] - \frac{1+3\nu}{8}\rho\omega^2 r^2 \tag{9-33b}$$

*where the primes stand for $d/dr$.

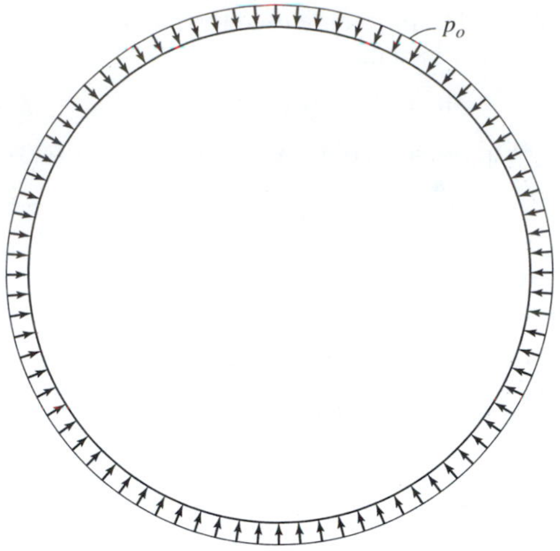


Figure 9-9

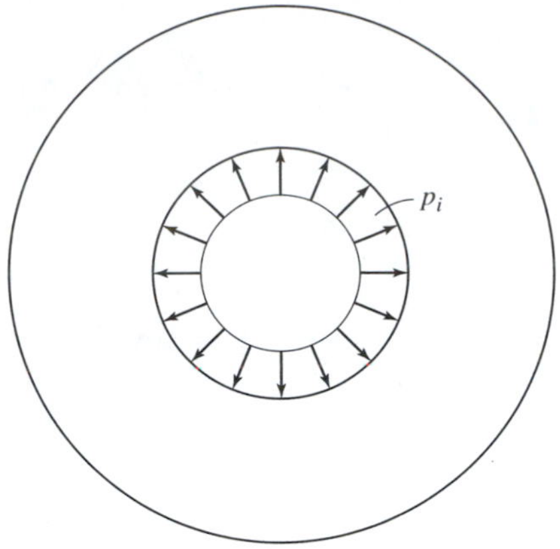


Figure 9-10

Two unknown constants $C_1$ and $C_2$ are determined by applying the appropriate boundary conditions. The most common ones are listed next:

1. If the disk is subject to the external pressure $p_o$ (Fig. 9-9), then

$$\sigma_{rr} = -p_o, \qquad \text{at } r = r_o \tag{9-34}$$

2. If the disk has a hole subject to internal pressure $p_i$ (Fig. 9-10), then the corresponding boundary condition is

$$\sigma_{rr} = -p_i, \qquad \text{at } r = r_i \tag{9-35}$$

   On the other hand, if the disk is solid (no hole), then instead of Eq. (9-35) the condition of finiteness of displacements and stresses requires that

$$C_2 = 0 \tag{9-36}$$

   This reduces to 1 the number of unknown constants so that condition (9-34) at the outer perimeter of the disk is sufficient to solve the problem.

3. Still another possibility consists in restricting the radial displacement at either of the perimeters of the disks. Let, for instance, the inner perimeter be attached to a rigid shaft via uniformly distributed springs (Fig. 9-11) with stiffness $k$ per unit of the arc of the perimeter. Now neither the displacement nor the radial stress are known at $r = r_i$. There is, however, a known relation between them, namely,

$$\sigma_{rr} = -ku \tag{9-37}$$

We now are able to solve a boundary value problem, that is, we can find the integration constants by applying the appropriate boundary conditions. First consider the state of stress in a stationary disk with a hole, with both edges of the disk subject to

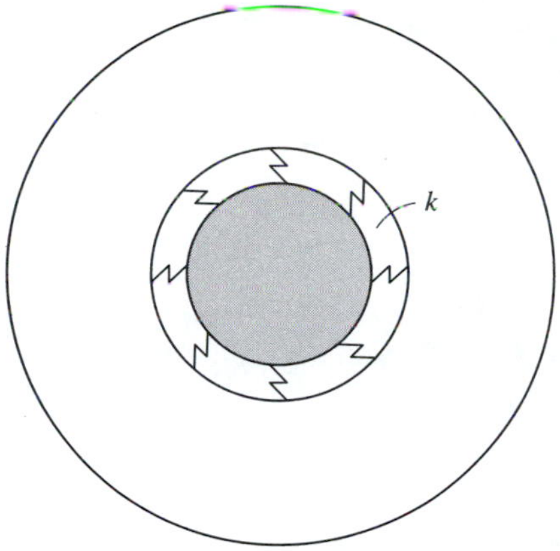


**Figure 9-11** Inner perimeter of the disk attached to a rigid shaft via uniformly distributed springs

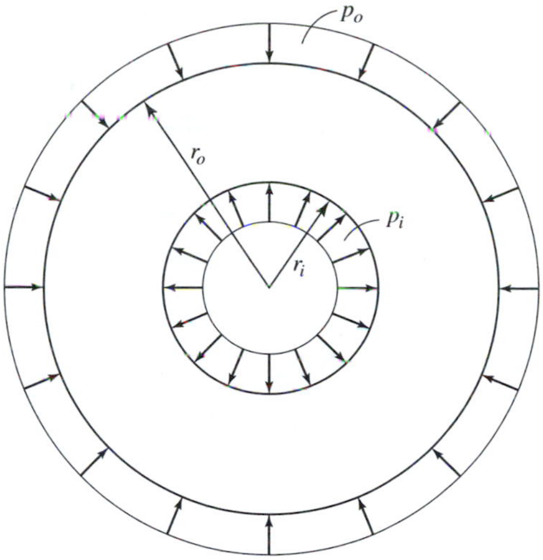


**Figure 9-12**

pressure (Fig. 9-12). The boundary conditions are

$$\begin{aligned} \sigma_{rr} &= -p_i, \qquad \text{at } r = r_i \\ \sigma_{rr} &= -p_o, \qquad \text{at } r = r_o \end{aligned} \tag{9-38}$$

We now substitute Eqs. (9-38) with $\omega = 0$ into Eqs. (9-33) and obtain the following system of equations:

$$\begin{aligned} \frac{E}{1-\nu^2}\left[(1+\nu)C_1 - \frac{1-\nu}{r_i^2}C_2\right] &= -p_i \\ \frac{E}{1-\nu^2}\left[(1+\nu)C_1 + \frac{1-\nu}{r_o^2}C_2\right] &= -p_o \end{aligned} \tag{9-39}$$

By solving these for the unknown constants $C_1$ and $C_2$ we have

$$C_1 = \frac{1-\nu}{E}\frac{r_i^2 p_i - r_o^2 p_o}{r_o^2 - r_i^2}, \qquad C_2 = \frac{1+\nu}{E}\frac{r_i^2 r_o^2(p_i - p_o)}{r_o^2 - r_i^2} \tag{9-40}$$

The next step is to substitute expressions (9-40) into Eqs. (9-33), which gives the following formulas for the stresses in a stationary disk:

$$\begin{aligned} \sigma_{rr} &= \frac{1}{r_o^2 - r_i^2}\left[r_i^2 p_i\left(1 - \frac{r_o^2}{r^2}\right) - r_o^2 p_o\left(1 - \frac{r_i^2}{r^2}\right)\right] \\ \sigma_{\theta\theta} &= \frac{1}{r_o^2 - r_i^2}\left[r_i^2 p_i\left(1 + \frac{r_o^2}{r^2}\right) - r_o^2 p_o\left(1 + \frac{r_i^2}{r^2}\right)\right] \end{aligned} \tag{9-41}$$

Finally Eq. (9-32) together with Eqs. (9-41) and $\omega = 0$ give the corresponding radial displacement,

$$u(r) = \frac{1}{E\left(r_o^2 - r_i^2\right)}\left[(1-\nu)\left(r_i^2 p_i - r_o^2 p_o\right)r + \frac{1}{r}(1+\nu)(p_i - p_o)r_i^2 r_o^2\right] \tag{9-42}$$

## EXAMPLE 9-3

Determine the stresses and the displacement in a steel cylinder with an inner radius of 1.5 in and an outer radius of 2.5 in. The cylinder is subject to the inside pressure $p_i = 3000$ lb/in$^2$ whereas the outside pressure $p_o$ is zero.

## SOLUTION

We will use Eqs. (9-41) and (9-42), noting that for a cylinder we must replace $E$ with $E/(1-\nu^2)$ and $\nu$ with $\nu/(1-\nu)$. Since for steel $E = 30 \times 10^6$ psi and $\nu = 0.3$, they must be replaced with $E = 30 \times 10^6/(1-0.3^2) = 32.97 \times 10^6$ psi and $\nu = 0.3/(1-0.3) = 0.4286$. Now using Eqs. (9-41) we obtain

$$\sigma_{rr} = \frac{1}{2.5^2 - 1.5^2}\left[1.5^2 \times 3000\left(1 - \frac{2.5^2}{r^2}\right)\right] = 1687 - \frac{10550}{r^2}$$
$$\sigma_{\theta\theta} = 1687 + \frac{10550}{r^2} \tag{a}$$

Maximum radial and tangential stresses appear at the perimeter of the hole,

$$\max \sigma_{rr} = 1687 - \frac{10550}{1.5^2} = -3002 \text{ psi}, \quad \max \sigma_{\theta\theta} = 1687 + \frac{10550}{1.5^2} = 6376 \text{ psi} \tag{b}$$

The displacement is obtained from Eq. (9-42),

$$u(r) = \frac{1}{32.97 \times 10^6(2.5^2 - 1.5^2)}$$
$$\times \left[(1-0.4286)1.5^2 \times 3000r + \frac{1}{r}(1+0.4286) \times 3000 \times 1.5^2 \times 2.5^2\right]$$
$$= \left(29.24r + \frac{456.9}{r}\right) \times 10^{-6} \tag{c}$$

To find the extremum of $u(r)$ we differentiate Eq. ($c$) with respect to $r$, and equating the result to zero we obtain $r = 3.95$ in, which is beyond the cylinder. Thus the maximum of $u(r)$ is local and appears at either of the perimeters of the cross section. Substituting, we find that max $u(r)$ occurs at the perimeter of the hole, $r = 1.5$,

$$\max u(r) = (43.86 + 304.6) \times 10^{-6} \text{ in} = 0.348 \times 10^{-3} \text{ in}$$

Now consider the special case, $p_i = p_o = p$. From Eqs. (9-41) and (9-42) it follows that

$$\sigma_{rr} = -p, \qquad \sigma_{\theta\theta} = -p, \qquad u(r) = -\frac{1-\nu}{E}pr \tag{9-43}$$

Thus we have a homogeneous state of stress.

Another case of interest occurs when the disk is solid (no hole), that is, when $r_i = 0$. Here we assume, according to Eq. (9-36), that $C_2 = 0$. The constant $C_1$ is obtained from the condition

$$\sigma_{rr} = -p_o, \qquad \text{at } r = r_o \tag{9-44}$$

which yields

$$C_1 = -\frac{(1-\nu)p_o}{E} \tag{9-45}$$

so that

$$\sigma_{rr} = -p_o, \qquad \sigma_{\theta\theta} = -p_o, \qquad u(r) = -\frac{(1-\nu)p_o}{E}r \tag{9-46}$$

The state of stress and displacement is the same for both cases. This could have been predicted by solving first the case of a solid disk, cutting in this disk a central hole, and replacing the action of the removed part by $p_i = p$. Because the resulting state of stress is homogeneous, the previous case would result.

Two other special cases deserve our attention. First consider a thin ring. To this end let $r_o - r_1 = d \ll 1$. With this $r_o + r_i \approx 2r_o$, and the stresses [Eqs. (9-41)] become

$$\begin{aligned}
\sigma_{rr} &= \frac{1}{2r_o d}\left\{(r_o - d)^2\left[p_i\left(1 - \frac{r_o^2}{r^2}\right) + p_o\frac{r_o^2}{r^2}\right] - p_o r_o^2\right\} \\
\sigma_{\theta\theta} &= \frac{1}{r_o d}\left[(r_o - d)^2 p_i - p_o r_o^2\right]
\end{aligned} \tag{9-47}$$

Since $r_o \gg d$ and $r \approx r_o$, we have approximately

$$\sigma_{rr} \approx 0, \qquad \sigma_{\theta\theta} \approx \frac{(p_i - p_o)r_o}{d} \tag{9-48}$$

Next assume that $r_o \gg r_i$ and $p_o = 0$. This gives

$$\sigma_{rr} = \sigma_{\theta\theta} = \frac{p_i}{r_o^2}\left[r_i^2\left(1 - \frac{r_o}{r^2}\right)\right] \approx -\frac{p_i r_i^2}{r^2} \tag{9-49}$$

Now let us examine the case known as *press-fit*. It occurs when we press one disk, here disk 1, into a hole in another disk, called disk 2 (see Fig. 9-13, where a dashed line depicts disk 1). It is assumed that the outer diameter of disk 1 is slightly larger than the hole diameter of disk 2. We denote

$$\Delta = R_{(1)o} - R_{(2)i} \tag{9-50}$$

where

$$\Delta \ll R_{(2)i} \tag{9-51}$$

is termed the *radial interference*. In order to insert disk 1 into disk 2 we have to make

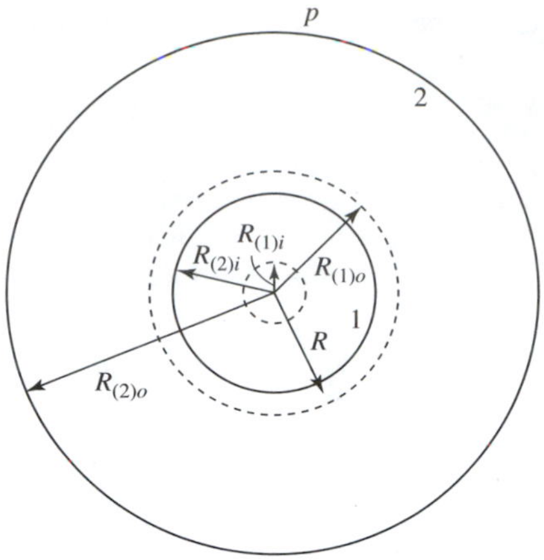


**Figure 9-13** The press-fit; the dashed line depicts disk 1 pressed into the hole of disk 2

$R_{(1)o}$ equal temporarily to $R_{(2)i}$ by, say, heating disk 2. Then, after insertion and removal of the heat source, disk 2 will have the tendency to shrink back to its original size. By doing so it will encounter the resistance of disk 1 in the form of unknown pressure $p$ uniformly distributed along the interface of the two disks. Ultimately a state of equilibrium of the two-disk system is established, with the interface radius $R$ being such that

$$R_{(1)o} < R < R_{(2)i} \tag{9-52}$$

Our goal is to determine the interface pressure $p$ and the state of stress and the displacements of the system. The first step in this development will be to relate the displacements of both mating surfaces. We note that the final interface radius $R$ can be expressed as

$$R = R_{(2)i} + u_{(2)}|_{r=R_{(2)i}}, \qquad \text{or} \qquad R = R_{(1)o} + u_{(1)}|_{r=R_{(1)o}} \tag{9-53}$$

where $u_{(1)}$ and $u_{(2)}$ are the displacements of disks 1 and 2, respectively. Comparing the two Eqs. (9-53) and using Eq. (9-50), we get

$$u_{(2)}(r)|_{r=R_{(2)i}} - u_{(1)}(r)|_{r=R_{(1)o}} = \Delta \tag{9-54}$$

Since for disk 1,

$$r_i = R_{(1)i}, \qquad r_o = R_{(1)o}, \qquad p_i = 0, \qquad p_o = -p \tag{9-55}$$

and for disk 2,

$$r_i = R_{(2)i}, \qquad r_o = R_{(2)o}, \qquad p_i = -p, \qquad p_o = 0 \tag{9-56}$$

and since the following quantities differ only by an infinitesimal amount,

$$R_{(1)o} \approx R_{(2)i} \approx R \tag{9-57}$$

therefore, with Eq. (9-42), the displacements appearing on the left-hand side of Eq. (9-54) become

$$\begin{aligned} u_{(2)}(r)|_{r=R_{(2)i}\approx R} &= \frac{pR}{E_2\left(R_{(2)o}^2 - R^2\right)}\left[(1-\nu_2)R^2 + (1+\nu_2)R_{(2)o}^2\right] \\ u_{(1)}(r)|_{r=R_{(1)o}\approx R} &= \frac{-pR}{E_1\left(R^2 - R_{(1)i}^2\right)}\left[(1-\nu_1)R^2 + (1+\nu_1)R_{(1)i}^2\right] \end{aligned} \tag{9-58}$$

We then find that with Eq. (9-54) the interface pressure is

$$p = \frac{\Delta}{R}\left\{\frac{1}{E_2}\left[\frac{R^2 + R_{(2)o}^2}{R_{(2)o}^2 - R^2} + \nu_2\right] + \frac{1}{E_1}\left[\frac{R^2 + R_{(1)i}^2}{R^2 - R_{(1)i}^2} - \nu_1\right]\right\}^{-1} \tag{9-59}$$

When both elements are made of the same material, $E_1 = E_2$, $\nu_1 = \nu_2$, and

$$p = \frac{\Delta E}{2R^3}\frac{\left(R_{(2)o}^2 - R^2\right)\left(R^2 - R_{(1)i}^2\right)}{R_{(2)o}^2 - R_{(1)i}^2} \tag{9-60}$$

## EXAMPLE 9-4

Find the stress distribution in a disk made by press-fitting the disk shown in Fig. 9-14*a* on the disk in Fig. 9-14*b*. Assume that both disks are made of steel with $E = 30 \times 10^6$ psi and $\nu = 0.3$, and that the smaller disk is subject to the inside pressure $p_i = 5000$ psi. Determine how the stress distribution will change when this two-disk system is replaced by a single disk with the same overall dimensions, made of the same material, and subject to the same inside pressure.

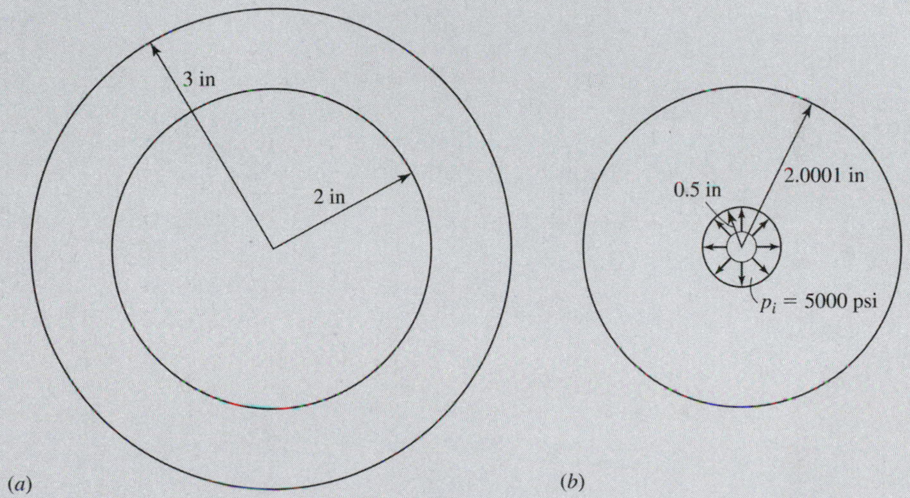


**Figure 9-14**

## SOLUTION

From Eq. (9-60), with $R = 2.0$ in, $R_{(1)i} = 0.5$ in, $R_{(2)o} = 3.0$ in, and $\Delta = 0.0001$ in we obtain

$$p = 0.0001 \times 30 \times 10^6 \frac{(3^2 - 2^2)(2^2 - 0.5^2)}{2 \times 2^3(3^2 - 0.5^2)} = 401.8 \text{ psi} \tag{a}$$

In order to find the stresses in the outside disk we use Eqs. (9-41). First we note that now

$$r_o = R_{(2)o} = 3.0 \text{ in}, \qquad r_i = R = 2.0 \text{ in}, \qquad p_o = 0, \qquad p_i = p = 401.8 \text{ psi} \qquad (b)$$

Substituting Eqs. (*b*) into Eqs. (9-41) we get

$$\sigma_{rr} = \frac{1}{3^2 - 2^2}\left[2^2 \times 401.8\left(1 - \frac{3^2}{r^2}\right)\right] = 321.4\left(1 - \frac{9}{r^2}\right)$$

$$\sigma_{\theta\theta} = 321.4\left(1 + \frac{9}{r^2}\right) \qquad (c)$$

Similarly, the stresses in the inside disk are found by substituting into Eqs. (9-41),

$$r_o = R = 2.0 \text{ in}, \qquad r_i = R_{(1)i} = 0.5 \text{ in}$$
$$p_o = p = 401.8 \text{ psi}, \qquad p_i = 5000 \text{ psi} \qquad (d)$$

which gives

$$\sigma_{rr} = \frac{1}{2^2 - 0.5^2}\left[0.5^2 \times 5000\left(1 - \frac{2^2}{r^2}\right) - 2^2 \times 401.8\left(1 - \frac{0.5^2}{r^2}\right)\right]$$

$$= 333.3\left(1 - \frac{4}{r^2}\right) - 428.6\left(1 - \frac{0.25}{r^2}\right) \qquad (e)$$

$$\sigma_{\theta\theta} = 333.3\left(1 + \frac{4}{r^2}\right) - 428.6\left(1 + \frac{0.25}{r^2}\right)$$

Figure 9-15 shows the distribution of the stresses along the radius.

Finally consider the disk shown in Fig. 9-16*a*. Now

$$r_i = 0.5 \text{ in}, \qquad r_o = 3.0 \text{ in}, \qquad p_i = 5000 \text{ psi}, \qquad p_o = 0$$

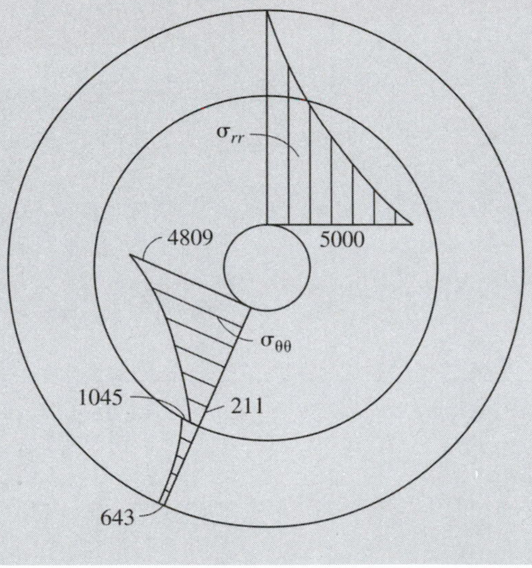


**Figure 9-15**

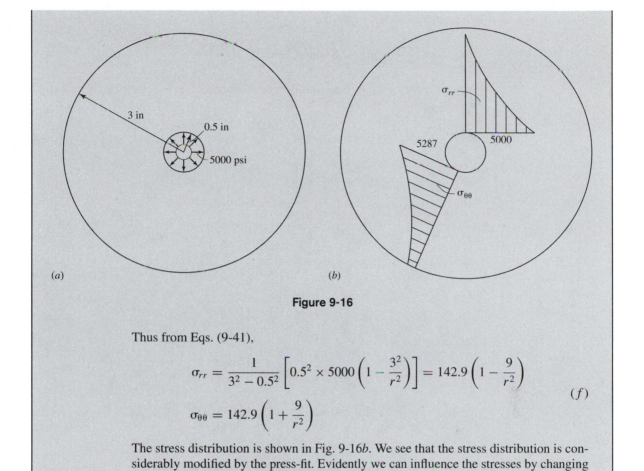

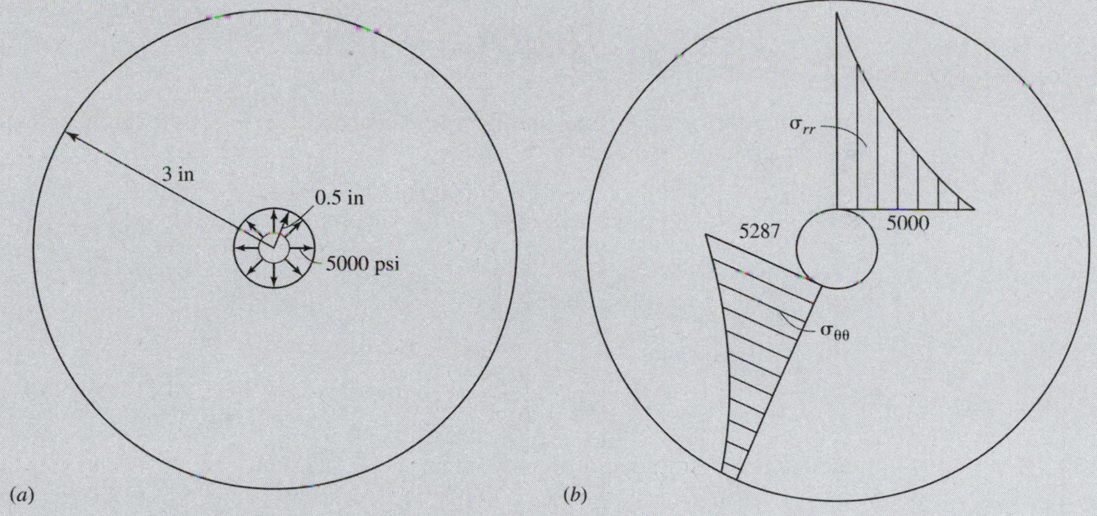


**Figure 9-16**

Thus from Eqs. (9-41),

$$\sigma_{rr} = \frac{1}{3^2 - 0.5^2}\left[0.5^2 \times 5000\left(1 - \frac{3^2}{r^2}\right)\right] = 142.9\left(1 - \frac{9}{r^2}\right)$$

$$\sigma_{\theta\theta} = 142.9\left(1 + \frac{9}{r^2}\right) \qquad (f)$$

The stress distribution is shown in Fig. 9-16*b*. We see that the stress distribution is considerably modified by the press-fit. Evidently we can influence the stresses by changing the material properties, the magnitude of the radial interference, and so on, while maintaining the same geometry.

Finally let us consider stresses in a spinning disk when neither an outside nor an inside pressure is applied. By using the boundary conditions

$$\sigma_{rr} = 0 \quad \text{when} \quad r = r_o, \qquad \sigma_{rr} = 0 \quad \text{when} \quad r = r_i$$

we obtain from Eq. (9-33*a*) the following relations:

$$\frac{E}{1-\nu^2}\left[(1+\nu)C_1 - \frac{1-\nu}{r_o^2}C_2\right] - \frac{3+\nu}{8}\rho\omega^2 r_o^2 = 0$$

$$\frac{E}{1-\nu^2}\left[(1+\nu)C_1 - \frac{1-\nu}{r_i^2}C_2\right] - \frac{3+\nu}{8}\rho\omega^2 r_i^2 = 0$$

Solving these we get

$$C_1 = \frac{(3+\nu)(1-\nu)}{8E}\left(r_i^2 + r_o^2\right)\rho\omega^2, \qquad C_2 = \frac{(3+\nu)(1+\nu)}{8E} r_i^2 r_o^2 \rho\omega^2$$

Substituting these expressions into Eqs. (9-32) and (9-33) gives the formulas for the displacements and the stresses in a spinning disk in the absence of outside pressure,

$$u = \frac{\rho\omega^2}{8Er}\left[r_i^2 r_o^2(1+\nu)(3+\nu) + (1-\nu)(3+\nu)\left(r_i^2 + r_o^2\right)r^2 - (1-\nu^2)r^4\right] \quad (9\text{-}61a)$$

$$\sigma_{rr} = \frac{3+\nu}{8}\left[r_i^2 + r_o^2 - \frac{r_i^2 r_o^2}{r^2} - r^2\right]\rho\omega^2 \quad (9\text{-}61b)$$

$$\sigma_{\theta\theta} = \frac{3+\nu}{8}\left[r_i^2 + r_o^2 + \frac{r_i^2 r_o^2}{r^2} - \frac{1+3\nu}{3+\nu}r^2\right]\rho\omega^2 \quad (9\text{-}61c)$$

To determine the location of the maximum radial stress and the maximum tangential stress we differentiate each of Eqs. (9-61) with respect to $r$ and equate to zero the resulting expressions. Consequently, in the case of radial stress we obtain

$$r = \sqrt{r_i r_o}$$

Substituting this into Eq. (9-61$b$),

$$\max \sigma_{rr} = \frac{3+\nu}{4}\rho\omega^2(r_o - r_i)^2 \quad (9\text{-}62)$$

Note that since $\sigma_{rr} = 0$ at both $r = r_o$ and $r = r_i$, in view of Eq. (9-62), it is positive for all values of $r_i < r < r_o$. Next we find that the shear stress is a monotonic function and that the maximum occurs at the inside perimeter,

$$\max \sigma_{\theta\theta} = \frac{\rho\omega^2}{4}\left[(3+\nu)r_o^2 + (1-\nu)r_i^2\right] \quad (9\text{-}63)$$

## EXAMPLE 9-5

A steel disk 20 in in diameter with a 4-in-diameter hole is press-fitted onto a steel shaft 4.006 in in diameter. At what rotational speed $\omega$ will the press-fit loosen? Here $E = 30 \times 10^6$ psi, $\nu = 0.3$, and the weight density $w = 0.284$ lb/in$^3$.

### SOLUTION

We begin by noting that radial stresses due to rotation are always positive and increasing with $\omega^2$. Thus the press-fit will loosen when the radial stresses due to rotation will balance, at the interface, the negative radial stresses due to radial interference. Also note that it is inaccurate to use the plane stress assumption for the shaft. In fact, inasmuch as the shaft is locally squeezed by the press-fitted disk, the stresses appearing in it depend also on the axial variable. This is, however, a small inaccuracy, and we will treat the portion of the shaft in contact with the disk as if it were also in a state of plane stress.

First calculate the radial interference $\Delta = 0.5(4.006 - 4.0) = 0.003$ in and, using Eq. (9-60), the interface pressure,

$$p = \frac{0.003 \times 30 \times 10^6}{2 \times 2^3} \frac{(10^2 - 2^2)2^2}{10^2} = 21{,}600 \text{ psi} \tag{a}$$

It is now necessary to apply Eq. (9-61$a$), requiring that the total radial stress at the interface be zero. Since prior to loosening the press-fit the disk and the shaft rotate together, we should substitute $r_i = 0$. Thus,

$$\frac{3+\nu}{8}\left(r_o^2 - R^2\right)\rho\omega^2 - p = 0 \tag{b}$$

where $R$ is the interface radius and the mass density $\rho = w/386 = 0.284/386 = 7.36 \times 10^{-4}$ lb $\cdot$ s$^2$/in$^4$. Solving Eq. ($b$) for $\omega$ we get

$$\begin{aligned}\omega &= \sqrt{\frac{8p}{(3+\nu)\left(r_o^2 - R^2\right)\rho}} \\ &= \sqrt{\frac{8 \times 21600}{(3+0.3)(10^2 - 2^2)7.36 \times 10^{-4}}} \\ &= 861 \text{ rad/s} = 8220 \text{ rpm}\end{aligned}$$

## 9-3-3 Thermal Effects

The thermal effects can be readily incorporated into the previous analysis by changing Hooke's law, Eq. (9-22), into the thermoelastic constitutive equation (see Section 4-15),

$$\begin{aligned}\sigma_{rr} &= \frac{E}{1-\nu^2}\left(\frac{du}{dr} + \nu\frac{u}{r}\right) - \frac{ET\alpha}{1-\nu} \\ \sigma_{\theta\theta} &= \frac{E}{1-\nu^2}\left(\frac{u}{r} + \nu\frac{du}{dr}\right) - \frac{ET\alpha}{1-\nu}\end{aligned} \tag{9-64}$$

Here $T$ is the known temperature change, assumed to be axisymmetric, and $\alpha$ is the coefficient of thermal expansion. Substituting Eq. (9-64) into the equation of equilibrium (9-18), this becomes

$$\frac{d^2u}{dr^2} + \frac{1}{r}\frac{du}{dr} - \frac{u}{r^2} = -\frac{1-\nu^2}{E}\rho\omega^2 r + (1+\nu)\alpha\frac{dT}{dr} \tag{9-65}$$

By using the superposition principle we consider the effects of temperature independently of rotation. In order to obtain the solution to Eq. (9-65) for arbitrary $T = T(r)$, we represent Eq. (9-65) in the form

$$\frac{d}{dr}\left[\frac{1}{r}\frac{d}{dr}(ru)\right] = (1+\nu)\alpha\frac{dT}{dr} \tag{9-66}$$

This can be easily integrated twice to obtain

$$u = C_1 r + \frac{C_2}{r} + \frac{(1+\nu)\alpha}{r}\int_{r_i}^{r} T\rho\, d\rho \tag{9-67}$$

Consequently the stresses are [see Eqs. (9-64)]

$$\begin{aligned}
\sigma_{rr} &= \frac{E}{1-\nu^2}\left[(1+\nu)C_1 - \frac{1-\nu}{r^2}C_2\right] - \frac{E\alpha}{r^2}\int_{r_i}^{r} T\rho\, d\rho \\
\sigma_{\theta\theta} &= \frac{E}{1-\nu^2}\left[(1+\nu)C_1 + \frac{1-\nu}{r^2}C_2\right] + \frac{E\alpha}{r^2}\int_{r_i}^{r} T\rho\, d\rho - ET\alpha
\end{aligned} \tag{9-68}$$

The constants $C_1$ and $C_2$ are then obtained from the boundary conditions at $r = r_i$ and $r = r_o$. In the absence of outside pressure they are

$$\sigma_{rr} = 0, \quad \text{at } r = r_i, \qquad \sigma_{rr} = 0, \quad \text{at } r = r_o \tag{9-69}$$

Thus we get

$$\begin{aligned}
\frac{E}{1-\nu^2}\left[(1+\nu)C_1 - \frac{1-\nu}{r_i^2}C_2\right] &= 0 \\
\frac{E}{1-\nu^2}\left[(1+\nu)C_1 - \frac{1-\nu}{r_o^2}C_2\right] - \frac{E\alpha}{r_o^2}\int_{r_i}^{r_o} T\rho\, d\rho &= 0
\end{aligned} \tag{9-70}$$

This gives

$$C_1 = \frac{(1-\nu)\alpha}{r_o^2 - r_i^2}\int_{r_i}^{r_o} T\rho\, d\rho$$

$$C_2 = \frac{(1+\nu)r_i^2\alpha}{r_o^2 - r_i^2}\int_{r_i}^{r_o} T\rho\, d\rho$$

Substituting these expressions into Eqs. (9-67) and (9-68), we obtain

$$u = \frac{\alpha}{\left(r_o^2 - r_i^2\right)r}\left[(1-\nu)r^2 + (1+\nu)r_i^2\right]\int_{r_i}^{r_o} T\rho\, d\rho + \frac{(1+\nu)\alpha}{r}\int_{r_i}^{r} T\rho\, d\rho \tag{9-71}$$

and

$$\begin{aligned}
\sigma_{rr} &= \frac{E\alpha}{r_o^2 - r_i^2}\left(1 - \frac{r_i^2}{r^2}\right)\int_{r_i}^{r_o} T\rho\, d\rho - \frac{E\alpha}{r^2}\int_{r_i}^{r} T\rho\, d\rho \\
\sigma_{\theta\theta} &= \frac{E\alpha}{r_o^2 - r_i^2}\left(1 + \frac{r_i^2}{r^2}\right)\int_{r_i}^{r_o} T\rho\, d\rho + \frac{E\alpha}{r^2}\int_{r_i}^{r} T\rho\, d\rho - E\alpha T
\end{aligned} \tag{9-72}$$

From Eq. (9-71) we obtain

$$u = \frac{2r_i \alpha}{r_o^2 - r_i^2} \int_{r_i}^{r_o} T\rho \, d\rho, \qquad \text{at } r = r_i \tag{9-73a}$$

$$\begin{aligned} u &= \left\{ \frac{\alpha}{\left(r_o^2 - r_i^2\right) r_o} \left[(1-\nu) r_o^2 + (1+\nu) r_i^2\right] + \frac{(1+\nu)\alpha}{r_o} \right\} \int_{r_i}^{r_o} T\rho \, d\rho \\ &= \frac{\alpha}{\left(r_o^2 - r_i^2\right) r_o} \left(2r_o^2\right) \int_{r_i}^{r_o} T\rho \, d\rho = \frac{2r_o \alpha}{r_o^2 - r_i^2} \int_{r_i}^{r_o} T\rho \, d\rho, \qquad \text{at } r = r_o \end{aligned} \tag{9-73b}$$

It is evident from here that $u(r_o) > u(r_i)$. Similarly, with Eqs. (9-72) we get

$$\begin{aligned} \sigma_{\theta\theta} &= \frac{2\alpha E}{r_o^2 - r_i^2} \int_{r_i}^{r_o} T\rho \, d\rho - \alpha E T(r_i), \qquad \text{at } r = r_i \\ \sigma_{\theta\theta} &= \frac{2\alpha E}{r_o^2 - r_i^2} \int_{r_i}^{r_o} T\rho \, d\rho - \alpha E T(r_o), \qquad \text{at } r = r_o \end{aligned} \tag{9-74}$$

Thus when $T(r_i) = T(r_o)$, then the stress $\sigma_{\theta\theta}$ must have extremum for $r_i < r < r_o$.

For a solid disk ($r_i = 0$) all static quantities must be finite at $r = 0$. Hence the expression

$$\lim_{r=0} \left[ \frac{C_2}{r} + \frac{(1+\nu)\alpha}{r} \int_0^r T\rho \, d\rho \right]$$

appearing in Eqs. (9-67) and (9-68) must be finite. First we find the limit of the second term,

$$\begin{aligned} \lim_{r=0} \frac{(1+\nu)\alpha}{r} \int_0^r T\rho \, d\rho &= (1+\nu)\alpha \lim_{r=0} \frac{\dfrac{d}{dr} \displaystyle\int_0^r T\rho \, d\rho}{\dfrac{d}{dr}(r)} \\ &= (1+\nu)\alpha \lim_{r=0} \frac{T(r) r}{1} = 0 \end{aligned} \tag{9-75}$$

where we have used the L'Hospital formula.* Similarly, the expression

$$\frac{1}{r^2} \int_0^r T\rho \, d\rho$$

can be shown to be finite at $r = 0$ as long as $T(r)$ is finite at $r = 0$. This implies that for finiteness of the results it is necessary that

$$C_2 = 0$$

*See any text on calculus.

Therefore now

$$u = C_1 r + \frac{(1+\nu)\alpha}{r} \int_0^r T\rho \, d\rho \tag{9-76a}$$

$$\sigma_{rr} = \frac{E}{1-\nu} C_1 - \frac{\alpha E}{r^2} \int_0^r T\rho \, d\rho \tag{9-76b}$$

$$\sigma_{\theta\theta} = \frac{E}{1-\nu} C_1 + \frac{\alpha E}{r^2} \int_0^r T\rho \, d\rho - \alpha E T \tag{9-76c}$$

The remaining unknown constant $C_1$ is obtained using the boundary condition at $r = r_o$. If this condition is, for example,

$$\sigma_{rr} = 0, \qquad \text{at } r = r_o$$

then with Eq. (9-76*b*) we obtain

$$\frac{E}{1-\nu} C_1 - \frac{\alpha E}{r_o^2} \int_0^{r_o} T\rho \, d\rho = 0$$

Therefore,

$$C_1 = \frac{\alpha(1-\nu)}{r_o^2} \int_0^{r_o} T\rho \, d\rho$$

and finally,

$$u = \frac{\alpha(1-\nu)r}{r_o^2} \int_0^{r_o} T\rho \, d\rho + \frac{(1+\nu)\alpha}{r} \int_0^r T\rho \, d\rho$$

$$\sigma_{rr} = \frac{E\alpha}{r_o^2} \int_0^{r_o} T\rho \, d\rho - \frac{E\alpha}{r^2} \int_0^r T\rho \, d\rho \tag{9-77}$$

$$\sigma_{\theta\theta} = \frac{E\alpha}{r_o^2} \int_0^{r_o} T\rho \, d\rho + \frac{E\alpha}{r^2} \int_0^r T\rho \, d\rho - \alpha E T$$

## EXAMPLE 9-6

Determine the temperature change $T$ needed to make the hole in the disk in Example 9-5 large enough to fit on the shaft from the same example.

### SOLUTION

The data from Example 9-5 are $r_i = 2.0$ in, $r_o = 4.0$ in, and $\Delta = 0.003$ in. The coefficient of thermal expansion for steel $\alpha = 0.0000117/°\text{C}$. From Eq. (9-73*a*), assuming

$T = $ constant, we get the displacement at $r = r_i$,

$$u(r_i) = \frac{2r_i\alpha T}{r_o^2 - r_i^2}\frac{r_o^2 - r_i^2}{2} = \alpha T r_i = \Delta$$

This gives

$$T = \frac{\Delta}{\alpha r_i} = \frac{0.003}{0.0000117 \times 4} = 64°\text{C} = 115°\text{F}$$

as the temperature change needed to increase the radius of the hole in the disk by $\Delta = 0.003$ in.

## 9-3-4 Plastic Cylinder

Although the solutions obtained in previous sections apply to the most commonly occurring situations, we must be able to deal with materials other than linearly elastic ones. We know from Chapter 4 that even an originally elastic material will behave plastically when an appropriate stress level (yield stress) is reached. This state may actually be imposed deliberately to use a material more efficiently by generating residual stresses (*prestress*), which cause the material to flow plastically (*autofrettage*), then reducing the applied pressure. We note that the equilibrium equation (9-18)

$$\frac{d\sigma_{rr}}{dr} + \frac{\sigma_{rr} - \sigma_{\theta\theta}}{r} + \rho\omega^2 r = 0 \tag{9-78}$$

was derived without any reference to material properties. Therefore it is valid whether a material is elastic, as was assumed in Section 9-4-2, or elastic–perfectly plastic (see Chapter 4), as we assume here. Let us consider the static case ($\omega = 0$). To proceed further we need another relation between stresses, which, while replacing the previously used Hooke's law, will show the plastic properties of a material and also, by relating $\sigma_{rr}$ to $\sigma_{\theta\theta}$, will provide a missing second equation, needed to solve the problem. This relation is the yield condition

$$\sigma_{\theta\theta} - \sigma_{rr} = kY \tag{9-79}$$

which has the same form (see Chapter 4 and also [2]), in both the case of Tresca's and that of von Mises' yield theory. Here $Y$ is the *yield strength* and

$$k = 1 \qquad \text{or} \qquad k = \sqrt{2}/3$$

the first value being applicable for Tresca's theory, the second for von Mises' yield theory. We substitute now Eq. (9-79) into Eq. (9-78) and obtain

$$\frac{d\sigma_{rr}}{dr} = \frac{kY}{r} \tag{9-80}$$

Integration gives

$$\sigma_{rr} = kY \ln r + C_1 \tag{9-81}$$

where $C_1$ is the constant of integration. Next assume that the cylinder is subjected to internal pressure $p_i$, and that $p_o = 0$. Thus the condition

$$\sigma_{rr} = -p_i, \qquad \text{at } r = r_i \tag{9-82}$$

gives

$$C_1 = -p_i - kY \ln r_i$$

and with Eqs. (9-81) and (9-79),

$$\begin{aligned} \sigma_{rr} &= -p_i + kY \ln\left(\frac{r}{r_i}\right) \\ \sigma_{\theta\theta} &= -p_i + kY\left[1 + \ln\left(\frac{r}{r_i}\right)\right] \end{aligned} \tag{9-83}$$

Examining Eqs. (9-83) we note that the absolute values of both stresses decrease as we move away from the hole. At a certain point $r = c$ the magnitudes of these stresses will be sufficiently small for the state of stress to change from plastic to elastic (Fig. 9-17). Beyond this point, that is, for $r > c$, the stresses are calculated from Eqs. (9-42), yielding, with $p_o = 0$,

$$\sigma_{rr} = \frac{p_c c^2}{r_o^2 - c^2}\left(1 - \frac{r_o^2}{r^2}\right), \qquad \sigma_{\theta\theta} = \frac{p_c c^2}{r_o^2 - c^2}\left(1 + \frac{r_o^2}{r^2}\right) \tag{9-84}$$

where $p_c$ is the interface pressure. As this is being calculated we require that at $r = c$. The stresses described by Eqs. (9-84) must also satisfy the yield condition, Eq. (9-69). Thus,

$$p_c = kY\frac{r_o^2 - c^2}{2r_o^2} \tag{9-85}$$

Finally, substituting Eq. (9-85) into Eqs. (9-84) we obtain the stresses in the elastic region,

$$\sigma_{rr} = kY\frac{c^2}{2r_o^2}\left(1 - \frac{r_o^2}{r^2}\right), \qquad \sigma_{\theta\theta} = kY\frac{c^2}{2r_o^2}\left(1 + \frac{r_o^2}{r^2}\right) \tag{9-86}$$

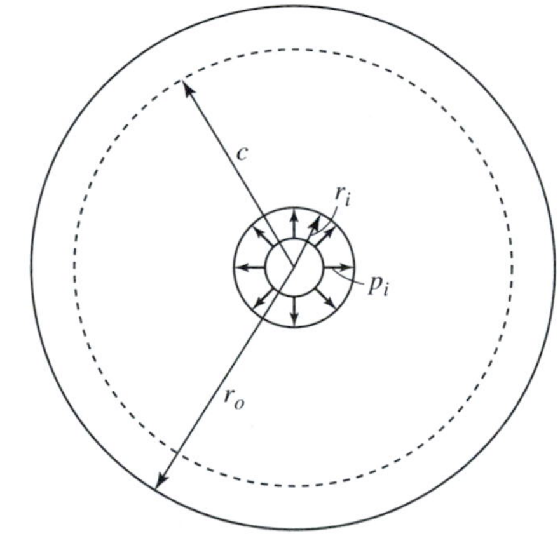


**Figure 9-17** Stresses outside of the dashed line are elastic

Next let us find the depth $c$ of the plastic zone corresponding to the given pressure $p_i$. We begin by calculating the range of values limiting $p_i$ in a manner that will ensure the presence of the plastic zone. From Eq. (9-85) we see that $c = r_i$ indicates the onset of yielding. Thus if

$$p_i \le kY \frac{r_o^2 - r_i^2}{2r_o^2}$$

then there is no plastic zone. On the other hand when the plastic zone extends to the outer edge ($r = r_o$), then the radial stress obtained from Eqs. (9-83) must be zero at $r = r_o$. Hence,

$$p_i = kY \ln \frac{r_o}{r_i}$$

Finally we have to check the depth of the plastic zone when

$$kY \frac{r_o^2 - r_i^2}{2r_o^2} \le p_i \le kY \ln \frac{r_o}{r_i} \tag{9-87}$$

If the inequalities (9-87) are satisfied by the applied pressure $p_i$, we find the depth $c$ by comparing the radial stress calculated from Eqs. (9-83) with the stress obtained from Eqs. (9-84). They must be equal at the interface $r = c$. Thus,

$$p_i = kY \left( \ln \frac{c}{r_i} + \frac{r_o^2 - c^2}{2r_o^2} \right)$$

This can be put in the form

$$\frac{c^2}{2r_o^2} - \ln \frac{c}{r_i} = \frac{1}{2} - \frac{p_i}{kY} \tag{9-88}$$

The quantity $c$ is then calculated numerically.

## 9-3-5 Composite Disks and Cylinders

Consider now a structure composed of two concentric disks (Fig. 9-18), each made of a different orthotropic material. The orthotropy of a composite material results as a rule

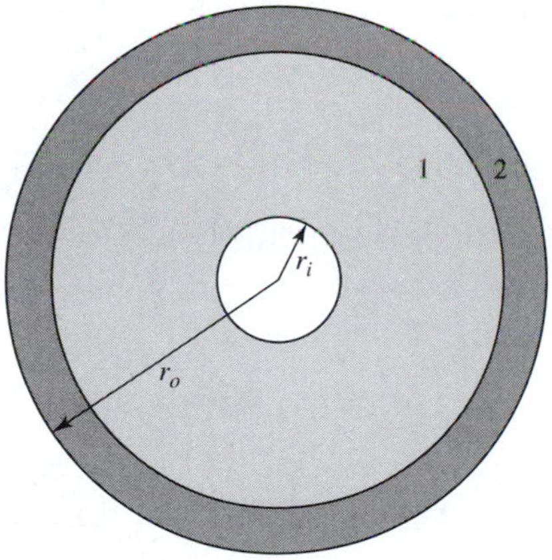


**Figure 9-18** Two concentric disks each made of a different orthotropic material.

from fiber reinforcement. This type of structure, modeling a more complicated disk formed of several layers, is of practical importance in turbines. However, the method shown here is valid for any number of layers, although an increase in the number of layers increases the algebraic complications. We begin by assuming a perfect bond between layers, and then examine the equations derived and used for homogeneous disks to see what must be changed. The equilibrium equation (9-18) is still valid, and so are the relations (9-19) and (9-20). A distinction must be made, however, between materials 1 and 2 by assuming that their physical properties differ. To do so, instead of using $\rho$ in Eq. (9-18), we use $\rho^{(1)}$ or $\rho^{(2)}$ when dealing with material 1 or 2. For the same reason we attach the index (1) or (2) to all stress, strain, and displacement components. The stress-strain relations, Eqs. (9-21), must, however, be modified for an orthotropic material. Solving the relations (see Chapter 4)

$$\varepsilon_{rr} = \frac{1}{E_r}(\sigma_{rr} - \nu_r \sigma_{\theta\theta}), \qquad \varepsilon_{\theta\theta} = \frac{1}{E_\theta}(\sigma_{\theta\theta} - \nu_\theta \sigma_{rr})$$

for $\sigma_{rr}$ and $\sigma_{\theta\theta}$, we get

$$\sigma_{rr} = \frac{1}{1 - \nu_r \nu_\theta}(E_r \varepsilon_{rr} + \nu_r E_\theta \varepsilon_{\theta\theta}), \qquad \sigma_{\theta\theta} = \frac{1}{1 - \nu_r \nu_\theta}(E_\theta \varepsilon_{\theta\theta} + \nu_\theta E_r \varepsilon_{rr}) \tag{9-89}$$

When Eqs. (9-19) and (9-20) are now substituted, we obtain the expressions for the stresses in terms of the displacements,

$$\sigma_{rr} = \frac{1}{1 - \nu_r \nu_\theta}\left(E_r \frac{du}{dr} + \nu_r E_\theta \frac{u}{r}\right) \tag{9-90$a$}$$

$$\sigma_{\theta\theta} = \frac{1}{1 - \nu_r \nu_\theta}\left(E_\theta \frac{u}{r} + \nu_\theta E_r \frac{du}{dr}\right) \tag{9-90$b$}$$

Next we substitute Eqs. (9-90) into the equilibrium equation (9-18). This results in the following differential equation for the displacement $u$:

$$E_r r^2 \frac{d^2u}{dr^2} + [(1 - \nu_\theta)E_r + \nu_r E_\theta] r \frac{du}{dr} - E_\theta u = -(1 - \nu_\theta \nu_r)\rho\omega^2 r^3 \tag{9-91}$$

This more complicated version of Eq. (9-23) is also Euler's equation, solvable using the same procedure that was used in Section 9-4-2. Thus we substitute

$$u = r^n$$

obtaining

$$E_r r^2 n(n-1) r^{n-2} + [(1 - \nu_\theta)E_r + \nu_r E_\theta] r n r^{n-1} - E_\theta r^n = 0$$

Dividing this by $r^n$ results in the indicial equation

$$E_r n(n-1) + [(1 - \nu_\theta)E_r + \nu_r E_\theta] n - E_\theta = 0 \tag{9-92}$$

from which it follows that

$$E_r n^2 + (\nu_r E_\theta - \nu_\theta E_r) n - E_\theta = 0 \tag{9-93}$$

Its roots are

$$n_{1,2} = \frac{\nu_\theta E_r - \nu_r E_\theta \pm \sqrt{(\nu_r E_\theta - \nu_\theta E_r)^2 + 4 E_r E_\theta}}{2E_r} \tag{9-94}$$

It is evident from the preceding that $n_1$ is always positive and $n_2$ always negative. It was determined in Chapter 4 that

$$\frac{\nu_r}{E_\theta} = \frac{\nu_\theta}{E_r} \tag{9-95}$$

When Eq. (9-94) is used with Eq. (9-95), it becomes

$$n_{1,2} = \frac{\nu_r(E_r^2 - E_\theta^2) \pm \sqrt{\nu_r^2(E_r^2 - E_\theta^2)^2 + 4E_r E_\theta^3}}{2E_r E_\theta} \tag{9-96}$$

For the isotropic case, $\nu_r = \nu_\theta = \nu$, $E_r = E_\theta = E$, and Eq. (9-94) reduces to Eqs. (9-27) obtained previously for a homogeneous disk,

$$n_{1,2} = \frac{\pm\sqrt{4E^2}}{2E} = \pm 1$$

Therefore the general solution to the homogeneous equation is

$$u(r) = C_1 r^{n_1} + C_2 r^{n_2} \tag{9-97}$$

By using the results from Section 9-4-2 we seek the particular solution to the nonhomogeneous equation in the form $u = Ar^3$. Substituting this into Eq. (9-91) we obtain

$$6E_r r^3 A + 3Ar^3[(1 - \nu_\theta)E_r + \nu_r E_\theta] - E_\theta A r^3 + (1 - \nu_\theta \nu_r)\rho\omega^2 r^3 = 0$$

or

$$A[6E_r + 3(1 - \nu_\theta)E_r + (3\nu_r - 1)E_\theta] = -(1 - \nu_\theta\nu_r)\rho\omega^2$$

so that

$$A = \frac{-(1 - \nu_r\nu_\theta)\rho\omega^2}{3(3 - \nu_\theta)E_r + (3\nu_r - 1)E_\theta} \tag{9-98}$$

Finally then

$$u(r) = C_1 r^{n_1} + C_2 r^{n_2} + Ar^3 \tag{9-99}$$

Such an expression must be written separately for each region. In order to distinguish between both materials we again attach the index (1) or (2) to the constants, the stresses, the diplacements, and so on. First let us calculate the stresses. Using Eq. (9-90*a*), we get

$$\begin{aligned}\sigma_{rr} &= \frac{1}{1 - \nu_\theta\nu_r}\left(E_r u' + \nu_r E_\theta \frac{u}{r}\right) \\ &= \frac{1}{1 - \nu_\theta\nu_r}[E_r(C_1 n_1 r^{n_1 - 1} + C_2 n_2 r^{n_2 - 1} + 3Ar^2) \\ &\quad + \nu_r E_\theta(C_1 r^{n_1 - 1} + C_2 r^{n_2 - 1} + Ar^2)]\end{aligned}$$

or

$$\begin{aligned}\sigma_{rr} = &\frac{1}{1 - \nu_\theta\nu_r}[C_1 r^{n_1 - 1}(E_r n_1 + \nu_r E_\theta) + C_2 r^{n_2 - 1}(E_r n_2 + \nu_r E_\theta) \\ &+ Ar^2(3E_r + \nu_r E_\theta)]\end{aligned} \tag{9-100}$$

Also from Eq. (9-90*b*),

$$\sigma_{\theta\theta} = \frac{1}{1-\nu_\theta \nu_r}\left(E_\theta \frac{u}{r} + \nu_\theta E_r u'\right)$$

$$= \frac{1}{1-\nu_\theta \nu_r}[E_\theta(C_1 r^{n_1-1} + C_2 r^{n_2-1} + Ar^2) + \nu_\theta E_r(C_1 n_1 r^{n_1-1} + C_2 n_2 r^{n_2-1} + 3Ar^2)]$$

or

$$\sigma_{\theta\theta} = \frac{1}{1-\nu_\theta \nu_r}[C_1 r^{n_1-1}(E_\theta + n_1 \nu_\theta E_r) + C_2 r^{n_2-1}(E_\theta + n_2 \nu_\theta E_r) + Ar^2(E_\theta + 3\nu_\theta E_r)] \tag{9-101}$$

Now the boundary conditions are

$$\sigma_{rr}^{(1)} = -p_o, \qquad \text{at } r = r_o$$
$$\sigma_{rr}^{(2)} = -p_i, \qquad \text{at } r = r_i \tag{9-102}$$

and the continuity conditions have the form

$$\sigma_{rr}^{(1)} = \sigma_{rr}^{(2)}, \quad \text{at } r = r_m, \qquad u^{(1)} = u^{(2)}, \quad \text{at } r = r_m \tag{9-103}$$

Representing the stresses in an abbreviated form,

$$\sigma_{rr}^{(1)} = C_1^{(1)} D_1^{(1)} r^{n_1^{(1)}-1} + C_2^{(1)} D_2^{(1)} r^{n_2^{(1)}-1} + D_3^{(1)} r$$
$$\sigma_{rr}^{(2)} = C_1^{(2)} D_1^{(2)} r^{n_1^{(2)}-1} + C_2^{(2)} D_2^{(2)} r^{n_2^{(2)}-1} + D_3^{(2)} r \tag{9-104}$$

and the displacements in the form

$$u^{(1)}(r) = C_1^{(1)} r^{n_1^{(1)}} + C_2^{(1)} r^{n_2^{(1)}} + A^{(1)} r^3 \tag{9-105a}$$

$$u^{(2)}(r) = C_1^{(2)} r^{n_1^{(2)}} + C_2^{(2)} r^{n_2^{(2)}} + A^{(2)} r^3 \tag{9-105b}$$

where

$$D_k^{(i)} \equiv \frac{1}{1-\nu_\theta^{(i)}\nu_r^{(i)}}\left[E_\theta^{(i)}\nu_r^{(i)} + n_k^{(i)} E_r^{(i)}\right], \qquad k = 1, 2$$
$$D_3^{(i)} = \frac{A^{(i)}}{1-\nu_\theta^{(i)}\nu_r^{(i)}}\left[3E_r^{(i)} + \nu_r^{(i)} E_\theta^{(i)}\right] \tag{9-106}$$

and using the boundary conditions (9-102) and the continuity conditions (9-103), a system of algebraic equations follows from which one can find $C_1^{(1)}$, $C_1^{(2)}$, $C_2^{(1)}$, and $C_2^{(2)}$.

For a solid disk $r_i = 0$, and the second boundary condition, Eq. (9-102*b*), must be replaced by another condition. Making use of the previous remark that $n_2 < 1$, we see by inspection of Eq. (9-105*b*) that when $r = 0$, the term

$$C_2^{(2)} r^{n_2^{(2)}}$$

tends to infinity and so does $u^{(2)}$. As this is physically unacceptable, we assume that

$$C_2^{(2)} = 0$$

eliminating in this way one of the unknown constants. The remaining three constants, $C_1^{(1)}$, $C_2^{(1)}$, and $C_1^{(2)}$, are readily calculated from the conditions (9-102*a*) and (9-103).

## EXAMPLE 9-7

Determine the stresses in a two-layer composite cylinder subject to an internal pressure of 5000 psi (see Fig. 9-19). The outside layer is made of T300/5208 graphite epoxy in the θ (circumferential) direction. The inside layer is made of E-glass/epoxy with fibers in the $z$ (axial) direction. The material properties are [10, p. 160] as follows:

- *T300/5208:*

| | |
|---|---|
| Young's modulus in circumferential direction | $E_\theta = 22.2 \times 10^6$ psi |
| Young's moduli in axial and radial directions | $E_z = E_r = 1.58 \times 10^6$ psi |
| Shear moduli of elasticity | $G_{z\theta} = G_{r\theta} = 0.81 \times 10^6$ psi |
| Poisson's ratios | $\nu_\theta = 0.30$, $\nu_z = 0.021$, $\nu_r = 0.021$ |

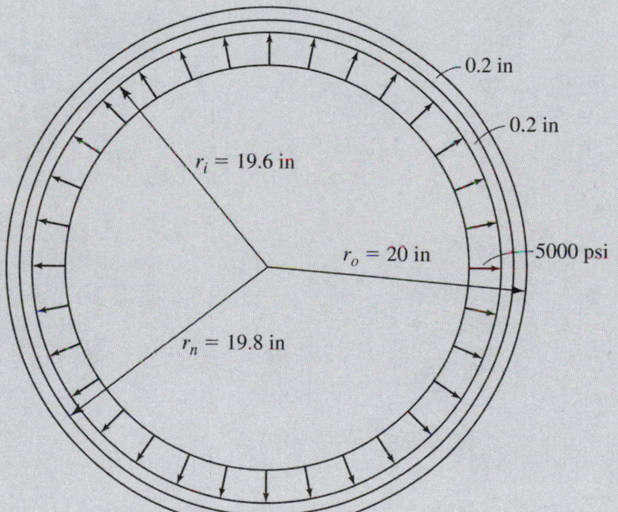


**Figure 9-19**

- *E-glass/epoxy:*

| | |
|---|---|
| Young's modulus in axial direction | $E_z = 8.8 \times 10^6$ psi |
| Young's moduli in circumferential and radial directions | $E_\theta = E_r = 3.6 \times 10^6$ psi |
| Shear moduli of elasticity | $G_{z\theta} = G_{r\theta} = 1.74 \times 10^6$ psi |
| Poisson's ratios | $\nu_\theta = 0.0941,\ \nu_z = 0.23,\ \nu_r = 0.0941$ |

## SOLUTION

Assume that the equations derived in Section 9-3-5 are also valid for cylinders. We begin by calculating the indices $n_1$ and $n_2$ for both layers. With Eqs. (9-94) we have for layer 1,

$$n_1^{(1)} = \frac{A}{2 \times 1.58 \times 10^6} = 3.751$$

$$n_2^{(1)} = \frac{B}{2 \times 1.58 \times 10^6} = -3.746 \qquad (a)$$

where

$$A = 0.30 \times 1.58 \times 10^6 - 0.021 \times 22.2 \times 10^6 + 10^6\sqrt{(0.021 \times 22.2 - 0.30 \times 1.58)^2 + 4 \times 1.58 \times 22}$$

$$B = 0.30 \times 1.58 \times 10^6 - 0.021 \times 22.2 \times 10^6 - 10^6\sqrt{(0.021 \times 22.2 - 0.30 \times 1.58)^2 + 4 \times 1.58 \times 22}$$

Since layer 2 (the inside cylinder) behaves like an isotropic material,

$$n_1^{(2)} = 1, \qquad n_2^{(2)} = -1 \qquad (b)$$

Next, using Eqs. (9-104), we determine the radial stresses. First note that [see Eqs. (9-106)]

$$D_1^{(1)} = \frac{10^6}{1 - 0.30 \times 0.021}(22.2 \times 0.021 + 3.751 \times 1.58) = 6.433 \times 10^6 \text{ psi}$$

$$D_2^{(1)} = \frac{10^6}{1 - 0.30 \times 0.021}(22.2 \times 0.021 - 3.746 \times 1.58) = -5.487 \times 10^6 \text{ psi}$$

$$D_1^{(2)} = \frac{10^6}{1 - 0.0941^2}(3.6 \times 0.0941 + 1 \times 3.6) = 3.974 \times 10^6 \text{ psi} \qquad (c)$$

$$D_2^{(2)} = \frac{10^6}{1 - 0.0941^2}(3.6 \times 0.0941 - 1 \times 3.6) = -3.290 \times 10^6 \text{ psi}$$

and that, with Eqs. (9-98) and (9-106),

$$A^{(1)} = A^{(2)} = 0, \qquad D_3^{(1)} = D_3^{(2)} = 0 \qquad (d)$$

Now substituting Eqs. (9-104) into the boundary conditions (9-102),

$$C_1^{(1)} \times 6.433 \times 10^6 \times 20^{3.751-1} - C_2^{(1)} \times 5.487 \times 10^6 \times 20^{-3.746-1} = 0$$

$$C_1^{(2)} \times 3.974 \times 10^6 \times (20 - 2 \times 0.20)^{1-1} - C_2^{(2)} \times 3.290 \times 10^6 \times (20 - 2 \times 0.20)^{-1-1} = -5000 \qquad (e)$$

At the interface $r = r_m = 19.8$ in, the continuity condition for the stresses, Eqs. (9-103), becomes

$$C_1^{(1)} \times 6.433 \times 10^6 \times (20 - 0.20)^{3.751-1} - C_2^{(1)} \times 5.487 \times 10^6 \times (20 - 0.20)^{-3.746-1} = C_1^{(2)} \times 3.974 \times 10^6 \times (20 - 0.20)^{1-1} - C_2^{(2)} \times 3.290 \times 10^6 \times (20 - 0.20)^{-1-1} \qquad (f)$$

Finally the continuity condition for the displacements, Eqs. (9-103), is

$$C_1^{(1)} \times (20 - 0.20)^{3.751} + C_2^{(1)} \times (20 - 0.20)^{-3.746} = C_1^{(2)} \times (20 - 0.20) + C_2^{(2)} \times (20 - 0.20)^{-1} \qquad (g)$$

When simplified, Eqs. (*e*), (*f*), and (*g*) become

$$\begin{aligned} 2.441 \times 10^4 C_1^{(1)} - 3.670 \times 10^{-6} C_2^{(1)} &= 0 \\ 3.974 \times 10^6 C_1^{(2)} - 8.564 \times 10^3 C_2^{(2)} &= -5000 \\ 2.374 \times 10^4 C_1^{(1)} - 3.849 \times 10^{-6} C_2^{(1)} &= 3.974 C_1^{(2)} - 8.392 \times 10^{-3} C_2^{(2)} \\ 7.308 \times 10^4 C_1^{(1)} + 1.389 \times 10^{-5} C_2^{(1)} &= 19.8 C_1^{(2)} + 0.05051 C_2^{(2)} \end{aligned} \qquad (h)$$

and their solution is

$$\begin{aligned} C_1^{(1)} &= 2.287 \times 10^{-6}, \qquad & C_1^{(2)} &= 0.008071 \\ C_2^{(1)} &= 15210, \qquad & C_2^{(2)} &= 4.32 \end{aligned} \qquad (i)$$

With Eqs. (*c*) and (*i*) we can calculate the stresses and the displacements everywhere in the composite cylinder. For instance, the radial stress at the interface is obtained from Eqs. (9-104),

$$\sigma_{rr}^{(1)} = 0.008071 \times 3.974 \times 10^6 \times 19.6^{1-1} - 4.3289 \times 3.290 \times 10^6 \times 19.6^{-1-1} = -4245 \text{ psi} \qquad (j)$$

## 9-3-6 Rotating Disks of Variable Thickness

Let us now assume that $t$ is a function of $r$, $t = t(r)$. We keep, however, in mind that the analysis that follows is only valid when max $t$ is still small compared to the other dimensions of the disk. We begin by deriving the equilibrium equation for an infinitesimal element. Because $t$ varies, becoming $t + dt$ at $r + dr$, we get

$$(\sigma_{rr} + d\sigma_{rr})(t + dt)(r + dr)\,d\theta - \sigma_{rr} tr\,d\theta$$
$$- 2\sigma_{\theta\theta}\left(t + \frac{dt}{2}\right) dr \sin\frac{d\theta}{2} + \rho r^2\omega^2\left(t + \frac{dt}{2}\right) dr\,d\theta = 0$$

Simplifying in a manner similar to the case $t$ = constant, we obtain

$$\frac{d\sigma_{rr}}{dr} + \frac{1}{t}\frac{dt}{dr}\sigma_{rr} + \frac{\sigma_{rr} - \sigma_{\theta\theta}}{r} + \rho\omega^2 r = 0 \tag{9-107}$$

The previous expressions for the strain are still valid and so are, for an elastic solid, the stress-strain relations. Therefore, with Eqs. (9-22), Eq. (9-107) becomes

$$\frac{d^2u}{dr^2} + \frac{1}{r}\frac{du}{dr} - \frac{u}{r^2} + \frac{1}{t}\left(\frac{du}{dr} + \nu\frac{u}{r}\right)\frac{dt}{dr} = -\frac{1-\nu^2}{E}\rho\omega^2 r \tag{9-108}$$

It is difficult to find the general solution to this equation analytically. There is, however, at least one case of variability of $t$ when the solution is readily obtained. It is of interest because it gives some additional insight and a possibility of comparing the accuracy of numerical solutions. An inspection of Eq. (9-108) indicates that if

$$\frac{1}{t}\frac{dt}{dr} = \frac{A}{r} \tag{9-109}$$

where $A$ is an arbitrary constant, then Eq. (9-108) becomes again an Euler's equation, which we have discussed previously. By solving Eq. (9-109) we see that for any variable thickness in the form

$$t = r^A \tag{9-110}$$

it can be solved exactly. Proceeding as before [see Eq. (9-24) and the subsequent equations], we find that the indicial equation becomes now

$$\tau(\tau - 1) + \tau - 1 + A\tau + \nu A\tau = 0$$

from which we get

$$\begin{aligned} \tau_1 &= \tfrac{1}{2}[-A(1+\nu) + \sqrt{A^2(1+\nu)^2 + 4}] > 0 \\ \tau_2 &= \tfrac{1}{2}[-A(1+\nu) - \sqrt{A^2(1+\nu)^2 + 4}] < 0 \end{aligned} \tag{9-111}$$

These replace $n_1$ and $n_2$ appearing in Eqs. (9-27). Consequently, using Eqs. (9-28), (9-32), and (9-22), we obtain the following expressions:

$$\begin{aligned} u(r) &= C_1 r^{\tau_1} + \frac{C_2}{r^{-\tau_2}} - \frac{1-\nu^2}{8E}\rho\omega^2 r^3 \\ \sigma_{rr} &= \frac{E}{1-\nu^2}[(\tau_1+\nu)C_1 r^{\tau_1 - 1} + (\tau_2+\nu)C_2 r^{\tau_2 - 1}] - \frac{3+\nu}{8}\rho\omega^2 r^2 \\ \sigma_{\theta\theta} &= \frac{E}{1-\nu^2}[(\tau_1+\nu)C_1 r^{\tau_1 - 1} - (\tau_2+\nu)C_2 r^{\tau_2 - 1}] - \frac{1+3\nu}{8}\rho\omega^2 r^2 \end{aligned} \tag{9-112}$$

## 9-4 AIRY'S STRESS FUNCTION

An inspection of the equations of equilibrium for the case of plane stress, Eqs. (2-55), indicates that we have two differential equations with three unknown functions, $\sigma_{xx}, \sigma_{xy}$, and $\sigma_{yy}$. Normally we would replace the stresses by strains via Hooke's law, and then express the strains as functions of the displacement components $u$ and $v$ (in the $x$ and $y$ directions). Finally, a system of two differential equations in two unknowns, $u$ and $v$, would result.

Occasionally it is, however, convenient to introduce, not unlike in Section 9-5-2, a single unknown function $\Phi(x, y)$ related to the unknown stresses, so that the equilibrium equations

$$\frac{\partial \sigma_{xx}}{\partial x} + \frac{\partial \sigma_{xy}}{\partial y} = 0, \qquad \frac{\partial \sigma_{xy}}{\partial x} + \frac{\partial \sigma_{yy}}{\partial y} = 0 \tag{9-113}$$

are identically satisfied. Indeed, if we assume that

$$\sigma_{xx} = \frac{\partial^2 \Phi}{\partial y^2}, \qquad \sigma_{xy} = -\frac{\partial^2 \Phi}{\partial x \, \partial y}, \qquad \sigma_{yy} = \frac{\partial^2 \Phi}{\partial x^2} \tag{9-114}$$

we find that Eqs. (9-113) are satisfied. The function $\Phi(x, y)$ is called *Airy's stress function*.* What remains to be done is to satisfy also the boundary conditions. This is not difficult when these conditions are imposed on stresses and when the geometry of the solid is sufficiently simple. Still, even in such cases we would end with an infinite number of possibilities in selecting an appropriate function $\Phi(x, y)$. We must remember, however, that the stress fields obtained from $\Phi(x, y)$ cannot be arbitrary, but they must satisfy the conditions of compatibility (Section 9-3),

$$\nabla^2(\sigma_{xx} + \sigma_{yy}) = 0 \tag{9-115}$$

or

$$\left(\frac{\partial^2}{\partial x^2} + \frac{\partial^2}{\partial y^2}\right)(\sigma_{xx} + \sigma_{yy}) = 0 \tag{9-116}$$

Therefore the stress function also must satisfy Eq. (9-116). Hence substituting here expressions (9-114), we obtain

$$\left(\frac{\partial^2}{\partial x^2} + \frac{\partial^2}{\partial y^2}\right)\left(\frac{\partial^2 \Phi}{\partial y^2} + \frac{\partial^2 \Phi}{\partial x^2}\right) = 0$$

which can also be represented in either of the equivalent forms

$$\nabla^2 \nabla^2 \Phi = 0, \qquad \frac{\partial^4 \Phi}{\partial x^4} + 2\frac{\partial^4 \Phi}{\partial x^2 \, \partial y^2} + \frac{\partial^4 \Phi}{\partial y^4} = 0 \tag{9-117}$$

This type of partial differential equation is called *biharmonic,* and its solutions are known as *biharmonic functions*. When the function $\Phi$ is found and adjusted to satisfy appropriate boundary conditions, the stresses result from applying Eqs. (9-114).

Since solving partial differential equations is beyond the scope of this text,† we limit ourselves to listing some biharmonic functions and determining the kinds of

*George B. Airy (1801–1892), English astronomer.

†A general method for solving a biharmonic equation is given, for example, in [8, p. 259].

problems for which they are the solutions. This type of "guessing" is known as *inverse method*. Although its usefulness is limited, it can give us insight into some simple problems of practical importance. First we note that any linear second- or third-order homogeneous polynomial

$$\Phi(x, y) = C_1 x + C_2 y + C_3 \tag{9-118a}$$

$$\Phi(x, y) = C_1 x^2 + C_2 xy + C_3 y^2 \tag{9-118b}$$

$$\Phi(x, y) = C_1 x^3 + C_2 x^2 y + C_3 xy^2 + C_4 y^3 \tag{9-118c}$$

is a biharmonic function. Also the functions

$$\Phi(x, y) = \sin ax(C_1 \sinh ay + C_2 \cosh ay + C_3 y \sinh ay + C_4 y \cosh ay) \tag{9-119a}$$

$$\Phi(x, y) = \cos ax(C_1 \sinh ay + C_2 \cosh ay + C_3 y \sinh ay + C_4 y \cosh ay) \tag{9-119b}$$

are biharmonic, as can be verified by substitution. Because the stresses are second derivatives of $\Phi$, they are zero everywhere in the solid when Eq. (9-118*a*) is selected as a solution. Therefore the boundary of the solid must be stress-free, which combined with the assumption of the absence of the body forces, makes this a trivial problem (no load = no stresses).

## EXAMPLE 9-8

Assuming Airy's stress function in the form of Eq. (9-118*a*) or (9-118*b*), find the corresponding stresses.

### SOLUTION

Choosing first Eq. (9-118*b*) as Airy's stress function, we find

$$\begin{aligned} \sigma_{xx} &= 2C_3 \\ \sigma_{yy} &= 2C_1 \\ \sigma_{xy} &= -C_2 \end{aligned} \tag{a}$$

This stress distribution corresponds to the problem illustrated in Fig. 9-20. Here we have a system of self-equilibrating uniformly distributed forces (any values of $C_1$, $C_2$, $C_3$ are acceptable) applied to the boundary of an elastic rectangle. Now assume the solution in the form of a cubic polynomial, Eq. (9-118*c*). The stresses become

$$\begin{aligned} \sigma_{xx} &= 2C_3 x + 6C_4 y \\ \sigma_{yy} &= 2C_2 y + 6C_1 x \\ \sigma_{xy} &= -2(C_2 x + C_3 y) \end{aligned} \tag{b}$$

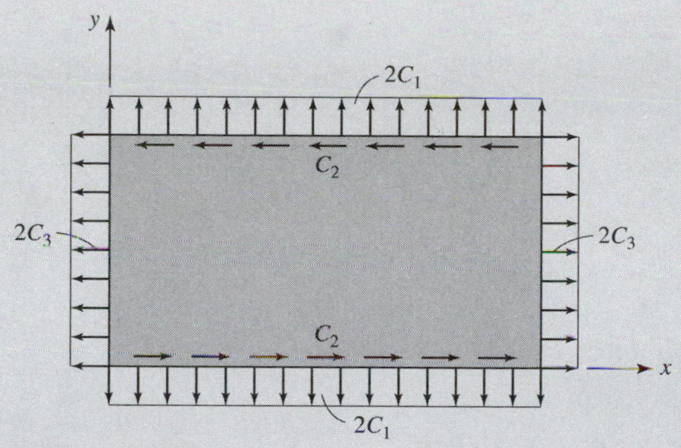


**Figure 9-20**

Considering again an elastic rectangle $a \times b$, we find that the following stresses appear at the boundaries:

$$\begin{aligned}
\sigma_{xx}(0, y) &= 6C_4 y \\
\sigma_{xy}(0, y) &= -2C_3 y \\
\sigma_{xx}(a, y) &= 2C_3 a + 6C_4 y \\
\sigma_{xy}(a, y) &= -2(C_3 a + C_3 y) \\
\sigma_{yy}(x, 0) &= 6C_1 x \\
\sigma_{xy}(x, 0) &= -2C_2 x \\
\sigma_{yy}(x, b) &= 2C_2 b + 6C_1 x \\
\sigma_{xy}(x, b) &= -2(C_2 x + C_3 b)
\end{aligned} \qquad (c)$$

The constants appearing here must be adjusted so that the stresses satisfy the boundary conditions. Specifically, the stresses applied to the boundary must ensure global equilibrium of the solid. Therefore the following conditions result:

$$-\int_0^a \sigma_{yy}(x, 0)\,dx + \int_0^a \sigma_{yy}(x, b)\,dx + \int_0^b \sigma_{xy}(0, y)\,dy - \int_0^b \sigma_{xy}(a, y)\,dy = 0$$

$$-\int_0^b \sigma_{xx}(0, y)\,dy + \int_0^b \sigma_{xx}(a, y)\,dy + \int_0^a \sigma_{xy}(x, 0)\,dx - \int_0^a \sigma_{xy}(x, b)\,dx = 0 \qquad (d)$$

Substituting Eqs. ($c$), we obtain

$$2C_2 ab - 2C_3 b^2 + 2C_2 ab = 0, \qquad 2C_3 ab - 2C_2 a^2 + 2C_3 ab = 0 \qquad (e)$$

which can only be satisfied if $C_2 = C_3 = 0$. When these conditions are satisfied, then the system of applied loads shown in Fig. 9-21$a$ corresponds to this Airy's function. In particular, when $C_1 = 0$, and when $b/a$ is small, this solid becomes equivalent to a thin

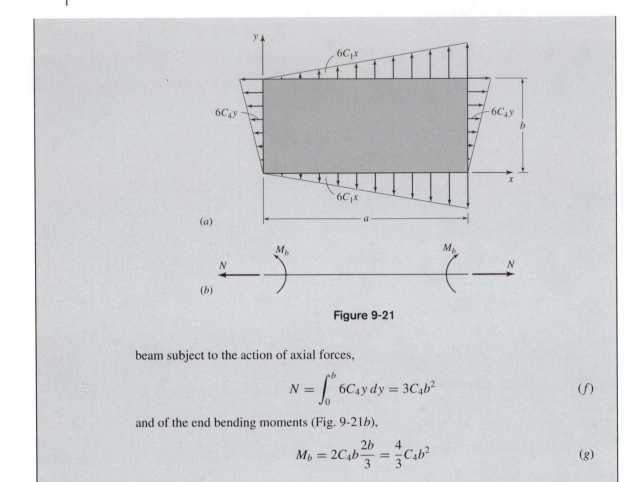

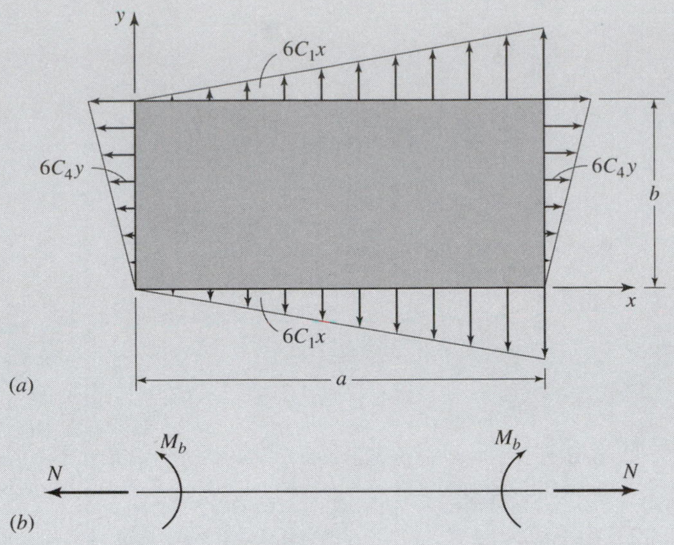


**Figure 9-21**

beam subject to the action of axial forces,

$$N = \int_0^b 6C_4 y\, dy = 3C_4 b^2 \tag{f}$$

and of the end bending moments (Fig. 9-21*b*),

$$M_b = 2C_4 b \frac{2b}{3} = \frac{4}{3} C_4 b^2 \tag{g}$$

## 9-5 TORSION

In this section we examine the torsion of noncircular prismatic elements. After the introduction of the membrane analogy, the torsion of tubes and other elements with hollow cross sections is discussed.

### 9-5-1 Circular Cross Section

We begin by recalling some concepts from elementary theory. When a cylindrical shaft of linearly elastic material, and of circular or annular cross section, is fixed at one end and subject to a torque $T$ at the other (Fig. 9-22), then the only stresses present at an

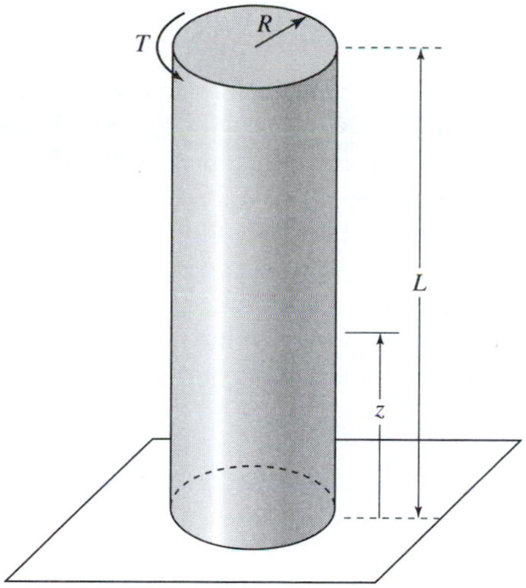


Figure 9-22 A cylindrical shaft of circular cross-section is fixed at one end and subjected to a torque $T$

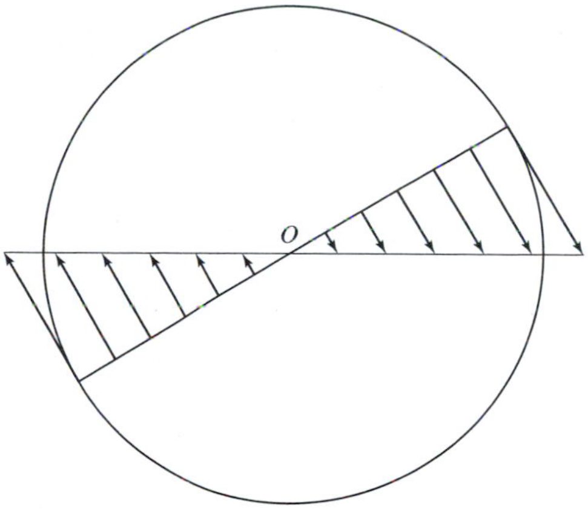


Figure 9-23 Shear stress along any diameter of the cross-section

arbitrary point $M$ are the shear stresses $\tau$, given by

$$\tau = \frac{Tr}{I_0} \tag{9-120}$$

Here $r$ is the distance of point $M$ from the axis of the shaft and $I_0$ is the polar moment of inertia of the area of cross section,

$$I_0 = \frac{\pi R^4}{2} \tag{9-121}$$

The distribution of the shear stress along any diameter of the cross section is shown in Fig. 9-23. This result is obtained assuming that plane cross sections remain plane, and that their shape is unaffected by the deformation. It is also assumed that the angle of twist $\theta$, that is, the angle of rotation of a radius, varies linearly with the distance $z$ from the fixed end,

$$\theta = \frac{Tz}{GI_0} \tag{9-122}$$

Therefore,

$$\theta_{\max} = \frac{TL}{GI_0} = \frac{TL}{C_T} \tag{9-123}$$

where $G$ is the shear modulus of elasticity and $C_T = GI_0$ is called the torsional stiffness of a circular shaft. We now generalize these results by considering the torsion of noncircular prismatic shafts. It will be shown that the more refined theory of torsion confirms the results obtained for a circular shaft.

## 9-5-2 Noncircular Prisms—Saint-Venant's Theory

Since for a circular cross section the stresses are perpendicular to a given diameter, they are tangential to concentric circles, $r =$ constant. Now consider a small element of the shaft next to the free outer surface (Fig. 9-24). According to the results of the elementary theory, this element is in equilibrium under the shear stresses acting on its four flat surfaces (Fig. 9-25). Since the curved surface is stress-free, these are the only stresses acting on the element. Saint-Venant's theory is based on the assumption that each cross section rotates about the $z$ axis as a rigid body and that all cross sections deform out of plane (warp) in the same way. A more rigorous approach of the linear theory of elasticity would be based on solving three differential equations of equilibrium [see Eqs. (9-1)]

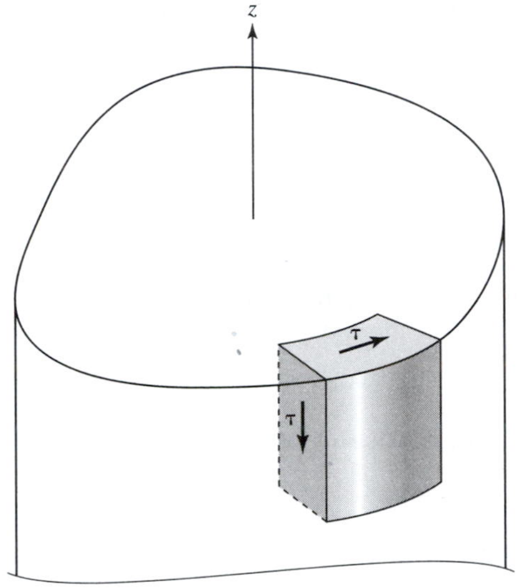


Figure 9-24

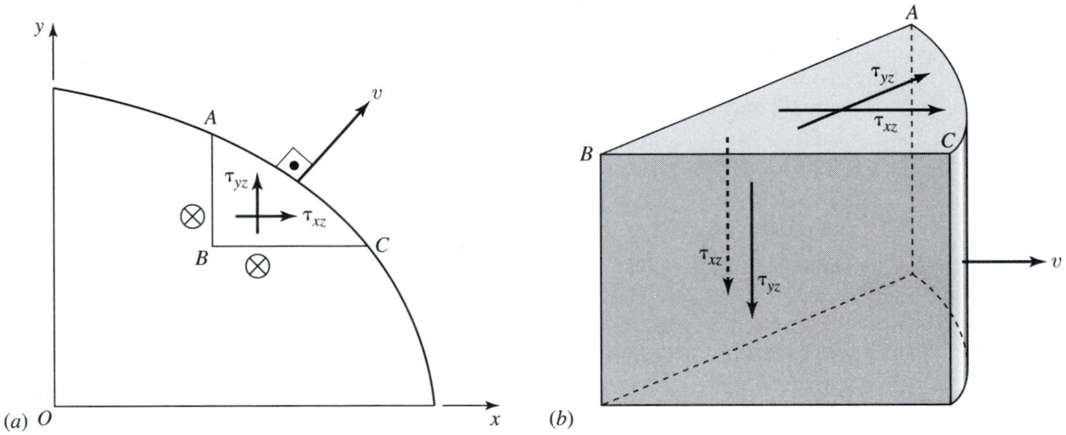


Figure 9-25 Element in equilibrium under the shear stresses acting on its four flat surfaces

without making assumptions regarding the deformation. Here it is further assumed that the angle of rotation of each cross section is proportional to the distance $z$ from a reference cross section. Thus Saint-Venant's theory bridges the gap between the strength of materials approach, familiar from the solution of torsion of circular bars, and the solution to the three-dimensional problem of the theory of elasticity, accurate in the mathematical sense but much more complicated. We first write the expression for the axial displacement $w$,

$$w = \beta\psi(x, y) \tag{9-124}$$

where the constant $\beta$ is the angle of twist per unit length. Evidently this is consistent with the assertion that each cross section warps in the same way. Although we still do not know what is $\psi(x, y)$, called the *warping function,* we now turn our attention to the displacement components $u$ and $v$ in the plane of each cross section. Let us examine the arbitrary cross section shown in Fig. 9-26, in which the origin $O$ is placed in the *center of twist*. Recall that after the deformation, the contour does not change, but only rotates about the $z$ axis, assuming the position indicated by the dashed line. The point $A$, like every other point of the cross section, moves around the origin $O$ by a small angle $\theta$. The projections of the displacement $AA'$ on $Ox$ and $Oy$ are

$$\begin{aligned} u(x, y) &= OA' \cos(\alpha - \theta) - OA\cos\alpha \\ &= OA' \cos\alpha\cos\theta + OA'\sin\alpha\sin\theta - OA\cos\alpha \\ v(x, y) &= OA'\sin(\alpha - \theta) - OA\sin\alpha \\ &= OA'\sin\alpha\cos\theta - OA'\cos\alpha\sin\theta - OA\sin\alpha \end{aligned}$$

Because $\theta$ is small, the following relationships result:

$$\sin\theta \approx \theta, \qquad \cos\theta \approx 1, \qquad OA\cos\alpha = OA'\cos\alpha = x, \qquad OA\sin\alpha = OA'\sin\alpha = y$$

Thus,

$$u = \theta y, \qquad v = -\theta x \tag{9-125}$$

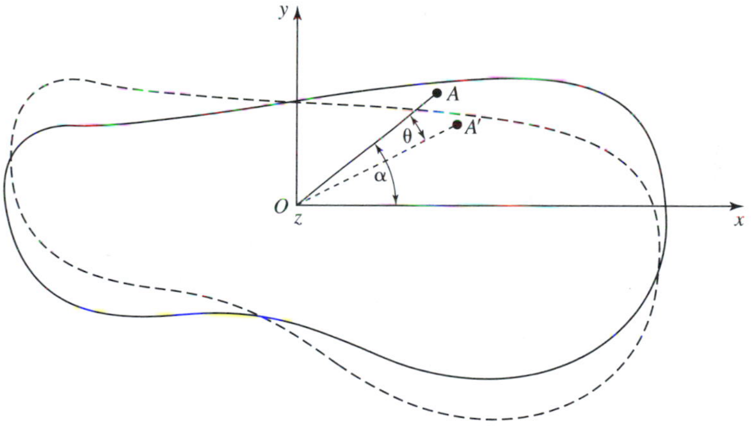


**Figure 9-26** An arbitrary cross section in which the origin $O$ is placed in the center of twist

But according to Saint-Venant's theory, $\theta$ is proportional to $z$, so that

$$\theta = \beta z \tag{9-126}$$

and finally,

$$u = \beta yz, \qquad v = -\beta xz \tag{9-127}$$

We now find the strain components, Eqs. (3-11),

$$\begin{aligned}
\varepsilon_{xx} &= \frac{\partial u}{\partial x} = 0, \qquad \varepsilon_{yy} = \frac{\partial v}{\partial y} = 0, \qquad \varepsilon_{zz} = \frac{\partial w}{\partial z} = 0 \\
\gamma_{xy} &= \frac{\partial u}{\partial y} + \frac{\partial v}{\partial x} = 0 \\
\gamma_{xz} &= \frac{\partial u}{\partial z} + \frac{\partial w}{\partial x} = \beta y + \beta \frac{\partial \psi}{\partial x} \\
\gamma_{yz} &= \frac{\partial v}{\partial z} + \frac{\partial w}{\partial y} = -\beta x + \beta \frac{\partial \psi}{\partial y}
\end{aligned} \tag{9-128}$$

From Hooke's law [Eqs. (4-115) and (4-116)] we see that

$$\begin{aligned}
\sigma_{xx} &= \sigma_{yy} = \sigma_{zz} = \sigma_{xy} = 0 \\
\sigma_{xz} &= G\gamma_{xz} = G\beta \left( y + \frac{\partial \psi}{\partial x} \right) \\
\sigma_{yz} &= G\beta \left( -x + \frac{\partial \psi}{\partial y} \right)
\end{aligned} \tag{9-129}$$

When the stresses, Eqs. (9-129), are substituted into the equilibrium equations (2-53),

$$\begin{aligned}
\frac{\partial \sigma_{xx}}{\partial x} + \frac{\partial \sigma_{xy}}{\partial y} + \frac{\partial \sigma_{xz}}{\partial z} &= 0 \\
\frac{\partial \sigma_{xy}}{\partial x} + \frac{\partial \sigma_{yy}}{\partial y} + \frac{\partial \sigma_{yz}}{\partial z} &= 0 \\
\frac{\partial \sigma_{xz}}{\partial x} + \frac{\partial \sigma_{yz}}{\partial y} + \frac{\partial \sigma_{zz}}{\partial z} &= 0
\end{aligned} \tag{9-130}$$

the first two equations are satisfied identically and the last one becomes

$$\frac{\partial^2 \psi}{\partial x^2} + \frac{\partial^2 \psi}{\partial y^2} = 0 \tag{9-131}$$

Therefore, in order to determine $\psi$ we have to solve the differential equation (9-131), known as *Laplace equation,** with the appropriate boundary conditions. Assuming that

*Pierre S. Laplace (1749–1827), French astronomer, mathematician, and physicist.

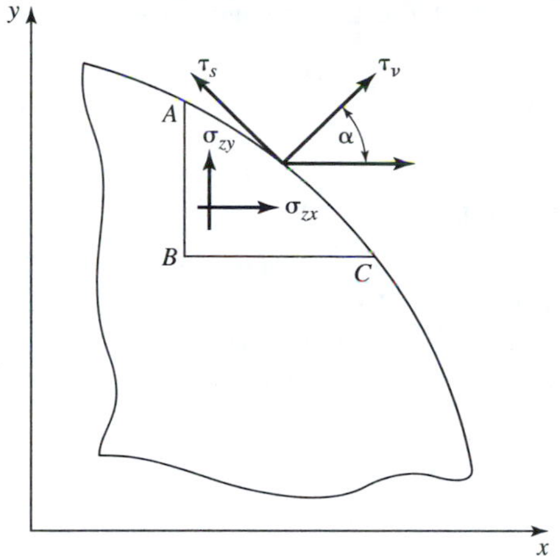


Figure 9-27

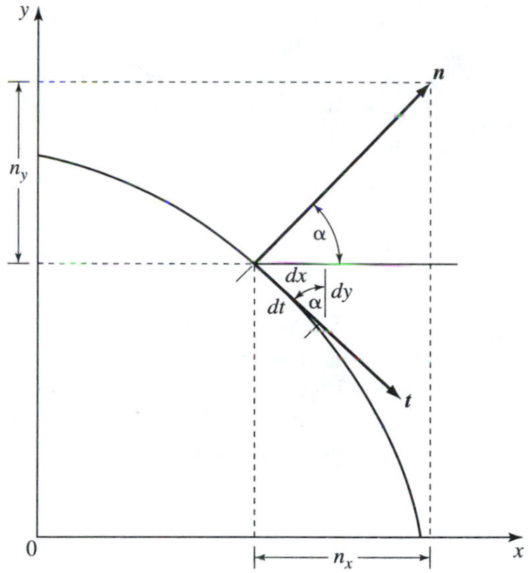


Figure 9-28 The projections of the normal vector $\boldsymbol{n}$

the lateral surfaces are stress-free, we consider the equilibrium of a small element of a twisted rod located next to a lateral surface (Fig. 9-27). We find that two out of three stress boundary conditions [Eqs. (2-14)] are identically satisfied while the third one becomes

$$\sigma_{xz} n_x + \sigma_{yz} n_y = 0 \tag{9-132}$$

Here $n_x$ and $n_y$ are projections of the normal vector $\boldsymbol{n}$ (Fig. 9-28) on $Ox$ and $Oy$, respectively. Hence

$$n_x = \cos(\boldsymbol{n}, x), \qquad n_y = \cos(\boldsymbol{n}, y) \tag{9-133}$$

Substituting expressions (9-129) and (9-133) into Eq . (9-132),

$$G\beta \left( y + \frac{\partial \psi}{\partial x} \right) \cos(\boldsymbol{n}, x) + G\beta \left( -x + \frac{\partial \psi}{\partial y} \right) \cos(\boldsymbol{n}, y) = 0 \tag{9-134}$$

But

$$\frac{\partial \psi}{\partial x} \cos(\boldsymbol{n}, x) + \frac{\partial \psi}{\partial y} \cos(\boldsymbol{n}, y) = \frac{\partial \psi}{\partial n} \tag{9-135}$$

Therefore,

$$\frac{\partial \psi}{\partial n} = -y \cos(\boldsymbol{n}, x) + x \cos(\boldsymbol{n}, y) \tag{9-136}$$

is the required boundary condition to be satisfied by the warping function $\psi$ on the contour $C$ of the cross section. When the function $\psi$ is known, then the stresses, except for the still unknown angle of twist per unit length $\beta$, can be obtained from expressions (9-129). Finding $\beta$ is not very difficult and we are going to do it later. The following approach leads to a differential equation with simpler boundary conditions (see, for example, [7]). We assume, following Prandtl [6, pp. 758–770], that a function $\varphi$ exists such that

$$\sigma_{xz} = -\frac{\partial\varphi}{\partial y}, \qquad \sigma_{yz} = \frac{\partial\varphi}{\partial x} \tag{9-137}$$

while, as before,

$$\sigma_{xx} = \sigma_{yy} = \sigma_{zz} = \sigma_{xy} = 0 \tag{9-138}$$

The function $\varphi$ is usually termed the *stress function*. First let us verify whether the equilibrium equations are satisfied. Indeed if Eqs. (9-137) and (9-138) are substituted into Eqs. (9-130), we see that they are identically satisfied, independently of the expression for $\varphi(x, y)$. Next we replace $\sigma_{xz}$ and $\sigma_{yz}$ in expressions (9-129) by their representations (9-137), obtaining

$$-\frac{\partial\varphi}{\partial y} = G\beta\left(y + \frac{\partial\psi}{\partial x}\right) \tag{9-139a}$$

$$\frac{\partial\varphi}{\partial x} = G\beta\left(-x + \frac{\partial\psi}{\partial y}\right) \tag{9-139b}$$

We recall from Chapter 2 and from Section 9-3 that when strains or stresses are given, they must satisfy the compatibility equations. On the other hand, when displacements are given and strains and stresses are obtained by proper differentiation, compatibility equations are identically satisfied. Here we have the former case. Compatibility is ensured in the following way. We want the warping function and its derivatives to be continuous. Therefore,

$$\frac{\partial^2\psi}{\partial x\,\partial y} = \frac{\partial^2\psi}{\partial y\,\partial x}$$

is a satisfactory test. Then, differentiating Eq. (9-139*a*) with respect to $x$ and Eq. (9-139*b*) with respect to $y$, and subtracting the first differentiated equation from the second, we obtain

$$\frac{\partial^2\varphi}{\partial x^2} + \frac{\partial^2\varphi}{\partial y^2} = -2G\beta$$

or

$$\nabla^2\varphi = -2G\beta \tag{9-140}$$

which is the desired differential equation. Substituting Eqs. (9-137) into the boundary condition (9-132), we get

$$-\frac{\partial\varphi}{\partial y}n_x + \frac{\partial\varphi}{\partial x}n_y = 0 \tag{9-141}$$

But we see in Fig. 9-28 that

$$\frac{d\varphi}{dt} = \frac{\partial\varphi}{\partial x}\frac{\partial x}{\partial t} + \frac{\partial\varphi}{\partial y}\frac{\partial y}{\partial t} = \frac{\partial\varphi}{\partial x}\sin\alpha - \frac{\partial\varphi}{\partial y}\cos\alpha = \frac{\partial\varphi}{\partial x}n_y - \frac{\partial\varphi}{\partial y}n_x$$

and therefore, according to Eq . (9-141),

$$\frac{d\varphi}{dt} = 0 \tag{9-142}$$

is the appropriate boundary condition to be satisfied on the contour of the twisted bar. Equation (9-142) implies that $\varphi =$ constant on the contour, and since the magnitude of this constant does not affect the stress that includes only derivatives of $\varphi$, we may as well take

$$\varphi = 0 \tag{9-143}$$

to be the boundary condition. An important relation between the shear stress and the derivative of the stress function $\varphi$ exists at the contour of the cross section. Let us examine the element $ABC$ close to the boundary of the cross section, as seen in Fig. 9-27. We decompose the shear stress acting on this element into two components, $\sigma_{zx}$ and $\sigma_{zy}$. However, it can also be decomposed into components $\tau_\nu$ and $\tau_s$, normal and tangential to the segment $AC$ of the contour. To ensure equilibrium, the component $\tau_\nu$ must have an equal counterpart acting on the outside surface of the twisted rod in the direction of the $z$ axis. Since we assumed that this surface is stress-free, it follows that $\tau_\nu = 0$, indicating that the shear stress at the contour is tangent to the contour. We now relate this stress $\tau_s$ to the components $\sigma_{zx}$ and $\sigma_{zy}$. Denoting by $\alpha$ the angle between the normal $\boldsymbol{n}$ to the contour and the $x$ axis, we have

$$\tau_s = \sigma_{zx} \sin\alpha + \sigma_{zy} \cos\alpha$$

Substituting here Eqs. (9-137) gives

$$\tau_s = \frac{\partial\varphi}{\partial y} \sin\alpha + \frac{\partial\varphi}{\partial x} \cos\alpha \tag{9-144}$$

Now we calculate $\partial\varphi/\partial n$ in the same manner as we calculated $\partial\varphi/\partial t$, that is

$$\frac{\partial\varphi}{\partial n} = \frac{\partial\varphi}{\partial x}\frac{\partial x}{\partial n} + \frac{\partial\varphi}{\partial y}\frac{\partial y}{\partial n} = \frac{\partial\varphi}{\partial x}\cos\alpha + \frac{\partial\varphi}{\partial y}\sin\alpha \tag{9-145}$$

Comparing expressions (9-144) and (9-145) we find that

$$\tau_s = \frac{\partial\varphi}{\partial n} \tag{9-146}$$

Thus, at the contour the shear stress equals the slope of the stress function $\varphi$. It can be shown [8, p. 117] that $\tau_s$ attains its maximum at the contour of the twisted bar. Next we want to relate the still unknown angle $\beta$ to the applied torque $T_0$. Considering the cross section of the twisted bar (Fig. 9-29) we determine the moment of all the forces with respect to 0. Those forces are due to the presence of the shear stresses $\sigma_{zy}$ and $\sigma_{zx}$ whose moment is

$$(\sigma_{zx}\, dA)y - (\sigma_{zx}\, dA)x$$

The resultant moment is obtained by integrating this expression over the entire area $A$ so that

$$M = \int_A\int (\sigma_{zx} y - \sigma_{zy} x)\, dA = -\int_A\int \left(\frac{\partial\varphi}{\partial y} y + \frac{\partial\varphi}{\partial x} x\right) dA \tag{9-147}$$

where the representations for shear stresses [Eqs. (9-137)] were used. In order to

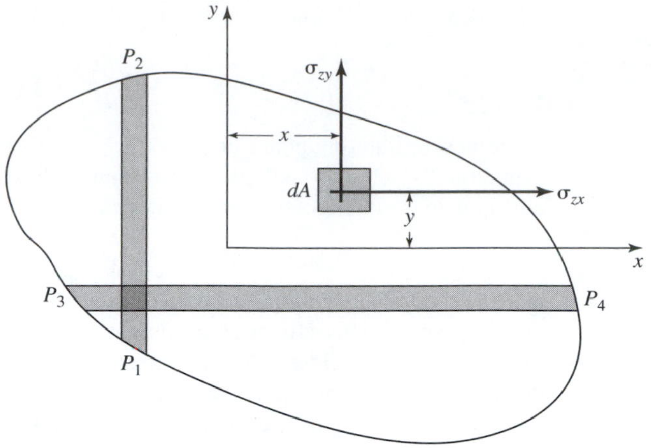


**Figure 9-29** Forces acting on the cross section of the twisted bar

eliminate the derivatives from the integrand in Eq. (9-147), each term will be integrated separately by parts. Consequently,

$$I_1 = \int_A\int \frac{\partial\varphi}{\partial y} y\, dA = \int_A\int \frac{\partial\varphi}{\partial y} y\, dx\, dy = \int dx \int_{P_1}^{P_2} \frac{\partial\varphi}{\partial y} y\, dy$$

$$= \int dx \left[ \varphi y \Big|_{P_1}^{P_2} - \int_{P_1}^{P_2} \varphi\, dy \right]$$

But $\varphi = 0$ at the contour points $P_1$ and $P_2$. Therefore,

$$I_1 = -\int dx \int_{P_1}^{P_2} \varphi\, dy = -\int_A\int \varphi\, dA \tag{9-148}$$

and similarly,

$$I_2 = \int_A\int \frac{\partial\varphi}{\partial x} x\, dA = \int_A\int \frac{\partial\varphi}{\partial x} x\, dx\, dy = \int dy \int_{P_3}^{P_4} \frac{\partial\varphi}{\partial x} x\, dx$$

$$= \int dy \left[ \varphi x \Big|_{P_3}^{P_4} - \int_{P_3}^{P_4} \varphi\, dx \right] = -\int_A\int \varphi\, dA \tag{9-149}$$

Finally then

$$M = -I_1 - I_2 = 2\int_A\int \varphi\, dA \tag{9-150}$$

and since $M$ must be in equilibrium with the external torque $T_0$,

$$T_0 = 2\int_A\int \varphi\, dA \tag{9-151}$$

The function $\varphi$ is presumably known, except for the constant $\beta$. Therefore Eq. (9-151) relates $\beta$ to $T_0$.

## 9-5-3 Membrane Analogy

It is always advantageous to be able to estimate the state of stress by inspection rather than only after solving a differential equation. In the case of torsion, we have the opportunity of achieving this by using the so-called *membrane analogy*. It allows us to consider a thin-walled tube with the same cross section as the twisted bar closed with a very thin membrane and subject to inside pressure. The shape of the membrane attached to this fictitious tube determines the stresses in the twisted bar. The more complicated the cross section (such as one with holes or notches), the more difficult it is to solve the differential equation, and the more useful becomes such a simple inspection.

Let us consider the torsion of a bar subject to a torque $T_0$, with the cross section shown in Fig. 9-30*a*. The contour of the cross section is now used to model the thin-walled, but nondeformable tube of Fig. 9-30*b*. It is assumed to be rigidly closed at the bottom while a thin flexible membrane is attached to its top by stretching it uniformly with tension $T$. Now we inflate the tube with air pressure. Since the tube is rigid, this will cause only the membrane to deform. The side view of the inflated membrane is shown in Fig. 9-30*c*. The deflections are assumed to be small, so that Fig. 9-30*c* is greatly exaggerated. We claim that the shape of the surface $w(x, y)$ is proportional to the shape of the surface of the stress function $\varphi(x, y)$ corresponding to the torsion problem of Fig. 9-30*a*. To prove this we consider the equilibrium of an infinitesimal element by projecting all forces in the direction of $w$. We get

$$T\,dy \sin\alpha + T\,dy \sin\beta - T\,dx \sin\delta + T\,dx \sin\gamma + p\,dx\,dy \qquad (9\text{-}152)$$

But

$$\sin\alpha \approx \tan\alpha = \frac{\partial w}{\partial x}, \qquad \sin\beta \approx \tan\beta = \frac{\partial w}{\partial x} + \frac{\partial}{\partial x}\left(\frac{\partial w}{\partial x}\right) dx$$

$$\sin\delta \approx \tan\delta = \frac{\partial w}{\partial y}, \qquad \sin\gamma \approx \tan\gamma = \frac{\partial w}{\partial y} + \frac{\partial}{\partial y}\left(\frac{\partial w}{\partial y}\right) dy$$

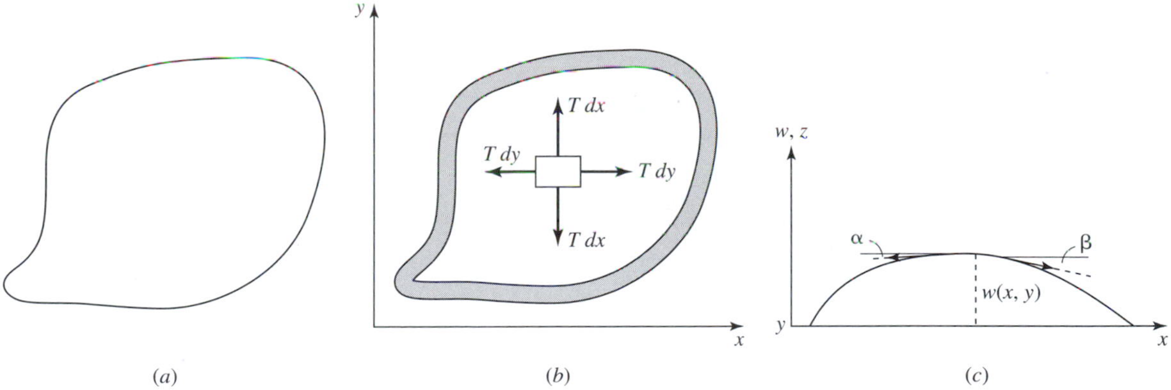


Figure 9-30

Plugging this into Eq. (9-152) and dividing the results by $T\,dx\,dy$ yields

$$\frac{\partial^2 w}{\partial x^2} + \frac{\partial^2 w}{\partial y^2} = -\frac{p}{T} \tag{9-153}$$

Noticing that $w(x, y) = 0$ at the contour and comparing Eq. (9-153) with Eq. (9-140),

$$\frac{\partial^2 \varphi}{\partial x^2} + \frac{\partial^2 \varphi}{\partial y^2} = -2G\beta \tag{9-154}$$

We see that if we make the constant $p/T$ equal to $2G\beta$, then $w(x, y) \equiv \varphi(x, y)$, as expected. If we can easily calculate $w(x, y)$, then we can obtain the torque $T_0$ using the analogy

$$T_0 = 2\int_A\!\!\int \varphi\, dA \equiv 2\int_A\!\!\int w\, dA \tag{9-155}$$

Hence the torque equals twice the volume between the inflated membrane and the $xy$ plane.

## 9-5-4 Rectangular and Related Cross Sections

The application of the membrane analogy will be illustrated using several examples. Consider first a thin rectangular section (Fig. 9-31*a*). Visualizing a membrane inflated over an identically shaped tube, we can predict that the shape of $w(x, y)$ will be similar to that shown in Fig. 9-31*b*. If the rectangle were extended to infinity, then $w(x, y)$ would certainly not depend on $x$. In the present case, the dependence of $w(x, y)$ on $x$ is basically restricted to regions close to the short ends.

Assuming therefore that

$$w = w(y)$$

we reduce Eq. (9-153) to

$$\frac{d^2 w}{dy^2} = -\frac{p}{T} \tag{9-156}$$

Two integrations give

$$\frac{dw}{dy} = -\frac{p}{T}y + C_1, \qquad w = -\frac{py^2}{2T} + C_1 y + C_2 \tag{9-157}$$

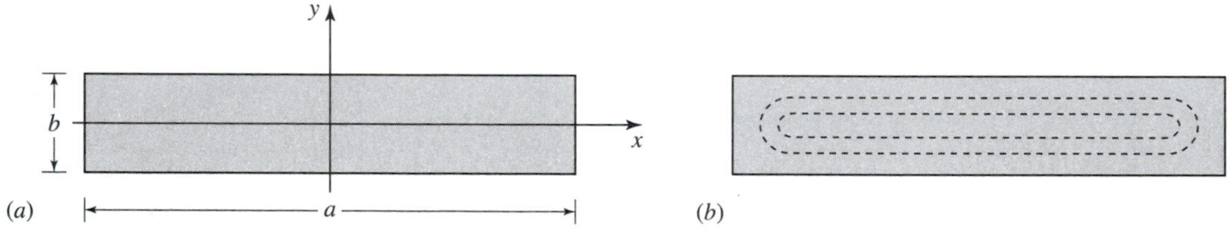


**Figure 9-31**

The constants $C_1$ and $C_2$ are determined from the boundary conditions

$$w\left(\frac{b}{2}\right) = 0, \qquad w\left(-\frac{b}{2}\right) = 0 \tag{9-158}$$

and they are found to be

$$C_1 = 0, \qquad C_2 = \frac{pb^2}{8T} \tag{9-159}$$

so that ultimately

$$w(y) = -\frac{p}{8T}(4y^2 - b^2) \tag{9-160}$$

and

$$\frac{dw}{dy} = -\frac{p}{T}y \tag{9-161}$$

Now we replace $p/T$ by $2G\beta$ and $w(y)$ by $\varphi(y)$, getting

$$\varphi(y) = -\frac{1}{4}G\beta(4y^2 - b^2), \qquad \frac{d\varphi}{dy} = -2G\beta y \tag{9-162}$$

As we know, maximum shear stress equals max $d\varphi/dx$, and since this obviously occurs at $x = \pm b/2$,

$$\max \tau = G\beta b \tag{9-163}$$

We can relate $\beta$ to the applied torque $T_0$ by using Eq. (9-155),

$$\begin{aligned} T_0 &= 2\int_A\int \varphi\, dA \approx 2\int_{-a/2}^{a/2} dx \int_{-b/2}^{b/2} \varphi(y)\, dy = 2a\int_{-b/2}^{b/2} \varphi(y)\, dy \\ &= -\frac{1}{4}G\beta \cdot 2a \int_{-b/2}^{b/2} (4y^2 - b^2)\, dy = -\frac{G\beta a}{2}\left(\frac{4}{3}y^3 - b^2 y\right)_{-b/2}^{b/2} \\ &= -\frac{G\beta a}{2}\left(2\frac{4}{3}\frac{b^3}{8} - 2b^2\frac{b}{2}\right) = -\frac{G\beta a}{2}\left(-\frac{2}{3}b^3\right) = \frac{G\beta ab^3}{3} \end{aligned}$$

From here,

$$\beta = \frac{3T_0}{Gab^3} = \frac{T_0}{GJ_T} = \frac{T_0}{C_T} \tag{9-164}$$

where $C_T = GJ_T = Gab^3/3$ is the *torsional stiffness* of a narrow rectangle. Thus with Eq. (9-153),

$$\max \tau = \frac{3T_0}{ab^2} = \frac{T_0 b}{J_T} \tag{9-165}$$

The situation will be similar for any of the shapes shown in Fig. 9-32, except that for the cases shown in Fig. 9-32*a* and *c* the stress patterns are distorted near the reentrant corners.

Considering, as an example, the case shown in Fig. 9-32*a* and disregarding the disturbance caused by the corner, we use the membrane analogy to find that $\varphi(y)$ has still the previous form, in which $b$ must be replaced by $b_1$ for the part $AB$ and by $b_2$ for the part $BC$. Since the total volume under the inflated membrane is the sum of the volumes over $AB$ and $BC$, calculating the relation between $T_0$ and $\beta$ (all those quantities are

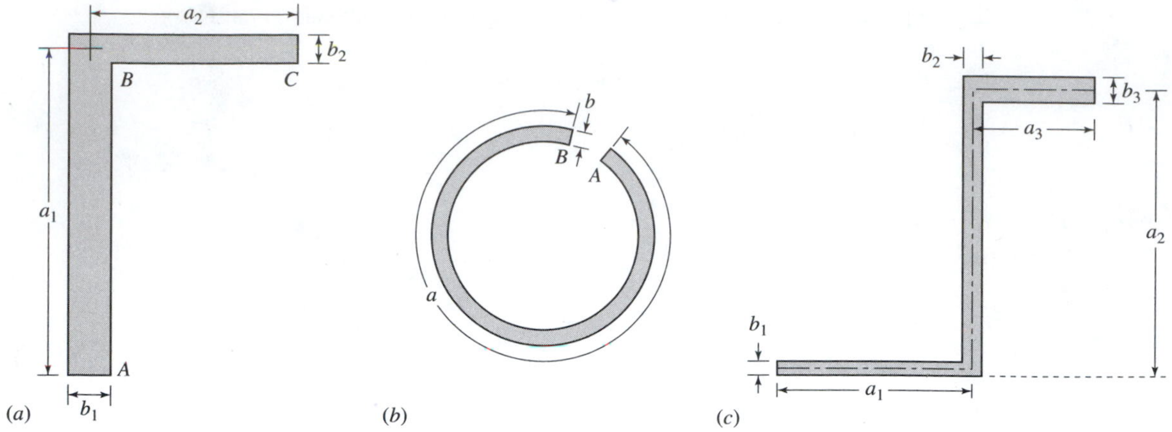


Figure 9-32

constant for the entire cross section) we obtain

$$T_0 = \frac{G\beta}{3}\left(a_1 b_1^3 + a_2 b_2^3\right) = \frac{G\beta}{3}\sum_{i=1}^{2} a_i b_i^3 \tag{9-166}$$

This result can be easily generalized to include cross sections composed of $n$ thin rectangular parts,

$$T_0 = G\beta J_T \tag{9-167}$$

where

$$J_T = \frac{1}{3}\sum_{i=1}^{n} a_i b_i^3 \tag{9-168}$$

Hence,

$$\beta = \frac{T_0}{GJ_T}$$

and using Eq. (9-163),

$$\max \tau = G\beta b_{i\,\max} \tag{9-169}$$

Expressions (9-166)–(9-169) can also be used for an approximate analysis of the torsion of open narrow cross sections with variable thickness. To this end the cross-sectional area must be split into several small elements, each of length $\Delta a_i$ and of variable average thickness $b_i$. After these preliminary steps, further analysis can be accomplished using previously derived expressions.

## 9-5-5 Torsion of Hollow Single-Cell and Multiple-Cell Members

We consider in this section the torsion of thin-walled, hollow single-cell sections as shown in Fig. 9-33$a$ and multiple-cell sections as shown in Fig. 9-33$b$. It is assumed that (1) the thickness of the wall is small compared to the dimensions of the cross section as

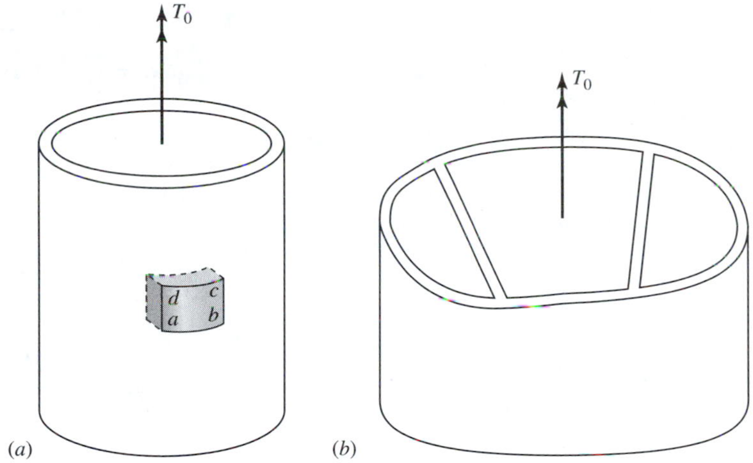


**Figure 9-33** Examples of hollow single-cell and multiple-cell sections

a whole, and (2) the twisted member is stiffened periodically by appropriate elements (rings or membranes) to ensure the integrity of the cross section. The collapse of the cross section, which might otherwise occur, would violate Saint-Venant's theory. On the other hand, we allow for the wall thickness to be variable. For the application of the membrane analogy to the solution of this problem the reader should consult, for example, DenHartog [3, p. 24]. Here another, quite general method is applied. We begin by analyzing the torsion of the tube shown in Fig. 9-33*a*. This tube has a single-cell cross section of wall thickness $t$ (see Fig. 9-34*a*). First we relate the torque $T_0$ transmitted by the cross section to the shear stress $\tau$ appearing at the wall. To this end we select an arbitrary point $O$ within the cross section. Next we note that the moment about $O$ of the stress resultant acting on an elementary area $t\ ds$ equals

$$dm_0 = \underbrace{(\tau t\, ds)}_{\text{resultant}}\ \underbrace{n}_{\text{distance}}$$

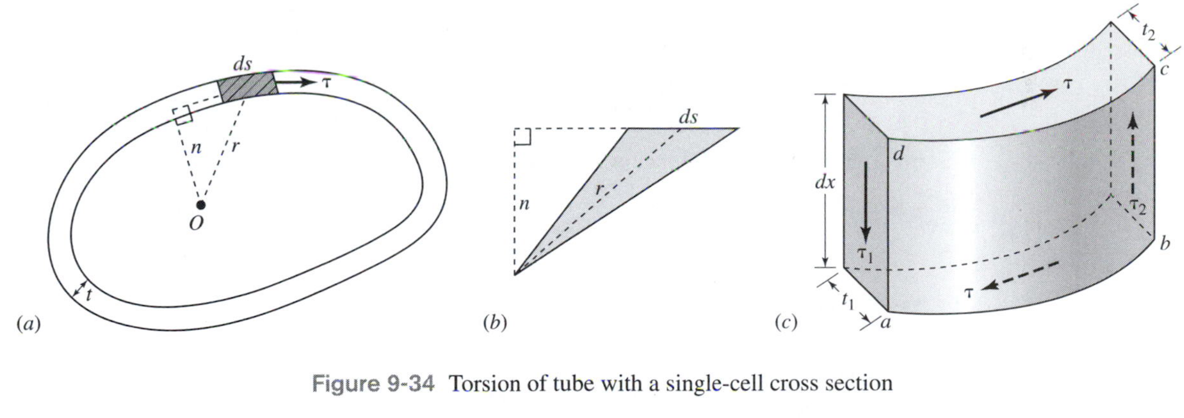


**Figure 9-34** Torsion of tube with a single-cell cross section

Hence,

$$T_0 = \int dm_0 = \int n\tau t \, ds \tag{9-170}$$

where the integration is along the periphery of the cross section. It is evident that $n\,ds$ is twice the area of the small triangle shown in Fig. 9-34*b*. To put this result in a simpler form, we consider the equilibrium of the tube element *abcd* shown in Fig. 9-33*a* and, enlarged, in Fig. 9-34*c*. Projecting all forces in the axial direction $x$, we obtain

$$-\tau_1 t_1 \, dx + \tau_2 t_2 \, dx = 0$$

or

$$\tau_1 t_1 = \tau_2 t_2 \equiv \tau t \equiv q \tag{9-171}$$

indicating that the quantity $q$, called the *shear flow,* is constant around the tube. With Eq. (9-171), Eq. (9-170) becomes

$$T_0 = \tau t \int n \, ds \tag{9-172}$$

Finally, we observe that the integral in Eq. (9-172) is the sum of the areas of the triangles from Fig. 9-34*b* obtained when the sum of all the elements $ds$ equals the length of the median line of the cross section. Thus,

$$\int n \, ds = 2A \tag{9-173}$$

where $A$ is the area enclosed within the median line of the cross section. With Eq. (9-173), Eq. (9-172) becomes

$$T_0 = 2A\tau t = 2Aq \tag{9-174}$$

and thus,

$$\tau = \frac{T_0}{2At} \tag{9-175}$$

Let us now use Castigliano's theorem (Chapter 5) to express the angle of twist per unit length $\beta$ in terms of the applied torque $T_0$. According to a special case of Eq. (5-72*a*), the strain energy per unit volume of a twisted rod is

$$u = \frac{\tau^2}{2G}$$

Integrating over the volume of the twisted rod (see Fig. 9-35), we obtain

$$U = \oint \frac{\tau^2}{2G} Lt \, ds = \oint \frac{T_0^2}{4A^2 t^2} \frac{1}{2G} Lt \, ds = \frac{T_0^2 L}{8A^2 G} \oint \frac{ds}{t} \tag{9-176}$$

To arrive at Eq. (9-176), Eq. (9-175) has been used. By Castigliano's principle,

$$\beta = \frac{\theta}{L} = \frac{1}{L} \frac{\partial U}{\partial T_0} = \frac{T_0}{4A^2 G} \oint \frac{ds}{t} \tag{9-177a}$$

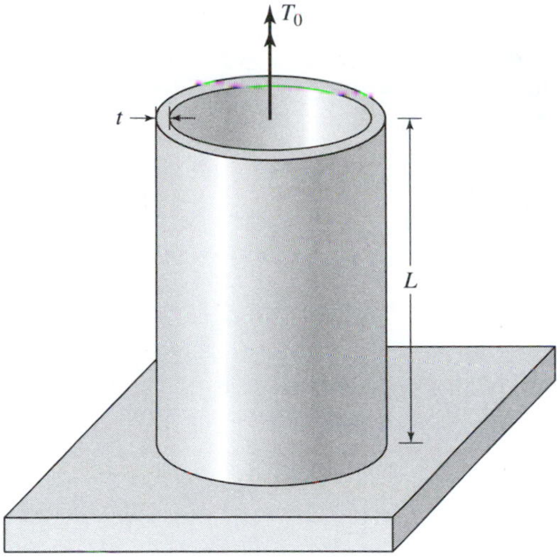


Figure 9-35

or, replacing $T_0$ by Eq. (9-174), we find

$$\beta = \frac{q}{2GA} \oint \frac{ds}{t} \tag{9-177b}$$

Both Eqs. (9-177*a*) and (9-177*b*) are known as Bredt's formulas [1].

## EXAMPLE 9-9

A torque $T_0 = 2000$ lb · in is applied to the tube shown in Fig. 9-36. Let $G = 11.6 \times 10^6$ psi. Find the maximum shear stress $\tau_s$ and the angle of twist per unit length $\beta$.

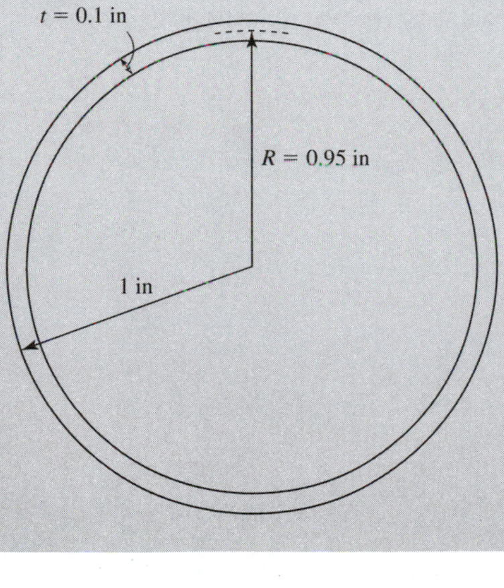


Figure 9-36

## SOLUTION

First evaluate the area $A$. From Fig. 9-36 we find that $A = \pi R^2 = \pi \times 0.95^2 = 2.8353\ \text{in}^2$. Now from Eq. (9-175) we get

$$\tau = \frac{T_0}{2At} = \frac{2000}{2 \times 2.8353 \times 0.1} = 3527\ \text{psi}$$

In order to obtain $\beta$ we begin by evaluating the integral appearing in Eqs. (9-177),

$$\oint \frac{ds}{t} = \frac{2\pi R}{t} = \frac{2\pi \times 0.95}{0.1} = 59.69$$

Thus from Eqs. (9-177) it follows that

$$\beta = \frac{\tau t}{2GA} \oint \frac{ds}{t} = \frac{3527 \times 0.1 \times 59.69}{2 \times 11.6 \times 10^6 \times 2.8353} = 320.1 \times 10^{-6}\ \text{rad/in}$$

Now consider the torsion of multiple-cell sections (see Fig. 9-33*b*). We begin by analyzing, as an example, the torsion of a tube with the cross section shown in Fig. 9-37. This cross section has two cells. It is assumed, as before, that the twisted tube has stiffeners, which will prevent a change in the cross-sectional geometry. Thus the cross section deforms as a rigid body. To obtain the expressions for the shear stress $\tau$ and for the angle of twist per unit length $\beta$, we proceed in the same manner as for a single-cell tube. Consequently Eq. (9-170) becomes

$$T_0 = \int_{BCA} n\tau t\, ds + \int_{ADB} n\tau t\, ds + \int_{AB} n\tau t\, ds \tag{9-178}$$

It is now necessary to examine Eq. (9-171). The presence of the web affects the continuity of the shear stress $\tau$ along the boundaries of the cross section. However, the shear

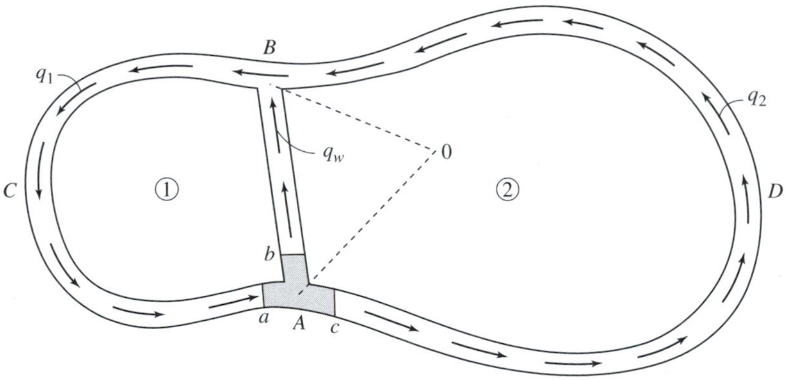


Figure 9-37

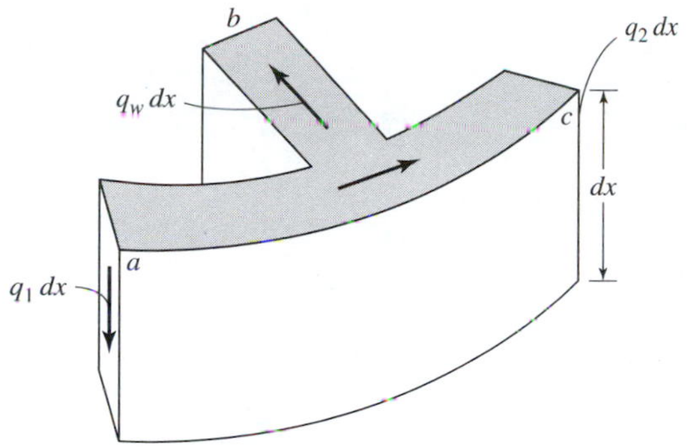


Figure 9-38 A tube element

flow is still constant for each portion of the wall. Thus (see Fig. 9-37, where we assumed arbitrarily that the shear flow $q_w$ runs from $A$ to $B$),

$$q_1 = \text{constant}, \qquad q_2 = \text{constant}, \qquad q_w = \text{constant} \tag{9-179}$$

Finally, using Eqs. (9-179), Eq. (9-178) becomes

$$T_0 = q_1 \int_{BCA} n\,ds + q_2 \int_{ADB} n\,ds - q_w \int_{AB} n\,ds \tag{9-180}$$

In order to relate $q_w$ to $q_1$ and $q_2$, we follow the derivation given in the case of a single-cell section. Thus we cut a tube element limited by the points $a$, $b$, and $c$, and by two horizontal cross sections at a distance $dx$ from each other (Fig. 9-38). The equilibrium of this element requires that

$$-q_1\,dx + q_2\,dx + q_w\,dx = 0$$

or

$$q_w = q_1 - q_2 \tag{9-181}$$

With this, Eq. (9-180) becomes

$$T_0 = q_1 \int_{BCA} n\,ds - (q_1 - q_2) \int_{AB} n\,ds + q_2 \int_{ADB} n\,ds$$

or

$$T_0 = q_1 \left( \int_{BCA} n\,ds - \int_{AB} n\,ds \right) + q_2 \left( \int_{ADB} n\,ds + \int_{AB} n\,ds \right) \tag{9-182}$$

By inspection of Fig. 9-39 we see that the first integral in Eq. (9-182) equals $2A_1 + 2A$, the second $2A$, the third $2A_2 - 2A$, and the fourth $2A$. Consequently,

$$T_0 = 2q_1 A_1 + 2q_2 A_2 \tag{9-183}$$

We note that in Eq. (9-183) there are two unknowns, $q_1$ and $q_2$. We thus need an additional equation. Since the entire cross section of the tube rotates as a rigid body, the angle $\beta$ is the same for every cell. Keeping this in mind, let us determine $\beta$ by applying

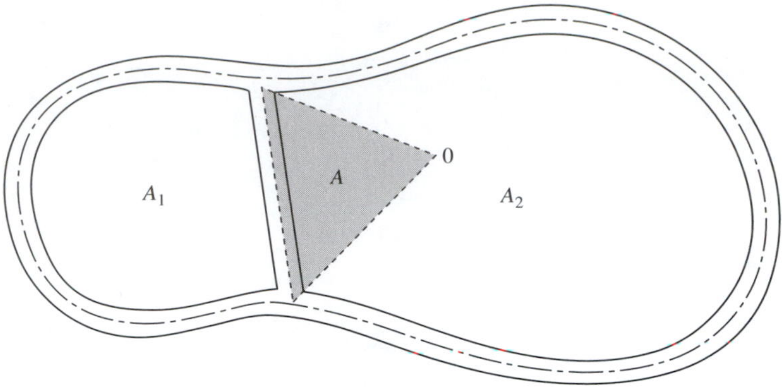


Figure 9-39

Castigliano's theorem. Now the strain energy can be put in the form

$$U = \sum_{n=1}^{2} U_i \tag{9-184}$$

where

$$U_1 = \oint_{(1)} \frac{\left[\tau_{(1)}\right]^2}{2G_{(1)}} L t_{(1)}\, ds = \frac{1}{A_1^2} \oint_{(1)} \frac{\left[T_{0,(1)}\right]^2}{8G_{(1)} t_{(1)}} L\, ds = \frac{L}{8A_1^2} \oint_{(1)} \frac{\left[T_{0,(1)}\right]^2}{G_{(1)} t_{(1)}}\, ds \tag{9-185}$$

and

$$U_2 = \oint_{(2)} \frac{\left[\tau_{(2)}\right]^2}{2G_{(2)}} L t_{(2)}\, ds = \frac{1}{A_2^2} \oint_{(2)} \frac{\left[T_{0,(2)}\right]^2}{8G_{(2)} t_{(2)}} L\, ds = \frac{L}{8A_2^2} \oint_{(2)} \frac{\left[T_{0,(2)}\right]^2}{G_{(2)} t_{(2)}}\, ds \tag{9-186}$$

are the strain energies accumulated in cells 1 and 2, respectively. The quantities $T_{0,(1)}$ and $T_{0,(2)}$ are the torques carried by these cells [see Eq. (9-183)],

$$T_{0,(1)} = 2q_1 A_1, \qquad T_{0,(2)} = 2q_2 A_2 \tag{9-187}$$

The indices (1) and (2) appearing below the integrals and as subscripts of $G$ and $t$ indicate that the respective quantities refer to cells 1 and 2. Finally, the line integrals appearing in Eqs. (9-185) and (9-186) are each taken around the perimeter of the corresponding cell. Now applying Castigliano's principle to each cell separately, we get the following system of equations:

$$\begin{aligned} \beta &= \frac{\theta}{L} = \frac{\partial U_1}{\partial T_{0,(1)}} = \frac{1}{4A_1^2} \oint_{(1)} \frac{T_{0,(1)}\, ds}{G_{(1)} t_{(1)}} = \frac{1}{2A_1} \oint_{(1)} \frac{q_{(1)}\, ds}{G_{(1)} t_{(1)}} \\ \beta &= \frac{\theta}{L} = \frac{\partial U_2}{\partial T_{0,(2)}} = \frac{1}{4A_2^2} \oint_{(2)} \frac{T_{0,(2)}\, ds}{G_{(2)} t_{(2)}} = \frac{1}{2A_2} \oint_{(2)} \frac{q_{(2)}\, ds}{G_{(2)} t_{(2)}} \end{aligned} \tag{9-188}$$

These results can be readily generalized to apply to an $n$-cell section, where $n$ is an arbitrary number. Thus we have

$$T_0 = 2 \sum_{i=1}^{n} q_i A_i \tag{9-189}$$

where the index $i$ applies to cell $i$. Similarly,

$$T_{0,(i)} = 2q_i A_i = 2\tau_{(i)} t_{(i)} A_i \tag{9-190}$$

Now the strain energy is

$$U = \sum_{n=1}^{n} U_i \tag{9-191}$$

where

$$U_i = \frac{L}{8A_i^2} \oint_{(i)} \frac{\left[T_{0,(i)}\right]^2}{G_{(i)} t_{(i)}} ds \tag{9-192}$$

is the strain energy accumulated in cell $i$. Thus applying Castigliano's principle individually to each cell gives a system of $n$ equations,

$$\beta = \frac{\theta}{L} = \frac{\partial U_i}{\partial T_{0,(i)}} = \frac{1}{4A_i^2} \oint_{(i)} \frac{T_{0,(i)}\, ds}{G_{(i)} t_{(i)}} = \frac{1}{2A_i} \oint_{(i)} \frac{q_i\, ds}{G_{(i)} t_{(i)}}, \qquad i = 1, \ldots, n \tag{9-193}$$

Equations (9-193) together with Eq. (9-183) constitute a system of $n + 1$ equations with $n$ unknowns $q_i$ and the unknown $T_0$.

## EXAMPLE 9-10

A torque $T_0 = 2000$ lb · in is applied to the tube shown in Fig. 9-40*a*. Let $G = 11.6 \times 10^6$ psi. Find the shear stresses $\tau_{(i)}$ and the angle of twist per unit length $\beta$. Compare the results with those of Example 9-9. Repeat the calculations for the two-cell section shown in Fig. 9-40*b*.

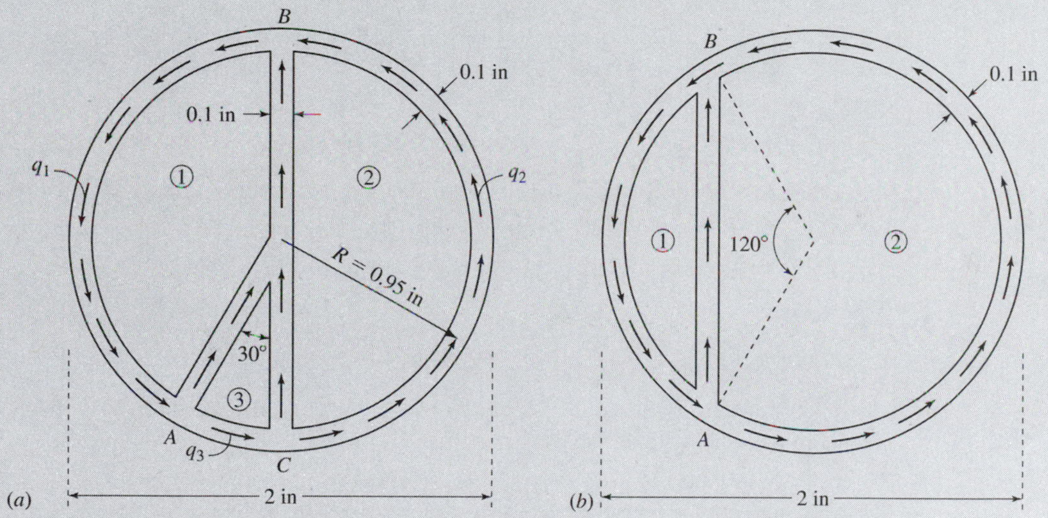


**Figure 9-40**

## SOLUTION

This is a three-cell section with constant wall thickness. Thus $n = 3$ and $t_{(i)} = t =$ constant. Also $G_{(i)} = G =$ constant. First apply Eq. (9-193) to cells 1, 2, and 3,

$$\begin{aligned}\beta &= \frac{1}{2GA_1t}\left[\frac{2\pi R}{3}q_1 + R(q_1 - q_2) + R(q_1 - q_3)\right]\\ \beta &= \frac{1}{2GA_2t}\left[\frac{2\pi R}{2}q_2 + R(q_3 - q_1) + R(q_2 - q_3)\right]\\ \beta &= \frac{3}{2GA_3t}\left[\frac{2\pi R}{6}q_3 + R(q_3 - q_1) + R(q_3 - q_2)\right]\end{aligned} \quad (a)$$

Next evaluate the geometric quantities appearing in Eq. (9-193),

$$\begin{aligned}A_1 &= \frac{R^2}{2}\frac{150^\circ - 120^\circ}{360^\circ}2\pi = \frac{\pi R^2}{3} = 0.9451 \text{ in}^2\\ A_2 &= \frac{\pi R^2}{2} = 1.4176 \text{ in}^2\\ A_3 &= \frac{\pi R^2}{6} = 0.4725 \text{ in}^2\end{aligned} \quad (b)$$

Using Eqs. ($b$), Eqs. ($a$) become

$$\begin{aligned}2G\beta &= \frac{0.95}{0.09451}(4.0944q_1 - q_2 - q_3)\\ 2G\beta &= \frac{0.95}{0.14176}(-q_1 + 5.1416q_2 - q_3)\\ 2G\beta &= \frac{0.95}{0.04725}(-q_1 - q_2 + 3.0472q_3)\end{aligned} \quad (c)$$

Solving Eqs. ($c$) we obtain

$$q_1 = 0.047498(2G\beta), \qquad q_2 = 0.047498(2G\beta), \qquad q_3 = 0.047498(2G\beta) \quad (d)$$

Hence the radial walls are stress-free and thus have no influence on the shear stress distribution in the tube. Substituting these results into Eq. (9-189) gives

$$2000 = 2(2G\beta) \times 0.047498(0.9451 + 1.4176 + 0.4725)$$

Thus,

$$\beta = \frac{2000}{4 \times 11.6 \times 0.13467 \times 10^6} = 320.1 \times 10^{-6} \text{ rad/in}$$

which, as expected, is the same result as for the single-cell tube solved in Example 9-9. The maximum shear stress equals

$$\begin{aligned}\max \tau_s &= \frac{q_1}{t} = \frac{0.047498(2G\beta)}{t} = \frac{2 \times 0.047498 \times 11.6 \times 10^6 \times 320.1 \times 10^6}{0.1}\\ &= 3527 \text{ psi}\end{aligned}$$

Now we repeat the calculations for the two-cell section of Fig. 9-40*b*. Here we find

$$A_2 = \frac{4\pi}{3}\frac{R^2}{2} + \frac{1}{2}(AB)h = R^2\left(\frac{2\pi}{3} + \frac{\sqrt{3}}{4}\right) = 2.2810 \text{ in}^2$$

$$A_1 = \pi R^2 - A_2 = 0.5543 \text{ in}^2$$

and

$$\beta = \frac{R}{2A_1Gt}\left[\frac{2\pi}{3}q_1 + \sqrt{3}(q_1 - q_2)\right]$$

$$\beta = \frac{R}{2A_2Gt}\left[\frac{4\pi}{3}q_2 + \sqrt{3}(q_2 - q_1)\right]$$

Substituting numerical data and solving for $q_1$ and $q_2$, we get

$$q_1 = 0.03874(2G\beta), \qquad q_2 = 0.05188(2G\beta)$$

Using Eq. (9-189), we get

$$2000 = 2(0.03874 \times 0.5543 + 0.05188 \times 2.2810)2G\beta$$

Thus,

$$\beta = \frac{2000}{4 \times 11.6 \times 0.1398 \times 10^6} = 308.3 \times 10^{-6} \text{ rad/in}$$

Hence the present configuration results in stresses in the internal wall. The maximum shear stress appears in the outside wall of cell 2 and equals

$$\max \tau_s = \frac{q_2}{t} = \frac{0.05188(2G\beta)}{t} = \frac{2 \times 0.05188 \times 11.6 \times 10^6 \times 308.3 \times 10^{-6}}{0.1}$$

$$= 3711 \text{ psi}$$

## 9-5-6 Pure Plastic Torsion

The application of Prandtl's stress function $\varphi$, introduced in Section 9-5-2, is not limited to torsion of elastic materials. However, the differential Eq. (9-140) is based on Hooke's law, and as such it is no longer valid. It has to be replaced by the yield condition,

$$\tau_{xz}^2 + \tau_{yz}^2 = \left(\frac{\sigma_0}{\alpha}\right)^2 \tag{9-194}$$

where $\sigma_0$ is the yield stress and where $\alpha = 2$ or $\alpha = \sqrt{3}$, depending on the usage of Tresca's or von Mises' conditions. Substituting here Eqs. (9-137), we get

$$\left(\frac{\partial\varphi}{\partial x}\right)^2 + \left(\frac{\partial\varphi}{\partial y}\right)^2 = \left(\frac{\sigma_0}{\alpha}\right)^2$$

or

$$\nabla\varphi \cdot \nabla\varphi = \left(\frac{\sigma_0}{\alpha}\right)^2$$

and hence,

$$|\nabla\varphi|^2 = \left(\frac{\sigma_0}{\alpha}\right)^2 \tag{9-195}$$

This nonlinear differential equation is the equivalent of the linear differential equation (9-140) obtained for the case of an elastic solid. The boundary condition $\varphi = 0$ [Eq. (9-143)] is still valid. Since $\partial\varphi/\partial x$ and $\partial\varphi/\partial y$ are Cartesian components of the gradient of $\varphi(x, y)$, Eq . (9-195) implies that the magnitude of the gradient is a constant that equals $\sigma_0$.

## 9-6 APPLICATION OF NUMERICAL METHODS TO SOLUTION OF TWO-DIMENSIONAL ELASTIC PROBLEMS

In this section we briefly discuss the application of finite differences to the solution to a two-dimensional elastic problem. Navier's equations of equilibrium in the case of plane stress [Eqs. (9-6)],

$$(1-\nu)\nabla^2 u + (1+\nu)\frac{\partial}{\partial x}\left(\frac{\partial u}{\partial x} + \frac{\partial v}{\partial y}\right) + \frac{2(1-\nu^2)}{E}X = 0$$

$$(1-\nu)\nabla^2 v + (1+\nu)\frac{\partial}{\partial y}\left(\frac{\partial u}{\partial x} + \frac{\partial v}{\partial y}\right) + \frac{2(1-\nu^2)}{E}Y = 0$$

will be solved for the case $X = Y = 0$.

### EXAMPLE 9-11

Consider the weightless beam shown in Fig. 9-41. The beam, clamped at both ends, carries a uniform load $q$.

#### SOLUTION

The symmetry of the beam and of the load with respect to the line $EF$ ensures the symmetry of the displacement $u$ and of the stress components $\sigma_{xx}$ and $\sigma_{yy}$. It also ensures antisymmetry of the displacement $v$ and the stress $\sigma_{xy}$. Thus it is sufficient to consider only one-half of the beam bounded by the polygon $AEFC$.

First we note the boundary conditions,

$$\begin{aligned} u &= 0, & v &= 0, & &\text{at } AC \\ \sigma_{xx} &= -q, & \sigma_{xy} &= 0, & &\text{at } AE \\ \sigma_{xx} &= 0, & \sigma_{xy} &= 0, & &\text{at } CF \end{aligned} \tag{a}$$

and the conditions resulting from symmetry,

$$v = 0, \quad \sigma_{xy} = 0, \qquad \text{at } EF \tag{b}$$

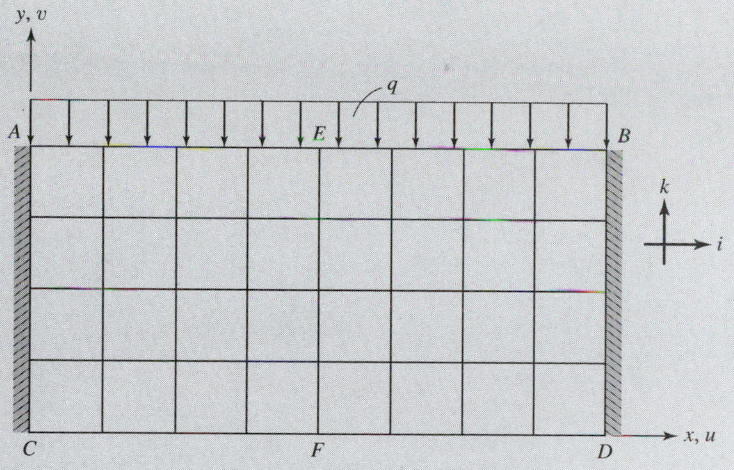


**Figure 9-41**

We begin by replacing derivatives in Eq. (9-6) by finite differences using formulas given in Chapter 6,

$$(\nabla^2 u)_{i,k} = \frac{1}{h^2}[-2(1+\beta^2)U_{i,k} + U_{i+1,k} + U_{i-1,k} + \beta^2 U_{i,k+1} + U_{i,k-1}]$$

$$(\nabla^2 v)_{i,k} = \frac{1}{h^2}[-2(1+\beta^2)V_{i,k} + V_{i+1,k} + V_{i-1,k} + \beta^2 V_{i,k+1} + V_{i,k-1}]$$

$$\left(\frac{\partial^2 u}{\partial x\,\partial y}\right)_{i,k} = \frac{1}{4hl}[U_{i+1,k+1} - U_{i+1,k-1} - U_{i-1,k+1} + U_{i-1,k-1}]$$

$$\left(\frac{\partial^2 v}{\partial x\,\partial y}\right)_{i,k} = \frac{1}{4hl}[V_{i+1,k+1} - V_{i+1,k-1} - V_{i-1,k+1} + V_{i-1,k-1}]$$

$$\left(\frac{\partial^2 u}{\partial x^2}\right)_{i,k} = \frac{1}{h^2}[U_{i+1,k} - 2U_{i,k} + U_{i-1,k}]$$

$$\left(\frac{\partial^2 v}{\partial y^2}\right)_{i,k} = \frac{1}{h^2}[V_{i+1,k} - 2V_{i,k} + V_{i-1,k}]$$

where $\beta = h/l$. We thus obtain

$$4(1-\nu)[-2(1+\beta^2)U_{i,k} + U_{i+1,k} + U_{i-1,k} + \beta^2(U_{i,k+1} + U_{i,k-1})] + 4(1+\nu)$$
$$\times (U_{i+1,k} - 2U_{i,k} + U_{i-1,k}) + \beta(1+\nu)(V_{i+1,k+1} - V_{i+1,k-1} - V_{i-1,k+1} + V_{i-1,k-1}) = 0$$

$$4(1-\nu)[-2(1+\beta^2)V_{i,k} + V_{i+1,k} + V_{i-1,k} + \beta^2(V_{i,k+1} + V_{i,k-1})] + 4\beta^2(1+\nu)$$
$$\times (V_{i+1,k} - 2V_{i,k} + V_{i-1,k}) + \beta(1+\nu)(U_{i+1,k+1} - U_{i+1,k-1} - U_{i-1,k+1} + U_{i-1,k-1}) = 0$$

which reduce to

$$\begin{aligned}-8[(1-\nu)\beta^2+2]U_{i,k}+4\beta^2(1-\nu)(U_{i,k-1}+U_{i,k+1})+8(U_{i-1,k}+U_{i+1,k})\\+\beta(1+\nu)(V_{i+1,k+1}-V_{i+1,k-1}-V_{i-1,k+1}+V_{i-1,k-1})=0\\-8[(1-\nu)+2\beta^2]V_{i,k}+8\beta^2(V_{i,k-1}+V_{i,k+1})+4(1-\nu)(V_{i-1,k}+V_{i+1,k})\\+\beta(1+\nu)(U_{i+1,k+1}-U_{i+1,k-1}-U_{i-1,k+1}+U_{i-1,k-1})=0\end{aligned}\tag{c}$$

Next, using Eqs. (4-21), (9-2), and (9-3), we express the stresses in terms of the displacements. Thus the boundary and symmetry conditions (*a*) and (*b*) become

$$\begin{aligned}\frac{E}{1-\nu^2}\left(\frac{\partial u}{\partial x}+\nu\frac{\partial v}{\partial y}\right)=-q,\quad&\frac{\partial u}{\partial y}+\frac{\partial v}{\partial x}=0,\quad\text{at } AE\\\frac{\partial u}{\partial x}+\nu\frac{\partial v}{\partial y}=0,\quad&\frac{\partial u}{\partial y}+\frac{\partial v}{\partial x}=0,\quad\text{at } CF\\v=0,\quad&\frac{\partial u}{\partial y}+\frac{\partial v}{\partial x}=0,\quad\text{at } EF\end{aligned}\tag{d}$$

Finally, with Eqs. (6-15) and (6-16) these become

$$\begin{aligned}U_{i,k}=0,\quad V_{i,k}=0,\quad&\text{at } AC\\\frac{E}{2h(1-\nu^2)}[U_{i+1,k}-U_{i-1,k}+\nu\beta(V_{i,k+1}-V_{i,k-1})]=-q\quad&\\\beta(U_{i,k+1}-U_{i,k-1})+V_{i+1,k}-V_{i-1,k}=0,\quad&\text{at } AE\\U_{i+1,k}-U_{i-1,k}+\nu\beta(V_{i,k+1}-V_{i,k-1})]=0\quad&\\\beta(U_{i,k+1}-U_{i,k-1})+V_{i+1,k}-V_{i-1,k}=0,\quad&\text{at } CF\\V_{i,k}=0,\quad\beta(U_{i,k+1}-U_{i,k-1})+V_{i+1,k}-V_{i-1,k}=0,\quad&\text{at } EF\end{aligned}\tag{e}$$

At this stage we make the problem more specific by assuming a square mesh, that is, $\beta=1$, and we select $\nu=0.3$. Consequently Eqs. (*a*) are

$$\begin{aligned}-21.6U_{i,k}+8(U_{i+1,k}+U_{i-1,k})+2.8(U_{i,k+1}+U_{i,k-1})\\+1.3(V_{i+1,k+1}-V_{i+1,k-1}-V_{i-1,k+1}+V_{i-1,k-1})=0\\-21.6V_{i,k}+8(U_{i,k+1}+U_{i,k-1})+2.8(V_{i+1,k}+V_{i-1,k})\\+1.3(U_{i+1,k+1}-U_{i+1,k-1}-U_{i-1,k+1}+U_{i-1,k-1})=0\end{aligned}\tag{f}$$

and the final form of the boundary conditions (*e*) is

$$\begin{aligned}U_{i,k}=0,\quad V_{i,k}=0,\quad&\text{at } AC\\U_{i+1,k}-U_{i-1,k}+0.3(V_{i,k+1}-V_{i,k-1})=-Q\quad&\\U_{i,k+1}-U_{i,k-1}+V_{i+1,k}-V_{i-1,k}=0,\quad&\text{at } AE\\U_{i+1,k}-U_{i-1,k}+0.3(V_{i,k+1}-V_{i,k-1})=0\quad&\\U_{i,k+1}-U_{i,k-1}+V_{i+1,k}-V_{i-1,k}=0,\quad&\text{at } CF\\V_{i,k}=0,\quad U_{i,k+1}-U_{i,k-1}+V_{i+1,k}-V_{i-1,k}=0,\quad&\text{at } EF\end{aligned}\tag{g}$$

where $Q\equiv1.82qh/E$.

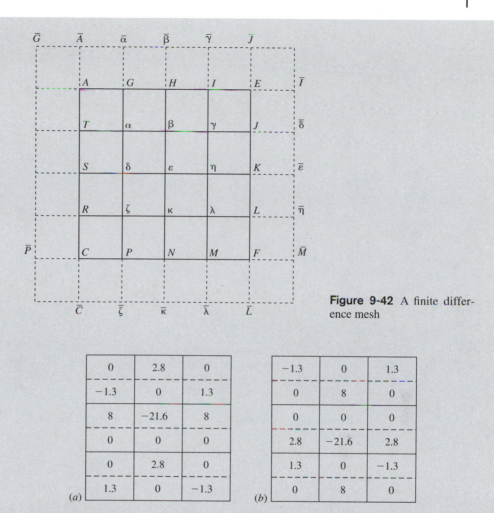

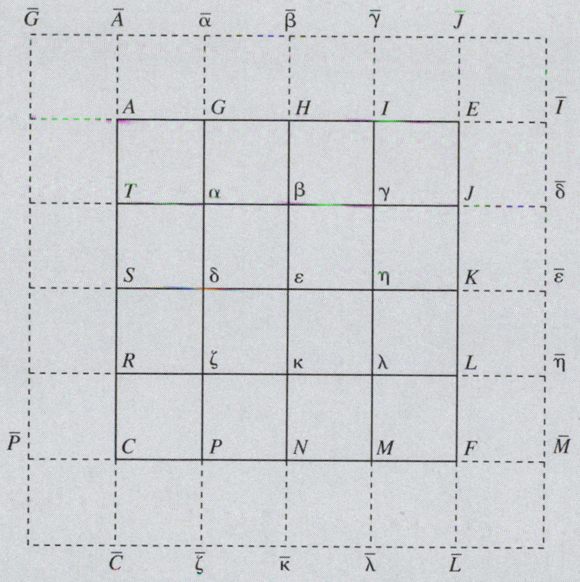


**Figure 9-42** A finite difference mesh

(*a*)

| | | |
|---|---|---|
| 0 | 2.8 | 0 |
| −1.3 | 0 | 1.3 |
| 8 | −21.6 | 8 |
| 0 | 0 | 0 |
| 0 | 2.8 | 0 |
| 1.3 | 0 | −1.3 |

(*b*)

| | | |
|---|---|---|
| −1.3 | 0 | 1.3 |
| 0 | 8 | 0 |
| 0 | 0 | 0 |
| 2.8 | −21.6 | 2.8 |
| 1.3 | 0 | −1.3 |
| 0 | 8 | 0 |

**Figure 9-43** Two stencils are needed, one for each set of the unknowns

At this point we label the internal pivots $\alpha, \beta, \ldots, \lambda$ and the boundary pivots (including pivots on the symmetry line) $A, B, \ldots, T$ (Fig. 9-42). Next we create a stencil (see Chapter 6) suitable for the system of the difference equations ($f$). Since there are two difference equations with two sets of unknowns, two stencils are needed, each with two cells within every box. The upper cell includes the coefficient of $U_{i,k}$ and the lower cell the coefficient of $V_{i,k}$. These stencils are shown in Fig. 9-43. Now we place the stencils consecutively at the inside pivots. Fig. 9-44 shows the center boxes of the stencils located at pivot $\alpha$. The first of Eqs. ($f$) becomes

$$2.8U_G + 8U_T - 21.6U_\alpha + 8U_\beta + 2.8U_\delta - 1.3V_A + 1.3V_H + 1.3V_S - 1.3V_\varepsilon = 0 \quad (h)$$

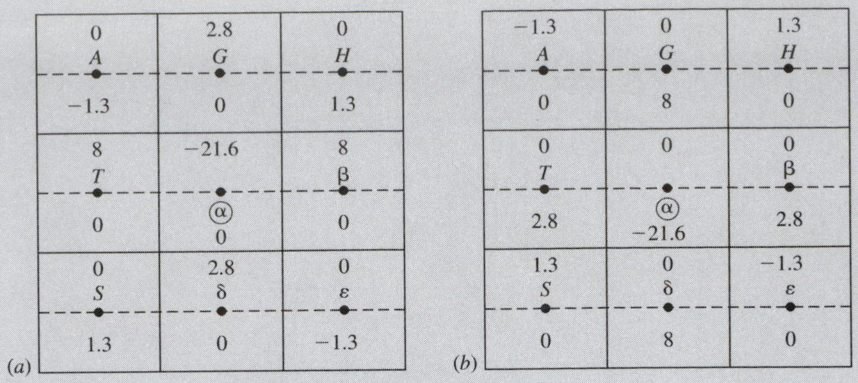


**Figure 9-44**

and the second,

$$-1.3U_A + 1.3U_H + 1.3U_S - 1.3U_\varepsilon + 8V_G + 2.8V_T - 21.6V_\alpha + 2.8V_\beta + 8V_\delta = 0 \qquad (i)$$

We continue in this manner, placing the centers of the stencils at the pivots β, γ, δ, ε, η, ζ, κ, and λ. We thus obtain 18 equations. Since the outside boxes of the stencils cover also the boundaries, these equations include 50 unknowns (two unknowns for each inside and boundary pivot). Additional equations are obtained by applying the boundary and symmetry conditions, Eqs. (*g*). We find that along *AC*,

$$U_A = V_A = U_T = V_T = U_S = V_S = U_R = V_R = U_C = V_C = 0 \qquad (j)$$

and along *GI*,

$$\begin{aligned} 0.3(U_H - U_A) + V_{\bar\alpha} - V_\alpha &= -Q & U_{\bar\alpha} - U_\alpha + V_H - V_A &= 0 \\ 0.3(U_I - U_G) + V_{\bar\beta} - V_\beta &= -Q & U_{\bar\beta} - U_\beta + V_I - V_G &= 0 \\ 0.3(U_E - U_H) + V_{\bar\gamma} - V_\gamma &= -Q & U_{\bar\gamma} - U_\gamma + V_E - V_H &= 0 \end{aligned} \qquad (k)$$

Along the symmetry line *EF* we have

$$\begin{aligned} &U_{\bar I} + U_I = 0, \quad U_{\bar\delta} + U_\gamma = 0, \quad U_{\bar\varepsilon} + U_\eta = 0 \\ &U_{\bar\eta} + U_\lambda = 0, \quad U_{\bar M} + U_M = 0 \end{aligned} \qquad (l)$$

and

$$V_{\bar I} = -V_I, \qquad V_{\bar\delta} = -V_\gamma, \qquad V_{\bar\varepsilon} = -V_\eta, \qquad V_{\bar\eta} = -V_\lambda, \qquad V_{\bar M} = -V_M \qquad (m)$$

At the edge *CF*,

$$\begin{aligned} 0.3(U_N - U_C) + V_\zeta - V_{\bar\zeta} &= 0, & U_\zeta - U_{\bar\zeta} + V_N - V_C &= 0 \\ 0.3(U_M - U_P) + V_\kappa - V_{\bar\kappa} &= 0, & U_\kappa - U_{\bar\kappa} + V_M - V_P &= 0 \\ 0.3(U_F - U_N) + V_\lambda - V_{\bar\lambda} &= 0, & U_\lambda - U_{\bar\lambda} + V_F - V_N &= 0 \end{aligned} \qquad (n)$$

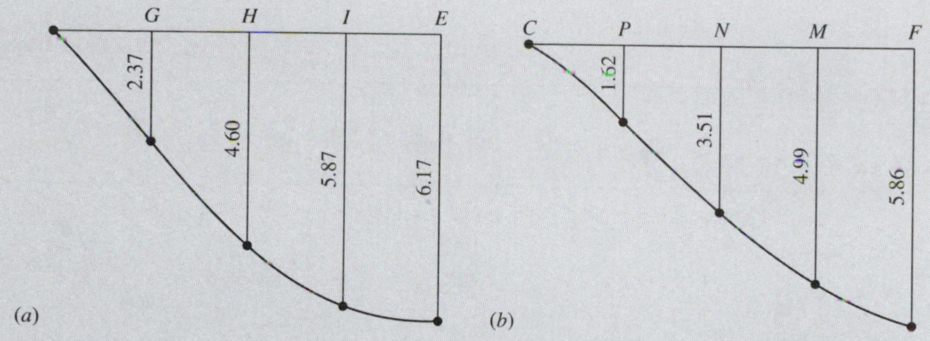


**Figure 9-45**

Now we notice the appearance of additional unknowns at fictitious outside pivots. To handle this we require that the equilibrium equations be also satisfied at the boundary pivots along the top and bottom edges. Thus placing the stencils at $G, H, I, P, N$, and $M$ we obtain additional equations. For instance, with the stencils at $H$ we get

$$
\begin{aligned}
2.8U_{\bar{\alpha}} + 8U_A - 21.6U_G + 8U_H + 2.8U_\alpha - 1.3V_{\bar{A}} + 1.3V_{\bar{\beta}} + 1.3V_T - 1.3V_\beta &= 0 \\
-1.3U_{\bar{\alpha}} + 8V_{\bar{\beta}} + 1.3U_{\bar{\gamma}} + 2.8V_G - 21.6V_H + 2.8V_I + 1.3U_\alpha + 8V_\beta - 1.3U_\gamma &= 0
\end{aligned} \tag{o}
$$

Another set of unknowns appears,

$$V_{\bar{A}},\quad U_{\bar{A}},\quad V_{\bar{J}},\quad U_{\bar{J}},\quad V_{\bar{C}},\quad U_{\bar{C}},\quad V_{\bar{L}},\quad U_{\bar{L}}$$

which are eliminated using the conditions

$$V_{\bar{A}} = V_{\bar{J}} = V_{\bar{C}} = V_{\bar{L}} = 0, \qquad U_{\bar{A}} = U_{\bar{C}} = 0, \qquad U_{\bar{J}} \approx U_E, \qquad U_{\bar{L}} \approx U_F \tag{p}$$

The system of linear algebraic equations thus obtained has been solved using Mathcad7, professional edition. The resulting deflections along the top and bottom surfaces of the beam are plotted in Fig. 9-45. The maximum deflection in the $y$ direction is

$$v_{\max} = -6.17Q = -6.17\frac{2qh(1-\nu^2)}{E} = -2.80\frac{qH}{E} \tag{r}$$

where $H$ is the depth of the beam. This compares to

$$v_{\max} = -2.19\frac{qH}{E} \tag{s}$$

obtained using a finite element program.* The relatively large error is due to an insufficiently fine mesh used in the finite difference analysis.

*Courtesy of Professor Herbert A. Koenig, Department of Mechanical Engineering, University of Connecticut. Prof. Koenig applied commercially available software MARC 5.1, developed by MARC Analysis Research Corporation, 150 North Main St., Manchester, CT.

In the next example we will solve the problem of torsion of a bar with a square cross section.

## EXAMPLE 9-12

A bar of square cross section is subjected to a torque $T_0$ (Fig. 9-46). Find the maximum stress.

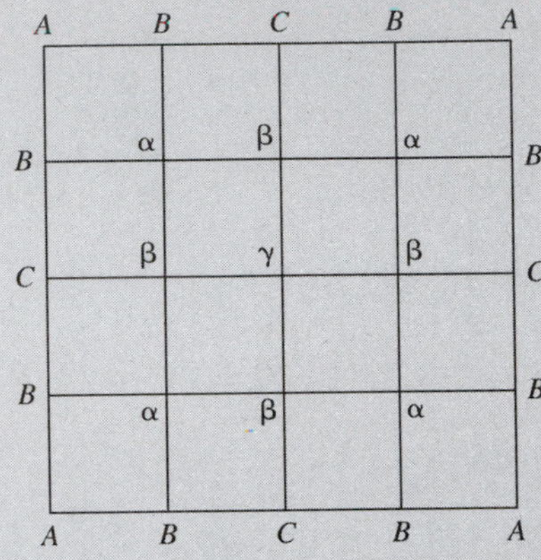


**Figure 9-46**

## SOLUTION

In terms of Prandtl's stress function φ (Section 9-5-2) the equilibrium equation becomes [see Eq. (9-140)]

$$\nabla^2\varphi = -2G\beta \qquad (a)$$

with the boundary condition [Eq. (9-143)]

$$\varphi = 0 \qquad (b)$$

After solving for φ we will use Eq. (9-151) to determine β in terms of the torque $T_0$. First we replace the right-hand side of Eq. (*a*) by −1. Since this is a linear equation, the principle of superposition applies and thus the solution to Eq. (*a*) is proportional to the solution to

$$\nabla^2\varphi = -1 \qquad (c)$$

Next, applying Eq. (6-27), we replace Eq. (*c*) by a finite differences equation,

$$-4\varphi_{i,k} + \varphi_{i+1,k} + \varphi_{i-1,k} + \varphi_{i,k+1} + \varphi_{i,k-1} = -h^2 \qquad (d)$$

with

$$\varphi_{i,k} = 0 \quad \text{at the boundary} \qquad (e)$$

| | | |
|---|---|---|
| 0 | 1 | 0 |
| 1 | −4 | 1 |
| 0 | 1 | 0 |

**Figure 9-47**

The stencil corresponding to Eq. (*d*) is shown in Fig. 9-47. Because of the symmetry of the function $\varphi$ it is sufficient to place the central box of the stencil at pivots $\alpha$, $\beta$, and $\gamma$. This results in the following system of algebraic equations:

$$\begin{aligned} -4\varphi_\alpha + \varphi_B + \varphi_\beta + \varphi_B + \varphi_\beta &= -h^2 \\ -4\varphi_\beta + \varphi_\alpha + \varphi_\alpha + \varphi_C + \varphi_\gamma &= -h^2 \\ -4\varphi_\gamma + \varphi_\beta + \varphi_\beta + \varphi_\beta + \varphi_\beta &= -h^2 \end{aligned} \qquad (f)$$

Simplifying Eqs. ($f$) and assuming that the dimensions of the square are $4 \times 4$, thus $h = 1$, we obtain

$$-4\varphi_\alpha + 2\varphi_\beta = -1, \qquad 2\varphi_\alpha - 4\varphi_\beta + \varphi_\gamma = -1, \qquad 4\varphi_\beta - 4\varphi_\gamma = -1 \qquad (g)$$

where $\varphi_B = \varphi_C = 0$ was also used. Solving Eqs. ($g$) gives

$$\varphi_\alpha = 0.6875, \qquad \varphi_\beta = 0.8750, \qquad \varphi_\gamma = 1.1250 \qquad (h)$$

Maximum shear stress appears at the location of the maximum slope of the surface $\varphi$ [see Eq. (9-146)]. As this occurs at the contour, we readily identify the pivot $C$ as the location of maximum stress,

$$\max \tau_s = \max \frac{\partial \varphi}{\partial n} \cong \frac{\varphi_\beta}{h} = 0.875 \qquad (i)$$

Remembering that the values of $\varphi$ have been obtained under the assumption that the right-hand side of the Eq. (9-154) is replaced by $-1$, we have to multiply Eq. ($i$) by $2G\beta$, getting finally

$$\max \tau_s = -1.750G\beta \qquad (j)$$

To find $\beta$ we use Eq. (9-151),

$$T_0 = 2 \int_A \int \varphi \, dA$$

First we notice that the double integral represents the volume enclosed between the surface $\varphi$ and the cross section of the bar. Since we do not have an analytical expression for $\varphi$, we calculate this volume $V_\varphi$ numerically, adding the volumes of parallelepipeds constructed over all boxes. The average height of a parallelepiped is taken as the average of the values of $\varphi$ calculated at all four corners of the box. Thus,

$$V_\varphi \approx 4\left(\frac{\varphi_\alpha + 2\varphi_\beta + \varphi_\gamma}{4} + 2\frac{\varphi_\alpha + \varphi_\beta}{4} + \frac{\varphi_\alpha}{4}\right) \approx 7.375 \tag{k}$$

and writing Eq. (9-151) as

$$T_0 = 2V_\varphi \cdot 2G\beta \tag{l}$$

we obtain

$$\beta = \frac{T_0}{29.5G} \tag{m}$$

When this is substituted into Eq. ($j$), we get

$$\max \tau = 0.0593 T_0 \tag{n}$$

The exact result is [9, p. 313, eq. 172]

$$\tau_{\max} = \frac{T_0}{0.208 \times 4^3} = 0.0751 T_0 \tag{o}$$

Using Eq. (9-162), which is valid for elongated rectangles, we have

$$\max \tau_s = \frac{3T_0}{a^3} = 0.0469 T_0 \tag{p}$$

To improve accuracy we have to use a finer mesh. The calculations were repeated for a 6 × 6 mesh, yielding

$$\max \tau_s = 0.0623 T_0 \tag{q}$$

## PROBLEMS

### Section 9-1

**9-1** Determine the boundary conditions for a two-dimensional solid supported by uniformly distributed springs of stiffness $k$ per unit length (Fig. P9-1).

**9-2** Determine the boundary conditions for a circular disk with a cutout $ABCD$ (Fig. P9-2).

**9-3** Determine the boundary conditions for a two-dimensional solid resting on an another two-dimensional solid (Fig. P9-3). Assume that there is no friction between solids I and II.

**9-4** Derive Eqs. (9-5).

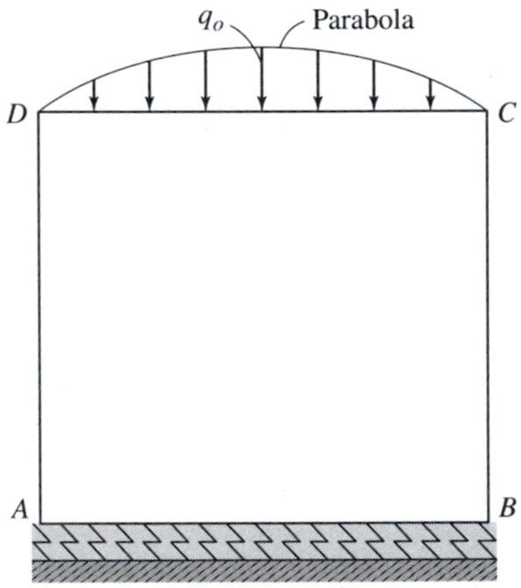


Figure P9-1

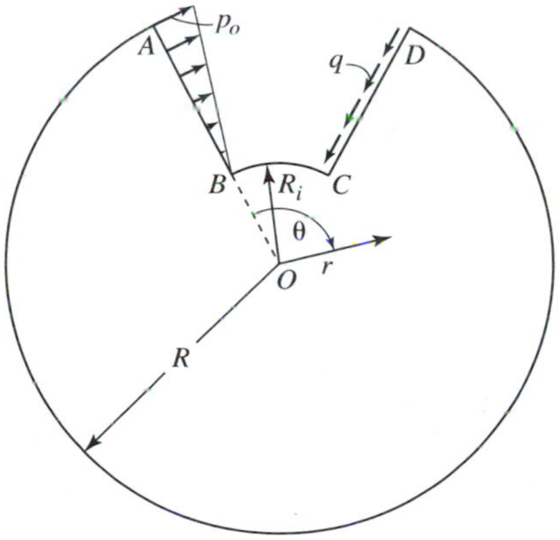


Figure P9-2

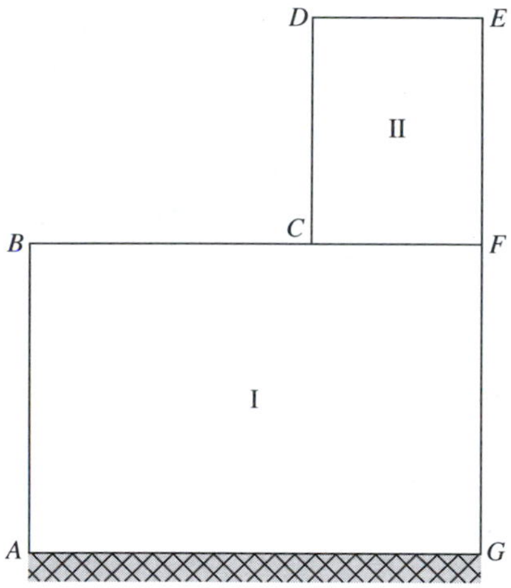


Figure P9-3

## Section 9-2

**9-5** Determine whether the following stress components are possible solutions of the equilibrium equations (9-14) in the case of the plane state of stress.

(a) $\sigma_{xx} = Ay^2, \quad \sigma_{yy} = Bx^2, \quad \sigma_{xy} = Cxy$
where $A$, $B$, and $C$ are constant.

(b) $\sigma_{xx} = Ax^2, \quad \sigma_{yy} = By^2, \quad \sigma_{xy} = Cxy$

**9-6** Question and data from Problem 9-5 but no body forces are present.

## Section 9-3

**9-7** Show all the details of the derivation of Eq. (9-23).

**9-8** What must be the ratio $p_i/p_o$ to ensure that $\sigma_{rr} = 0$ [Eqs. (9-41)] at $r_i < r < r_o$?

**9-9** Find a ratio $p_i/p_o$ such that it makes the radial displacement $u(r)$ [Eq. (9-42)] vanish at a point $r$, where $r_i < r < r_o$.

**9-10** Show that when $p_i = p_o$, both $\sigma_{rr}$ and $\sigma_{\theta\theta}$ are constant.

**9-11** A rigid (undeformable) inclusion of radius $r_m = 5.001$ in is inserted into a circular disk with an outer radius $r_o = 20$ in and an inner radius $r_i = 5$ in. Determine the resulting displacement and the stresses.

**9-12** A deformable disk is attached to a rigid shaft such that $r_i \ll r_o$. Derive expressions for the displacement and for the stresses in the disk.

**9-13** A steel cylinder ($E = 30 \times 10^6$ psi) of radii $r_o = 10$ in and $r_i = 5$ in is press-fitted on a steel shaft of radius $r_m = 5.002$ in.

(a) Calculate the outside pressure $p_o$ needed to loosen the fit.

(b) Calculate the tangential stresses occurring at $p_o$ calculated in part (a).

**9-14** A steel cylinder of the same dimensions and properties as in Problem 9-13 is press-fitted on a steel shaft of unknown radius. Assuming that the interface pressure is 6000 psi, find the speed $\omega$ needed to loosen the fit.

**9-15** Use Eqs. (9-61) to determine the displacement and stresses in a rotating disk with $r_o = r_i + \delta$, where $\delta/r_i \ll 1$.

**9-16** Determine the change in radius of a circular disk with $r_o = 10$ in and $r_i = 5$ in subjected to a temperature distribution

$$T(r) = T_0\left(\frac{r}{r_i} - 1\right)$$

where $T_0$ = constant, $\nu = 0.3$ psi, and $\alpha = 0.00001117/°C$.

**9-17** A rotating disk with a hole is subjected to an inside pressure $p_i$ while the outside pressure is zero. Determine the total stresses resulting from the temperature change $\Delta T(r)$.

**9-18** A steel cylinder with a Young's modulus $E = 30 \times 10^6$ psi and a yield stress $Y = 48 \times 10^6$ psi is subjected to an internal pressure $p_i = 12 \times 10^6$ psi. Determine the depth of the plastic zone $c$ assuming that $r_o = 20$ in and $r_i = 15$ in. Find the radial and tangential stress distributions.

**9-19** For the cylinder described in Problem 9-18 determine the internal pressure needed to make the depth of the plastic zone $c = 18$ in.

## Section 9-4

**9-20** Assume Airy's stress function in the form

$$\Phi(x, y) = 3C_1 bxy^2 - 2C_1 xy^3$$

and let $x \geq 0$, $b \geq y \geq 0$ (Fig. P9-20). Find the stresses at the boundaries.

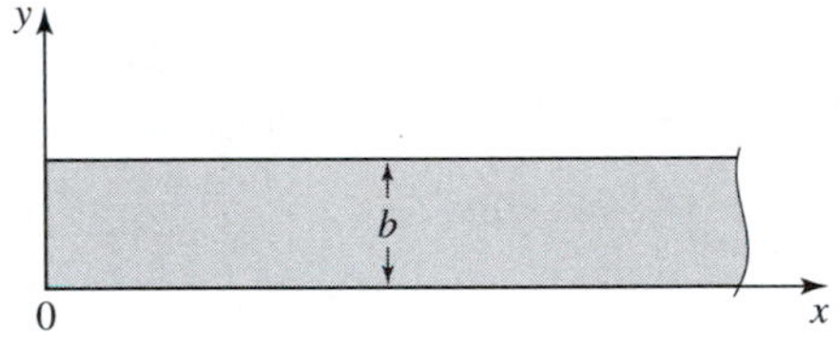


Figure P9-20

**9-21** Determine the stresses corresponding to Airy's stress function in the form of:

(a) Eq. (9-119*a*). (b) Eq. (9-119*b*).

## Section 9-5

**9-22** Let a function $\varphi$ be given in the form

$$\varphi = \frac{G\beta}{2}\left(-x^2 - y^2 + b^2 + 2ax - \frac{2ab^2 x}{x^2 + y^2}\right)$$

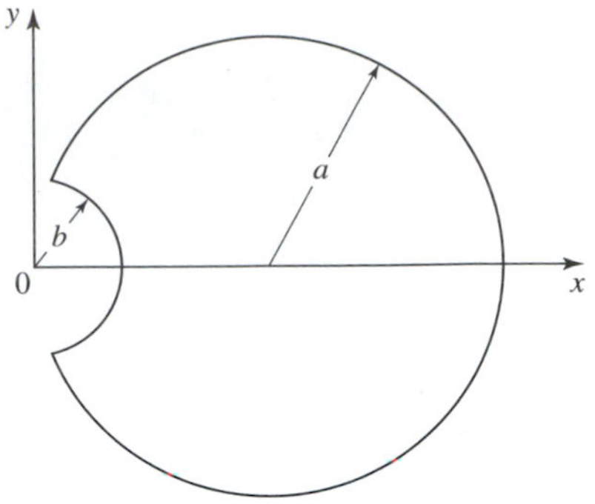


Figure P9-22

(a) Verify that it satisfies the boundary condition, Eq. (9-143), at the contour of the cross section shown in Fig. P9-22 and that it thus is the stress function.

(b) Calculate the corresponding stresses.

(c) Find the maximum stress.

**9-23** Let a function $\varphi$ be given in the form

$$\varphi = \frac{G\beta}{2}\left(-x^2 - y^2 + \frac{y^3}{a} - \frac{3x^2 y}{a} + \frac{4a^2}{27}\right)$$

(a) Verify that it satisfies Eqs. (9-130) and the boundary condition, Eq. (9-132), at the contour of the cross section shown in Fig. P9-23 and that it thus is the stress function.

(b) Calculate the corresponding stresses.

(c) Find the maximum stress.

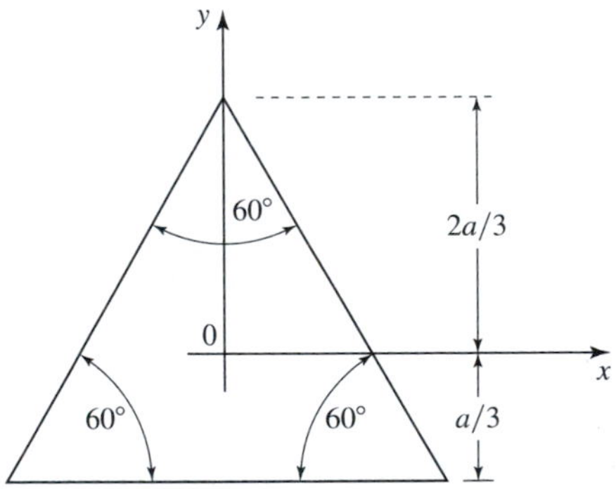


Figure P9-23

**9-24** Verify that

$$\varphi = \frac{G\beta a^2 b^2}{2(a^2 + b^2)} \left( \frac{x^2}{a^2} + \frac{y^2}{b^2} - 1 \right)$$

is the stress function for the elliptic cross section shown in Fig. P9-24.

(a) Use Eq. (9-151) to find the angle $\beta$.
(b) Find the stress components in terms of the applied torque $T_0$.

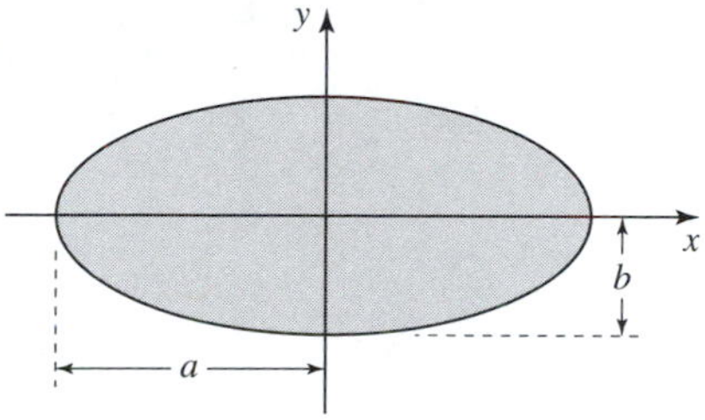


Figure P9-24

**9-25** The square thin-walled tube shown in Fig. P9-25*a* is subjected to a twisting moment $T_0 = 20{,}000$ psi.

(a) Find the maximum shear stress.
(b) Assuming that the maximum shear stress in the annular tube shown in Fig. P9-25*b* is the same as in (a), find the radius *R*.

**9-26** Install a stiffening element *AB* in the square tube shown in Fig. P9-26. The element is made of the same material as the rest of the tube. Now change the angle $\alpha$ by rotating *AB* around the center point *O*. Find such values of $\alpha\,(45^\circ < \alpha < 90^\circ)$ that ensure

(a) Maximum torsional stiffness.
(b) Minimum value of maximum shear stress.

**9-27** Find the minimum wall thickness *t* for the tube shown in Fig. P9-27 when the applied torque is $T_0 = 5000$ lb and the admissible stress is $\tau_0 = 5000$ psi. Assume $G = 11.6 \times 10^6$ psi.

**9-28** Using the data from Problem 9-27 and assuming that $t = 0.5$ in, find the smallest admissible radius *R* of the tube shown in Fig. P9-28. What is the corresponding angle $\beta$?

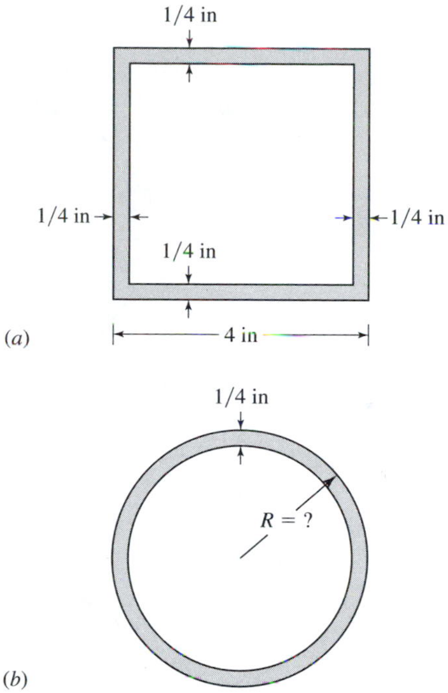


Figure P9-25

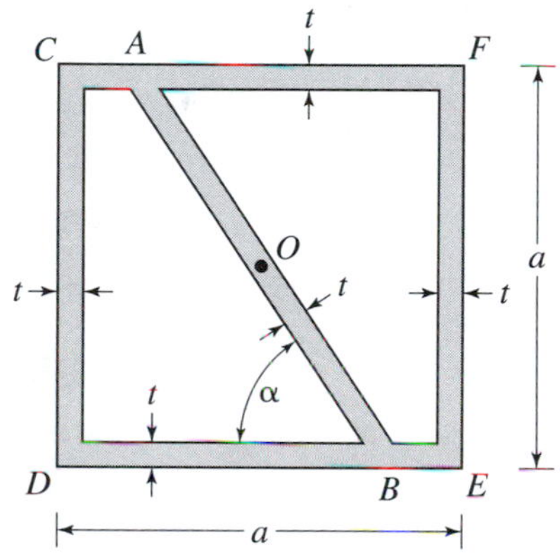


Figure P9-26

**9-29** A bar with a triangular cross section (Fig. P9-29) is subjected to a twisting moment $T_0 = 10{,}000$ lb · in. Find the maximum shear

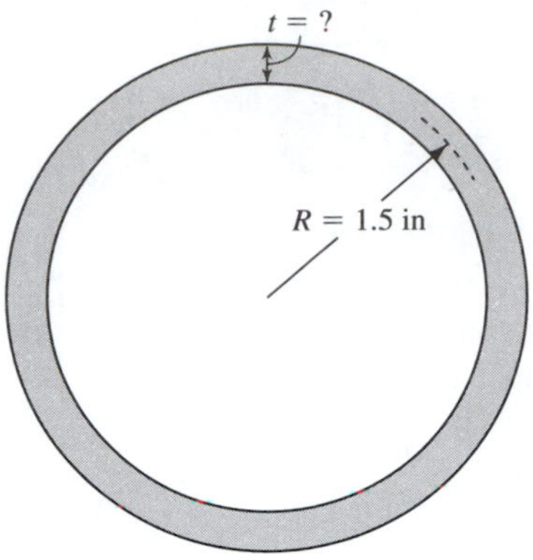


Figure P9-27

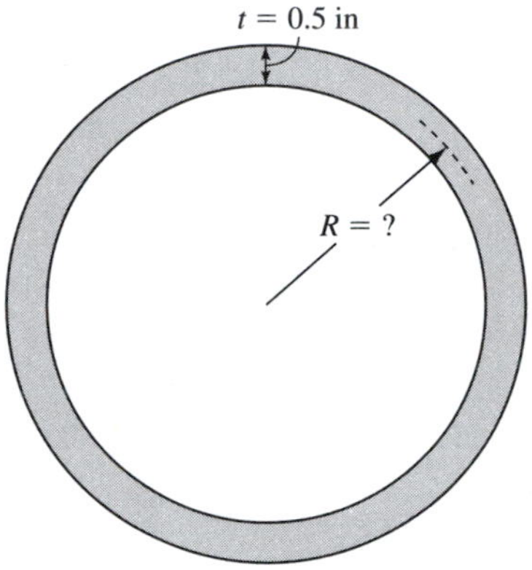


Figure P9-28

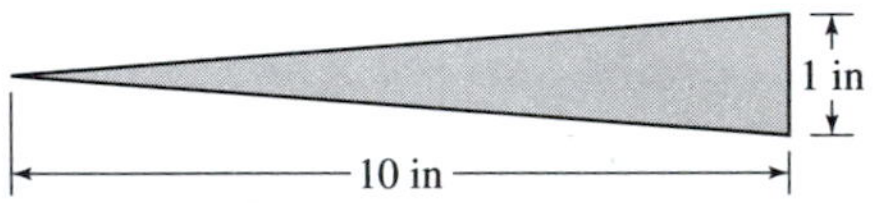


Figure P9-29

stress $\tau_s$ and the angle of twist per unit length $\beta$. *Hint:* use Eq. (9-168) and find the approximate value of $J_T$ by subdividing the cross section into five elements of equal length. Assume $G = 12 \times 10^6$ lb/in$^2$.

**9-30** Design a square tube carrying the same torque and made of the same material as the tube in Example 9-9. Assume the maximum shear stresses to be also the same.

**9-31** Three bars with the cross sections shown in Fig. P9-31 are subjected to the same torques $T_0$ and are made of the same material. Compare the angles of twist per unit length $\beta$.

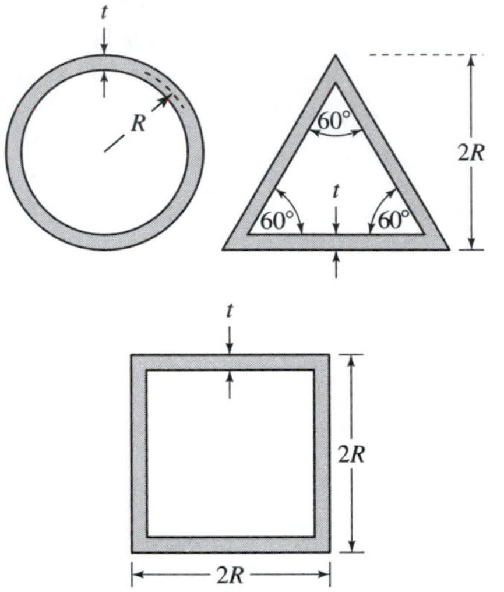


Figure P9-31

**9-32** Find the maximum shear stress and the angle of twist per unit length for the tube shown in Fig. 9-32*b* with the curved wall *AB* slit open.

**9-33** Find the allowable torque for the two-cell section of the steel shaft shown in Fig. P9-33. Assume the allowable shear stress to be 10,000 psi.

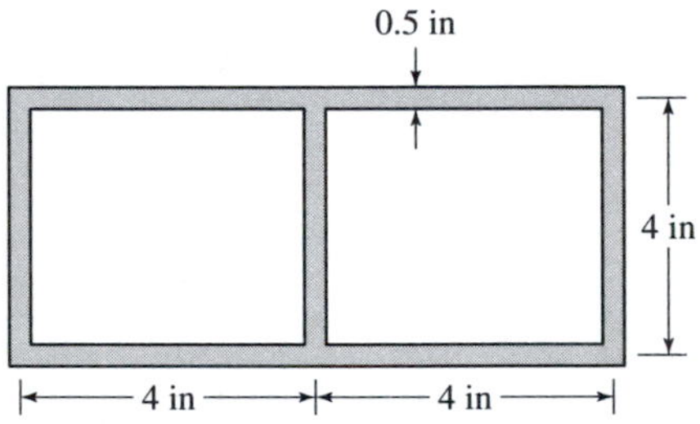


Figure P9-33

## REFERENCES

[1] Bredt, R., *VDI,* 40, p. 815 (1896).
[2] Cook, R. D., and W. C. Young, *Advanced Mechanics of Materials* (Macmillan, New York, 1985).
[3] DenHartog, J. P., *Advanced Strength of Materials* (McGraw-Hill, New York, 1952).
[4] Fung, Y. C., *Foundations of Solid Mechanics* (Prentice-Hall, Englewood Cliffs, NJ, 1965).
[5] Hildebrand, F. B., *Methods of Applied Mathematics,* 2nd ed. (Prentice-Hall, Englewood Cliffs, NJ, 1965).
[6] Prandtl, L., "Zur Torsion von prismatischen Stäben," *Z. Physik,* 4, pp. 758–770 (1903).
[7] Saada, A. S., *Elasticity, Theory and Applications* (Pergamon, New York, 1974).
[8] Sokolnikoff, I. S., *Mathematical Theory of Elasticity* (Wiley, New York, 1956).
[9] Timoshenko, S. P., and J. N. Goodier, *Theory of Elasticity,* 3rd ed. (McGraw-Hill, New York, 1970).
[10] Vinson, J. R., and R. L. Sierakowski, *The Behavior of Structures Composed of Composite Materials,* 1st ed. (Martinus Nijhoff, Dordrecht, The Netherlands, 1986).

# 10 PLATES

In this chapter we derive equations for the deflection of a thin elastic plate. The bulk of the chapter is devoted to the solution of the problem of axisymmetric bending of solid and annular circular plates. Symbolic manipulation packages are used extensively to obtain solutions for various combinations of the boundary conditions as well as for various loads, with an emphasis on Green's functions. The method of Fourier series is applied to analyze the deflection of a simply supported rectangular plate, a plate on an elastic foundation, and a rectangular membrane. Approximate methods (finite differences and Rayleigh–Ritz method) are discussed in some detail, and worked-out examples provide a comparison with exact results. Interactive Fortran programs are accessible on a diskette.

## 10-1 INTRODUCTION

A thin plate is a three-dimensional structure bounded by two parallel planes, such that the distance between them is small compared to the other two dimensions (Fig. 10-1). This geometric assumption allows for an important simplifying hypothesis. Let us call the plane halving the distance between the two bounding planes the *middle surface of the plate,* and define this to be the $z = 0$ plane. The following assumptions are made.

1. The normal to the middle surface remains normal after deformation. This is analogous to the hypothesis of the simple beam theory, which states that a plane, normal to the neutral axis, remains normal after deformation. The consequences of both assumptions are also similar: linear distribution of stress across the thickness of the beam or of the plate ($Oz$ direction in Fig. 10-1).
2. The middle surface is unstrained.
3. The normal stress in the $Oz$ direction is negligible.

We begin by deriving again the equilibrium equation for the plate in a Cartesian coordinate system. To this end we cut an infinitesimal through-element *ABCD-EFGH*

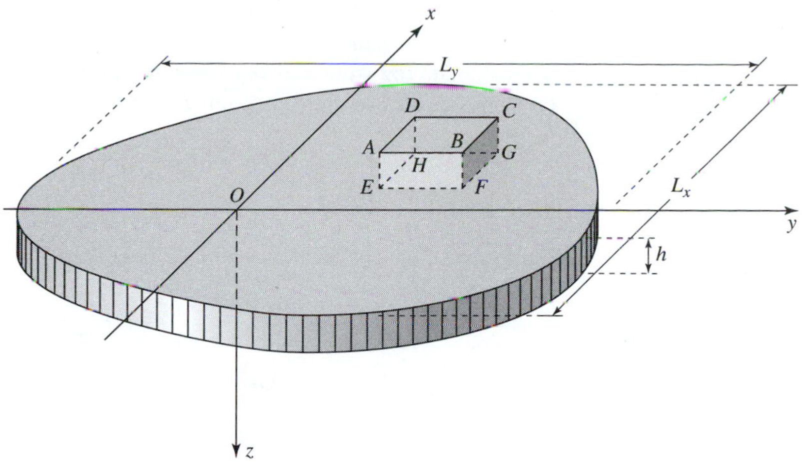


Figure 10-1 A thin plate

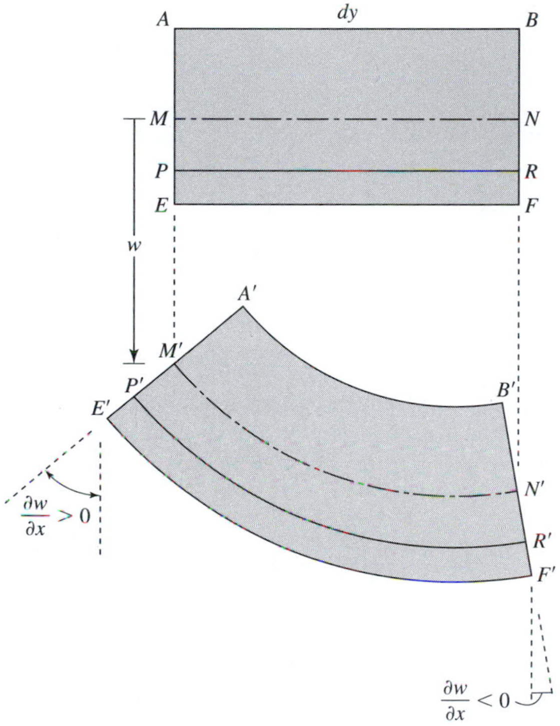


Figure 10-2 An infinitesimal element before and after deformation

from the plate in Fig. 10-1 and consider its deformation. Looking at face *ABFE* of this element (Fig. 10-2) we denote by $A'B'F'E'$ the positions of its corners after deformation. The deformed shape is consistent with the previously made assumptions of the normal to the middle surface remaining normal. This implies perpendicularity of $A'E'$ to $M'N'$, and so on. Denoting by $w = w(x, y)$ the vertical displacement (the $z$ component of the displacement) of the middle surface, we observe that any point throughout the thickness of the plate with the same $x$, $y$ coordinates will have the same displacement $w(x, y)$. Next we calculate the strains of the fibers parallel to $Oy$ caused by this displacement. If $v(x, y)$ is the displacement of point $P$ in the $y$ direction, then for small slopes we find that

$$v(x, y) = -z\frac{\partial w}{\partial y}$$

Thus using the strain-displacement relation from Chapter 3, Eq. (3-11), we get

$$\varepsilon_{yy} = \frac{\partial v}{\partial y} = -z\frac{\partial^2 w}{\partial y^2} \tag{10-1}$$

Similarly, looking at the face $BCGF$ of the same element, we first find that the horizontal displacement in the $x$ direction is

$$u(x, y) = -z\frac{\partial w}{\partial x}$$

and therefore,

$$\varepsilon_{xx} = \frac{\partial u}{\partial x} = -z\frac{\partial^2 w}{\partial x^2} \tag{10-2}$$

Now the expression for shear strain $\gamma(x, y)$ is readily obtained from Eq. (3-11),

$$\gamma_{xy} = \frac{\partial u}{\partial y} + \frac{\partial v}{\partial x} = -z\left(\frac{\partial^2 w}{\partial x\,\partial y} + \frac{\partial^2 w}{\partial y\,\partial x}\right) = -2z\frac{\partial^2 w}{\partial x\,\partial y} \tag{10-3}$$

The remaining shear strains $\gamma_{xz}$ and $\gamma_{yz}$ are considered to be negligible, which is consistent with assumption 1,

$$\gamma_{xz} \approx 0, \qquad \gamma_{yz} \approx 0 \tag{10-4}$$

We observe that, as a consequence of the first hypothesis, all stress components are linear functions of the distance from the middle surface $z = 0$. Assume now that the plate material is elastic and isotropic. By making use of Hooke's law [Eq. (4-21)] we find the stresses

$$\begin{aligned}
\sigma_{xx} &= \frac{E}{1-\nu^2}(\varepsilon_{xx} + \nu\varepsilon_{yy}) = -\frac{Ez}{1-\nu^2}\left(\frac{\partial^2 w}{\partial x^2} + \nu\frac{\partial^2 w}{\partial y^2}\right) \\
\sigma_{yy} &= \frac{E}{1-\nu^2}(\varepsilon_{yy} + \nu\varepsilon_{xx}) = -\frac{Ez}{1-\nu^2}\left(\frac{\partial^2 w}{\partial y^2} + \nu\frac{\partial^2 w}{\partial x^2}\right) \\
\sigma_{xy} &= G\gamma_{xy} = -\frac{Ez}{1+\nu}\frac{\partial^2 w}{\partial x\,\partial y}
\end{aligned} \tag{10-5}$$

The linearity of the stresses throughout the thickness results both from the linearity of normal strains and from the assumption that the material is linearly elastic. Because in our analysis shear strains [Eqs. (10-4)] are neglected, we cannot use Hooke's law to obtain shear stresses $\sigma_{xz}$ and $\sigma_{yz}$. Instead we use the first two of the equations of equilibrium for an infinitesimal element [Eqs. (2-53)], which in the absence of body forces become

$$\frac{\partial \sigma_{xx}}{\partial x} + \frac{\partial \sigma_{xy}}{\partial y} + \frac{\partial \sigma_{xz}}{\partial z} = 0, \qquad \frac{\partial \sigma_{xy}}{\partial x} + \frac{\partial \sigma_{yy}}{\partial y} + \frac{\partial \sigma_{yz}}{\partial z} = 0 \tag{10-6}$$

Substituting here expressions (10-4), integrating the resulting equations with respect to $z$, and taking into account that $\sigma_{xz}$ and $\sigma_{yz}$ are zero at both $z = h/2$ and $z = -h/2$, gives

$$\begin{aligned} \sigma_{xz} &= -\frac{E}{2(1-\nu^2)}\left(\frac{h^2}{4} - z^2\right)\frac{\partial}{\partial x}(\nabla^2 w) \\ \sigma_{yz} &= -\frac{E}{2(1-\nu^2)}\left(\frac{h^2}{4} - z^2\right)\frac{\partial}{\partial y}(\nabla^2 w) \end{aligned} \tag{10-7}$$

where

$$\nabla^2 w \equiv \frac{\partial^2 w}{\partial x^2} + \frac{\partial^2 w}{\partial y^2} \tag{10-8}$$

In analogy to simple beam theory, the plate theory makes use of resultant moments and resultant shear forces. To be another look at the plate element (Fig. 10-3). At this point we replace the moment of the normal stresses $\sigma_{xx}$ with respect to a line parallel to the $Oy$ axis by the resultant moment per unit length,

$$\begin{aligned} M_x &= \int_{-h/2}^{h/2} \sigma_{xx} z\, dz = -\frac{E}{1-\nu^2}\left(\frac{\partial^2 w}{\partial x^2} + \nu\frac{\partial^2 w}{\partial y^2}\right)\int_{-h/2}^{h/2} z^2\, dz \\ &= -\frac{Eh^3}{12(1-\nu^2)}\left(\frac{\partial^2 w}{\partial x^2} + \nu\frac{\partial^2 w}{\partial y^2}\right) = -D\left(\frac{\partial^2 w}{\partial x^2} + \nu\frac{\partial^2 w}{\partial y^2}\right) \end{aligned} \tag{10-9}$$

where

$$D = \frac{Eh^3}{12(1-\nu^2)} \tag{10-10}$$

The quantity $D$ is called the *flexural stiffness,* or *flexural rigidity,* of the elastic plate.

Proceeding in a similar manner with another face of the element, parallel to $Oyz$, we obtain the bending moment per unit length corresponding to the normal stress $\sigma_{yy}$,

$$M_y = \int_{-h/2}^{h/2} \sigma_{yy} z\, dz = -D\left(\frac{\partial^2 w}{\partial y^2} + \nu\frac{\partial^2 w}{\partial x^2}\right) \tag{10-11}$$

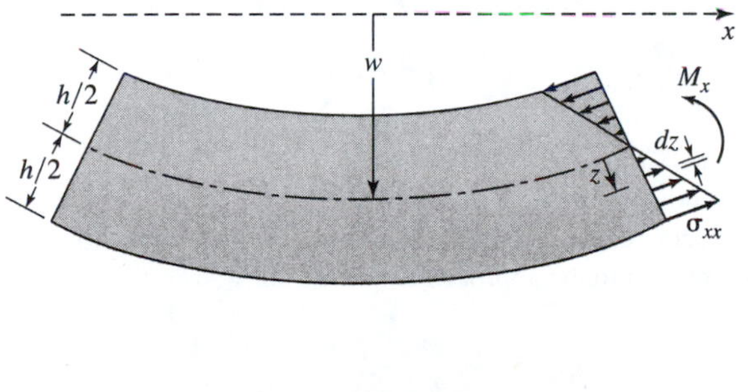


**Figure 10-3** Resultant bending moment

Finally, the resultant moment of the shear stresses $\sigma_{xy}$, known as the *twisting moment* per unit length, is defined in an analogous manner as

$$M_{xy} = \int_{-h/2}^{h/2} \sigma_{xy} z\, dz = -\frac{E}{1+\nu}\frac{\partial^2 w}{\partial x\, \partial y}\int_{-h/2}^{h/2} z^2\, dz = -D(1-\nu)\frac{\partial^2 w}{\partial x\, \partial y} \tag{10-12}$$

We observe that the symmetry of the shear stresses, $\sigma_{yx} = \sigma_{xy}$, implies that also $M_{yx} = M_{xy}$. The *shear forces* $Q_x$ and $Q_y$ are defined as the resultants of the shear stresses $\sigma_{xz}$ and $\sigma_{yz}$, respectively,

$$Q_x = \int_{-h/2}^{h/2} \sigma_{xz}\, dz = -\frac{E}{2(1-\nu^2)}\frac{\partial}{\partial x}(\nabla^2 w)\int_{-h/2}^{h/2}\left(\frac{h^2}{4} - z^2\right) dz = -D\frac{\partial}{\partial x}(\nabla^2 w)$$

$$Q_y = \int_{-h/2}^{h/2} \sigma_{yz}\, dz = -\frac{E}{2(1-\nu^2)}\frac{\partial}{\partial y}(\nabla^2 w)\int_{-h/2}^{h/2}\left(\frac{h^2}{4} - z^2\right) dz = -D\frac{\partial}{\partial y}(\nabla^2 w) \tag{10-13}$$

where we used Eqs. (10-7). A comparison of Eqs. (10-5) with Eqs. (10-9), (10-11), and (10-12) yields the following relations for stresses:

$$\sigma_{xx} = \frac{12M_x z}{h^3}, \qquad \sigma_{yy} = \frac{12M_y z}{h^3}, \qquad \sigma_{xy} = \frac{12M_{xy} z}{h^3} \tag{10-14}$$

Similarly, comparing Eqs. (10-7) with Eqs. (10-13) gives

$$\sigma_{xz} = \frac{3Q_x}{2h}\left[1 - \left(\frac{2z}{h}\right)^2\right], \qquad \sigma_{yz} = \frac{3Q_y}{2h}\left[1 - \left(\frac{2z}{h}\right)^2\right] \tag{10-15}$$

Consequently we find from Eqs. (10-14) and (10-15) that the maximum stresses are

$$\max \sigma_{xx} = \frac{6M_x}{h^2}, \qquad \max \sigma_{yy} = \frac{6M_y}{h^2}, \qquad \max \sigma_{xy} = \frac{6M_{xy}}{h^2}$$

$$\max \sigma_{xz} = \frac{3Q_x}{2h}, \qquad \max \sigma_{yz} = \frac{3Q_y}{2h}$$

Equations (10-9), (10-11), and (10-12) are used to derive the differential equation for the bending of plates in the Cartesian coordinate system. In order to obtain this equation, examine the equilibrium of the infinitesimal plate element shown in Fig. 10-4.* In addition to the resultant moments $M_x$, $M_y$, and $M_{xy}$, shear forces $Q_x$ and $Q_y$ also act on the vertical faces of this element (Fig. 10-5). First we note that using Eqs. (10-4), the equilibrium in the $x$ and $y$ directions is identically satisfied.

Next, projecting onto the $z$ axis all forces acting on the element, and including the load of intensity $q(x, y)$ per unit area perpendicular to the top surface of the plate, we obtain

$$-Q_y\, dx + \left(Q_x + \frac{\partial Q_x}{\partial x}dx\right)dy + \left(Q_y + \frac{\partial Q_y}{\partial y}dy\right)dx - Q_x\, dy + q\, dx\, dy = 0$$

*When we move from one face of the element to a parallel one, the moments change. This is taken into account by expanding the moment into a Taylor series and retaining only the portion linear with respect to the appropriate differential.

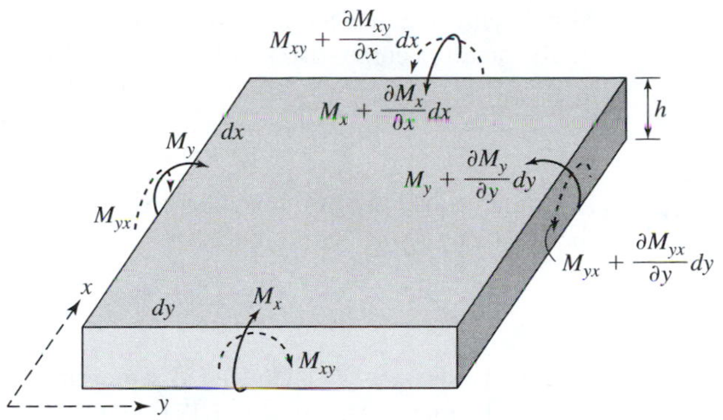


Figure 10-4 Moments acting on an infinitesimal plate element

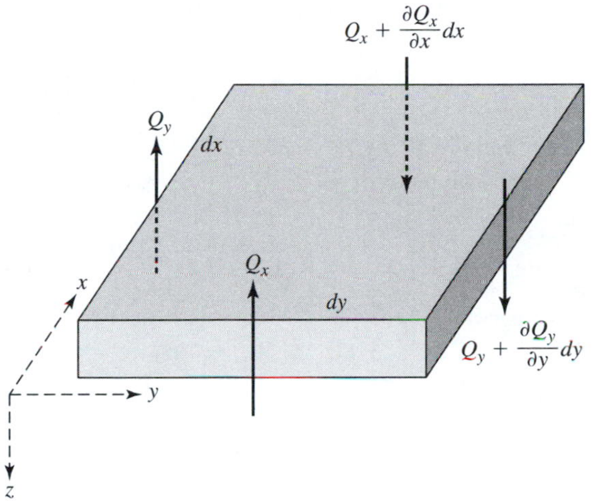


Figure 10-5 Shear forces acting on an infinitesimal plate element

Thus,

$$\frac{\partial Q_x}{\partial x} + \frac{\partial Q_y}{\partial y} + q = 0 \tag{10-16}$$

is the resulting expression, analogous to the relation $dQ_x/dx + q = 0$ known from simple beam theory. Now we calculate the sum of the moments of all forces acting on the element with respect to the $Ox$ axis. This gives

$$M_y\,dx - \left(M_{xy} + \frac{\partial M_{xy}}{\partial x}\right) dy - \left(M_y + \frac{\partial M_y}{\partial y}\,dy\right) dx + M_{xy}\,dy$$

$$+ \left(Q_x + \frac{\partial Q_x}{\partial x}\,dx\right) dy\,\frac{dy}{2} + \left(Q_y + \frac{\partial Q_y}{\partial y}\,dy\right) dx\,dy - Q_x\,dy\frac{dy}{2} + q\,dx\,dy\frac{dy}{2} = 0$$

Simplifying and neglecting all terms, including three differentials, considered infinitesimally small in comparison with the remaining terms , this becomes

$$\frac{\partial M_{xy}}{\partial x} + \frac{\partial M_y}{\partial y} - Q_y = 0 \tag{10-17}$$

A similar equation is obtained when taking the sum of the moments with respect to the $Oy$ axis,

$$\frac{\partial M_{xy}}{\partial y} + \frac{\partial M_x}{\partial x} - Q_x = 0 \tag{10-18}$$

Note that Eqs. (10-17) and (10-18) represent an integrated version of Eqs. (10-6). Replacing the moments in Eqs. (10-17) and (10-18) by the expressions (10-9), (10-11), and (10-12), we obtain again the relations between the shear forces $Q_x$, $Q_y$ and the displacement $w$, Eqs. (10-13).

Now, replacing $Q_x$ and $Q_y$ in Eqs. (10-16) by the expressions obtained from Eqs. (10-17) and (10-18) in the result, we arrive at

$$\frac{\partial^2 M_x}{\partial x^2} + 2\frac{\partial^2 M_{xy}}{\partial x\,\partial y} + \frac{\partial^2 M_y}{\partial y^2} = -q \tag{10-19}$$

Finally, using expressions (10-9), (10-11), and (10-12) gives the differential equation for the deflection,

$$-D\left(\frac{\partial^4 w}{\partial x^4} + \nu\frac{\partial^4 w}{\partial x^2\,\partial y^2}\right) - 2D(1-\nu)\frac{\partial^4 w}{\partial x^2\,\partial y^2} - D\left(\frac{\partial^4 w}{\partial y^4} + \nu\frac{\partial^4 w}{\partial x^2\,\partial y^2}\right) = -q$$

or

$$D\left(\frac{\partial^4 w}{\partial x^4} + 2\frac{\partial^4 w}{\partial x^2\,\partial y^2} + \frac{\partial^4 w}{\partial y^4}\right) = q \tag{10-20}$$

Using Eq. (10-8) this can be written as

$$D\nabla^2\nabla^2 w = q \tag{10-21}$$

We note that the expression (10-8) represents the sum of the curvatures of the surface $w(x, y)$ in the two mutually perpendicular directions $x$ and $y$, and that the quantity $\nabla^2 w(x, y)$ is also a surface. The sum of its curvatures,

$$\frac{\partial^2}{\partial x^2}(\nabla^2 w) + \frac{\partial^2}{\partial y^2}(\nabla^2 w) \tag{10-22}$$

is proportional to the load $q$, as is evident from Eq. (10-20). Since $q = q(x, y)$ depends only on the coordinates of the point, but not on the coordinate system used to represent it, it follows that the sum of the curvatures of $\nabla^2 w$ in two mutually perpendicular but otherwise arbitrary directions is invariant. This observation will help in establishing the equilibrium equation of plates in the polar coordinate system. The solution to Eq. (10-20) will be derived later for a simply supported rectangular plate. At this point we can analyze a simplified problem of a plate subject to *cylindrical bending*. This occurs when the load depends only on one variable, say $x$, while extending to infinity in the $y$ direction. Consequently the plate is forced to bend only in the $x$ direction (Fig. 10-6).

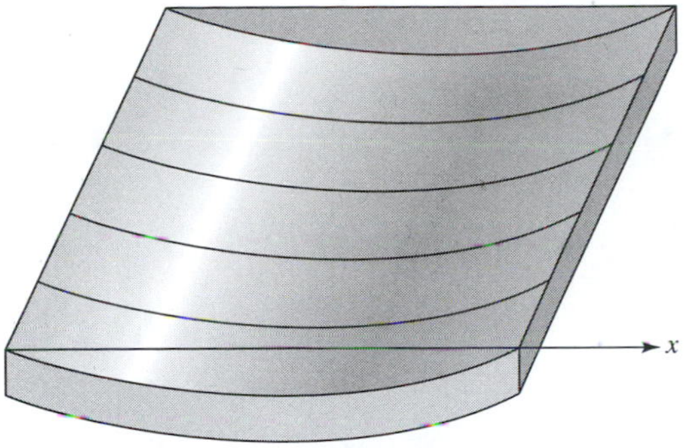


Figure 10-6

Since now $w = w(x)$, Eq. (10-20) becomes

$$D\frac{\partial^4 w}{\partial x^4} = q \tag{10-23}$$

This equation is similar to the equation for the deflection of a beam, except for the difference between the stiffness $D$ of the plate and the stiffness $EI$ of the beam,

$$D = \frac{Eh^3}{12(1-\nu^2)}, \qquad EI = \frac{Ebh^3}{12}$$

This difference is the result of using stresses and their resultants per unit length, that is, $b = 1$, when deriving Eq. (10-23). The term $1 - \nu^2$ appearing in $D$ reflects the actions of neighboring elements on the element under consideration.

## 10-2 AXISYMMETRIC BENDING OF CIRCULAR PLATES

### 10-2-1 General Expressions

In this section we analyze solid and annular circular plates with axisymmetric boundary conditions and subject to an axisymmetric load. It is convenient to use a cylindrical polar coordinate system (Fig. 10-7). Because of the symmetry of the shape and the load, all statical quantities are also symmetric. This signifies that neither the deflection $w$ nor its derivatives or their combinations, such as the bending moment or the shear force, depend on the angle $\theta$,

$$w = w(r) \tag{10-24}$$

Note that when the deformation of a circular plate is not axisymmetric, then the following procedure is no longer valid and a more general approach is needed [12, pp. 282–293]. We will now derive the differential equation for the deflection of the plate in a polar coordinate system. This can be accomplished either directly by analyzing the equilibrium of an infinitesimal plate element in polar coordinates, or indirectly by employing the results from Section 10-1. In the latter case, as noted in the paragraph following Eq. (10-21), $\nabla^2 w$ represents the sum of the curvatures of the plate in two

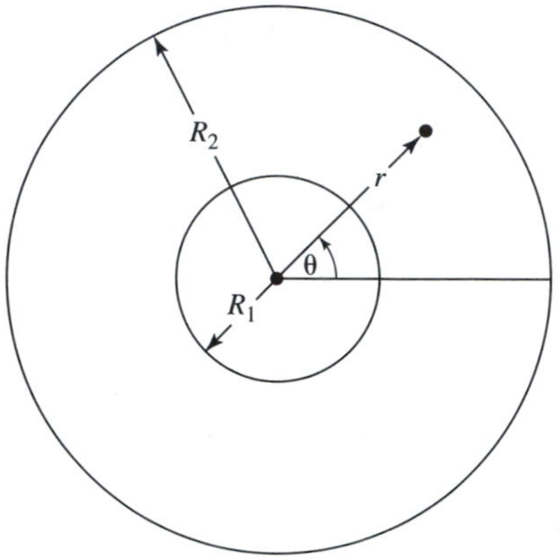


Figure 10-7

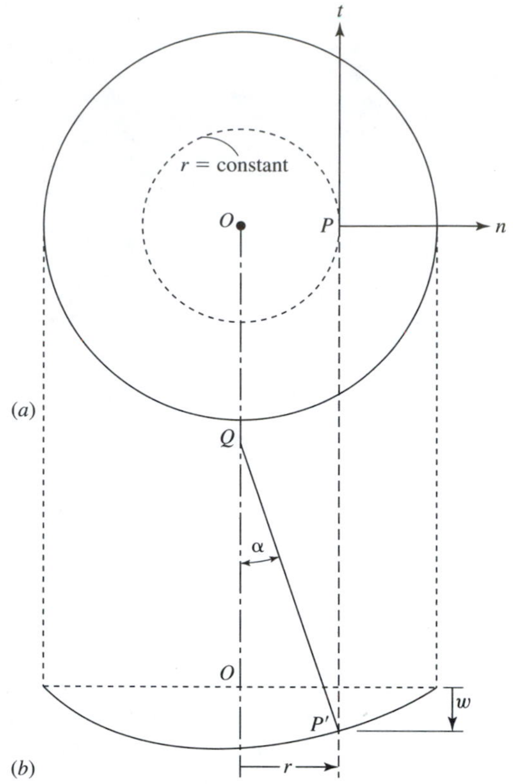


Figure 10-8 Axisymmetric deformation of a circular plate

mutually perpendicular directions, namely, the radial direction $n$ and the direction $t$ tangent to the circle $r$ = constant (Fig. 10-8*a*). Figure 10-8*b* shows a side view of the deformed plate. Since $w = w(r)$, the curvature $K_n$ in the direction $n$ is given by

$$K_n = \frac{d^2w}{dr^2} \equiv w'' \tag{10-25}$$

In order to calculate the curvature of the surface $w(r)$ in the $t$ direction, we first draw a normal to the deformed plate through the point $P$ intersecting the axis of symmetry at $Q$. The distance between $P$ and $Q$ represents the radius of curvature $R_t$ in the $t$ direction. The angle $\angle OQP = \alpha$ equals the angle between the tangent to $w(r)$ at $P$ and the horizontal line,

$$\alpha \approx \tan\alpha = \frac{dw}{dr} = w'$$

Therefore,

$$R_t = QP = \frac{r}{\sin\alpha} \approx \frac{r}{\alpha} = \frac{r}{w'}$$

so that the curvature $K$ is

$$K_t = \frac{1}{R_t} = \frac{w'}{r} \tag{10-26}$$

Consequently the sum of the curvatures in two mutually perpendicular directions equals

$$K_n + K_t = w'' + \frac{w'}{r} \tag{10-27}$$

This expression represents a surface $f(r)$. It is evident from Eq. (10-22) that the sum of its curvatures in the directions $n$ and $t$ is proportional to $q$,

$$D\left(\frac{d^2 f}{dr^2} + \frac{1}{r}\frac{df}{dr}\right) = q(r)$$

or

$$D\left(\frac{d^2}{dr^2} + \frac{1}{r}\frac{d}{dr}\right)\left(\frac{d^2 w}{dr^2} + \frac{1}{r}\frac{dw}{dr}\right) = q(r) \tag{10-28}$$

This is the differential equation for the axisymmetric bending of circular plates. We can obtain the bending moments by observing that

$M_x = -D \times$ (curvature in $x$ direction $+ \nu \times$ curvature in direction perpendicular to it, i.e., $y$ direction)

$M_y = -D \times$ (curvature in $y$ direction $+ \nu \times$ curvature in direction perpendicular to it, i.e., $x$ direction)

Thus, by analogy,

$M_r = -D \times$ (curvature in $r$ direction $+ \nu \times$ curvature in $\theta$ direction)

$M_\theta = -D \times$ (curvature in $\theta$ direction $+ \nu \times$ curvature in $r$ direction)

Also,

$$Q_x = D\frac{\partial}{\partial x}(\nabla^2 w)$$

implies that

$$Q_r = D\frac{\partial}{\partial r}(\nabla^2 w) \tag{10-29}$$

The same equation can be obtained, as indicated previously, by using a direct method. To this end we examine the equilibrium of a wedge-like infinitesimal element in a polar coordinate system (Fig. 10-9). We found in Chapter 3, Eq. (3-82), that for axially symmetric problems,

$$\varepsilon_{rr} = \frac{du}{dr}, \qquad \varepsilon_{\theta\theta} = \frac{u}{r} \tag{10-30}$$

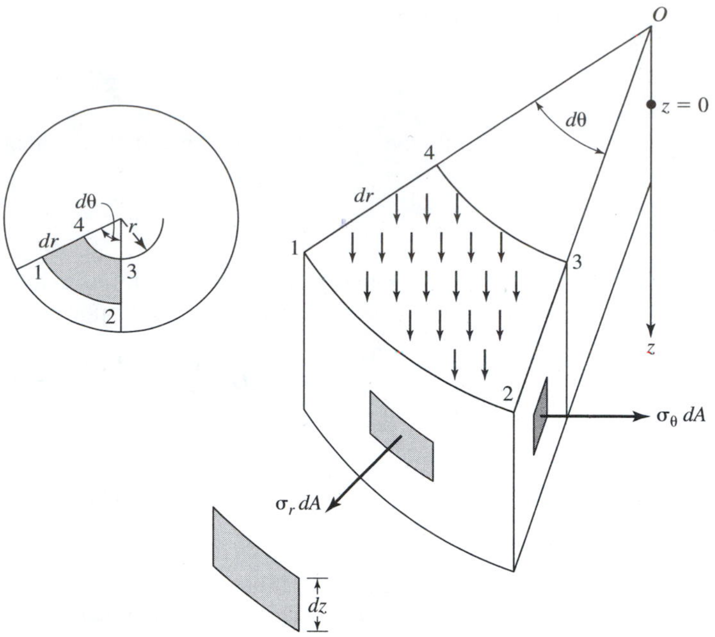


Figure 10-9

where $u$ is the displacement component in the $r$ direction. Recalling the procedure used for the Cartesian components of strain, we find

$$\varepsilon_{rr} = -z\frac{d^2w}{dr^2}, \qquad \varepsilon_{\theta\theta} = -\frac{z}{r}\frac{dw}{dr} \tag{10-31}$$

where $w$ is the displacement component in the $z$ direction. Thus the radial and tangential stresses in an elastic plate are given by

$$\begin{aligned} \sigma_{rr} &= \frac{E}{1-\nu^2}(\varepsilon_{rr} + \nu\varepsilon_{\theta\theta}) = -\frac{Ez}{1-\nu^2}\left(w'' + \frac{\nu}{r}w'\right) \\ \sigma_{\theta\theta} &= \frac{E}{1-\nu^2}(\varepsilon_{\theta\theta} + \nu\varepsilon_{rr}) = -\frac{Ez}{1-\nu^2}\left(\frac{1}{r}w' + \nu w''\right) \end{aligned} \tag{10-32}$$

while $\sigma_{r\theta} = 0$ because of the assumed axial symmetry. Next we define the bending moments in the $r$ and $\theta$ directions,

$$\begin{aligned} M_r &= \int_{-h/2}^{h/2} \sigma_{rr} z\, dz = -D\left(w'' + \frac{\nu}{r}w'\right) \\ M_\theta &= \int_{-h/2}^{h/2} \sigma_{\theta\theta} z\, dz = -D\left(\frac{1}{r}w' + \nu w''\right) \end{aligned} \tag{10-33}$$

The torque $M_{r\theta} = 0$ as the result of $\sigma_{r\theta} = 0$.

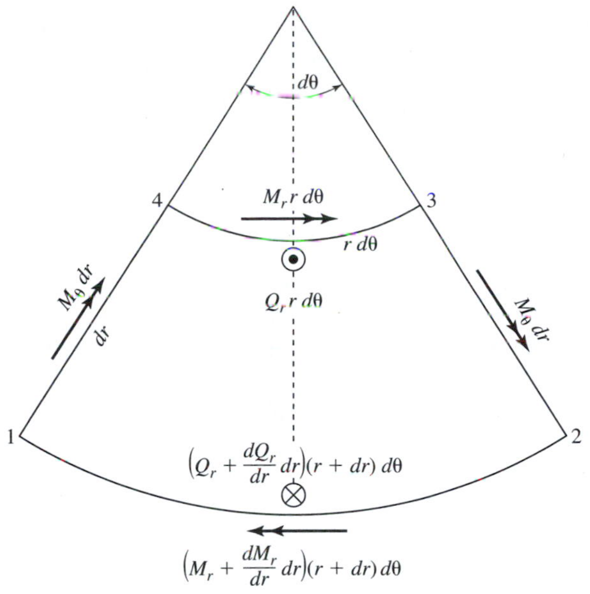


Figure 10-10

Comparing these with Eqs. (10-32) we find the relation between the stresses and the bending moments,

$$\sigma_{rr} = \frac{12M_r}{h^3}z, \qquad \sigma_{\theta\theta} = \frac{12M_\theta}{h^3}z \tag{10-34}$$

It follows that at any given point, the maximum stress occurs at $z = h/2$, where it becomes

$$\sigma_{rr} = \frac{6M_r}{h^2}, \qquad \sigma_{\theta\theta} = \frac{6M_\theta}{h^2} \tag{10-35}$$

Now let us consider the equilibrium of the element shown in Fig. 10-10,* where the stresses have been replaced by bending moments, and where the resultants of vertical shear stresses have been added in the form of the shear forces $Q_r$. Using the fact that the bending moments $M_\theta$ do not change with $\theta$, and taking the sum of the moments with respect to the tangent to line 1-2 attached at its center, we obtain

$$M_r r\, d\theta - \left(M_r + \frac{dM_r}{dr}\, dr\right)(r + dr)\, d\theta + M_\theta\, dr \sin\frac{d\theta}{2}$$
$$+ M_\theta\, dr \sin\frac{d\theta}{2} + (Q_r r\, d\theta)\, dr + qr\, dr\, d\theta\, dr = 0$$

*The symbols ⊙ and ⊗ appearing in this and other figures indicate a force pointed toward the reader and toward the paper, respectively.

Simplifying this expression, deleting higher order terms, and using the relationship

$$\sin\frac{d\theta}{2} \approx \frac{d\theta}{2}$$

we get

$$\frac{M_r - M_\theta}{r} + \frac{dM_r}{dr} - Q_r = 0 \tag{10-36}$$

Using Eq. (10-33) this becomes

$$-D\left(\frac{1}{r}w'' + \frac{\nu}{r^2}w' - \frac{1}{r^2}w' - \frac{\nu}{r}w''\right) - D\left(w''' + \frac{\nu}{r}w'' - \frac{\nu}{r^2}w'\right) = Q_r$$

Thus, summing up similar terms we obtain $Q_r$ as a function of $w$,

$$Q_r = -D\left(w''' + \frac{1}{r}w'' - \frac{1}{r^2}w'\right) \tag{10-37}$$

which can also be written as

$$Q_r = -D\frac{d}{dr}\left[\frac{1}{r}\frac{d}{dr}(rw')\right] \tag{10-38}$$

Expanding expression (10-38), Eq. (10-37) is readily obtained. The second equation of equilibrium is found by projecting on the $z$ axis all forces acting on this element,

$$-Q_r r\,d\theta + \left(Q_r + \frac{dQ_r}{dr}dr\right)(r + dr)\,d\theta + qr\,dr\,d\theta = 0$$

or

$$Q_r + r\frac{dQ_r}{dr} = -qr \tag{10-39}$$

This can also be put in the form

$$\frac{d}{dr}(rQ_r) = -qr \tag{10-40}$$

Integrating we obtain

$$rQ_r = -\int_R^r qr\,dr \tag{10-41}$$

where $R$ is the radius of the inside hole. Next Eq. (10-36) is multiplied by $r$, Eq. (10-41) is substituted, and the resulting equation is differentiated to give

$$rM_r'' + 2M_r' - M_\theta' + qr = 0$$

Finally expressions (10-33) are substituted for $M_r$ and $M_\theta$, and in this manner we derive the differential equation

$$D(r^4 w^{\text{iv}} + 2r^3 w''' - r^2 w'' + rw') = r^4 q(r) \tag{10-42}$$

which can be put into the form

$$D\frac{1}{r}\frac{d}{dr}\left\{r\frac{d}{dr}\left[\frac{1}{r}\frac{d}{dr}\left(r\frac{dw}{dr}\right)\right]\right\} = q(r) \tag{10-43}$$

or into the form of Eq. (10-28). Equation (10-42) has the same structure as Eq. (9-23), which was identified as Euler's differential equation. Consequently we seek the solution to the reduced equation in the form

$$w_h(r) = r^n \tag{10-44}$$

This leads to the indicial equation

$$n(n-1)(n-2)(n-3) + 2n(n-1)(n-2) - n(n-1) + n = 0$$

which can be expressed as

$$n^2(n-2)^2 = 0 \tag{10-45}$$

Thus we obtain two double roots,

$$n_1 = n_2 = 0, \qquad n_3 = n_4 = 2 \tag{10-46}$$

According to the theory of differential equations [5, p. 142], the presence of double roots introduces logarithmic solutions so that the solution to the reduced equation becomes finally

$$w_h(r) = C_1 + C_2 \ln\frac{r}{R_2} + C_3 r^2 + C_4 r^2 \ln\frac{r}{R_2} \tag{10-47}$$

Differentiating Eq. (10-47) we find the slope,

$$w_h'(r) = \frac{C_2}{r} + 2C_3 r + C_4 r\left(1 + 2\ln\frac{r}{R_2}\right) \tag{10-48}$$

the bending moments [see Eq. (10-33)],

$$M_{rh}(r) = -D\left\{-\frac{1-\nu}{r^2}C_2 + 2(1+\nu)C_3 + \left[2(1+\nu)\ln\frac{r}{R_2} + 3 + \nu\right]C_4\right\} \tag{10-49}$$

$$M_{\theta h}(r) = -D\left\{\frac{1-\nu}{r^2}C_2 + 2(1+\nu)C_3 + \left[2(1+\nu)\ln\frac{r}{R_2} + 1 + 3\nu\right]C_4\right\} \tag{10-50}$$

and the shear force [see Eq. (10-36)],

$$\begin{aligned} Q_{rh}(r) &= \frac{M_{rh} - M_{\theta h}}{r} + M_{rp}' \\ &= -4D\frac{C_4}{r} \end{aligned} \tag{10-51}$$

The quantities (10-47)–(10-51) do not depend on the load applied directly to the plate because we have solved only the reduced equation. In order to obtain the actual deflection $w(r)$ and other static quantities, we have to add the particular solution $w_q(r)$ of

Eq. (10-42) corresponding to a given load distribution $q$. Thus,

$$\begin{aligned} w(r) &= w_h(r) + w_q(r), & w'(r) &= w_h'(r) + w_q'(r) \\ M_r(r) &= M_{rh}(r) + M_{rq}(r), & M_\theta(r) &= M_{\theta h}(r) + M_{\theta q}(r) \\ Q_r(r) &= Q_{rh}(r) + Q_{rq}(r) \end{aligned} \tag{10-52}$$

where the relation between $M_{rq}(r)$ and $w_q(r)$ has the same form as the relation between $M_{rh}(r)$ and $w_h(r)$, and so on. Some examples of evaluating $w_q(r)$ are given in the next section. Note that when the loads are applied only at the boundary, then $w_q(r)$ and all of its derivatives equal zero. When the particular solutions are known, they are substituted into Eqs. (10-52), which still depend on four unknown constants. These must be determined using boundary conditions, as will be shown in later sections.

## 10-2-2 Particular Solutions for Selected Types of Axisymmetric Loads

### Case 1

Consider a uniform load distribution $q(r) = q_0 =$ constant. In this case the particular solution to Eq. (10-42) is assumed in the polynomial form

$$w_q(r) = Ar^4 \tag{10-53}$$

To find the unknown constant $A$, $q(r) = q_0$ is substituted together with Eq. (10-53) into Eq. (10-42), yielding

$$A = \frac{q_0}{64D}$$

Therefore,

$$\begin{gathered} w_q(r) = \frac{q_0 r^4}{64D}, \qquad w_q'(r) = \frac{q_0 r^3}{16D} \\ M_{rq}(r) = -\frac{(3+\nu)q_0 r^2}{16}, \qquad M_{\theta q}(r) = -\frac{(1+3\nu)q_0 r^2}{16}, \qquad Q_{rq}(r) = -\frac{q_0 r}{2} \end{gathered} \tag{10-54}$$

### Case 2

When the load distribution is a linear function of $r$ [10, p. 513] (see Fig. 10-11), then

$$q(r) = q_0 \frac{r}{R} \tag{10-55}$$

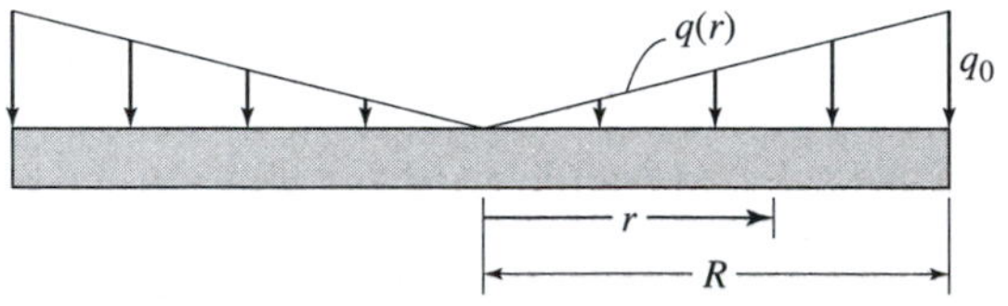


Figure 10-11

Now the right-hand side of the equation for deflection [Eq. (10-42)] is in the form of a monomial of fifth order. Therefore we seek the particular solution in the form

$$w_q(r) = Ar^5 \tag{10-56}$$

With Eqs. (10-55) and (10-56) substituted into Eq. (10-42) this becomes

$$D(r^4 \cdot 5 \cdot 4 \cdot 3 \cdot 2Ar + 2r^3 \cdot 5 \cdot 4 \cdot 3Ar^2 - r^2 \cdot 5 \cdot 4Ar^3 + r \cdot 5Ar^4) = r^4 q_0 \frac{r}{R}$$

so that

$$A = \frac{q_0}{225DR} \tag{10-57}$$

Thus (see also [10, p. 513]),

$$w_q(r) = \frac{q_0 r^4}{225D}\frac{r}{R}, \qquad w_q'(r) = \frac{5q_0 r^3}{225D}\frac{r}{R}$$

$$M_{rq}(r) = -\frac{5(4+\nu)q_0 r^2}{225}\frac{r}{R}, \quad M_{\theta q}(r) = -\frac{5(1+4\nu)q_0 r^2}{225}\frac{r}{R}, \quad Q_{rq}(r) = -\frac{q_0 r}{3}\frac{r}{R} \tag{10-58}$$

### Case 3

When the load distribution is a general linear function of $r$ (Fig. 10-12), then

$$q(r) = q_c + \frac{q_2 - q_c}{R_2} r \tag{10-59}$$

In order to obtain the particular solution $w_q(r)$ and its derivatives, the results of cases 1 and 2 are superimposed. Therefore [see Eqs. (10-54) and (10-58)],

$$\begin{aligned}
w_q(r) &= \frac{q_c r^4}{64D} + \frac{(q_2 - q_c)r^4}{225D}\frac{r}{R_2} \\
w_q'(r) &= \frac{q_c r^3}{16D} + \frac{5(q_2 - q_c)r^3}{225D}\frac{r}{R_2} \\
M_{rq}(r) &= -\frac{(3+\nu)q_c r^2}{16} - \frac{5(4+\nu)(q_2 - q_c)r^2}{225}\frac{r}{R_2} \\
M_{\theta q}(r) &= -\frac{(1+3\nu)q_c r^2}{16} - \frac{5(1+4\nu)(q_2 - q_c)r^2}{225}\frac{r}{R_2} \\
Q_{rq}(r) &= -\frac{q_c r}{2} - \frac{(q_2 - q_c)r}{3}\frac{r}{R_2}
\end{aligned} \tag{10-60}$$

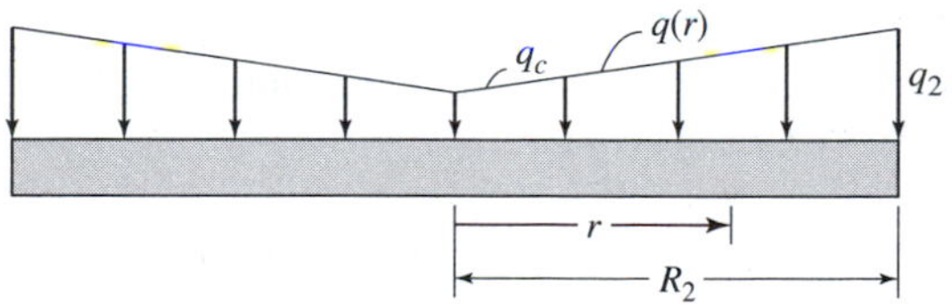


Figure 10-12

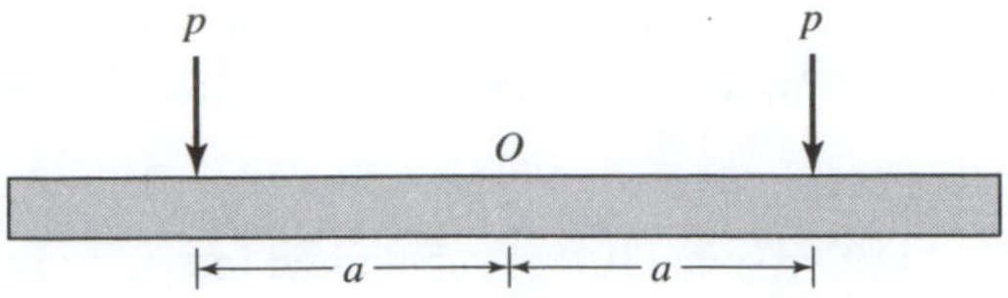


Figure 10-13

Now assume that we know that $q(r) = q_1$ at $r = R_1$ and $q(r) = q_2$ at $r = R_2$, as is the case for the annular plates (see Section 10-2-6). We then find from geometry that

$$q_c = \frac{q_1 R_2 - q_2 R_1}{R_2 - R_1} = q_2 \frac{q_1 R_2/q_2 - R_1}{R_2 - R_1} \tag{10-61}$$

This is the quantity to be used in expressions (10-60).

### Case 4

Still another type of load is shown in Fig. 10-13. It represents a circularly distributed line load of intensity $p$ per unit of arc over a circle of radius $a$. The importance of this case lies in that the resulting deflection can be used as an influence function (Green's function) to generate expressions for arbitrarily distributed continuous loads (see also Chapter 8). This problem can be solved in various ways [12, p. 63]. To find the particular solution $w_q(r)$ we consider Eq. (10-38),

$$\frac{d}{dr}\left[\frac{1}{r}\frac{d}{dr}(rw_q')\right] = -\frac{Q_r}{D}$$

Designating by $P$ the total force applied to the plate, we obtain the load intensity $p$ from the relation

$$p = \frac{P}{2\pi a} \tag{10-62}$$

Considering the overall equilibrium of a disk of radius $r$ cut out from the plate, we find

$$\begin{aligned} Q_r \times 2\pi r &= 0, \qquad & r < a \\ Q_r \times 2\pi r &= -p \times 2\pi a, \qquad & r > a \end{aligned} \tag{10-63}$$

Expressions (10-63) can be written as

$$2\pi r Q_r = -2\pi a p H(r - a)$$

or

$$Q_r = -\frac{paH(r-a)}{r} \tag{10-64}$$

where $H$ is a Heaviside (or step) function (see Chapter 8). Substituting Eq. (10-64) into Eq. (10-38), this becomes

$$\frac{d}{dr}\left[\frac{1}{r}\frac{d}{dr}(rw_q')\right] = \frac{paH(r-a)}{Dr}$$

Consecutive integrations with respect to $r$, where $0 \le r$, give the following results:*

$$\frac{d}{dr}(r w_q') = \frac{par}{D} \ln \frac{r}{a} H(r-a)$$

and

$$w_q' = \frac{pa}{4Dr}\left[\left(2\ln\frac{r}{a} - 1\right)r^2 + a^2\right]H(r-a) \tag{10-65}$$

$$w_q = \frac{pa}{4D}\left[(r^2+a^2)\ln\frac{r}{a} - (r^2-a^2)\right]H(r-a) \tag{10-66}$$

Therefore the expressions for the bending moments and the shear force become

$$M_{rq} = -\frac{pa}{4}\left[2(1+\nu)\ln\frac{r}{a} - (1-\nu)\frac{a^2}{r^2} + 1 - \nu\right]H(r-a) \tag{10-67}$$

$$M_{\theta q} = -\frac{pa}{4}\left[2(1+\nu)\ln\frac{r}{a} + (1-\nu)\frac{a^2}{r^2} - (1-\nu)\right]H(r-a) \tag{10-68}$$

$$Q_{rq} = -\frac{pa}{r}H(r-a) \tag{10-69}$$

## 10-2-3 Solid Plate: Boundary Conditions and Examples

We now have all the building blocks necessary to determine the deflections and the stresses in a symmetrically loaded and symmetrically supported circular plate. In order to determine four constants of integration, $C_1$, $C_2$, $C_3$, and $C_4$, we need four additional equations, that is, the *boundary conditions*. There are three basic cases of boundary supports:

1. Simply supported edge, $r = R$
2. Clamped edge, $r = R$
3. Free edge, $r = R$

These are shown symbolically in Fig. 10-14.† At each of the supports there are two known quantities:

| | | | |
|---|---|---|---|
| *Case 1* | $w(R) = W,$ | $M_r(R) = M$ | (10-70*a*) |
| *Case 2* | $w(R) = W,$ | $w'(R) = \Theta$ | (10-70*b*) |
| *Case 3* | $M_r(R) = M,$ | $Q_r(R) = Q$ | (10-70*c*) |

Here the quantities appearing on the right-hand sides are known constants, which may or may not be equal to zero. For example, $W \neq 0$ and $\Theta \neq 0$ in case 2 indicates

*Vanishing of the integrals at $r = a$ has been taken into account.

†In cases 1 and 2 the edges of the plate are free to move as necessary to prevent straining of the middle surface.

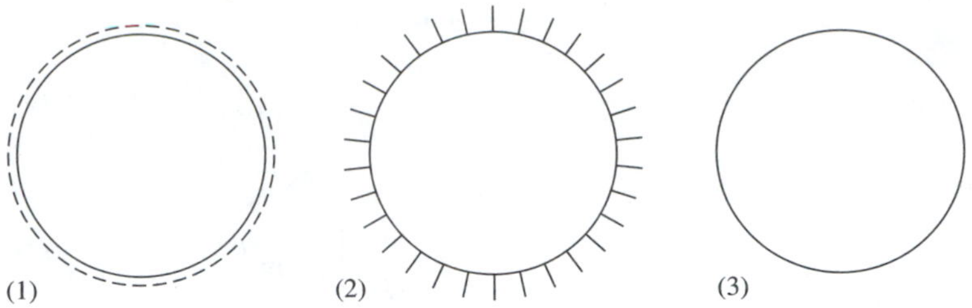


Figure 10-14

that the clamped edge is subject to a displacement $W$ and a rotation $\Theta$. $W$ alone will not generate stresses in the plate for the case of axial symmetry, but $\Theta$ will. Consequently even if no load is applied to the surface of the plate in case 2, the plate will be stressed when $\Theta \neq 0$. Similar reasoning applies to cases 1 and 3.

Now the question arises: how can we determine four constants for a solid plate that has only one edge and, hence, only two boundary conditions? The answer is found by inspection of expressions (10-47)–(10-50). We see that even when no load is applied, all those quantities become infinite at $r = 0$. But this is physically impossible, and the only way to remove this contradiction is by requiring that

$$C_2 = 0, \qquad C_4 = 0 \tag{10-71}$$

Using Eqs. (10-47)–(10-52), we now obtain the following expressions for the deflection, the slope, the bending moments, and the shear force in a solid plate:

$$\begin{aligned} &w(r) = C_1 + C_3 r^2 + w_q(r), && w'(r) = 2C_3 r + w'_q \\ &M_r(r) = -2D(1+\nu)C_3 + M_{rq}(r), && M_\theta(r) = -2D(1+\nu)C_3 + M_{\theta q}(r) \\ &Q_r(r) = Q_{qr}(r) \end{aligned} \tag{10-72}$$

## EXAMPLE 10-1

Find the deflection, the bending moments, and the maximum stresses for a simply supported circular plate of radius $R$ carrying a uniform load of intensity $q_0 =$ constant.

### SOLUTION

Using Eqs. (10-72) and (10-54) we have

$$w(r) = C_1 + C_3 r^2 + \frac{q_0 r^4}{64D} \tag{10-73}$$

Now the boundary conditions (10-70$a$) take the form

$$w(R) = 0, \qquad M_r(R) = 0 \tag{10-74}$$

Thus if use is made of the third of Eqs. (10-72), Eq. (10-73), and the third of Eqs. (10-54), a system of algebraic equations follows:

$$C_1 + C_3 R^2 + \frac{q_0 R^4}{64D} = 0, \qquad -2D(1+\nu)C_3 - \frac{(3+\nu)q_0 R^2}{16} = 0$$

Solving this system we obtain the constants of integration,

$$C_3 = -\frac{3+\nu}{1+\nu}\frac{q_0 R^2}{32D}, \qquad C_1 = \frac{5+\nu}{1+\nu}\frac{q_0 R^4}{64D} \tag{10-75}$$

and then, substituting into Eq. (10-73),

$$w(r) = \frac{q_0 R^4}{64(1+\nu)D}\left[(1+\nu)\left(\frac{r}{R}\right)^4 - 2(3+\nu)\left(\frac{r}{R}\right)^2 + 5 + \nu\right] \tag{10-76}$$

Maximum deflection occurs evidently at the center of the plate,

$$w_{\max} = \frac{(5+\nu)q_0 R^4}{64(1+\nu)D} \tag{10-77}$$

The bending moments are obtained by substituting $C_3$ from Eq. (10-75) into the third and fourth of Eqs. (10-72) and then replacing $M_{rq}$ and $M_{\theta q}$ by their expressions (10-54). Thus the following formulas are deduced:

$$\begin{aligned} M_r(r) &= \frac{(3+\nu)q_0 R^2}{16}\left[1 - \left(\frac{r}{R}\right)^2\right] \\ M_\theta(r) &= \frac{q_0 R^2}{16}\left[3 + \nu - (1+3\nu)\left(\frac{r}{R}\right)^2\right] \end{aligned} \tag{10-78}$$

Maximum bending moments are found at the center of the plate, where

$$\max M_r = \max M_\theta = \frac{(3+\nu)q_0 R^2}{16} \tag{10-79}$$

From Eqs. (10-35) we obtain the corresponding maximum stress,

$$(\sigma_{rr})_{\max} = (\sigma_{\theta\theta})_{\max} = \frac{6M_r}{h^2} = \frac{3(3+\nu)q_0 R^2}{8h^2} \tag{10-80}$$

## EXAMPLE 10-2

We are using the data from Example 10-1, but for a clamped* plate.

*It is assumed that the plate edges can move horizontally (thus the strains in the middle surface are avoided).

**SOLUTION**

Here again

$$w(r) = C_1 + C_3 r^2 + \frac{q_0 r^4}{64D} \tag{10-81}$$

but now the applicable boundary conditions are (10-70*b*), that is,

$$w(R) = 0, \qquad w'(R) = 0 \tag{10-82}$$

so that using Eqs. (10-72*b*), (10-81), and (10-54) we derive the following system of algebraic equations for the unknown constants:

$$C_1 + C_3 R^2 + \frac{q_0 R^4}{64D} = 0, \qquad 2C_3 R + \frac{q_0 R^3}{16D} = 0$$

Its solution has the form

$$C_1 = \frac{q_0 R^4}{64D}, \qquad C_3 = -\frac{q_0 R^2}{32D} \tag{10-83}$$

Substituting this into Eq. (10-81) we obtain*

$$w(r) = \frac{q_0 R^4}{64D}\left[1 - \left(\frac{r}{R}\right)^2\right] \tag{10-84}$$

with maximum deflection at the center,

$$w_{\max} = \frac{q_0 R^4}{64D} \tag{10-85}$$

about four times smaller than the maximum deflection for a simply supported plate. The bending moments are determined using expressions (10-83), (10-72), and (10-54), from which we find

$$\begin{aligned} M_r(r) &= \frac{q_0 R^2}{16}\left[1 + \nu - (3 + \nu)\left(\frac{r}{R}\right)^2\right] \\ M_\theta(r) &= \frac{q_0 R^2}{16}\left[1 + \nu - (1 + 3\nu)\left(\frac{r}{R}\right)^2\right] \end{aligned} \tag{10-86}$$

Inspection of the bracketed expressions in Eqs. (10-86) shows that neither of the moments has a local extremum. However, when evaluating the boundary values of the moments we find that the bending moments at the edge $r = R$ are

$$M_r = -\frac{q_0 R^2}{8}, \qquad M_\theta = -\frac{\nu q_0 R^2}{8} \tag{10-87}$$

*It is interesting to note that the deflection is independent of Poisson's ratio.

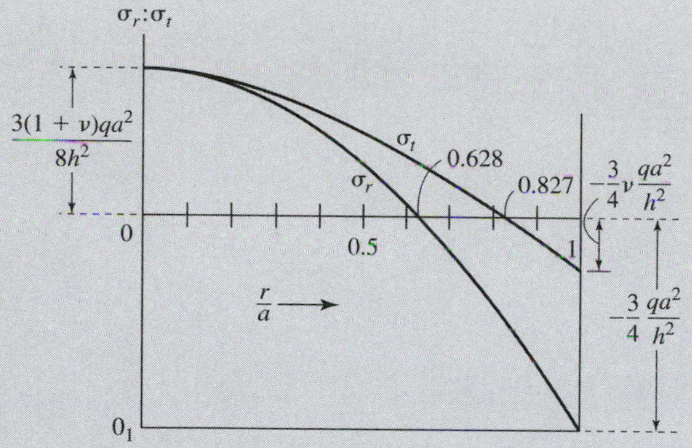


**Figure 10-15***

while their value at the center, $r = 0$, is

$$M_r = M_\theta = \frac{(1+\nu)q_0R^2}{16} \tag{10-88}$$

Because $1 + \nu < 2$, the maximum bending moments appear at the boundary of the plate, and thus the maximum stress is

$$(\sigma_{rr})_{\max} = -\frac{6M_r}{h^2} = \frac{3q_0R^2}{4h^2} \tag{10-89}$$

The variation of $\sigma_{rr}$ and $\sigma_{\theta\theta}$ along $r$ is shown in Fig. 10-15 (see [12, p. 56, fig. 29]).

## EXAMPLE 10-3

We want to show that the boundary deformation alone can generate stresses in a plate.

### SOLUTION

Consider a plate with a built-in edge that is subject to a small rotation by an angle $\alpha \ll 1$ (Fig. 10-16). The boundary conditions are now

$$w(R) = 0, \qquad w'(R) = -\alpha \tag{10-90}$$

Since no load is directly applied to the plate, both $w_q(r)$ and $w_q'(r)$ are zero. By using Eqs. (10-72) we obtain, from Eqs. (10-90),

$$C_1 + C_3R^2 = 0, \qquad 2C_3R + \alpha = 0$$

*Reprinted by permission of McGraw-Hill, New York.

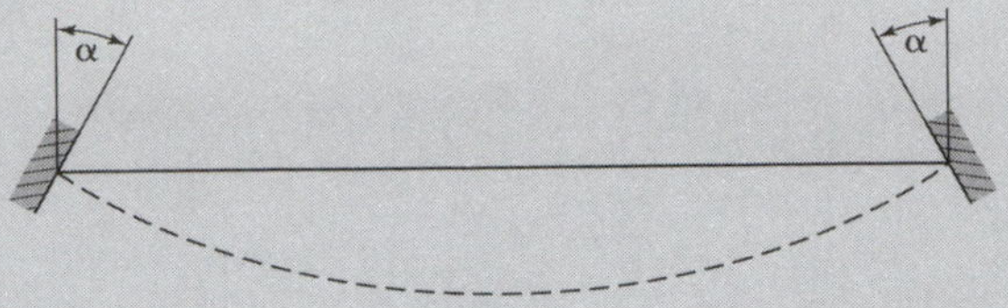


**Figure 10-16** A circular plate with built-in edge subject to small rotation by an angle $\alpha \ll 1$

from which it follows that

$$C_1 = \frac{\alpha R}{2}, \qquad C_3 = -\frac{\alpha}{2R} \tag{10-91}$$

Therefore the displacement obtained from Eqs. (10-72) becomes

$$w(r) = \frac{\alpha R}{2}\left[1 - \left(\frac{r}{R}\right)^2\right] \tag{10-92}$$

with the maximum deflection at the center equal to

$$w_{\max} = \frac{\alpha R}{2} \tag{10-93}$$

The bending moments resulting from Eqs. (10-72) and (10-91) are

$$M_r(r) = M_\theta(r) = \frac{(1+\nu)D\alpha}{R} = \text{constant} \tag{10-94}$$

## 10-2-4 Solid Plate: Influence Functions (Green's Functions)

The particular solution obtained when a line load is applied along a circle of radius $a$ will now be used to generate influence functions.

First we consider a simply supported plate of radius $R$ (Fig. 10-17). By using Eqs. (10-72) and (10-66) we get

$$w(r) = C_1 + C_3 r^2 + \frac{pa}{4D}\left[(r^2 + a^2)\ln\frac{r}{a} - (r^2 - a^2)\right]H(r-a) \tag{10-95}$$

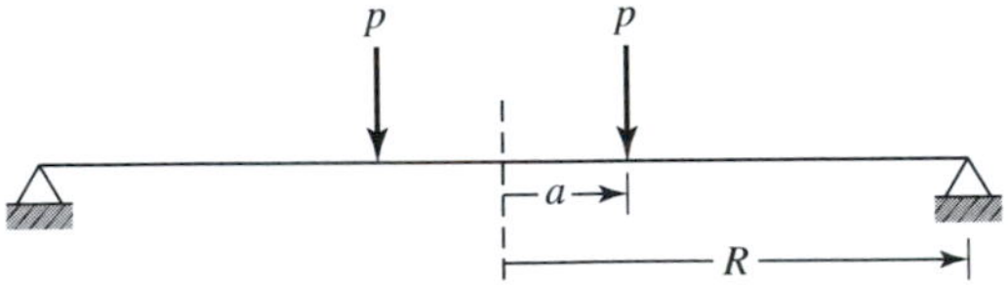


Figure 10-17

and with Eqs. (10-72$c$) and (10-67),

$$M_r(r) = -2D(1+\nu)C_3 - \frac{pa}{4}\left[2(1+\nu)\ln\frac{r}{a} - (1-\nu)\frac{a^2}{r^2} + 1 - \nu\right]H(r-a) \tag{10-96}$$

Boundary conditions (10-70$a$) are again applicable,

$$w(R) = 0, \qquad M_r(R) = 0$$

and after substituting here Eqs. (10-95) and (10-96) we obtain

$$C_1 + C_3R^2 + \frac{pa}{4D}\left[(R^2+a^2)\ln\frac{R}{a} - (R^2-a^2)\right] = 0$$

$$-2D(1+\nu)C_3 - \frac{pa}{4}\left[2(1+\nu)\ln\frac{R}{a} - (1-\nu)\frac{a^2}{R^2} + 1 - \nu\right] = 0$$

from which it follows that

$$\begin{aligned} C_1 &= \frac{pa}{4D}\left[-a^2\ln\frac{R}{a} + \frac{1-\nu}{2(1+\nu)}R^2 - \frac{1-\nu}{2(1+\nu)}a^2 + R^2 - a^2\right] \\ C_3 &= -\frac{pa}{8(1+\nu)D}\left[2(1+\nu)\ln\frac{R}{a} + (1-\nu)\left(1 - \frac{a^2}{R^2}\right)\right] \end{aligned} \tag{10-97}$$

By substituting expressions (10-97) into Eq. (10-95) we get

$$\begin{aligned} w(r) &= \frac{pa}{4D}\left[(a^2+r^2)\ln\frac{a}{R} + \left(1-\frac{a^2}{R^2}\right)\frac{(3+\nu)R^2-(1-\nu)r^2}{2(1+\nu)}\right], \quad 0 \le r \le a \\ w(r) &= \frac{pa}{4D}\left[(a^2+r^2)\ln\frac{r}{R} + \left(1-\frac{a^2}{R^2}\right)\frac{(3+\nu)R^2-(1-\nu)r^2}{2(1+\nu)} + a^2 - r^2\right] \\ &\qquad a \le r \le R \end{aligned} \tag{10-98}$$

In arriving at the last expression, the identity

$$\ln\frac{r}{a} = \ln\frac{r}{R} + \ln\frac{R}{a} \tag{10-99}$$

has been used. Expressions (10-98) give the deflection of the plate at a distance $r$ from the center due to a load $p$ at a distance $a$ from the center. It is, in other words, the influence of an arbitrarily located load (at $r = a$) on the deflection at an arbitrary place. When $p = 1$ this deflection is often called the *influence function* or *Green's function,** and it is usually denoted by

$$w(r) \equiv G(r, a)$$

*George Green (1793–1830), English mathematician.

Thus,

$$G(r,a) = G_1(r,a)$$
$$= \frac{a}{4D}\left[(a^2+r^2)\ln\frac{a}{R} + \left(1-\frac{a^2}{r^2}\right)\frac{(3+\nu)R^2-(1-\nu)r^2}{2(1+\nu)}\right] \qquad 0 \le r \le a$$

$$G(r,a) = G_2(r,a)$$
$$= \frac{a}{4D}\left[(a^2+r^2)\ln\frac{r}{R} + \left(1-\frac{a^2}{R^2}\right)\frac{(3+\nu)R^2-(1-\nu)r^2}{2(1+\nu)} + a^2 - r^2\right] \qquad a \le r \le R$$

(10-100)

In order to see how this solution can be used, consider an arbitrarily loaded simply supported plate, as shown in Fig. 10-18. We begin the analysis by representing the load as a collection of very thin cylindrical tubes, each of the thickness $dr$. The problem we discuss is that of determining the deflection $w(r)$ due to such a tubular load located at an arbitrary distance $a$ from the center. Next the effects of other similar tubular loads are added on. This sum will represent the total deflection due to the given load. Figure 10-19 shows the same plate, but carrying only the tubular load shaded in Fig. 10-18. Since $da$ is infinitesimal, the effect of this load does not differ from the effect of a line load considered to derive Eqs. (10-100), except that its intensity is $q(a)\,da$ times larger. Thus the resulting deflection $w_1$ is obtained by multiplying the deflection Eq. (10-100) by $q(a)\,da$,

$$w_1(r) = G(r,a)q(a)\,da$$

Now varying $a$ we add other loads represented in Fig. 10-18, thus covering the entire loaded area $a(r)$. Since this addition is simply integration, we can write

$$w(r) = \int_0^R G(r,a)q(a)\,da \tag{10-101}$$

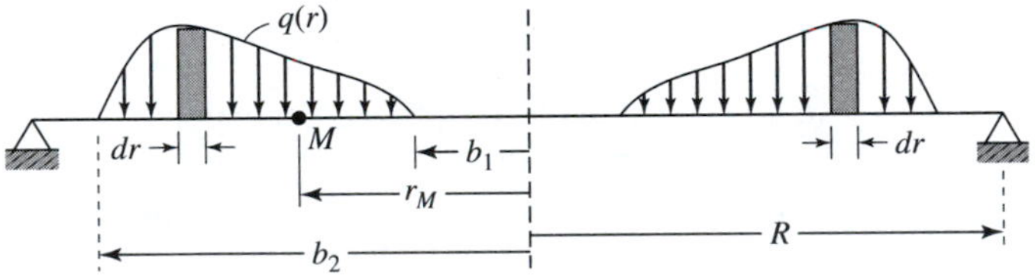


Figure 10-18 Using Green's function to determine the deflection caused by an arbitrary load

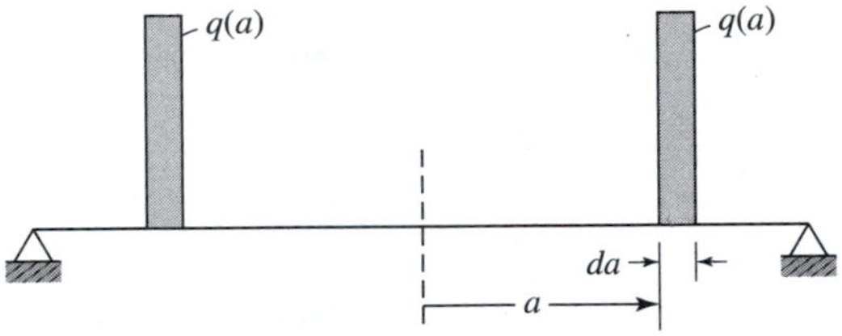


Figure 10-19

where $q(a) = 0$ on some portions of the interval $(0, R)$. Let us calculate the deflection at a point $M$, whose location is shown in Fig. 10-18. First note that the load $q(a)$ appearing under the integral in Eq. (10-101) is zero unless $b_2 \geq r_M \geq b_1$. Consequently Eq. (10-101) can be written as

$$w(r_M) = \int_{b_1}^{b_2} G(r_M, a)q(a)\, da$$

Next we see in Fig. 10-18 that the variable of integration $a$ is either less or greater than $r_M$. Thus $G_2(r_M, a)$ and $G_1(r_M, a)$ must be substituted for $G(r_M, a)$ appearing under the integral. Finally, using Eqs. (10-100) and (10-99), where $\ln r/R$ is written as

$$\ln\frac{r}{R} = \ln\frac{a}{R} + \ln\frac{r}{a} \tag{10-102}$$

we arrive finally at the following expression for the deflection of a simply supported plate under an arbitrary load,

$$w(r) = \frac{1}{4D}\int_{b_1}^{b_2} q(a)a\left[(a^2 + r^2)\ln\frac{a}{R} + \left(1 - \frac{a^2}{R^2}\right)\frac{(3+\nu)R^2 - (1-\nu)r^2}{2(1+\nu)}\right] da$$
$$+ \frac{1}{4D}\int_{b_1}^{r} q(a)a\left[a^2 - r^2 + (a^2 + r^2)\ln\frac{r}{a}\right] da, \qquad b_1 \leq r \leq b_2 \tag{10-103}$$

Similar expressions can be obtained for the deflection at a point $r$ located beyond the interval $(b_1, b_2)$.

## EXAMPLE 10-4

Let us use the Green's function to determine the deflection of a simply supported plate under a uniformly distributed load of intensity $q_0$ = constant (see Example 10-1).

### SOLUTION

Replacing $q(a)$ by $q_0$ in Eq. (10-103) and using appropriate values for $b_1$ and $b_2$, we obtain

$$w(r) = \frac{q_0}{4D}\int_0^R a\left[(a^2 + r^2)\ln\frac{a}{R} + \left(1 - \frac{a^2}{R^2}\right)\frac{(3+\nu)R^2 - (1-\nu)r^2}{2(1+\nu)}\right] da$$
$$+ \frac{q_0}{4D}\int_0^r a\left[a^2 - r^2 + (a^2 + r^2)\ln\frac{r}{a}\right] da$$
$$= \frac{q_0 R^4}{64D(1+\nu)}\left[(1+\nu)\left(\frac{r}{R}\right)^4 - 2(3+\nu)\left(\frac{r}{R}\right)^2 + 5 + \nu\right]$$

which is identical to the result in Example 10-1.

The integration becomes more cumbersome for loads that are more complicated than the load in this example. For many cases however, a symbolic manipulation package can remove the tedium from the calculation, as shown in the next example.

## EXAMPLE 10-5

We use a software package Mathcad7, Professional Edition [9] to determine the deflection of a simply supported plate carrying a load (Fig. 10-20) $q(r) = q_0(r/R)^2$. The following is the printout of a Mathcad session.

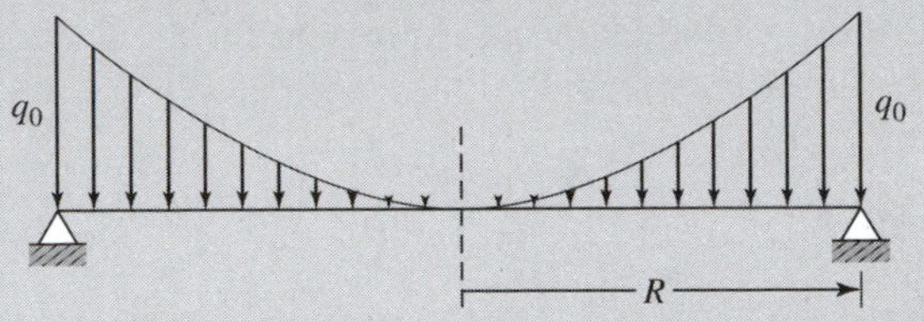


**Figure 10-20**

## SOLUTION

This is the integrand of the first integral

$$\left(\frac{a}{R}\right)^2 a\left[(a^2+r^2)\ln\left(\frac{a}{R}\right)+\left(1-\frac{a^2}{R^2}\right)\frac{(3+\nu)\cdot R^2-(1-\nu)r^2}{2(1+\nu)}\right]$$

Denoting $a = aR * R$ we get:

$$\left(\frac{a}{R}\right)^2 R\,aR\left[(aR^2\,R^2+r^2)\ln(aR)+(1-aR^2)\cdot\frac{((3+\nu)R^2-(1-\nu)r^2)}{2(1+\nu)}\right]$$

This is integrated using the Calculus Palette

$$\int_0^1 \left(\frac{a}{R}\right)^2 R^2\,aR\left[(aR^2\,R^2+r^2)\cdot\ln(aR)+(1-aR^2)\cdot\frac{((3+\nu)R^2-(1-\nu)r^2)}{2(1+\nu)}\right]daR$$

yielding

$$\frac{-1}{144}R^2\frac{15r^2-14R^2-2R^2\nu+3r^2\nu}{1+\nu}$$

This is the integrand of the second integral

$$\left(\frac{a}{R}\right)^2 a\left[a^2-r^2+(a^2+r^2)\ln\left(\frac{r}{a}\right)\right]$$

or

$$(aR)^2 R\,aR\left[R^2\,aR^2-r^2+(R^2\,aR^2+r^2)\left(\ln\left(\frac{r}{R}\right)-\ln(aR)\right)\right]$$

Integrating this we get

$$\int_0^{rR} (aR)^2 R^2\, aR(R^2\, aR^2 - r^2 + (R^2\, aR^2 + r^2)\cdot(\ln(rR)$$

$$-\ln(aR)))daR - \frac{3}{16}R^2 r^2 rR^4 + \frac{7}{36}R^4\, rR^6 \qquad \text{where } rR = r/R$$

Simplifying this result we obtain

$$\frac{1}{144}\cdot\frac{r^6}{R^2}$$

Adding both integrals and multiplying the result by $q_0/4D$ yields

$$\frac{q_0}{4D}\left(\frac{-1}{144}R^2\frac{15r^2 - 14R^2 - 2R^2\nu + 3r^2\nu}{1+\nu} + \frac{1}{144}\cdot\frac{r^6}{R^2}\right)$$

Hence

$$w(r) = \frac{1}{576}\cdot\frac{q_0}{D}\cdot\frac{2(7+\nu)R^6 - 3(5+\nu)R^4r^2 + (1+\nu)r^6}{(1+\nu)R^2} \tag{10-104}$$

In the cases when only few numerical results are needed, or when symbolic integration does not give closed-form expressions, or when no analytical (but only a numerical) expression is available for $q(r)$, we can use the Fortran program 10-1 to calculate the deflections.

Next let us determine the influence function for a clamped plate of radius $R$ (Fig. 10-21). We have

$$w(r) = C_1 + C_3r^2 + \frac{pa}{4D}\left[(r^2+a^2)\ln\frac{r}{a} - (r^2 - a^2)\right]H(r-a) \tag{10-105}$$

Now the boundary conditions (10-70*b*) are applicable,

$$w(R) = 0, \qquad w'(R) = 0 \tag{10-106}$$

Because [see Eqs. (10-65) and (10-105)]

$$w'(r) = 2C_3r + \frac{pa}{4Dr}\left[\left(2\ln\frac{r}{a} - 1\right)r^2 + a^2\right]H(r-a) \tag{10-107}$$

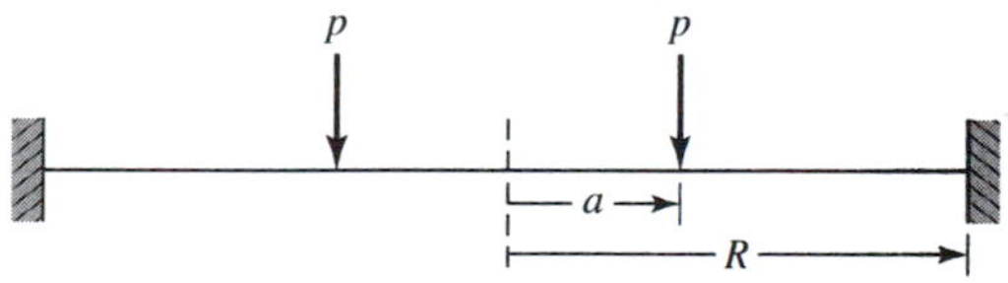


Figure 10-21

boundary conditions (10-106) become

$$C_1 + C_3 R^2 + \frac{pa}{4D}\left[(R^2 + a^2)\ln\frac{R}{a} - (R^2 - a^2)\right] = 0$$

$$2C_3 R + \frac{pa}{4DR}\left[\left(2\ln\frac{R}{a} - 1\right)R^2 + a^2\right] = 0$$

from which it follows that

$$C_1 = \frac{pa}{8D}\left(-2a^2\ln\frac{R}{a} + R^2 - a^2\right)$$
$$C_3 = -\frac{pa}{8DR^2}\left[\left(2\ln\frac{R}{a} - 1\right)R^2 + a^2\right] \tag{10-108}$$

Finally this is substituted into Eq. (10-105), and $p = 1$ is then used in the result,

$$w(r) = \frac{a}{8D}\left\{-2(r^2 + a^2)\ln\frac{R}{a} + (R^2 - a^2)\left[\left(\frac{r}{R}\right)^2 + 1\right]\right\}, \qquad r \le a$$

$$w(r) = \frac{a}{8D}\left\{-2(r^2 + a^2)\ln\frac{R}{a} + (R^2 - a^2)\left[\left(\frac{r}{R}\right)^2 + 1\right]\right.$$
$$\left. + 2(r^2 + a^2)\ln\frac{r}{a} - 2(r^2 - a^2)\right\}, \qquad r \ge a \tag{10-109}$$

As in the case of a simply supported plate, the deflection (10-109) is the Green's function $G(r, a)$ for a clamped plate. We can use Eq. (10-101) to obtain the deflection of a clamped plate carrying an arbitrary load (Fig. 10-22). The first step in this development will be to subdivide the interval $(0, R)$ into several subintervals. This subdivision is dictated by the nature of the applied load $q$. In our example $q(a) \equiv 0$ for $0 \le r \le b_1$ and $b_2 \le r \le R$. Thus using Eq. (10-101) we get

$$\begin{aligned} w(r) &= \int_0^R G(r, a)q(a)\,da \\ &= \int_0^{b_1} G(r, a)q(a)\,da + \int_{b_1}^{b_2} G(r, a)q(a)\,da + \int_{b_2}^{R} G(r, a)q(a)\,da \\ &= \int_{b_1}^{b_2} G(r, a)q(a)\,da \end{aligned} \tag{10-110}$$

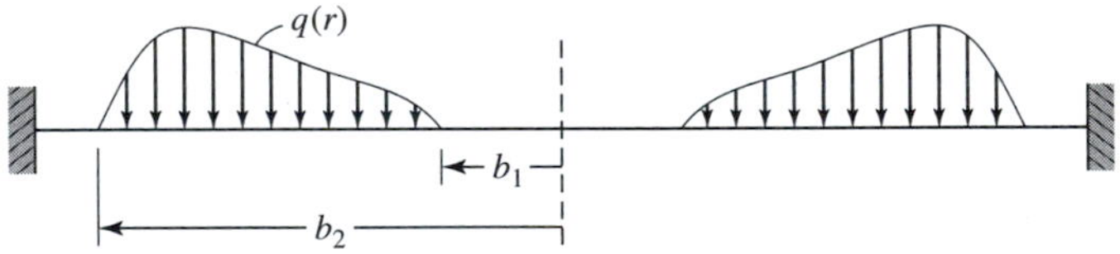


Figure 10-22

The form of $w(r)$ depends on the position of point $r$.

- When $r \leq b_1$, then $r$ is also always less than $a$ (because $b_1 \leq a$). Thus we replace $G(r, a)$ in Eq. (10-110) by the first of Eqs. (10-109),

$$w(r) = \frac{a}{8D} \int_{b_1}^{b_2} q(a)a \left\{ -2(r^2 + a^2) \ln \frac{R}{a} + (R^2 - a^2) \left[ \left( \frac{r}{R} \right)^2 + 1 \right] \right\} da \tag{10-111}$$

- When $r \geq b_2$, then $r$ is always greater than $a$ (because $a \leq b_2$). $G(r, a)$ appearing in Eq. (10-110) must now be replaced by the second of Eqs. (10-109) to give

$$w(r) = \frac{a}{8D} \int_{b_1}^{b_2} \left\{ -2(r^2 + a^2) \ln \frac{R}{a} + (R^2 - a^2) \left[ \left( \frac{r}{R} \right)^2 + 1 \right] + 2(r^2 + a^2) \ln \frac{r}{a} - 2(r^2 - a^2) \right\} da \tag{10-112}$$

- Finally, when $b_1 \leq r \leq b_2$, then the second of Eqs. (10-109) is substituted into Eq. (10-110) for $G(r, a)$ when $b_1 \leq a \leq r$, and the first of Eqs. (10-109) is substituted when $r \leq a \leq b_2$,

$$w(r) = \frac{1}{8D} \int_{b_1}^{b_2} q(a)a \left\{ -2(r^2 + a^2) \ln \frac{R}{a} + (R^2 - a^2) \left[ \left( \frac{r}{R} \right)^2 + 1 \right] \right\} da + \frac{1}{4D} \int_{b_1}^{r} q(a)a \left[ (r^2 + a^2) \ln \frac{r}{a} - r^2 + a^2 \right] da, \qquad b_1 \leq r \leq b_2 \tag{10-113}$$

## 10-2-5 Solid Plate with Additional Support

The results from Section 10-2-4 can also be used to determine deflections and other quantities for a plate with additional concentric supports. Consider, for example, the plate shown in Fig. 10-23$a$ with a circular support at a radius $r = c$. We first remove the

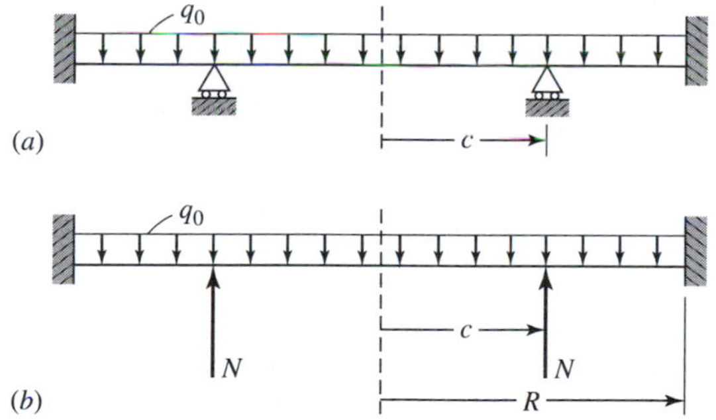


Figure 10-23

additional support, replacing it by an unknown uniformly distributed reaction force $N$ (Fig. 10-23*b*). In order to make the systems in Fig. 10-23*a* and *b* equivalent, we must require that the sum of the deflections at $r = c$, due to the load $q_0$ and to the reactions $N$, be zero. With Eqs. (10-109) and (10-84) this becomes

$$w(c) = \frac{q_0 R^4}{64D}\left(1 - \frac{c}{R}\right)^2 - \frac{Nc}{8D}\left\{-2(c^2 + c^2)\ln\frac{R}{c} + (R^2 - c^2)\left[\left(\frac{c}{R}\right)^2 + 1\right]\right\} = 0$$

Solving this for the unknown reaction $N$, we obtain

$$N = -\frac{q_0 R\left(1 - \dfrac{c}{R}\right)^2}{8\dfrac{c}{R}\left[4\left(\dfrac{c}{R}\right)^2 \ln\dfrac{c}{R} - 1 + \left(\dfrac{c}{R}\right)^4\right]}, \qquad 0 < c < R \qquad (10\text{-}114)$$

Now using Eqs. (10-84) and (10-109), we can determine the deflection of the plate,

$$w(r) = \frac{q_0 R^2}{64D}\left\{(R - r)^2 + \frac{(R - c)^2}{4R^2c^2\ln\dfrac{R}{c} - R^4 + c^4}\left[-2R^2(r^2 + c^2)\ln\frac{R}{c} + (R^2 - c^2)(r^2 + R^2)\right]\right\}, \qquad r \le c$$

$$w(r) = \frac{q_0 R^2}{64D}\left\{(R - r)^2 + \frac{(R - c)^2}{4R^2c^2\ln\dfrac{R}{c} - R^4 + c^4}\left[-2R^2(r^2 + c^2)\ln\frac{R}{c} + (R^2 - c^2)(r^2 + R^2) + 2R^2(r^2 + c^2)\ln\frac{r}{c} - 2R^2(r^2 - c^2)\right]\right\}, \qquad r \ge c$$

(10-115)

The dependence of $N$ on the location $r = c$ of the support and the deflection at $c = R/2$ are shown in Figs. 10-24 and 10-25 and in Tables 10-1 and 10-2, respectively.

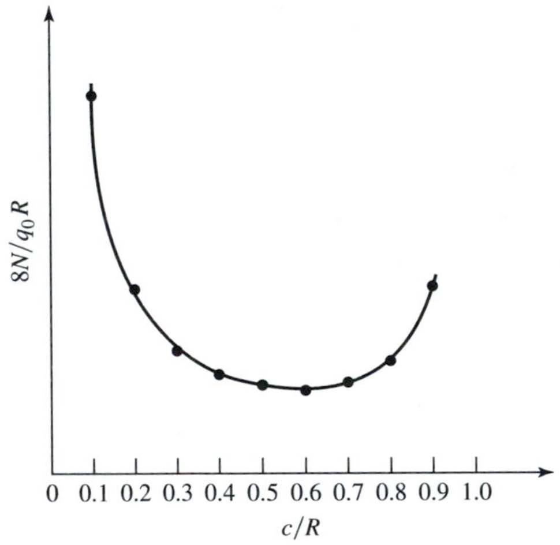


**Figure 10-24** The dependence of the reaction $N$ on the position $r = c$ of the internal support

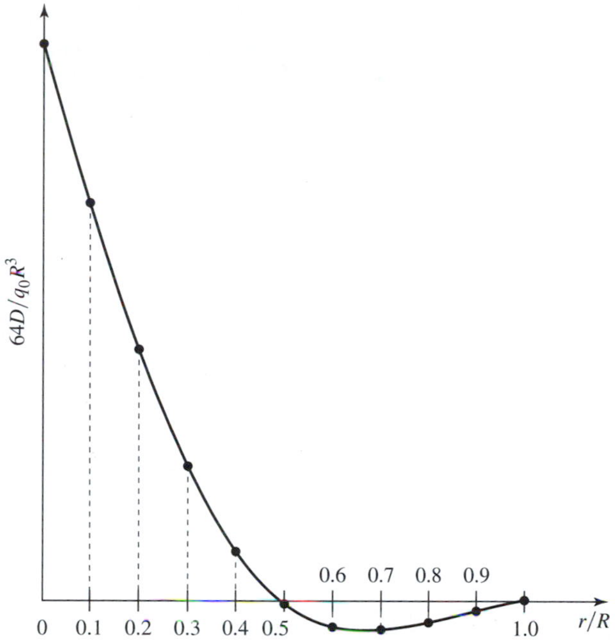


**Figure 10-25** The deflection of the plate at $r = R/2$ as a function of the position of the support $r = c$

**Table 10-1**

| $\frac{c}{R}$ | $\frac{8N}{q_0R}$ |
|---|---|
| 0.1 | 8.9227 |
| 0.2 | 4.3191 |
| 0.3 | 2.9247 |
| 0.4 | 2.3197 |
| 0.5 | 2.0462 |
| 0.6 | 1.9781 |
| 0.7 | 2.1141 |
| 0.8 | 2.6106 |
| 0.9 | 4.3884 |

**Table 10-2**

| $\frac{r}{R}$ | $\frac{64D}{q_0R^2}w$ | $\frac{r}{R}$ | $\frac{64D}{q_0R^2}w$ |
|---|---|---|---|
| 0.00 | 0.5872 | 0.55 | −0.0136 |
| 0.05 | 0.4914 | 0.60 | −0.0209 |
| 0.10 | 0.4038 | 0.65 | −0.0232 |
| 0.15 | 0.3244 | 0.70 | −0.0222 |
| 0.20 | 0.2533 | 0.75 | −0.0187 |
| 0.25 | 0.1904 | 0.80 | −0.0140 |
| 0.30 | 0.1358 | 0.85 | −0.0090 |
| 0.35 | 0.8950 | 0.90 | −0.0044 |
| 0.40 | 0.5141 | 0.95 | −0.0012 |
| 0.45 | 0.2158 | 1.00 | 0.0000 |
| 0.50 | 0.0000 | | |

The cases of other applied loads, of different boundary conditions, and of larger numbers of additional supports may be treated in an analogous way.

## 10-2-6 Annular Plate: Boundary Conditions and Examples

An annular plate has two concentric circular boundaries: the inner edge $r = R_1$ and the outer edge $r = R_2$ (Fig. 10-26). Since two boundary conditions of the type given by Eqs. (10-70) are available at each boundary, we have used a total of four conditions to

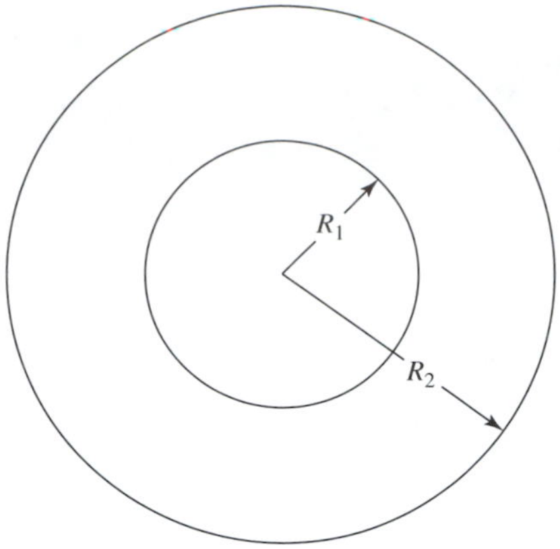


Figure 10-26

determine four constants of integration. The deflection, the slope, the bending moments, and the shear force take the forms [see Eqs. (10-47)–(10-51)]

$$w(r) = C_1 + C_2 \ln \frac{r}{R_2} + C_3 r^2 + C_4 r^2 \ln \frac{r}{R_2} + w_q(r) \tag{10-116a}$$

$$w'(r) = \frac{C_2}{r} + 2C_3 r + C_4 r \left(1 + 2\ln \frac{r}{R_2}\right) + w'_q(r) \tag{10-116b}$$

$$M_r(r) = -D\left\{-\frac{1-\nu}{r^2}C_2 + 2(1+\nu)C_3 + \left[2(1+\nu)\ln \frac{r}{R_2} + 3 + \nu\right]C_4\right\} + M_{rq}(r) \tag{10-116c}$$

$$M_\theta(r) = -D\left\{\frac{1-\nu}{r^2}C_2 + 2(1+\nu)C_3 + \left[2(1+\nu)\ln \frac{r}{R_2} + 1 + 3\nu\right]C_4\right\} + M_{\theta q} \tag{10-116d}$$

$$Q_r(r) = -4D\frac{C_4}{r} + Q_{qr}(r) \tag{10-116e}$$

## EXAMPLE 10-6

Find the deflection, the bending moments, and the maximum stress for a "cantilevered" plate carrying a uniform load of intensity $q_0 =$ constant (Fig. 10-27).

### SOLUTION

Now the boundary conditions are

$$w(R_1) = 0, \qquad w'(R_1) = 0, \qquad M_r(R_2) = 0, \qquad Q_r(R_2) = 0 \tag{10-117}$$

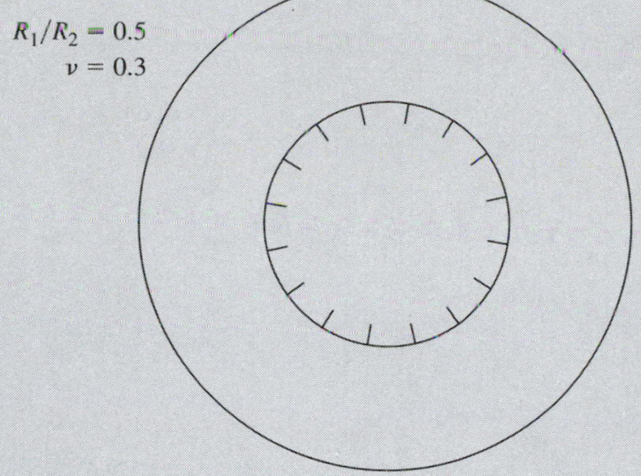


**Figure 10-27**

Employing Eqs. (10-116), (10-117), and (10-54) we obtain the following system of algebraic equations,

$$C_1 + C_2 \ln \frac{R_1}{R_2} + C_3 R_1^2 + C_4 R_1^2 \ln \frac{R_1}{R_2} + \frac{q_0 R_1^4}{64D} = 0$$

$$\frac{1}{R_1} C_2 + 2C_3 R_1 + C_4 R_1 \left(1 + 2\ln \frac{R_1}{R_2}\right) + \frac{q_0 R_1^3}{16D} = 0$$

$$-D\left[-\frac{1-\nu}{R_2^2} C_2 + 2(1+\nu)C_3 + (3+\nu)C_4\right] - \frac{(3+\nu)q_0 R_2^2}{16} = 0$$

$$-4D\frac{C_4}{R_2} - \frac{q_0 R_2}{2} = 0$$

This system has been solved applying the symbolic manipulation package Mathcad7, professional edition [9], yielding

$$C_1 = \frac{q_0 R_1^2}{64D\left[(1-\nu)R_1^2 + (1+\nu)R_2^2\right]} \left\{(1-\nu)R_1^4 + \left[-5 + 3\nu + 4(1+\nu)\ln \frac{R_1}{R_2}\right] R_1^2 R_2^2 + \left[-2(3+\nu)\left(1 - 2\ln \frac{R_1}{R_2}\right) - 16(1+\nu)\ln^2 \frac{R_1}{R_2}\right] R_2^4\right\}$$

$$C_2 = \frac{q_0 R_1^2 R_2^2}{16D\left[(1-\nu)R_1^2 + (1+\nu)R_2^2\right]} \left\{-(1+\nu)R_1^2 + \left[-(1-\nu) + 4(1+\nu)\ln \frac{R_1}{R_2}\right] R_2^2\right\}$$

$$C_3 = \frac{q_0}{32D\left[(1-\nu)R_1^2 + (1+\nu)R_2^2\right]} \left[-(1-\nu)R_1^4 + 2(1-\nu)\left(1 + 2\ln \frac{R_1}{R_2}\right) R_1^2 R_2^2 + (3+\nu)R_2^4\right]$$

$$C_4 = -\frac{q_0 R_2^2}{8D} \tag{10-118}$$

For $R_1/R_2 = 0.5$ and $\nu = 0.3$ these become

$$C_1 = -0.073160\frac{q_0 R_2^4}{D}, \qquad C_2 = -0.049049\frac{q_0 R_2^4}{D}$$
$$C_3 = 0.066124\frac{q_0 R_2^2}{D}, \qquad C_4 = -0.125\frac{q_0 R_2^2}{D} \tag{10-119}$$

Hence from Eq. (10-116*a*) we obtain the deflection,

$$w(r) = \frac{q_0 R_2^4}{D}\left[-0.073160 - 0.049040 \ln\frac{r}{R_2} + 0.066124\left(\frac{r}{R_2}\right)^2 - 0.125\left(\frac{r}{R_2}\right)^2 \ln\frac{r}{R_2} + 0.015625\left(\frac{r}{R_2}\right)^4\right]$$

The deflected shape of the plate diameter is shown in Fig. 10-28. Expression (10-116*c*)

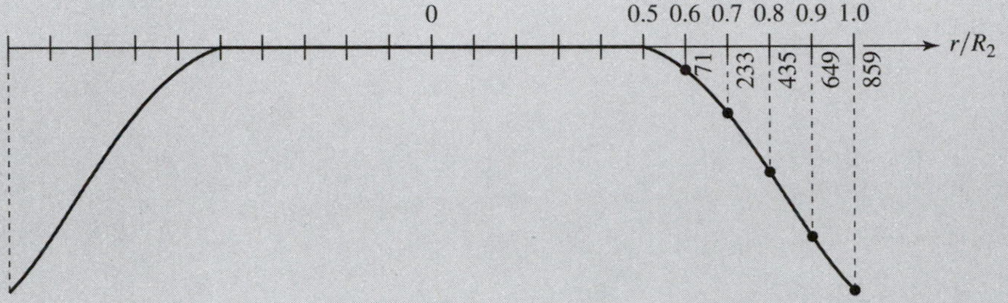


**Figure 10-28** The deflection along a radius of an annular plate carrying a uniform load

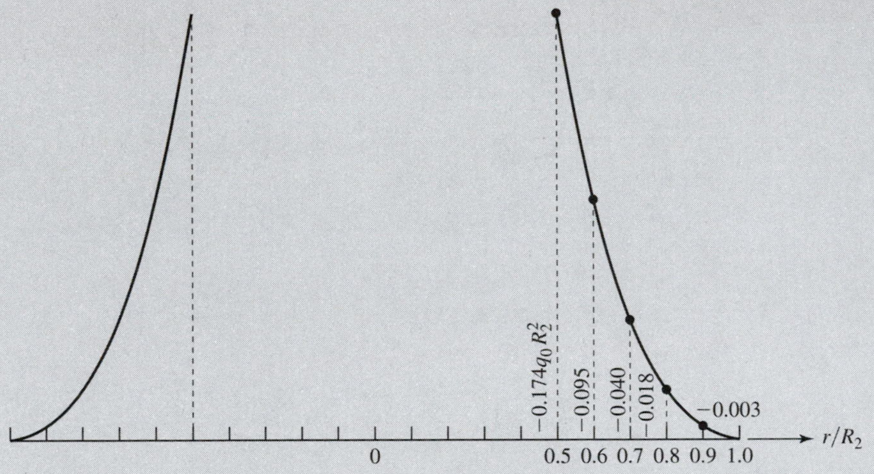


**Figure 10-29** The bending moment along a radius of an annular plate carrying a uniform load

for the bending moment $M_r(r)$ becomes

$$M_r(r) = -q_0 R_2^2 \left[ 0.034328 \left( \frac{R_2}{r} \right)^2 - 0.24058 - 0.325 \ln \frac{r}{R_2} + 0.20625 \left( \frac{r}{R_2} \right)^2 \right]$$

The graph of $M_r(r)$ is shown in Fig. 10.29. Maximum bending moment $M_r(r)$ occurs at $r = R_1 = 0.5R_2$ where it equals:

$$\max M_r(R_1) = -0.174\, q_0 R_2^2$$

Maximum stress is obtained from Eq. (10-35)

$$(\sigma_{rr})_{\max} = \frac{6M_r}{h^2} = \pm 1.04 \frac{q_0 R_2^2}{h^2}$$

The constants of integration for other types of boundary conditions for annular plates carrying a uniform load are collected in Table 10-3 (see also [8, p. 174]).

The deflections, the slopes, the bending moments, and the shear forces of an annular plate with various combinations of boundary conditions on both edges are calculated by the interactive Fortran program 10-2 for three types of load: (1) uniform load, (2) load varying linearly but of zero intensity at the center of the plate, and (3) general linearly varying load.

## 10-2-7 Annular Plate: Influence Functions (Green's Functions)

In order to determine the influence function for the circular plate shown in Fig. 10-30 (see also Section 10-2-4) we use the general solution, Eqs. (10-116), where we substitute Eqs. (10-65)–(10-69) for the particular solution and its derivatives.* Applying the boundary conditions (10-117) from Example 10-6, we get

$$w(R_1) = C_1 + C_2 \ln \frac{R_1}{R_2} + C_3 R_1^2 + C_4 R_1^2 \ln \frac{R_1}{R_2} = 0$$

$$w'(R_1) = \frac{1}{R_1} C_2 + 2C_3 R_1 + C_4 R_1 \left( 1 + 2 \ln \frac{R_1}{R_2} \right) = 0$$

$$\begin{aligned} M_r(R_2) = -D\left[ -\frac{1-\nu}{R_2^2} C_2 + 2(1+\nu)C_3 + (3+\nu)C_4 \right] \\ - \frac{pa}{4}\left[ 2(1+\nu)\ln\frac{R_2}{a} - (1-\nu)\frac{a^2}{R_2^2} + 1 - \nu \right] = 0 \end{aligned} \tag{10-120}$$

$$Q_r(R_2) = -4D \frac{C_4}{R_2} - \frac{pa}{R_2} = 0$$

*Vanishing of the integrals at $r = a$ has been taken into account.

**TABLE 10-3 Integration Constants for Annular Plate Carrying a Uniform Load $q_0$ with Various Boundary Conditions**

| Schematic | Constants of Integration |
|---|---|

1.

$$C_1 = \frac{q_0 R_1^2}{64DN}\left\{(1-\nu)R_1^4 + \left[-5+3\nu+4(1+\nu)\ln\frac{R_1}{R_2}\right]R_1^2R_2^2 + \left[-2(3+\nu)(1-2\ln-16(1+\nu)\left(\ln\frac{R_1}{R_2}\right)^2\right]R_2^4\right\}$$

$$C_2 = \frac{q_0 R_1^2 R_2^2}{16DN}\left\{-(1-\nu)R_1^2 + \left[-1+\nu+4(1+\nu)\ln\frac{R_1}{R_2}\right]R_2^2\right\}$$

$$C_3 = \frac{q_0}{32DN}\left\{-(1-\nu)R_1^4 + 2(1-\nu)\left(1+2\ln\frac{R_1}{R_2}\right)R_1^2R_2^2 + (3+\nu)R_2^4\right\}$$

$$C_4 = -\frac{q_0 R_2^2}{8D}$$

where

$$N = (1-\nu)R_1^2 + (1+\nu)R_2^2$$

2.

$$C_1 = \frac{q_0 R_1^2 R_2^2}{64DN}\left\{\left[-1+\nu+(2-\nu)\ln\frac{R_1}{R_2}\right]R_1^4 + 2\left[-1-\nu+2(3+\nu)\ln\frac{R_1}{R_2}\right]R_1^2R_2^2 + \left[(3+\nu)\left(1-2\ln\frac{R_1}{R_2}\right) - 4(5+\nu)\left(\ln\frac{R_1}{R_2}\right)^2\right]R_2^4\right\}$$

$$C_2 = \frac{q_0 R_1^2}{64DN\ln\frac{R_1}{R_2}}\left\{\left(1-\nu+\nu\ln\frac{R_1}{R_2}\right)R_1^6 + \left[8+3\nu-4(1+\nu)\ln\frac{R_1}{R_2}\right]R_1^4R_2^2\ln\frac{R_1}{R_2} - 8(3+\nu)R_1^2R_2^4\ln\frac{R_1}{R_2} + 4\left[\nu+4+(\nu+5)\ln\frac{R_1}{R_2}\right]R_2^6\ln\frac{R_1}{R_2}\right\}$$

$$C_3 = \frac{q_0}{64DN}\left\{\left[1-\nu-(2-\nu)\ln\frac{R_1}{R_2}\right]R_1^6 + \left[\nu+3-4(3+\nu)\ln\frac{R_1}{R_2}\right]R_1^4R_2^2 + \left[-1+\nu-2(1-\nu)\ln\frac{R_1}{R_2} + 8(3+\nu)\left(\ln\frac{R_1}{R_2}\right)^2\right]R_1^2R_2^4 - (3+\nu)R_2^6\right\}$$

$$C_4 = \frac{q_0}{32DN}\left\{(1-\nu)R_1^6 + \left[-3+\nu+4(1+\nu)\ln\frac{R_1}{R_2}\right]R_1^4R_2^2 + \left[7+\nu-4(3+\nu)\ln\frac{R_1}{R_2}\right]R_1^2R_2^4 - (5+\nu)R_2^6\right\}$$

where

$$N = (1-\nu)R_1^4 + 2\left[-1-\nu+4\ln\frac{R_1}{R_2} - 2(1+\nu)\left(\ln\frac{R_1}{R_2}\right)^2\right]R_1^2R_2^2 + (3+\nu)R_2^4$$

3.

$$C_1 = \frac{-q_0 R_1^2 R_2^2}{64DN}\left\{\left(2\ln\frac{R_1}{R_2}-1\right)R_1^4 + 2\left(1-2\ln\frac{R_1}{R_2}\right)R_1^2R_2^2 + \left(-1+2\ln\frac{R_1}{R_2}+4\left(\ln\frac{R_1}{R_2}\right)^2\right)R_2^4\right\}$$

$$C_2 = \frac{q_0 R_1^2}{64DN \ln \dfrac{R_1}{R_2}} \left\{ \left(1 - 2\ln\frac{R_1}{R_2}\right) R_1^6 + 2\left(-1 - \ln\frac{R_1}{R_2} + 2\left(\ln\frac{R_1}{R_2}\right)^2\right) R_1^4 R_2^2 \right.$$

$$\left. + 2\left(-1 + \ln\frac{R_1}{R_2} - 4\left(\ln\frac{R_1}{R_2}\right)^2\right) R_1^2 R_2^4 + \left(1 + 2\ln\frac{R_1}{R_2} + 4\left(\ln\frac{R_1}{R_2}\right)^2\right) R_2^6 \right\}$$

$$C_3 = \frac{q_0}{64DN} \left\{ \left(-1 + 2\ln\frac{R_1}{R_2}\right) R_1^6 + \left(1 - 4\ln\frac{R_1}{R_2}\right) R_1^4 R_2^2 \right.$$

$$\left. + \left(1 + 2\ln\frac{R_1}{R_2} + 8\left(\ln\frac{R_1}{R_2}\right)^2\right) R_1^2 R_2^4 - R_2^6 \right\}$$

$$C_4 = \frac{q_0}{32DN \ln \dfrac{R_1}{R_2}} \left\{ -R_1^6 + \left(1 + 4\ln\frac{R_1}{R_2}\right) R_1^4 R_2^2 + \left(1 - 4\ln\frac{R_1}{R_2}\right) R_1^2 R_2^4 - R_2^6 \right\}$$

where

$$N = R_1^4 - 2\left(1 + 2\left(\ln\frac{R_1}{R_2}\right)^2\right) R_1^2 R_2^2 + R_2^4$$

4.

$$C_1 = \frac{q_0 R_2^2}{64DN} \left\{ -2R_1^4 \left(3 + \nu + 4\nu \ln\frac{R_1}{R_2} + 4\left(\ln\frac{R_1}{R_2}\right)^2\right) + (-5 + 3\nu) R_1^2 R_2^2 + (1 - \nu) R_2^4 \right\}$$

$$C_2 = \frac{q_0 R_1^2 R_2^2}{16DN} \left\{ -4(1 + \nu) R_1^2 \ln\frac{R_1}{R_2} - (1 - \nu) R_1^2 - (1 + \nu) R_2^2 \right\}$$

$$C_3 = \frac{q_0}{32DN} \left\{ \left(3 + \nu - 4\nu \ln\frac{R_1}{R_2} + 4\ln\frac{R_1}{R_2}\right) R_1^4 + 2(1 - \nu) R_1^2 R_2^2 + (-1 + \nu) R_2^4 \right\}$$

$$C_4 = -\frac{q_0 R_1^2}{8D}$$

where

$$N = (1 + \nu) R_1^2 + (1 - \nu) R^2$$

5.

$$C_1 = \frac{q_0 R_2^2}{64DN} \left\{ \left[ -4 - \nu + 2(5 + \nu)\ln\frac{R_1}{R_2} + 4(1 + \nu)\left(\ln\frac{R_1}{R_2}\right)^2 \right] R_1^2 R_2^4 \right.$$

$$\left. + \left[ 2 + 2\nu - 4(3 + \nu)\ln\frac{R_1}{R_2} \right] R_1^4 R_2^2 - \left[ 3 + \nu - 2(5 + \nu)\ln\frac{R_1}{R_2} \right] R_1^6 \right\}$$

$$C_2 = -\frac{q_0 R_1^2 R_2^2}{32DN} \left\{ \left[ 9 + 2\nu + 2(1 + \nu)\ln\frac{R_1}{R_2} \right] R_2^4 - 4(3 + \nu) R_1^2 R_2^2 \right.$$

$$\left. + 2\left[ 4 + \nu - (5 + \nu)\ln\frac{R_1}{R_2} \right] R_1^4 \right\}$$

$$C_3 = \frac{q_0}{64DN} \left\{ \left[ 2 - \nu - 2(9 + \nu)\ln\frac{R_1}{R_2} - 8(1 + \nu)\left(\ln\frac{R_1}{R_2}\right)^2 \right] R_1^2 R_2^4 \right.$$

$$\left. + \left[ 1 - \nu + 4(3 + \nu)\ln\frac{R_1}{R_2} \right] R_1^4 R_2^2 + \left[ 3 + \nu - 2(5 + \nu)\ln\frac{R_1}{R_2} \right] R_1^6 - (1 - \nu) R_2^6 \right\}$$

$$C_4 = \frac{q_0}{32DN} \left\{ \left[ 3 - \nu + 4(1 + \nu)\ln\frac{R_1}{R_2} \right] R_1^2 R_2^4 - \left[ 7 + \nu + 4(3 + \nu)\ln\frac{R_1}{R_2} \right] R_1^4 R_2^2 \right.$$

$$\left. + (5 + \nu) R_1^6 - (1 - \nu) R_2^6 \right\}$$

*(continued)*

TABLE 10-3 ***Continued***

| Schematic | Constants of Integration |
|---|---|

where

$$N = 2\left[1+\nu+4\ln\frac{R_1}{R_2}+2(1+\nu)\left(\ln\frac{R_1}{R_2}\right)^2\right]R_1^2R_2^2-(3+\nu)R_1^4+(1-\nu)R_2^4$$

6.

$$C_1 = -\frac{q_0R_1^2R_2^2}{64DN}\left\{\left[-(1-\nu)(3+\nu)-2(23+8\nu+\nu^2)\ln\frac{R_1}{R_2}-4(5+6\nu+3\nu^2)\left(\ln\frac{R_1}{R_2}\right)^2\right]R_2^4\right.$$
$$\left.+2\left[(1-\nu)(3+\nu)+2(3+\nu)^2\ln\frac{R_1}{R_2}\right]R_1^2R_2^2+(1-\nu)\left[-3-\nu+2(5+\nu)\right]R_1^6\ln\frac{R_1}{R_2}\right\}$$

$$C_2 = -\frac{q_0R_1^2R_2^2}{16DN}\left[(3+\nu)^2\left(R_1^2-R_2^2\right)^2+(1+\nu)(5+\nu)\left(R_1^4-R_2^4\right)\ln\frac{R_1}{R_2}\right]$$

$$C_3 = \frac{q_0}{64DN}\left\{\left[3-2\nu-\nu^2-2(23+8\nu+\nu^2)\ln\frac{R_1}{R_2}-8(1+\nu)(3+\nu)\left(\ln\frac{R_1}{R_2}\right)^2\right]R_1^2R_2^4\right.$$
$$+\left[3-2\nu-\nu^2+4(3+\nu)^2\ln\frac{R_1}{R_2}\right]R_1^4R_2^2+\left[-(1-\nu)(3+\nu)+2(1-\nu)(5+\nu)\ln\frac{R_1}{R_2}\right]R_1^6$$
$$\left.-(1-\nu)(3+\nu)R_2^6\right\}$$

$$C_4 = \frac{q_0}{32DN}\left\{\left[(1-\nu)(5+\nu)+4(1+\nu)(3+\nu)\ln\frac{R_1}{R_2}\right]R_1^2R_2^4\right.$$
$$\left.+\left[(1-\nu)(5+\nu)-4(1+\nu)(3+\nu)\ln\frac{R_1}{R_2}\right]R_1^4R_2^2-(1-\nu)(5+\nu)\left(R_1^6+R_2^6\right)\right\}$$

where

$$N = 2\left[-(1-\nu)(3+\nu)+2(1+\nu)^2\left(\ln\frac{R_1}{R_2}\right)^2\right]R_1^2R_2^2+(1-\nu)(3+\nu)\left(R_1^4+R_2^4\right)$$

7.

$$C_1 = \frac{q_0R_2^2}{64DN_1}\left\{-4(4+\nu)R_1^2R_2^4+\left[17+5\nu+8(1+\nu)\ln\frac{R_1}{R_2}\right]R_1^4R_2^2\right.$$
$$\left.-2\left[3+\nu+4(1+\nu)\ln\frac{R_1}{R_2}\right]R_1^6+(5+\nu)R_2^6\right\}$$

$$C_2 = \frac{q_0R_1^2R_2^2}{32DN_2}\left\{(3+\nu)R_2^2-\left[3+\nu-4(1+\nu)\ln\frac{R_1}{R_2}\right]R_1^2\right\}$$

$$C_3 = q_0\left[(3+\nu)\left(R_1^2-R_2^2\right)^2+4(1+\nu)R_1^4\right]\Big/32DN_3$$

$$C_4 = -\frac{q_0R_2^2}{8D}$$

where

$$N_1 = (1+\nu)\left(R_1^2-R_2^2\right)^2$$

$$N_2 = (1-\nu)\left(R_1^2-R_2^2\right)$$

$$N_3 = (1+\nu)\left(R_1^2-R_2^2\right)$$

8.

$$C_1 = \frac{q_0 R_1^2}{64DN}\left\{(5+\nu)R_1^4 + \left[-11 - 3\nu + 4(3+\nu)\ln\frac{R_1}{R_2}\right]R_1^2 R_2^2 + \left[6 + 2\nu - (11+9\nu)\ln\frac{R_1}{R_2} - 16(1+\nu)\left(\ln\frac{R_1}{R_2}\right)^2\right]R_2^4\right\}$$

$$C_2 = \frac{q_0 R_1^2 R_2^2}{16DN}\left\{-(3+\nu)R_1^2 + \left[3 + \nu + 4(1+\nu)\ln\frac{R_1}{R_2}\right]R_2^2\right\}$$

$$C_3 = -\frac{q_0}{32DN}\left[(3+\nu)\left(R_1^2 - R_2^2\right) - 4(1+\nu)R_1^2 R_2^2 \ln\frac{R_1}{R_2}\right]$$

$$C_4 = -\frac{q_0 R_2^2}{8D}$$

where

$$N = (1+\nu)\left(R_1^2 - R_2^2\right)$$

---

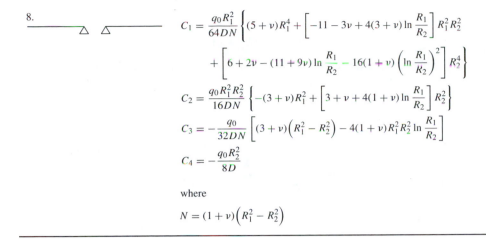

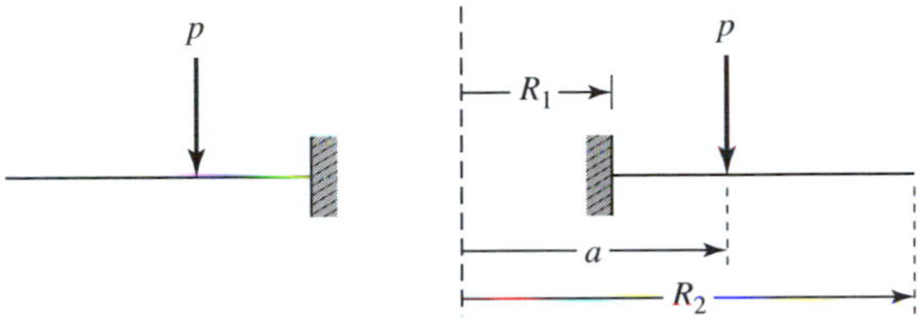


Figure 10-30 A radial cut through an annular plate with clamped inner edge and free outer edge

Solving this system of equations we obtain

$$C_1 = -\frac{paR_1^2}{8D\Delta_1}\left[-4R_2^2(1+\nu)\ln\frac{R_1}{R_2} + 4(1+\nu)R_2^2\ln\frac{R_2}{a}\ln\frac{R_1}{R_2} - 2a^2(1-\nu)\ln\frac{R_1}{R_2} + 4(1+\nu)R_2^2\left(\ln\frac{R_1}{R_2}\right)^2 + (1-\nu)R_1^2 + 2(1+\nu)R_2^2 - 2(1+\nu)R_2^2\ln\frac{R_2}{a} + (1-\nu)a^2\right]$$

$$C_2 = \frac{paR_1^2}{4D\Delta_1}\left[-(1-\nu)a^2 + 2(1+\nu)R_2^2\ln\frac{R_1}{a} - (1+\nu)R_2^2\right] \tag{10-121}$$

$$C_3 = \frac{paR_1^2}{8D\Delta_1}\left[(1-\nu)R_1^2 + 2(1-\nu)R_1^2\ln\frac{R_1}{R_2} - 2(1+\nu)R_2^2\ln\frac{R_2}{a} + (1-\nu)a^2 + 2(1+\nu)R_2^2\right]$$

$$C_4 = -\frac{pa}{4D}$$

where

$$\Delta_1 = (1-\nu)R_1^2 + (1+\nu)R_2^2$$

The constants of integration $C_1, \ldots, C_4$ have been calculated in a similar way for other combinations of boundary conditions using the symbolic manipulation package *Macsyma* [7]. The results are given in Table 10-4.

**TABLE 10-4 Constants of Integration for Green's Functions for Annular Plates with Various Boundary Conditions**

| Schematic | Constants of Integration |
|---|---|
| 1.  | $C_1 = -\frac{pa}{8D\Delta_2}\left\{2(1+\nu)R_1^2\left[\left(R_2^2+a^2\right)\ln\frac{R_2}{a} - R_2^2 + a^2\right] + (1-\nu)R_2^2\left(2a^2\ln\frac{R_2}{a} - R_2^2 + a^2\right)\right\}$<br>$C_2 = -\frac{paR_1^2}{4D\Delta_2}(1+\nu)\left(2R_2^2\ln\frac{R_2}{a} - R_2^2 + a^2\right)$<br>$C_3 = -\frac{pa}{8D\Delta_2}(1-\nu)\left(2R_2^2\ln\frac{R_2}{a} - R_2^2 + a^2\right)$<br>$C_4 = 0$<br>where<br>$\Delta_2 = (1+\nu)R_1^2 + (1-\nu)R_2^2$ |
| 2. 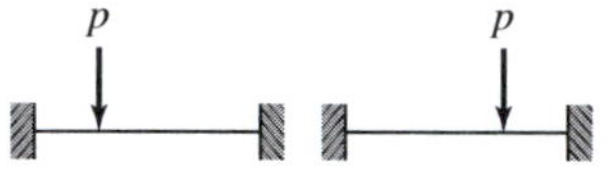 | $C_1 = -\frac{paR_1^2}{4D\Delta_3}\left\{\left(R_2^2 - 2R_2^2\ln\frac{R_1}{R_2} - R_1^2\right)\left[\left(R_2^2+a^2\right)\ln\frac{R_2}{a} - R_2^2 + a^2\right] + 2R_2^2\ln^2\frac{R_1}{R_2}\left(2a^2\ln\frac{R_2}{a} - R_2^2 + a^2\right)\right\}$<br>$C_2 = \frac{paR_1^2}{4D\Delta_3}\left[\left(R_1^2 - R_2^2\right)\left(2R_2^2\ln\frac{R_2}{a} - R_2^2 + a^2\right) + 2R_2^2\ln\frac{R_1}{R_2}\left(2a^2\ln\frac{R_2}{a} - R_2^2 + a^2\right)\right]$<br>$C_3 = -\frac{pa}{4D\Delta_3}\left\{\left(2R_1^2\ln\frac{R_1}{R_2} + R_1^2 - R_2^2\right)\left[\left(R_2^2+a^2\right)\ln\frac{R_2}{a} - R_2^2 + a^2\right] + 2R_1^2\ln^2\frac{R_1}{R_2}\left(2R_2^2\ln\frac{R_2}{a} - R_2^2 + a^2\right)\right\}$<br>$C_4 = \frac{pa}{4D\Delta_3}\left[4R_1^2\ln\frac{R_1}{R_2}\left(2R_2^2\ln\frac{R_2}{a} - R_2^2 + a^2\right) + \left(R_1^2 - R_2^2\right)\left(2a^2\ln\frac{R_2}{a} - R_2^2 + a^2\right)\right]$<br>where<br>$\Delta_3 = -R_2^4 + 4R_1^2R_2^2\ln^2\frac{R_1}{R_2} + 2R_1^2R_2^2 - R_1^4$ |

3.

$$C_1 = -\frac{paR_1^2}{8D(1-\nu^2)\left(R_1^2-R_2^2\right)}\left[4(1+\nu)^2R_2^2\ln\frac{R_1}{R_2}\ln\frac{R_1}{a}+(1-\nu^2)R_2^2\right.$$
$$\left.\times\left(-2+4\ln\frac{R_1}{R_2}+2\ln\frac{R_2}{a}\right)-(1-\nu)^2a^2+(3+\nu)(1-\nu)R_1^2-2(1-\nu^2)a^2\ln\frac{R_1}{R_2}\right]$$

$$C_2 = \frac{paR_1^2}{4D(1-\nu)\left(R_1^2-R_2^2\right)}\left[(1-\nu)(R_2^2-a^2)+2(1+\nu)R_2^2\ln\frac{R_1}{a}\right]$$

$$C_3 = \frac{pa}{8D(1+\nu)\left(R_1^2-R_2^2\right)}\left[2(1+\nu)\left(R_1^2\ln\frac{R_1}{R_2}-R_2^2+R_2^2\ln\frac{R_2}{a}\right)\right.$$
$$\left.+(3+\nu)R_1^2-(1-\nu)a^2\right]$$

$$C_4 = -\frac{pa}{4D}$$

4.

$$C_1 = -\frac{pa}{8D(1+\nu)\left(R_1^2-R_2^2\right)}\left\{(1+\nu)R_1^2\left[2\ln\frac{R_2}{a}\left(R_2^2+a^2\right)-2\left(R_2^2-a^2\right)\right]\right.$$
$$\left.-2R_2^2a^2(1+\nu)\ln\frac{R_2}{a}+(3+\nu)R_2^2\left(R_2^2-a^2\right)\right\}$$

$$C_2 = -\frac{paR_1^2}{4D(1-\nu)\left(R_2^2-R_1^2\right)}\left[2R_2^2\ln\frac{R_2}{a}(1+\nu)+\left(R_2^2-a^2\right)(1-\nu)\right]$$

$$C_3 = \frac{pa}{8D(1+\nu)\left(R_1^2-R_2^2\right)}\left[2R_2^2\ln\frac{R_2}{a}(1+\nu)+\left(R_2^2-a^2\right)(1-\nu)\right]$$

$$C_4 = 0$$

5.

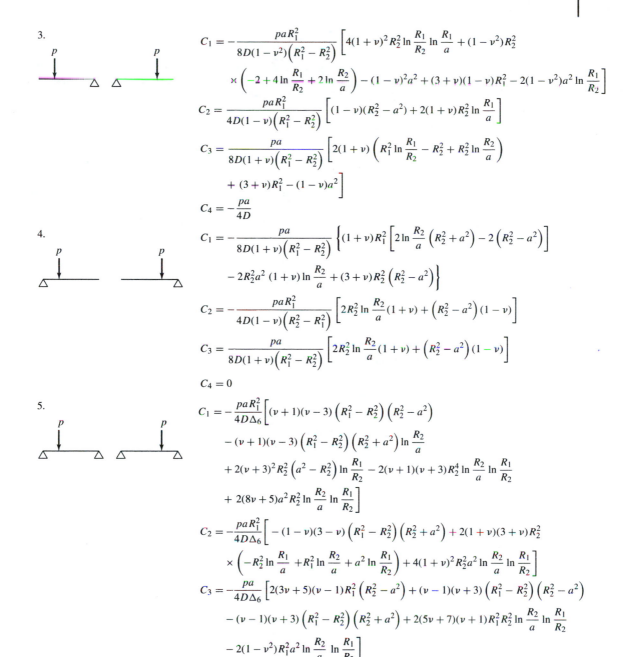

$$C_1 = -\frac{paR_1^2}{4D\Delta_6}\left[(\nu+1)(\nu-3)\left(R_1^2-R_2^2\right)\left(R_2^2-a^2\right)\right.$$
$$-(\nu+1)(\nu-3)\left(R_1^2-R_2^2\right)\left(R_2^2+a^2\right)\ln\frac{R_2}{a}$$
$$+2(\nu+3)^2R_2^2\left(a^2-R_2^2\right)\ln\frac{R_1}{R_2}-2(\nu+1)(\nu+3)R_2^4\ln\frac{R_2}{a}\ln\frac{R_1}{R_2}$$
$$\left.+2(8\nu+5)a^2R_2^2\ln\frac{R_2}{a}\ln\frac{R_1}{R_2}\right]$$

$$C_2 = -\frac{paR_1^2}{4D\Delta_6}\left[-(1-\nu)(3-\nu)\left(R_1^2-R_2^2\right)\left(R_2^2+a^2\right)+2(1+\nu)(3+\nu)R_2^2\right.$$
$$\left.\times\left(-R_2^2\ln\frac{R_1}{a}+R_1^2\ln\frac{R_2}{a}+a^2\ln\frac{R_1}{R_2}\right)+4(1+\nu)^2R_2^2a^2\ln\frac{R_2}{a}\ln\frac{R_1}{R_2}\right]$$

$$C_3 = -\frac{pa}{4D\Delta_6}\left[2(3\nu+5)(\nu-1)R_1^2\left(R_2^2-a^2\right)+(\nu-1)(\nu+3)\left(R_1^2-R_2^2\right)\left(R_2^2-a^2\right)\right.$$
$$-(\nu-1)(\nu+3)\left(R_1^2-R_2^2\right)\left(R_2^2+a^2\right)+2(5\nu+7)(\nu+1)R_1^2R_2^2\ln\frac{R_2}{a}\ln\frac{R_1}{R_2}$$
$$\left.-2(1-\nu^2)R_1^2a^2\ln\frac{R_2}{a}\ln\frac{R_1}{R_2}\right]$$

$$C_4 = -\frac{pa}{4D\Delta_6}\left[-2(1-\nu^2)R_1^2\left(R_2^2-a^2\right)\ln\frac{R_1}{R_2}-2(1-\nu^2)a^2\left(R_2^2-R_1^2\right)\ln\frac{R_2}{a}\right.$$
$$\left.+4(\nu+1)^2R_1^2R_2^2\ln\frac{R_2}{a}\ln\frac{R_1}{R_2}-(\nu+3)(\nu-1)\left(R_2^2-R_1^2\right)\left(R_2^2-a^2\right)\right]$$

$$\Delta_6 = -8(1+\nu)^2R_1^2R_2^2\ln\frac{R_1}{R_2}-(1-\nu)(3+\nu)\left(R_1^2-R_2^2\right)^2$$

## 10-3 BENDING OF RECTANGULAR PLATES

In this section we discuss briefly the general boundary conditions and then determine the deflections of a simply supported rectangular plate. Bending of rectangular plates with other boundary conditions is treated in specialized texts (see, for example, [12], [14]).

### 10-3-1 Boundary Conditions

The differential equation for bending, Eq. (10-20), requires two boundary conditions at each edge to be satisfied. The boundary conditions are physical constraints, thus they are the same in any coordinate system. However, their mathematical description depends on the coordinate system. As we have seen in Section 10-2-3, there are three basic types of boundary conditions: (1) simply supported, (2) clamped, and (3) free edge.

#### Simply Supported

Both the deflection $w$ and the bending moment in the direction normal to the edge must be zero. Thus for $x$ = constant,

$$w(x, y) = 0, \qquad M_x(x, y) = -D\left(\frac{\partial^2 w}{\partial x^2} + \nu\frac{\partial^2 w}{\partial y^2}\right) = 0 \tag{10-122}$$

while for $y$ = constant,

$$w(x, y) = 0, \qquad M_y(x, y) = -D\left(\frac{\partial^2 w}{\partial y^2} + \nu\frac{\partial^2 w}{\partial x^2}\right) = 0 \tag{10-123}$$

However, $w = 0$ along a line implies that its second derivative in the direction tangent to this line is also zero. Therefore conditions (10-122) and (10-123) reduce to

$$\begin{aligned} w = 0, \qquad \frac{\partial^2 w}{\partial x^2} = 0, \qquad \text{at } x = \text{constant} \\ w = 0, \qquad \frac{\partial^2 w}{\partial y^2} = 0, \qquad \text{at } y = \text{constant} \end{aligned} \tag{10-124}$$

#### Clamped

The deflection and the slope in the normal direction are zero. Thus,

$$\begin{aligned} w = 0, \qquad \frac{\partial w}{\partial x} = 0, \qquad \text{at } x = \text{constant} \\ w = 0, \qquad \frac{\partial w}{\partial y} = 0, \qquad \text{at } y = \text{constant} \end{aligned} \tag{10-125}$$

#### Free Edge

Both the bending moment and the shear force vanish. The conditions for the bending moment to vanish are given mathematically by Eqs. (10-122) and (10-123) [however,

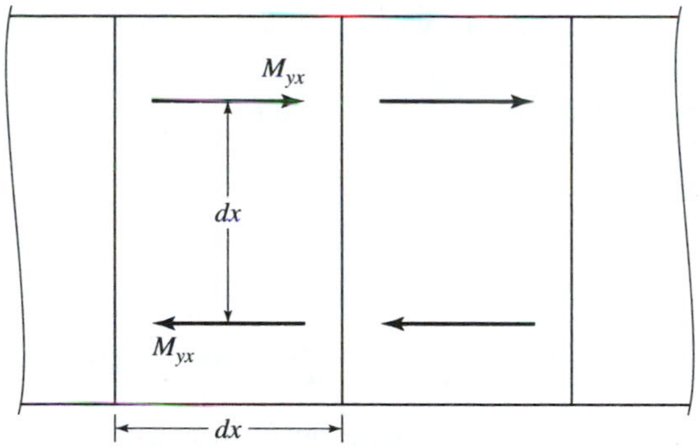


Figure 10-31 Elementary moments $M_{yx}\,dx$ acting on an infinitesimal element $dx$ replaced by couples of horizontal forces

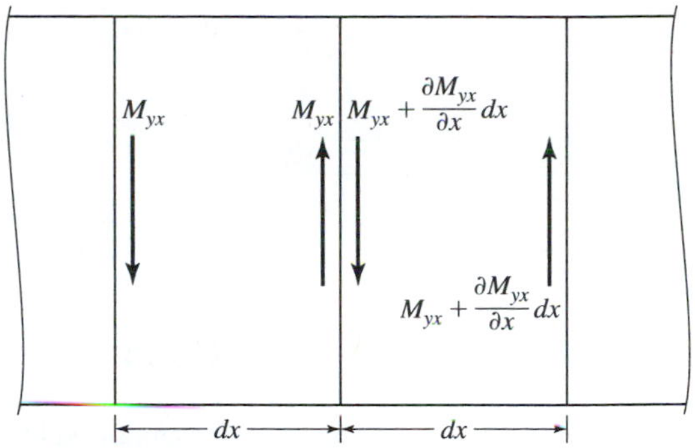


Figure 10-32

without the simplifications leading to Eqs. (10-124)]. Thus,

$$\frac{\partial^2 w}{\partial x^2} + \nu\frac{\partial^2 w}{\partial y^2} = 0, \qquad \text{at } x = \text{constant}$$
$$\frac{\partial^2 w}{\partial y^2} + \nu\frac{\partial^2 w}{\partial x^2} = 0, \qquad \text{at } y = \text{constant} \tag{10-126}$$

The conditions for the shear force to vanish differ from those for a circular plate (Section 10-2). In Section 10-2 we considered only axisymmetric bending. Thus the torque $M_{r\theta}$ was zero. Here the torque $M_{yx}$ is, in general, not zero. Therefore it appears on the free edge as a third unknown while we have only two boundary conditions available. We note, however, that the elementary moments $M_{yx}\,dx$ acting on an infinitesimal element $dx$ can be replaced by couples of horizontal forces (Fig. 10-31). These moments will not change if the forces are redirected, as seen in Fig. 10-32. This replacement causes only local changes in the stress distribution. Thus two forces act on each element $dx$, that is, $M_{yx}$ and $M_{yx} + (\partial M_{yx}/\partial x)\,dx$. Their resultant is

$$Q'_y = -\left(\frac{\partial M_{yx}}{\partial x}\right)_{y=\text{const}} \tag{10-127}$$

Similarly, at the boundary $x$ = constant,

$$Q'_x = -\left(\frac{\partial M_{xy}}{\partial y}\right)_{x=\text{const}} \tag{10-128}$$

Adding $Q'_x$ to $Q_x$, and $Q'_y$ to $Q_y$, Eqs. (10-12) give with Eqs. (10-13)

$$V_x = Q_x + \frac{\partial M_{xy}}{\partial y} = -D\left[\frac{\partial^3 w}{\partial x^3} + (2-\nu)\frac{\partial^3 w}{\partial x\,\partial y^2}\right]$$
$$V_y = Q_y + \frac{\partial M_{yx}}{\partial x} = -D\left[\frac{\partial^3 w}{\partial y^3} + (2-\nu)\frac{\partial^3 w}{\partial x^2\,\partial y}\right] \tag{10-129}$$

The forces $V_x$ and $V_y$ are known as *reduced,* or *Kirchhoff's,* or *effective shear forces* [14].

Consequently the conditions that the shear forces vanish at the free edge become

$$\frac{\partial^3 w}{\partial x^3} + (2-\nu)\frac{\partial^3 w}{\partial x\,\partial y^2} = 0, \qquad \text{at } x = \text{constant}$$
$$\frac{\partial^3 w}{\partial y^3} + (2-\nu)\frac{\partial^3 w}{\partial x^2\,\partial y} = 0, \qquad \text{at } y = \text{constant} \tag{10-130}$$

## 10-3-2 Bending of a Simply Supported Rectangular Plate

For a simply supported rectangular plate (Fig. 10-33) we select the Cartesian coordinate system, and consequently the differential equation for bending has the form of Eq. (10-20). Let us determine the boundary conditions before deciding on the method of solution. First we note that the differential equation (10-20)

$$D\left(\frac{\partial^4 w}{\partial x^4} + 2\frac{\partial^4 w}{\partial x^2\,\partial y^2} + \frac{\partial^4 w}{\partial y^4}\right) = q \tag{10-131}$$

is fourth-order equation with respect to $x$ and to $y$. Therefore it requires four boundary conditions for fixed $x$ and four for fixed $y$. Since the boundary is composed of four segments ($x$ is fixed at two of them and $y$ at the other two), we have two boundary conditions at each of these segments. The deflection is zero and the bending moment in the direction perpendicular to the edge is zero, that is,

$$w(0, y) = 0, \qquad w(a, y) = 0, \qquad w(x, 0) = 0, \qquad w(x, b) = 0$$
$$M_x(0, y) = 0, \qquad M_x(a, y) = 0, \qquad M_y(x, 0) = 0, \qquad M_y(x, b) = 0 \tag{10-132}$$

By using the relations between the bending moments and the second derivatives of the deflection $w(x, y)$ [Eqs. (10-9) and (10-11)], we write the last four of the boundary conditions (10-132) in the following form:

$$\frac{\partial^2 w}{\partial x^2} + \nu\frac{\partial^2 w}{\partial y^2} = 0, \qquad \text{at } x = 0 \text{ and } x = a$$
$$\frac{\partial^2 w}{\partial y^2} + \nu\frac{\partial^2 w}{\partial x^2} = 0, \qquad \text{at } y = 0 \text{ and } y = b \tag{10-133}$$

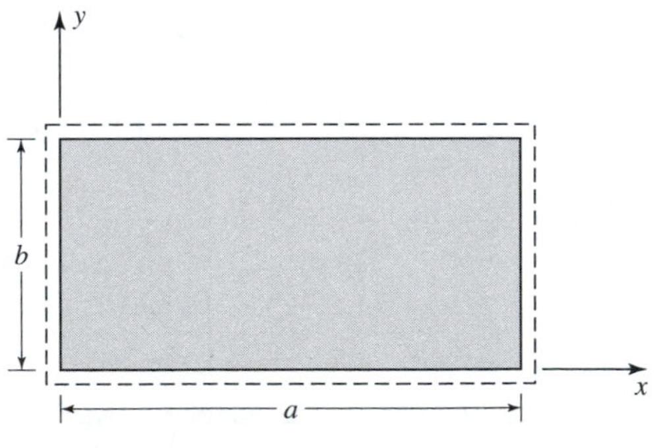


**Figure 10-33** A simply supported rectangular plate

Recalling the boundary conditions (10-132), we note that

$$\frac{\partial w}{\partial y} = \frac{\partial^2 w}{\partial y^2} = 0, \qquad \text{at } x = 0 \text{ and } x = a$$

$$\frac{\partial w}{\partial x} = \frac{\partial^2 w}{\partial x^2} = 0, \qquad \text{at } y = 0 \text{ and } y = b$$

so that the boundary conditions (10-133) reduce to

$$\begin{aligned} \frac{\partial^2 w}{\partial x^2} &= 0, \qquad \text{at } x = 0 \text{ and } x = a \\ \frac{\partial^2 w}{\partial y^2} &= 0, \qquad \text{at } y = 0 \text{ and } y = b \end{aligned} \tag{10-134}$$

The easiest way to obtain the solution to the differential equation (10-131) with the boundary conditions (10-132) and (10-134) is to use the Fourier series.* The expansion of a function of a single variable into a Fourier series was discussed briefly in Section 8-9. The expansion of a function of two variables, such as $w(x, y)$, into a Fourier series is similar. Here we restrict ourselves to the double sine series. (Other types of two-dimensional expansion of a function into a Fourier series are the double cosine series and the mixed sine-cosine series.) We write

$$w(x, y) = \sum_{m=1}^{\infty} \sum_{n=1}^{\infty} w_{mn} \sin \frac{m\pi x}{a} \sin \frac{n\pi y}{b} \tag{10-135}$$

where the Fourier coefficients $w_{mn}$ are given by

$$w_{mn} = \frac{4}{ab} \int_0^a \int_0^b w(x, y) \sin \frac{m\pi x}{a} \sin \frac{n\pi y}{b}\, dx\, dy, \qquad m, n = 1, \ldots, \infty \tag{10-136}$$

This result can be verified by substituting the series (10-135) for $w(x, y)$ into Eq. (10-136) and using the orthogonality property (see Section 8-9). Moreover, the known function $q(x, y)$ appearing in Eq. (10-131) is also expanded into a double sine series,

$$q(x, y) = \sum_{m=1}^{\infty} \sum_{n=1}^{\infty} q_{mn} \sin \frac{m\pi x}{a} \sin \frac{n\pi y}{b} \tag{10-137}$$

where

$$q_{mn} = \frac{4}{ab} \int_0^a \int_0^b q(x, y) \sin \frac{m\pi x}{a} \sin \frac{n\pi y}{b}\, dx\, dy \tag{10-138}$$

Now we substitute Eq. (10-135) into the boundary conditions, Eqs. (10-134), and also, using Eq. (10-137), into the differential equation (10-131) and keep in mind that the series (10-135) can be differentiated term by term because both $w(x, y)$ and its second derivatives vanish at all edges. Here the boundary conditions are identically satisfied,

*Joseph Fourier (1768–1830), French mathematician.

whereas for different boundary conditions other, more complicated, differentiation formulas must be used (see [1, pp. 348–356] and [13, pp. 137–143]; see also Section 8-9). The differential equation (10-131) becomes

$$D\sum_{m=1}^{\infty}\sum_{n=1}^{\infty}\left(\frac{m^2\pi^2}{a^2}+\frac{n^2\pi^2}{b^2}\right)^2 w_{mn}\sin\frac{m\pi x}{a}\sin\frac{n\pi y}{b}$$

$$-\sum_{m=1}^{\infty}\sum_{n=1}^{\infty} q_{mn}\sin\frac{m\pi x}{a}\sin\frac{n\pi y}{b}=0$$

By combining both terms we get an equation of the form

$$\sum_{m=1}^{\infty}\sum_{n=1}^{\infty} A_{mn}\sin\frac{m\pi x}{a}\sin\frac{n\pi y}{b}=0, \qquad 0\le x\le a, \quad 0\le y\le b$$

implying that $A_{mn}=0$ and thus

$$D\left(\frac{m^2\pi^2}{a^2}+\frac{n^2\pi^2}{b^2}\right)^2 w_{mn}=q_{mn}, \qquad m,n=1,\ldots,\infty$$

or

$$w_{mn}=\frac{q_{mn}}{D\left(\dfrac{m^2\pi^2}{a^2}+\dfrac{n^2\pi^2}{b^2}\right)^2}, \qquad m,n=1,\ldots,\infty \tag{10-139}$$

Recalling the expansion (10-135), we express the deflection of the plate at an arbitrary point in the form

$$w(x,y)=\frac{a^4}{D\pi^4}\sum_{m=1}^{\infty}\sum_{n=1}^{\infty}\frac{q_{mn}}{(m^2+n^2(a^2/b^2))^2}\sin\frac{m\pi x}{a}\sin\frac{n\pi y}{b} \tag{10-140}$$

## EXAMPLE 10-7

Find the maximum deflection of a square plate carrying a uniform load $q(x,y)=q_0=$ constant.

### SOLUTION

We first assume that $b=a$ and calculate $q_{mn}$,

$$q_{mn}=\frac{4}{a^2}\int_0^a\int_0^a q_0\sin\frac{m\pi x}{a}\sin\frac{n\pi y}{a}\,dx\,dy=\frac{4q_0}{a^2}\frac{a^2}{mn\pi^2}\cos\frac{m\pi x}{a}\bigg|_0^a\cos\frac{n\pi y}{a}\bigg|_0^a$$

$$=\frac{4q_0}{mn\pi^2}[1-(-1)^m][1-(-1)^n], \qquad m,n=1,\ldots,\infty$$

where we used the result $\cos m\pi = (-1)^m$. Putting this in a more convenient form,

$$q_{mn} = \begin{cases} 0, & m \text{ or } n \text{ even} \\ \dfrac{16q_0}{mn\pi^2}, & m \text{ and } n \text{ odd} \end{cases} \tag{10-141}$$

and substituting this into the expression (10-140), we obtain

$$w(x, y) = \frac{16q_0a^4}{D\pi^6} \sum_{m=1,3}^{\infty} \sum_{n=1,3}^{\infty} \frac{1}{mn(m^2+n^2)^2} \sin\frac{m\pi x}{a} \sin\frac{n\pi y}{a} \tag{10-142}$$

Because of symmetry, the maximum deflection occurs at the center of the plate,

$$\begin{aligned} \max w(x, y) = w\left(\frac{a}{2}, \frac{a}{2}\right) &= \frac{16q_0a^4}{D\pi^6} \sum_{m=1,3}^{\infty} \sum_{n=1,3}^{\infty} \frac{\sin\dfrac{m\pi}{2}\sin\dfrac{n\pi}{2}}{mn(m^2+n^2)^2} \\ &= \frac{16q_0a^4}{D\pi^6}\left[\frac{1}{2^2} - \frac{1}{3\times 10^2} - \frac{1}{3\times 10^2} + \frac{1}{9\times 18^2} + \frac{1}{5\times 26^2} \right. \\ &\qquad \left. + \frac{1}{5\times 26^2} - \frac{1}{15\times 34^2} - \frac{1}{15\times 34^2} + \frac{1}{25\times 50^2} + \cdots\right] \\ &= \frac{16q_0a^4}{D\pi^6}0.244 \end{aligned} \tag{10-143}$$

It is easily seen that taking into account only the first term ($m = 1, n = 1$) of the series results in a 2.5% error. Considerably more terms would be needed to ensure sufficient accuracy of the series representing the bending moments, whereas the series representing the shear forces are very slowly converging. This is seen by inspection of Eq. (10-142): each differentiation with respect to $x$ or $y$ results in multiplying the general term in the series by $m$ or $n$. Thus the numerator of each term increases, making its influence on the total sum more important. These difficulties can be overcome, however, by replacing the double series by a single series. This can be accomplished in the following way:

$$\begin{aligned} S(x, y) &= \sum_{m=1,3}^{\infty} \sum_{n=1,3}^{\infty} \frac{1}{mn(m^2+n^2)^2} \sin\frac{m\pi x}{a} \sin\frac{n\pi y}{a} \\ &\equiv \sum_{m=1,3}^{\infty} \frac{\sin\dfrac{m\pi x}{a}}{m} \sum_{n=1,3}^{\infty} \frac{\sin\dfrac{n\pi y}{a}}{n(n^2+m^2)^2} \end{aligned} \tag{10-144}$$

The rightmost series can be represented in closed form [11], yielding

$$S(x, y) = \sum_{m=1,3}^{\infty} \frac{\sin\dfrac{m\pi x}{a}}{m} f_m(y) \tag{10-145}$$

where $f_m(y)$ is

$$f_m(y) = \frac{\pi}{4m^4} - \frac{\pi}{16m^4 \cosh^2 \dfrac{m\pi}{2}} \left[ 4 \cosh \frac{m\pi[1-2(x/a)]}{2} \cosh \frac{m\pi}{2} - m\pi\left(1 - 2\frac{x}{a}\right) \right.$$

$$\left. \times \sinh \frac{m\pi[1-2(x/a)]}{2} \cosh \frac{m\pi}{2} + m\pi \cosh \frac{m\pi[1-2(x/a)]}{2} \sinh \frac{m\pi}{2} \right]$$

It is evident that for large $m$, $f_m(y)$ behaves like $1/m^3$, making the series rapidly converging. However, in the present example the series appearing in Eq. (10-143) is also converging rapidly, so there is little numerical advantage of replacing it by a closed-form expression. On the other hand, thousands of terms may be needed to obtain sufficiently accurate sums of series converging more slowly. In such cases closed-form expressions are very important. Note that the double series, such as $S(x, y)$ or its derivatives, can be further simplified by replacing hyperbolic functions by their asymptotic expressions.

## 10-4 PLATES ON ELASTIC FOUNDATIONS

In Section 8-4 we examined beams resting on Winkler's foundation (see Section 8-4). Now we will analyze plates on Winkler's foundation, noting that the differential equation for the deflection of such plates changes only because of the presence of the foundation's reaction. This is assumed to be proportional to the deflection of the plate. Thus the distributed load $q$ appearing on the right-hand side of Eq. (10-20) is now replaced by $q - kw$, where $k$ is the *foundation modulus*. Therefore in the Cartesian coordinate system, Eq. (10-20) becomes

$$D\left(\frac{\partial^4 w}{\partial x^4} + 2\frac{\partial^4 w}{\partial x^2\,\partial y^2} + \frac{\partial^4 w}{\partial y^4}\right) = q - kw \tag{10-146}$$

By analyzing here a rectangular plate, which, in addition to resting on an elastic foundation, is simply supported along its boundary, we find that the boundary conditions obtained in Section 10-3 are still valid. Thus applying the method of expansion into a Fourier series and assuming

$$w(x, y) = \sum_{m=1}^{\infty}\sum_{n=1}^{\infty} w_{mn} \sin\frac{m\pi x}{a} \sin\frac{n\pi y}{b} \tag{10-147}$$

we repeat the steps outlined in Section 10-3, arriving at the following expression for $w_{mn}$:

$$w_{mn} = \frac{q_{mn}}{D\left(\dfrac{m^2\pi^2}{a^2} + \dfrac{n^2\pi^2}{b^2}\right)^2 + k} \tag{10-148}$$

where $q_{mn}$ is the Fourier coefficient of $q(x, y)$ given by Eq. (10-138). Therefore the expression for the deflection becomes

$$w(x, y) = \sum_{m=1}^{\infty}\sum_{n=1}^{\infty} \frac{q_{mn}}{D\left(\dfrac{m^2\pi^2}{a^2} + \dfrac{n^2\pi^2}{b^2}\right)^2 + k} \sin\frac{m\pi x}{a} \sin\frac{n\pi y}{b} \tag{10-149}$$

By examining Eq. (10-149) we note that for $k = 0$ we recover the result (10-140), that is, the entire load $q(x, y)$ is then transmitted to the supports. With the increase of $k$ from 0 to $\infty$ the load transmitted to the foundation also increases. Finally, as $k$ reaches $\infty$, the deflection $w(x, y)$ becomes zero everywhere, the support reactions vanish, and the entire load is transmitted to the foundation. The solutions for plates of other shapes and with other boundary conditions can be found in [4].

## 10-5 STRAIN ENERGY OF AN ELASTIC PLATE

The derivation of the expression for the strain energy of a plate resembles that for a beam (see Chapter 5). We begin again with Eq. (5-83), expressing the total strain energy stored in the body,

$$U = \frac{1}{2}\int_V (\sigma_{xx}\varepsilon_{xx} + \sigma_{yy}\varepsilon_{yy} + \sigma_{zz}\varepsilon_{zz} + \sigma_{xy}\gamma_{xy} + \sigma_{xz}\gamma_{xz} + \sigma_{yz}\gamma_{yz})\,dV \tag{10-150}$$

and substitute for the strains $\varepsilon_{xx}$, $\varepsilon_{yy}$, $\gamma_{xy}$ and for the stresses $\sigma_{xx}$, $\sigma_{yy}$, $\sigma_{xy}$ their expressions (10-1)–(10-3) and (10-5), respectively. Equating the remaining stresses to zero, we obtain

$$U = \frac{E}{2(1-\nu^2)}\int_V \left[\left(\frac{\partial^2 w}{\partial x^2} + \nu\frac{\partial^2 w}{\partial y^2}\right)\frac{\partial^2 w}{\partial x^2} + \left(\frac{\partial^2 w}{\partial y^2} + \nu\frac{\partial^2 w}{\partial x^2}\right)\frac{\partial^2 w}{\partial y^2} + 2(1-\nu)\left(\frac{\partial^2 w}{\partial x\,\partial y}\right)^2\right] z^2\,dV$$

But in the Cartesian coordinate system $dV = dx\,dy\,dz$, and none of the terms in parentheses depends on $z$. Thus we can integrate this expression with respect to $z$ independently of the other variables, which then becomes

$$\int_{-h/2}^{h/2} z^2\,dz = \frac{h^3}{12}$$

Rearranging terms and recalling that

$$\frac{Eh^3}{12(1-\nu^2)} \equiv D$$

we get

$$U = \frac{D}{2}\int_A \left\{\left(\frac{\partial^2 w}{\partial x^2} + \frac{\partial^2 w}{\partial y^2}\right)^2 - 2(1-\nu)\left[\frac{\partial^2 w}{\partial x^2}\frac{\partial^2 w}{\partial y^2} - \left(\frac{\partial^2 w}{\partial x\,\partial y}\right)^2\right]\right\} dx\,dy \tag{10-151}$$

where $A$ is the area of the plate. In order to use another coordinate system, we replace the Cartesian element of the area, $dx\,dy$, by an expression appropriate for the new system, and we also transform the partial derivatives appearing in the integrand. Expression (10-151) is particularly useful when employing methods based on the principles of minimum energy (see Section 10-8-2).

## 10-6 MEMBRANES

A membrane is a very thin plate with a negligibly small bending stiffness $D$. If a membrane, supported along its edge, were subject to a load $p = p(x, y)$, then, in view of its negligible stiffness, it would not be able to carry this load but would collapse. On the other hand, if we assume that a self-equilibrating system of tensile forces $T$ is applied to the edges of the membrane, then, as we have seen in Chapter 9 in connection with "membrane analogy," a state of equilibrium exists. It is described, in Cartesian coordinates, by the differential equation

$$\nabla^2 w \equiv \frac{\partial^2 w}{\partial x^2} + \frac{\partial^2 w}{\partial y^2} = -\frac{p}{T} \tag{10-152}$$

Note that if the deformation of the membrane were not very small, we would have to take into account its effect on the magnitude and on the distribution of the tensile forces $T$. This would lead to a nonlinear problem considerably more difficult to analyze than the linearized version described by Eq. (10-152). We note that such idealized structures hardly exist, because every plate has some, however small, bending stiffness. However, the value of discussing membrane problems lies in an analogy [14] between Eq. (10-152) and an equation that we will now develop. Adding Eqs. (10-9) and (10-11) we get

$$M_x + M_y = -D(1+\nu)\left(\frac{\partial^2 w}{\partial x^2} + \frac{\partial^2 w}{\partial y^2}\right)$$

or

$$\nabla^2 w = -\frac{M}{D} \tag{10-153}$$

where

$$M = \frac{M_x + M_y}{1+\nu}$$

is called the *moment sum*. Next Eq. (10-153) is substituted into Eq. (10-21),

$$\nabla^2 M = -q \tag{10-154}$$

Both Eq. (10-153) and Eq. (10-154) are of the same form as the membrane equation (10-152). We also note that $M$ is proportional to the sum of the curvatures of $w(x, y)$ in two mutually perpendicular directions. Therefore on an edge not parallel to the $x$, $y$ axes,

$$M = \frac{M_n + M_s}{1+\nu} = -D\left(\frac{\partial^2 w}{\partial n^2} + \frac{\partial^2 w}{\partial s^2}\right)$$

where $n$ and $s$ are directions normal and tangent to this edge, respectively. When $w$ is zero at the edge, so are its derivatives in the tangent, that is, the $s$ direction. Thus $M_n$ and $M$ reduce to

$$M_n = -D(1+\nu)\frac{\partial^2 w}{\partial n^2}, \qquad M = -D\frac{\partial^2 w}{\partial n^2}$$

Consequently, when the edge is simply supported,

$$w = 0, \qquad M_n = 0 > M = 0$$

Now assume that $T = T_0$, a constant, and let us find the deflection of a rectangular membrane carrying a uniform load $p_0$. Because of the shape of the membrane, the method of expansion into the Fourier series (see Section 10-3) is particularly suitable. We begin by expanding the unknown deflection into a Fourier sine series,

$$w(x, y) = \sum_{m=1}^{\infty}\sum_{n=1}^{\infty} w_{mn} \sin\frac{m\pi x}{a} \sin\frac{n\pi y}{b} \tag{10-155}$$

This choice is dictated by the boundary conditions

$$w(x, 0) = w(x, b) = w(0, y) = w(a, y) = 0 \tag{10-156}$$

identically satisfied by expression (10-155). By expanding also the right-hand side of Eq. (10-152) into a Fourier sine series we get

$$-\frac{p_0}{T_0} = \sum_{m=1}^{\infty}\sum_{n=1}^{\infty} p_{mn} \sin\frac{m\pi x}{a} \sin\frac{n\pi y}{b} \tag{10-157}$$

where

$$p_{mn} = -\frac{4p_0}{abT_0}\int_0^a\int_0^b \sin\frac{m\pi x}{a} \sin\frac{n\pi y}{b}\, dx\, dy = \begin{cases} -\dfrac{16p_0}{mn\pi^2 T_0}, & m \text{ and } n \text{ odd} \\ 0, & m \text{ or } n \text{ even} \end{cases} \tag{10-158}$$

Substituting this, and the expression for $w(x, y)$, into the differential equation (10-152), gives

$$-\left(\frac{m^2\pi^2}{a^2} + \frac{n^2\pi^2}{b^2}\right) w_{mn} = \begin{cases} -\dfrac{16p_0}{mn\pi T_0}, & m \text{ and } n \text{ odd} \\ 0, & m \text{ or } n \text{ even} \end{cases} \tag{10-159}$$

In arriving at Eq. (10-159), the transformations explained in Section 10-3 have been used. Evidently the deflection of the membrane is inversely propotional to the applied tension $T_0$,

$$w(x, y) = \frac{16p_0 a^2}{\pi^3 T_0} \sum_{m=1,3}^{\infty}\sum_{n=1,3}^{\infty} \frac{1}{[m^2 + n^2(a^2/b^2)]mn} \sin\frac{m\pi x}{a} \sin\frac{n\pi y}{b} \tag{10-160}$$

## 10-7 COMPOSITE PLATES

In this section we consider plates composed of several layers made of different materials and known as *laminated plates*. For a more general analysis see [16], [6, pp. 239–259], [15, pp. 63–78], and [13].

### 10-7-1 Laminated Plates with Isotropic Layers

Let us analyze a plate composed of $N$ perfectly bonded layers (*laminae*) made of different isotropic materials (Fig. 10-34). All the assumptions made in reference to homogeneous plates are still valid. We also make an additional assumption of a perfect bond between the layers. This, combined with the assumption of the normal remaining normal after deformation, ensures that the in-plane displacements $u$ and $v$ of the laminated element remain linear functions of the $z$ coordinate. With Hooke's law this implies that the strains remain also linear functions of $z$. Thus the expressions (10-1)–(10-3) derived in Section 10-1, relating strains to the vertical displacement $w$, are still valid,

$$\varepsilon_{yy} = -z\frac{\partial^2 w}{\partial y^2}, \qquad \varepsilon_{xx} = -z\frac{\partial^2 w}{\partial x^2}, \qquad \gamma_{xy} = -2z\frac{\partial^2 w}{\partial x\,\partial y} \tag{10-161}$$

The assumed validity of Hooke's law also ensures that the stress varies linearly within each lamina, but the differences in material properties of adjacent laminae lead to discontinuities in the overall stress diagram (see Fig. 10-35). We then have the following

Figure 10-34 A laminated plate with isotropic layers

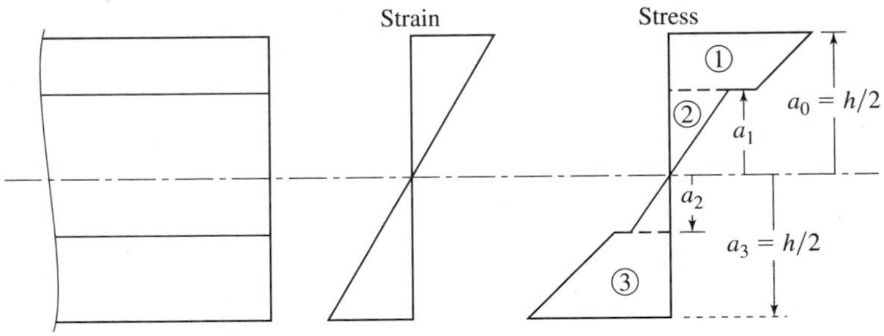


Figure 10-35 The differences in material properties of the layers lead to discontinuities in the stress diagram

set of expressions for stresses within each of the $k$ laminae [see Eqs. (10-5)]:

$$\begin{aligned}
\sigma_{xx}^{(k)} &= -\frac{E_k z}{1-\nu_k^2}\left(\frac{\partial^2 w}{\partial x^2} + \nu_k \frac{\partial^2 w}{\partial y^2}\right) = -z\left(A_k \frac{\partial^2 w}{\partial x^2} + B_k \frac{\partial^2 w}{\partial y^2}\right) \\
\sigma_{yy}^{(k)} &= -\frac{E_k z}{1-\nu_k^2}\left(\frac{\partial^2 w}{\partial y^2} + \nu_k \frac{\partial^2 w}{\partial x^2}\right) = -z\left(A_k \frac{\partial^2 w}{\partial y^2} + B_k \frac{\partial^2 w}{\partial x^2}\right) \qquad (10\text{-}162) \\
\sigma_{xy}^{(k)} &= -\frac{E_k z}{1+\nu_k}\frac{\partial^2 w}{\partial x\, \partial y} = -zC_k \frac{\partial^2 w}{\partial x\, \partial y}, \qquad k = 1, \ldots, N
\end{aligned}$$

where

$$A_k = \frac{E_k}{1-\nu_k^2}, \qquad B_k = \frac{E_k \nu_k}{1-\nu_k^2}, \qquad C_k = \frac{E_k}{1+\nu_k} \qquad (10\text{-}163)$$

In order to derive the expressions for shear stresses $\sigma_{xz}$ and $\sigma_{yz}$, we first modify Eqs. (10-6) to read

$$\frac{\partial \sigma_{xx}^{(k)}}{\partial x} + \frac{\partial \sigma_{xy}^{(k)}}{\partial y} + \frac{\partial \sigma_{xz}^{(k)}}{\partial z} = 0, \qquad \frac{\partial \sigma_{xy}^{(k)}}{\partial x} + \frac{\partial \sigma_{yy}^{(k)}}{\partial y} + \frac{\partial \sigma_{yz}^{(k)}}{\partial z} = 0 \qquad (10\text{-}164)$$

where the superscript $k$ indicates the position of the infinitesimal element. Next Eqs. (10-162) are substituted into Eqs. (10-164) to give

$$\begin{aligned}
\frac{\partial \sigma_{xz}^{(k)}}{\partial z} &= z\left[A_k \frac{\partial^3 w}{\partial x^3} + (B_k + C_k)\frac{\partial^3 w}{\partial x\, \partial y^2}\right] \\
\frac{\partial \sigma_{yz}^{(k)}}{\partial z} &= z\left[A_k \frac{\partial^3 w}{\partial y^3} + (B_k + C_k)\frac{\partial^3 w}{\partial x^2\, \partial y}\right]
\end{aligned} \qquad (10\text{-}165)$$

Integrating Eq. (10-165) with respect to $z$ and taking into account that both $\sigma_{xz}$ and $\sigma_{yz}$ are zero at $z = \pm h/2$, we get

$$\begin{aligned}
\sigma_{xz}^{(k)}(z) &= \frac{1}{2}\left(\frac{h^2}{4} - z^2\right)\left[A_k \frac{\partial^3 w}{\partial x^3} + (B_k + C_k)\frac{\partial^3 w}{\partial x\, \partial y^2}\right] \\
\sigma_{yz}^{(k)}(z) &= \frac{1}{2}\left(\frac{h^2}{4} - z^2\right)\left[A_k \frac{\partial^3 w}{\partial y^3} + (B_k + C_k)\frac{\partial^3 w}{\partial x^2\, \partial y}\right]
\end{aligned} \qquad (10\text{-}166)$$

When defining the global quantities $M_x$, $M_y$, and $M_{xy}$ [see Eqs. (10-9), (10-11), and (10-12)] we must take into account the changes in stresses between various layers. Therefore by using as an example the arrangement of Fig. 10-35, we get

$$\begin{aligned}
M_x &= \int_{-h/2}^{h/2} \sigma_{xx} z\, dz \\
&= -\frac{\partial^2 w}{\partial x^2}\left(\int_{a_3=-h/2}^{a_2} A_3 z^2\, dz + \int_{a_2}^{a_1} A_2 z^2\, dz + \int_{a_1}^{a_0=h/2} A_1 z^2\, dz\right) \\
&\quad - \frac{\partial^2 w}{\partial y^2}\left(\int_{a_3}^{a_2} B_3 z^2\, dz + \int_{a_2}^{a_1} B_2 z^2\, dz + \int_{a_1}^{a_0} B_1 z^2\, dz\right)
\end{aligned}$$

Integrating and simplifying the results, this becomes

$$M_x = -D_{11}\frac{\partial^2 w}{\partial x^2} - D_{12}\frac{\partial^2 w}{\partial y^2} \tag{10-167}$$

where

$$D_{11} = \frac{1}{3}\sum_{k=1}^{3} A_k\left(a_{k-1}^3 - a_k^3\right), \qquad D_{12} = \frac{1}{3}\sum_{k=1}^{3} B_k\left(a_{k-1}^3 - a_k^3\right) \tag{10-168}$$

Proceeding in a similar manner, we get

$$\begin{aligned} M_y &= -D_{12}\frac{\partial^2 w}{\partial x^2} - D_{22}\frac{\partial^2 w}{\partial y^2} \\ M_{xy} &= -D_{66}\frac{\partial^2 w}{\partial x\,\partial y} \end{aligned} \tag{10-169}$$

where $D_{22} = D_{11}$ and

$$D_{66} = \frac{1}{3}\sum_{k=1}^{3} C_k\left(a_{k-1}^3 - a_k^3\right) \tag{10-170}$$

Finally, using Eqs. (10-166) the shear forces $Q_x$ and $Q_y$ can be written,

$$\begin{aligned} Q_x &= \int_{-h/2}^{h/2}\sum_{k=1}^{3}\sigma_{xz}^{(k)}\,dz \\ &= \sum_{k=1}^{3}\left[\frac{a_k^3 - a_{k-1}^3}{6} - \frac{h^2(a_k - a_{k-1})}{8}\right]\left[A_k\frac{\partial^3 w}{\partial x^3} + (B_k + C_k)\frac{\partial^3 w}{\partial x\,\partial y^2}\right] \\ Q_y &= \int_{-h/2}^{h/2}\sum_{k=1}^{3}\sigma_{yz}^{(k)}\,dz \\ &= \sum_{k=1}^{3}\left[\frac{a_k^3 - a_{k-1}^3}{6} - \frac{h^2(a_k - a_{k-1})}{8}\right]\left[A_k\frac{\partial^3 w}{\partial y^3} + (B_k + C_k)\frac{\partial^3 w}{\partial x^2\,\partial y}\right] \end{aligned} \tag{10-171}$$

Thus the reduced shear forces $V_x$ and $V_y$ [Eqs. (10-129)] become

$$\begin{aligned} V_x &= Q_x + \frac{\partial M_{xy}}{\partial y} \\ &= \sum_{k=1}^{3}\left[\frac{a_k^3 - a_{k-1}^3}{6} - \frac{h^2(a_k - a_{k-1})}{8}\right]\left[A_k\frac{\partial^3 w}{\partial x^3} + (B_k + C_k)\frac{\partial^3 w}{\partial x\,\partial y^2}\right] - D_{66}\frac{\partial^2 w}{\partial x\,\partial y} \\ V_y &= Q_y + \frac{\partial M_{yx}}{\partial x} \\ &= \sum_{k=1}^{3}\left[\frac{a_k^3 - a_{k-1}^3}{6} - \frac{h^2(a_k - a_{k-1})}{8}\right]\left[A_k\frac{\partial^3 w}{\partial y^3} + (B_k + C_k)\frac{\partial^3 w}{\partial x^2\,\partial y}\right] - D_{66}\frac{\partial^2 w}{\partial x^2\,\partial y} \end{aligned} \tag{10-172}$$

It is evident that for $N$ layers the preceding expressions must be modified by replacing 3 by $N$ as the sum limit. The derivation of the equilibrium equations is identical with that shown for homogeneous plates until we obtain Eq. (10-19),

$$\frac{\partial^2 M_x}{\partial x^2} + 2\frac{\partial^2 M_{xy}}{\partial x\, \partial y} + \frac{\partial^2 M_y}{\partial y^2} = -q \tag{10-173}$$

At this point we substitute the expressions (10-167), (10-170), and (10-171) for the moments into Eq. (10-173). This gives the following differential equation for bending of a laminated, thin plate composed of isotropic layers,

$$D_{11}\frac{\partial^4 w}{\partial x^4} + 2(D_{12} + D_{66})\frac{\partial^4 w}{\partial x^2\, \partial y^2} + D_{11}\frac{\partial^4 w}{\partial y^4} = q \tag{10-174}$$

If we now examine the definitions of $D_{12}$ and $D_{66}$, we find that

$$D_{12} + D_{66} = D_{11} \tag{10-175}$$

Therefore Eq. (10-170) becomes

$$D_{11}\nabla^2\nabla^2 w = q \tag{10-176}$$

which is identical to Eq. (10-21) derived for the homogeneous plate, except that the bending stiffness $D$ must be replaced by the bending stiffness of a laminated plate, Eq. (10-168). Obviously the differential equation (10-176) cannot be solved without prescribing boundary conditions. If these involve bending moments or torques, then the stiffnesses $D_{12}$ and $D_{66}$ will appear in the final solution of the problem, unless the plate is supported all around. For example, for a rectangular plate with dimensions $a \times b$ the boundary conditions will be as follows.*

### Simply Supported Edge

$$\begin{aligned} w = 0, &\qquad M_x = -D_{11}\frac{\partial^2 w}{\partial x^2} - D_{12}\frac{\partial^2 w}{\partial y^2} = 0, &\qquad \text{at } x = \text{constant} \\ w = 0, &\qquad M_y = -D_{12}\frac{\partial^2 w}{\partial x^2} - D_{22}\frac{\partial^2 w}{\partial y^2} = 0, &\qquad \text{at } y = \text{constant} \end{aligned} \tag{10-177}$$

which reduce to

$$\begin{aligned} w = 0, &\qquad \frac{\partial^2 w}{\partial x^2} = 0, &\qquad \text{at } x = \text{constant} \\ w = 0, &\qquad \frac{\partial^2 w}{\partial y^2} = 0, &\qquad \text{at } y = \text{constant} \end{aligned} \tag{10-178}$$

These are identical to the boundary conditions, Eqs. (10-124), valid for a homogeneous plate.

*We used here Eqs. (10-122)–(10-130) with modified definitions of moments and shear forces, Eqs. (10-167), (10-169), and (10-171).

### Clamped Edge

$$
\begin{aligned}
w &= 0, \qquad \frac{\partial w}{\partial x} = 0, \qquad \text{at } x = \text{constant} \\
w &= 0, \qquad \frac{\partial w}{\partial y} = 0, \qquad \text{at } y = \text{constant}
\end{aligned}
\tag{10-179}
$$

are again the same as the conditions for a homogeneous plate, Eqs. (10-125).

### Free Edge

The conditions at a free edge, requiring that the appropriate bending moment and the effective shear force be zero, become

$$
\begin{aligned}
&-D_{11}\frac{\partial^2 w}{\partial x^2} - D_{12}\frac{\partial^2 w}{\partial y^2} = 0 \\
&\sum_{k=1}^{3}\left[\frac{a_k^3 - a_{k-1}^3}{6} - \frac{h^2(a_k - a_{k-1})}{8}\right]\left[A_k\frac{\partial^3 w}{\partial x^3} + (B_k + C_k)\frac{\partial^3 w}{\partial x\,\partial y^2}\right] \\
&\qquad - D_{66}\frac{\partial^3 w}{\partial x\,\partial y^2} = 0, \qquad \text{at } x = \text{constant}
\end{aligned}
\tag{10-180}
$$

and

$$
\begin{aligned}
&-D_{12}\frac{\partial^2 w}{\partial x^2} - D_{22}\frac{\partial^2 w}{\partial y^2} = 0 \\
&\sum_{k=1}^{3}\left[\frac{a_k^3 - a_{k-1}^3}{6} - \frac{h^2(a_k - a_{k-1})}{8}\right]\left[A_k\frac{\partial^3 w}{\partial y^3} + (B_k + C_k)\frac{\partial^3 w}{\partial x^2\,\partial y}\right] \\
&\qquad - D_{66}\frac{\partial^3 w}{\partial x^2\,\partial y} = 0, \qquad \text{at } y = \text{constant}
\end{aligned}
\tag{10-181}
$$

For axisymmetric bending of a laminated circular plate we repeat the analysis presented in Section 10-2 for homogeneous plates. This gives the following differential equation for deflection:

$$
D_{11}\left(\frac{d^2}{dr^2} + \frac{1}{r}\frac{d}{dr}\right)\left(\frac{d^2 w}{dr^2} + \frac{1}{r}\frac{dw}{dr}\right) = q(r)
\tag{10-182}
$$

The expressions for the bending moments and the shear force are

$$
\begin{aligned}
M_r &= -D_{11}w'' - D_{12}\frac{1}{r}w' \\
M_\theta &= -D_{11}\frac{1}{r}w' - D_{12}w'' \\
Q_r &= \frac{M_r - M_\theta}{r} + M_r'
\end{aligned}
\tag{10-183}
$$

The last equation can be expressed as

$$Q_r = -D_{11}\frac{d}{dr}\left[\frac{1}{r}\frac{d}{dr}(rw')\right] \tag{10-184}$$

Note that both $w(r)$ and $w'(r)$ depend only on $D_{11}$, that is, for the results obtained previously for a clamped homogeneous plate to be valid, we have to replace $D$ by $D_{11}$.

## EXAMPLE 10-8

(a) Find the maximum deflection for a circular simply supported laminated two-ply plate, carrying a uniform load $q_0$. It is assumed that the bottom ply has a thickness equal to $1/4$ of the total thickness of the plate (Fig. 10-36), and that its Young's modulus is ten times that of the upper ply. It is further assumed that $\nu$ is constant.

(b) Compare the result with that obtained for an identical plate but made totally of the material of the bottom or of the top ply.

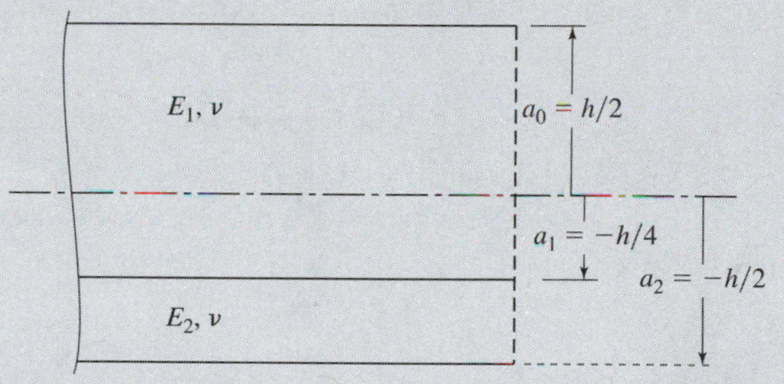


**Figure 10-36**

## SOLUTION

(a) Using the previously obtained general solution (see Example 10-1) we have

$$w(r) = C_1 + C_3r^2 + \frac{q_0r^4}{64D_{11}} \tag{10-185}$$

The boundary conditions are

$$w(R) = 0, \qquad M_r(R) = 0 \tag{10-186}$$

Thus using Eqs. (10-185) and (10-183), and recalling that the particular solution has the form

$$w_q(r) = \frac{q_0r^4}{64D_{11}}$$

we get the following system of algebraic equations:

$$C_1 + C_3R^2 + \frac{q_0R^4}{64D_{11}} = 0$$

$$-2D_{11}C_3 - 2D_{12}C_3 - \frac{(3D_{11} + D_{12})q_0R^2}{16D_{11}} = 0$$

The solution to this system yields

$$C_1 = \frac{5D_{11} + D_{12}}{D_{11}(D_{11} + D_{12})}\frac{q_0R^4}{64}, \qquad C_3 = -\frac{3D_{11} + D_{12}}{D_{11}(D_{11} + D_{12})}\frac{q_0R^2}{32} \tag{10-187}$$

Therefore the deflection is

$$w(r) = \frac{q_0R^4}{64(D_{11} + D_{12})D_{11}}\left[5D_{11} + D_{12} - (3D_{11} + D_{12})\left(\frac{r}{R}\right)^2 + (D_{11} + D_{12})\left(\frac{r}{R}\right)^4\right] \tag{10-188}$$

The maximum deflection is at the center of the plate,

$$w_{\max} = \frac{(5D_{11} + D_{12})q_0R^4}{64D_{11}(D_{11} + D_{12})} \tag{10-189}$$

In order to calculate $D_{11}$ and $D_{12}$, we need the quantities $A_k$ and $B_k$. The distribution of the materials throughout the thickness of the plate is shown in Fig. 10-36.

By assuming that $E_2 = 10E_1$ we obtain

$$A_1 = \frac{E_1}{1 - \nu^2}, \qquad A_2 = \frac{E_2}{1 - \nu^2} = \frac{10E_1}{1 - \nu^2}$$

$$B_1 = \frac{E_1\nu}{1 - \nu^2}, \qquad B_2 = \frac{E_2\nu}{1 - \nu^2} = \frac{10E_1\nu}{1 - \nu^2}$$

$$D_{11} = \frac{1}{3}\sum_{k=1}^{2} A_k\left(a_{k-1}^3 - a_k^3\right) = \frac{E_1h^3}{3(1 - \nu^2)}\left[\frac{1}{8} + \frac{1}{64} + 10\left(-\frac{1}{64} + \frac{1}{8}\right)\right]$$

$$= \frac{E_1h^3}{3(1 - \nu^2)}\left(\frac{9}{64} + \frac{70}{64}\right) = \frac{79}{192}\frac{E_1h^3}{1 - \nu^2}$$

$$D_{12} = \frac{79}{192}\frac{\nu E_1h^3}{1 - \nu^2} = \nu D_{11}$$

$$w_{\max} = \frac{(5 + \nu)q_0R^4}{64(1 + \nu)D_{11}} \tag{10-190}$$

(b) If the plate were made totally from material 1 of the top ply, then the deflection would be

$$w_{\max} = \frac{(5+\nu)q_0 R^4}{64D(1+\nu)}$$

where

$$D = \frac{E_1 h^3}{12(1-\nu^2)} \tag{10-191}$$

Now

$$D_{11} = \frac{12 \times 79D}{192} = 4.9375D$$

so that the deflection of the laminated plate is about five times smaller than the deflection of a homogeneous plate made from material 1. If, on the other hand, the plate were made totally from material 2 of the bottom ply, then

$$D = \frac{E_2 h^3}{12(1-\nu^2)} = \frac{10E_1 h^3}{12(1-\nu^2)}$$

Hence,

$$D_{11} = \frac{12 \times 79}{192 \times 10} D = 0.4938D$$

so that the deflection of the laminated plate would be about two times greater than the deflection of a plate made from material 2.

## 10-7-2 Laminated Plates with Orthotropic Layers

Let us now analyze a plate composed of orthotropic layers (see Section 4-8-2) with axes of orthotropy parallel to either the $x$ or the $y$ coordinate axis. Such materials are often made of an isotropic *matrix* with embedded, equally spaced *fibers*. Assumptions made in Section 10-7-1 are still valid, and so are the expressions (10-161) for strains. With the stress-strain relations (4-61) the stress expression for each of the $k$ laminae becomes

$$\begin{aligned}
\sigma_{xx}^{(k)} &= -\frac{E_{kx} z}{1-\nu_{kxy}\nu_{kyx}}\left(\frac{\partial^2 w}{\partial x^2} + \nu_{kyx}\frac{\partial^2 w}{\partial y^2}\right) = -z\left(A_{kx}\frac{\partial^2 w}{\partial x^2} + B_{ko}\frac{\partial^2 w}{\partial y^2}\right) \\
\sigma_{yy}^{(k)} &= -\frac{E_{ky} z}{1-\nu_{kxy}\nu_{kyx}}\left(\frac{\partial^2 w}{\partial y^2} + \nu_{kxy}\frac{\partial^2 w}{\partial x^2}\right) = -z\left(A_{ky}\frac{\partial^2 w}{\partial y^2} + B_{ko}\frac{\partial^2 w}{\partial x^2}\right) \\
\sigma_{xy}^{(k)} &= -G_{kxy} z \frac{\partial^2 w}{\partial x\, \partial y}, \qquad k = 1, \ldots, N
\end{aligned} \tag{10-192}$$

where

$$A_{kx} = \frac{E_{kx}}{1 - \nu_{kxy}\nu_{kyx}}, \qquad A_{ky} = \frac{E_{ky}}{1 - \nu_{kxy}\nu_{kyx}}$$
$$B_{ko} = \frac{E_{kx}\nu_{kyx}}{1 - \nu_{kxy}\nu_{kyx}} \equiv \frac{E_{ky}\nu_{kxy}}{1 - \nu_{kxy}\nu_{kyx}} \tag{10-193}$$

Here $E_{kx}$ signifies Young's modulus of layer $k$ in the $x$ direction, and so on. Next Eqs. (10-192) are substituted into Eqs. (10-165), and the resulting expressions are then integrated to give (see Section 10-7-1 for details)

$$\sigma_{xz}^{(k)}(z) = \frac{1}{2}\left(\frac{h^2}{4} - z^2\right)\left[A_{kx}\frac{\partial^3 w}{\partial x^3} + (B_{ko} + G_{kxy})\frac{\partial^3 w}{\partial x\,\partial y^2}\right]$$
$$\sigma_{yz}^{(k)}(z) = \frac{1}{2}\left(\frac{h^2}{4} - z^2\right)\left[A_{ky}\frac{\partial^3 w}{\partial y^3} + (B_{ko} + G_{kxy})\frac{\partial^3 w}{\partial x^2\,\partial y}\right] \tag{10-194}$$

By using again the arrangement of Fig. 10-35 for a three-layer plate we get

$$M_x = \int_{-h/2}^{h/2} \sigma_{xx} z\, dz = -D_{11}\frac{\partial^2 w}{\partial x^2} - D_{12}\frac{\partial^2 w}{\partial y^2} \tag{10-195}$$

where

$$D_{11} = \frac{1}{3}\sum_{k=1}^{3} A_{kx}\left(a_{k-1}^3 - a_k^3\right), \qquad D_{12} = \frac{1}{3}\sum_{k=1}^{3} B_{ko}\left(a_{k-1}^3 - a_k^3\right) \tag{10-196}$$

and where 3, the upper limit of the sum, should be replaced by $N$ when an $N$-layer plate is analyzed. In a similar manner we obtain

$$M_y = -D_{12}\frac{\partial^2 w}{\partial x^2} - D_{22}\frac{\partial^2 w}{\partial y^2}, \qquad M_{xy} = -D_{66}\frac{\partial^2 w}{\partial x\,\partial y} \tag{10-197}$$

where

$$D_{22} = \frac{1}{3}\sum_{k=1}^{N} A_{ky}\left(a_{k-1}^3 - a_k^3\right), \qquad D_{66} = \frac{1}{3}\sum_{k=1}^{N} G_{kxy}\left(a_{k-1}^3 - a_k^3\right) \tag{10-198}$$

It is evident that the expressions for shear forces are similar to Eqs. (10-171), with the constants being appropriately replaced. Thus,

$$Q_x = \sum_{k=1}^{3}\left[\frac{a_k^3 - a_{k-1}^3}{6} - \frac{h^2(a_k - a_{k-1})}{8}\right]\left[A_{kx}\frac{\partial^3 w}{\partial x^3} + (B_{ko} + G_{kxy})\frac{\partial^3 w}{\partial x\,\partial y^2}\right]$$
$$Q_y = \sum_{k=1}^{3}\left[\frac{a_k^3 - a_{k-1}^3}{6} - \frac{h^2(a_k - a_{k-1})}{8}\right]\left[A_{ky}\frac{\partial^3 w}{\partial y^3} + (B_{ko} + G_{kxy})\frac{\partial^3 w}{\partial x^2\,\partial y}\right] \tag{10-199}$$

and

$$V_x = \sum_{k=1}^{3} \left[ \frac{a_k^3 - a_{k-1}^3}{6} - \frac{h^2(a_k - a_{k-1})}{8} \right] \times \left[ A_{kx} \frac{\partial^3 w}{\partial x^3} + (B_{ko} + G_{kxy}) \frac{\partial^3 w}{\partial x\, \partial y^2} \right] - D_{66} \frac{\partial^2 w}{\partial x\, \partial y^2}$$

$$V_y = \sum_{k=1}^{3} \left[ \frac{a_k^3 - a_{k-1}^3}{6} - \frac{h^2(a_k - a_{k-1})}{8} \right] \times \left[ A_{ky} \frac{\partial^3 w}{\partial y^3} + (B_{ko} + G_{kxy}) \frac{\partial^3 w}{\partial x^2\, \partial y} \right] - D_{66} \frac{\partial^2 w}{\partial x^2\, \partial y} \tag{10-200}$$

The subsequent procedure is again identical with that of Section 10-7-1 and leads to the following differential equation for the deflection of a laminated plate composed of orthotropic layers:

$$D_{11} \frac{\partial^4 w}{\partial x^4} + 2(D_{12} + D_{66}) \frac{\partial^4 w}{\partial x^2\, \partial y^2} + D_{22} \frac{\partial^4 w}{\partial y^4} = q \tag{10-201}$$

where $D_{11}$, $D_{12}$, and $D_{66}$ are given by Eqs. (10-196) and (10-198). The boundary conditions are identical to those described in Section 10-7-1 by Eqs. (10-178) and (10-179). The only exception are the boundary conditions at the free edge, Eqs. (10-180) and (10-181). There $B_k$ and $C_k$ must be replaced by $B_{ko}$ and $G_{kxy}$, while $A_k$ must be replaced by $A_{kx}$ in Eq. (10-180) and by $A_{ky}$ in Eq. (10-181).

## EXAMPLE 10-9

Consider a simply supported square plate of 10 by 10 in, composed of three laminae, each 0.01 in thick, and made of T300/5208 unidirectional graphite-epoxy, with a stacking sequence* 0°/90°/0°. The properties of T300/5208 are as follows [15]: $E_1 = 22.2 \times 10^6$ psi, $E_2 = 1.58 \times 10^6$ psi, $G_{12} = 0.65 \times 10^6$ psi, $\nu_{12} = 0.30$, and $\nu_{21} = 0.021$, where 1 refers to the direction of the fibers and 2 to the direction perpendicular to the fibers. The plate carries a uniformly distributed load $q_0$ in lb/in$^2$. Determine the maximum deflection.

### SOLUTION

We begin by solving Eq. (10-201),

$$D_{11} \frac{\partial^4 w}{\partial x^4} + 2(D_{12} + D_{66}) \frac{\partial^4 w}{\partial x^2\, \partial y^2} + D_{22} \frac{\partial^4 w}{\partial y^4} = q \tag{a}$$

*The stacking sequence shows the relative orientation of the fibers in adjacent laminae.

with the boundary conditions [Eqs. (10-178)]

$$
\begin{aligned}
&\frac{\partial^2 w}{\partial x^2} = 0, \qquad \text{at } x = 0 \text{ and } x = a \\
&w = 0, \qquad \frac{\partial^2 w}{\partial y^2} = 0, \qquad \text{at } y = 0 \text{ and } y = a
\end{aligned}
\tag{b}
$$

The first step will be to assume the solution to Eq. (*a*) in the form of a Fourier series (see Section 10-3 for details),

$$
w(x, y) = \sum_{m=1}^{\infty} \sum_{n=1}^{\infty} w_{mn} \sin \frac{m\pi x}{a} \sin \frac{n\pi y}{b} \tag{c}
$$

Next Eq. (*c*) is substituted into Eq. (*a*) to give

$$
\begin{aligned}
&\sum_{m=1}^{\infty} \sum_{n=1}^{\infty} \left[ D_{11} \frac{m^4\pi^4}{a^4} + 2(D_{12} + D_{66}) \frac{m^2 n^2 \pi^4}{a^4} + D_{22} \frac{n^4 \pi^4}{a^4} \right] w_{mn} \sin \frac{m\pi x}{a} \sin \frac{n\pi y}{a} \\
&\quad - \sum_{m=1}^{\infty} \sum_{n=1}^{\infty} q_{mn} \sin \frac{m\pi x}{a} \sin \frac{n\pi y}{a} = 0
\end{aligned}
\tag{d}
$$

Satisfaction of Eq. (*d*) for arbitrary *x* and *y* requires that

$$
\left[ D_{11} \frac{m^4\pi^4}{a^4} + 2(D_{12} + D_{66}) \frac{m^2 n^2 \pi^4}{a^4} + D_{22} \frac{n^4 \pi^4}{a^4} \right] w_{mn} = q_{mn}, \quad \text{for } m, n = 1, \ldots, \infty \tag{e}
$$

By using the fact that the load is constant and following the development shown in Example 10-7, we find that

$$
q_{mn} = \begin{cases} 0, & m \text{ or } n \text{ even} \\ \dfrac{16 q_0}{mn\pi^2}, & m \text{ and } n \text{ odd} \end{cases} \tag{f}
$$

When Eq. (*f*) is used in Eq. (*e*) and the latter is solved for $w_{mn}$, we get

$$
w_{mn} = \frac{16 q_0 a^4}{mn\pi^6 [D_{11} m^4 + 2(D_{12} + D_{66}) m^2 n^2 + D_{22} n^4]}, \qquad \text{for } m \text{ and } n \text{ odd} \tag{g}
$$

When Eq. (*g*) is substituted into Eq. (*a*), this becomes

$$
w(x, y) = \frac{16 q_0 a^4}{\pi^6} \sum_{m=1,3,\ldots}^{\infty} \sum_{n=1,3,\ldots}^{\infty} \frac{\sin \dfrac{m\pi x}{a} \sin \dfrac{n\pi y}{a}}{mn[D_{11} m^4 + 2(D_{12} + D_{66}) m^2 n^2 + D_{22} n^4]} \tag{h}
$$

It is now necessary to calculate the coefficients $D_{11}$, $D_{12}$, $D_{22}$, and $D_{66}$. Let directions 1 and 2 be collinear with the *x* and *y* axes, respectively (Fig. 10-37). Then the fibers in the

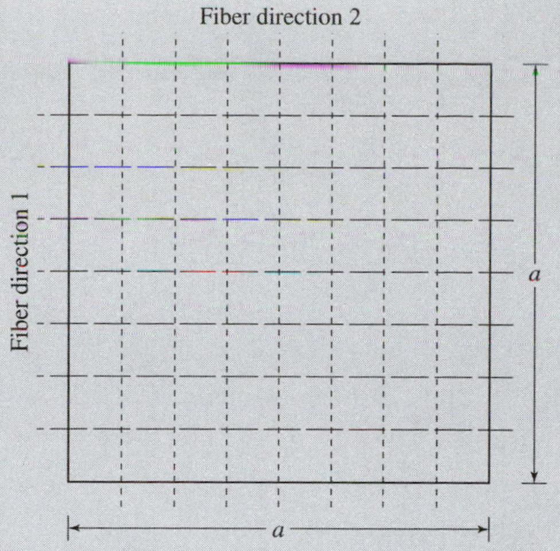


**Figure 10-37**

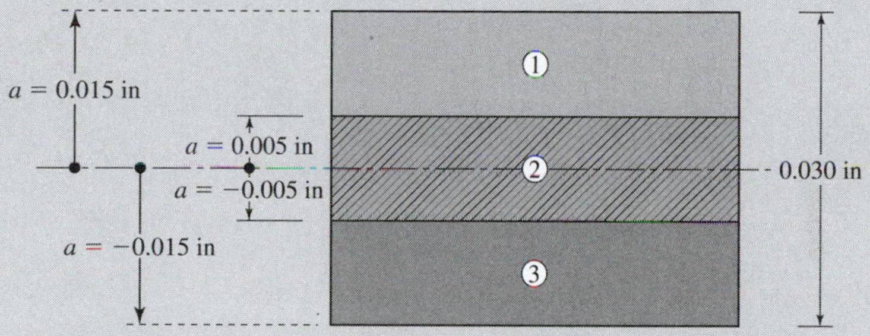


**Figure 10-38**

outer laminae have the direction $x$ and those in the inner layers the direction $y$. Next we calculate the constants $A_{kx}$, $A_{ky}$, and $B_{ko}$ from Eq. (10-193). Assigning $k = 1, 2$, or 3 to the laminae shown in Fig. 10-38 we have, for laminae 1 and 3,

$$
\begin{aligned}
A_{1x} = A_{3x} &= \frac{E_{1x}}{1 - \nu_{1xy}\nu_{1yx}} = \frac{E_1}{1 - \nu_{12}\nu_{21}} = \frac{22.2}{1 - 0.3 \times 0.021} = 22.53 \times 10^6 \text{ psi} \\
A_{1y} = A_{3y} &= \frac{E_{1y}}{1 - \nu_{1xy}\nu_{1yx}} = \frac{E_2}{1 - \nu_{12}\nu_{21}} = 1.604 \times 10^6 \text{ psi} \\
B_{1o} = B_{3o} &= \frac{E_{1x}\nu_{1yx}}{1 - \nu_{1xy}\nu_{1yx}} = \frac{E_1\nu_{21}}{1 - \nu_{12}\nu_{21}} = 0.473 \times 10^6 \text{ psi} \\
G_{12} &= 0.65 \times 10^6 \text{ psi}
\end{aligned}
\qquad (i)
$$

For lamina 2 the fiber direction is parallel to the $y$ axis and we get

$$
\begin{aligned}
A_{2x} &= \frac{E_{2x}}{1-\nu_{2xy}\nu_{2yx}} = \frac{E_2}{1-\nu_{12}\nu_{21}} = 1.604 \times 10^6 \text{ psi} \\
A_{2y} &= \frac{E_{2y}}{1-\nu_{2xy}\nu_{2yx}} = \frac{E_1}{1-\nu_{12}\nu_{21}} = 22.53 \times 10^6 \text{ psi} \\
B_{2o} &= \frac{E_{2x}\nu_{2yx}}{1-\nu_{2xy}\nu_{2yx}} = \frac{E_2\nu_{12}}{1-\nu_{21}\nu_{12}} = 0.481 \times 10^6 \text{ psi} \\
G_{12} &= 0.65 \times 10^6 \text{ psi}
\end{aligned} \tag{j}
$$

Using Eqs. (10-196) and (10-198) the constants $D_{11}$, $D_{22}$, $D_{12}$, and $D_{66}$ are readily evaluated,

$$
\begin{aligned}
D_{11} &= \tfrac{1}{3}\left[A_{1x}(a_0^3 - a_1^3) + A_{2x}(a_1^3 - a_2^3) + A_{3x}(a_2^3 - a_3^3)\right] \\
&= \tfrac{2}{3}(73.22 + 0.20) = 49.08 \text{ lb}\cdot\text{in} \\
D_{22} &= \tfrac{1}{3}\left[A_{1y}(a_0^3 - a_1^3) + A_{2y}(a_1^3 - a_2^3) + A_{3y}(a_2^3 - a_3^3)\right] \\
&= \tfrac{2}{3}(5.413 + 2.816) = 5.486 \text{ lb}\cdot\text{in} \\
D_{12} &= \tfrac{1}{3}\left[B_{1o}(a_0^3 - a_1^3) + B_{2o}(a_1^3 - a_2^3) + B_{3o}(a_2^3 - a_3^3)\right] \\
&= \tfrac{2}{3}(1.596 + 0.060) = 1.104 \text{ lb}\cdot\text{in} \\
D_{66} &= \tfrac{1}{3}\left[G_{12}(a_0^3 - a_1^3) + G_{12}(a_1^3 - a_2^3) + G_{12}(a_2^3 - a_3^3)\right] \\
&= \tfrac{2}{3}0.65(3.375 + 0.125) = 1.517 \text{ lb}\cdot\text{in}
\end{aligned} \tag{k}
$$

In arriving at these results we have used the notation of Fig. 10-38. Examining expressions ($k$) we note that, except for $D_{22}$, the influence of the inner lamina (the second term in the parantheses on the right) is relatively small. It is evident that maximum deflection occurs at $x = y = a/2$. Thus when Eq. ($k$) is substituted into Eq. ($h$) we obtain

$$
\begin{aligned}
\max w(x, y) &= w\left(\frac{a}{2}, \frac{a}{2}\right) \\
&= \frac{16q_0a^4}{\pi^6} \sum_{m=1,3,\ldots}^{\infty} \sum_{n=1,3,\ldots}^{\infty} \frac{\sin\dfrac{m\pi}{2}\sin\dfrac{n\pi}{2}}{mn[49.08m^4 + 2(1.104 + 1.517)m^2n^2 + 5.486n^4]}
\end{aligned} \tag{l}
$$

At this point we use only the first four terms of the infinite series in Eq. ($h$), that is, $m = n = 1$; $m = 1$ and $n = 3$; $m = 3$ and $n = 1$; and $m = n = 3$. This gives the normalized maximum deflection,

$$
\frac{\max w(x, y)}{q_0} \approx \frac{16a^4}{\pi^6} \times 0.01368 = 2.277 \text{ in}^3/\text{lb} \tag{m}
$$

## 10-8 APPROXIMATE METHODS IN THE ANALYSIS OF PLATES AND MEMBRANES

So far in this chapter we have worked out exact solutions for some of the plate bending problems. Because of various limiting assumptions, such as axial symmetry for a circular plate, the solutions were of a closed form. Analytical solutions in closed form or expanded form are very useful because they are parametric and often allow for a relatively simple analysis of the effect of changing the geometry or load. Unfortunately we do not often have the luxury of analytical solutions. Many technical problems deal with plates whose shapes, materials, and other properties are such that the differential equations describing bending either are impossible to solve analytically or can only be solved at the expense of an excessive amount of time. With the development and accessibility of computers, numerical methods have gained in popularity. At present, there are numerous finite element and other packages available for easy handling of complicated problems. In this chapter we discuss the application of finite differences (Chapter 6) and of the Rayleigh–Ritz method (Chapter 5) to the bending of the plates.

### 10-8-1 Application of Finite Differences

The application of the finite differences method to the solution to partial differential equations was discussed in Chapter 6 (see also [2], [12], and [4]). It was also shown that the differential equation for bending an elastic plate of constant stiffness $D$, which in the Cartesian coordinate system is (see Section 10-1)

$$D\left(\frac{\partial^4 w}{\partial x^4} + 2\frac{\partial^4 w}{\partial x^2\,\partial y^2} + \frac{\partial^4 w}{\partial y^4}\right) = q \tag{10-202}$$

becomes, at the pivot $i$, $k$ and with a rectangular mesh (Fig. 10-39),

$$\begin{aligned}\frac{D}{h^4}[2(3+4\beta^2+3\beta^4)W_{i,k} - 4(1+\beta^2)(W_{i+1,k}+W_{i-1,k}) - 4\beta^2(1+\beta^2)(W_{i,k+1}+W_{i,k-1})\\ + 2\beta^2(W_{i+1,k+1}+W_{i+1,k-1}+W_{i-1,k+1}+W_{i-1,k-1}) + W_{i+2,k}+W_{i-2,k}+\beta^4(W_{i,k+2}+W_{i,k-2})] = q_{i,k}\end{aligned} \tag{10-203}$$

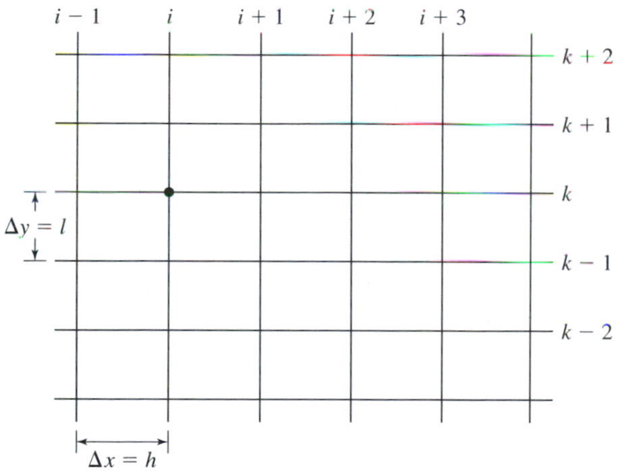


Figure 10-39 A rectangular mesh

where $\beta = h/l$. The concept of stencils was then introduced and applied for a rectangular simply supported plate carrying a uniform load $q$ (see Example 6-2). Let us now consider a more complicated case of bending of a plate.

## EXAMPLE 10-10

Determine the deflection of the L-shaped plate shown in Fig. 10-40.

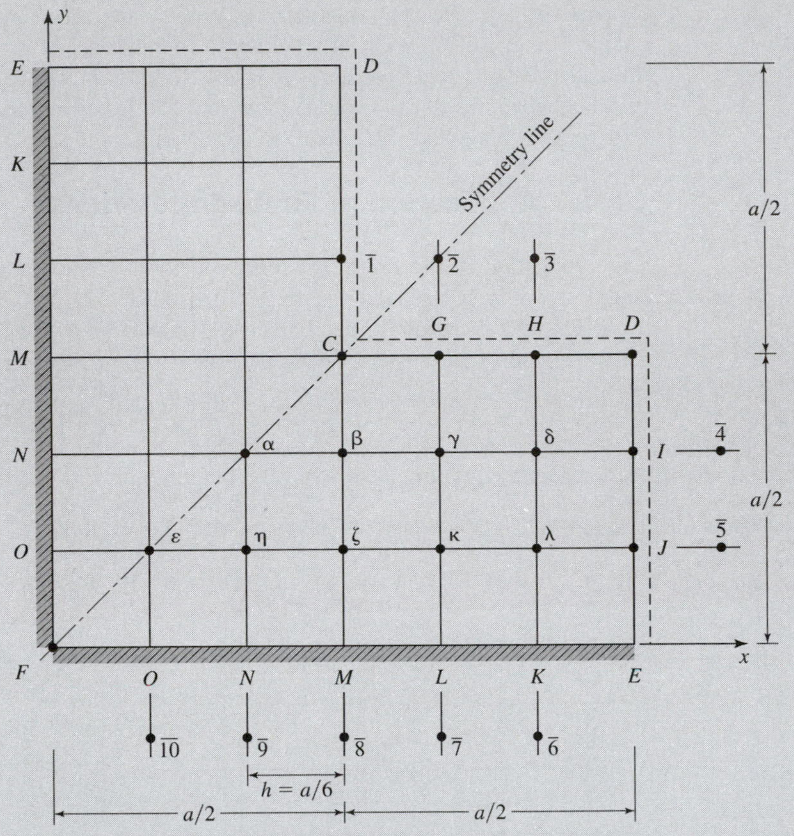


**Figure 10-40**

## SOLUTION

Recalling the differential equation for bending, Eq. (10-202),

$$D\left(\frac{\partial^4 w}{\partial x^4} + 2\frac{\partial^4 w}{\partial x^2\,\partial y^2} + \frac{\partial^4 w}{\partial y^4}\right) = q \tag{a}$$

we note that the plate is symmetric and subject to a symmetric load. Thus it is sufficient to analyze only the portion *CDEF* of the plate. The boundary conditions are

$$w = 0 \qquad \text{and} \qquad \frac{\partial^2 w}{\partial y^2} = 0 \tag{b}$$

at the edge *CD* of the plate,

$$w = 0 \qquad \text{and} \qquad \frac{\partial^2 w}{\partial x^2} = 0 \tag{c}$$

at the edge *DE*, and

$$w = 0 \qquad \text{and} \qquad \frac{\partial w}{\partial y} = 0 \tag{d}$$

at the edge *EF*. We now introduce a square mesh $h = l = a/6$. Thus $\beta = h/l = 1$ and, at an arbitrary pivot $i$, $k$, the finite difference equation (10-203) becomes (see also Example 6-2)

$$\begin{aligned} 20W_{i,k} &- 8(W_{i+1,k} + W_{i,k+1} + W_{i-1,k} + W_{i,k-1}) + 2(W_{i+1,k+1} + W_{i-1,k+1} \\ &+ W_{i+1,k-1} + W_{i-1,k-1}) + W_{i+2,k} + W_{i-2,k} + W_{i,k+2} + W_{i,k-2} = Q \end{aligned} \tag{e}$$

We now label the internal pivots $\alpha, \ldots, \lambda$ and the boundary pivots $C, \ldots, F$, as seen in Fig. 10-40. This change of indices from a double array $i$, $k$ to single characters enables the use of matrices (see also Chapter 6). Next the boundary conditions (*b*) are transformed into

$$\begin{aligned} W_C = W_G = W_H = W_D = 0, \qquad W_\beta + W_{\bar{1}} = 0, \\ W_\gamma + W_{\bar{2}} = 0, \qquad W_\delta + W_{\bar{3}} = 0 \end{aligned} \tag{f}$$

at the edge *CD*. Note that the pivots $\bar{1}$, $\bar{2}$, and $\bar{3}$ are located outside the plate, and the boundary conditions (*f*) will help to eliminate them from Eq. (*e*). We also note that the pivot $\bar{1}$ is located symmetrically to $G$, and that consequently $W_{\bar{1}} = 0$. This implies

$$W_\beta = 0 \tag{g}$$

Similarly, for the edge *DE* the boundary conditions (*c*) become

$$W_I = W_J = 0, \qquad W_\delta + W_{\bar{4}} = 0, \qquad W_\lambda + W_{\bar{5}} = 0 \tag{h}$$

Finally the boundary conditions (*d*) at the edge *EF* are

$$\begin{aligned} W_F = W_O = W_N = W_M = W_L = W_K = W_E = 0 \\ W_{\overline{10}} = W_\varepsilon, \quad W_{\bar{9}} = W_\eta, \quad W_{\bar{8}} = W_\zeta, \quad W_{\bar{7}} = W_\kappa, \quad W_{\bar{6}} = W_\lambda \end{aligned} \tag{i}$$

We now use the stencil developed in Section 6-1-2 and place it consecutively at $\alpha, \beta, \ldots$. Specifically, with the stencil at $\alpha$, we get

$$W_\gamma + 2W_\zeta - 8W_\beta - 8W_\eta + 20W_\alpha - 8W_\beta + W_\gamma + 2W_\varepsilon - 8W_\eta + 2W_\zeta = Q \tag{j}$$

in which $Q = qh^4/D$. Consider also the case of the stencil at δ,

$$W_{\bar{3}} + 2W_G - 8W_H + 2W_D + W_N + W_\beta - 8W_\gamma + 20W_\delta - 8W_I + W_{\bar{4}} + 2W_\kappa - 8W_\lambda + 2W_J + W_K = Q \qquad (k)$$

Next Eqs. (*g*), (*h*), and (*i*) are substituted into Eqs. (*j*) and (*k*), which become

$$20W_\alpha + 2W_\gamma + 2W_\varepsilon - 16W_\eta + 4W_\zeta = Q \qquad (l)$$

and

$$-8W_\gamma + 18W_\delta + 2W_\kappa - 8W_\lambda = Q \qquad (m)$$

Proceeding in a similar manner with all other internal pivots, except for β where we already found the deflection to be zero, we get the difference equations in matrix form,

$$[\boldsymbol{C}]\{\boldsymbol{W}\} = \{\boldsymbol{Q}\} \qquad (n)$$

where

$$[\boldsymbol{C}] = \begin{bmatrix} 20 & 2 & 0 & 2 & -16 & 4 & 0 & 0 \\ 1 & 19 & -8 & 0 & 0 & 2 & -8 & 2 \\ 0 & -8 & 18 & 0 & 0 & 0 & 2 & -8 \\ 2 & 0 & 0 & 18 & -16 & 2 & 0 & 0 \\ -8 & 0 & 0 & -8 & 23 & -8 & 1 & 0 \\ 2 & 2 & 0 & 1 & -8 & 21 & -8 & 1 \\ 0 & -8 & 1 & 0 & 1 & -8 & 21 & -8 \\ 0 & 2 & -8 & 0 & 0 & 1 & -8 & 20 \end{bmatrix} \qquad (o)$$

and

$$\{\boldsymbol{W}\} = \begin{Bmatrix} W_\alpha \\ W_\gamma \\ W_\delta \\ W_\varepsilon \\ W_\eta \\ W_\zeta \\ W_\kappa \\ W_\lambda \end{Bmatrix}, \qquad \{\boldsymbol{Q}\} = \begin{Bmatrix} Q \\ Q \\ Q \\ Q \\ Q \\ Q \\ Q \\ Q \end{Bmatrix} \qquad (p)$$

The solution to the system of Eq. (*n*) is

$$\begin{aligned} W_\alpha &= 0.1356Q = 0.0001046\frac{qa^4}{D}, & W_\gamma &= 0.1953Q = 0.0001507\frac{qa^4}{D} \\ W_\delta &= 0.2048Q = 0.0002048\frac{qa^4}{D}, & W_\zeta &= 0.1646Q = 0.0001270\frac{qa^4}{D} \\ W_\varepsilon &= 0.1580Q = 0.0001219\frac{qa^4}{D}, & W_\eta &= 0.1923Q = 0.0001538\frac{qa^4}{D} \\ W_\kappa &= 0.2425Q = 0.0001871\frac{qa^4}{D}, & W_\lambda &= 0.2012Q = 0.0001552\frac{qa^4}{D} \end{aligned} \qquad (q)$$

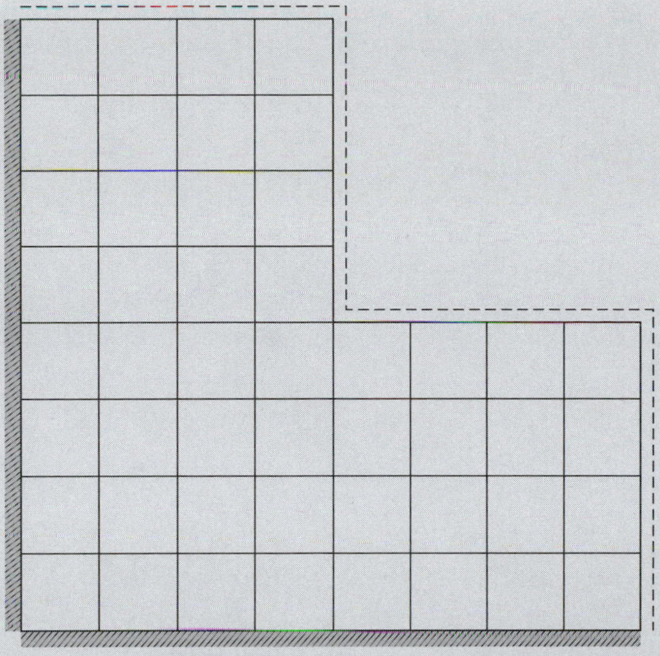

**Figure 10-41** A finer mesh, $h = a/8$

Consequently the maximum deflection occurs in the neighborhood of pivot $\zeta$. Now let us apply a finer mesh, with $h = a/8$, as seen in Fig. 10-41. Using again a stencil, we arrive at a system of 17 linear algebraic equations ($n$), where

$$[C] = \begin{bmatrix}
20 & 2 & 0 & 0 & 2 & -16 & 4 & 0 & 0 & 0 & 0 & 0 & 2 & 0 & 0 & 0 & 0 \\
1 & 19 & -8 & 1 & 0 & 0 & 2 & -8 & 2 & 0 & 0 & 0 & 0 & 0 & 1 & 0 & 0 \\
0 & -8 & 19 & -8 & 0 & 0 & 0 & 2 & -8 & 2 & 0 & 0 & 0 & 0 & 0 & 1 & 0 \\
0 & 1 & -8 & 18 & 0 & 0 & 0 & 0 & 2 & -8 & 0 & 0 & 0 & 0 & 0 & 0 & 1 \\
2 & 0 & 0 & 0 & 20 & -16 & 2 & 0 & 0 & 0 & 2 & -16 & 4 & 0 & 0 & 0 & 0 \\
-8 & 0 & 0 & 0 & -8 & 22 & -8 & 1 & 0 & 0 & 0 & 3 & -8 & 2 & 0 & 0 & 0 \\
2 & 2 & 0 & 0 & 1 & -8 & 20 & -8 & 1 & 0 & 0 & 0 & 2 & -8 & 2 & 0 & 0 \\
0 & -8 & 2 & 0 & 0 & 1 & -8 & 20 & -8 & 1 & 0 & 0 & 0 & 2 & -8 & 2 & 0 \\
0 & 2 & -8 & 2 & 0 & 0 & 1 & -8 & 20 & -8 & 0 & 0 & 0 & 0 & 2 & -8 & 2 \\
0 & 0 & 2 & -8 & 0 & 0 & 0 & -1 & -8 & 19 & 0 & 0 & 0 & 0 & 0 & 2 & -8 \\
0 & 0 & 0 & 0 & 2 & 0 & 0 & 0 & 0 & 0 & 22 & -16 & 2 & 0 & 0 & 0 & 0 \\
0 & 0 & 0 & 0 & -8 & 3 & 0 & 0 & 0 & 0 & -8 & 23 & -8 & 1 & 0 & 0 & 0 \\
1 & 0 & 0 & 0 & 2 & -8 & 2 & 0 & 0 & 0 & 1 & -8 & 21 & -8 & 1 & 0 & 0 \\
0 & 0 & 0 & 0 & 0 & 2 & -8 & 2 & 0 & 0 & 0 & 1 & -8 & 21 & -8 & 1 & 0 \\
0 & 1 & 0 & 0 & 0 & 0 & 2 & -8 & 2 & 0 & 0 & 0 & 1 & -8 & 21 & -8 & 1 \\
0 & 0 & 1 & 0 & 0 & 0 & 0 & 2 & -8 & 2 & 0 & 0 & 0 & 1 & -8 & 21 & -8 \\
0 & 0 & 0 & 1 & 0 & 0 & 0 & 0 & 0 & -8 & 0 & 0 & 0 & 0 & 1 & -8 & 20
\end{bmatrix} \qquad (r)$$

At this point it would be worthwhile for the reader to examine the correctness of the matrix $[\boldsymbol{C}]$. Solving the system of equations ($n$), with Eq. ($r$) substituted for $[\boldsymbol{C}]$, we get

$$
\begin{aligned}
W_\alpha &= 0.2967Q = 0.0000724\frac{qa^4}{D}, & W_\gamma &= 0.4057Q = 0.0000990\frac{qa^4}{D} \\
W_\delta &= 0.5903Q = 0.0001441\frac{qa^4}{D}, & W_\varepsilon &= 0.4424Q = 0.0001080\frac{qa^4}{D} \\
W_\eta &= 0.7094Q = 0.0001732\frac{qa^4}{D}, & W_\zeta &= 0.6391Q = 0.0001560\frac{qa^4}{D} \\
W_\kappa &= 0.5373Q = 0.0001312\frac{qa^4}{D}, & W_\lambda &= 0.7449Q = 0.0001819\frac{qa^4}{D} \\
W_\mu &= 0.8253Q = 0.0002015\frac{qa^4}{D}, & W_\nu &= 0.5820Q = 0.0001421\frac{qa^4}{D} \\
W_o &= 0.2562Q = 0.0000625\frac{qa^4}{D}, & W_\Omega &= 0.4355Q = 0.0001063\frac{qa^4}{D} \\
W_\psi &= 0.3597Q = 0.0000878\frac{qa^4}{D}, & W_\rho &= 0.4565Q = 0.0001115\frac{qa^4}{D} \\
W_\sigma &= 0.4430Q = 0.0001082\frac{qa^4}{D}, & W_\tau &= 0.5094Q = 0.0001244\frac{qa^4}{D}
\end{aligned}
\qquad (s)
$$

$$
W_\theta = 0.5175Q = 0.0001263\frac{qa^4}{D}
$$

Here the maximum deflection occurs in the neighborhood of the pivot κ, which is near but not at the same location as the previously found pivot ζ. The difference in the maximum deflections obtained for both meshes is about 7.7%.

## 10-8-2 Application of the Rayleigh–Ritz Method

The general idea of the Rayleigh–Ritz method was presented, in Chapter 5 and then used in Chapter 8 for the solution of some problems of beam bending. Its application to the solution of problems of bending of plates is very similar to that for beams. First we write the expression for the potential energy of an elastic isotropic plate. By using Eq. (10-151) for the strain energy we have

$$
\Pi = \frac{D}{2}\int_A\!\!\int \left\{\left(\frac{\partial^2 w}{\partial x^2} + \frac{\partial^2 w}{\partial y^2}\right)^2 - 2(1-\nu)\left[\frac{\partial^2 w}{\partial x^2}\frac{\partial^2 w}{\partial y^2} - \left(\frac{\partial^2 w}{\partial x\,\partial y}\right)^2\right]\right\} dx
$$
$$
- \int_A\!\!\int qw\,dx\,dy \qquad (10\text{-}204)
$$

Next the unknown deflection $w(x, y)$ is expressed in the form [12, p. 344], [2, p. 432], [4, p. 181]

$$w(x, y) = a_1 f_1(x, y) + a_2 f_2(x, y) + \cdots + a_n f_n(x, y) \tag{10-205}$$

Since the functions $f_i(x, y)$, $i = 1, \ldots, n$, are required to satisfy the kinematic boundary conditions, their selection for irregularly shaped plates can be quite complicated. Here we limit ourselves to rectangular plates for which the choice of trial functions does not pose such difficulties. Once the functions $f_i(x, y)$ are adopted, Eq. (10-205) is substituted into Eq. (10-204) for the potential energy $\Pi$, which is now dependent on the unknown parameters $a_i$, $i = 1, \ldots, n$. Thus the conditions for minimum $\Pi$ become

$$\frac{\partial \Pi}{\partial a_1} = 0, \quad \frac{\partial \Pi}{\partial a_2} = 0, \ldots, \quad \frac{\partial \Pi}{\partial a_n} = 0 \tag{10-206}$$

yielding a system of $n$ algebraic equations in $n$ unknowns $a_1, a_2, \ldots, a_n$.

## EXAMPLE 10-11

Find the maximum deflection of a simply supported square plate carrying a uniform load $q_0$ (see Fig. 10-42).

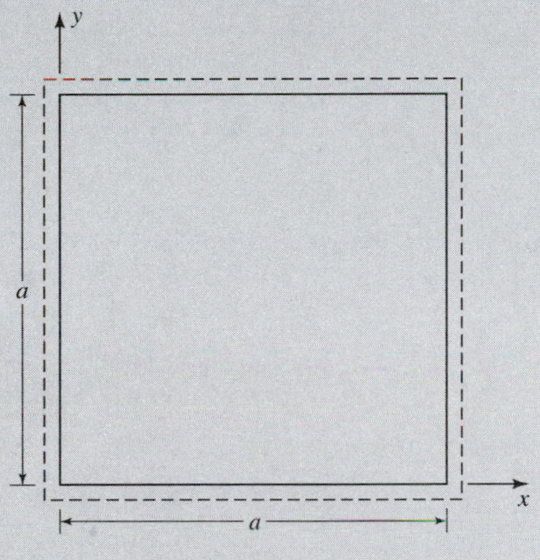


Figure 10-42

### SOLUTION

By inspection of the boundary conditions given in Section 10-3 the kinematic boundary conditions are easily found to be

$$w(0, y) = w(a, y) = w(x, 0) = w(x, a) = 0 \tag{10-207}$$

Let us assume as trial functions

$$\sin\frac{m\pi x}{a}\sin\frac{n\pi y}{a}, \qquad m=1,\ldots,M, \quad n=1,\ldots,N \tag{10-208}$$

obviously satisfying boundary conditions (10-207). It is convenient to assign two subscripts to the unknown parameters and to put the deflection $w(x, y)$ in the form

$$w(x,y) = a_{11}\sin\frac{\pi x}{a}\sin\frac{\pi y}{a} + a_{12}\sin\frac{\pi x}{a}\sin\frac{2\pi y}{a} + a_{21}\sin\frac{2\pi x}{a}\sin\frac{\pi y}{a} + a_{22}\sin\frac{2\pi x}{a}\sin\frac{2\pi y}{a} + \cdots \tag{10-209}$$

which can easily be represented as a double series,

$$w(x,y) = \sum_{m=1}^{M}\sum_{n=1}^{N} a_{mn}\sin\frac{m\pi x}{a}\sin\frac{n\pi y}{a} \tag{10-210}$$

Substituting this into Eq. (10-204) gives

$$\begin{aligned}\Pi = \frac{D}{2}\int_0^a\int_0^a \Bigg\{ &\left[\sum_{m=1}^{M}\sum_{n=1}^{N}\left(\frac{m^2\pi^2}{a^2}+\frac{n^2\pi^2}{a^2}\right)a_{mn}\sin\frac{m\pi x}{a}\sin\frac{n\pi y}{a}\right] \\ &\times\left[\sum_{r=1}^{M}\sum_{s=1}^{N}\left(\frac{r^2\pi^2}{a^2}+\frac{s^2\pi^2}{a^2}\right)a_{rs}\sin\frac{r\pi x}{a}\sin\frac{s\pi y}{a}\right] \\ &-2(1-\nu)\Bigg[\left(\sum_{m=1}^{M}\sum_{n=1}^{N}\frac{m^2\pi^2}{a^2}a_{mn}\sin\frac{m\pi x}{a}\sin\frac{n\pi y}{a}\right) \\ &\times\left(\sum_{r=1}^{M}\sum_{s=1}^{N}\frac{s^2\pi^2}{a^2}a_{rs}\sin\frac{r\pi x}{a}\sin\frac{s\pi y}{a}\right) \\ &-\left(\sum_{m=1}^{M}\sum_{n=1}^{N}\frac{mn\pi^2}{a^2}a_{mn}\cos\frac{m\pi x}{a}\cos\frac{n\pi y}{a}\right) \\ &\times\left(\sum_{r=1}^{M}\sum_{s=1}^{N}\frac{rs\pi^2}{a^2}a_{rs}\cos\frac{r\pi x}{a}\cos\frac{s\pi y}{a}\right)\Bigg]\Bigg\}\,dx\,dy \\ &-q_0\int_0^a\int_0^a\sum_{m=1}^{M}\sum_{n=1}^{N}a_{mn}\sin\frac{m\pi x}{a}\sin\frac{n\pi y}{a}\,dx\,dy\end{aligned} \tag{10-211}$$

We now differentiate $\Pi$ with respect to each of the unknown parameters $a_{mn}$, $m=1,\ldots,M, n=1,\ldots,N$, postponing the integration until later. We begin by differentiating with respect to $a_{11}$ and notice that it appears under every double series. We thus obtain

$$\begin{aligned}\frac{\partial \Pi}{\partial a_{11}} = \frac{D}{2}\int_0^a\int_0^a \Bigg\{&\left[\left(\frac{\pi^2}{a^2}+\frac{\pi^2}{a^2}\right)\sin\frac{\pi x}{a}\sin\frac{\pi y}{a}\right]\left[\sum_{r=1}^{M}\sum_{s=1}^{N}\left(\frac{r^2\pi^2}{a^2}+\frac{s^2\pi^2}{a^2}\right)a_{rs}\sin\frac{\pi x}{a}\sin\frac{\pi y}{a}\right]\\
&+\left[\sum_{m=1}^{M}\sum_{n=1}^{N}\left(\frac{m^2\pi^2}{a^2}+\frac{n^2\pi^2}{a^2}\right)a_{mn}\sin\frac{m\pi x}{a}\sin\frac{n\pi y}{a}\right]\left[\left(\frac{\pi^2}{a^2}+\frac{\pi^2}{a^2}\right)\sin\frac{\pi x}{a}\sin\frac{\pi y}{a}\right]\\
&-2(1-\nu)\Bigg[\left(\frac{\pi^2}{a^2}\sin\frac{\pi x}{a}\sin\frac{\pi y}{a}\right)\left(\sum_{r=1}^{M}\sum_{s=1}^{N}\frac{s^2\pi^2}{a^2}a_{rs}\sin\frac{r\pi x}{a}\sin\frac{s\pi y}{a}\right)\\
&+\left(\sum_{m=1}^{M}\sum_{n=1}^{N}\frac{m^2\pi^2}{a^2}a_{mn}\sin\frac{m\pi x}{a}\sin\frac{n\pi y}{a}\right)\left(\frac{\pi^2}{a^2}\sin\frac{\pi x}{a}\sin\frac{\pi y}{a}\right)\\
&-\left(\frac{\pi^2}{a^2}\cos\frac{\pi x}{a}\cos\frac{\pi y}{a}\right)\left(\sum_{r=1}^{M}\sum_{s=1}^{N}\frac{rs\pi^2}{a^2}a_{rs}\cos\frac{r\pi x}{a}\cos\frac{s\pi y}{a}\right)\\
&-\left(\sum_{m=1}^{M}\sum_{n=1}^{N}\frac{mn\pi^2}{a^2}a_{mn}\cos\frac{m\pi x}{a}\cos\frac{n\pi y}{a}\right)\left(\frac{\pi^2}{a^2}\cos\frac{\pi x}{a}\cos\frac{\pi y}{a}\right)\Bigg]\Bigg\}\,dx\,dy\\
&-q_0\int_0^a\int_0^a \sin\frac{\pi x}{a}\sin\frac{\pi y}{a}\,dx\,dy = 0 \end{aligned} \tag{10-212}$$

Integrating we find that, for example,

$$\int_0^a \sin\frac{r\pi x}{a}\sin\frac{s\pi x}{a}\,dx = 0, \qquad \int_0^a \cos\frac{r\pi y}{a}\cos\frac{s\pi y}{a}\,dy = 0 \tag{10-213}$$

unless $r = s$, in which case these integrals equal $a/2$. This property, known as orthogonality (see Chapter 8), occurs each time a summation sign appears in Eq. (10-212), and it leads to considerable simplifications. Thus,

$$\begin{aligned}\frac{\partial \Pi}{\partial a_{11}} = \frac{Da^2}{8}\Bigg\{&\frac{2\pi^2}{a^2}\frac{2\pi^2}{a^2}a_{11} + \frac{2\pi^2}{a^2}\frac{2\pi^2}{a^2}a_{11}\\
&-2(1-\nu)\left[\frac{\pi^4}{a^4}a_{11}+\frac{\pi^4}{a^4}a_{11}-\frac{\pi^4}{a^4}a_{11}-\frac{\pi^4}{a^4}a_{11}\right]\Bigg\} - \frac{a^2}{\pi^2}4q_0 = 0\end{aligned}$$

from which we have

$$a_{11} = \frac{4q_0a^4}{D\pi^6} \tag{10-214}$$

Proceeding in a similar manner for the conditions

$$\frac{\partial \Pi}{\partial a_{12}} = 0, \qquad \frac{\partial \Pi}{\partial a_{21}} = 0, \qquad \frac{\partial \Pi}{\partial a_{22}} = 0, \text{ etc.} \tag{10-215}$$

we find that in general

$$a_{mn} = \begin{cases} \dfrac{16 q_0 a^4}{D\pi^4 mn(m^2+n^2)^2}, & m \text{ and } n \text{ odd integers} \\ 0, & m \text{ or } n \text{ even integer} \end{cases} \tag{10-216}$$

so that the deflection is

$$w(x, y) = \frac{16 q_0 a^4}{D\pi^4} \sum_{m=1}^{M} \sum_{n=1}^{N} \frac{1}{mn(m^2+n^2)} \sin\frac{m\pi x}{a} \sin\frac{n\pi y}{a} \tag{10-217}$$

For $M \to \infty$ and $N \to \infty$ this result matches the exact solution, Eq. (10-142).

## EXAMPLE 10-12

Solve the problem in Example 10-11 using a different set of trial functions.

### SOLUTION

We satisfy all kinematic boundary conditions, assuming, for example,

$$w(x, y) = xy(a-x)(a-y)(a_1 + a_2 x^2 + a_3 y^2) \tag{10-218}$$

The calculations have been performed with the help of the software package Mathcad7, professional edition [9], and the following is adapted from the Mathcad7 printout. We define the trial function $w(x, y, u, v, z)$, where $u$, $v$, and $z$ are constants to be determined,

$$w(x, y) = xy(a-x)(a-y)(u + vx^2 + zy^2)$$

Next we calculate the second derivative using using the calculus palette,

$$\frac{\partial^2 w}{\partial x^2} = \frac{d^2}{dx^2} w(x, y)$$

Selecting the right-hand side of the preceding, we apply SYMBOLICS > FACTOR to get

$$2y(-a+y)(zy^2+u+6vx^2-3vxa)$$

In a similar way we calculate the second derivative with respect to $y$,

$$\frac{\partial^2 w}{\partial y^2} = 2x(a-x)(3zya - vx^2 - 6zy^2 - u)$$

The mixed second derivative is obtained by two consecutive differentiations,

$$\frac{\partial^2 w}{\partial x\,\partial y} = 3a^2zy^2 + a^2u + 3a^2vx^2 - 6axzy^2 - 2axu - 4ax^3v - 4azy^3 - 2yau - 6yavx^2 + 8xzy^3 + 4yxu + 8yx^3v$$

Next we calculate the energy, Eq. (10-204). Denoting by FIRST the first part of the integrand in Eq. (10-204), by SECOND its second part, and by THIRD its last part, we get

$$\text{FIRST} = \left(\frac{\partial^2 w}{\partial x^2} + \frac{\partial^2 w}{\partial y}\right)^2$$

We now integrate this with respect $x$, then with respect to $y$ from 0 to $a$, using the symbolic palette. It yields the following result:

$$\text{INTXYFIRST} = \frac{367}{1575}z^2a^{10} + \frac{12}{35}zua^8 + \frac{12}{35}vua^8 + \frac{61}{525}vza^{10} + \frac{367}{1575}v^2a^{10} + \frac{22}{45}u^2a^6$$

Using the symbolic palette again to calculate the derivatives of this expression with respect to the constants $u$, $v$, and $z$, we obtain

$$\text{EQ1A} = \frac{12}{35}za^8 + \frac{12}{35}va^8 + \frac{44}{45}ua^6$$

$$\text{EQ2A} = \frac{12}{35}ua^8 + \frac{61}{525}za^{10} + \frac{734}{1575}va^{10}$$

and

$$\text{EQ3A} = \frac{734}{1575}za^{10} + \frac{12}{35}ua^8 + \frac{61}{525}va^{10}$$

The same steps are repeated with respect to the second part,

$$\text{SECOND} = -2(1-\nu)\left[\frac{\partial^2 w}{\partial x^2}\frac{\partial^2 w}{\partial y^2} - \left(\frac{\partial^2 w}{\partial x\,\partial y}\right)^2\right]$$

All the derivatives of the resulting integral are found to be zero. The last term is

$$\text{THIRD} = \left(\frac{2}{D}\right)[xy(a-x)(a-y)(u+vx^2+zy^2)]$$

Integrating this, calculating its derivatives with respect to $u$, $v$, and $z$, and adding to the previous results leads to a system of algebraic equations in $u$, $v$,and $z$, which is solved.

*Given*:

$$\frac{12}{35}za^8 + \frac{12}{35}va^8 + \frac{44}{45}ua^6 - \left(\frac{2}{36D}qa^6\right) = 0$$

$$\frac{12}{35}ua^8 + \frac{61}{525}za^{10} + \frac{734}{1575}va^{10} - \left(\frac{2}{120D}qa^8\right) = 0$$

$$\frac{734}{1575}za^{10} + \frac{12}{35}ua^8 + \frac{61}{525}va^{10} - \left(\frac{2}{120D}qa^8\right) = 0$$

*Solution:*

$$\text{MINERR}(u, v, z) = \begin{bmatrix} \dfrac{0.0625920}{D}q \\ \dfrac{-0.00823301}{(a^2D)}q \\ \dfrac{-0.00823301}{(a^2D)}q \end{bmatrix}$$

With $\xi = x/a$ and $\eta = y/a$, the deflection becomes

$$w(\xi, \eta) = \frac{q_0a^4}{D}\xi\eta(1-\xi)(1-\eta)(0.0626 - 0.00823\xi^2 - 0.00823\eta^2) \quad (10\text{-}219)$$

The maximum deflection occurs at $x = h = 0.5$ and equals

$$\max w(x, y) = w\left(\frac{a}{2}, \frac{a}{2}\right) = 0.00365\frac{q_0a^4}{D} \quad (10\text{-}220)$$

while the exact result, obtained in Section 10.3, is

$$w\left(\frac{a}{2}, \frac{a}{2}\right) = \frac{16q_0a^4}{D\pi^6}0.244 = 0.00406\frac{q_0a^4}{D} \quad (10\text{-}221)$$

The 11% error would have been reduced had we taken more terms in our initial assumption.

# PROBLEMS

## Section 10-1

**10-1** Derive Eqs. (10-15).

**10-2** Derive Eq. (10-18).

**10-3** Derive Eq. (10-19).

## Sections 10-2-1 through 10-2-3

**10-4** Derive Eq. (10-42).

**10-5** Prove that Eqs. (10-42) and (10-43) are identical.

**10-6** Derive Eq. (10-45).

**10-7** Show that Eq. (10-47) is the solution to the differential equation for bending [Eq. (10-42)] when $q(r) \equiv 0$.

**10-8** Find a particular solution to Eq. (10-42) for the case $q(r) = q_0(r/R)^2$.

**10-9** Derive Eqs. (10-54).

**10-10** Derive Eqs. (10-65) and (10-66).

**10-11** Derive Eqs. (10-67), (10-68), and (10-69).

**10-12** Find the deflection of a simply supported circular plate carrying the load $q(r) = q(r/R)^2$ (see Problem 10-8). Compare this result with Eq. (10-104).

**10-13** Find the deflection, the bending moments, and the shear force for a simply supported circular plate carrying distributed moments of intensity $M_0$ along its perimeter (Fig. P10-13).

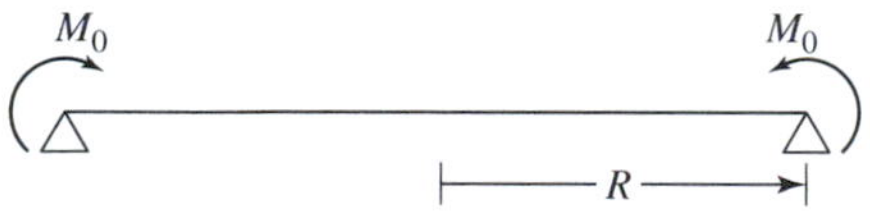


Figure P10-13

## Sections 10-2-4 and 10-2-5

**10-14** Derive expression (10-102).

**10-15** Derive expression (10-103).

**10-16** Use the Green's function to determine the deflection of a simply supported circular plate carrying an axisymmetric load, as shown in Fig. P10-16

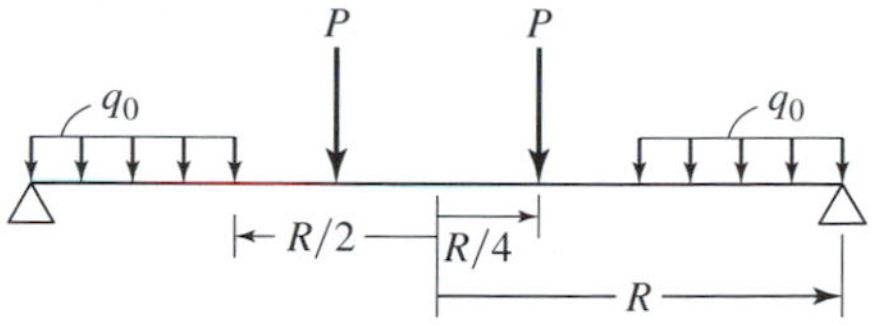


Figure P10-16

**10-17** Use the Green's function to determine the deflection of a simply supported circular plate carrying an axisymmetric load, as shown in Fig. P10-17.

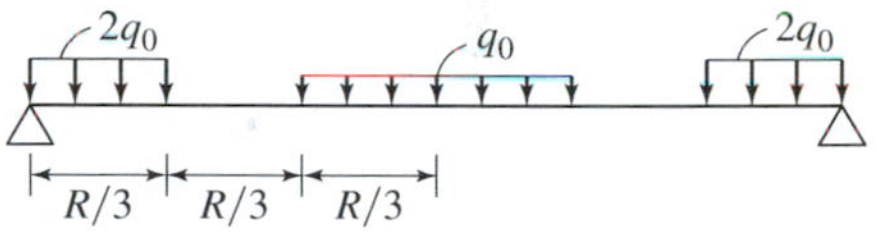


Figure P10-17

**10-18** Use the Green's function to determine the deflection of a circular plate clamped along its perimeter and carrying an axisymmetric load, as shown in Fig. P10-18.

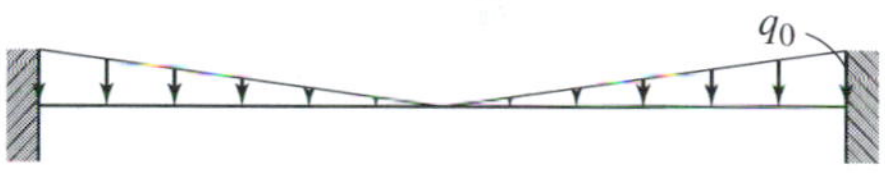


Figure P10-18

**10-19** Determine the reaction force at an additional support, located at a distance $c$ from the center, for a simply supported circular plate carrying a uniform load

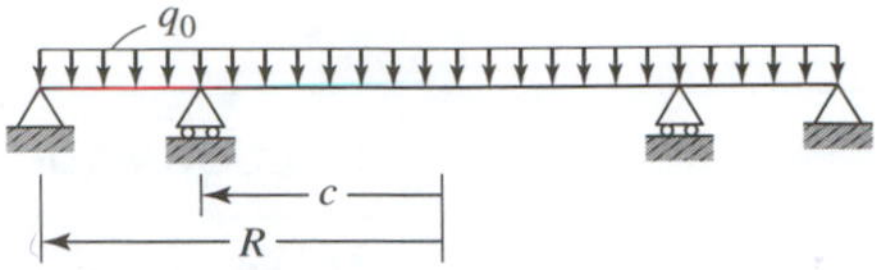


Figure P10-19

(Fig. P10-19). Draw a diagram of $N/(q_0 R)$ as a function of $c/R$.

**10-20** Find such a position $c$ of the additional support in Problem 10-19 that the deflection $w(r)$ at $r = 0$ equals the maximum deflection at $c < r < R$.

**10-21** A circular plate, clamped at the perimeter and carrying a uniform load $q$, is located slightly above a thin rigid circular cylinder of radius $c$ (Fig. P10-21). Determine the intensity of the load $q = q_{\lim}$ needed to just eliminate the clearance $\Delta$ such that $\Delta/R \ll 1$.

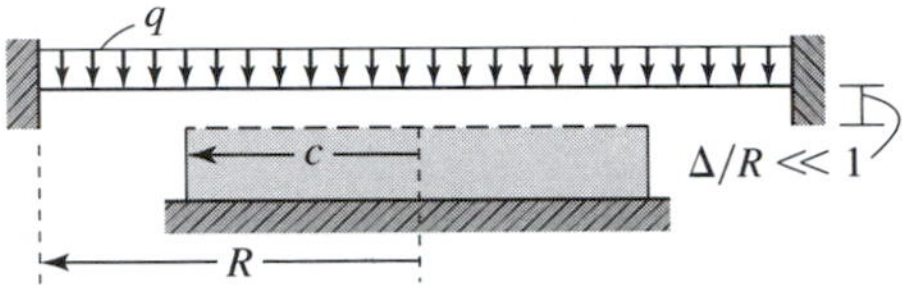


Figure P10-21

## Sections 10-2-6 and 10-2-7

**10-22** The internal perimeter of the annular plate shown in Fig. P10-22 is subject to a small rotation $\alpha \ll 1$. Determine the deflection, the slope, the bending moments, and the shear force.

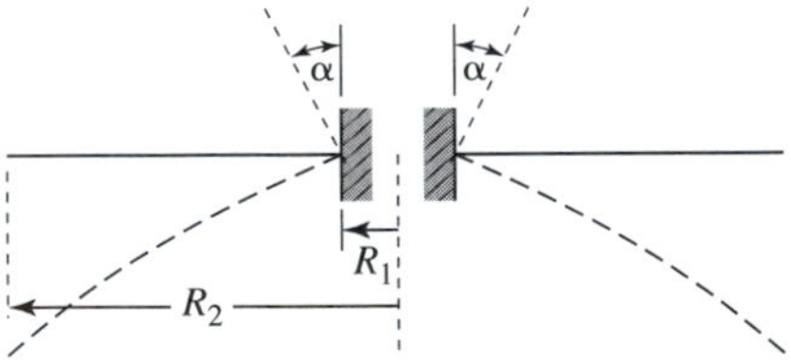


Figure P10-22

**10-23** The annular plate shown in Fig. P10-23 is carrying a uniform load $q$.

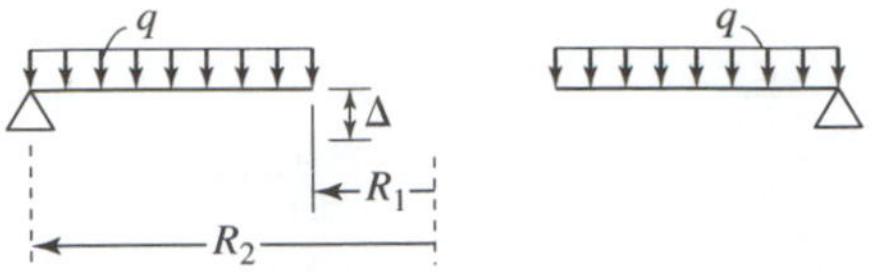


Figure P10-23

Determine the magnitude of $q$ needed to produce a deflection $\Delta$ at $r = R_1$.

**10-24** The annular plate shown in Fig. P10-24 carries a uniform load of intensity $q_0$ such that the total load on the plate is constant equal to $Q = \pi(R_2^2 - R_1^2)q_0$ pounds. Determine the ratio $R_1/R_2$ needed to minimize the weight of the plate. *Note:* Assume that the material properties are fixed.

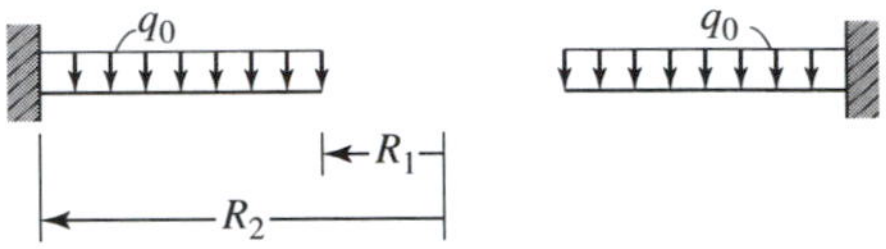


Figure P10-24

## Sections 10-3 and 10-4

**10-25** Find the bending moments in a simply supported rectangular plate carrying a uniform load $q_0$.

**10-26** Find the bending moments in a rectangular plate on an elastic foundation with a simply supported perimeter, carrying a uniform load $q_0$.

## Section 10-7

**10-27** Show that for an isotropic laminated plate, $D_{12} + D_{66} = D_{11}$.

**10-28** Derive Eqs. (10-169) for $M_y$ and $M_{xy}$ in an isotropic laminated plate.

**10-29** Derive Eq. (10-182) for the deflection of a laminated circular plate.

**10-30** Derive Eqs. (10-183) and (10-184).

**10-31** A laminated circular simply supported plate of 12-in radius is composed of two layers and is supposed to carry a uniform load of an intensity of 10 psi. Assuming that one layer is made of steel and the other one of aluminum with the following properties:

| | |
|---|---|
| Young's modulus | $30 \times 10^6$ psi, $10.5 \times 10^6$ psi |
| Ultimate strength | $60 \times 10^3$ psi, $42 \times 10^3$ psi |
| Density | $0.284$ lb/in$^3$, $0.098$ lb/in$^3$ |

Design a plate of minimum weight. Assume Poisson's ratio to equal 0.3 for both materials.

**10-32** Design the plate in Problem 10-31 assuming that it is clamped.

## Section 10-8

**10-33** Write finite differences equations and the boundary conditions for a square membrane under uniform tension $T$ and carrying a uniform load $q_0$.

**10-34** Apply the Rayleigh–Ritz method to find the deflection of the rectangular plate shown in Fig. P10-34. *Hint:* Assume trial functions of the form $w(x, y) = (1 - x^2)^2 \times (1 - y^2)(a_1 + a_2x^2 + a_3y^2)$. Verify first whether the kinematic boundary conditions are satisfied.

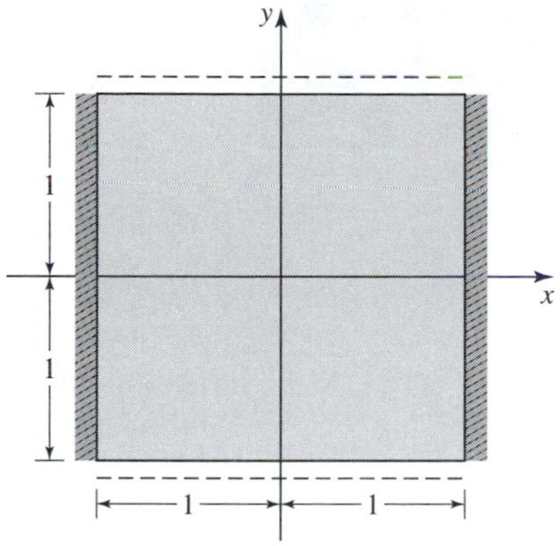


Figure P10-34

## REFERENCES


[1] Churchill, R. V., *Operational Mathematics,* 3rd ed. (McGraw-Hill, New York, 1972).
[2] Collatz, L.,*The Numerical Treatment of Differential Equations,* 3rd ed. (Springer, Berlin, 1960).
[3] Den Hartog, J. P., *Advanced Strength of Materials* (McGraw-Hill, New York, 1952).
[4] Hildebrand, F. B., *Methods of Applied Mathematics,* 2nd ed. (Prentice-Hall, Englewood Cliffs, NJ, 1965).
[5] Ince, E. L., *Ordinary Differential Equations* (Dover Publications, New York, 1956).
[6] Jones, R. M., *Mechanics of Composite Materials* (Scripta Book Co., Washington, DC, 1975).
[7] *Macsyma,* version 13 (Symbolics, Inc., Burlington, MA, 1988).
[8] Marguerre, K., and H. T. Woernle, *Elastic Plates* (Blaisdell Publishing, Waltham, MA, 1969).
[9] *Mathcad7*, professional ed. (MathSoft, Cambridge, MA, 1997).
[10] Panc, V., *Theories of Elastic Plates* (Noordhoff International, Leyden, 1975).
[11] Solecki, R., and G. Zhao, *Industrial Mathematics,* vol. 38 (1988).
[12] Timoshenko, S., and S. Woinowsky-Krieger, *Theory of Plates and Shells*, 2nd ed. (McGraw-Hill, New York, 1959).
[13] Tolstov, G. P., *Fourier Series* (Dover Publications, New York, 1962).
[14] Ugural, A. C., *Stresses in Plates and Shells* (McGraw-Hill, New York, 1981).
[15] Vinson, J. R., and R. L. Sierakowski, *The Behavior of Structures Composed of Composite Materials* (Martinus Nijhoff, Dordrecht, 1986).
[16] Whitney, J. M., *Structural Analysis of Laminated Anisotropic Plates* (Technomic Publishing, Lancaster, PA, 1987).

# BUCKLING AND VIBRATION

Stability and vibration are important components of structural and machine design. Since they are related physically (vibration ceases when the buckling stage is attained) and also mathematically, we discuss them together in this chapter. We deal specifically with the elastic stability and vibration of two-dimensional structures (beams, rings, arches) and three-dimensional structures (rectangular plates).

## 11-1 BUCKLING AND VIBRATION OF BEAMS AND COLUMNS

### 11-1-1 Equation of Motion and Its Solution

Consider an element of a laterally vibrating beam subjected at the ends to time-independent, self-equilibrating axial forces $P$. We will see that the presence of such forces influences the frequencies of vibration. We will also prove that when a compressive force reaches the critical value, vibration ceases completely and buckling occurs. To begin let us derive the differential equation describing this problem.* To this end we analyze the element of Fig. 11-1 shown in the deformed configuration in Fig. 11-2,† where $M = M(x, t)$, $N = N(x, t)$, and $V = V(x, t)$ are the bending moment, the axial force, and the shear force, respectively. The projection of all forces on the $Ox$ axis gives

$$N \cos\beta - (N + dN)\cos(\beta + d\beta) - V \sin\beta + (V + dV)\sin(\beta + d\beta) = 0 \qquad (11\text{-}1)$$

Since $\beta \ll 1$, then $\sin\beta \approx \beta$ and $\cos\beta \approx 1$. Thus Eq. (11-1) becomes

$$-dN + \beta dV + V d\beta = 0 \qquad (11\text{-}2)$$

---

*We neglect the rotary inertia.

†Only in this configuration do the axial forces contribute to bending.

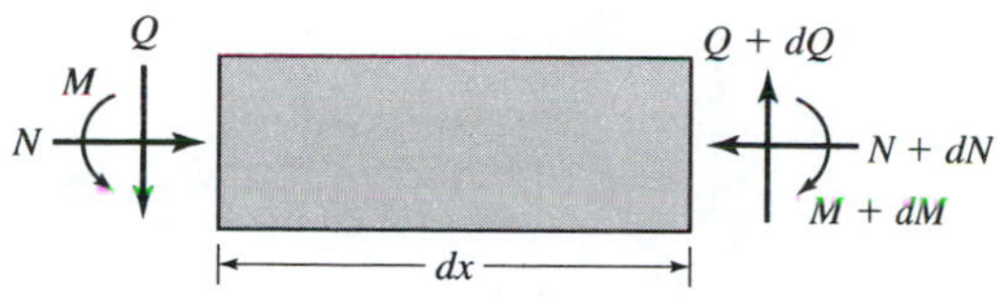


**Figure 11-1** The undeformed element

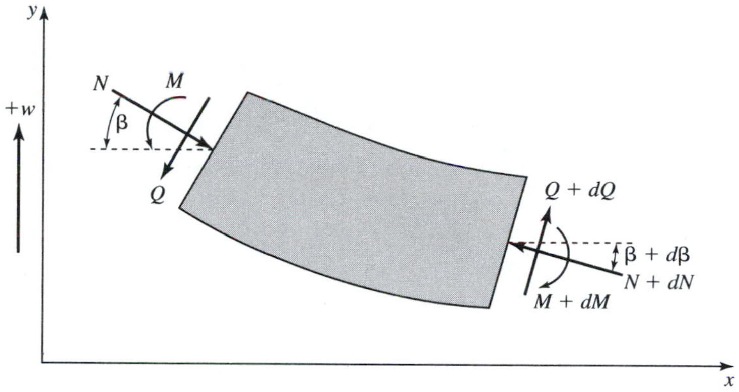


**Figure 11-2** The deformed element

But since $V$ is usually much smaller than $N$ [1, 8], the last two terms in Eq. (11-2) are much smaller than the first, implying that

$$dN \equiv \frac{\partial N}{\partial x}\, dx = 0 \qquad \text{or} \qquad \frac{\partial N}{\partial x} = 0 \tag{11-3}$$

We continue projecting all forces on $Oy$,

$$-N \sin\beta + (N + dN)\sin(\beta + d\beta) - V\cos\beta$$
$$+ (V + dV)\cos(\beta + d\beta) = \rho \frac{\partial^2 w}{\partial t^2}\, dx$$

Deleting all quantities of the second order and remembering that $dN = 0$ we get, upon division by $dx$,

$$N\beta' + V' = \rho \frac{\partial^2 w}{\partial t^2} \tag{11-4}$$

where the prime indicates a partial derivative with respect to $x$. Next, considering the balance of the moment of momentum, we obtain

$$-M - (V + dV)\, dx + (M + dM) = 0$$

Thus,

$$M' = V \tag{11-5}$$

Substituting

$$\beta = -\frac{\partial w}{\partial x} = -w'$$

we represent Eqs. (11-3)–(11-5) in the form

$$N' = 0, \qquad -Nw'' + V' = \rho \frac{\partial^2 w}{\partial t^2}, \qquad M' = V \tag{11-6}$$

or, eliminating $V$,

$$N' = 0, \qquad -Nw'' + M'' = \rho \frac{\partial^2 w}{\partial t^2} \tag{11-7}$$

We know from simple beam theory that

$$M = -EIw'' \tag{11-8}$$

Thus Eqs. (11-7) become

$$N' = 0, \qquad EIw^{\text{iv}} + Nw'' = -\rho \frac{\partial^2 w}{\partial t^2} \tag{11-9}$$

Integrating the first of these equations results in

$$N = \text{constant}$$

Projecting all forces acting on the beam between its end and the current section $x$ on $Ox$ results in $P - N = 0$. Thus the axial force $N$ equals the applied force $P$ so that the second of Eq. (11-9) becomes

$$EIw^{\text{iv}} + Pw'' = -\rho \frac{\partial^2 w}{\partial t^2} \tag{11-10}$$

This is the differential equation for a vibrating beam subjected to the action of a constant axial force $P$. We note that when the displacement is time-independent, $w = w(x)$, Eq. (11-10) reduces to a pure buckling equation for a beam.

On the other hand, if $P = 0$, then Eq. (11-10) represents the differential equation for a vibrating beam. Now we assume that the motion of the beam is harmonic so that the deflection can be represented in the form

$$w(x, t) = W(x) \sin \omega t \tag{11-11}$$

Substituting this into Eq. (11-10) we obtain

$$EIW^{\text{iv}}(x) + PW''(x) - \rho\omega^2 W(x) = 0$$

Dividing this by $EI$ we get

$$W^{\text{iv}}(x) + \lambda^2 W''(x) - \gamma^2 W(x) = 0 \tag{11-12}$$

where

$$\lambda^2 = \frac{P}{EI}, \qquad \gamma^2 = \frac{\rho\omega^2}{EI} \tag{11-13}$$

It can be verified* that the general solution to Eq. (11-12) has the form

$$W(x) = C_1 \cosh ax + C_2 \sinh ax + C_3 \cos bx + C_4 \sin bx \tag{11-14}$$

where

$$\begin{aligned} a &= \sqrt{-\frac{\lambda^2}{2} + \sqrt{\frac{\lambda^4}{4} + \gamma^2}} \\ b &= \sqrt{\frac{\lambda^2}{2} + \sqrt{\frac{\lambda^4}{4} + \gamma^2}} \end{aligned} \tag{11-15}$$

The arbitrary constants $C_1$, $C_2$, $C_3$, and $C_4$ must be found from the boundary conditions.

To understand better the mechanics of pure buckling, consider a column that is initially straight and unloaded. The column is bent and constrained to remain in the bent position. A compressive load $P$ is applied to the end of the column in its bent configuration and then the constraint is removed. If $P$ is small enough then, on removal of the constraint, the column will simply return to its undeformed configuration. On the other hand, if $P$ is large enough then, on removal of the constraint, the column will collapse. This implies that there is a certain value of $P$, say $P_{\text{cr}}$, at which when the constraint is removed, the column will remain in its bent configuration. If $P$ is decreased slightly, the column will return to its undeformed configuration, whereas if $P$ is increased slightly, the column will collapse. Thus $P_{\text{cr}}$ is the maximum compressive load that can be applied to the column without causing collapse or buckling. Note that with the application of $P_{\text{cr}}$ to the column in its bent position and subsequent removal of the constraint, the column is in static equilibrium, and therefore Eq. (11-12) with $\gamma = 0$ applies. Now the modified Eq. (11-12) together with appropriate boundary conditions constitute a homogeneous differential equation and homogeneous boundary conditions. On first sight, the only solution is the trivial one: $w(x) = 0$. On the other hand, based on the physical arguments put forth in the preceding, it seems reasonable that a nontrivial solution exists, at least for certain values of $P$. The question we wish to address, then, is whether we can find values of $P$ for which a nontrivial solution exists to this homogeneous differential equation and the homogeneous boundary conditions. Those solutions, if they exist, represent the static equilibrium configuration for the deformed (bent) column subjected to a compressive load and no lateral load. The search for those solutions is achieved, as the reader remembers from an introductory strength of materials course, by employing the solution

$$W(x) = C_1 \frac{x}{\lambda^2} + C_2 \frac{1}{\lambda^2} + C_3 \sin \lambda x + C_4 \cos \lambda x \tag{11-16}$$

When the boundary conditions are imposed, a system of homogeneous linear algebraic equations in $C_1$, $C_2$, $C_3$, $C_4$ results. Next the determinant of the coefficients of those constants is equated to zero. The possible values of $\lambda$ (and thus of $P$) are obtained by solving this *characteristic equation*.

---

*The standard procedure results in a fourth-order algebraic equation with two real and two imaginary roots.

## 11-1-2 Frequencies and Critical Loads for Various Boundary Conditions

The basic types of boundary conditions for a beam are as follows:

- Hinged support $W = 0, \quad W'' = 0$
- Clamped support $W = 0, \quad W' = 0$
- Free end $W'' = 0, \quad \lambda^2 W' + W''' = 0$*

For future reference we also list the expressions for the derivatives of $W(x)$ obtained from Eq. (11-14),

$$\begin{aligned} W'(x) &= C_1 a \sinh ax + C_2 a \cosh ax - C_3 b \sin bx + C_4 b \cos bx \\ W''(x) &= C_1 a^2 \cosh ax + C_2 a^2 \sinh ax - C_3 b^2 \cos bx - C_4 b^2 \sin bx \\ W'''(x) &= C_1 a^3 \sinh ax + C_2 a^3 \cosh ax + C_3 b^3 \sin bx - C_4 b^3 \cos bx \end{aligned} \tag{11-17}$$

### EXAMPLE 11-1

Consider the simply supported beam shown in Fig. 11-3.

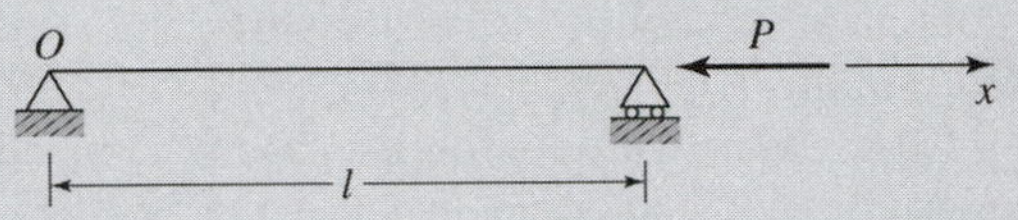


**Figure 11-3**

### SOLUTION

Here the boundary conditions are

1. $W(0) = 0$
2. $W''(0) = 0$
3. $W(l) = 0$
4. $W''(l) = 0$

Using conditions 1 and 2 with Eq. (11-16) we get

$$C_1 + C_3 = 0, \qquad C_1 a^2 - C_3 b^2 = 0 \tag{a}$$

*The requirement that the $Oy$ projection of all forces acting at an end be zero (see Fig. 11-2) yields: $N \sin\beta + V \cos\beta = 0$. We note that since $N$ is usually much larger than $V$, the first term is of the same order of magnitude as the second. Then, using $\beta = -w'$ and Eqs. (11-5), (11-8), and (11-13), we get the desired boundary condition.

These can be satisfied only when $C_1 = 0$ and $C_3 = 0$. The remaining boundary conditions, 3 and 4, with Eqs. (11-17), give

$$C_2 \sinh al + C_4 \sin bl = 0, \qquad C_2 a^2 \sinh al - C_4 b^2 \sin bl = 0 \tag{b}$$

The only possibility for this system of homogeneous linear algebraic equations to have a nontrivial solution exists if the determinant of the matrix composed of the coefficients of the unknowns in system ($b$), a *characteristic matrix,* equals zero. Thus,

$$\det \begin{pmatrix} \sinh al & \sin bl \\ a^2 \sinh al & -b^2 \sin bl \end{pmatrix} = 0$$

Expanding this determinant yields

$$(a^2 + b^2) \sinh al \sin bl = 0 \tag{c}$$

This equation implies that since $\sinh al \neq 0$ for $a > 0$, either

$$a^2 + b^2 = 0 \tag{d}$$

or

$$\sin bl = 0 \tag{e}$$

But condition ($d$) is never satisfied because by Eqs. (11-15), always $a^2 + b^2 > 0$. On the other hand, condition ($e$) implies that

$$bl = n\pi \tag{f}$$

Using the definition of $b$, Eq. (11-15), we get

$$\frac{\lambda^2}{2} + \sqrt{\frac{\lambda^4}{4} + \gamma^2} = \frac{n^2\pi^2}{l^2}$$

Solving this for $\gamma$ yields

$$\gamma = \frac{n\pi}{l}\sqrt{\frac{n^2\pi^2}{l^2} - \lambda^2} \tag{g}$$

In terms of the original variables this becomes

$$\omega_n = \frac{n\pi}{l}\sqrt{\frac{EI}{\rho}\sqrt{\frac{n^2\pi^2}{l^2} - \frac{P}{EI}}}, \qquad n = 1, 2, \ldots, \infty \tag{h}$$

We note that when $P$ attains its critical value causing buckling,

$$P = P_{\text{cr}} = \frac{EIn^2\pi^2}{l^2}, \qquad n = 1, 2, \ldots, \infty \tag{i}$$

then $\omega_n = 0$. We also note that a tensile force will have a sign opposite to $P$, thus increasing the frequency of vibrations.

TABLE 11-1

| Schematic | Boundary Condition | Characteristic Equation |
|---|---|---|
| 1.  | $S+S$ | $\frac{\overline{\lambda}^2}{2}+\sqrt{\frac{\overline{\lambda}^4}{4}+\overline{\gamma}^2}=n^2\pi^2$ |
| 2.  | $C+F$ | $(\overline{\lambda}^4+2\overline{\gamma}^2)\cosh\overline{a}\cos\overline{b}-\overline{\gamma}\overline{\lambda}^2\sinh\overline{a}\sin\overline{b}+2\overline{\gamma}^2=0$ |
| 3.  | $C+C$ | $2\overline{a}(1-\cosh\overline{a}\cos\overline{b})+\frac{\overline{a}^2-\overline{b}^2}{\overline{b}}\sinh\overline{a}\sin\overline{b}=0$ |
| 4.  | $C+S$ | $\overline{a}\cosh\overline{a}\sin\overline{b}-\overline{b}\sinh\overline{a}\cos\overline{b}=0$ |

$S=$ simply supported, $C=$ clamped, $F=$ free.

The characteristic equations for other boundary conditions are obtained in a similar manner. These are, as a rule, transcendental equations. Thus the expressions for frequencies or for critical loads cannot be obtained in closed form. A summary of these characteristic equations is given in Table 11-1 in terms of dimensionless quantities,* where

$$\overline{a}=al,\qquad \overline{b}=bl,\qquad \overline{\lambda}=\lambda l,\qquad \overline{\gamma}=\gamma l^2 \tag{11-18}$$

(For a more general type of support (spring) see also [4]).

When there is no axial force ($P=0$), we obtain the characteristic equations by setting

$$\overline{\lambda}=0,\qquad \overline{a}=\overline{b}=\sqrt{\overline{\gamma}} \tag{11-19}$$

into the expressions listed in Table 11-1. This results in the characteristic equations for frequencies shown in Table 11-2.

This procedure can be adopted for multispan beams with only slight modifications. Let us consider a vibrating two-span beam subjected to compression or tension (Fig. 11-4). The deflection and its derivatives have, for each span, the same form as Eq. (11-14),

$$\begin{aligned}
W_1(x_1)&=C_1\cosh ax_1+C_2\sinh ax_1+C_3\cos bx_1+C_4\sin bx_1\\
W_1'(x_1)&=C_1a\sinh ax_1+C_2a\cosh ax_1-C_3b\sin bx_1+C_4b\cos bx_1\\
W_1''(x_1)&=C_1a^2\cosh ax_1+C_2a^2\sinh ax_1-C_3b^2\cos bx_1-C_4b^2\sin bx_1\\
W_1'''(x_1)&=C_1a^3\sinh ax_1+C_2a^3\cosh ax_1+C_3b^3\sin bx_1-C_4b^3\cos bx_1
\end{aligned} \tag{11-20}$$

*The equations were obtained with the help of the software package Mathcad7, professional edition [2].

TABLE 11-2

| Schematic | Boundary Condition | Characteristic Equation |
|---|---|---|
| 1.  | $S + S$ | $\overline{\gamma} = n^2\pi^2, \quad n = 1, 2, \ldots, \infty$ |
| 2.  | $C + F$ | $\cosh\overline{\gamma}\cos\overline{\gamma} + 1 = 0$ |
| 3.  | $C + C$ | $\cosh\overline{\gamma}\cos\overline{\gamma} - 1 = 0$ |
| 4.  | $C + S$ | $\cosh\overline{\gamma}\sin\overline{\gamma} - \sinh\overline{\gamma}\cos\overline{\gamma} = 0$ |

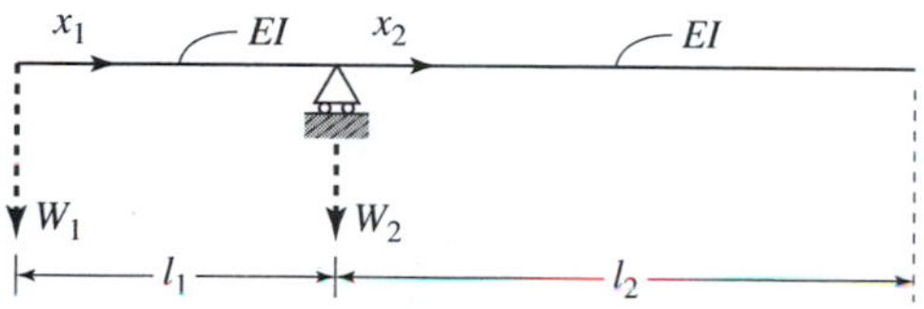


**Figure 11-4** A vibrating two-span beam with unspecified end conditions

$$
\begin{aligned}
W_2(x_2) &= C_5 \cosh ax_2 + C_6 \sinh ax_2 + C_7 \cos bx_2 + C_8 \sin bx_2 \\
W_2'(x_2) &= C_5 a \sinh ax_2 + C_6 a \cosh ax_2 - C_7 b \sin bx_2 + C_8 b \cos bx_2 \\
W_2''(x_2) &= C_5 a^2 \cosh ax_2 + C_6 a^2 \sinh ax_2 - C_7 b^2 \cos bx_2 - C_8 b^2 \sin bx_2 \\
W_2'''(x_2) &= C_5 a^3 \sinh ax_2 + C_6 a^3 \cosh ax_2 + C_7 b^3 \sin bx_2 - C_8 b^3 \cos bx_2
\end{aligned}
\tag{11-21}
$$

Since there are eight constants of integration, we need eight conditions. We have four conditions at the outside ends of the beam (they depend on the manner in which the beam is supported). The remaining four are the conditions of continuity of deflection, slope, and bending moment at the intermediate support, and the condition that the deflection at this support is zero. Thus,

$$
W_1(l) = 0, \qquad W_2(0) = 0, \qquad W_1'(l) = W_2'(0), \qquad EIW_1''(l) = EIW_2''(0) \tag{11-22}
$$

which in terms of expressions (11-20) and (11-21) become

$$
\begin{aligned}
&C_1 \cosh al + C_2 \sinh al + C_3 \cos bl + C_4 \sin bl = 0, \qquad C_5 + C_7 = 0 \\
&C_1 a \sinh al + C_2 a \cosh al - C_3 b \sin bl + C_4 b \cos bl = C_6 a + C_8 b \\
&EI(C_1 a^2 \cosh al + C_2 a^2 \sinh al - C_3 b^2 \cos bl - C_4 b^2 \sin bl) = EI(C_5 a^2 - C_7 b^2)
\end{aligned}
\tag{11-23}
$$

## EXAMPLE 11-2

Derive the characteristic equation for a two-span beam simply supported at both ends.

### SOLUTION

Besides Eqs. (11-23) we have the following boundary conditions:

$$W_1(0) = 0, \qquad W_1''(0) = 0, \qquad W_2(l) = 0, \qquad W_2''(l) = 0 \tag{a}$$

Substituting Eqs. (11-20) and (11-21) we get

$$\begin{aligned} &C_1 + C_3 = 0, \qquad C_1a^2 - C_3b^2 = 0 \\ &C_5 \cosh al + C_6 \sinh al + C_7 \cos bl + C_8 \sin bl = 0 \\ &C_5a^2 \cosh al + C_6a^2 \sinh al - C_7b^2 \cos bl - C_8b^2 \sin bl = 0 \end{aligned} \tag{b}$$

Evidently $C_1 = C_3 = 0$. Thus the characteristic equation is obtained from the condition

$$\det \begin{pmatrix} \sinh \bar{a} & \sin \bar{b} & 0 & 0 & 0 & 0 \\ 0 & 0 & 1 & 0 & 1 & 0 \\ \bar{a} \cosh \bar{a} & \bar{b} \cos \bar{b} & 0 & -\bar{a} & 0 & -\bar{b} \\ 0 & 0 & \cosh \bar{a} & \sinh \bar{a} & \cos \bar{b} & \sin \bar{b} \\ \bar{a}^2 \sinh \bar{a} & -\bar{b}^2 \sin \bar{b} & -\bar{a}^2 & 0 & \bar{b}^2 & 0 \\ 0 & 0 & \bar{a}^2 \cosh \bar{a} & \bar{a}^2 \sinh \bar{a} & -\bar{b}^2 \cos \bar{b} & -\bar{b}^2 \sin \bar{b} \end{pmatrix} = 0 \tag{c}$$

Using the software package Mathcad7, professional edition [2], we simplified this to

$$\sin \bar{b}(\bar{b} \sinh \bar{a} \cos \bar{b} - \bar{a} \cosh \bar{a} \sin \bar{b}) = 0 \tag{d}$$

The first factor corresponds to antisymmetric modes and is identical to the characteristic equation for a single-span beam simply supported on both ends. The second factor, representing symmetric modes, is identical to the characteristic equation for a single-span beam simply supported at one end and clamped at the another.

Table 11-3 lists the characteristic equations for beams with two equal-length spans and various boundary conditions.

The characteristic equations for the special case $P = 0$ (no axial force) are listed in the Table 11-4. They are obtained from the results in Table 11-3 by setting

$$\bar{\lambda} = 0, \qquad \bar{a} = \bar{b} = \sqrt{\bar{\gamma}} \tag{11-24}$$

The results for the other special case—no vibration but only axial force present, thus causing pure buckling—could also be obtained from the general solution, Eq. (11-14 ). However, this would require a limit analysis, $a \to 0$. It is simpler to set $\gamma = 0$ in

TABLE 11-3

| Schematic | Boundary Condition | Characteristic Equation |
|---|---|---|
| 1.  | $S + S$ | $\sin\overline{b} = 0$ or $\overline{b}\sinh\overline{a}\cos\overline{b} - \overline{a}\cosh\overline{a}\sin\overline{b} = 0$ |
| 2. 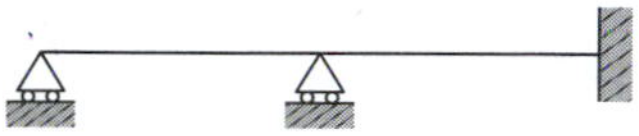 | $S + C$ | $\overline{a}^2\sin^2\overline{b}\cosh 2\overline{a} - 2\overline{b}^2\sinh^2\overline{a}\left(\frac{1}{2} - \cos^2\overline{b}\right)$<br>$+ 2\overline{a}\overline{b}\sin\overline{b}\sinh\overline{a}(1 - 2\cosh\overline{a}\cos\overline{b}) = 0$ |
| 3.  | $C + C$ | $\overline{a}(\overline{a}^2 - 3\overline{b}^2)\sinh 2\overline{a}\cos^2\overline{b} + \frac{1}{2}\overline{b}(3\overline{a}^2 - \overline{b}^2)\cosh 2\overline{a}\sin 2\overline{b}$<br>$- \overline{a}(\overline{a}^2 - \overline{b}^2)\sinh 2\overline{a} + 4\overline{a}\overline{b}^2\sinh\overline{a}\cos\overline{b}$<br>$- 4\overline{a}^2\overline{b}\cosh\overline{a}\sin\overline{b} + \overline{b}(\overline{a}^2 + \overline{b}^2)\sin\overline{b}\cos\overline{b} = 0$ |
| 4. 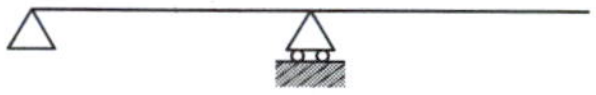 | $S + F$ | $2\lambda^2\sinh\overline{a}\sin\overline{b} - \lambda^2\sinh 2\overline{a}\sin 2\overline{b}$<br>$+ \overline{a}\overline{b}\sinh 2\overline{a}(4\sin^2\overline{b} - 1) + \overline{a}\overline{b}(2\cos^2\overline{b} - 1) = 0$ |
| 5. 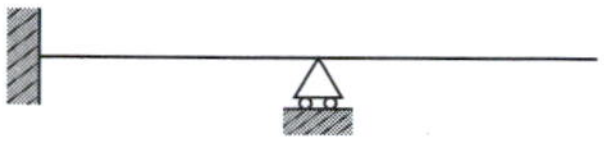 | $C + F$ | $\overline{a}\overline{b}^2(2\overline{a}^4 - \overline{a}^2\overline{b}^2 + \overline{b}^4)\sinh 2\overline{a}\cos^2\overline{b}$<br>$- \frac{1}{2}\overline{a}^2\overline{b}(\overline{a}^4 - \overline{a}^2\overline{b}^2 + 2\overline{b}^4)\cosh 2\overline{a}\sin 2\overline{b}$<br>$- \frac{1}{2}\overline{a}\overline{b}^2\lambda^4\sinh 2\overline{a} + 2\overline{a}^3\overline{b}^2\lambda^2\sinh\overline{a}\cos\overline{b}$<br>$+ 2\overline{a}^2\overline{b}^3\lambda^2\cosh\overline{a}\sin\overline{b} - \overline{a}^2\overline{b}^3(\overline{a}^2 + \overline{b}^2)\sin\overline{b}\cos\overline{b} = 0$ |

TABLE 11-4

| Schematic | End Condition | Characteristic Equation |
|---|---|---|
| 1.  | $S + S$ | $\sin\overline{\gamma} = 0$<br>or<br>$\sinh\overline{\gamma}\cos\overline{\gamma} - \cosh\overline{\gamma}\sin\overline{\gamma} = 0$ |
| 2. 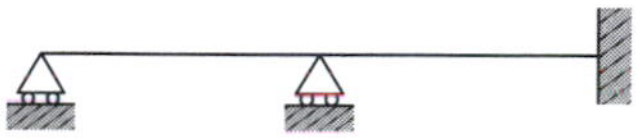 | $S + C$ | $2\sinh\overline{\gamma}\sin\overline{\gamma}(1 - 2\cosh\overline{\gamma}\cos\overline{\gamma}) + \cosh^2\overline{\gamma} - \cos^2\overline{\gamma} = 0$ |
| 3. 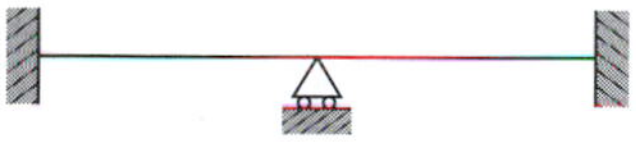 | $C + C$ | $\sinh\overline{\gamma}\cos\overline{\gamma} - \cosh\overline{\gamma}\sin\overline{\gamma} = 0$<br>or<br>$1 + \cosh\overline{\gamma}\cos\overline{\gamma} = 0$ |

Eq. (11-12), thus obtaining

$$W^{iv}(x) + \lambda^2 W''(x) = 0 \tag{11-25}$$

and solve it independently. This gives

$$W(x) = C_1\frac{x}{\lambda^2} + C_2\frac{1}{\lambda^2} + C_3\sin\lambda x + C_4\cos\lambda x \tag{11-26}$$

TABLE 11-5

| Schematic | Boundary Condition | Characteristic Equation |
|---|---|---|
| 1.  | $S+S$ | $\sin\bar{\lambda}=0$ |
| 2.  | $S+C$ | $\bar{\lambda}\cos\bar{\lambda}-\sin\bar{\lambda}=0$ |
| 3.  | $C+C$ | $2-2\cos\bar{\lambda}-\bar{\lambda}\sin\bar{\lambda}=0$ |
| 4.  | $C+F$ | $\cos\bar{\lambda}=0$ |

A list of the characteristic equations for the case of single-span beams with various boundary conditions is given in Table 11-5, where

$$\bar{\lambda}=\lambda l \tag{11-27}$$

The results obtained for two-span beams with unequal spans of lengths $l_1$, $l_2$ are shown in Table 11-6.

## 11-1-3 Application of the Rayleigh–Ritz Method

In Section 5-11 we presented the fundamentals of the Rayleigh–Ritz method. Here it will be applied to problems of buckling of straight beams. We note that the potential energy of a beam in bending without the transverse load is given by Eq. (5-123) in the form

$$\Pi=\int_0^l \frac{EI}{2}\left(\frac{d^2w}{dx^2}\right)^2 dx-\sum_{i=1}^{n} w_i P_i \tag{11-28}$$

Now it becomes

$$\Pi=\int_0^l \frac{EI}{2}\left(\frac{d^2w}{dx^2}\right)^2 dx-P\Delta \tag{11-29}$$

where $P$ is the applied axial force and $\Delta$ is the axial displacement of one end relative the other (Fig. 11-5). To calculate this displacement, we consider bending alone, disregarding the effect of the shortening of the axis of the beam [5, p. 211] due to compression. We see from Fig. 11-6 that

$$ds^2=dx^2+dw^2$$

$$ds=\sqrt{1+\left(\frac{dw}{dx}\right)^2}\,dx$$

The assumed inextensibility of the beam leads to the following relation (see Fig. 11-6):

$$L+\Delta=\int_0^l ds\equiv\int_0^L\sqrt{1+\left(\frac{dw}{dx}\right)^2}\,dx\cong\int_0^L\left[1+\frac{1}{2}(w')^2\right]dx \tag{11-30}$$

**TABLE 11-6**

| Schematic | End Condition | Characteristic Equation: Two-Span Beams | Characteristic Equation: Spans of Equal Length ($l_1 = l_2 = l$) |
|---|---|---|---|
| 1. 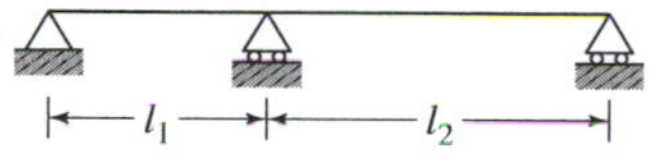  | $S + S$ | $\sin\bar{\lambda}_1[(1+\beta)\sin\bar{\lambda}_1\beta - \bar{\lambda}_1\cos\bar{\lambda}_1\beta]$ $- \bar{\lambda}_1\cos\bar{\lambda}_1\sin\bar{\lambda}_1\beta = 0$ | $\sin\bar{\lambda} - \bar{\lambda}\cos\bar{\lambda} = 0$ or $\sin\bar{\lambda} = 0$ |
| 2. 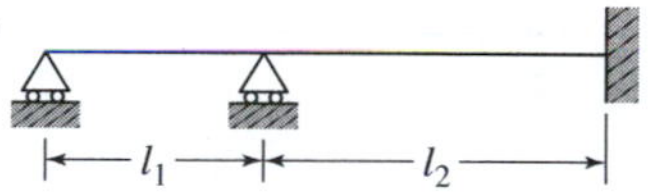  | $S + C$ | $\sin\bar{\lambda}_1[(1+\bar{\lambda}_1)\sin\bar{\lambda}_1\beta - \bar{\lambda}_1\cos\bar{\lambda}_1\beta]$ $- \bar{\lambda}_1\cos\bar{\lambda}_1\sin\bar{\lambda}_1\beta = 0$ | $\sin\bar{\lambda} - \bar{\lambda}\cos\bar{\lambda} = 0$ or $\sin\bar{\lambda} = 0$ |
| 3. 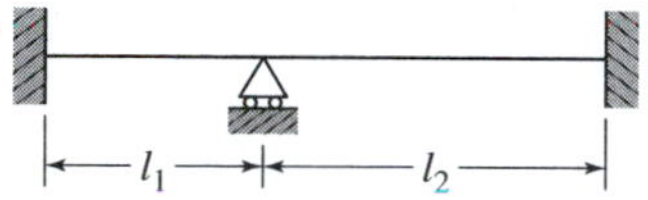  | $S + F$ | $\bar{\lambda}_1\cos\bar{\lambda}_1\sin\bar{\lambda}_1\beta$ $+ \sin\bar{\lambda}_1(\bar{\lambda}_1\cos\bar{\lambda}_1\beta - \sin\bar{\lambda}_1\beta) = 0$ | $2\bar{\lambda}\cos\bar{\lambda} - \sin\bar{\lambda} = 0$ or $\sin\bar{\lambda} = 0$ |
| 4. 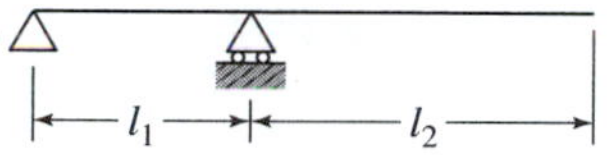  | $C + C$ | $\left(2 - \bar{\lambda}_1^2\beta\right)\sin[\bar{\lambda}_1(1+\beta)] + \bar{\lambda}_1(1+\beta)$ $\times(\sin\bar{\lambda}_1\sin\bar{\lambda}_1\beta - 2\cos\bar{\lambda}_1\cos\bar{\lambda}_1\beta)$ $+ 2\bar{\lambda}_1(\cos\bar{\lambda}_1 + \beta\cos\bar{\lambda}_1\beta)$ $- 2\sin\bar{\lambda}_1 - 2\sin\bar{\lambda}_1\beta = 0$ | $(2 - \bar{\lambda}^2)\sin 2\bar{\lambda} + 2\bar{\lambda}$ $\times(\sin^2\bar{\lambda} - 2\cos^2\bar{\lambda})$ $+ 4\bar{\lambda}\cos\bar{\lambda}_1 - 4\sin\bar{\lambda}_1 = 0$ |
| 5. 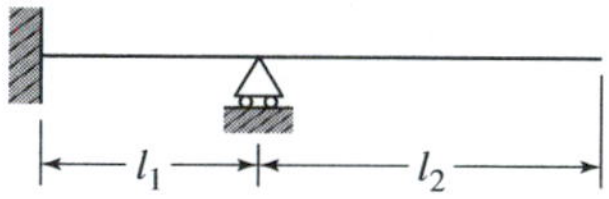  | $C + F$ | $\bar{\lambda}_1\cos[\bar{\lambda}_1(1+\beta)] - 2\cos\bar{\lambda}_1\sin\bar{\lambda}_1\beta$ $- \sin\bar{\lambda}_1\cos\bar{\lambda}_1\beta + 2\sin\bar{\lambda}_1\beta = 0$ | $2\bar{\lambda}\cos^2\bar{\lambda} - 3\sin\bar{\lambda}\cos\bar{\lambda}$ $+ 2\sin\bar{\lambda} - \bar{\lambda} = 0$ |

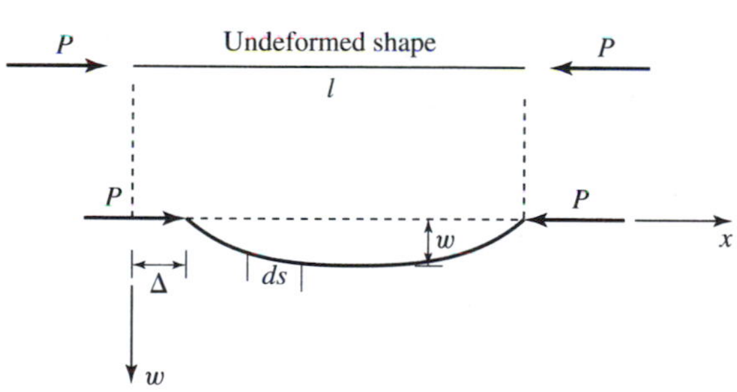


Figure 11-5

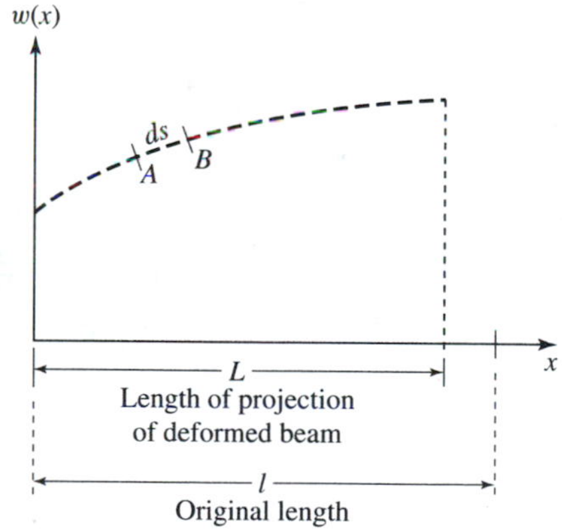


Figure 11-6

where we used a Taylor series expansion in which we neglected higher order terms. Thus,

$$\Delta = \frac{1}{2}\int_0^L [w'(x)]^2\,dx \cong \frac{1}{2}\int_0^l [w'(x)]^2\,dx \tag{11-31}$$

Substituting this into Eq. (11-29), we get the final expression for potential energy,

$$\Pi = \frac{1}{2}\int_0^l [EI(w'')^2 - P(w')^2]\,dx \tag{11-32}$$

## EXAMPLE 11-3

Use the Rayleigh–Ritz method to find the critical load for a beam clamped at both ends (Fig. 11-7).

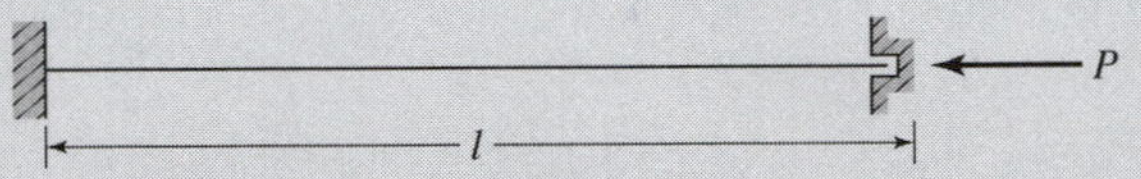


**Figure 11-7**

### SOLUTION

The boundary conditions are

$$w(0) = w(l) = 0, \qquad w'(0) = w'(l) = 0 \tag{$a$}$$

The assumed deflection $w(x)$ must satisfy all conditions ($a$) because they are all geometric. We take, for instance,

$$w(x) = a_1\left(1 - \cos\frac{2\pi x}{l}\right) \tag{$b$}$$

It is easy to check that this function satisfies all conditions ($a$) and is therefore an appropriate candidate for the Rayleigh–Ritz method. Substituting the derivatives of $w(x)$, Eq. ($b$), into the expression for potential energy, Eq. (11-32), we get

$$\Pi = \frac{1}{2}\int_0^l \left[\frac{16EI\pi^4}{l^4}a_1^2\cos^2\frac{2\pi x}{l} - \frac{4P\pi^2}{l^2}a_1^2\sin^2\frac{2\pi x}{l}\right]dx \tag{$c$}$$

Applying the condition

$$\frac{\partial\Pi}{\partial a_1} = 0 \tag{$d$}$$

results in

$$P_{\text{cr}} = \frac{4EI\pi^2}{l^2} \tag{$e$}$$

We notice that this equals the exact value. Evidently our initial "guess," Eq. (*b*), satisfies not only the boundary conditions but also the differential equation for buckling.

Now repeat this procedure using a polynomial as an approximation to $w(x)$,

$$w(x) = a_0 + a_1\left(\frac{x}{l}\right) + a_2\left(\frac{x}{l}\right)^2 + a_3\left(\frac{x}{l}\right)^3 + a_4\left(\frac{x}{l}\right)^4 + a_5\left(\frac{x}{l}\right)^5 \qquad (f)$$

To begin with, impose the kinematic boundary conditions (*a*). This leads to

$$w(0) = 0 \Rightarrow a_0 = 0, \qquad w'(0) = 0 \Rightarrow a_1 = 0$$

$$w(l) = 0 \Rightarrow a_2 + a_3 + a_4 + a_5 = 0, \qquad w'(l) = 0 \Rightarrow 2a_2 + 3a_3 + 4a_4 + 5a_5 = 0 \qquad (g)$$

Using Eqs. (*g*) we eliminate $a_4$ and $a_5$,

$$a_4 = -3a_2 - 2a_3, \qquad a_5 = 2a_2 + a_3$$

Thus the assumed deflection, Eq. (*f*), becomes

$$w(x) = a_2\left[\left(\frac{x}{l}\right)^2 - 3\left(\frac{x}{l}\right)^4 + 2\left(\frac{x}{l}\right)^5\right] + a_3\left[\left(\frac{x}{l}\right)^3 - 2\left(\frac{x}{l}\right)^4 + \left(\frac{x}{l}\right)^5\right] \qquad (h)$$

This example was completed using the software package Mathcad7, professional edition [2]. First the potential energy $\Pi$ was evaluated. Next the conditions

$$\frac{\partial \Pi}{\partial a_2} = 0, \qquad \frac{\partial \Pi}{\partial a_3} = 0 \qquad (i)$$

were applied, leading to a system of homogeneous linear algebraic equations. Finally the requirement that the characteristic determinant of this system vanish leads to a quadratic equation in $P$,

$$P^2 - 132\frac{EI}{l^2}P + 3780\left(\frac{EI}{l^2}\right)^2 = 0 \qquad (j)$$

Its smallest root is the critical value. Thus,

$$P_{\text{cr}} = \frac{42EI}{l^2} \qquad (k)$$

which is about 6% higher than the exact value, Eq. (*e*).

We noted in Chapter 5, Section 5-11 that a structure constrained by the assumptions of the Rayleigh–Ritz method is stiffer than the actual one. This implied that the displacements calculated by the Rayleigh–Ritz method are underestimated. This conclusion is also valid for cases of buckling and vibration. In the former case the critical loads predicted by the Rayleigh–Ritz method are larger than the corresponding exact values (see also [5, p. 223]). In the latter case the frequencies resulting from the application of the Rayleigh–Ritz method are larger than the exact ones (see [3, sec. 5.8]).

## 11-2 BUCKLING OF RINGS, ARCHES, AND THIN-WALLED TUBES

### 11-2-1 Equations of Motion and Their Solution

Consider a thin circular ring subjected to uniform outside pressure $p$* (Fig. 11-8). This problem is related to that of a column compressed by an axial force, despite geometric differences. In both cases a state of equilibrium exists in which the original geometry is preserved—a slightly shorter straight line for the column and a slightly smaller circle for the ring. There is no bending present in either case. Increasing the magnitudes of the applied forces also leads to similar phenomena in both cases, that is, when the magnitude of the axial force applied to the column, or the magnitude of the pressure applied to the disk, reach certain critical values, then even the smallest deviation from the original geometry leads to bending and ultimately to buckling. To find the critical pressure $p_{cr}$ we analyze the equilibrium of a ring in its deformed, noncircular state (Fig. 11-9). We begin by establishing the relation between the bending moment and the change of curvature caused by this deformation. Let $u(\theta)$ be the radial positive outward displacement. The original curvature of the ring is $1/R$. The curvature of the deformed ring is, according to the definition of the curvature of a curve at a point, the limit of the change of the slope at points adjacent to the given point. At a point $P$ at an angle $\theta$ the slope about the local tangent is

$$\frac{du}{ds} = \frac{du}{R\,d\theta} = \frac{1}{R}u' \tag{11-33}$$

whereas at the neighboring point $P'$ at an angle $\theta + d\theta$ the slope about the tangent is incrementally larger,

$$\frac{1}{R}u' + \frac{d}{d\theta}\left(\frac{u'}{R}\right) d\theta = \frac{1}{R}(u' + u''\,d\theta) \tag{11-34}$$

Thus the change in slope is the difference between the slopes at $P'$ and $P$ plus the slope $d\theta$ due to the curvature of the undeformed axis,

$$-\frac{1}{R}(u' + u''\,d\theta) + \frac{1}{R}u' + d\theta = \left(1 - \frac{1}{R}u''\right) d\theta$$

Therefore the curvature of the deformed ring at any point is

$$\frac{(1 - (1/R)u'')\,d\theta}{r\,d\theta} = \frac{(1 - (1/R)u'')\,d\theta}{(R + u)\,d\theta}$$

Expanding the denominator into a Taylor series and retaining in this expansion only two terms, we get

$$\frac{1}{R} - \frac{1}{R^2}u - \frac{1}{R^2}u''$$

*This is, strictly speaking, the pressure per unit length of the ring. Thus it is measured in pounds per inch.

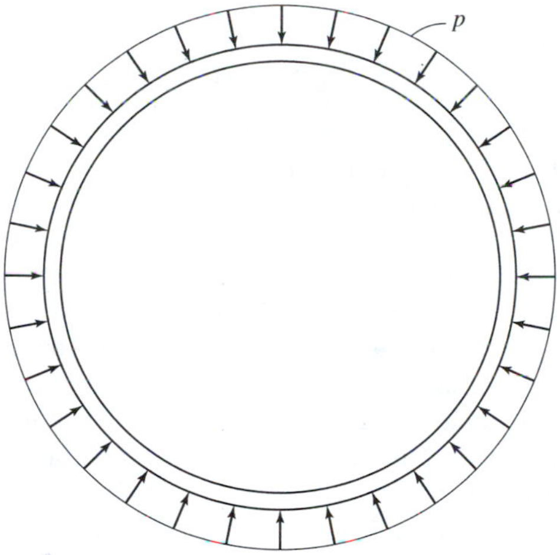


Figure 11-8

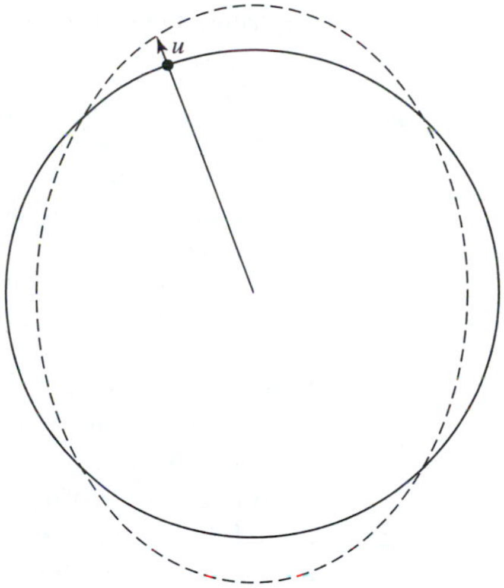


Figure 11-9 Undeformed and deformed state of a circular ring

Thus the change in curvature resulting from the deformation is the difference between this result and the original curvature $1/R$,

$$-\frac{1}{R^2}(u + u'') \tag{11-35}$$

Finally we relate the bending moment $M$ to the change in curvature. To this end recall Eq. (8-109), which can be written in the form

$$\frac{\Delta\, d\varphi}{d\varphi} = \frac{M}{EAe} \tag{11-36}$$

Next, denoting by $\varrho$ the radius of curvature of the deformed bar, we see from Fig. 8-31 that

$$ds = \rho(d\varphi - \Delta\, d\varphi) = R\, d\varphi \tag{11-37}$$

Thus dividing by $d\varphi$, we get

$$\frac{\Delta\, d\varphi}{d\varphi} = R\left(\frac{1}{R} - \frac{1}{\rho}\right) \tag{11-38}$$

But from the Section 8-3-3-2 we also find that

$$e = \frac{1}{A}\int_A \frac{y^2\, dA}{R - y} \approx \frac{I}{AR} \tag{11-39}$$

where $I$ is the moment of inertia of the cross section. Substituting this and Eq. (11-38) into Eq. (11-36), we get the relation between the bending moment and the change in curvature,

$$M = EI\left(\frac{1}{R} - \frac{1}{\rho}\right) \tag{11-40}$$

Replacing the change in curvature by the expression (11-35) results in

$$M = \frac{EI}{R^2}(u + u'') \tag{11-41}$$

where it has been assumed that the positive bending moment straightens the ring (therefore decreases the curvature). Next we derive the differential equation for the equilibrium in the deformed state. To this end we calculate the bending moment appearing in an arbitrary cross section.

Consider a circular ring of radius $R$, subjected to a uniform external pressure $p$. If the ring maintains its circular shape, $p$ can be increased until the yield stress is reached. However, if the ring deviates from its circular shape, buckling can occur. To find the internal forces and moments in this case, consider a slightly deformed ring (Fig. 11-10). If $u$ is the radial displacement, note that at points $A'$ and $B'$ the slope $u' = \partial u/\partial\theta$ is zero. Cut the deformed ring at these points and draw the free-body diagram, as shown in Fig. 11-11. No internal horizontal forces are present because $A'B'$ lies in a plane of symmetry and because the horizontal resultant of the pressure is zero. Consider the shape in a pressure field $p$ shown in Fig. 11-12. Force equilibrium in the vertical direction requires that the vertical resultant of the pressure on the curved surface be $p[2(R - u_A)]$. Thus,

$$Q_A = p(R - u_A) \tag{11-42}$$

Now cut the ring at some location $\theta$, measured from $A$ (see Fig. 11-13). Then, summing moments about $C$,

$$M(\theta) = M_A + M_Q - M_p(\theta) \tag{11-43}$$

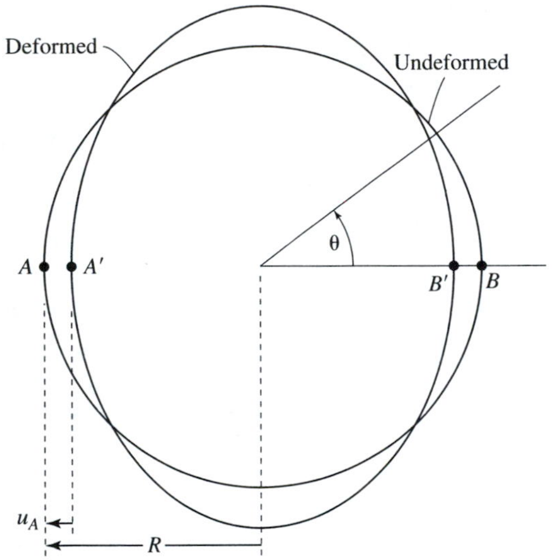


Figure 11-10

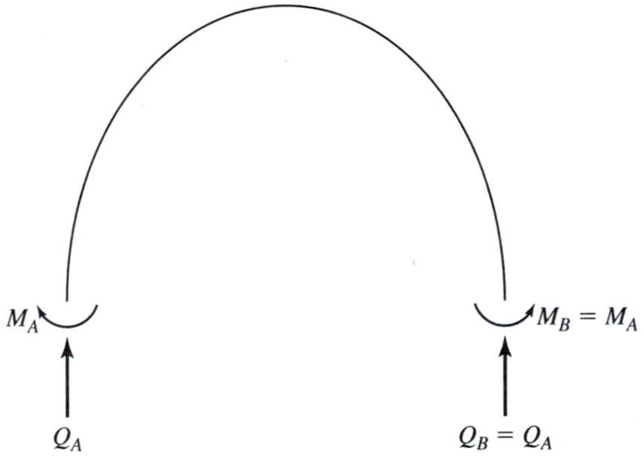


Figure 11-11 A free body diagram of the deformed ring cut at the points with zero slopes

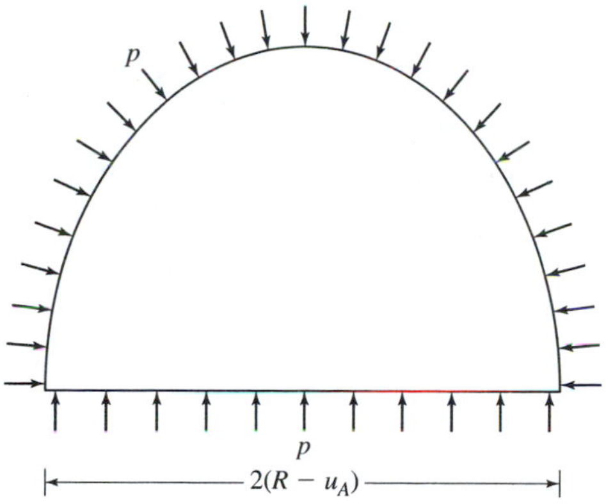


Figure 11-12

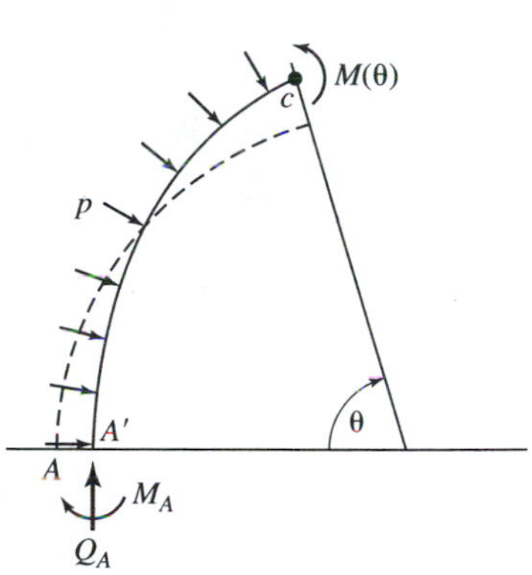


Figure 11-13

where $M_Q$ is the moment due to $Q_A$ and $M_p(\theta)$ is the moment due to the pressure. From Fig. 11-14*a* we see that the moment arm for $Q_A$ is $(R - u_A) - (R + u)\cos\theta$ where $u$ is the length of the vector $\boldsymbol{u}$. Thus, using Eq. (11-42),

$$\begin{aligned} M_Q &= p(R - u_A)[R - u_A - (R + u)\cos\theta] \\ &= pR^2\left(1 - \frac{u_A}{R}\right)\left[1 - \frac{u_A}{R} - \left(1 + \frac{u}{R}\right)\cos\theta\right] \\ &= pR^2\left[1 - \frac{u_A}{R} - \left(1 + \frac{u}{R}\right)\cos\theta - \frac{u_A}{R} + \left(\frac{u_A}{R}\right)^2 + \frac{u_A}{R}\left(1 + \frac{u}{R}\right)\cos\theta\right] \end{aligned}$$

Note that $(u_A/R)^2 \ll 1$ and $u\, u_A/R^2 \ll 1$. Neglecting these terms, we get

$$M_Q = pR^2\left[1 - 2\frac{u_A}{R} - \left(1 + \frac{u - u_A}{R}\right)\cos\theta\right] \tag{11-44}$$

To find $M_p$ consider again Fig. 11-14. Figure 11-14*a* shows the basic geometry, Fig. 11-14*b* shows the polygon $A'C'CC''$ in a constant-pressure field $p$, and Fig. 11-14*c* shows the polygon with the pressure resultants on the straight sides. We notice that the hydrostatic pressure applied to $A'C'$ is equivalent to the sum of the pressures applied to $C'C$ and $CC''$, and to the secant $A'C''$. To prove this we remove the dashed polygon in Fig. 11-14*a* and place it in a pressure field $p$, as shown in Fig. 11-14*b*. Under the action of the pressure, the polygon in Fig. 11-14*b* must be in static equilibrium. Thus we replace the arc $A'C'$ by the polygon $A'C''CC'$, making the calculation much easier. Next we note that the sum of the moments about $C'$ due to the pressure must be zero. Hence if $M_p(\theta)$ is the moment about $C'$ due to the pressure along the arc $A'C'$, then from Fig. 11-14*c*,

$$\sum M = -M_p(\theta) + F_1a_1 + F_2a_2 + F_3a_3 = 0 \tag{11-45}$$

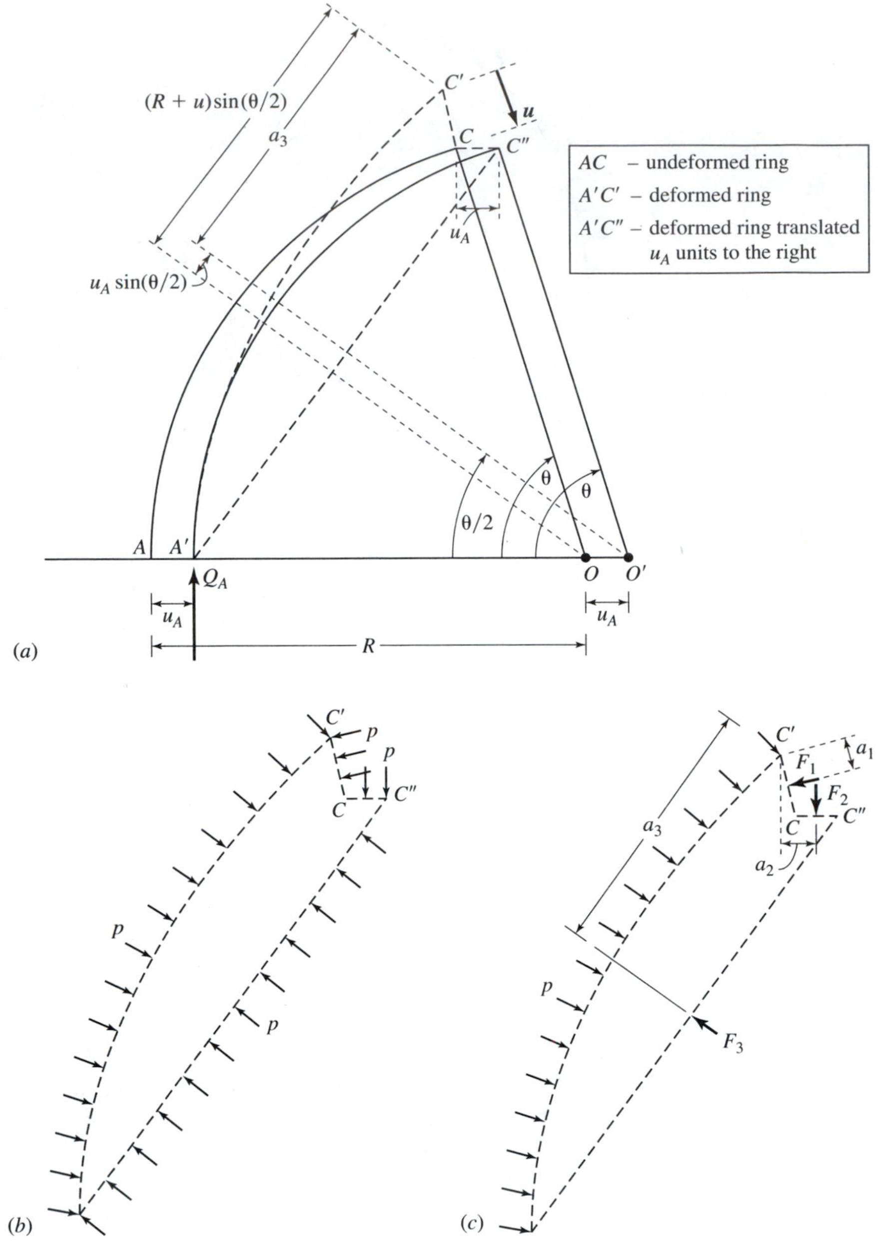

$(R + u)\sin(\theta/2)$
$a_3$
$C'$
$u$
$C$
$C''$
$AC$ – undeformed ring
$A'C'$ – deformed ring
$A'C''$ – deformed ring translated $u_A$ units to the right
$u_A$
$u_A \sin(\theta/2)$
$\theta$
$\theta$
$\theta/2$
$A$
$A'$
$Q_A$
$O$
$O'$
$u_A$
$u_A$
$R$
(a)
$C'$
$p$
$p$
$C''$
$C$
$p$
$p$
(b)
$C'$
$a_1$
$F_1$
$F_2$
$C''$
$C$
$a_3$
$a_2$
$p$
$F_3$
(c)


Figure 11-14

Referring to Fig. 11-14 we find

$$F_1 = pu, \quad F_2 = pu_A, \quad F_3 = p\left(2R\sin\frac{\theta}{2}\right)$$
$$a_1 = \frac{1}{2}u, \quad a_2 = \frac{1}{2}u_A + u\cos\theta, \quad a_3 = (R+u)\sin\frac{\theta}{2} - u_A\sin\frac{\theta}{2} \tag{11-46}$$

Therefore, using Eq. (11-45),

$$M_p(\theta) = pR^2\left\{\frac{1}{2}\left(\frac{u}{R}\right)^2 + \left[\frac{1}{2}\left(\frac{u_A}{R}\right)^2 + \frac{u_A u}{R^2}\cos\theta\right] + 2\sin\frac{\theta}{2}\left[\left(1+\frac{u}{R}\right)\sin\frac{\theta}{2} - \frac{u_A}{R}\sin\frac{\theta}{2}\right]\right\}$$

Now, following the derivation of the expression for $M_Q$, we neglect the first three terms as vanishingly small and obtain

$$M_p(\theta) = 2pR^2\sin^2\frac{\theta}{2}\left(1+\frac{u-u_A}{R}\right) \tag{11-47}$$

Using Eqs. (11-43), (11-44), and (11-47),

$$M(\theta) = M_A + pR^2\left[1 - 2\frac{u_A}{R} - \left(1+\frac{u-u_A}{R}\right)\cos\theta\right] - 2pR^2\sin^2\frac{\theta}{2}\left(1+\frac{u-u_A}{R}\right)$$

which, thanks to the trigonometric identity

$$\cos\theta = 1 - 2\sin^2\frac{\theta}{2}$$

simplifies, to

$$M(\theta) = M_A - pR(u_A + u) \tag{11-48}$$

Substituting this into Eq. (11-41) we get

$$\frac{EI}{R^2}(u + u'') = M_A - pR(u + u_A)$$

Thus the differential equation for buckling results,

$$u'' + u\left(1 + \frac{pR^3}{EI}\right) = \frac{M_A R^2}{EI} - \frac{pR^3 u_A}{EI} \tag{11-49}$$

The right-hand side of Eq. (11-49) is constant. Thus its particular solution is $u(\theta) =$ constant, that is, axisymmetric. An axisymmetric deformation does not cause buckling. It can therefore be set equal to zero. We finally have

$$u'' + \left(1 + \frac{pR^3}{EI}\right)u = 0 \tag{11-50}$$

The solution to this equation has the form

$$u(\theta) = C_1 \sin k\theta + C_2 \cos k\theta \qquad (11\text{-}51)$$

where

$$k^2 = 1 + \frac{pR^3}{EI} \qquad (11\text{-}52)$$

One boundary condition results from the requirement that $u'(0) = (0)$ (that is how we selected point $A$). The second condition is supplied by the periodicity of the displacement, which implies that $u(2\pi) = u_A = u(0)$. Since

$$u'(\theta) = C_1 k \cos k\theta - C_2 k \sin k\theta$$

the boundary conditions lead to the following system of equations:

$$C_1 k = 0, \qquad C_2 = C_2 \cos 2k\pi$$

The solution is

$$C_1 = 0, \qquad \cos 2k\pi = 1, \qquad k = 1, 2, \ldots, \infty$$

The critical pressure $p_{cr}$ is now obtained from Eq. (11-52),

$$p_{cr} = \frac{EI}{R^3}(k^2 - 1), \qquad k = 1, 2, \ldots, \infty \qquad (11\text{-}53)$$

Buckling occurs at the smallest value of $p_{cr}$, which corresponds to $k = 2$,

$$p_{cr} = \frac{3EI}{R^3} \qquad (11\text{-}54)$$

The associated mode shape

$$u(\theta) = C_2 \cos 2\theta$$

is shown in Fig. 11-15$a$. We also note that for $k = 1$ the radial displacement is given by

$$u(\theta) = C_2 \cos \theta$$

That this represents a rigid body translation of the ring (Fig. 11-15$b$) is also evident from Eqs. (11-40) and (11-41). It is seen that now the change of curvature $EI(u + u'')/R^2 = 0$.

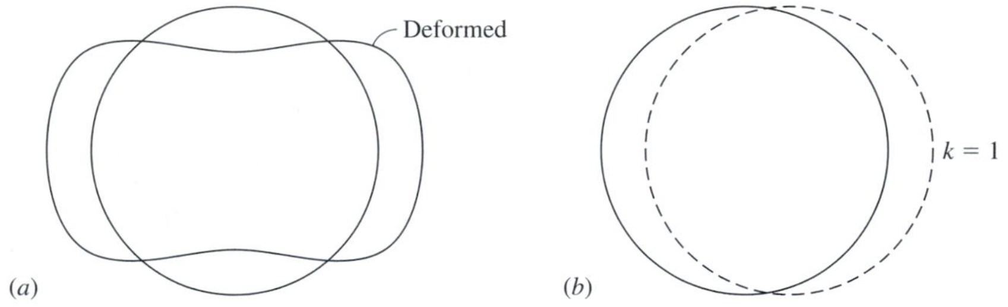


**Figure 11-15** Modes shapes corresponding to $k = 2$ (buckling) and $k = 1$ (rigid body displacement)

## 11-3 BUCKLING OF THIN RECTANGULAR PLATES

In this section we analyze a simply supported rectangular plate subjected to in-plane compression by uniformly distributed edge loads $p_x$ and $p_y$ (Fig. 11-16). In analogy to a beam, when these loads attain certain critical values, the initially flat plate buckles, that is, it bends out of plane. Our objective is to find these critical values. We start with finding the relationship between the edge loads and the internal in-plane forces per unit length, $N_{xx}$, $N_{yy}$, and $N_{xy}$. Evidently they must satisfy the equations of equilibrium (2-55) derived in Chapter 2 for the case of a two-dimensional state of stress. In the absence of body forces these equations become

$$\frac{\partial N_{xx}}{\partial x} + \frac{\partial N_{xy}}{\partial y} = 0, \qquad \frac{\partial N_{xy}}{\partial x} + \frac{\partial N_{yy}}{\partial y} = 0 \tag{11-55}$$

They must also satisfy the boundary conditions

$$\begin{aligned} N_{xx} &= -p_x, && \text{at } x = 0 \text{ and } x = a \\ N_{yy} &= -p_y, && \text{at } y = 0 \\ N_{xy} &= 0 && \text{all around} \end{aligned} \tag{11-56}$$

The solution to Eqs. (11-55) satisfying the conditions (11-56) is

$$N_{xx} = -p_x, \qquad N_{yy} = -p_y, \qquad N_{xy} = 0; \qquad 0 \le x \le a, \quad 0 \le y \le b \tag{11-57}$$

Next consider the equilibrium of an infinitesimal element of the deformed plate. Its top view and its two side views are shown in Fig. 11-17. We start by projecting on the $Oz$ axis all forces acting on this element. This leads to the following equation:

$$\begin{aligned} &Q_x \cos\beta\, dy - \left(Q_x + \frac{\partial Q_x}{\partial x} dx\right)\cos\beta\, dy + Q_y \cos\gamma\, dx - \left(Q_y + \frac{\partial Q_y}{\partial y} dy\right)\cos\gamma\, dx \\ &+ N_{xx}\sin\beta\, dy - \left(N_{xx} + \frac{\partial N_{xx}}{\partial x} dx\right)\sin\left(\beta + \frac{\partial \beta}{\partial x} dx\right) dy + N_{yy}\sin\gamma\, dx \\ &- \left(N_{yy} + \frac{\partial N_{yy}}{\partial y} dy\right)\sin\left(\gamma + \frac{\partial \gamma}{\partial y} dy\right) dx = 0 \end{aligned} \tag{11-58}$$

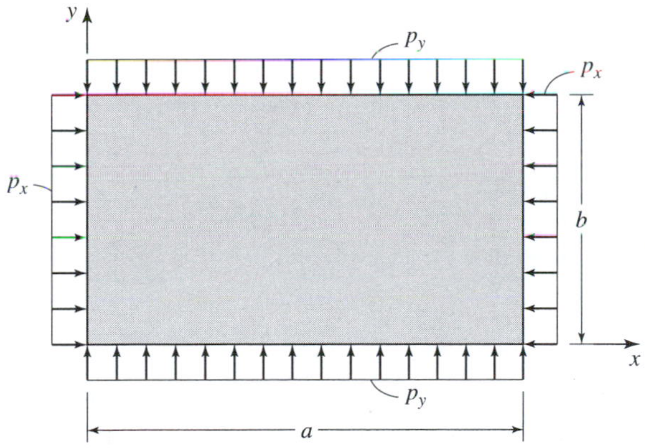


**Figure 11-16** A simply-supported rectangular plate under biaxial compression

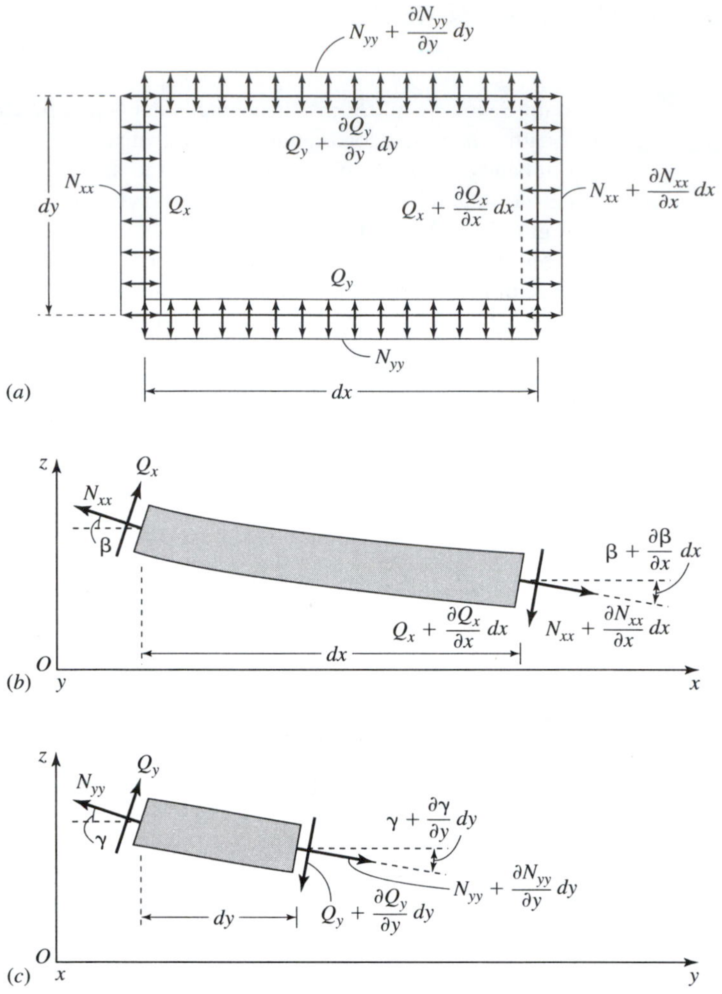


**Figure 11-17** Top and side views of an infinitesimal element

But the derivatives of the axial forces vanish in view of Eq. (11-57), while

$$\cos\beta \simeq 1, \qquad \cos\gamma \simeq 1, \qquad \sin\beta \simeq \beta \simeq \frac{\partial w}{\partial x}, \qquad \sin\gamma \simeq \gamma \simeq \frac{\partial w}{\partial y} \tag{11-59}$$

Substituting these into Eq. (11-58) and dividing by $dx\,dy$, we obtain

$$\frac{\partial Q_x}{\partial x} + \frac{\partial Q_y}{\partial y} + N_{xx}\frac{\partial^2 w}{\partial x^2} + N_{yy}\frac{\partial^2 w}{\partial y^2} = 0 \tag{11-60}$$

Next we calculate the sums of the moments of all forces (including the bending moments shown in Fig. 11-18) with respect to the lines parallel to the $Ox$ and $Oy$ axes. This

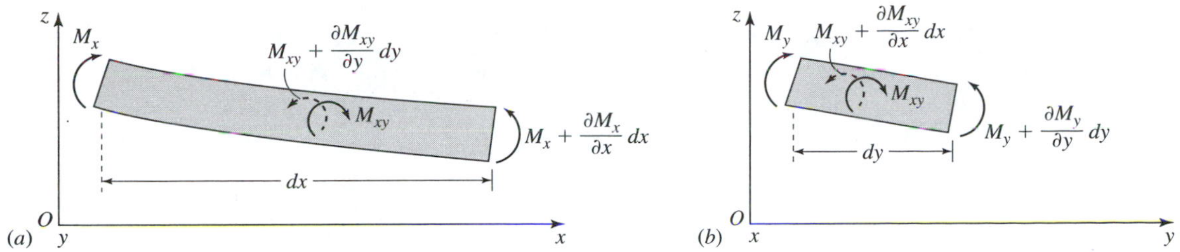


Figure 11-18

leads to Eqs. (10-17) and (10-18), which are repeated here,

$$\frac{\partial M_{xy}}{\partial x} + \frac{\partial M_y}{\partial y} - Q_y = 0, \qquad \frac{\partial M_x}{\partial x} + \frac{\partial M_{xy}}{\partial y} - Q_x = 0 \tag{11-61}$$

Now we substitute $Q_x$ and $Q_y$ obtained from Eqs. (11-61) into Eq. (11-60) and use the definitions of $M_x$, $M_y$, and $M_{xy}$ [Eqs. (10-9), (10-11), and (10-12)] to get

$$D\nabla^2\nabla^2 w - N_{xx}\frac{\partial^2 w}{\partial x^2} - N_{yy}\frac{\partial^2 w}{\partial y^2} = 0 \tag{11-62}$$

Finally, using Eqs. (11-57) this becomes

$$D\nabla^2\nabla^2 w + p_x\frac{\partial^2 w}{\partial x^2} + p_y\frac{\partial^2 w}{\partial y^2} = 0 \tag{11-63}$$

## EXAMPLE 11-4

Find the critical load for a simply supported rectangular plate.

### SOLUTION

We start by introducing a dimensionless parameter α [1, p. 103] such that

$$p_y = \alpha p_x \tag{a}$$

Now Eq. (11-63) becomes

$$D\nabla^2\nabla^2 w + p_x\left(\frac{\partial^2 w}{\partial x^2} + \alpha\frac{\partial^2 w}{\partial y^2}\right) = 0 \tag{b}$$

The boundary conditions are (see Chapter 10)

$$\begin{aligned} w = 0, \quad & \frac{\partial^2 w}{\partial x^2} = 0 \quad \text{at } x = 0, \quad x = a \\ w = 0, \quad & \frac{\partial^2 w}{\partial y^2} = 0 \quad \text{at } y = 0, \quad y = b \end{aligned} \tag{c}$$

A double Fourier sine series (see Section 10-3-2) satisfies both the differential equation ($b$) and the boundary conditions ($c$). Since we are interested in the smallest value of the buckling load, it is sufficient to assume the solution as only one term,

$$w = w_{11} \sin \frac{m\pi x}{a} \sin \frac{n\pi y}{b} \qquad (d)$$

and later select the values of the integers $m$ and $n$. Substituting this into Eq. (b) we get

$$D\left[\left(\frac{m\pi}{a}\right)^2 + \left(\frac{n\pi}{b}\right)^2\right]^2 - p_x\left[\left(\frac{m\pi}{a}\right)^2 + \alpha\left(\frac{n\pi}{b}\right)^2\right] = 0 \qquad (e)$$

Let

$$P_x = p_x \frac{b^2}{\pi^2 D} \qquad (f)$$

Now Eq. ($e$) can be put in the form

$$\left[\left(\frac{mb}{a}\right)^2 + n^2\right]^2 - P_x\left[\left(\frac{mb}{a}\right)^2 + \alpha n^2\right] = 0 \qquad (g)$$

Thus,

$$P_x = \frac{\left[\left(\frac{mb}{a}\right)^2 + n^2\right]^2}{\left(\frac{mb}{a}\right)^2 + \alpha n^2} \qquad (h)$$

For any given $b/a$ and $\alpha$ we select the integers $m$ and $n$ in Eq. ($h$) to find the minimum $P_x$. Let, for instance, $\alpha = 1$ and $b = a$. Then minimum $P_x$ occurs for $m = n = 1$,

$$\min P_x = 2$$

The critical loads follow then from Eq. ($f$) and Eq. ($a$).

## PROBLEMS

### Section 11-1

**11-1** Derive the equation for buckling of a cantilever bar subjected to a uniform continuous load along its axis (see Fig. P11-1).

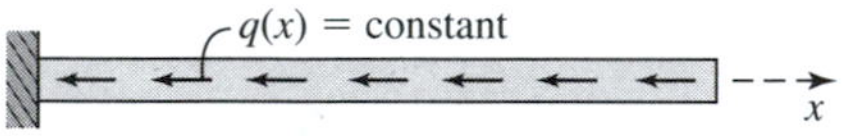


Figure P11-1

**11-2** Explain when a column of length $l$ and cross section $A$ will collapse even though $P < P_{cr}$.

**11-3** Find the approximate value for minimum $P_{cr}$ for pure buckling of a single-span beam with the following boundary conditions (see also Table 11-5):

(a) Clamped–free

(b) Clamped–clamped

(c) Clamped–simply supported

**11-4** Derive the characteristic equation for a beam with one end clamped and the other supported by a spring with stiffness $S$ (Fig. P11-4).

Figure P11-4

**11-5** Use the finite difference method to solve Problem 11-3(a).

**11-6** A straight column 10 in long has a rectangular cross section 1 in wide and 0.3 in thick. If $E = 29 \times 10^6$ psi, determine the elastic buckling load for:

(a) Simply supported column ends

(b) Clamped ends

**11-7** Derive an expression for the critical buckling force for a simply supported beam on an elastic foundation (see Chapter 8).

**11-8** Use any symbolic manipulation package to derive the characteristic equation for buckling of the three-span beam shown in Fig. P11-8.

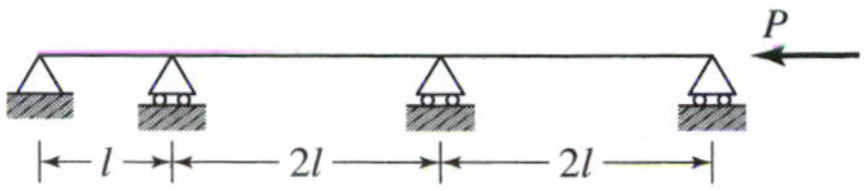


Figure P11-8

**11-9** Use the Rayleigh–Ritz method to find an approximate value of the buckling load for the beam described in Problem 11-8.

## Section 11-2

**11-10** Find the minimum critical pressure $P_x$ for a rectangular plate when $\alpha = -1$ and $a = 2b$.

## REFERENCES


[1] Brush, D. O., and B. O. Almroth, *Buckling of Bars, Plates and Shells* (McGraw-Hill, New York, 1975).

[2] *Mathcad7, professional ed.* (MathSoft, Cambridge, MA, 1997).

[3] Meirovitch, L., *Elements of Vibration Analysis* (McGraw-Hill, New York, 1975).

[4] Pedersen, P., *Bjaelkesojler og stabilitet af simple rammer* (Technical University of Denmark, Lyngby, 1988).

[5] Tauchert, T. R., *Energy Principles in Structural Mechanics* (McGraw-Hill, New York, 1974).

# INTRODUCTION TO FRACTURE MECHANICS

This chapter deals with the influence of cracks on the behavior of a linearly elastic solid. We consider solutions of some idealized problems in which the crack is assumed to be straight, finite, or semi-infinite. Of particular interest is the stress distribution near the tip of the crack. Thus the concept of a stress intensity factor (SIF) is introduced. The stress intensity factor helps in predicting the possibility of fracture damage. The concept of the $J$ integral is also introduced, and its relation to the energy release rate in the case of linearly elastic materials is determined.

## 12-1 INTRODUCTORY CONCEPTS

All materials crack due to such conditions as material imperfections, overload, or drying out. Fracture mechanics deals with the effects of cracking on the load-bearing ability of structural elements. In this chapter we merely introduce certain fundamental concepts of fracture mechanics. We consider only two-dimensional brittle fracture (with small exceptions) and assume that the cracks are rectilinear. We also assume that the cracked solid is made of a linearly elastic material. Let us not overlook the fact that a crack can be beneficial. Take, for instance, wood chopping: we introduce a crack artificially by hitting wood with an ax. We can then enlarge the crack thus formed by inserting a wedge and hammering this wedge down. This operation will cause crack propagation and ends with splitting the piece of wood. Consider another example: tearing a perforated sheet of paper starts with a crack forming at one end, and propagating to the nearest hole. The "bridge" between this hole and the next one is subject to even more tension than the original "bridge," thus a new crack appears, and so on. These are examples of "beneficial" cracks. On the other hand, when a hole is made for structural reasons, then unwanted stress concentration occurs at the hole. Any flaw or imperfection in the material combined with this stress concentration may now lead to the formation of a crack. Note that any singularity such as a notch or inclusion may cause an elevation of the stress level, which may lead to a crack. The mere presence of a crack, or even of several cracks, does not indicate that the element must be replaced. The concept of the stress intensity factor, usually denoted by $K$, together with the concept of *plane strain structure toughness* $K_{Ic}$, help in making the decision about the load-bearing capability of a cracked element. If the calculated quantity $K$ reaches the empirical quantity $K_{Ic}$, the crack is expected to propagate. Note that structure toughness, crack size, and stress level are the three most important factors controlling the susceptibility of a structure to brittle fracture [1, p. 15].

In the following we will distinguish between three basic *modes* characterizing crack behavior. For convenience let us call the *crack front* the straight line connecting points $A$ and $B$, as shown in Fig. 12-1. In *mode I,* called the *opening mode,* the crack walls separate so that the displacements occur only in the direction normal to the crack surfaces (Fig. 12-2). This mode (think again of wood splitting) is the most likely cause of crack propagation. In *mode II,* referred to as the *sliding mode,* the crack walls move in a shearing fashion in the direction normal to the crack front (Fig. 12-3). There is also shearing in *mode III,* known as *tearing mode,* but this movement is in the direction parallel to the crack front (Fig. 12-4). This mode brings to mind the tearing of a sheet of paper.

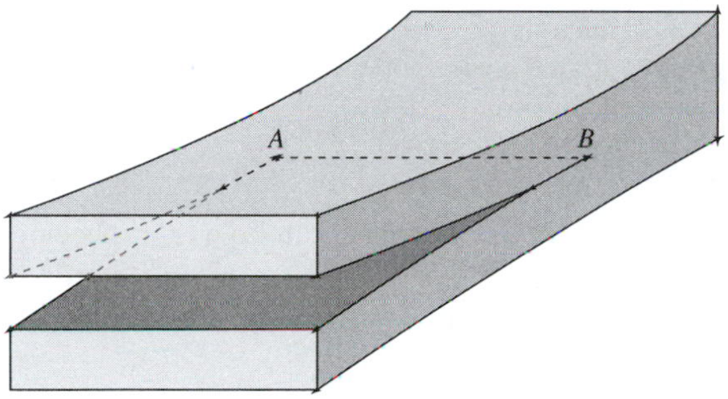

Figure 12-1 The crack front *AB*

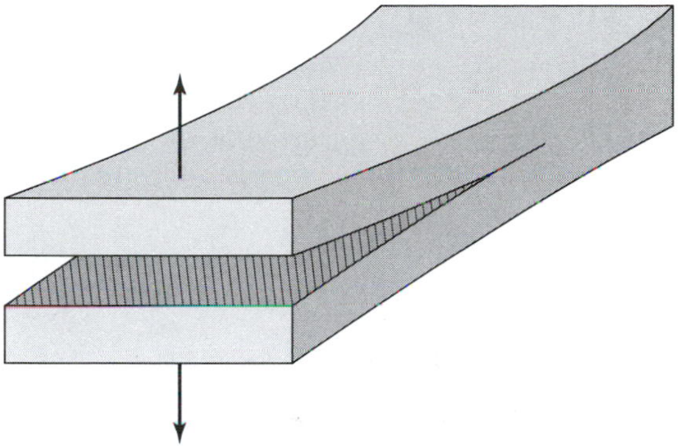
Figure 12-2 Mode I, opening

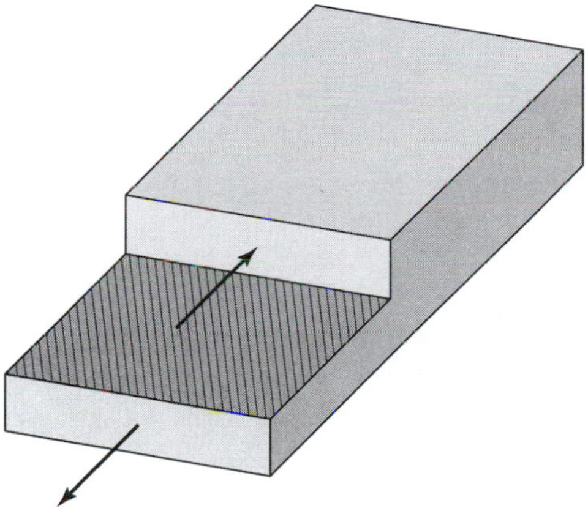
Figure 12-3 Mode II, sliding

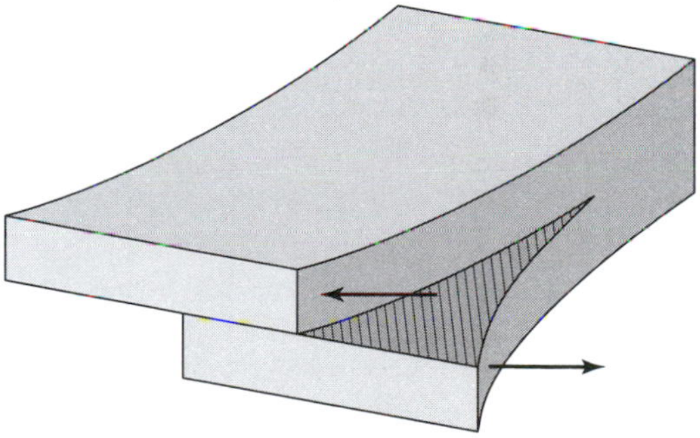
Figure 12-4 Mode III, tearing

## 12-2 LINEAR CRACKS IN TWO-DIMENSIONAL ELASTIC SOLIDS—WILLIAMS'S SOLUTION, STRESS SINGULARITY

Consider a two-dimensional elastic solid with a crack. In general the crack will be curvilinear, as shown in Fig. 12-5. We would like to know how its presence affects the stress distribution and the displacements within a solid. We suspect, intuitively, that the crack is not beneficial from the strength of materials point of view. It seems to weaken the solid. Using reasoning from elementary strength of materials we note that the normal stress in the cross section weakened by the crack will increase considerably. This is evident when studying, for instance, a simply supported beam with a centrally located crack (Fig. 12-6)—the cracked beam will deflect more than the solid beam, and the maximum stress will increase. We can actually perform a simplified elementary calculation, which will give us some idea of what to expect. We know from elementary strength of materials that the maximum stress in the unflawed beam occurs at the middle and equals

$$\sigma_{\max} = \frac{6M_{\max}}{bh^2} = \frac{3Pl}{2bh^2} \tag{12-1}$$

Assume now that we know the crack length $c$, the fracture toughness $K_{Ic}$, and the safety factor $k$. Let us determine the allowable magnitude of the applied load $P$. The stress intensity factor for a small edge crack is found to be approximately [see Table 12-1 (Section 12-6), 2, for $c/b \ll 1$]

$$K = 1.12\sigma_{\max}\sqrt{\pi c} \tag{12-2}$$

For safety we want $K < K_{Ic}/k$, and thus Eq. (12-2) together with Eq. (12-1) becomes

$$\frac{K_{Ic}}{k} > 1.12\frac{3Pl}{2bh^2}\sqrt{\pi c}$$

from which we determine the maximum allowable load $P$,

$$P < \frac{bh^2 K_{Ic}}{1.68kl\sqrt{\pi c}} \tag{12-3}$$

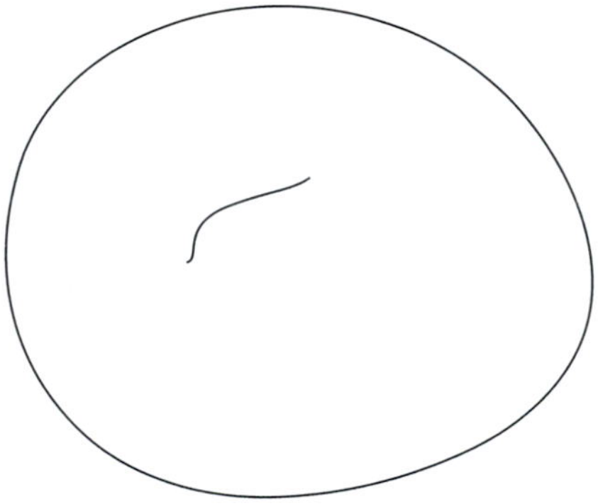

Figure 12-5

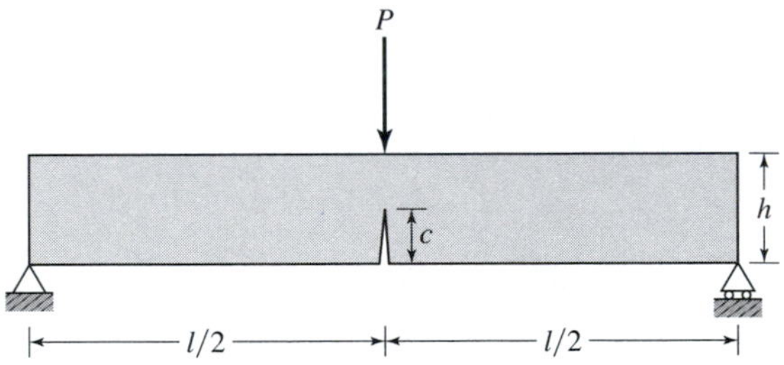


Figure 12-6

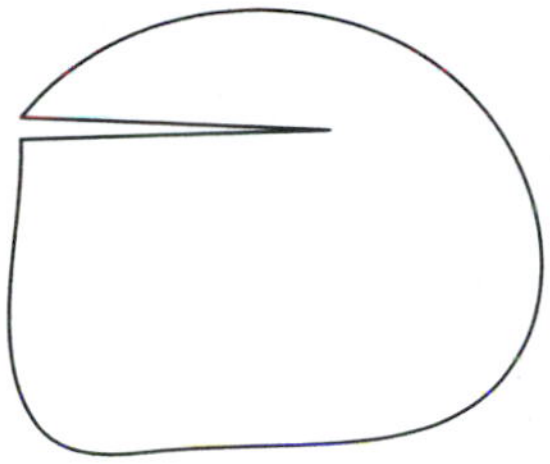

**Figure 12-7** An infinite solid with a semi-infinite rectilinear crack

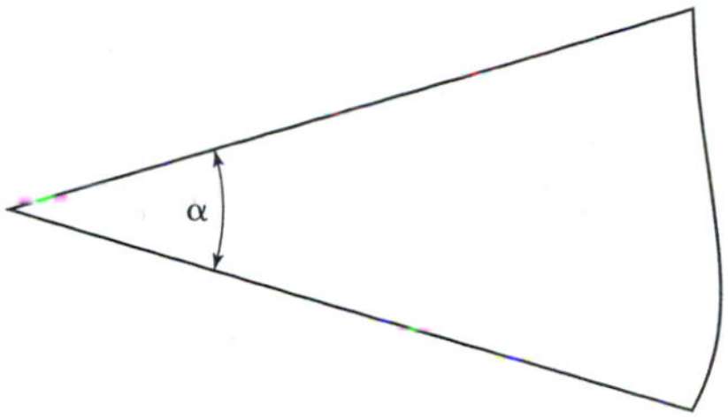


**Figure 12-8**

Hence the maximum value of $P$ obtained from Eq. (12-3) represents the safe operating load. Now we will determine the stress in an infinite elastic solid with a semi-infinite rectilinear crack (see Fig. 12-7). The assumption of infinite dimensions simplifies the analysis considerably without affecting the stress behavior in close vicinity to the crack tip. This is due to Saint Venant's principle: if "close vicinity" means "infinitesimal distance," then the crack length does not affect the behavior of the stress at the tip. We will attempt to determine the anticipated *singular* (that is, tending to infinity) behavior of the stress at the tip. Thus the question is whether the stresses tend to infinity at the tip, and if so how? For instance, the quantities $r^{-0.5}, r^{-1}, r^{-2}, \ldots$ all become infinite when $r \to 0$ while the integrals of the last two quantities do not.

First we will analyze an elastic wedge with a vertex angle $\alpha$, as shown in Fig. 12-8. This wedge is a generalization of the infinite cracked solid in Fig. 12-7. Indeed, setting $\alpha = 2\pi$, we turn the infinite wedge into an infinite solid with a semi-infinite crack. We will solve this two-dimensional problem using Airy's stress function.

We have seen in Chapter 9 that the application of Airy's stress function $\Phi$, where the stresses are given by Eqs. (9-114),

$$\sigma_{xx} = \frac{\partial^2 \Phi}{\partial y^2}, \qquad \sigma_{yy} = \frac{\partial^2 \Phi}{\partial x^2}, \qquad \sigma_{xy} = -\frac{\partial^2 \Phi}{\partial x\, \partial y}$$

sometimes reduces a two-dimensional elastic problem to a single partial differential equation, Eq. (9-117),

$$\nabla^2 \nabla^2 \Phi = 0 \tag{12-4}$$

which in a Cartesian coordinate system becomes

$$\frac{\partial^2 \Phi}{\partial x^2} + \frac{\partial^2 \Phi}{\partial y^2} = 0 \tag{12-5}$$

Looking, however, at Fig. 12-8, we see that a polar coordinate system $r$, $\theta$ is more suitable than a Cartesian system because in this system one of the independent variables, $\theta$, is constant along the sides of the wedge. Assuming that

$$\sigma_{rr} = \frac{1}{r}\frac{\partial \Phi}{\partial r} + \frac{1}{r^2}\frac{\partial^2 \Phi}{\partial \theta^2}, \qquad \sigma_{\theta\theta} = \frac{\partial^2 \Phi}{\partial r^2}, \qquad \delta_{r\theta} = -\frac{\partial}{\partial r}\left(\frac{1}{r}\frac{\partial \Phi}{\partial \theta}\right) \tag{12-6}$$

we satisfy identically the two-dimensional equilibrium equations in the cylindrical coordinate system [see Eqs. (2-60)],

$$\begin{aligned}\frac{\partial\sigma_{rr}}{\partial r}+\frac{1}{r}\frac{\partial\sigma_{r\theta}}{\partial\theta}+\frac{1}{r}(\sigma_{rr}-\sigma_{\theta\theta})+b_r=0\\ \frac{\partial\sigma_{r\theta}}{\partial r}+\frac{1}{r}\frac{\partial\sigma_{\theta\theta}}{\partial\theta}+\frac{2}{r}\sigma_{r\theta}+b_\theta=0\end{aligned} \tag{12-7}$$

But [Eq. (3-77)]

$$\begin{aligned}\frac{\partial}{\partial x}&=\frac{\partial}{\partial r}\frac{\partial r}{\partial x}+\frac{\partial}{\partial\theta}\frac{\partial\theta}{\partial x}\\ \frac{\partial}{\partial y}&=\frac{\partial}{\partial r}\frac{\partial r}{\partial y}+\frac{\partial}{\partial\theta}\frac{\partial\theta}{\partial y}\end{aligned} \tag{12-8}$$

Therefore,

$$\begin{aligned}\frac{\partial^2}{\partial x^2}&=\left(\cos\theta\frac{\partial}{\partial r}-\frac{1}{r}\sin\theta\frac{\partial}{\partial\theta}\right)\left(\cos\theta\frac{\partial}{\partial r}-\frac{1}{r}\sin\theta\frac{\partial}{\partial\theta}\right)\\ &=\cos^2\theta\frac{\partial^2}{\partial r^2}+\frac{1}{r^2}\sin^2\theta\frac{\partial^2}{\partial\theta^2}-\frac{2}{r^2}\sin\theta\cos\theta\frac{\partial}{\partial\theta}\\ &\quad+\frac{2}{r}\sin\theta\cos\theta\frac{\partial^2}{\partial r\,\partial\theta}+\frac{1}{r}\cos^2\theta\frac{\partial}{\partial r}\\ \frac{\partial^2}{\partial y^2}&=\left(\sin\theta\frac{\partial}{\partial r}+\frac{1}{r}\cos\theta\frac{\partial}{\partial\theta}\right)\left(\sin\theta\frac{\partial}{\partial r}+\frac{1}{r}\cos\theta\frac{\partial}{\partial\theta}\right)\\ &=\sin^2\theta\frac{\partial^2}{\partial r^2}+\frac{1}{r^2}\cos^2\theta\frac{\partial^2}{\partial\theta^2}+\frac{2}{r^2}\sin\theta\cos\theta\frac{\partial}{\partial\theta}\\ &\quad-\frac{2}{r}\sin\theta\cos\theta\frac{\partial^2}{\partial r\,\partial\theta}+\frac{1}{r}\sin^2\theta\frac{\partial}{\partial r}\end{aligned}$$

Hence, adding both, we get

$$\nabla^2=\frac{\partial^2}{\partial x^2}+\frac{\partial^2}{\partial y^2}=\frac{\partial^2}{\partial r^2}+\frac{1}{r}\frac{\partial}{\partial r}+\frac{1}{r^2}\frac{\partial^2}{\partial\theta^2} \tag{12-9}$$

Finally, repeating this operation, we obtain the following partial differential equation defining $\Phi$ in the cylindrical coordinate system,

$$\begin{aligned}\nabla^2\nabla^2\Phi&=\frac{\partial^4\Phi}{\partial r^4}+\frac{2}{r}\frac{\partial^3\Phi}{\partial r^3}-\frac{1}{r^2}\frac{\partial^2\Phi}{\partial r^2}+\frac{1}{r^3}\frac{\partial\Phi}{\partial r}+\frac{1}{r^4}\frac{\partial^4\Phi}{\partial\theta^4}\\ &\quad+\frac{4}{r^4}\frac{\partial^2\Phi}{\partial\theta^2}+\frac{2}{r^2}\frac{\partial^4\Phi}{\partial\theta^2\,\partial r^2}-\frac{2}{r^3}\frac{\partial^3\Phi}{\partial\theta^2\,\partial r}=0\end{aligned} \tag{12-10}$$

The appropriate boundary conditions for an infinite wedge are (see Fig. 12-8)

$$\sigma_{\theta\theta}=\frac{\partial^2\Phi}{\partial r^2}=0,\qquad \sigma_{r\theta}=-\frac{\partial}{\partial r}\left(\frac{1}{r}\frac{\partial\Phi}{\partial\theta}\right)\qquad \theta=\pm\alpha \tag{12-11}$$

Now, following Williams [13], we assume that $\Phi(r, \theta)$ can be represented as a product of two functions,

$$\Phi(r, \theta) = r^{\lambda+1} f(\theta) \tag{12-12}$$

where $\lambda$ is an unknown real parameter. Substituting Eq. (12-12) into the differential equation (12-10) and into the boundary conditions (12-11), we get

$$\begin{aligned} &\frac{d^4 f}{d\theta^4} + 2(\lambda^2 + 1)\frac{d^2 f}{d\theta^2} + (1 - \lambda^2)^2 f = 0 \\ &f = 0, \qquad \frac{df}{d\theta} = 0, \qquad \theta = \pm\alpha \end{aligned} \tag{12-13}$$

Solving the ordinary differential equation with constant coefficients, Eq. (12-13), and substituting the result into Eq. (12-12), we have

$$\begin{aligned} \Phi(r, \theta) &= r^{\lambda+1} f(\theta) \\ &= r^{\lambda+1}[C_1 \sin(\lambda + 1)\theta + C_2 \cos(\lambda + 1)\theta + C_3 \sin(\lambda - 1)\theta + C_4 \cos(\lambda - 1)\theta] \end{aligned} \tag{12-14}$$

Substituting Eq. (12-14) into the boundary conditions (12-11) results in the following homogeneous system of linear algebraic equations:

$$\begin{aligned} &C_1 \sin(\lambda + 1)\alpha + C_2 \cos(\lambda + 1)\alpha + C_3 \sin(\lambda - 1)\alpha + C_4 \cos(\lambda - 1)\alpha = 0 \\ &-C_1 \sin(\lambda + 1)\alpha + C_2 \cos(\lambda + 1)\alpha - C_3 \sin(\lambda - 1)\alpha + C_4 \cos(\lambda - 1)\alpha = 0 \\ &(\lambda + 1)C_1 \cos(\lambda + 1)\alpha - (\lambda + 1)C_2 \sin(\lambda + 1)\alpha \\ &\qquad + (\lambda - 1)C_3 \cos(\lambda - 1)\alpha - (\lambda - 1)C_4 \sin(\lambda - 1)\alpha = 0 \\ &(\lambda + 1)C_1 \cos(\lambda + 1)\alpha + (\lambda + 1)C_2 \sin(\lambda + 1)\alpha \\ &\qquad + (\lambda - 1)C_3 \cos(\lambda - 1)\alpha + (\lambda - 1)C_4 \sin(\lambda - 1)\alpha = 0 \end{aligned} \tag{12-15}$$

We note that adding or subtracting the first two and the last two equations leads to two independent systems of homogeneous algebraic equations, each containing only two arbitrary constants and the unknown parameter $\lambda$. They have the following matrix forms:

$$\begin{aligned} &\begin{bmatrix} \cos(\lambda + 1)\alpha & \cos(\lambda - 1)\alpha \\ (\lambda + 1)\sin(\lambda + 1)\alpha & (\lambda - 1)\sin(\lambda - 1)\alpha \end{bmatrix} \begin{Bmatrix} C_2 \\ C_4 \end{Bmatrix} = \begin{Bmatrix} 0 \\ 0 \end{Bmatrix} \\ &\begin{bmatrix} \sin(\lambda + 1)\alpha & \sin(\lambda - 1)\alpha \\ (\lambda + 1)\cos(\lambda + 1)\alpha & (\lambda - 1)\cos(\lambda - 1)\alpha \end{bmatrix} \begin{Bmatrix} C_1 \\ C_3 \end{Bmatrix} = \begin{Bmatrix} 0 \\ 0 \end{Bmatrix} \end{aligned} \tag{12-16}$$

Each of these systems of homogeneous equations has a nontrivial solution if and only if its characteristic determinant is zero. Thus we get two characteristic equations,

$$\begin{aligned} (\lambda - 1)\sin(\lambda - 1)\cos(\lambda + 1) - (\lambda + 1)\sin(\lambda + 1)\cos(\lambda - 1) = 0 \\ (\lambda - 1)\sin(\lambda + 1)\cos(\lambda - 1) - (\lambda + 1)\sin(\lambda - 1)\cos(\lambda + 1) = 0 \end{aligned}$$

which can be simplified to yield

$$\lambda \sin 2\alpha + \sin 2\alpha\lambda = 0, \qquad \lambda \sin 2\alpha - \sin 2\alpha\lambda = 0 \tag{12-17}$$

For each $\alpha$, defining half of the apex angle of the wedge, we get a numerical value of $\lambda$. We are, however, only interested in the value $\alpha = \pi$, because in this case both edges of the wedge become infinitesimally close to each other, thus forming a semi-infinite crack. Hence substituting $\alpha = \pi$ into Eqs. (12-17) we obtain

$$\sin 2\pi\lambda = 0$$

from which it follows that

$$\lambda = \tfrac{1}{2}m, \qquad m = 1, 2, 3, \ldots \tag{12-18}$$

Substituting this value into the matrix equations (12-16), we get either

$$\begin{bmatrix} \sin\left(\frac{m}{2}+1\right)\pi & \sin\left(\frac{m}{2}-1\right)\pi \\ \left(\frac{m}{2}+1\right)\cos\left(\frac{m}{2}+1\right)\pi & \left(\frac{m}{2}-1\right)\cos\left(\frac{m}{2}-1\right)\pi \end{bmatrix} \begin{Bmatrix} C_{1m} \\ C_{3m} \end{Bmatrix} = \begin{Bmatrix} 0 \\ 0 \end{Bmatrix}$$

or

$$\begin{bmatrix} \cos\left(\frac{m}{2}+1\right)\pi & \cos\left(\frac{m}{2}-1\right)\pi \\ \left(\frac{m}{2}+1\right)\sin\left(\frac{m}{2}+1\right)\pi & \left(\frac{m}{2}-1\right)\sin\left(\frac{m}{2}-1\right)\pi \end{bmatrix} \begin{Bmatrix} C_{2m} \\ C_{4m} \end{Bmatrix} = \begin{Bmatrix} 0 \\ 0 \end{Bmatrix}$$

We note now that solving those equations for $m = 1, 3, \ldots$ results in

$$C_{2m} = -\frac{m-2}{m+2}C_{4m}, \qquad C_{1m} = -C_{3m} \tag{12-19}$$

whereas for $m = 2, 4, \ldots$ we get

$$C_{1m} = -\frac{m-2}{m+2}C_{3m}, \qquad C_{2m} = -C_{4m} \tag{12-20}$$

For each of the infinite number of values of $m$ we determine an appropriate Airy's stress function from Eq. (12-14), where we replace the constants $C_1, C_2, \ldots$ by $C_{1m}, C_{2m}, \ldots,$ thus obtaining the following result:

$$\begin{aligned} \Phi(r,\theta) &= \sum_{m=1,2,\ldots} r^{1+(m/2)} f_m(\theta) \\ &= \sum_{m=1,3,\ldots} r^{1+(m/2)} \left[\left(\sin\frac{m-2}{2}\theta - \sin\frac{m+2}{2}\theta\right) C_{3m}\right. \\ &\quad \left. + \left(\cos\frac{m-2}{2}\theta - \frac{m-2}{m+2}\cos\frac{m+2}{2}\theta\right) C_{4m}\right] \\ &\quad + \sum_{m=2,4,\ldots} r^{1+(m/2)} \left[\left(\sin\frac{m-2}{2}\theta - \frac{m-2}{m+2}\sin\frac{m+2}{2}\theta\right) C_{3m}\right. \\ &\quad \left. + \left(\cos\frac{m-2}{2}\theta - \cos\frac{m+2}{2}\theta\right) C_{4m}\right] \end{aligned} \tag{12-21}$$

We notice that in the expression for $f(\theta)$ the coefficients of $C_{3m}$ are skew symmetric with respect to $\theta$ whereas the coefficients of $C_{4m}$ are symmetric.

Substituting Eq. (12-14) into the stress expressions, Eqs. (12-6), we have

$$\sigma_{rr} = r^{\lambda-1}[(\lambda+1)f(\theta) + f''(\theta)], \qquad \sigma_{\theta\theta} = \lambda(\lambda+1)r^{\lambda-1}f(\theta)$$
$$\sigma_{r\theta} = -\lambda r^{\lambda-1}f'(\theta) \tag{12-22}$$

We observe that the normal stresses also have a symmetric part composed of the coefficients of $C_{4m}$ and a skew-symmetric part composed of the coefficients of $C_{3m}$. The opposite is true for the shear stress, as it depends on $f'(\theta)$. This observation implies that if the stresses are symmetric with respect to the crack line, so are the deformations. Hence the coefficients of $C_{4m}$ correspond to the state of deformation shown in Fig. 12-2, where the side plane is the $r$, $\theta$ plane, which we called mode I. The coefficients of $C_{3m}$ on the other hand correspond to the state of deformation shown in Fig. 12-3, which we called mode II.

It is seen that with $\lambda = m/2$, $m = 1, 3, \ldots$, all stress components start with the term $r^{-1/2}$, which tends to infinity when $r$ tends to 0. All other powers of $r$ have positive exponents. We say that stresses have a *square-root singularity* at the tip of a linear crack. Since we are now interested in stress behavior near the tip, let us disregard all the terms except for the first one in the series (12-21).

Substituting this into the stress expressions, Eqs. (12-22), we get for mode I,

$$\sigma_{rr} = \frac{C_{41}}{4}r^{-1/2}\left(5\cos\frac{\theta}{2} - \cos\frac{3\theta}{2}\right) + \text{regular terms}$$
$$\sigma_{\theta\theta} = \frac{C_{41}}{4}r^{-1/2}\left(3\cos\frac{\theta}{2} + \cos\frac{3\theta}{2}\right) + \text{regular terms} \tag{12-23}$$
$$\sigma_{r\theta} = \frac{C_{41}}{4}r^{-1/2}\left(\sin\frac{\theta}{2} + \sin\frac{3\theta}{2}\right) + \text{regular terms}$$

and for mode II,

$$\sigma_{rr} = \frac{C_{31}}{4}r^{-1/2}\left(-5\sin\frac{\theta}{2} + 3\sin\frac{3\theta}{2}\right) + \text{regular terms}$$
$$\sigma_{\theta\theta} = \frac{C_{31}}{4}r^{-1/2}\left(-3\sin\frac{\theta}{2} - 3\sin\frac{3\theta}{2}\right) + \text{regular terms} \tag{12-24}$$
$$\sigma_{r\theta} = \frac{C_{31}}{4}r^{-1/2}\left(\cos\frac{\theta}{2} + 3\cos\frac{3\theta}{2}\right) + \text{regular terms}$$

Simple trigonometric transformations (see [7, p. 53]) put these into equivalent forms. For mode I,

$$\sigma_{rr} = C_{41}r^{-1/2}\cos\frac{\theta}{2}\left(1 + \sin^2\frac{\theta}{2}\right) + \text{regular terms}$$
$$\sigma_{\theta\theta} = C_{41}r^{-1/2}\cos^3\frac{\theta}{2} + \text{regular terms} \tag{12-25}$$
$$\sigma_{r\theta} = C_{41}r^{-1/2}\sin\frac{\theta}{2}\cos^2\frac{\theta}{2} + \text{regular terms}$$

and for mode II,

$$
\begin{aligned}
\sigma_{rr} &= C_{31} r^{-1/2} \sin\frac{\theta}{2}\left(1 - 3\sin^2\frac{\theta}{2}\right) + \text{regular terms} \\
\sigma_{\theta\theta} &= -3C_{31} r^{-1/2} \sin\frac{\theta}{2}\cos^2\frac{\theta}{2} + \text{regular terms} \\
\sigma_{r\theta} &= C_{31} r^{-1/2} \cos\frac{\theta}{2}\left(1 - 3\sin^2\frac{\theta}{2}\right) + \text{regular terms}
\end{aligned}
\tag{12-26}
$$

Using Eqs. (2-52) we can also find easily the Cartesian stress components in terms of polar coordinates. For mode I,

$$
\begin{aligned}
\sigma_{xx} &= C_{41} r^{-1/2} \cos\frac{\theta}{2}\left(1 - \sin\frac{\theta}{2}\sin\frac{3\theta}{2}\right) + \text{regular terms} \\
\sigma_{yy} &= C_{41} r^{-1/2} \cos\frac{\theta}{2}\left(1 + \sin\frac{\theta}{2}\sin\frac{3\theta}{2}\right) + \text{regular terms} \\
\sigma_{xy} &= C_{41} r^{-1/2} \sin\frac{\theta}{2}\cos\frac{\theta}{2}\cos\frac{3\theta}{2} + \text{regular terms}
\end{aligned}
\tag{12-27}
$$

and for mode II,

$$
\begin{aligned}
\sigma_{xx} &= -C_{31} r^{-1/2} \sin\frac{\theta}{2}\left(2 + \cos\frac{\theta}{2}\cos\frac{3\theta}{2}\right) + \text{regular terms} \\
\sigma_{yy} &= C_{31} r^{-1/2} \sin\frac{\theta}{2}\cos\frac{\theta}{2}\cos\frac{3\theta}{2} + \text{regular terms} \\
\sigma_{xy} &= C_{31} r^{-1/2} \cos\frac{\theta}{2}\left(1 - \sin\frac{\theta}{2}\sin\frac{3\theta}{2}\right) + \text{regular terms}
\end{aligned}
\tag{12-28}
$$

In order to derive the expressions for the displacements $u$, $v$, and $w$ along the crack edges we use the formulas for strain from Chapter 3 [Eqs. (3-82)],

$$
\begin{aligned}
\varepsilon_{rr} &= \frac{\partial u_r}{\partial r} \\
\varepsilon_{\theta\theta} &= \frac{u_r}{r} + \frac{1}{r}\frac{\partial u_\theta}{\partial \theta} \\
\gamma_{r\theta} &= \frac{1}{r}\frac{\partial u_r}{\partial \theta} + \frac{\partial u_\theta}{\partial r} - \frac{u_\theta}{r}
\end{aligned}
\tag{12-29}
$$

as well as Hooke's law in the polar coordinate system, obtained from Eqs. (4-8) and (4-11),

$$
\varepsilon_{xx} = \frac{1}{E}[\sigma_{xx} - \nu(\sigma_{yy} + \sigma_{zz})], \qquad \varepsilon_{yy} = \frac{1}{E}[\sigma_{yy} - \nu(\sigma_{xx} + \sigma_{zz})], \qquad \gamma_{xy} = \frac{1}{G}\sigma_{xy}
$$

where we replace the Cartesian components by the polar components,

$$\begin{aligned} &\varepsilon_{xx} = \varepsilon_{rr}, \qquad \varepsilon_{yy} = \varepsilon_{\theta\theta}, \qquad \gamma_{xy} = \gamma_{r\theta} \\ &\sigma_{xx} = \sigma_{rr}, \qquad \sigma_{yy} = \sigma_{\theta\theta}, \qquad \sigma_{xy} = \sigma_{r\theta}, \qquad \sigma_{zz} = 0 \end{aligned} \tag{12-30}$$

to obtain

$$E\varepsilon_{rr} = \sigma_{rr} - \nu\sigma_{\theta\theta}, \qquad E\varepsilon_{\theta\theta} = \sigma_{\theta\theta} - \nu\sigma_{rr}, \qquad G\gamma_{r\theta} = \sigma_{r\theta} \tag{12-31}$$

Substitution of Eqs. (12-25) and (12-29) into Eq. (12-31) leads to the system of equations needed to determine the displacement components corresponding to mode I,

$$E\frac{\partial u_r}{\partial r} = C_{41}r^{-1/2}\cos\frac{\theta}{2}\left(1 + \sin^2\frac{\theta}{2} - \nu\cos^2\frac{\theta}{2}\right) \tag{12-32$a$}$$

$$E\frac{\partial u_\theta}{\partial \theta} = -Eu_r + C_{41}r^{1/2}\cos\frac{\theta}{2}\left[\cos^2\frac{\theta}{2} - \nu\left(1 + \sin^2\frac{\theta}{2}\right)\right] \tag{12-32$b$}$$

Next we integrate Eq. (12-32$a$) with respect to $r$, getting

$$Eu_r = 2C_{41}r^{1/2}\cos\frac{\theta}{2}\left(1 + \sin^2\frac{\theta}{2} - \nu\cos^2\frac{\theta}{2}\right) \tag{12-33}$$

With this substituted into Eq. (12-32$b$) we obtain

$$E\frac{\partial u_\theta}{\partial \theta} = C_{41}r^{1/2}\cos\frac{\theta}{2}\left[\cos^2\frac{\theta}{2} - \nu\left(1 + \sin^2\frac{\theta}{2}\right) - 2 - 2\sin^2\frac{\theta}{2} + 2\nu\cos^2\frac{\theta}{2}\right] \tag{12-34}$$

When this is integrated with respect to $\theta$ we obtain, using the symbolic manipulation package Maple [8],

$$Eu_\theta = 2C_{41}r^{1/2}\sin\frac{\theta}{2}\left[(1+\nu)\cos^2\frac{\theta}{2} - 2\right] \tag{12-35}$$

The integration constants have been disregarded as contributing to the rigid body motion only.

Simple manipulations lead to the following expressions for the displacement components in mode I for plane stress,

$$\begin{aligned} u_r &= C_{41}\frac{r^{1/2}}{4G}\left(\frac{5-3\nu}{1+\nu}\cos\frac{\theta}{2} - \cos\frac{3\theta}{2}\right) \\ u_\theta &= C_{41}\frac{r^{1/2}}{4G}\left(-\frac{7-\nu}{1+\nu}\sin\frac{\theta}{2} + \sin\frac{3\theta}{2}\right) \end{aligned} \tag{12-36}$$

Repeating the preceding steps for mode II we get

$$\begin{aligned} u_r &= C_{31}\frac{r^{1/2}}{4G}\left(-\frac{5-3\nu}{1+\nu}\sin\frac{\theta}{2} + 3\sin\frac{3\theta}{2}\right) \\ u_\theta &= C_{31}\frac{r^{1/2}}{4G}\left(-\frac{7-\nu}{1+\nu}\cos\frac{\theta}{2} + 3\cos\frac{3\theta}{2}\right) \end{aligned} \tag{12-37}$$

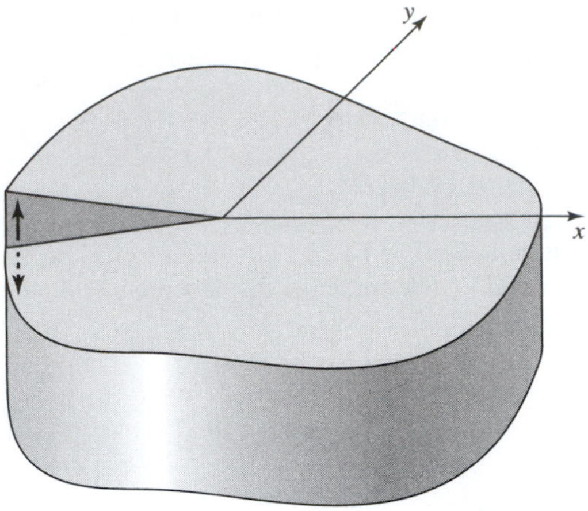


Figure 12-9

Next we will study the case of one-half of the solid above the crack moving upward without changing its shape in the $xy$ plane while the other half moves downward in the same way (Fig. 12-9). This implies that the solid remains undeformed along the line $y = 0$ to the right of the crack tip. This kind of deformation, called *antiplane* (see also [5, p. 11] and Fig. 12-4) is described by the following states of stress and displacement, called *mode III*,

$$\sigma_{xx} = \sigma_{yy} = \sigma_{xy} = \sigma_{zz} = u = v = 0$$
$$u_z = w(r, \theta), \qquad \sigma_{xz} = \sigma_{xz}(r, \theta), \qquad \sigma_{yz} = \sigma_{yz}(r, \theta) \tag{12-38}$$

Substituting two Hooke's law equations [Eqs. (4-10) and (3-11), where we replace $x$ by $r$, $y$ by $r\theta$],

$$\sigma_{rz} = G\gamma_{rz} = G\frac{\partial w}{\partial r}, \qquad \sigma_{\theta z} = G\gamma_{\theta z} = G\frac{1}{r}\frac{\partial w}{\partial \theta} \tag{12-39}$$

into the third of the equilibrium equations (2-58) (other equilibrium equations are identically equal to zero),

$$\frac{\partial}{\partial r}(r\sigma_{rz}) + \frac{\partial \sigma_{\theta z}}{\partial \theta} = 0 \tag{12-40}$$

we get

$$\frac{\partial}{\partial r}\left(r\frac{\partial w}{\partial r}\right) + \frac{1}{r}\frac{\partial^2 w}{\partial \theta^2} = 0 \qquad \text{or} \qquad \nabla^2 w = 0 \tag{12-41}$$

We solve this equation in a way analogous to solving Eq. (12-10), that is, we assume the solution in the form of a product of a function of $r$ alone by a function of $\theta$ alone,

$$w(r, \theta) = r^{\zeta} g(\theta) \tag{12-42}$$

Substituting Eq. (12-42) into Eq. (12-41) and dividing the result by $r^{\zeta-1}$ we obtain

$$\frac{d^2g(\theta)}{d\theta^2} + \zeta^2 g(\theta) = 0 \tag{12-43}$$

Solving this gives

$$g(\theta) = C_1 \cos(\zeta\theta) + C_2 \sin(\zeta\theta) \tag{12-44}$$

This expression is subject to the boundary condition at the symmetry line where, due to the antisymmetry of $w$,

$$w = 0 \tag{12-45}$$

and at the crack edges ($\theta = \pi$),

$$\sigma_{\theta z} = 0, \qquad \text{at } \theta = \pi \tag{12-46}$$

Equation (12-45) implies that $C_1 = 0$ while in view of expressions (12-39) and (12-42), Eq. (12-46) becomes

$$g'(\pi) = 0 \tag{12-47}$$

Substituting here Eq. (12-44), we get the characteristic equation

$$\cos(\zeta\pi) = 0 \tag{12-48}$$

Its characteristic values are

$$\zeta = n/2, \qquad n = 1, 3, 5, \ldots \tag{12-49}$$

For each value of $n$ we get an appropriate solution $g(\theta)$. Replacing in Eq. (12-44) $C_1$ and $C_2$ by $C_{1n}$ and $C_{2n}$ we obtain the following result:

$$g(\theta) = \sum_{n=1,3,5,\ldots} C_{2n} \sin\frac{n\theta}{2} \tag{12-50}$$

Thus using Eq. (12-42) we find the dominant term at the crack tip,

$$w(r,\theta) = C_{21} r^{1/2} \sin\frac{\theta}{2} + \text{terms vanishing at crack tip} \tag{12-51}$$

Next, using Eqs. (12-39) we calculate the corresponding shear stresses $\sigma_{\theta z}$,

$$\sigma_{\theta z} = \frac{G}{2} C_{21} r^{-1/2} \cos\frac{\theta}{2} + \text{regular terms} \tag{12-52}$$

Let us consider yet another example. Look at an elliptical hole in an infinite two-dimensional elastic solid (Fig. 12-10) subject to a uniform tension $p$ in infinity. This case is important to us because in the limit, when the length $2b$ of the ellipse's minor axis tends to zero, the hole degenerates to a rectilinear finite crack. (The process of changing the elliptical hole into a crack is shown in Fig. 12-10 in the form of a sequence of narrowing ellipses drafted with dashed lines.) But since we can solve the original problem analytically, this solution includes as a special case the stress distribution around the limiting rectilinear finite crack. The stress analysis of a two-dimensional

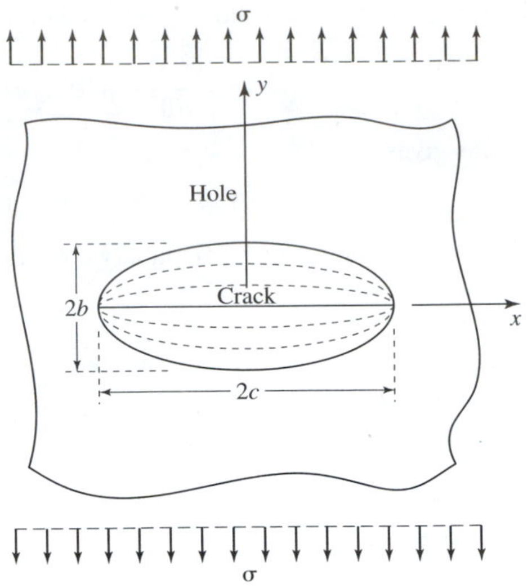


Figure 12-10

solid with an elliptical crack is straightforward but tedious because it requires the application of an elliptical coordinate system. Hence we will give only the final results. Thus with $b = 0$ we get

$$\sigma_{xx} = \sigma\left(\frac{x}{\sqrt{x^2 - c^2}} - 1\right), \qquad \sigma_{yx} = \frac{\sigma x}{\sqrt{x^2 - c^2}}, \qquad x \geq 0 \tag{12-53}$$

where $\sigma$ is the applied tension and $2c$ is the crack length (see [10, p. 567]). To determine the stress behavior in the vicinity of the crack tip, we replace $x$ by $c + r$, where $r$ is the radial distance from the crack tip, expand Eqs. (12-53) into a power series, and consider only the dominant term, which is the first term in the resulting series. This yields

$$\sigma_{xx} = \sigma_{yy} = \sigma\sqrt{\frac{c}{2r}} \tag{12-54}$$

## 12-3 STRESS INTENSITY FACTOR

We pointed out previously that it is important to know the stress intensity factor $K$ because when it reaches a certain critical value, the crack may start to grow, causing damage to the structure.

Considering the same problem as in Section 2-2, we calculate the stresses at $\theta = 0$ and the displacements at $\theta = \pm\pi$. In the Cartesian coordinate system these values of $\theta$ correspond to $y = 0$, that is, to the line of the crack and its extension (Fig. 12-11). We start by introducing a new notation,

$$K_{\text{I}} = \sqrt{2\pi}C_{41}, \qquad K_{\text{II}} = \sqrt{2\pi}C_{31}, \qquad K_{\text{III}} = G\sqrt{\pi/2}C_{21} \tag{12-55}$$

called the *stress intensity factors* (SIFs), corresponding to modes I, II, and III, respectively.

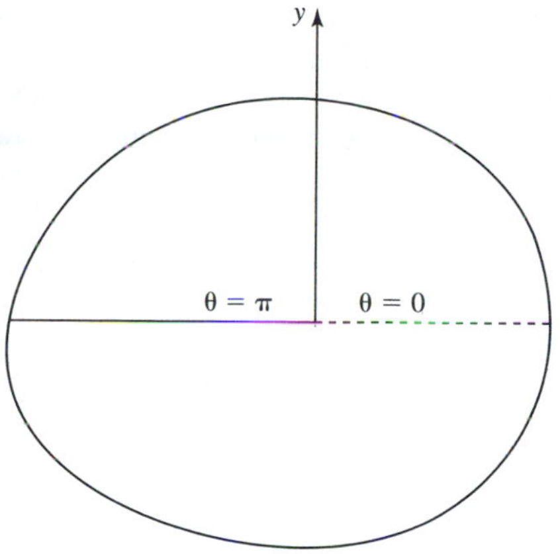


Figure 12-11

Thus for mode I, using Eq. (12-27) and $\theta = 0$, we have

$$\sigma_{yy} = \sigma_{\theta\theta} = \frac{K_{\mathrm{I}}}{\sqrt{2\pi x}}, \qquad x \geq 0 \tag{12-56}$$

and, using Eq. (12-36) and $\theta = \pm\pi$,

$$v = -u_\theta = \pm\frac{2}{G(1+\nu)}K_{\mathrm{I}}\sqrt{\frac{-x}{2\pi}} \tag{12-57}$$

where we took into account that for $\theta = \pm\pi$ the polar distance $r$ becomes $-x$.

For mode II we obtain [see Eqs. (12-28) with $\theta = 0$]

$$\sigma_{xy} = \sigma_{r\theta} = \frac{K_{\mathrm{II}}}{\sqrt{2\pi x}}, \qquad x \geq 0 \tag{12-58}$$

and [see Eqs. (12-37) with $\theta = \pm\pi$]

$$u = -u_r = \pm\frac{2}{G(1+\nu)}K_{\mathrm{II}}\sqrt{\frac{-x}{2\pi}} \tag{12-59}$$

Finally, for mode III, using Eqs. (12-52) and (12-55) with $\theta = 0$, we have

$$\sigma_{yz} = \sigma_{\theta z} = \frac{K_{\mathrm{III}}}{\sqrt{2\pi x}}, \qquad x \geq 0 \tag{12-60}$$

and, using Eqs. (12-51) and (12-55) and $\theta = \pm\pi$,

$$w = \pm\frac{2}{G}K_{\mathrm{III}}\sqrt{\frac{-x}{2\pi}} \tag{12-61}$$

Note that all the expressions for stress in the neighborhood of the crack tip can be put in the form

$$\sigma = K(2\pi r)^{-1/2}F(\theta) \tag{12-62}$$

where $K$ stands for any of the stress intensity factors defined in Eqs. (12-55). We see that stress is a product of a function of $\theta$ alone by a function of $r$ alone and by the stress intensity factor, which depends only on the crack geometry and on the applied load. In other words, the stress distribution around the crack has always the same shape, and the stress intensity factor is just a proportionality factor describing the intensity of the stress field (see [7, p. 55]). It remains to be decided how to determine the stress intensity factor for a given geometry and load. Some results are presented in Section 12-6. Here we recall only that the principle of superposition is valid for linearly elastic problems. This implies that given a cracked solid subject to a certain system $(a)$ of forces and the same solid subject to another system $(b)$ of forces, both systems causing (say) mode I crack behavior, and the stress intensity factors being $K_a$ and $K_b$, respectively, then the system $(a) + (b)$ results in a stress intensity factor $K_a + K_b$. Hence the stress intensity factors are additive. We also note that in order to make use of the stress intensity factor, we compare—as mentioned earlier—the value of the stress intensity factor determined analytically or numerically with the value of the fracture toughness $K_{Ic}$. As a rule this is done experimentally. For convenience we give here just a couple of values. Others may be found in the literature (see, for instance, [10, p. 572]). Hence from [1, p. 89] we have

- Aluminum 7001-T75 $\quad K_{Ic} = 19.8 \text{ ksi} \cdot \text{in}^{1/2} = 1.099 \times 19.8$
  $= 21.8 \text{ MPa} \cdot \text{m}^{1/2}$
- Steel 18Ni, maraging $\quad K_{Ic} = 113 \text{ ksi} \cdot \text{in}^{1/2} = 1.099 \times 113$
  $= 124 \text{ Mpa} \cdot \text{m}^{1/2}$

## 12-4 CRACK DRIVING FORCE AS AN ENERGY RATE

First we determine the *energy release rate* $\Gamma$, also called *crack driving force* (see [1, p. 33], [6, p. 38], [7, p. 49]) during the growth of the crack. Our interest in this entity stems from the Griffith *energy-balance* concept according to which the potential energy is maximum at the onset of catastrophic crack propagation. In the next section this entity, $\Gamma$, will be connected to the stress intensity factor. Let a two-dimensional solid carry some arbitrary surface loads $\boldsymbol{p}$ and the body forces $\boldsymbol{b}$. Let the initial length of the crack be $c$ (state 1), with a tip at $C_1$, and that of the extended crack, $c + \Delta c$ (state 2), with a tip at $C_2$ (see Fig. 12-12). Consider the work done by traction appearing at the surface, corresponding to the path of the extending crack. During the growth process these tractions change from some unknown values $\boldsymbol{p}^{(1)}$ to the final values $\boldsymbol{p}^{(2)} = 0$. Let the corresponding infinitesimal displacements be $d\boldsymbol{u}$.

In Fig. 12-13$a$, depicting state 1, we see both the crack $AC_1B$ and the dashed line $C_1C_2$ representing the expected and the idealized lines along which the crack will grow. Just before the growing process begins, that is, at $t = t^{(1)}$, the unknown stresses $\boldsymbol{p}^{(1)}$ act on both sides of the line $C_1C_2$. During the crack growing process they change to reach the value $\boldsymbol{p}^{(2)} = \boldsymbol{0}$ (Fig. 12-13$b$), when the crack opens along $C_1C_2$ at $t = t^{(2)}$. The work done on the crack extension zone by traction exerted on this zone during this process (that is, from $t = t^{(1)}$ to $t = t^{(2)}$) is

$$-\int_{C_1C_2}\int_{t^{(1)}}^{t^{(2)}} \boldsymbol{p}\, d\boldsymbol{u}\, dA \tag{12-63}$$

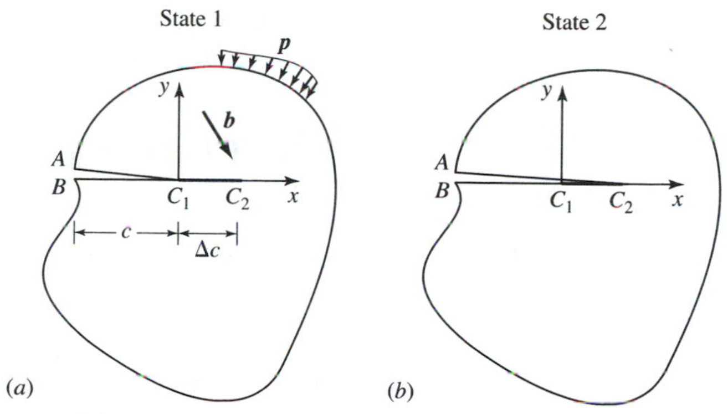


Figure 12-12

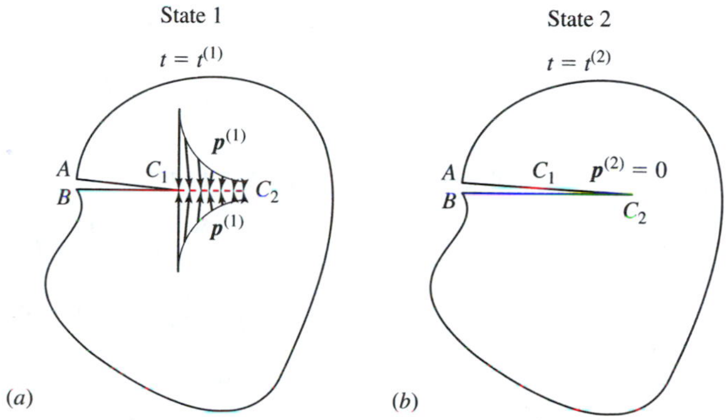


Figure 12-13

Thus to determine the strain energy release rate we divide this expression by the crack length increment $\Delta c$ and then let $\Delta c$ approach 0, thus obtaining

$$\Gamma = -\lim_{\Delta c \to 0} \frac{1}{\Delta c} \int_{C_1 C_2} \int_{t^{(1)}}^{t^{(2)}} \boldsymbol{p}\, d\boldsymbol{u}\, dA \tag{12-64}$$

Note that this expression is valid for a two-dimensional solid of unit thickness. Otherwise we have to replace the denominator on the right-hand side of the Eq. (12-64) by $B\Delta c$, where $B$ is the actual thickness. Next we find the expression for the work associated with the extension of the crack. To this end we apply the principle of virtual work [see Eq. (5-112)] [6, pp. 113, 159],

$$\int_{\overline{V}} \boldsymbol{b} \cdot \delta \boldsymbol{u}\, d\overline{V} + \int_A \boldsymbol{T} \cdot \delta \boldsymbol{u}\, dA = \int_{\overline{V}} (\sigma_{xx}\delta\varepsilon_{xx} + \sigma_{yy}\delta\varepsilon_{yy} + \sigma_{xy}\delta\gamma_{xy})\, d\overline{V} \tag{12-65}$$

where $A$ is the external area of the body plus the area of the crack faces,

$$A = A_{\text{ext}} + A_{\text{crack}} \tag{12-66}$$

Now we integrate Eq. (12-65) with respect to the time between $t = t^{(1)}$ and $t = t^{(2)}$, corresponding to states 1 and 2, respectively. This results in

$$\int_V \left( \int_{t^{(1)}}^{t^{(2)}} \boldsymbol{b} \cdot d\boldsymbol{u} \right) dV + \int_{A_{\text{ext}}} \left( \int_{t^{(1)}}^{t^{(2)}} \boldsymbol{T} \cdot d\boldsymbol{u} \right) dA$$
$$- \int_V \left[ \int_{t^{(1)}}^{t^{(2)}} (\sigma_{xx}\delta\varepsilon_{xx} + \sigma_{yy}\delta\varepsilon_{yy} + \sigma_{xy}\delta\varepsilon_{xy}) \right] dV = - \int_{A_{\text{crack}}} \left( \int_{t^{(1)}}^{t^{(2)}} \boldsymbol{T} \cdot d\boldsymbol{u} \right) dA \tag{12-67}$$

Assuming that the changes of the traction $\boldsymbol{T}$ and the body forces $\boldsymbol{b}$ are negligible between $t^{(1)}$ and $t^{(2)}$ and denoting

$$\Delta\boldsymbol{u} = \boldsymbol{u}(t^{(2)}) - \boldsymbol{u}(t^{(1)}), \qquad \Delta\boldsymbol{b} = \boldsymbol{b}(t^{(2)}) - \boldsymbol{b}(t^{(1)}), \qquad \Delta\boldsymbol{F} = \boldsymbol{F}(t^{(2)}) - \boldsymbol{F}(t^{(1)}) \tag{12-68}$$

we put the left-hand side of Eq. (12-67) into the form

$$\int_V \boldsymbol{b} \cdot d\boldsymbol{u}\, dV + \int_{A_{\text{ext}}} \boldsymbol{T} \cdot d\boldsymbol{u}\, dA - \Delta\Phi \tag{12-69}$$

where $\Phi$ is the total *elastic strain energy*,

$$\Phi = \int_V \varphi\, dV \tag{12-70}$$

and where $\varphi$ is the strain energy density, that is, the strain energy per unit volume,

$$d\varphi = \sigma_{xx}\, d\varepsilon_{xx} + \sigma_{yy}\, d\varepsilon_{yy} + \sigma_{xy}\, d\varepsilon_{xy} \tag{12-71}$$

On the right-hand side of Eq. (12-67) we recognize $\Gamma$ [Eq. (12-64)], and hence using Eq. (12-69) we get

$$W - \Phi' = \Gamma \tag{12-72}$$

where

$$\Phi' = \lim_{\Delta c \to 0} \frac{\Delta\Phi}{\Delta c} \tag{12-73}$$

Let $U$ be the total potential energy,

$$U = \Phi - \int_A \boldsymbol{p}\, d\boldsymbol{u} \tag{12-74}$$

Considering the increment of both sides of Eq. (12-74) and dividing by the increment of $c$ yields

$$U' = \Phi' - \int_A \boldsymbol{p}\, d\boldsymbol{u}' \tag{12-75}$$

where the prime has the same meaning as in Eq. (12-73).

But the integral on the right-hand side of Eq. (12-75) equals $W$, hence Eq. (12-72) becomes

$$\Gamma = \frac{\partial U}{\partial c} \tag{12-76}$$

which means that the crack driving force equals the potential energy release rate. Note that for a solid of thickness $B$ we must multiply by $B$ the denominator on the right-hand side of Eq. (12-76).

## 12-5 RELATION BETWEEN Γ AND THE STRESS INTENSITY FACTORS

Let us look again at Eq. (12-64),

$$\Gamma = -\lim_{\Delta c \to 0} \frac{1}{\Delta c} \int_{C_1 C_2} \int_{t^{(1)}}^{t^{(2)}} \boldsymbol{p} \cdot d\boldsymbol{u}\, dA$$

Since we are considering a two-dimensional problem and we anticipate the crack to grow along the straight line $AC_1C_2$, it is expedient to use a Cartesian coordinate system $x$, $y$, $z$ with an $Ox$ axis collinear with the crack. We assume also that the crack depth equals 1. Now we note that the scalar components of the surface traction vector $\boldsymbol{p}$ are the stresses acting on both sides of the edge $C_1C_2$. Hence,

$$\boldsymbol{p} = s_{yx}\boldsymbol{i} + s_{yy}\boldsymbol{j} + s_{yz}\boldsymbol{k} \tag{12-77}$$

We also note that if $\boldsymbol{p}$ is positive on the upper edge of the crack then, by Newton's law, it must be negative on the lower edge. Let the scalar components of the displacement vector $\boldsymbol{u}$ be $u$, $v$, $w$, so that

$$\boldsymbol{u} = u\boldsymbol{i} + v\boldsymbol{j} + w\boldsymbol{k} \tag{12-78}$$

Finally let the magnitudes of the displacements at the same points on the crack, but belonging to the upper edge or the lower edge be labeled + or −, respectively. Keeping all this in mind, we represent Eq. (12-64) in the following form:

$$\begin{aligned}\Gamma &= -\lim_{\Delta c \to 0} \frac{1}{2\Delta c} \int_0^{\Delta c} (\sigma_{yx}\boldsymbol{i} + \sigma_{yy}\boldsymbol{j} + \sigma_{yz}\boldsymbol{k}) \cdot \int_{t^{(1)}}^{t^{(2)}} (du\,\boldsymbol{i} + dv\,\boldsymbol{j} + dw\,\boldsymbol{k})\, dx \\ &= -\lim_{\Delta c \to 0} \frac{1}{2\Delta c} \int_0^{\Delta c} (\sigma_{yx}\boldsymbol{i} + \sigma_{yy}\boldsymbol{j} + \sigma_{yz}\boldsymbol{k}) \cdot [(u^+ - u^-)\boldsymbol{i} \\ &\quad + (v^+ - v^-)\boldsymbol{j} + (w^+ - w^-)\boldsymbol{k}]\, dx \\ &= -\lim_{\Delta c \to 0} \frac{1}{2\Delta c} \int_0^{\Delta c} [\sigma_{yx}(u^+ - u^-) + \sigma_{yy}(v^+ - v^-) + \sigma_{yz})(w^+ - w^-)]\, dx\end{aligned} \tag{12-79}$$

We replace now the stress components $\sigma_{yx}$, $\sigma_{yy}$, and $\sigma_{yz}$ by Eqs. (12-58), (12-56), and (12-60), and for the displacement components $u$, $v$, and $w$ we use Eqs. (12-59), (12-57), and (12-61). In the latter case we note that the crack tip is at the origin of the coordinate system. Now, for the growing crack of length $\Delta c$, the crack tip is at $x = \Delta c$. Hence to use the previous expressions we must replace the argument $-x$ by $\Delta c - x$. This transforms the integral on the right-hand side of Eq. (12-79) into the following expression:

$$\int_0^{\Delta c} \left\{ \frac{K_{\text{II}}}{\sqrt{2\pi x}} \left[ \frac{2}{(1+\nu)G} K_{\text{II}} \sqrt{\frac{\Delta c - x}{2\pi}} + \frac{2}{(1+\nu)G} K_{\text{II}} \sqrt{\frac{\Delta c - x}{2\pi}} \right] \right.$$
$$+ \frac{K_{\text{I}}}{\sqrt{2\pi x}} \left[ \frac{2}{(1+\nu)G} K_{\text{I}} \sqrt{\frac{\Delta c - x}{2\pi}} + \frac{2}{(1+\nu)G} K_{\text{I}} \sqrt{\frac{\Delta c - x}{2\pi}} \right] \quad (12\text{-}80)$$
$$\left. + \frac{K_{\text{III}}}{\sqrt{2\pi x}} \left[ \frac{2}{G} K_{\text{III}} \sqrt{\frac{\Delta c - x}{2\pi}} + \frac{2}{G} K_{\text{III}} \sqrt{\frac{\Delta c - x}{2\pi}} \right] \right\} dx$$

which after simplification becomes

$$\left[ \frac{2}{\pi G(1+\nu)} \left( K_{\text{II}}^2 + K_{\text{I}}^2 \right) + \frac{2}{\pi G} K_{\text{III}}^2 \right] \int_0^{\Delta c} \sqrt{\frac{\Delta c - x}{x}}\, dx \qquad (12\text{-}81)$$

Substituting this and

$$\int_0^{\Delta c} \sqrt{\frac{\Delta c - x}{x}}\, dx = \frac{\pi \Delta c}{2}$$

into Eq. (12-79) we get

$$\Gamma = \frac{1}{E} \left( K_{\text{II}}^2 + K_{\text{I}}^2 \right) + \frac{1+\nu}{E} K_{\text{III}}^2 \qquad (12\text{-}82)$$

which is the relation between the crack driving force and the stress intensity factors. It shows that, unlike the stress intensity factors, the crack driving forces are additive for different modes.

## 12-6 SOME SIMPLE CALCULATIONS OF STRESS INTENSITY FACTORS

Consider the specimen shown in Fig. 12-14, known as a double-cantilever beam (see [5, p. 56]). It is a partly split beam, modeling a beam with a crack. We will calculate the crack driving force $\Gamma$ and the stress intensity factor $K_{\text{I}}$ at the tip of the crack. We notice that each beam formed by splitting behaves like a cantilever of length $c$. Thus the crack opening $w$ equals twice the deflection of such a beam at its tip,

$$w = 2\frac{Pc^3}{3EI} \qquad (12\text{-}83)$$

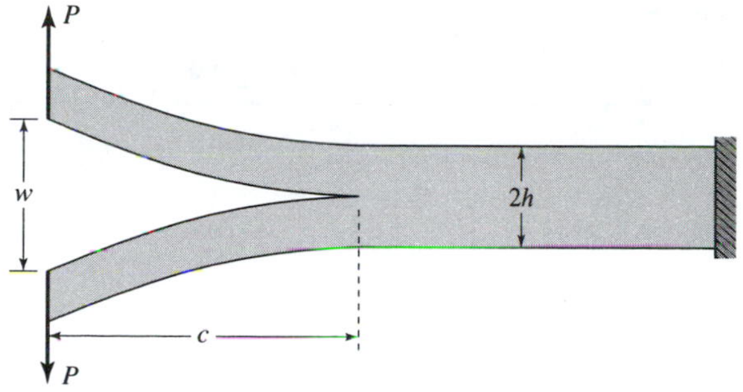


Figure 12-14

where for a rectangular beam cross section,

$$I = \frac{bh^3}{12}$$

Next we find the strain energy $U$,

$$U = \frac{1}{2}Pw = \frac{P^2c^3}{3EI} \tag{12-84}$$

Hence, using Eq. (12-76), we get

$$\Gamma = \frac{1}{B}\frac{\partial U}{\partial c} = \frac{P^2c^2}{BEI} \tag{12-85}$$

But this problem considers only the crack opening. Thus $K_{\text{II}}$ and $K_{\text{III}}$ are equal to zero, and from Eq. (12-82) we get

$$\Gamma = \frac{K_{\text{I}}^2}{E} \tag{12-86}$$

Comparing this with Eq. (12-85),

$$\frac{P^2c^2}{BEI} = \frac{K_{\text{I}}^2}{E}$$

we obtain

$$K_{\text{I}} = \frac{Pc}{\sqrt{BI}} \tag{12-87}$$

The problem just solved is a simple one. The problem solved in Section 12-3 is also relatively simple. The evaluation of the stress intensity factors is, generally speaking, considerably easier than the analysis of cracked solids with finite dimensions. Often, however, we can make use of Saint-Venant's principle, solving a finite problem by treating it as infinite. This is justified as long as we are only interested in the stress distribution near the crack tip. In other cases either we use a more sophisticated type of analysis or we solve the problem numerically. The stress intensity factors for some more common cases are listed in Table 12-1.

**TABLE 12-1 Stress Intensity Factors**

| Schematics | Stress Intensity Factor | Source |
|---|---|---|
| 1. 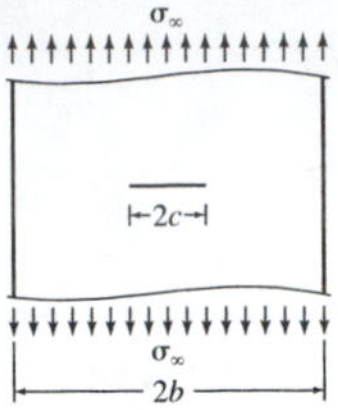  | $K_I = \sigma_\infty\sqrt{\pi c}\dfrac{1 - \dfrac{c}{2b} + 0.326\dfrac{c^2}{b^2}}{\sqrt{1 - \dfrac{c}{b}}}$ | [2] |
| 2. 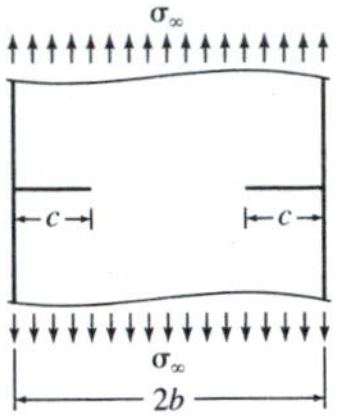  | $K_I = \sigma_\infty\sqrt{\pi c}\dfrac{1.12 - 0.61\dfrac{c}{b} + 0.13\dfrac{c^3}{b^3}}{\sqrt{1 - \dfrac{c}{b}}}$ | [2] |
| 3. 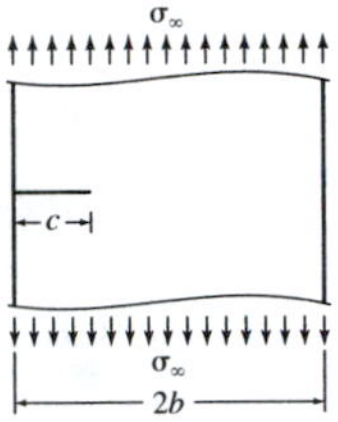  | $K_I = \sigma_\infty\sqrt{\pi c}\left(1.12 - 0.23\dfrac{c}{b} + 10.6\dfrac{c^2}{b^2} - 21.7\dfrac{c^3}{b^3} + 30.4\dfrac{c^4}{b^4}\right), \quad \dfrac{c}{b} < 0.7$ | [4] |
| 4. 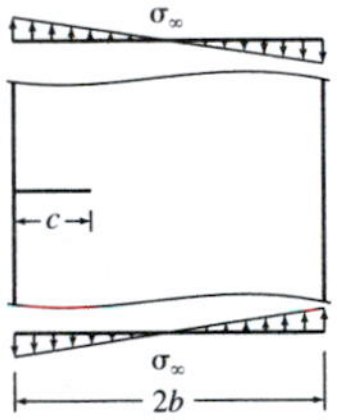  | $K_I = \sigma_\infty\sqrt{\pi c}\left(1.12 - 1.39\dfrac{c}{b} + 7.3\dfrac{c^2}{b^2} - 13\dfrac{c^3}{b^3} + 14\dfrac{c^4}{b^4}\right), \quad \dfrac{c}{b} < 0.7$ | [4] |
| 5. 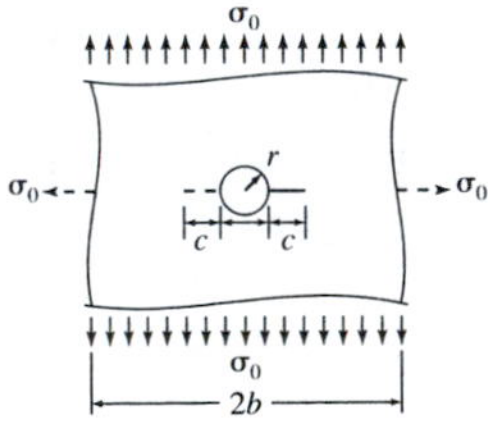  | $K_I = \sigma_0\sqrt{\dfrac{\pi}{c}}F\left(\dfrac{c}{r}\right)^*$ | [3], [9] |

6.

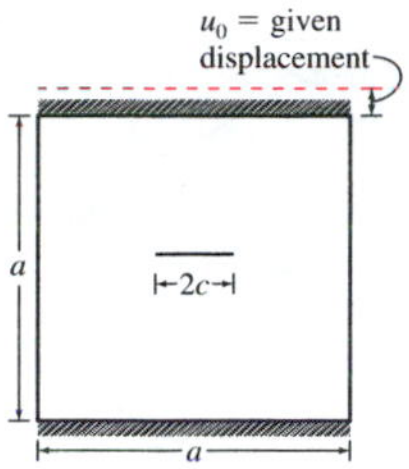


$$K_\text{I} = \alpha \frac{\mu u_0 \sqrt{c}}{a}^{\dagger}$$

[12, pp. 233–243]

7.

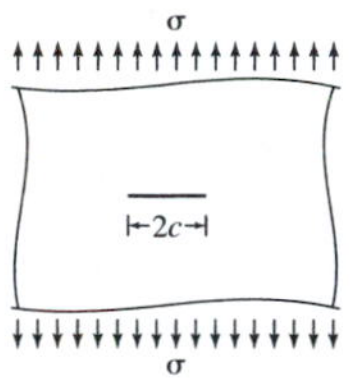


$$K_\text{I} = \sigma\sqrt{\pi c}$$

[6, p. 16]

*$F(c/r)$ is given in Table 12-2.

†where α is given in Table 12-3 and $\mu = \dfrac{E}{2(1+\nu)}$

**TABLE 12-2 Values of $F(c/r)$**

| | One Crack | | Two Cracks | |
|---|---|---|---|---|
| $c/r$ | Uniaxial | Biaxial | Uniaxial | Biaxial |
| 0.0 | 3.36 | 2.24 | 3.39 | 2.24 |
| 0.1 | 2.73 | 1.98 | 2.73 | 1.98 |
| 0.2 | 2.30 | 1.82 | 2.41 | 1.83 |
| 0.3 | 2.04 | 1.67 | 2.15 | 1.70 |
| 0.4 | 1.86 | 1.58 | 1.96 | 1.61 |
| 0.5 | 1.73 | 1.49 | 1.83 | 1.57 |
| 0.6 | 1.64 | 1.42 | 1.71 | 1.52 |
| 0.8 | 1.47 | 1.32 | 1.58 | 1.43 |
| 1.0 | 1.37 | 1.22 | 1.45 | 1.38 |
| 1.5 | 1.18 | 1.06 | 1.29 | 1.26 |
| 2.0 | 1.06 | 1.01 | 1.21 | 1.20 |
| 3.0 | 0.94 | 0.93 | 1.14 | 1.13 |
| 5.0 | 0.81 | 0.81 | 1.07 | 1.06 |
| 10.0 | 0.75 | 0.75 | 1.03 | 1.03 |
| ∞ | 0.707 | 0.707 | 1.00 | 1.00 |

**TABLE 12-3 Values of α (see Table 12-1, pos. 6)**

| $2c/a$ | $\nu = 0.2$ | $\nu = 0.3$ |
|---|---|---|
| 0.1 | 2.60029 | 3.15135 |
| 0.2 | 2.49602 | 3.00550 |
| 0.4 | 2.17958 | 2.78474 |
| 0.5 | 2.00829 | 2.34322 |

## 12-7 THE J INTEGRAL

A path-independent integral, known as *J integral,* which was introduced and applied to fracture mechanics by Rice [11], is defined as

$$J = \int_C \left( \varphi n_x - T_x \frac{\partial u}{\partial x} - T_y \frac{\partial v}{\partial x} \right) ds \tag{12-88}$$

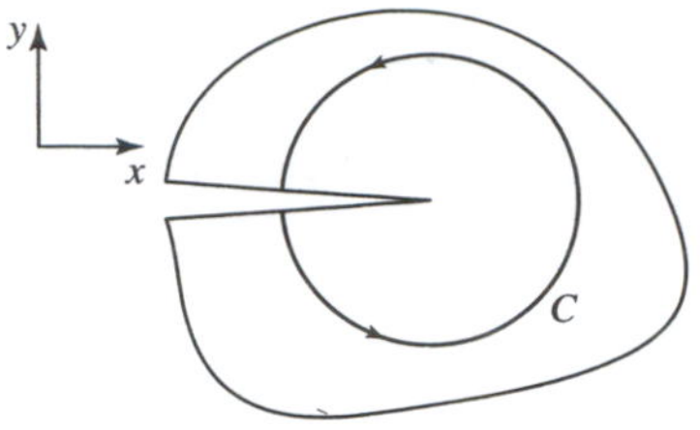


Figure 12-15 The integration contour $C$

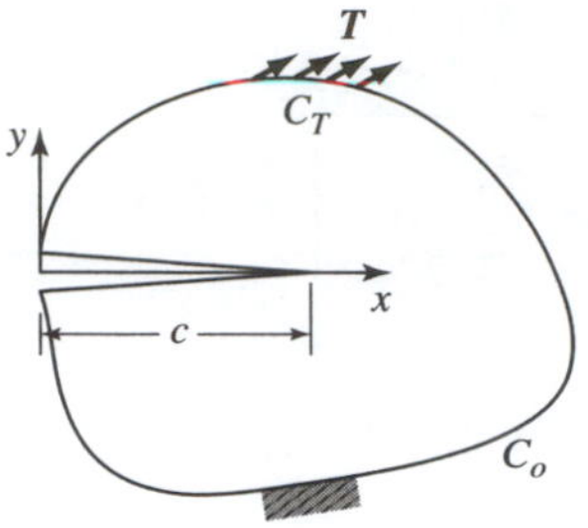


Figure 12-16

where the integration is around an arbitrary contour starting at the lower face of the crack, extending around the crack tip, and ending on the upper face of the crack (see Fig. 12-15). Here $\varphi$ is the strain energy density and $T_x$ and $T_y$ are the Cartesian components of the surface traction. The $J$ integral is used to describe the stresses in the vicinity of the crack by choosing a path of integration such that this integral can be evaluated relatively easily, even when the material behavior near the crack tip is not linearly elastic.

Let us relate the $J$ integral to the energy release rate. To this end we consider the two-dimensional linearly elastic solid shown in Fig. 12-16. Its geometry and the boundary conditions are assumed to be independent of the crack, which is of variable length $c$. The potential energy of this solid of outer contour $C_o$ and area $A$ is

$$\Pi(c) = \int_A \varphi\, dA - \int_{C_T} (T_x u + T_y v)\, ds \tag{12-89}$$

where $\Pi(c)$ indicates the dependence of the potential energy on the length of the crack, and $C_T$ is the part of the outer contour that is subject to the given traction $\boldsymbol{T}$. Note that the second integral, representing the work of the outside forces, equals zero over the part of the outside boundary where there is no traction or where there are no displacements (fixed boundary). Hence, without changing anything we can replace the integral over $C_T$ by the integral over the entire contour $C_o$ of the body. Now we differentiate Eq. (12-89) with respect to $c$ getting

$$\frac{d\Pi(c)}{dc} = \int_A \frac{d\varphi}{dc}\, dA - \int_{C_T} \left( T_x \frac{du}{dc} + T_y \frac{dv}{dc} \right) ds \tag{12-90}$$

Next we introduce a coordinate system $X, Y$ attached to the crack tip, such that

$$X = x - c, \qquad Y = y \tag{12-91}$$

Now all quantities appearing in the integrands are functions of $X$, $Y$, and $c$. Thus by the rule of differentiation of compound functions, we have

$$\frac{d}{dc} = \frac{\partial}{\partial c} + \frac{\partial}{\partial X}\frac{\partial X}{\partial c} + \frac{\partial}{\partial Y}\frac{\partial Y}{\partial c} = \frac{\partial}{\partial c} - \frac{\partial}{\partial X} = \frac{\partial}{\partial c} - \frac{\partial}{\partial x} \tag{12-92}$$

as $dX = dx$ and $\partial/\partial X = \partial/\partial x$ whereas $\partial Y/\partial c = 0$ and $\partial X/\partial c = -1$. Using these results we represent Eq. (12-90) in the form

$$\frac{d\Pi(c)}{dc} = \int_A \left(\frac{\partial\varphi}{\partial c} - \frac{\partial\varphi}{\partial x}\right) dA - \int_{C_o}\left[T_x\left(\frac{\partial u}{\partial c} - \frac{\partial u}{\partial x}\right) + T_y\left(\frac{\partial v}{\partial c} - \frac{\partial v}{\partial x}\right)\right] ds \tag{12-93}$$

However, according to Eq. (12-71),

$$d\varphi = \sigma_{xx}\,d\varepsilon_{xx} + \sigma_{yy}\,d\varepsilon_{yy} + \sigma_{zz}\,d\varepsilon_{zz} + \sigma_{xy}\,d\gamma_{xy} + \sigma_{xz}\,d\gamma_{xz} + \sigma_{yz}\,d\gamma_{yz} \tag{12-94}$$

Hence, reducing the problem to two dimensions, we get

$$\frac{\partial\varphi}{\partial\varepsilon_{xx}} = \sigma_{xx}, \qquad \frac{\partial\varphi}{\partial\varepsilon_{yy}} = \sigma_{yy}, \qquad \frac{\partial\varphi}{\partial\gamma_{xy}} = \sigma_{xy} \tag{12-95}$$

On the other hand,

$$\frac{\partial\varphi}{\partial c} = \frac{\partial\varphi}{\partial\varepsilon_{xx}}\frac{\partial\varepsilon_{xx}}{\partial c} + \frac{\partial\varphi}{\partial\varepsilon_{yy}}\frac{\partial\varepsilon_{yy}}{\partial c} + \frac{\partial\varphi}{\partial\varepsilon_{xy}}\frac{\partial\varepsilon_{xy}}{\partial c} + \frac{\partial\varphi}{\partial\varepsilon_{yx}}\frac{\partial\varepsilon_{yx}}{\partial c} \tag{12-96}$$

and since

$$\frac{\partial\varepsilon_{xx}}{\partial c} = \frac{\partial}{\partial c}\left(\frac{\partial u}{\partial x}\right) = \frac{\partial}{\partial x}\left(\frac{\partial u}{\partial c}\right), \qquad \frac{\partial\varepsilon_{xy}}{\partial c} = \frac{\partial}{\partial c}\left(\frac{\partial u}{\partial y}\right) = \frac{\partial}{\partial y}\left(\frac{\partial u}{\partial c}\right), \ldots \tag{12-97}$$

therefore,

$$\frac{\partial\varphi}{\partial c} = \sigma_{xx}\frac{\partial}{\partial x}\left(\frac{\partial u}{\partial c}\right) + \sigma_{yy}\frac{\partial}{\partial y}\left(\frac{\partial v}{\partial c}\right) + \sigma_{xy}\left[\frac{\partial}{\partial x}\left(\frac{\partial v}{\partial c}\right) + \frac{\partial}{\partial y}\left(\frac{\partial u}{\partial c}\right)\right] \tag{12-98}$$

Now let us return to the principle of virtual work [Eq. (5-112)] whose two-dimensional form is

$$\int_0^C \boldsymbol{T}\cdot\delta\boldsymbol{u}\,dC = \int_{\bar{A}}(\sigma_{xx}\,\delta\epsilon_{xx} + \sigma_{yy}\,\delta\epsilon_{yy} + \sigma_{xy}\,\delta\gamma_{xy})\,dA \tag{12-99}$$

Replacing $\delta\boldsymbol{u}$ by $\partial\boldsymbol{u}/\partial c$ and using Eq. (12-97), we put this into the form

$$\int_C\left(T_x\frac{\partial u}{\partial c} + T_y\frac{\partial v}{\partial c}\right) dC = \int_A\left\{\sigma_{xx}\frac{\partial}{\partial x}\left(\frac{\partial u}{\partial c}\right) + \sigma_{yy}\frac{\partial}{\partial y}\left(\frac{\partial v}{\partial c}\right) + \sigma_{xy}\left[\frac{\partial}{\partial y}\left(\frac{\partial u}{\partial c}\right) + \frac{\partial}{\partial x}\left(\frac{\partial v}{\partial c}\right)\right]\right\} dA \tag{12-100}$$

When the right-hand side of Eq. (12-98) is replaced by the left-hand side of Eq. (12-100) and the resulting equation substituted into Eq. (12-90), the latter takes the following form:

$$\frac{d\Pi(c)}{dc} = -\int_A \frac{\partial\varphi}{\partial x}\,dA + \int_{C_o}\left(T_x\frac{\partial u}{\partial x} + T_y\frac{\partial v}{\partial x}\right) ds \tag{12-101}$$

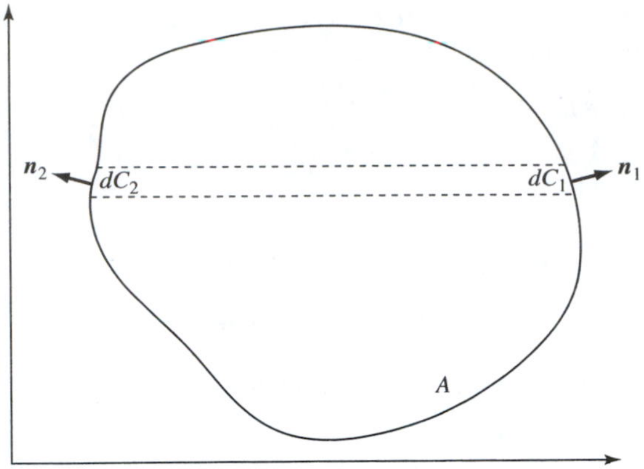


**Figure 12-17** Proof of Gauss's theorem in two dimensions

But by Gauss's theorem [see also Eq. (5-45)],

$$\int_A \frac{\partial \varphi}{\partial x}\, dA = \int_C \varphi n_x\, dC \tag{12-102}$$

To prove this, consider a closed area $A$ with the boundary $C$ (Fig. 12-17). Let the function $\varphi(x, y)$ be defined there, with $x$, $y$ being a Cartesian coordinate system. We now draw a strip parallel to the $x$ axis, with its ends touching the boundary $C$ along $dC_1$ at the right end and $dC_2$ at the left. Let the unit vectors normal to $dC_1$ and $dC_2$ be $\boldsymbol{n}_1$ and $\boldsymbol{n}_2$, respectively, and let $\varphi_1$ and $\varphi_2$ be the values of the function $\varphi$ at the right and left ends of the strip. Thus performing the integration we get

$$\int_A \frac{\partial \varphi}{\partial x}\, dx\, dy = \int_C (\varphi_1 - \varphi_2)\, dy \tag{12-103}$$

But at the right end we have $dy = n_{1x} dC_1$ and at the left end, $dy = -n_{2x} dC_2$, where the subscripts $x$ denote the $x$ components of the vectors $\boldsymbol{n}_1$ or $\boldsymbol{n}_2$. Substituting these results into the right-hand side of Eq. (12-103) we get

$$\int_A \frac{\partial \varphi}{\partial x}\, dx\, dy = \int_C \varphi n_x\, dC \tag{12-104}$$

Thus,

$$\int_C (\varphi_1 - \varphi_2)\, dy = \int_C (\varphi n_{1x}\, dC_1 + \varphi n_{2x}\, dC_2) = \int_C \varphi n_x\, dC \tag{12-105}$$

and Eq. (12-101) becomes now

$$\frac{d\Pi(c)}{dc} = \int_{C_0} \left( -\varphi n_x + T_x \frac{\partial u}{\partial x_1} + T_y \frac{\partial v}{\partial x_1} \right) ds \tag{12-106}$$

This expression differs from the $J$ integral defined in Eq. (12-88) in that here the integration is performed around the boundary of the solid, whereas the integration with the $J$ integral is around an arbitrary contour starting at the lower face of a crack, extending around the crack tip, and ending on the upper face of the crack. It is necessary to prove

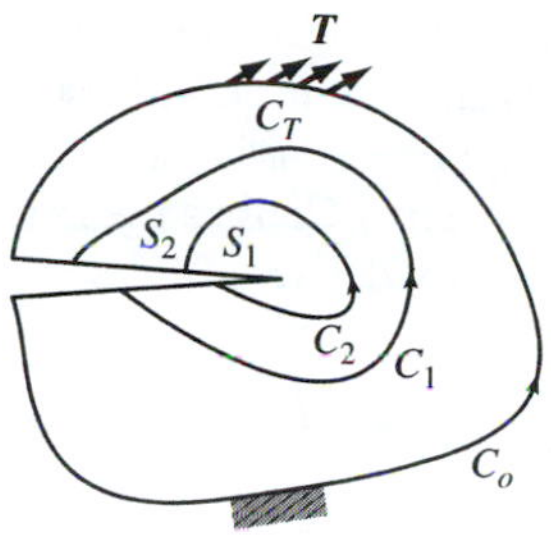


Figure 12-18

that the $J$ integral is path-independent, and thus that the contour of the solid may also be chosen as the path of integration. Let $C_1$ and $C_2$ be two arbitrarily selected paths of integration (each starting at the lower face of a crack and continuing counterclockwise until it meets the upper face of the crack) as they appear in the $J$ integral (Fig. 12-18), and let $J_1$ and $J_2$ be the corresponding $J$ integrals;

$$J_1 = \int_{C_1} \left( \varphi n_x - T_x \frac{\partial u}{\partial x} - T_y \frac{\partial v}{\partial x} \right) dC$$
$$J_2 = \int_{C_2} \left( \varphi n_x - T_x \frac{\partial u}{\partial x} - T_y \frac{\partial v}{\partial x} \right) dC \qquad (12\text{-}107)$$

Now we generate a closed contour by including the portions $S_1$ and $S_2$ of the crack faces, extending from the tip of the crack to both ends of $C_1$ and $C_2$, respectively. As all $n_x$, $T_x$, and $T_y$ equal zero at the crack faces, the values of the integrals are not affected by this addition. Hence the value of this integral along the closed contour is

$$J_1 - J_2 = \int_{C_1+C_2+S_1+S_2} \left( \varphi n_x - T_x \frac{\partial u}{\partial x} - T_y \frac{\partial v}{\partial x} \right) dC \qquad (12\text{-}108)$$

Using again Gauss's theorem, Eq. (12-102), we obtain

$$\int_{C_{1+}+C_2+S_1+S_2} \varphi n_x \, dC = \int_A \frac{\partial \varphi}{\partial x} dA \qquad (12\text{-}109)$$

where $A$ is the area included within the closed contour. Using Eqs. (2-14) we get

$$T_x = \sigma_{xx} n_x + \sigma_{xy} n_y, \qquad T_y = \sigma_{yx} n_x + \sigma_{yy} n_y \qquad (12\text{-}110)$$

and thus we obtain

$$\begin{aligned}
&\int_{C_1+C_2+S_1+S_2} \left( T_x \frac{\partial u}{\partial x} + T_y \frac{\partial v}{\partial x} \right) dC \\
&= \int_{C_1+C_2+S_1+S_2} \left[ (\sigma_{xx} n_x + \sigma_{xy} n_y) \frac{\partial u}{\partial x} + (\sigma_{yx} n_x + \sigma_{yy} n_y) \frac{\partial v}{\partial x} \right] dC \\
&= \int_{C_1+C_2+S_1+S_2} \left[ \left( \sigma_{xx} \frac{\partial u}{\partial x} + \sigma_{yx} \frac{\partial v}{\partial x} \right) n_x + \left( \sigma_{xy} \frac{\partial u}{\partial x} + \sigma_{yy} \frac{\partial v}{\partial x} \right) n_y \right] dC \qquad (12\text{-}111)
\end{aligned}$$

Now we can use Gauss's theorem again to represent the integral on the right-hand side of Eq. (12-111) as an area integral. First we notice that we can use this theorem directly by replacing the $\varphi$ appearing in Eq. (12-109) by the coefficient of $n_x$ in the integrand of Eq. (12-111). To apply Gauss's theorem to the second parentheses of the integrand in Eq. (12-111) we easily derive a formula analogous to Eq. (12-109),

$$\int_{C_1+C_2+S_1+S_2} \varphi n_y \, ds = \int_A \frac{\partial \varphi}{\partial y} \, dA \tag{12-112}$$

Finally we put Eq. (12-111) into the following form:

$$\int_{C_1+C_2+S_1+S_2} \left( T_x \frac{\partial u}{\partial x} + T_y \frac{\partial v}{\partial x} \right) dC$$
$$= \int_A \left[ \frac{\partial}{\partial x} \left( \sigma_{xx} \frac{\partial u}{\partial x} + \sigma_{yx} \frac{\partial v}{\partial x} \right) + \frac{\partial}{\partial y} \left( \sigma_{xy} \frac{\partial u}{\partial x} + \sigma_{yy} \frac{\partial v}{\partial x} \right) \right] dA \tag{12-113}$$

Hence substituting Eqs. (12-109) and (12-113) into Eq. (12-108), we obtain

$$J_1 - J_2 = \int_A \left[ \frac{\partial \varphi}{\partial x} - \frac{\partial}{\partial x} \left( \sigma_{xx} \frac{\partial u}{\partial x} + \sigma_{yx} \frac{\partial v}{\partial x} \right) - \frac{\partial}{\partial y} \left( \sigma_{xy} \frac{\partial u}{\partial x} + \sigma_{yy} \frac{\partial v}{\partial x} \right) \right] dA \tag{12-114}$$

By invoking the equilibrium equations (2-55), which in the absence of body forces take the form

$$\frac{\partial \sigma_{xx}}{\partial x} + \frac{\partial \sigma_{xy}}{\partial y} = 0, \qquad \frac{\partial \sigma_{yx}}{\partial x} + \frac{\partial \sigma_{yy}}{\partial y} = 0$$

and using Eqs. (12-98) with $c$ replaced by $x$,

$$\frac{\partial \varphi}{\partial x} = \sigma_{xx} \frac{\partial}{\partial x} \left( \frac{\partial u}{\partial x} \right) + \sigma_{yy} \frac{\partial}{\partial y} \left( \frac{\partial v}{\partial x} \right) + \sigma_{xy} \left[ \frac{\partial}{\partial x} \left( \frac{\partial v}{\partial x} \right) + \frac{\partial}{\partial y} \left( \frac{\partial u}{\partial x} \right) \right]$$

we conclude that

$$J_1 = J_2 \tag{12-115}$$

that is, the value of the $J$ integral is path-independent. Thus the right-hand side of Eq. (12-90) is also a $J$ integral, namely,

$$J = -\frac{\partial \Pi}{\partial c} \tag{12-116}$$

Hence the $J$ integral, like the energy release rate, reaches its critical value simultaneously with the stress intensity factor. This implies that any of these parameters can be used in establishing a fracture criterion.

## PROBLEMS

### Section 12-2

**12-1** Assuming that the beam shown in Fig. 12-6 carries a load $P$ at a distance $l/4$ from the support, determine the approximate value of the allowable load.

**12-2** Find the stress function for a crack with one surface free and the other fixed so that the displacement component normal to it equals zero.

*Hint:* Radial displacement $u_r$ is given by

$$2Gu_r = -\frac{\partial \Phi}{\partial r} + \left(1 - \frac{\nu}{1+\nu}\right) r \frac{\partial \psi}{\partial \theta}$$

where $\psi(r, \theta) = r^m f(\theta)$ is the solution of the differential equation

$$\frac{\partial^2 \psi}{\partial r^2} + \frac{1}{r}\frac{\partial \psi}{\partial r} + \frac{1}{r^2}\frac{\partial^2 \psi}{\partial \theta^2} = 0$$

and where $\Phi$ and $\psi$ are related

$$\nabla^2 \Phi = \frac{\partial}{\partial r}\left(r \frac{\partial \psi}{\partial \theta}\right)$$

**12-3** Determine the stresses corresponding to the conditions described in Problem 12-2.

**12-4** Show the details of the derivation of the following equations. Use a symbolic manipulation package if available.

(a) Eqs. (12-23).

(b) Eqs. (12-24).

(c) Eqs. (12-27).

(d) Eqs. (12-28).

**12-5** Use Eqs. (12-23) to determine the stresses $\sigma_{rr}$ at $r = 0.1$ in, 0.5 in, 1.0 in for $\theta = 0°, 30°, 45°, 90°$.

### Section 12-6

**12-6** A large structural element with a central crack of width $2c$ is subject to 1000-MPa tension. Determine the critical crack size when the material of the element is made from: (a) 18 Ni maraging steel; (b) aluminum 7001-T75.

**12-7** A long steel strip of width $b = 10$ cm has an edge crack of critical length $c =$ 20 mm. What is the value of the applied stress $\sigma_\infty$ if the fracture toughness $K_{Ic}$ equals 120 MPa $\cdot$ m$^{1/2}$? *Hint:* Use Table 12-1 #3.

**12-8** In Problem 12-7 assume that there are two edge cracks and find $\sigma_\infty$ (see Table 12-1 #2)

**12-9** Make a diagram of $K_I$ as a function of $c/b$ for the case:

(a) Table 12-1 #1;

(b) Table 12-1 #2;

(c) Table 12-1 #3.

## REFERENCES

[1] Barsom, J. M., and S. T. Rolfe, *Fracture and Fatigue Control in Structures,* 2nd ed. (Prentice-Hall, Englewood Cliffs, NJ, 1987).

[2] Benthem, J. P., and W. T. Koiter, "Asymptotic Approximations to Crack Problems," in G. C. Sih, Ed., *Mechanics of Fracture,* vol. 1 (Noordhoff, Leyden, 1973).

[3] Bowie, O. L., "Analysis of an Infinite Plate Containing Radial Cracks Originating at the Boundary of an Internal Circular Hole," *Journal of Mathematics and Physics,* **35**, pp. 60–71 (1956).

[4] Gross, B., J. E. Srawley, and W. F. Brown, Jr., "Stress Intensity Factors for a Single Edge Notch Tension Specimen by Boundary Collocation of a Stress Function," NASA Tech. Note D-2395 (1964).

[5] Hellan, K., *Introduction to Fracture Mechanics* (McGraw-Hill, New York, 1984).

[6] Kanninen, M. F., and C. H. Popelar, *Advanced Fracture Mechanics* (Oxford University Press, New York, 1985).

[7] Lawn, B. R., and T. R. Wilshaw, *Fracture of Brittle Solids* (Cambridge University Press, Cambridge, MA, 1975).

[8] *Maple V,* release 5 (Waterloo Maple Inc., Waterloo, ON, Canada, 1998).

[9] Paris, C. P., and G. C. Sih, "Stress Analysis of Cracks," in *Fracture Toughness Testing and Its Applications,* ASTM STP 381 (American Society for Testing and Materials, Philadelphia, PA, 1965).

[10] Ragab, A. R., and S. E. Bayoumi, *Engineering Solid Mechanics* (CRC Press, Boca Raton, FL, 1998).

[11] Rice, J. R., "A Path Independent Integral and the Approximate Analysis of Strain Concentration by Notches and Cracks," *Journal of Applied Mechanics,* 35, pp. 379–386 (1968).

[12] Tang, R., and F. Erdogan, "A Clamped Rectangular Plate Containing a Crack," *Theoretical and Applied Fracture Mechanics,* 4, pp. 233–243 (1985).

[13] Williams, M. L., "On the Stress Distribution at the Base of a Stationary Crack," *Journal of Applied Mechanics,* 24, pp. 109–114 (1957).

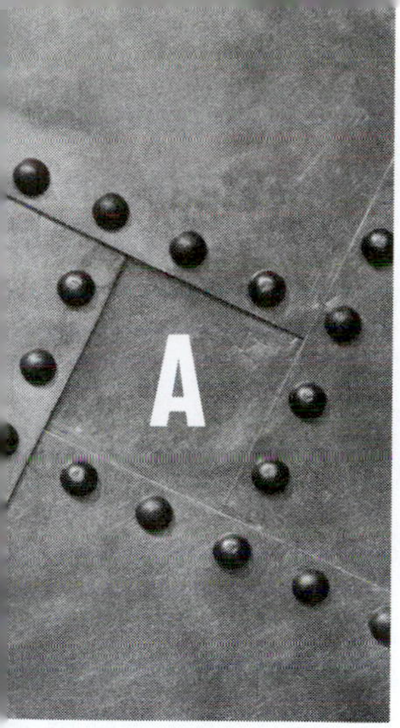


# APPENDIX

## MATRICES

A *matrix* is simply a rectangular array of elements (numbers or symbols) arranged into rows and columns. It is written as follows:

$$\begin{bmatrix} a_{11} & a_{12} & \cdots & a_{1n} \\ a_{21} & a_{22} & \cdots & a_{2n} \\ \vdots & \vdots & \vdots & \vdots \\ a_{m1} & a_{m2} & \cdots & a_{mn} \end{bmatrix}$$

The array of elements is usually enclosed in either brackets [ ] or braces { }. Symbolically a matrix is often designated by a boldface letter enclosed in brackets or braces,

$$[\boldsymbol{a}] = \begin{bmatrix} a_{11} & a_{12} & \cdots & a_{1n} \\ a_{21} & a_{22} & \cdots & a_{2n} \\ \vdots & \vdots & \vdots & \vdots \\ a_{m1} & a_{m2} & \cdots & a_{mn} \end{bmatrix}$$

A matrix has no numerical value. It is simply a convenient device for displaying arrays of numbers.

Each number appearing in a matrix is called an *element*. Since a matrix is, in general, a two-dimensional array of numbers, the elements of the matrix are identified by double subscripts. The first subscript designates the *row* in which an element is located and the second designates the *column*. Thus $a_{21}$ refers to the element in the second row, first column.

A matrix that contains only one column is referred to as a *column matrix,* and one that contains only one row is called a *row matrix*. Usually a column matrix is enclosed in braces. A square matrix contains the same number of rows and columns. The *leading diagonal* (also called the *main* or *principal diagonal,* or simply the *diagonal*) of a square matrix extends from the upper left corner to the lower right corner. Elements along the diagonal are referred to as *diagonal elements*. All other elements are *off-diagonal elements*. The diagonal extending from the upper right to the lower left has no special significance in dealing with matrices.

A *diagonal matrix* is one in which all off-diagonal elements are zero. For example,

$$\begin{bmatrix} -5 & 0 & 0 \\ 0 & 3 & 0 \\ 0 & 0 & 7 \end{bmatrix}, \qquad \begin{bmatrix} 2 & 0 & 0 & 0 \\ 0 & 0 & 0 & 0 \\ 0 & 0 & -3 & 0 \\ 0 & 0 & 0 & 8 \end{bmatrix}$$

A square matrix whose diagonal elements are 1's and whose off-diagonal elements are 0's is a *unit* or *identity matrix*. It is usually designated as $[\boldsymbol{I}]$.

If all elements of a matrix are zero, the matrix is called a *null* or *zero matrix*. Such matrices need not be square.

A square matrix for which $a_{ji} = a_{ij}$ for all $i$ and $j$ is a *symmetric matrix*. The matrix

$$[\boldsymbol{a}] = \begin{bmatrix} a_{11} & a_{12} & a_{13} \\ a_{12} & a_{22} & a_{23} \\ a_{13} & a_{23} & a_{33} \end{bmatrix}$$

is symmetric.

## Algebraic Operations

Two matrices can be added (or subtracted) only if each contains the same number of rows and the same number of columns. If this condition is met, then

$$\begin{bmatrix} a_{11} & a_{12} \\ a_{21} & a_{22} \\ a_{31} & a_{32} \end{bmatrix} \pm \begin{bmatrix} b_{11} & b_{12} \\ b_{21} & b_{22} \\ b_{31} & b_{32} \end{bmatrix} = \begin{bmatrix} a_{11} \pm b_{11} & a_{12} \pm b_{12} \\ a_{21} \pm b_{21} & a_{22} \pm b_{22} \\ a_{31} \pm b_{31} & a_{32} \pm b_{32} \end{bmatrix} \quad \text{(A-1)}$$

Clearly, the operations of matrix addition and subtraction are commutative.

Two matrices can be multiplied only if the number of columns in the first matrix and the number of rows in the second matrix are the same. If this condition is met, then the matrices are said to be *conformable for multiplication*. The matrix that results from multiplication contains the same number of rows as the first matrix and the same number of columns as the second matrix. Thus if a $2 \times 3$ matrix is to be multiplied by a $3 \times 5$ matrix, the matrices are conformable for multiplication and the resulting matrix is $2 \times 5$. If a matrix $[\boldsymbol{a}]$ is to be multiplied by a matrix $[\boldsymbol{b}]$ to yield a product $[\boldsymbol{c}]$, then $[\boldsymbol{a}][\boldsymbol{b}] = [\boldsymbol{c}]$. The element $[\boldsymbol{c}]$ located in row $i$, column $j$ is obtained through term-by-term multiplication of the $i$th row of $[\boldsymbol{a}]$ by the $j$th column of $[\boldsymbol{b}]$. Thus, for example,

$$c_{ij} = \sum_k a_{ik} b_{kj} \quad \text{(A-2)}$$

Note that matrix multiplication is not, in general, commutative. Consequently a distinction is made between *postmultiplication* of $[\boldsymbol{a}]$ by $[\boldsymbol{b}]$ (that is, $[\boldsymbol{a}][\boldsymbol{b}]$) and *premultiplication* of $[\boldsymbol{a}]$ by $[\boldsymbol{b}]$ (that is, $[\boldsymbol{b}][\boldsymbol{a}]$).

When a row matrix $(1 \times m)$ is postmultiplied by a column matrix $(m \times 1)$, the result is a scalar, whereas when a row matrix $(1 \times m)$ is premultiplied by a column matrix $(m \times 1)$, the result is a square matrix $(m \times m)$.

It is easy to verify that when a matrix $[\boldsymbol{a}]$ is pre- or postmultiplied by the appropriately sized identity matrix, the result is simply $[\boldsymbol{a}]$.

The transpose of $[\boldsymbol{a}]$ is obtained by interchanging the rows and columns of $[\boldsymbol{a}]$. For example, if

$$[\boldsymbol{a}] = \begin{bmatrix} a_{11} & a_{12} & a_{13} \\ a_{21} & a_{22} & a_{23} \end{bmatrix}$$

then

$$[\boldsymbol{a}]^T = \begin{bmatrix} a_{11} & a_{21} \\ a_{12} & a_{22} \\ a_{13} & a_{23} \end{bmatrix}$$

where the superscript $T$ indicates the transpose of $[\boldsymbol{a}]$. Note that $[\boldsymbol{a}]$ need not be a square matrix. It is easily demonstrated that

$$([\boldsymbol{a}]^T)^T = [\boldsymbol{a}] \tag{A-3}$$

If $[\boldsymbol{a}]$ is *symmetric,* then

$$[\boldsymbol{a}]^T = [\boldsymbol{a}] \tag{A-4}$$

The transpose of a product is the product of the individual transposes taken in reverse order,

$$([\boldsymbol{a}][\boldsymbol{b}][\boldsymbol{c}]\cdots[\boldsymbol{h}])^T = [\boldsymbol{h}]^T\cdots[\boldsymbol{c}]^T[\boldsymbol{b}]^T[\boldsymbol{a}]^T \tag{A-5}$$

Suppose $[\boldsymbol{a}]$ is a square matrix. If a matrix $[\boldsymbol{b}]$ exists such that

$$[\boldsymbol{b}][\boldsymbol{a}] = [\boldsymbol{a}][\boldsymbol{b}] = [\boldsymbol{I}]$$

then $[\boldsymbol{b}]$ is called the *inverse* of $[\boldsymbol{a}]$ and designated as $[\boldsymbol{a}]^{-1}$. Thus,

$$[\boldsymbol{a}]^{-1}[\boldsymbol{a}] = [\boldsymbol{a}][\boldsymbol{a}]^{-1} = [\boldsymbol{I}] \tag{A-6}$$

If the determinant of $[\boldsymbol{a}]$ is zero, then $[\boldsymbol{a}]$ is said to be a *singular matrix* and its inverse does not exist; otherwise the inverse does exist.

If $[\boldsymbol{a}], [\boldsymbol{b}], [\boldsymbol{c}], \ldots, [\boldsymbol{h}]$ are all $m \times m$ square matrices, then

$$([\boldsymbol{a}][\boldsymbol{b}][\boldsymbol{c}]\cdots[\boldsymbol{h}])^{-1} = [\boldsymbol{h}]^{-1}\cdots[\boldsymbol{c}]^{-1}[\boldsymbol{b}]^{-1}[\boldsymbol{a}]^{-1} \tag{A-7}$$

If the inverse of $[\boldsymbol{a}]$ exists, then by taking the inverse of $[\boldsymbol{a}]^{-1}[\boldsymbol{a}] = [\boldsymbol{I}]$, premultiplying the result by $[\boldsymbol{a}]$, noting that $[\boldsymbol{I}]^{-1} = [\boldsymbol{I}]$, and using Eq. (A-6), it follows that

$$([\boldsymbol{a}]^{-1})^{-1} = [\boldsymbol{a}] \tag{A-8}$$

The inverse of a symmetric matrix is also symmetric.

The inverse of a diagonal matrix is a diagonal matrix whose elements are the reciprocals of the corresponding elements of the original matrix. Thus,

$$[\boldsymbol{a}] = \begin{bmatrix} a_{11} & 0 & \cdots & 0 \\ 0 & a_{22} & \cdots & 0 \\ \vdots & \vdots & \ddots & \vdots \\ 0 & 0 & 0 & a_{nn} \end{bmatrix}, \qquad [\boldsymbol{a}]^{-1} = \begin{bmatrix} \dfrac{1}{a_{11}} & 0 & \cdots & 0 \\ 0 & \dfrac{1}{a_{22}} & \cdots & 0 \\ \vdots & \vdots & \ddots & \vdots \\ 0 & \cdots & 0 & \dfrac{1}{a_{nn}} \end{bmatrix} \tag{A-9}$$

The inverse of a transpose equals the transpose of the inverse,

$$([\boldsymbol{a}]^T)^{-1} = ([\boldsymbol{a}]^{-1})^T \tag{A-10}$$

If it happens that, for a square matrix $[\boldsymbol{a}]$,

$$[\boldsymbol{a}]^{-1} = [\boldsymbol{a}]^T \tag{A-11}$$

then $[\boldsymbol{a}]$ is said to be an *orthogonal matrix*. The rotation matrix given by Eq. (7.28) is an example of an orthogonal matrix.

Consider the following system of equations:

$$\begin{aligned} a_{11}x_1 + a_{12}x_2 + a_{13}x_3 &= b_1 \\ a_{21}x_1 + a_{22}x_2 + a_{23}x_3 &= b_2 \\ a_{31}x_1 + a_{32}x_2 + a_{33}x_3 &= b_3 \end{aligned} \tag{A-12}$$

Following the procedure for matrix multiplication, these can be written in matrix form as

$$\begin{bmatrix} a_{11} & a_{12} & a_{13} \\ a_{21} & a_{22} & a_{23} \\ a_{31} & a_{32} & a_{33} \end{bmatrix} \begin{Bmatrix} x_1 \\ x_2 \\ x_3 \end{Bmatrix} = \begin{Bmatrix} b_1 \\ b_2 \\ b_3 \end{Bmatrix}$$

or, symbolically, as

$$[\boldsymbol{a}]\{\boldsymbol{x}\} = \{\boldsymbol{b}\} \tag{A-13}$$

where

$$\begin{aligned} [\boldsymbol{a}] &= \begin{bmatrix} a_{11} & a_{12} & a_{13} \\ a_{21} & a_{22} & a_{23} \\ a_{31} & a_{32} & a_{33} \end{bmatrix} \\ \{\boldsymbol{x}\} &= [\,x_1 \quad x_2 \quad x_3\,]^T \\ \{\boldsymbol{b}\} &= [\,b_1 \quad b_2 \quad b_3\,]^T \end{aligned} \tag{A-14}$$

Suppose that $[\boldsymbol{a}]$ and $\{\boldsymbol{b}\}$ are known and it is desired to determine $\{\boldsymbol{x}\}$. This can be accomplished by premultiplying Eq. (A-13) by $[\boldsymbol{a}]^{-1}$ and using Eq. (A-6) to get

$$\{\boldsymbol{x}\} = [\boldsymbol{a}]^{-1}\{\boldsymbol{b}\} \tag{A-15}$$

Thus the solution of Eq. (A-13) can be obtained formally by premultiplying $\{\boldsymbol{b}\}$ by the inverse of $[\boldsymbol{a}]$. It should be pointed out that numerical computation of the inverse of $[\boldsymbol{a}]$ requires many operations. Consequently unless $[\boldsymbol{a}]$ is very small (on the order of, say, $3 \times 3$), it is computationally more efficient to obtain a numerical solution to Eq. (A-13) using a technique such as Gaussian elimination.

## Partitioning of Matrices

Consider the following matrix:

$$[\boldsymbol{c}] = \begin{bmatrix} c_{11} & c_{12} & c_{13} & c_{14} & c_{15} \\ c_{21} & c_{22} & c_{23} & c_{24} & c_{25} \\ c_{31} & c_{32} & c_{33} & c_{34} & c_{35} \\ c_{41} & c_{42} & c_{43} & c_{44} & c_{45} \\ c_{51} & c_{52} & c_{53} & c_{54} & c_{55} \end{bmatrix}$$

From time to time it is desirable to *partition* a matrix into segments by drawing horizontal and vertical dividing lines,

$$[\boldsymbol{c}] = \left[\begin{array}{cc:ccc} c_{11} & c_{12} & c_{13} & c_{14} & c_{15} \\ c_{21} & c_{22} & c_{23} & c_{24} & c_{25} \\ c_{31} & c_{32} & c_{33} & c_{34} & c_{35} \\ \hdashline c_{41} & c_{42} & c_{43} & c_{44} & c_{45} \\ c_{51} & c_{52} & c_{53} & c_{54} & c_{55} \end{array}\right]$$

There can be any number of segments, and they can be of any size. Each segment is itself a matrix, and therefore the original matrix may be written in terms of *submatrices,*

$$[\boldsymbol{c}] = \begin{bmatrix} [\boldsymbol{C}_{11}] & [\boldsymbol{C}_{12}] \\ [\boldsymbol{C}_{13}] & [\boldsymbol{C}_{14}] \end{bmatrix}$$

where the submatrices are

$$[\boldsymbol{C}_{11}] = \begin{bmatrix} c_{11} & c_{12} \\ c_{21} & c_{22} \\ c_{31} & c_{32} \end{bmatrix}, \qquad [\boldsymbol{C}_{12}] = \begin{bmatrix} c_{13} & c_{14} & c_{15} \\ c_{23} & c_{24} & c_{25} \\ c_{33} & c_{34} & c_{35} \end{bmatrix}$$

$$[\boldsymbol{C}_{21}] = \begin{bmatrix} c_{41} & c_{42} \\ c_{51} & c_{52} \end{bmatrix}, \qquad [\boldsymbol{C}_{22}] = \begin{bmatrix} c_{43} & c_{44} & c_{45} \\ c_{53} & c_{54} & c_{55} \end{bmatrix}$$

For example, the stiffness matrix for an orthotropic material has the following form:

$$[\boldsymbol{c}] = \begin{bmatrix} c_{11} & c_{12} & c_{13} & 0 & 0 & 0 \\ c_{12} & c_{22} & c_{23} & 0 & 0 & 0 \\ c_{13} & c_{23} & c_{33} & 0 & 0 & 0 \\ 0 & 0 & 0 & c_{44} & 0 & 0 \\ 0 & 0 & 0 & 0 & c_{55} & 0 \\ 0 & 0 & 0 & 0 & 0 & c_{66} \end{bmatrix}$$

This matrix can be partitioned into four 3 × 3 submatrices,

$$[\boldsymbol{c}] = \left[\begin{array}{ccc:ccc} c_{11} & c_{12} & c_{13} & 0 & 0 & 0 \\ c_{12} & c_{22} & c_{23} & 0 & 0 & 0 \\ c_{13} & c_{23} & c_{33} & 0 & 0 & 0 \\ \hdashline 0 & 0 & 0 & c_{44} & 0 & 0 \\ 0 & 0 & 0 & 0 & c_{55} & 0 \\ 0 & 0 & 0 & 0 & 0 & c_{66} \end{array}\right]$$

or

$$[\boldsymbol{c}] = \begin{bmatrix} [\boldsymbol{C}_{11}] & [\boldsymbol{0}] \\ [\boldsymbol{0}] & [\boldsymbol{C}_{22}] \end{bmatrix}$$

where

$$[\boldsymbol{C}_{11}] = \begin{bmatrix} c_{11} & c_{12} & c_{13} \\ c_{12} & c_{22} & c_{23} \\ c_{13} & c_{23} & c_{33} \end{bmatrix}, \qquad [\boldsymbol{C}_{22}] = \begin{bmatrix} c_{44} & 0 & 0 \\ 0 & c_{55} & 0 \\ 0 & 0 & c_{66} \end{bmatrix}$$

and $[\boldsymbol{0}]$ is a 3 × 3 null matrix.

## Differentiation of Matrices

Suppose we have a scalar $A = A(x_1, x_2, \ldots, x_m)$ that we wish to differentiate successively with respect to $x_1, x_2, \ldots, x_m$. This leads to

$$\frac{\partial A}{\partial x_1}, \frac{\partial A}{\partial x_2}, \ldots, \frac{\partial A}{\partial x_m}$$

The result can be displayed as a matrix,

$$\begin{Bmatrix} \dfrac{\partial A}{\partial x_1} \\ \dfrac{\partial A}{\partial x_2} \\ \vdots \\ \dfrac{\partial A}{\partial x_m} \end{Bmatrix}$$

The following notation can be adopted:

$$\frac{\partial A}{\partial \{\boldsymbol{x}\}} = \begin{Bmatrix} \dfrac{\partial A}{\partial x_1} \\ \dfrac{\partial A}{\partial x_2} \\ \vdots \\ \dfrac{\partial A}{\partial x_m} \end{Bmatrix} \tag{A-16}$$

where

$$\{\boldsymbol{x}\} = [x_1 \quad x_2 \quad \cdots \quad x_m]^T \tag{A-17}$$

and $\partial A/\partial\{\boldsymbol{x}\}$ implies differentiation of $A$ successively with respect to each element of $\{\boldsymbol{x}\}$ and arranging the result in a column matrix. Now suppose we have

$$\begin{aligned} A_1 &= A_1(x_1, x_1, \ldots, x_m) \\ A_2 &= A_2(x_1, x_1, \ldots, x_m) \\ &\vdots \\ A_n &= A_n(x_1, x_1, \ldots, x_m) \end{aligned}$$

Each of these can be differentiated with respect to $\{\boldsymbol{x}\}$ according to Eq. (A-16) to give

$$\frac{\partial A_1}{\partial\{\boldsymbol{x}\}} = \begin{Bmatrix} \dfrac{\partial A_1}{\partial x_1} \\ \dfrac{\partial A_1}{\partial x_2} \\ \vdots \\ \dfrac{\partial A_1}{\partial x_m} \end{Bmatrix}, \quad \frac{\partial A_2}{\partial\{\boldsymbol{x}\}} = \begin{Bmatrix} \dfrac{\partial A_2}{\partial x_1} \\ \dfrac{\partial A_2}{\partial x_2} \\ \vdots \\ \dfrac{\partial A_2}{\partial x_m} \end{Bmatrix}, \ldots, \frac{\partial A_n}{\partial\{\boldsymbol{x}\}} = \begin{Bmatrix} \dfrac{\partial A_n}{\partial x_1} \\ \dfrac{\partial A_n}{\partial x_2} \\ \vdots \\ \dfrac{\partial A_n}{\partial x_m} \end{Bmatrix}$$

By defining $\{\boldsymbol{A}\}$ as

$$\{\boldsymbol{A}\} = [A_1 \quad A_2 \quad \cdots \quad A_n]^T \tag{A-18}$$

these results can be arranged in matrix form as

$$\frac{\partial\{\boldsymbol{A}\}^T}{\partial\{\boldsymbol{x}\}} = \begin{bmatrix} \dfrac{\partial A_1}{\partial x_1} & \dfrac{\partial A_2}{\partial x_1} & \cdots & \dfrac{\partial A_n}{\partial x_1} \\ \dfrac{\partial A_1}{\partial x_2} & \dfrac{\partial A_2}{\partial x_2} & \cdots & \dfrac{\partial A_n}{\partial x_2} \\ \vdots & \vdots & \vdots & \vdots \\ \dfrac{\partial A_1}{\partial x_m} & \dfrac{\partial A_2}{\partial x_m} & \cdots & \dfrac{\partial A_n}{\partial x_m} \end{bmatrix} \tag{A-19}$$

Thus the symbolic operator to the left of the equal sign implies the $m \times n$ matrix on the right. Note that the rules for matrix multiplication apply in that the $m \times 1$ symbolic operator $\partial/\partial\{\boldsymbol{x}\}$ operates on (or "multiplies") the $1 \times n$ matrix $\{\boldsymbol{A}\}^T$. Thus the matrices are conformable for multiplication and the result is an $m \times n$ matrix.

It can be shown by direct substitution that

$$\frac{\partial\{\boldsymbol{x}\}^T}{\partial\{\boldsymbol{x}\}} = [\boldsymbol{I}] \tag{A-20}$$

Also, if $\{\boldsymbol{A}\}$ and $\{\boldsymbol{B}\}$ are $m \times 1$ column matrices such that $\{\boldsymbol{A}\} = \{\boldsymbol{A}(x_1, x_2, \ldots, x_n)\}^T$, $\{\boldsymbol{B}\} = \{\boldsymbol{B}(x_1, x_2, \ldots, x_n)\}^T$, and $\boldsymbol{C} = \{\boldsymbol{A}\}^T\{\boldsymbol{B}\}$, then

$$\frac{\partial \boldsymbol{C}}{\partial\{\boldsymbol{x}\}} = \frac{\partial}{\partial\{\boldsymbol{x}\}}(\{\boldsymbol{A}\}^T\{\boldsymbol{B}\}) = \frac{\partial\{\boldsymbol{A}\}^T}{\partial\{\boldsymbol{x}\}}\{\boldsymbol{B}\} + \frac{\partial\{\boldsymbol{B}\}^T}{\partial\{\boldsymbol{x}\}}\{\boldsymbol{A}\} \tag{A-21}$$

If $\{\boldsymbol{A}\}$ is independent of $\{\boldsymbol{x}\}$, then

$$\frac{\partial \boldsymbol{C}}{\partial \{\boldsymbol{x}\}} = \frac{\partial}{\partial \{\boldsymbol{x}\}}(\{\boldsymbol{A}\}^T\{\boldsymbol{B}\}) = \frac{\partial \{\boldsymbol{B}\}^T}{\partial \{\boldsymbol{x}\}}\{\boldsymbol{A}\} \tag{A-22}$$

whereas if $\{\boldsymbol{B}\} = \{\boldsymbol{x}\}$, then

$$\frac{\partial \boldsymbol{C}}{\partial \{\boldsymbol{x}\}} = \frac{\partial}{\partial \{\boldsymbol{x}\}}(\{\boldsymbol{A}\}^T\{\boldsymbol{x}\}) = \frac{\partial \{\boldsymbol{A}\}^T}{\partial \{\boldsymbol{x}\}}\{\boldsymbol{x}\} + \{\boldsymbol{A}\} \tag{A-23}$$

Finally, if $\boldsymbol{C} = \{\boldsymbol{A}\}^T[\boldsymbol{D}]\{\boldsymbol{A}\}$, where $[\boldsymbol{D}]$ is an $m \times m$ matrix whose elements are independent of $x_1, x_2, \ldots, x_n$, then

$$\frac{\partial \boldsymbol{C}}{\partial \{\boldsymbol{x}\}} = \frac{\partial}{\partial \{\boldsymbol{x}\}}(\{\boldsymbol{A}\}^T[\boldsymbol{D}]\{\boldsymbol{A}\}) = \frac{\partial \{\boldsymbol{A}\}^T}{\partial \{\boldsymbol{x}\}}([\boldsymbol{D}] + [\boldsymbol{D}]^T)\{\boldsymbol{A}\} \tag{A-24}$$

If $[\boldsymbol{D}]$ is a symmetric matrix, then this becomes

$$\frac{\partial \boldsymbol{C}}{\partial \{\boldsymbol{x}\}} = \frac{\partial}{\partial \{\boldsymbol{x}\}}(\{\boldsymbol{A}\}^T[\boldsymbol{D}]\{\boldsymbol{A}\}) = 2\frac{\partial \{\boldsymbol{A}\}^T}{\partial \{\boldsymbol{x}\}}[\boldsymbol{D}]\{\boldsymbol{A}\} \tag{A-25}$$

and if, in addition, $\{\boldsymbol{A}\} = \{\boldsymbol{x}\}$,

$$\frac{\partial \boldsymbol{C}}{\partial \{\boldsymbol{x}\}} = \frac{\partial}{\partial \{\boldsymbol{x}\}}(\{\boldsymbol{x}\}^T[\boldsymbol{D}]\{\boldsymbol{x}\}) = 2[\boldsymbol{D}]\{\boldsymbol{x}\} \tag{A-26}$$

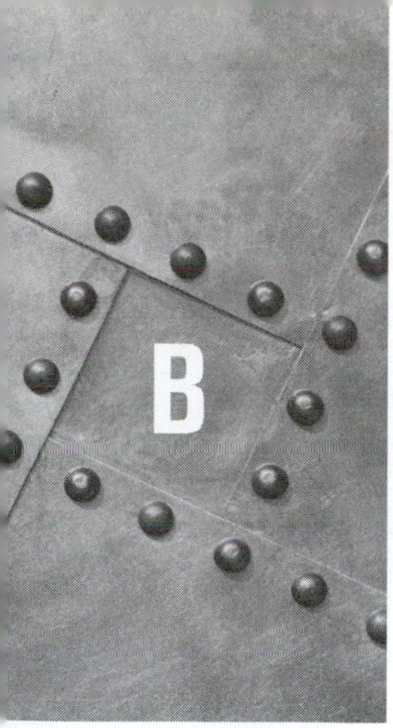


# APPENDIX

## COORDINATE TRANSFORMATIONS

Let us consider a vector $\boldsymbol{A}$ connecting two points in space, $O$ and $P$. We establish a right-handed orthogonal coordinate system $x$, $y$, $z$ with origin at point $O$ and unit vectors $\boldsymbol{i}, \boldsymbol{j}$, and $\boldsymbol{k}$ directed along the $x$, $y$, and $z$ axes, respectively, as shown in Fig. B-1. We can write $\boldsymbol{A}$ in terms of its components $A_x$, $A_y$, $A_z$ and the unit vectors as

$$\boldsymbol{A} = A_x\boldsymbol{i} + A_y\boldsymbol{j} + A_z\boldsymbol{k} \tag{B-1}$$

$\boldsymbol{A}$ can, of course, be described just as easily in another right-handed orthogonal coordinate system, say, the $x'$, $y'$, $z'$ coordinate system shown in Fig. B-2 whose unit vectors in the $x'$, $y'$, and $z'$ coordinate directions are $\boldsymbol{i}'$, $\boldsymbol{j}'$, and $\boldsymbol{k}'$, respectively. In terms of its components $A_x$, $A_y$, and $A_z$, the vector $\boldsymbol{A}$ is given by

$$\boldsymbol{A} = A_x\boldsymbol{i}' + A_y\boldsymbol{j}' + A_z\boldsymbol{k}' \tag{B-2}$$

Supposing that the components of $\boldsymbol{A}$ are known in the $x$, $y$, $z$ coordinate system and that the relationship between the primed and unprimed coordinate systems is known, the components of $\boldsymbol{A}$ in the primed coordinate system can be found.

To begin with, note that $A_{x'}$ is simply the projection of $\boldsymbol{A}$ on the $x'$ axis. Thus with the nomenclature of Fig. B-3,

$$A_{x'} = |\boldsymbol{A}|\cos(\theta) = \boldsymbol{i}' \cdot \boldsymbol{A} \tag{B-3}$$

where $|\boldsymbol{A}|$ is the magnitude of $\boldsymbol{A}$ and $\theta$ is the angle between $\boldsymbol{A}$ and the $x'$ axis measured in the plane defined by $\boldsymbol{A}$ and $\boldsymbol{i}'$.

When Eq. (B-1) is used in Eq. (B-3), it yields, after collecting terms,

$$A_{x'} = (\boldsymbol{i}' \cdot \boldsymbol{i})A_x + (\boldsymbol{i}' \cdot \boldsymbol{j})A_y + (\boldsymbol{i}' \cdot \boldsymbol{k})A_z \tag{B-4}$$

We now define the relationship between the two coordinate systems in terms of the angles between the coordinate axes in the unprimed system and those in the primed system. In order to do this we will agree that the angles are measured *from* the unprimed

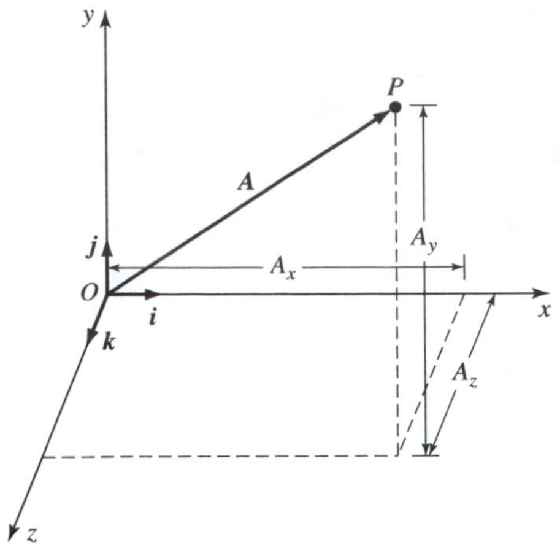


Figure B-1

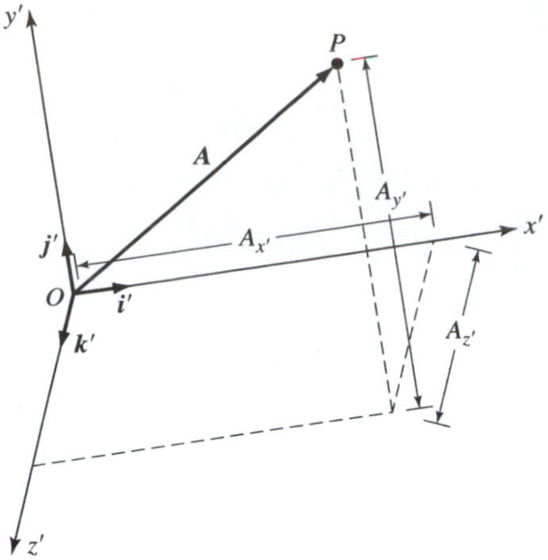


Figure B-2

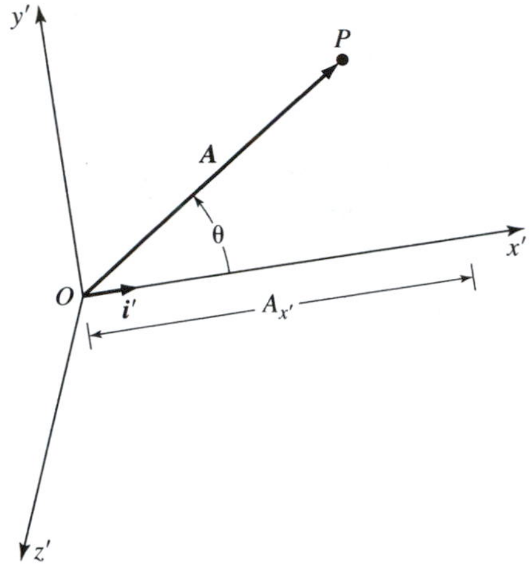


Figure B-3

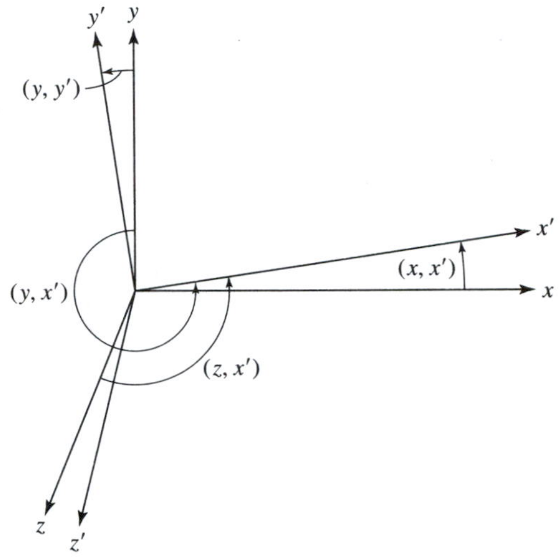


Figure B-4

axes *to* the primed axes in the counterclockwise direction in the plane defined by the two axes. The term counterclockwise as used here refers to the counterclockwise direction when the coordinate system in three-dimensional space is projected onto a flat plane such as a piece of paper. For example, in Fig. B-4 the angle from the $y$ axis to the $y'$ axis is in the counterclockwise direction, as is the angle from the $z$ axis to the $x'$ axis. Angles are designated by two letters enclosed in parentheses and separated by commas. The first letter identifies the "from" axis and the second the "to" axis. Thus, as shown in Fig. B-4, the angle $(x, x')$ is the angle from the $x$ axis to the $x'$ axis, measured in the counterclockwise direction in the plane defined by the $x$ and $x'$ axes. Similarly, the angle $(y, x')$ is the angle from the $y$ axis to the $x'$ axis measured in the counterclockwise direction in the plane defined by the $y$ and $x'$ axes.

With the angles so defined, the dot products of the unit vectors in Eq. (B-4) become

$$\boldsymbol{i}' \cdot \boldsymbol{i} = |\boldsymbol{i}'||\boldsymbol{i}| \cos(x, x') = \cos(x, x')$$
$$\boldsymbol{i}' \cdot \boldsymbol{j} = |\boldsymbol{i}'||\boldsymbol{j}| \cos(y, x') = \cos(y, x')$$
$$\boldsymbol{i}' \cdot \boldsymbol{k} = |\boldsymbol{i}'||\boldsymbol{k}| \cos(z, x') = \cos(z, x')$$

so that

$$A_{x'} = A_x \cos(x, x') + A_y \cos(y, x') + A_z \cos(z, x') \tag{B-5}$$

Similarly, $A_{y'}$ is the projection of $\boldsymbol{A}$ on the $y'$ axis so that

$$A_{y'} = (\boldsymbol{j}' \cdot \boldsymbol{i})A_x + (\boldsymbol{j}' \cdot \boldsymbol{j})A_y + (\boldsymbol{j}' \cdot \boldsymbol{k})A_z$$

Evaluating the dot products we get

$$A_{y'} = A_x \cos(x, y') + A_y \cos(y, y') + A_z \cos(z, y') \tag{B-6}$$

Finally, since $A_{z'}$ is the projection of $\boldsymbol{A}$ on the $z'$ axis,

$$A_{z'} = A_x \cos(x, z') + A_y \cos(y, z') + A_z \cos(z, z') \tag{B-7}$$

Equations (B-5)–(B-7) can be written in matrix form,

$$\begin{Bmatrix} A_{x'} \\ A_{y'} \\ A_{z'} \end{Bmatrix} = \begin{bmatrix} \cos(x, x') & \cos(y, x') & \cos(z, x') \\ \cos(x, y') & \cos(y, y') & \cos(z, y') \\ \cos(x, z') & \cos(y, z') & \cos(z, z') \end{bmatrix} \begin{Bmatrix} A_x \\ A_y \\ A_z \end{Bmatrix} \tag{B-8}$$

This can be written symbolically as

$$\{\boldsymbol{A}'\} = [\boldsymbol{R}]\{\boldsymbol{A}\} \tag{B-9}$$

where

$$\{\boldsymbol{A}'\} = [A_{x'} \quad A_{y'} \quad A_{z'}]^T \tag{B-10}$$

$$\{\boldsymbol{A}\} = [A_x \quad A_y \quad A_z]^T \tag{B-11}$$

$$[\boldsymbol{R}] = \begin{bmatrix} \cos(x, x') & \cos(y, x') & \cos(z, x') \\ \cos(x, y') & \cos(y, y') & \cos(z, y') \\ \cos(x, z') & \cos(y, z') & \cos(z, z') \end{bmatrix} \tag{B-12}$$

where $[\boldsymbol{R}]$ is the *rotation matrix*. Its elements are *direction cosines*.

The direction cosines are often expressed using the letters $l$, $m$, and $n$,

$$\begin{bmatrix} l_1 & m_1 & n_1 \\ l_2 & m_2 & n_2 \\ l_3 & m_3 & n_3 \end{bmatrix} = \begin{bmatrix} \cos(x, x') & \cos(y, x') & \cos(z, x') \\ \cos(x, y') & \cos(y, y') & \cos(z, y') \\ \cos(x, z') & \cos(y, z') & \cos(z, z') \end{bmatrix} \tag{B-13}$$

The square of the length of $\boldsymbol{A}$ is given by

$$|\boldsymbol{A}|^2 = \boldsymbol{A} \cdot \boldsymbol{A} = A_x^2 + A_y^2 + A_z^2 = \{\boldsymbol{A}\}^T\{\boldsymbol{A}\} \tag{B-14}$$

Since the length is unchanged by a coordinate transformation, we have

$$\{A'\}^T\{A'\} = \{A\}^T\{A\} \tag{B-15}$$

Substituting Eq. (B-9), we get

$$\{A\}^T[R]^T[R]\{A\} = \{A\}^T\{A\} \tag{B-16}$$

Comparing the two sides of this equation it is clear that

$$[R]^T[R] = [I] \tag{B-17}$$

and by expanding the left-hand side of Eq. (B-17) and then equating the corresponding elements on the right and left sides, we get the following relationships among the direction cosines:

$$\begin{aligned} l_1^2 + l_2^2 + l_3^2 &= 1 \\ m_1^2 + m_2^2 + m_3^2 &= 1 \\ n_1^2 + n_2^2 + n_3^2 &= 1 \\ l_1 m_1 + l_2 m_2 + l_3 m_3 &= 0 \\ l_1 n_1 + l_2 n_2 + l_3 n_3 &= 0 \\ m_1 n_1 + m_2 n_2 + m_3 n_3 &= 0 \end{aligned} \tag{B-18}$$

These equations give relationships that must be satisfied by the nine direction cosines so that only three direction cosines can be chosen arbitrarily. The remaining six are determined by Eqs. (B-18). Thus three angles are needed to define a general coordinate transformation in three-dimensional space.

We now postmultiply Eq. (B-17) by the inverse of $[R]$. This yields

$$[R]^T = [R]^{-1} \tag{B-19}$$

that is, the inverse of the rotation matrix and its transpose are equal. Thus, as noted in Appendix A, the rotation matrix is an *orthogonal matrix*.

Now consider a special case of the general three-dimensional transformation, namely, the *plane transformation*. This is a transformation in which one axis (say, the $z$ axis) remains fixed. As seen in Fig. B-5, only one angle is needed to define the transformation. If this angle is designated $\theta$, then

$$\begin{aligned} \cos(x, x') &= l_1 = \cos(\theta) \\ \cos(x, y') &= l_2 = \cos(\pi/2 + \theta) = -\sin(\theta) \\ \cos(x, z') &= l_3 = \cos(\pi/2) = 0 \\ \cos(y, x') &= m_1 = \cos(3\pi/2 + \theta) = \sin(\theta) \\ \cos(y, y') &= m_2 = \cos(\theta) \\ \cos(y, z') &= m_3 = \cos(\pi/2) = 0 \\ \cos(z, x') &= n_1 = \cos(\pi/2) = 0 \\ \cos(z, y') &= n_2 = \cos(\pi/2) = 0 \\ \cos(z, z') &= n_3 = \cos(0) = 1 \end{aligned}$$

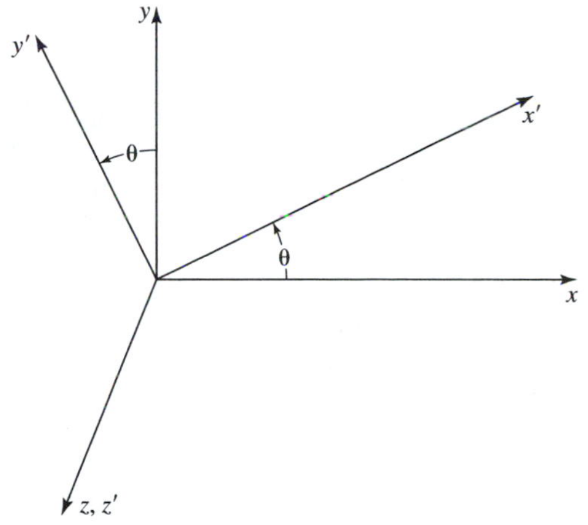


Figure B-5

With these, Eqs. (B-18) reduce to

$$
\begin{aligned}
l_1^2 + l_2^2 &= 1 \\
m_1^2 + m_2^2 &= 1 \\
l_1 m_1 + l_2 m_2 &= 0
\end{aligned}
\tag{B-20}
$$

The rotation matrix then becomes

$$
[\boldsymbol{R}] = \begin{bmatrix} \cos(\theta) & \sin(\theta) & 0 \\ -\sin(\theta) & \cos(\theta) & 0 \\ 0 & 0 & 1 \end{bmatrix} \tag{B-21}
$$

Since in a plane transformation the $z$ component of the vector remains unchanged, as indicated by the third row of the rotation matrix, Eq. (B-21), the rotation matrix is often written as

$$
[\boldsymbol{R}] = \begin{bmatrix} \cos(\theta) & \sin(\theta) \\ -\sin(\theta) & \cos(\theta) \end{bmatrix}
$$

# INDEX